Moscow International Symposium on Magnetism (MISM 2017)

EPJ Web of Conferences Volume 185 (2018)

Moscow, Russia
1-5 July 2017

Editors:

N. Perov **A. Semisalova**

ISBN: 978-1-5108-6596-9

Printed from e-media with permission by:

Curran Associates, Inc.
57 Morehouse Lane
Red Hook, NY 12571

proceedings.com

Some format issues inherent in the e-media version may also appear in this print version.

Printed by Curran Associates, Inc. (2018)

For additional information, please contact EDP Sciences – Web of Conferences at the address below.

EDP Sciences – Web of Conferences
17, Avenue du Hoggar
Parc d'Activité de Courtabœuf
BP 112
F-91944 Les Ulis Cedex A
France

Phone: +33 (0) 1 69 18 75 75
Fax: +33 (0) 1 69 28 84 91

contact-edps@webofconferences.org

Additional copies of this publication are available from:

Curran Associates, Inc.
57 Morehouse Lane
Red Hook, NY 12571 USA
Phone: 845-758-0400
Fax: 845-758-2633
Email: curran@proceedings.com
Web: www.proceedings.com

TABLE OF CONTENTS

MODELING OF MAGNETIC LABEL FIELD SENSING BY GMR STRUCTURE 1
Bukanov Kirill, Babaizev Georgy, Chechenin Nikolay

SIZE EFFECT IN THE ELECTRONIC TRANSPORT OF THIN FILMS OF BI_2SE_3 5
Marchenkov V.V., Chistyakov V.V., Huang J.C.A., Perevozchikova Y.A., Domozhirova A.N., Eisterer M.

EFFECTS OF AVERAGE GRAIN SIZE ON THE MAGNETIC PROPERTIES OF PERMALLOY FILMS 8
Djuzhev Nikolay, Iurov Aleksey, Mazurkin Nikita, Chinenkov Maksim, Trifonov Aleksey, Pushkina Marina

IMAGE RECONSTRUCTION ALGORITHMS FOR THE MICROWAVE HOLOGRAPHIC VISION SYSTEM WITH RELIABLE GAP DETECTION AT THEORETICAL LIMITS 12
Leshchiner Dmitry, Zvezdin Konstantin, Popkov Anatoly, Chepkov Grigory, Perlo Pietro

INVERSE SPIN HALL EFFECT IN HETEROSTRUCTURES "NANOSTRUCTURED FERROMAGNET/TOPOLOGICAL INSULATOR" 16
Petrov P.N., Davydova M.D., Skirdkov P.N., Zvezdin K.A., Lin J.G., Huang J.C.A.

TOP AND BOTTOM SPIN VALVES WITH NI-FE-MN ANTIFERROMAGNETIC LAYER 20
Blinov Ilya, Krinitsina Tatiana, Milyaev Mikhail, Popov Vladimir, Ustinov Vladimir

MAGNETIC INTERACTION AT THE INTERFACE OF EPITAXIAL MANGANITE FILM AND $(TBCO_2/FECO)_N$ SUPERLATTICE 23
Grishin Alexander S., Ovsyannikov Gennady A., Klimov Alexey A., Demidov Viktor V., Constantinian Karen Y., Borisenko Igor V., Preobrazhensky Vladimir L., Tiercelin Nicolas, Pernod Phillipe

BIFURCATION ANALYSIS OF THE DYNAMICAL SYSTEM FOR A THREELAYERED VALVE WITH PERPENDICULAR ANISOTROPY 27
Ostrovskaya Natalia, Skidanov Vladimir, Iusipova Iulia, Skvortsov Maxims

CALCULATION OF CPP-AND CIP-MAGNITORESISTANCE IN MULTILAYER MAGNETIC STRUCTURES 31
Romanovskii Dmitry E., Prudnikov Vladimir V., Prudnikov Pavel V., Mamonova Marina V.

MAGNETICALLY INTERCALATED MULTILAYER SILICENE 35
Tokmachev Andrey M., Averyanov Dmitry V., Karateev Igor A., Parfenov Oleg E., Kondratev Oleg A., Taldenkov Alexander N., Storchak Vyacheslav G.

MODELING OF GMR LOOPS ASYMMETRY IN SPIN VALVES 39
Kurenkov Alexander, Babaizev Georgy, Chechenin Nikolay

CRITICAL SWITCHING CHARACTERISTICS OF THREE-LAYERED SPIN VALVE FOR DIFFERENT MATERIALS AND ALLOYS WITH UNIAXIAL ANISOTROPY 43
Iusipova Iuliia

HIGH FIELD MAGNETORESISTANCE OF NANOCOMPOSITES $(CO_{84}NB_{14}TA_2)_X(AL_2O_3)_{100-X}$ NEAR THE PERCOLATION THRESHOLD 47
Blinov Mikhail, Zakharchuk Ivan, Lähderanta Erkki, Sitnikov Alexander, Rodionov Igor, Prudnikov Valerii, Rylkov Vladimir, Granovsky Alexander

MAGNETOTRANSPORT PROPERTIES OF $CO_X(MGF_2)_{100-X}$ OXYGEN-FREE NANOCOMPOSITES 51
Tregubova Tatyana, Stognei Oleg, Kirpan Veronika, Sitnikov Aleksandr

TUNNEL MAGNETORESISTANCE AND SPIN TRANSFER TORQUE IN MAGNETIC TUNNEL JUNCTION WITH EMBEDDED NANOPARTICLES 55
Useinov Niazbeck

TUNNEL MAGNETORESISTIVE EFFECT IN NANOSTRUCTURED COMPOSITE SYSTEMS 59
Stognei O.V., Sitnikov A.V., Kalinin Yu.E.

THERMODYNAMIC ASPECTS OF DESCRIBING THE CONTRIBUTION OF SPONTANEOUS MAGNETISM OF ELECTRONS TO THE HALL RESISTANCE 64
Okulov Vsevolod, Pamyatnykh Evgeny

INVESTIGATION OF INSB:MN THIN FILMS MAGNETORESISTANCE BY I-V CHARACTERISTICS METHOD 66
Novodvorsky Oleg, Mikhalevsky Vladimir, Parshina Liubov, Khramova Olga, Lotin Andrey, Cherebylo Elena, Marenkin Sergey, Aronov Aleksey, Kochura Aleksey, Panchenko Vladislav

SPIN-FILTER STATE OF AU-CO NANOWIRES 69
Smelova Ekaterina, Tsysar Kseniya, Saletsky Alexander

SPIN-TORQUE QUANTIZATION AND MICROWAVE SENSITIVITY OF A NANO-SIZED SPIN DIODE 74
Demin Gleb, Popkov Anatoly

NUMERICAL SIMULATION OF SPIN TRANSPORT IN SYSTEMS WITH COMPLEX GEOMETRY 78
Andrianov Timofey, Vedyayev Anatoly

THE INFLUENCE OF "INJECTED" AND "THERMAL" MAGNONS ON A SPIN WAVE CURRENT AND DRAG EFFECT IN HYBRID STRUCTURES 82
Lyapilin Igor, Okorokov Mikhail

INVESTIGATION OF ANGULAR AND TEMPERATURE DEPENDENCE OF SPIN-WAVE RESONANCE SPECTRA IN MULTILAYER MAGNETIC FILMS 86
Bazhanov Andrey

EXPERIMENTAL STUDY OF MICROWAVE PROPERTIES OF COMPOSITES FILLED WITH PERMALLOY POWDER 90
Bobrovskii Stanislav Y., Garanov Victor A., Naboko Andrei S., Osipov Alexei V., Rozanov Konstantin N.

A NOVEL METHOD OF ESR OSCILLATING MAGNETIZATION VALUE DETERMINATION IN STRONGLY CORRELATED METALS 94
Gilmanov Marat, Semeno Alexey, Samarin Alexander, Demishev Sergey
CONDITIONS FOR DIVISION AND MULTIPLICATION OF FREQUENCIES IN FERRITE 98
Pleshev Dmitriy, Vlasov Vladimir, Asadullin Fanur, Kotov Leonid, Poleshchikov Sergey, Shavrov Vladimir, Shcheglov Vladimir
DETERMINATION OF MAGNETOELASTIC OSCILLATIONS IN FERRITE PLATE WITH DIFFERENT VALUES OF ANISOTROPY CONSTANT 102
Kotov Leonid, Severin Pavel, Vlasov Vladimir, Beznosikov Dmitry
SUPERDIRECTED BEAM OF THE BACKWARD VOLUME SPIN WAVE 106
Annenkov Alexander Yu., Gerus Sergey V., Lock Edwin H.
TEMPERATURE DEPENDENCE OF GILBERT DAMPING IN MANGANITE/NORMAL METAL HETEROSTRUCTURE 109
Shcaihulov Timur A., Demidov Victor V., Borisenko Igor V., Ovsyannikov Gennady A.
STUDY OF TORQUE ACTING ON MAGNETIZATION DURING 180° PULSED REVERSAL PROCESS IN FERRITE-GARNET FILMS WITH BIAXIAL ANISOTROPY 113
Kolotov Oleg, Matyunin Andrey, Nikoladze Georgiy, Polyakov Petr
THE GRANULE SIZE DISTRIBUTION INFLUENCE IN NANOCOMPOSITES ON OPTICAL AND MAGNETOOPTICAL SPECTRA 115
Yurasov Alexey, Gan'shina Elena, Sokolov Alexey, Granovsky Nikita, Zazymkina Daria
MEASUREMENT OF MAGNETIC NOISE IN MAGNETOIMPEDANCE SENSING ELEMENT 119
Antonov Anatoliy S., Buznikov Nikita A., Sultan-Zade Timur T., Shkliaev Alexey A.
MANIFESTATIONS OF THE EFFECTS OF A "WEAK" SUBLATTICE IN AN YIG FILM WITH THE REPLACEMENT OF YTTRIUM BY GD IONS 123
Poimanov Vladislav, Shkar Vladiimir, Nepochatykh Yuriy, Sampath Vedamanickam, Shavrov Vladiimir, Koledov Victor
MAGNETOSTATIC WAVES IN A MEDIUM WITH DAMPING 127
Makarov Pavel, Maltceva Lidiya, Kotov Leonid, Shcheglov Vladimir
WAVEGUIDE MAGNETOPLASMONIC STRUCTURE BASED ON FERRITE GARNET FILM 130
Basiladze Georgy, Dolgov Aleksandr, Berzhansky Vladimir, Karavainikov Andrey, Prokopov Anatoly
INVESTIGATION OF WELDS BY THE METHOD OF THE MAGNETO-OPTICAL EDDY CURRENT FLAW DETECTION 134
Lugovskoy Nazar, Berzhansky Vladimir, Filippov Dmitry, Prokopov Anatoliy, Shuyskyy Alexandr
DISPERSION OF THE MAGNETOSTATIC VOLUME WAVES IN A MEDIUM WITH DAMPING 137
Maltceva Lidiya, Makarov Pavel, Kotov Leonid, Shcheglov Vladimir
TAMM PLASMON-POLARITONS STRUCTURES WITH BI-SUBSTITUTED GARNET LAYERS 141
Mikhailova Tatyana, Shaposhnikov Alexander, Prokopov Anatoly, Karavainikov Andrey, Tomilin Sergey, Lyashko Sergey, Berzhansky Vladimir
HYPERSOUND EXCITATION OF MAGNETIZATION AND ELASTIC DISPLACEMENT IN CASE OF MAGNETIZATION REVERSAL 145
Asadullin Fanur, Pleshev Dmitriy, Vlasov Vladimir, Kotov Leonid, Poleshchikov Sergey, Shavrov Vladimir, Shcheglov Vladimir
ROTATIONAL REMAGNETIZATION AND ORIENTATIONAL TRANSITIONS IN SUBMICRON YIG FILMS 148
Vladyslav Poimanov, Yurii Nepochatykh, Victor Koledov, Vladimir Shavrov
FREQUENCY DEPENDENCE OF MICROWAVE COMPLEX PERMEABILITY UNDER MAGNETIC BIAS 151
Shiryaev Artem O., Bobrovskii Stanislav Y., Osipov Aleksey V., Petrov Dmitry A., Rozanov Konstantin N., Starostenko Sergey N.
ELECTRICAL PROPERTIES OF FERROMAGNETIC-INSULATOR NANOCOMPOSITES 155
Kalinin Yu.E., Makagonov V.A., Sitnikov A.V., Granovsky A.B.
THE ORIGIN OF FINE STRUCTURE IN MAGNETIZATION CURVE OF α-COV_2O_6 158
Kudasov Yuri, Kozabaranov Roman
COEXISTENCE OF THE MAGNETICALLY ORDERED AND HALDANE STATES IN $(Y_{1-X}ND_X)_2BANIO_5$ 162
Popova Elena, Klimin Sergei, Popova Marina, Klingeler Rudiger, Buchner Bernd, Vasiliev Alexander
CRITICAL PHENOMENA IN 1D ISING MODEL WITH ARBITRARY SPIN 166
Proshkin Alexey, Kassan-Ogly Felix, Zarubin Alexander, Ponomareva Tatyana, Menshikh Ivan
MAGNETIZATION CURVES AND HYSTERESIS LOOPS OF NANOSTRUCTURED RARE EARTH TITANATES 170
Rinkevich Anatoly B., Korolev Alexander V., Samoilovich Mikhail I., Demokritov Sergej O., Perov Dmitry V.
APPEARANCE OF ITINERANT ELECTRONS DETECTED BY IR SPECTROSCOPY AND ITS CORRELATION WITH SURFACE MAGNETISM IN CO-DOPED TIO_2 NANOPOWDERS 174
Mostovshchikova E.V., Yermakov A.Ye., Uimin M.A., Minin A.S.
FIRST-PRINCIPLE CALCULATIONS OF THE EXCHANGE INTERACTION CHARACTERISTICS FOR MULTILAYER MAGNETIC STRUCTURES 177
Mamonova Marina, Prudnikov Vladimir
MONTE CARLO SIMULATION OF AGING PHENOMENA IN MULTILAYER MAGNETIC STRUCTURES 181
Purtov Alexandr N., Mamonova Marina V., Prudnikov Vladimir V., Prudnikov Pavel V.
STRUCTURAL AND MAGNETIC CHARACTERISTICS OF THE CO/CU/CO THIN-FILM SYSTEMS 185
Shalygina Elena, Kharlamova Anna, Makarov Andrey, Kurlyandskaya Galina, Svalov Andrey
PHASE DYNAMICS IN ARRAYS OF COUPLED VORTEX SPIN-TORQUE NANO-OSCILLATORS 189
Katkova Olga, Safin Ansar, Udalov Nikolay, Kapranov Mikhail
MICROWAVE AND MAGNETO-OPTIC PROPERTIES OF ε-FE_2O_3 NANOPARTICLES ARISING IN BORATE GLASSES DOPED WITH FE AND GD 193
Ivanova Oxana, Kliava Janis, Edelman Irina, Ivantsov Ruslan, Petrakovskaja Eleonora

ON NATURE OF THE INDUCED ANISOTROPY IN THE CHEMICALLY DEPOSITED CO-P FILMS 197
Chzhan Anatoly V., Podorozhnyak Sergey A., Volochaev Mikhail N., Patrin Gennady S.

MAGNETO-OPTICAL RESPONSE AMPLIFICATION IN MULTI-LAYER NANOCOMPOSITE-SEMICONDUCTOR STRUCTURES 200
Buravtsova Victoria, Gan'shina Elena, Kalinin Yuri, Sitnikov Alexander

EFFECT OF CHEMICAL ORDERING ON OPTICAL PROPERTIES OF FE_3SI EPITAXIAL FILMS 204
Tarasov Ivan, Popov Zakhar, Visotin Maxim, Yakovlev Ivan, Varnakov Sergey

ATOMIC ORDERING AND COERCIVE FORCE OF NANOCRYSTALLINE FE-CO ALLOY FILMS 208
Repina Nataliya

HEAT TREATMENT EFFECT ON MAGNETIC MICROSTRUCTURE OF $FE_{73.9}CU_1NB_3SI_{13.2}B_{8.9}$ THIN FILMS 211
Mikhalitsyna Evgeniya, Zakharchuk Ivan, Soboleva Ekaterina, Geydt Pavel, Kataev Vasiliy, Lepalovskij Vladimir, Lähderanta Erkki

MAGNETO-OPTICAL AND MAGNETIC RESONANCE PROPERTIES OF NANO-SCALED GRANULAR FILMS $(COFEB)_X(SIO_2)_{100-X}$ AND $(COFEB)_XC_{100-X}$ 215
Vyzulin Sergey, Gan'shina Elena, Garshin Vladimir, Perova Natalia, Syr'ev Nikolaj

INFLUENCE OF HYDROGENATION ON MAGNETORESONANCE CHARACTERISTICS OF NANOCOMPOSITE $(COFEB)_MC_{100-M}$ FILMS 219
Vyzulin Sergei, Kevraletin Alexandr, Syr'ev Nikolaj

FERROMAGNETIC RESONANCE AND RANDOM MAGNETIC ANISOTROPY IN NANOCRYSTALLINE FE-ZR-N FILMS 222
Sheftel Elena, Harin Eugene, Tedzhetov Valentin, Koksharov Yuri

MAGNETIC PROPERTIES OF POLY(P-XYLYLENE)-NI NANOCOMPOSITES 226
Vdovichenko Artem, Streltsov Dmitry, Orekhov Anton, Vasiliev Alexander, Yan Zubavichus, Grigoriev Evgenii, Zavyalov Sergei, Oveshnikov Leonid, Sergei Chvalun

INTERACTION OF MAGNETIC ELEMENTS IN METALLIC FERROMAGNETIC NANOFILMS 230
P Panaetov V.

MAGNETIC PROPERTIES OF THE NATURAL AND ISOTOPE-MODIFIED DIAMOND AND SILICON CARBIDE 232
Taldenkov A.N., Inyushkin A.V., Chistotina E.A., Ralchenko V.G., Bolshakov A.P., Mokhov E.N.

EFFECTS OF SPIN POLARIZATION ON RESONANT PHOTOEMISSION FROM D-F STATES IN $TBNI_2MN_X$ COMPOUNDS 235
Grebennikov Vladimir I., Kuznetsova Tatyana V., Gerasimov Evgenii.G., Mushnikov Nikolai.V.

MAGNETIC STRUCTURES OF $(GD_{1-X}Y_X)MN_2GE_2$ AND $(TB_{1-X}Y_X)MN_2GE_2$ STUDIED BY NEUTRON POWDER DIFFRACTION 239
Granovsky Sergey, Gaidukova Irina, Ritter Clemens

STRUCTURAL AND MAGNETIC PROPERTIES OF COPPER SUBSTITUTED MG-FERRITES 243
Kiseleva Tatiana, Kabanov Vladislav, Ilyushin Alexander, Markov Gennadiy, Sanga Deleg, Hirazawa H.

THE INVESTIGATION OF CRYSTAL AND MAGNETIC STRUCTURES OF SOLID SOLUTIONS $BAFE_{12-X}D_XO_{19}$ (D= IN AND GA\; X= 0.1 – 1.2) 248
Turchenko Vitalii, Trukhanov Alex V., Trukhanov Sergei V., Kazakevich I.S.

MAGNETISM OF PURIFIED AMORPHOUS CARBON 252
Komlev Anton, Lähderanta Erkki, Shevchenko Evgeniy, Vorob'ev-Desyatovskii Nikolay

AB INITIO STUDY OF MAGNETIC AND STRUCTURAL PROPERTIES OF FE-GA ALLOYS 256
Matyunina Mariya, Zagrebin Mikhail, Sokolovskiy Vladimir, Buchelnikov Vasiliy

OPTICAL AND MAGNETIC PROPERTIES OF AL/NIFE AND AL/GE/NIFE NANOSIZED FILMS 260
Hashim H., Singh S.P., Panina L.V., Pudonin F.A., Sherstnev I.A., Podgornaya S.V., Shpetnyy I.

GIANT FORCED MAGNETOSTRICTION IN $TB_{0.2}GD_{0.8}$ SINGLE CRYSTAL 264
Nikitin Sergey, Ivanova Tatiana, Zvonov Alexey, Rogacki Krzysztof, Koshkid\'ko Yurii, Cwik Jacek

EFFECT OF SPIN-ORBITAL INTERACTION ON CONTINUOUS AND JUMPWISE REVERSAL MAGNETIZATION IN $[MN(II)(HL)(H_2O)][MN(III)(CN)_6]$ $2H_2O$ MOLECULAR FERRIMAGNET 267
Kirman Marina, Morgunov Roman

MONTE CARLO STUDY OF MAGNETIC STRUCTURES IN RARE-EARTH AMORPHOUS ALLOYS 271
Bondarev Alexey, Bataronov Igor

MAGNETIC PROPERTIES OF TM_2FE_{16} UNDER PRESSURE 275
Kuchin Anatoly, Arnold Zdenek, Kamarád Jiri, Platonov Sergey

MAGNETIC INVESTIGATION OF LOW TEMPERATURE PHASE TRANSITION IN IRON SELENIDES 279
Tigran Baluyan, Alla Novakova, Marat Khairullin

EFFECT OF AGING ON THE MAGNETIC PROPERTIES OF RAPIDLY QUENCHED ND-RICH ND-FE ALLOYS 282
Wiafe-Akenteng Dennis, Menushenkov Vladimir, Shchetinin Igor, Savchenko Alexander

MAGNETIC PROPERTIES OF THE NON-STOICHIOMETRIC $TBCO_2NI_X$ ALLOYS 286
Inishev Alexander, Gerasimov Evgeny, Mushnikov Nikolay, Terentev Pavel, Gaviko Vasily

STRUCTURAL AND MAGNETIC ANISOTROPY OF DIRECTIONALLY-CRYSTALLIZED FERROMAGNETIC MICROWIRES 290
Evstigneeva Svetlana, Morchenko Alexander, Trukhanov Alexey, Panina Larissa, Larin Vladimir, Volodina Natalia, Yudanov Nikolay, Nematov Makhsudsho, Hashim Hisham, Ahmad Haseeb

COMPETING ANISOTROPY IN THE $(TM_XPR_{1-X})_2FE_{17}$ SYSTEM 294
Platonov Sergey, Kuchin Anatoly, Volegov Alexey, Korolev Alexander, Neznakhin Dmitriy, Voronin Vladimir, Proskurnina Natal'ya, Kolodkin Denis

STATIC AND HIGH-FREQUENCY MAGNETIC PROPERTIES OF NANOCRYSTALLINE FE-ZR-N FILMS 298
Sheftel Elena, Harin Eugene, Tedzhetov Valentin, Rozanov Konstantin, Zezyulina Polina, Bobrovskii Stanislav, Kiryukhantsev-Korneev Philipp

ELECTOMAGNETIC ABSORPTION OF COMPOSITES BASED ON EPOXY RESIN AND METALLIC IRON NANOPARTICLES 302
Krekhno Roman V., Safronov Alexander P., Beketova Anna I., Beketov Igor V.

EFFECT OF ALLOYING ELEMENTS (ZR, HF, CO), HEAT AND MECHANICAL TREATMENT CONDITIONS ON THE PHASE COMPOSITION AND MAGNETIC PROPERTIES OF $SMFE_{11}TI$ COMPOUNDS WITH $THMN_{12}$ STRUCTURE 306
Urzhumtsev Andrey, Anikin Maksim, Tarasov Evgeniy, Semkin Mikhail, Cherepkov Maksim, Kudrevatykh Nikolay, Zinin Aleksandr, Moskalev Vladimir

STRESS-INDUCED MAGNETIC DOMAIN STRUCTURE IN $DYFE_{11}TI$ COMPOUND 310
Semenova Elena, Lyakhova Marina, Karpenkov Dmitriy, Kuznetsova Yulia, Karpenkov Alexey, Skokov Konstantin

SPONTANEOUS MAGNETIZATION REVERSAL CAUSED BY MAGNETIC NOISE IN ε-$IN_{0.24}FE_{1.76}O_3$ NANOPARTICLES 314
Dmitriev Alexei

STUDY OF REVERSIBILITY OF COERCIVITY UNDER HEAT TREATMENT IN PERMANENT MAGNETS 317
Lileev Alexey, Pinkas Victoria, Voronchihina Kristina, Gunbin Anton

STRESS EFFECTS ON MAGNETIC PROPERTIES OF AMORPHOUS MICROWIRES SUBJECTED TO CURRENT ANNEALING 320
Nematov Makhsudsho G., Yudanov Nicolay A., Yakubtsev Rodion A., Panina Larissa V., Beklemisheva Anna V., Morchenko Alexander T., Atul Thakur

INFLUENCE OF ADDITIONAL COMPOUNDS ON COERCIVITY OF SINTERED ND-FE-B MAGNETS 324
Pham Thanh, Nguyen Duong, Nguyen Yen, Nguyen Lam, Dinh Quang, Nguyen Ngoc, Victor Koledov, Alexander Kamantsev, Alexey Mashirov, Nguyen Dan

INVESTIGATION OF MAGNETIC PHASE TRANSITION AND MAGNETOCALORIC EFFECT OF (NI,CO)-MN-AL MELT-SPUN RIBBONS 328
Nguyen Yen, Nguyen Mai, Vu Quang, Pham Thanh, Koledov Victor, Kamantsev Alexander, Mashirov Alexey, Tran Thanh, Kieu Hau, Seong Yu, Nguyen Dan

PHASE FORMATION AND MAGNETOCALORIC EFFECT IN (PR,ND)-FE ALLOYS PREPARED BY RAPIDLY QUENCHED METHOD 332
Nguyen Dan, Nguyen Ha, Nguyen An, Nguyen Yen, Pham Thanh, Koledov Victor, Kamantsev Alexander, Mashirov Alexey, Tran Thanh, Kieu Hau, Yu Seong

DOUBLE MAGNETOCALORIC PEAK FEATURE OBSERVED IN QUATERNARY NI-MN-IN BASED HEUSLER ALLOYS 336
Rodionov I. D., Dubenko I. S., Prudnikov V. N., Granovsky A. B., Titov I. S., Saletsky A. M.

VISUALIZATION OF MAGNETOSTRUCTURAL TRANSITION IN HEUSLER ALLOYS BY MAGNETIC FORCE MICROSCOPY 339
Geydt Pavel, Rodionov Igor D., Granovsky Alexander B., Soboleva Ekaterina, Fadeev Egor, Titov Ivan S., Dubenko Igor, Lähderanta Erkki

INVESTIGATION OF STRUCTURAL AND MAGNETIC PROPERTIES OF FE-RH-(Z) (Z = CO, PT) ALLOYS BY FIRST PRINCIPLES METHOD 343
Pavlukhina Oksana, Sokolovskiy Vladimir, Zagrebin Mikhail, Buchelnikov Vasily

THE TRANSFORMATION OF THE MAGNETOSTRUCTURAL PHASE TRANSITION WITH TI ADDITION IN $GD_5SI_2GE_2$ 347
Nikitin Sergey, Smirnov Andrey, Bogdanov Anatoly, Ovchenkova Ioulia

STRUCTURAL AND MAGNETIC PROPERTIES OF HEUSLER ALLOYS PD_2MNZ (Z=GA, GE, AS): AB INITIO STUDY 351
Miroshkina Olga, Zagrebin Mikhail, Sokolovskiy Vladimir, Buchelnikov Vasiliy

PECULIARITIES OF THE MAGNETOCALORIC EFFECT IN FERH-BASED ALLOYS IN THE VICINITY OF THE FIRST ORDER MAGNETIC PHASE TRANSITION 355
Gimaev Radel, Zverev Vladimir, Spichkin Yury, Tishin Alexander, Miyanaga Takafumi

MAGNETOCALORIC PROPERTIES OF $GD(CO_{1-X}FE_X)_2$ COMPOUNDS, WITH $X \leq 0.60$ 359
Anikin Maksim, Tarasov Evgeniy, Kudrevatykh Nikolay, Zinin Aleksander

GIANT MAGNETOCALORIC EFFECT IN COMPOSITES BASED ON POLYMERIC MATRIX AND MANGANESE ARSENIDE 363
Pankratov Nikolay, Mitsiuk Victor, Vetcher Alexander, Krokhotin Andrei, Govor Gennadiy, Nikitin Sergey

MAGNETIC PHASE DIAGRAMS OF GD-H, TB-H, DY-H SYSTEMS 367
Chzhan Victoria, Tereshina Irina, Burkhanov Gennady, Politova Galina, Drulis Henryk

TERNARY PHASE DIAGRAM OF NI-MN-GA: INSIGHTS FROM AB INITIO CALCULATIONS 371
Sokolovskaya Yulia, Zagrebin Mikhail, Buchelnikov Vasiliy, Zayak Alexey

SHUBNIKOV-DE HAAS OSCILLATIONS IN DILUTED MAGNETIC SEMICONDUCTORS $(CD_{1-X-Y}ZN_XMN_Y)_3AS_2$ 375
Zakhvalinskii Vasilii S., Nikulicheva Tatyana B., Lähderanta Erkki, Kochura Aleksey V., Nikitovskaia Ekaterina A., Pilyk Evgeny A., Pogrebneak Mikhail A.

EFFECT OF AGEING ON MAGNETIC AND ELECTRICAL PROPERTIES OF TI–SUBSTITUTED LA-SR MANGANITES 377
Badelin A., Datskaya Z., Estemirova S., Karpasyuk V., Merkulov D.

MAGNETOTRANSPORT PROPERTIES OF INSB-MNSB NANOSTRUCTURED FILMS 381
Kochura Aleksey, Marenkin Sergey, Novodvorsky Oleg, Mikhalevsky Vladimir, Davidov Alexander, Shakhov Mikhail, Lahderanta Erkki, Zakhvalinskii Vasilii, Kuzmenko Alexander

EFFECT OF OXYGEN NONSTOICHIOMETRY ON MAGNETIC PHASE TRANSITIONS IN FRUSTRATED COBALTITES $YBACO_4O_{7+X}$ (X = 0, 0.1, 0.2) 385
Kazei Zoya, Snegirev Vyacheslav, Kozeeva Ludmila, Kameneva Margarita, Lavrov Alexander

VIBRONIC INTERACTION AS MAIN REASON OF MAGNETIC ORDERING IN INSULATING MANGANITES $R_{1-X}A_XMNO_3$ 389
Gonchar Liudmila, Nikiforov Anatoliy

STRAIN EFFECT ON RESISTIVITY OF $LA_{0.7}BA_{0.3}MNO_3$ THIN FILM 393
Shaikhulov Timur A., Shakhunov Valery A., Demidov Victor V., Ovsyannikov Gennady A., Andreev Nikolay V., Pestun Anna E., Preobrazhensky Vladimir L.

QUANTUM EFFECTS IN MAGNETOTRANSPORT OF INGAAS QUANTUM WELLS WITH REMOTE MN IMPURITIES 397
Oveshnikov Leonid, Morgun Leonid, Nekhaeva Elena, Kulbachinskii Vladimir, Aronzon Boris

AB-INITIO STUDY OF ELECTRONIC AND MAGNETO-OPTICAL PROPERTIES OF INAS:MN 401
Gan'shina Elena, Kulatov Erkin, Golik Leonard, Kun'kova Zoya, Uspenskii Yurii, Zykov Georgy, Yuan Ye, Zhou Shengqiang

SUPERPARAMAGNETIC BEHAVIOR OF MOCVD GROWN ZNO:CO FILMS 405
Kytin Vladimir G., Maximova Olga V., Kulbachinskii Vladimir A., Burova Lidia I., Kaul Andrey R., Bandyopadhya Sudipta, Ahmed Azaharuddin, Banerjee Aritra

MAGNETIC PROPERTIES OF $PB_{1-X-Y}SN_XV_YTE$ ALLOYS 409
Skipetrov Evgeny, Konstantinov Nikita, Skipetrova Liudmila, Knotko Alexander, Slynko Vasily

COLOSSAL MAGNETORESISTANCE AND ANOMALOUS HALL EFFECT IN NONMAGNETIC SEMICONDUCTORS 413
Obukhov Sergei, Panysheva Tatiana

ELECTRONIC AND MAGNETO-OPTICAL PROPERTIES OF ZNO:CO 417
Kulatov Erkin, Novodvorskii Oleg, Lotin Andrey, Rylkov Vladimir, Gan'shina Elena, Shevyakov Dmitrii, Tugushev Victor, Uspenskii Yurii

INFLUENCE OF OXYGEN VACANCIES ON MAGNETIC AND TRANSPORT PROPERTIES OF THIN $ZN_{1-X}CO_XO_Y$ FILMS 421
Lotin Andrey A., Kuz'mina Alina S., Novodvorsky Oleg A., Parshina Liubov S., Khramova Olga D., Mikhalevsky Vladimir A., Cherebilo Elena A., Shneider Alexander G., Perov Nikolay S., Makarova Liudmila A.

CONNECTION OF THERMOPOWER, MAGNETOTHERMOPOWER WITH RESISTIVITY AND MAGNETORESISTANCE IN MANGANITES WITH ND AND SM 424
Koroleva Luidmila, Batashev Ivan, Morozov Artem, Balbashov Anatolii, Szymczak Henryk, Slavska-Wanniewska Anna

DILUTED MAGNETIC SEMICONDUCTORS INFESB PREPARED BY LASER ABLATION: SPECTROSCOPIC AND MICROSCOPIC INVESTIGATIONS 426
Gan'shina Elena, Golik Leonard, Kun'kova Zoya, Zykov Georgy, Rukovishnikov Alexander, Temiryazeva Marina, Lesnikov Vladimir, Markin Yuri

ELECTROINDUCED MAGNETIC BUBBLE DOMAIN NUCLEATION 430
Kulikova Daria, Gareev Timur, Nikolaeva Elena, Pyatakov Alexander

OPTICAL SPECTROSCOPY OF $ERFE_3(BO_3)_4$: DETECTION OF PHASE TRANSITIONS AND CRYSTAL-FIELD LEVELS OF THE ER^{3+} GROUND MULTIPLET 433
Erofeev Dmitry, Jablunovskis Artjoms, Chukalina Elena

POLARIZABILITY OF ELECTRICALLY INDUCED MAGNETIC VORTEX "ATOMS" 437
Karpov P.I., Mukhin S.I.

TEMPERATURE CHARACTERISTICS OF THE NONLINEAR MAGNETOELECTRIC EFFECT 441
Dmitry, Burdin, Nikolay Ekonomov, Dmitry Chashin, Yury Fetisov

MAGNETOELECTRIC EFFECT IN FERROMAGNETIC-SEMICONDUCTOR LAYERED COMPOSITE STRUCTURES 445
Fetisov Leonid, Chashin Dmitri, Saveliev Dmitri, Plekhanova Daria, Makarova Ludmila, Stognii Alexandr

SUPPRESSION OF THE MAGNETOELECTRIC HYSTERESIS IN A COMPOSITE LANGATATE-METGLAS STRUCTURE 449
Burdin D.A., Chashin D.V., Ekonomov N.A., Fetisov L.Y., Fetisov Y.K.

INFLUENCE OF MAGNETIC ORDER PARAMETER SYMMETRY AND SPIN FLUCTUATIONS ON FORMING OF FERROELECTRIC ORDER IN MN_2O_5 OXIDES 452
Men'shenin Vladimir

COMPOSITE MULTIFERROIC MATERIALS CONSISTING OF NDFEB AND PZT PARTICLES EMBEDDED IN ELASTIC MATRIX: THE APPEARANCE OF ELECTRICAL POLARIZATION IN A CONSTANT MAGNETIC FIELD 455
Makarova Liudmila, Alekhina Yuliya, Kramarenko Elena, Omelyanchik Alexander, Rodionova Valeria, Malyshkina Olga, Perov Nikolai

MFM STUDY OF THE DOMAIN STRUCTURE OF PERMALLOY MICROPARTICLES UNDER MECHANICAL STRESS 460
Biziyaev Dmitry, Bukharaev Anastas, Nurgazizov Niyaz

TEMPERATURE MÖSSBAUER STUDY OF THE SPATIAL SPIN-MODULATED STRUCTURE IN THE MULTIFERROIC $BIFEO_3$ 463
Rusakov Vyacheslav, Pokatilov Vyacheslav, Sigov Alexander, Matsnev Mikhail, Pyatakov Alexander

ISOLATION OF PROXIMITY-INDUCED TRIPLET PAIRING CHANNEL IN A SUPERCONDUCTOR/FERROMAGNET SPIN VALVE 467
Kamashev Andrey, Validov Aidar, Garif'yanov Nadir, Fominov Yakov, Leksin Pavel, Schumann Joachim, Thomas Jürgen, Kataev Vladislav, Büchner Bernd, Garifullin Ilgiz

SUPERCONDUCTING MESA-STRUCTURES WITH A SPIN FILTERING MANGANITE INTERLAYER 471
Constantinian Karen, Ovsyannikov Gennady, Kislinskii Yulii, Demidov Victor, Shadrin Anton, Petrzhik Andrey

INTERPLAY BETWEEN MAGNETISM AND SUPERCONDUCTIVITY IN METALLIC HYDROGEN AND HYDRIDES AT HIGH PRESSURE 475
Mazov Lev

INHOMOGENEITY AND IRREVERSIBILITY FIELD OF SUPERCONDUCTING NB-TI TAPES 479
Guryev Valentin, Shavkin Sergey, Kruglov Vitaly

NONLINEAR RESPONSE OF A SUPERCONDUCTOR OF THE BI-SR-CA-CU-O SYSTEM TO AN ALTERNATING MAGNETIC FIELD IN THE SUPERCONDUCTING TRANSITION TEMPERATURE RANGE 483
Sergeev Alexander, Golev Igor, Gvozdevskaya Victoria, Barkalova Anastasia

UNCONVENTIONAL PHASE TRANSITIONS IN STRONGLY ANISOTROPIC 2D (PSEUDO)SPIN SYSTEMS 486
Konev Vitaly, Vasinovich Evgeny, Ulitko Vasily, Panov Yury, Moskvin Alexander

FEATURES OF MICROSTRUCTURE AND MAGNETIC FLUX DYNAMICS IN SUPERCONDUCTING NB-TI TAPES WITH STRONG ANISOTROPIC PINNING 491
Shavkin Sergey, Guryev Valentin, Kruglov Vitaly, Ovcharov Alexey, Likhachev Igor, Vasiliev Alexandre, Veligzhanin Alexey, Zubavichus Yan

THERMOMAGNETIC CONVECTION IN A CYLINDRICAL ENCLOSURE. PART 1: ZERO GRAVITY 495
Krakov Mikhail S., Nikiforov Ivan V.

EFFECTS OF MAGNETIC FIELDS ON CAPILLARY-GRAVITY WAVES IN THE PRESENCE OF MAGNETIC SURFACTANTS 501
Zhukov Alexander V.

FREQUENCY DEPENDENCE OF MAGNETOTHERMAL PROPERTIES FOR MAGNETIC FLUID AND MAGNETICALLY FUNCTIONALIZED IMPLANTS 504
Salakhova R.T., Vylegzhanin A. G., Kashtanov E.A., Zverev V.I., Müller R., de Sena Pereira F.D., Parfenov V.A., Mironov V.A. V.A., Kritskaya E.A., Pyatakov A.P., Markov V.K., Malyshev A.Yu, Kamilov K.I, Tishin A.M

MICROSTRUCTURE FORMATION AND MACROSCOPIC DYNAMICS OF FERROFLUID EMULSION IN ROTATING MAGNETIC FIELD 508
Kolesnikova Anastasia, Zakinyan Arthur, Dikansky Yuri

MECHANICS OF MAGNETIC FLUID ACTIVE ELEMENT IN STRONG MAGNETIC FIELD 511
Polunin V.M., Ryapolov P.A., Sheldeshova E.V.

DEFORMATION OF A MAGNETIC LIQUID DROP IN AN APPLIED NON-STATIONARY UNIFORM MAGNETIC FIELD 516
Tyatyushkin Alexander

COMPLEX REFRACTIVE INDEX OF CONCENTRATED MAGNETIC FLUIDS 520
Yerin Constantine, Lykhmanova Victoria, Erina Marina

LEVITATION AND MOTION OF A MAGNET PARTIALLY IMMERSED INTO A MAGNETIC FLUID 524
Pelevina Daria, Kobzev Mikhail, Kalmykov Sergey, Merkulov Dmitrii, Naletova Vera

RESEARCH OF THE DEFORMATION OF A THIN BODY WITH A MAGNETIZABLE ELASTOMER IN THE MAGNETIC FIELD OF A COIL 528
Merkulov Dmitrii

MAGNETIC FLUID BRIDGE BETWEEN TWO CONES AND A CYLINDER IN THE MAGNETIC FIELD OF A CURRENT-CARRYING WIRE 531
Vinogradova Alexandra S., Turkov Vladimir A., Naletova Vera A.

ON THE TEMPERATURE DEPENDENCE OF FERROFLUID SUSCEPTIBILITY 535
Dikansky Yuri, Ispiryan Anna, Kunikin Stanislav

MAGNETIC PROPERTIES AND GIANT MAGNETOIMPEDANCE OF SURFACE MODIFIED CO-BASED AMORPHOUS RIBBONS WITH CARBON COVERING 539
Elizaveta V. Golubeva, Anna A. Chlenova, Elena A. Stepanova, Galina V. Kurlyandskaya

HETEROGENEITY OF POPULATION OF MICROORGANISMS GROWN IN PRESENCE OF IRON OXIDE MAGHEMITE NANOPARTICLES 543
Denisova Tatyana P., Simonova Elena V., Lubov A. Kokorina, Maximova Evgenia N., Samatov Oleg M., Safronov Alexander P., Kurlyandskaya Galina V.

MAGNETIC PROPERTIES AND GIANT MAGNETOIMPEDANCE EFFECT FOR COFEMOSIB SURFACE MODIFIED AMORPHOUS RIBBONS COVERED BY WATER BASED FERROFLUID 547
Lotfollahi Zahra, Amirabadizadeh Ahmad, Safronov Alexander P., Kurlyandskaya Galina V.

SELECTED FEATURES OF MORPHO-FUNCTIONAL REACTIONS OF EUKARYOTIC MICROORGANISMS GROWN IN THE PRESENCE OF MAGHEMITE IRON OXIDE NANOPARTICLES OBTAINED BY LASER TARGET EVAPORATION 551
Simonova Elena V., Denisova Tatyana P., Khandukhanov Roman T., Melnikov Grigory Yu., Medvedev Anatoly I., Beketov Igor V., Andrade Ricardo

MODELLING OF THIN FILM MAGNETOIMPEDANCE SENSITIVE ELEMENT DESIGNED FOR BIODETECTION 555
Volchkov S.O., Chlenova A.A., Lepalovskij V.N.

VOLUMETRIC REGISTRATION OF MAGNETIC NANOPARTICLES FOR OPTIMIZATION OF QUANTITATIVE IMMUNOCHROMATOGRAPHIC ASSAYS FOR DETECTION OF SMALL MOLECULES 559
Guteneva Natalia V., Znoyko Sergey L., Orlov Alexey V., Nikitin Maxim P., Nikitin Petr I.

METHODOLOGICAL ASPECTS OF SMALL IRON CONCENTRATIONS DETERMINATION IN BLACK YEASTS GROWN IN THE PRESENCE OF IRON OXIDE NANOPARTICLES 563
Melnikov Grigory Yu., Denisova Tatyana P., Safronov Alexander P., Samatov Oleg M., Khandukhanov Roman T., Kulesh Nikita A., Ricardo Andrade, Kurlyandskaya Galina V.

POSSIBILITY TO USE OF THE FE_3O_4/TA_2O_5 CORE-SHELL NANOPARTICLES IN RADIOTHERAPY 567
Lukyanenko Kseniya, Afremov Leonid, Apanasevich Vladimir, Shmykova Mariya, Medkov Mikhail, Lukyanov Pavel, Tarakova Olga, Kustov Vladimir
CRITICAL PROPERTIES OF 2D DISORDERED 3-STATE ANTIFERROMAGNETIC POTTS MODEL ON TRIANGULAR LATTICE 571
Murtazaev A.K., Babaev A.B., Ataeva G.Y.
FRUSTRATIONS AND PHASE TRANSITIONS IN MAGNETS OF VARIOUS DIMENSIONALITY 575
Kassan-Ogly Felix, Proshkin Alexey
LAWS OF VOLUME ELASTICITY IN PHYSICAL PROCESSES OF THE EFFECT OF T, P, H PARAMETERS ON THE LATTICE MODIFICATIONS, PHASE TRANSITIONS, CONDUCTIVITY AND MAGNETISM 579
Polyakov Petr
CURRENT-DRIVEN QUANTUM DYNAMICS OF TOROIDAL MOMENT IN RARE-EARTH NANOCLUSTERS 582
Plokhov D.I., Popov A.I., Zvezdin A.K.
NON-EQUILIBRIUM CRITICAL BEHAVIOR OF HEISENBERG THIN FILMS 586
Shlyakhtich Maria A., Prudnikov Pavel V.
UNCONVENTIONAL PHASE SEPARATION IN THE MODEL 2D SPIN-PSEUDOSPIN SYSTEM 590
Budrin K.S., Panov Yu.D., Moskvin A.S., Chikov A.A.
MAGNETIC EXCITATIONS OF CARBON NANOTUBES IN CHIRAL MODEL OF GRAPHENE 594
Rybakov Yu.P.
PHASE TRANSITIONS AND THERMODYNAMIC PROPERTIES OF THE ANTIFERROMAGNETIC POTTS MODEL ON A FACE-CENTERED CUBIC LATTICE 597
Babaev Albert, Murtazaev Akai, Kassan-Ogly Felix, Proshkin Alexey
NON-EQUILIBRIUM CRITICAL DYNAMICS OF LOW-DIMENSIONAL MAGNETICS AND MULTILAYER STRUCTURES 601
Prudnikov Pavel V., Prudnikov Vladimir V., Danilova Alena Yu., Borzilov Vadim O., Baksheev Georgy G.
THE MONTE CARLO SIMULATION OF 2D ANNNI-MODEL 605
Akai Kurbanovich Murtazaev, Zhavrail Gadzhievich Ibaev
SPIN-DENSITY CORRELATIONS AND SHORT-RANGE ORDER IN METALS ABOVE THE CURIE TEMPERATURE 609
Melnikov Nikolai B., Reser Boris I., Paradezhenko Georgii V.
QUASI-CLASSICAL LOCALIZED STATES IN THE 2D FERRIMAGNET 614
Panov Yury, Moskvin Alexander
Author Index

edp sciences

EPJ Web of Conferences

Statement of Peer review

In submitting conference proceedings to EPJ Web of Conferences, I certify to the Publisher that I adhere to the **Policy on Publishing Integrity** of the journal in order to safeguard good scientific practice in publishing.

1. All articles have been subjected to peer review administered by the proceedings editors.
2. Reviews have been conducted by expert referees, who have been requested to provide unbiased and constructive comments aimed, whenever possible, at improving the work.
3. Proceedings editors have taken all reasonable steps to ensure the quality of the materials they publish and their decision to accept or reject a paper for publication has been based only on the merits of the work and the relevance to the journal.

Title, date and place of the conference

Moscow International Symposium on Magnetism, 1-5 July, 2017, Moscow, Russia

Proceedings editor(s): Nikolai Perov, Anna Semisalova

Date and editor's signature 08.06.2018

MODELING OF MAGNETIC LABEL FIELD SENSING BY GMR STRUCTURE

Kirill Bukunov[1,2], *Georgy* Babaizev[1,2,*], and *Nikolay* Chechenin[1,2]

[1]Skobeltsyn Institute of Nuclear Physics, Lomonosov Moscow State University, Leninskie Gory, 1/2, 119991 Moscow, Russia
[2]Faculty of Physics, Lomonosov Moscow State University, Leninskie Gory,1/2, 119991 Moscow, Russia

Abstract. An analytical model is developed to calculate 3D spatial distribution of vector magnetic field ***H(r)*** around a reference magnetic label of an arbitrary shape. A reliable method of ***H(r)*** modelling is important for location and precise positioning of a labelled object using GMR-based magnetic field sensors.

1 Introduction

Giant Magnetoresistance (GMR) sensors are widely used in countless applications, where sensors of linear or rotation motion are ones of them [1]. The main idea is a sensing of the magnetic field created by a reference magnetic label, which is normally a kind of permanent magnet placed on the moving part to be localized or positioned. The configuration of the field of the label is important to know. The magnitude and orientation of the magnetic field vector ***H(r)*** determines the sensing ability of GMR sensor placed at a remote point ***r*** with respect to the reference label position. For precise positioning and displacement sensing, a reliable method of modelling of magnetic field spatial configuration, produced by the reference labels, is highly important. Here we report on our method of calculation of the spatial configuration of magnetic field produced by a reference magnetic label.

2 The model

Considering the problem of magnetostatics to determine the spatial configuration of stationary magnetic field around the magnetic label at absence of electric currents, we can write Maxwell's equations together with additional equations closing the system in SI system as:

$$[\nabla \times \boldsymbol{H}] = 0 \tag{1}$$

$$(\nabla \cdot \boldsymbol{B}) = 0 \tag{2}$$

$$\boldsymbol{B} = \mu_0 (\boldsymbol{H}+\boldsymbol{M}) \tag{3}$$

where ∇ – the Helmholtz operator, $\boldsymbol{B} = \boldsymbol{B}(\boldsymbol{r})$, $\boldsymbol{H} = \boldsymbol{H}(\boldsymbol{r})$ – vectors of magnetic induction and magnetic field, respectively, $\boldsymbol{M} = \boldsymbol{M}(\boldsymbol{r})$ – the magnetization vector, μ_0 – magnetic constant.

Eq. (1) gives right to introduce a scalar magnetic potential $\phi = \phi(\boldsymbol{r})$ which satisfy the condition:

$$\boldsymbol{H} = -\nabla\phi \tag{4}$$

Substituting (3) into (2) and using (4), we obtain the Poisson equation for the scalar potential:

$$\Delta\phi = (\nabla \cdot \mathbf{M}), \tag{5}$$

where Δ is the Laplace operator. We denote the spatial region of localization of the magnetic substance (volume) as Ω, and the surface bounding this volume as Σ. The problem arrives to find a sufficiently smooth function ϕ(r) in the outer region $\boldsymbol{r}\in\Omega$, satisfying Eq. (5) with the given distribution $\boldsymbol{M}(\boldsymbol{r})$ and such that $\phi(\boldsymbol{r}) \xrightarrow[|r|\to\infty]{} 0$. The solution of this problem exists, is unique and stable with respect to the input data. In general form the solution for the case of a uniform volume distribution of the magnetization is written via the Green's function of the Laplace operator:

$$\phi(\boldsymbol{r}) = -\frac{1}{4\pi}\int \frac{(\nabla' \cdot \boldsymbol{M}(\boldsymbol{r}'))}{|\boldsymbol{r}-\boldsymbol{r}'|} d\boldsymbol{r}' \tag{6}$$

Applying the Ostrogradskii-Gauss theorem and taking into account the fact that the sources of the magnetic field are localized in the volume of the label and are absent at infinity, Eq. (6) transforms into:

$$\phi(\boldsymbol{r}) = \frac{1}{4\pi}\iiint_\Omega \left(\boldsymbol{M}(\boldsymbol{r}') \cdot \nabla' \frac{1}{|\boldsymbol{r}-\boldsymbol{r}'|}\right) dV' \tag{7}$$

To find the explicit form of the scalar potential $\phi(\boldsymbol{r})$ it is necessary to determine the choice of the coordinate system, and hence the choice of the shape of the magnetic label, since it will be associated with the symmetry of a certain coordinate system. Due to the fact that the mark should give sufficient amplitude for the magnetic field to detect the sensor, we choose a flat label shape, with a microsize in one spatial dimension and macro-dimensions for the other two. In order to simplify the model and not to take into account the evolution of the initial state of magnetization in time, we choose a sufficiently elongated shape of the label along one of the directions, thereby reducing the edge effects on the magnetization direction. We arrange the label, which is a rectangular parallelepiped with the sides Δx, Δy, Δz along the corresponding axes to the centre of the Cartesian coordinate system and with the ratio of the linear dimensions of the label $t{:}w{:}l = 1{:}\ 10^2{:}\ 10^3$. We

* Corresponding author: kvyvg-george93@mail.ru

consider the simplest case of the distribution $\boldsymbol{M}(\boldsymbol{r})$, which has an analytic solution - the case of a homogeneous magnetization over the entire volume of the label:

$$\boldsymbol{M} = \begin{cases} M_0 \boldsymbol{e}_z, & \boldsymbol{r} \in \Omega, \\ 0, & \boldsymbol{r} \notin \Omega \end{cases} \tag{8}$$

Thus, we have

$$\phi(\boldsymbol{r}) = \frac{M_0}{4\pi} \iiint_\Omega (\boldsymbol{e}_z \cdot \boldsymbol{\nabla}' \frac{1}{|r-r'|}) dV' = \frac{M_0}{4\pi} \iiint_\Omega \boldsymbol{div}(\frac{\boldsymbol{e}_z}{|r-r'|}) dV' = \frac{M_0}{4\pi} \oiint_\Sigma (\frac{\boldsymbol{e}_z \boldsymbol{n}'}{|r-r'|}) d\sigma' \tag{9}$$

Here we have used the Ostrogradskii-Gauss theorem; $\boldsymbol{n}'$ is the field of surface normals. The resulting expression must be integrated over the surface Σ of a rectangular parallelepiped. Given that

$$(\boldsymbol{e}_z \boldsymbol{n}') = \begin{cases} 1, & z = \frac{l}{2} \\ -1, & z = -\frac{l}{2} \end{cases} \quad \begin{cases} -\frac{t}{2} \le x \le \frac{t}{2} \\ -\frac{w}{2} \le y \le \frac{w}{2} \end{cases} \tag{10}$$

we obtain

$$\phi(\boldsymbol{r}) = \phi(x,y,z) = \frac{M_0}{4\pi} \iint \frac{dx'dy'}{\sqrt{(x-x')^2+(y-y')^2+(z-z')^2}} \Bigg|_{z'=-\frac{l}{2}}^{z'=\frac{l}{2}} =$$

$$\frac{M_0}{4\pi} \int_{-\frac{t}{2}}^{\frac{t}{2}} dx' \int_{-\frac{w}{2}}^{\frac{w}{2}} dy' \left(\frac{1}{\sqrt{(x-x')^2+(y-y')^2+\left(z-\frac{l}{2}\right)^2}} - \frac{1}{\sqrt{(x-x')^2+(y-y')^2+\left(z+\frac{l}{2}\right)^2}} \right) \tag{11}$$

We introduce the function

$$F(x,y,z) \overset{def}{=} \iint \frac{dxdy}{\sqrt{x^2+y^2+z^2}} =$$

$$-zarctg\left(\frac{xy}{z\sqrt{x^2+y^2+z^2}}\right) + xln\left(y + \sqrt{x^2+y^2+z^2}\right) +$$

$$yln\left(x + \sqrt{x^2+y^2+z^2}\right) - x - y + zarctg\left(\frac{x}{z}\right) +$$

$$zarctg\left(\frac{y}{z}\right) + const \tag{12}$$

It expresses the value of the indefinite integral (11) in terms of elementary functions. For this integral a substitution is performed in symmetric limits and the last five terms in (12) can be omitted, since they are odd functions for at least one argument and the constant term does not play a role. As a result, we obtain an explicit expression for the scalar potential of the magnetic field of the label, written in Cartesian coordinates through the function (12) with the omitted five last members:

$$\phi(x,y,z) = \frac{M_0}{4\pi} \left\{ \begin{matrix} F(x-\frac{t}{2}, y-\frac{w}{2}, z) - F(x-\frac{t}{2}, y+\frac{w}{2}, z) \\ -F(x+\frac{t}{2}, y-\frac{w}{2}, z) + F(x+\frac{t}{2}, y+\frac{w}{2}, z) \end{matrix} \right\} \Bigg|_{z'=-\frac{l}{2}}^{z'=\frac{l}{2}} \tag{13}$$

To find the required magnetic field vector $\boldsymbol{H} = \{H_x, H_y, H_z\}$ we need to take the gradient from the scalar potential, as it is written in (1)

$$\begin{cases} H_x(x,y,z) = -\frac{\partial\phi(x,y,z)}{\partial x} \\ H_y(x,y,z) = -\frac{\partial\phi(x,y,z)}{\partial y} \\ H_z(x,y,z) = -\frac{\partial\phi(x,y,z)}{\partial z} \end{cases} \tag{14}$$

Expressions for each of the components have a complicated and cumbersome analytical form, so their graphic interpretation is presented below.

3 Spatial configuration of magnetic field of particular reference label shapes

Figure 1 shows a three-dimensional picture of the distribution of the magnetic field vector $\boldsymbol{H}$ in the form of discrete arrows regularly located in 3D space. The colour gradation reflects the modulus of the $\boldsymbol{H}$ vector at each particular point in space and the length of the arrows is

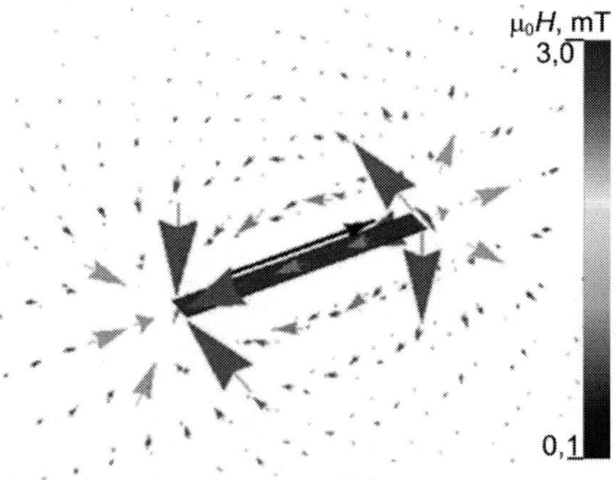

Fig. 1. 3D picture of $\boldsymbol{H}$ field configuration around the reference label.

proportional to the projection of $\boldsymbol{H}$ on the plane of drawing. Here and in the following figures the magnetic strip is shown in red. The black arrow above indicates the direction of magnetization (the positive direction of the z axis). It is worth to note that the change in absolute label sizes does not affect the field configuration, but causes only a proportional change in the magnetic field strength.

In Figs. 2,3 distribution of components of magnetic field vector are over the y-z-plane with fixed x coordinate (x_0=1 mm). The third dimension of the plot is the projection of the magnetic field strength on a particular axis, x (Fig. 2a) or y (Fig. 3a). The spatial coordinates are normalized on the length of the label (2 cm).

The choice of the fixed coordinate x perpendicular to the main surface of the label was made because the main relative label-sensor displacements are assumed in y-z plane. It is obvious that the closer to zero the x coordinate value, i.e. the closer the sensor set to the label, the stronger is the field.

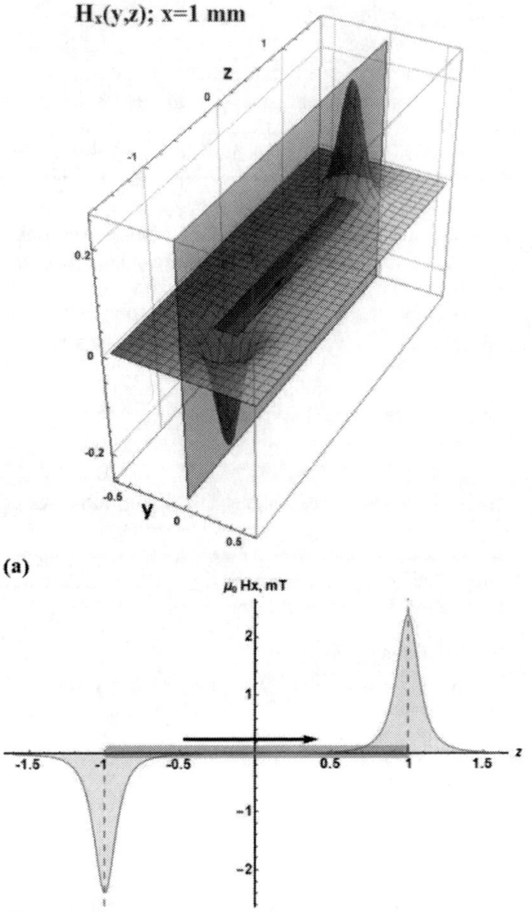

Fig.2. Distribution of x-component of the magnetic field $\mu_0 H_x(x_0,y,z)$ at x_0 = 1 mm (a) and cross-section of this distribution at $y = 0$ (b).

The cross-sections of the field spatial configuration over the strongest field variations are shown in Figs. 2b and 3b,c. The cross-section planes are highlighted in the same colour in Figs. 2a and 3a as the plots in Figs. 2b and 3b,c. The numerical values of the components of vector $\boldsymbol{H}$ in the plots are given as multiplied by the magnetic constant and correspond to the components of magnetic induction in tesla, which coincide with magnetic field strength, Eq. (3), in the region outside the label where magnetisation is zero. The electromagnetic potential in Eq. (6 to 13) and, hence, the magnetic field strength depends on the uniform magnetization M_0 which was obtained as was obtained by:

$$M_0 = \alpha n \mu_B, \tag{15}$$

where n is the concentration of ferromagnetic atoms in the volume, μ_B is the Bohr magneton, and α is a dimensionless empirical constant, indicating the number of Bohr magnetons per 1 atom of the magnetic substance. In our estimations we take $\alpha \cong 1$. In this case estimated field strength fits the experimental values for cobalt label. The magnetic field from a cobalt strip of 20 μm x 2 mm x 2 cm (thickness x width x length) at a

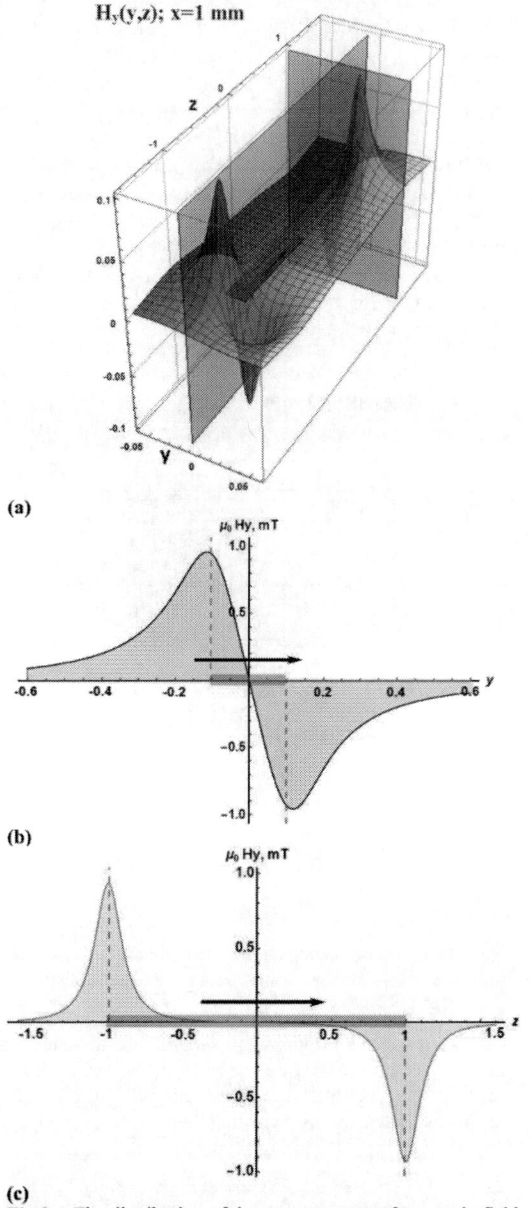

Fig.3. The distribution of the y-component of magnetic field $\mu_0 H_y(x_0,y,z)$ at x_0 = 1 mm (a) and cross-section $\mu_0 H_y(x_0,y)$ of this distribution at x_0 = 1mm, $z=l/2$ (b) and $\mu_0 H_y(x_0,z)$ at x_0 = 1mm, $y = -w/2$, (c).

distance of 1 mm in the region of the poles gives a field

larger by order of magnitude than that of the Earth, which is about 50 μT.

One can note a common feature of the cross-sections in Figs. 2b, 3b,c: narrow peaks, in comparison with the size of the label, localized in the regions of the magnet poles. The collection of information about the magnetic field occurs only along the selected direction in space, therefore only the projections of the field strength for this selected direction are read. Of the possible combination of the field components $\{H_x(i), H_y(i), H_z(i)\}$, where $i=x,y,z$, the choice was made in favor of $H_z(z)$, Fig. 4, since it was in this graph that several useful, in practical terms, properties are combined:

the presence of a "double drop" in the function graph in each constant-sign region of variation of its argument, this facilitates unambiguous determination of the position of the magnetic label relative to the sensor;

the field amplitude not diminishing to zero in the region between the poles: it also gives an advantage in the uniqueness of position determination;

a large gradient of the function in the direction of displacement (in areas close to the edges of the label): this improves the space resolution of the label location.

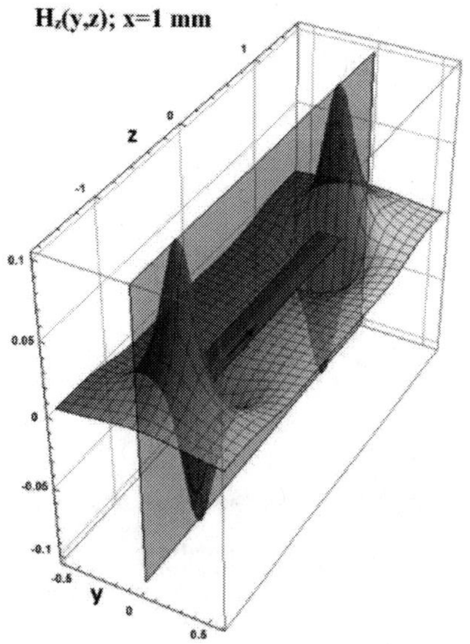

Fig. 4. The configuration of z-component of the field, $\mu_0 H_z(x_0,y,z)$ at distant $x_0 = 1$ mm from the magnetic label.

Based on the foregoing, it is possible to determine the relative arrangement of sensors in the sensor block, formulate the conditions for the minimum and maximum distances between the sensors, their number and give estimates of the positioning accuracy.

To follow the $H_z(z)$ component, as a possible solution, the sensor block can be a series of sensors in a row along the forward direction, which coincides with the magnetization direction in the label, that is, with the direction of its long side. The sensor unit should be positioned as close to the center of the label as possible to capture the maximum field amplitude. Obviously, one sensor in the sensor block will not be enough to uniquely identify the position in space over the field amplitude. The arrangement of the sensors in the row can be of regular or irregular spaced to optimize the accuracy of label displacement.

As an example, a detailed analysis which includes the mapping of the field derivative over the displacement direction, $\partial H_z/\partial z$, shows that for regular equidistant spacing along the label of l=20 mm, 3 sensors is enough to obtain the label positioning accuracy of 2 μm.

The magnetic field space distribution is used to estimate the effect of the field on the magnetoresistance of the spin-valve (SV) structure displaced by a certain distance x_0 from the magnetic stripe. The efficiency of the magnetoresistive sensing effect of the GMR SV structure strongly depends not only from the strength of magnetic field exerted, but also on magnetic properties ferromagnetic and antiferromagnetic layers, composing the SV. In a simple approximation, the component of the magnetic field from the reference magnetic label oriented at 90° to the uniaxial anisotropy direction in the of the free layer in the SV structure rotates the magnetic moment by the angle [2]

$$\varphi = \arccos(M_F H_F / 2K_u), \qquad (16)$$

where M_F and H_F are magnetic moment and magnetic field, acting in the free layer of SV structure, and K_u is the uniaxial anisotropy energy of the free ferromagnetic layer. The magnetoresistance of the SV in this case is determined from approximate relation

$$R_{GMR} = R_{GMR}^0(1-\cos(\varphi))/2 \qquad (17)$$

Using the described technique we can estimate the signal response to variable field strength and, hence, sensitivity and space resolution of the device based on the array of GMR sensors to detect position and movement of the object.

4 Conclusion

An analytical approach is developed to calculate the 3D configuration of the magnetic field near the magnetic label. It allows modeling the response of the magnetic sensor or arrays of the sensors and optimization the sensor-to-sensor arrangement within an array of GMR sensors in order to increase the sensitivity and spatial resolution of the displacement transducer.

References

1. Reig C., Cardoso C., Mukhopadhyay S., *Giant Magnetoresistance (GMR) Sensors. From Basis to State-of-the-Art Applications.* (Springer-Verlag, Berlin Heideberg, 2013) 131p.
2. E. C. Stoner, E.P. Wohlfarth. Phil. Trans. Roy. Soc. (London) **A240,** 599 (1948)

Size effect in the electronic transport of thin films of Bi_2Se_3

V.V. Marchenkov[1,2], *V.V.* Chistyakov[1], *J.C.A.* Huang[3], *Y.A.* Perevozchikova[1], *A.N.* Domozhirova[1,2], and *M.* Eisterer[4]

[1]M.N. Mikheev Institute of Metal Physics, UB RAS, 620137 Ekaterinburg, Russia
[2]Ural Federal University, 620002 Ekaterinburg, Russia
[3]National Cheng Kung University, Tainan, Taiwan
[4]TU Wien Atominstitut, 1020 Vienna, Austria

Abstract. Thin films of a topological insulator (TI) Bi_2Se_3 of various thicknesses from 20 nm to 75 nm were obtained. The resistivity measurements were carried out according to the conventional 4-contact DC technique. This allows to "separate" the bulk and surface conductivities at different temperatures and magnetic fields. It was suggested that similar effects should be observed in other TIs and systems with inhomogeneous distribution of dc-current on sample cross section.

1 Introduction

The new functional materials with the unique physical properties are needed for spintronic devices. One of such promising materials are the topological insulators (TIs) [1], which have a nontrivial topological band structure, arising from strong spin-orbital interaction [2]. The TI has an energy gap and, hence, is the insulator or semiconductor, in a bulk and has the protected gapless conduction states on its surface. A rigid connection between the directions of the momentum and the electron spin leads to the emergence of spin polarization of charge carriers and the possibility of a spin-polarized current flowing near the TI surface with practically no loss [3]. This spin-polarized surface current can be used for spintronic devices.

It is known that Bi_2Se_3 compound is the TI [4, 5] with a metallic conductivity near surface and the gapless semiconductive one [6] in its bulk. Since the electroconductivity value in a bulk and near surface of such materials can differ substantially, it is of interest to "divide" them experimentally. For this, we can use the results of Ref. [7], where we studied the size effect in the conductivity of pure tungsten single crystals under conditions of the static skin effect (SSE) [8], i.e. a predominant flow of direct electric current near a sample surface. The aim of this paper is to search for and study the size effect in the electronic transport of thin films of TI Bi_2Se_3.

2 Samples and measurements

Thin films of Bi_2Se_3 were grown by the molecular beam epitaxy method on Al_2O_3 substrates [4, 5] with thickness from 20 to 75 nm. The XRD data and the atomic content of elements analysis showed the synthesized films have Bi_2Se_3 composition (see Refs. [4, 5]). The atomic content of elements was measured by a scanning electron microscope equipped with an EDAX X-ray microanalysis attachment. Our examination showed that the deviations from a stoichiometric composition were insignificant in all samples. The measurements of the electroresistivity ρ_0 and magnetoresistivity ρ_{xx} were carried out by the conventional 4-points method at dc-current in the temperature range from 4.2 to 80 K and in magnetic fields of up to 10 T. The results are presented in units of conductivities $\sigma_0 \approx 1/\rho_0$ and $\sigma_{xx} \approx 1/\rho_{xx}$.

3 Results and discussion

Schematic view of the thin film of TI is shown in Fig.1. The electric current I is passed through the film of thickness d and width c (Fig. 1), and the voltage U is measured between the potential leads located at a distance L. In the near-surface layer of thickness δ, a region with high surface conductivity σ^{sur} appears.

Total conductivity of such a system contains two terms, namely, the surface conductivity σ^{sur} and the bulk conductivity σ^{bulk}. It is quite similar to Refs. [7, 8], where the dc-current is concentrated near a sample surface due to the static skin effect (SSE) [8]. It is quite easy to show that

* Corresponding author: wchist@imp.uran.ru

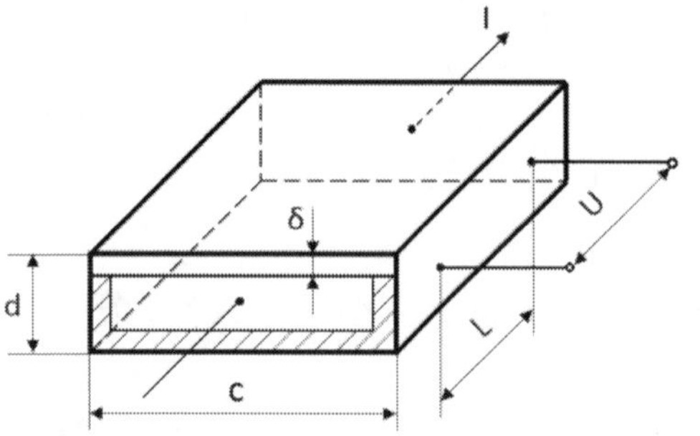

Fig. 1. The schematic view of experiment. Where *c* is the width of the sample; *d* is the thickness of the sample; *δ* is the thickness of the near-surface layer; *L* is the distance between potential contacts, *I* is dc current, *U* is measured voltage.

$$\sigma \approx \sigma^{sur} \cdot (\delta/d) + \sigma^{bulk} \quad (1)$$

Thus, a linear dependence of the conductivity on the film thickness should be observed

$$\sigma = f(d^{-1}) \quad (2)$$

The first term in Eq.(1) is proportional to surface conductivity σ^{sur} and second one is the bulk conductivity σ^{bulk}, i.e. we can "separate" σ^{sur} and σ^{bulk}. To do this the films of different thickness *d* were synthesized. Fig. 2 presents the experimental results for conductivity of Bi_2Se_3 films without magnetic field (Fig. 2a) and in a field of 10T at T=4.2 K (Fig. 2b). One can see that there is a linear dependence on d^{-1} both for σ_0 and σ_{xx}. It allowed us to separate the bulk and surface conductivities. Taking into account Refs [11], it was assumed that delta is not more than 1 nm*.

According to estimations (evaluations) at T = 4.2 K, $\sigma_0^{bulk} \sim 0.5 \cdot 10^3$ and $\sigma_0^{sur} \sim 9.8 \cdot 10^4$ without magnetic field and $\sigma_{xx}^{bulk} \sim 0.2 \cdot 10^3$ and $\sigma_{xx}^{sur} \sim 4.8 \cdot 10^4$ in a field of 10T. That means the surface conductivity σ^{sur} is almost 200 times higher than σ^{bulk}, σ_0^{sur}, $\sigma_{xx}^{sur} >> \sigma_0^{bulk}$, σ_{xx}^{bulk}. The obtained results are in good qualitative agreement with the Ref. [6].

Fig. 3a and Fig. 4a show the temperature dependence of σ_0^{sur} and σ_0^{bulk} without field. σ_0^{sur} decreases with temperature as it should be for protected conductive states on TI surface. σ_0^{bulk} also decreases with T as it was observed for Bi_2Se_3 in Ref. [6] since Bi_2Se_3 has relatively small gap in its electron energy spectrum at Fermi level. It leads to that fact the scattering of charge carriers begin to play the main role in formation of its conductivity. As a result, σ_0^{bulk} decreases with temperature due to the scattering processes.

Magnetic field should change a type of bulk conductivity, as it should be in semiconductors with small gaps and in gapless semiconductors. Fig. 3b, 4b present the temperature dependence of the magnetoconductivities σ_{xx}^{sur} and σ_{xx}^{bulk}. It is seen that σ_{xx}^{sur} decreases with temperature, and it is typical for systems with metallic states. σ_{xx}^{bulk} increases with temperature, which can be explained by changing the character of current carriers scattering for systems in which the scattering processes play the main role in conductivity.

It should be noted that a similar change in a type of temperature dependence for surface and bulk conductivities was observed in Refs. [9, 10], where dc-current was concentrated near a samples surface under the SSE [7, 8].

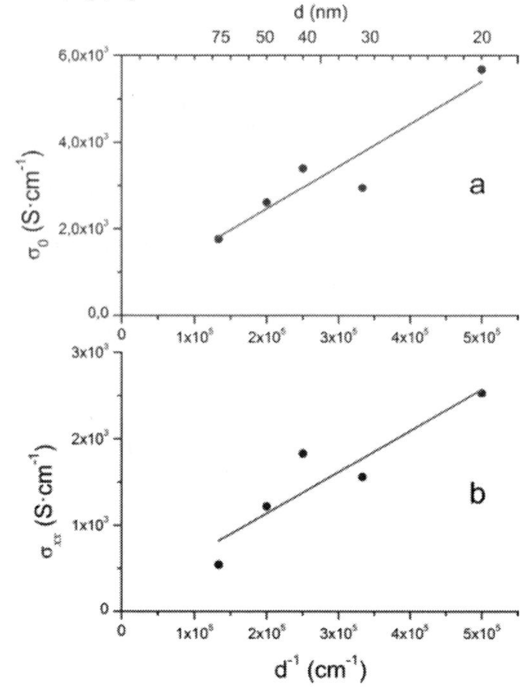

Fig.2. Size effect in the conductivity of thin films of Bi_2Se_3 in magnetic fields 0T (a) and 10T (b) at 4.2K.

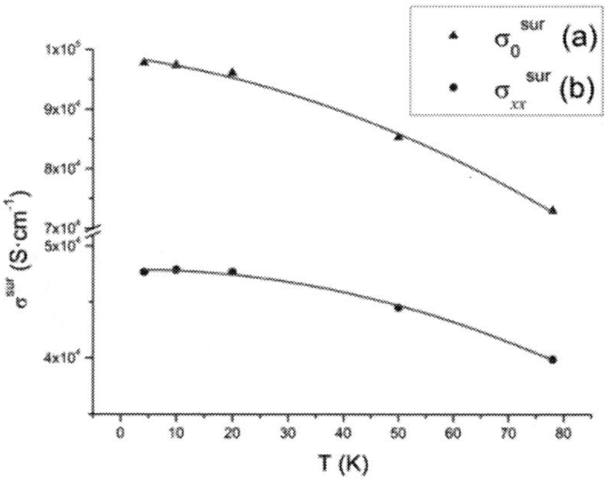

Fig. 3. Temperature dependences of surface conductivity of thin films Bi_2Se_3 in magnetic fields 0T (a) and 10T (b).

* It is an estimation from "above", which gives a smaller value of σ^{sur}.

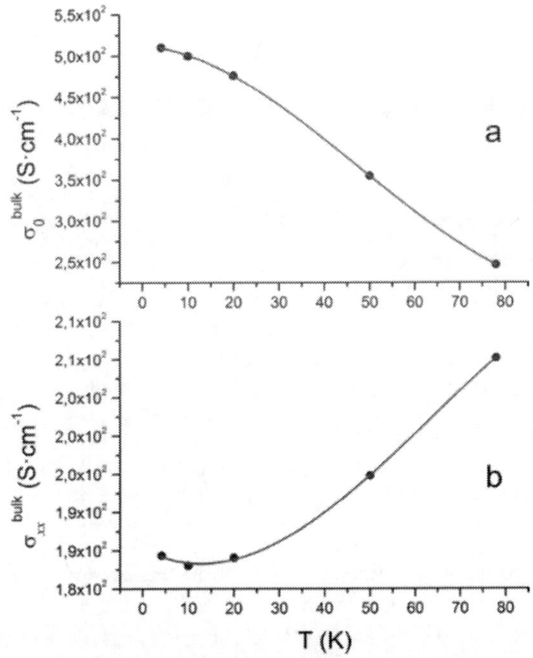

Fig. 4. Temperature dependences of bulk conductivity of thin films Bi_2Se_3 in magnetic fields 0T (a) and 10T (b).

4 Conclusions

The size effect, i.e. a dependence of the conductivity on the film inverse thickness, was observed in thin films of TI Bi_2Se_3. This allows us to "separate" the bulk and surface conductivities at different temperatures and magnetic fields. In apparently, similar effects should be observed in other TIs and systems with inhomogeneous distribution of dc-current on sample cross section.

This work was partly supported by the state assignment of FASO of Russia (theme "Spin" No. AAAA-A18-118020290104-2), by the RFBR (project No. 17-52-52008), by the Government of the Russian Federation (state contract No. 02.A03.21.0006), and by grant of Russian Ministry of Education and Science No 14.Z50.31.0025.

References

1. H. Zhang, C.X. Liu et al., Nat. Phys. **5**, 438 (2009)
2. S.Y. Xu, I. Belopolski, et al., Science **349**, 6248 (2015)
3. Z.K. Liu, L.X. Yang et al., Nat. Mat. **15**, 27 (2016)
4. Y.H. Liu, C.W. Chong et al., Appl. Phys. Lett. **107**, 012106 (2015)
5. Y. Liu, C. Chong, W. Chen et al., Jpn. J. Appl. Phys. **56**, 070311 (2017)
6. B. Bhattacharyya, A. Sharma et al., J. Phys.: Condens. Matter **29**, 115602 (2017)
7. V.V. Marchenkov, H.W. Weber, J. Low Temp. Phys. **132**, 135 (2003)
8. N.V. Volkenshtein, M. Glinski et al., Sov. Phys. JETP **68**, 1216 (1989)
9. V.V. Marchenkov, A.N. Cherepanov et al., J. Low Temp. Phys. 98, 425 (1995)
10. V.V. Marchenkov, H.W. Weber et al., J. Low Temp. Phys. **102**, 133 (1996)
11. X.L. Qi, S.C. Zhang, Rev. Mod. Phys. **83**, 1057 (2011)

Effects of average grain size on the magnetic properties of permalloy films

Nikolay Djuzhev[1], *Aleksey* Iurov[1], *Nikita* Mazurkin[1], *Maksim* Chinenkov[1,*], *Aleksey* Trifonov[2] and *Marina* Pushkina[1]

[1]National Research University of Electronic Technology (MIET), 124498, Moscow, Russia
[2]F.V. Lukin State Scientific-Reseach Institute of Physical Problems, 124460, Moscow, Russia

Abstract. In this paper the results of substrate temperature effects on average grain size in FeNi 20:80 are shown. It was found that with an increase of the substrate temperature from 270 to 390 °C, the anisotropic magnetoresistive (AMR) effect increases from 1.2 to 2.3% and the coercive force from 1.6 to 5.3 Oe. The presence of two conflicting processes: the task of increasing AMR effect to achieve high sensitivity of AMR-sensor and to decrease the coercive force to increase the precision of conversion, leads to the necessity to find the optimum deposition temperature of the permalloy film. The dependence of the AMR effect and the coercive force on the substrate temperature during deposition was obtained. This dependence shows that the substrate temperature increasing above 320 °C leads to coercive force increasing and does not lead to a substantial AMR increasing. In this regard, the substrate temperature 320 °C was determined as optimal.

1 Introduction

The ability of the magnetic thin films to change its resistance under the influence of an external magnetic field is applied in the creation of effective magnetic field measurement devices [1, 2, 3, 4, 5]. Permalloy films FeNi widely used to create anisotropic magnetoresistive sensors (AMR sensors), which convert magnetic field into an electrical signal due to a change in conductivity with a change of the angle between the current vector and the magnetization vector in the permalloy film. The AMR effect is expressed by the formula:

$$\rho(\theta) = \rho_0 + \Delta\rho\cos^2\theta \qquad (1)$$

where θ - the angle between the magnetization vector and the electrical current direction, $\Delta\rho/\rho_0$ – the magnetoresistivity coefficient.

To ensure a high level of output characteristics of AMR sensors it is needed to improve the magnetic properties of the magnetoresistive material. In particular it is needed to maximize the AMR effect, which determines the output signal amplitude, and to prevent the strong growth of the coercive force and anisotropy field for maximum sensitivity and minimum hysteresis achievement. However, it should be taken into account that the magnetoresistive material parameters are determined not only by its internal characteristics, but significantly by the composition and parameters of the film growth, particularly average grain size.

2 Experiment

Permalloy films were obtained by magnetron deposition on a Phase II AJA unit. The target diameter is 50 mm, the distance to the substrate is 112 mm, the thickness of the working and dielectric layers is given in Table 1.

Table 1. Test structure.

Layer	Estimated layer thickness, nm
FeNi 20:80	30
Si_3N_4	120
SiO_2	600
Si	substrate

To create the anisotropy and the AMR in the magnetron deposition camera magnetic field was generated at least 40 Oe.

In this article an experiment was conducted to determine the impact of process parameters on the deposition rate of the material and the grain size in films of FeNi 20:80 during magnetron deposition. To establish the relationship between the characteristics of the films of permalloy and output parameters AMR sensors, primarily sensitivity a complex of studies showing a significant influence of the substrate temperature on the average grain size in the film of permalloy (fig. 1) at a constant magnetron power of 150 W and a working gas pressure of 0.5 Pa was conducted. The average grain size measurement was conducted on the multifunctional x-ray analytical systems "RIKOR-8", "X-Ray MiniLab-6".

* Corresponding author: chinenkov@inbox.ru

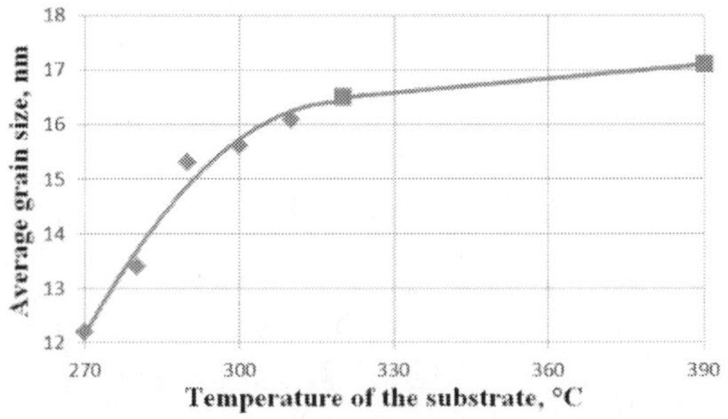

Fig. 1. The dependence of the average grain size on the substrate temperature in the FeNi film.

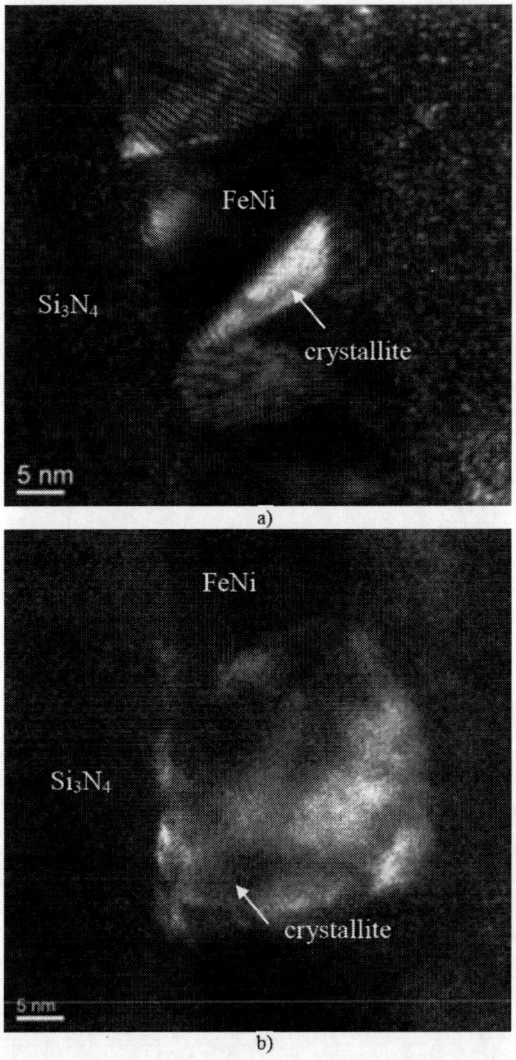

Fig. 2. The image of the grain in the dark field obtained on a transmission electron microscope. a) the substrate temperature during the deposition was 270 °C; b) the substrate temperature during the deposition was 390 °C.

Figure 2 shows photographs taken on TEM grown at different substrate temperatures. It can be seen from the figure that when the substrate temperature increases from 270 a) to 390 °C, the size of crystallite and the surface roughness increases. The increase in the surface roughness is confirmed by atomic force microscope, Fig. 3. It can be seen from the figure that the average height of particles increased from 0.75 to 5.85 nm (on the sidebar to the figures).Figure 1 shows that average grain size increases significantly to a temperature of 300-320 °C. Above this temperature grain growth is slowing sharply. We assume that this transition is associated with the achievement of the average grain size of the film thickness and further growth leads to the development of the surface topography, as evidenced by research on transmission electron microscope and atomic force microscope, fig.2 and fig.3.

Fig. 3. The surface image of the permalloy films at different deposition temperatures taken with AFM AIST-NT Smart SPM-1000. a) the substrate temperature during the deposition was 270 °C, Ra is equal to 0.16 nm, and b) the substrate temperature during the deposition was 390 °C, Ra is equal to 0.97 nm.

It is shown that grain size increasing lead to increase of the magnitude of the AMR and the coercivity (fig.4). The presence of two conflicting processes, namely the AMR increasing for high sensitivity achievement and a coercivity decreasing to increase the precision of conversion, leads to the necessity of finding the optimum deposition temperature of the permalloy film.To determine the optimal deposition temperature based on the data presented in fig. 1 and fig. 4, the dependence of the AMR effect and the coercive force on the substrate temperature during the deposition was obtained (fig. 5). This dependence shows that the substrate temperature increasing above 320 °C leads to coercivity increasing and does not lead to a substantial AMR increasing. In this regard, the temperature 320 °C was determined as optimal.

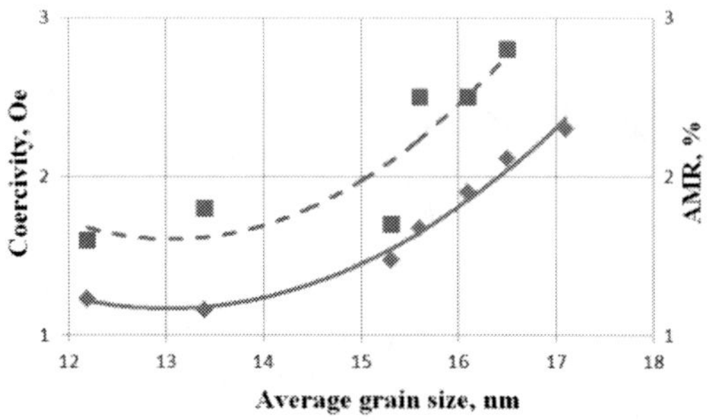

Fig. 4. The dependence of the coercivity (dashed line) and the AMR (solid line) on the average grain size.

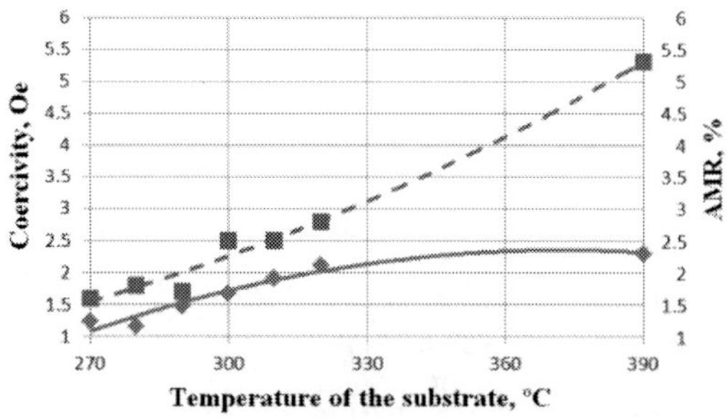

Fig. 5. The dependence of the coercivity (dashed line) and the AMR (solid line) on the substrate temperature.

Other magnetron deposition parameters affect the average grain size and output signal of AMR sensors much weaker. The results of the experimental study parameters of the permalloy film from the substrate temperature are given in table 2. The diffraction patterns were measured in a standard symmetrical scanning scheme with Bragg-Brentano focusing, Fig. 6. The size distribution of the crystallites is approximated by the log-normal distribution, estimated by the Warren-Averbach method.

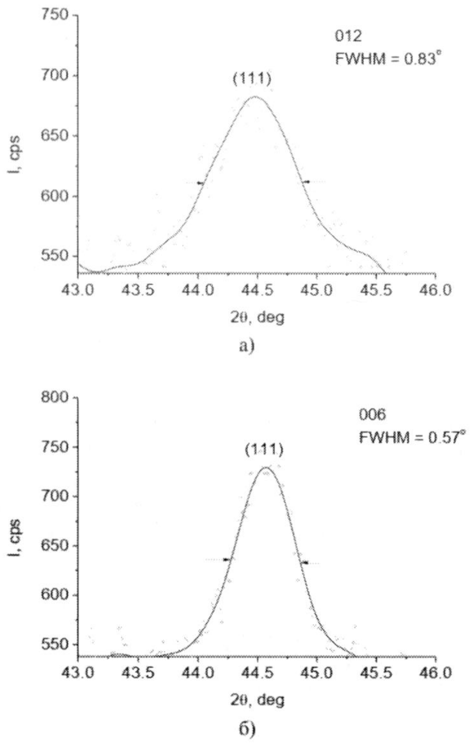

Fig. 6. The diffraction peak of permalloy film for the substrate temperature: a) 270 °C; b) 390 °C.

It is known that the AMR increases with the permalloy film thickness increasing. However, with the thickness increasing the conductivity of the layer grows, and the resistance of the diagonal of the bridge AMR sensor decreases, which leads to significant heating and temperature characteristics deterioration. Therefore, the magnetoresistive layer thickness is selected maximum on the basis of the requirements of achieving the resistance of the bridge diagonal, the values of 1.5 kOhm in an existing topology (fig.7).

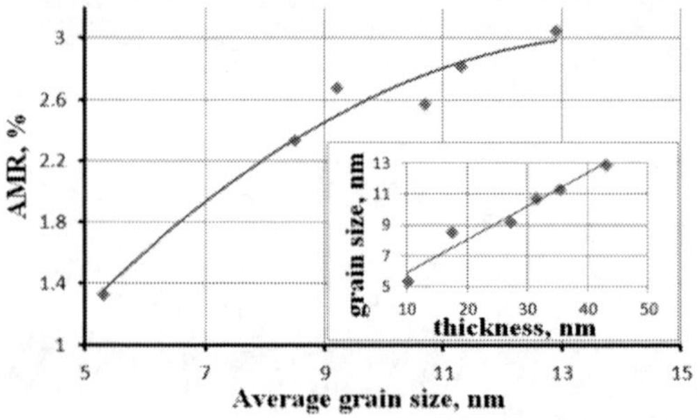

Fig. 7. The dependence of the AMR on the average grain size and film thickness.

Table 2. The dependence of the parameters of the FeNi 20:80 permalloy film on the substrate temperature.

Substrate temperature, °C	AMR, %	Coercivity, Oe	Anisotropy, Oe	Average grain size, nm	film thickness, nm	Surface resistance, Ohm
390	2.30	5.3	7	17.1	26.6	8.25
320	2.01	2.8	6	16.5	26.6	8.21
310	1.90	2.5	5	16.1	26.5	8.48
300	1.82	2.5	6	15.6	25.5	8.92
290	1.47	1.7	6	15.3	26.6	8.57
280	1.16	1.8	7	13.4	26.7	8.88
270	1.23	1.6	7	12.2	26.7	9.29

3 Conclusion

Based on the results of experiments the substrate temperature 320 °C was determined as optimal for the installation of Phase II AJA 150. The magnetic field sensor based on the investigated permalloy films were made. This sensor had a sensitivity value 9.4 (mV/V)/(kA/m) in the bias field value of 1 kA/m instead of 3.0 (mV/V)/(kA/m).

The work was supported by the Minobrnauki RF (Goszadanie № 16.2475.2017/4.6) and conducted using the equipment of Multi-access center «Microsystem technics and electronic component base» MIET.

References

1. S. Tumanski, *Thin Film Magnetoresistive Sensors* (IOP Publishing Ltd, 2001).
2. Dibbern, U. *Magnetoresistive Sensors, in: W.Gopel, J. Hesse, J. N. Zemel, Sensors vol. 5,* Copyright VCH Verlagsgesellschaft mbH, Hannover, 1989, pp. 341 – 380.
3. N.A. Djuzhev, N.S. Mazurkin, V.S. Pozdnyakov, A.S. Iurov, M.Yu. Chinenkov. Semiconductors, Vol. **49**, № 13, pp. 1739-1742 (2015).
4. V. Bespalov, N. Djuzhev, A. Iurov, N. Mazurkin, M. Chinenkov, R. Preobrazhensky, Solid State Phenomena, **249** (2016).
5. V. Bespalov, N. Djuzhev, A. Iurov, N. Mazurkin, M. Chinenkov, R. Preobrazhensky Journal of Surface Investigation: X-ray, Synchrotron and Neutron Techniques, Vol. **11**, No. 2, pp. 343-345 (2017).

Image reconstruction algorithms for the microwave holographic vision system with reliable gap detection at theoretical limits

Dmitry Leshchiner[1,*], *Konstantin* Zvezdin[1], *Anatoly* Popkov[1], *Grigory* Chepkov[2], and *Pietro* Perlo[3]

[1]Moscow Institute of Physics and Technology, Moscow Region, Russia
[2]Higher School of Economics, Moscow, Russia
[3]IFEVS & Torino e-District, Torino, Italy

Abstract. We present a reliable image reconstruction algorithm suitable for a microwave holographic vision system with several sensors coupled to the spin-diode based microwave detector and a single emission source. An objective is, by reconstructing the spatial microwave scattering density on the scene, to detect the presence and the nature of road obstacles impeding driving in the near vehicle zone. The idea of holographic visualization is to reconstruct the spatial microwave scattering density of an object by detecting an amplitude and phase of a reflected signal by lattice of sensors. We discuss versions of an algorithm, determine and analyse its resolution limits for various distances with different number of sensors for a one-dimensional test problem of detecting two walls (or posts) separated by a gap at a fixed distance. The maximal interval between sensors needed for a reliable reconstruction equals approximately Fresnel zone width. We show that maximal resolution achieved by our algorithm with an appropriate number of sensors was about 40% of Fresnel zone width for wall detection and about 30% of zone width for gap detection.

1 Introduction

The growing interest in automated and connected vehicle technology prompted the need for efficient traffic and obstacle detection techniques in large variety of settings. For inner-city mobility, as well as for vehicle parking setting, critical issues are control of obstacles and object movement in the near-vehicle zone, about 0.3 to 4.5 meters distance to detection point. That would allow for speed control and adjustment in dense environment, vehicle-to-vehicle and pedestrian automatic emergency braking (with reaction time < 0.1 sec) at speeds less than 30 km/h, mitigating crashes in most inner-city environments. A promising technology direction in this area is the development of microwave-based detection techniques. Microwaves can penetrate various media – such as stone and water – so it can work at every weather conditions and sense distant or inaccessible objects.

Recently, there suggested an alternative approach to close-distance imaging [1] based on digital microwave holography, using high sensitivity spin diodes [2] as wave sensors. That could provide compact and cost effective devices for dynamic detection of surrounding objects in different driving scenarios and weather conditions. We could see the full 3D density of the objects – not just the surface. Different densities indicate different materials – stone, flesh, metal, wood etc.

At short distances this technology combines the advantages of radars (can work at every weather conditions, low cost, can see behind obstacles) and of LIDAR (doing 3D scenario reconstruction), adding also an extra capacity for object type detection, based on accurate determination of object material. In this article, we discuss the potential of that technique for accurate obstacle recognition in near-vehicle zone from the image reconstruction and algorithmic point of view. We show that there exists a capacity to use it as a component in an integrated realistic object recognition automotive system.

2 Image resolution and object detection

Digital holography techniques were initially proposed in 1960-ies [3,4]. They typically use an array of sensors comparable in size to the object studied and a wide parallel imaging beam. In our application, however, there is no practical way to provide a parallel imaging beam comparable in size to the imaging field. We need to rely on just one or few compact radiation sources, emitting each within a certain angle. That has serious consequences for spatial resolution achievable. Indeed, looking at Fresnel zones size at the object edges, we would not normally expect spatial resolution much better than the half of the zone width. It remains unclear though if even that resolution is reliably achievable given the limited number of sensors available for wave detection. As provided by Kotelnikov–Whittaker–Nyquist–Shannon sampling theorem, the maximal interval between sensors needed for a reliable reconstruction equals about Fresnel zone width. Same consideration implies that using more frequent sampling would not actually increase the amount of useful information available. With that in mind, we would be

* Corresponding author: dmitry_1111@inbox.ru

looking at positioning no more than few dozens of sensors at the vehicle's front. We will need effective parameters regularization techniques applied to reduce the dimensionality of the problem and still get the meaningful object detection outcome.

2.1. The density reconstruction setting

We will consider a task with one monochromatic radiation emitter, and several sensors. Given is the vector **P** of K sensor signals (both phase and amplitude) and the positions of both the emitter and the sensors. **Fig. 1** gives an overall scheme of the corresponding system.

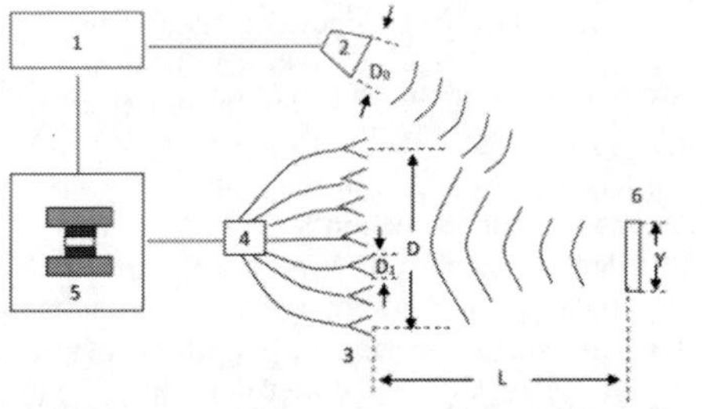

Fig. 1. The block scheme for spin-diode based microwave holographic registration: 1 – microwave oscillator, 2 – horn antenna, 3 – sensors array, 4 – channel switch, 5 - spin-diode based registration unit, 6 – the registered object.

The article [1] gave the following expression for the scattered signal **P** in terms of the functions expressing objects density in a point **r** within the imaging area, $f(\mathbf{r})$, and the emission intensity in direction to **r**, $g(\mathbf{r})$:

$$\mathbf{P}(p)=\int f(\mathbf{r})g(\mathbf{r})/R\, exp(-i\omega R)\, d\mathbf{r} \qquad (1)$$

Here p is a position of a sensor, ω is signal frequency and R is the length of the two-segment broken line from a position of the emitter to the point **r**, and then to the p. That expression, however, takes no account of the signal adsorption within the object. Considering the adsorption intensity to be equal to the scattering density, the total adsorption coefficient is then given by $exp(-\oint f(x)dx)$, where the integral in the exponent is taken along the same two-segment broken line from emitter to **r** and then to the p, of the length R. We obtain a non-linear integral equation for the density function $f(\mathbf{r})$:

$$\int f(\mathbf{r})(g(\mathbf{r})/R\, exp(-\oint(f(x)+i\omega)dx)d\mathbf{r} = \mathbf{P}(p) \qquad (2)$$

What makes that equation non-linear is the adsorption term. However, given a density function, we may compute it, and fixing the term will make the equation linear. It could then be readily solved and its solution will provide a correction to a previously given density. We may use that solution in its turn to obtain a newer correction, and so on.

The iterative matrix inversion task could still become too costly to perform in real time. The solution, however, may lie in ignoring the adsorption term altogether. Indeed, ignoring the term, we would still get accurate scattering picture of the front parts of the obstacles - the most critical ones to detect - and the image of distant parts slightly distorted by scattering adsorption. If we ignore the term, the task of inverting the matrix is still there, but the matrix no longer depends on the density function $f(\mathbf{r})$, so its inversion could just be precomputed. Moreover, what one needs to obtain the result is simply a basis of density functions corresponding to each individual sensor – so the result is simply a sum of these functions multiplied by each sensor's signal: a map from a signal space to a density space. That gives us an easy and effective algorithm for object density reconstruction, suitable for the practical purposes.

To obtain a linear resolution procedure, one could represent the regularization preferences by a quadratic weight functional. The solution procedure (Tikhonov regularization [5]) is to find a solution with a minimal weight – or, more generally, a solution that minimizes the sum of a penalty for deviation from exact solution and of a regularization weight penalty. The weight is defined by a matrix $\mathbf{\Gamma}$ and is given by $\|\mathbf{\Gamma} \boldsymbol{f}\|^2$, where $\boldsymbol{f}$ is a proposed solution, $\|*\|$ is a standard (Euclidean or Hermitian) norm, and $\mathbf{\Gamma}$ is a matrix mapping the vector $\boldsymbol{f}$ into a norm space. The form for the total penalty is then $\|\mathbf{A}\boldsymbol{f}-\mathbf{P}\|^2+\|\mathbf{\Gamma}\boldsymbol{f}\|^2$, where **P** is the vector of sensor signals and **A** is a matrix for linearized integral equation. The regularized solution is given by $\boldsymbol{f} = (\mathbf{A}^T\mathbf{A}+\mathbf{\Gamma}^T\mathbf{\Gamma})^{-1}\mathbf{A}^T\mathbf{P}$, where $\mathbf{A}^T$ and $\mathbf{\Gamma}^T$ are Hermitian conjugate matrices to **A** and $\mathbf{\Gamma}$. So this is a result of multiplication of vector **P** by a rectangular matrix, which maps K-dimensional vector **P** into N-dimensional vector $\boldsymbol{f}$ – or, in other words, a collection of K basic functions f_i, each corresponding to its own sensor. Please note that it depends on sensor configuration only, so one may precompute this. To express a preference for solution smoothness (in Fourier basis) is to define a diagonal matrix $\mathbf{\Gamma}$ of elements increasing by harmonics number. We found in numerical experiments that choosing the weight proportional to a harmonics number gives about the best performance.

2.2 The 1D object shape extraction

We performed numerical experiments corresponding to a task of imaging and detecting a 1D object structure (a slot – two walls separated by a gap) at a given distance. For that task, we used a particular object detection algorithm. Assume that object density could only take two values 0 and 1 (all objects are of the same density - as in, of the same material) so we would need to distinguish between the two cases. The way to do it is to set a density threshold and assume zero density for all reconstructed density values below the threshold and density one for all the values above. We take the point lattice with a period equal to the wave length, take all the reconstructed values in these points, and put a threshold in the middle of the largest interval between the values (to the point of "lowest values density" – corresponding to the point between the high and low density values).

Applying the chosen threshold to the reconstructed density, we then get back a reconstructed object shape. There was a series of experiments detecting the 1D slots (two walls divided by a gap) at a given distance to determine the spatial resolution limits for the algorithm described above. Spatial resolution refers to the ability of the imaging to differentiate two objects. In our case, objects are the slot walls and the closeness of the objects is the width of the gap. We determine the gap resolution limits depending on the object distance and the configuration of walls, for several configuration of sensors. The imaging field size was fixed to be of 64 wavelengths (for a realistic case of spin diodes set to 10 GHz frequency, that would be about 2 m), and the size of sensors lattice was fixed to be the same (that would roughly correspond to a width dimension of a car). We assumed the emitter was monochromatic, wide-beam, positioned at the centre of sensors field, and sensors positioned at a lattice with approximately equal spacing (affected by a small random noise). The configurations tested had 4, 8, 16, 32 and 64 sensors.

The object distances varied from 10 to 500 wavelengths (in physical units, it is from 0.3 to 15 meters). Two types of slots, positioned to the centre of the imaging field, were tested - one with walls of equal width, another with one wall twice as wide as the other. We did not add any noise to the sensor signals modelled, assuming them exact. **Fig. 2** gives an illustration of how the algorithm worked for a particular case (32 sensors, object distance is 50 wavelengths). A reconstructed 1D density, for a given slot shape: two unequal wall separated by a gap. X-axis is the object coordinate measured in wavelength units ($x=D/\lambda$). Y-axis is an object density (original object density is in red – equal 0 or 1; the reconstructed density is in blue). The thin horizontal line shows the density threshold obtained.

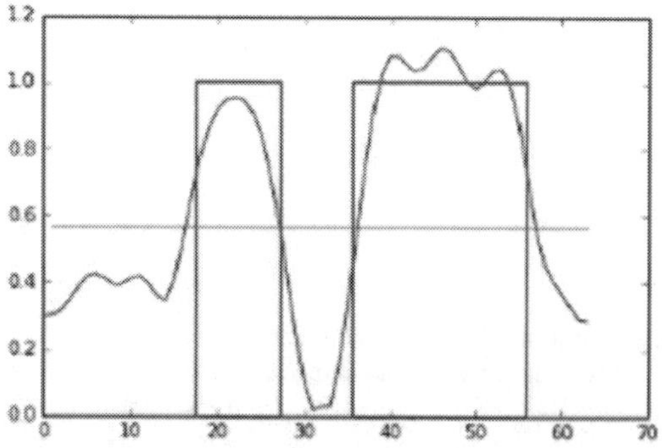

Fig. 2. An object shape given applying the density threshold.

2.3. Resolution limits obtained

We summarize the results of detection experiments in terms of minimal object dimensions required for reliable slot detection. It is of interest to quantify the resolution quality by introducing the quality parameter as the ratio: $\kappa=L/r_2$, where L is given minimal critical dimension (i.e. minimal gap or wall size etc.) and $r_2=\sqrt{(2D+1)}$ is the first Fresnel zone width (from zone center to second ring boundary in zone plate) at the distance D. One may note that the maximal possible resolution at first Fresnel zone, $\Delta_l=1.22\Delta r_2$, equals about $0.36r_2$. Hence, we may at best obtain 0.36 as quality parameter value. Here is the resolution ratio values graph, obtained for unequal walls:

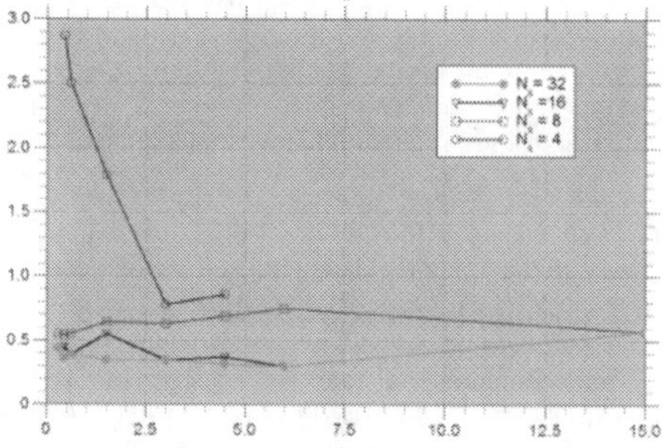

Fig. 3. A resolution ratio, by distance and number of sensors.

We may see that for 16 or more sensors, the resolution well approaches the theoretical limit. Given below are the tables of minimal wall and gap widths available for reliable recognition by object distance and number of sensors (for both the unequal walls and equal walls configuration). For the equal walls slots, detection is quite reliable (detection failure rate is about 1% to 3%) at distances of up to 15 meters. Failure rate is measured as percentage of configurations for which gap detection fails, among all tested configurations within obtained minimal wall and gap dimensions limits. For the slots with unequal walls, it is quite reliable (in low single digit percentage rate) at distances up to 5 m, which is enough for near vehicle detection. One may also note there is approximately six times (for the wall) and 10-12 times (for the gap) gain in detection distance compared to the worst-case scenario of resolution equal to the first Fresnel zone width. That is, e.g. the gap of 18 cm was possible to detect at a distance of 6 m instead of worst-case 52 cm distance (which would render it unusable).

Table 1. Minimal sum of wall size and the gap, and minimal gap value (in centimetres), required for slot shape with equal walls for reliable recognition, depending on the distance of sensors to the object.

Distance (m)	4 sensors		8 sensors		16 sensors	
	Sum	Gap	Sum	Gap	Sum	Gap
0.6 meters	78	33	18	9	18	9
1.5 meters	81	48	60	9	26	11
3.0 meters	50	29	45	20	41	11
4.5 meters	60	33	54	27	50	15
6.0 meters	74	45	57	30	44	17
15 meters			62	36	68	36

Table 2. Minimal smaller wall size and gap (in centimetres) required for slot shape with unequal walls for reliable recognition, depending on the distance of sensors to the object.

Distance (m)	4 sensors		8 sensors		16 sensors	
	Wall	Gap	Wall	Gap	Wall	Gap
0.6 meters	21	48	21	11	9	8
1.5 meters	30	54	20	20	12	17
3.0 meters	24	33	21	27	18	15
4.5 meters	8	45	23	36	20	20
6.0 meters			24	45	24	18
15 meters			24	54	24	54

3 Discussion

To reach level 4 of truly autonomous vehicles (Levels of Driving Automation, SAE international standard J3016) the standard is putting emphasis on the automotive vision systems. Self-driving vehicles currently developed by Alphabet, Uber, or Toyota rely heavily on LIDAR. LIDAR can scan more than 100 meters in all directions around the car, generating a precise 3D map of the car's surroundings. The main issues related to LIDAR in automotive industry are that they generate a large amount of data and that they are still too expensive for mass-market implementation. Much cheaper solutions, in terms of price and computations, are those based on radars and cameras. The cheapest and most available sensors are the cameras, the best for classification and texture interpretation. The cameras, however, use massive amounts of data, making processing a computational intense and algorithmically complex job.

Radars are computationally much lighter than cameras and LIDAR. Radars can work in every weather condition and even use reflection to see behind obstacles; however, they are much less accurate than LIDAR. Radar is a proven technology increasingly becoming more efficient for the autonomous car. This will also make radar more complementary to the cameras as the "dynamic duo" to "cross validate" the potential reconstructed scenario. The microwave holographic technology with its ability to detect the object material and volume (e.g. helping to tell a bicycle apart from a motorcycle) could become an additional useful complement to the emerging picture of synergetic collaborative and cross-validating automotive obstacles detection and classification environment. There exist wide varieties of signal processing approaches with the given algorithm, defined by the choice of functional space and a regularization metrics, allowing for adaptive treatment of sensor data based on the output of the other parts of the collaborative object recognition system.

4 Conclusions

We present, by combining the well-known building blocks (generalized Fourier-basis deconvolution, a type of Tikhonov regularization in Hermitian space, and a threshold-based object detection technique), a version of microwave holographic imaging and obstacle detection algorithm for an environment with a single coherent emitter and few detection sensors. The version is computationally efficient, requiring just $O(N*K)$ operations per image frame, where K is the number of signals detected (sensors) and N is the number of image elements (pixels). The detection distances and spatial resolution achieved (better than 20 cm with distances up to 4.5 meters) are sufficient for the near-vehicle object detection purposes. An important direction for future research is to focus on distance measurement precision with object detection algorithms (the experiments presented in this article concentrated on a width measurement task only). The numerical experiments on 2D and 3D shapes detection, systematic treatment for signal noise effects, and the testing with the actual physical objects all are the subject for the future work.

We thank the Russian Science Foundation, project number 16-19-00181, for supporting this work.

References

1. L. Fu, Y. S. Gui, L. H. Bai, H. Guo, H. Abou-Rachid, and C.-M. Hu, J. Appl. Phys. **117**, 213902 (2015)
2. Miwa, S. et al., Nature Materials **13**, 50–56 (2014)
3. J.W. Goodman, R.W. Lawrence, Appl. Phys. Lett. **11**, 77–79 (1967)
4. T. Huang, *Digital holography*. Proc. IEEE. **59**, 1335–1346 (1971)
5. Tikhonov, A. N. Doklady Akademii Nauk SSSR **39**, 195–198 (1943)

Inverse spin Hall effect in heterostructures "nanostructured ferromagnet/topological insulator"

P.N. Petrov[1,*], *M.D.* Davydova[1], *P.N.* Skirdkov[1,2,3], *K.A.* Zvezdin[1,2,3], *J.G.* Lin [4], *J.C.A.* Huang[5]

[1] Moscow Institute of Physics and Technology, 141700 Dolgoprudny, Russia
[2] Russian Quantum Center, 143025 Skolkovo, Moscow Region, Russia
[3] A.M. Prokhorov General Physics Institute, RAS, 119991 Moscow, Russia
[4] National Taiwan University, Taipei, 106 Taiwan, Republic of China
[5] National Cheng Kung University, Tainan, 701 Taiwan, Republic of China

Abstract. Interaction between magnetization dynamics and spin polarized electronic transport has been studied for ferromagnet nanodisk situated upon a 3D topological insulator (TI) film. Resonant magnetization dynamics leads to generation of spin current, which flows into the topological insulator, where spin to charge conversion occurs. Using micromagnetic simulations for magnetization dynamics we estimate the dc voltage, which is created due to this process in topological insulator. Contribution from different modes, which are characteristic for nanodisks, to the voltage was calculated.

1 Introduction

Interaction between magnetization dynamics and spin polarized electronic transport has been first studied by Slonczewski [1] and Berger [2]. It has been predicted and later experimentally proven that the spin polarized current can drive magnetization dynamics via a spin torque mechanism, and that in an inverse process, resonant magnetization dynamics can lead to generation of spin current which is related to the spin pumping mechanism [3]. When spin polarized current flows into the material with strong spin-orbit interaction, the spin-to-charge conversion takes place. One of the possible mechanisms of such conversion is inverse spin Hall effect (ISHE) [4,5], which occurs in volume, and the other is the inverse Rashba-Edelstein effect (IREE) [5], connected with the spin-momentum locking for surface currents. We consider the heterostructure with thin metallic ferromagnet situated upon the topological insulator film. Due to the ferromagnet layer the bulk conductance in TI may be restored and the origin of the charge current in the experiment often lies in ISHE [6,7]. We consider the dc voltage, which arises in the system due to such mechanism. On the other hand, in work by Wang *et al.* [8] in YIG-TI bilayers with Bi_2Se_3 the IREE effect was found to be dominant over ISHE (the assumption of negligence of ISHE mechanism was made due to the smallness of the fitted spin diffusion length λ_{SD}). However, the interface between metallic FM like Py and nonmetallic YIG and TI may have very different properties, which may explain the discrepancy in the experimental results.

Recently, spin pumping and dc voltage generation was studied in bilayer systems with metallic ferromagnet and three-dimensional topological insulator Bi_2Se_3 at room temperature [7], which has proved the possibility of efficient spin-to-charge conversion in such systems. Three-dimensional (3D) topological insulators are known for their strong spin-orbit coupling and the existence of spin-textured topological surface states which could be potentially used for spintronics applications. The measure of the spin-to-charge conversion rate in considered mechanism corresponds to spin Hall angle θ_{SH}, which for Bi_2Se_3 has the giant value, equal to 0.38 [9], which makes it a perfect candidate material for topological spintronics [10].

We consider structure, which consists of a ferromagnetic (FM) nanodisk with diameter either 200 or 400 nm and thickness 4 nm, which is situated upon the 50 nm thick topological insulator film, as shown in Fig. 1. The necessity for investigation of nanostructured system is connected with the need for miniaturization of future spintronic devices. It is known that in structured ferromagnets shape modes arise [11], which have been studied previously [12]. The existence of additional modes comparing to sheet film structures opens a possibility for frequency tuning and increasing the number of varying parameters in the system, which may be important for practical applications. Moreover, such structures can be arranged into 2D arrays, in which coupling occurs and new properties arise [13].

In this work we use micromagnetic modelling to study properties of spin pumping in heterostructures, consisting of FM nanodot upon a topological insulator.

* Corresponding author: petr.petrov@phystech.edu

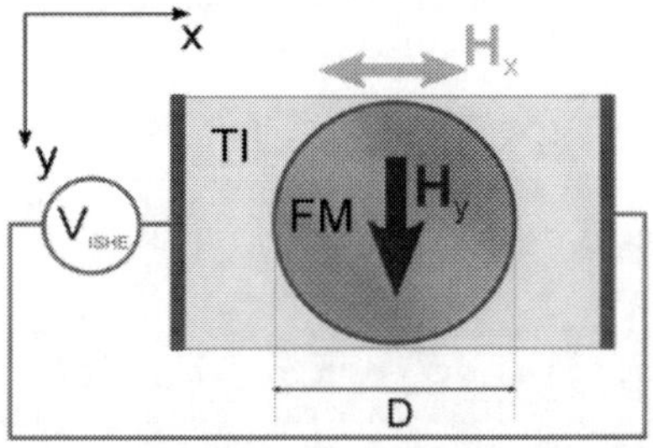

Fig. 1. Schematic structure of the considered bilayer system. D = 200/400 nm denotes the diameter of the ferromagnet nanodisk and V_{ISHE} is the measured dc current due to spin pumping from ferromagnetic resonance in ferromagnet.

We use permalloy (FM) and Bi_2Se_3 (TI) as modelling materials. Short investigation of the sensitivity of the second ('end') mode on the ferromagnet properties, such as exchange constant, g-factor and net magnetization was performed. We then calculate spin current density in each cell and obtain dc voltage due to spin-to-charge conversion. The second mode reveals itself as an additional peak in calculated dc voltage. Such peak was not observed in recent experimental results with spin pumping between dielectric ferromagnet and bismuth selenide [8] but seen clearly in experiments on ferromagnetic resonance in permalloy nanodisks [14].

2 Micromagnetic modelling and ISHE voltage

Magnetization dynamics in the nanodisk is caused by the ferromagnetic resonance in the presence of large constant in-plane magnetic field H_y, which is applied along y-axis (see Fig. 1), and a small oscillating perpendicular in-plane component H_x. The magnetization dynamics was obtained using our micromagnetic finite-difference code 'SpinPM', which is based on the fourth-order Runge-Kutta method with an adaptive timestep control for the time integration and mesh size of 2.5 nm × 2.5 nm × 4 nm..

The results for the average net magnetization of the nanodisk with D = 200 nm with enhanced Gilbert damping $\alpha = 0.0275$ (this damping corresponds to effective spin-mixing constant g^{eff} at the interface, equal to 3.7×10^{19} m^{-2}, which is characteristic value for similar heterostructures [7]) taken into account are shown in Figs. 2 and 3. Enhanced Gilbert damping, which can be expressed as $\Delta\alpha = \left(g\mu_B g^{eff}\right)/\left(4\pi M_s h_{FM}\right)$ (h_{FM} is the thickness of the ferromagnetic layer, M_S is the saturation magnetization) is the result of angular momentum transfer from ferromagnet into attached layer of nonmagnetic material [15].

The results for micromagnetic simulations of *x*- and *z*- components of magnetization are shown in Figs. 2a and 2b. The *y*-component is approximately equal to saturation magnetization due to strong bias magnetic field in this direction.

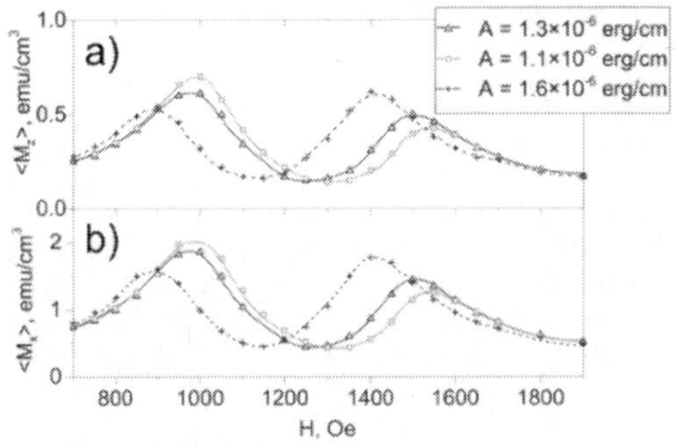

Fig. 2. Dependence of average in period for a) – *x*-component and b) – *z*-component of net magnetization in nanodisk with D = 200 nm for different exchange constants on bias magnetic field. M_s = 800 emu/cm^3, g = 2. Lines are guides to the eyes.

In Fig. 2 dependence of the behavior of both resonance peaks on the exchange constant in ferromagnet is illustrated. The lower field mode corresponds to Kittel homogeneous resonance, whereas the higher mode is 'end' mode investigated in ref. [14]. However, the behavior of this mode in presence of varying material parameters was not considered there due to absence of analytical investigations, which is also an important question from practical point of view, which we address for further thorough research. Here we provide only a short insight into the problem. As it is seen from Fig. 2, the distance between the two peaks remains almost constant, whereas ratio of resonance height varies a lot depending on the exchange constant.

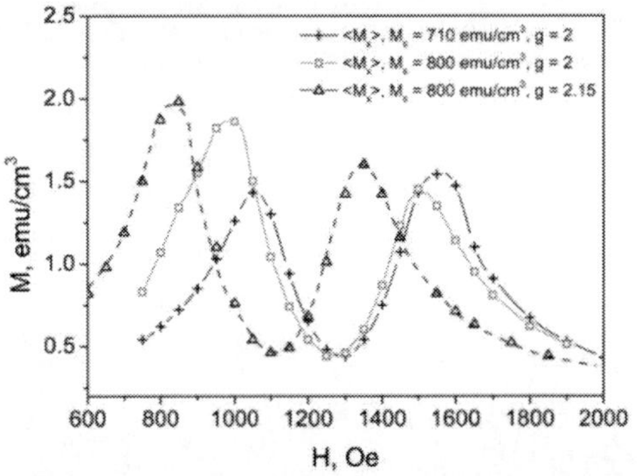

Fig. 3. Dependence of average in period net magnetization in nanodisk with D = 200 nm for different saturation magnetizations and g-factor on bias magnetic field. Exchange constant is taken to be A = 1.3×10^{-6} erg/cm. Lines are guides to the eyes.

In Fig. 3 the ferromagnetic resonance is simulated for different saturation magnetizations and different g-factors, which are the parameters which also vary between different experiments for the same materials. The homogeneous resonant peak is naturally sensitive to the saturation magnetization, whereas 'end' mode is much less sensitive, which may allow additional method for controlling their ratio.

The magnetization dynamics data with enhanced Gilbert damping taken into account was used to obtain spin current flow due to spin pumping in each cell. According to the conventional spin pumping theory spin current appears as a result of oscillations of chemical potential in nonmagnetic layer [15]. It was shown that bulk ISHE pumping signal may be described with the same expressions as in theory for the ferromagnet-normal metal systems [16]. However, it is important to note that the material coefficients in the theory are effective constants with difference, which lies in large spin orbit coupling on the FM-TI interface. The spin current density flow to the TI in z-direction (perpendicular to the interface) is expressed in the following way:

$$(\mathbf{j}_s)_z = \frac{\hbar}{4\pi} g^{eff} \left[\mathbf{m}, \frac{d\mathbf{m}}{dt} \right], \quad (1)$$

where g^{eff} is the effective spin-mixing constant, $\mathbf{m}$ is unit magnetization vector in the current cell and the vector components $\mathbf{j}_z$ correspond to spin polarization. Equation (1) describes the spin current at the interface, whereas in the bulk it decays according to [16]:

$$\mathbf{j}_s(z) = \mathbf{j}_s(0) \sinh\left(\frac{h-z}{\lambda_{SD}} \right) / \sinh\left(\frac{h}{\lambda_{SD}} \right), \quad (2)$$

where h is the thickness of the TI film and $\lambda_{SD} \approx 6$ nm is the spin diffusion length in Bi_2Se_3. Due to the inverse spin Hall effect in TI the spin current flow gives the rise to charge current using empirical spin Hall angle parameter θ_{SH} [17]:

$$\mathbf{j}_c = \frac{2e}{\hbar} \theta_{SH} \left[j_s \mathbf{e}_z, \boldsymbol{\sigma} \right]. \quad (3)$$

Here we take that inverse to spin Hall effect process is reciprocal in terms of the spin-to-charge conversion constant $\theta_{SH} \approx 0.4$, which lies in agreement with experimental results [7,17], and $\boldsymbol{\sigma}$ is the polarization of the spin current.

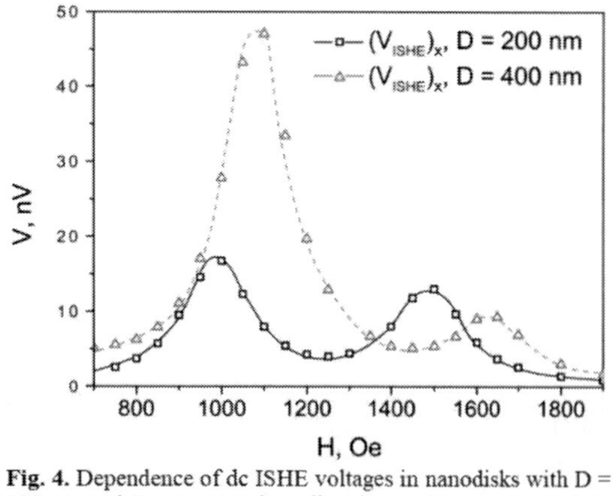

Fig. 4. Dependence of dc ISHE voltages in nanodisks with D = 200 nm and D = 400 nm in x-direction on bias magnetic field at parameters $M_s = 800$ emu/cm^3, g = 2, A = 1.3×10^{-6} erg/cm. Lines are guides to the eyes.

Finally, the voltage, which is found on the edges of topological insulator due to the charge current flow, can be expressed as

$$V_{ISHE} = \langle j_c \rangle SR = \langle j_c \rangle DhR, \quad (4)$$

where R is the resistance of the film in x-direction, which we take to be roughly equal to 240 Ω [18]. The average spin current polarization is nonzero for y-direction and from the structure of eq. (3) it is seen that the voltage will arise in x-direction. The results for the x-components of ISHE voltage calculations are present in Fig. 4. The peak values of this dc voltage are approximately equal to 47 nV for 400 nm disk and 17 nV for 200 nm disk, which is consistent with results from experiments with sheet films [6,7], if scaled.

The main difference between voltage in such nanostructure from sheet samples is that the existence of the second peak (see Fig. 4). It is seen from that input to the ISHE current from the 'end' mode is much smaller in comparison to homogeneous resonance in case of larger disk diameter, and in limit of sheet films this resonance will vanish [7]; whereas for smaller disks both peaks are approximately same in height. A resonance with lower frequency corresponds to the higher peak in terms of magnetic field, which will possibly provide frequency tunability for future prospective spintronic devices.

3 Conclusion

In conclusion, we have investigated the ISHE voltage in heterostructures, which consist of thin nanodisk, situated upon topological insulator film. Two resonances with respect to bias magnetic field were found and investigated. The ratio between the two resonances was shown to be sensitive to the properties of ferromagnet. Verification of the existence of this second resonance in ISHE voltage with respect to magnetic field and further study of ISHE in nanostructured systems is a matter for future experimental verification.

The grant RSF No.17-12-01333 is acknowledged.

References

1. Slonczewski, J. C., *J. Magn. Magn. Mater* **159**, 1-2 (1996).
2. Berger L., *Phys. Rev. B.* **54**, 9533 (1996).
3. Ando, K., Takahashi, S., Ieda, J., Kajiwara, Y., Nakayama, H., Yoshino, T., Harii, K., Fujikawa, Y., Matsuo, M., Maekawa, S. and Saitoh, E., *J. Appl. Phys.* **109**, 10, (2011).
4. Saitoh, E., Ueda M., Miyajima H., and Tatara G., *Appl. Phys. Lett.* **88**, 18 (2006)
5. Edelstein, V. M., *Solid State Commun.* **73**, 3 (1990)
6. Deorani, P., Son, J., Banerjee, K., Koirala, N., Brahlek, M., Oh, S. and Yang, H., *Phys. Rev. B* **90**, 9 (2014)

7. Jamali, M., Lee, J.S., Jeong, J.S., Mahfouzi, F., Lv, Y., Zhao, Z., Nikolić, B.K., Mkhoyan, K.A., Samarth, N. and Wang, J.P., *Nano Letters* **15**, 10 (2015)
8. Wang, H., Kally, J., Lee, J.S., Liu, T., Chang, H., Hickey, D.R., Mkhoyan, K.A., Wu, M., Richardella, A. and Samarth, N., *Phys. Rev. Lett.* **117**, 7 (2016).
9. Mellnik, A.R., Lee, J.S., Richardella, A., Grab, J.L., Mintun, P.J., Fischer, M.H., Vaezi, A., Manchon, A., Kim, E.A., Samarth, N. and Ralph, D.C., *Nature* **511**, 7510 (2014)
10. Wang, K.L., Lang, M. and Kou, X., *Handbook of Spintronics*, pp. 431-462. (2016)
11. Dillon Jr, J.F., 1960, *J. Appl. Phys.,* **31**, 9 (1960)
12. Shaw, J.M., Silva, T.J., Schneider, M.L. and McMichael, R.D., *Phys. Rev. B* **79**, 18 (2009)
13. Velten, S., Streubel, R., Farhan, A., Kent, N., Im, M.Y., Scholl, A., Dhuey, S., Behncke, C., Meier, G. and Fischer, P., *Appl. Phys. Lett.* **110**, 26 (2017)
14. Shaw, J.M., Silva, T.J., Schneider, M.L. and McMichael, R.D., *Phys. Rev. B* **79**, 18 (2009)
15. Tserkovnyak, Y., Brataas, A., Bauer, G.E. and Halperin, B.I., *Rev. Mod. Phys* **77**, 4 (2005)
16. Chen, K. and Zhang, S., *Phys. Rev. Lett.* **114**, 12 (2015)
17. Saitoh, E., Ueda, M., Miyajima, H. and Tatara, G., *Appl. Phys. Lett.* **88**, 18 (2006)
18. Stordeur, M., Ketavong, K.K., Priemuth, A., Sobotta, H. and Riede, V., *Phys. Status Solidi B* **169**, 2 (1992)

Top and Bottom Spin Valves With Ni-Fe-Mn Antiferromagnetic Layer

Ilya Blinov*, *Tatiana* Krinitsina, *Mikhail* Milyaev, *Vladimir* Popov and *Vladimir* Ustinov

M.N. Miheev Institute of Metal Physics, 620990 Ekaterinburg, Russia

Abstract. Structure, magnetic and magnetoresistive properties of spin valves with Ni-Fe-Mn antiferromagnet as a pinning layer have been studied. A technique of fabrication of spin valves with an enhanced thermal stability and improved hysteretic characteristics has been elaborated.

1 Introduction

Bilayers of permalloy/antiferromagnetic triple Ni-Fe-Mn alloy have been studied, as well as magnetic and magnetoresistive properties of spin valves (SVs) with Ni-Fe-Mn antiferromagnet (AF) as a pinning layer. A technique of fabrication of bottom spin valves based on Ni-Fe-Mn ordered AF with an enhanced thermal stability and improved hysteretic characteristics has been elaborated.

To apply the ordered Ni-Fe-Mn AF phase in spin valves, a definite deposition order of permalloy and manganese layers is necessary. Notably, the permalloy layer should be deposited on manganese or the manganese containing alloy [1]. With this deposition order, the authors of [2] fixed the high value H_{ex} = 110 Oe after the annealing of the FeMn(15nm)/NiFe(15nm) bilayers. It should be noted that the formation of this ordered system is only possible with the thermal treatment of the films that contain the permalloy and manganese layers or the layers of the manganese containing alloy. The annealing of $Ni_{48}Fe_{12}Mn_{40}$ and $Ni_{32}Fe_{8}Mn_{60}$ films which after magnetron sputtering from the targets of corresponding compositions, were in the state of a homogeneous ternary solid solution did not result in the formation of an ordered antiferromagnetic phase. After annealing at 400°C, the decomposition of a ternary solid solution into the nickel based solid solution and almost pure manganese was revealed. A similar result was found in [3].

3 Experimental details

The samples were made by DC magnetron sputtering and by electron-beam evaporation on glass (Corning) substrates and single-crystalline sapphire (1012). To form a unidirectional anisotropy the magnetic field of 110 Oe was applied in the process of nanostructures growth, and the thermal-magnetic treatment was performed at pressure of 10^{-4} Pa in permanent magnetic field of 2 kOe applied in the sample plane at 260 °C for 4 h. The exchange bias field temperature dependence was measured in the temperature range of 20–260 °C. Etching of samples for preparation of bottom SVs was carried out in a device for reactive ion-plasmic etching PlasmaPro NGP 80 RIE Oxford.

2 Results and discussion

According to the results of the studies carried out, the bilayers with antiferromagnetic triple alloy $Ni_{14}Fe_{6}Mn_{80}$ are characterized by medium blocking temperature T_b= 150 °C (Fig. 1) and moderate energy of exchange interaction J_{ex} = 0.05 erg/cm^2 [4].

Higher T_b = 270 °C (Fig. 1) and J_{ex} = 0.27 erg/cm^2 were obtained for the annealed manganese/permalloy bilayers. In this case, an ordered AF phase of Ni-Fe-Mn is formed [5]. The ordered AF phase formation is testified by an appearance of super-structural Debye rings (100), (110), (210), (211) (indicated by arrows in Fig. 2a) in electron diffraction patterns of sample Al_2O_3/Mn(50 nm)/$Ni_{77}Fe_{23}$(30 nm)/Ta(5 nm) after its annealing in the magnetic field at T_{ann} = 260 °C for 4 h. In this case in the electron-microscope images one can see the columnar structure, and an intermediate layer between manganese and permalloy layers is absent (Fig. 2b).

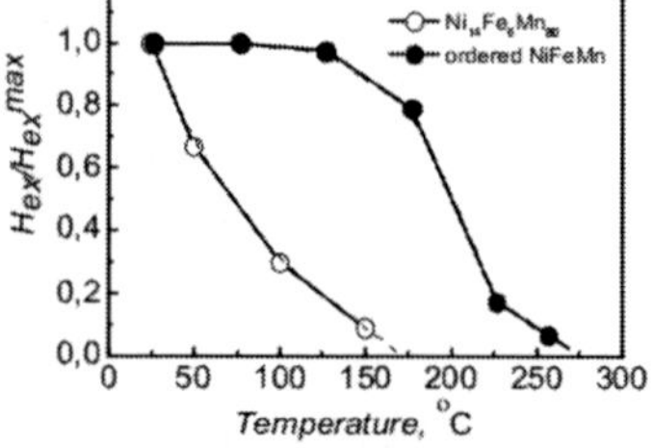

Fig. 1. Temperature dependence of the exchange bias field (H_{ex}) for two types of AF used.

* Corresponding author: blinov@imp.uran.ru

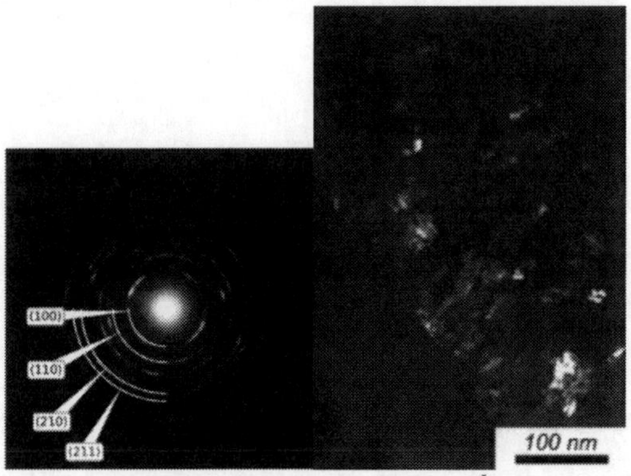

a b

Fig. 2. Electron diffraction pattern (a) and dark-field image in (100) reflection (b) of a cross section TEM image of Al_2O_3/Mn(50 nm)/$Ni_{77}Fe_{23}$(30 nm)/Ta(5 nm) after thermal-magnetic treatment at 260 °C for 4 h. Super-structural reflections are shown by arrows.

Based on the ordered AF Ni-Fe-Mn phase the spin valves with an enhanced thermal stability can be fabricated.

As shown by the studies of the effect of the AF layer $Ni_{14}Fe_6Mn_{80}$ thickness on the magnetoresistance and shift of the pinned layer $Co_{90}Fe_{10}$ loop of the top spin valve glass/Ta/$Ni_{80}Fe_{20}$/$Co_{90}Fe_{10}$/Cu/$Co_{90}Fe_{10}$/$Ni_{14}Fe_6Mn_{80}$/Ta, the combination of maximal values of magnetoresistive effect and pinned layer $Co_{90}Fe_{10}$ loop shift is observed at t_{AF} = 25 nm. At this thickness ($\Delta R/R_s$) = 6.95 %.

To study the magnetoresistance dependence of a spin valve on the copper layer thickness the samples with t_{AF} = 25 nm were prepared.

With increasing copper layer thickness the magnetoresistance $\Delta R/R_s$ at first increases and then decreases [4]. The dependence obtained is non-monotone and demonstrates qualitative agreement with the data published in [6].

The maximal value of $\Delta R/R_s$ = 7.30 % corresponds to t_{Cu} = 2.8 nm (Fig. 3). The magnetoresistance sensitivity determined as an average value for the ascending and descending hysteresis loop of the free layer $Ni_{80}Fe_{20}$/$Co_{90}Fe_{10}$ is $\Delta(\Delta R/R_s)/\Delta H$ = 0.75 %/Oe. The data obtained demonstrate possibility of using disordered alloy $Ni_{14}Fe_6Mn_{80}$ as a pinning layer in top spin valves.

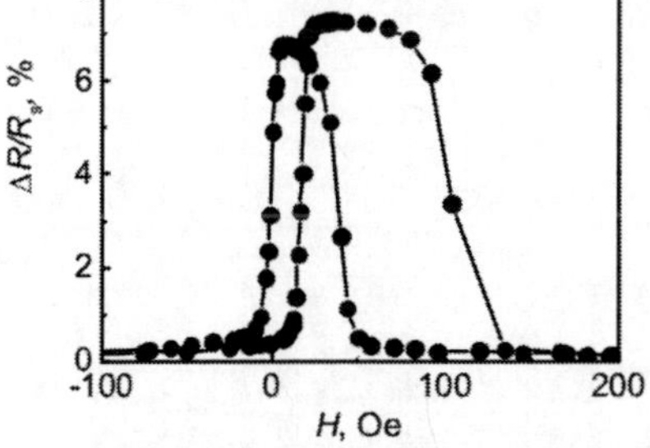

Fig. 3. Field dependence of magnetoresistance of a top spin valve glass/Ta(5 nm)/$Ni_{80}Fe_{20}$(2 nm)/$Co_{90}Fe_{10}$(5.5 nm)/ Cu(2.8 nm)/$Co_{90}Fe_{10}$(5.5 nm)/$Ni_{14}Fe_6Mn_{80}$(25 nm)/Ta(3 nm).

To fabricate a bottom SV with the ordered AF phase Ni-Fe-Mn a technological cycle has been worked out, including the following operations:

1) Formation of the ordered AF phase Ni-Fe-Mn in a sample Al_2O_3/$Ni_{77}Fe_{23}$(5 nm)/Mn(50 nm)/$Ni_{77}Fe_{23}$(30 nm)/Ta(5 nm) by the thermal- magnetic treatment at 260 °C for 4 h.

2) Ion etching for 20 min of the annealed sample for the surface layer removing.

The ion etching duration was chosen to assure the removal of the contaminated surface layer and retaining the ferromagnetic area together with the ordered AF phase required for the formation of the unidirectional anisotropy in the FM layer sputtered on the sample after etching.

3) Magnetron sputtering of the layered structure consisting of ferromagnetic $Co_{90}Fe_{10}$ layers separated by Cu on the as-prepared sample Al_2O_3/Ni-Fe-Mn.

4) Annealing of the as-prepared structure in a magnetic field at T_{ann} = 300 °C for 15 min, the annealing temperature T_{ann} = 300 °C being higher than T_b = 270 °C for Al_2O_3/$Ni_{77}Fe_{23}$(5 nm)/Mn(50 nm)/$Ni_{77}Fe_{23}$(30 nm)/ Ta(5 nm).

The magnetoresistance of the as-fabricated spin valve Al_2O_3/Ni-Fe-Mn/$Co_{90}Fe_{10}$(5.5 nm)/Cu(3.6 nm)/$Co_{90}Fe_{10}$(5.5 nm)/Ta(5 nm) is $\Delta R/R_s$ = 3.8 % (Fig. 4). This value is significantly higher than the effect obtained in [1], since the replacement of permalloy in the free and pinning layer by Co or the $Co_{90}Fe_{10}$ alloy leads to an increase in spin-dependent scattering and an increase in the giant magnetic resistance effect (GMR) in the spin valves [7].

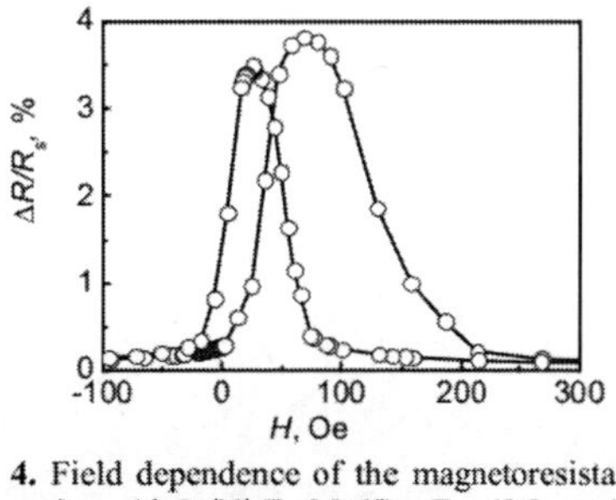

Fig. 4. Field dependence of the magnetoresistance of bottom spin valve Al_2O_3/Ni-Fe-Mn/$Co_{90}Fe_{10}$(5.5 nm)/Cu(3.6 nm)/ $Co_{90}Fe_{10}$(5.5 nm)/Ta(5 nm).

4 Conclusion

A possibility of application of the Ni-Fe-Mn AF for SV devices is demonstrated on a sample of composition Ta/$Ni_{80}Fe_{20}$/$Co_{90}Fe_{10}$/Cu/$Co_{90}Fe_{10}$/$Ni_{14}Fe_6Mn_{80}$/Ta. A technology of fabrication of a spin valve based on the ordered AF phase Ni-Fe-Mn has been worked out. A spin valve with giant magnetic resistance exceeding the GMR of previously known structures has been created, and the possibility of using the ordered AF phase Ni-Fe-Mn as a pinning layer in a spin valve has been proved.

The research has been done within the state assignment of FASO of Russia (theme "Spin", with partial support by UB RAS (project No. 18-10-2-3) and RFBR (project No. 16-32-00128).

References

1. A.V. Svalov, V. O. Vas'kovskii, and Yu . M. Yarmoshenko, Fiz. Met. Metalloved. **79**, 270 (1995)
2. K.C. Chen, C.T. Yang, Y. H. Wu, C.H. Huang, K.M. Wu, J. C. Wu, S. L. Young, and L. Horng, Phys. Status Solidi **C 4**, 4372 (2007).
3. C. S. Yoon, S. J. Kim, S. J. Kim, and C. R. Kim, J. Appl. Phys. **94**, 539 (2003)
4. I. V. Blinov, T. P. Krinitsina, M. A. Milyaev, V. V. Popov, V. V. Ustinov, Sol. Stat. Phenomena **233-234**, 517 (2015)
5. I. V. Blinov, T. P. Krinitsina, S. A. Matveev, M. A. Milyaev, P. N. Sedova, V. V. Popov, V. V. Ustinov, J. Nanosci. Nanotech. **12**, 7562 (2012)
6. V.V. Ustinov, M.A. Milyaev, L.I. Naumova, V.V. Proglyado, N.S. Bannikova, T.P. Krinitsina, Physics of Metals and Metallography **113,** 341 (2012)
7. Coehoorn, R. Giant Magnetoresistance and Magnetic Interactions in Exchange-Biased Spin-Valves, in: K.H.J. Buschow (Ed.), Handbook of magnetic materials **15** – Elsevier B.V., Amsterdam, P.1 – 199 (2003)

Magnetic interaction at the interface of epitaxial manganite film and $(TbCo_2/FeCo)_n$ superlattice

Alexander S. Grishin[1,2*], *Gennady* A. Ovsyannikov[1], *Alexey* A. Klimov[1,2], *Viktor* V. Demidov[1], *Karen* Y. Constantinian[1], *Igor* V. Borisenko[1], *Vladimir* L. Preobrazhensky[3,4], *Nicolas* Tiercelin[4], *Phillipe* Pernod[4]

[1]Kotel'nikov Institute of Radio Engineering and Electronics Russian Academy of Sciences, 11 Mokhovaya str., 125009 Moscow, Russia
[2]Moscow Technological University MIREA, 86 Vernadsky av., 119571 Moscow, Russia
[3]The Wave Research Center, Prokhorov General Physics Institute Russian Academy of Sciences, 38 Vavilova str., 119991 Moscow, Russia
[4]Univ. Lille, CNRS, Centrale Lille, ISEN, Univ. Valenciennes, UMR 8520 - IEMN, F-59000 Lille, France

Abstract. Hybrid magnetic heterostructures made of epitaxial manganite $La_{0.7}Sr_{0.3}MnO_3$ thin film and intermetallic superlattice $(TeCo_2/FeCo)_n$ were prepared on orthorhombic $NdGaO_3$ substrates and characterized by means of magneto-optical Kerr effect and magnetoresistance. Experimental data show that magnetic interaction between $La_{0.7}Sr_{0.3}MnO_3$ and $(TeCo_2$ $/FeCo)_n$ is of the antiferromagnetic type. The features observed in magnetotransport characteristics are caused by the magnetization reversal at the interface between manganite thin film and intermetallic superlattice.

1 Introduction

The development of spintronic devices based on nanoscale interfaces of magnetic materials is a challenging problem. Stacks of complex materials (manganites, multiferroics, rare earth compounds and others) for non-volatile memory devices were reviewed in [1]. Manipulation of phase transition in manganite $La_{0.7}Sr_{0.3}MnO_3$ (LSMO) thin film was achieved by applying positive/negative voltage over an additional gate electrode. Large tunneling resistance effect, accompanied with a moderately magnetoresistance was reported for $LSMO/BaTiO_3/Co$ junctions [2].

Recently a promising solution was proposed for a memory cells providing reliable spin manipulation and switching of magnetization between two stable positions in intermetallic structures [3]. The solution is based on a superlattice of exchange-coupled rare-earth compound $TbCo_2$ and 3d transition metal FeCo nanolayers. The structure is characterized by a giant magnetostriction and demonstrates spin reorientation transitions by means of an external magnetic field, and/or elastic strain [4,5]. In transition metal oxides $R_{1-x}A_xMnO_3$ (manganites) where R is one of the rare earth elements La or Pr and A is one of the alkaline earth metals Sr or Ca, new electronic states and magnetic phases can arise under electric fields and strains in thin film or at the interfaces with dielectrics or other oxides.

However, the magnetic interaction and electronic and spin states at the interface between manganites and intermetallic structures were not studied yet. In this paper, we fabricated manganite/intermetallic heterostructures in order to determine magnetization behavior at interface and to reveal the influence of spin-polarized current on magnetotransport characteristics.

2 Sample fabrication and experimental

The heterostructures under consideration consist of two layers. The top layer is the rare earth intermetallic superlattice, consisting of an exchange-coupled multilayered $(TbCo_2$ (4nm)/FeCo(4 nm))n (TCFC) stack with $n = 25$. The bottom layer is the lanthanum - strontium manganite $La_{0.7}Sr_{0.3}MnO_3$ (LSMO) 30 nm thick film which was epitaxially deposited on an orthorhombic $NdGaO_3$ (NGO) substrate.

The LSMO films were deposited onto $5\times5\times0.5$ mm^3 (110) NGO substrates by magnetron sputtering at 700-750°C temperature and partial oxygen pressure between 0.1 and 0.3 mbar. The desired crystal structure of the LSMO film was ensured by the NGO substrate that promoted the dominance of uniaxial in-plane magnetic anisotropy [3]. The intermetallic superlattice was obtained by the sequentially sputtering of $TbCo_2$ and FeCo nano-layers under applied magnetic field directed in the plane of the substrate, which defined the direction of the uniaxial easy axis magnetic anisotropy. The strength of the magnetic anisotropy in the TCFC superlattice is controlled by the ratio of Tb, Co, Fe elements in the layers as well as by the thickness and number n of $TbCo_2$ and FeCo layers. The heterostructures were fabricated with directions of easy axis magnetization of the manganite and intermetallic superlattice being in parallel. The LSMO film covers the whole 5×5 mm^2 area of substrate, whereas the TCFC is a 3×3 mm^2 square at the center of the sample (Fig. 1).

Magnetic parameters, the coercive force (H_c) and saturation field (H_s) of heterostructures were studied using magneto-optical Kerr effect (MOKE). The magneto-optical setup involved a semiconductor laser operating at $\lambda = 0.63$ μm wavelength and power $P = 5$ μW, a beam-splitter glass plate for forming the reference and signal

* alex1995@hitech.cplire.ru

https://doi.org/10.1051/epjconf/201818501007

beams, a $\lambda/2$ thick phase-shifter plate for selecting the *s*- or *p*-type polarization of the incident beam, and a polarizer/analyzer for selecting the signal caused by the meridional Kerr effect. The laser spot diameter at the measuring point is about 1 mm. The investigated sample was placed on a rotary table in the gap of electromagnet. The angle (δ) of polarization plane rotation of the beam reflected from the surface of the magnetized sample was determined by the compensation method using the polarizer-analyzer. The dependence $\delta(H)/\delta_s \propto M(H)/MS$ was measured, where δ_s is the magneto-optical Kerr effect (MOKE) signal value at M = MS (M_S is the saturation magnetization of the sample). A detailed description of the experimental setup is given in [3]. For heterostructures characterization, we also used the vibrating sample magnetometer (VSM). Magnetoresistance $R(H)$ dependencies were obtained by measuring the differential resistance $R=dV/dI$ as a function of the applied magnetic field H. Most of measurements were carried out at room temperature T=300 K.

The Curie temperatures (T_C) for both TCFC and LSMO films were well above 300 K. Nevertheless, some of the resistive characteristics for higher accuracy were measured at 77 K by a 4-point method.

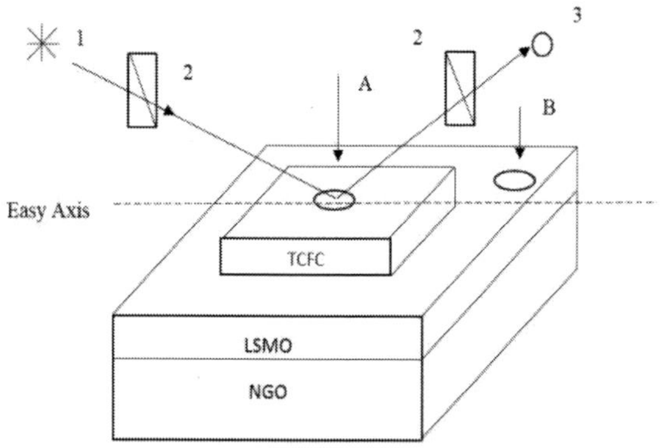

Fig. 1. Schematic view of the investigated heterostructures and Kerr-effect geometry: He-Ne laser (1), polarizers (2), detector (3). "A" and "B" are respectively the spot positions where TCFC/LSMO structure and LSMO film only were probing.

3 Results and discussion

3.1 Kerr effect measurements

Figure 2 shows the magnetic field dependence of magnetization that was obtained by MOKE measurements when an external magnetic field was directed along the hard axis of magnetization. The magnetization behavior for the TCFC/LSMO heterostructure (the dashed line in Fig. 2b) was obtained when the probing laser at the central part of heterostructure (spot A in Fig.1) whereas the LSMO film was probed when beam was focused at spot B (the solid line in Fig.2b).

It is seen that the saturation field for TCFC/LSMO, $H_S \approx 1000$ Oe was much higher than for the LSMO film, $H_S \approx 20$ Oe, and that the saturation magnetization of TCFC/LSMO heterostructure is also much higher than the one of LSMO film. Note, the influence of the underlying LSMO film cannot be observed by MOKE at the spot A, since the laser beam is screened by the TCFC film. When probing the LSMO film only (spot B), a change of magnetization sign was observed that points to an antiferromagnetic magnetic ordering at the TCFC/LSMO interface of the heterostructure.

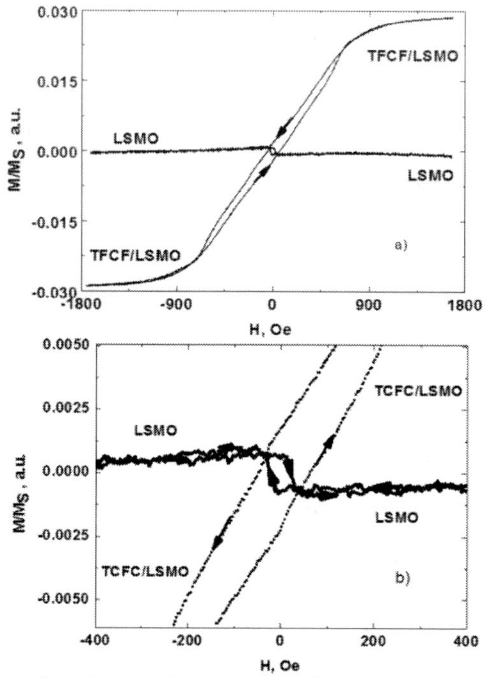

Fig. 2. Magnetization curves of the TCFC/LSMO heterostructure. a) Full-scale plot of magnetization curve for a ±2kOe sweeping range. Spot position (A) TCFC/LSMO and (B) LSMO film only (see Fig.1). The external magnetic field is directed along the hard magnetization axis (ha) as shown on inset; b) Magnified central part of the magnetization curves for both TCFC and LSMO film with a±400 Oe sweeping range. The arrows indicate the magnetic field evolution.

Figure 3 shows hysteresis loops measured at H-field directed along the easy axis. Figure 3a shows the hysteresis loop for LSMO film and LSMO/TCFC heterostructure when remagnetization range was larger than 2 kOe. The widths of the hysteresis loop for both the TCFC/LSMO heterostructure and the LSMO film approximately correspond to a coercive force of about 200 Oe. From Fig. 3b it is seen in a shorter remagnetization range, less than 400 Oe, the hysteresis loop of the TCFC/LSMO heterostructure exhibits a $M(H)$ dependence with an inverted sign as in the case when magnetic field was directed along the hard axis. The hysteresis loop width is approximately equal to the coercitivity of the LSMO film obtained before TCFC deposition on top. When H-field sweeping amplitude was larger than 400 Oe, the magnetization curve for LSMO film repeats the shape of the hysteresis loop of the TCFC superlattice. Increasing the magnetic field sweeping range over 400 Oe leads to a magnetization curve of the LSMO film that coincides with the one shown in Figure 3a.

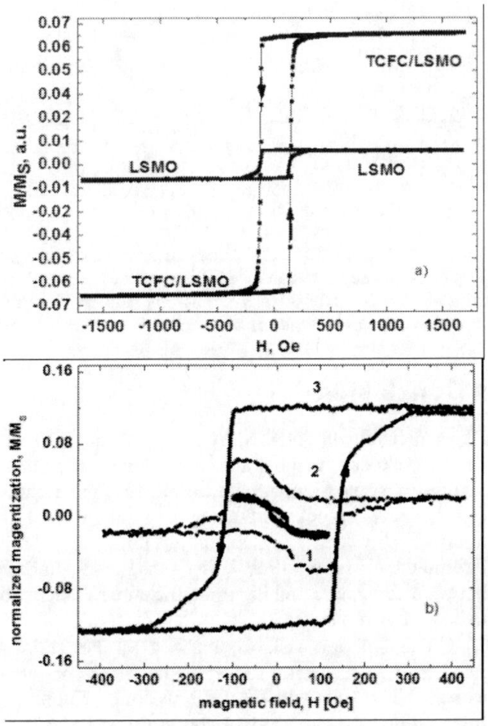

Fig. 3. Magnetization versus magnetic field for the TCFC/LSMO heterostructure. Magnetization normalized for TCFC/LSMO part of the heterostructure. a) Full-scale magnetization curve for ±2 kOe sweep ranges. The external magnetic field is directed along the easy magnetization axis (ea). b) Magnified central part of the magnetization curve for the LSMO film for various sweep ranges:(1) ±100 Oe, (2) ±400 Oe and (3)±2 kOe. The arrows indicate the direction of the magnetic field evolution.

3.2 Magnetization

Fig. 4a displays VSM-measured the hysteresis loops of single LSMO film deposited on NGO. Hysteresis loops are given for two orthogonal orientations of magnetic field. The saturation magnetic field and magnetization are equal to H_S= 20 Oe, M_0=198 emu/cm^3, correspondingly. The coercive force and saturation magnetic field are close to each other.

The VSM-measured hysteresis loops of TCFC/LSMO heterostructure for two orthogonal orientations of magnetic field are shown in Fig. 4b. Along the easy axis, the hysteresis curve exhibits two steps corresponding to the reversal of magnetization in the soft (LSMO) and the hard (TCFC) magnetic layers. A similar behavior of magnetization was obtained in the [6] where computer simulations were performed for trilayer magnetic structures taking into account the coupling between layers. In our case, the third layer providing the interaction of two ferromagnetics is the interface of LSMO/TCFC. The thickness of the active layers can be estimated using the approach in the work [7]. As it follows from the magnetization curves, (Fig. 4b) a stable antiferromagnetic (AFM) configuration takes place at the certain range of the external magnetic field. Taking into account the magnetization distributions discussed in [6], we can conclude that the reversal process proceeds more uniformly in the hard magnetic layer than in the soft one. In the former case, the magnetization undergoes a discontinuous change, with its distribution being relatively uniform before the switching. In contrast, the soft magnetic layer is switched very non-uniformly. The non-uniformity expands gradually, moving from the edges of the layer towards its center.

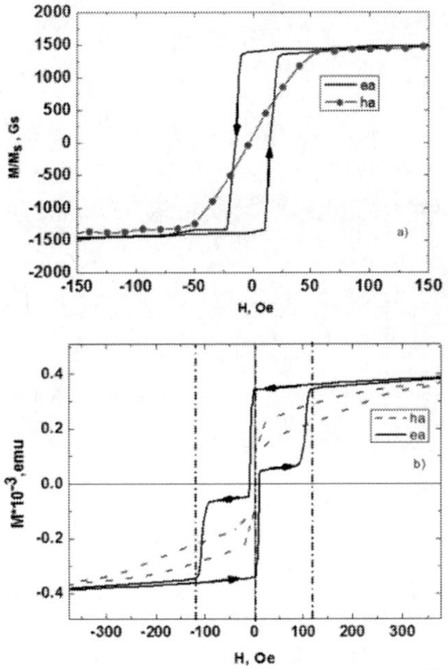

Fig. 4. Hysteresis loops for two structures: a) LSMO film, b) TCFC /LSMO heterostructure. The external magnetic field is directed along the hard axis (dashed lines ha) and along the easy axis (solid lines ea).

3.3 Magnetoresistance

The magnetoresistance of the TCFC/ LSMO heterostructure was measured at T=77 K for two directions of the applied magnetic field. The variation of magnetoresistance was about 0.2% at T=77 K but it was smaller at T=300K. The resistance was measured over the diagonal direction of the substrate, which corresponds to the magnetic hard axis. The magnetic field was applied either along hard magnetization axis (Fig.5a) or along easy magnetization axis (Fig.5 b) of TCFC layer.

At low field there is a shift of the maximum of $R(H)$ which indicates the presence of ferromagnetism. The distance between the peaks of $R(H)$ allows to estimate the coercive force of the heterostructure. The slope of the *R(H)* curves indicates the presence of colossal magnetoresistance in the manganite film. The additional features of magnetization are observed in the magnetic field ranged 250- 500 Oe applied along hard axis. They are probably caused by the switch of magnetic moment in the TCFC superlattice. When the magnetic field is applied along the easy axis *R(H)* maximum is also shifted about 80 Oe from *H*=0. But there are no features in the range from 250 to 500 Oe of external magnetic field.

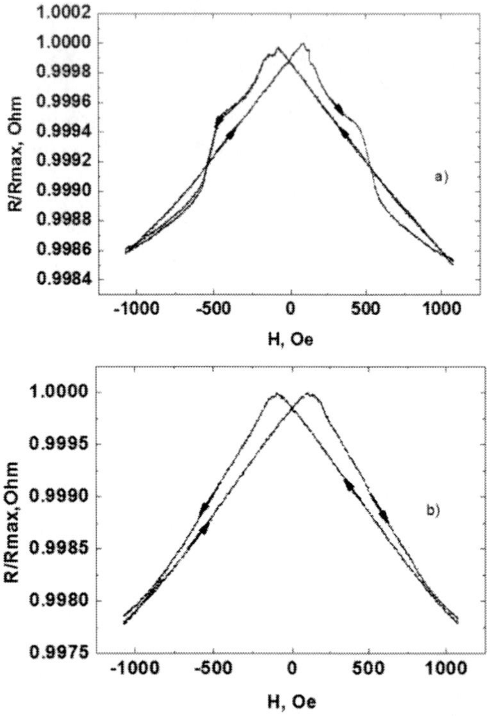

Fig. 5. Dependence of the resistance on magnetic field applied along the hard (a) and easy (b) axes.

Figure 6 shows the difference between magnetoresistances for magnetic field oriented along hard and easy axes. The shape of the functions shown in Fig.6 reflects the contribution of the interlayer interaction to magnetoresistance. Giant magnetoresistance (GMR) of the heterostructure and anisotropic magnetoresistance (AMR) are proportional to $(\cos\theta)^2$ where θ is the angle between magnetizations of the layers [3]. The features of the magnetoresistance observed for magnetic field applied along the hard axis and caused by switching of the TCFC film magnetization are similar to ones predicted in [4].

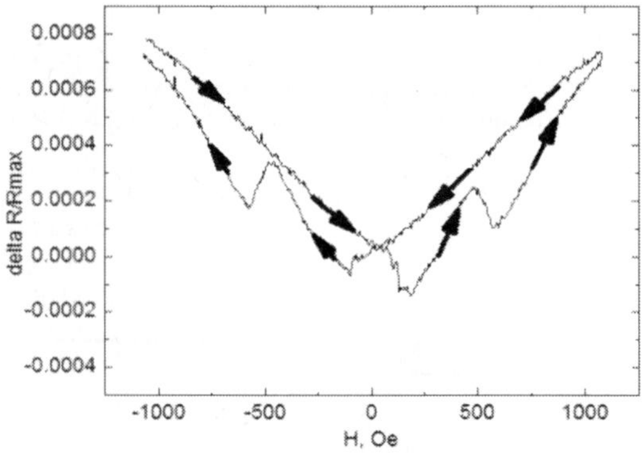

Fig. 6. Difference between magnetoresistance presented on Fig. 5a and Fig 5b. The arrows point. The arrows indicate the direction of magnetic field change.

4 Conclusion

Experimental data obtained by means of magneto-optical Kerr effect and magnetoresistance and show that the features observed in magnetotransport characteristics are caused by the process of magnetization reversal at the interface between a manganite thin film and an intermetallic superlattice. The magnetic interaction between manganite and intermetallic superlattice is found to be antiferromagnetic.

The authors wish to thank A.M. Petrzhik, A.V. Shadrin, Y.V. Kislinskii and T.A. Shayhulov for help with experiment and useful discussion. Support by International Associated Laboratory LEMAC-LICS, Russian Foundation for Basic Research 16-29-14022, and Scientific School NSh-8168.2016.2 is acknowledged. Samples were partly processed with the support of the RENATECH technological network.

References

1. D. S. Jeong, R. Thomas, R. S. Katiyar, J. F. Scott, H. Kohlstedt, A. Petraru and Cheol Seong Hwang, Rep. Prog. Phys.75, 076502 (2012).

2. H.J. Mao, P. X. Miao, J. Z. Cong, C. Song, B. Cui, J. J. Peng, F. Li, G. Y. Wang, Y. G. Zhao, Y. Sun, L. R. Xiao, and F. Pan, J. Appl. Phys. 116, 053703 (2014).

3. Y. Dusch, N. Tiercelin, A. Klimov, S. Giordano,V. Preobrazhensky, and Ph. Pernod, J. Appl. Phys. **113**, 17C719 (2013).

4. A. Klimov, N. Tiercelin, V. Preobrazhensky, and Ph. Pernod, IEEE Transactions on Magnetics, **42**, 3090 (2006).

5. Y. Dusch, V. Rudenko, N. Tiercelin, S. Giordano, V. Preobrazhensky, and P. Pernod, Nanomater. Nanostruct. 2, 44 (2012).

6. K.A. Zvezdin, J. Physics of the Solid State 1, 120-125 (2000)

7. A.Aharoni, J. Appl. Phys. **76**, 6977 (1994).

BIFURCATION ANALYSIS OF THE DYNAMICAL SYSTEM FOR A THREE-LAYERED VALVE WITH PERPENDICULAR ANISOTROPY

Natalia Ostrovskaya[1,*], *Vladimir* Skidanov[1], *Iulia* Iusipova[1,2], and *Maxims* Skvortsov[1,2]

[1]Institute for Design Problems in Microelectronics, Department of Microelectronic Components, Soviet Str. 3, Zelenograd, 124365 Moscow, Russia
[2]National Research University of Electronic Technology, Department of Applied Mathematics, Zelenograd, Bld.1, Shokin Square, 124498, Moscow, Russia

Abstract. The features of switching dynamics in a model of a three-layered valve have been investigated theoretically and numerically. For this purpose, the system of ordinary differential equations in the approximation of the uniform magnetization distribution for the magnetization dynamics in the valve with perpendicular anisotropy was derived. It was shown that in such a system, in contrast with the system for the in-plane anisotropy, there are only two equilibrium positions of the magnetization vector. The stability analysis of the stationary points of the system has been carried out. With its help, the classification of types of dynamics versus field and current values was performed. The regions of limit cycles existence and the regions of optimal magnetization switching were revealed.

The magnetic random access memory (MRAM) attracts the attention of electronics engineers due to its high speed of magnetic switching, low energy consumption, high data storage density, and reliability. In the basic publication by J. Slonczewski (1996) [1], the first suggested model is composed of two ferromagnetic layers with in-plane anisotropy separated by a nonmagnetic interlayer. However, both calculations and experiments showed that switching currents in the classical ferromagnetics, such as cobalt and iron, are too high. It was found experimentally that the most optimal materials for the ferromagnetic layers are the ferromagnetic alloys FeCoB, and the most promising design solution is a memory cell with perpendicular magnetic anisotropy of the ferromagnetic layers [3–7].

1 Model

The object under study is a three-layered valve structure consisting of two ferromagnetic layers separated by nonmagnetic one. The cross-section of the structure is the $d_s \times d_s$ square; the thickness of the thin layer is d_1. These dimensions assumed to be small enough to use the uniform distribution of the magnetization in the layer. The magnetization of the reference (thick) layer is fixed and directed from the thick to the thin layer. This direction is taken as positive for the OZ-axis, which is perpendicular to the layers. The current passes in parallel to OZ-axis. Its density J lies in the interval from 0 to 10^{13} A/m^2. The structure is placed into the external magnetic field parallel to OZ. This field can be either positive or negative (Fig. 1).

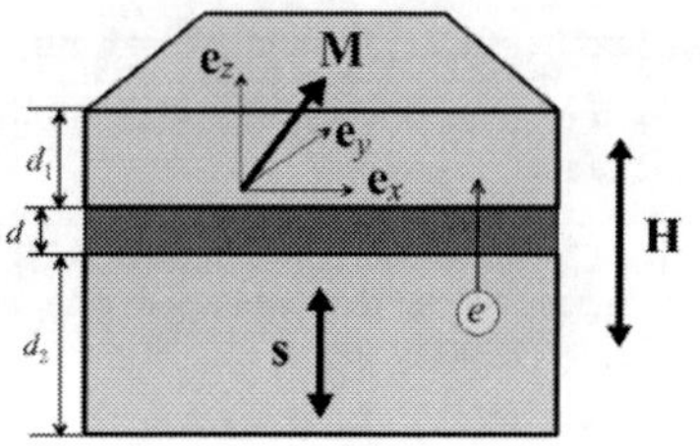

Fig. 1. Geometry of the model.

1.1 Basic Equations

The theory of the phenomenon is proposed by J. Slonczewski [1]. The model is based on the fundamental Landau–Lifshits–Gilbert equation that describes the dynamics of the magnetization vector $\mathbf{M}$ in the free ferromagnetic layer:

$$\frac{\partial \mathbf{M}}{\partial t} = -|\gamma|\mu_0\left[\mathbf{M}\times\mathbf{H}^{\text{eff}}\right] + \frac{\alpha}{M_s}\left[\mathbf{M}\times\frac{\partial \mathbf{M}}{\partial t}\right]. \qquad (1)$$

Here $\mu_0 = 4\pi\cdot 10^{-7}$ N/A^2 is the magnetic constant, $\gamma = 1.76\cdot 10^{11}$ rad$\cdot$T$^{-1}\cdot$s^{-1} is the gyromagnetic ratio, α is the dimensionless damping parameter, M_s is the saturation magnetization, $\mathbf{H}^{\text{eff}}$ is the effective magnetic field, which is the equivalent of those physical interactions that are taken into account in the model. In the case under consideration, it is

$$\mathbf{H}^{\text{eff}} = \mathbf{H} + \mathbf{H}_a + \mathbf{H}_f + \mathbf{H}_c, \qquad (2)$$

* Corresponding author: n.ost@ippm.ru

where $\mathbf{H}$ is the external magnetic field, $\mathbf{H}_a$ is the effective field of the magnetic anisotropy, $\mathbf{H}_f$ is the effective demagnetizing field that appears because of the finite sizes of the valve, $\mathbf{H}_c$ is the effective field originated due to the spin-polarized injection current. The exchange interaction is ignored due to the small sizes of the structure (the approximation of uniform magnetization). We supposed the anisotropy effective field and the magnetization of the fixed layer to be perpendicular to the valve layers. At that, the magnetization of the fixed layer is supposed to be directed to the free layer of the valve (the vector $\mathbf{s}$ in Fig. 1, $|\mathbf{s}|=1$). The magnetization of the free layer at the first moment is taken to be co-directed with the magnetization of the fixed layer (parallel orientation). After the simultaneous field and current engaging, it can pass to another stationary position. It is natural to name the parallel orientation of the magnetization vectors as "zero", and the anti-parallel orientation as "one" (or vice versa).

Equation (1) can be written in the dimensionless form

$$\frac{\partial \mathbf{m}}{\partial \tilde{\tau}} = -\left[\mathbf{m}\times\mathbf{h}^{\text{eff}}\right] + \alpha\left[\mathbf{m}\times\frac{\partial \mathbf{m}}{\partial \tilde{\tau}}\right], \tag{3}$$

where $\mathbf{m}=\dfrac{\mathbf{M}}{M_s}$, $\mathbf{h}^{\text{eff}}=\dfrac{\mathbf{H}^{\text{eff}}}{M_s}$, $|\mathbf{m}|=1$, time $\tilde{\tau}$ is measured in the units $(\gamma\mu_0 M_s)^{-1}$, and $\mathbf{h}^{\text{eff}}=\mathbf{h}+\mathbf{h}_a+\mathbf{h}_f+\mathbf{h}_c$.

1) In the general case, the external field $\mathbf{h}$ can be presented as

$$\mathbf{h} = h_x\mathbf{e}_x + h_y\mathbf{e}_y + h_z\mathbf{e}_z.$$

In the case under consideration, the external field is parallel to OZ-axis, i.e.

$$\mathbf{h} = h\mathbf{e}_z.$$

2) The anisotropy field is also oriented along OZ-axis

$$\mathbf{h}_a = k(\mathbf{m},\mathbf{e}_z)\mathbf{e}_z = km_z\mathbf{e}_z,$$

where $k = 2K_a\mu_0^{-1}M_s^{-2}$, and K_a is the anisotropy constant.

3) The demagnetizing field $\mathbf{h}_f$ is defined as $\mathbf{h}_f = -\hat{\mathbf{q}}\mathbf{m}$, where $\hat{\mathbf{q}}$ is a tensor (form-factor). In the geometry of this model we can put (see [2], for example)

$$\mathbf{h}_f = -m_z\mathbf{e}_z.$$

4) According to the Slonczewski–Berger theory, the contribution into effective field caused by the spin-polarized current is

$$\mathbf{h}_c = G\frac{J}{J_n}[\mathbf{s}\times\mathbf{m}],$$

where $\mathbf{s}$ is the unit vector of the spin polarization, the direction of which is coincident with the direction of the magnetization in the thick layer (in this geometry $\mathbf{s}\equiv\mathbf{e}_z$), J is the dimensional density of the spin-polarized current, J_n is the normalization coefficient, which is equal to

$$J_n = \frac{d_1 e\mu_0 M_s^2}{\hbar}, \tag{4}$$

($\hbar$ is the Plank constant, μ_0 is the magnetic constant, e is the charge of the electron, d_1 is the thickness of the free ferromagnetic layer). Hence, the dimensionless current density is $j = J/J_n$. According to [1], the scalar dimensionless function $G(\mathbf{m})$ is as follows:

$$G = \frac{4P^{3/2}}{(1+P)^3(3+(\mathbf{s},\mathbf{m}))-16P^{3/2}},$$

where P is the parameter of the spin polarization in the interface of the valve.

Taking into account that, in this case, $\mathbf{s}\equiv\mathbf{e}_z$, we obtain $(\mathbf{s},\mathbf{m}) = m_z$. Hence,

$$G = \frac{4P^{3/2}}{(1+P)^3(3+m_z)-16P^{3/2}} = \frac{c}{b+m_z},$$

($c = 4P^{3/2}/(1+P)^3$, $b = 3-4c$). Note that, in our geometry, $[\mathbf{s}\times\mathbf{m}] = -m_y\cdot\mathbf{e}_x + m_x\mathbf{e}_y$. Therefore, for the valve with the perpendicular anisotropy, the effective magnetic field takes the following form:

$$\mathbf{h}^{\text{eff}} = \begin{pmatrix} h_x^{\text{eff}} \\ h_y^{\text{eff}} \\ h_z^{\text{eff}} \end{pmatrix} = \begin{pmatrix} -gm_y \\ gm_x \\ km_z - m_z + h \end{pmatrix}, \tag{5}$$

where $g = Gj$.

After certain algebraic transformations of (3), we obtain

$$\frac{\partial \mathbf{m}}{\partial \tau} = -\left[\mathbf{m}\times\mathbf{h}^{\text{eff}}\right] + \alpha\mathbf{h}^{\text{eff}} - \alpha\mathbf{m}(\mathbf{m},\mathbf{h}^{\text{eff}}), \tag{6}$$

where $\tau = \dfrac{\tilde{\tau}}{1+\alpha^2} = |\gamma|\dfrac{\mu_0 M_s}{1+\alpha^2}t$. In the coordinate form, the vector equation (6) is equivalent to the system

$$\begin{aligned}
\frac{dm_x}{dt} &= j\frac{c}{b+m_z}m_x m_z - m_y(km_z - m_z + h) - \alpha j\frac{c}{b+m_z}m_y - \\
&\quad -\alpha m_x m_z(km_z - m_z + h), \\
\frac{dm_y}{dt} &= m_x(km_z - m_z + h) + j\frac{c}{b+m_z}m_y m_z + \alpha j\frac{c}{b+m_z}m_x - \\
&\quad -\alpha m_y m_z(km_z - m_z + h), \\
\frac{dm_z}{dt} &= -j\frac{c}{b+m_z}m_y^2 - j\frac{c}{b+m_z}m_x^2 + \alpha(km_z - m_z + h) - \\
&\quad -\alpha m_z^2(km_z - m_z + h).
\end{aligned} \tag{7}$$

The parameters of the Co/Cu/Co structure, which were used in the numerical calculations, were $\alpha = 0.02$, $P = 0.35$, $\mu_0 M_s = 1.76$ T, $K_a = 0.53\cdot10^6$ J/m^3, the thickness of the free layer was $d_1 = 4$ nm. The normalization coefficients were as follows:

$$t = \frac{1+\alpha^2}{\gamma\mu_0 M_s}\tilde{\tau} \cong 2\cdot10^{-10}\tau \text{ (s)},$$

$$K_a = \frac{\mu_0 M_s^2}{2}k \cong 1.2\cdot10^6 k \text{ (J/m}^3\text{)},$$

$$H = M_s h \cong 1.4\cdot10^6 h \text{ (A/m)},$$

$$J = \frac{d_1 e\mu_0 M_s^2}{\hbar}j \cong 1.9\cdot10^{13} j \text{ (A/m}^2\text{)}.$$

2 Stationary points

The first step to the qualitative classification of the dynamical regimes in the system (7) is to find the equilibrium states of the magnetization in the free layer of the valve. With this purpose, the right-hand parts of (7) must be equated to zero:

$$\begin{cases}(m_z h_y - m_y h_z) + \alpha h_x - \alpha m_x L = 0, \\ (m_x h_z - m_z h_x) + \alpha h_y - \alpha m_y L = 0, \\ (m_y h_x - m_x h_y) + \alpha h_z - \alpha m_z L = 0,\end{cases} \tag{8}$$

where $L = (\mathbf{m}, \mathbf{h})$. As a result, we obtain the algebraic fractionally-rational set of three equations for three variables m_x, m_y, m_z. Parameters α, P, k are thought to be fixed (internal), whereas the field h and the current density j are scalable (in other words, external or control ones). Solutions m_x, m_y, m_z of (8) at the current values h, j are nothing else but the stationary points of (7). At $h = 0$, $j = 0$, the system (8) degenerates to the form

$$\begin{cases}m_y m_z (1-k) - \alpha m_x L = 0, \\ (k-1) m_x m_z - \alpha m_y L = 0, \\ \alpha m_z (k-1) - \alpha m_z L = 0.\end{cases}$$

Here, $L = (k-1) m_z^2$, and the system (7) has two stationary points on the surface of the unit sphere with coordinates $T_{1,2}(0,0,\pm 1)$, and the singular line, which coincides with the equator of the sphere.

To find the singular points at the nonzero external parameters, it is necessary to solve (8) in the general form. It is not difficult to show that it has no other roots, except $T_{1,2}(0,0,\pm 1)$.

2.1.1 Type and stability of singular points

1) *Point* $T_1(0,0,+1)$.

The matrix of the system (7) linearized in the vicinity of the singular point $T_1(0,0,+1)$ has the form

$$\mathbf{A}|_{T_1} = \begin{pmatrix} \dfrac{jc - \alpha(b+1)(h-1+k)}{b+1} & -\dfrac{\alpha jc + (b+1)(h-1+k)}{b+1} & 0 \\ \dfrac{\alpha jc + (b+1)(h-1+k)}{b+1} & \dfrac{jc - \alpha(b+1)(h-1+k)}{b+1} & 0 \\ 0 & 0 & -2\alpha(h-1+k) \end{pmatrix}.$$

Its eigenvalues are

$$\lambda_{1,2}^+ = \frac{jc - \alpha(b+1)(h-1+k)}{b+1} \pm i \frac{\alpha jc + (b+1)(h-1+k)}{b+1},$$

$$\lambda_3^+ = -2\alpha(h-1+k).$$

The product of the eigenvalues λ_1^+ and λ_2^+ is

$$\lambda_1^+ \lambda_2^+ = \frac{(1+\alpha^2)}{(b+1)^2}\left[(b+1)^2 (h-1+k)^2 + j^2 c^2\right] \geq 0,$$

therefore, this singular point cannot be of a saddle-type, but is only a focus or a node.

Analyzing the eigenvalues, we see two threshold relations for the point $T_1(0,0,+1)$

$$L_1: \ \alpha cj + (b+1)h - (b+1)(1-k) = 0,$$

$$L_2: \ cj - \alpha(b+1)h + \alpha(b+1)(1-k) = 0.$$

In the plane of the control parameters (h, j), line L_1 separates the regions II and III, where the directions of the trajectories rotation in the vicinity of the singular point are opposite. Exactly in the line L_1, the singular point degenerates into a node.

Line L_2 separates the regions of the focus stability (I) and instability (II), therefore, according to the Andronov–Hopf theorem, in the vicinity of this line, a limit cycle must exist. It really does exist under the line L_2 and is unstable. The bifurcation diagram for the singular point $T_1(0,0,+1)$ is shown in Fig. 2a.

2) *Point* $T_2(0,0,-1)$

The matrix of the system linearized in the vicinity of this point is

$$\mathbf{A}|_{T_2} = \begin{pmatrix} \dfrac{jc + \alpha(1-b)(h+1-k)}{1-b} & \dfrac{\alpha jc - (1-b)(h+1-k)}{1-b} & 0 \\ -\dfrac{\alpha jc - (1-b)(h+1-k)}{1-b} & \dfrac{jc + \alpha(1-b)(h+1-k)}{1-b} & 0 \\ 0 & 0 & 2\alpha(h+1-k) \end{pmatrix}.$$

The eigenvalues are as follows:

$$\lambda_{1,2}^- = \frac{-jc + \alpha(b-1)(h+1-k)}{b-1} \pm i \frac{\alpha jc + (b-1)(h+1-k)}{b-1},$$

$$\lambda_3^- = 2\alpha(h+1-k).$$

As in the previous case, the product of λ_1^- and λ_2^- is always positive, because

$$\lambda_1^- \lambda_2^- = \frac{(1+\alpha^2)}{(b-1)^2}\left[(b-1)^2 (h+1-k)^2 + j^2 c^2\right] \geq 0.$$

That is why, the same as the singular point $T_1(0,0,+1)$, the point $T_2(0,0,-1)$ cannot be of the saddle type. The equations of the critical lines, which separate the (h, j)-plane into the regions of the identical magnetization dynamics in the vicinity of $T_2(0,0,-1)$, are

$$L_3: \ \alpha cj + (b-1)h + (b-1)(1-k) = 0,$$

$$L_4: \ cj - \alpha(b-1)h - \alpha(b-1)(1-k) = 0.$$

Line L_3 separates the areas with opposite directions of the rotation of the trajectories around the point $T_2(0,0,-1)$ on the sphere. The points that belong to the line correspond to nodes. Line L_4 separates the regions of foci stability and instability. As above, this is the line of the appearance/disappearance of the limit cycle. The limit cycles exist below this line. The bifurcation diagram for the point $T_2(0,0,-1)$ is shown in Fig. 2b.

In the center of figure 3, the superposition of the phase diagrams for the points T_1 and T_2 is displayed.

Critical lines L_1, L_2, L_3, L_4 divide the (h, j)-plane into the regions of the total identical magnetization dynamics and the equivalent phase portraits. Note that, in the case of the opposite magnetization **s** of the reference layer, the common bifurcation diagram would be reflected from the vertical axis j.

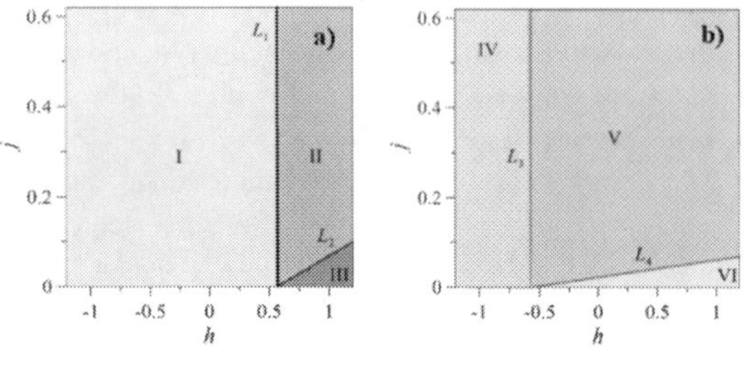

Fig. 2. Bifurcation diagrams: **a)** for the point $T_1(0,0,+1)$, **b)** for the point $T_2(0,0,-1)$.

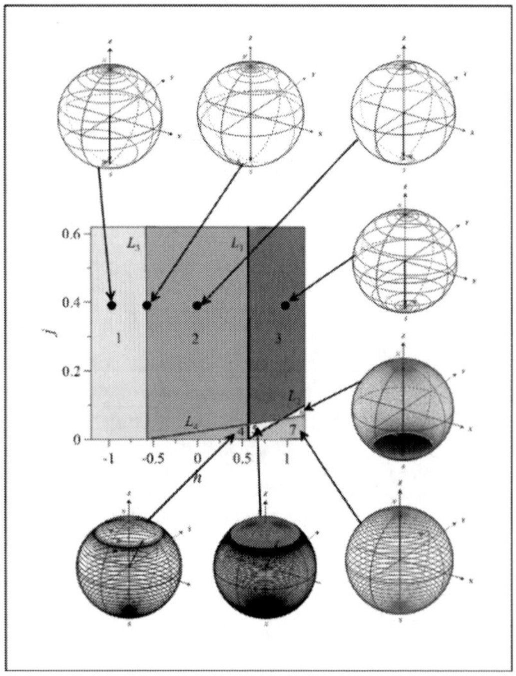

Fig. 3. Superposition of the bifurcation diagrams for the points $T_{1,2}(0,0,\pm1)$ (in the center) and the examples of the magnetization hodographs for each region in the plane (h, j).

3 Results

If the control parameters h and j belong to regions 1,2,3, the singular point T_1 will be an unstable focus, although the direction of the vector rotation in these regions will be different. At the same time, the singular point T_2 will be a stable focus at these parameters. Therefore, at the parameters taken from these regions, the stable switching from parallel to anti-parallel configuration takes place, which, for the parameters from region 2, happens with the change of the rotation direction (see the corresponding hodographs in Fig. 3). It should be noted that, at the same values of the current density, the speed of switching is higher in region 1, than in 2 and 3.

For the parameters from region 4, the points T_1 and T_2 are the unstable foci with the opposite direction of the trajectory rotation. These points on the surface of the unit sphere are separated by the stable limit cycle. The trajectories started from T_1 and T_2 are attracted by this cycle. Let us note that the trajectories started from the point T_1 have the turning point, where they change the direction of rotation. Therefore, at the parameters belonging to region 4, switching is impossible. This is true also for region 5 — the difference with region 4 is only in the absence of the turning point and, therefore, the unidirectional rotation of the magnetization for all the trajectories (Fig. 3). For the values of the field and the current from region 6, the singular points T_1 and T_2 are the stable foci with the same direction of the trajectory rotation. They are separated by an unstable limit cycle (Fig. 3). Hence, in this region of the parameters, switching is impossible.

At the control parameters that belong to region 7, the point T_1 is a stable focus, whereas the point T_2 is an unstable one with the constant direction of the trajectory rotation (Fig. 3). Here the reverse switching of the magnetization is possible, i.e., from the anti-parallel to parallel configuration. This switching occurs mainly under the influence of the magnetic field.

References

1. J. Slonczewski, J.Magn.Magn.Matter, **159**, L1–L7 (1996).
2. R. Skomski, *Simple models of Magnetism* (Oxford University Press, 2012).
3. Yiming Huai, US PATENT No US 6,967,863 B2 Nov. 22 (2005).
4. Desin Wang, Haiwen Xi, Yuankai Zheng, D. Dimitar,// US PATENT Application Publication, Pub. No. US 2010/0109110 AI, May 6 (2010).
5. S. N. Piramanayagam, Chong, C. Tow, *Developments in Data Storage: Materials and Perspective* (A. John Wiley & Sons, Inc., 2011)
6. S. Mangin, D. Ravelosona, Y. Henry, J. A. Katine, E. E. Fullerton, AAPPS Bulletin, **18**, No.6, 41 (2008).
7. P. Kh. Amiri, Z. M. Zeng, J. Langer, H. Zhao et al., Appl.Phys.Lett. **98**, 112507 (2011).

Calculation of CPP- and CIP-magnitoresistance in multilayer magnetic structures

Dmitry E. Romanovskii[1*], *Vladimir* V. Prudnikov[1], *Pavel* V. Prudnikov[1], and *Marina* V. Mamonova[1]

[1] Omsk State University, Mira 55A, Russia

Abstract. A Monte Carlo simulation of magnetic properties for structures Fe/Cr/Fe and Co/Cu/Co, constructed from two ferromagnetic films divided by nonmagnetic film are carried out. The calculation of the magnetoresistance is carried out in case of CPP and CIP geometry for different thicknesses of the ferromagnetic films with the use of the anisotropic Heisenberg model for determination of magnetic properties. It was shown, that the obtained temperature dependence for the magnetoresistance agrees very well with experimental results.

1 Introduction

The artificial created multilayer magnetic superlattices has become of great interest for wide range of applications based on the phenomena of the giant magnetoresistance (GMR) and the tunneling magnetoresistance. Devices based on the GMR effect are widely used as read heads of hard disks, memory devices, sensors, etc [1]. Achievements in the development of technology give possibility now to receive a high-quality ultrathin films and multilayer coatings on the basis of the magnetic transition metals Fe, Co, and Ni. The magnetic properties of ultrathin films and superstructures are sensitive to the effects of anisotropy generated by the crystal field of a substrate or nonmagnetic layers. Therefore, the physical properties of ultrathin films based on Fe, Co, and Ni can be described by the anisotropic Heisenberg model [2, 3].

We developed in [4, 5] a new methodology for determination of the magnetoresistance (MR) with the use of the Monte Carlo method for measurements of MR with the current perpendicular to plane (CPP – MR). The magnetoresistance had been calculated for three-layer and spin-valve magnetic structures at various thicknesses of ferromagnetic films. It had been shown that the calculated temperature dependence of the magnetoresistance is in good agreement with experimental data obtained for the Fe(001)/Cr(001) multilayer structure and the CFAS/Ag/CFAS/IrMn spin valve based on the $Co_2FeAl_{0.5}Si_{0.5}$ (CFAS) Heusler alloy.

In this paper, we expand our methodology [4] for determination of MR in case of measurements with the current in plane (CIP–MR). We introduce for calculation of the CIP-MR a phenomenological parameter of electron scattering at F/N interfaces. The parameter of scattering can be determined from comparison of calculated and experimentally measured CIP-MR values for Fe/Cr(100)/Fe and Co/Cu(100)/Co multilayer structures [6].

2 Model and methods

We use the Hamiltonian of the anisotropic Heisenberg model [4, 5] for the Monte Carlo statistical description of the magnetic properties of Fe films for Fe/Cr(100)/Fe multilayer structure with the out of plane magnetization in Fe films

$$H = -\sum_{<i,j>} J_{ij}\{(1-\Delta_1(N))(S_i^x S_j^x + S_i^y S_j^y) + S_i^z S_j^z\}, \quad (1)$$

and for Co films in Co/Cu(100)/Co multilayer structure with the in plane magnetization in Co films

$$H = -\sum_{<i,j>} J_{ij}\{(S_i^x S_j^x + S_i^y S_j^y) + (1-\Delta_2(N))S_i^z S_j^z\}. \quad (2)$$

Therein, J_{ij} characterizes the short-range exchange interaction between the spins **S***i*, fixed at the sites of a body-centered cubic (bcc) lattice for Fe films and face-centered cubic (fcc) lattice for Co films. $S_i = (S_i^x, S_i^y, S_i^z)$ is introduced as a three-dimensional unit vector. $\Delta_{1,2}(N)$ are an anisotropy parameters. The dependence of the anisotropy parameters $\Delta_{1,2}(N)$ on the film thickness N was selected as proportional to the experimental critical temperatures for Fe/Cr(100) and Co/Cu(100) systems [7, 8], and the curves of $\Delta_{1,2}(N)$ are given in Figure 1.

We considered a multilayer structure, which consists of two ferromagnetic films separated by a nonmagnetic metal film (Figure 2). The simulations were performed for the films with linear sizes $L\times L\times N$ with applied of the periodic boundary conditions in the film plane. $\nu L \times L$ gives the number of spins in each layer, where ν is the number of atoms in unit cell, and N is the number of monolayers in the thin film. We considered Fe/Cr system as bcc-structure with $\nu = 2$ and Co/Cu as fcc-

* Corresponding author: romanovskii.dm@mail.ru

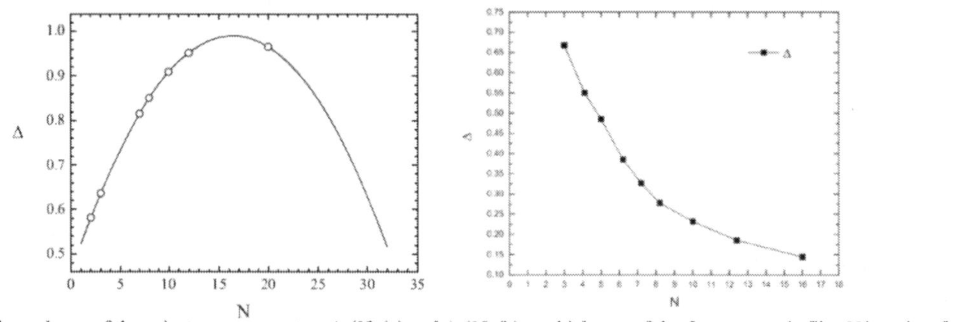

Fig. 1. The dependence of the anisotropy parameters $\Delta_1(N)$ (a) and $\Delta_2(N)$ (b) on thickness of the ferromagnetic film N in units of monatomic layers (ML). The circles are experimental data for Fe/Cr(100) (a) and Co/Cu(100) (b) systems.

structure with $\nu = 4$. The value of the exchange integral that determines the interaction of neighbouring spins inside the ferromagnetic film is specified as $J_1 > 0$ and the interaction between the films is taken as $J_2 = -0{:}1J_1$. The temperature T of the system is measured in units of the exchange integral J_1/k_B. Since $J_2 < 0$, the two films in the structure are magnetized in the opposite direction (one up and the other down). The spin configurations of the Heisenberg ferromagnetic films in the structures are updated using the Swendsen-Wang cluster algorithm.

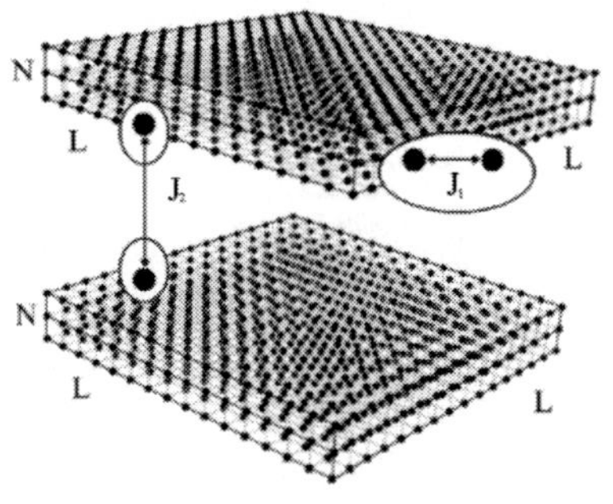

Fig. 2. Model of a multilayer structure consisting of two ferromagnetic films separated by the nonmagnetic metal film; L and N are the linear sizes of the films, J_1 and J_2 are the exchange integrals.

We calculated the magnetoresistance for a multilayer structures which is introduced as

$$\delta_h = \frac{R_{AP} - R_P}{R_P}, \tag{3}$$

where R_{AP} is the resistance of the structure when the magnetizations of adjacent ferromagnetic layers are aligned antiparallel, and R_P is the resistance of the structure for parallel orientation of the magnetization of ferromagnetic layers. The multilayer structure can connect two ways in the current line for measurements of the resistance: first case with the current in plane (CIP), when the current conduction is realized along layers and the electrodes are situated on one side of the structure, and second case with the current perpendicular to plane (CPP), when the current conduction is realized perpendicular to the layers of the superstructure and the electrodes are situated on both sides of the structure. It was shown in [6] that the CPP-MR is characterized by larger values than the CIP-MR, approximately twice as high, and becomes more competitive as the device size shrinks. Now this method has great practical interest because the CPP-MR sensors demonstrate more sensitivity than CIP-MR sensors.

3 Methodology and results of calculation of CPP- and CIP-MR

We present in this section the methodology which was developed for the calculation of magnetoresistence in multilayered structures in case of CPP and CIP geometry. We denote the resistance of an ferromagnetic film for two groups of electrons with spins up $R_\uparrow$ and spin down $R_\downarrow$. Here we use the simple two-current Mott model for the description of the resistance of different conductance channels. This model suggests conservation of orientation of electron spin moments during penetration of the structure and better corresponds to measurements of the CPP-MR. The total resistance of the trilayer structure (Fig. 2) for the antiparallel configuration, which is realized in the absence of a magnetic field, is determined by the relation $R_{AP} = (R_\uparrow + R_\downarrow)/2$. The parallel configuration of the trilayer structure for the magnetic field $H \geq H_S$, where H_S is saturation field, is characterized by the resistance in the form $R_P = (2\ R_\uparrow\ R_\downarrow)/(R_\uparrow + R_\downarrow)$. Consequently, the magnetoresistance of the trilayer structure is determined by the relation

$$\delta_h = \frac{(R_\uparrow - R_\downarrow)^2}{4R_\uparrow R_\downarrow} = \frac{(J_\uparrow - J_\downarrow)^2}{4J_\uparrow J_\downarrow}, \tag{4}$$

where $\mathbf{J}_{\uparrow,\downarrow} = en_{\uparrow,\downarrow}\,\mathbf{V}_{\uparrow,\downarrow}$ is the current density. Here, $n_{\uparrow,\downarrow}$ is the density of electrons with z components of spin moment equal to $+1/2$ and $-1/2$, $n = n_\uparrow + n_\downarrow$ is the total electron density and $\mathbf{V}_{\uparrow,\downarrow}$ are the averaged velocities of electrons with corresponding spin projections. The electron densities with spin up and down can be expressed through the magnetization of the film $n_{\uparrow,\downarrow}/n = (1 \pm m)/2$ determined in the process of the Monte Carlo simulation. The averaged electron velocity $\mathbf{V}_{\uparrow,\downarrow}$ can be expressed through the electron mobility and the external electric field intensity $\mathbf{E}$, and after that through the probability of electron displacement in unit time (corresponding to one Monte Carlo step per spin) from unit cell i to a neighbouring unit cell in the direction of the electric field with averaging over all film unit cells:

$$< V_{\uparrow,\downarrow} >= \mu_{\uparrow,\downarrow} E = \frac{e}{T} E e^{-\frac{\Delta E_{i,\uparrow,\downarrow}}{T}}, \tag{5}$$

 https://doi.org/10.1051/epjconf/201818501009

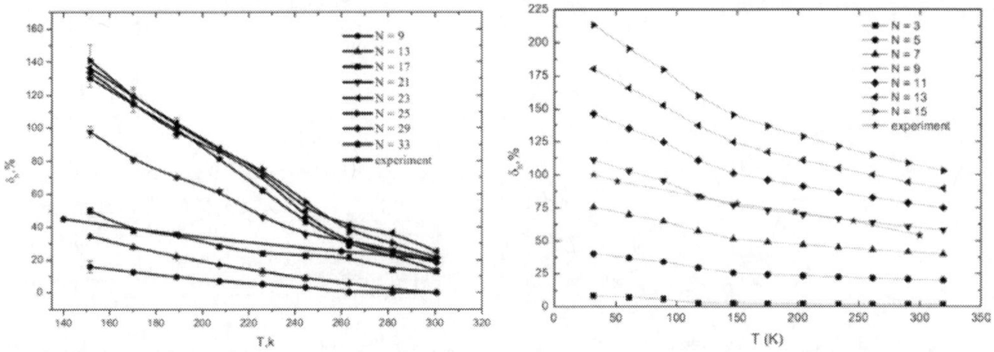

Fig. 3. The dependence of the CPP-MR of a trilayer structures Fe/Cr(100)/Fe (a) and Co/Cu(100)/Co (b) on the temperature for different values N of the thickness of ferromagnetic films. The comparison of the calculated and experimentally measured in [6] temperature dependences was made for structures with the thickness of the Fe film equal to 4 nm ($N = 21$ ML) and with the thickness of the Co film equal to 1.2 nm ($N = 9$ ML).

where $\mu_{\uparrow,\downarrow}$ is the electron mobility, ΔE_i characterizes the change of system energy connected with electron jump from i-cell to a neighbouring cell. $\Delta E_{i,\uparrow,\downarrow}$ is determined by the relation in case of CPP-MR

$$E_{i,\uparrow,\downarrow} = \mp J\left[\sum_{j\neq i} S_j^z(n_{j\uparrow} - n_{j\downarrow}) - S_i^z(n_{i\uparrow} - n_{i\downarrow})\right], \quad (6)$$

and following relation for CIP-MR

$$E_{i,\uparrow,\downarrow} = \mp J\left[\sum_{j\neq i} S_j^z(n_{j\uparrow} - \alpha n_{j\downarrow}) - S_i^z(n_{i\uparrow} - \alpha n_{i\downarrow})\right]. \quad (7)$$

The phenomenological parameter $0 < \alpha < 1$ is introduced in (7) which characterizes the spin flip scattering process in case of CIP geometry.

On the basis of the above presented relations, we calculated the temperature dependence of the CPP- and CIP-MR for the trilayer structure with different thicknesses N of ferromagnetic films. The procedure of calculation consists of the following steps: the first step is connected with Monte Carlo simulations of a magnetic structure in the equilibrium state at temperature T with determination of the magnetization of ferromagnetic films m_1 and m_2, which gives the possibility of calculating the electron densities $n_{\uparrow,\downarrow}$ for film cells; in the second step the average electron velocities $\mathbf{V}_{\uparrow,\downarrow}$ and the current densities $\mathbf{J}_{\uparrow,\downarrow}$ are calculated under relations (5)–(7) subject to the spin configuration realized at a given time of the simulation and averaged over Monte Carlo steps at times of the equilibrium state simulation; in the last step the calculation of the CPP- or CIP-MR is carried out under relation (4).

Graphs of the calculated temperature dependence of the CPP-MR are presented in Figure 3 for the multilayer structures Fe/Cr(100)/Fe and Co/Cu(100)/Co with different thicknesses N of the magnetic films. The values of exchange integral was calculated using Curie-Weiss theory, $J = 1.7\cdot10^{-14}$ erg for Fe and $J = 4.4\cdot10^{-14}$ erg for Co. The magnetoresistance for the Fe/Cr(100)/Fe system is characterized by increase for films with thicknesses in the interval $N = 9 \div 25$, but the magnetoresistance begins to decrease with a further increase in the thickness. For the Co/Cu(100) system, it can be seen that the magnetoresistance increases with increasing thickness of Co film in whole considered thickness range. We carried out the comparison of the calculated and experimentally measured in [6] temperature dependences of the CPP-MR for structures with the thickness of the Fe films equal to 4 nm, corresponding to $N = 21$ ML and with the thickness of the Co films equal to 1.2 nm, corresponding to $N = 9$ ML. One can see in Figure 3 that the calculated temperature dependence of the CPP-MR agrees very well with experimental data.

Calculation of the CIP-MR is begun at the first step from determination of the phenomenological parameter of scattering α as a value for which the best agreement of the CIP-MR with experimental results will be achieved. For the Fe/Cr(100)/Fe system, the best agreement of calculated CIP-MR with experimental data [6] is observed with $\alpha_{FeCr} = 0.36$ (see Figure 4(a)) and with $\alpha_{CoCu} = 0.20$ for Co/Cu(100)/Co system (Figure 4(b)). The results of calculation presented in Figure 4 show that the CIP-MR is decreased with increasing the parameter of scattering α.

At the second step, we calculated the CIP-MR with appointed values of α and determined the temperature dependence of the CIP-MR on thickness N of the magnetic films (Figure 5). For the Fe/Cr(100)/Fe system, the CIP-MR is characterized by by increase for films with thicknesses in the interval $N = 9 \div 25$, but the CIP-MR begins to decrease with a further increase in the thickness. For the Co/Cu(100)/Cu system, the CIP-MR increases with increasing thickness N ferromagnetic film in the considered range of thicknesses. Results of our calculations demonstrate that the magnetoresistance in the case of CPP geometry is higher than the CIP-MR, that is confirmed by experimental data [6].

4 Conclusions

In this paper, we presented the basic statements of our methodology for determination of the magnetoresistance in case of measurements both with the current perpendicular to plane (CPP-MR) and with the current in plane (CIP-MR). We introduced for calculation of the CIP-MR a phenomenological parameter of electron scattering at F/N interfaces. We realized the methodology for calculation of the CPP- and CIP-MR in Fe/Cr(100)/Fe and Co/Cu(100)/Co multilayer structures. The parameter of scattering was determined from comparison of calculated and experimentally measured CIP-MR values for Fe/Cr(100)/Fe and Co/Cu(100)/Co structures. We study the temperature dependence of the CPP- and CIP-MR in

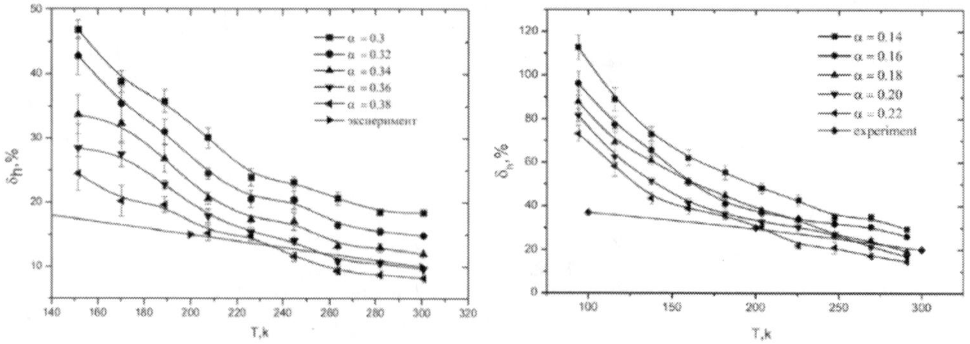

Fig. 4. The dependence of the CPP-MR of a trilayer structures Fe/Cr(100)/Fe (a) and Co/Cu(100)/Co (b) on the temperature for different values of the spin flip scattering parameter α. The comparison of the calculated and experimentally measured in [6] results was carried out with the thickness of the Fe films equal to 4 nm (*N* = 21 ML) and with the thickness of the Co films equal to 1.2 nm (*N* = 9 ML).

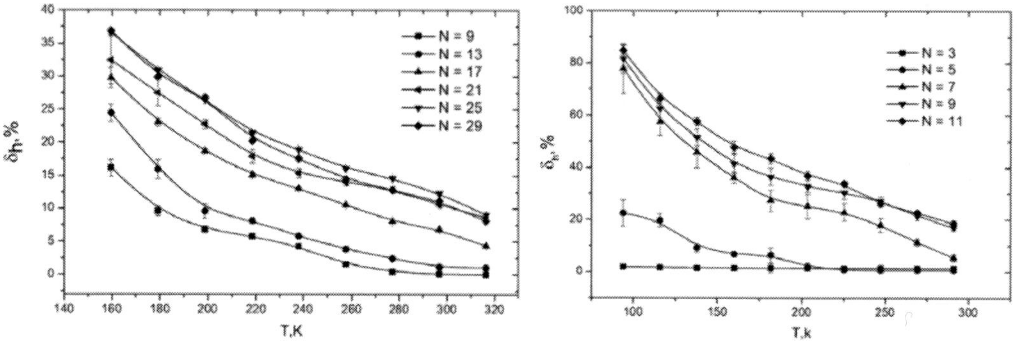

Fig. 5. The dependence of the CPP-MR of a trilayer structures Fe/Cr(100)/Fe (a) and Co/Cu(100)/Co (b) on the temperature for different values *N* of the thickness of ferromagnetic films.

Fe/Cr(100)/Fe and Co/Cu(100)/Co structures on thickness *N* of the magnetic films.

We carried out the comparison of the calculated and experimentally measured in [6] temperature dependences of the MR for these structures. It was shown that the calculated temperature dependence of the CPP- and CIP-MR agrees very well with experimental data.

This work was supported in part by the RFBR through Grant 17-02-00279 and by Grant MD-3815.2010.2 of Russian Federation President. Simulations were carried out on the SKIF-MSU of the Moscow State University and with use of computing resources of the "Applied theoretical physics and parallel computing laboratory" of Omsk State University.

References

1. C. Chappert, A. Fert, F.N. Van Dau, Nature Mater. **6,** 813 (2007)
2. P.V. Prudnikov, V.V. Prudnikov, M. A. Medvedeva, JETP Lett. **100**, 446 (2014)
3. P.V. Prudnikov, V.V. Prudnikov, M.A. Menshikova, N.I. Piskunova, J. Magn. Magn. Mater. **387**, 77 (2015)
4. V.V. Prudnikov, P.V. Prudnikov, D.E. Romanovskii, JETP Lett. **102**, 668 (2015)
5. V.V. Prudnikov, P.V. Prudnikov, D.E. Romanovskii, J. Phys. D: Appl. Phys. **49**, 235002 (2016)
6. M.A. Gijs, S.K.J. Lenczowski, J.B. Giesbers, et al., J. Appl. Phys. **75**, 6709 (1994)
7. A. Berger, H. Hopster, Phys. Rev. Lett. **73**, 193 (1994)
8. F. Huang, M.T. Kief, G.J. Mankey, R.F. Willis, Phys. Rev. B **49**, 3962 (1994)

Magnetically intercalated multilayer silicene

Andrey M. Tokmachev[1,*], *Dmitry V.* Averyanov[1], *Igor A.* Karateev[1], *Oleg E.* Parfenov[1], *Oleg A.* Kondratev[1], *Alexander N.* Taldenkov[1], and *Vyacheslav G.* Storchak[1,*]

[1]National Research Centre "Kurchatov Institute", Kurchatov Sq. 1, 123182 Moscow, Russia

Abstract. Silicene, a Si-based analogue of graphene, is predicted to exhibit topological electronic phases with exotic properties capable to revolutionize electronics. In particular, the silicene structure is highly advantageous for spintronics. However, lack of synthetic routes to free-standing and magnetically functionalized silicene compounds prevents experimental corroboration of the predictions. Here we synthesize $EuSi_2$, multilayer silicene intercalated with inherently magnetic Eu atoms, on $SrSi_2$/Si(001) templates. The resulting films are formed by crystallites of two mutually orthogonal orientations. The structure is firmly established with electron diffraction, X-ray diffraction and electron microscopy. The compound $EuSi_2$ exhibits non-trivial magnetic and transport properties. The data are compared with those for $EuSi_2$ films grown on $SrSi_2$/Si(111) templates.

1 Introduction

The exceptional rise of graphene to prominence in condensed-matter physics has driven a surge in studies of two-dimensional (2D) materials. Buckled 2D Xene sheets formed by group IVA atoms arranged in a honeycomb lattice are of particular interest to nanotechnology due to diversity and high tunability of their electronic states [1]. The most important member of the family of 2D Xenes is silicene, a Si-based analogue of graphene [2, 3]. Similar to graphene, physical properties of silicene are largely determined by massless Dirac fermions. The difference is that the band gap in silicene is easily tuned by chemical functionalization and electric fields. Despite technological difficulties, silicene is demonstrated as a promising material for field-effect transistors [4]. Electric and exchange fields induce different topological states in silicene [3, 5] leading to unconventional properties including quantum anomalous Hall effect, quantum spin Hall effect and valley polarized quantum Hall effect.

Silicene is thought to be especially valuable for application in spintronics [3]. Unlike graphene, silicene benefits from relatively strong spin-orbit interaction, large spin-coherence length and spin-diffusion time. Especially useful are valley-spin coupling and accessible topological states with non-trivial spin transport. However, to make use of these functionalities magnetic states in silicene should be induced, for example by chemical modification. In this respect, adsorption or intercalation of metal atoms are considered as a prospective way to bring diverse magnetic properties into the silicene physics [3].

A lot of predictions are made about unique and exotic properties of silicene compounds but their experimental realization is well behind the theory and electronic structure calculations. The problem is in synthesis of silicene and its derivatives. Synthesis of 2D materials is usually tricky [6] but in the case of silicene it is vastly aggravated by its high chemical reactivity caused by significant buckling of silicene layers. Free-standing silicene is yet to be produced. 2D silicene structures are formed by adsorption of Si on metal substrates; however, a strong hybridization of silicene with the substrate destroys the Dirac states [2, 3, 7].

It is suggested that intercalation of silicene with active atoms restores Dirac cones [7]. This hypothesis is experimentally corroborated in multilayer silicene intercalated with Ca, a polymorph of $CaSi_2$ [8]. Recently we succeeded in synthesis of Sr-intercalated multilayer silicene, a new polymorph of $SrSi_2$ [9]. It is engineered by epitaxial stabilization on Si(111) and Si(001). In contrast to $CaSi_2$, layered $SrSi_2$ enjoys a regular stacking of silicene layers, thus avoiding the momentum shift of Dirac cones from high-symmetry points [9].

Both $CaSi_2$ and $SrSi_2$ are nonmagnetic. To induce magnetic properties into silicene lattice we proposed a synthesis of multilayer silicene intercalated with inherently magnetic atoms [10]. We chose Eu because it is the most active among magnetic metals, the half-filled *f*-shell of Eu^{2+} provides strong magnetic properties, and the isomorphism of Eu(II) and Sr compounds provides an opportunity for successful synthesis. In Ref. [10] we thoroughly explored synthesis and properties of EuSi2 grown on $SrSi_2$/Si(111) with silicene layers parallel to the surface. However, many unique properties of silicene lattice come from electric fields acting normal to silicene surface [3, 5]. Here, we explore transport across magnetic silicene layers by studying in detail $EuSi_2$ synthesized on Si(001) with a $SrSi_2$ template layer.

* Corresponding authors: tokm@rambler.ru, mussr@triumf.ca

2 Methods

The films of silicides are grown by molecular beam epitaxy in the Riber Compact 12 system. The UHV system consists of Gamma Vacuum Titan Ion Pump, Cryo-Torr 8 cryopump, cryopanels and a titanium sublimation pump. The base pressure in the growth chamber is less than 10^{-10} Torr. The metals (Sr and Eu) of 4N quality are effused from Knudsen cell sources. The intensity of molecular beams is measured with a Bayard-Alpert ionization gauge. The substrate temperature is determined with a PhotriX ML-AAPX/090 infrared pyrometer. The surface is controlled *in situ* with a RHEED diffractometer.

X-ray diffraction (XRD) study of the films employs Rigaku SmartLab diffractometer with a 9 kW copper anode. Electron microscopy studies are carried out for cross-sections prepared in Helios scanning electron microscope/focused ion beam (FIB) dual system. The surface of the samples is covered by a layer of Pt. Ga^+ ions are used for FIB milling, thinning and cleaning cross-sections. The specimens are protected from oxidation by carbon deposition. The structure of the films is studied with a Titan 80-300 microscope.

Magnetic properties of $EuSi_2$ are studied with MPMS XL-7 Superconducting Quantum Interference Device employing the reciprocating sample option. Transport properties of the films are determined in the van der Pauw configuration with a Lake Shore 9709A measurement system. Ohmic contacts to metallic films are produced by placing a Ga-Sn-Ag alloy.

3 Results

Multilayer silicene intercalated with Eu atoms (trigonal $EuSi_2$) is less stable than the tetragonal $EuSi_2$. The latter is formed in the reaction of Eu atoms with Si surface, irrespective of its orientation [11]. We employ epitaxial stabilization to produce trigonal $EuSi_2$. The first step is to prepare a template of layered $SrSi_2$ on a clean Si(001) surface following the recipe proposed in Ref. [9]. The synthesis goes through a number of surface reconstructions of Sr on Si(001) [12]. Due to natural incommensurability of the hexagonal lattice of $SrSi_2$ and the cubic lattice of Si(001), the resulting film is formed by crystallites of two orientations differing by lateral rotations by 90°. Silicene layers are orthogonal to the Si(001) surface for both orientations.

$EuSi_2$ is then grown by directing a beam of Eu atoms (10^{-8} Torr) on top of the $SrSi_2$/Si(001) template kept at 450 °C. Si atoms go through $SrSi_2$ via the well-known vacancy mechanism. The structure is capped with SiO_x to prevent its degradation by atmosphere. RHEED images (Fig. 1a) show that the film is formed by two types of $EuSi_2$ crystallites with orthogonal orientations repeating the morphology of the underlying $SrSi_2$. Regretfully, the epitaxial stabilization is short-range: the film adopts the most stable form of $EuSi_2$ [11] when the thickness of $EuSi_2$ exceeds 7 nm. It becomes evident in RHEED images (Fig. 1b) revealing appearance of the pattern corresponding to tetragonal $EuSi_2$.

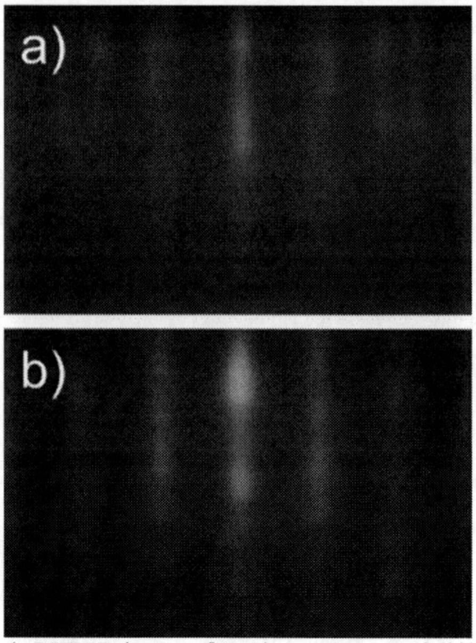

Fig. 1. RHEED image of $EuSi_2$ grown on $SrSi_2$/Si(001) viewed along azimuth [110] of the substrate. a) thickness less than 7 nm; b) thickness exceeds 7 nm.

The structure of the films is further characterized *ex situ*. XRD θ-2θ scan for the $SiO_x/EuSi_2/SrSi_2$/Si(001) film is shown in Fig. 2. Although peaks from $SrSi_2$ and $EuSi_2$ are not resolved due to isomorphism of the compounds and closeness of the corresponding lattice constants, the scan is consistent with the proposed orientation of crystallites and the absence of any alien phases.

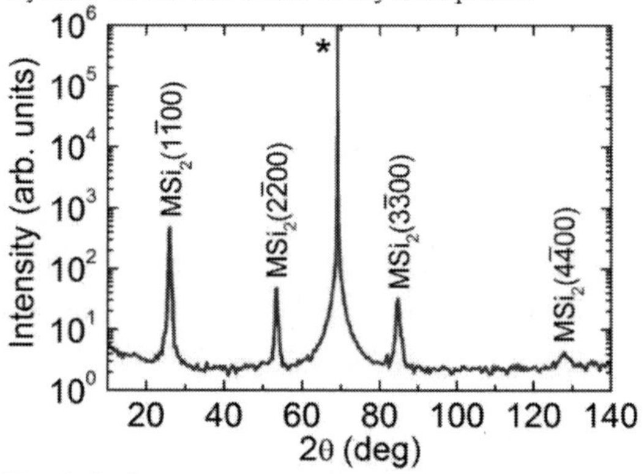

Fig. 2. θ-2θ XRD scan for the $SiO_x/EuSi_2/SrSi_2$/Si(001) film. Peaks from SrSi2 and EuSi2 are not resolved, a peak from Si is marked *.

The atomic structure of the films is revealed by electron microscopy. Fig. 3 shows HAADF-STEM image for one of the crystallites with silicene layers running normal to the cross-section. The image confirms the orientational relation between $SrSi_2$ and $EuSi_2$. It also reveals the presence of edge dislocations: extra half-planes of atoms are introduced at the $EuSi_2/SrSi_2$ boundary to account for some mismatch between the lattices. Fig. 4 shows a crystallite of the other orientation, with silicene layers parallel to the cross-section. We employ the bright-field mode with reduced Z-contrast to show the honeycomb structure of silicene. The boundary between $EuSi_2$ and $SrSi_2$ is formed by a single layer of mixed $Eu_xSr_{1-x}Si_2$ silicide marked as red rectangle.

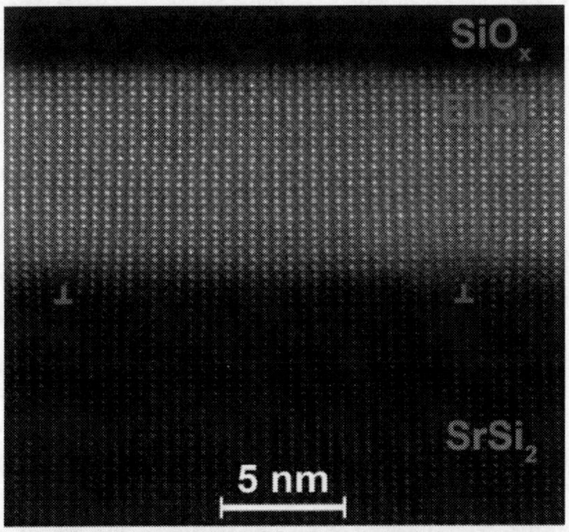

Fig. 3. Cross-sectional HAADF-STEM image of the $SiO_x/EuSi_2/SrSi_2/Si(001)$ film. Edge dislocations in $EuSi_2$ are marked by red pointers.

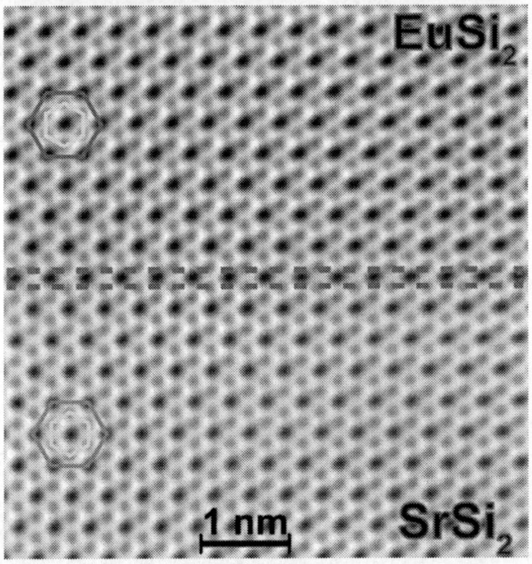

Fig. 4. Cross-sectional BF-STEM image of the $SiO_x/EuSi_2/SrSi_2/Si(001)$ film. It shows the boundary between $EuSi_2$ and $SrSi_2$ phases as well as hexagons of Eu (red), Sr (blue) and Si (yellow).

The aim of making trigonal $EuSi_2$ is to get a magnetic (spin) response from a silicene-based system. Fig. 5 shows temperature dependence of the magnetic moment measured in magnetic field of 1 T directed along 3 axes. The film demonstrates antiferromagnetic (AFM) response with moments oriented normal to the plane and parallel to silicene layers.

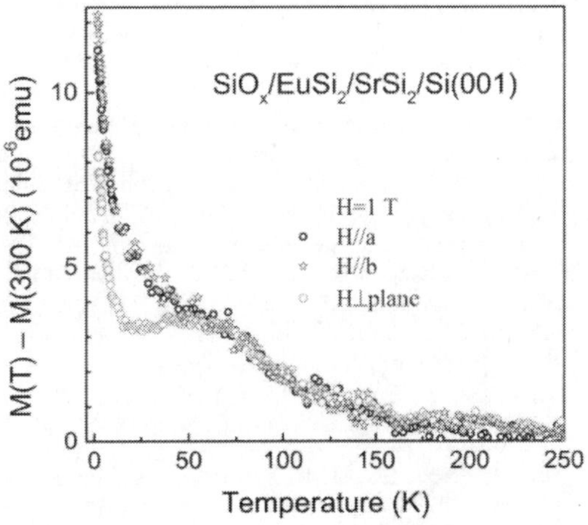

Fig. 5. Temperature dependence of magnetic moment of the $SiO_x/EuSi_2/SrSi_2/Si(001)$ film for different directions of the external field (1 T) with respect to the substrate: (001) – red, (110) – black and $(1\bar{1}0)$ – blue.

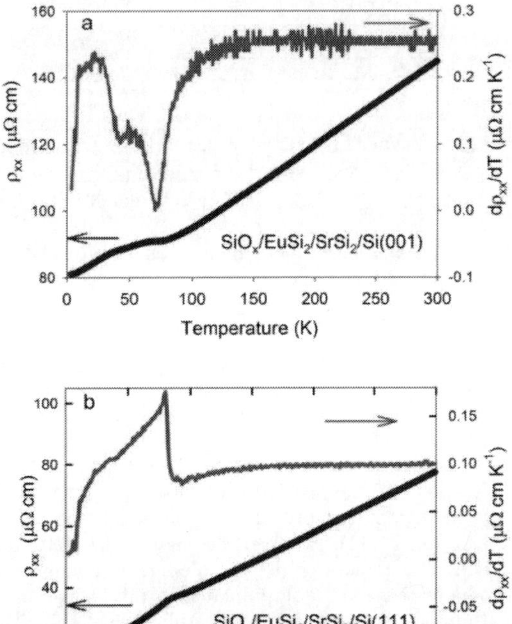

Fig. 6. Temperature dependence of resistivity ρ_{xx} (blue) and its temperature derivative $d\rho_{xx}/dT$ (red) for thin films: a) $SiO_x/EuSi_2/SrSi_2/Si(001)$ and b) $SiO_x/EuSi_2/SrSi_2/Si(111)$.

Thus, we succeeded in inducing magnetism into a system based on silicene layers. It is reasonable to expect that the proximity of magnetic Eu^{2+} ions can result in non-trivial transport properties of the films, especially for the transport across a stack of silicene layers. Temperature dependence of resistivity ρ_{xx} and its temperature derivative for $EuSi_2$ grown on Si(001) and Si(111) are shown in Fig. 6. In the latter case the transport is carried out along silicene layers, while in the former case carriers move partly along the layered structure and partly across it. Naturally, $SrSi_2$ also contributes into the film conductivity but this transport channel shows no anomalies [9]. The resistivity of a magnetic metal is dominated by local spin fluctuations; the temperature dependence of $d\rho_{xx}/dT$ should follow that of the magnetic specific heat. The opposite sign of derivative changes at the AFM transition for the $EuSi_2/SrSi_2/Si(001)$ and $EuSi_2/SrSi_2/Si(111)$ structures is a consequence of different direction of magnetic moment with respect to the current in the two films.

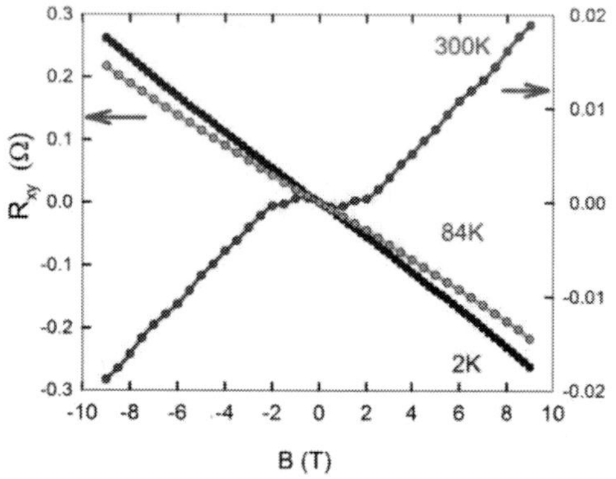

Fig. 7. Magnetic field dependence of Hall resistance R_{xy} for $SiO_x/EuSi_2/SrSi_2/Si(001)$ structure measured at three temperatures: 2 K (blue), 84 K (green) and 300 K (red).

In Ref. [10] we found an unconventional magnetoresistance and its temperature dependence in transport across silicene layers. Here, we study Hall resistance R_{xy} for the films. In the case of the $EuSi_2/SrSi_2/Si(111)$ structure, Hall response is unexceptional – dependence of R_{xy} on B is linear at all T; the sign of the derivative corresponds to electron carriers. The situation is quite different for the $EuSi_2/SrSi_2/Si(001)$ structure, as shown in Fig. 7. The behaviour of the Hall resistance at low temperatures (including those above T_N) is quite similar to that of $EuSi_2$ on Si(111). However, the sign of the Hall effect changes from electron-like to hole-like as T approaches room temperature. More importantly, this change in the electronic structure is accompanied by anomalous contribution to the Hall resistivity, which is highly unusual for a paramagnet. This behaviour may be caused by interplay of more than one band.

4 Conclusions

Silicene is one of the most promising materials for nanoelectronics. Its structure ensures efficient transport properties which can be manipulated by external stimuli. A major promise of silicene is for spin-dependent phenomena and transport. However, the bottleneck of silicene studies – synthesis – precludes the theoretical predictions from experimental verification. Here, we attempted to bridge the gap between the theory and experiment by studies of stacks of silicene layers intercalated with magnetic Eu atoms. We described the complex synthesis procedure and verified the atomic structure of the resulting layered polymorph of $EuSi_2$ grown on Si(001) with a $SrSi_2$ template. The film exhibits a complex magnetic structure dominated by antiferromagnetism. This conclusion is confirmed by studies of transport across silicene layers. The electronic structure of the films changes as temperature goes up from 2 K to 300 K, signified by changes in the carrier type and unexpected room-temperature anomaly in Hall measurements. The results call for additional studies, both experimental and theoretical.

This work is partially supported by NRC "Kurchatov Institute", the Russian Foundation for Basic Research (grants 16-07-00204, 16-29-03027, and 17-07-00170), and the Russian Science Foundation (grant 14-19-00662). The measurements are partially carried out using the equipment of the resource centres of electrophysical, laboratory X-ray, and electron microscopy techniques of NRC "Kurchatov Institute".

References

1. A. Molle, J. Goldberger, M. Houssa, Y. Xu, S.-C. Zhang, D. Akinwande, Nature Mat. **16**, 163 (2017).
2. C. Grazianetti, E. Cinquanta, A. Molle, 2D Mat. **3**, 012001 (2016).
3. J. Zhao, H. Liu, Z. Yu, R. Quhe, S. Zhou, Y. Wang, C.C. Liu, H. Zhong, N. Han, J. Lu, Y. Yao, K. Wu, Prog. Mat. Sci. **83**, 24 (2016).
4. L. Tao, E. Cinquanta, D. Chiappe, C. Grazianetti, M. Fanciulli, M. Dubey, A. Molle, D. Akinwande, Nature Nanotech. **10**, 227 (2015).
5. M. Ezawa, Phys. Rev. Lett. **109**, 055502 (2012).
6. A.J. Mannix, B. Kiraly, M.C. Hersam, N.P. Guisinger, Nature Rev. Chem. **1**, 0014 (2017).
7. R. Quhe, Y. Yuan, J. Zheng, Y. Wang, Z. Ni, J. Shi, D. Yu, J. Yang, J. Lu, Sci. Rep. **4**, 5476 (2014).
8. E. Noguchi, K. Sugawara, R. Yaokawa, T. Hitosugi, H. Nakano, T. Takahashi, Adv. Mat. **27**, 856 (2015).
9. A.M. Tokmachev, D.V. Averyanov, I.A. Karateev, O.E. Parfenov, A.L. Vasiliev, S.N. Yakunin, V.G. Storchak, Nanoscale **8**, 16229 (2016).
10. A.M. Tokmachev, D.V. Averyanov, I.A. Karateev, O.E. Parfenov, O.A. Kondratev, A.N. Taldenkov, V.G. Storchak, Adv. Funct. Mat. **27**, 1606603 (2017).
11. D.V. Averyanov, A.M. Tokmachev, C.G. Karateeva, I.A. Karateev, E.F. Lobanovich, G.V. Prutskov, O.E. Parfenov, A.N. Taldenkov, A.L. Vasiliev, V.G. Storchak, Sci. Rep. **6**, 25980 (2016).
12. D.V. Averyanov, C.G. Karateeva, I.A. Karateev, A.M. Tokmachev, M.V. Kuzmin, P. Laukkanen, A.L. Vasiliev, V.G. Storchak, Mater. & Design **116**, 616 (2017).

Modeling of GMR loops asymmetry in spin valves

Alexander Kurenkov [1,2], *Georgy* Babaizev[1,2], and *Nikolay* Chechenin[1,2,*]

[1]Skobeltsyn Institute of Nuclear Physics, Lomonosov Moscow State University, Leninskie Gory, 1/2, 119991 Moscow, Russia
[2]Faculty of Physics, Lomonosov Moscow State University, Leninskie Gory,1/2, 119991 Moscow, Russia

Abstract. The regular shape of the magnetoresistance as a function of applied magnetic field, *R(H)*, is important for numerous spin valves applications. Nevertheless, in many practical studies the shape of the *R(H)* curve is unpredictably complicated. A simple technique, proposed in this paper, allows interpreting the *R(H)* features and calculating *R(H)* curves from known hysteresis loops of ferromagnetic layers, composing spin valves. And vice versa, the shape of the hysteresis loops of the composing ferromagnetic layers can be obtained from *R(H)* curves.The method does not give an insight into the physical factors determining the shape of hysteresis loops, but it is suitable for prompt selection of promising spin valve stacks. A series of spin valve structures was prepared and their magnetoresistance curves were measured. The analysis of the experimental data demonstrates the capacity of the proposed method.

1 Introduction

Spin valves (SV) are widely applied in spintronic devices. Active part of SVs normally consists of ferromagnetic (F) – spacer (S) – ferromagnetic (F) – antiferromagnetic (AF). In such structures the magnetization in the first ferromagnetic layer (FF) is free to be influenced by an external magnetic field, while in the magnetization in the second ferromagnetic (PF) is pinned by exchange bias field exerted from the adjacent AF layer. Various applications require SVs with different properties, which means different giant magnetoresistance (GMR) characteristics. Characteristics of linear region of magnetoresistance (MR) as a function of external magnetic field, *H*, like the amplitude of the resistance *R(H)* and position of the MR peak on the *H* scale, are of a practical interest. In SV-based magnetic field sensors, the tilt of MR in the linear region defines the precision, and, together with the position of linear region along the *H* axis, the operating range of the sensor. A high amplitude is desirable for both the high precision and wider operating range, while the tilt compromises between the resolution and the field range. In case of magnetoresistive random-access memory, the switching thresholds are defined by linear region position, while high amplitude is important for sustainable state readout. That is why MR curve shape control and mastering is crucial for development of devices with desirable properties.
The switching field in the MR curves can be estimated by minimizing system's energy in Stoner-Wolfarth model [1, 2] or considering macrospin dynamics with Landay-Lifshitz-Gilbert equation [3,4]. However, to date, these approaches cannot describe some important features of MR curves, such as hysteresis in the switching field in the increasing and decreasing scans and other additional features of the MR loops, emerging in some cases as illustrated below. Here we propose an estimation method, which interprets MR curve as a consequence of the composition of hysteresis loops of individual F layers in the SV structures. Method reveals their properties, though, does not provide a deep insight into mechanisms which determine the shape of the hysteresis in the individual layers, like tilt of the PF layer's hysteresis loop, which partly results from ferromagnetic-antiferromagnetic exchange interaction, and must be a subject of separate considerations.

2 Sample preparation

The samples were prepared by DC magnetron sputtering (Ar atmosphere 2,9*10^{-3} Torr, base pressure 10^{-6} Torr) in the presence of permanent magnetic field of about 400 Oe for inducing uniaxial anisotropy in F-layers. The stripe-like structures were deposited through gaps in the mask with lateral dimensions of 200 x 500 μm and with long side of the mask oriented perpendicular to the field. Unidirectional anisotropy appeared due to antiferromagnetic (AF) layer growth on the top of the saturated F layer. Stacks had Ta[22]/$Ni_{80}Fe_{20}$[12]/Cu[3]/$Ni_{80}Fe_{20}$[7]/$Ir_{80}Mn_{20}$[t_{AF}]/Ta[22] structure where the figures in the brackets are the layers thicknesses in nm with t_{AF} = 1 to 15 nm. MR curves were measured at room temperature using the four-point contact method with current flowing perpendicular to the swept external magnetic field aligned parallel to the easy axis in the F layers The contribution of the anisotropic magnetoresistance is negligible in given setup.

3 Coupling between magnetization and magnetoresistance

As a rough approximation, GMR can be described as a function of the angle ϕ between magnetization directions in successive ferromagnetic layers [5]:

$$R_{GMR} = R^0_{GMR} (1 - \cos(\phi(H)))/2, \quad (1)$$

where R^0_{GMR} is the amplitude of GMR effect. Dependence $\phi(H)$ can be derived from hysteresis loops assuming macrospin model. Here hysteresis loops describe projections of magnetizations on the axis of external magnetic field and macrospin assumption implies constant magnetization vector length during magnetization reversal. Thus:

$$\phi(H) = \arccos(m_{FF}(H)) - \arccos(m_{PF}(H)), \quad (2)$$

$$R(H) = R^0 + R^0_{GMR}(1 - \cos(\arccos(m_{FF}(H)) + \arccos(m_{PF}(H))))/2, \quad (3)$$

where $m_{FF} = M_{FF}^{ext}/M_s$, $m_{PF} = M_{PF}^{ext}/M_s$ are relative magnetizations and M_{FF}^{ext}, M_{PF}^{ext} are projections of magnetization vector of FF and PF, respectively, on the direction of the external field.

Eq. (3) was applied to construct MR curves, using predefined FF and PF hysteresis loops, and to fit the experimental MR curves. The fitting parameters were coercivity, H_C^{PF}, exchange bias, H_{EB}, and tilt, dM/dH, for PF hysteresis loop (Fig.1a).

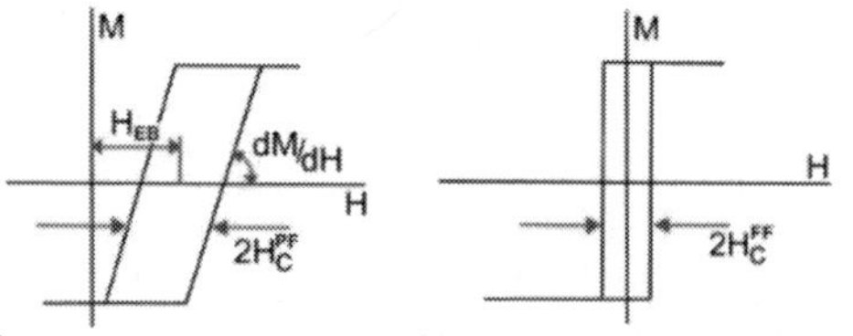

a) b)

Fig.1 Typical hysteresis loops for pinned (a) and free (b) F layers and their varied parameters coercivity, H_C^{PF}, exchange bias, H_{EB}, and tilt, dM/dH, for PF hysteresis loop (Fig.1a) and solely coercivity, H_C^{FF}, for the FF hysteresis loops.

Since the magnetic field during the MR measurements was aligned parallel to the easy axis in the samples, the hysteresis loop for FF layer can be approximated by a rectangular shape, while the exchange interaction on the F-AF interface can originate the tilt in the hysteresis loop for PF layer, as schematically illustrated in Fig.1. For the FF hysteresis loops in practical cases in our study, it was found sufficient to vary the coercivity H_C^{FF} solely (Fig.1b). If AF had a perfect structure, magnetization reversal of F would occur in the narrow range of external magnetic field, like in the case of F itself. However, due to variation of AF grain size and orientation, some of F domains experience larger exchange bias then the others. This leads to a spreading of magnetization reversal, widening the switching range of the external magnetic field, which manifests in the PF hysteresis loop.

4 Illustrations of magnetization and magnetoresistance loops coupling

Fig.2 shows hysteresis loops (a,c,e,g) and MR curves

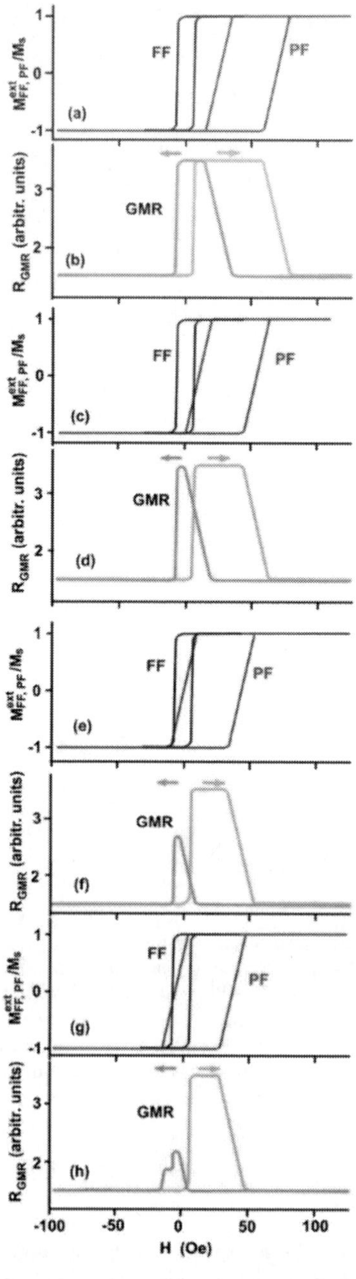

Fig.2 Hysteresis loops (blue – free F, red – pinned F) and corresponding calculated MR curves (light blue for increasing external magnetic field, light-green – for decreasing). Exchange bias decreases from (a) to (h).

(b,d,f,h) for increasing and decreasing external fields, calculated from them. Let us vary solely the exchange bias first (exchange bias decreasing from top to bottom in Fig.2). Estimation shows, that when exchange bias, H_{EB}, is large enough, compared to coercive field, H_c, in both FF and PF layers, both forward and backward MR scans (measured with increasing and decreasing external magnetic field) reach GMR effect with a plateau and maximal

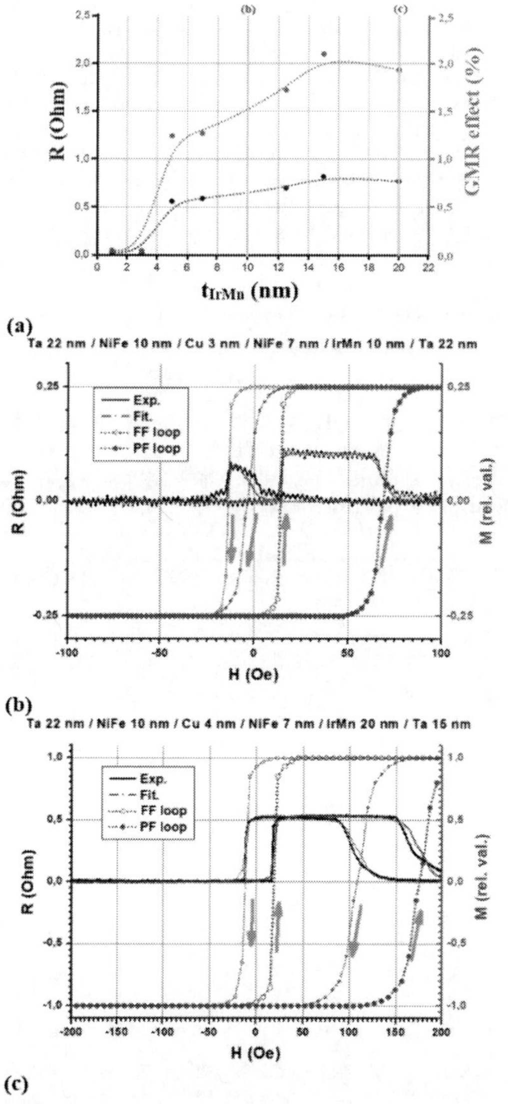

Fig.3 (a) AF thickness GMR dependence, absolute (black, Ohm) and relative (red, percent, the same scale) values. Experimental MR curves (black) with fitting (red) and derived hysteresis loops for FF and PF layers (green) for cases of 10nm- (b) and 20nm-thick (c) AF layer. Insufficient exchange bias (b) leads to superposition of hysteresis loops thus to GMR decrease.

magnitude, Fig. 2a. If exchange bias decreases, the plateaus shorten and, at some point, one of MR curves loses its plateau, Fig. 2d, and then deforms to a narrow form with reduced amplitude, Fig. 2f,h. This evolution of MR curves with a reduction of H_{EB} is experimentally illustrated in Fig. 3. Fig.3a shows experimental dependence of GMR on AF layer thickness. The value of 2% is rather typical for room temperature for this type of system [6]. Here GMR monotonically increases with the thickness due to increasing exchange bias and saturates at the AF thickness of 15 nm. At t_{AF} = 10 nm, the MR is not at its maximum, Fig. 3a, and the backward (decreasing) field scan shows a lower and narrower MR effect than in the forward (increasing) field scan, Fig.3b. The forward-backward asymmetries in SV structures have been discussed in more details in [7]. Basing on the shapes of the MR scans the FF and PF hysteresis loops can be constructed, as shown in Fig. 3b. The insufficient exchange bias does not allow a complete separation of the FF and PF hysteresis loop causing the reduced (but not zero) GMR. In contrast, the saturated GMR effect at t_{AF} = 20 nm, Fig. 2a, leads to full-magnitude MR curves with plateaus, Fig.3c, evidencing on separated hysteresis FF and PF. By fitting the experimental MR curves the hysteresis loops of composing FM layers can be easily constructed, as shown on Fig.3b,c.

These examples show a critical condition of nonintersecting FF and PF hysteresis loops with large enough exchange bias in PF layer for MR curves with a large amplitude and regular shape. At lower exchange bias, FF and PF hysteresis loops are not separated and intersection causes decrease of GMR. Below, we illustrate that the tilt of the hysteresis loops, dM/dH, plays also important role in interpretation of the MR curves shape.

The experimental MR curves, shown in Fig. 4, are of a triangular shape with sharp edge in low-field side, slow decrease at high-field side and asymmetry in the amplitude in the forward-backward scans of the MR measurement. Besides, there is a small additional peak in the negative field region of the backward scan. Fitting of the MR curves with a variation of the FF and PF hysteresis loops parameters can describe

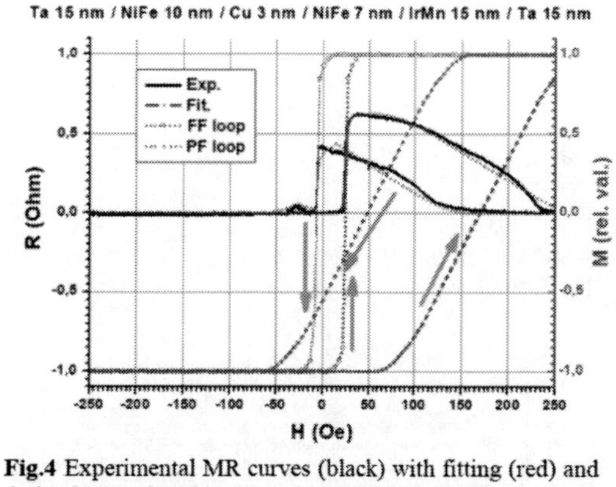

Fig.4 Experimental MR curves (black) with fitting (red) and derived FF and PF hysteresis loops (green). Small extra peak belongs to the MR curve, corresponding to decreasing external magnetic field.

reasonably well the MR curves with their irregular shape at sharp and smooth edges, the asymmetry of amplitude and even the small extra peak. The MR curves features originate from the specific combinations of exchange bias, coercivities and tilts of the FF and PF layers. In particular, the additional peak emerges when corresponding hysteresis loops' branches (i.e. both for increasing or both for decreasing field) intersect. From a physical point of layers become equal, reducing to zero the magnetoresistance. At further decrease of the field, the F-layer with smaller anisotropy (smaller tilt of the loop) sharper reverses the magnetization in the rest domains than that in domains in the counterpart F-layer, spiking the MR, as observed in backward scan *R(H)* around *H*~-25 Oe in Fig.4. Evidently, this effect is connected with a local variation of the exchange energy, holding a fraction of the PF domains in the unreversed state and leading to a significant tilt of the *M(H)* curve. The additional peak can be eliminated not only by a decreasing the tilt, but also by a decreasing the coercive force, or by an increasing exchange bias of PF hysteresis loop.

In practice, each MR curve is a combined effect of a number hysteresis loops' parameters which are not taken into account in our simple approach. Nevertheless, the suggested method, in principle is simple and, at least, can be useful for qualitative interpretation of the MR curves. The method can also be used for inverse task, i.e., to predict the shape of the MR curve, basing on the hysteresis loops of the FF and PF layers. This is very important for optimization of spintronic devices, like spin valves, magnetic field, sensors with required characteristics, reducing the time and resource consuming experimental development.

5 Conclusions

A method for analysis of shapes of GMR curves in spin valves of a simple configuration FF/S/PF/AF is proposed. The method is based on the connection between MR curves and properties of the SV composing ferromagnetic layers. This insight enables constructing SVs with desirable properties by view it means, that magnetization of the FF layer reverses when magnetization of the PF layer still partly keeps the original orientation. Decreasing the external field to the point of intersection, the fractions of domains with reversed magnetizations in FF and PF selecting ferromagnetic layers with proper characteristics. The functionality of the method is illustrated on specific cases of MR measurements. It is proved that it can be helpful in understanding the origin of a series of experimental features in MR curves and can be useful for diagnosing undesirable features of MR characteristics. The method does not give an impact on the nature of the magnetic properties of the components of the SV structure, which make the features of the MR curves, assuming that these issues are to be the subject of further studies of mechanisms and processes, influencing the hysteresis loops in the SV stack components.

References

1. K. Nishioka, S. Gangopadhyay, H. Fujiwara, M. Parker, IEEE Trans. Magn., **31,** 3949 (1995)
2. M.R. Parker, H. Fujiwara, S. Hossain, IEEE Trans. Magn., **31**, 2618 (1995)
3. C. Yin, Z. Jia, W.-c. Ma, Y.-l. Ren, Int. J. Miner. Metall. Mater., **20**, 700 (2013)
4. J.-S. Park, S.-R. Lee, Y.K. Kim, S.H. Lim, J. Magn. Magn. Mater., **250**, 25 (2002)
5. B. Dieny, J. Magn. Magn. Mater., **136**, 335 (1994)
6. S. Bae, W. F. Egelhoff, Jr., P. J. Chen, and J. H. Judy, IEEE Trans. Magn. **36**(5), 2617 (2000)
7. N. G. Chechenin, P. N. Chernykh, S. A. Dushenko, I. O. Dzhun, A. Y. Goikhman, V. V. Rodionova,. J. Supercond. and Novel Magn, **27**, 1547 (2014)

Critical switching characteristics of three-layered spin valve for different materials and alloys with uniaxial anisotropy

Iuliia Iusipova [1,2,*]

[1]MIET, General Physics Department, Bld. 1, Shokin Square, Zelenograd, Moscow, Russia, 124498
[2]IPPM RAS, Department of design of microelectronic components for nanotechnology, 3, Sovetskaya Street, Zelenograd, Moscow, Russia, 124365

Abstract. We analyze the dependence of the current density and magnetic field switching on the magnetic parameters of the material of the ferromagnetic layers of the spin valve. Comparison of critical characteristics of the spin valve with longitudinal anisotropy of ferromagnetic layers fabricared of different materials showed that the promising materials for the fabrication of spin valve are cobalt, iron, their alloys, ferroborates of cobalt and alloys of cobalt with gadolinium. For these materials we produced and analyzed the bifurcation diagrams of equations describing the switching process of the spin valve. Based on the study of the dynamics of the magnetization vector we obtained the numerical evaluation of time switching.

1 Introduction

Currently, the existing types of memory are approaching to the limits of their possibilities, and intensive development of new, particularly non-volatile types of memory is observed. In 2016, the IBM Corporation, in cooperation with the company Samsung has demonstrated a new magnetoresistive random access memory (MRAM), the diameter of the cells of which is 11 nanometers. The switching time of this cell is superior to all existing types of memory and is only 10 nanoseconds [1]. Magnetoresistive random access memory (MRAM), has many advantages over other types of memory, however, has one significant drawback: the values of current density and magnetic field that must be applied to switch the free layer of the spin valve, which is included in a memory cell of MRAM, are too large

The present work investigates the influence of the magnetic characteristics of the material of the ferromagnetic layers of the spin valve on the dynamics of the magnetization vector of the free layer. The aim of this study is the search of ferromagnetic material, providing better switching characteristics of the spin valve.

2. The mathematical model of a spin valve

The transverse section of a spin valve is a square with side 11 nm. The thickness of the free layer d is 2 nm, the thickness of the pinned layer is 5 nm, the thickness of the non-magnetic interlayer is 1.2 nm. Physical basics of spin valve under the action of spin–polarized current was first described in the work of J.C.Slonczewski [2].

The dynamic equation of the magnetization in the free layer of the spin valve in the form of the canonical dynamical system has the form (see e.g. [3-5]):

$$\frac{\partial \mathbf{m}}{\partial \tau} = -\left[\mathbf{m}\times\mathbf{h}^{eff}\right] + \alpha\mathbf{h}^{eff} - \alpha\mathbf{m}(\mathbf{m},\mathbf{h}^{eff}), \quad (1)$$

where $\mathbf{m}$ is the magnetization of the free layer and the effective field $\mathbf{h}^{eff}$ is determined by the relation (2).

$$\mathbf{h}^{eff} = \begin{pmatrix} h_x \\ h_y \\ h_z \end{pmatrix} = \begin{pmatrix} h + km_x \\ -jGm_z \\ jGm_y - m_z \end{pmatrix}. \quad (2)$$

We denote in (1), (2) $h = \frac{H}{M_s}$ $\tau = \frac{t\gamma\mu_0 M_s}{(1+\alpha^2)}$, $j = \frac{J\hbar}{de\mu_0 M_s^{\,2}}$, $k = \frac{2K}{\mu_0 M_s^2}$, where H is the external magnetic field, M_s is the saturation magnetization, t is the time, μ_0 is the magnetic constant, α is the coefficient of dissipation, γ is the gyromagnetic ratio, d is the thickness of the free layer, e is the elementary charge, $\hbar$ is the Planck constant, K is the coefficient of anysotropy.

The impact of the spin–polarized current on the dynamics of the magnetization in such model is described by a current term in the form of Slonczewski – Berger. The coefficient of the current polarization included in the expression for that term is

$$G = \frac{4P^{3/2}}{(1+P)^3(3+\mathbf{m}\cdot\mathbf{s}) - 16P^{3/2}} = \frac{c}{b+m_x},$$

* Corresponding author: linda_nike@mail.ru

here $c=\frac{4P^{3/2}}{(1+P)^3}$, $b=3-4c$, $P=\frac{n_\uparrow - n_\downarrow}{n_\uparrow + n_\downarrow}$ is a parameter of the polarization, **s** is the magnetization of the the magnetization of the pinned layer [2]. We consider in this work a case of the longitudinal anysotropy where $\mathbf{s}=(1,0,0)$.

In the coordinate record the system (1) has a view

$$\begin{aligned}
\dot{m}_x &= -j(m_y^2+m_z^2)G+\alpha(h+km_x)- \\
&-\alpha m_x L + m_y m_z, \\
\dot{m}_y &= jm_x m_y G - m_x m_z - hm_z - km_x m_x - \\
&-\alpha jm_z G - \alpha m_y L, \\
\dot{m}_z &= m_y(h+km_x)+\alpha(jm_y G - m_z)- \\
&-\alpha m_z L + jm_x m_z G,
\end{aligned} \quad (3)$$

where $L=hm_x+km_x^2-m_z^2$, m_x, m_y, m_z are projections of the vector of magnetization of the free layer of spin velve on the axis *OX*, *OY* and *OZ* respectively.

The important characteristics of the dynamics of the magnetization are singular points of system (3), i.e. the equilibrium points of the magnetization of the valve structure. In works [3-5] the following expressions for calculation of the coordinates of singular points were obtained

$$m_x^2-1=0, \quad (4a)$$

$$\begin{aligned}
&m_x{}^4(k^2+k)+m_x{}^3(h+2bk^2+2hk+2bk)+ \\
&+m_x{}^2(h^2+4hbk+2hb+b^2k^2+b^2k+c^2j^2)+ \\
&+m_x(hb^2+2h^2b+2hb^2k)+h^2b^2=0.
\end{aligned} \quad (4b)$$

Equation (4A) corresponds to two singular points that are in the system for all currents and fields, equation (4B) allows to calculate the coordinates of other singular points, depending on the current and field and to determine areas of their existence. In our cosideration, we were interested in the existence in the system (3) trajectory connecting the equilibrium points $T_{1,2}(\pm 1,0,0)$ — in other words, the switching mode. Such a trajectory can exist only if the instability of the initial position of the magnetization vector corresponding to the point $T_1(+1,0,0)$, and stability of the final position of the magnetization vector – point $T_2(-1,0,0)$.

In works [3-5] expressions for determining the stability of the point $T_1(+1,0,0)$ were obtained :

$$\frac{\alpha de M_S \mu_0}{\hbar G}H - J + \frac{\alpha de}{\hbar G}\left(\frac{4K+M_S^2\mu_0^2}{2\mu_0}\right)=0 \quad (5a)$$

$$\left(\frac{H+\frac{4K+M_S^2\mu_0}{2M_S}}{\frac{M_S}{2}}\right)^2+\left(\frac{J}{\frac{deM_S^2\mu_0}{2G\hbar}}\right)^2-1=0 \quad (5b)$$

Expressions (5a) and (5b) are responsible for the minimum switching current of the spin valve. Expression (5a) is the canonical form of the equation of the straight line separating the regions in which the point $T_1(+1,0,0)$ is a stable or an unstable focus, and which intersects the axis at the point

$$J_{\min}=\frac{\alpha de}{\hbar G}c\text{, где } c=\left(\frac{4K+M_S^2\mu_0^2}{2\mu_0}\right). \quad (6)$$

The expression (5b) is the canonical form of the equation of an ellipse within which the point $T_1(+1,0,0)$ has the form of a saddle, and which crosses the *H*-axis at the points with coordinates $H_{\min}=\frac{-2K}{M_s\mu_0}$ and $H_{\max}=-\frac{2K+M_S^2\mu_0}{M_s\mu_0}$. Details of the bifurcation analysis was presented in works [3-5].

The decrease of the switching current can be done in several ways :
1) applying a magnetic field directed opposite to the magnetization of the pinned layer;
2) reducing the thickness of the free layer;
3) choosing the manufacturing material of the free layer of the spin valve, the magnetic properties of which require minimal stability region of point $T_1(+1,0,0)$.

3. The influence of the magnetic properties of ferromagnetic materials on threshold fields and on switching currents of spin valve

Consider an influence of the magnetic properties of materials constituting the ferromagnetic layers of the spin valve, on the threshold switching current for zero magnetic field $J_{\min}$ and minimum magnetic switching field $H_{\min}$ for zero current. For this we have chosen the following materials:
1) ferroborates of cobalt (they possess the best magnetic properties for reducing the switching current due to the high value of the polarization parameter): $Fe_{60}Co_{20}B_{20}$ which was annealed at the temperature 450°C, and $Fe_{40}Co_{40}B_{20}$ and $Fe_{70}Co_{30}$ which were annealed at the temperature 300°C ;
2) alloys of cobalt with gadolinium (they possess the best magnetic properties for reducing the switching magnetic field: $Co_{80}Gd_{20}$ and $Co_{93}Gd_7$ which were annealed at the temperature 200°C ;
3) materials, production of mono-crystal membrans of which has the lowest complexity and is less expensive: cobalt, iron.

Magnetic properties of the considered materials are given in Table 1.

Table 1. The saturation magnetization $M_s\mu_0$, the coefficient of anisotropy K, the spin polarization P and the coefficient of dissipation α for different materials.

Material*)	$M_s\mu_0$, T	$K, \frac{\text{J}}{\text{m}^3}$	P	α
Co (cobalt) [6]	1,76	$5,3\cdot10^5$	0,35	0,020
Fe (iron) [6,7]	2,15	$4,8\cdot10^4$	0,40	0,008
$Fe_{70}Co_{30}$ [7-9]	2,40	$3,5\cdot10^4$	0,55	0,015
$Fe_{60}Co_{20}B_{20}$ [10-12]	1,96	$2,1\cdot10^5$	0,53	0,040
$Fe_{40}Co_{40}B_{20}$ [13,14]	1,30	$3,4\cdot10^{-2}$	0,52	0,010
$Co_{93}Gd_7$ [15]	1,21	$1,88\cdot10^3$	0,30	0,020
$Co_{80}Gd_{20}$ [15]	0,10	$1,38\cdot10^3$	0,10	0,020

*) In square brackets references to the data sources are given

Figure 1 shows the line of stability of the point $T_1(+1,0,0)$ for various materials. Switching of spin valve is possible only in the region above the line whose equation is has the form (5a), and within semi-ellipse determined by formula (5b). Table 2 shows the values of the minimum switching currents for zero magnetic field $J_{\min}$ and minimum magnetic field for zero current $H_{\min}$ for various materials. Note that, from physical considerations, there is a number of restrictions on the limiting value of the density of the injection current J and an external magnetic field H, associated with electromigration and heating of the sample, which we did not consider in our calculation.

As can be seen from Figure 1, the line of stability of $Co_{80}Gd_{20}$ has the minimum slope, in this case, accordingly, there is a minimum threshold current $J_{\min}$ at zero magnetic field (see Table 2). The memory cells MRAM produced by technology with 10 nm by IBM in partnership with Samsung, use current of density $J = 7,89\cdot10^{10}\ \frac{\text{A}}{\text{m}^2}$ [1] in the switching, which is significantly more than the estimated values $J_{\min}$ for $Co_{80}Gd_{20}$.

Based on figure 1 and table 2 we can conclude that the smallest value of the magnetic field switching at zero current has alloy $Fe_{40}Co_{40}B_{20}$.

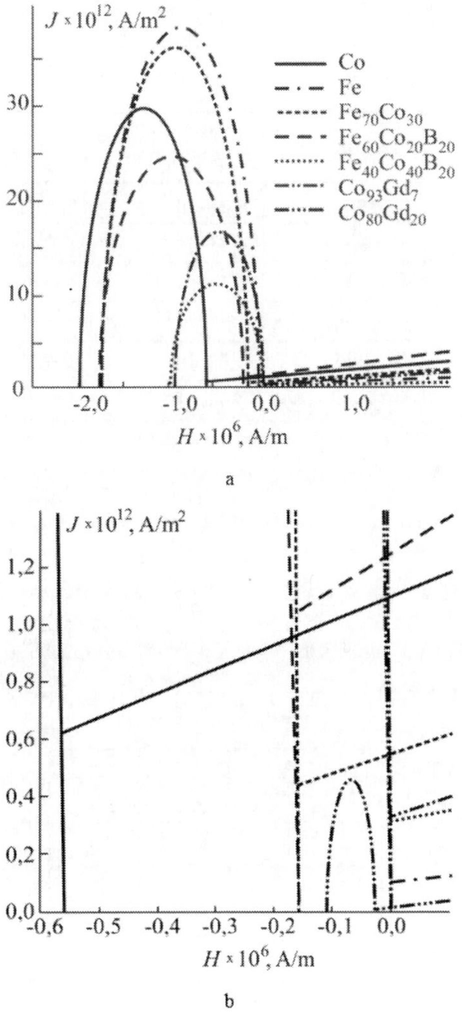

Fig. 1. The stability line of singular point $T_1(+1,0,0)$ for cobalt (solid line), iron (dash–dotted line), $Fe_{70}Co_{30}$ (dashed line with short strokes), $Fe_{60}Co_{20}B_{20}$ (the dashed line with long dashes), $Fe_{40}Co_{40}B_{20}$ (dotted line), $Co_{93}Gd_7$ (dash–dotted line with two dots), $Co_{80}Gd_{20}$ (dash–dotted line with three dots); a —in the range $H[-2,5;2]\times10^6\ \frac{\text{A}}{\text{m}}$ and in the current range $J[0;40]\times10^{12}\ \frac{\text{A}}{\text{m}^2}$; b — in the range $H[-0,6;0,1]\times10^6\ \frac{\text{A}}{\text{m}}$ and in the current range $J[0;1,4]\times10^{12}\ \frac{\text{A}}{\text{m}^2}$

Table 2. Threshold currents J_{min} and magnetic fields H_{min} of switching of the spin valve for various materials

Material	$J_{min}, \frac{A}{m^2}$	$H_{min}, \frac{A}{m}$
Co(cobalt)	$1{,}12 \cdot 10^{12}$	$-6{,}02 \cdot 10^{5}$
Fe(iron)	$3{,}38 \cdot 10^{11}$	$-4{,}47 \cdot 10^{4}$
$Fe_{70}Co_{30}$	$5{,}79 \cdot 10^{11}$	$-2{,}92 \cdot 10^{4}$
$Fe_{60}Co_{20}B_{20}$	$1{,}31 \cdot 10^{12}$	$-2{,}14 \cdot 10^{5}$
$Fe_{40}Co_{40}B_{20}$	$1{,}15 \cdot 10^{11}$	$-5{,}17 \cdot 10^{-1}$
$Co_{93}Gd_{7}$	$3{,}89 \cdot 10^{11}$	$-3{,}13 \cdot 10^{3}$
$Co_{80}Gd_{20}$	$2{,}66 \cdot 10^{9}$	$-2{,}76 \cdot 10^{4}$

4. The calculation of the switching time

To estimate the time of switching of the MRAM cell, the dynamics of the process of switching the spin valve from parallel to antiparallel state was simulated. Table 3 presents the values of the time of switching of the spin valve which is included in the MRAM memory cell and contains the spin valve of different materials.

Table 3. The switching time of spin valve (in nano-seconds) for different materials

Material	$H = 0$	$J = 0$
Co (cobalt)	187,2	14,5
Fe (iron)	339,2	91,9
$Fe_{70}Co_{30}$	171,5	21,1
$Fe_{60}Co_{20}B_{20}$	52,6	79,4
$Fe_{40}Co_{40}B_{20}$	564,0	4690,9
$Co_{93}Gd_{7}$	79,38	63,85
$Co_{80}Gd_{20}$	960,6	105,5

As Table 3 shows, the lowest value of the switching time corresponds to the spin valve of cobalt at zero current. However, switching modes, using a magnetic field, suggest the complication of the production of the chip of magnetoresistive memory MRAM, since the application of the magnetic field requires further massive tire. Therefore, the priority mode of switching by the electric current (which gives the most high-speed) is a mode for the spin valve based on $Fe_{60}Co_{20}B_{20}$.

5. Conclusion

Thus, in the present work, we conducted an analysis of the dependence of the current switching on characteristics of the materials of magnetic layers of the spin valve. Based on this analysis it was found that the promising materials for magnetic layers are cobalt, iron, and compounds such as $Fe_{60}Co_{20}B_{20}$ (which were annealed at the temperature 450°C), $Fe_{40}Co_{40}B_{20}$ and $Fe_{70}Co_{30}$ (which were annealed at the temperature 300°C); $Co_{80}Gd_{20}$ and $Co_{93}Gd_{7}$ (which were annealed at the temperature 200°C). We investigated the dynamics of the magnetization vector of the free layer of the spin valve fabricated on the basis of these materials. As a result of the calculation of the critical currents and fields, as well as the switching time for each mode without accounting physical restrictions on the values of current density and magnetic field, it is concluded that the most suitable for the manufacture of spin valve contained in the memory cell of MRAM, the materials can serve two alloys, namely $Fe_{60}Co_{20}B_{20}$ and $Co_{80}Gd_{20}$ which were annealed at the temperatures 300°C and 200°C respectively.

References

1. J.J. Nowak, R.P. Robertazzi, J.Z. Sun et. all., IEEE Magnetics Letters, **7**, 3102604 (2016)
2. J.C. Slonczewski, Journal of Magnetism and Magnetic Materials, **159**, L1 (1996)
3. N.V. Ostrovskaya, V.A. Skidanov, Iu.A. Iusipova, Solid State Phenomena, **233-234**, 431 (2015)
4. N.V. Ostrovskaya, V.A. Skidanov, Iu.A. Iusipova, Proceedings of IPPM RAS Ed. Stempkovskiy A.L., Moscow: IPPM RAS, 199 (2016)
5. N.V. Ostrovskaya, V.A. Skidanov, Iu.A. Iusipova, Computer Research and Modeling, **8**, 4, 605 (2016)
6. R. Skomski, *Simple Models of Magnetism* (Oxford university press, New York, 2008)
7. T. Kawai1, Y. Asai1, M. Ohtake et. all., EPJ Web of Conferences, **40**, 13001 (2013)
8. J.S. Moodera, C. Tanaka et. all., Philosophical Magazine B, **80**, 2, 195 (2000)
9. A.P. Babichev, N.A. Babushkina, et. al, *Physical quantities: Reference book Ed. Grigor'ev I.S., Meylihov E.Z.* (Energoatomizdat, Moscow, 1991)
10. S.X. Huang, T.Y. Chen, C.L. Chien, Appl. Phys. Lett., **92**, 242509 (2008)
11. M.H. Nguyen, C.F. Pai, K.X. Nguyen et. all., Appl. Phys. Lett., **106**, 222402 (2015)
12. S. Ikeda, K. Miura, H. Yamamoto et. all., Nature Materials, **9**, 721 (2010)
13. J. Hayakava, S. Ikeda, Y.M. Lee, et. all., Jap. J. Appl. Phys., **44(2)**, 37–41, L1267 (2005)
14. H. Almasi, D.R. Hickey, T.Newhouse-Illige et. all., Appl. Phys. Lett., **106**, 182406 (2015)
15. C. Kaiser, S.S. PapWorth Parkin United States Patent № US007230265B2 (2007)

EPJ Web of Conferences **185**, 01013 (2018) https://doi.org/10.1051/epjconf/201818501013
MISM 2017

High field magnetoresistance of nanocomposites $(Co_{84}Nb_{14}Ta_2)_x(Al_2O_3)_{100-x}$ near the percolation threshold

Mikhail Blinov[1], *Ivan Zakharchuk*[2], *Erkki Lähderanta*[2], *Alexander Sitnikov*[3], *Igor Rodionov*[1], *Valerii Prudnikov*[1], *Vladimir Rylkov*[4,5], *and Alexander Granovsky*[1,5]

[1] Faculty of Physics, Lomonosov Moscow State University, 119991 Moscow, Russia
[2] Lappeenranta University of Technology, 53851 Lappeenranta, Finland
[3] Voronezh State Technical University, 394026 Voronezh, Russia
[4] National Research Centre "Kurchatov Institute", 123182 Moscow, Russia
[5] Institute of Applied and Theoretical Electrodynamics RAS, 127412 Moscow, Russia

Abstract. We present results of experimental studies of magnetic properties, resistivity and magnetoresistance (MR) of $(Co_{84}Nb_{14}Ta_2)_x(Al_2O_3)_{100-x}$ films deposited onto a glass-ceramic substrate by the ion-beam sputtering, focusing on MR in high magnetic fields for compositions close to the percolation threshold (x=47-57 at.%). The samples consist on Co-Nb-Ta metallic nanogranules size of 2-5 nm which are embedded into the non-stoichiometric Al-O matrix. Magnetization was measured by SQUID magnetometer at T=4.2-350 K. MR was studied in the pulsed magnetic fields μ_0H up to 20 T at T=70-300 K in three geometries: magnetic field in plane parallel and perpendicular to current, magnetic field perpendicular to plane. The pulse duration was 11-12 ms. For the sample with x=57 at.% the temperature dependence of conductivity follows the ln$T$ behavior that matches a strong tunnel coupling between nanogranules. With decreasing metal volume fraction ln$T$ behavior gradually changes to the $T^{1/2}$ dependence at 47 at.%. For all samples MR is small (<1%) and negative. For x<57 at.% it is slightly anisotropic at μ_0H<1.0 T and almost saturates with increasing magnetic field. There is an evidence of small positive contribution to MR at μ_0H=20T. Accordingly to structural and magnetic data a large amount of metallic atoms are located between magnetic nanogranules that diminish the tunnel barrier height and make tunnel MR small and weakly dependent on temperature.

1 Introduction

Recent discovery of memristor effect in metal-insulator nanostructures, in particular in those based on non-stoichiometric oxides AlOz (z<1.5) [1,2], triggered renewed interest to structural, magnetic and magnetotransport properties of magnetic nanocomposites [3,4]. In the case of $(CoFeB)_x$ $(Al\text{-}O)_{1-x}$ nanogranular thin films with metal content x=49-56 at.% it was found a large number of Co, Fe and B atoms dispersed in Al-O matrix [3], logarithmic temperature dependence of conductivity [3, 4] and tunnel anomalous Hall effect [3]. These observations raised numerous questions on dominant transport mechanisms in nanocomposites. Particularly for mechanisms of magnetoresistance (MR), which may be due to tunneling, hopping, Lorentz force, spin-orbit interaction, weak localization and so on. The granular size about 2-5 nm and presence of isolated magnetic atoms in matrix requires for saturation of magnetization the use of high magnetic fields in MR measurements. In this paper we focus on MR in nanocomposites $(Co_{84}Nb_{14}Ta_2)$x(Al-O)1-x in the pulsed magnetic fields μ_0H up to 30 T.

* Corresponding author: gran60@mail.ru

2 Experiment

The nanocomposites $(Co_{84}Nb_{14}Ta_2)_x(Al\text{-}O)_{1-x}$ with x=47-57 at.% were obtained by the ion-beam sputtering of targets, which consisted of a metal base of the $Co_{84}Nb_{14}Ta_2$ alloy with several Al_2O_3 plates on its surface. Details of this method are presented in [3, 5]. The thickness of samples was about d=2.7μm.

The magnetization measurements were performed with a SQUID magnetometer (Cryogenic S700X) between 4.2 and 350 K at in-plane and out-of-plane magnetic fields μ_0H up to 3 T.

The resistivity was measured with a standard dc method, and the pulsed magnetic field B up to 30 T was applied to measure the MR. The pulse duration was 11–12 ms. The magnetic field inhomogeneity did not exceed 3% within the size of the investigated area (5 mm^2).

3 Results and discussion

3.1. Magnetization

The field and temperature dependencies of magnetization for studied samples (Fig.1) are quite

https://doi.org/10.1051/epjconf/201818501013

typical for nanogranular thin films with compositions in the vicinity of percolation threshold [3]. The saturation field for in-plane geometry is less than that for out-of-plane case, increases gradually when x decreases from 57 to 47 at.%, and does not exceed 0.6-0.7 T. This behavior can be easily understood taking into account demagnetization effects. For example, the saturation field of 0.6-0.7 T is well coincided with the demagnetization of spherical Co particles. The data shown on Fig.1b indicates on superparamagnetic behavior. The temperature dependence of magnetization at 2.0 T (not shown) is similar to that reported in Ref. [3] (see Fig. 8c), namely, magnetization strongly increases at T<25 K due to presence of a large amount of metallic atoms located between magnetic nanogranules and is almost constant at T=100-300 K.

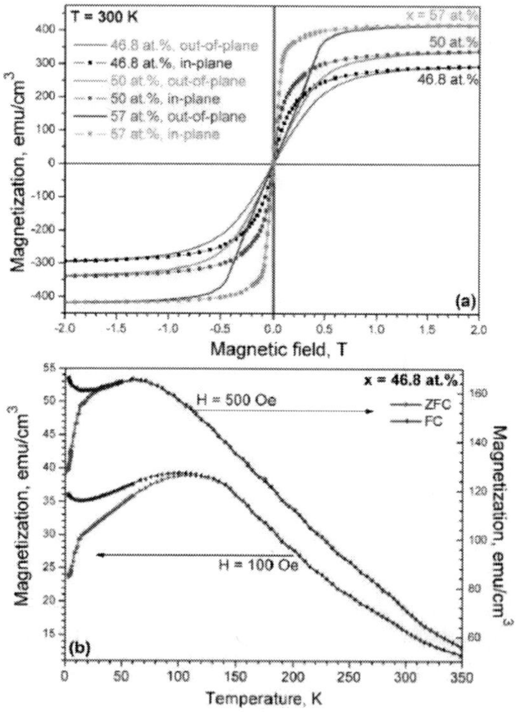

Fig. 1. (a) Magnetic field dependences of magnetization of samples with x = 46.8 at.%, 50 at.%, and 57 at.% at room temperature. The dash-dot lines present the in-plane curves and solid lines denote out-of-plane curves. (b) Magnetization temperature dependences of sample with x = 46.8 at.% measured in H = 100 Oe and 500 Oe. The red curves are zero-field magnetizations (ZFC) and blue curves are field-cooled magnetizations (FC).

3.2. Resistivity

Fig. 2 shows that resistivity of the sample with x=46.8 at. % is at least one order of magnitude larger than for x=57 at.%. It confirms that the considered samples composition is close to the percolation threshold. The temperature dependence of resistivity of the sample with x=46.8 at.% follows quite well the $T^{1/2}$ law but it is not the case for samples with x=57 at.% (Fig.2). For the sample with x=57 at.% the temperature dependence of conductivity follows the $\ln T$ behavior that accordingly to the theory of Efetov et al. [6,7] matches a strong tunnel coupling between nanogranules. Therefore the main transport mechanism in the studied compositions falls into the transition region between strong and weak tunnel coupling.

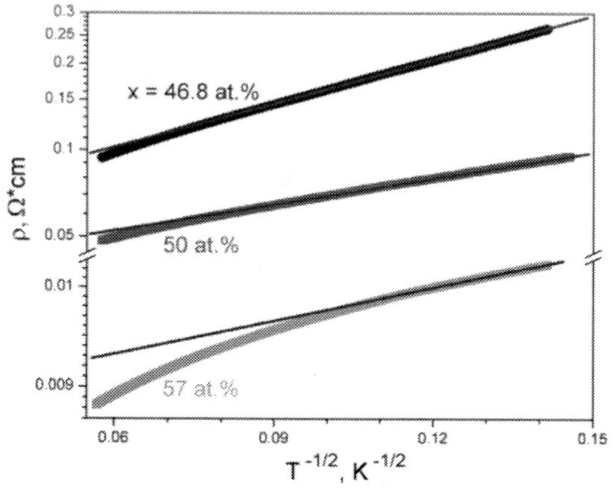

Fig. 2. Temperature dependences of resistivity of samples with x = 46.8 at.%, 50 at.%, and 57 at.% plotted as ln(ρ) versus $T^{-1/2}$.

3.3. Magnetoresistance

3.3.1. Negative magnetoresistance

For all samples MR is small and negative in weak magnetic fields (Figs. 3-5). Negative MR can be connected with spin-dependent tunnelling, suppression of spin disorder inside ferromagnetic granules and spontaneous anisotropy (see [8]). It is quite reasonable for nanocomposites to neglect two last mechanisms. Indeed, since in studied samples granules are small the second mechanism unlikely contributes to MR. Negative MR in our case is approximately the same for magnetic field applied parallel and perpendicular to the current direction and this observation allows to completely exclude anisotropic MR. Therefore we can conclude that negative MR in the samples under investigation is due to tunneling.

Let us now discuss why tunnel MR is small, less than 1% (in comparison with record values about 10-13% at low temperatures [9,10]), and almost negligible for the sample with x=57 at. %. There are several factors which can diminish or even suppress tunnel MR: not optimal composition, wide size distribution of granules, spin-flip processes, weak spin polarization of current carriers, rough interfaces of granular surfaces, scattering by impurities or metallic atoms located inside tunnel barriers. We think that in our case all these factors play role but the most important is the presence of a large amount of metallic atoms in the Al-O. The composition of the sample with x=57 at.% is quite far from the percolation threshold and that is the reason for its extremely small negative MR.

Tunnel MR weakly depends on temperature in the Inoue-Maekawa model [11] and shows the distinctive $1/T$ decay of the Helman-Abeles model [12]. Fig. 5 demonstrates that our data follows qualitatively the Inoue-Maekawa model, but MR decreases in the temperature range 200-300 K stronger than magnetization. Perhaps, it is connected with low tunnel barrier height in our case [3].

3.3.2. Positive magnetoresistance

MR reaches its greatest negative value in magnetic fields about 1- 3 T (Figs. 3 and 4) and after that instead of saturation starts to increase with magnetic field (Figs. 3 and 5). It is clearly indicate on the positive contribution to MR in high magnetic fields. Positive MR can be due to Lorentz force, MR in hopping regime, or the influence of magnetic field on e-e contribution to MR [13].

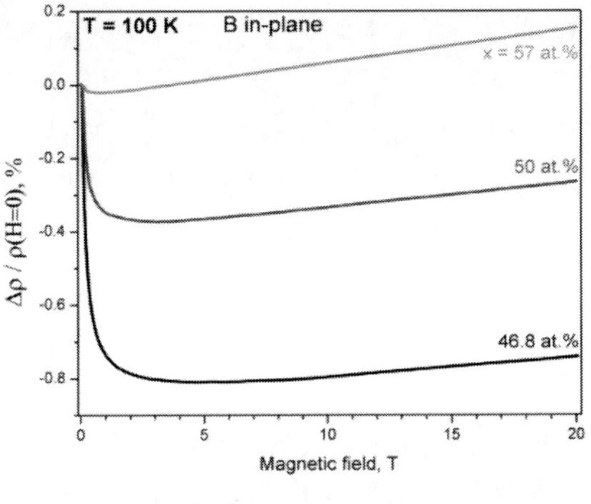

Fig. 3. Magnetoresistance of samples with x = 46.8 at.%, 50 at.%, and 57 at.% at temperature of 100 K. The measurements were carried out with magnetic field applied parallel to the sample plane along the current.

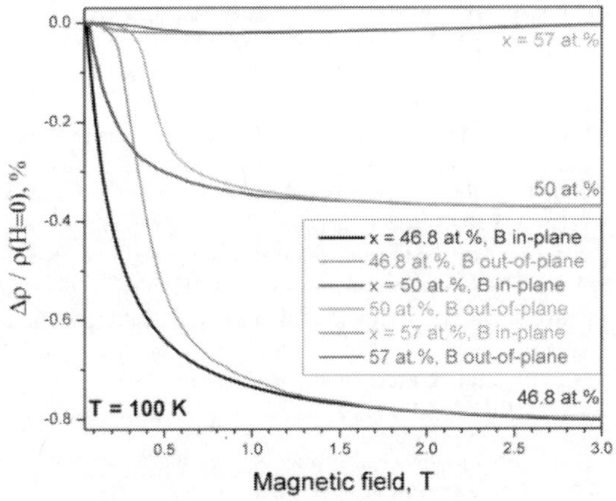

Fig. 4. The low magnetic field part of the magnetoresistance of samples with x = 46.8 at.%, 50 at.%, and 57 at.% at temperature of 100 K for both orientations of magnetic field: in-plane and out-of-plane.

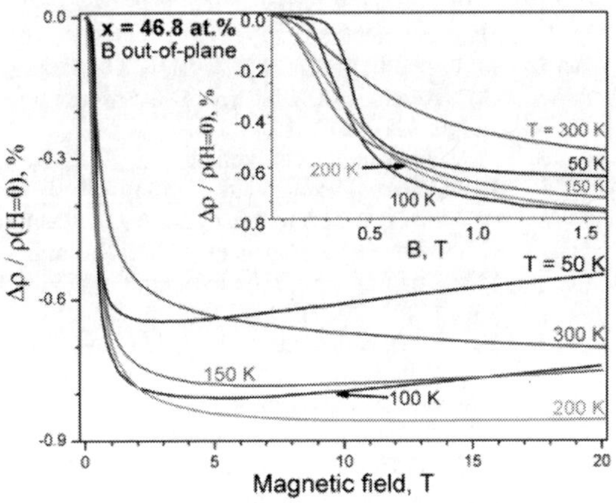

Fig. 5. The out-of-plane magnetoresistance of the sample with x = 46.8 at.% at various temperatures T = 50, 100, 150, 200, and 300 K. Inset shows the low magnetic field part of the main plot.

The Lorentz MR is described by the well-known relation $\frac{\Delta\rho}{\rho} = \frac{1}{24ne^2}\left(\frac{B}{\rho}\right)^2$, i.e. depends quadratically on the magnetic field and is inversely proportional to the square of resistivity. However, it is not the case for our samples (Fig. 4). The positive contribution is approximately the same for samples with quite different resistivity and does not follow B^2 law.

The influence of magnetic field on resistivity in hopping regime can be represented by the general form $\rho(B)=\rho(B=0)\exp(ConstB^m)$, where m varies between 0.5 and 2 depending on the model. Our data do not correspond to this field relation. In spite of significantly different resistivity of samples with x=46.8 and 57 at. %, the positive contribution to MR is almost the same. Therefore in spite of a large amount of metallic atoms in the matrix there is no evidence of hopping transport both in resistivity and MR. The positive MR in high magnetic fields in thin films and nanogranular alloys was observed in [13] and was explained by the influence of magnetic field on e-e contribution to MR. By contrast to [13] in our case the positive MR slightly depends on resistivity and temperature. There is no evidence of significant e-e contribution to resistivity and MR.

Since the basic transport mechanism in the samples under investigations is tunneling we can suppose that the observed positive contribution to MR in high magnetic field is due to influence of magnetic field on tunnel barrier height. This hypothesis requires further experimental and theoretical researches.

This work was financially supported by the Russian Science Foundation, Grant No. 16-19-10233.

References

1. J. J. Yang, D. B. Strukov, and D.R.Stewart, Nat. Nanotechnol. **8**,13 (2013)
2. M. Prezioso, F. Merrikh-Bayat, B. D. Hoskins, G.C. Adam, K. K. Likharev, D.B.Strukov, Nature 521, **61** (2015)
3. V. V. Rylkov, S. N. Nikolaev, K.Yu.Chernoglazov, V. A. Demin, A.V.Sitnikov, M.Yu. Presnyakov, A.L.Vasiliev, N.S. Perov, A. S. Vedeneev, Yu. E. Kalinin, V.V. Tugushev, A. B. Granovsky, Phys. Rev. B **95**, 144202 (2017)
4. Yu. O. Mikhailovskii, V. N. Prudnikov, V.V.Ryl'kov, K. Yu. Chernoglazov, A.V.Sitnikov, Yu. E. Kalinin, and A.B.Granovskii, Physics of the Solid State **58**, 444 (2016)
5. Yu. E. Kalinin, A. N. Remizov, A.V. Sitnikov, Physics of the Solid State **46**, 2146 (2004)
6. K. B. Efetov and A. Tschersich, Phys. Rev. B **67**, 174205 (2003).
7. I. S. Beloborodov, A. V. Lopatin, V.M.Vinokur, and K.B. Efetov, Rev. Mod. Phys. **79**, 469 (2007).
8. A. Milner, A Gerber, B. Groisman, M. Karpovsky, and A. Gladkikh, Phys. Rev. Lett. 76,475 (1996)
9. Y.A. Vovk, V.O. Golub, L. Malkinski, A.F.Kravetz, A.M. Pogorely, and O.V. Shypil, JMMM **272-276**, e1403 (2004).
10. N. kobayashi, S. Ohnuma, T. Masumoto, and H. Fujimori, J. of Appl. Phys., 90, 4159 (2001)
11. J. Inoue and S. Maekawa, Phys. Rev. B **53**, R11927 (1996).
12. J. S. Helman and B. Abeles, Phys. Rev. Lett. **37**, 1429 (1976).
13. A. Gerber, I. Kishon, I. Ya. Korenblit, O. Riss, A. Segal, and M. Karpovski, Phys. Rev. Lett. 99, 027201 (2007)

Magnetotransport properties of $Co_x(MgF_2)_{100-x}$ oxygen-free nanocomposites

Tatyana Tregubova*, *Oleg* Stognei, *Veronika* Kirpan, and *Aleksandr* Sitnikov

Voronezh State Technical University, 394026 Voronezh, Russia

Abstract. Electric and magnetoresistive properties of $Co_x(MgF_2)_{100-x}$ thin films ($14 \leq x$, ат. % ≤ 62) were investigated. It is established that the received films have a composite structure. Magnetoresistive and electric properties of the films in initial state and after heat treatment have been investigated. The magnetoresistive effect of the studied samples in initial state reaches 7,5% in magnetic field of 10 kOe. It is established that the composites structure is steady against thermal heating up to 250 °C. The change of composites magnetoresistive effect after heat treatment depends on the Co concentration. The particularities of the magnetoresistive changes after annealing are discussed.

Introduction

One of the most characteristic properties of nanogranular ferromagnetic metal-insulator composites is the tunneling magnetoresistance (TMR). High values of TMR have been observed in cobalt based composites [1]: in Co-Al-O composites the magnetoresistance reaches 7-8 % [2] and in some cases it reaches 10 % [3]. Despite the high values of the TMR observed in the composites the presence of an oxide matrix leads to some undesirable phenomena. In particular the metal phase can be oxidized during deposition and subsequent storage of the composites. The oxidizing affects to the properties of the formed samples usually worsening them. The aim of the present work was to obtain a two-phase metal-insulator nanocomposite based on an oxygen-free matrix capable to exhibit high values of the magnetoresistive effect and save these values under thermal action.

Experimental procedure

The Co-MgF_2 system was chosen as an object of investigation. Magnesium fluoride was selected as the dielectric phase since the formation of MgF_2 is much more energetically favorable than the formation of CoF_2. That ensures the production of a composite structure in the chosen system. All samples have been prepared by ion-beam sputter of mosaic target made from a Co base and MgF_2 plates nonsymmetrically located on its surface. The samples were deposited out on glass (for structure investigations) and glassceramic substrates (for resistive measurements).

The elemental composition of the composites was determined by electron-probe energy-dispersive microanalysis. The measurements were carried out on a scanning X-ray microanalyzer JXA-840 with an error not exceeding 1.5% of the content of the measured element. The thickness of the obtained samples was monitored with the help of the MII-4 interferometer. A BRUKER D2 Phaser diffractometer was used to perform the phase analysis of the samples.

Investigation of the magnetoresistive effect was carried out by direct measurement of the electrical resistivity of the samples during the change of the external magnetic field. The resistance of the samples was measured by a potentiometric method using a two-probe DC method. The error of the electrical resistivity measurements did not exceed 0.1%. Annealing of the samples was carried out in a vacuum with a residual pressure not less than 10^{-3} Torr.

Results

As a result of the deposition the samples of $Co_x(MgF_2)_{100-x}$ fine-dispersed thin-films were obtained. The thickness of the investigated films is 1-3 μm depending on the composition. The concentration of the metal phase in the obtained samples smoothly changes from 14 to 62 atomic %.

Analysis of the diffractograms obtained from the samples in the initial state indicates the presence of a two-phase (composite) structure for all obtained compositions (fig. 1). There are peaks corresponding only to two phases: magnesium fluoride and cobalt on the diffractograms. As can be seen from Figure 1 the peaks corresponding only to the MgF_2 phase are observed on the diffractograms corresponding to the lower metal concentration. As the concentration of the metal increases the intensity of the peaks of the dielectric component decreases and the peaks from the crystalline Co with hexagonal structure become more pronounced.

* Corresponding author: ttv1507@ya.ru

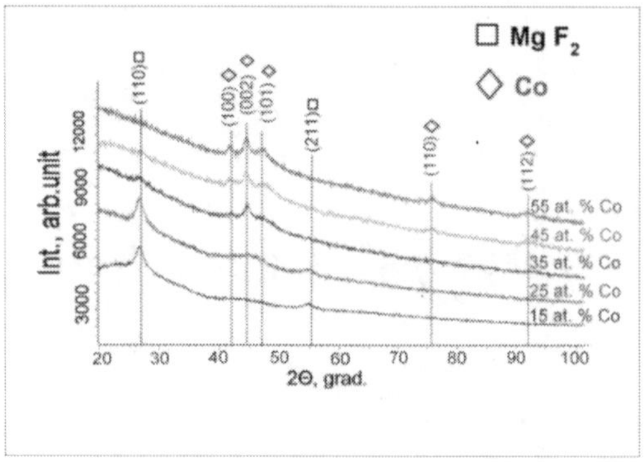

Fig. 1. Diffractograms of $Co_x(MgF_2)_{100-x}$ samples. The cobalt concentration (in at. %) is indicated by the numbers

It is wellknown that the electrical properties of any nanogranular composite system change radically with increasing of metal phase concentration [4]. In particular, the concentration dependence of the resistivity of $Co_x(MgF_2)_{100-x}$ composites in the initial state is typical for composite metal-insulator systems (fig. 2).

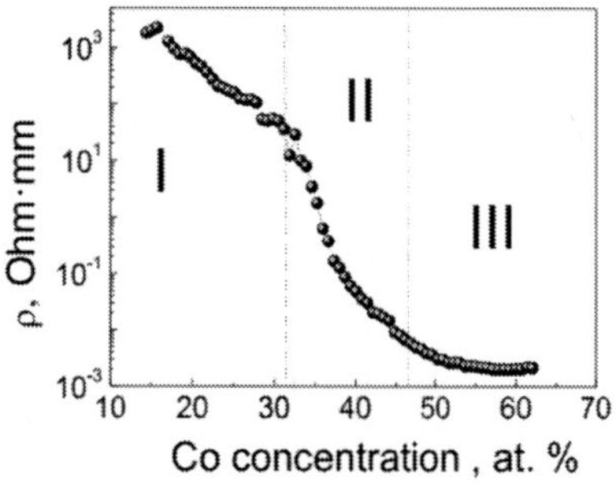

Fig. 2. Concentration dependence of the electrical resistivity of $Co_x(MgF_2)_{100-x}$ composites. The data are obtained at room temperature

The whole studied concentration range can be divided into three regions, differing from each other by the angle of inclination of the ρ dependence. The second region is characterized by a sharp decrease of resistance against the background of a relatively small change in Co concentration. Thus, it can be concluded that this region of concentration dependence corresponds to the percolation threshold [5]. An estimation of the concentration position of the percolation threshold for the $Co_x(MgF)_{100-x}$ system corresponds to an interval of 30-36 at.% Co (fig. 2).

In the vicinity of the threshold the $Co_x(MgF)_{100-x}$ samples exhibit magnetoresistive effect. It reaches quite high values even at relatively small applied field (fig. 3). In a case of larger magnetic field (up to 10 kOe) maximum values of the tunnel magnetoresistance reaching 7.5% (fig.3, incert). However, it should be noted that even in the region of the TMR maximum the saturation of the effect is not achieved (fig. 3, inset) and can be larger in a case of the field increasing. One should note that the presence of the magnetoresistive effect unambiguously confirms the compositeness of the samples.

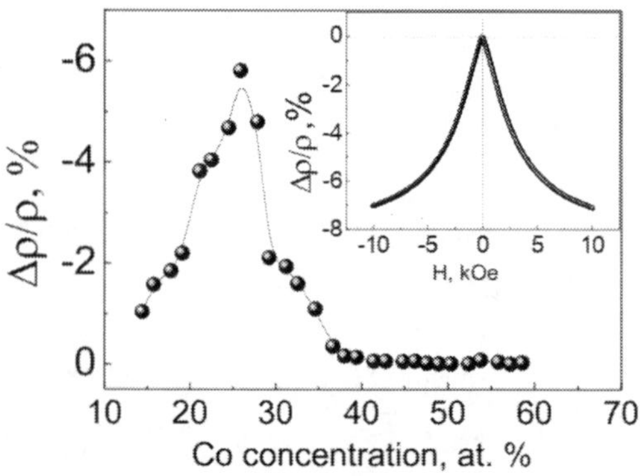

Fig. 3. The concentration dependence of the magnetoresistive effect of $Co_x(MgF_2)_{100-x}$ composites in the initial state. The TMR values were determined in a field of 5 kOe. The inset shows the field dependence of TMR for $Co_{25}(MgF_2)_{75}$. The data are obtained at room temperature

Nanogranular composites prepared by deposition technique has a nonequlibrium structure and any thermal influence should change physical properties of the material. To determine the thermal stability of the prepared composites the electrical resistance versus temperature has been measured in some samples (fig. 4). As follows from the presented data the heating of the samples to a temperature of 250 °C affects their electrical resistance in different ways. The resistance of the samples with a concentration corresponding to the pre-threshold interval increases with temperature increasing and after cooling it is higher than in the initial state (for example $Co_{24}(MgF)_{76}$, fig. 4). In samples with a higher cobalt concentration (31 at.% Co and above) the resistance begins to decrease during heating and after cooling it is lower than in the initial state (fig. 4).

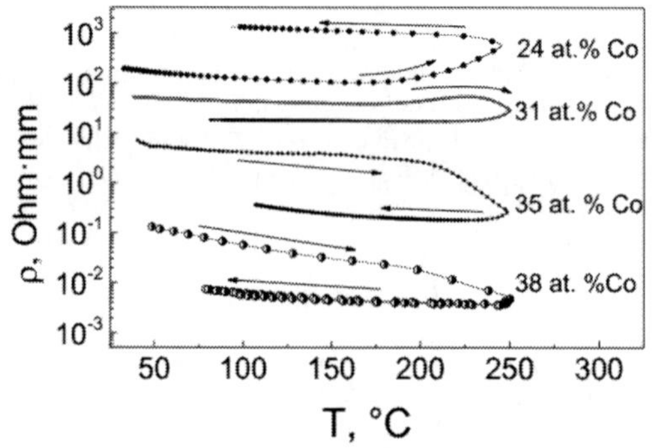

Fig. 4. Temperature dependence of the electrical resistivity of $Co_x(MgF_2)_{100-x}$ composites

The analysis of the obtained data also indicates that no structural changes occur in composites up to 200 °C. This follows from the fact that the resistance of the samples varies practically linearly in this temperature interval without any anomalies.

A more significant argument in favor of the fact that the structure of composites does not undergo any changes in this temperature interval is the absence of any change in the

 https://doi.org/10.1051/epjconf/201818501014

TMR values (fig. 5). If the samples annealing was carried out at 180 °C it did not lead to any changes in the TMR values (fig. 5). This fact testifies the stability of both the samples dielectric matrix and the composite structure as a whole. The invariance of the TMR values may also indicate a low level of defectiveness of the fluoride matrix. In the opposite case the defect concentration should decrease during annealing and as a consequence the TMR values should increase [6] but such result is not observed in our experiment.

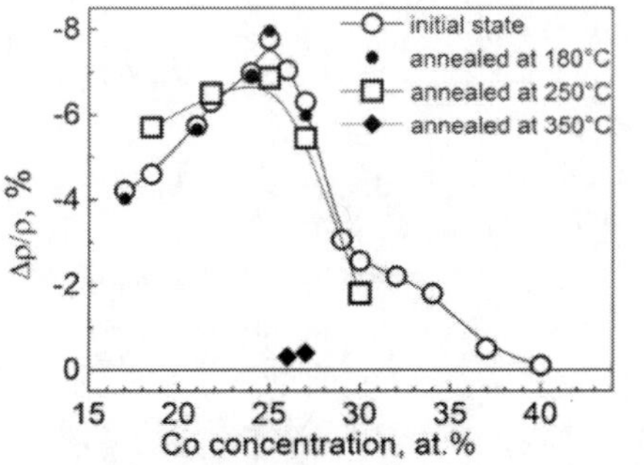

Fig. 5. The concentration dependence of the magnetoresistance of $Co_x(MgF_2)_{100-x}$ composites in the initial state and after heat treatment. All TMR values (for as-deposited and annealed samples) were determined in a field of 10 kOe at room temperature

Heating the samples to 200 - 250 °C initiates the beginning of structural changes which result in a significant change of the resistance (at room temperature the difference of the samples resistance in the initial state and after heat treatment at 250 °C is almost one order or larger, see Figure 4). Moreover, this annealing procedure changes the TMR values. The change of the TMR is systematic: the TMR value increases in the composites with low Co concentration (before the percolation threshold) but the TMR decreases in the composites located beyond the threshold (fig. 5).

A similar reaction of composites on the thermal action was observed in oxide-based composites ($CoFeB-SiO_2$) [7]. The magnetoresistance of the composites decreased after annealing if the concentration of metal phase was larger than the percolation threshold. On the contrary the TMR values increased after heat treatment if the metal concentration was lower than the percolation.

An increase of the heating temperature up to 350 °C leads to a significant change of the $Co_x(MgF_2)_{100-x}$ composites resistance. The resistance of the composites lying before the percolation threshold increases by several orders of magnitude and reaches hundreds of MΩ which does not allow to fulfill the potentiometric measurements (fig. 6). The resistance decreases sharply and the sign of the temperature coefficient of electrical resistance (α) changes from negative to positive if the composites located behind the percolation threshold. The positive α sign is typical for the metallic conductivity mechanism and indicates the formation of the percolation cluster after treatment at 350 °C. Such significant changes of electrical resistance indicate either interfacial interaction or significant changes in the properties of the phases themselves, provided that the heterogeneity (compositness) of the samples is preserved.

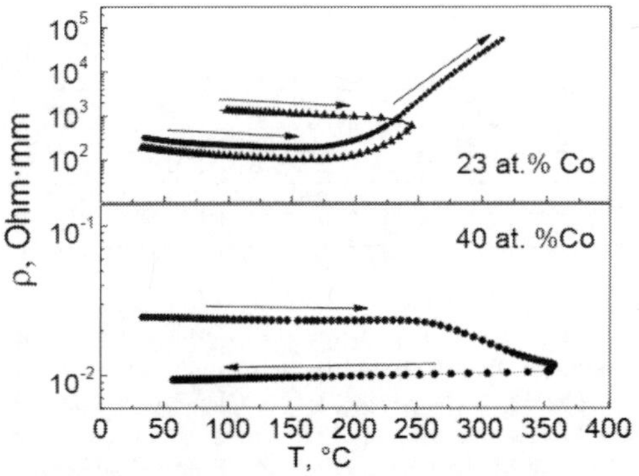

Fig. 6. Temperature dependence of electrical resistance for two $Co_x(MgF_2)_{100-x}$ composites located before and after percolation threshold

To detect the possible changes in the structure of the composites the X-ray structural studies of samples annealed at 350 ° C for 10 min were performed Figure 7 shows the diffraction patterns of the samples in the initial state and after heating at 350 °C. Obviously, the temperature influence does not affect the shape of the diffractograms: only an increase in the intensity of the peaks and a change in their shape are observed. The diffraction peaks became higher and narrower which indicates an increase in the size of the coherent scattering.

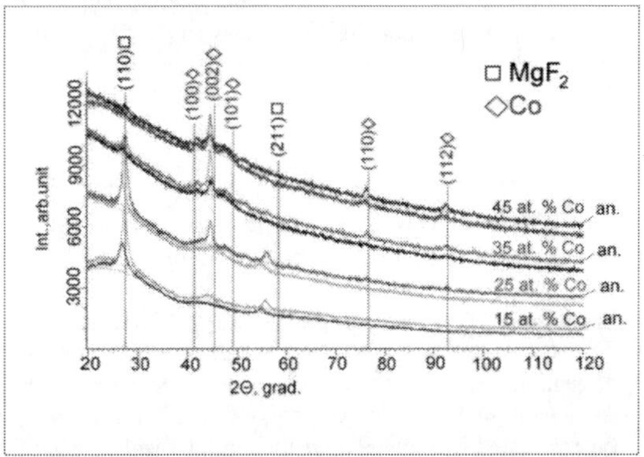

Fig. 7. Diffractograms of $Co_x(MgF_2)_{100-x}$ samples in the initial state (bottom curve for each sample) and after annealing at T = 350 °C (upper curve, indicated by symbol "an."). The Co concentration (in at. %) is indicated by the numbers

Using the Sherrer method, which takes into account the half-width of the diffraction maximum, the source wavelength, and also the Bragg diffraction angle, the values of the coherent scattering regions (L) were calculated for the cobalt phase. It seems quite reasonable to assume that the regions of coherent scattering in the metallic phase are the cobalt nanogranules. The results of calculations are presented in Table 1.

The growth of cobalt granules is obvious and for the composite located on the percolation threshold the increase reaches more than 100%. Such increase of the metal granules size explains the sharp decrease of the TMR to practically zero in composites located beyond the percolation threshold after heat treatment at 350 °C (fig. 5).

Table 1. The size of the cobalt grains in the initial state and after annealing at T = 350 °C.

	L, nм	
Composition	Initial state	After annealing at T = 350 °C
$Co_{25}(MgF_2)_{75}$	4,5	10,6
$Co_{35}(MgF_2)_{65}$	8,5	13,3
$Co_{45}(MgF_2)_{55}$	11,3	15,6

First, as the size of the metal granules increases, the width of the dielectric barrier should decreases. Thus, interaction can take place between the granules that leads to the formation of large magnetic domains (magnetic clusters). The magnetic moments of individual granules do not depend on the external magnetic field within this domain size. This local region of the composite does not contribute to the formation of the magnetoresistive effect (the conductivity through such a domain does not depend on the magnetic field). The second reason for the reduction of the TMR after annealing is the formation of electric percolation clusters, where electrical conductivity is due to the usual metallic conductivity while the tunneling does not occur. The formation of percolation clusters and the change of the conductivity mechanism are confirmed by the temperature dependences of the electrical resistivity of the composites with large cobalt content (behind the percolation threshold. fig. 6).

In pre-percolation composites the situation is different. Their morphology (and, consequently, their nanogranular structure) remains practically unchanged (fig. 7). However, in this case, the resistance increased to such high values that it did not allow to correctly determine the magnitude of the TMR effect. A possible explanation for this fact is a significant reduction in the degree of defectiveness of the dielectric matrix as a result of annealing at 350 C. It is known that in composites in addition to direct tunneling of electrons between the granules the hopping conductivity along the defects of the dielectric matrix also exists [6]. In our case, such defects are the broken bonds of magnesium atoms and cobalt atoms dissolved in fluoride. During the annealing the cobalt atoms diffuse to the granules increasing their size and the broken magnesium bonds can be connected by oxygen contained in the residual gas of the vacuum chamber.

As a result of these processes, the electrical resistivity of composites should increase, which is observed in our experiments (fig. 6). Also, the result of these processes should be the growth of the magnetoresistive effect. It should be noted that heating of the pre-percolation composites to a lower temperature (fig. 4, fig 5) leads precisely to this result: the growth of resistance and the growth of the magnetoresistive effect.

Conclusion

The nanocrystalline composites $Co_x(MgF_2)_{100-x}$ (14 ≤x, at.% Co ≤ 62) were prepared by ion-beam sputtering of a composite target in an argon atmosphere. Composites exhibit a tunneling magnetoresistance, reaching 7.5% in fields of 10 kOe. Composites exhibit thermal stability with respect to heating to 250 °C, while maintaining high TMR values.

Acknowledgment

The work has been partly supported by RFBR grant № 15-02-05920_a

References

1. A.E.Berkowitz, J.R.Mitchell, M.J.Carey et al., Phys.Rev.Lett., **68**, 3745, (1992)
2. S. Mitani, H. Fujimori, S. Ohnuma., JMMM, **177**, 919, (1998)
3. K. Yakushiji, S. Mitani, K. Takanashi et al, JMMM,.**212**, 75, (2000)
4. B.Abeles, P. Sheng M.D.Coutts and Y. Arie, Advances in Physics, **24,** 407, (1975)
5. Yu.Yu. Tarasevich, *Perkolyztsiya, teoriya, prilozheniya, algoritmi*, URSS, (2002)
6. Yu.E. Kalinin, A.V. Sitnikov, O.V. Stognei et al, Mater. Sci & Eng., **304**, 941, (2001)
7. O.V. Stognei, V.A. Slyusarev, Yu.E. Kalinin et al, Microelectronics Engineering, **69**, 476,(2003)

Tunnel magnetoresistance and spin transfer torque in magnetic tunnel junction with embedded nanoparticles

Niazbeck Useinov[1,*]

[1] *Institute of Physics, Kazan Federal University, Kazan 420008, Russia*

Abstract. The theoretical model of spin-dependent transport in magnetic tunnel junctions (MTJ) containing magnetic or non-magnetic nanoparticle is developed. The dependences of tunnel magnetoresistance (TMR) and in-plane component of spin transfer torque (STT) on the applied voltage for various sizes of nanoparticles of the order of the mean free path of the conduction electron are calculated. The calculation is performed in the approximation of the ballistic transport of conduction electrons through the insulating layers of the MTJ and the nanoparticles.

1 Introduction

Currently the magnetic tunnel junction (MTJ) are widely studied in spintronics. The effects of tunnel magnetoresistance (TMR) and spin transfer torque (STT) in such nanostructures find their application in magnetic field sensors, spin current filters, non-volatile magnetoresistive random-access memory (MRAM, STT-MRAM), resonant tunnel diodes, spin transistors, etc.

The main goal of this work is to highlight the important conditions for dimension and voltage induced thresholds as well as demonstrate TMR and STT behaviors due to presence of nanoparticles (NPs) in insulator of MTJ. The MTJ-NPs (MTJ with embedded NPs) characteristics are similar to those in double barrier magnetic tunnel junctions (DMTJs), since the basic tunneling phenomena is the same. Double barrier MTJs are promising structures for MRAM applications due to better thermal stability, low noise and less critical current value for the switching in comparison to single barrier tunnel junctions (SMTJs) [1]. Moreover, both DMTJ and MTJ-NPs show a different spin-dependent quantum effects on ferromagnet/barrier and metallic (or magnetic) NP/barrier interfaces, which can be related to Coulomb blockade (CB), quantum well (QW) states, ballistic tunneling, resonant spin filtering and Kondo effect depending on middle layer thickness and applied voltage [2–7].

The basic tunneling approach has similar origin with the theory of giant magnetoresistance in point-like contacts [8] reproducing the limit of the point-like contact when NP touches top and bottom ferromagnetic (FM) layers. However, many aspects of the problem related to variation of barrier asymmetry, size distribution, quantization of the states and temperature dependence were not completely studied.

*e-mail: nuseinov@mail.ru

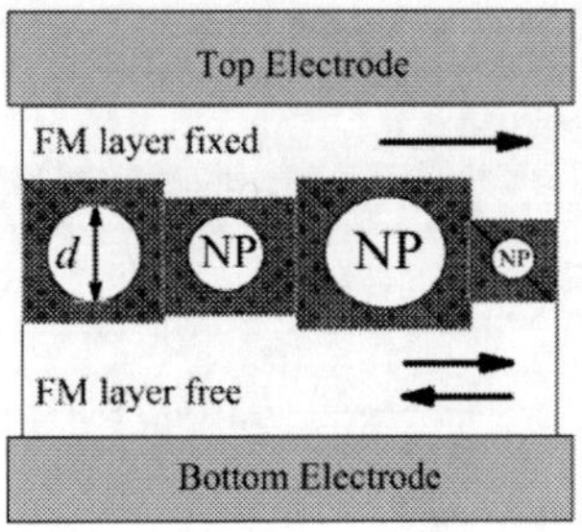

Figure 1. Schematic view of a magnetic tunnel junction with tunnel cells and nanoparticles (NP) of different sizes in an isolating middle layer (shaded). The arrows show the parallel (P) and antiparallel (AP) direction of the magnetizations of the free and fixed FM layers.

The present work is based on the assumption of the ballistic transport in MTJ-NPs, which were considered as a chain of the double barrier tunneling cells (TCs) connected in parallel, as shown in Fig. 1. Experimental works [6, 7] were considered as a data sources for the comparison with our results. In particular the MTJ-NPs was characterized by the average diameters of nanoparticles from $d = 1.0 \pm 0.4$ nm to $d = 3.2 \pm 0.7$ nm.

2 Ballistic transport of electrons through insulating layers and nanoparticles

Magnetic tunnel junction with embedded nanoparticles was simulated as one hundred TCs connected in parallel ($N_{\text{tot}} = 100$). A few TCs are shown schematically in Fig. 1, each TC contains one NP. Ten fractions $f = 1..10$ were considered with diameter $d^{(f)}$, and $\sum_f i_f = 100$,

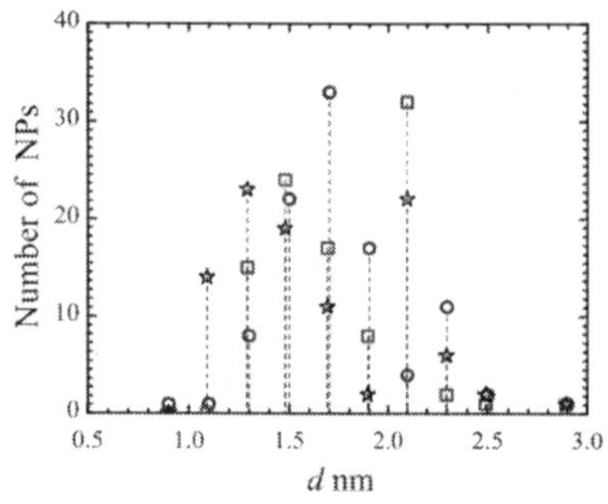

Figure 2. Distribution nanoparticles in MTJ with the set diameter d. The number nanoparticles with the given diameter in fraction is designated by corresponding signs with a line to an axis of abscisses. These numbers and signs correspond to calculations and the schedules presented further in drawings. The rings, rectangles and stars correspond to the simulation solid lines in Fig. 4, curve 1, 2 and 3.

where i is a number of NPs. The *fraction* is defined as a set of TCs with the same diameter of NPs. The distribution of nanoparticles and their separation into fractions in MTJ, depending on their diameter is shown schematically in Fig. 2. Note that this distribution corresponds to the size of the NPs studied in [6].

The ballistic tunneling was considered as a basic transport approach. The spin-polarized current for each TC was defined as follows [9, 10]:

$$I_{s,f}^{\mathrm{P(AP)}} = \frac{e^2}{h} \frac{\sigma_f k_{F,s}^2 V}{2\pi} \left\langle \cos(\theta_s) D_s^{\mathrm{P(AP)}}(V) \right\rangle, \qquad (1)$$

where V is the applied voltage, θ_s is the incident angle which arises from conservation of the longitudinal projections of the k-vectors to the interfaces, $s = \uparrow, \downarrow$ is the spin index, σ_f is an area of TC, $k_{F,s}$ is defined as a Fermi wavenumber of the top FM layer for positive voltage. The Eq. (1) consists integrals over the top incident electron trajectory angles θ_s and φ and it is derived more accurate than in case WKB method, since the transmission coefficient D_s was obtained as exact analytical solution for the double barrier system [11]. In our case D_s is a function of $k_{F,s}$, θ_s and wavenumber through of the nanoparticle, $k_{n,s}^{(f)}$.

The wavenumber values for majority and minority electron bands were assumed as initial parameters of FM layers at zero voltage for all considered TCs as $k_{F\uparrow} = 1.09\,\text{Å}^{-1}$ and $k_{F\downarrow} = 0.421\,\text{Å}^{-1}$ respectively. In present work the k-values are assumed as effective parameters which reflect the barrier transparency to filter out an electron wavefuctions and properties of spin bands at FM/barrier interfaces [12, 13].

To consider the positive voltage impact and spin filtering effect of NPs, electron wavenumbers in NPs and bottom FM layer were modified as follows:

$$\begin{aligned} k_{n,s}^{(f)}(V) &= \left[\left(k_n^{(f)}\right)^2 \pm \frac{m_{\mathrm{eff}}\Delta}{\hbar^2} + c\frac{m_{\mathrm{eff}}}{m_0}\frac{V}{2}\right]^{1/2}, \\ k_{F,s}(V) &= \left(k_{F,s}^2 + cV\right)^{1/2}, \end{aligned} \qquad (2)$$

where $c = 2m_0 e/\hbar^2 \cdot 10^{-18} \approx 26.25\,\mathrm{nm^{-2}eV^{-1}}$ is the dimensional factor, m_{eff} and m_0 is the effective and free electron mass, Δ is the exchange energy which determines the difference between electron spin states, and the plus or minus sign corresponds to $s = \uparrow$ or $s = \downarrow$. In the case of a nonmagnetic NP, electronic states are degenerate in spin.

The transverse wavenumber $k_{n,s}^{(f)}$ in NP is the key parameter, since the result of integration (1) is very sensitive to its value. The related choice of initial $k_{n,s}^{(f)}$ was varied from $0.1\,\text{Å}^{-1}$ to $0.6\,\text{Å}^{-1}$. The lowest $k_{n,s}^{(f)}$ values corresponds to NPs distribution with greatest averaged NP diameter: $d_{\mathrm{av}} = \frac{1}{N_{\mathrm{tot}}}\sum i_f d^{(f)}$. The quantum well solution, which was used as initial approach of $k_n^{(f)}$ distribution over d, is presented below:

$$k_n^{(f)} = \pi n/d^{(f)} - \Phi^{(f)}, \qquad (3)$$

where correction term $\Phi^{(f)}$ that places in $k_n^{(f)}$ aside from ideal QW solution corresponding to the real systems; index n is obtained as an integer number corresponding to QW discrete levels. The Eq. (3) is simplified quantization rule, $2k_n d + \Phi_B + \Phi_T = 2\pi n$, and term $\Phi_{B(T)}$ is the phase change of the electron wave function on interfaces [3, 4]. However, tunneling behavior at $d > 1.8$ nm can be different due to increased NP volume, and then NP reveals more bulk-assisted properties, whereas degree of quantization is vanished.

In our study tunnel magnetoresistance in the MTJ-NPs is determined by the ratio: $\mathrm{TMR} = \left(I^{\mathrm{P}} - I^{\mathrm{AP}}\right)/I^{\mathrm{AP}} \times 100\,\%$, where

$$I^{\mathrm{P(AP)}} = \sum i_f \cdot \left(I_{\uparrow,f}^{\mathrm{P(AP)}} + I_{\downarrow,f}^{\mathrm{P(AP)}}\right), \qquad (4)$$

and each term $I_{s,f}^{\mathrm{P(AP)}}$ was calculated by (1). The tunneling current are passing through insulator between NPs, was considered as negligible impact. The barriers thicknesses in symmetric MTJ-NPs were fixed as $L = 1.0$ nm, the height of potential barriers $U_B = 1.2$ eV, the effective electron mass $m_{\mathrm{eff}} = 0.8\,m_0$ for NPs, $m_{\mathrm{eff}} = 0.4\,m_0$ for barriers and $m_{\mathrm{eff}} = m_0$ for FM layers are assumed.

One of important dynamic MTJ properties of the magnetic distortion is spin transfer torque (STT) effect [14, 15]. The STT is determined by vector which consists of parallel $T_{\parallel}$ and perpendicular $T_{\perp}$ components in relation to interfaces. The parallel component $T_{\parallel}$ vanishes at zero voltage, while at finite voltage it lies in the plane of FM layer interfaces triggering additional magnetization fluctuation. The value of $T_{\perp}$ is determined by interlayer exchange interaction, which isn't vanished at zero voltage, $T_{\perp} \neq 0$. Only $T_{\parallel}$ component will be considered below, since we assumed in our case $T_{\parallel} >> T_{\perp}$. Parallel STT component is calculated as difference between spin current densities $\mathfrak{J}^{\mathrm{P(AP)}}$, Ref.[14, 15]:

$$T_{\parallel} = \frac{\sin(\gamma)}{2}\left(\mathfrak{J}^{\mathrm{P}} - \mathfrak{J}^{\mathrm{AP}}\right), \qquad (5)$$

where $\mathfrak{J}^{\mathrm{P(AP)}} = \hbar\left(J_{\uparrow}^{\mathrm{P(AP)}} - J_{\downarrow}^{\mathrm{P(AP)}}\right)/2e$, γ is an angle between FM layers magnetizations; $J_s^{\mathrm{P(AP)}} = I_s^{\mathrm{P(AP)}}/S$ is the charge current density. Here $S = \frac{\pi}{4}\sum_{f=1..10} i_f\left[d^{(f)}\right]^2$ is the total active cross-section area of the MTJ-NPs.

https://doi.org/10.1051/epjconf/201818501015

3 Results and discussion

3.1 Tunnel magnetoresistance

Now we will consider several effects that characterize MTJ-NPs, when the magnetizations of the FM layers are localized in the contact plane. Abnormal behavior of TMR shown in Fig. 3, which reproduces peak-like TMR dependencies and their suppression at zero voltages $V \simeq 0$. We shall denote such domains of TMR with the index zero TMR_0. The calculations show that MTJ-NPs increase the electron transmission coefficient and create resonance conditions providing a quantization of the conductance. This is shown by curve 1 in Fig. 3 which was obtained for MTJ with non-magnetic NPs at values of $k_n = 0.4508$ Å^{-1} and $d_{\mathrm{av}} = 2.6$ nm. The Fig. 3 shows that TMR_0 can take even negative values. Similar TMR dependencies were previously demonstrated in the experimental works [6, 7]. In our calculations this corresponds to states with small wavenumber: $k_n \approx 0.1 \div 0.36$ Å^{-1}. For example, TMR_0

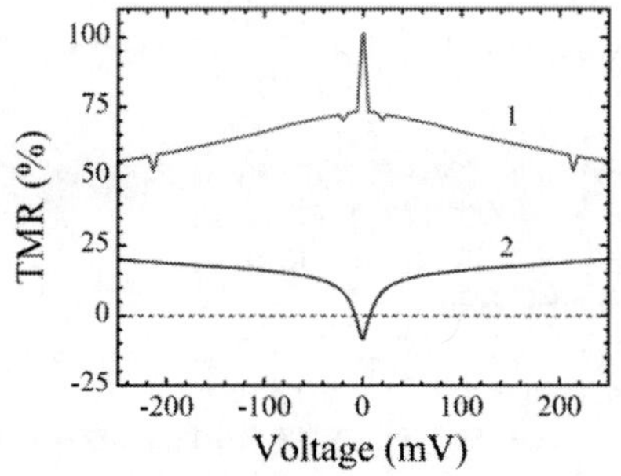

Figure 3. Bias voltage dependence of TMR for the magnetic tunnel junctions with embedded non-magnetic nanoparticles: curve 1 shows peak-like TMR for the case of wavenumber $k_n = 0.4508$ Å^{-1} of the nanoparticle, curve 2 shows negative TMR_0=-8% for $k_n = 0.115$ Å^{-1}.

becomes negative for $k_n = 0.115$ Å^{-1} and $d_{\mathrm{av}} = 2.6$ nm, see Fig. 3, curve 2. Such values of TMR_0 can also be observed experimentally due to the influence of interfaces and the uneven thickness of the barriers, as well as due to the influence of the anisotropic shape of the nanoparticles, when the wavenumber is described by the formula (3). The calculation of the transmission coefficient $D_s^{\mathrm{P(AP)}}$ for small $k_{n,s}$ also includes solutions in which the tunnel transparency at the expense of open trajectories in the allowed angles θ_s. The influence of temperature on the value of TMR in our model is not discussed, since in the derivation of the expression for the current $I_{s,f}^{\mathrm{P(AP)}}$ is assumed that the operating temperatures are much less than the characteristic Fermi temperature E_F/k_B, where k_B is the Boltzmann constant.

The tunneling magnetoresistance as a function of the applied voltage is strongly modified at $n = 1, 2$ in Fig. 4. Also the Fig. 4 show the experimental data fitting which was made according to the presented theoretical approach (1) and (2) for magnetic NPs. The TMR_0 curve corresponds to $n = 1$ and $\Phi = 0.1$ nm^{-1}. The diameter distribution, Fig. 2 (rings), is close to experimental data [6],

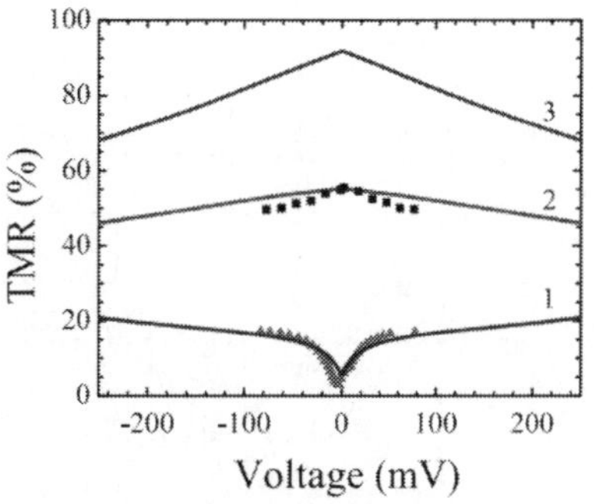

Figure 4. TMR-V dependencies with related diameter distributions: the curve 1 with reduced TMR_0 is shown for $n = 1$, $\Phi = 0.1$ nm^{-1} in comparison to exp. data [6] with diameter of NPs distribution similar to the Fig. 2 (rings); curve 2 and 3 at $n = 2$, $\Phi = 0$ for distributions rectangles and stars of the Fig. 2, respectively.

where dip-like TMR behavior was observed only at low temperatures, $T \approx 2.5$ K. The gradual changes TMR variations show the initially suppressed TMR at low voltages (curve 1), and then the dome-like behavior (curve 3), depending on the distribution of diameters shown in Fig. 2 (see rings and stars). All initial $k_n^{(f)}$ are different for each fraction according to (3). The formation of the high TMR_0, with dome-like behavior, follows to the fraction with lowest $k_n^{(f)}$ (curve 2) that is similar to the case $n = 1$. The lowest $k_n^{(f)}$ values corresponds to NPs distribution with greatest averaged NP diameter d_{av}. The calculated TMR curves satisfactorily reproduces experimental data at this case (see Fig. 4 and Fig. S6 in Ref.[6]), however, TMR behavior in relation to parameter d is opposite to experimental observations for $d_{\mathrm{av}} > 1.8$ nm. Thus, this can be more rational explanation when $k_{n,s}^{(f)}$ also increases with d_{av}. It is assumed, that there is a size threshold at $d_{\mathrm{av}} \approx 1.8$ nm between quasi-1D and bulk-assisted states in the middle conducting layer, and thus, quantization rule (3) have to be changed.

3.2 Spin transfer torque

The parallel component $T_{\parallel}$ was calculated for the case when the magnetization directions of the upper and lower FM layers are perpendicular to each other, i.e. $\gamma = \pi/2$. Note that $T_{\parallel}$ has the maximum value when the direction of the magnetic moment of the nanoparticle follows the direction (or coincides with the direction) of the magnetization of the lower FM layer. Due to the identical determination of the spin-polarized current in single-barrier MTJ and in MTJ with nanoparticles during ballistic tunneling of electrons, in both cases the parallel STT component was calculated in a similar way using the equation (5).

Figure 5 shows the dependence of $T_{\parallel}$ on the applied voltage for MTJ with nanoparticles in different states (see curves 1-3) and for a single-barrier MTJ (see curve 4). The values of the parameters at which the dependences were obtained correspond to Fig. 3. One can see a fast

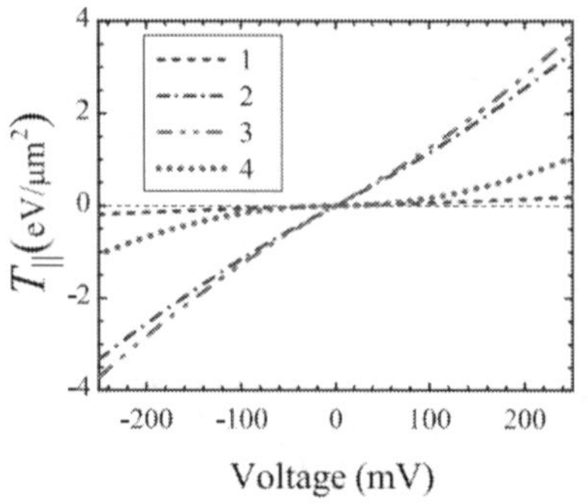

Figure 5. Dependences of the parallel component of the STT on the applied voltage for MTJ with nanoparticles (curves 1 and 2) and single-barrier MTJ (curve 4). The values of the parameters for which the curves were obtained correspond to Fig. 3. Curve 3 shows the dependence of the $T_{\|}$ on V for magnetic nanoparticles, when $k_{n,\uparrow(\downarrow)} = 0.5026 \pm 0.0025$ Å^{-1}.

and almost linear increase of $T_{\|}$ for MTJ with nanoparticles having averaged diameter $d_{av} = 1.25$ nm (see curve 2). For MTJ with magnetic nanoparticles, the values of $T_{\|}$ show the greatest increase with the voltage (see curve 3). In addition, the calculations show that if the thickness of the MTJ with non-magnetic NPs is $t = 3,25$ nm, where $t = 2L + d$, then $T_{\|}$ is comparable to its value in a single-barrier MTJ with a barrier thickness of 1.0 nm (see curve 4). This value of $T_{\|}$ can be even greater in the case of a magnetic NP, see curve 3 in Fig. 5.

Thus, we come to the conclusion that in the case of MTJ with magnetic nanoparticles, the large value of the parallel component STT can increase the fluctuations of two quantities: the magnetic moment of the nanoparticle and the magnetization of the soft magnetic FM layer (without taking into account the influence of temperature). As a result, this effect can also reduce the switching value of the critical current in the memory cell operating on the spin moment transfer effect, which agrees with the experimental data; see, for example, the work [16].

4 Conclusion

The development of the theory of spin-polarized electron transport in the process of quantization in magnetic NPs is based on the use of quantum mechanical and quasiclassical methods. Using the notion of two-barrier magnetic tunnel junctions with quantum wells made it possible to calculate the conductivity and reproduce qualitatively the TMR anomalies demonstrated in the experiments [6, 7]. Low electronic states in quantum wells and small values of wave vectors in NPs play an important role in observing the quantized conductivity regime.

The TMR anomalies such as peak behavior or suppression of their maximum value (negative TMR values) at low voltages are obtained and explained under the assumption of the ballistic tunneling mode based on the values of the wave vectors $k_{n,s}$ of the NPs and in view of their distribution along the diameters.

The above theoretical model also predicts a rather large effect of the spin transfer torque in magnetic tunnel junctions with built-in NPs. This indicates that in such systems it is possible to reduce the critical current density for switching the magnetization of the FM layers. The calculations showed that the contacts studied are complex systems in which, for certain parameters, various dependences (in particular, resonance) of the TMR on the applied voltage can be realized, as well as an energetically effective STT leading to a switching of the magnetization of the FM layer.

References

[1] G.Q. Yu, Z. Diao, J.F. Feng, H. Kurt, X.F. Han, J. M. D. Coey, Appl. Phys. Let. **98**, 112504 (2011)

[2] T.V. Pham, S. Miwa, D. Bang, T. Nozaki, et al., Solid State Commun. **183**, 18 (2014)

[3] Y. Wang, Z.Y. Lu, X.G. Zhang, X.F. Han, Phys. Rev. Let. **97**, 087210 (2006)

[4] T. Nozaki, N. Tezuka, K. Inomata, Phys. Rev. Let. **96**, 027208 (2006)

[5] P. Sheng, F. Bonell, S. Miwa, T. Nakamura, et al., Appl. Phys. Rev. **102**, 032406 (2013)

[6] H. Yang, S. Yang, S.S.P. Parkin, Nano Lett. **8**, 340 (2008)

[7] D. Ciudad, Z.C. Wen, A.T. Hindmarch, E. Negusse, et al., Phys. Rev. B. **85**, 214408 (2012)

[8] L.R. Tagirov, B.P. Vodopyanov, K.B. Efetov, Phys. Rev. B. **63**, 104428 (2001)

[9] N.Kh. Useinov, L.R. Tagirov, Physics Procedia **75**, 995 (2015)

[10] A.N. Useinov, L.X. Ye, N.K. Useinov, T.H. Wu, C.H. Lai, Sci. Rep. **5**, 18026 (2015)

[11] A.N. Useinov, J. Kosel, N.K. Useinov, L.R. Tagirov, Phys. Rev. B. **84**, 085424 (2011)

[12] X.G. Zhang, W.H. Butler, Phys. Rev. B. **70**, 172407 (2004)

[13] W.H. Butler, Sci. Technol. Adv. Mater. **9**, 014106 (2008)

[14] I. Theodonis, N. Kioussis, A. Kalitsov, M. Chshiev, W. Butler, Phys. Rev. Lett.**97**, 237205 (2006)

[15] A. Kalitsov, W. Silvestre, M. Chshiev, J Velev, Phys. Rev. B. **88**, 104430 (2013)

[16] Z. Diao, A. Panchula, Y. Ding, et al., Appl. Phys. Lett. **90**, 132508 (2007).

Tunnel Magnetoresistive Effect in Nanostructured Composite Systems

Stognei O.V., Sitnikov A.V., Kalinin Yu.E.*

Voronezh State Technical University, Voronezh, Russia

Abstract. The results of the investigation of tunnel magnetoresistive effect in a large number of metal-dielectric nanocomposites obtained by the same technology and under identical conditions are presented. The composites with multi-elemental amorphous metal phase were considered. It has been shown that the large values of the tunnel magnetoresistive effect in these composites are ensured when the value of the density of electronic states on the Fermi surface is large while other conditions are being equal. The conditions for the appearance of an anomalous positive tunneling magnetoresistance are formulated based on the study of composites with a mono-elemental ferromagnetic phase from cobalt.

Introduction

Nanogranular composites are artificial materials consisting of two different phases (as a rule – from metal and dielectric materials). In a case of relatively low metal concentration (less than 50 at. %), the metal phase has the form of nano-sized granules randomly distributed in the dielectric matrix. Such morphology determines unusual physical properties of the composites and one of this unique phenomenon is a tunneling magnetoresistance (TMR) [1].

The TMR is due to tunneling of spin-polarized electrons between ferromagnetic nanogranules through dielectric barriers [1]. Owing to this complicated mechanism the TMR is very sensitive to a large number of different factors [2-4]: granule's size distribution, the distance between granules, presence or absence of clusters coexisting together with separated granules, the magnetic anisotropy of granules, parameters of the tunnel barrier, co-tunneling and interface effects etc. All these features can be observed within one composite system when the ratio between metal and dielectric phases is being changed. Being dependent on mentioned factors the tunneling magnetoresistance shows complicated concentration and temperature dependence as well as sensitivity to external influence (heat or electric field treatment) and preparation conditions [5-7].

On the other hand if one compares the maximum values of the TMR in composite systems with different metal phases (Ni, Fe or Co) it is possible to see that the TMR differs by more than an order of magnitude [8-11]. The maximum TMR values in Ni-based composites always do not exceed 0.1-0.2 % [8,9], while in Co-based composites the maximum TMR values usually reach 7-8 % [10,11]. Why does it happen? According to the base theories, the value of TMR effect depends on the polarised coefficient of electrons of the metal phase [12] and mutual orientation of magnetic moments of neighbour granules [13]. However experimental data give a more complicated picture. Some experimental results concerning the influence of the metal phase electronic properties as well as composite's morphology on the TMR values are presented in the paper.

1 Methods

The presented results have been obtained at the investigation of different composites systems prepared by one deposition equipment and under the same deposition conditions (residual pressure, operating gas pressure, sputter power and so on). All composites have been prepared by an ion-beam sputter of mosaic targets made from a metal base and dielectric plates nonsymmetrically located on its surface. Nonsymmetrical targets give the possibility to obtain samples with the different ratio of metal and dielectric phase in one deposition process and under the same conditions. All studied composites were thin films (3-4 mkm) deposited on glassceramic substrates (NaCl in a case of HRTM investigation). Magnetoresistance of the samples was measured by the two-probe potentiometric method under a sweeping magnetic field. All measurements were carried out at room temperature.

2 Results

2.1. Composites with amorphous metal granules

The TMR effect is determined by reorientation of granules magnetic moments under applied magnetic field (it depends on the average angle between the granule magnetic moments) [13]. As a rule, the granules in

* Corresponding author: sto@sci.vrn.ru

metal-dielectric composites are crystalline and therefore can have different crystalline anisotropy. One can suppose that the anisotropy determines the remagnetization process and influences on the TMR values (for example anisotropy of Co is much larger than anisotropy of Ni). To check this point the composites with amorphous metal granules were studied.

The composites containing metallic phases which can be easily produced in the amorphous state ($Co_{84}Nb_{14}Ta_2$, $Co_{41}Fe_{39}B_{20}$ and $Co_{45}Fe_{45}Zr_{10}$) were used for investigation. According to the TEM data, the composites morphology does not depend on the metal phase and is similar in different systems. See the example of TEM photographs for $(Co_{86}Ta_{12}Nb_2)_{34}(SiO_n)_{66}$ and $(Co_{41}Fe_{39}B_{20})_{30}(SiO_n)_{70}$ composites having the close concentration of the metal phase (fig. 1).

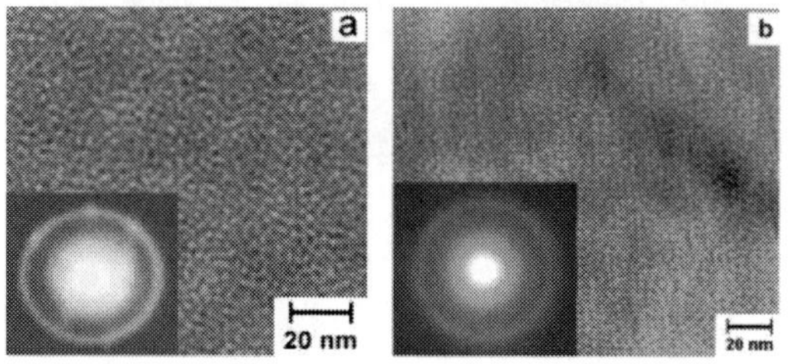

Fig.1. TEM photographs of the composites with different metal phases: a - $(Co_{86}Ta_{12}Nb_2)_{34}(SiO_n)_{66}$; b - $(Co_{41}Fe_{39}B_{20})_{30}(SiO_n)_{70}$

Besides the absence of the crystalline anisotropy, the metal phases of the studied composites should have similar polarization coefficient of electrons. Really, two metal phases have very close element composition ($Co_{41}Fe_{39}B_{20}$ and $Co_{45}Fe_{45}Zr_{10}$), the third metal phase contains only one ferromagnetic element – cobalt. If we compare the known values of the polarization coefficients for $Co_{50}Fe_{50}$ alloy and pure cobalt we will see that the values are close: 50 % for the alloy and 45 % for cobalt [14]. Therefore magnetoresistance of composites with amorphous metal phase should be equal. However, the experiment gives a completely different result (see fig. 2).

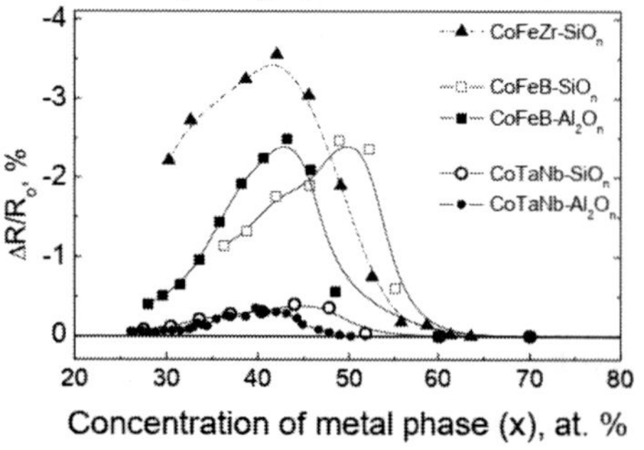

Fig.2. Concentration dependence of the magnetoresistive effect in amorphous granular composites with various combination of metallic [$(Co_{86}Ta_{12}Nb_2)$, $(Co_{41}Fe_{39}B_{20})$ and $(Co_{45}Fe_{45}Zr_{10})$] and dielectric [SiO_n and Al_2O_n] phases. The magnetoresistance values were determined in a magnetic field of 10 kOe.

One should note that all compared composites were obtained by the same equipment and under the same conditions. Maximum TMR values were observed in composites with $Co_{45}Fe_{45}Zr_{10}$ metal phase and minimum values are shown by composites with $Co_{86}Ta_{12}Nb_2$ in spite of the fact that composites with metal phase from pure cobalt (the Co-Al-O composites) have high values of the TMR effect (7-8 %) [10,11]. Another important fact is that the TMR values do not depend on the dielectric matrix: composites with different matrix (silicon or alumina oxide) but the same metal phase have equal magnetoresistance (see fig. 2).

Besides the observed difference in TMR values the very unusual correlation was also found in the amorphous nanocomposites: the maximum TMR values, observed in the composites with different metal phase, correlate with the magnitude of saturation magnetostriction of this metal phase (see fig. 3). The magnetostriction in all alloys is positive and in the simplest approach one can suppose that granules are stretched along the applied magnetic field and this makes the distance between granules smaller. The probability of electrons tunnel throw the dielectric barrier exponentially depends on the barrier thickness [1], therefore larger value of magnetostriction leads to the smaller distance between neighbour granules and more intensive electron tunneling when a magnetic field is applied to composite. However, the calculations of the changes in the distance between neighbour granules due to magnetostriction have shown that this effect is negligible (less than 10-13 m) and cannot influence the electron tunneling. Direct estimates of the change of the composite's electrical resistivity based on the tunnel model [1] and taking into account the change in the intergranular distance caused by magnetostriction, also give a negligible value by several orders of magnitude less than the actual TMR values. It is obvious that the mechanism of the observed correlation should have a more fundamental reason.

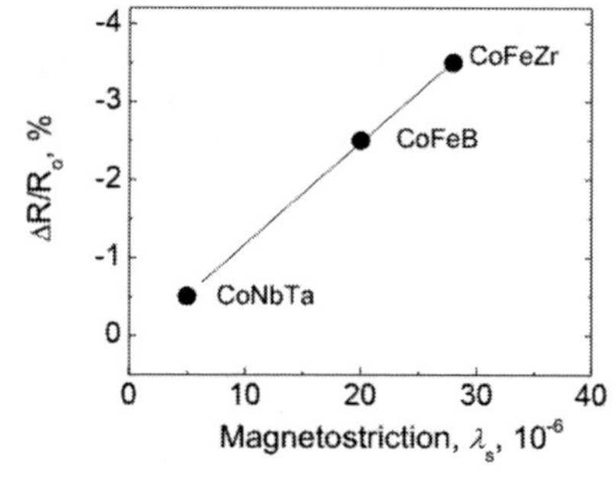

Fig.3. Maximum value of the magnetoresistive effect vs. values of saturation magnetostriction of the metal phase.

Properties of condensed matters associated with charge transport and electronic interaction are determined by the electrons which have energy close to the Fermi level [15] (even the value of magnetostriction effect depends on the density of electronic states [16]). Therefore, the density of electronic states at the Fermi level ($g(E_F)$) was estimated for the studied composites to explain the obtained result.

The determination was carried out using the « Mott's model » for conductivity in non-metallic systems [15]. The calculation was based on the temperature dependence of composites resistance in the low-temperature region. In a case of realization of hopping conductivity in the composites one can apply the «Mott's model» and based on the recalculating of the experimental dependencies get the $g(E_F)$ values [17]. The obtained results are presented at the fig.4.

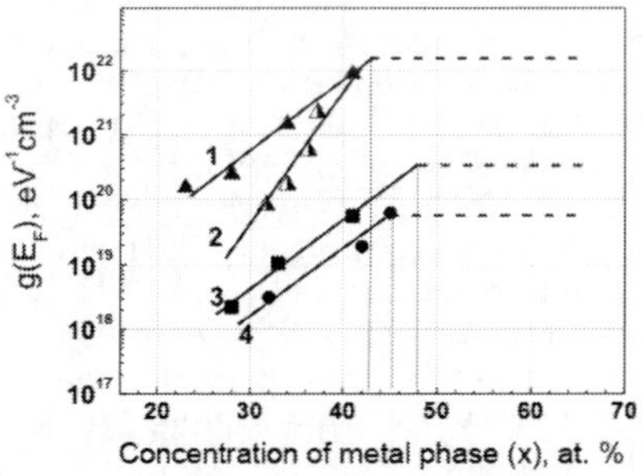

Fig.4. Concentration dependences of the density of electronic states at the Fermi level for different nanocomposites:
1 - $(Co_{45}Fe_{45}Zr_{10})_x(SiO_n)_{100-x}$; 2 - $(Co_{45}Fe_{45}Zr_{10})_x(Al_2O_n)_{100-x}$;
3 - $(Co_{41}Fe_{39}B_{20})_x(SiO_n)_{100-x}$; 4 - $(Co_{84}Nb_{14}Ta_2)_x(SiO_n)_{100-x}$.
Vertical lines show the concentration of the percolation threshold for each composite

The calculated values indicate the $g(E_F)$ in the dielectric matrix of the composites, that is why the $g(E_F)$ values for $(Co_{45}Fe_{45}Zr_{10})_x(SiO_n)_{100-x}$ and $(Co_{45}Fe_{45}Zr_{10})_x(Al_2O_n)_{100-x}$ are different. As the concentration of the metal increases, the values of $g(E_F)$ also increase and reach values corresponding to the pure metallic phase when metal concentration approaches the percolation threshold. The concentration of the percolating threshold is been pointed by vertical lines in the fig. 4. Since we consider the magnetoresistive effect the only electrons which are in metal phase and directly tunnel from one granule to another one has to be taken into account. Therefore only $g(E_F)$ values of metal phase are important and these values have to be compared if one compare TMR of different composites.

It is obvious, that the density of states determined from the experiments directly correlates to the magnetoresistivity values of the composites (compare fig.2 and fig.4). Thus, to obtain large values of the magnetoresistive effect, not only maximum values of the electron polarization coefficient are needed. The large values of the density of electronic states at the Fermi level are also required.

2.2. Cobalt based composites and positive TMR

The technique which has been used to produce the composites allows obtaining samples with very small step in composition. This makes it possible to accurately investigate effects that appear in narrow concentration intervals (for example, near the percolation threshold). It is known that in addition to the usual negative magnetoresistive effect the anomalous positive magnetoresistive effect is observed in the composites with mono-element ferromagnetic phase: Fe-SiO_2 [18, 19], Co-SiO_2 [20], Fe- Al_2O_3 [21], Co- Al_2O_n [22] (see fig. 5).

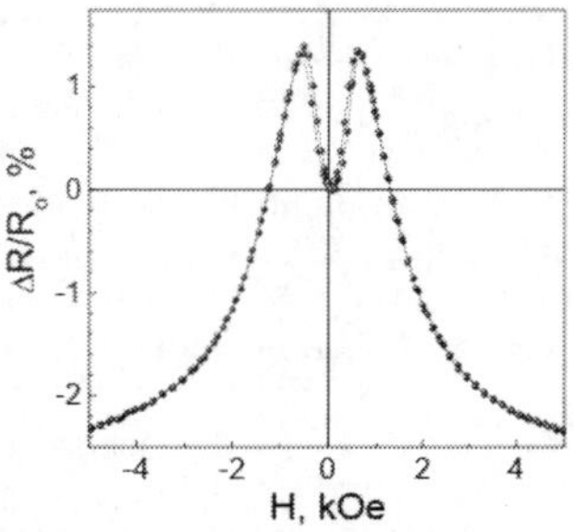

Fig. 5. Example of the positive tunneling magnetoresistance observed in $Co_{60}(Al_2O_n)_{40}$ nanocomposite.

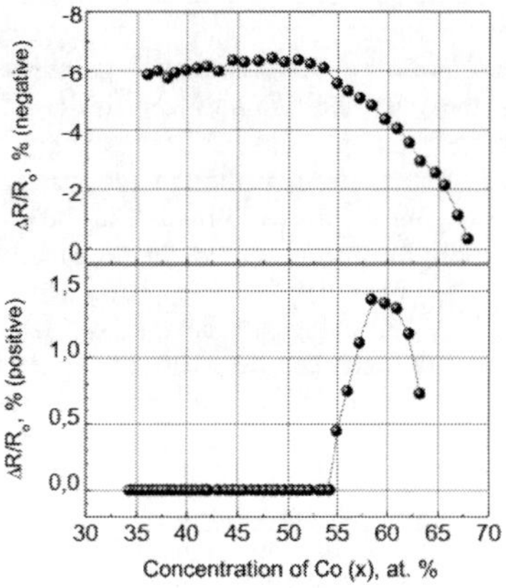

Fig. 6. Concentration dependence of the tunneling magnetoresistance (positive and negative) in $Co_x(Al_2O_n)_{100-x}$ nanocomposites.

The positive effect is observed only in the vicinity of the percolation threshold (see fig. 6) and has the tunnel origin like the usual negative TMR [20, 22]. The tunnel nature of the positive effect is confirmed by the nanogranular structure and tunneling conductivity mechanism observed in these composites [20]. Moreover, the positive effect is isotropic: its value and sign do not depend on the mutual orientation of electric current and magnetic field [22], as it could be if the origin of the effect would be different. For example, in a case of anisotropic magnetoresistance, where the mechanism is not a tunnel one, the positive effect is observed only when the current is parallel to the field and the effect is negative when the current is perpendicular to the field [23,24].

The positive isotropic effect is due to the complicated morphology of the composites: simultaneous coexistence of relatively large clusters and individual nanogranules [20, 22]. From one hand the clusters and granules have very different anisotropy and remagnetize under different magnetic fields (clusters need much higher field due to specific chain-like shape [20, 22]). On the other hand, there is a dipole-dipole interaction between clusters and neighbour granules, therefore the field strength where granules are depends on the granules position (how close they are to the cluster). Individual granules remagnetized at the relatively low applied field, but granules need higher field if they are close to clusters and interact with it. As a result, the maximum disorder of the granules magnetic moments (condition for the realisation of maximum resistivity of the composite) takes place at a non-zero magnetic field but at some non zero value. Such situation can take place only when the metal concentration is large i.e. at the threshold.

The question is why the positive magnetoresistance is not observed in every crystalline composite. For example, in Co-CaFn composites, obtained under the same conditions as all other systems, the positive magnetoresistive effect is not observed [25]. Keeping in mind that all composites showing the positive effect have an oxide matrix it could be assumed that positive magnetoresistance is associated with oxygen and partial oxidation of the ferromagnetic granules. Oxidation occurs unevenly and this can lead to the formation of a complicated morphology in the composites. In the absence of an oxide shell the granules can join and form clusters during the growth process, in the presence of an oxide shell on the granules the clusters will not be formed at a given metal concentration. However, studies of the Co-MgFn composite have shown that, despite the absence of the oxygen, the positive magnetoresistance is observed in the samples at the percolation threshold. It is characteristic that in the oxygen-free composites the percolation threshold is shifted toward compositions with the larger concentration of a dielectric phase. This is due to the fact that atoms of the metal phase do not interact with the dielectric and, accordingly, a dielectric layer (the shell) is not formed on the surface of the granules giving the possibility of percolation cluster formation.

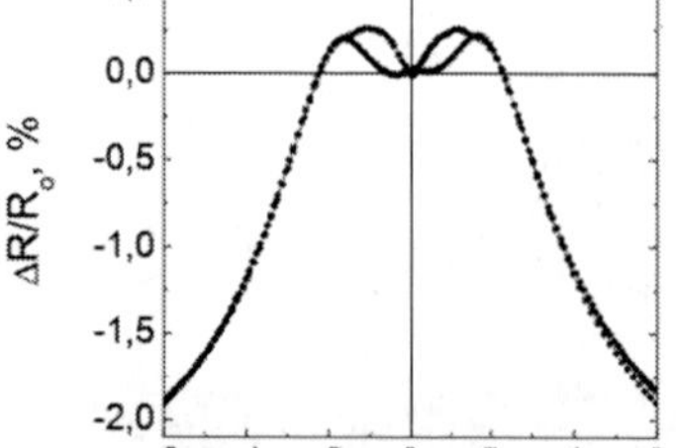

Fig. 7. Example of the positive tunneling magnetoresistance observed in $Co_{30}(MgF_n)_{70}$ nanocomposite

We believe that one of the factors affecting the morphology of composites is the surface energies of the phases forming the composite. The ratio of the surface energies of dielectric (γ_d) and metallic (γ_m) phases affects the granules size, their mutual arrangement and the possibility of clusters formation from several neighbouring granules. This is analogous to the difference in the mechanisms of thin film formation (Frank - Van der Merve or Vollmer - Weber growth mode [26]), depending on the ratio of the surface energies of the substrate and the film. It should be noted that values of the surface energy for metals are always greater than for dielectrics, therefore the dielectric does not form granules in composites with a large content of the metal phase as it occurs with the metal phase at low metal concentration. The difference between the surface energy values of metal and dielectric can vary widely and this affects the morphology of the composite. In the case when the difference γ_m-γ_d is small, it is advantageous to form a morphology where the growing granules can be joined into clusters having the form of chains from the contacting granules. This mechanism of composite formatting is similar to the Vollmer - Weber mechanism. If the difference γ_m-γ_d is large, that is the surface energy of the dielectric phase is small, then it is more advantageous to form mutually-isolated granules and this case is similar to the Frank - Van der Merve mechanism for thin films. The study of the structure of $Co_{55}(Al_2O_n)_{45}$ and $Co_{49}(CaF_n)_{51}$ composites is the experimental proof of this assumption (see fig.8).

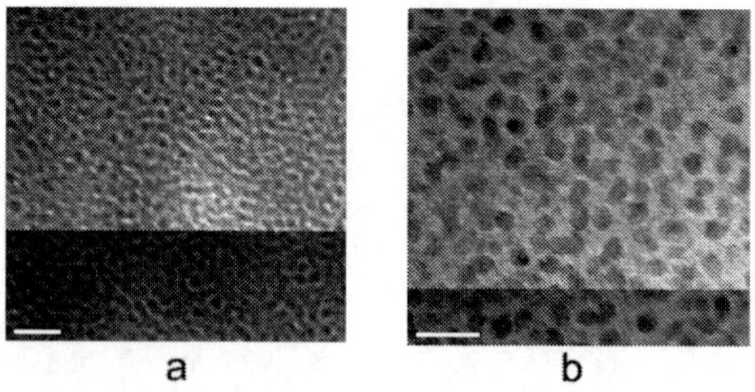

Fig.8. TEM photographs of composites with different dielectric phases: a - $Co_{55}(Al_2O_n)_{45}$; b - $Co_{49}(CaF_n)_{51}$. The white tab corresponds to 10 μm.

The metal phase of both composites is the same and is characterized by a large surface energy (γ_{Co}=2,7 J/m^2) [27], while the values of γ_d of the dielectric phases differ significantly: (γ_{CaF}=0,45 J/m^2 [27], γ_{AlO}=1,4 J/m^2 [27]. It is obvious that the morphology of these two composites is significantly different despite the close composition (the cobalt concentration is similar in the composites). A complex labyrinth-like structure with clusters in the form of chains of contacting granules is observed in the $Co_{55}(Al_2O_n)_{45}$ composite. A similar structure was observed in composites with SiO_2 matrix [20]. In the $Co_{49}(CaF_n)_{51}$ composite the morphology is different: the granules are larger in size and, most importantly, they are isolated from each other and do not form any clusters. We believe that the observed difference in morphology is a consequence of the low value of the surface energy of magnesium fluoride. Such an unrelated arrangement of the granules does not create the conditions for the appearance of the positive magnetoresistance. It should be noted that there is no positive magnetoresistance in the $Co_{49}(CaF_n)_{51}$ composites.

Conclusion

The value of tunnel magnetoresistive effect in metal-dielectric nanocomposites is determined not only by the polarizing coefficient but also by the density of electronic states on the Fermi level as well.

Difference between the surface energy of metallic and dielectric phases determines the nanocomposites morphology. The large difference (in a case of small surface energy of the dielectric phase) provides the mutual separation of metal granules even at big metal concentration.

References

1. B.Abeles, P. Sheng M.D.Coutts and Y. Arie Advances in Physics **24**, 407 (1975)
2. V.I.Kozub, A.V.Shumilin Solid State Communications 171, .55 (2013)
3. A.Ya.Vovk, V.O.Golub et al. JMMM **272–276**, E1403 (2004)
4. L.Kraus, O. Chayka et al. JMMM **226–230**, 669 (2001)
5. T.N. Koltunowicz, P. Zukowski et al. Vacuum 120 44 (2015)
6. A.Svito, A. K.Fedotov et al. J. Alloys and Comp. **699**, 818 (2017)
7. O.V. Stognei, V.A. Slyusarev et al. Microelectronics Engineering **69**, 476 (2003)
8. A.Gerber, A.Milner, B.Groisman et al. PR B **55**, 6446 (1997)
9. O.V.Stognei, A.J. Al-Maliki et al. Semiconductors **50**, 709 (2016)
10. S.Sankar, D.Dender, J.A.Borchers et al. JMMM **221**, 1 (2000)
11. S.Mitani, H.Fujimori, S.Ohnuma JMMM **165**, 141 (1997)
12. M.Julliere Phys.Let. **54A**, 225 (1975)
13. J.C.Slonczewski PR B, **39**, 6995 (1989)
14. J.S.Moodera, G.Mathon JMMM 200, 248 (1999)
15. N. F. Mott, E.A. Davis, *Electronic Processes in Non-Crystalline Materials* (Clarendon Press Oxford, 1971)
16. I.Yoshikazu, N. **Miura** *Physics and Engineering Applications of Magnetism* (Springer, 1991)
17. Y.E.Kalinin, A.N.Remizov, A.V.Sitnikov, Physics of the Solid State **46**. 2176 (2004)
18. A. E. Varfolomeev, M. V. Sedova Physics of the Solid State **45**, 529 (2003)
19. B. A. Aronzon, D. Yu. Kovalev, A. E. Varfolomeev Physics of the Solid State **41**, 857 (1999)
20. S. Sankar, A.E. Berkowitz, D.J. Smith. Phys.Rev. **B62**, 14273 (2000)
21. M.A.S. Boff, J. Geshev, J.E. Scmidt. J. Appl.Phys. **91**, 12, 9909 (2002)
22. O. V. Stognei, A. V. Sitnikov, Yu. E. Kalinin, S. F. Avdeev, M. N. Kopytin Physics of the Solid State 49, 164 (2007)
23. K. Y. Wang, K. W. Edmonds et al. PR B **72**, 085201 (2005)
24. A. Grebennikov, O. Stognei Solid State Phenomena **233-234,** 501 (2015)
25. O.V. Stognei, A.V. Sitnikov, S.F. Avdeev, Yu.E Kalinin CNM Int. Conf. 114 (2006)
26. Pimpinelli, Alberto; Jacques Villain. *Physics of Crystal Growth*. (Cambridge: Cambridge University Press, 1998)
27. F.J.Himpsel, J.E.Ortega et. al. Advances in Physics 47, 511 (1998)

Thermodynamic aspects of describing the contribution of spontaneous magnetism of electrons to the Hall resistance

Vsevolod Okulov[1,2], and *Evgeny* Pamyatnykh[2]

[1]M.N. Miheev Institute of Metal Physics of Ural Branch of RAS 620990 Ekaterinburg, Russia
[2]Ural Federal University, 620083, Ekaterinburg, Russia

Abstract. On the base of the analysis of quantum-statistical description of the magnetization of electron system containing the spontaneous spin polarization contribution there were found the magnetization and conduction current densities in equilibrium state. It has been shown that equilibrium surface conduction current ensures realization of demagnetization effects but in local equilibrium state determines local equilibrium part of the Hall conductivity. As a result one is given the justification of influence of the spontaneous magnetization on galvanomagnetic effects, which is not related to spin-orbital interaction.

1 Introduction

Now the study of manifesting the spontaneous magnetization of electrons in the Hall effect and other galvanomagnetic phenomena is actively developed. In the present report we show that for describing the phenomena relating to anomalous Hall effect there can be used naturally the methods of equilibrium and nonequilibrium thermodynamics. The term "thermodynamic anomalous Hall effect" is introduced by us to designate anomalous Hall effect which can be described by thermodynamic methods. Next we consider the connection of thermodynamic anomalous Hall effect with equilibrium magnetization.

Necessity of such researches has been explicit after revealing in [1] the contribution to the Hall resistance of semiconductor from spontaneous magnetization of donor electrons of iron impurities of very low concentrations. Such low concentration does not admit ordinary explanation of the given effect by the manifestation of spin-orbital interaction. It is the search of other explanations that has resulted in revealing other physical mechanisms of the anomalies, which are not related to spin-orbital interaction.

2 Equilibrium conduction current

From the beginning of considering we choose as initial a well known formula for the density vector of equilibrium magnetization $\mathbf{M}$ of quantum system of electrons

$$\mathbf{M} = \left(\frac{\partial \Omega}{\partial \mathbf{H}}\right)_{T\zeta} = m\mathbf{h} + \frac{1}{H} P\mathbf{h} \qquad (1)$$

It is obtained from thermodynamic potential Ω in variables of temperature, chemical potential ζ and magnetic field strength $\mathbf{H}$. Taking into account real inhomogeneity of a system, this formula can be transformed to the expression for varying the potential Ω with respect to vector potential of magnetic field $\mathbf{A}(\mathbf{r})$:

$$\delta\Omega = -\frac{1}{c}\int d\mathbf{r}\,\delta\mathbf{A}(\mathbf{r})\,\mathbf{j}_m(\mathbf{r}) = -\frac{1}{c}\int d\mathbf{r}\,\delta\mathbf{A}(\mathbf{r})\left[\mathbf{j}_{eq}(\mathbf{r}) - \mathbf{j}_c(\mathbf{r})\right] \qquad (2)$$

In this equality $\mathbf{j}_m(\mathbf{r}) = c\; rot\,\mathbf{M}$ is the magnetization current density, $\mathbf{j}_{eq}(\mathbf{r}) = c\; rot\,\mathbf{m}$ is the density of mean magnetic moment flux of an electron and we assume that the vectors $\mathbf{H}, \mathbf{M}$ and $\mathbf{m}$ are directed along the symmetry axis. Then the vector $\mathbf{j}_c(\mathbf{r})$ is directed perpendicularly to $\mathbf{H}$ and defined by the equality:

$$\frac{1}{c}\left[\mathbf{j}_c(\mathbf{r}) \cdot \mathbf{H}\right] = grad\,P \qquad (3)$$

Here $P = -(\Omega/V)$ is the pressure of electron system. This equality is the condition of the equilibrium of the external force, with which magnetic field acts the system with current, and the pressure force of the system. It means that the quantity $\mathbf{j}_c(\mathbf{r})$ should be considered as the conduction current density, due to existence of the external force mentioned above. If the system has a spontaneous magnetization M_0, the pressure in (3) includes the energy $M_0 H$. In homogeneous equilibrium system the currents considered are concentrated at the boundaries. According to the equation (3) the magnitude and distribution of equilibrium conduction current depend on the structure and shape of the boundaries and it is this current that is responsible for the existence of so-called demagnetizing field. Simple definition of such a field can be represented by Maxwell equations inside the boundaries of a system:

$$rot\,\mathbf{B}_{eq} - 4\pi \cdot rot\,\mathbf{M} = rot\,\mathbf{H}_{eq} = \frac{4\pi}{c}\mathbf{j}_c(\mathbf{r})$$

$$div\,\mathbf{H}_{eq} = 0\,, \qquad (4)$$

($\mathbf{B}_{eq}$, $\mathbf{H}_{eq}$ are the induction and strength vectors of magnetic field) and outside the boundaries of a system one has:

$$rot\,\mathbf{H}_e = \frac{4\pi}{c}\mathbf{j}_e(\mathbf{r}) + \frac{4\pi}{c}\mathbf{j}_c(\mathbf{r})$$

$$div\,\mathbf{H}_e = 0\ , \qquad (5)$$

($\mathbf{H}_e$ and $\mathbf{j}_e(\mathbf{r})$ are the magnetic field strength and current density of the field source). Equations (4) and (5) describe a modification of the initial magnetic field strength $\mathbf{H}$ inside the system by demagnetizing coefficients, which is specified by the current of a source $\mathbf{j}_e(\mathbf{r})$. As a whole the role of conduction current, determined by us, can been characterized by averaged value of $\mathbf{J}_c$, which is obtained by integrating the equation (4) over the layer, adjoining to the boundary, and in the plane of cross-section, perpendicular to the magnetic field strength one has:

$$H\cdot\mathbf{J}_c = -\frac{c}{s}\int d\mathbf{r}\,rot(P\mathbf{h})\ , \qquad (6)$$

s is the area of integrating.

3 The Hall current

The presented ideas about equilibrium conduction current can be applied to studying the Hall resistance, as far as it contains the contribution from the current of local equilibrium. Local equilibrium in the field of current source with potential $\varphi(\mathbf{r})$ is described by adding the energy $e\varphi(\mathbf{r})$ to chemical potential. Extending the results, quoted above, to local equilibrium state of electron system and replacing $P(\zeta)$ by $P(\zeta + e\varphi(\mathbf{r}))$ in the right-hand part of (3), one obtains additional term

$$-en\mathbf{E} - e\mathbf{E}\frac{\partial M_0}{\partial\zeta}\ , \qquad (7)$$

where $\mathbf{E} = -\,grad(\varphi(\mathbf{r}))$ is the electric field strength, $n = \partial P/\partial\zeta$ is the electron concentration.

Then averaged equilibrium conduction current, defined by equation (6), is replaced by local equilibrium one

$$\mathbf{J}_c^L = \mathbf{J}_c + \left\{\frac{ecn}{H} + ec\frac{\partial M_0}{\partial\zeta}\right\}[\mathbf{E}\cdot\mathbf{h}] = \\ = \mathbf{J}_c + \left(\sigma_H^L + \sigma_m\right)[\mathbf{E}\cdot\mathbf{h}] = \mathbf{J}_c + \mathbf{J}_H^L \qquad (8)$$

In this expression as a term of local equilibrium current there enters well known contribution of local equilibrium to the Hall current with conductivity σ_H^L, which is predominant in high magnetic field and also the contribution of spontaneous magnetization with the conductivity σ_m revealed first in work [1], and the contribution of equilibrium conduction current $\mathbf{J}_c$ represented by us in the present work. The terms reflecting the manifestation of spontaneous magnetization of electron system in galvanomagnetic phenomena are of fundamental importance.

4 Final remarks

It should be noted that the Hall conductivity, considered by us, includes also the diagonal element of electrical resistivity tensor in magnetic field. Thus in the given work we have shown that spontaneous magnetization of electrons has more strong effect on galvanomagnetic properties than the phenomena, which are due to spin-orbital interaction. The specific behaviors revealed due to spontaneous magnetization must be taken into account in studying the magnetism of semiconductor systems with the transition element impurities what is confirmed by the results of the works [1-3].

Acknowledgements

The work was performed as a part of the state task FASO Russia (subject "Electron», № 01201463326) with partial support by the RFBR (project №15-02-08909), Fundamental Research Program of UB RAS 2015-2017 (project № 15-17-2-32), by Act 211 Government of the Russian Federation, agreement № 02.A03.21.0006, by the Ministry of Education and Science of the Russian Federation, projects 5719.

References

1. A.T.Lonchakov, V.I.Okulov, T.E. Govorkova, M.D. Andriychuk, L.D.Paranchich, JETP Letters **96**, 405 (2012).

2. A.T. Lonchakov, V.I. Okulov, E.A. Pamyatnykh, T.E. Govorkova, M.A.Andriichuk, L.D. Paranchich, S.B. Bobin and V.V. Deryushkin , Solid State Phenomena **233-234**, 456 (2015).

3. Lonchakov, A.T., Okulova, K.A., Pamyatnykh, E.A., Podgornykh, S.M., Andrichuk, M.D., Paranchich, L.D. , Low Temperature Physics **39** , 384 (2013).

https://doi.org/10.1051/epjconf/201818501018

Investigation of InSb:Mn thin films magnetoresistance by I-V characteristics method

Oleg Novodvorsky[1,*], *Vladimir* Mikhalevsky[1], *Liubov* Parshina[1], *Olga* Khramova, *Andrey* Lotin[1], *Elena* Cherebylo[1], *Sergey* Marenkin[2], *Aleksey* Aronov[2], *Aleksey* Kochura[3], *and Vladislav* Panchenko[1]

[1]ILIT RAS – Branch of FSRC "Crystallography and Photonics" of RAS, Shatura, Russia
[2] Kurnakov Institute of the general and inorganic chemistry of the RAS, Moscow, Russia
[3] South – West State University, Kursk, Russia

Abstract. We have investigated the effect of magnetoresistance in the thin films of InSb:Mn using the data of I-V curves. The thin films were produced by pulsed laser deposition (PLD) method from the InSb+MnSb composite alloy targets with the concentrations of MnSb to 20%. The temperature dependence of magnetoresistance of the films has been studied in the temperature range from 10K to 300K.

1. Introduction

The fabrication of In1−xMnxSb thin films with a uniform distribution of magneto-active component presents a challenge [1]. The investigation of thin films of ferromagnetic semiconductors (AIII-BV):Mn is of great interest in connection with a possibility of their use in the magnetoelectronic and spintronic devices [2]. A bulk sample of the InSb-MnSb composite eutectic represents the narrow band semiconductor of p-type that has the lattice parameters approaching those of InSb and is characterized by high mobility of carriers. The Curie temperature in In1−xMnxSb alloys reaches 600 K. Low solubility of manganese in InSb is responsible for formation of nanoclusters of MnSb in bulk sample with concentrations higher than 3,5% [3]. Though solubility of manganese in InSb is small, the pulsed laser deposition method (PLD) makes possible obtaining homogeneous thin films of InSb:Mn with concentration of manganese to 20% from the targets of InSb-MnSb alloys. The PLD method allows producing thin films of various materials on single-crystal, amorphous and flexible organic substrates [4]. The presence of drops is the main factor of reduction in the quality of the films obtained by traditional PLD. Application of the droplet-free mode of PLD provides the production of the high-quality smooth uniform films on the substrate of room temperature owing to removing of the drops which take off from the target and fall on the substrate in the process of film deposition. In the process of film growth the mechanical separator permits the passage of quickly flying atoms and ions to the substrate, but presents falling of large drops on the growing film. The droplet-free mode of film deposition provides smooth homogeneous high-quality films even at the ambient temperature of the substrate thanks to elimination of the drops which are emitted from the target onto the substrate in the deposition process [5]. The films of the diluted magnetic semiconductors with high Curie temperature and high extent of spin polarization of carriers [6] were obtained by this method. The magnetoresistance in thin films of InSb:Mn can have influence on the characteristics of the magneto-operated p-(InSb+MnSb)/n-InSb diodes in which the effect of giant magnetoresistance is observed [7-9].

2.Experimental procedure

The thin InSb:Mn films were grown on sapphire (0001) substrates by ablation with an excimer KrF laser using the droplet-free pulsed laser deposition (PLD) method [4]. The size of substrates was 10x10 mm. The InSb+MnSb composite alloy target contained up to 20% of MnSb. The composition homogeneity and the morphology of the films were investigated by the X-ray spectroscopy, the SEM method with an attachment for energy-dispersive X-ray spectroscopy, by X-ray phase analysis (XPS), and by the AFM method with a magnetic cantilever. We used Hall method to measure the conductivity type of the charge carriers in the films. The films exhibited p-type conductivity. The thickness of the films was determined with the aid of the MII-4 interferometer.

3. Results and discussion

The investigation of the films with SEM showed the homogeneous distribution of the In, Sb and Mn elements in the surface layer of the film. According to the results of energy-dispersive X-ray spectroscopy, the concentration of these elements corresponded to the composition of the initial target. We deposited 10 films with concentration of manganese from 3,5% (eutectic) up to 20%. Thickness of the films was 50-80 nm. The results of the AFM analysis using a magnetic cantilever point to the presence of ferromagnetic clusters of the size of 4 nm of our thin films of InSb:Mn with concentration

* Corresponding author: onov@mail.ru

of manganese 20%. The InSb: Mn films with MnSb concentration of 3,5% (eutectic) were examined with the help of SEM and AFM, which showed the homogeneity of the distribution of elements on the surface of the InSb:Mn film. X-ray diffraction studies of the InSb:Mn film on a Bruker D8 Advance diffractometer showed the presence of very weak MnSb peak (Fig.1).

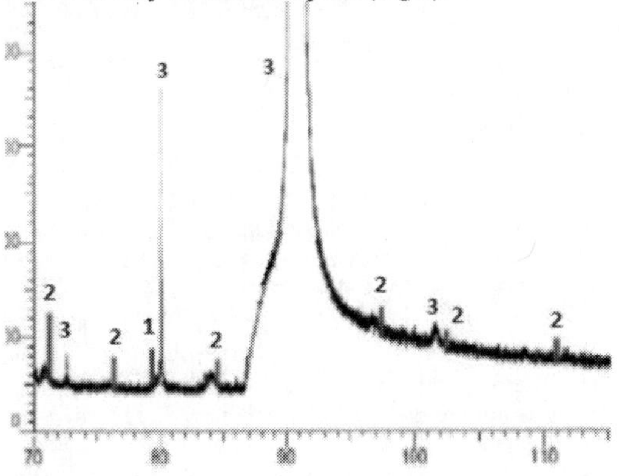

Fig.1. X-ray pattern of an InSb:Mn film with MnSb concentration of 3,5% (eutectic) on a sapphire substrate: 1 – MnSb phase, 2 – InSb phase, 3 – InMnSb phase (eutectic).

3.1. Magnetoresistance of thin InSb:Mn films

In this paper, we have for the first time investigated the magnetoresistive effect in the thin InSb:Mn films using the *I-V* curve method. Thin films of InSb:Mn ferromagnetic semiconductors are widely investigated [1,7]. However, the papers does not show the dependence of the magnetoresistance of films on the voltage applied in the plane of the film. Such dependencies can be obtained from the I-V characteristic. The magnetoresistance of the *p*-(InSbMn) films with MnSb concentration of 3,5% (eutectic) was measured under the constant magnetic field of two values, 0.15 T and 1 T. The measurements were made under room temperature. The field of 0.15 *T* was applied collinearly to the current flowing in the film (Fig.2), and the field of 0.15 T or the field of 1 T was applied perpendicular to the film plane and to the current flowing in the film (Fig.3). The indium ohmic contacts were formed on the film surface before measuring of the I-V curve using the Keithley 2612 device. The *I-V* curve was measured in the range of applied voltage from -10 V to +10 V in steps of 1 mV. Switching of the voltage polarity allowed reversing the polarity of the electric field. Figures 2, 3 illustrates the dependence of the magnetoresistance of the *p*-InSb:Mn thin films on the voltage for collinear and perpendicular electric and magnetic fields under room temperature.
The magnetoresistance MR (U) was measured on the basis of *I-V* curve in the temperature range from 10 K to 300 K using the formula $(R(H) - R(0)) / R(0) * 100\%$, where $R(0)$ and $R(H)$ are the resistances in the absence and presence of a magnetic field, respectively. It was observed that the magnetoresistance of the sample was negative and depended on the magnetic field orientation with respect to the current direction at room temperature.

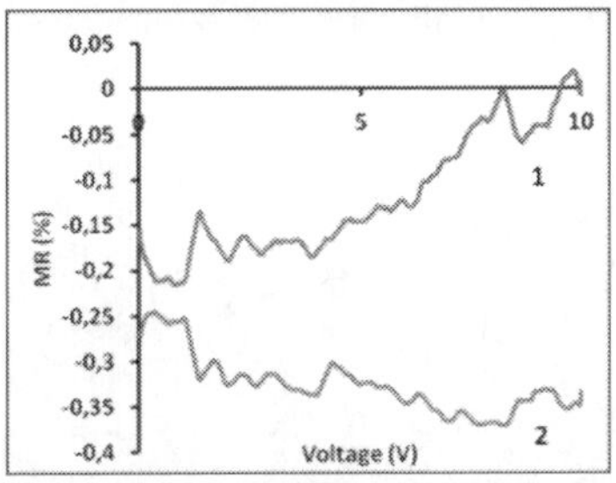

Fig.2. The magnetoresistance of the *p*-InSb:Mn thin films vs the voltage for collinear electric and magnetic fields located in the plane of the film under room temperature (1 - $E\uparrow\downarrow B$, 2 - $E\uparrow\uparrow B$), B=0.15T.

The observed different behavior of the magnetoresistance of the films depending on mutual orientation of the collinear magnetic and electric fields can be caused by different mean free paths of the charge carriers for the parallel and antiparallel directions of the fields E and H in the thin InSb:Mn films caused by the presence of surface magnetic nanoclusters, which are registrated by the AFM with a magnetic cantilever.
The measurements were made of the dependence of the film magnetoresistance on the voltage for the magnetic field perpendicular to the film surface for two magnetic field values B = 0.15 T and B = 1 T. The electric field is localized within the film plane (Fig.3).

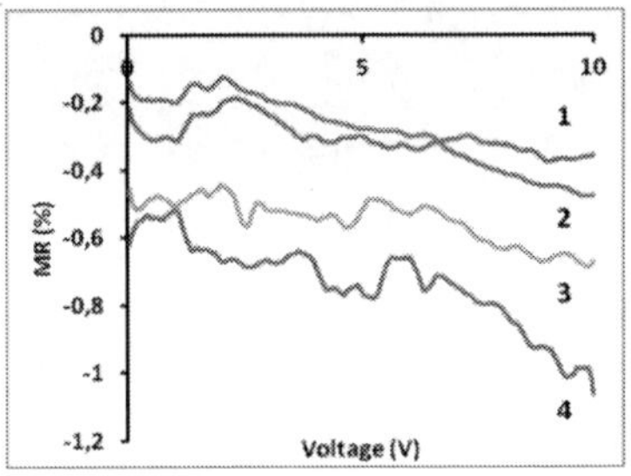

Fig.3. The magnetoresistance of the film vs the voltage for the magnetic field perpendicular to the film under room temperature (1 – B = 0,15 T ↑, 2 – B = 0,15 T ↓, 3 – B = 1 T ↑, 4 – B = 1 T ↓).

For the magnetic field normal to the film a similar dependence on the magnitude of the field is disclosed [7].
The temperature dependence of the magnetoresistance of p-InSb:Mn films with the MnSb concentration of 20% was studied in the temperature range from 10 K to 300 K. The investigations were carried out by measuring and analyzing their I-V curves in the temperature range from 10 K to 300 K. The measurement wires were connected to the indium ohmic contacts which were formed on the surface of each film. The films were placed in a cryostat. The measurements of the I-V curves were performed in the field with permanent magnet and without it. The dependences of the magnetoresistance for a fixed temperature and magnetic field were calculated on the

onov@mail.ru

basis of the I-V curve. The results of measurements R and calculations MR are shown in Fig. 4, 5.

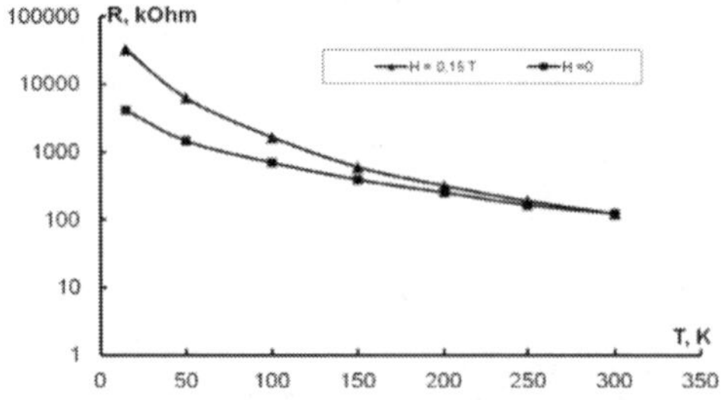

Fig.4. The dependence of the film InSb:Mn resistance on the temperature in the absence (squares) and in the presence of a magnetic field of 0.15 T (triangles). The voltage applied to the sample is 10V.

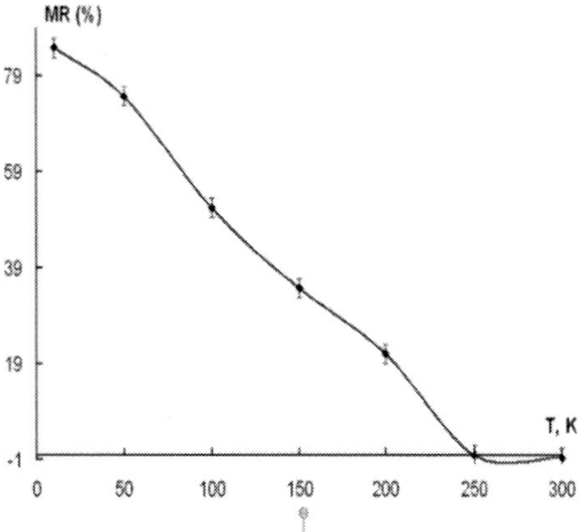

Fig. 5. The dependence of the thin InSb:Mn film magnetoresistance on the temperature at an applied voltage of 10V and under magnetic field of 0.15 T in the plane of the film.

Figure 5 shows that the crossover from positive to negative giant magnetoresistance (GMR) is observed at the temperature of 250 K during cooling. As the sample cooled down, the value of GMR increased and reached 80% at 15 K in the magnetic field of 0.15 T directed perpendicular to the plane of the film. Thus, it has been shown that the magnetoresistance of the InSb:Mn thin films is registered even at room temperature under small magnetic fields, which opens up considerable prospects for using this material to create various spintronic devices.

Conclusion

The temperature dependence of the magnetoresistance of InSb: Mn films is studied. In the temperature range of the film from room temperature to 250 K, a negative magnetoresistance was observed. At 250 K, a transition from a negative giant magnetoresistance (GMR) to a positive one was observed. As the sample cooled down, the GMR increased and reached 80% at 15 K in a magnetic field of 0.15 T directed perpendicular to the plane of the film. The magnetoresistance of InSb: Mn thin films is recorded even at room temperature in small magnetic fields, which opens the prospect of using this material to create various spintronics devices.

The work was supported by the Federal Agency of Scientific Organizations (Agreement No 007-ГЗ/Ч3363/26) in part of films deposition and by the Russian Foundation for Basic Research (projects No. 15-07-03580,15-29-01171, 16-29-05385, 16-07-00657, 16-07-00842) in part films analysis, pr. of ONIT RAS (code IV.3.3) and Presidium RAS (Code I.III) in part of numerical simulation.

References

1. N. D. Parashar, N. Rangaraju, V. K. Lazarov, S. Xie, B. W. Wessels, Phys. Rev. B **81**, 115321 (2010)
2. A.V. Kochura, B. A. Aronzon, K. G. Lisunov, A. V. Lashkul, A. A. Sidorenko, R. De Renzi, S. F. Marenkin, M. Alam, A. P. Kuzmenko, and E. Lahderanta, JAP **113**, 083905 (2013)
3. S.F. Marenkin, A.V. Kochura, I.V. Fedorchenko, A.D. Izotov, M.G.Vasil'ev, V.M.Trukhan, T.V.Shelkovaya, O.A. Novodvorsky, A.L. Zheludkevich, Inorganic materials **52**, № 3, 268 (2016)
4. O.A. Novodvorskiy, V.Ya. Panchenko, in Collective monograph *"Modern laser-information technologies"*, ed. Acad. V.Ya. Panchenko and prof. F.V. Lebedev – Moscow: Publishing house Intercontact Science, - 959 p. (2015)
5. L.S. Parshina, O.A. Novodvorsky, O.D. Khramova, I.A. Petukhov, V.A. Mikhalevsky, A.A. Lotin, E.A. Cherebilo, V.Ya. Panchenko, Optical and Quantum Electronics **48**, 316(9) (2016)
6. V.V. Rylkov, A.S. Bugaev, O.A. Novodvorskii, V.V. Tugushev, E.T.Kulatov, A.V. Zenkevich, A.S. Semisalova, S.N. Nikolaev, A.S. Vedeneev, A.V. Shorokhova, D.V. Aver'yanov, K.Yu. Chernoglazov, E.A. Gan'shina, A.B. Granovsky, Y. Wang, V.Ya. Panchenko, S. Zhou, JMMM **383**, 39 (2015)
7. J.A. Peters, N. Rangaraju, C. Feeser and B. W. Wessels, Appl. Phys. Lett, **98**, 193506 (2010)
8. O.D. Khramova, V.A. Mikhalevsky, L.S. Parshina, O.A. Novodvorsky, S.F.Marenkin, A.A. Lotin, E.A. Cherebilo, B.A. Aronzon, A.N. Aronov, V.Ya. Panchenko, Optical and Quantum Electronics. **48**, 361(8) (2016)
9. Peters *et al.* PHYSICAL REVIEW B **82**, 205207 (2010)

onov@mail.ru

EPJ Web of Conferences **185**, 01019 (2018) https://doi.org/10.1051/epjconf/201818501019
MISM 2017

Spin-filter state of Au-Co nanowires

Ekaterina *Smelova*[1,*], Kseniya *Tsysar*[1] *and* Alexander *Saletsky*[1]

[1]Lomonosov Moscow State University, Faculty of Physics, (*Lomonosov MSU*), 119991, Moscow, Russia

Abstract. Our theoretical study reveals the dependence of quantum conductance of Au-Co nanowires on their atomic structure. The results show the emergence of spin-filter state in one-dimensional Au-Co bimetallic nanowires. We found the existence of two transmission regime in Au-Co nanowires with low and high conductivity $1G_0$ and $2G_0$ for "zig-zag" and linear nanowire correspondingly. The study of transmission spectra of Au-Co nanowires reveals the control capability of spin transport regime by changing of bias voltage between bulk electrodes.

1 Introduction

Bimetallic atomic wires are the most promising objects for construction of new nanoelectronics and spintronics devices [1, 2]. Bimetallic atomic nanowires possess the unique magnetic [3, 4, 8, 11] and conductive [2, 5] properties, i.e. spin polarized tunnel current, spin-filtering properties [5], giant magnetoresistance [2, 12] and giant magnetic anisotropy [11]. These properties of one-dimensional structures used now in leading laboratories of the world for the development of new storage or recording nanodevices. Recently IBM's labs (Yorktown, New York) by using CMOS technology produced the first prototype of racetrack memory based on "giant magnetoresistance" in vertical magnetic nanowires. By passing "spin polarized" currents through the data, the magnetic data can be moved up and down the skyscrapers which are vertical racetracks [6]. Thus the main problem of spintronics is the search for new spin control capabilities in one dimensional nanostructures [7].

2 Method and model

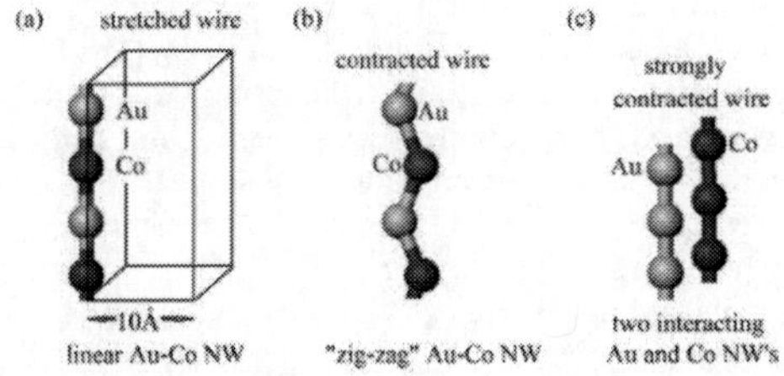

Fig.1 Model of Au-Co nanowire.

In the present paper the model of one dimensional atomic Au-Co nanowire is built in a full agreement with the result of experimental works on NC's formation in MCBJ technic [2,31] (Figure 1(a)). Under nanojunction formation with help of MCBJ technic at the last stage only one atomic chain formed between two massive electrodes with ballistic regime of quantum electron conductance through suspended atomic chain [1, 2]. Our recent studies revealed the most stable configurations of bimetallic nanowires as evenly mixed bimetallic atomic chain [9, 10]. We found that dimer formation in nanojunction (NJ) chain leads to the destabilization of atomic bonds and as result to early breaking of NJ under stretching process [9,13]. Thus at the last stage of MCBJ or STM formation process only evenly mixed atomic nanowires form long conductive atomic bridges between two alloy Au/Co electrodes, which emphasizes the quantum character of conductivity of nanocontacts, studied in the present work. Recent studies are focused on the dependence of transport properties of nanocontacts on the electrodes structure, atomic composition or magnetic ordering of bulk electrodes [1, 2]. The atomic structure and magnetic properties of suspended atomic chain consciously have not been investigated however they can have the decisive impact on the electronic transport in the ballistic regime of bimetallic NJ. Moreover the electronic structure and magnetic properties of NJ chain strongly affect the spin polarised electron transport through a NJ. We offer an explanation of transport properties of bimetallic one-dimentional structures with help of detail investigation of "inter-chain" interactions.

The study of the magnetic properties of the mixed Au-Co nanowires was carried out using the program Vienna Ab-initio Simulation Package (VASP) which is based on the density functional theory (DFT) [14]. The calculations presented in this paper were based on the self-consistent Kohn-Sham equations [15]. Electronic states were described using the basis of plane waves [16]. The cutoff energy for plane wave basis of 250 eV was used. All calculations presented in this paper were performed using the generalized gradient approximation (GGA) and also local density approximation (LDA) for exchange-correlation

* Corresponding author: smelova_e_m@mail.ru

functional and PAW (Projector Augmented-Wave) pseudopotentials (Perdew-Wang '91 treatment) [17, 18]. A structural relaxation was performed via a quasi-Newton algorithm, using the exact Hellmann-Feynman forces acting on each atom [19]. The total energies of the system were converged up to 1 meV/atom, while the residual force acting on each atom was less than 0.01 eV/A. The integration over Brillouin Zone (BZ) was performed using the tetrahedron method with Blöchl corrections. To calculate the total energy of the system we carried out the integration in reciprocal space by a special k-points grid of 1 × 1 × 7 constructed according to k-point mesh in Monkhorst-Pack scheme [20]. The structure of mixed Au-Co nanowires (NWs) for magnetic calculations has been represented by means of three dimension supercell with periodic boundary conditions. Au-Co wire was represented by an atomic chain aligned along the z axis separated from its images along the x and y directions (~10Å) to avoid spurious interactions with them. A one-dimensional Brillouin zone was used, along the z axis of the wire. Mixed Au-Co nanowire was modelled as a four-atomic chain of two Au and two Co atoms, located along the z axis. To compare the calculated results for different bimetallic NWs (areas of existence, transmission, energies, etc.) for stretched or contracted mixed NWs, we consider them as the functions of an average interatomic distance. Average interatomic distance (d) for bimetallic NW represents the sum of all bond lengths in the wire projected on the wire axis (z axis) divided by the number of atoms in the wire.

Transport calculations were carried out in the elastic limit, ballistic transport conductivity calculations were conducted within the Landauer–Buttiker formalism, where the (spin-dependent) conductance is given by (1) [21, 22]:

$$G_\sigma = \frac{e^2}{h}\sum_\sigma \sum_{ij} T_{ij}^\sigma = \frac{e^2}{h}\sum_\sigma T^\sigma(E) = \frac{e^2}{h}\sum_\sigma Tr[t_\sigma(E)t_\sigma^+(E)] = \frac{G_0}{2}\sum_\sigma T^\sigma(E) \quad (1)$$

Where $\sum_{ij}$ indicates that the sum is performed over all channels Fermi energy (the open channels) and we have introduced the spin index σ (σ=↑, ↓, spin direction), $T^\sigma(E)$-transmission coefficients($T(E) = \sum_\sigma T^\sigma(E)$), $t_\sigma(E)$-transmission coefficients of scattering matrix S, which relates the wavefunction of the incoming and outgoing electrons with respect to the step potential, where $S = \begin{pmatrix} r & t' \\ t & r' \end{pmatrix}$ and t and r the transmission and reflection coefficients correspondingly for incoming waves from the left, whereas t', and r' are the counter parts for incoming waves from the right. We can clearly see that the conductance is written in terms of the conductance quantum G_0 ($1G_0 = \frac{2e^2}{h} = \frac{1}{12.9}$ kΩ). The NEGF (non-equilibrium Green function) formalism for transport calculation [23–27] combined with density-functional theory and implemented in SMEAGOL, [17–19] is used here to calculate the transmission coefficients (2):

$$G = \frac{2e^2}{h}\sum_{Tr}[\Gamma_L G_M^{R+}\Gamma_R G_M^R] \quad (2)$$

Where G -is total conductance, where transmission coefficients represented by (3):

$$T(E) = \mathrm{Tr}[\Gamma_L G_M^{R+}\Gamma_R G_M^R](E) \quad (3)$$

Where $G_M^R(E)$ –retarded Green's function [22]. All quantities with the index L and R refer to the left and right electrodes respectively.

The CA (Ceperley/Alder) approximation (LDA) for the exchange-correlation potential is employed [28]. Troullier-Martins norm-conserving nonrelativistic and nonlocal pseudopotentials (PPs) are used for Au and Co. The atomic configurations for generating the PPs are $4s^2 3d^7$ (Co) and $5d^{10}6s^1$ (Au). The cutoff radii (in atomic units) for s, p, and d components of the pseudopotential generation are respectively 2.6 a.u., 2.73 a.u. and 2.6 a.u. for Au and 2.35 a.u., 2.47 a.u. and 2.235 a.u. for Co. The ab initio engine behind SMEAGOL is the numerical implementation of DFT contained in the SIESTA (Spanish Initiative for Electronic Simulations with Thousands of Atoms) code [29, 30] which uses finite-ranged numerical orbitals as basis set. Split-valence bases for Co are of double zeta type for s, p and d orbitals with the largest cutoff radii being 6.0 a.u. The electronic temperature is 300K. A mesh cutoff of 250 Ry, which specifies the real-space grid for defining the potential and the electron density, is used. Because Au atom has a large spin-orbit interaction we actually use calculation method which includes the effect of spin-orbit coupling. This amendment to Hamiltonian is taken into account in the used software codes SIESTA and VASP.

3 RESULTS

We present in this paper results of our theoretical calculations of conducting properties of Au-Co nanowire and also the bias effect on the wire conductivity. Our atomic structure calculations predicted that mixed Au-Co nanowires are stable in large area of average interatomic distances (1.6Å- 3.0Å) and may possess a high structural stability at room temperature. Our ab-initio calculations have shown the transition of the Au-Co nanowire from linear configuration to a "zig-zag" one under wire contraction up to 2.4Å (Fig.1 (b)). Further contraction of bimetallic nanowire leads to formation of two single component Au and Co nanowires strongly interacting with each other (Fig.1(c)). We calculated the conductance of Au-Co nanowires in the whole range of average interatomic distances, corresponding to the area of the wire existence (1.6Å - 3.0Å) including the region of two interacting Au and Co nanowires. These calculations simulate the real processes taking place in the chain of mixed Au-Co nanocontact during its formation in experiment by using MKBJ [2, 31] or STM [3, 4, 32] methods (Fig.1). We found the strong dependence of the total conductance of the wire on the "stretching-contraction" deformations emerging during the wire formation (see Fig.2, solid line). The total transmission of linear stretched nanowires (with average interatomic distance >2.4Å) was calculated about $2.0G_0$ and remains practically unchanged in the range (2.4Å -3.0Å) (Fig.2, solid line). Under contraction wire goes into "zig-zag" configuration at average interatomic distance ~2.4Å. This structural transition of the wire leads to an abrupt decreasing of transmission value to $1G_0$ (Fig.2, solid line). When the system passes to two interacting wires (1.6 Å-2.0 Å), the conductivity in each of the wires is $1G_0$, the total conductivity is $2G_0$ (distance between the two wires ~2Å).

By the fact that Au-Co nanowires are magnetic and their magnetic properties strongly correlate with wire geometry we study in present paper the spin dependent electron transport in Au-Co nanowire [33, 34]. We found spin polarization of electron transport in contracted "zig-zag" nanowires. However our study revealed the vanishing of spin polarization of electron transport in linear stretched Au-Co nanowire. Similar changes of conductivity were observed in our recent work in the Pt-Fe nanowires [35] and in work of Nemnes et al. in graphene nanoribbons with magnetic impurities [36,37]. For quantitative estimation of the spin polarization of the electron transport through nanowire, we use a term - degree of polarization [35] (4):

$$g = \frac{G_\uparrow - G_\downarrow}{G_\uparrow + G_\downarrow} \quad (4)$$

The value of g drastically decreases under wire stretching to ~0 for linear nanowire (Fig.2, solid green line).

To study in detail the conductivity of bimetallic Au-Co nanowires we calculated the transmission spectra of linear and contracted "zig-zag" nanowires. Figure 3a illustrates the transmission spectra of the linear wire at average interatomic distance 2.5Å. Our transmission spectra calculations have shown the absence of spin polarizations of electron transport in linear Au-Co nanowires. We can see that transmission through spin-up and spin-down channels at Fermi level (Fermi energy of left and right electrodes) is identical and both about

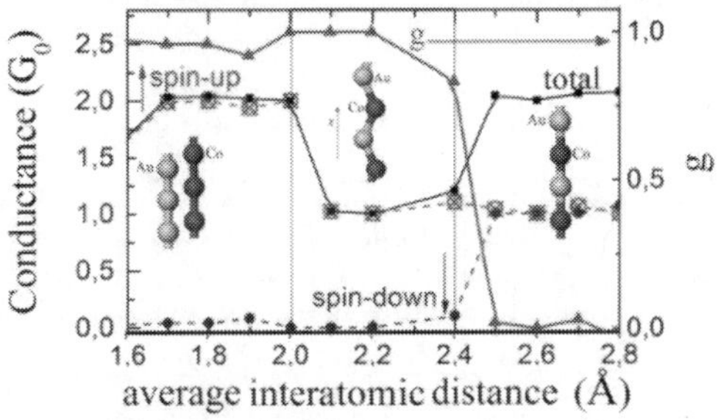

Fig. 2 Total quantum conductance of Au-Co nanowire (solid black line), spin-polarized conductance of spin-up (dashed red line) and spin-down (dashed blue line) channels and also the degree of spin polarization (g) of quantum electron transport (solid green line) as a function of average interatomic distance. Insight figure represents the model of evenly mixed "zig-zag" Au-Co nanowire.

~$1G_0$. Transmission pattern differs greatly for the contracted Au-Co nanowire. Figure 3b illustrates the partial transmission through spin-up and spin-down channels for average interatomic distance equals 2.2Å. We can see that only transmission through spin-up channel is presented at Fermi level (~$1G_0$), the transmission through spin-down channel equals zero. The degree of spin polarization (g) increases to 0.99 in "zig-zag" nanowire (see Fig.2, solid green line). Thus our transmission spectra calculations revealed that contracted "zig-zag" Au-Co nanowires represent the spin-filter system. We found also the dependence of spin conductance on the bias between bulk electrodes. Our transmission spectra calculations have shown that spin filtering state of Au-Co nanowire is metastable with narrow plateau of zero transmission spectra for one spin state at the Fermi level thus bias voltage about 0.06eV can destroy the spin filter state (Fig.3(b)) and put the system in non-polarized state (Fig.3(a)). As a result of our calculations we found the ability to control the transition point to the *spin-filter* state by changing of the bias between bulk electrodes. Our electronic structure calculations revealed the complex hybridized s-d_{z^2}-d_{xz}-d_{yz} band formation between atoms in "zig-zag Au-Co nanowire (Fig. 4). This leads to the emergence of

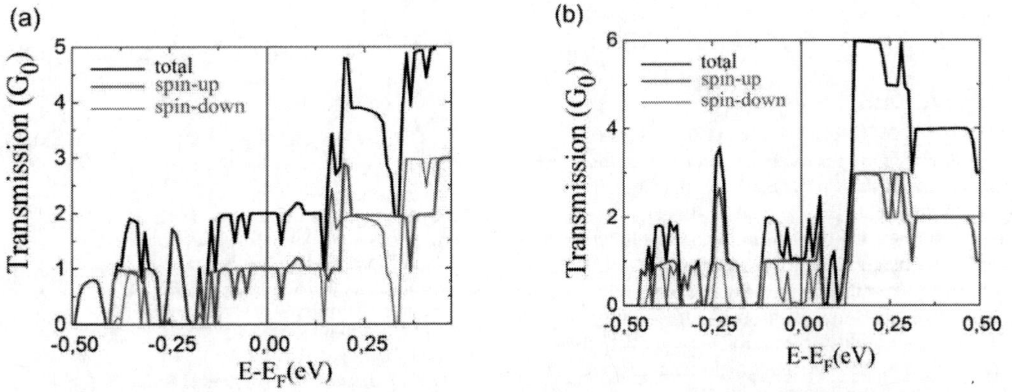

Fig. 3 Transmission $T(E)$ through spin-up and spin-down channel and total transmission at zero-bias of (a) linear Au-Co nanowire at average interatomic distance 2.5Å and (b) zig-zag nanowires at 2.2Å in spin filtering regime. E_F is the Fermi energy of left and right hand side electrodes.

coherent one-electron resonant tunnelling regime into a wire with two type of spin directions of electrons – spin up and spin down ones. The ratio of a length of a tunnelling area and the mean free path of an electron indicates that quantum interference of electrons exists in these systems (Au-37.7nm, Co-11.8nm) [38]. In case of linear nanowire each electron sequentially interact with magnetic and gold atoms and loss its original spin polarization as a result the quantum transport through the wire become non spin-polarised. In "zig-zag"

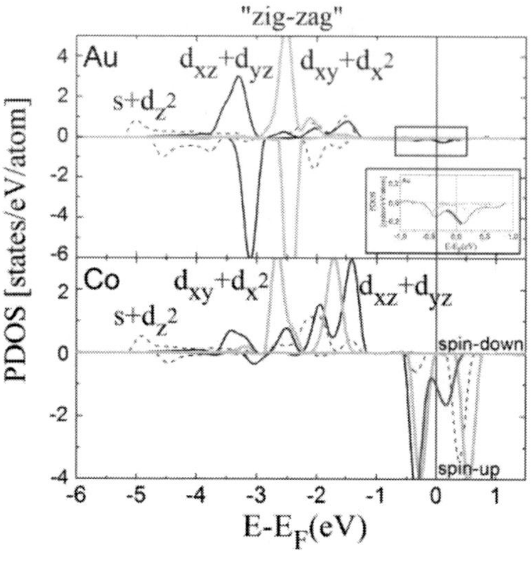

Fig. 4 Partial density of states for Au and Co atoms in "zig-zag" Au-Co nanowire in *spin-filter* state at average interatomic distance 2.2Å. Red line- d_{xz}-d_{yz} band at the Fermi level for Au and Co atoms

configuration electrons propagate through the wire without the loss of spin polarisation thus electron may interact simultaneously with magnetic Co atoms (with spin up polarisation) and also with Au atoms in tunnel area. The strong exchange leads to appearance of nonzero phase difference for electrons with spin up and spin down polarisation as the result of interaction with magnetized cobalt atoms. The interference of these electron waves leads to the formation of a *spin -filter* state in one-dimensional "zig-zag" nanowires with preferred spin up direction of electron polarization.

Conclusions

The study revealed the dependence of quantum conductance of Au-Co nanowire on the elongation of the wire. We found the new spin-filter state in one-dimensional Au-Co bimetallic nanowires. Moreover we found the existence of two transmission regimes in Au-Co nanowires with low and high conductivity $1G_0$ and $2G_0$ in "zig-zag" and linear nanowire correspondingly. We found the ability to control the conductivity of the wire by changing the geometry and length of the wire, as well as by applying an external voltage.

Acknowledgements

This work was supported by RFBR. The reported study was supported by the Supercomputing Center of Lomonosov Moscow State University and cluster MVS50K of the Interagency Supercomputer Center of the Russian Academy of Sciences.

References

1. A. Sokolov, Ch. Zhang, E. Y. Tsymbal, J. Redepending, B. Doudin, Nature Nanotechnology Lett. **2007**, 2, 171 – 175 Quantum Wires Shik A.Ya. 1997 SOROSOVSKY EDUCATIONAL JOURNAL, **1997**, No. 5,87-92.
2. H.-D. Chopra and S. Z. Hua, Phys. Rev. B **2002**, 66, 020403. M. J. Lagos, F. Sato, P. A. S. Autreto, D. S. Galvao, V. Rodrigues, D. Ugarte, Nanotechnology **2010**, 21, 485702.
3. J.Velev, R. F. Sabirianov, S. S. Jaswal, E.Y. Tsymbal, Phys. Rev. Lett. **2005**, 94, 127203; J.C.Tung and G.Y.Guo, Phys. Rev. B **2010**, 81, 094422.
4. V.S. Stepanyuk, P. Bruno, A.L. Klavsyuk et al. Phys. Rev. B **2004**, 69, 3, 033302 ; T. Haug, K. Perzlmaier, and C. H. Back, Phys. Rev. B **2009**, 79, 024414; K. M. Smelova, D. I. Bazhanov V.S. Stepanyuk et al. Phys. Rev. B **2008**, 77, 3, 033408.
5. K. M. Tsysar, D. I. Bazhanov and A. M. Saletsky, Phys. Rev. B **2011**, 84, 08545.
6. S.P. Parkin, M. Hayashi, L. Thomas, Science **2008**, 320, 5873, 190-194.
7. N. Müller, W. Eckstein, W. Heiland, and W. Zinn, Phys. Rev. Lett. **1972**, 29, 1651.
8. J.Bettini, F.Sato, P.Z.Coura, S.O.Danatas, D.S.Galvao, D.Ugarte, Nature Nanotechnology **2006**, 1, 182 .
9. E. M. Smelova, K. M. Tsysar', D. I. Bazhanov and A. M. Saletsky, JETP Lett. **2011**, 93, 3, 129-132; K. M. Tsysar', E. M. Smelova, D. I. Bazhanov and A. M. Saletsky, JETP Lett. **2011**, 94, 3, 228-232.
10. K.M.Tsysar, D.I. Bazhanov, A.M. Saletsky, V.S. Stepanyuk, and W.Hergert, Phys. Solid State 2010, 52, 3, 641.
11. E.M. Smelova, A.L. Klavsyuk, K.M. Tsysar, A.M. Saletsky, Moscow University Physics Bulletin, **2013**, 68, 1, 92–95.
12. S. Egle, C. Bacca, H. – Fr. of Pernau, M. Hafner, D. Hinzke, U. Nowak, E. Scheer, Phys. Rev. B **2010**, 81, 134402
13. K.M.Tsysar, D.I.Bazhanov, E.M.Smelova, and A.M.Saletsky Appl. Phys. Lett. **2012**, 101, 043108.
14. G. Kresse and J. Furthmuller, Phys. Rev. B **1996**, 54 , 11169.
15. W. Kohn, L.J. Sham, Phys. Rev. A **1965**, 140, 1133-1138.
16. J. Hafner, G. Kresse, A. Eichler, R. Lorentz, R. Hirschl, M. Marsman, Vasp workshop, http://cms.mpi.univie.ac.at/vasp-workshop/slides/documentation.htm.
17. J.P. Perdew, J.A. Chevary, S.H. Vosko, K.A. Jackson, M.R. Pederson, D.J.Singh, C. Fiolhais, Phys. Rev. B **1992**, 46, 6671-6687; M. C. Payne, M. P.Teter, D. C. Allan , T.A. Arias, J.D. Joannopoulos, Rev. Mod. Phys. **1992**, 64, 4, 1045-1097; M. C. Payne, et al., Rev. Mod. Phys. **1992**, 64, 1045-1077.
18. P P. Blochl, Phys. Rev. B **1994**, 50, 17953.
19. P. N. Molin, Solid State Storage of Hydrogen in Magnesium Alanate - a Density Functional Study, Department of Physics University of Oslo, (2005).
20. H. J. Monkhorst and J. D. Pack, Phys. Rev. B **1976**, 13, 5188.
21. R. Landauer, Philos. Mag. **1970**, 21, 863.
22. Theoretical and Computational Aspects of Electronic Transport at the Nanoscale, Alexandre Reily Rocha 2007, Trinity College.

23. L. P. Kadanoff and G. Baym, Quantum Statistical Mechanics_Benjamin/Cummings, **1962**, New York.
24. L. V. Keldysh, Sov. Phys. JETP **1965**, 20, 1018.
25. Y. Meir and N. S. Wingreen, Phys. Rev. Lett. **1992**, 68, 2512.
26. S. Datta, Electronic Transport in Mesoscopic Systems Cambridge University Press, **1995**,Cambridge, England.
27. H. Haug and A. P. Jauho, Quantum Kinetics in Transport and Optics of Semiconductors Springer-Verlag, **1996**, Berlin.
28. N. Troullier and J. L. Martins, Phys. Rev. B **1991**, 43.
29. P. Ordejon, E. Artacho, and J. M. Soler, Phys. Rev. B **1996**, 53, R10441.
30. J. Soler, E. Artacho, J. D. Gale, A. Garcia, J. Junquera, P. Ordejon, and D. Sanchez-Portal, J. Phys.: Condens. Matter **2002**, 14, 2745.
31. M. Czerner, A. Bagrets, V. S. Stepanyuk, A. L. Klavsyuk, and I. Mertig Phys. Rev. B, **2006**, 74, 115108.
32. A.L. Klavsyuk, A.M. Saletsky, Physics-Uspekhi 58, 933 (2015); A.L. Klavsyuk, S.V. Kolesnikov, E.M. Smelova, A.M. Saletsky, JETP Lett., **2010**, 91, 158.
33.Tsysar K.M., Bazhanov D.I., Smelova E.M., Saletsky A.M., Applied Physics Letters, American Institute of Physics (United States), **2012**,101, 043108-043112.
34. Smelova E.M., Tsysar K.M., Bazhanov D.I., Saletsky A.M., Letters to the "Journal of Experimental and Theoretical Physics", **2011**, 93, № 3, 139-142.
35. E.M. Smelova, K.M. Tsysar, A.M. Saletsky, Physical Chemistry Chemical Physics, **2014**, 16, 8360 .
36.G.A.Nemnes, S.Antohe, Materials Science and Engeniring B, **2013**,178, 1347.
37. G.A.Nemnes, Journal of Nanomaterials, **2013**, 408475
38. Daniel Gall, JOURNAL OF APPLIED PHYSICS, **2013**, 119, 085101.

Spin-torque quantization and microwave sensitivity of a nano-sized spin diode

Gleb Demin[1,2*] and *Anatoly* Popkov[1,2]

[1]Moscow Institute of Physics and Technology (MIPT), 141701 Moscow region, Dolgoprudny, Russia
[2]National Research University of Electronic Technology (MIET), 124498 Moscow, Russia

Abstract. Rectification of microwave signal by the spin-torque diode is very promising for its practical applications in microwave imaging. This is due to a very high sensitivity of magnetic tunnel junction under the bias current, which was previously demonstrated in a number of works [1-3]. The decreasing of cross-sectional area of the spin-torque diode up to the nano-sized dimensions below 10 nm allows one to reach high sensitivity without any bias current. Transverse quantization of electron states in the magnetic nanowire based on nano-sized metallic spin valves and magnetic tunnel junctions can create an additional impact not only on the magnetoresistance, but also on the spin-transfer torque in such structures. In this work we present an analysis of the quantization effect of conductance and spin-transfer torques on the microwave sensitivity of nano-sized spin-torque diodes during the reduction of its cross-sectional area. It was found that the magnetoresistance values up to 130 % can be achieved in a magnetic nanowire containing spin-valve diode with the nonmagnetic metal spacer. As a result, the maximum microwave sensitivity of spin-torque diodes based on these structures can be increased several times that opens the way for the further development of highly sensitive microwave detectors.

1 Introduction

Nowadays much attention in the field of spintronics is paid to the consideration of so-called spin-torque diodes based on magnetic tunnel junctions (MTJ), which have a large magnetoresistive effect and demonstrate very high sensitivity to the microwave signal. Microwave alternating current in MTJ generates dc voltage $V_{dc} = < R(t)I(t) >$ having maximum value at the resonant frequency of spin oscillations induced by the spin-transfer torque [1]. The sensitivity of the spin-torque diode $\varepsilon = V_{dc} / P_{in}$, where P_{in} is the input microwave power, can exceed the sensitivity of a semiconductor Schottky diode by an order of value in the presence of the bias current [2-4]. High effect of the rectification of the microwave signal in MTJ makes spin-torque diode attractive for practical applications in the microwave vision techniques [5]. However, the threshold current density, in the vicinity of which there is a sharp increase in the spin-torque diode sensitivity, is generally high. For that reason there are some problems of the usage of spin-torque diode shifted by bias current. In this connection, it is of interest to consider an alternative way of increasing the sensitivity of a spin-torque diode. This can be realized by reducing its lateral dimensions, since the spin-torque diode sensitivity increases inversely with the area of MTJ [6]. On the other hand, such scaling effects can crucially influence both magnetoresistance and spin-transfer torques in nano-sized region, where the effects of transverse quantization play an important role as it was shown for the conductivity of nano-point magnetic contacts [7, 8]. In this work we performed an analysis of the microwave sensitivity of magnetic nano-sized spin-torque diode based on magnetic tunnel structures depending on the lateral dimensions of its cross-sectional area. We also consider possible effects of quantization of the magnetoresistance and spin-transfer torques in a magnetic nanowire consisting of the metallic spin-valve structure.

2 Model and basic equations

2.1 Charge and spin currents

We analyze the spin-transfer torque effect in a magnetic nanowire consisting of two semi-infinite ferromagnetic (FM) layers separated by a thin nonmagnetic spacer (S) having the thickness d_S and rectangular cross section $a \times b$. Considered trilayer structure «FM-S-FM» of the magnetic nanowire is shown in Fig.1c. The magnetization unit vector $\mathbf{m_P}$ in the reference (left) FM layer is assumed to be fixed, while the magnetization of the free (right) FM layer $\mathbf{m}$ is initially directed at the angle θ to the direction of magnetization in the reference layer. We use the Sommerfeld model for the description of the ballistic spin transport of free electrons. Within this model the charge and spin fluxes are determined by spinor wave functions $\Psi_{\sigma i}$ of

* Corresponding author: gddemin@gmail.com

electrons with a given spin polarization according to well-known relations:

$$\begin{cases} j_{\sigma i} = -\dfrac{i\hbar}{2m_{i*}}\left(\Psi^*_{\sigma i}\cdot\dfrac{\partial\Psi_{\sigma i}}{\partial x} - \Psi_{\sigma i}\cdot\dfrac{\partial\Psi^*_{\sigma i}}{\partial x}\right) \\ j^s_{\mu,\sigma i} = \dfrac{i\hbar}{2m_{i*}}\left(\left(\dfrac{\partial\Psi^*_{\sigma i}}{\partial x}\right)\sigma_\mu\Psi_{\sigma i} - \Psi^*_{\sigma i}\sigma_\mu\left(\dfrac{\partial\Psi_{\sigma i}}{\partial x}\right)\right) \end{cases}, \quad (1)$$

where $\hbar$ is the Planck constant, m_{i*} is the effective electron mass in the $i-$th layer, σ_μ are the Pauli matrices ($\mu = x, y, z$). To determine the spinor wave functions $\Psi_{\sigma i}$ in each $i-$th layer ($i = L, S, R$) for a given spin polarization $\sigma = \uparrow, \downarrow$ of incident electrons, it is necessary to solve the Schrödinger equation:

$$-\frac{\hbar^2}{m_{i*}}\nabla^2\Psi_{\sigma i}(x,y,z) = (E - U(x))\Psi_{\sigma i}(x,y,z), \quad (2)$$

where $E = E_{\sigma xi} + E_{\perp mni}$ is the total energy of the electron including the longitudinal ($E_{\sigma xi}$) and transverse ($E_{\perp mni}$) energy terms. In the nano-sized magnetic nanowire the transverse energy is quantized and for the case of its rectangular cross section it can be expressed as $E_{\perp mni} = \hbar^2 k^2_{mn}/2m_{i*} = \beta_i E_{mn}$, where $\beta_i = m_e/m_{i*}$, $k_{mn} = \pi\sqrt{m^2/a^2 + n^2/b^2}$, $E_{\sigma xi} = \hbar^2 k^2_{\sigma xi}/.2m_{i*}$, $k_{\sigma xi}$ is the spin-dependent longitudinal wave vector in the $i-$th layer. The spinor wave function $\Psi_{\sigma i}(x,y,z)$ in this case can be presented as the combination of plane waves $\Psi_{\sigma i}(x,y,z) = \psi_{\sigma i}(x)\cdot\Phi_i(y,z)$, where $\psi_{\sigma i}(x) = (\psi_{\sigma\uparrow_i}(x), \psi_{\sigma\downarrow_i}(x))$ corresponds to the longitudinal component of the spinor, whereas $\Phi_i(y,z) = C_\varphi \sin(k_y y)\sin(k_z z)$ is the function of transverse plane wave with the normalization factor $C_\varphi = 2/\sqrt{ab}$, where $k_y = \pi m/a$, $k_z = \pi n/b$ with quantum numbers $m, n = 1,2,3,\ldots$.

We consider the potential profile $U(x)$ in the limit of low bias voltage $eV \ll E_F$ which is shown in Fig. 1a, b for the two types of «FM-S-FM» structure. In general, the x-dependent potential $U(x)$ takes the form:

$$U(x) = \begin{cases} -\sigma_L\cdot\Delta_L + \sigma_V^{(-1)}(\delta E_{FL,R} + eV), x \le 0 \\ U_S, 0 < x < d_S \\ -\sigma_R\cdot\Delta_R + (\sigma_V^{(-1)} - 1)\delta E_{FL,R} - \sigma_V^{(+1)}eV, x \ge d_S \end{cases}, \quad (3)$$

where $U_S = E_F + \varphi_L - (\Delta\varphi + eV)x/d_S + \sigma_V^{(-1)}eV$ in the case of tunnel spacer (Fig. 1a), and $U_S = \delta E_{F,NM}$ in the case of metal spacer (Fig. 1b), $\delta E_{F,NM} = E_F - E_{F,NM}$, $E_{F,NM}$ is the Fermi level in the metal spacer, $\sigma_V^{(\pm1)} = 0.5\sigma_V(\sigma_V \pm 1)$, $\sigma_V = \pm1$ is the sign of the bias voltage ($\sigma_V = 0$ at zero voltage), e is the electron charge, $\delta E_{FL,R} = E_{FR} - E_{FL}$, $E_{FL(R)}$, $\Delta_{L(R)}$, $\varphi_{L(R)}$ are the Fermi level, spin splitting and the work function in the left (right) FM electrode, $E_F = E_{FL} + \sigma_V^{(-1)}\delta E_F$, $\Delta\varphi = \varphi_L - \varphi_R$, and $\sigma_{L(R)} = \pm1$ corresponds to the spin direction $\sigma_{L(R)} = \uparrow, \downarrow$ in the left (right) FM layer.

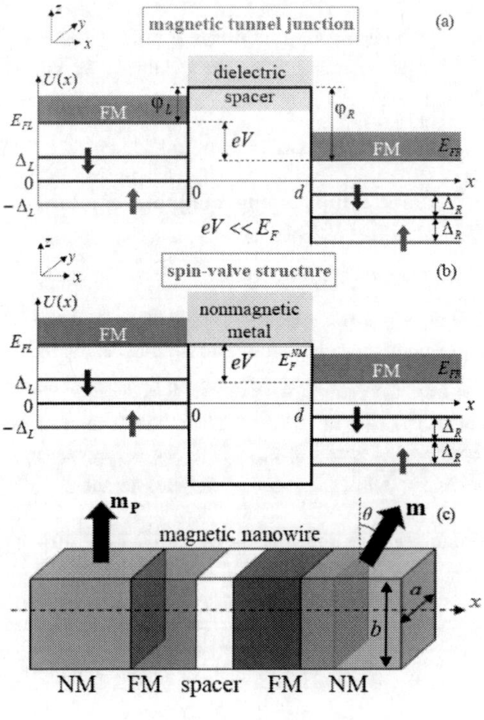

Fig. 1. The schematic potential diagram for the case of (a) magnetic tunnel junction, (b) spin-valve structure. (c) A view of magnetic nanowire with noncollinear magnetization alignment, where NM denotes the «normal metal».

The solution of the quantum-mechanical problem (2) with the potential (3) allows us to find both the charge ($<I_{\sigma\sigma'(e)}>$) and spin ($<I_{\mu,\sigma\sigma'(s)}>$) currents in the spin channel $\sigma \leftrightarrow \sigma'$ of the proposed structures of magnetic nanowire according to formulas (1) for spin fluxes. Thus, the spin current $<I_{\mu,\sigma\sigma'(s)}> = <I^{L\to R}_{\mu,\sigma\sigma'(s)}> - <I^{R\to L}_{\mu,\sigma'\sigma(s)}>$ can be represented as the difference between the thermodynamically averaged electron spin currents «from the left to the right electrode» ($<I^{L\to R}_{\mu,\sigma\sigma'(s)}$), which can be and «from the right to the left electrode» ($<I^{R\to L}_{\mu,\sigma'\sigma(s)}>$), which take the following form:

$$\begin{cases} <I^{L\to R}_{\mu,\sigma\sigma'(s)}> = \dfrac{m_{L*}}{2\pi\hbar^2}\cdot\sum\limits_{m,n}\int\limits_0^{+\infty}\dfrac{1}{k^\sigma_{xL}}\,j^{(TR)}_{\mu,\sigma\sigma'LR}F_{LR}dE \\ <I^{R\to L}_{\mu,\sigma'\sigma(e)}> = \dfrac{m_{R*}}{2\pi\hbar^2}\cdot\sum\limits_{m,n}\int\limits_0^{+\infty}\dfrac{1}{k^{\sigma'}_{xR}}\,j^{(TR)}_{\mu,\sigma'\sigma RL}F_{RL}dE \end{cases}, \quad (4)$$

where $\sigma,\sigma'=\uparrow,\downarrow$ is the spin direction of electrons before and after entering the free FM layer, $F_{LR}=f_L^{(-1)}\left(1-f_R^{(+1)}\right)$, $F_{RL}=f_R^{(+1)}\left(1-f_L^{(-1)}\right)$, $f_{L(R)}^{(-1)(+1)}=f_{L(R)}\left(E_V^{(-1)(+1)},T\right)$ is the Fermi distribution function in the left (right) FM electrode, $E_V^{(\pm1)}=E+\sigma_V^{(\pm1)}e|V|$, T is the temperature. $j_{\mu,\sigma\sigma'LR}^{(TR)}=j_{\mu,\sigma\sigma'LR}^{(TR)}(E,E_{mn})$, $j_{\mu,\sigma'\sigma RL}^{(TR)}=j_{\mu,\sigma'\sigma RL}^{(TR)}(E,E_{mn})$ are the transmitted spin fluxes for the $\sigma(\sigma')\leftrightarrow\sigma'(\sigma)$ conduction channel, E_{mn} are quantized energy levels chosen in accordance with the condition $E_F-e|V|<E=E_{\sigma(\sigma')xL(R)}+E_{\perp mnL(R)}<E_F$.

2.2 Quantization of the magnetoresistance and spin-transfer torque

Thus, the total charge current $<I_{(e)}>$ and the total spin current $<I_{\mu(s)}>$ can be calculated as the sum of corresponding contributions of spin-dependent currents $<I_{\sigma\sigma'(e)}>$ and $<I_{\mu,\sigma\sigma'(s)}>$ over all possible conducting channels $\sigma(\sigma')\leftrightarrow\sigma'(\sigma)$. Then the conductance of magnetic nanowire can be expressed as $<G_S>=\sum_{\sigma,\sigma'}<I_{\sigma\sigma'(e)}>/V$ that goes to the Landauer's formula in the limit of $T\to0$ and $V\to0$ [9]. In this case, it leads to the following relationships for the charge and spin current:

$$\begin{cases} <I_{(e)}>^{T=0K}=\dfrac{e^2V}{h}\sum_{\sigma,\sigma'}\sum_{m,n}P_{T\sigma\sigma'(e)}(E_F,E_{mn})\Big|_{E_{mn}\le E_F-U_{\sigma\sigma'}} \\ <I_{\mu(s)}>^{T=0K}=\dfrac{eV}{h}\sum_{\sigma,\sigma'}\sum_{m,n}P_{T\mu,\sigma\sigma'(s)}(E_F,E_{mn})\Big|_{E_{mn}\le E_F-U_{\sigma\sigma'}} \end{cases}, \quad (5)$$

where V is the voltage, $P_{T\sigma\sigma'(e)}=\left|k_{xR}^{\sigma'}/k_{xL}^{\sigma}\right|\cdot|T_{\sigma\sigma'}|^2$, $P_{Tx,\sigma\sigma'(s)}=\mathrm{Re}\left[\left|k_{xR}^{\sigma'}/k_{xL}^{\sigma}\right|\cdot\left(T_{\sigma\uparrow}^{*}T_{\sigma\downarrow}\right)e^{-i\left(k_{xR}^{\uparrow}-k_{xR}^{\downarrow}\right)x}\right]$, $P_{Ty,\sigma\sigma'(s)}=\mathrm{Im}\left[\left|k_{xR}^{\sigma'}/k_{xL}^{\sigma}\right|\cdot\left(T_{\sigma\uparrow}^{*}T_{\sigma\downarrow}\right)e^{-i\left(k_{xR}^{\uparrow}-k_{xR}^{\downarrow}\right)x}\right]$, $P_{Tz,\sigma\sigma'(s)}=\mathrm{Im}\left[is_{\sigma'}\left|k_{xR}^{\sigma'}/k_{xL}^{\sigma}\right|\cdot|T_{\sigma\sigma'}|^2\right]$, $s_{\sigma'=\uparrow,\downarrow}=\pm1$, $k_{xR}^{\sigma'}=k_{xR}^{\sigma'}(E,E_{mn})$, $T_{\sigma\sigma'}=T_{\sigma\sigma'}(E,E_{mn})$ is the transmission coefficient for the $\sigma\leftrightarrow\sigma'$ conduction channel. The amplitudes of in-plane ($\tau_{\parallel}$) and out-of-plane ($\tau_{\perp}$) spin-transfer torques are proportional to the corresponding spin currents transmitted in the magnetic nanowire:

$$\begin{cases} \tau_{\parallel}=\dfrac{0.5\sigma_V\hbar}{2}\{[\sigma_V+1]\cdot I_{x(s)}(x=d)+[\sigma_V-1] \\ \cdot\left[I_{x(s)}(x=0)\cos\theta-I_{z(s)}(x=0)\sin\theta\right]\} \\ \tau_{\perp}=\dfrac{\sigma_V\hbar}{2}\{[\sigma_V+1]\cdot I_{y(s)}(x=d)+[\sigma_V-1]\cdot I_{y(s)}(x=0)\} \end{cases}, \quad (6)$$

It follows from the expressions (4) - (6) that the quantization of the energy levels for each conduction channel lead both to the conductance quantization $<G_S>$ (proportional to e^2/h) and to the quantization of spin-transfer «torquance» $\partial\tau_{\parallel(\perp)}/\partial V$ (proportional to $e/4\pi$), which should affect the microwave sensitivity of a spin-torque diode.

2.3 Microwave sensitivity of spin-torque diode based on magnetic nanowire

In [4] it was shown that the maximum microwave sensitivity ε_{MAX} of the spin-torque diode in the absence of magnetic field for the case of mutually perpendicular magnetization geometry can be presented:

$$\varepsilon_{MAX}\approx\varepsilon_P\rho(\eta_{\parallel}+\alpha\eta_{\perp})/(\alpha\cdot(2k_{eff}-k_b)), \quad (7)$$

where $\varepsilon_P=v/J_PS$, $v=4Z_0\bar{R}/(Z_0+\bar{R})^2$ is the matching ratio coefficient, $J_p=2ed_FM_S^2/\hbar$, Z_0 is the impedance of the line, S is the cross-sectional area of the spin-torque diode, $\bar{R}=2R_{\uparrow\uparrow}(1+\delta_{MR})/(2+\delta_{MR})$, $R_{\uparrow\uparrow}=\left(<G_S>_{V\to0}^{\uparrow\uparrow}\right)^{-1}$ is the diode resistance for the parallel magnetization alignment of FM layers, d_F is the thickness of the free FM layer, M_S is the saturation magnetization, α is the Gilbert damping factor, $k_{eff}=2K_u/M_s^2-4\pi N_z+4\pi N_y$, $k_b=4\pi(N_y-N_x)$, K_u is the constant of induced magnetic anisotropy, $N_{x,y,z}$ are the demagnetization factors. In this case $\rho=\delta_{MR}/(2+\delta_{MR})$ is the modified value of the magnetoresistance, $\eta_{\parallel(\perp)}$ is the in-plane (out-of-plane) spin-transfer torque efficiency. It should be noted that for the magnetic nanowire the value of the spin-transfer torque efficiency $\eta_{\parallel(\perp)}=(2e/\hbar)\tau_{\parallel(\perp)}(\theta)/<I_{(e)}(\theta)>,\theta=\pi/2$, which follows from the definition of the in-plane spin-transfer torque $\boldsymbol{\tau}_{\parallel}\propto m\times m\times m_P$ and out-of-plane spin-transfer torque $\boldsymbol{\tau}_{\perp}\propto m\times m_P$ correspondingly.

3 Simulation results

Numerical calculations of the conductance and spin-transfer torques in the magnetic nanowire were carried out for the case of its rectangular cross section ($\varsigma=b/a=2$), where Co was chosen as the material of FM layers with the parameters: $m_{L*}=m_{R*}=m_e$, $E_{FL}=E_{FR}=3.24eV$, $\Delta_L=\Delta_R=1.37eV$ (these values correspond to Fermi wave vectors for Co $k_F^{\uparrow}=1.1\text{A}^{-1},k_F^{\downarrow}=0.7\text{A}^{-1}$ estimated for nano-sized magnetic point contact in [8]). The width of the nanowire cross section a was varied in the range from 1 to 3 nm, while the aspect ratio ς was assumed to be constant. When considering the magnetic nanowire consisting of a magnetic tunnel junction with MgO tunnel spacer (the work function $\varphi_L = \varphi_R = 1.2eV$, the thickness of MgO dielectric layer $d_S = 0.8nm$), the effects of the quantization of conductance and spin-transfer torques were weakly expressed for the chosen parameters of the tunnel structure. This can be caused by the small transmission probability for various dimensions of the nanowire's cross section. However, in the case of the spin-valve structure (Co-Au-Co) having $d_S = 0.942nm$, $E_F^{NM} = 5.51eV$ the quantization of the magnetoresistance $\delta_{MR} = (<G_S^{\uparrow\uparrow}>/<G_S^{\uparrow\downarrow}>)-1$ was obtained with the increase of its amplitude up to 130 % as it is shown in Fig. 2.

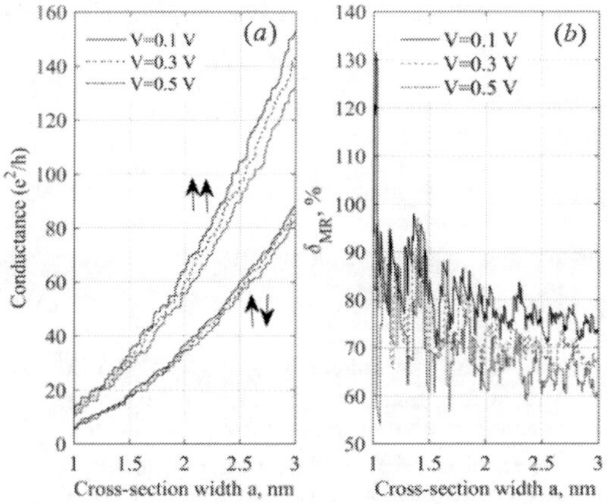

Fig. 2. The dependence of (a) conductance and (b) magnetoresistance of the magnetic nanowire based on spin-valve structure (Co-Au-Co) on the cross-section width for the different bias voltages, where ($\uparrow\uparrow$ ($\uparrow\downarrow$)) is the parallel (antiparallel) magnetization alignment of FM layers.

Fig. 3 shows that the quantization of the transverse energy levels in magnetic nanowire also affects the amplitudes of spin-transfer torques. It was found from Fig. 3a that the frequency and the number of quantum jumps will be growing up with increasing the bias voltage. Like in a bulk magnetic tunnel junction, the amplitudes of in-plane ($\tau_{||}$) and perpendicular ($\tau_{\perp}$) spin-transfer torques have opposite sign in the case of positive bias voltage, but they are additionally quantized due to the confined geometry of magnetic nanowire. The dependence of maximum microwave sensitivity ε_{MAX} of the spin-torque diode based on the spin-valve structure (Co-Au-Co) from the cross-section width a is shown in Fig. 3b. The following parameters were used: $\varepsilon_P = 85.5mV/mW$, $\alpha = 0.01$, $k_{eff} = 0.6$, $k_b = 0.3$. From Fig.3b it can be seen that with the decrease of the characteristic width of the spin-torque diode a the sensitivity of the diode rises and begins to oscillate, starting at approximately 3 nm.

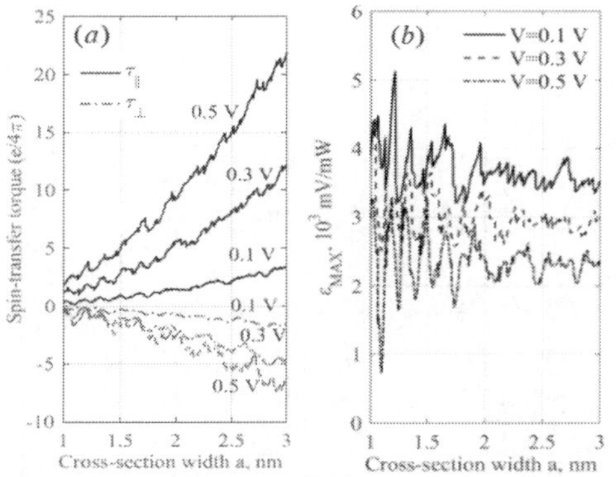

Fig. 3. The quantization of (a) spin-transfer torques and (b) maximum microwave sensitivity in dependence from the cross-section width of magnetic nanowire with the metallic spin-valve structure (Co-Au-Co) for the different bias voltages.

Calculations show that a decrease to lateral sizes less 10 nm makes it possible to obtain an increase in the microwave sensitivity of the spin-torque diode in the absence of a bias magnetic field by more than two order of value higher than in a nano-pillar spin-torque diode of 100 nm cross-sectional size. These results can be useful for future development of high-sensitivity microwave detectors. The considered features of the quantization of the spin-transfer torques are very important for the understanding of the physics of magnetoresistive memory elements during its scalability to technological nodes less than 10 nm [10]. Scaling less than 1 nm requires new technological and theoretical approaches related to monoatomic physics, which is beyond the scope of this work.

The work was supported by the Russian Science Foundation (project № 16-19-00181).

References

1. S. Miwa, Nature materials, **13**, 50-56 (2014)
2. X. Li, Y. Zhou, P.W.T. Pong, Journal of Nanotechnology, **8347280**, 1-11 (2016)
3. B. Fang, M. Carpentieri, X. Hao, Nature Communications, **7**, 11259 (2016)
4. A.F. Popkov, N.E. Kulagin, G.D. Demin, Solid State Communications, **248**, 140-143 (2016)
5. L. Fu, Y. S. Gui, L. H. Bai, H. Guo, H. Abou-Rachid, J. Appl. Phys., **117**, 213902 (2015)
6. A.F. Popkov, N.E. Kulagin, G.D. Demin, K.A. Zvezdin, Proc.of Universities. Electronics, **22**, 101-119 (2017)
7. A.K. Zvezdin, A.F. Popkov, J. Exp. Theor. Phys., **71**, 209-212 (2000)
8. L.R. Tagirov, B.P. Vodopyanov, B.M. Garipov, J. Magn. Magn. Mater. **258-259**, 61-66 (2003)
9. R. Landauer et.al., Phys. Rev. B., **31**, 6207 (1985)
10. R. Requist, Nature Nanotech. **11**, 499-508 (2016)

Numerical simulation of spin transport in systems with complex geometry

Timofey Andrianov[1,*], and *Anatoly* Vedyayev[1,2]

[1]M. V. Lomonosov Moscow State University, Faculty of Physics, Leninskie Gory 1, 1199991 Moscow, Russia
[2]SPINTEC, UMR CEA/CNRS/UJF-Grenoble1/Grenoble-INP, INAC, 38054 Grenoble, France

Abstract. The spin diffusion and charge equations in Levy-Fert and Waintal models were numerically solved, using finite element method in complex non-collinear geometry with strongly inhomogeneous current flow. As an illustration, spin-dependent transport through a magnetic pillar and nonmagnetic spacer separating two magnetic layers was investigated. It is shown, that the structure with number of pillars gives a higher value of Giant Magnetoresistance (GMR) effect rather than a structure with one pillar of equivalent diameter. The inhomogeneity of spin currents, which has one of the strongest impacts on GMR effect value leads to the occurrence of spin-current vortices. Introduction of l_T and l_L lengths in Waintal model gives a better description of angular dependence of GMR effect rather than Levy-Fert model.

1 Introduction

Discovery of the effect of giant magnetoresistance (GMR) was a first step which allowed spintronics to become a very promising field of studies [1,2]. The scrutinized investigation of this effect allowed creating a new revolutionary type of reading heads for hard disc drives, which gave a huge leap for computer technologies development. The GMR effect was explained in terms of spin-dependent scattering phenomena, which takes place at the interfaces and in the adjacent magnetic layers. Later a few theoretical models of GMR effect were offered for geometries current-in-plane (CIP) [3,4] and current-perpendicular-plane (CPP) [5–8]. One of the ground theories introduces the concept of spin accumulation and spin diffusion length as key parameters to describe the diffusive transport in CPP metallic multilayers, which were generalized for both collinear and non-collinear cases. Recently another model was proposed by *Waintal et al* [9] which generalizes Valet-Fert theory, involving two new coherence and Larmor lengths. However, all these theoretical models were applied for simple geometries with homogeneous current flow so far. However, many spintronic devices involve inhomogeneous current flows. In our paper, we numerically investigate the structures with geometries that cause non-uniform current flows, using Levy-Fert [6] and Waintal [9] transport models for comparison. We show that inhomogeneity of current flows, appearing in GMR structures leads to creation of vortices in distribution of spin currents and compare angular dependence of GMR for Levy-Fert and Waintal models.

2 Method

Here we investigate and compare two type of structures. The first one consists of two 5-nm thick ferromagnetic layers connected with paramagnetic spacer of variable size sandwiched between two 100-nm thick paramagnetic electrodes (Fig. 1).

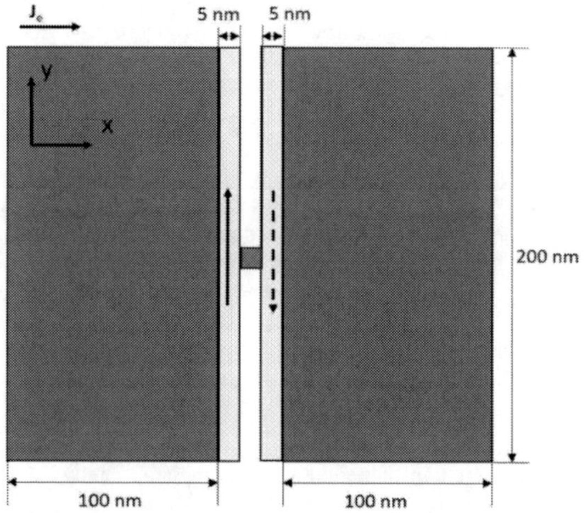

Fig.1. Structure with metallic spacer used for the finite element calculation of CPP spin transport through a nanoconstricted spin valve. In case of 2D model, all quantities are calculated per 1 nm depth.

The second structure consists of so-called magnetic pillars sandwiched between two 100-nm thick paramagnetic electrodes (Fig. 2).

* Timofey Andrianov: timofey.andrianov@gmail.com

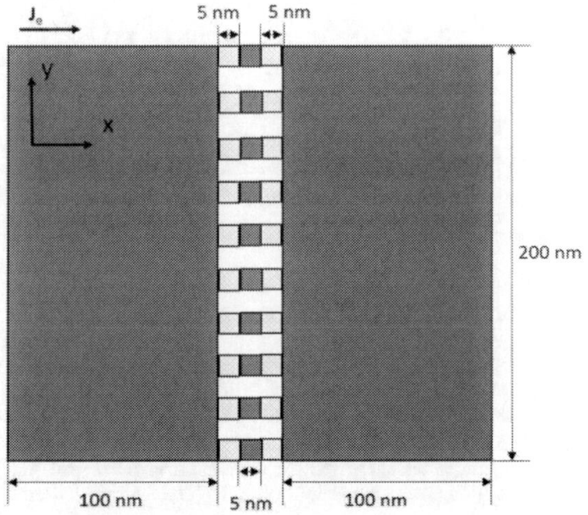

Fig.2. Structure with magnetic pillar used for the finite element calculation of CPP spin transport through a nanoconstricted spin valve. In case of 2D model, all quantities are calculated per 1 nm depth.

The pillar is a thin tube corresponding to a simple spin valve with number of repeating ferromagnetic and paramagnetic layers. In both structures, one of the ferromagnetic layers has a fixed magnetization directed perpendicular to current flow. Magnetization in the other layer rotates from parallel to antiparallel state relatively to the first one.

In our numerical simulation, we use and compare two models which describe spin transport in the structures. The diffusive spin transport equations proposed by Levy-Fert extended on non-collinear geometry look as follows [6]:

$$\mathbf{J_e} = -\sigma\nabla\varphi - \beta\frac{\sigma}{\nu}\left(\mathbf{M},\nabla\mathbf{m}\right) \quad (1)$$

$$\mathbf{J_m} = -\sigma\beta\left(\mathbf{M},\nabla\varphi\right) - \frac{\sigma}{\nu}\nabla\mathbf{m} \quad (2)$$

$$\mathrm{div}\mathbf{J_e} = 0 \quad (3)$$

$$\mathrm{div}\mathbf{J_m} = -\frac{\sigma}{\nu l_{sf}^2}\mathbf{m} - \frac{\sigma}{\nu l_J^2}\left[\mathbf{M},\mathbf{m}\right] \quad (4)$$

Where φ – electric potential, σ – conductivity, β – parameter of spin asymmetry of conductivity, **M** – unit vector of magnetization; **m** – spin accumulation vector, which has 3 components in spin dimension, l_{sf} – spin diffusion length, l_J – precession length, ν – density of states at Fermi level. Unless electric current, spin current is not conserved and can vary either because of spin relaxation or because of the local spin-transfer torque, which induces a precession of the spin accumulation around the local magnetization due to *s-d* exchange interaction.

Spin transport equations presented by *Waintal et al* derived directly from Valet-Fert [8] theory and introduces two new characteristic lengths: the transverse spin coherence length l_T and the (Larmor) spin precession length l_L. The first length is responsible for decaying and the second one for precession. Introduction of these two lengths give a better description of the angular dependence of GMR in studied structures.

Here we take into account only bulk spin-dependent scattering, which would not change qualitative description of the phenomena presented in paper.

The numerical simulation was provided with finite element method on the basis of Comsol Multiphysics® software. For all simulations, we used the following values of the key parameters: σ^F=0.001 (Ω · nm)$^{-1}$; σ^P=0.01 (Ω · nm)$^{-1}$; ν=0.1 (eV)$^{-1}$; l_{sf}^P=100 nm; β=0.7; J_e=10^8 A/cm^2; l_J=1 nm, where indexes *F* and *P* refer to ferromagnetic and paramagnetic layers accordingly.

3 Results

First, we compared two different geometries. The first one has long paramagnetic electrodes with long and thin ferromagnetic layers connected with a thin paramagnetic spacer. The second has two paramagnetic electrodes connected with one or several thin pillars in form of F/P/F. In the Fig.3 dependence of GMR on diameter of pillar and spacer is shown.

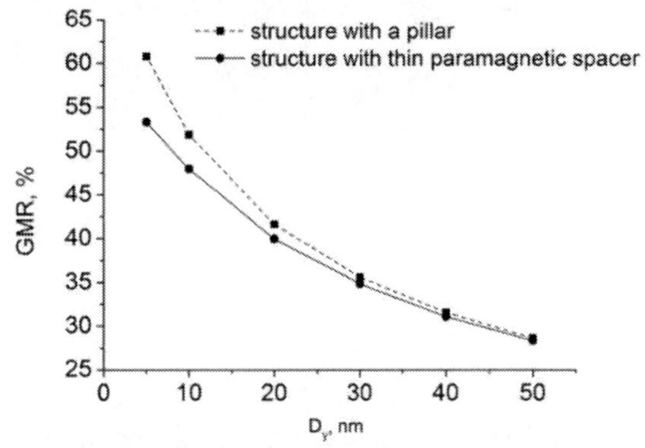

Fig.3. Dependence of GMR from diameters of pillar and spacer. l_{sf}^F=10 nm.

The result of numerical simulation showed that in structure with F/P/F pillar of small diameter the value of GMR effect is higher than in structure with paramagnetic spacer of the same diameter.

As the next step, we investigated the system where electrodes connected with different number of F/P/F pillars. For this geometry GMR effect was calculated and compared with dependence of GMR for the geometry with one pillar of equivalent diameter (Fig.4). Naturally, due to decreasing of total resistance of system with pillars the value of GMR effect decreases with the increasing of diameter or number of pillars. Fig.4 shows that geometry with number of pillars gives significantly higher GMR effect rather than geometry with one pillar of equivalent diameter. It means that one of the most important impacts on GMR effect makes a non-uniform distribution of spin currents in the investigated system [10].

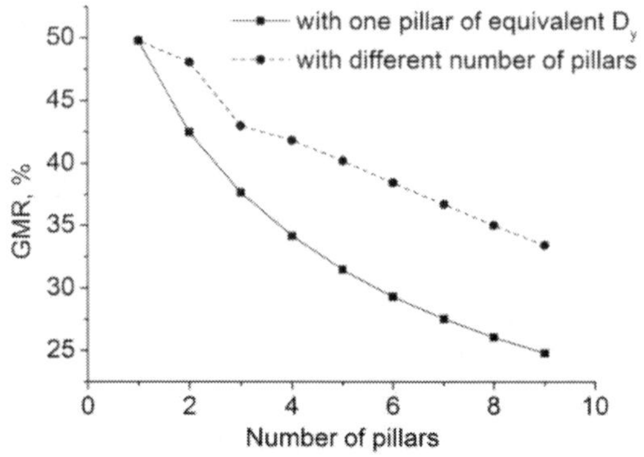

Fig.4. Dependence of GMR from number of pillars and one pillar with equivalent diameter. l_{sf}^F=5 nm.

To demonstrate non-uniform behavior of spin current in the structure we plot spin current streamlines. One can notice that in some regions of the structure the streamlines started to curl forming spin current vortices, which can be considered as inhomogeneity of spin current distribution (Fig. 5).

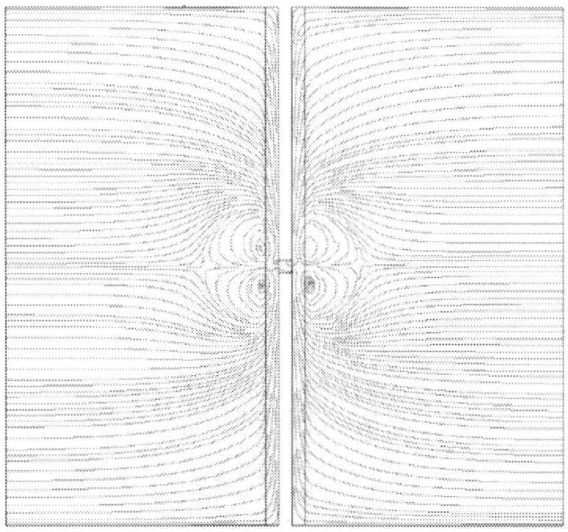

Fig.5. Distribution of spin current in the system with paramagnetic spacer. l_{sf}^F=50 nm.

For comparison of the angular dependence of GMR effect in Levy-Fert and Waintal models we used a reduced magnetoresistance introduced by Slonczewski [11] :

$$r = \frac{1-\cos^2(\theta/2)}{1+\chi\cos^2(\theta/2)} \quad (5)$$

Where χ is a parameter directly connected with material properties and which describes the deviation of the r from linearity in respect to $cos^2(\theta/2)$. We calculated parameter χ for different numerical simulation data obtained using Waintal model for different l_T and l_L values and also for data obtained using Levy-Fert model when l_J = 1 nm (Fig. 6).

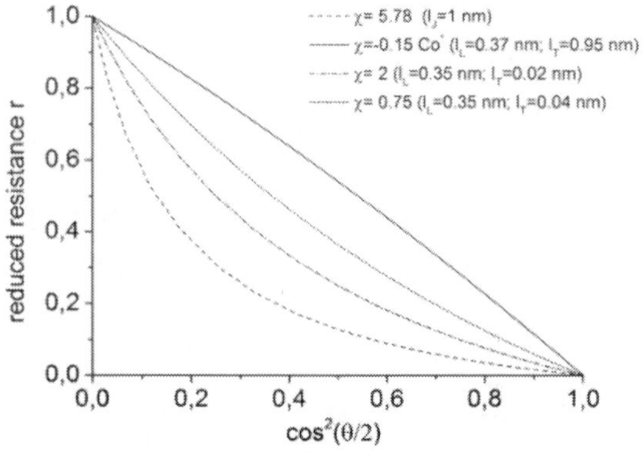

Fig.6. Angular dependence of reduced magnetoresistance for different values of l_T , l_L and l_J obtained from numerical simulation data using Levy-Fert and Waintal models. l_{sf}^F=50 nm.

For values of l_T =0.95 nm and l_L=0.37 nm corresponding to Co taken from *Waintal et al* the value of χ parameter equals to -0.15. However, the experimental data [12] shows that the value of χ is in the range of 1.2 to 1.6. The values of calculated χ close to ones observed in experiment correspond to the curves with χ equals to 2 and 0.75. These values are obtained in case if l_T has a rather small value. It means that transverse component of spin accumulation is destroyed by strong spin flip scattering on the interface.

4 Conclusions

Finite element approach was applied to investigate spin transport properties of structures with complex geometry. It was shown, that application of magnetic pillars is more efficient from the point of GMR effect value than application of a thin paramagnetic spacer. As non-uniformity of spin currents has a significant impact on the GMR effect value, it was shown that application of number of pillars is more effective than application of one pillar of equivalent diameter. Also non-uniform distribution of spin currents leads to occurrence of spin current vortices. Despite the fact that spin transport equations presented by *Waintal et al* give the same GMR effect as spin transport equations presented by Levy-Fert, they give a better description of angular dependence of the GMR effect.

References

1. M. N. Baibich, J. M. Broto, A. Fert, F. N. Van Dau, F. Petroff, P. Etienne, G. Creuzet, A. Friederich, and J. Chazelas, Phys. Rev. Lett. **61**, 2472 (1988).
2. G. Binasch, P. Grünberg, F. Saurenbach, and W. Zinn, Phys. Rev. B **39**, 4828 (1989).

3. R. E. Camley and J. Barnaś, Phys. Rev. Lett. **63**, 664 (1989).
4. P. M. Levy, S. Zhang, and A. Fert, Phys. Rev. Lett. **65**, 1643 (1990).
5. A. Vedyayev, M. Chshiev, N. Ryzhanova, B. Dieny, C. Cowache, and F. Brouers, J. Magn. Magn. Mater. **172**, 53 (1997).
6. S. Zhang, P. M. Levy, and A. Fert, Phys. Rev. Lett. **88**, (2002).
7. A. Vedyayev, B. Dieny, and N. Ryzhanova, Europhys. Lett. EPL **19**, 329 (1992).
8. T. Valet and A. Fert, Phys. Rev. B **48**, 7099 (1993).
9. C. Petitjean, D. Luc, and X. Waintal, Phys. Rev. Lett. **109**, (2012).
10. N. Strelkov, A. Vedyayev, N. Ryzhanova, D. Gusakova, L. D. Buda-Prejbeanu, M. Chshiev, S. Amara, N. de Mestier, C. Baraduc, and B. Dieny, Phys. Rev. B **84**, (2011).
11. J. . Slonczewski, J. Magn. Magn. Mater. **247**, 324 (2002).
12. L. Giacomoni, B. Dieny, W. P. Pratt, R. Loloee, and M. Tsoi, (n.d.).

https://doi.org/10.1051/epjconf/201818501022

THE INFLUENCE OF "INJECTED" AND "THERMAL" MAGNONS ON A SPIN WAVE CURRENT AND DRAG EFFECT IN HYBRID STRUCTURES

Igor Lyapilin[1,2] and *Mikhail* Okorokov[1]e-mail: Okorokovmike@gmail.com

[1]*Institute of Metal Physics, Ural Branch of the Russian Academy of Sciences, ul. S. Kovalevskoi 18, Yekaterinburg, 620990 Russia*

[2]*Ural Federal University, ul. Mira 19, Yekaterinburg, 620002 Russia*

Abstract. The formation of the two: injected and thermally excited, different in energies magnon subsystems and the influence of its interaction with phonons and between on drag effect under spin Seebeck effect conditions in the magnetic insulator part of the metal/ferromagnetic insulator/metal structure is studied. The analysis of the macroscopic momentum balance equations of the systems of interest conducted for different ratios of the drift velocities of the magnon and phonon currents show that the injected magnons relaxation on the thermal ones is possible to be dominant over its relaxation on phonons. This interaction will be the defining in the forming of the temperature dependence of the spin-wave current under spin Seebeck effect conditions, and inelastic part of the magnon-magnon interaction is the dominant spin relaxation mechanism.

1 Introduction

Recently, a great deal of attention is devoted to the investigation of thermally excited magnons, particularly in studies of the spin Seebeck effect (SSE) in Pt/YIG/Pt structures. The propagation of magnons in a magnetic insulator is described by two characteristic quantities: mean free path and spin diffusion length that are governed, in turn, by various magnon relaxation mechanisms. A series of experiments determine the range of the diffusion lengths as being quite wide: from $4\mu m$ to $120\mu m$ [1–5]. To explain so large values of the spin diffusion lengths, the number of papers has put forward several concepts of appearance, along with thermal magnons, of long-wave subthermal magnons in a magnetic insulator. For interpreting the experimental results, the works [4, 5] have adopted the hypothesis of the existence of the two magnon subsystems with different energies: thermal and subthermal magnons in a magnetic insulator. As to the temperature dependence of the Seebeck coefficient, it is non-monotonic and reaches its maximum within the range of 50 - 100K. And as the investigations have shown, it is affected by strength of a magnetic field, dimensions of the samples, and quality of the interface [5]. To explain the low temperature enhancement proposed the phonon-drag SSE scenario based on a theoretical model [6–8]. However, the observed transport of magnons over a long distance of up to millimeters in magnetic insulators implies a relative weak interaction with phonons and impurities, and the measurements of the temperature dependent thermal conductance of YIG single crystals show that the phonon contribution to the thermal conductivity reaches its maximum at around 25 K, which is 50 K lower than the observed peak in the SSE.

We have developed an "enhanced model" of the spin current and drag effect, which is based on the formation of two interacting magnon flows. Indeed, when SSE occurs in such structures, there are two magnon groups with different energies. The first group consists of magnons produced by an inhomogeneous temperature field applied to the magnetic insulator- "thermal" magnons. The energy of the magnons is of the order of a temperature ($k_B T$). Along with them, there are the magnons injected into the magnetic insulator due to inelastic scattering of spin-polarized electrons of the metal by localized spins located in the vicinity of the interface. The energy of the "injected" magnons is of the order of spin accumulation energy of conduction electrons of the metal $\Delta_s > k_B T$. Thus, it can be said that the magnetic system of the insulator forms another subsystem of "injected" magnons that are actually responsible for the SSE. As a consequence, in the presence of a non-uniform temperature field, there are three flows inside the magnetic insulator, namely, phonon and two magnon ones. The evolution of the magnon and phonon subsystems to equilibrium occurs due to the relaxation of both their energy and their moment. Obviously, the interaction between the flows gives rise to the drag effect.

https://doi.org/10.1051/epjconf/201818501022

2 Macroscopic momentum balance equations

Under the influence of a non-uniform temperature field (a temperature gradient) applied to the system, the magnons and phonons begin travelling; their macroscopic drift affects the propagation of the spin-wave current. The problem to be solved reduces to constructing and analyzing a set of macroscopic momentum balance equations for the magnon ($i = 1, 2$) and phonon ($i = p$) subsystems. To derive the macroscopic equations

$$\left\langle \dot{P}_i(\mathbf{r}) \right\rangle^t = Sp\{\dot{P}_i(\mathbf{r})\,\rho(t)\},\ (i = m_1, m_2, p), \qquad (1)$$

The expression for the non-equilibrium statistical operator (NSO) needs to be sought. According to [9], for $\rho(t)$, in the linear approximation in deviation from equilibrium, we have:

$$\rho(t) = \rho_q(t) - \int\limits_{-\infty}^{0} dt'\, e^{\epsilon t'} e^{it'L} \int\limits_{0}^{1} d\tau\, \rho_0^{\tau} \dot{S}(t+t') \rho_0^{-\tau}\, \rho_0. \qquad (2)$$

$\dot{S}(t) = \partial S(t)/\partial t + (i\hbar)^{-1}[S(t), H]$ is the entropy production operator. $e^{itL}A = e^{-itH/\hbar}\,A\,e^{itH/\hbar}$, $\rho_0 = \exp\{-S_0\}$. S_0 is the entropy operator for the equilibrium system. Thus, the problem boils down to finding the entropy production operator. We can write the entropy operator as

$$\begin{aligned} S(t) = \Phi(t) + \int d\mathbf{r}\{\beta_{m_i}(\mathbf{r},t)[H_{m_i}(\mathbf{r}) + H_{m_i m_i}(\mathbf{r}) +\\ + H_{m_i m_j}(\mathbf{r})] - \beta\mu_{m_i}(\mathbf{r},t)N_{m_i}(\mathbf{r}) +\\ + \beta_p(\mathbf{r},t)[H_p(\mathbf{r}) + H_{pp}(\mathbf{r}) + H_{pm_i}(\mathbf{r})] -\\ -\beta_{m_i}(\mathbf{r},t)V_{m_i}(\mathbf{r},t)P_{m_i}(\mathbf{r}) + \beta_p(\mathbf{r},t)V_p(\mathbf{r},t)P_p(\mathbf{r})\}. \end{aligned} \qquad (3)$$

Here

$$I^{*}_{m_i}(\mathbf{r}) = [I_{H_{m_i}}(\mathbf{r}) + I_{P_{m_i}}(\mathbf{r})V_{m_i}(\mathbf{r})],$$

$$I^{*}_{p}(\mathbf{r}) = [I_{H_p}(\mathbf{r}) + V_p(\mathbf{r})I_{P_p}(\mathbf{r})]$$

the density flows of the energy and momentum magnons and phonons. We also have taken into account that $\nabla\left(\beta_k(\mathbf{r},t)V_k(\mathbf{r},t)\right) \sim V_k(t)\nabla\beta_k(\mathbf{r},t)$. Here $\beta_{m_i}(\mathbf{r},t)$ are local-equilibrium values of the inverse temperatures of the magnon ($i = 1, 2$) and phonon subsystems $\beta_p(\mathbf{r},t)$. $\mu_{m_i}(\mathbf{r},t)$ is a local equilibrium value of the chemical potential of the magnons. $N(\mathbf{r}) = N_{m_1}(\mathbf{r}) + N_{m_2}(\mathbf{r})$ is the magnon number density operator. V_{m_i}, V_p are the drift velocities of the magnons ($i = 1, 2$) and the phonons respectively.

Inserting the entropy production operator (3) into the expression for the NSO (2), we average the operator equations (1) for momenta of the subsystems under discussion. Then we have

$$\begin{aligned} \left\langle \dot{P}_{m_1}(\mathbf{r}) \right\rangle^t = -\int\limits_{-\infty}^{0} dt'\, e^{\epsilon t'} \int d\mathbf{r}'\beta\{D_{m_1 m_1}(\mathbf{r},\mathbf{r}',t')\,\nabla\mu_{m_1}(\mathbf{r}',\bar{t}) +\\ + L_{(m_1,m_1 p)}(\mathbf{r},\mathbf{r}',t')\,\delta V_{m_1,p}(\mathbf{r}',\bar{t}) +\\ + L_{(m_1,m_1 m_2)}(\mathbf{r},\mathbf{r}',t')\,\delta V_{m_1,m_2}(\mathbf{r}',\bar{t})\}, \end{aligned} \qquad (4)$$

$$\begin{aligned} \left\langle \dot{P}_{m_2}(\mathbf{r}) \right\rangle^t = -\int\limits_{-\infty}^{0} dt'\, e^{\epsilon t'} \int d\mathbf{r}'\beta\{D_{m_2 m_2}(\mathbf{r},\mathbf{r}',t')\,\nabla\mu_{m_2}(\mathbf{r}',\bar{t}) +\\ + L_{(m_2,m_2 p)}(\mathbf{r},\mathbf{r}',t')\,\delta V_{m_2,p}(\mathbf{r}',\bar{t}) +\\ + L_{(m_2,m_1 m_2)}(\mathbf{r},\mathbf{r}',t')\,\delta V_{m_2,m_1}(\mathbf{r}',\bar{t})\}, \end{aligned} \qquad (5)$$

here $\bar{t} \equiv t + t'$. The first summands in the right-hand side of (4), (5) describe magnon-diffusion processes due to inhomogenius magnon chemical potential. The last two - magnon-phonon and "thermal" - "coherent" magnons scattering respectively

$$\begin{aligned} \left\langle \dot{P}_{p}(\mathbf{r}) \right\rangle^t = -\int\limits_{-\infty}^{0} dt'\, e^{\epsilon t'} \int d\mathbf{r}'\beta\{-D_{pp}(\mathbf{r},\mathbf{r}',t')\,\nabla\beta_p(\mathbf{r}',\bar{t}) +\\ + L_{(p,m_1 p)}(\mathbf{r},\mathbf{r}',t')\,\delta V_{p,m_1}(\mathbf{r}',\bar{t}) +\\ + L_{(p,m_2 p)}(\mathbf{r},\mathbf{r}',t')\,\delta V_{p,m_2}(\mathbf{r}',\bar{t}) +\\ + L_{(p,pp)}(\mathbf{r},\mathbf{r}',t')V_p(\mathbf{r}',\bar{t})\}. \end{aligned} \qquad (6)$$

The expression (6) describes the diffusive processes, processes of non-magnon relaxation of phonons and the magnon-phonon scattering. Here we introduce the notations: $\delta V_{ik} = V_i - V_k$, $D_{pp}(\mathbf{r},\mathbf{r}',t') = \beta(\nabla I_{P_p}(\mathbf{r}), I^{*}_p(\mathbf{r}',t'))$ and

$$(A,B) = \int\limits_{0}^{1} d\lambda Sp\{A\rho_0^{\lambda}\,\Delta B\,\rho_0^{1-\lambda}\}.$$

The momentum balance equations allow conducting the analysis of how the interaction between the subsystems at hand affects the implementation of the drag effect. We introduce the average values of the forces induced by the chemical potential and temperature gradients

$$F_{m_i}(\mathbf{r}) = \int\limits_{-\infty}^{0} dt'\, e^{\epsilon t'} \int d\mathbf{r}'\, D_{m_i m_j}(\mathbf{r},\mathbf{r}',t')\,\nabla\mu_{m_j}(\mathbf{r}',\bar{t}),$$

$$F_p(\mathbf{r}) = \int\limits_{-\infty}^{0} dt'\, e^{\epsilon t'} \int d\mathbf{r}' D_{pp}(\mathbf{r},\mathbf{r}',t')\,\nabla\beta_p(\mathbf{r}',\bar{t}).$$

Besides, introduce the inverse times of the magnon and phonon momentum relaxation caused by interaction with phonons processes of non-magnon relaxation of phonons. Let us designate them as $\omega_{(mp)}$, and $\omega_{(pp)}$, respectively [9]

$$\omega_{(\gamma,v)} = (P_\gamma, P_\gamma)^{-1} \int\limits_{-\infty}^{0} dt'\, e^{\epsilon t'} (\dot{P}_{(\gamma,v)}, \dot{P}_{(\gamma,v)}(t')),$$

$\gamma = m_1, m_2, p$. We restrict ourselves to the discussion of a stationary case. For this purpose, we average the balance equations over time t. To start the analysis, we consider the simplest case when the drift velocities of the magnon systems are equal: $V_{m_1} = V_{m_2} \equiv V_m$ and $\beta_{m_1} = \beta_{m_2}$. The balance equation for the magnon momentum acquires the form

$$F_m = V_m \mathcal{P}_m \frac{\omega_{(p,pp)}\,\omega_{(m,mp)}}{\omega_{(m,mp)} + \omega_{(p,pp)}} \equiv V_m \mathcal{P}_m \Omega. \quad (7)$$

where $\mathcal{P}_m \equiv (P_m, P_m)$, $\mathcal{P}_p \equiv (P_p, P_p)$. Finally, the dragging leads to the change in frequency of the magnon-phonon collisions, and the quantity Ω is the inverse relaxation time of the magnon momentum by non-equilibrium phonons. From the expression (7) it follows that the drag effect has an influence on the magnon-phonon collision frequency. The criterion of realizing the drag effect consists in the requirement $\omega_{(m,mp)} > \omega_{(p,pp)}$ that coincides with the solution of the kinetic equation [10].

Another limiting case corresponds to the situation when the drift velocities of thermal magnons and phonons are equal to $V_{m_2} = V_p$ and $(\beta_{m_2} = \beta_p)$. In this case, thermal magnons and phonons form one subsystem. From balance equations we obtain

$$F_{m_1} = V_{m_1} \frac{\omega_{(p,pp)}\,[\omega_{(p,mp)} + \omega_{(m,m_1m_2)}]}{\omega_{(p,mp)} + \omega_{(p,pp)}}. \quad (8)$$

From the expression (8) it follows that if $\omega_{(p,pp)} \gg \omega_{(p,mp)}$ then $F_1 \sim \omega_{(p,mp)}$ and $F_1 \sim \omega_{(m,m_1m_2)}$ if $\omega_{(p,mp)} \ll \omega_{(m,m_1m_2)}$. If the opposite inequality, when $\omega_{(p,pp)} \ll \omega_{(p,mp)}$ then $F_1 \sim \omega_{(p,pp)}\,[1 + \omega_{(m,m_1m_2)}/\omega_{(p,mp)}]$ and $F_1 \sim \omega_{(p,pp)}$ if $\omega_{(m,m_1m_2)} \ll \omega_{(p,mp)}$ and $F_1 \sim \omega_{(p,pp)}\,\omega_{(m,m_1m_2)}/\omega_{(p,mp)}$ when $\omega_{(m,m_1m_2)} \gg \omega_{(p,mp)}$.

Now we look into the drag effect in the event of two magnon and one phonon systems. Then, the momentum balance equations can be written as follows. The set of the equations (5), (6) implies

$$F_{m_1} = \{\omega_{(m_1,mp)} + \omega_{(m,m_1m_2)} - \frac{\omega_{(m_1,mp)}\omega_{(m_1,mp)}}{\Omega}\} V_{m_1} - \{\frac{\omega_{(m_1,mp)}\omega_{(m_2,mp)}}{\Omega} + \omega_{(m,m_1m_2)}\} \cdot \{\frac{F_{m_2} + (\omega_{(m,m_1m_2)} + \omega_{(m_1,mp)}\omega_{(m_2,mp)}/\Omega) V_{m_1}}{\omega_{(m_2,mp)} + \omega_{(m,m_1m_2)} - \omega_{(m_2,mp)}\omega_{(m_2,mp)}/\Omega}\}, \quad (9)$$

where $\Omega = \omega_{(m_1,mp)} + \omega_{(m_2,mp)} + \omega_{(p,pp)}$.

Let the energy transfer channels from the magnon subsystems to the phonon subsystem be equal $\omega_{(m_1,mp)} = \omega_{(m_2,mp)} = \omega_{(m,mp)}$, $V_{m_1} = V_m$. In this case we have

$$F_{m_1} = \{\omega_{(m,mp)} + \omega_{(m,m_1m_2)} - \omega_{(m,mp)}/\Omega\} V_m - \{\omega_{(m,mp)}/\Omega + \omega_{(m,m_1m_2)}\} \times \{\frac{F_{m_2} + (\omega_{(m,m_1m_2)} + \omega_{(m,mp)}/\Omega) V_m}{\omega_{(m,mp)} + \omega_{(m,m_1m_2)} - \omega_{(m,mp)}/\Omega}\}. \quad (10)$$

Here $\Omega = 2 + \omega_{(p,pp)}/\omega_{(m,mp)}$. If $\omega_{(p,pp)} \gg \omega_{(m,mp)}$, then

$$F_{m_1} = \{\omega_{(m,mp)} + \omega_{(m,m_1m_2)}\} V_m - \omega_{(m,m_1m_2)} \cdot \{\frac{F_{m_2} + \omega_{(m,m_1m_2)}}{\omega_{(m,mp)} + \omega_{(m,m_1m_2)}}\}. \quad (11)$$

The expression (11) claims that the spin-wave current $\sim F_1$ is determined by the relations between the correlation functions $\omega_{(m,m_1m_2)}$ and $\omega_{(m,mp)}$. As it follows from the expression (11) that if $\omega_{(m,m_1m_2)} \ll \omega_{(m,mp)}$ then $F_1 \sim \omega_{(m,mp)}$. In this case magnon-phonon interaction is the dominant channel of a magnon relaxation. If we have the opposite inequality $\omega_{(m,m_1m_2)} \gg \omega_{(m,mp)}$ then $F_1 \sim \omega_{(m,m_1m_2)}$. In this case the interaction between "injected" and "thermal" magnons is the dominant channel of a magnon relaxation. Moreover, the inelastic scattering of the "injected" magnons by "thermal" ones can be regarded as scattering by impurity centers whose concentration is temperature-varied. This interaction will determine the temperature-field behavior of the spin-wave current under the conditions of the Seebeck spin effect.

Conclusion. The analysis of the macroscopic momentum balance equations of the systems of interest conducted for different ratios of the drift velocities of the magnon and phonon currents show that the "injected" magnons relaxation on the "thermal" ones is possible to be dominant over its relaxation on the phonons. This interaction will be the defining in the forming of the temperature dependence of the spin-wave current under SSE conditions, and inelastic part of the magnon-magnon interaction is the dominant spin relaxation mechanism.

Acknowledgments. The given work has been done as the part of the state task on the theme "Electron" 01201463330 (project 12-T-2-1011) with the support of the Ministry of Education of the Russian Federation (Grant 14.Z50.31.0025, 16-02-00044) and project no. 15- 17-2-17).

References

[1] S. R. Boona, J. P. Heremans, Phys. Rev. B **90**, 064421 (2014).
[2] L. J. Cornelissen, J. Liu, R. A. Duine, J. B. Youssef, B. J. van Wees, Natur. Phys. **11**, 1022 (2015).
[3] B. L. Giles, Z. Yang, J. S. Jamison, R. C. Myers, Phys. Rev. B **92**, 224415 (2015).
[4] A. Kehlberger, U. Ritzmann, D. Hinzke, E. J. Guo, J. Cramer, G. Jakob, M. C. Onbasli, D. H. Kim, C. A. Ross, M. B. Jungfleisch, B. Hillebrands, U. Nowak, M. Kleaui, Phys, Rev. Lett. **115**, 096602 (2015).
[5] T. Kikkawa, K. Uchida, S. Daimon, Y. Shiomi, H. Adachi, Z. Qiu, D. Hou, X. F. Jin, S. Maekawa, E. Saitoh, Phys. Rev. B **88**, 214403 (2013).
[6] L. E. Gurevich and G. A. Roman, Sov. Phys. Solid State **8**, 2102 (1967).

[7] F. J. Blatt, P. A. Schroeder, C.L. Foiles and D. Greig, Thermoelectric Power of Metals (New York: Plenum 1976).
[8] M. Bailyn, Phys. Rev. **126**, 2040 (1962).
[9] I.I. Lyapilin, M.S. Okorokov, V. V. Ustinov, Phys.Rev. B 91, 195309 (2015).
[10] R. J. Kubo, Peport on Progr in Phys. **29**, 1 (1966).

INVESTIGATION OF ANGULAR AND TEMPERATURE DEPENDENCE OF SPIN-WAVE RESONANCE SPECTRA IN MULTILAYER MAGNETIC FILMS

Andrey Bazhanov

Mordovian State University, Saransk, Russia

Abstract. In this letter we study the dependences of the resonance field H_n and the linewidth $2\Delta H_n$ of the spin-wave resonance (SWR) modes in two-layer ferrite-garnet films with dissipative and mixed spin-pinning mechanisms. It is established that for films with a dissipative mechanism of spin pinning, the increase in the linewidth at $\theta_H = 35^0$ is associated with the maximum misorientation between the external magnetic field and the direction of the equilibrium magnetization of the sample. For films with a mixed spin-pinning mechanism, an increase in the linewidth $2\Delta H_n$ with increasing angle θ_H is associated with an increase in the depth of penetration of the spin wave into the pinning layer for such films. On the temperature dependence of the linewidth $2\Delta H_n$ for samples with a mixed fixation mechanism at a perpendicular orientation, an increase in the width of the absorption peak lines is observed. And with an increase in the peak number, the maximum of the linewidth shifts to the low-temperature region. With parallel orientation, the linewidth $2\Delta H_n$ in a wide temperature range from 20°C to 200°C remains practically unchanged.

1 INTRODUCTION

The essence of ferromagnetic resonance is the selective absorption of the energy of a high-frequency electromagnetic field by a ferromagnet. The presence of an exchange interaction in magnetically ordered substances leads to a strong correlation between the orientations of the spins of neighboring atoms and the existence of exchange spin waves in such systems. Under certain boundary conditions, a uniform variable field can excite spin waves with a discrete set of wave numbers. This appearance was called the spin-wave resonance (SWR).

Despite the large number of papers devoted to spin-wave resonance (SWR) in thin films, the angular and temperature dependences of the SWR spectra remain far from being studied. Almost in all works, analysis and experimental studies of the SWR spectra are carried out for perpendicular or parallel orientations of a constant magnetic field relative to the plane of the film. At the same time, an analysis of the angular and temperature dependences of the resonance field H_n and the linewidth $2\Delta H_n$ makes it possible to better understand the properties and features of the excitation of spin waves in magnetic films. For example, a change in the angle between $\vec{H}$ and the film as well as an increase in temperature can be used for certain values of the layer parameters as one of the ways of smoothly changing the degree of pinning of the spins.

Therefore, the purpose of this paper was to study the angular and temperature dependences of the SWR spectra in multilayer magnetic films.

2 EXPERIMENT

The investigations were carried out on double-layer single-crystal films of ferrit-garnets. The first type of samples had the following parameters: the first layer close to the substrate (the excitation layer of harmonic standing spin-wave (SW) modes) of the composition $Y_{2.98}Sm_{0.02}Fe_5O_{12}$ had thickness $h = 1.0$ μm, saturation magnetization $4\pi M_1 = 1740$ Gs, Gilbert damping parameter $\alpha_1 = (\Delta H\gamma/\omega) = 0.0009$, gyromagnetic ratio $\gamma_1 = 1.76\cdot10^7$ Oe^{-1}s^{-1}, exchange interaction constant $A = 3.7\cdot10^{-7}$ erg/cm, effective uniaxial anisotropy field $H_k^{eff} = (2K_u/M) - 4\pi M_s = -1715$ Oe, the second layer of the composition $(LaEr)_3(FeGa)_5O_{12}$ had $h = 1.2$ μm, $4\pi M_2 = 450$ Gs, $\alpha_2 = (\Delta H\gamma/\omega) = 0.84$, $\gamma_2 = 1.76\cdot10^7$ Oe^{-1}s^{-1}, $A = 3.7\cdot10^{-7}$ erg/cm, $H_k^{eff} = (2K_u/M) - 4\pi M_s = -78$ Oe. Here ΔH - the half-width of the absorption line, ω is the circular frequency

* Corresponding author: bajanovag@mail.ru

of the microwave field. In this sample, the dominant mechanism for the spin pinning was the dissipative mechanism. Such a pinning mechanism arises in multilayer films with widely differing values of the Gilbert damping parameter in the layers. The presence of an exchange coupling between the layers leads to the appearance of a node of a standing spin wave at the interface of the layers or near it. One of the qualitative differences between the dissipative mechanism for the spin pinning from the dynamic one lies in the fact that it does not depend on the orientation of the external magnetic field relative to the film, which is related to the isotropy of the damping parameter. For any orientation, the excitation region of standing harmonic spin waves is localized in the layer with a small one.

The second sample had the following parameters: the first layer had a composition $Y_{2.98}Sm_{0.02}Fe_5O_{12}$, $h = 0.46$ μm, $4\pi M_s = 1740$ Gs, $\gamma = 1.76\times10^7$ Oe^{-1}s^{-1}, $H_k^{eff} = -1715$ Oe, $\alpha = (\Delta H\gamma/\omega) = 0.003$, $A = 3.7\cdot10^{-7}$ erg / cm. The second layer (the spin-pinning layer) had composition $(SmEr)_3Fe_5O_{12}$ and material parameters: $h = 1.2$ mkm, $4\pi M_s = 1330$ Gs, $\gamma = 1.38\cdot10^7$ Oe^{-1}s^{-1}, $H_k^{eff} = 96$ Oe, $\alpha = 0.2$, $A = 3.7\cdot10^{-7}$ erg/cm.

The third sample the first layer close to the substrate had the composition $Y_3Fe_{4.97}Ge_{0.03}O_{12}$, $h = 0.343$ mkm, $4\pi M_s = 1680$ Gs, $\gamma = 1.76\times10^7$ Oe^{-1}s^{-1}, $H_k^{eff} = (2K_u/M) - 4\pi M_s = -1620$ Oe, $\alpha = (\Delta H\gamma/\omega) = 0.0009$, $A = 3.7\cdot10^{-7}$ erg/cm. The second layer (the spin-pinning layer) had composition $(YSmLuCa)_3(FeGe)_5O_{12}$ and material parameters: $h = 2.0$ mkm, $4\pi M_s = 560$ Gs, $\gamma = 1.74\cdot10^7$ Oe^{-1}s^{-1}, $H_k^{eff} = 980$ Oe, $\alpha = 0.12$, $A = 3.7\cdot10^{-7}$ erg/cm. In the second and third samples the dominant mechanism for pinning the spins was the mixed mechanism of spin pinning (dissipative and dynamic).

In films with different fields of homogeneous resonance, the dynamic mechanism of spin pinning is realized in the layers. With a dynamic mechanism, a microwave field excites localized modes that are harmonic in one layer (a layer with a large value of the field of a homogeneous resonance $\vec{H}_{01}$) and exponentially fall off in another layer, which in external fields larger than its homogeneous resonance field $\vec{H}_{02}$ is for spin waves by a reactive (elastic) medium and thereby ensures the pinning of the spins. The interval of fields in which spin waves are intensively excited, with dynamic pinning mechanism is limited by the values $\vec{H}_{01}$, $\vec{H}_{02}$. In our films different fields of homogeneous resonance are realized and the damping parameters in the layers vary greatly; therefore, both the dissipative and the dynamic mechanism of spin pinning are appear, i.e. a mixed mechanism for the spin pinning is realized.

The registration of the SWR spectra was carried out on a RE-1301 radio spectrometer at the frequency of the microwave field $9.34\cdot10^9$ Hz, and the magnetic field was measured with the help of the Hall's magnetic induction meter RSh1-10. The SWR spectra were recorded at different angles θ_H of the external magnetic field relative to the film plane. For carrying out the temperature studies, a special thermostat attachment was used, which allows setting the sample temperature in the range from -100^{0}C to 300^{0}C.

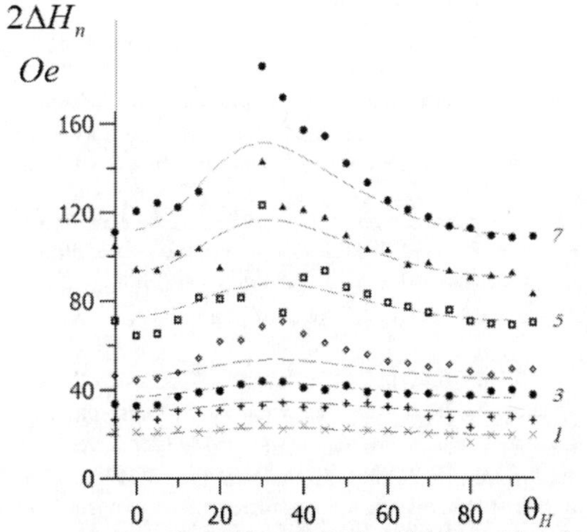

Fig.1. Dependence of the linewidth in the film with the dissipative mechanism of spin pinning on the angle between the external magnetic field and the film plane (the points - the experiment, the dashed line - the calculation). Figures for curves - numbers of spin-wave modes.

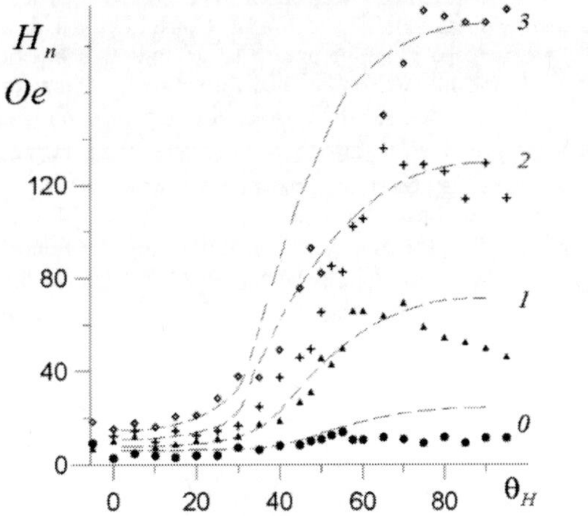

Fig.2. Dependence of the linewidth in a film with a mixed mechanism of spin pinning on the angle between the external magnetic field and the film plane (the points - the experiment, the dashed line - the calculation). Figures for curves - numbers of spin-wave modes.

3 RESULTS AND DISCUSSION

As a result of experimental studies the following was established. In a sample with a dissipative mechanism of spin pinning on the anguar dependences of the linewidth a maximum of the linewidth $2\Delta H_n$ is observed for $\theta_H = 35^0$ (Fig. 1).

In a sample with a mixed mechanism of spin pinning with an increase in the angle between the external magnetic field and the film plane θ_H the fields of uniform resonance in the layers gradually approach each other. The number of spin-wave modes in the excitation layer decreases. On the angular dependences of the linewidth, a monotonous increase in the width of the line is observed. In this case, the maximum width of the lines is achieved with a parallel orientation of the external field relative to the film plane $\theta_H = 90^0$ (Fig. 2).

On the temperature dependences of the resonance fields H_p in films with a mixed spin-pinning mechanism for a perpendicular orientation of the external magnetic field relative to the film plane a monotonic decrease in the resonance fields is observed. In this case, the width of the line $2\Delta H_n$ of the zero- and first- modes with increasing temperature remains practically unchanged. However, the linewidth of the second mode increases monotonically at a temperature T = 100 ° C reaches a maximum at T = 150 ° C and decreases monotonically to the previous values with a further increase in temperature. An analogous behavior of the linewidth is also observed for subsequent lines of spin-wave modes. In this case, the maxima on the dependences $2\Delta H_n$ shift towards smaller temperatures with increasing absorption peak number (Fig. 3, 4).

The linewidth was calculated as follows. According to [1-4], the width of the absorption line of each SW-mode $2\Delta H_n$ can be expressed in terms of the effective damping parameter of the n-th spin-wave mode:

$$2\Delta H_n = 2\alpha_n^{eff}\frac{\omega}{\gamma_i}, \qquad (1)$$

$$\alpha_n^{eff} = \frac{\dfrac{\alpha_1}{\gamma_1 M_1}\displaystyle\int_0^{h_1} m_{1n}^2\,dz + \dfrac{\alpha_2}{\gamma_2 M_2}\displaystyle\int_{h_1}^{h_1+h_2} m_{2n}^2\,dz}{\dfrac{1}{\gamma_1 M_1}\displaystyle\int_0^{h_1} m_{1n}^2\,dz + \dfrac{1}{\gamma_2 M_2}\displaystyle\int_{h_1}^{h_1+h_2} m_{2n}^2\,dz}, \qquad (2)$$

The distributions of the variable magnetization in the excitation layer $m_1(z)$ and in the pinning layer $m_2(z)$ were found as a result of solving the equations of the equilibrium orientation of the magnetization and the dispersion relations written for each of the layers of a two-layer magnetic film and also taking into account the boundary conditions at the outer boundaries of the film and at the interlayer boundary.

The calculation carried out showed that in films with a dissipative mechanism of spin pinnng an increase in the linewidth $2\Delta H_n$ at $\theta_H = 35^0$ is observed [3-4]. It is established that this increase is associated with an increase in the misorientation angle between the external magnetic field and the direction of the equilibrium orientation of the magnetization of the sample $\theta_H - \theta_M$.

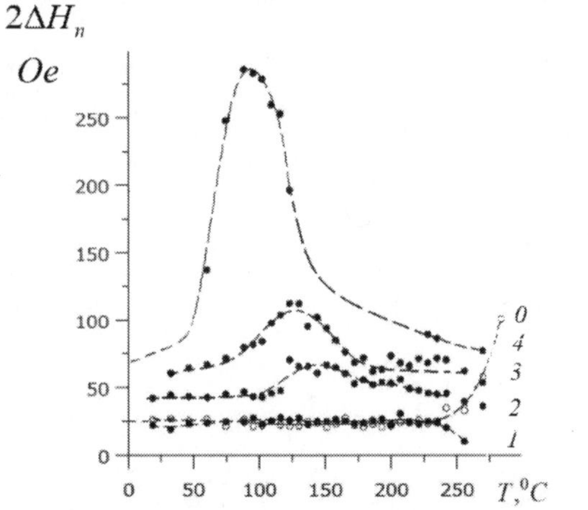

Fig.3. Dependence of the linewidth in a film with a mixed mechanism of spin pinning on temperature at a perpendicular orientation (points - experiment, dashed line - calculation). Figures for curves - numbers of spin-wave modes.

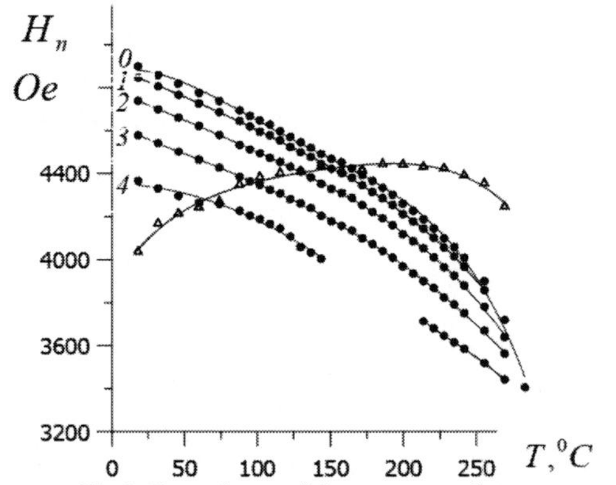

Fig.4. Dependence of the resonance field in a film with a mixed mechanism of spin pinning on temperature at a perpendicular orientation (points - experiment, solid line - calculation). Figures for curves - numbers of spin-wave modes.

Approximately equal values $2\Delta H_n$ for the perpendicular and parallel orientations are related to the same values of the depth of penetration of the spin wave into the pinning layers for both orientations.

In films with a mixed spin-pinning mechanism for a perpendicular orientation of the external field relative to the film plane ($\theta_H = 0^0$), the linewidth of all modes have values from 0 to 20 Oe. With an increase in the angle θ_H, the linewidth for all modes monotonically increases. The maximum width of the lines is achieved with a parallel orientation of the external field relative to the film plane ($\theta_H = 90^0$), which is due to an increase in the depth of penetration of the spin wave into the pinning layer for this orientation.

To explain the dependences of the linewidth on temperature, the additional calculation was carried out. The decrease in the magnetization of the excitation layer of spin waves with increasing temperature is described by the Bloch law:

$$M(T) = M(0)\left(1 - \left(\frac{T}{T_k}\right)^{3/2}\right), \quad (3)$$

where $M(0)$ is the saturation magnetization at 0 K, $M(T)$ is the saturation magnetization at a given temperature, T_k is the Curie temperature of the given layer.

It can be seen from expression (3) that as the temperature increases, the magnetization of the substance decreases monotonically, as a result the resonance fields decrease with perpendicular orientation of the external field relative to the film, and increase with parallel.

The calculation showed that at a perpendicular orientation there is an intersection of the resonance field of the zero mode of the pinning layer H_{02} and the resonance field of the corresponding SW-mode of the excitation layer H_n (fig. 4). As a result, the depth of penetration of this SW-mode into the pinning layer increases, and accordingly, in the same temperature range, the linewidth of this SW-mode increases. For example, the intersection of the resonance fields H_{02} and H_3 for the third SW-mode occurs in the interval $T \approx 100 \div 120^0C$, in the same interval the linewidth increases, for the second SW-mode - $T \approx 125 \div 150^0C$, etc. And with an increase in the number of the SW-modes, the maximum of the linewidth shifts to the low temperature region, which is associated with an increase in the penetration depth of the spin wave into the anchoring layers at these temperatures (fig.3, 4).

For a parallel orientation, the intersection of resonance fields of peaks corresponding to different layers does not occur (figs 5, 6). There is also no increase in the linewidth of the corresponding peaks.

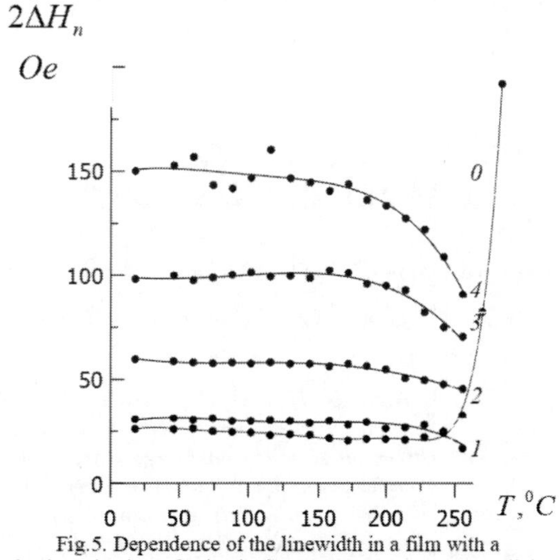

Fig.5. Dependence of the linewidth in a film with a mixed mechanism of spin pinning on temperature at a parallel orientation (points - experiment, solid line - calculation). curves - numbers of spin-wave modes.

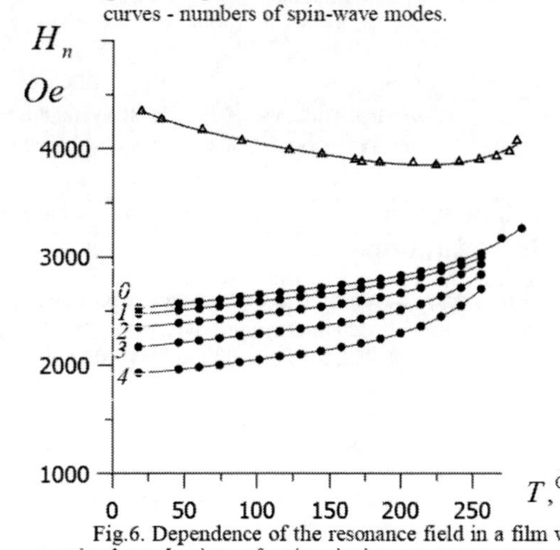

Fig.6. Dependence of the resonance field in a film with a mixed mechanism of spin pinning on temperature at a parallel orientation (points - experiment, solid line - calculation). Figures for curves - numbers of spin-wave modes.

4 CONCLUSION

Thus, we can draw the following conclusions:

1. It is established that for films with a dissipative mechanism of spin pinning, the increase in the linewidth $2\Delta H_n$ at $\theta_H = 35^0$ is associated with the maximum misorientation between the external magnetic field and the direction of the equilibrium magnetization of the sample $\theta_H - \theta_M$. Approximately equal values $2\Delta H_n$ for the perpendicular and parallel orientations are related to the same values of the depth of penetration of the spin wave into the pinning layer for both orientations.

2. It is established that for films with a mixed spin-pinning mechanism, an increase in the linewidth $2\Delta H_n$ with increasing angle θ_H is associated with an increase in the depth of penetration of the spin wave into the pinning layer for such films.

3. On the temperature dependence of the linewidth for samples with a mixed pinning mechanism at a perpendicular orientation, an increase in the linewidth of the absorption peak lines is observed. And with an increase in the peak number, the maximum linewidth shifts to low temperatures.

4. With parallel orientation, the linewidth of the SW-modes in a wide temperature range from 20^{0}C to 200^{0}C remains practically unchanged.

1. A.M. Zyuzin, A.G. Bazhanov, etc, Physic Solid State, **42**, 7 (2000).
2. B. Hoekstra, R.P. Stapele, J.M. Robertson. J. Appl. Phys., **48**, 1 (1977).
3. A.M. Zyuzin, A.G. Bazhanov, Bul. of Russian Academian. Phys. series, **77**, 10 (2013).
4. A.M. Zyuzin, A.G. Bazhanov, Tech. Phys. Letters, **32**, 10 (2006).
5. A.M. Zyuzin, A.G. Bazhanov, JETP Letters, **63**, 7 (1996).

Experimental study of microwave properties of composites filled with Permalloy powder

Stanislav Y. Bobrovskii*, *Victor A.* Garanov, *Andrei S.* Naboko, *Alexei V.* Osipov *and Konstantin N.* Rozanov

Institute for Theoretical and Applied Electromagnetics RAS

Abstract. Frequency dependencies of permittivity and permeability of composites filled with Permalloy powder with various concentrations and shapes of particles have been measured and discussed. The powder particles of various shapes were obtained by mechanical milling of industrial Permalloy powder in organic media. Paraffin wax was used as a matrix for producing the composites. A shape data of the particles was obtained with an electron microscope. The permittivity and permeability were measured in the frequency range of 0.13 to 10 GHz. The shape of the particles changes from sphere to platelet with increase of milling time. The change leads to a shift of the ferromagnetic resonance (FMR) frequency and magnetic loss peak. It is shown that for larger particles, a decrease of the Q-factor of the ferromagnetic resonance is observed, which is explained by the strong influence of the skin effect. With increase of the concentration of the Permalloy particles in composite, the magnetic loss peak shifts to low frequencies. When the spherical particles are flattened, the resonance frequency shifts to high frequencies, which indicates that the shift to high frequencies due to the decrease of the skin effect and the change in the magnetic structure is stronger than the shift to the low frequencies by changing the shape and reducing the demagnetization.

1 Introduction

The frequency dependencies of microwave permeability have a complex shape in many materials. Analysis of such dependencies is a challenging problem [1-4], that allows, in particular, obtaining estimates of the largest feasible values of microwave permeability. In this regard, the relations between the parameters of the magnetic loss peak are important, because such relations allows the number of independent variables used in describing the magnetic spectra to be reduced.

In the simplest case, the dependence of the complex permeability $\mu=\mu'+i\mu''$ on frequency f obeys the Lorentzian dispersion law:

$$\mu(f) = 1 + \frac{\mu_s - 1}{1 - \frac{if}{f_1} - (\frac{f}{f_0})^2}, \qquad (1)$$

where μ_s is the static permeability, f_0 is the resonant frequency, f_1 is the relaxation frequency. According to (1), the real permeability is close to μ_s at frequencies below f_0, and falls sharply at higher frequencies. In this way, f_0 can be considered as the cut-off frequency of the permeability. For many magnetic materials μ_s and f_0 are related by Snoek's law [5]:

$$(\mu_s - 1)f_0 = (2/3)\gamma\, 4\pi M_0, \qquad (2)$$

where $4\pi M_0$ is the saturation magnetization of the material, $\gamma\approx3$ GHz/kOe. Therefore, the higher the static permeability of the material, the lower frequencies where dynamic magnetic properties starts to disappear. A measure of the high-frequency magnetic properties of a material is the quantity on the right-hand side of (2), is Snoek's parameter S.

Snoek's law relates the parameters of the permeability spectrum – the amplitude and position of the absorption peak, which depend substantially on the crystalline and magnetic structure of the material, with the saturation magnetization determined solely by the composition. Snoek's law follows from the Landau-Lifshitz equation that describes the ferromagnetic resonance in a single-domain spherical particle.

If the magnetic particles are in shape of thin plates and their magnetic moment is in the plane of the plate, then[6]:

$$(\mu_s - 1)f_0^2 = \left(\frac{1}{2}\right)(\gamma\, 4\pi M_0)^2, \qquad (3)$$

where the factor 1/2 corresponds to the uniform orientation of the magnetic moments in the plane.

It is seen from (3) that thin ferromagnetic films can possess the highest values of microwave permeability of all magnetic materials. The right-hand side of (3) is a measure of the high-frequency properties of materials based on thin films and is called Acher's parameter A.

Composites with conducting particles are of interest for a number of practical applications. This study is devoted to the composites filled with Permalloy powder. The effect of concentration, size and shape of inclusions on the properties of composites is investigated.

* Corresponding author: sbobrowskiy@yandex.ru

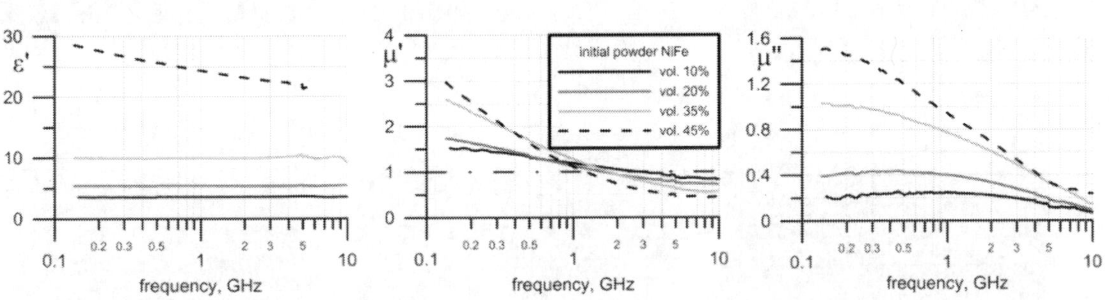

Fig. 1. The measured frequency dependencies of the permittivity and permeability of the composites with initial Permalloy powder with various volume concentration.

2 Experimental details

The shape of the particles of the powders under study was transformed by high-energy mechanical milling of the Permalloy powder. The initial powder consisted of spherical particles and was separated into four fractions by particle size: less than 40 μm, from 40 to 63 μm, 63 to 90
μm and above 90 μm. Milling time varied from 20 to 320 seconds. Images of powders were obtained using an electron microscope.

Samples for a measurement were prepared by manual stirring the necessary amount of Permalloy powder and molten paraffin wax to form a uniform distribution of inclusions in the composite. The permittivity ε of paraffin wax is 2.25. Then the mixture was placed in a mold to give the sample a shape corresponding to the measuring cell. The thickness of the samples was usually chosen in the range from 1 to 5 mm.

Measurements of material parameters (permittivity and permeability) were performed using a standard coaxial line of 7/3 mm in the frequency range of 0.13 to 10 GHz and HP8720A vector analyzer. The permittivity and permeability were calculated from the measures of the reflection coefficients, with the short-circuited and open-circuited sample [7].

3 Results

Figure 1 shows the measured frequency dependencies of material parameters of composites for different volume concentrations of initial Permalloy powder. The imaginary permittivity is close to zero and is of no interest. With an increase of the concentration of inclusions, the frequency of the resonance and the magnetic loss peak shifts to low frequencies, which is in agreement with the effective medium theory [8].

The permittivity of the composites under study does not depend on the frequency in the measured frequency range, and dielectric loss is small. The values of the real permeability at high frequencies drop below unity up to the level corresponding to the volume concentration of inclusions, which indicates a strong influence of the skin effect [9]. The magnetic loss peak is half the amplitude of the ferromagnetic resonance, so as the values of $\mu_s - 1$. The shift to low frequencies with increasing concentration can be associated with proximity to the percolation threshold. The ratio between the absorption maximum and the amplitude indicates a high damping of the resonance: the shape of the curve corresponds to the Debye formula, so the spectrum is very spread.

In Fig. 2, the measured frequency dependencies of the material parameters of composites with particles of different sizes are shown.

The permittivity of all powders is almost the same, which indicates the same shape of the particles. The permeability shifts to low frequencies, both in terms of the resonance frequency ($\mu' = 1$) and of the frequency of the maximum loss, which corresponds to the ferromagnetic resonance theory. For particles larger than 90 μm, a decrease in the amplitude without a shift in frequency is

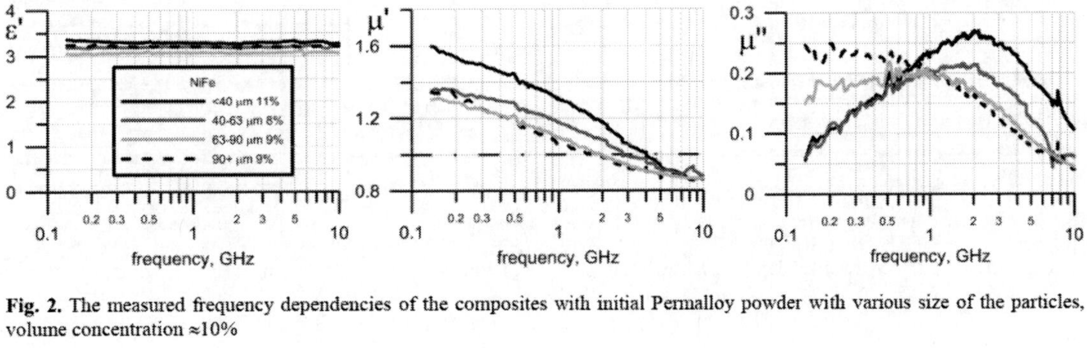

Fig. 2. The measured frequency dependencies of the composites with initial Permalloy powder with various size of the particles, volume concentration ≈10%

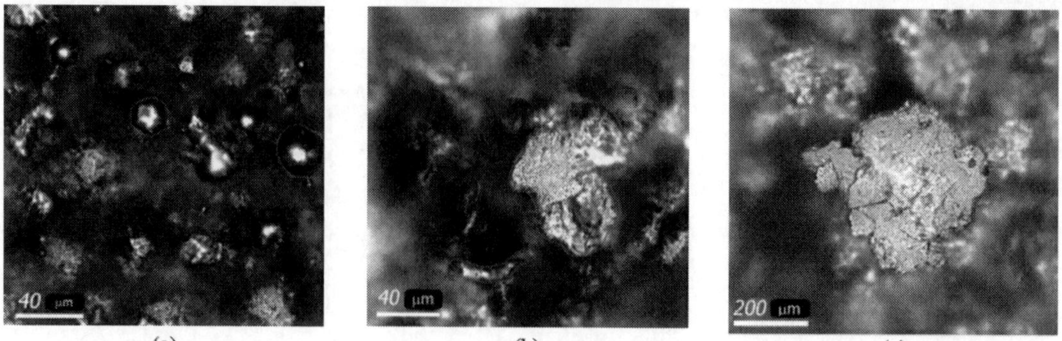

(a) (b) (c)

Fig.3. Optical representation of the initial powder with particle size up to 40 µm (a), after 80 s milling (b), after 160 s milling (c)

observed, as compared with 63-90 µm particles. The frequency of ferromagnetic resonance for the smallest particles is 4 GHz or higher. Since for conductive particles the permittivity is determined only by the shape, then the equality of ε implies the similarity of the shape. As is shown below, the powder particles of all fractions are in the form of spheres. Consequently, the cause of the change in permeability is the influence of size, i.e. skin effect. This is confirmed by a decrease in the Q-factor with increasing size, as can be seen from the broadening of the magnetic loss peak.

A method to avoid the influence of the skin effect is to change the shape of the particles by flattening. This should reduce the skin effect, which is determined by the least particle size, and the ferromagnetic resonance should shift to the lower frequencies by reducing the effective demagnetization factor.

To determine the shape of the particles under study, an optical image of the powder of the fraction up to 40 µm was obtained with a confocal microscope (Fig. 3a). The image was made with x50 lens. A sample of the powder consists of granules with a diameter of 10-30 µm (in the plane, i.e., in area). The variation in the focal length in a confocal microscope makes it possible to estimate the size of the particles in height, which is also 20-30 µm. Consequently, the considered particles have a near-spherical shape; the aspect ratio is close to 1:1.

After milling for t_m=80 s (Fig. 3b), the shape of the powder particles changed from spherical to platelet. The particle diameter in the plane is 50-80 µm, the thickness is 10 µm, while the particles have a slight bending in the plane. The aspect ratio is 1:5-1:8, therefore the particles are flattening during milling. In Fig. 3c there is an optical image of the powder milled for t_m=160 s. After milling for 160 seconds, the particle thickness is less than 2-3 µm. The diameter of the particles in the plane has a range from 50 to 500 µm, and many smaller fragments.

Frequency dependencies of permittivity and permeability of composites of the Permalloy powder with the various milling time were measured. The particle size before milling were in range of 63 to 90 µm with volume concentration of the inclusions p=0.1. The measured result is presented at the Fig. 4.

It is seen that the permittivity increases uniformly up to ε = 18, then decreases at t_m=320 s. The permeability grows in amplitude up to t_m=240 s, the resonance frequency and the frequency of maximum loss slowly shifts to higher frequencies. At t_m=320 s the amplitude decreases, the resonance frequency continues to increase, the magnetic loss peak shifts to high frequencies from 1 to 3 GHz and decreases in amplitude. For milling time t_m=240 s, Acher's and Snoek's parameters are A=0.6 GHz^2, S=2.4 GHz, this indicates that the magnetic structure is in some intermediate state between Snoek's law (spherical particles) and Acher's law (planar particles with inplane magnetization).

The magnitude of the absorption maximum is half the amplitude, hence, from the point of view of the frequency mechanism, relaxation is observed. The change in ε indicates that the particles are flattened up to t_m=240 s, then their milling occurs, so that a significant part of the powder is a number of small fragments. When the

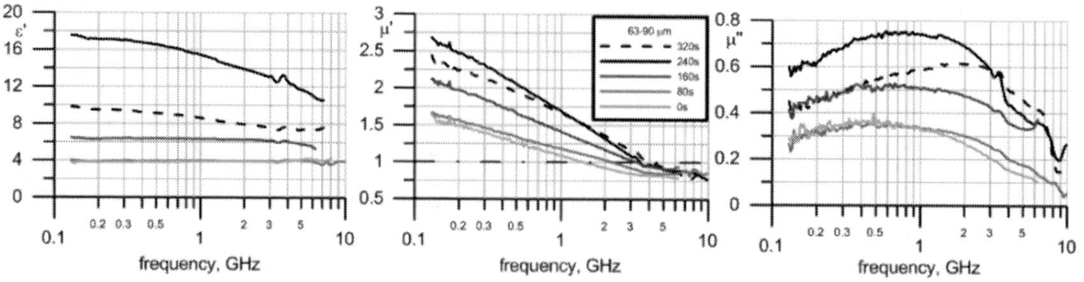

Fig.4. The measured frequency dependencies of composites with Permalloy powder 63–90 µm, volume concentration≈10%, with various milling time t_m

spherical particles are flattened, the resonance frequency should shift to lower frequencies. The shift in the opposite direction can be caused both by a decrease in skin effect and by a change in the magnetic structure of the particles. The decrease in skinning is confirmed by the increase of the Q-factor, based on the values of Acher's and Snoek's parameters. It may be concluded that the magnetic structure of the particles is predominantly closer to bulk polycrystalline that obeys Snoek's law.

4 Conclusions

The frequency dependencies of permittivity and permeability of composites for fractions of different sizes were measured. It is shown that for larger particles, it is a decrease of the Q-factor of the ferromagnetic resonance, which is explained by the strong influence of the skin effect. Frequency dependencies of permittivity and permeability were measured for different milling time. It is shown that as the milling time increases, the amplitude of the magnetic permeability increases. The maximum of magnetic loss and the frequency of the ferromagnetic resonance shifts to high frequencies. It is shown that due to milling, the shape of the particles changes from spherical to platelet, and the calculated values of Acher's and Snoek's parameters indicate that the magnetic structure of the resulting particles is closer to a bulk polycrystalline that satisfies Snoek's law. When the spherical particles are flattened, the resonance frequency shifts to high frequencies, which indicates that the shift to high frequencies due to the decrease of the skin effect and the change in the magnetic structure is stronger than the shift to low frequencies by changing the shape and reducing the demagnetization.

The authors acknowledge financial support from Russian Science Foundation (RSF), according to agreement no. 16-19-10490.

References

1. S.S.Maklakov, A.N.Lagarkov, S.A. Maklakov, Y.A. Adamovich, D.A. Petrov, K.N. Rozanov, I.A. Ryzhikov, A.Y. Zarubina, K.V. Pokholok, D.S. Filimonov, J Alloys Compounds, **706** (2017) 267-273.
2. S.F. Lomayeva, A.V. Syugaev, A.N. Maratkanova, A.A. Shakov, K.N. Rozanov, D.A. Petrov, C.A. Stergiou, J. Alloys Compounds, **721** (2017) 18-27.
3. S.N. Starostenko, K.N. Rozanov, A.O. Shiryaev, A.N. Shalygin, and A.N. Lagarkov, J. Appl. Phys. **121**, 245107 (2017).
4. X.G. Huang, J. Zhang, W.F. Rao, T.Y. Sang, B.Song, C.P. Wong, J. Alloys Compounds, **662**, 409 (2016)
5. J.L. Snoek, Physica, **14**, 207 (1948)
6. O. Acher, P.-M. Jacquart, J.-M. Fontaine, Ph. Baclet, and G. Perrin, IEEE Trans Magn., **30**, 4533 (1994)
7. K.N. Rozanov, N.A. Simonov, A.V. Osipov, J. Commun. Technol. Electron., **47**, 229 (2002)
8. A.N.Lagarkov, K.N.Rozanov, J. Magn. Magn. Mater., **321**, 2082 (2009)
9. L. Lewin, Theory of Waveguides, (John Wiley & Sons, Canada, 1975)

A novel method of ESR oscillating magnetization value determination in strongly correlated metals

Marat Gilmanov[1,2,*], *Alexey* Semeno[2], *Alexander* Samarin[2], and *Sergey* Demishev[1,3].

[1]Moscow Institute of Physics and Technology, Dolgoprudny, Russia
[2]Prokhorov General Physics Institute of RAS, Moscow, Russia
[3]National Research University "Higher School of Economics", Moscow, Russia

Abstract. We propose a powerful method of direct measurement of oscillating magnetization by the electron spin resonance, based on dependence of resonant conditions on geometry of the experiment. Theoretical consideration of the matter leads to a simple expression for oscillating magnetization. Approbation of this method is implemented by means of cavity ESR spectrometer (60 GHz) on two diverse metallic systems, where static magnetization at the resonance field varies by an order of magnitude. Quantitative values of oscillating magnetization (905 G for EuB_6 at T = 4.2 K and 94 G for CeB_6 at T = 1.8 K) are in appropriate agreement with the one obtained by the other methods.

1 Introduction

It was noticed quite some time ago that application of the concept of the permeability coefficient could not explain all experimental data concerning dynamic magnetic properties and permeability tensor should be used [1]. However, in experiment one could observe only the scalar values and therefore the correspondence between tensor and scalar quantities is not trivial, as it was demonstrated by Young and Uehling for two main cases of magnetization parallel and perpendicular to the surface of a metallic sample [2].

The parameters describing the electron spin resonance (ESR) line shape give information about magnetic state of the material, and while g-factor and line width could be directly obtained from the experiment, the integrated intensity of the resonance line is believed to be proportional to the static susceptibility of oscillating moments and requires significant additional efforts to be obtained in absolute values from the ESR data. Basically, the integrated intensity is defined by oscillating magnetization M_0, which may be introduced in the same way as in [3, 4]. The M_0 parameter defines magnitude of the ESR absorption, and in most cases it coincides with static magnetization M_{st} (EuB_6 [3], $YbRh_2Si_2$ [5]), but there are some materials (CeB_6 [4], [6]) that demonstrate a significant difference between the oscillating M_0 and total static magnetization M_{st} measured by standard technique for studying of magnetization. The comparison of M_0 and M_{st} may provide important information on different magnetic contributions or interactions within the system [6].

In addition, the problem of experimental determination of M_0 recently arose due to growing interest in strongly correlated metallic systems [4-8]. However, only a few methods have been developed in this area up to now.

One of the techniques that could be used to obtain M_0 is quite similar to the one being applied often to the dielectric samples. It consists in measuring the investigating material together with some reference sample with known magnetic properties and magnetic resonance response in similar conditions. Such procedure was performed for the heavy fermion compound $YbRh_2Si_2$ by comparison the ESR intensity with the samples YPd_3:Yb1% [5] and YPd_3:Yb0.6% [7], and it was discovered that M_0 and M_{st} in this system coincides. Nevertheless, this method has some major uncertainties which are due to (i) the assumption concerning comparable quality factors of the microwave cavity for both samples and from (ii) uncertainties in determining the sample size, mass, and penetration depth [5].

Another method was developed for the ferromagnetic strongly correlated metal EuB_6 [3] and been also applied to a helimagnet compound MnSi [8] and dense Kondo system CeB_6 [4]. This method exploits the fact that the absorption in the sample depends on the combination $(\mu_{eff}\rho)^{1/2}$ of effective permeability μ_{eff} and resistivity ρ. In the case of strong magnetoresistance, the dependence of resistivity on the magnetic field ρ(H) could be used to calibrate values of μ in the absolute values. Application of this method to the CeB_6 system showed a significant deviation of M_0 from M_{st}, which is a consequence of the key role of ferromagnetic correlations in the genesis of magnetic resonance in dense Kondo systems [4]. However, the main difficulty of this method is that it requires strong dependencies of magnetoresistance of investigated material to provide acceptable accuracy of absolute calibration of μ_{eff}.

* Corresponding author: gilmanov@lt.gpi.ru

We argue that the same parameter M_0 could be directly obtained from the difference in positions of the ESR line in experiments with different mutual orientations of magnetization with relation to the sample surface. Testing of this method was performed for two different systems with high (EuB_6) and low (CeB_6) oscillating magnetization to demonstrate a wide range of the applicability of proposed ansatz.

2 Method

2.1. Theoretical background

As it was mentioned above, the effective resonant function of magnetic permeability $\mu_{eff}(H)$ includes two elements of permeability tensor $\mu(H)$ and $\mu_a(H)$ [2] due to gyrotropic origin of ESR. While in the case **k** || **H** ESR response is defined by two circularly polarized functions $\mu_\pm = \mu \pm \mu_a$, the scheme with $\mathbf{k} \perp \mathbf{H}$ results in $\mu_{eff} = \mu - \mu_a^2/\mu$ [2]. We wish to emphasize that in these two cases the parameters of the ESR line should be strongly affected not only by the demagnetization fields, but also by the conditions of the experiment itself.

Exact equations describing the ESR line shape in these two cases could be found elsewhere [2]. For current consideration, we only would be interested in the difference of resonant conditions.

For the case of $\mathbf{k} \perp \mathbf{H}$, the position of resonance $H_\perp$ could be described by:

$$\omega_0^2 = \gamma^2 H_\perp (H_\perp + 4\pi M_0)(1 + a^2) \quad (1)$$

and

$$\omega_0 = \gamma H_{||} \quad (2)$$

for the case of **k** || **H**. Here $H_\perp$ and $H_{||}$ correspond to resonant fields in considered experimental geometries and α stands for dissipation parameter in Landau-Lifshitz damping term.

Although the equations (1)-(2) depend on M_0 only, the influence of the demagnetization fields should also be taken into account. This could be performed quite easily by modification of values of the external magnetic field for both cases as $H_{||}' = H_{||} - N_{z1}M_{st}$ and $H_\perp' = H_\perp - N_{z2}M_{st}$ where N_{z1} and N_{z2} denote the demagnetization factors for different directions of the magnetic field with respect to the sample. Such correction of the $H_\perp$ and $H_{||}$ should be performed precise enough for accounting of the significant impact of the N_{z1}, N_{z2} on the final result. This could be done by means of numerical simulation applicable to any particular form of the sample [9]. Values of N_{z1} and N_{z2} calculated in this way for disk shaped samples are given in Table 1. It also should be noted that accounting for the demagnetization factor would not lead to Kittel resonant conditions [10] due to the interaction of the sample with microwave radiation being limited by the skin-effect.

Summing up equations (1), (2) and taking into account that experiments in both geometries are performed at the same frequency, the following expression for oscillating magnetization M_0 could be derived:

$$4\pi M_0 = \frac{(H_{||} - N_{Z1}M_{st})^2}{(H_\perp - N_{Z2}M_{st})(1 + a^2)} - (H_\perp - N_{Z2}M_{st}) \quad (3)$$

2.2 Experimental details

Experimental method was tested with the help of high frequency (60 GHz) cavity ESR spectrometer based on Agilent E8261C Network Analyzer [11]. Sample is placed as a part of the cavity wall to evade field inhomogeneity effects. Two types of coupling between waveguides and resonator are used to establish different geometries of the experiment (Fig.1). Also, cavities are designed to be practically identical in both experiments to simplify the calculations.

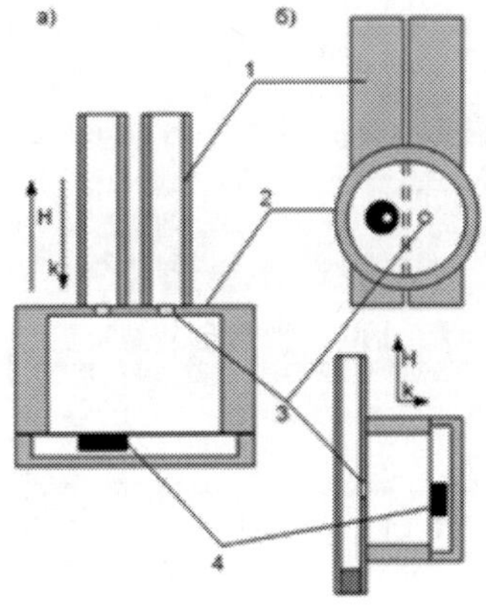

Figure 1. Experimental layouts corresponding to the cases of $\mathbf{k} \perp \mathbf{H}$ (a) and **k** || **H** (b) in ESR experiment. The direction of the vector **k** corresponds to its orientation at the sample surface. 1 – waveguides, 2 – cylindrical resonator, 3 – communication holes, 4 – sample on thin cupper foil aperture.

3 Approbation

Two rather different metallic systems are selected to demonstrate a wide range of applicability of the proposed method. First, EuB_6 is a ferromagnetic strongly correlated metal with $M_{st} = 912$ G at resonant field (31.6 kOe). Spectra obtained at T = 4.2 K for both magnetization conditions are presented at Fig.2. Narrow ESR line and a large magnetization leads to a significant difference in line shape and resonance fields as well as to considerable values of demagnetization fields. The obtained parameters are presented at the Table 1. The calculated value of M_0 coincides with the one obtained by different method [3] and, as expected, with the value of static magnetization within the error limits.

Dense Kondo system CeB_6 is on the other edge of experimental complexity. Small values of magnetization ($M_{st} = 96$ G at T = 1.8 K) cause the difference in resonant fields to be relatively low (see Fig.3). Moreover widening of the ESR line makes more difficult the resonance field determination. Nevertheless, even in this case, the calculated M_0 values agree well with the ones reported earlier [4] (see Table 1).

For the case of differing M_0 and M_{st} in CeB_6 (T = 2.6 K), considerable broadening simultaneously with decreasing in intensity of the line prevents the possibility of obtaining very accurate value of M_0. In spite of that it is still possible to obtain estimate the value of M_0 being significantly lower than M_{st} (Table 1) which also correlates with the results obtained earlier [6].

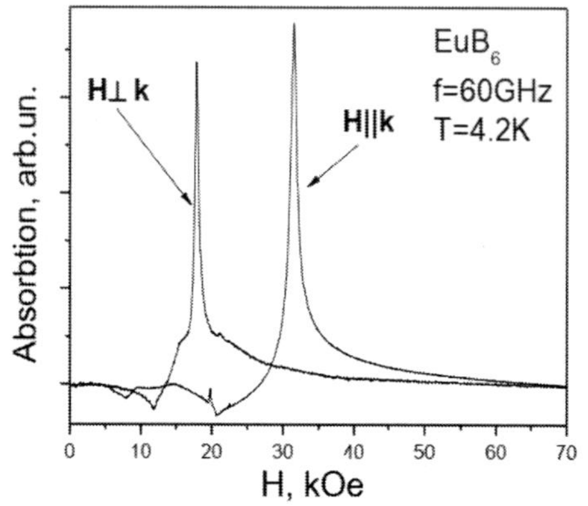

Figure 2. ESR spectra of EuB_6 system for two different conditions of magnetization

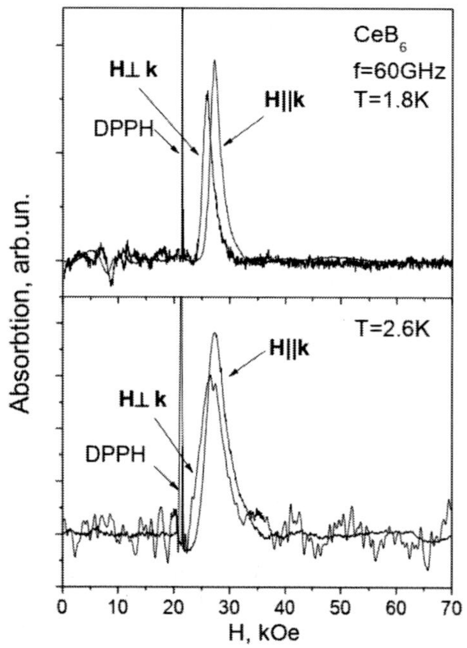

Figure 3. ESR spectra of CeB_6 system for two different temperatures and conditions of magnetization. DPPH is a reference signal from an α-diphenyl-β-picrylhydrazyl with g-factor = 2.0036.

Table 1. Values of demagnetization factors N_{z1}, N_{z2} calculated by the method described in [9], resonance fields $H_\perp$, $H_\parallel$ and static magnetization M_{st} measured in experiments, and oscillating magnetization M_0 values estimated from (3).

Sample	T, K	N_{Z1}, 4π	N_{Z2}, 4π	$H_\perp$, kOe	$H_\parallel$, kOe	M_{st}, G	M_0, G
EuB_6	4.2	0.86	0.07	17.6	31.6	911	905±16
CeB_6	1.8	0.84	0.08	25.7	27.2	96	94±7
CeB_6	2.6	0.84	0.08	26.3	27.1	71	21±10

4 Conclusion

In conclusion, we propose a method of direct measurement of ESR oscillating magnetization M_0 based on difference in resonant conditions for different cases of magnetization. It is shown that precise correction on demagnetization fields is required. Method approbation is performed on two rather distinct metallic systems, where static magnetization at the resonance field varies by an order of magnitude. The suggested technique gives appropriate results in agreement with the data collected by the other methods. The obtained results for CeB_6 system at temperature T = 2.6 K allows concluding that the proposed method could be applied to collect relevant information about oscillating magnetization M_0 deviating from the static magnetization M_{st}.

Acknowledgements. This work was supported by the Russian Scientific Foundation, Grant №17-12-01426

References

1. J.A. Young jr., E.A. Uehling, Phys. Rev. **90**, 990 (1953)
2. J.A. Young jr., E.A. Uehling, Phys. Rev. **94**, 3, 545 (1954)
3. A.V. Semeno, V.V. Glushkov, A.V. Bogach, N.E. Sluchanko, A.V. Dukhnenko, V.B. Filipov, N.Yu. Shitsevalova, S.V. Demishev, Phys. Rev. B **79**, 014423, (2009)
4. S.V. Demishev, A.V. Semeno, A.V. Bogach, N.A. Samarin, T.V. Ishchenko, V.B. Filipov, N.Yu. Shitsevalova, N.E. Sluchanko, Phys. Rev. B **80**, 245106 (2009)
5. J. Sichelschmidt, V.A. Ivanshin, J. Ferstl, C. Geibel, F. Steglich, Phys. Rev. Lett. **91**, 156401 (2003)
6. A.V. Semeno, M.I. Gilmanov, A.V. Bogach, V.N. Krasnorussky, A.N. Samarin, N.A. Samarin, N.E. Sluchanko, N.Yu. Shitsevalova, V.B. Filipov, S.V. Demishev, V.V. Glushkov, Sci. Rep. **6**, 39196 (2016).
7. J. Wykhoff, J. Sichelschmidt, G. Lapertot, G. Knebel, J. Flouquet, I.I. Fazlishanov, H.A. Krug von Nidda, C. Krellner, C. Geibel, F. Steglich, STAM **8**, 389 (2007)
8. S.V. Demishev, A.V. Semeno, A.V. Bogach, V.V. Glushkov, N.E. Sluchanko, N.A. Samarin, A.L. Chernobrovkin, Jetp Lett. **93**, 213 (2011)

https://doi.org/10.1051/epjconf/201818502003

9. S. Tandon, M. Beleggia, Y. Zhu, M. De Graef, JMMM **271**, 1, 27 (2004)
10. C. Kittel, Phys. Rev. **73**, 155 (1948)
11. A.N. Samarin, A.V. Semeno, M.I. Gilmanov, V.V. Glushkov, I.I. Lobanova, N.A. Samarin, I.I. Sannikov, N.M. Chubova, V.A. Dyadkin, S.V.Grigoriev, S.V. Demishev, Phys. Procedia **71**, 337 (2015).

EPJ Web of Conferences **185**, 02004 (2018) https://doi.org/10.1051/epjconf/201818502004
MISM 2017

Conditions for division and multiplication of frequencies in ferrite

Dmitriy Pleshev[1,2*], *Vladimir* Vlasov[2], *Fanur* Asadullin[1,2], *Leonid* Kotov[2], *Sergey* Poleshchikov[1], *Vladimir* Shavrov[3], and *Vladimir* Shcheglov[3]

[1]Saint-Petersburg state forest technical university named after S.M. Kirov, St-Petersburg, Russia
[2]Syktyvkar State University, Syktyvkar, Russia
[3]Institute of Radioengineering and Electronics of the Russian Academy of Sciences, Moscow, Russia

Abstract. The paper considers investigation of hypersound magnetic and elastic oscillations excitation by use of alternating magnetic field in normal magnetized plate. Distinctive features of frequency division and multiplication are studied in the paper. The relation between processes of frequency division and multiplication and structure of the plate as well as parameters of saturation magnetization, anisotropy constant, magnetic and elastic damping has been revealed.

1 Introduction

The magnetoelastic oscillations generation caused by magnetostriction is one of the interesting interaction mechanism of magnetic, elastic and magnetoelastic subsystems, that attracted attention of researchers for a long time [1-5]. This process can be executed in case of action on the plate by alternating magnetic field, hypersound oscillations as well as pulsed magnetisation reversal in a constant field. Dimensions and shape of a plate are fitted in such a way that avoid chance of spin waves excitation [6]. The present work deals with investigation of the features of the magnetization vector nonlinear precession and the elastic displacement excited by alternating magnetic field. For this study, sample with Ni parameters has been taken which is also of technical interest.

2 Geometry of the problem and basic equations

The plane-parallel ferrite plate has thickness d. The external DC magnetic field $\mathbf{H}_0$ is applied perpendicularly to the plane of the plate as shown in Fig. 1. Right-hand circularly polarized alternating field is applied in the plate plane.

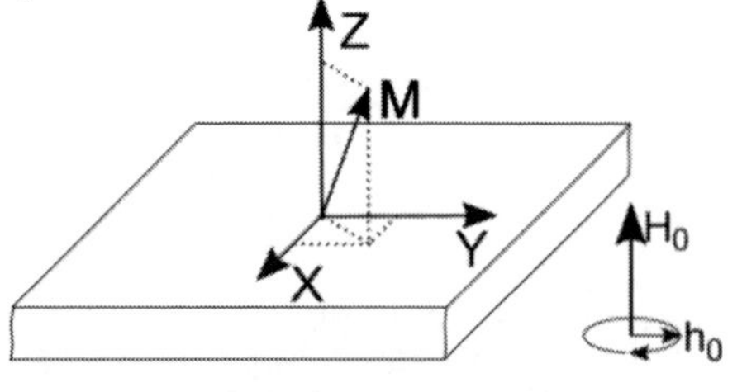

Fig. 1. Geometry of the problem.

The problem is solved in a Cartesian coordinate system Oxyz. The plane Oxy of the coordinate system coincides with the plane of the plate. The coordinate axes are parallel to the edges of the cube crystallographic cell. The center of the coordinate system O is in the center of the plate, so that the plate planes coordinates are $z=\pm d/2$. We consider only excitement of the shear elastic oscillations.

Assuming that the total energy density of the plate U in the field $\mathbf{H}=\{0;0;\mathbf{H}_0\}$ equals the sum of magnetic, elastic and magnetoelastic energy densities, we obtain:

$$U = -M_0 h_x m_x - M_0 h_y m_y - M_0 H_0 m_z + 2\pi M_0^2 m_z^2 + 2c_{44}(u_{xy}^2 + u_{yz}^2 + u_{zx}^2) + 2B_2(m_x m_y u_{xy} + m_y m_z u_{yz} + m_z m_x u_{zx}) \quad (1)$$

where $\mathbf{m}=\mathbf{M}/M_0$ - normalized magnetization vector, M_0 - the saturation magnetization of the magnetic plate, u_{ij} – the components of the strain tensor, c_{44} - the elastic constant and B_2 – the magnetoelastic coupling constant.

The initial equations for the system are the Landau-Lifshitz-Gilbert equation and the equations for the elastic displacement vector components $u_{x,y}$:

$$\frac{\partial \vec{m}}{\partial t} = -\gamma\left[\vec{m}\times\vec{H}_e\right] + \alpha\left[\vec{m}\times\frac{\partial \vec{m}}{\partial t}\right], \quad (2)$$

$$\frac{\partial^2 u_{x,y}}{\partial t^2} = -2\beta\frac{\partial u_{x,y}}{\partial t} + \frac{c_{44}}{\rho}\cdot\frac{\partial^2 u_{x,y}}{\partial z^2}, \quad (3)$$

* Corresponding author: dpleshev@gmail.com

where γ - the gyromagnetic ratio; α, β - damping constants of the magnetic and elastic subsystems.

The effective field is determined by the following equation:

$$\vec{\mathbf{H}}_e = -\frac{1}{M_0} \cdot \frac{\partial U}{\partial \vec{\mathbf{m}}} \cdot \quad (4)$$

The boundary conditions are:

$$c_{44} \frac{\partial u_{x,y}}{\partial z} \Big|_{z=\pm d/2} = -B_2 m_{x,y} m_z \cdot \quad (5)$$

The system of the equations [7] was solved numerically by the Runge-Kutta 7–8 orders method with control of the integration at every step length.

The material parameters used in the calculation: ρ=8.968 g·cm^{-3}, M_0=480 G, c_{44}=1.2·10^{11} erg·cm^{-3}, B_2=0.16·10^9 erg·cm^{-3}, α=0.035.

3 Frequency division

Consider the case when the frequency of elastic resonance was multiple of alternating field frequencies, i.e. thickness of the plane was selected multiple of resonance thickness i.e. $d=n \cdot d_r$, where n is multiplicity, d_r - the resonance thickness.

The process of frequency division is followed by two transient regimes connected with relaxations of elastic and magnetic oscillations. Optimum conditions for frequency division can be traced when relaxation of magnetic oscillations finishes whereas relaxation of elastic oscillations does not appear. The second necessary condition is nonlinear regime of magnetoelastic oscillations when $\mathbf{h}_0 \geq 10$ Oe.

Development of magnetization oscillations and elastic displacements in the nonlinear ($\mathbf{h}_0$=1000 Oe) case can be seen in the Fig. 2.

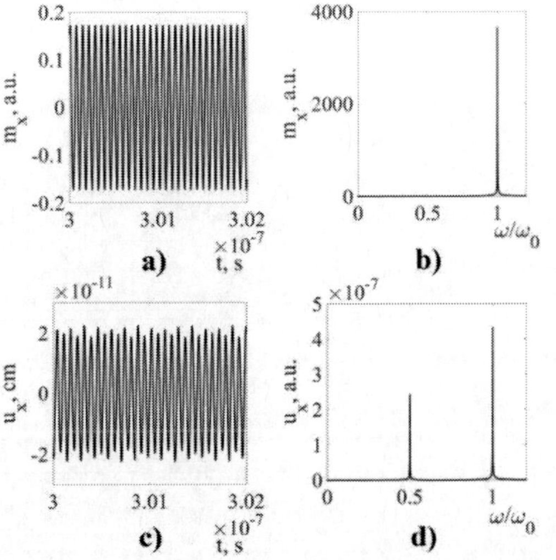

Fig. 2. Development of magnetization oscillations and elastic displacements (a, c) and their spectra (b, c) in case of $\mathbf{h}_0$=1000 Oe and n=2.

Fig.2b and Fig.2d illustrate that magnetization oscillations and elastic oscillations have a spectral peak on excitation frequency and elastic oscillations have a spectral peak on elastic resonance frequency of the plate in this case and the elastic resonance frequency occurs on one-half of excitation frequency. It should be noticed here, that the elastic displacement frequency is equal to excitation frequency after magnetic and elastic subsystems relaxation.

We consider the case integer frequency division when the elastic resonance frequency of the plate is a multiple of the frequency of the alternating field at a ratio of 1/2, 1/3 and 1/4, which corresponds to n = 2, 3 and 4 correspondingly.

Fig.3 displays that elastic oscillation spectra have two peaks. The one is on the excitation frequency and the other on the frequency of elastic resonance. The magnetic oscillations hold the shape while the elastic oscillations have the shape distortion due to undesired oscillation that occur on excitations frequency.

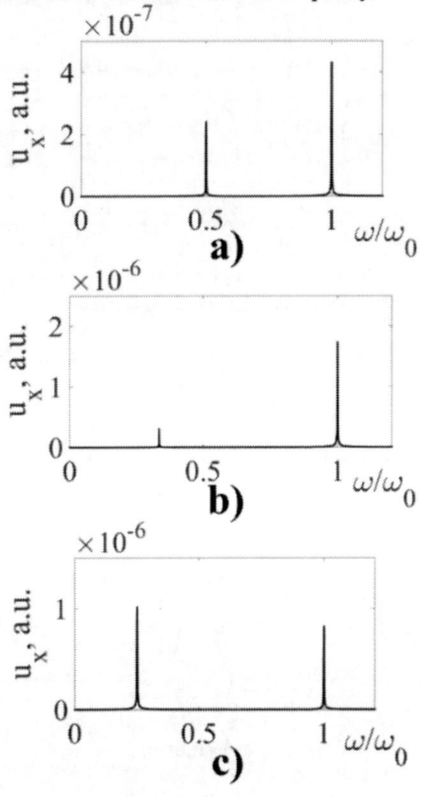

Fig. 3. Spectra of elastic displacements oscillations in case of $\mathbf{h}_0$=1000 Oe and (a) n=2, (b) n=3, (c) n=4.

In case of the fractional frequency division when n=2.2, n=2.5 and n=2.8, the spectra of elastic displacement get a strong change as is shown in Fig.4.

The paper [8] considers the situation where the multiplicity shifted to some integer, for example, for n = 2.2 or n = 2.8, the amplitude of magnetization oscillations on the frequency of the elastic resonance is two orders of magnitude higher if the division multiplicity corresponds to n = 2.5, 3.5, i.e. multiplicity is in the middle of the neighboring integer multiplicities. As opposed to [8] in this case the similar dependence is not available.

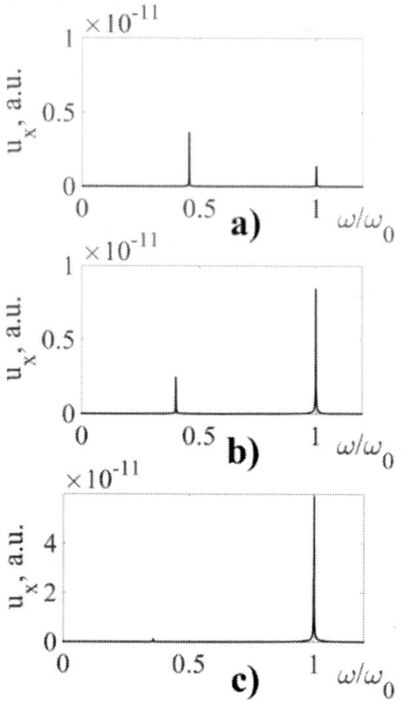

Fig. 4. Spectra of elastic displacements oscillations in case of h_0=1000 Oe and (a) n=2.2, (b) n=2.5, (c) n=2.8.

The spectrum analysis of elastic oscillations makes it possible to determine the distribution of amplitudes maxima on excitation frequency and frequency of elastic resonance in relation to division multiplicity that are shown in Fig. 5.

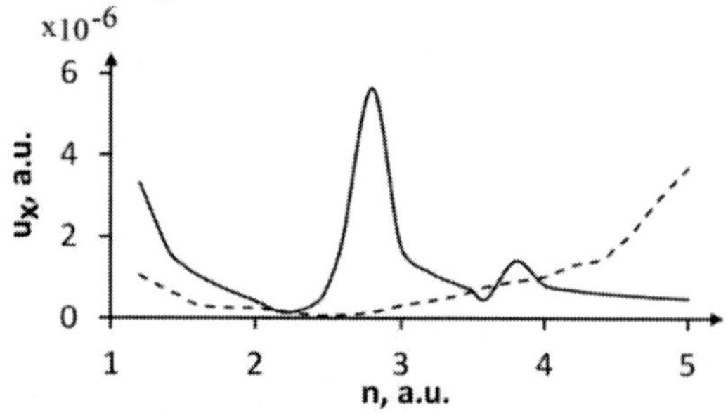

Fig. 5. Distribution of elastic oscillations amplitudes as a function of the division multiplicity on excitations frequency (solid line) and on the frequency of the elastic resonance (dash line).

Amplitudes of elastic oscillations decrease exponentially on frequency of excitation with increase of multiplicity and have peaks at multiplicity n=2.8 and n=3.8. Note that the amplitudes of the magnetic subsystem are not prone to considerable changes.

The distribution of the maximum amplitudes as a function of the fission multiplicity has a shape of an irregular-shaped parabola with a vertex at n = 2.5 on the frequency of the acoustic resonance. The efficient frequency division is possible in case of multiplicity value n<1.5 or n>4.

4 Frequency multiplication

Frequency multiplication is implemented by fitting of thickness of the plate in such a way that the frequency of elastic resonance was multiple bigger than excitation frequency. That is the thickness of the plate equals $d = d_r/n$, where n is multiplicity, d_r - the resonance thickness. The optimal conditions of frequency multiplications are analogous to case of frequency division. The excitation field should be linearly polarized in order the frequency multiplication is efficient. The amplitudes of excitation field components are equal to $h_x \neq 0$, $h_y = 0$.

Consider the case of the frequency multiplication when n=2, 3, 4, i.e. n is integer. The spectra of elastic oscillations contain frequency component of elastic resonance and excitation frequency component as is shown in Fig. 6. The oscillation amplitudes ratio on the excitation and elastic resonance frequencies changes subject to increasing multiplicity of frequency multiplication.

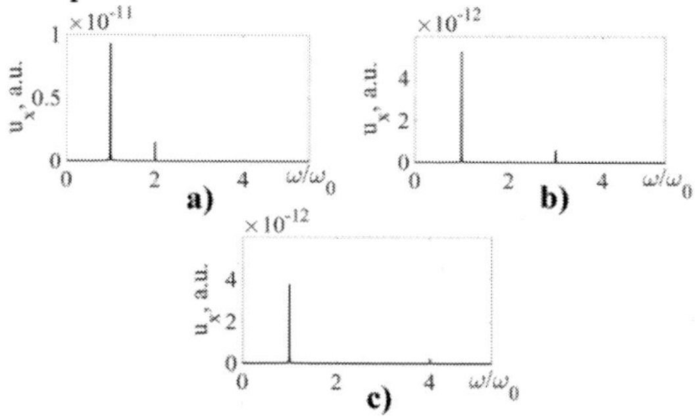

Fig. 6. Spectra of elastic displacements oscillations in case of h_0=1000 Oe and (a) n=2, (b) n=3, (c) n=4.

Fig.7 displays spectra of elastic oscillations in case of fractional frequency multiplication when n=2.2, n=2.5 and 3.5.

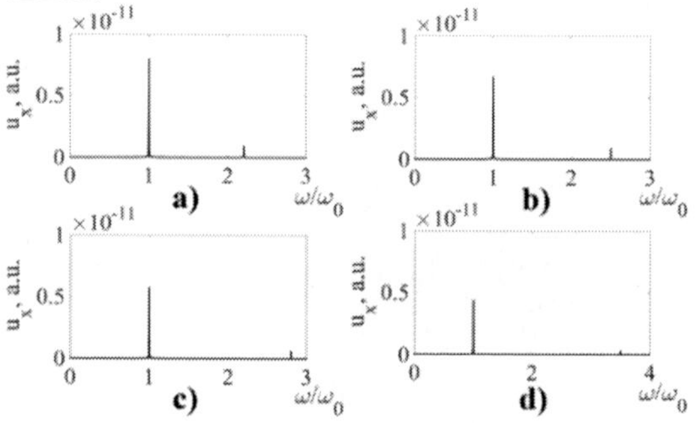

Fig. 7. Spectra of elastic displacements oscillations in case of h_0=1000 Oe and (a) n=2.2, (b) n=2.5, (c) n=2.8, (d) n=3.5.

It is important to note that magnitude of elastic oscillations on the frequency of elastic resonance decreases with increase of multiplicity. Considering the fact that the amplitude of elastic oscillations on frequency of elastic resonance is an order of magnitude less than the amplitude of elastic oscillations on excitation frequency, multiplication of frequency is possible, but it's of no practical interest at high multiplicity.

Fig.8 depicts the amplitudes maxima distribution of elastic oscillations on excitation frequency and frequency of elastic resonance in relation to multiplicity of multiplication. Amplitudes of elastic oscillations decrease monotonically on the frequency of excitation and the frequency of elastic resonance with increase of multiplicity. The elastic oscillations on the frequency of elastic resonance have peaks at multiplicity n=2.2 and n=2.5. The efficiency of frequency multiplication decreases with increase of multiplication multiplicity.

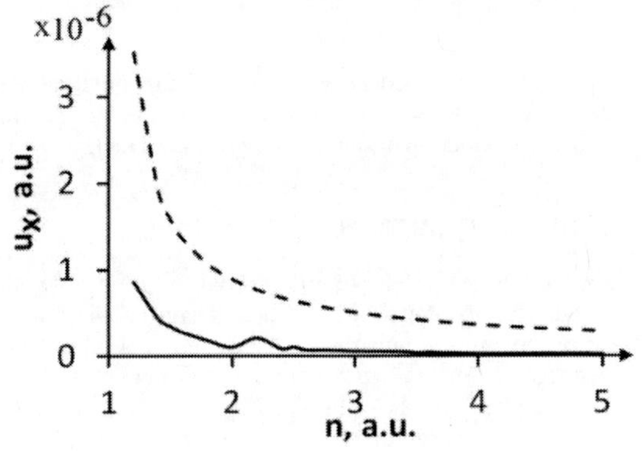

Fig. 8. Distribution of elastic oscillations amplitudes as a function of the multiplication multiplicity on excitations frequency (dash line) and on the frequency of the elastic resonance (solid line).

5 Conclusion

Thus the coupled oscillations of magnetization and elastic displacement in normally magnetized ferrite plate that possess magnetoelastic properties are considered in the paper.

The research shows that the efficient processes of frequency division and multiplication are possible only when regime of excitement is nonlinear. In case of multiplication, polarization of excitation alternating field should be linear.

Correlation of division and multiplication processes with time of magnetic and elastic relaxations are described. The paper reveals the fact that stable division and multiplication processes exist when relaxation time of elastic oscillations are bigger by an order of magnitude than relaxation time of magnetic oscillations.

The dependencies of effective division and multiplication conditions are defined. The spectrum analysis showed that components on exciting frequencies are dominant over components on multiply and division frequencies. However, in case of frequency division, components on frequency of elastic resonance are greater than components on exciting frequency if multiplicity is more than $n>4$.

This work was supported by RFBR (grant no. 17-02-01138-a, grant 17-57-150001 НЦНИ-а.).

References

1. R.L. Comstock, R.C. LeCraw, Generation of microwave elastic vibrations in a disk by ferromagnetic resonance, J. Appl. Phys., 34 (1963), 3022.
2. M. Ye, H. Dötsch. Magnetoelastic instabilities in the ferrimagnetic resonance of magnetic garnet films, Phys. Rev. B, 44 (1991), 9458.
3. A.M. Shutyĭ, D.I. Sementsov. Nonlinear effects of magnetization precession near ferromagnetic resonance, Physics of the Solid State. 42 (2000) 1306-1309.
4. L. Dreher et al. Surface acoustic wave driven ferromagnetic resonance in nickel thin films: Theory and experiment, Phys. Rev. B, 86 (2012), 134415(13).
5. L. Thevenard et al. Irreversible magnetization switching using surface acoustic waves, Phys. Rev. B, 87 (2013), 144402(9).
6. V.S. Vlasov, L.N. Kotov, V.G. Shavrov, V.I. Shcheglov, Nonlinear excitation of hypersound in a ferrite plate under the ferromagnetic-resonance conditions, J. Comm. Tech. El. 54(2009) 821-832.
7. M.S. Kirushev, V.S. Vlasov, D.A. Pleshev, F.F. Asadullin, L.N. Kotov, V.G. Shavrov, V.I. Shcheglov, Solid State Phenomena Vols 233-234 (2015) 73-78.
8. F.F. Asadullin, S.M. Poleshikov, D.A. Pleshev, L.N. Kotov, V.S. Vlasov, V.G. Shavrov, V.I. Shcheglov, Journal of Siberian Federal University. Mathematics & Physics 2017, 10(1), 36–39.

Determination of magnetoelastic oscillations in ferrite plate with different values of anisotropy constant

Leonid Kotov[1], *Pavel* Severin[1,*], *Vladimir* Vlasov[1] and *Dmitry* Beznosikov[1]

[1]Syktyvkar State University named after Pitirim Sorokoin, Komi Republic, Syktyvkar 167000, Russia

Abstract. The maximum amplitudes of magnetic and elastic oscillations are calculated at various material parameters and parameters of external fields. Normally and tangentially magnetized anisotropic magnetic plates were considered. To obtain the maximum amplitudes with additional constraints, the simulated annealing method was used. The behaviour of the magnetic and elastic components of the oscillations was considered. The changes in the position of the magnetization vector and the equilibrium displacements as a function of the first anisotropy constant were revealed.

1 Introduction

Analysis of parameters of differential equations with the account for the boundary conditions can be reduced to more general problem of finding the extrema of the cost function, which is a reflection of specific simulated characteristics (for example, the degree of randomness of the excited oscillations, the amplitude of the elastic and magnetic oscillations). For global optimization simulated annealing method is used to solve a large class of problems. The advantage of the method is that, even with limited computing resources, it allows the extremum to be found that is close to the global one [1].

The nonlinear magnetoelastic dynamics of films and plates is a rapidly developing region [2-6]. Thus, the use of optimization methods in the study of magnetoacoustic interactions is also an urgent task.

Since the experimental search for the maxima of the amplitudes of elastic oscillations is usually hampered by the lack of high-power radio frequency generators and sensitive receivers with low noise, then this calculation allows to narrow down the frequency and amplitude of the ranges of external magnetic fields, in which it is necessary to carry out the study. Thus the solution of this problem allows reducing financial and time expenses in nonlinear magnetoelastic tasks.

In this study, the simulated annealing method is used to find a set of parameters of a system of differential equations describing the excitation of oscillations by a radio-frequency magnetic field. For these parameters, the smallest change in the constants of the cubic magnetic anisotropy corresponds to the largest change in the amplitudes of the elastic oscillations. The regions with high sensitivity of the amplitude of the elastic oscillations to the changes of the anisotropy constants were observed.

Further we consider the behavior of the magnetic and elastic components at the founded points of the multidimensional parameter space of the model.

2 Basic equations

The complete system of differential equations for the problem of excitation of magnetoelastic oscillations of radio frequency magnetic field in the normally magnetized ferrite plate is written following [2].

The plane-parallel plate with a thickness d having magnetic, elastic and magnetoelastic properties is considered. External DC magnetic field $\mathbf{H_0}$ is perpendicular to the plane of the plate and the alternating magnetic field vector $\mathbf{h_0}$ is tangential to the surface of the plate. The problem is solved in a Cartesian coordinate system $Oxyz$. The plane Oxy of the coordinate system coincides with the plane of the plate. The coordinate axes are parallel to the edges of the cube crystallographic cell. The center of the coordinate system O is in the center of the plate, so that the plate planes coordinates are $z = \pm d / 2$, as shown in Figure 1.

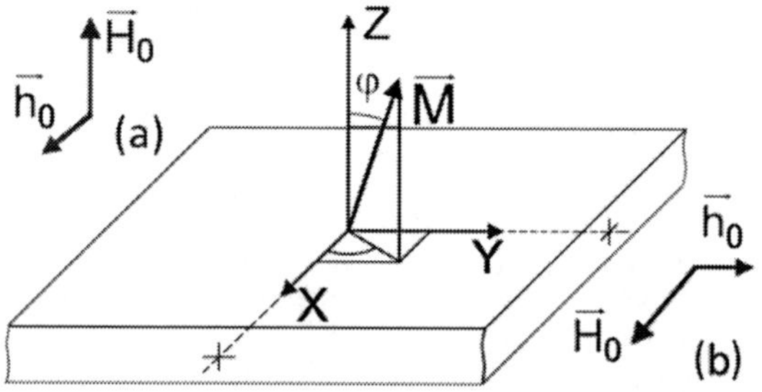

Fig. 1. Direction of magnetic fields and magnetic moment for a normally magnetized plate (a). Direction of magnetic fields (b) for a tangentially magnetized plate (the magnetic moments in this case are directed along the Ox axis)

* Corresponding author: pav9687@yandex.ru

Assuming that the total energy density of the plate U in the field $\mathbf{H} = (h_x;h_y;H_0)$ equals the sum of magnetic, elastic and magnetoelastic energy densities, we obtain [2]:

$$\begin{aligned} U = & -M_0 h_x m_x - M_0 h_y m_y - M_0 H_0 m_z + 2\pi M_0^2 m_z^2 \\ & + K_1(m_x^2 m_y^2 + m_x^2 m_z^2 + m_z^2 m_y^2) + K_2 m_x^2 m_y^2 m_z^2 \\ & + 2c_{44}\left(u_{xy}^2 + u_{yz}^2 + u_{zx}^2\right) + \\ & 2B_2\left(m_x m_y u_{xy} + m_y m_z u_{yz} + m_z m_x u_{zx}\right) \end{aligned} \quad (1)$$

where, $\mathbf{m} = \mathbf{M} / M_0$ is the normalized magnetization vector, M_0 is the saturation magnetization, K_1, K_2 – the first and second magnetocrystalline anisotropy constants, u_{ij} are components of the deformation tensor, c_{44} is the constant of elasticity and B_2 is the constant of magnetoelastic interaction. The system of the ordinary differential equations describing the magnetoelastic oscillations are obtained in [2].

The system describes the excitation of the transverse elastic oscillations. The initial equations for the obtained system are the Landau-Lifshitz-Gilbert equation and the equations for the elastic displacement vector components $u_{x,y}$:

$$\frac{\partial \vec{m}}{\partial t} = -\gamma \cdot \left[\vec{m} \times \vec{H}_e\right] + \alpha \cdot \left[\vec{m} \times \frac{\partial \vec{m}}{\partial t}\right], \quad (2)$$

$$\frac{\partial^2 u_{x,y}}{\partial t^2} = -2\beta \frac{\partial u_{x,y}}{\partial t} + \frac{c_{44}}{\rho}\frac{\partial^2 u_{x,y}}{\partial z^2}, \quad (3)$$

where effective fields are:

$$\begin{aligned} H_{ex} = & h_x + \frac{B_2^2}{M_0 c_{44}} m_x m_z^2 - \frac{2B_2}{M_0 d} m_z v_x - \\ & \frac{2K_1}{M_0} m_x (m_y^2 + m_z^2) - \frac{2K_2}{M_0} m_x m_y^2 m_z^2, \end{aligned} \quad (4)$$

$$\begin{aligned} H_{ey} = & h_y + \frac{B_2^2}{M_0 c_{44}} m_y m_z^2 - \frac{2B_2}{M_0 d} m_z v_y - \\ & \frac{2K_1}{M_0} m_y (m_x^2 + m_z^2) - \frac{2K_2}{M_0} m_y m_x^2 m_z^2, \end{aligned} \quad (5)$$

$$\begin{aligned} H_{ez} = & H_0 - 4\pi M_0 m_z + \frac{B_2^2}{M_0 c_{44}} m_z \left(m_x^2 + m_y^2\right) - \\ & -\frac{2B_2}{M_0 d} m_x v_x - \frac{2B_2}{M_0 d} m_y v_y - \\ & \frac{2K_1}{M_0} m_z (m_x^2 + m_y^2) - \frac{2K_2}{M_0} m_z m_x^2 m_y^2. \end{aligned} \quad (6)$$

The equations for the elastic displacement are [2]:

$$u_{x,y} = -\frac{B_2}{c_{44}} m_{x,y} m_z z + v_{x,y} \sin(\frac{\pi}{d} z). \quad (7)$$

Boundary conditions are:

$$c_{44} \frac{\partial u_{x,y}}{\partial z} \Big|_{z = \pm d/2} = -B_2 m_{x,y} m_z. \quad (8)$$

In case of a tangentially magnetized ferrite plate, the geometry of the problem is described as follows: a DC external magnetic field $\mathbf{H_0}$ is directed along Ox axis and the alternating magnetic field $\mathbf{h_0}$ is directed along Oy axis.

The system of the equations is solved numerically by 7th to 8th order Runge-Kutta-Fehlberg method with control of the length of the integration step.

Reliability of the results of modeling is provided by use of universally recognized methods and the correctness of the mathematical formulation of the problem [1-2, 7].

3 Parameters of manganese-zinc spinel

To investigate the processes with different values of anisotropy constants a manganese-zinc spinel (MZS) crystal grown by the Verneuil method with a specific composition $Mn_{0.61}Zn_{0.35}Fe_{2.04}O_4$ was chosen.

The choice is due to the availability of experimental works [7-8], as well as available characteristics for this and similar in composition crystals [7].

Dependences of the anisotropy constants and other parameters of MZS on temperature are taken from [7], as well as from experimental data (the velocity of a shear elastic wave) and are presented in Table 1. We used a linear extrapolation based on the available data to obtain the values of the parameters in the low-temperature region. The Curie temperature for the selected material is $T_c = 458$ K.

Table 1. Parameters of MZS in dependence on temperature.

T (K)	M_s (G)	$K_1\times10^{-3}$ (erg·cm^{-3})	$K_2\times10^{-3}$	$B_2\times10^{-6}$	$C_{44}\times10^{-11}$	$\alpha\times10^2$
450	18	0.35	-1.09	-3.43	10.88	14.8
400	125	1.41	-1.15	-11.96	10.58	12.9
350	225	2.82	-1.15	-20.42	10.45	11.3
300	318	3.32	-2.88	-30.21	10.27	10.1
263	383	-0.03	-6.26	-39.65	9.99	9.4
250	404	-2.94	-7.80	-44.56	10.01	9.2
200	482	-22.21	-15.07	-64.42	9.63	8.3
150	550	-64.77	-19.30	-90.30	9.72	7.5
100	608	-133.90	-24.21	-115.71	9.78	6.7
50	653	-229.59	-34.00	-140.97	9.80	6.0
30	666	-275.31	-39.80	-150.96	9.80	5.8
10	675	-325.28	-46.27	-160.96	9.80	5.5

The thickness of the plate and the density of the plate material was chosen as: $\rho = 5.4$ g cm^{-3}, $d = 7.69\times10^{-5}$ cm. The oscillations were excited by an alternating magnetic field pulses. The pulse duration was 10 periods of excitation frequency and the amplitude was 0.01 Oe.

The evolution of the oscillations was considered in a the time interval 10 times larger than pulse duration. The damping parameter of the elastic system is typical for MZS crystal: $\beta = 1.5\times10^{6}$ s^{-1}.

In order to reduce the oscillation development time to a stationary amplitude and to exclude short-term relaxation effects from the simulation, the damping parameter of the magnetic system was taken as $\alpha = 0.5$.

4 Using simulated annealing

We take $P = (M_s, f, \Delta H)$ as the set of parameters of a system of differential equations describing the excitation of oscillations by a radio-frequency magnetic field, at which the smallest change in the constants of the crystallographic anisotropy K_1 and K_2 would correspond to the largest change in the amplitudes of the elastic vibrations u_x.

We denote ΔK_1 and ΔK_2 as the desired changes in the anisotropy constants, and ΔK_1^{MAX} and ΔK_2^{MAX} as the maximum possible changes determined by the boundaries of the search area. The cost function was used in simulated annealing algorithm. The maximum of the function should be found by formula:

$$F = \left|u_x(P,K_1,K_2) - u_x(P,K_1+\Delta K_1,K_2+\Delta K_2)\right| \left(1-\sqrt{\frac{\Delta K_1^2+\Delta K_2^2}{\Delta K_1^{2MAX}+\Delta K_2^{2MAX}}}\right) \left(1-\frac{|K_1|+|K_1+\Delta K_1|}{|2\Delta K_1^{MAX}|}\right)\left(1-\frac{|K_2|+|K_2+\Delta K_2|}{|2\Delta K_2^{MAX}|}\right). \quad (9)$$

The cost function implicitly contains the maximization condition for u_x. The remaining factors describe the additional conditions of the extremum and are normalized. It is necessary to ensure the equilibrium of these conditions in the search for the maximum of the cost function.

The ranges of parameters changes are: frequency of alternating magnetic field f varied from 300 MHz to 3000 MHz; the DC field exceeded the saturation magnetization field $4\pi M_s$ by an amount ΔH from 100 Oe to 1000 Oe. The anisotropy constants and the saturation magnetization varied within the limits corresponding to a change in temperature from 10 to 450 K. We change these parameters to search the maximum of the elastic displacements.

The remaining parameters were taken from Table 1 for a temperature $T = 263$ K ($K_1 \approx 0$).

The values of the input parameters which correspond to two states of the model was found using the simulation annealing (Figure 2).

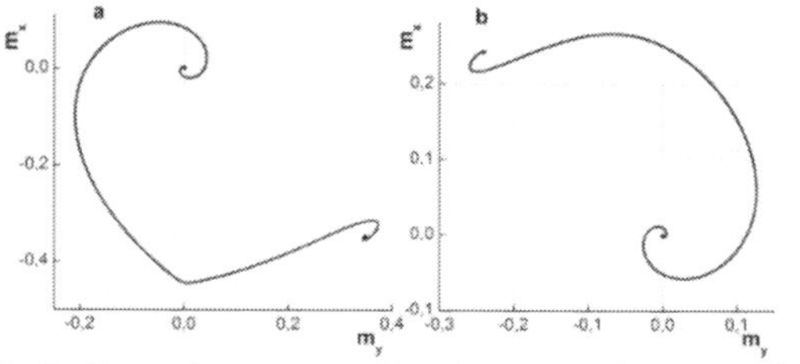

Fig. 2. Precession patterns for the case of a normally magnetized plate with $f = 1359$ MHz; $M_s = 163.05$ G, $\Delta H = 971.83$ Oe and $u_x = 5.12\times10^{-10}$ cm, $K_1 = -181398$ erg cm^{-3}, $K_2 = -1090$ erg cm^{-3} (a); $u_x = 6.96\times10^{-12}$ cm, $K_1 = -114143$ erg cm^{-3}, $K_2 = -1096$ erg cm^{-3} (b)

5 Numerical solution

Let us consider the behavior of the elastic component and the dependence of the precession angle φ on K_1 (In this geometry, changes in K_2 lead to changes in u_x in order of 10%).

It is seen from Figure 3 that the change in the anisotropy constant K_1 of about 40% entails a change in the final position of the magnetization vector during reorientation (Figure 2) and also a significant change the amplitude of the elastic oscillations.

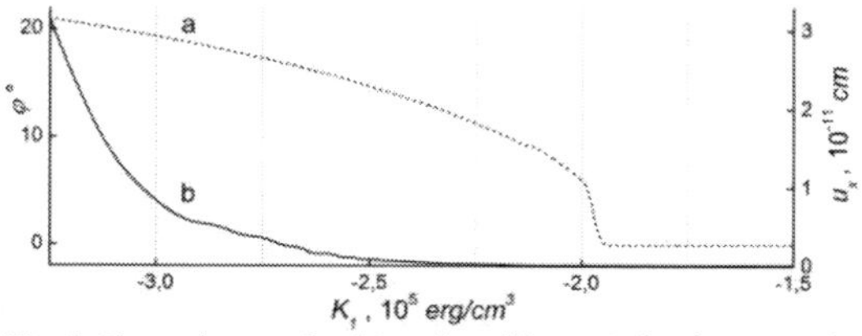

Fig. 3. Dependence of φ (a) and u_x (b) on K_1 for the case of a normally magnetized plate

The mean-square values of the oscillation amplitude were calculated without taking into account the constant component in the considered time interval (which included the region of excitation of the oscillations).

The threshold value K_1 for these values of M_s and ΔH is about -1.8×10^{5} erg cm^{-3}. The reorientation occurs at K_1 smaller than this value. The increase of the amplitude of the elastic oscillations is observed.

Figure 4 shows that the DC field H_0 and the saturation magnetization M_s influence to the threshold value K_1.

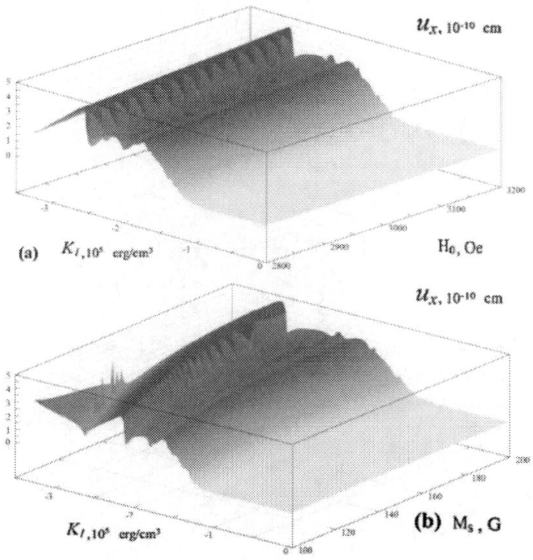

Fig. 4. Dependence of u_x amplitude on K_1 and H_0 (a) and on K_1 and M_s (b) for the case of a normally magnetized plate

Thus, the threshold character of the change in the amplitude of the elastic oscillations can be explained by the interaction of the anisotropy field with the magnetization field.

The calculation of the average value of the amplitude of the magnetic component (Figure 5) makes possible to estimate its final position at the end of the reorientation.

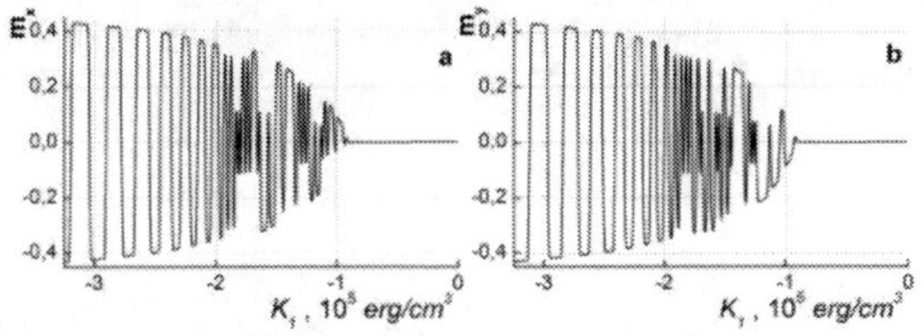

Fig. 5. Average value of the amplitude of the magnetic components m_x (a) and m_y (b) on K_1 for the case of a normally magnetized plate, $\alpha = 0.05$

It is seen that as the module K_1 increases, the final position of magnetization deviates from zero, while spontaneous changes in the direction of the projection of the magnetization vector onto the corresponding axis are observed. Changes in the direction are especially frequent for values of K_1 close to the threshold.

It should be noted that for the geometry of a normally magnetized plate considered above, the precession angle is 0°–20°.

Let us consider the case of tangentially magnetized plate. We use pulse excitation with short pulses duration 5.2×10^{-9} s and amplitude 1 Oe to avoid parametric decay processes. Parameters that differ from those used in the geometry of a normally magnetized plate are: the observation time 10^{-7} s, $d = 1.098\times10^{-4}$ cm, f = 1958 MHz, H_0 = 100 Oe.

The dependence of the mean-square amplitudes of the elastic oscillations and the precession angle φ on K_1 is shown in Figure 6.

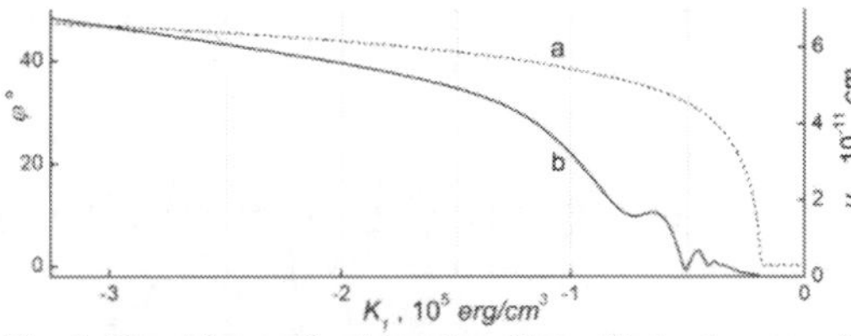

Fig. 6. Dependence of φ (a) and u_y (b) on K_1 for the case of a tangentially magnetized plate

The amplitude of the elastic oscillations is slightly larger than the amplitude in the geometry of the normally magnetized plate. But the amplitude of the alternating field 100 times larger than in the geometry of the normally magnetized plate.

Thus the efficiency of elastic oscillations excitation is far better in the geometry of the normally magnetized plate.

6 Conclusions

The behavior of the amplitudes of magnetic and elastic oscillations was described in the manganese-zinc spinel monocrystalline plate. The optimization method simulating annealing was used. The system of magnetoelastic equations was solved using 7th to 8th order Runge-Kutta-Fehlberg method. Precession patterns and other characteristics for the components of magnetization and elastic displacements were obtained as a function of the anisotropy constants. The value of the first anisotropy constant -1.8×10^5 erg·cm^{-3} was found at which a reorientation of the magnetization vector occurs under the action of a small amplitude alternating field. The changes in the position of the magnetization vector and the equilibrium displacements as the function of the first anisotropy constant were revealed.

The support from RFBR (grants ## 17-02-01138, 17-57-150001) is gratefully acknowledged.

References

1. L. Ingber, Control Cybern. **25**, 33 (1996)
2. V.S. Vlasov, L.N. Kotov, V.G. Shavrov, V.I. Shheglov, J. Communic. Technol. Electron. **54**, 863 (2009)
3. Nicolas Tiercelin et.al. Proc. SPIE, Spintronics X, 103571T (2017).
4. Kim J.W., Bigot J.Y. Phys. Rev. B **95**, 144422 (2017).
5. Kim J. W., Vomir M., Bigot J. Y. Scientific reports **5**, 8511 (2015).
6. L. Thevenard et al. Phys. Rev. B **87**, 144402 (2013).
7. B.A. Goldin, L.N. Kotov, L.K. Zarembo, S.N. Karpachov, *Spin-phonon interactions in crystals* (1991)
8. L. N. Kotov, S. N. Karpachev, Tech. Phys. Lett. **28**, 105 (2002).

Superdirected beam of the backward volume spin wave

Alexander Yu. Annenkov, Sergey V. Gerus and Edwin H. Lock *

Kotel'nikov Institute of Radio Engineering and Electronics of RAS (Fryazino branch), 141190 Vvedensky sq.1, Fryazino, Moscow region, Russia

Abstract. Visualized patterns of the backward volume spin wave (BVSW) excited by arbitrarily oriented linear transducer in tangentially magnetized ferrite film are investigated experimentally in the plane of ferrite film for the case where the transducer length D is much larger than the wavelength λ_0. Superdirected BVSW beam having zero angular width and minimal smearing of the beam energy along the film surface is observed experimentally. Thus, it is proved that such phenomenon as "superdirected propagation of the wave" exists in the nature.

1 Introduction

Some years ago it was investigated theoretically a two-dimensional diffraction patterns arising in the far-field region of a ferrite slab plane for the common geometry, where a plane surface spin wave with noncollinear group and phase velocities is incident on the wide slit in opaque screen with arbitrary orientation [1, 2]. Using magnetostatic approximation [3], it was found, that as a distinct from isotropic media the angular width $\Delta\psi$ of the diffracted spin wave beam is defined not only by the ratio λ/D (here λ is the incident wavelength and D is the slit length), but also by the mathematical properties of isofrequency dependence (ID)[a] for the diffracted wave. The general formula for the angular width $\Delta\psi$ of a diffracted beam was derived as a result of this investigation. It was shown that this formula is valid not only for various types of spin waves, but also for other waves, propagating in various anisotropic media and structures (including metamaterials). As follows from the obtained formula, the angular width of the beam in anisotropic media can not only take values greater or less than λ_0/D, but also can be equal to zero under certain conditions [2]. Thus, analyzing mathematical properties of the wave ID in the certain anisotropic medium or structure, one can find out the wave vector orientation for which superdirected beams (i.e. nonexpanding beams with zero angular width) can be observed experimentally.

Using the analysis of the spin wave ID it was shown in the work [4] that superdirected beams of the backward volume spin waves (BVSW) can be observed in the free ferrite slab too. Mention must be made, that the similar results for BVSW was predicted also in [5], where for superdirected propagation of the beam it was used the terms "focusing" and "caustic propagation".

In order to confirm validity of theoretical results described above, a visualized patterns of the surface spin wave excited by arbitrarily oriented linear transducer were investigated and the angular width of spin wave beams was measured [6]. It was found in the [6] that experimental results are in good agreement with theoretical calculations, basing on [1, 2]. Moreover, superdirected surface spin wave beam was observed experimentally [6].

Below we present the experimental visualized pattern of superdirected BVSW beam excited by arbitrarily oriented linear transducer.

2 Experimental and calculated results

The general scheme of experimental setup is shown in Fig. 1. Propagation of spin wave beams was studied in yttrium iron garnet (YIG) film *5*, grown on a gadolinium gallium garnet (GGG) substrate with a diameter of 76 mm and a thickness of 0.5 mm. YIG film with thickness s = 48.5 μm and magnetization $4\pi M_0$ = 1800 Gs was magnetized to saturation by the tangential uniform magnetic field $\mathbf{H_0}$ with a value H_0 = 500 Oe.

Microwave signal, propagating as a spin wave in the YIG film from input transducer *6* to output transducer *4*, was generated and received by the device *1*, measuring complex transmission coefficient. Then received analog signal was transformed by analog-to-digital converter and was analyzed by special computer *2*. Input transducer *6*, exciting spin wave beam, and output receiving transducer *4* were made of gold-plated tungsten wire of 12 μm thick. Exciting transducer *6* has the length D = 5 mm and receiving transducer *4* has the aperture L ~ 0.5 mm. Two identical systems had provided in-plane displacement of the receiving transducer in two orthogonally directions along YIG film surface. Thus, receiving transducer equipped also by the

[a] ID is also known as "section of wavevector surface", "section of the isoenergy surface" and "equifrequency line".

* Corresponding author: edwin@ms.ire.rssi.ru

special position sensors *3* was turned into movable scanning probe, directed by computer *2*.

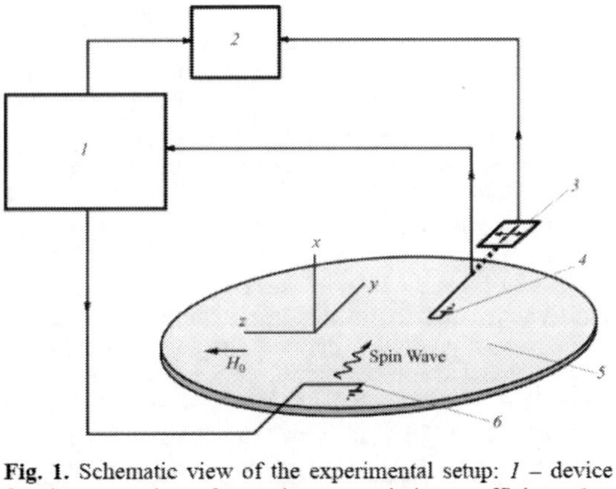

Fig. 1. Schematic view of the experimental setup: *1* – device for the measuring of complex transmission coefficient, *2* – computer, *3* – sensors of the transducer position, *4* – receiving transducer (probe), *5* – YIG film, *6* –linear transducer exciting spin wave.

The distribution of spin wave beam in the plane of the yttrium iron garnet film is studied experimentally by means of the probe scanning method, that was used earlier in [7 - 9]. Recently this method was improved cardinally [10] and now it is possible to visualise distribution of amplitude and phase of spin wave along the film surface by means of computer signal processing (for more detail see [10]). Mention must be made that dipole spin wave distribution may occupy a large area compared with YIG film surface of 76 mm in diameter. Therefore, this method is appear much more effective for dipole spin waves than the similar well known method based on the Brillouin scattering of light by the spin wave [11], while the last method is widely used to study spin wave distribution in the ferrite film plane [12-14].

Note, that the theory [2] is also valid for the case, where spin wave beam is excited by arbitrary oriented linear transducer (this assumption is justified in section 9 of [2]). In this case general formula (30) in [2] for angular beam width $\Delta\psi$ is simplified to the next form

$$\Delta\psi=(\lambda_0/D)\ |d\psi/d\varphi(\varphi_0)| \quad (1)$$

Here angle φ_0 is orientation of the wave vector $\mathbf{k_0}$ respect to the *y* axis (or angle φ_0 may be considered as orientation of transducer's aperture line respect to the vector $\mathbf{H_0}$); angles φ and ψ are orientations for an arbitrary wave vector **k** and a corresponding group velocity vector **V** respectively; $d\psi/d\varphi(\varphi_0)$ – the derivative value, corresponding to the point of ID in which the wave vector $\mathbf{k_0}$ is ended (for more detail see section 8 in [2]).

Characterizing spin wave beams it is convenient to determine both their *absolute* angular width $\Delta\psi$ and their *relative* angular width σ, which is equal to the ratio between the angular width $\Delta\psi$ (in radians) and $\Delta\psi_{isotr}$ = λ_0/D (the angular width of the similar diffracted beam in isotropic medium, in radians):

$$\sigma = \Delta\psi/\Delta\psi_{isotr} = \Delta\psi/(\lambda_0/D) = d\psi/d\varphi(\varphi_0) \quad (2)$$

From the physical standpoint, relative angular width σ shows how much $\Delta\psi$ is smaller (or greater) than the absolute angular width $\Delta\psi_{isotr}$ for the similar beam in isotropic media.

Formula (1) shows, that the angular beam width $\Delta\psi$ is determined not only by the ratio λ_0/D, but also by the derivative value $d\psi/d\varphi$ (φ_0) characterizing the curvature of ID. So if one can find the structure (medium) and geometry (excitation topology), at which $d\psi/d\varphi$ $(\varphi_0) = 0$, then one can observe superdirected (nonexpanding) beam with angular width $\Delta\psi = 0$, i.e. the beam retaining constant absolute width during its propagation!

As it is found in [4], the values $d\psi/d\varphi$ (φ_0) and σ can be equal to zero for certain wave vector orientations φ_0, corresponding to the inflection points of BVSW ID (Fig.2). In these points vectors **k** and **V** are noncollinear. For example, if $f_0 = 2800$ MHz (curve 4 in Fig.2), then $\sigma = 0$ at $\varphi_0 = \pm 17^{\circ}$ and at $\varphi_0 = 180^{\circ} \pm 17^{\circ}$ (for more detail see also fig. 2 - 4 in [4]).

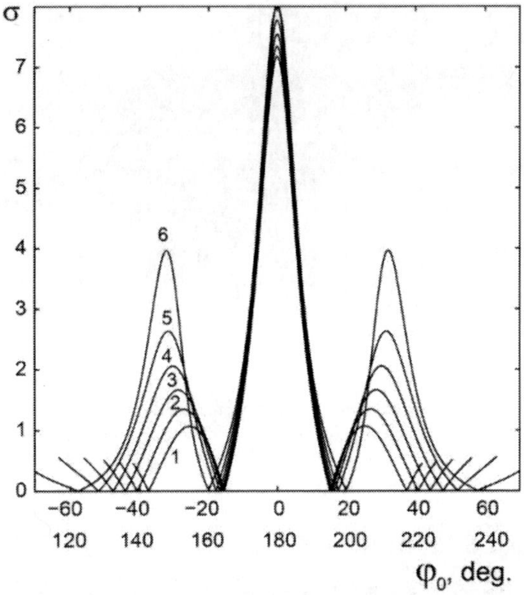

Fig. 2. Relative angular width σ of BVSW beams versus angle φ_0 for various frequencies: 1 – 2500, 2 – 2600, 3 – 2700, 4 – 2800, 5 – 2900, 6 – 3000 MHz.

So to observe superdirected BVSW beam corresponding to the case $\sigma = d\psi/d\varphi$ $(\varphi_0) = 0$, we had excited BVSW with frequency $f_0 = 2800$ MHz and transducer orientation $\varphi_0 = 180^{\circ} - 17^{\circ} = 163^{\circ}$ (at which the ID of BVSW had the inflection point with $d\psi/d\varphi$ (φ_0) $= 0$). At this ID point the wave had wave number k_0 = 118 cm^{-1}, wave length λ_0 = 532 μm and ratio λ_0/D = 0.106. Visualized pattern corresponding to the amplitude distribution of the beam for this BVSW in the YIG film plane is shown in Fig. 3.

The change of color in Fig.3 corresponds to the change of spin wave amplitude about 3 dB. It should be noted, that in isotropic media such beam (with ratio λ_0/D = 0.106) would have angular width about 6.1°, whereas in Fig.3 spin wave beam has absolute angular width $\Delta\psi_{exp}$ = 0.5° and relative angular width σ_{exp} = 0.08! These experimentally measured values $\Delta\psi_{exp}$ and σ_{exp} are slightly differ from calculated zero values, because of as a distinct from the light the spin wave can propagate only at a maximal distance[b] ~ 50 mm and so it is possible to measure absolute angular width with accuracy about 0.3° by means of a probe with aperture of 0.5 mm (minimal angle $\Delta\psi_{exp}$ = 0.5/50/2 = 0.005 radian ≈ 0.3°). Since the superdirected wave beam retains its absolute width during propagation, the direction ψ_0 of the energy transfer for this beam is measured with the greatest possible accuracy ψ_0 = 59.2° ± 0.3°. Mention must be made that the using of the probe with a smaller aperture does not increase the accuracy of measurement because at the large distances the probe sensitivity is decreased proportionally to the aperture size.

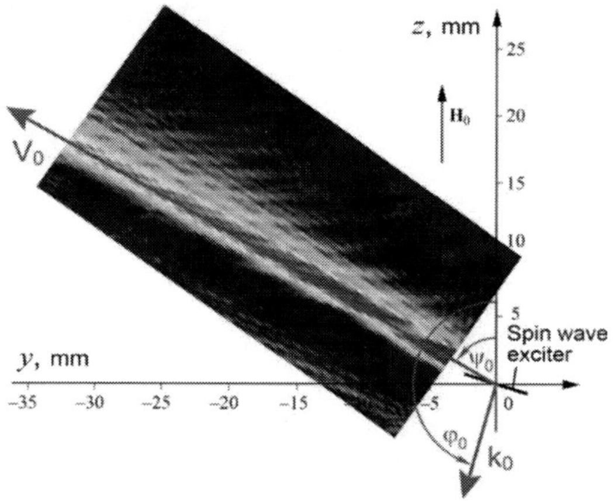

Fig. 3. Amplitude distribution of superdirected beam of BVSW in the YIG film plane (experiment) for frequency f_0 = 2800 MHz, transducer orientation φ_0 = 163°, k_0 = 118 cm^{-1} (wave length λ_0 = 532 μm), ratio λ_0/D = 0.106 and direction of the energy transfer ψ_0 = 59.2°.

3 Summary

Patterns of BVSW beams excited by arbitrarily oriented linear transducer in tangentially magnetized ferrite film are studied experimentally in the plane of ferrite film for the case, where the transducer length D is much larger than the wavelength λ_0. In the study there is used the probe scanning method, that give possibility to visualise the amplitude distribution and phase distribution of the spin wave along the film surface. The angular beam width of spin waves is measured experimentally and calculated theoretically by means of the general formula for the angular beam width in anisotropic media [2].

It was proved experimentally that as a distinct from the beams in isotropic media the angular beam width $\Delta\psi$ of BVSW is not *constant* value: it depends on transducer orientation φ_0 and can take values *greater* or *smaller* than the ratio λ_0/D (where λ_0 is exited wavelength and D is the exciter length). Moreover, it is found such transducer orientation, at which the angular beam width $\Delta\psi$ is about zero (it means physically, that the beam retains its absolute width during propagation). *Thus it was shown that such phenomenon as "superdirected propagation of the waves" exists in the nature.* This phenomenon takes place when the orientation φ_0 of excited wave vector $\mathbf{k_0}$ (or the transducer orientation) corresponds to the ID inflexion point. The direction ψ_0 of the energy transfer for this beam can be measured with the greatest possible accuracy. The comparison between calculated and measured values of angular beam widths both for BVSW and for surface spin waves (see [6]) demonstrates good agreement and gives us hope that the same diffractive phenomenon may take place in the other anisotropic media and structures.

The work is supported by Russian Foundation for Basic Research (project No. 17-07-00016).

References

1. E.H. Lock, Cornell University Library: https://arxiv.org/abs/1112.3929 (2011)
2. E.H. Lock, Physics-Uspekhi, **55**, 1239 (2012).
3. R.W. Damon, J.R. Eshbach. J. Phys. Chem. Solids, **19**, 308 (1961).
4. E.G. Lokk. J. of Communications Techn. Electron. **60**, 97 (2015).
5. V. Veerakumar, R.E. Camley, Phys. Rev. B. **74**, 214401 (2006).
6. A.Yu. Annenkov, S.V. Gerus, E.H. Lock, Cornell University Library: https://arxiv.org/abs/1611.09595 (2016)
7. V.I. Zubkov, E.G. Lokk, V.I. Shcheglov. Radiotekhnika i elektronika. **35**, 1617 (1990).
8. A.V. Vashkovsky, V.I. Zubkov, E.H. Lock, V.I. Shcheglov. IEEE Trans. on Magn. **26**, 1480 (1990).
9. A.Yu. Annenkov, I.V. Vasil'ev, S.V. Gerus, S.I. Kovalev. Tech. Phys. **40**, 330 (1995).
10. A.Yu. Annenkov, S.V. Gerus. J. of Communications Techn. Electron. **57**, 519 (2012).
11. S.O. Demokritov and V.E. Demidov. IEEE Trans. on Magn. **44**, 6 (2008).
12. A. V. Sadovnikov et al. Appl. Phys. Lett. **107**, 202405 (2015).
13. A. V. Sadovnikov et al. Appl. Phys. Lett. **108**, 172411 (2016).
14. A. V. Sadovnikov et al. Phys. Rev. B. **96**, 144428 (2017).

[b] The maximal distance value (~ 50 mm) is defined by the spin wave dissipation, that is much larger than for the light wave in the lot of transparent medium.

Temperature dependence of Gilbert damping in manganite/normal metal heterostructure

Timur A. Shcaihulov*, *Victor V.* Demidov, *Igor V.* Borisenko, *Gennady A.* Ovsyannikov

Kotel'nikov IRE RAS, Mokhovaya 11-7, Moscow 125009, Russia

Abstract. The temperature dependence of the spin-pumping effect on the Gilbert damping in a bilayer based on epitaxial manganite film grown on neodymium galate substrate was investigated by measuring of the linewidth of the ferromagnetic resonance spectrum (FMR). Ferromagnetic resonance in thin ferromagnetic manganite $La_{0.7}Sr_{0.3}MnO_3$ (LSMO) films is used to produce a spin current at the interface between a metallic and a ferromagnetic layer. Pt/LSMO/NGO, Au/LSMO/NGO and LSMO/NGO heterostrucutres were measured. An increase of the linewidth in FMR spectrum at the temperature lower liquid nitrogen temperature (77K) is observed .

1 Introduction

The operation of the most promising spintronics devices is based on a spin transport in magnetic heterostructures associated with the transfer of the spin moment when a spin-polarized current flows through them. Magnetic heterostructure, as a rule, consists of magnetic and nonmagnetic layers. Rare-earth manganite perovskites with the structure $Re_{1-x}A_xMnO_3$ (Re is-rare-earth materials of the type La or Nd), and A is alkaline earth metals such as Sr, Ca, Ba) exhibit a wide range of unusual electrical and magnetic properties, including the colossal magnetoresistance effect (see, review [1]). The parameters of epitaxial films of these materials very often differ substantially from the properties of single crystals. A significant influence on the magnetic and electrical properties of the films is caused by the strain in the films caused by a mismatch with the substrate on which the manganite film [1-3] is deposited, as well as the effects of phase separation and the presence of a nonmagnetic layer at the substrate/film interface [4], which are the strongest Appear in very thin films (less than 10 nm). The lanthanum-strontium manganites $La_{0.7}Sr_{0.3}MnO_3$ (LSMO) have high spin polarization (up to 100%) and can be used in magnetic tunnel junctions [5,6] and spin valves [7,8]. Manganite films for which the Curie temperature Tc is close to room temperature, especially attractive for practical applications. Although a number of studies have been carried out on the excitation of a spin current by ferromagnetic resonance in two LSMO/NM layer structures (NM is a normal metal, usually platinum) [9,10], there is no data on the temperature dependence of the width of the FMR line in the presence of spin current in ferromagnets.

2 Model for spin current

Due to phonon or magnon scattering an infinite precession of the magnetization around the external magnetic field reaches a stable equilibrium state parallel to the effective magnetic field. For a correct description of this phenomena Gilbert proposed introduced a phenomenological correction responsible for the dissipation in the magnetic dynamics [11], which leads to the Landau-Lifshitz-Gilbert equation (LLG):

$$\frac{d\vec{m}}{dt} = -\gamma_0 \vec{m} \times \vec{H}_{eff} + \alpha \vec{m} \times \frac{d\vec{m}}{dt}, \quad (1)$$

here α is the Hilbert damping coefficient, $m=M/|M|$ is unit magnetization vector of ferromagnetic layer, $\gamma=g\mu_B/h$ is the gyromagnetic ration, H_{eff} is s the local effective magnetic field including the external, demagnetization and crystal anisotropy field. The Gilbert damping α is the times scale for relaxation which can take place in two different ways: either the spin angular momentum is transferred to non-magnetic degrees of freedom, which is called direct, homogenous damping; or the spin angular momentum is transferred to other magnetic degrees of freedom such as spin waves, impurity, which is called indirect Inhomogeneous dampingm Significant relaxation processes in thin ferromagnetic films [12] are spin-orbit relaxation damping, two-magnon scattering and eddy current damping.

The increase in the Gilbert damping parameter in a heterostructure from a ferromagnetic and a normal (nonmagnetic) metal caused by precession of the magnetization in a ferromagnetic that causes the spin current to flow through the boundary into a normal metal was considered theoretically [13]. The theory explains the injection of a spin current from a ferromagnetic to a nonmagnetic metal by scattering of the spin at the

* Corresponding author: shcaihulov@hitech.cplire.ru

interface. The spin current from ferromagnetic to the non-magnetic layer moves perpendicular to the interface

$$j_s^0\vec{s} = \frac{\hbar}{8\pi}\mathrm{Re}(2g\uparrow\downarrow)\left[\vec{m}\times\frac{\partial m}{\partial t}\right] \qquad (3)$$

where Re(2g↑↓) is the spin mixing conductance which is additively added to the components of the Gilbert damping.

The Gilbert damping parameter can be written as $\alpha = \alpha_0 + \alpha'$, where α_0 is the intrinsic contribution and α' is the additional damping due to spin pumping [13,14]

$$\alpha' = \frac{\gamma\hbar g^{\uparrow\downarrow}}{4\pi M_0 t_{FM}} \qquad (2)$$

3 Samples and experimental technique

Epitaxial films $La_{0.7}Sr_{0.3}MnO_3$ (LIMO) were deposited by magnetron sputtering on the (110) plane of a single-crystal substrate of NdGaO3 (NGO). 10-20 nm of Pt or Au were sputtered ex situ just after cooling of the film. Contact areas were produced by sputtering of Pt through a metal mask. The resistance of the films was studied by a four-point method, which excludes the influence of contact resistance.

The magnetic characteristics were measured by magnetic resonance technique using ER-200 EPR spectrometer from Bruker (frequency 9.76 GHz). When recording the signals, a synchronous detection was used at the modulation frequency of an external magnetic field, the value of which was 100 kHz. The investigated samples were located in the microwave cavity of the spectrometer in such a way that the plane of the sample was always parallel to the direction of the constant external magnetic field and the magnetic component of the microwave field (parallel orientation). The parallelism of the plane of the samples to a constant magnetic field was controlled by the minimum of the resonance value of the field of the ferromagnetic resonance (FMR) lines). This arrangement of the samples eliminated the change in the magnetic resonance spectra due to shape anisotropy. Rotation of the samples was carried out around an axis perpendicular to the plane of the samples.

The films just after deposition (is designd as d-LSMO) and heated at T=720C during one hour (is designed as h-LSMO) were investigated

4 Magnetic parameters of the films

The method for determining the parameters of the magnetic anisotropy consists in processing the angular dependences of the resonant fields of the FMR spectra. The solution of the Landau-Lifshitz equation for the evolution of the magnetization M in an external constant magnetic field H under the action of the magnetic component of the radio-frequency field is used, which gives an analytic equation for the resonance field H_0 and the frequency ω [15]

$$\left(\frac{\omega}{\gamma}\right)^2 = \left(4\pi M_0 + H_0 + \frac{2K_u}{M_0}\cos^2\varphi_u + \frac{2K_c}{M_0}\frac{1+\cos^2 2\varphi_c}{2}\right)\left(H_0 + \frac{2K_u}{M_0}\cos 2\varphi_u + \frac{2K_c}{M_0}\cos 4\varphi_c\right) \qquad (3)$$

here γ is the gyromagnetic ratio, M_0 is equilibrium magnetization, K_u and K_c are the uniaxial strain and the biaxial cubic anisotropy constants correspondingly. tφ_u and φ_c are the angles between the earsy axis of the uniaxial and cubic anisotropies and the external magnetic field correspondingly. K_u, K_c, M_0 as well as angles φ_u and φ_c were obtained by the fitting of the angular dependence (Fig.1)

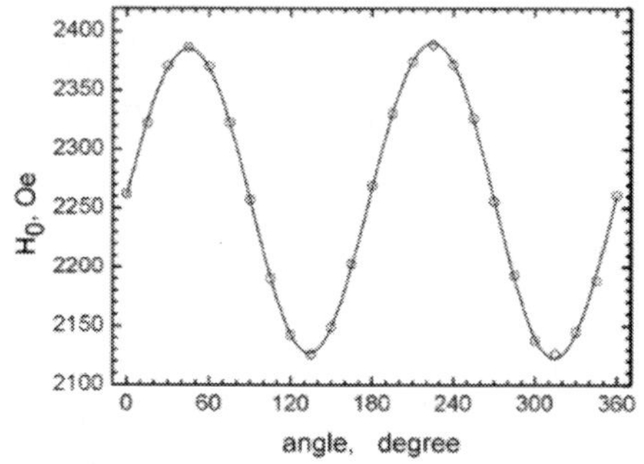

Fig. 1. Angular dependence of resonance field for h-LSMO film at T=300K. The fitting of the experimental data using eq. (3) give the following magnetic parameters of LSMO film: magnetization M_0=300 Oe, H_u=190 Oe, H_c=10 Oe

Fig. 1 shows the angular dependence of the resonance value of the magnetic field H_0 for the LSMO film taken at room temperature and which was rotated around the normal to the film plane.φthe sample by an angle φ. The angle was measured from one of the faces of the substrate. The constant magnetic field and the magnetic component of the microwave field were in the plane of the film. Since the substrate with the film was square 5x5 mm, the effect of the anisotropy of the shape of the sample is minimal, and the entire shift of the resonant field is due to the planar magnetic anisotropy of the LBMO film. The angular dependence was described by a resonance relation taking into account magnetic uniaxial and cubic plane anisotropies (3) [15]. It can be seen that the relationship used describes the experimental data quite well. As a result, the magnetization M_0, anisotropy constants Ku and Kc $H_{u,c} = 2K_{U,C} / M_0$ were determined.

The temperature dependences of magnetic parameter are shown on Fig.2. Magnetization of the film increases up to $2.2\mu_B$/Mn at T≤200K. Magnetic anisotropy fields increase at lower T<50K. Uniaxial magnetic anisotropy dominates over cubic anisotropy.

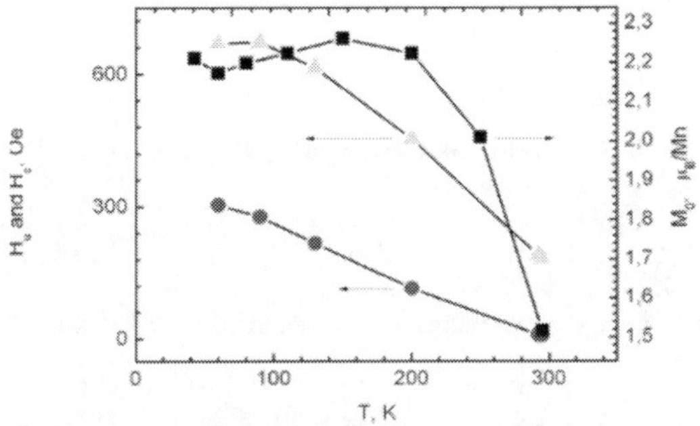

Fig.2. Temperature dependence of magnetization M_0, biaxial H_u and cubic H_c magnetic field anisotropy correspondingly for h-LSMO film

5 The linewidth of FMR

The width of the ferromagnetic resonance line (FMR), measured during the scanning of an external magnetic field H, is defined as the difference betweem the extrema of the spectrum.

$$\delta_{pp} = H_{p+} - H_{p-} \qquad (5)$$

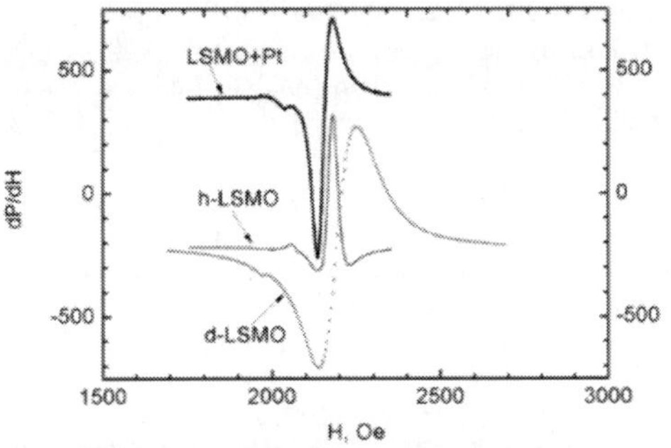

Fig. 3. FMR spectrum for d-LSMO and Pt/LSMO heterostructures. The increasing of linewidth is clearly observed. The spectra for d-LSMO and Pr/h-LSMO are moved on the axis dP/dH.

The initial and final points of which H_{p+} and H_{p-} constitute the position of the maximum and minimum of the first derivative dP/dH of the high-frequency field absorption signal P (H) (see Fig. 2). At this value of the resonance field, H_0, defined as the transition point of the signal dP/dH through zero, always lies in the range of $H_{p+} < H_0 < H_{p-}$. It is known that the linewidth δ_{pp} determined in the above way depends on the angle φ that determines the orientation of the resonant magnetic field with respect to axes of anisotropy.

The physical reason for the change in linewidth ΔH_{pp} films of anisotropic ferromagnets due to the change of the equilibrium orientation of the magnetization M with the registration of the absorption line. It is known that in manganite films contribution to the angular dependence of H_0 are making as uniaxial (substrate influence) and cubic (crystallographic) magnetic anisotropy. When the deviation of the magnetic field from the axis of the anisotropy of the linewidth of the FMR increases .

We deposited normal metal (Pt, Au) on LSMO to evaluate the efficiency of generation of spin current at the boundary of a ferromagnetic–normal metal. Using the following expressions, we calculated the spin mixing conductance in bilayers LSMO/Pt.

$$g_{eff}^{\uparrow\downarrow} = \frac{4\pi\gamma_g M_s t_{LSMO}}{g\mu_\beta\omega_f}\left(\delta H_{Pt/LSMO} - \delta H_{LSMO}\right) \qquad (6)$$

$\gamma_g = 17{,}605 \cdot 10^6$ is gyromagnetic ratio for electron , $\omega_f = 2\pi \cdot 9.51 \cdot 10^9 s^{-1}$ is the microwave angular frequency, $M_s = 300$ Oe is the magnetization LSMO film, $t_{LSMO} = 40$ nm is the thickness of the layer of platinum, $\mu_B = 9.274 \cdot 10^{-21}$ erg/Gs is the Bohr magneton, g=2 is Lande factor. We got in the case $\delta_{pp} = 130$ Oe at T=60K $g_{eff} = 1{,}5\ 10^{20}\ m^{-2}$. For comparison for interface Py/Pt it was obtained $g_{eff} = 2{,}1 \cdot 10^{19}$ м$^{-2}$ [16] and for YIG/Pt $g_{eff} = 4{,}8 \cdot 10^{20}$ м$^{-2}$ [17]. There is no spin conductance in bulk Pt taken into consideration. So the obtained value of spin-conductance is over estimated.

6 Temperature dependence

Temperature dependence of FMR spectrum and, in particular, the width of the resonance lines of magnetization, and other parameters in the heterostructure of Pt/LSMO and LSMO were measured. Possible sources of additional damping could be the sputtered film of normal metal (Pt and Au) on LSMO. The mechanism of spin pumping weakens the precession of the magnetization, creating a spin current from the ferromagnetic film to the layer of normal metal[16-18]. Several other processes contribute to the temperature dependence of damping [14].

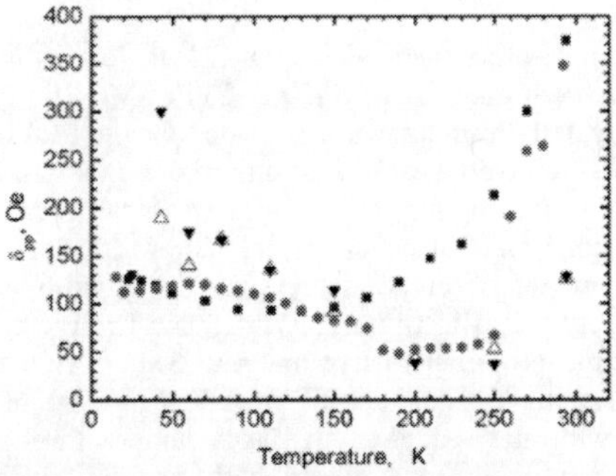

Fig.4. Temperature dependence of FMR linewidth for the following heterostuctures Au/d-LSMO, Pt/d-LSMO (triangle), d-LSMO (empty triangle, inverse triangle) for two angles of magnetic field directions

The temperature dependence of the FMR linewidth for the heterostructures based on d-LSMO are given in figure 4. It is seen that with decreasing temperature the linewidth of all the structures is reduced. The increase in linewidth in the heterostructure Au/LSMO/NGO at low temperatures T<150K can be attributed to existence the layer with the reduction of Curie temperature at the interface LSMO with substrate. It plays the role of a barrier which blocks the spin pumping interface at T low TCU of interfaced. The effect was observed for heterostructure Pt/SRO/LSMO. Here SRO is SrRuO3 film with T_{CU} =150K[19].

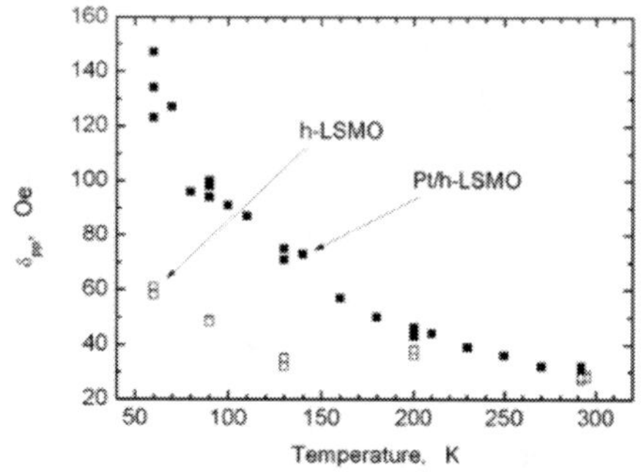

Fig.5. Temperature dependence of FMR linewidth for the films h-LSMO, and Pt/h-LSMO taken with an external magnetic field directed along the easy axis of uniaxial magnetic anisotropy

The temperature dependences of linewidth for the manganite film h-LSMO and for Pt covered Pt/h-LSMO are presented at Fig. 5. There is no reduction for linewidth at high T observed. Comparison data on Fig.4 and Fig.5 show that inhomogeneous increasing of line width clearly observed at high temperature T>150K. Increasing of linewidth at low T still exist even at h-LSMO which more homogeneous then d-LSMO. Here possible mechanism increasing of linewidth with T is connected with interface at the substrate and LSMO film.

7 Conclusion

The temperature dependence of FMR linewidth and magnetic parameters of LSMO and Pt/LSMO films were investigated. From linewidth obtained for the film before and after deposition of Pt film on top of epitaxial LSMO films the spin mixing conductance was determined. The non monotonic dependence of linewidth of FMR spectrum was observed; decreasing of linewidth at high temperature and increasing at T<150K. The increase in linewidth in the heterostructure Au/LSMO/NGO at low temperatures T<150K can be attributed to existence the layer with the reduction of Curie temperature at the interfaces of LSMO either with normal metal or substrate.

V.A. Atsarkin is acknowledged for fruitful discussion. Support by RFBR 17-02-00145, 16-29-14022 and Scientific school NSh -8168.2016.2.

References

1. A.-M. Haghiri-Cosnet and J.P. Renard, J. Phys D:Appl. Phys. **36**, R127 (2003).
2. Zh. Huang, G.Y. Gao, Zh.Zh. Yin, X.X. Feng, Y.Zh. Chen, X.R. Zhao, J.R. Sun, W.B. Wu, J. Appl. Phys. **105**, 113919 (2009).
3. G.A. Ovsyannikov, A.M. Petrzhik, I.V. Borisenko, A.A. Klimov, Yu.A. Ignatov, V.V. Demidov, S.A. Nikitov, JETP, **108**, 48 (2009).
4. N.D. Mathur, G. Burnell, S.P. Isaac, T.J. Jackson, B.S. Teo, J.L. MacManus-Driscoll, L.F. Cohen, J.E. Evetts, M.G. Blamire, Nature **387**, 266 (1997).
5. M. Bowen, M. Bibes, A. Barthélémy, J.P. Contour, A. Anane, Y. Lemaître, and A. Fert, Appl. Phys. Lett. **82**, 233 (2003).
6. Y. Ishii, H. Yamada, H. Sato, H. Akoh, M. Kawasaki, and Y. Tokura, Appl. Phys. Lett. **87**, 22509 (2005).
7. Z. H. Xiong, D. Wu, Z. V. Vardeny, and J. Shi, Nature **427**, 821 (2004).
8. V. Dediu, M. Murgia, F. C. Matacotta, C. Taliani, and S. Barbanera, Solid State Commun **122**, 181 (2002).
9. G. Y. Luo, C. R. Chang, and J. G. Lin, J . Appl. Phys **115-119**, 17C508 (2014).
10. V.A. Atsarkin, B.V. Sorokin, I.V. Borisenko, V.V. Demidov and G.A. Ovsyannikov, J. Phys. D: Appl. Phys. **49** (2016) 125003 (8pp).
11. T.L. Gilbert, IEEETrans.Magn. **40**, 3443 (1949)
12. B Heinrich, R Urban and G Woltersdorf, *J. Appl. Phys.* **91,** 7523 (2002)
13. A. Tserkovnyak, Brataas, E.W. Bauer, Phys. Rev. Lett., **88**, 117601 (2002).
14. A. Azevedo,L.H. Vilela-Leao,R. L. Rodrıguez-Suarez, A.F. Lacerda Santos, and S. M. Rezende Phys.Rev. **B83**, 144402 (2011).
15. V.V. Demidov, G.A. Ovsyannikov, A.M. Petrzhik, I.V. Borisenko, A.V. Shadrin, and R. Gunnarsson J. Appl. Phys. **113,** 163909 (2013).
16. O. Mosendz, V. Vlaminck, J. E. Pearson, F.Y. Fradin, G. E. W. Bauer, S. D. Bader, and A. Hoffmann, Phys. Rev. **82**, 214403 (2010).
17. M. Rezende, R. L. Rodr_ıguez-Suarez, M. M. Soares,1 L. H. Vilela-Le, D. Ley Dom_ınguez, and A. Azeved, Appl.Phys.Lett. **102**, 012402 (2013).
18. T.G.A. Verhagen, H.N. Tinkey, H.C. Overweg, M. van Son, M. Huber, J.M. van Ruitenbeek and J. Aarts, J. Phys. Condens. Matter **28,** 056004 (2016) (10pp).
19. S. Emori, U.S. Alaan, M.T. Gray, V.Sluka, Y. Chen, A.D.Kent, Y.Suzuki, Phys. Rev **B94**, 224423 (2016).

STUDY OF TORQUE ACTING ON MAGNETIZATION DURING 180° PULSED REVERSAL PROCESS IN FERRITE-GARNET FILMS WITH BIAXIAL ANISOTROPY

Oleg Kolotov[1], *Andrey* Matyunin[1,*], *Georgiy* Nikoladze[1], and *Petr* Polyakov[1]

[1]MSU, Faculty of Physics, 1-2 Leninskiye Gory, GSP-1, Moscow, 119991, Russia

Abstract. The mode of 180° pulsed magnetization reversal process of ferrite-garnet films with in-plane anisotropy in the external fields region, in which the mechanism of the homogeneous rotation of magnetization takes place, was firstly investigated. Analysis of calculated and experimental longitudinal signals showed that the presence of biaxial anisotropy in real ferrite-garnet film leads to the occurrence of nonlinear damped magnetization oscillations (with a period of the main harmonic of ~1.5 ns).

To calculate the torque acting on the magnetization during the 180° pulsed reversal process in real ferrite-garnet films, in which besides in-plane anisotropy biaxial anisotropy is present, the method based on the analysis of the trajectory of an operating point is used [1]. As the coordinates of the operating point the azimuthal angle φ and the torque component T_p caused by the switching field H_p are used. During the calculations we used the parameters of the real film: saturation magnetization $M_S = 14$ G, saturation magnetization field $H_{sat} = 2.5$ Oe and effective uniaxial, biaxial and in-plane anisotropy fields, respectively: $H_{K1} = 4$ Oe, $H_{K2} = 16K_2/M_S = 36$ Oe and $H_{Kp} = 1100$ Oe. Composition of the film — $(YLuBi)_3(FeGa)_5O_{12}$, thikness $d = 4$ μm. Landau-Lifshitz damping constant λ was equal to $2 \cdot 10^7$ Hz (in that way the best possible agreement of the calculated and experimental signals occurs). Fig.1 illustrates this

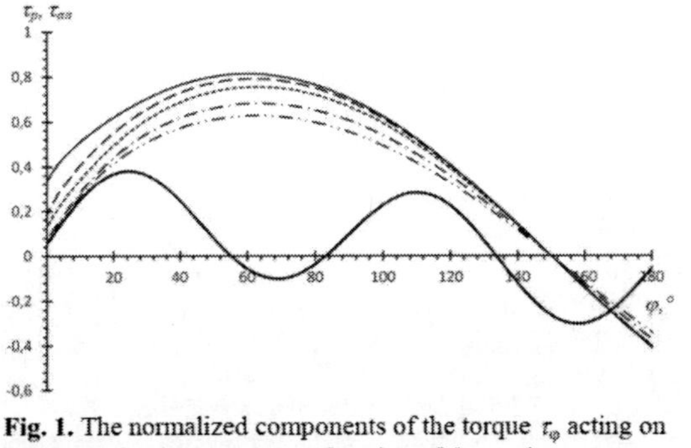

Fig. 1. The normalized components of the torque τ_φ acting on the magnetization vector as a function of the angle φ.

statement. Solid thick line corresponds to the braking torque $\tau_{an} = T_{an}/(M_S \cdot H_{K2})$ (here we use the normalized value of the torque) caused by the uniaxial H_{K1} and biaxial H_{K2} anisotropy and setup field H_0. Other lines correspond to the different values of accelerating torque $\tau_p = T_p/(M_S \cdot H_{K2})$ and applied switching field H_p (= 30 Oe): thin solid line corresponds to the H_p field pulse with rise time of 0.3 ns, dashed line – 0.9 ns, dotted line – 1.5 ns, dash-dotted line – 3.5 ns, dash-dot-dotted line – 6.5 ns. The behaviour of the magnetization vector in this regime is similar to the process of 90° pulsed magnetization of such ferrite-garnet films, which was investigated earlier [2,3]. Thus it has been shown that the «effect of delayed acceleration of the transient process» also takes place during this regime (at the time moment t^* - see Fig.2, where the longitudinal calculated

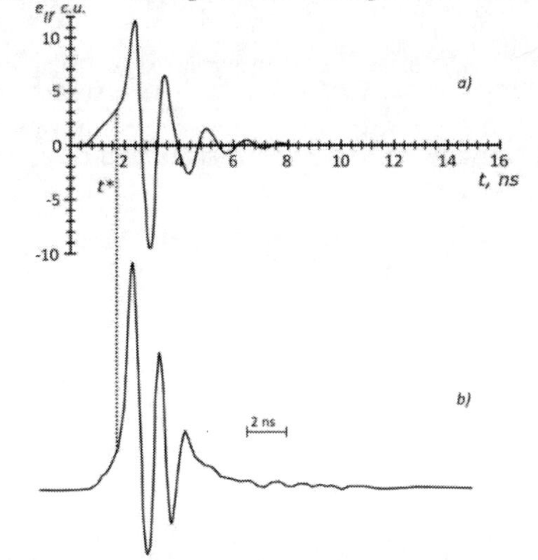

Fig. 2. Longitudinal calculated (*a*) and experimental (*b*) magnetization reversal signals obtained for the H_p field pulse with rise time of 1.5 ns and the value of 30 Oe.

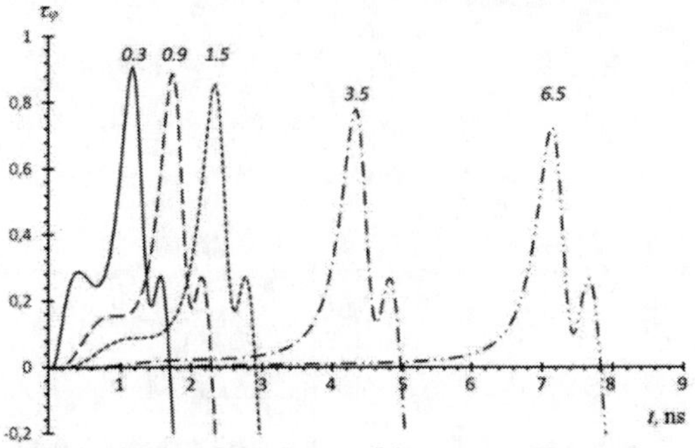

Fig. 3. The time dependences of the torque τ_φ. The numbers above the dependences represent the corresponding front duration values.

---

[*] Corresponding author: physphak@mail.ru

and experimental signals represented). Meanwhile an additional maximum on the resulting torque $\tau_\varphi = \tau_p - \tau_{an}$ is observed (Fig.3). It appears at the moment of excitation of nonlinear magnetization oscillations (with a period of the main harmonic of ~1.5 ns (see Fig.2)) [3], when magnetization reaches easy axis. Its influence is not significant and can be explained by the magnetization's inertia.

Conclusion

Thus, the 180° pulsed reversal process is similar to the 90° pulsed magnetization process in many manifestations: for example, the «effect of delayed acceleration of the transient process» is also observed. Also an additional maximum on the resulting torque is observed. Its influence is not significant and can be explained by the magnetization's inertia.

References

1. O.S. Kolotov, A.V. Matyunin, G.M. Nikoladze, P.A. Polyakov, *Sol. State Phenomena*, **233**, 490 (2015).
2. O.S. Kolotov, A.V. Matyunin, G.M. Nikoladze, P.A. Polyakov, *Phys. Solid State*, **54**, 2380 (2012).
3. O.S. Kolotov, A.V. Matyunin, G.M. Nikoladze, P.A. Polyakov, *Tech. Phys.*, **60**, 1809 (2015).

THE GRANULE SIZE DISTRIBUTION INFLUENCE IN NANOCOMPOSITES ON OPTICAL AND MAGNETOOPTICAL SPECTRA

Alexey Yurasov [1*], *Elena* Gan'shina [2], *Alexey* Sokolov[1], *Nikita* Granovsky[1] and Daria Zazymkina[1]

[1] Moscow Technological University (MIREA) , Physical technological institute, 119454, Moscow, Russia
[2] Moscow State University by M.V. Lomonosov (MSU) , Physical Department, 119991, Moscow, Russia

Abstract. We have investigated the size effect (quasi-classical size effect) in nanocomposites. It was shown that the size effect can change the amplitude, form and sign of the optical and magnetooptical spectra. We have deduced formulas for size effect and discussed the applications of the distributions for corrected description of optical and magnetooptical properties with regard to the granule size effect. It is very important to consider the distribution on the granule size in size effect. This fact allows to describe optical and magnetooptical spectra of nanocomposites better, especially in near IR due to intraband electron transitions. We have deduced formulas for size effect and discussed applications of the distributions for corrected description of optical and magnetooptical properties with regard to the effect of the granule size.

1 Introduction

Optical and magnetooptical features in nanocomposites are strictly connected with the size effects [1-3]. So as there is a very important problem to describe optical and magnetooptical spectra of these structures especially in near IR due to intraband electron transitions. It was shown that the size effect can change the amplitude, form and sign of the optical and magnetooptical spectra[2,4]. The calculations with consideration of the distribution on the granule size in size effect allow to describe magnetooptical spectra for layerwise sprayed $(Co_{40}Fe_{40}B_{20})_x(SiO_2)_{100-x}$ nanocomposites qualitatively. Nanocomposites are heterogeneous magnetic materials in which ferromagnetic particles are placed in a para - or diamagnetic matrix of the dielectric. Nanocomposites are interesting because of the presence of a percolation threshold near which the electrical, optical and magneto-optical properties change significantly. This structure is a good example for the consideration of this phenomenon[1,2].

To describe the properties of ferromagnetic nanocomposites effective medium approaches[1] are used. It is necessary to consider that the scattering on the surface of the granules, leading to quasi-classical size effect, modifies as a diagonal $\varepsilon_{xx}=\varepsilon=\varepsilon_1-i\cdot\varepsilon_2$, and the nondiagonal $\varepsilon_{xy}=\gamma$ components of the tensor of the dielectric granule permittivity , if they are of average size (radius r_0) comparable with the free path length of electron l. The size effect has a significant influence on the optical and magneto-optical properties of nanocomposites particularly in the near infrared region of the spectrum by changing the amplitude, shape and the sign of the spectra. The present work discusses the influence of granule size distribution on the optical and magneto-optical properties of nanocomposites.

2 Experimental data

Nanocomposite "amorphous ferromagnet – insulator" $(Co_{41}Fe_{39}B_{20})_x(SiO_2)_{100-x}$ was obtained in Voronezh state technical University by the method of ion-beam sputtering in argon atmosphere. The choice of such a complex granule composition $Co_{41}Fe_{39}B_{20}$ was made due to the fact that the amorphous structure of such a ferromagnetic is stable at room temperature. The specimen is produced by ion-beam sputtering in argon atmosphere. When using a vacuum chamber with three sources ion-beam [5].

*Corresponding author: alexey_yurasov@mail.ru

A composite alloy target was used for the deposition. Floatable target $Co_{41}Fe_{39}B_{20}$ is manufactured by an induction vacuum melting. Cobalt with the purity 99.98% is used for the preparation of alloys, the amount of the carbonyl iron and boron correspond to the alloy composition. The composite target was an alloy target attached to the surface plate of a single crystal quartz/alumina. The thickness of the obtained film samples was 0.15-6.5 μm.
The composition of the resulting composites was controlled by the electron probe x-ray microanalysis. The electrical and magnetoresistive properties of the obtained samples were studied by out two-probe potentiometric method.
The measurement of TKE was carried out by the dynamic method, consisting in the detection of the small changes in the intensity of the reflected light when the sample is remagnetized by an alternating magnetic field [see, e.g. 6]. All other experimental details are shown in [6].

3 Calculations and discussion

We'll find a given size granule distribution in the following way[1]. We considered the granules as sphere particles with size r_0. The free time path of electrons in the granule (τ_{part}) is less than its corresponding time in the massive sample (τ_{bulk}) due to the collisions with the surface of the granules [1]:

$$\frac{1}{\tau_{part}} = \frac{1}{\tau_{bulk}} + \frac{v_f}{r_0}, \qquad (1)$$

where v_f is Fermi velocity and

$$\varepsilon_{mod} = \varepsilon + \frac{\omega_p^2}{\omega(\omega+i/\tau_{bulk})} - \frac{\omega_p^2}{\omega(\omega+i/\tau_{part})} \qquad (2)$$

The mentioned above values are the diagonal components of the permittivity tensor with the size effect influence ε_{mod}, where ω is frequency of the light, ω_p is plasma frequency.
Then, taking into account that the frequency dependence of the intraband conductivity can be described by the Drude-Lorentz formulas, similar to the formula [1] is the following:

$$\gamma_{mod} = \gamma + \frac{4\pi\sigma_{xy}^{bulk}(0)/\tau_{bulk}^2}{\omega(\omega+i/\tau_{bulk})^2} - \frac{4\pi\sigma_{xy}^{gr}(0)/\tau_{part}^2}{\omega(\omega+i/\tau_{part})^2}, \qquad (3)$$

They are nondiagonal components of the permittivity tensor with the size effect influence,
where $\sigma_{xy}^{bulk}(0) = 4\pi M_s R_{bulk} / \rho_{bulk}^2$, $\sigma_{xy}^{gr}(0) = 4\pi M_s R_{gr} / \rho_{gr}^2$, M_s is saturation magnetization, R_{gr} is the coefficient of anomalous Hall effect, τ_{bulk} is the corresponding free time of the electrons in the bulk material, τ_{gr} is the corresponding free time of the electrons in the granule, ρ_{bulk} is the resistance of the bulk material, ρ_{gr} is the resistance of the granules[4].
The size effect has an impact on the coefficient of the extraordinary Hall effect, and resistivity. The latter is given by the expression $\rho_{gr} = \rho_{bulk}(1+l/r_0)$, and the influence of the size effect on factor of the extraordinary Hall effect granules can be written as:

$$R_{gr} = R_{bulk} + 0.2 R_s \frac{l}{r_0}(1 + \frac{l}{r_0}), \qquad (4)$$

where R_s is the value of the coefficient of the extraordinary Hall effect on the surface material of the granules [7]. The characteristic dependences of the optical and magneto-optical parameters from granule size are given in [1,2,4].
In the above formulas (1) to (3) a possible distribution of granules by size is not taken into account. At the same time, it is important for the description of optical and magneto-optical spectra of the nanocomposites, since the diagonal components of the dielectric permittivity tensor are responsible for the optical properties, and the nondiagonal for the magneto-optical properties. This is especially true in the near infrared region of the spectrum that is due to intraband transitions [2, 4]. (2)
Accounting for the size distribution of the granules allows to describe the spectra of the transverse Kerr effect (TKE) more accurately on the p-component is calculated using equations (2-3) according to the formula:

$$\delta_P = (A\gamma_1 + B\gamma_2)\frac{2\sin 2\varphi}{A^2 + B^2}, \qquad (5)$$

where $A = \varepsilon_2(2\varepsilon_1\cos^2\varphi - 1)$, $B = \cos^2\varphi(\varepsilon_2^2 - \varepsilon_1^2 + 1) + \varepsilon_1 - 1$, φ is the incidence angle of light, δ_p is the parameter of TKE [8]. We considered magnetic nanocomposite for which were optical and magneto-optical parameters experimentally measured [6]. The calculated theoretical spectra without size effect of TKE is shown in Fig. 1

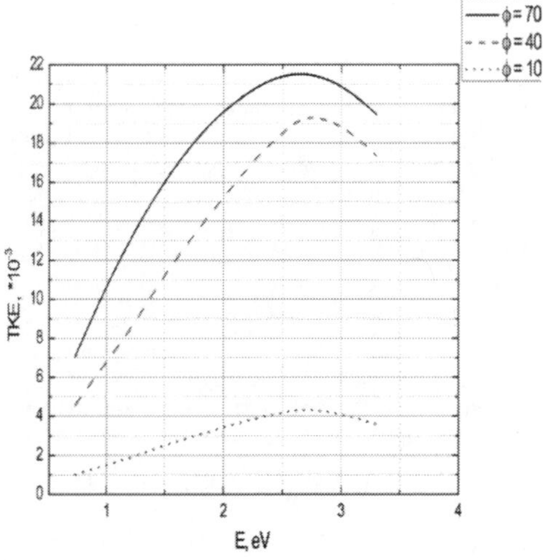

Fig. 1 The calculated spectra of the transversal Kerr effect $(Co_{41}Fe_{39}B_{20})x\ (SiO_2\ )_{100-x}$ nanocomposite: φ=70^0 - solid line ; φ=40 ^{0} - dash line; φ=10 ^{0} - dot line (x = 34%, without size effect).

It is clearly seen that the magnitude of this effect strongly depends on the incidence angle of light [Fig.1]. It was shown that the magnitude of the TKE varied more than an order of magnitude [Fig.1].

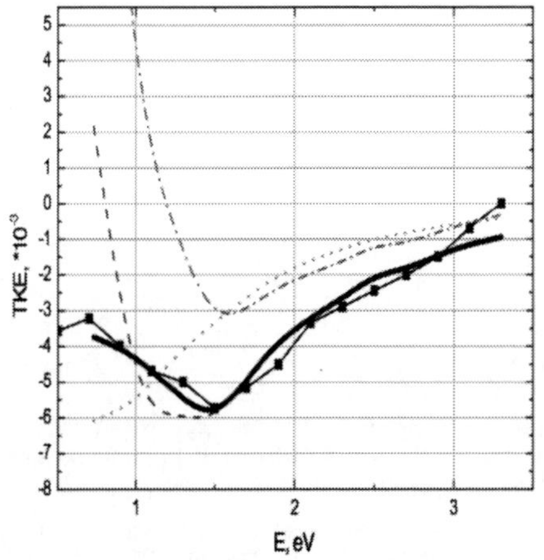

Fig. 2 The calculated spectra of the transversal Kerr effect with the size effect influence compared to the experiment: squares- experimental spectrum; dash dot line-calculated spectra at φ=70^0 ; dash line-calculated spectra taking at φ= 77 ^{0} ; dot line-calculated spectra at φ=83^0 (x = 34%, L=0.33, r_0=2 nm, R_s /R_{bulk}=-2.25) and solid line-calculated spectra taking into account the size effect and particle size distribution at φ=77 ^{0}.

Fig. 2 shows the calculated spectra of the TKE with the size effect influence compared to the experiment. It was proved that the size effect can change the amplitude, form and sign of the optical and magnetooptical spectra and allows to describe the experimental data well.

For a more accurate description we must take into account the granules distribution by size. Schematically, such a nanocomposite is shown in Fig. 3. Figure 3 shows one of the variants of layer-by-layer sputtered composites heterogeneity. It is shown that the ferromagnetic granules, which are closer to the previous composite layers have a smaller size compared with those that are in the scope of layers. It is due to the fact that after the deposition of each layer the films are cooled. When we started spraying a new layer, ferromagnetic and dielectric phase falling on the still cold of the previous layer and cooled immediately, which led to the formation of small granules. In proportion the growth of the layer thickness the sample is heated that leads to the formation of larger ferromagnetic grains in the layer. Thus, we can say that composites sputtered layers are not just bulk-composites with different randomly distributed size granules. In layer-by-layer sputtered composites we can observed a periodic layered structure, the characteristics of which strongly influence the magneto-optical properties of the films.

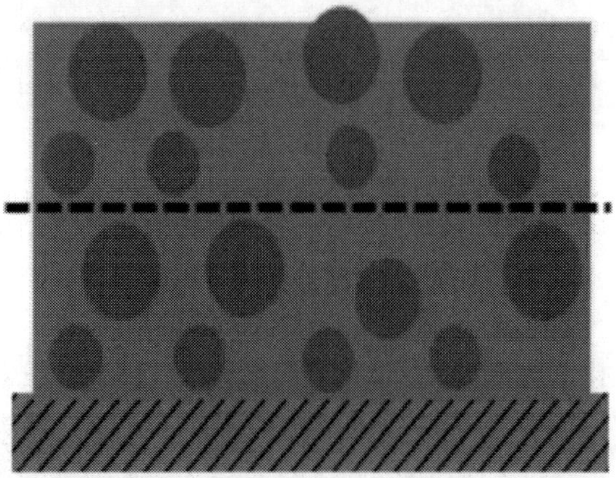

Fig. 3 The particle size distribution in the layers of deposited nanocomposite (dash line – the board between of layer-by-layer sputtered composites heterogeneity).

In the case of a uniform distribution the probability density has the form:

$$f_x(x) = \begin{cases} \dfrac{1}{b-a}, & x \in [a,b] \\ 0, & x \notin [a,b] \end{cases} \qquad (6)$$

And the distribution function is :

$$F_x(x) \equiv P(X \le x) = \begin{cases} 0, & x < a \\ \dfrac{x-a}{b-a}, & a \le x < b \\ 1, & x \ge b \end{cases} \qquad (7)$$

In our version, a is the smallest possible size of a granule, and b is the largest dimension. Now we denote by r_0 is the smallest size, r_1 is the highest size, and r is the current size, then the function size of the granules $R(r)$ can be written in the form:

$$R(r) = r_0(1+F(r)). \qquad (8)$$

If the size of the granules is changed two times compared to the original r_0, the function takes the form:

$$R(r) = r_0(1+\frac{r-r_0}{r_1-r_0}). \qquad (9)$$

It is important to note that the uniform distribution (6-7) is certainly important, but not the general case. Therefore, in consideration of nanocomposites of various compositions, it is important to find the magnitude of the $F(r)$. In figure 2 the solid line presents the spectrum of TKE based on the granules distribution by size and by calculations using the function (9) .This particle size distribution allows to describe the corresponding experimental data well (Fig.2).

Conclusion

Spectra of transversal Kerr effect in $(Co_{41}Fe_{39}B_{20})_x(SiO_2)_{100-x}$ nanocomposite have been calculated. Both experimental and theoretical spectra have a local minimum at E ~ 1.5 eV. This is due to the fact that the size effect in the IR spectral region is connected with intraband transitions. The good interpretation of an experimental data is possible only when the size effect influence is taken into account. Furthermore, for more accurate calculations a distribution function of granules by its size should be used. Thus, the obtained results concerning the size effect and size distribution of the granules allow to improve the description and to explain the amplitude, form and sigh of the optical and magnetooptical spectra in nanocomposites mainly in the near infrared region.

This research was supported by the Russian Foundation for Basic Researches №15-02-02077 and Grant of the RF President for Leading scientific schools SS - 8003.2016.2.

References

1. A.N. Yurasov , RTZ, **1(10)**, 25 (2016) .

2. G.A. Niklasson, C.G.Granqvist, J. Appl. Phys. **55**, 3382 (1984).

3. D. R. Stevens, E. W. Skau, L. N. Downen, M. P. Roman, and L. I. Clarke, Physical Review E **84**, 021126 (2011).

4. A.B. Granovsky, E.A. Gan'shina, A.N. Vinogradov, I.K.Rodin, A.N. Yurasov and H.R. Khan, Physics of Metals and Metallography, **91**, S52 (2001).

5. C. L. Chien, Xiao John Q., and J. Jiang Samuel , J. Appl. Phys., **73**, 5309 (1993).

6. V. Buravtsova, . E. Ganshina, E. Lebedeva, N. Syr'ev, I. Trofimenko, S. Vyzulin, I. Shipkova, S. Phonghirun, Yu. Kalinin, A. Sitnikov, Solid State Phenomena, **168-169**, 533 (2011).

7. E. Ganshina, A. Granovsky, V.Gushin, Kuzmichov M., P. Podrugin, A. Kravetz, E. Shipil, Physica A: Statistical Mechanics and its Applications., **241**, 45 (1997).

8. P. M. Xui, D.Stroud, Appl. Phys. Lett, **50**, 950 (1987).

Measurement of magnetic noise in magnetoimpedance sensing element

Anatoliy S. Antonov[1,*], *Nikita A.* Buznikov[2], *Timur T.* Sultan-Zade[1], and *Alexey A.* Shkliaev[3]

[1]IMPEDANCE JSC, 124527 Zelenograd, Moscow, Russia
[2]Scientific & Research Institute of Natural Gases and Gas Technologies – Gazprom VNIIGAZ, 142717 Razvilka, Leninsky District, Moscow Region, Russia
[3]Decima LLC, 124498, Zelenograd, Moscow, Russia

Abstract. A method for measurement of magnetic noise in magnetoimpedance sensing element is proposed. Glass-coated Co-based amorphous microwires with a slightly negative magnetostriction were used as the sensing element. The operation principle of the sensing element was based on the nonlinear off-diagonal magnetoimpedance, when field dependent higher harmonic components appeared in voltage in the pick-up coil wound around the microwire. The magnetic noise for the second harmonic in the pick-up coil voltage was studied. The dependences of the magnetic noise and the signal-to-noise ratio in the sensitive element on the current amplitude were analyzed. The noise performance of the second harmonic in the pick-up coil voltage was compared to that for a flux-gate sensing element.

1 Introduction

The magnetoimpedance (MI) implies a strong dependence of the impedance of a magnetic conductor on an external magnetic field. The MI is attracted considerable attention due to its potential for magnetic sensing applications (see, for example, [1–4]). The MI effect is usually observed as a change of the voltage across the magnetic sample ends related to the external field variation. Another approach to detect the field dependent signal consists in the measurement of the voltage in the pick-up coil wound around the magnetic conductor [5]. The latter effect is referred to as the off-diagonal MI.

At relatively high amplitudes of the excitation current, the generation of higher harmonics in voltage has been reported (the so-called nonlinear MI) [6–12]. The origin of the nonlinear voltage response is ascribed to the magnetization reversal of a conductor under the magnetic field generated by the current. The nonlinear MI may be promising for applications, since higher harmonics are very sensitive to the external field.

One of the most important features of magnetic sensors is the capability to detect the external field with high accuracy. In this connection, it is of importance to achieve low noise level. The noise performance in different sensors based on the linear MI effect has been studied previously (see, for example, [13–17]). A model to calculate the intrinsic magnetic noise of MI sensing element has been proposed in [18], and the contributions of the sensing element and electronic circuits to the noise have been analyzed in detail [14,15]. However, precise separation of noise sources is a difficult task.

In this work, we present a direct method for the magnetic noise measurement in MI sensing element. The noise in the sensing element was measured by using well-known method of the transferring of low-frequency range to the high-frequency range based on the principle of the heterodyne transformation. In contrast to the previous studies, we investigated the nonlinear off-diagonal MI [12], and the second harmonic amplitude in the pick-up coil voltage was used for the analysis of the magnetic noise performance. The magnetic noise in the sensitive element was analyzed as a function of the excitation current amplitude. The magnetic noise in the nonlinear off-diagonal MI sensing element was compared to the noise in a flux-gate sensing element.

2 Experimental

Glass-coated CoFeSiB amorphous microwires with a slightly negative magnetostriction were used as a MI sensing element. The microwires fabricated by Taylor–Ulitovsky method had the amorphous core diameter of 16 μm, and the glass coating thickness was 1 μm. The studied samples had the length of 6 mm.

The MI effect in the microwire was measured by means of a conventional procedure using Agilent Impedance Analyzer 4294A. To study the nonlinear off-diagonal MI, the pick-up coil was wound around the microwire. The coil had 125 turns and its inner diameter was of 2 mm. In experiments, the current amplitude was high enough to induce the magnetization reversal process in the microwire and was changed up to 30 mA. The pick-up coil voltage was measured by means of Rohde&Schwarz FSH8 Spectrum Analyzer.

* Corresponding author: asantonov@inbox.ru

https://doi.org/10.1051/epjconf/201818502010

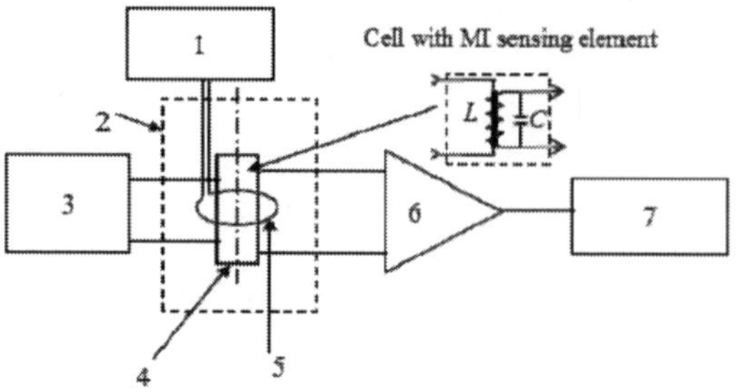

Fig. 1. A schematic of the experimental setup for noise measurements. 1, low-frequency generator; 2, magnetic shield; 3, high-frequency generator; 4, sensing element; 5, calibrated coil; 6, low-noise amplifier; 7, FSH8 Spectrum Analyzer.

The magnetic noise in the sensing element was measured by means of the frequency demodulation method. The schematic of the experimental setup is shown in Fig. 1. The setup allows one to study the different principles of the field detection in the sensing element, namely, the MI effect, off-diagonal MI and nonlinear off-diagonal MI. In this work, we focus on the latter effect and analyze the noise for the second harmonic in the pick-up coil voltage frequency spectrum.

The frequency of the excitation current was about 1 MHz. The pick-up coil wound around the microwire can be included in the resonant circuit with a capacitor, which was tuned to the second harmonic frequency. The studied microwire was placed into the magnetic shield in order to reduce external noise. The magnetic shield provided the protection at the level of 40 dB. The low-noise amplifier was used to match with the low-impedance input of FSH8 Spectrum Analyzer.

3 Results and discussion

Fig. 2 shows the field dependence of the microwire impedance at the frequency $f_0 = 2$ MHz. The measurements were carried out for the excitation current amplitude of 1 mA. Within this frequency range, the imaginary part of the impedance has the strong dependence on the external field. The dependence has a typical two-peak behavior, with the maxima being at the field of the order of ±1 Oe.

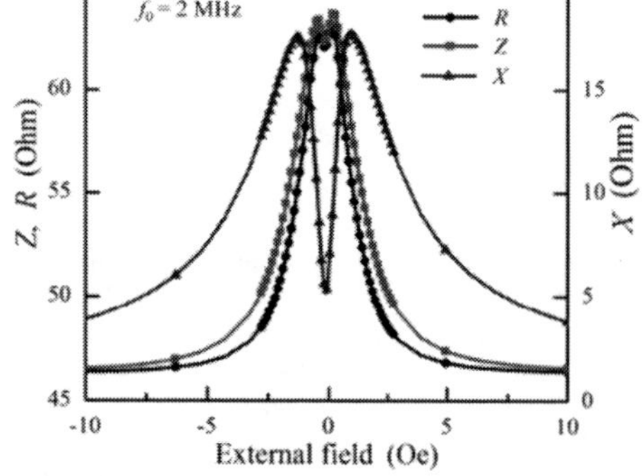

Fig. 2. Field dependences of the microwire impedance Z, its real R and imaginary X parts.

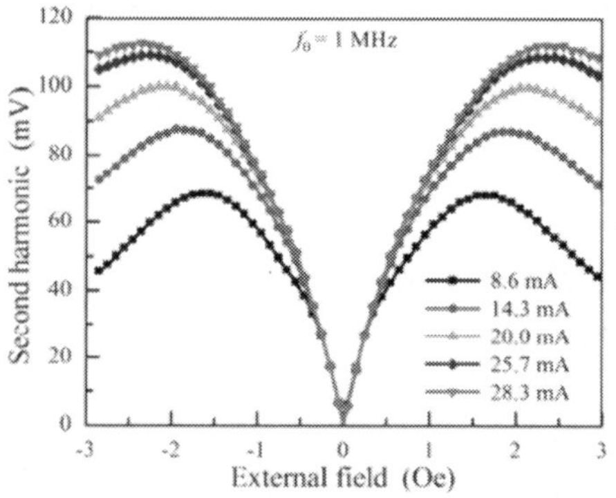

Fig. 3. Field dependence of the second harmonic amplitude in the pick-up coil voltage at different current amplitudes.

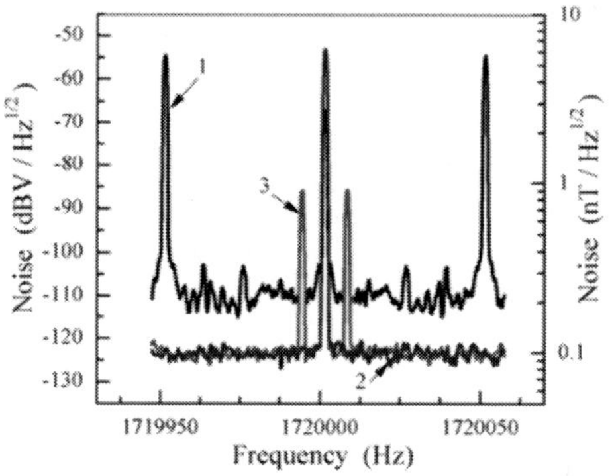

Fig. 4. An example of measured frequency spectrum of magnetic noise. Curve 1, noise measured in the absence of the magnetic shield; 2, noise measured in the presence of the magnetic shield for signal without modulation; 3, noise measured in the presence of the magnetic shield for signal with modulation.

At relatively high current amplitudes, the pick-up coil voltage response becomes nonlinear and consists of a number of harmonics due to the magnetization reversal in the microwire. It was shown previously that the second harmonic is dominant one in the pick-up voltage within a wide range of current amplitudes [12]. The measured field dependence of the second harmonic amplitude is presented in Fig. 3 for different values of the current amplitude for the excitation frequency $f_0 = 1$ MHz. The second harmonic exhibited high sensitivity to the external field of the order of 50 mV/Oe.

A typical frequency spectrum obtained in the measurements of the noise in the sensing element is presented in Fig. 4. The frequency range was selected as ±55 Hz with respect to the second harmonic frequency $2f_0 = 1.72$ MHz. Three curves in Fig. 4 correspond to the measurements of the noise without the magnetic shield and in the presence of the shield without modulation and with modulation by a signal of the frequency of 7 Hz. As a result of the modulation of the signal from the pick-up coil, two satellite peaks appeared in the noise frequency spectrum. The satellite peaks were symmetric with respect to the second harmonic frequency $2f_0$ and differed from that at 7 Hz.

The amplitude of the pick-up coil signal in the units of nT was determined using the calibrating coil with an additional measurement by means of a magnetometer. A sinusoidal signal of the frequency of 7 Hz with the amplitude of 60 nT was used. Using the calibration, the equivalent magnetic noise was recalculated in the units of $nT/Hz^{1/2}$.

Shown in Fig. 5 are the current amplitude dependences of the magnetic noise in the sensitive element based on the nonlinear off-diagonal MI and the signal-to-noise ratio for the excitation frequency of 0.86 MHz. The results are presented for both the resonant and non-resonant modes. In the resonant mode, the pick-up coil was connected to the resonance circuit tuned to the frequency $2f_0 = 1.72$ MHz. It follows from Fig. 5 that the magnetic noise and the signal-to-noise ratio depended slightly on the mode of the measurements. Note that the use of the capacitor in the circuit of the output signal can be of significance for applications, since it allows one to increase the transformation coefficient taking into account the resonance circuit quality factor. In the experiment, the quality factor was of the order of 10, which allows one to obtain the transformation coefficient at the level of 1 V/Oe. As a result, the requirements to the primary amplifier of electronic circuit can be reduced for the same signal-to-noise ratio.

It follows from Fig. 5 (a) that the magnetic noise increased sharply for the range of current amplitudes 5–7 mA. The range corresponds to the circular magnetic field at the microwire surface of 1.25–1.75 Oe, when the

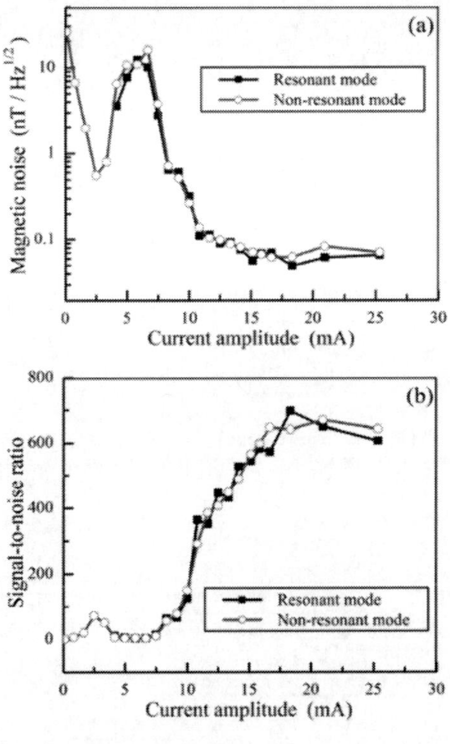

Fig. 5. Magnetic noise (a) and signal-to-noise ratio (b) versus current amplitude at $f_0 = 0.86$ MHz.

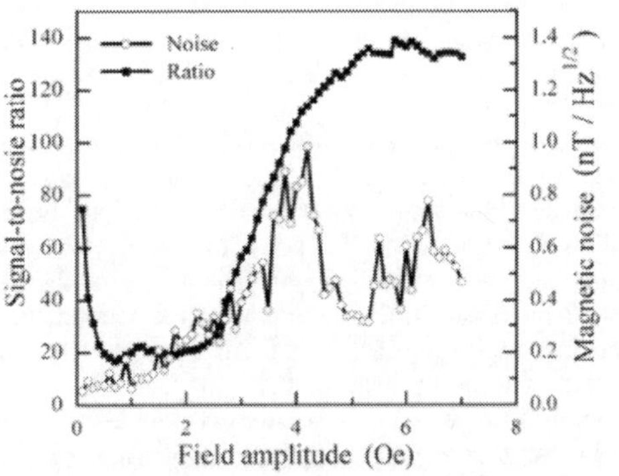

Fig. 6. The dependences of the magnetic noise and signal-to-noise ratio on the amplitude of the excitation magnetic field at $f_0 = 0.86$ MHz for the flux-gate sensing element.

transition of the pick-up coil voltage response to the nonlinear mode takes place. With a further increase of the current amplitude, the magnetic noise decreased and attained the level of 60 $pT/Hz^{1/2}$ at the current amplitude exceeding 15 mA. For the studied microwire, this range of the current amplitude seems to be optimal from the point of view of the signal-to-noise ratio.

Fig. 6 shows the results of the experimental study of the magnetic noise for the fluxgate configuration. The experiments were carried out for the same amorphous microwires. To obtain the signal with higher amplitude the volume of the sensing element was increased using twelve parallel microwires with length of 6 mm. For the measurements, the same pick-up coil having 125 turns, inner diameter of 2 mm and length of 6 mm was used.

It follows from Fig. 6 that the magnetic noise for the flux-gate sensitive element attained maximum at the excitation field amplitude of the order of 4 Oe. This value is approximately two times higher than the amplitude of the threshold magnetic field for the sensitive element based on the nonlinear off-diagonal MI. When the field amplitude exceeded 5 Oe, the signal-to-noise ratio stabilized at the level of 140. The magnetic noise for the flux-gate sensitive element was approximately 4.5 times higher than that for the nonlinear off-diagonal MI element, although the total length of the flux-gate element was twelve times higher.

This difference may be due to several reasons. Although the detailed explanation is beyond the scope of this work, note that the main reason of the difference in the magnetic noise level can be related to the principle of the excitation of the sensing element. In the flux-gate sensing element, the magnetization reversal appears due to the longitudinal field, and the alternating magnetic fields in the primary and output circuits have the collinear components. For the sensing element based on the nonlinear off-diagonal MI, the magnetization reversal is induced by the transverse magnetic field. The primary circuit forms the structure orthogonal to the output circuit, and there are no parallel components of the alternating magnetic field in these circuits. This fact results in the noise in the primary circuit of the sensing element is not transformed to the noise in the output circuit.

4 Conclusion

A direct method for the magnetic noise measurement in the MI sensing element is proposed. The method allows one to study the magnetic noise in the sensing element for different principles of the field detection. The advantage of the method is that the signal and noise can be measured in a single experiment. In addition, the method allows one to avoid the noise of the electronic circuit forming the detected signal. The noise in the sensing element based on the detection of the second harmonic in the nonlinear off-diagonal MI response was investigated. The dependences of the magnetic noise and the signal-to-noise ratio in the sensitive element on the current amplitude were analyzed. A minimum noise level for the studied sensing elements was about 60 $pT/Hz^{1/2}$ in the limit of low frequencies (at 1 Hz). It was demonstrated that the magnetic noise in the nonlinear off-diagonal MI sensing element is significantly less than the noise in the flux-gate for a comparable size of the sensing element.

References

1. M. Vázquez, J. Magn. Magn. Mater. **226–230**, 693 (2001)
2. M. Knobel, K.R. Pirota, J. Magn. Magn. Mater. **242–245**, 33 (2002)
3. L. Kraus, Sens. Actuators A **106**, 187 (2003)
4. M.-H. Phan, H.-X. Peng, Prog. Mater. Sci. **53**, 323 (2008)
5. A.S. Antonov, I.T. Iakubov, A.N. Lagarkov, J. Magn. Magn. Mater. **187**, 252 (1998)
6. R.S. Beach, N. Smith, C.L. Platt, F. Jeffers, A.E. Berkowitz, Appl. Phys. Lett. **68**, 2753 (1996)
7. A. Yelon, M. Britel, D. Ménard, P. Ciureanu, Physica A **241**, 439 (1997)
8. S. Iida, O. Ishii, S. Kambe, Jpn. J. Appl. Phys. **37**, L869 (1998)
9. P. Aragoneses, A.P. Zhukov, J. Gonzalez, J.M. Blanco, L. Dominguez, Sens. Actuators A **81**, 86 (2000)
10. C. Gómez-Polo, M. Knobel, K.R. Pirota, M. Vázquez, Physica B **299**, 322 (2001)
11. G.V. Kurlyandskaya, H. Yakabchuk, E. Kisker, N.G. Bebenin, H. García-Miquel, M. Vázquez, V.O. Vas'kovskiy, J. Appl. Phys. **90**, 6280 (2001)
12. A.S. Antonov, N.A. Buznikov, I.T. Iakubov, A.N. Lagarkov, A.L. Rakhmanov, J. Phys. D: Appl. Phys. **34**, 752 (2001)
13. S. Yabukami, H. Mawatari, N. Horikoshi, Y. Murayama, T. Ozawa, K. Ishiyama, K. Arai, J. Magn. Magn. Mater. **290**, 1318 (2005)
14. L. Ding, S. Saez, C. Dolabdjian, L.G.C. Melo, A. Yelon, D. Ménard, IEEE Sens. J. **9**, 159 (2009)
15. B. Dufay, S. Saez, C. Dolabdjian, A. Yelon, D. Ménard, IEEE Sens. J. **13**, 379 (2013)
16. M. Malatek, B. Dufay, S. Saez, C. Dolabdjiana, Sens. Actuators A **204**, 20 (2013)
17. E. Fernández, A. García-Arribas, M. Barandiarán, A.V. Svalov, G.V. Kurlyandskaya, C.P. Dolabdjian, IEEE Sens. J. **15**, 6707 (2015)
18. L.G.C. Melo, D. Ménard, A. Yelon, L. Ding, S. Saez, C. Dolabdjian, J. Appl. Phys. **103**, 033903 (2008)

Manifestations of the effects of a "weak" sublattice in an YIG film with the replacement of yttrium by Gd ions

Vladislav Poimanov[1]*, *Vladiimir* Shkar[2], *Yuriy* Nepochatykh[2], *Vedamanickam* Sampath[3], *Vladiimir* Shavrov[4] and *Victor* Koledov[4]
[1]Donetsk National University, Universitetskaya 24, Donetsk, Ukraine
[2]Donetsk Institute for Physics and Engineering named after A.A. Galkin, R. Luxemburg str. 72, Donetsk, Ukraine
[3]Department of Metallurgical and Materials Engineering, Indian Institute of Technology Madras, Chennai-600 036, India
[4]Kotelnikov IRE RAS, Moscow, Russia, 125009, Mokhovaya 11/7

Abstract. Ferrite-garnet films of submicron thickness were investigated by ferromagnetic resonance method and by the autodyne oscillator method. Compensation of the magnetization of such a film is detected on the curve of its temperature dependence, which is associated with the inclusion of gadolinium ions from the gadolinium-gallium garnet substrate in its composition. It was found that the magnetization of the sample in perpendicular to the surface of the film magnetic field showed an abnormal change in the magnetization and also rotation of the magnetic axis. This behaviour of a ferromagnetic material in the presence of a magnetic field can be explained by the expressions for the linear magnetostrictive effect, and the piezomagnetic effect arising from the para-processes operating within a "weak" rare-earth magnetic sublattice which is located inside a unidirectional exchange field of "strong" iron sublattice.

1. Introduction

Gadolinium ferrite-garnet shows the unique characteristic of unidirectional anisotropy of the exchange interaction, which influences the "weak" gadolinium sublattice. It is shown [1] that the presence of the 'weak' sublattice in rare-earth ferrite-garnets leads to manifestation of abnormal phenomena. In particular, in gadolinium ferrite-garnet a breach of magnetostriction "evennies" and also magnetoresistance and magnetocaloric effect are observed. Magnetostriction shows linear dependence of magnetization due to the influence of unidirectional exchange anisotropy. To the linear magnetostriction is associated a thermodynamically opposite effect, namely piezomagntetism (the occurrence of the magnetization under the action of elastic stresses in the absence of a magnetic field). The occurrence of these phenomena is solely attributed to the paraprocess [1]. For yttrium iron garnet, which is far from the point of magnetic ordering, magnetostriction is a quadratic function of the magnetization. The aim of the present study was to detect piezomagnetic and other related effects given rise to by the presence of a 'weak' sublattice in ferrite.

2. Samples

The samples were synthesized by the two-crucible liquid-phase epitaxy technique from wetting melt-solution [7] on (111)-oriented substrate of gadolinium-gallium garnet. In the first crucible containing a superheated melt-solution of yttrium iron garnet (YIG), overgrazing (dissolution) of the upper layer of the substrate forms so that the various surface defects can be gotten rid of. Then the substrate, which spent almost the entire time in the hot zone, was transferred to second pot. During the transfer time of the substrate a submicron YIG type thin film grows from the remaining on it solution-melt of YIG and partially etched and dissolved substrate. In second pot was grown a uniaxial film with the required thickness for bubbles magnetic domain devices. Both these and earlier investigations [5,6] were conducted on this film of submicron thickness, which was grown on the substrate during the transfer time of it from the first crucible to another. The width of ferrimagnetic resonance (FMR) line for these films was ~30-40 Oe, while that for pure YIG was $\leq$ 0.5 Oe. The saturation magnetization for these films was 1250 Gs, while that for YIG was 1760 Gs. The film thickness was in the range 0.1- 0.2 μm, and the cubic anisotropy field was ~36 Oe.

It is known [4] that if impurity ions are present in the rare-earth sublattice the width of the FMR line of pure YIG increases. In this case, such impurity ions are magnetic gadolinium cations, which have diffused from the dissolved substrate (non-magnetic gallium ions dilute iron sublattice) [8]. Increased width of the line by nearly two orders of magnitude and decreased magnetization of the sample indicate a significant concentration of these ions. Thus, the total magnetization of the films arises due not only to the "iron" sublattice, but also to "weak"

* Corresponding author: author@e-mail.org

(yttrium diluted by gadolinium) sublattice. Therefore, unusual phenomena may occur in them.

3. Experiment

The details pertaining to the FMR apparatus that was used to carry out studies on the thin films are given in [4,5]. Experiments were performed at the ambient temperature. Figure 1 shows the frequency-field dependence of the fundamental mode of the FMR for the negative (H-) and positive (H+) external magnetic fields that are normal to the surface of the film (Higher-order modes, detected for this sample in [6], were not generated in this experiment).

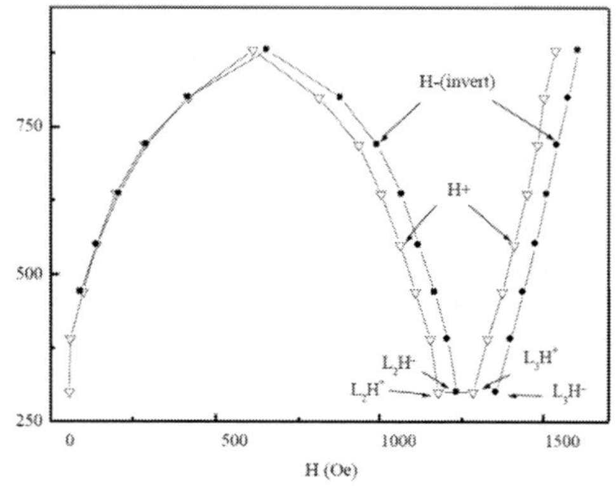

Fig.1. Experimental frequency-field dependence of FMR lines. The field is perpendicular to the surface of the film. (Dependence with a negative field is inverted to the positive axis).

The frequency-field dependence with a negative field is inverted to the positive axis H and is shown as curve H-(invert) in Fig. 1. Figure 1 shows that both for the negative and positive directions of the field the plots of frequency vs. field start from the zero value of the field, and with increasing absolute value of the field, the curves gradually diverge from each other to the field of orientation of the magnetization along the external field, which is ~1250 Oe. (In zero field the magnetization lies almost on the plane of the film due to the demagnetizing energy). Such behavior of the curves can be explained by the change in the demagnetizing field during the magnetization of the sample.

With increasing field along one direction, the demagnetizing energy grows greater, and in another increases less. Upon the sample reaching to the maximum in magnetization, the change of the demagnetizing energy is terminated. Thus, for one direction of the field, normal to the plane of the film, the saturation magnetization increased, while it decreased for the opposite direction.

Let us estimate the magnitude of change of magnetization in order to evaluate the differences in frequency-field dependence shown in Fig. 1. The difference is found to be 36 Oe near the saturation point. This means that for a change of magnetization of 1.43 Gs the demagnetizing field changes to 18 Oe along each direction.

In the next experiment, the external magnetic field was allowed to subtend an angle (θ) of 1 degree with the normal to the surface of the film. Consider the behavior of two FMR lines at a field strength that is closer to saturation at a frequency of 300 MHz and is dependent on the azimuthal angle, φ, lying on the surface of the film. This line is L2H-in the negative field up to the saturation point; and after the saturation point it changes to L3H$^-$, whose position in the frequency-field space (Fig. 1) is indicated by arrows. In the positive field, lines L2H$^+$ and L3H$^+$ correspond to this behavior.

Since the field is tilted from the normal, which is represented by a diagonal of a cubic crystal in a single-crystal film, i.e. the direction of type {111}, then at the azimuthal sample rotation, the cubic magnetic anisotropy is due to the presence of the other three diagonals (one of the methods for determining the cubic anisotropy [3]). For a given geometry and experimental conditions, the explanation for cubic magnetic anisotropy is as follows: during rotation of the sample and passing of the magnetic moment near one of the three edges of the cube, the distance between the FMR lines reaches its the maximum. At a distance from the edges, they approach and disappear. This fact is illustrated in Fig. 2, which shows the dependence of the derivatives of the absorption lines on the magnetic field for a number of azimuthal angles close to the direction {112}, which is representative of the projection of both the diagonal and the edge of the cube on the plane of the film [3].

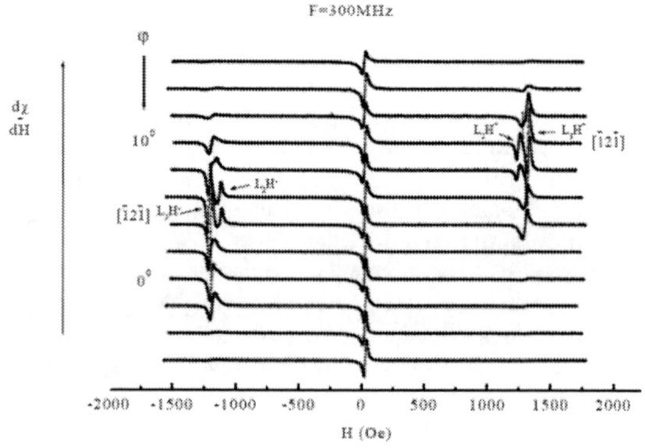

Fig. 2. Variation of the derivative of magnetic susceptibility with quasinormal external (θ_H = 1 degree) field for a range of azimuthal angles (φ_H) separated by an interval of 2 degrees. (F = 300 MHz and T= Ambient).

By rotating the azimuthal angle to 360 degrees, there are three such ranges of FMR line appearing at a distance of 120 degrees from one another. One of them is shown in Fig. 2. From this figure it follows that if we define the direction of the crystallographic axes in accordance with the FMR spectra, they are dependent on the direction of the field. This difference is about 4 degrees.

Consequently, there is an effect, which depends on the magnetic field, which changes the magnetization of the sample apart from rotating the magnetic axis.

To determine the form of the dependence of the saturation magnetization on temperature and the search for the compensation point, measurements of the values of the normal field of the transferring film to the saturation state were made for different temperatures. Such measurements were made to supplement and confirm the results and conclusions [6]. The measurements were performed from the records of the change in the magnetic susceptibility of the sample at a frequency of 5 MHz using the autodyne oscillator method [9]. The recording of the signal of absorption of high-frequency energy in a film with an increase in the normal field by an acute maximum (minimum in the inset in Fig. 3) fixes its transition to the saturation state. Such records were made for a sequence of stabilized temperatures. The obtained dependence (Fig. 3) on the temperature of the measured values of the transition field to the saturation magnetization has an acute deep minimum at a temperature of 57 °C. By recording the high-frequency energy absorption signal in such a film, as a function of temperature, the Neel point is determined to be 248 °C. The detected minimum is determined by the corresponding magnetic compensation point. Its appearance is due to the entry of gadolinium ions from the GGG substrate into a thin transition layer between the substrate and the YIG film and the formation of a mixed yttrium-gadolinium iron garnet in which a compensation point is possible. The magnetization of a very thin YIG film does not overlap this minimum by its monotonic decrease in magnetization with increasing temperature.

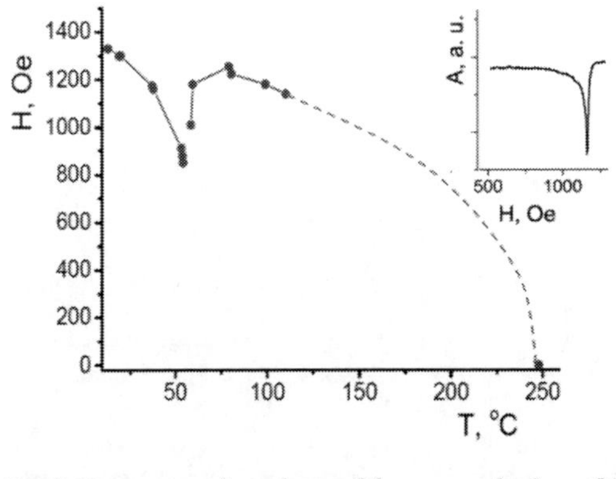

Fig. 3. Temperature dependence of the measured values of the transition field of the film to the saturation state.

4. Discussion

Based on the concepts developed in [1], the features observed during a study on the film using the FMR method can be explained as follows. The technique of fabrication of the ferromagnetic film involves a rare-earth sublattice incorporating ions of gadolinium (magnetic) and yttrium (nonmagnetic) with the iron sublattice being diluted with gallium ions. The exchange interaction between gadolinium ions at room temperature is so weak that it does not lead to their ordering. ("weak" sublattice). Their ordering occurs due to influence of strong (about 200 kOe) negative exchange field from the iron sublattice (paraprocess). The interaction of the rare-earth sublattice with iron is weak. This unidirectional exchange interaction leads to a linear rather than a quadratic dependence of magnetostriction on magnetization. Consequently, when the sample is magnetized in a direction that is parallel to the field, it gets elongated, while it gets compressed when it is aligned in a direction that is perpendicular to the field. As the symmetry analysis shows, the accounting of the linear magnetostrictive effect and piezomagnetic effect under cubic symmetry [1,8], in a coordinate system with the axis [111] along the normal leads to the appearance of non-diagonal components in the strain tensor. In turn, this may lead not only to the anisotropic stretching of the sample (without changing its volume), but also to a purely mechanical torsion of crystallographic axes. Moreover, with respect to the applied external field, this effect also shows unidirectionality. Since thermodynamically it is the opposite effect, piezomagnetism corresponds to the linear magnetostriction. Such deformation of the sample in a magnetic field causes an increase in magnetization due to tension and a decrease in it due to compression and also to the rotation of the magnetic axes by twisting. Thus, as determined from the experiment the change in magnetization and the rotation of the magnetic axes in the magnetic field are interrelated and are defined by the simultaneous manifestation of two effects, namely linear magnetostriction and piezomagnetic effect.

5. Conclusions

A deep minimum was found that marks the compensation point for the magnetization of thin YIG films on the curve of its temperature dependence. The presence of compensation is associated with the entry of gadolinium ions from the GGG substrate into the transition layer between the substrate and the YIG film during liquid-phase epitaxy from the wetting melt - solution of the YIG and from the GGG substrate dissolved in it, which results in the formation of a mixed yttrium-gadolinium iron garnet. The total magnetization of such a film is due to the "iron" and "weak" sublattices with a possible point of its compensation. It has been found that during magnetization in a normal magnetic field of such a film an abnormal change in the magnetization of the sample as well as rotation of the magnetic axes occur: i.e., in the straight (one) field direction the magnetization increases and the axis of the magnetic field rotates to align with it; on the other hand, in the opposite external field direction, magnetization decreases and the axis rotates to align with it in the opposite direction. This ferromagnetic behavior in a magnetic field can be explained by the simultaneous manifestation of the linear magnetostriction and the piezomagneticeffect that operate due to the paraprocess in a "weak" magnetic sublattice, wich is located in a unidirectional exchange field of the "strong" sublattice.

References

1. K.P. Belov, Physics-Uspekhi, **42** (1999)
2. A.S.Borovik-Romanov, JETP **11** (1960)
3. A.M. Balbashov, F.V. Lisovsky, V.K. Raevet al. Magnetic bubble elements and devices: Directory, Radio and Communications, Moscow, (1987)
4. A.G. Gurevich, Ferromagnetic resonance in ferrite and antiferromagnetics, Moscow, Nauka, 1973
5. V.F.Shkar', V.N.Varyukhin, JETP Letters 88 (2008)
6. V. Poimanov, V. Shkar, Yu. Nepochatykh al. JMMM **443**, 1 December (2017)
7. Y. I. Nikolaev, I. A. Krasin, Crystallography Reports **33** (1988)
8. S. I. Yushuk, Zhurnal Technicheskoi Fiziki **69** (12) (1999)
9. V.F.Shkar', V.N.Varyukhin, JETP Letters **92** (2010)

Magnetostatic waves in a medium with damping

Pavel Makarov[1,*], *Lidiya* Maltceva[1], *Leonid* Kotov[1], and *Vladimir* Shcheglov[2,**]

[1] *Pitirim Sorokin Syktyvkar State University, Oktyabrsky 55, Syktyvkar, 167001, Russia*
[2] *Institute of Radio-Engineering and Electronics of RAS, Mokhovaya 11-7, Moscow, 125009, Russia*

Abstract. By means of consistent solution of the motion equation for magnetization the consideration of energy dissipation for magnetostatic surface waves propagating on in-plane magnetized ferrite plate is regarded. The complex character of the wave number caused by the dissipation is revealed. In this case the dispersion relations for the real and imaginary parts of the wave number are obtained. It is shown that dispersion curves are limited both by the wave number and frequency, and these restrictions tighten with an increase in the damping parameter. Also, it is predicted that there is a new set of branches of the backward dissipative waves. Both for forward and backward waves there is a critical value of the damping parameter and propagation angle, beyond which it does not exist.

1 Introduction

The utilization of the ferromagnetic structures for the magnonic devices is one of the perspective field of magnon spintronics and magnonics. Magnetostatic surface waves (MSSW) propagating in magnetic films are the basis for the construction of various microwave devices for analog and digital data processing, information transfer and control of GHz and THz dynamics [1–4]. Among the novel spin-wave based devices the magnonic coupler can also be considered as a candidate to beyond-CMOS devices [5, 6].

Dispersion is an inherent property of any waveguide based on ferrite. The signal attenuation level, the delay time, the phase shift and all other important characteristics dramatically depends on the dispersion law. Minimization of wave losses is a prerequisite for waveguide correct work, because the dissipation plays a cruicial role for the functionality of spin-wave devices. However, the MSSW damping is understudied as the vast majority of research is performed without consideration of attenuation [4, 7, 8]. Though, the damping may significantly change the dispersion law [9]. In particular, it limits the maximum value of the wave number, which results in increasing the minimum value of the signal delay time.

In large number of invesigations [3, 4, 7–9], propagation of MSSW is studied in the special cases when wave is transferred perpendicularly to the direction of the permanent external magnetic field or along it. However, the wave propagation at arbitrary angles is essential for magnonic networks circuits [10–12], multi-channel filters [1, 2], etc. Therefore, the aim of this research is to study the impact of the damping effect on the dispersion of MSSW at arbitrary angles.

*e-mail: makarovpa@syktsu.ru
**e-mail: vshcheg@cplire.ru

2 Main equations

The geometry of the problem is the following. An unbounded ferrite film of thickness d is magnetized to saturation $4\pi\mathbf{M}$ by the constant field $\mathbf{H}$. The Cartesian coordinates $Oxyz$ are chosen so as the origin is at the middle of the film thickness, the Oyz plane is parallel to the plane of the film, and the axis Ox is perpendicular to latter. The Oz axis is oriented along the field $\mathbf{H}$. The wave vector $\mathbf{k}$ lies on the Oyz plane and is tilted by the angle φ with respect to the axis Oy.

It is assumed that the medium is homogeneous and isotropic. Waves are generated by a source with a predetermined frequency ω. Thus, the wave amplitude can only be reduced if the wave number is the complex value: $k = \eta - i\xi$.

The Landau — Lifshitz equation [13] with the dissipative term of the Gilbert form is used to describe the motion of the magnetization vector $\mathbf{m}$ in a lossy medium (with the damping coefficient α). In this geometry the dispersion equation for MSSW has the form:

$$\beta - 2\mu\vartheta\coth(kd\vartheta) - 1 = 0. \quad (1)$$

The equation above has been obtained by solving the Landau — Lifshitz — Gilbert equation with the boundary conditions. In (1) we have one constant parameter d, four variables: k, μ — the magnetic permeability of the medium of the form:

$$\mu = 1 + \frac{\Omega_H + i\alpha\Omega}{\Omega_H^2 - (1+\alpha^2)\Omega^2 + i2\alpha\Omega\Omega_H}, \quad (2)$$

and two ancillary quantities ϑ and β, which are expressed by formulae

$$\vartheta = \sqrt{\cos^2\varphi - \frac{\sin^2\varphi}{\mu}}, \quad \beta = (\nu^2 - \mu^2 + \mu)\cos^2\varphi - \mu, \quad (3)$$

where ν is the magnetic susceptibility of the medium having the form:

$$\nu = \frac{\Omega}{\Omega_H^2 - (1+\alpha^2)\Omega^2 + i2\alpha\Omega\Omega_H}. \quad (4)$$

The quantities Ω and Ω_H given in equations (2) and (4) are the normalized wave frequency and the magnetic field respectively:

$$\Omega = \frac{\omega}{4\pi\gamma M}, \quad \Omega_H = \frac{H}{4\pi M}, \quad (5)$$

where γ is the gyromagnetic ratio.

3 Calculation results

Figure 1 demonstrates the assemblage of dispersion curves for the real and imaginary parts of the wave number obtained by the numerical solution of equation (1). To construct figure 1 were used parameters for YIG-films: $4\pi M = 1750$ Gs, $\gamma = 1.76 \cdot 10^7$ rad/(s · Gs), $H = 437.5$ Oe, $d = 15\,\mu$m.

We neglect the uniaxial anisotropy field H_u in YIG due to its extreme smallness for film with thickness $d = 15\,\mu$m, which may be considered as a bulk sample [14].

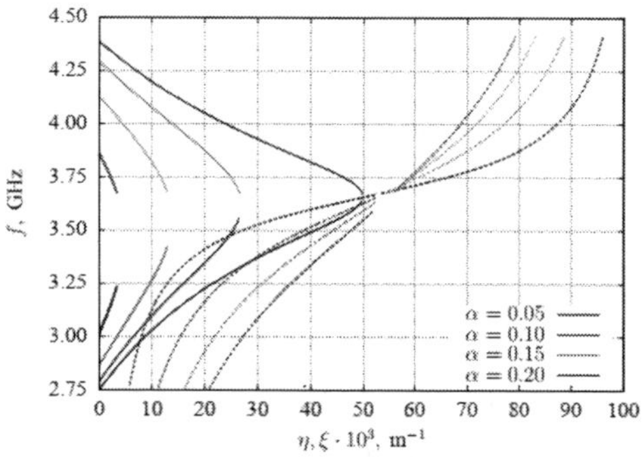

(a) $\varphi = 0°$

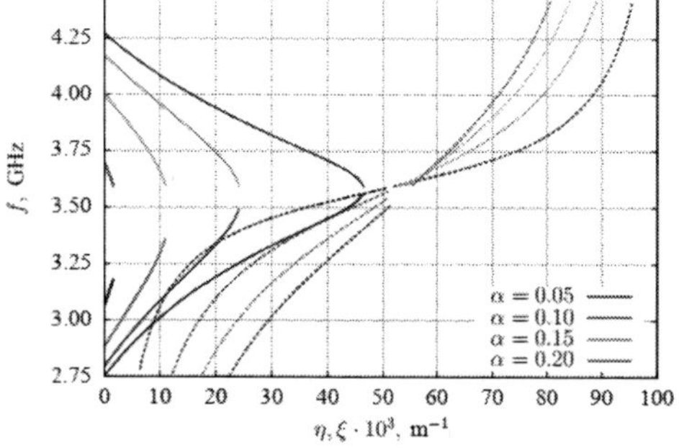

(b) $\varphi = 15°$

Figure 1. Dispersion law of the MSFSW and MSBSW with different values of α and φ. Solid curves — $f(\eta)$; dashed — $f(\xi)$.

Figure 1(a) illustrates the case when the wave propagates directly perpendicularly to magnetic field direction ($\varphi = 0°$). A set of solid curves represents the dependence of the wave frequency from the real part of the wave number $f(\eta)$; the family of dashed curves displays the dependence of the frequency from the imaginary part of the wave number $f(\xi)$. Each of the four lines in the set is constructed for different values of the damping parameter α. As figure 1(a) indicates, each set of curves is separated into two subset branches corresponding to dispersion of forward (MSFSW) and backward (MSBSW) waves.

Between the MSFSW and MSBSW branches of the dispersion curves $f(\eta)$ there is a bandgap of width Δf that is the increasing function of the damping coefficient. For instance, with α ranging from 0.05 to 0.20 Δf increases from about 32 MHz to 440 MHz. The frequency gap Δf between forward and backward waves is associated with a change in the nature of the dispersion law from normal to anomalous. The transition between these two types of dispersion is associated with a slowing down of the energy transfer of the forward waves due to the dissipation.

Therewith, the frequency range of MSFSW and MSBSW significantly narrows with the increase of α. Furthermore, there is a limitation to the maximum of η for a given α. For example, when $\alpha = 0.05$, the value of η_{max} is approximately equal to $50 \cdot 10^3$ m^{-1}, whereas when $\alpha = 0.20$, $\eta_{max} \approx 3.4 \cdot 10^3$ m^{-1} (figure 1(a)). From the physical point of view, the reduction of the MSFSW and MSBSW frequency range with the increase of damping constant is related with the following feature. The speed of the wave energy transfer is reduced stronger with the growth of the dissipation. Therefore, the transition from the normal to the anomalous dispersion law occurs earlier both in terms of the wave number and frequency.

As we can see from the figure 1, for all given damping parameters, $\eta < 50 \cdot 10^3$ m^{-1} (i.e., the wavelength $\lambda > 125\,\mu$m). It is known, that the effect of the exchange interaction is noticeable only at a wavelength of $\lambda < 1\,\mu$m and in this case this leads to a change in the dispersion frequency by less than one thousandth of the normal frequency. Thus, under the conditions described in this paper, the effect of the exchange is negligible.

The dispersion curves for the dissipation $f(\xi)$ show that ξ is a monotonically increasing function of the wave frequency, and with an increase of α this dependence becomes more linear. For this reason, the increase of α from 0.05 to 0.20 for MSFSW with frequency 3.1 GHz leads to a growth of ξ from $9.5 \cdot 10^3$ m^{-1} to $30.5 \cdot 10^3$ m^{-1}.

Figure 1(b) displays the case of wave propagation in the direction of straight tilted by the angle $\varphi = 15°$ with respect to the axis Oy. The basic regularities in this case are the same as for case $\varphi = 0°$ (figure 1(a)), however the restrictions of wave number and frequency are more enhanced.

Also, we obtained boundary frequences of magnetostatic waves by solving equation (1) in case when there is no damping. This frequencies, defined from the limit transition $\eta \longrightarrow 0$ and $\eta \longrightarrow \infty$ have the form:

$$\begin{cases} f_1 = \gamma\sqrt{H(H+4\pi M)}, \\ f_2 = \gamma(H+2\pi M), \\ f_3 = \gamma\sqrt{(H+2\pi M)^2 + (2\pi M)^2}. \end{cases} \quad (6)$$

Thus, the MSFSW propagate in a strictly defined frequency range $f_1 \leqslant f \leqslant f_2$ and MSBSW in range from f_2 to f_3. The curves of the boundary frequencies as functions of a constant magnetic field are shown on figure 2(a). In addition, this figure shows the field dependences of width of the MSSW frequency range (dashed and dotted lines). The boundary frequencies of the MSSW are determined by the value of the saturation magnetization and can vary over a wide range.

On figure 2(b) we have dependence of the boundary frequencies of the magnetostatic waves, propagating in YIG film (saturation magnetization $4\pi M = 1750$ Gs) on angle φ when $\alpha = 0$ and magnetic field $H = 437.5$ Oe. We can see that higher frequency of forward and bacward surface waves is a deacreasing function of angle φ. Moreover, there is critical value of angle φ, above which surface waves cannot propagate.

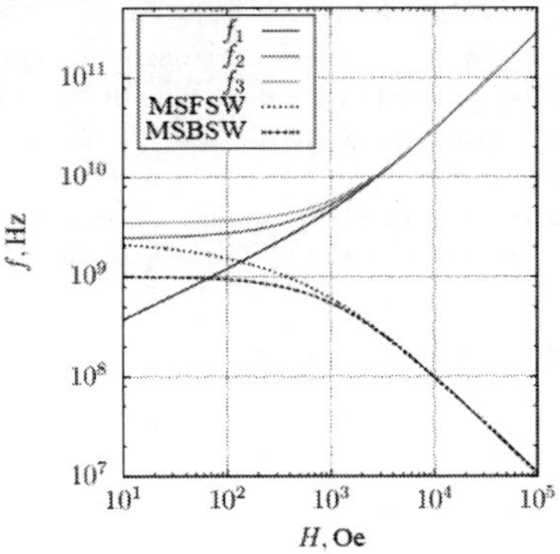

(a)

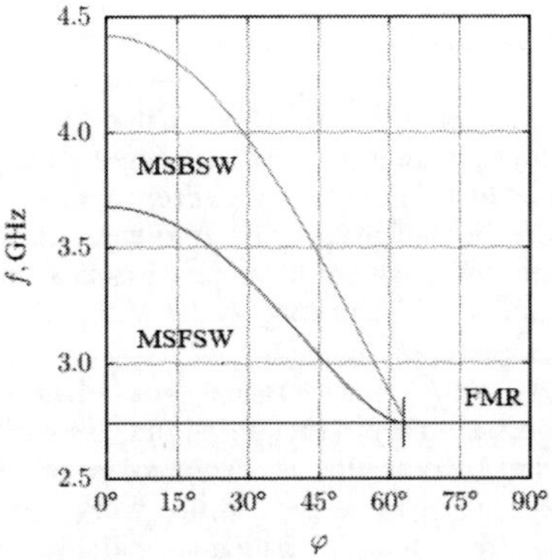

(b)

Figure 2. MSSW frequency spectrum without consider damping

4 Conclusions

In this study dispersion relations for the real and imaginary parts of the wave number were obtained. It was shown that dispersion curves are limited both by the wave number and frequency, and these restrictions tighten with an increase in the damping parameter.

A new branch of the backward dissipative waves was theoreticaly predicted. This solution doesn't contradict with the basic principles of electrodynamics, because it corresponds to the anomalous dispersion law.

MSFSW and MSBSW are characterised by existence of a critical value of the damping parameter and propagation angle, beyond which it does not exist. With the increase of the angle φ, limitation of dispersion curves tightens, the branches of the dispersion curves are shifted towards lower wave numbers and the damping parameter limits them from below and above.

The support from RFBR (grants #17-02-01138, #17-57-150001) are gratefully acknowledged.

References

[1] W.S. Ishak, Proc. of the IEEE **76** 171–187 (1988)
[2] Y. Kajiwara et al., Nature **464** 262–266 (2010)
[3] O. Kolokoltsev, N. Qureshi, E. Mejía-Uriarte, C.L. Ordóñez-Romero, J. Appl. Phys. **112** 013902 (2012)
[4] S. Jun, S.A. Nikitov, R. Marcelli, P. De Gasperis, J. Appl. Phys. **81**, 1341 (1997)
[5] A.V. Sadovnikov et al. Appl. Phys. Lett. **107**, 202405 (2015)
[6] A.V. Sadovnikov et al. Phys. Rev. Appl. **7**, 014013 (2017)
[7] R.W. Damon, J.R. Eshbach, J. Phys. Chem. Solids **19** 308–320 (1961)
[8] K.L. Wong et al., Appl. Phys. Lett. **105** 232403 (2014)
[9] J.H. Kwon, S.S. Mukherjee, P. Deorani, M. Hayashi, H. Yang Appl. Phys. A **111** 369–378 (2013)
[10] V.E. Demidov et al. Sci. Rep. **5**, 8578 (2015)
[11] A.V. Sadovnikov et al. Appl. Phys. Lett. **108**, 172411 (2016)
[12] A.V. Sadovnikov et al. Phys. Rev. B **96**, 060401(R) (2017)
[13] L.D. Landau, E.M. Lifshitz *Statistical physics Part 2: Theory of the condensed state* (Pergamon Press, Oxford, 1981) 387
[14] D. Stancil, A. Prabhakar *Spin Waves. Theory and Applications* (Springer, New York, 2009) 364

Waveguide magnetoplasmonic structure based on ferrite garnet film

Georgy Basiladze, *Aleksandr* Dolgov, *Vladimir* Berzhansky, *Andrey* Karavainikov, and *Anatoly* Prokopov

V. I. Vernadsky Crimean Federal University, 295007 Simferopol, Russia

Abstract. The magnetoplasmonic structure, which is a planar multimode magnetooptical waveguide, 10 μm thick with plasmon resonance SiO_2 / Cu coating on the surface is experimentally realized. With the total length of the magnetooptical waveguide 4.6 mm and the length of the SiO_2 / Cu coating equal to 3.9 mm, the insertion loss of the structure for TM- and TE-polarized light was 22 and 4 dB, respectively, at a wavelength of 1550 nm. Linearly polarized light was launched into the magneto-optical waveguide using a SMF-28 single-mode optical fiber coupled to the polished input edge of the waveguide. The structure is of interest for use as miniature magnetically controlled modulators of light intensity.

1 Introduction

Planar waveguide structures based on dielectric magnetic films are of considerable interest due to the unique property of magneto-optical nonreciprocity and the ability to control the characteristics of the light propagating through them using an external magnetic field. The study of such structures is widely represented in the literature, where their properties are studied in detail and the possibility of using them as various planar optical devices, such as isolators [1], circulators [2], modulators [3], has been shown. At the same time, similar structures in which the actions of the Faraday effect in a dielectric magnetic waveguide and of surface plasmon resonance (SPR) in a metal layer on waveguide surface are combined are not sufficiently studied. This is especially true for structures based on sufficiently thick (8-10 μm) dielectric magnetic waveguides, which can be relatively easily coupled to a telecomunication single-mode optical fiber. As follows from modern studies [4], the combination of magneto-optical and plasmon effects opens new prospects for the use of magnetic materials in sensor systems and miniature light-control devices. As an example of waveguide structures using dielectric magnetic materials and photon-plasmon interaction for controlling the parameters of an optical wave by means of an external magnetic field, devices theoretically developed in [5, 6] can serve. So, in [5] the possibility of creating Faraday rotator based on a hybrid ridge-plasmonic waveguide, including the yttrium iron garnet (Bi: YIG). In [6] the possibility of amplifying the nonreciprocity effect in a structure based on a silicon waveguide with a lattice plasmon structure on a surface covered with a dielectric magnetic material Bi: GIG is shown.

The aim of this work was the experimental creation a planar structure based on a waveguide from an liquid-phase epitaxial ferrite garnet film, about 10 μm in thickness, with a plasmon resonance coating on the surface that is capable of absorbing a TM-polarized near-infrared wavelength. As is well known [7-9] similar structures on the basis of not magnetic material well perform the function of the polarizer. In our case, when a magnetic material is used instead of a dielectric waveguide layer, the structure assumes the possibility of being used as an active element for various planar devices, for example, a light intensity modulator [10].

2 Model of the structure

It is well known [7-9] that the effect of the optical waveguide TE polarizer with a plasmon structure on the surface of the waveguide is based on the resonant coupling of the waveguide TM mode to the plasmon mode, which is supported by the metal layer due to the excitation of the surface plasmon resonance. Such a resonance arises under the condition of phase synchronization of the waveguide and plasmon TM modes. This condition is achieved due to the dielectric buffer layer, which is located between the dielectric waveguide and the metal layer. From the correctness of the choice of thicknesses and refractive indices of the metal and buffer layers for a given optical waveguide, the efficiency of quenching of the optical TM wave in the structure largely depends.

Fig. 1 shows the model of the waveguide magnetoplasmonic structure, the possibility of experimental realization and obtaining of a plasmon effect in which was verified in this work. Here are the substrate from the nonmagnetic gadolinium gallium garnet (GGG) $Gd_3Ga_5O_{12}$ (1), the dielectric waveguide layer from the epitaxial film of ferrite garnet $(Bi,Lu)_3(Fe,Ga)_5O_{12}$ (2), SiO_2 buffer layer (3), metal layer (4).

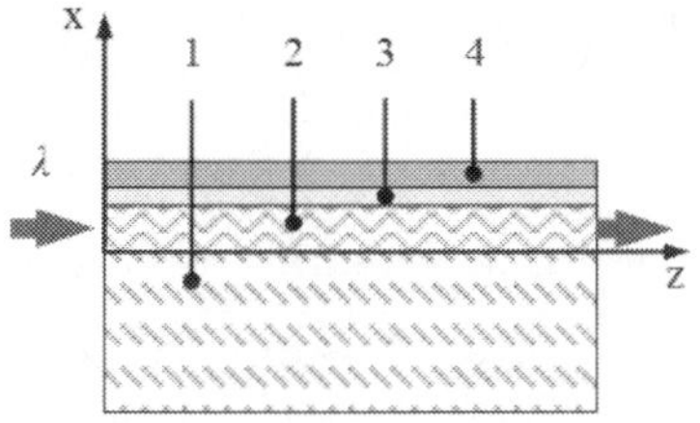

Fig. 1. Magnetoplasmonic structure: 1 – GGG substrate, 2 – ferrite garnet film, 3 – buffer layer 4 – metal layer.

3 Fabricating the structure

To realize the model, we used a dielectric magnetic film with a thickness $d_f = 10$ µm (refractive index $n_f \approx 2.22$ at $\lambda = 1550$ nm), which had a easy-plane magnetic anisotropy with relatively small external magnetic fields for its saturation along the hard (~ 4 Oe) and easy (~ 0,6 Oe) of the magnetization axes, with a specific Faraday rotation of 55.3° / mm at a wavelength of 1550 nm. The film was grown by liquid-phase epitaxy from a melt solution on a substrate of GGG, which has a thickness of $d_S \approx 500$ µm ($n_S \approx 1.92$ at $\lambda = 1550$ nm) end the orientations of the crystallographic axes (111)-plane. According to the AFM measurement, the root mean square roughness of the surface of the magnetic film is 1.682 nm. The input and output edge of the magnetic film were polished along the hard magnetization axis so that light could be launched into film in the direction of the axis of easy magnetization. The distance between the polished ends was $L_f = 4.6$ mm, the size along the polished ends was 8 mm. It should be noted that, in contrast to the plasmon structures presented in [7, 8], where the waveguide is single-mode, the magneto-optical waveguide used in this work, according to the above optical and geometric parameters, is a multimode.

We used easily accessible and relatively inexpensive materials such as fused quartz (SiO_2) and copper (Cu – 99.9 %) to deposit the buffer and metal layers on the surface of the magnetic film. Deposition was carried out by reactive ion-beam sputtering of the target in a vacuum chamber. Wherein, the possibilities of the equipment used to maintain the vacuum in the sputtering chamber were limited to a pressure of 1.6×10^{-6} kPa. SiO_2 and Cu were deposited on a sample of a magnetic film sequentially, each in its cycle – first a SiO_2 buffer coating then a Cu layer.

SiO_2 was deposited by sputtering a target from fused quartz in the presence of a gas mixture of argon and oxygen at a pressure of 9.6×10^{-5} kPa, an accelerating voltage of 5 kV and a current of 0.16 A. The deposition rate was 8 nm / min. The gas mixture in the sputtering chamber, which was previously evacuated to a vacuum depth of 1.6×10^{-6} kPa, was formed by the injection of argon to a pressure of 1.1×10^{-5} kPa, then of oxygen to the working pressure in the chamber of 9.6×10^{-5} kPa.

The buffer coating on the surface of the magnetic film was formed in the form of steps along the large side of the sample, each of which had its own thickness. This technique was used to increase the probability of achieving resonance in the fabricated structure. The stepped coating was formed with the aid of a special tooling having a movable shutter adjacent to the surface of the sample. This shutter, after the deposition of one of the SiO_2 layers, moved a distance of ~ 1 mm along the large side of the sample to deposit the next layer. As a result, a buffer coating was obtained from SiO_2 with a thickness of steps 8, 13, 18 and 23 nm varying along the large side of the sample. The dimensions of these steps along the large side of the sample were 1.2 mm, 1.3 mm, 1.3 mm and 1 mm, respectively, and 3.9 mm in the direction of light propagation.

Cu was deposited over the buffer coating, as well as on the adjacent non-coated portion of the magnetic film surface. Precipitation was carried out by sputtering a copper target in the presence of argon at a pressure of 7.7×10^{-5} kPa (residual gas partial pressure 1.6×10^{-6} kPa), accelerating voltage 3 kV and current 75 mA. The sample was located at a distance of 20 cm from the target. Wherein, the deposition rate of copper was 10 nm/min. The deposition time was 3 min, the obtained copper layer thickness $d_{Cu} = 30$ nm. The size of the copper coating on the surface of the magnetic film along its large side was 6.2 mm, and in the direction of light propagation $L_{Cu} = 3.9$ mm.

Fig. 2 shows a photograph of the surface of a structure with a dimensional partition into areas. Here are the area of the uncoated film (a), the area with a copper layer without buffer (b), SiO_2 / Cu heterostructures with SiO_2 thicknesses, respectively, 8 nm (c), 13 nm (d), 18 nm (e) and 23 nm (f) at a thickness of Cu – $d_{Cu} = 30$ nm.

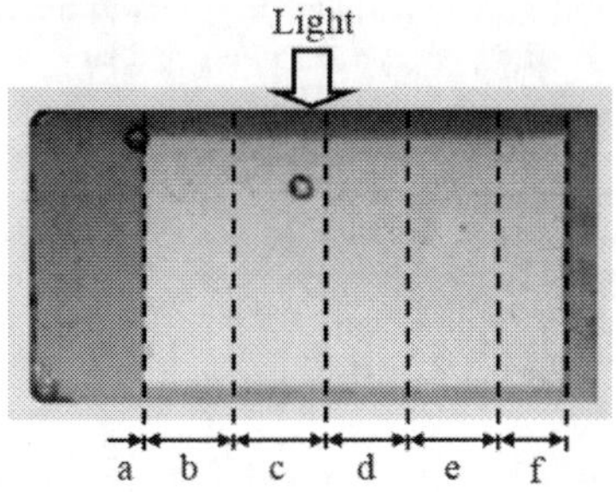

Fig. 2. The image of the surface of the structure (the size of the structure is 8 × 4.6 mm).

Thus, on the surface of the magneto-optical film for plasmon resonance studies, we had four plasmon SiO_2/Cu heterostructures with different SiO_2 thicknesses.

4 Structure study

Fig. 3 shows in a simplified form the design of the waveguide magnetoplasmon structure, realized in this work, and the scheme of the experiment for its investigation.

As a light source, a semiconductor laser with a wavelength of 1550 nm was used, to the output of which a fiber optic (FO) polarizer (not shown in the figure), made of a SMF-28 single-mode optical fiber was connected. Linearly polarized light from the FO

https://doi.org/10.1051/epjconf/201818502013

polarizer is launched into the magnetic film 2 through its polished end-face connected to the fiber terminal 6 of the polarizer. Thus, by establishing the corresponding polarization direction at the input of the structure, it was possible to excite only the TE or only the TM mode in the magneto-optical waveguide. Wherein, to study different areas of the structure, the sample was moved in a horizontal plane by means of a precision positioner.

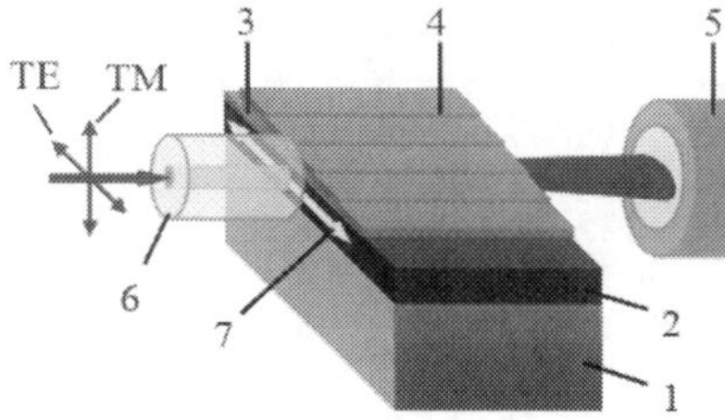

Fig. 3. Scheme of the experiment: 1 – GGG substrate ($d_S \approx 500$ μm), 2 – ferrite garnet film (d_f =10 μm), 3 – SiO_2 buffer layer with a step profile (d_{SiO2} = 8-23 nm), 4 – Cu layer (d_{Cu}=30 nm), 5 – Ge photodiode, 6 – optic fiber, 7 – sample scanning direction with respect to the fiber.

Measurements of the signals at the output of the structure were performed using a germanium photodiode 5 and a photocurrent detector (not shown in Fig. 3). Wherein, the light attenuation in the structure was calculated for each of the polarizations at the input by the formula:

$$\alpha = -10\times\log(I_0/I_{out}),$$

where I_0 and I_{out} are the light intensities at the input and output of the magnetic film. When performing measurements, the film was magnetized by an external magnetic field of strength > 50 Oe in the plane, transverse to the direction of light. For this purpose, permanent magnets located on the sides of the sample (not shown in Fig. 3) were used.

5 Results and discussion

Fig. 4 shows the optical loss of TM- and TE-polarized light beams in the magnetoplasmonic structure as a function of the coordinate of a launch into the input edge of the waveguide. The coordinate of the launch of the light beam into the waveguide was changed by scanning the sample in a horizontal plane relative to the fiber. Vertical dashed lines show the boundaries between areas of the surface with different coating compositions (See Fig. 2). The noise level in these dependencies does not exceed ± 0.9 dB.

It can be seen from Fig. 4 the attenuation of light for both polarizations during propagation in a magnetic film along a-area, where any coating is absent does not exceed 4 dB. Then, in the transition to b-area, where the copper layer does not have a buffer, for TM-polarized light, the attenuation increases to a level of 7.5 dB and continues to increase in the c (d_{SiO2} = 8 nm) and d (d_{SiO2} = 13 nm) regions, reaching in the d-area of the largest value ≈ 22 dB. With a further thickening of the buffer layer in e (d_{SiO2} = 18 nm) and f (d_{SiO2} = 23 nm) regions, a certain decrease in the curve for the TM polarization is observed. Simultaneously, for TE-polarized light, the attenuation level within the experimental accuracy can be considered independent of the thickness of the buffer layer. According to the figure, this attenuation level on the average corresponds to a value of about 4 dB.

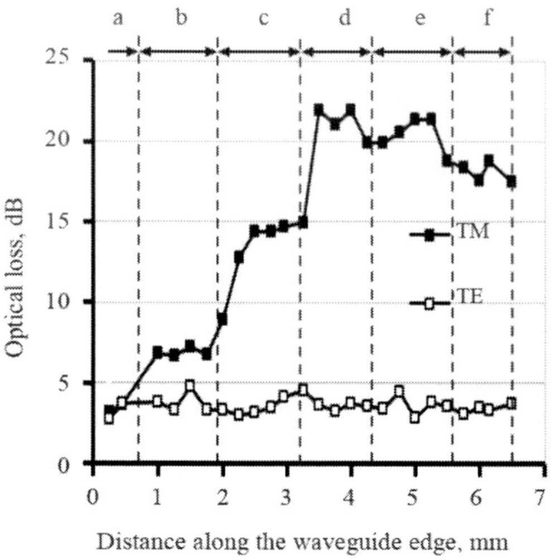

Fig. 4. TM, TE mode loss vs. SiO_2 layer thickness, where d_{SiO2}: 0 nm (a, b), 8 nm (c), 13 nm (d), 18 nm (e), 23 nm (f); d_{Cu}: 0 nm (a), 30 nm (b-f).

The explanation of the physical mechanism of damping of TM-polarized light in the structure, in our opinion, logically follows from the behavior of the obtained dependence on the thickness of the buffer layer (see Fig. 4). In this regard, one should note a certain similarity of the above dependence with the theoretically obtained [11, 12] and experimentally [8] curves for structures in which the resonance interaction of the waveguide and plasmon-polariton waves is the dominant damping mechanism. The similarity is that, as in our case, with a thickening of the buffer layer to a thickness where the resonance is maximal, the damping of the TM mode increases. If there was no such mechanism in our structure, then at an increase in the thickness of the buffer layer, an increase in the optical signal would be observed, since the dominant mechanism of ohmic losses would undergo a decrease due to a decrease in the fraction of the light energy penetrating into the metal.

Thus, we come to the conclusion that the nature of the attenuation of TM-polarized light in the resulting structure is directly related to the resonance interaction of the waveguide and plasmon-polariton waves. Such interaction is first amplified by equalizing the phase velocities of these waves with an increase in the thickness of the buffer layer from 0 to 13 nm. Then, with further thickening of the buffer layer and caused by this attenuation of the interaction between the waveguide and plasmon-polariton waves, the attenuation of the TM polarized waveguide mode decreases. The comparatively small efficiency of the observed resonance can be explained by the insufficient optimization of the parameters of the buffer and metal layers, as well as by the multimode nature of the optical waveguide system. The latter factor can lead to a significant transformation of the transverse distribution of the light wave field as it propagates in the waveguide. Therefore, initially the high degree of localization of the field at the metal interface, which is inherent in the lowest order TM mode in the metallized waveguide [12, 13], can be gradually broken, resulting in a disruption of the connection with the plasmon polariton.

With regard to the possibility of modulating the intensity of light by means of the proposed structure, it can be realized by the action of an external alternating magnetic field directed parallel to the light propagation path. The Faraday rotation of the plane of light polarization caused by such an action will lead to an angular vibration of the direction of light polarization in the magnetic film. At those times when the direction of polarization is perpendicular to the plane of the film, the output of the device will exhibit a minimum of light intensity caused by the transformation of the energy of the waveguide TM mode into a plasmon mode. When the light is polarized in the plane of the magnetic film, the intensity will be maximum, since the TE wave propagates without exciting plasmon oscillations in the structure. Thus, be efficiency of transforming the TM-polarized component of the electromagnetic field of the waveguide mode into plasmon mode will determine such an important modulator characteristic as the modulation depth. Based on the maximum difference between the attenuation values for TM- and TE-polarized signals ($\approx$ 22 dB and $\approx$ 4 dB, respectively), this parameter for a modulator based on the sample considered in this paper can be about 96 %.

A certain feature of the technology used to make the investigated sample of the magnetoplasmonic structure is a relatively low level of vacuum and a relatively low purity of the sputtered materials that form a plasmon heterostructure on the surface of the magnetic film. At the same time, in accordance with the current recommendations [14], to obtain high-quality plasmon coatings, it is necessary to adhere to the more stringent requirements for the purity of plasmon materials ($\geq$ 99.99 %) and the pressure in the vacuum chamber of $< 10^{-5}$ Torr ($\approx 1.333 \times 10^{-6}$ kPa). It is logical to assume that the fulfillment of these conditions can lead to a significant increase of the efficiency of plasmon resonance in the considered structure.

6 Conclusion

The possibility of fabricating a waveguide the magnetoplasmonic structure on the basis of a planar multimode waveguide, which is an epitaxial film of a garnet ferrite with a SiO_2/Cu resonant heterostructure on the surface, is demonstrated experimentally. A feature of the technology of deposition of a resonant heterostructure using the method of reactive ion beam sputtering of the target is a relatively high baseline pressure in the vacuum chamber ($> 10^{-5}$ torr) and a relatively low purity of the initial metal for sputtering (~ 99.9 %). The thus obtained plasmon resonance SiO_2/Cu coating (d_{SiO2} = 13 nm, d_{Cu} = 30 nm), L_{Cu} = 3.9 mm in length with a total waveguide length of L_f = 4.6 mm and its thickness of d_f = 10 μm, provided quenching of the introduced TM-polarized wave (λ = 1550 nm) about 22 dB and insertion loss for the TE-polarized wave about 4 dB. The efficiency of the resonance coupling of the optical and plasmon modes in such a structure can be substantially increased by optimizing the thickness of the buffer and metallic layers, as well as the technology of their deposition.

The waveguide magnetoplasmonic structure presented in this paper can be used as an active element of a miniature magnetooptical light intensity modulator in planar design [10].

Acknowledgement

This work is support by the RF Ministry of Education and Science in the framework of the state task (project no. 3.7126.2017/8.9).

References

1. J.P. Castera, G. Hapner, Appl. Opt. **16**, 8, 2031-2034, (1977)
2. R. Wolfe, R.A. Lieberman, V.J. Fratello, R.E. Scotti, N. Kopylov, Appl. Phys. Lett. **56**, 5, 426-428 (1990)
3. S.E. Irvine, A.Y. Elezzabi, Opt. Commun, **220**, 325-329 (2003)
4. G. Armelles, A. Cebollada, A. García-Martín, M.U. González, Adv. Opt. Mater. **1**, 10-35 (2013)
5. C. J. Firby, PoHan Chang, Amr S. Helmy, and A. Y. Elezzabi. ACS Phot. 3, 2344 (2016)
6. L. Halagacka, K. Postava, M. Vanwolleghem, B. Dagensb, and J. Piˇstora. Proc. of SPIE Vol. 9516, 95161 (2015)
7. M.N. Zervas, I.P. Giles, Opt. Lett. **15**, 9, 513-515 (1990)
8. I.V. Il'ichev, N.V. Toguzov, A.V. Shamray, Tech. Phys. Lett. **35**, 9, 831-833 (2009)
9. A. Kumar, Rakesh, M. Kumar, G. Bhatt, A. Kapoor, Intern. J. Eng. & Appl. Sciences, **7**, 2, 31-36 (2015)
10. G.D. Basiladze, V.N. Berzhansky, A.I. Dolgov, Patent RU 161388 (2016)
11. J.N. Polky, G.L. Mitchell, J. Opt. Soc. Am. **64**, 3, 274-279 (1974)
12. Y. Yamamoto, T. Kamiya, H. Yanai, IEEE J. Quant. Electron. **QE-11**, 9, 729-736 (1975)
13. S.C. Rashleigh, Opt. & Quant. Electronics **8**, 49-60 (1976)
14. K.M. McPeak, S.V. Jayanti, S.J.P. Kress, S. Meyer, S. Iotti, A. Rossinelli, D.J. Norris, ACS Photonics, 2, 326-333 (2015)

 https://doi.org/10.1051/epjconf/201818502014

Investigation of welds by the method of the magneto-optical eddy current flaw detection

Nazar Lugovskoy*, *Vladimir* Berzhansky, *Dmitry* Filippov, *Anatoliy* Prokopov, *and Alexandr* Shuyskyy

V.I. Vernadsky Crimean Federal University, 95007, Academician Vernadsky Ave. 4, Simferopol, Russia

Abstract. The paper deals with the magneto-optical eddy current method of flaw detection of conducting materials, in which epitaxial films of ferrite garnet are used as sensors. In particular, the possibilities of visualization of welded seams in magnetic and non-magnetic samples, as well as the defectoscopy of the seams themselves, are considered. The second part of the work is devoted to mathematical modeling of the distribution of the magnetic fields of eddy currents near similar defects.

1 Experiment

One of the prospective flaw detection technologies is the magneto-optical eddy current (MOEC) method, in which surface, subsurface and fatigue defects can be recorded in products from both magnetic and non-magnetic metals [1-4].

1.1 Introduction

The principle of operation of the magneto-optical eddy current flaw detector is based on the reaction of the mobile domain structure (DS) of the sensor on the distribution of magnetic fields produced by the EC in the conductive material under study. The presence of defects in the metal object leads to a change in the trajectories of the eddy currents and to a corresponding change in the configuration of the magnetic fields generated by them. Visualization of the DS is carried out with the help of the Faraday Effect, as a result of which the rotation of the plane of polarization of the transmitted light occurs as a function of the direction of the local magnetization. The scheme and principle of operation are described in more detail in [3].

The main advantage of MOEC defectoscopy is the direct visualization of defects in real time, small dimensions and relative ease of manufacturing of the main parts of the flaw detector.

1.2 Experimental technique and samples

In [3,4] the influence of different characteristics of the flaw detector, such as the frequency of the excitation current, the magnitude and sign of the acting magnetic fields, on the MOEC image was studied.

In this paper, we investigated the possibility of observing welds and hidden defects in them, in products made from magnetic and nonmagnetic metal alloys.

In our device we used alternating field inductors of two configurations that created a magnetic field either in the plane of the sample (parallel inductor) or normal to it (normal inductor).

During the measurements, single-crystal epitaxial films based on substituted bismuth ferrite-garnet with the general formula $(R, Bi)_3(Fe, M)_5O_{12}$ [5] with the type of magnetic crystallographic anisotropy "easy axis" were used as a magneto-optical sensor. The thickness of the films varied in the range from 1.9 to 7.2 μm, the anisotropy field from 500 to 3000 Oe, the period of the domain structure from 5.5 to 32 μm, and the saturation field from 22 to 77 Oe.

1.3 Non-defective welds

Model non-defective welds made from both magnetic and non-magnetic materials were investigated. Figure 1 shows optical and magneto-optical images of welds in aluminum and stainless steel samples measured at a frequency f = 25 kHz, alternating magnetic field H_m = 150 Oe.

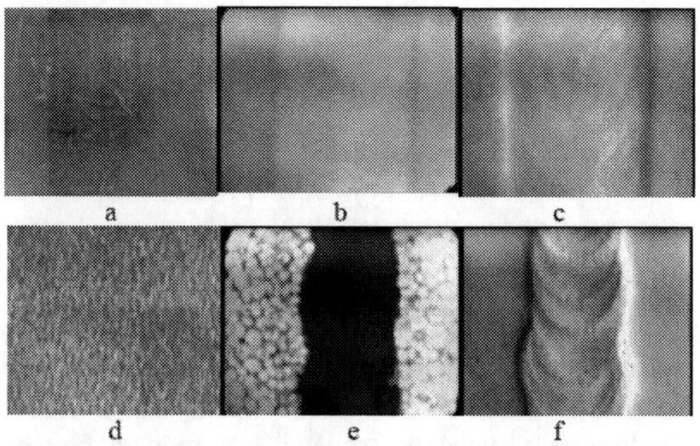

Fig. 1. Optical (a, d) and magneto-optical (b, c, e, f) images of welds: aluminum (a-c), stainless steel (d-f); orthogonal (b, e) and longitudinal (c, f) excitation of eddy currents.

It can be seen that welded seams are well registered in model samples of various types. In stainless steel samples,

* Corresponding author: lugovskoynv@mail.ru

the sensitivity of the MO sensor is higher than in samples made of aluminum. Figure 1e shows a nonstationary cylindrical domain structure that fills the entire film space along the seam. The sensor above the seam is in a saturated state, which allows obtaining a clear and contrasting image of the welded seam. On MO images 1c and 1f we see an "analog picture" of the structure of the seam, especially on stainless steel, all the inhomogeneities (influxes) formed during the welding process and invisible during normal visual contact are visible. In the same pictures, we see clear welds boundaries represented by a black and white narrow frame of the defect. This indicates the opposite direction of the magnetic field of the eddy currents at the boundaries of the seam with longitudinal EC excitation.

The effect of the frequency of the alternating magnetic field on the MO imaging of the seams was investigated. It was found that the efficiency of MO imaging decreases with increasing frequency (Fig. 2) and to increase the efficiency it is required to increase the amplitude of the alternating magnetic field.

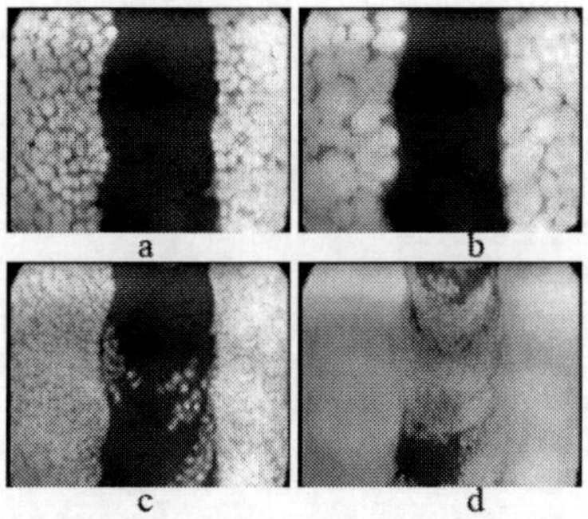

c d

Fig. 2. Magnetooptical images of a weld in a stainless steel sample with orthogonal excitation of eddy currents with frequencies: a - 8, b - 15, c - 30, d – 45 kHz.

Analysis of MO images in magnetic and nonmagnetic samples shows that the magnitude of the EC magnetic fields is much larger in stainless steel samples. This is due to the much greater magnetic permeability of these samples. Therefore saturation of epitaxial ferrite-garnet films at the locations of defect mapping caused by EC magnetic fields occurs in a magnetic sample at lower values of H_s = 120 Oe than in a nonmagnetic sample H_s = 300 Oe. Such a state leads to a contrasting and clear image of defects. The use of the orthogonal constant bias field H_ leads to a significant decrease of these values. For example, a constant bias field H_ = 24 Oe reduces saturation fields up to 80 Oe and 220 Oe for magnetic and nonmagnetic samples, respectively. The effect of the constant bias field and the frequency of the alternating field on the MO imaging of defects in magnetic samples is more pronounced than in the nonmagnetic ones.

1.4. Defective welds

To test the operation of the MOEC method under real conditions, samples of welds of steel pipes with a diameter of 50 and 70 mm using electric and gas welding were made. MO defectoscopy was carried out at frequencies from 8 to 60 kHz, in alternating magnetic fields from 40 to 440 Oe. It was shown that the optimal conditions in this case are the following: an alternating current of 25 kHz, an alternating magnetic field of 80 Oe, a constant magnetic field of 0 Oe. Optical and MO images obtained under these conditions are shown in Figure 3.

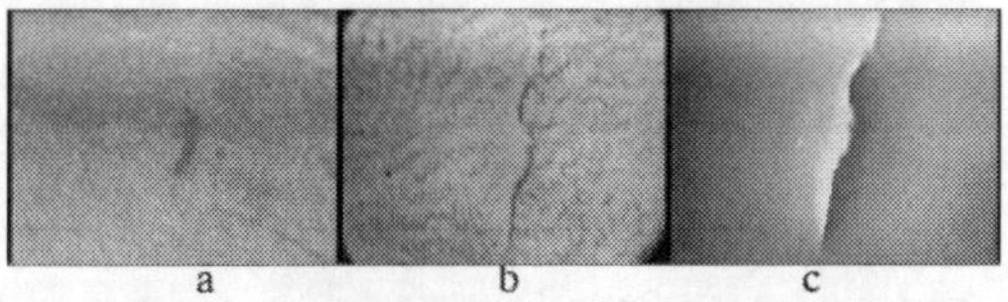

a b c

Fig. 3. The defects in welds of steel pipes: optical (a) and magneto-optical (b, c) images; orthogonal (b) and longitudinal (c) excitation of eddy currents.

MO images in Figure 3 were obtained with equal parameters of alternating and constant fields, the difference is only in the direction of generation of the excitation field. It is clearly seen that the MO image of the defective area contains much more information than a simple optical snapshot. The use of the MOEC method allows detecting defects that lie in the depth of the seam.

An interesting result was received the use of ferrite-garnet film with an easy-plane anisotropy as a sensor. It made it possible to obtain an "analog picture" of the welds on a stainless steel sample using an inductor of a normal field.

2 Simulation of magnetic field of eddy currents

2.1. Mathematic model

A mathematical model of the magnetic field of a conducting block consisting of three parts with different values of electric conductivity and magnetic permeability is constructed. This mathematical model is a system of integro-differential equations for the densities of secondary sources. These sources include: eddy currents (determined by the instantaneous value of the bulk current density), electric charges (specified by the instantaneous value of the surface charge density), and magnetization currents (given by the instantaneous value of the surface current density). We note that the derivation of this system of integro-differential equations is carried out using the Ohm's law in differential form, as well as the boundary conditions for the normal component of the eddy current density and the tangential component of the magnetic field strength vector. [6]

The magnitude of the magnetic field acting on the magneto-optical sensor is superposition fields of the inductor and eddy current of the defect sample and can be represented in the form:

$$\vec{H}(M,t)=\vec{H}_0(M,t)+\frac{1}{4\pi\mu_0}\sum_{k=1}^{n}\mu_k\int_{V_k}\frac{\left[\vec{\delta}(N,t),\vec{r}_{MN}\right]}{r_{MN}^3}dV_N+$$
$$+\frac{1}{4\pi}\sum_{k=1}^{m}\oint_{S_k}\frac{\left[\vec{j}(P,t),\vec{r}_{MP}\right]}{r_{MP}^3}dS_P \qquad (1)$$

where M — an arbitrary point of space (the point of observation); N — point belonging to the volume of the conductor; P — point belonging to the surface of a conductor; $\vec{H}_0(M,t)$ – the magnetic field of the inductor; $\vec{\delta}(N,t)$ — instantaneous value of the bulk density of eddy currents; $\vec{j}(P,t)$ — instantaneous value of the surface density of magnetization currents; μ_k — magnetic permeability of the k conductor; n — number of volumes of conductors; m — number of surfaces separating regions with different electrical conductivities.

2.2 Non-ferromagnetic material

We will give an example of modeling a magnetic field above the surface of a conducting massive block containing a defect in the form of a welded seam. First, consider the case where the block is made of a non-ferromagnetic material (aluminum).

When manufacturing a welded seam, a change in the structure of the material inevitably occurs, which leads to a decrease in the conductivity of the conductor in the weld region. On figure 4 show are the distributions of the normal component of the intensity of the magnetic field in the region above the block.

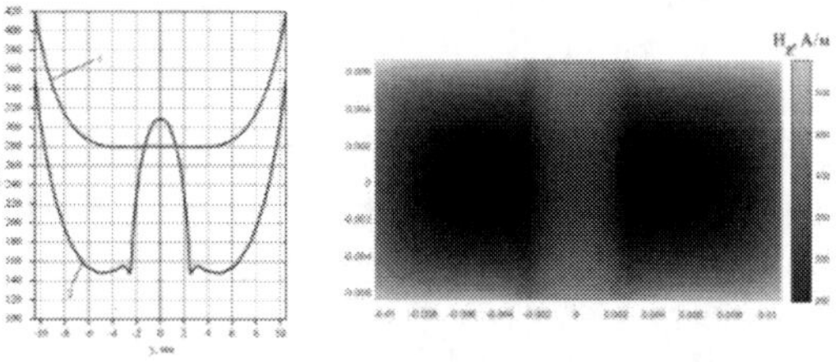

Fig.4. Distribution of strength magnetic field in the area above the block (a): block without weld (1); block with weld (2); topogram of the magnetic field in the region over a block with a welded seam (b).

In calculations, the conductivity of the weld was assumed to be 1.5 × 10^7 (S/m), which is 40% of the conductivity of pure aluminum. As can be seen, the presence of a weld with a lower conductivity than the conductivity of adjacent regions of the conductor leads to a significant increase in the contrast of the topogram of the magnetic field. The boundaries of the seam are clearly traced. It is natural to expect that the higher the difference in conductivity of are conductors, the more contrast the picture of the field, and, consequently, the MO image of varies part of sample will also be more contrast.

2.3 Ferromagnetic material

The presence of ferromagnetic properties in the sample from stainless steel leads to the appearance of magnetization currents. In contrast to eddy currents, the effect of which leads to a decrease in the resulting magnetic field, the magnetization currents tend to increase this field. Therefore, in the case of a non-ferromagnetic sample, the magnetic field is always smaller than in the case of a ferromagnetic sample, even with a relatively low magnetic permeability. The distribution of the normal component of the magnetic field, which allow one to compare the cases of a ferromagnetic and nonferromagnetic samples represented in figure 5.

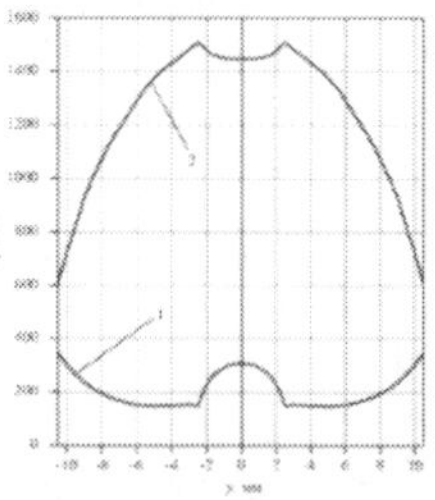

Fig. 5. Distribution of the normal component of the magnetic field intensity above the samples with a welded seam: non-ferromagnetic block (1); ferromagnetic block (2).

The last figure shows that the field in the central region in the case of a ferromagnetic conductor exceeds the analogous value for a non-ferromagnetic conductor by more than 4.5 times. Note that in calculations it was assumed that the magnetic permeability of stainless steel is $30\mu_0$. A significantly increased field in the case of the use of a ferromagnetic conductor can lead to a rapid saturation of the film, which is confirmed by the experimental results.

3 Conclusions

The possibility of using ferrite-garnet films to visualization and control the quality of defects of the "welded seam" type by the method of magneto-optical eddy current flaw detection has been shown.

The efficiency of visualization of welds seams in ferromagnetic samples is higher than in nonmagnetic samples and depends on the eddy current frequency and the constant bias field.

By the method of integro-differential equations, a model is obtained for the distribution of magnetic fields of eddy currents in the vicinity of welded seams, which is in satisfactory agreement with the experimental data.

Acknowledgements

The author Lugovskoy N.V. thanks for support the Project of Development Program of V.I. Vernadsky Crimean Federal University Network "Academic mobility of Young Scientists of Russia - AMMUR" in 2016

References

1. U. Radtke, R. Zielke, H. Rademacher, Opt. Laser. Eng. 36 (2001)
2. Y. Deng, X. Lin, Y. Fan, IEEE Trans. Magn. 42(2006)
3. V.Vishnevskiy, V.Berzhansky, N. Lugovskoy, A. Prokopov, F. Pankov, SSP, 230 (2014)
4. V.N. Berzhansky, D.M. Filippov, N.V. Lugovskoy, Phys. Procedia, 82C (2016)
5. V.N. Berzhansky, A.V. Karavainikov, T.V. Mikhailova, V.I. Belotelov, N.V. Lugovskoy, JMMM, 440 (2017)
6. D.M. Filippov, G.P. Kozik, A.V. Fursenko, V.N. Fedorovsky In Proc. ICIEAM 2 (2016)

Dispersion of the magnetostatic volume waves in a medium with damping

Lidiya Maltceva[1,*], *Pavel* Makarov[1,**], *Leonid* Kotov[1], and *Vladimir* Shcheglov[2,***]

[1] *Pitirim Sorokin Syktyvkar State University, Oktyabrsky 55, Syktyvkar, 167001, Russia*
[2] *Institute of Radio-Engineering and Electronics of RAS, Mokhovaya 11-7, Moscow, 125009, Russia*

Abstract. In magnetostatic approximation of electrodynamics the dispersion relation for magnetostatic backward volume waves (MSBVW) in a ferrite film, placed in a constant longitudinal and transverse alternating magnetic field is obtained. The magnetic field, damping and temperature influence on the spectrum and dispersion characteristics of the MSBVW is regarded.

1 Introduction

It is well known that spin waves have an extremely rich and peculiar dispersion that is nonlinear, anisotropic, and nonreciprocal [1, 2]. The spin wave dispersion is very sensitive to the sample's magnetic properties and micromagnetic state, including both the internal magnetic field and magnetization. Nowadays, creating devices on magnetostatic surface (MSSW) and volume (MSVW) waves is one of the most actively developing areas of spintronics [3]. The use of monocrystalline ferro- and ferrimagnetic structures, such as iron-yttrium garnet (YIG) films, in systems based on MSVW allows the creation of compact devices with unique magnetic properties [4–9].

One of the main requirements for spintronics devices is the thermal stabilization of their magnetic parameters in the operating temperature range. At the same time, it is well known that YIG films have a serious drawback — the strong temperature dependence of the saturation magnetization.

Minimization of the energy losses is the key to the successful operation of spintronics devices, but MSVW attenuation and temperature instability has not been studied enoung, because most of the research are carried out without taking into account the attenuation and the effect of temperature [10].

2 Model and basic equations

The geometry of the problem is shown in Figure 1. Let's consider a boundless ferritic film of thickness d, magnetized to saturation. The plane yOz corresponds to the plane of the film, the positions of the boundaries are determined by the coordinates $x = \pm d/2$. The film is magnetized to saturation by a constant field $\mathbf{H}$ applied along the Oz axis. The angle between the Oy axis and the wave vector $\mathbf{k}$ is denoted as φ.

*e-mail: lidija.malceva@rambler.ru
**e-mail: makarovpa@syktsu.ru
***e-mail: vshcheg@cplire.ru

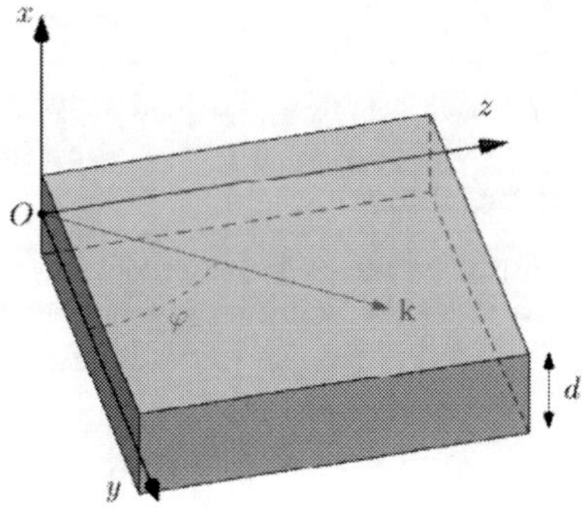

Figure 1. Geometry of the problem

Despite the fact that magnetostatics is only an approximation of electrodynamics, it can be used to model the work of a variety of spintronics devices. In the magnetostatic approximation Maxwell's magnetic equations have the form [11]

$$\begin{cases} \nabla \times \mathbf{H} = 0, \\ \nabla \cdot (\mathbf{H} + 4\pi \mathbf{M}) = 0. \end{cases} \tag{1}$$

If we introduce the magnetostatic potential ψ:

$$\mathbf{H} = \nabla \psi, \tag{2}$$

then from (1) we obtain the Walker equation

$$\Delta \psi + 4\pi \nabla \cdot \mathbf{M} = 0. \tag{3}$$

The Landau-Lifshitz equation with a dissipative term in the Gilbert form was used to calculate the magnetization dynamics

$$\frac{\partial \mathbf{M}}{\partial t} = -\gamma \left[\mathbf{M} \times \mathbf{H}\right] + \frac{\alpha}{M_0} \left[\mathbf{M} \times \frac{\partial \mathbf{M}}{\partial t}\right]. \tag{4}$$

In (4) M_0 is the constant length of the magnetization vector, γ is the gyromagnetic ratio, and α is the dissipation parameter. The dispersion equation for the magnetostatic volume waves, obtained from (3) and (4), has the form:

$$\tan(kd\vartheta) = \frac{2\mu\vartheta}{\beta - 1}, \tag{5}$$

where ϑ, β — ancillarity quantities:

$$\vartheta = \sqrt{\cos^2\varphi + \frac{\sin^2\varphi}{\mu}}, \quad \beta = \nu^2\cos\varphi - \mu^2\vartheta^2. \tag{6}$$

In (6) μ is the permeability and ν is the magnetic susceptibility of the medium

$$\mu = 1 + \frac{\Omega_H + i\alpha\Omega}{\Omega_H^2 - (1+\alpha^2)\Omega^2 + i2\alpha\Omega\Omega_H}, \quad \nu = \frac{\Omega}{\Omega_H^2 - (1+\alpha^2)\Omega^2 + i2\alpha\Omega\Omega_H}, \tag{7}$$

which depend of the normalized magnetic field Ω_H and normalized frequency Ω:

$$\Omega_H = \frac{H}{4\pi M_0}, \quad \Omega = \frac{\omega}{4\pi\gamma M_0}. \tag{8}$$

3 Main results

The dispersion curves for the first five modes of the magnetostatic volume waves in the case when the angle $\varphi = \pi/2$ without damping ($\alpha = 0$) are shown in Figure 2.

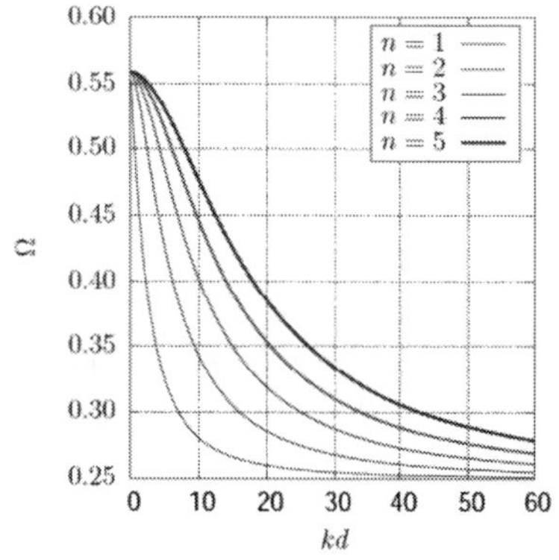

Figure 2. Dispersion curves for first five modes of MSBVW without damping in case $\varphi = \pi/2$.

The analysis of dispersion curves (Figure 2) shows, that in our geometry (Figure 1) only the magnetostatic backward volume waves (MSBVW) can exist.

The boundary frequencies of the MSBVW, defined from the limit transition $k \longrightarrow 0$ and $k \longrightarrow \infty$ of the dispersion relation (5) have the form:

$$\begin{cases} f_1 = \gamma H, \\ f_2 = \gamma\sqrt{H(H + 4\pi M_0)}. \end{cases} \tag{9}$$

Thus, the volume waves propagate in a strictly defined frequency range $f_1 \leqslant f \leqslant f_2$. The boundary frequencies of the MSBVW are determined by the value of the saturation magnetization and can vary over a wide range. At the same time, the saturation magnetization of the material substantially depends on the temperature.

3.1 Influence of the constant magnetic field value and the direction of propagation on the dispersion of MSBVW

The curves of the boundary frequencies as functions of a constant magnetic field are shown in Figure 3. In addition, this figure shows the field dependences of the width of the MSBVW frequency range (dashed line).

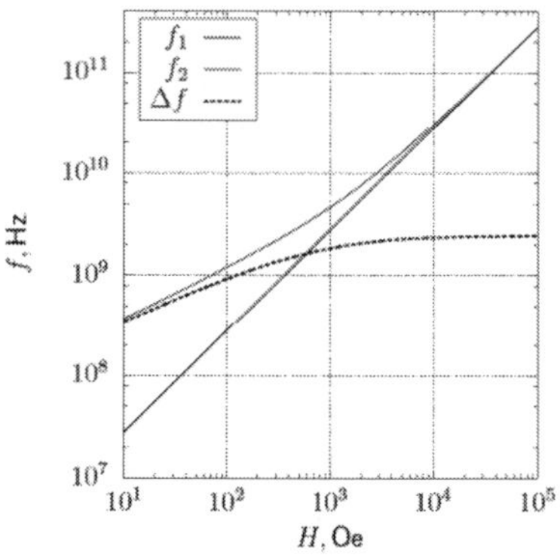

Figure 3. Magnetic field effect on the MSBVW spectrum

Figure 3 illustrates that in weak fields the lower boundary frequency of the MSBVW, in contrast to the upper one, linearly depends on the field. However, in the limiting transition $H \rightarrow \infty$ both of f_1 and f_2 are linearly dependent on H. Thus, with an increase in the magnetic field from 10 Oe to about 10 kOe, the frequency spectrum of MSBVW expands from about 0.5 GHz to 2 GHz. In fields H more than 10 kOe, the width of the spectrum does not change and depends on the saturation magnetization.

The influence of the angle φ on the boundary frequencies is shown in Figure 4. From this figure we can see that the high frequency of MSBVW spectrum f_2 equals to the FMR frequency and does not depend on angle φ. The low boundary frequency of MSBVW spectrum f_1 is a decreasing function of angle between magnetic field and direction of wave propagation. Therefore, the MSBVW spectrum decreases with the reduction of angle φ.

Figure 5 shows the frequency dependence of the first mode of the MSBVW for the real part of the wave number at different angles in case when damping is negligible. The angle change leads to the alteration of the spectrum width of the MSBVW.

It is evident that the frequency spectrum of the MSBVW expands with the increasing angle φ. The region of existence of volume waves is limited by the angle $\varphi \neq 0$.

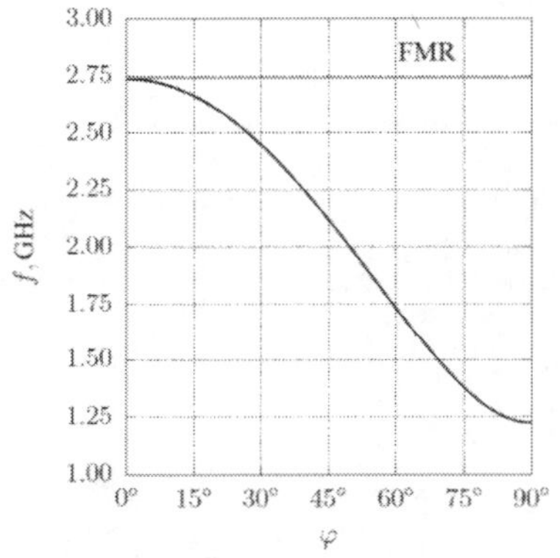

Figure 4. Angle effect on the MSBVW spectrum

3.2 Temperature influence

It is well-known that YIG films exhibit strong temperature instability, which can be caused by a number of factors [12]. In our model, the effect of temperature reduces to the fact that the normalized frequencies of the waves Ω and the magnetic fields Ω_H, determined by the relations (8), shift up with changing T.

In the temperature range T from 213 K to 353 K the function $4\pi M(T)$ can be approximated by a linear dependence, and tangent of the slope angle is equal to −2.9 G/K [12]. Thus, with the temperature decreasing from $T = 300$ K to $T = 273$ K, the saturation magnetization increases from 1760 G to 1840 G.

Dispersion curves for the first five modes of the MSBVW at different temperatures ($T - 273$ K and $T = 300$ K) are shown in Figure 6.

It is clear from Figure 6(a) and Figure 6(b) that changing of the temperature alterates the dispersion curves slightly, in particular, with increasing T the frequency spectrum shifts to a lower frequency region.

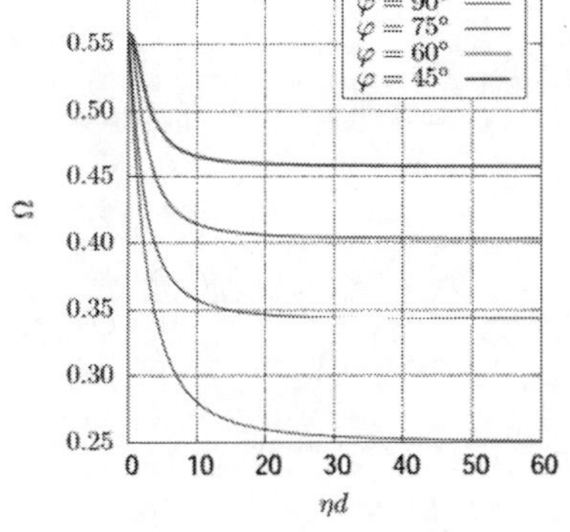

Figure 5. The angle influence on the dispersion of first mode of MSBVW

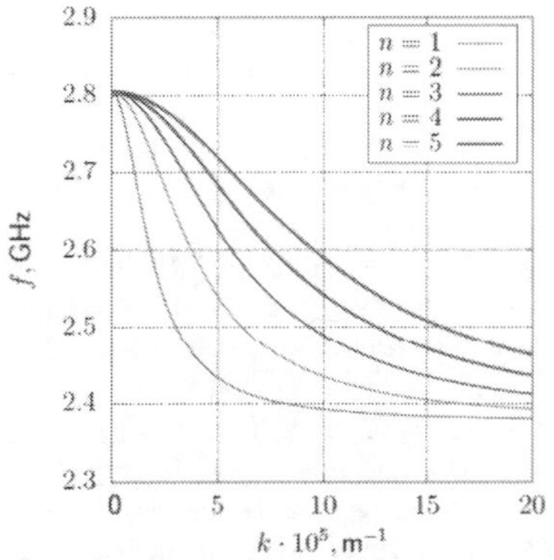

(a) $T = 273$ K

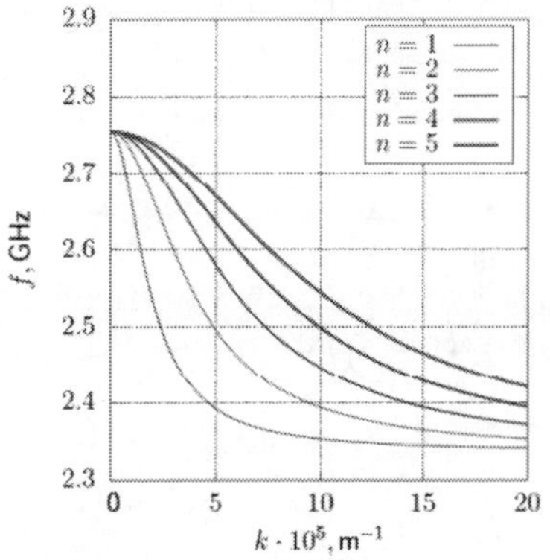

(b) $T = 300$ K

Figure 6. Dispersion curves for the first five modes of MSBVW

In Figures 6(a), and 6(b) the follows parameters were used: film thickness $d = 15$ mkm, gyromagnetic ratio $\gamma = 2$, magnetic field value $H = 440$ Oe, angle $\varphi = 40°$. The magnetic field H and angle φ values were chosen as intermediate numbers between minimum and maximum quantities. This choice approximately corresponds to the average point of the dashed curve in Figure 3 and middle value of the spectrum width in Figure 4.

3.3 The influence of dissipation

Let's consider the case when $\alpha \neq 0$. In this case parameters ν, μ, β, ϑ, as well as the wave vector $k = \eta - i\xi$ are complex values. The imaginary part ξ of the wave vector determines the logarithmic damping decrement.

The left-hand side of the equation (5) is a function of both the wave vector k and the frequency Ω. However, with help of the auxiliary notation (10), it is easy to separate the variables.

$$\mathcal{Z} = \arctan \frac{2\mu\vartheta}{\beta - 1} \quad (10)$$

In this way, the dispersion relation (5) can be written as follows:

$$\eta d = \frac{Z'\vartheta' + Z''\vartheta''}{(\vartheta')^2 + (\vartheta'')^2}, \quad \xi d = \frac{Z'\vartheta'' - Z''\vartheta'}{(\vartheta')^2 + (\vartheta'')^2}. \qquad (11)$$

In case of arbitrary geometry (for various angles φ) and when damping taking into account the dispersion of the MSBVW is complex.

Figure 7 illustrates the effect of damping on the dispersion characteristics of the first five modes of the MSBVW. This effect leads to the fact that the wave number is limited both from above (in the low-frequency region) and from below (in the high-frequency region). Also, it is clear that damping, in addition to limiting the possible values of the wave number (and hence also the wavelength of the MSBVW), affects the shape of the dispersion curves. This means that the presence of attenuation changes the propagation velocity of the waves.

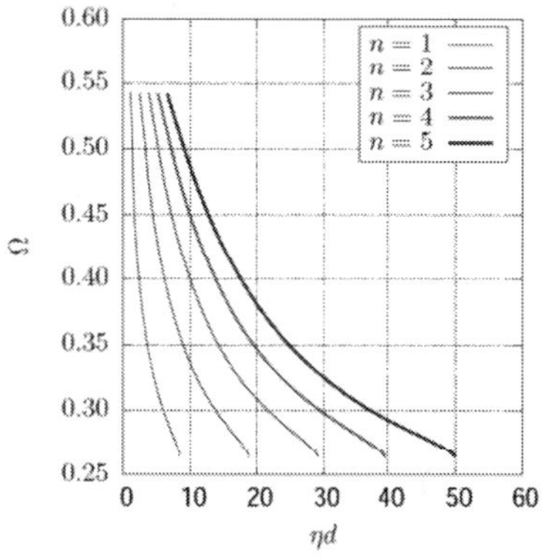

Figure 7. Dispersion curves of the first five modes of the MSBVW when $\alpha = 0.05$ for $\varphi = \pi/2$

4 Conclusion

Thus, the analysis of dispersion curves with and without of damping, showed that dissipation leads to the limitation of the spectrum of MSBVW not only in frequency, but also in wavelength.

Moreover, field and angular dependences of the boundary frequencies and spectrum width for magnetostatic backward volume waves are obtained. In addition, the effect of temperature on the dispersion curves of the MSBVW is determined: with the increase of the temperature, the frequency spectrum shifts to a lower frequency region.

This research was supported by Russian Scientific Fund (projects 17-02-01138 and 17-57-150001).

References

[1] S.O. Demokritov *Spin Wave Confinement: Propagating Waves (2nd Edition)* (Pan Stanford Publishing Pte. Ltd., Singapore, 2017) 449

[2] D. Stancil, A. Prabhakar *Spin Waves. Theory and Applications* (Springer, New York, 2009) 364

[3] A.V. Sadovnikov et al. Phys. Rev. B **96**, 060401(R) (2017)

[4] J.P. Parekh, K.W. Chang, H.S. Tuan, Circuits, Systems and Signal Processing **4**, 9–39 (1985)

[5] W.S. Ishak, Proceedings of the IEEE **76**, 171–187 (1988)

[6] K.L. Wang, Z. Zhao, A. Khitun, Thin Solid Films **517**, 184–190 (2008)

[7] A.A. Serga, A.V. Chumak, B. Hillebrands, J. Phys. D: Appl. Phys. **43**, 264002 (2010)

[8] A. Khitun, M.Q. Bao, K.L. Wang, J. Phys. D: Appl. Phys., **43**, 264005 (2010)

[9] A.V. Chumak, V.I. Vasyuchka, A.A. Serga, B. Hillebrands, Nat. Phys., **11**, 453 (2015)

[10] R.W. Damon, J.R. Eshbach, J. Phys. Chem. Solids **19**, 308 (1961)

[11] L.D. Landau, E.M. Lifshitz *The Classical Theory of Fields* (Pergamon Press, Oxford, 1971) 430

[12] S.I. Yushchuk, P.S. Kostyuk, I.E. Lopatinskii, Tech. Phys. The Russian J. of Appl. Phys. **43**, 1051–1054 (1998)

Tamm plasmon-polaritons structures with Bi-substituted garnet layers

Tatyana Mikhailova*, *Alexander* Shaposhnikov, *Anatoly* Prokopov, *Andrey* Karavainikov, *Sergey* Tomilin, *Sergey* Lyashko, and *Vladimir* Berzhansky

V. I. Vernadsky Crimean Federal University, 295007 Simferopol, Russia

Abstract. New original Tamm plasmon-polaritons (TPP) structures with Bi-substituted iron garnet and Au plasmonic layers were proposed, synthesized and investigated. The structures with single and double garnet layers were modelled to form a TPP mode at the center of photonic band gab. The top Au layer has the gradient thickness varied in the range from 0 to 65.2 nm. It was found the features of TPP resonances as a function of the thickness of metal coating. The resonances on TPP have the maximum optical quality factor and transmission at the vicinity of Au thickness of 30 nm. These configurations are optimum to form the highest intensity of electric field of light wave in the area of the magnetic layers. It was found the spectral blue and red shifts of TPP mode with increasing of Au thickness. The blue and red shifts can be explained respectively by structural and thickness changes of Au coating. The maximum resonant values of Faraday rotation were −2.1° at 664 nm and −12.3° at 645 nm for structures with single and double garnet layers, respectively, and thickness of Au coating of 65.2 nm.

1 Introduction

The one-dimensional nanostructures of photonic crystals with bismuth-substituted iron garnet (Bi: IG) layers are perspective for magnetooptical (MO) sensors in optoelectronic devices allowed significant rotation of the polarization plane of the transmitted light. One of the first proposed structures in magnetophotonics was the microcavity based on nonmagnetic dielectric Bragg mirrors (BM) and a Bi: IG layer [1-5]. In such structures, Faraday and Kerr rotation enhancement is achieved most effectively. Later, the photonic and magnetophotonic crystals limited on the one side by a metal were investigated [6-10]. The amplification of MO effects in these structures are result of the excitation of the optical Tamm state or Tamm plasmon-polaritons (TPP). In addition the TPP mode can be combined with resonances of another nature [11-15]. In our previous works, we proposed and investigated the new type of microcavity structure on basis of nonmagnetic dielectric BM and double iron garnet layer with low and high Bi content [15-17]. In this paper, the new effective magnetophotonic structures with TPP resonance mode are proposed and realised.

2 Experimental technique and modelling

The investigated TPP structures consist of a seven-pair dielectric BM of titanium and silicon oxides, on the surface of which single or double garnet layer, silicon oxide buffer layer and Au plasmonic coating are placed consistently. The top Au plasmonic coating has the gradient thickness. Formulas of structures are listed below

No. 1 – GGG / $[TiO_2/SiO_2]^7$ / M1 / SiO_2 / Au,
No. 2 – GGG / $[TiO_2/SiO_2]^7$ / M1 / M2 / SiO_2 / Au.

There GGG is substrate of gadolinium gallium garnet with crystallographic orientation of (111). M1 and M2 are Bi: IG of compositions of $Bi_{1.0}Y_{0.5}Gd_{1.5}Fe_{4.2}Al_{0.8}O_{12}$ and $Bi_{1.5}Gd_{1.5}Fe_{4.5}Al_{0.5}O_{12}$, respectively. Schematic diagrams of configurations are shown in Figure 1. The buffer layer of silicon oxide was used to reduce the absorption of light at the magnetic layer-metal interface.

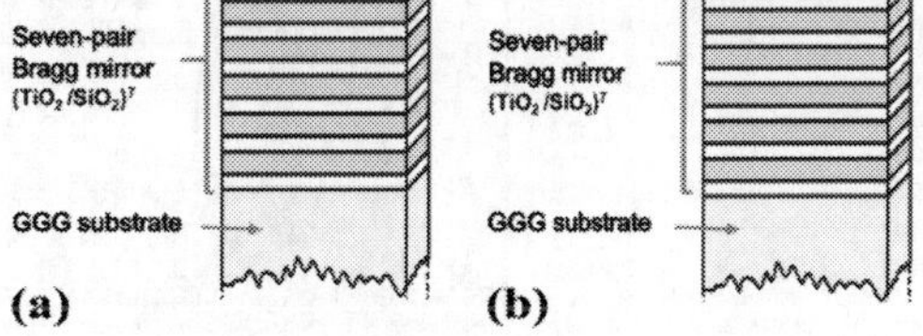

Fig. 1. Schematic diagram of synthesized structures No. 1 (a) and No. 2 (b).

2.1 Modelling

These structures were modelled to form a TPP mode at the center of photonic band gab (PBG) of BM. The

* Corresponding author: taciamikh@gmail.com

parameters of the structure were previously calculated using the 4×4 transfer matrix method [18] for numerical solution of Maxwell's equations. The components of permittivity tensors of layers obtained by fitting the optical and magneto-optical spectra of previously synthesized single-layer films and magnetophotonic crystals were taken into account [15-17]. The same components of permittivity tensor were used for Au coating with thicknesses in the range from 0 to 70 nm [15]. The following thicknesses were used in presented in section 3 calculations of the spectra of single BM, BM with magnetic layers and structures without and with the top Au coating:

h_{TiO2} = 73 nm and h_{SiO2} = 115 nm for BM layers;

h_{M1} = 108 nm and h_{bSiO2} = 140 nm for M1 and SiO_2 buffer layers of structure No. 1;

h_{M1} = 55 nm, h_{M2} = 178 nm and h_{bSiO2} = 80 nm for M1, M2 and SiO_2 buffer layers of structure No. 2.

The PBG center of both structures is placed at λ_0 = 655 nm.

2.2 Synthesis

Synthesis of the structures occurred in several stages. Dielectric BM were synthesized by electron beam evaporation on hot (400 °C) substrate. The thickness of layers was optically controlled during deposition. Iron garnet layers were fabricated by reactive ion beam sputtering of corresponding ceramic targets in argon-oxygen mixture and crystallized in the annealing process at the air. Detailed description of the methods and conditions of garnet synthesis can be found in [15-17].

Au thin film with a gradient of thickness h_{Au} from 0 to 70 nm along the chosen direction on the sample surface was deposited by thermal evaporation in vacuum using a technique described in [19, 20]. The length of sample along the gradient of Au thickness was 12 mm. Therefore, the gradient of Au thickness was 5.8 nm/mm. Garnet and Au thickness was determined by MII-4 microinterferometer.

Investigation of transmittance was carried out using an automated spectrophotometer KFK-3. Measurements of Faraday rotation (FR) were performed using hand-made computer-control spectropolarimeter by compensation method in saturation fields. The beam aperture and the gradient of Au thickness at the scale of beam aperture were respectively 0.1 mm and 0.6 nm.

3 Results and discussion

3.1 Structure with single Bi: IG layer

Spectral dependencies of transmittance and magnetooptical FR as a function of the thickness of Au coating for structure No. 1 are shown in Figure 2. The spectra of BM, BM with magnetic layer and structure without Au coating are also presented.

A characteristic feature of the spectra is the presence of a pronounced resonant peak, the position of which depends on the thickness of Au layer. In the calculated spectra with increasing of Au thickness h_{Au} from 6.1 to 65.2 nm, the blue shift of resonance peak of TPP mode from 710 to 664 nm takes place. Nevertheless, the red shift of resonant mode from 640 to 664 nm is observed in experimental spectra for h_{Au} in the range from 6.1 to 30 nm. Blue shift of resonance from 666 to 664 nm occurs in the experiment for h_{Au} in the range from 30 to

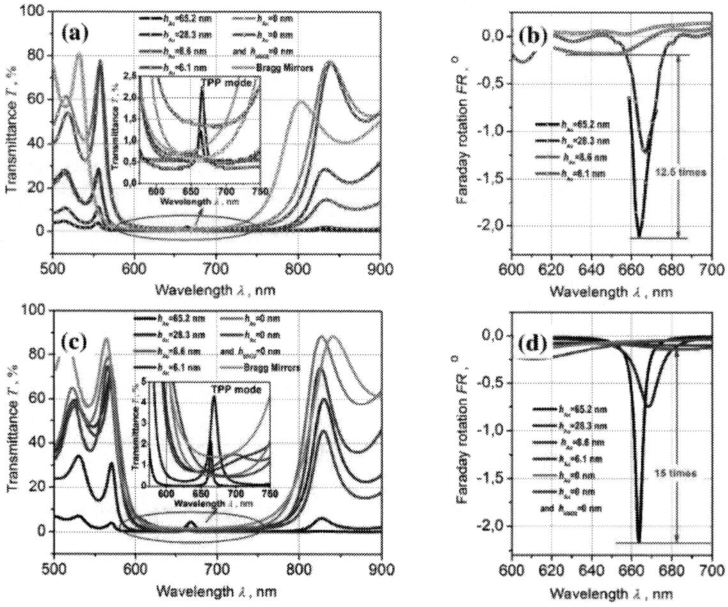

Fig. 2. Measured (a, b) and calculated (c, d) transmittance and FR spectra of synthesized seven-pair Bragg mirror, Bragg mirror with single garnet layer and TPP structure No.1 with single garnet layer as a function of the thickness of Au coating.

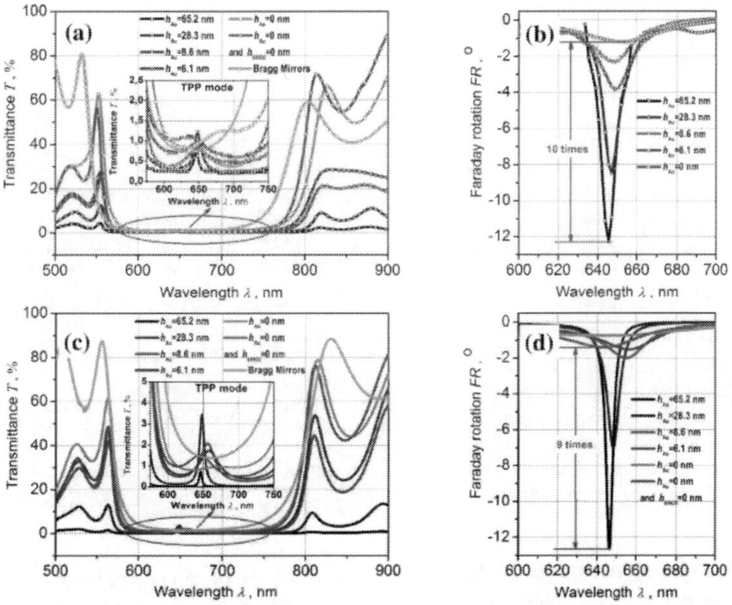

Fig. 3. Measured (a, b) and calculated (c, d) transmittance and FR spectra of synthesized seven-pair Bragg mirror, Bragg mirror with double garnet layer and TPP structure No.2 with double garnet layer as a function of the thickness of Au coating.

65.2 nm. A reason of red shift of the mode is the changes of granularity, surface roughness and continuity of Au coating (structural changes) with increase of its thickness, which leads to a change of optical properties of coating. Blue shift of TPP modes is caused by the change in the coating thickness itself, since in theory we assume smooth boundaries, continuity and non-gradient thickness of Au layer. The surface roughness of layers, discontinuity and thickness gradient of Au coating in the beam aperture in the experiments lead to increasing of half-width of the resonant peaks.

The maximum resonant FR value are –2.1° at 664 nm for configuration with h_{Au} = 65.2 nm.

3.2 Structure with double Bi: IG layer

Measured and calculated transmittance and FR spectra of synthesized structure No. 2 are shown in Figure 3. In the spectra the shifts of TPP mode are observed analogous to the shifts in the spectra of structure No. 1. In the calculated spectra with increasing of h_{Au} from 0 to 65.2 nm, the blue shift from 675 to 645 nm takes place. In the experiment, the red shift is from 638 to 649 nm for h_{Au} in the range from 0 to 30 nm. Blue shift is from 649 to 645 nm for h_{Au} in the range from 30 to 65.2 nm. Experimental and calculated spectra of structure without Au coating are identical and have the peak at 675 nm.

The maximum resonant FR value are –12.3° at 645 nm for configuration with h_{Au} = 65.2 nm.

Calculated spatial distribution of electric field intensity in structure No. 2 with Au coating of different thickness, structure No. 2 without Au coating, BM with magnetic layers and single BM is shown in Figure 4.

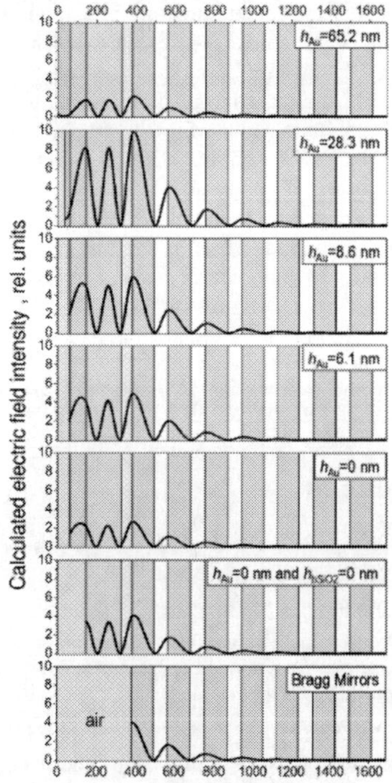

Fig. 4. Calculated spatial distribution of electric field intensity in structure No. 2 with different Au thickness, structure No. 2 without Au coating, BM with magnetic layers and single BM.

The distribution is classical for TPP mode. Distribution in magnetic double layer is characterized by asymmetry. The maximum intensity is observed in the layers adjacent to the plasmonic coating and for configuration with thickness of Au layer of 28.3 nm. This is in good agreement with the experimental results and indicates that a coating thickness is optimal for such type of structure. Consequently, the resonances on TPP has the maximum optical quality factor and transmittance at the vicinity of Au thickness of 30 nm for both structures No. 1 and No. 2.

4 Conclusions

In the work, the new original magnetophotonic structures $[TiO_2/SiO_2]^7/M1/SiO_2/Au$ and $[TiO_2/SiO_2]^7/M1/M2/SiO_2/Au$ with Bi-substituted iron garnets of composition $Bi_{1.0}Y_{0.5}Gd_{1.5}Fe_{4.2}Al_{0.8}O_{12}$ (M1) and $Bi_{1.5}Gd_{1.5}Fe_{4.5}Al_{0.5}O_{12}$ (M2) and Au plasmonic layer were proposed and synthesized. The structures were modelled to form a TPP mode at the center of photonic band gab.

For the first time, the investigation of the features of TPP resonances for TPP structures as a function of the thickness of metal coating was done experimentally and theoretically. The resonances on TPP have the maximum optical quality factor and transmittance at the vicinity of Au thickness of 30 nm. Calculations show that these configurations are optimum to form the highest intensity of electric field of light wave in the area of the magnetic layers. It was found the spectral shift of TPP mode with increasing of Au thickness. Spectral red and blue shifts are caused by different factors. A reason of blue shift of the mode in the experiment is the structural changes in Au coating with increase of its thickness. The red shift of TPP modes is caused by the change in the coating thickness itself.

It was achieved the maximum resonant values of FR −2.1° at 664 nm and −12.3° at 645 nm for structures with single and double garnet layers, respectively, and thickness of Au coating of 65.2 nm.

The authors acknowledge support by the RF Ministry of Education and Science in the framework of the state task (project no. 3.7126.2017/8.9).

References

1. M. Inoue, K. Arai, T. Fujii, M. Abe, J. Appl. Phys. **85**, 5768 (1999)
2. M. Inoue, A.V. Baryshev, T. Goto, S.M. Baek, S. Mito, H. Takagi, and P.B. Lim, *Magnetophotonics* (Springer-Verlag Berlin Heidelberg, 2013)
3. I.L. Lyubchanskii, N.N. Dadoenkova, M.I. Lyubchanskii, E.A. Shapovalov, Th. Rasing, Journal of Physics D: Applied Physics **36**, R277 (2003)
4. S. Kahl and A.M. Grishin, Appl. Phys. Lett. **84**, 1438 (2004)
5. S.I. Khartsev and A.M. Grishin, J. Appl. Phys. **101**, 053906 (2007)
6. A.P. Vinogradov, A.V. Dorofeenko, S.G. Erokhin, M. Inoue, A.A. Lisyansky, A.M. Merzlikin and A.B. Granovsky, Phys. Rev. B **74**, 045128 (2006)
7. M. Kaliteevski, I. Iorsh, S. Brand, R.A. Abram, J.M. Chamberlain, A.V. Kavokin, and I.A. Shelykh, Phys. Rev. B **76**, 165415 (2007)
8. T. Goto, A.V. Dorofeenko, A.M. Merzlikin, A.V. Baryshev, A.P. Vinogradov, M. Inoue, A.A. Lisyansky and A.B. Granovsky, Physical Review Letters **101** 113902 (2008)
9. M. Kaliteevski, S. Brand, R.A. Abram, I. Iorsh, A.V. Kavokin and I.A. Shelykh, Applied Physics Letters **95** 251108 (2009)
10. A.P. Vinogradov, A.V. Dorofeenko, A.M. Merzlikin, Uspekhi Fizicheskikh Nauk **180** 249 (2010)
11. R. Brückner, M. Sudzius, S.I. Hintschich, H. Fröb, V.G. Lyssenko and K. Leo, Phys. Rev. B **83** 033405 (2011)
12. Hai-Chun Zhou, Guang Yang, Kai Wang, Hua Long, Pei-Xiang Lu, Chin. Phys. Lett. **29**, 067101 (2012)
13. B.I. Afinogenov, V.O. Bessonov, A.A. Nikulin and A.A. Fedyanin, Applied Physics Letters **103**, 061112 (2013)
14. S.K. S.-U. Rahman, T. Klein, S. Klembt, J. Gutowski, D. Hommel and K. Sebald, Sci. Rep. **6**, 34392 (2016)
15. T.V. Mikhailova, V.N. Berzhansky, A.V. Karavainikov, A.N. Shaposhnikov, A.R. Prokopov and S.D. Lyashko, J. Phys.: Conf. Ser. **741**, 012121 (2016)
16. V.N. Berzhansky, T.V. Mikhailova, A.V. Karavainikov, A.R. Prokopov, A.N. Shaposhnikov, I.N. Lukienko, Yu.N. Kharchenko, O.V. Miloslavskaya, N.F. Kharchenko, J. Magn. Soc. Jpn. **36**, 42 (2012)
17. V. Berzhansky, A. Shaposhnikov, A. Prokopov, A. Karavainikov, T. Mikhailova, I. Lukienko, Yu. Kharchenko, V. Golub, O. Salyuk, V. Belotelov, JETP **123**, 744 (2016)
18. C.P. Yin, T.B. Wang and H.Z. Wang, Eur. Phys. J. B **85**, 104 (2012)
19. S.V. Tomilin, A.S. Yanovsky, J. of Nano- and Electronic Physics **5**, 03014 (2013)
20. S.V. Tomilin, V.N. Berzhansky, A.N. Shaposhnikov, A.R Prokopov, E.T. Milyukova, A.V. Karavaynikov and O.A. Tomilina, J. Phys.: Conf. Ser. **741**, 012113 (2016)

Hypersound excitation of magnetization and elastic displacement in case of magnetization reversal

Fanur Asadullin[1,2,*], *Dmitriy* Pleshev[1,2], *Vladimir* Vlasov[2], *Leonid* Kotov[2], *Sergey* Poleshchikov[1], *Vladimir* Shavrov[3], and *Vladimir* Shcheglov[3]

[1]Saint-Petersburg state forest technical university named after S.M. Kirov, St-Petersburg, Russia
[2]Syktyvkar State University, Syktyvkar, Russia
[3]Institute of Radioengineering and Electronics of the Russian Academy of Sciences, Moscow, Russia

Abstract. The paper gives a description of hypersound magnetoelastic oscillations excitation along magnetisation reversal of normal magnetized single-layer and multi-layers ferrite plates, also provides definition of distinctive features of magnetic and elastic oscillations development. The relation of oscillations to the parameters of saturation magnetization, magnetic and elastic damping has been revealed.

1 Introduction

The magnetoelastic oscillations generation caused by magnetisation reversal is one of the interesting interaction mechanism of magnetic, elastic and magnetoelastic subsystems. This process can be executed in case of action on the plate by alternating magnetic field, hypersound oscillations as well as pulsed magnetisation reversal in a DC field. Dimensions and shape of a plate are fitted in such a way that avoid chance of uniform precession parametric decay into exchange spin wave [1].

2 Geometry of the problem and basic equations

The plane-parallel ferrite plate has thickness *d*. The external DC magnetic field $\mathbf{H}_0$ is applied perpendicularly to the plane of the plate as shown in Fig. 1. The problem is solved in a Cartesian coordinate system Oxyz. The plane Oxy of the coordinate system coincides with the plane of the plate. The coordinate axes are parallel to the edges of the cube crystallographic cell. The center of the coordinate system O is in the center of the plate, so that the plate planes coordinates are $z=\pm d/2$. We consider only excitement of the shear elastic oscillations.

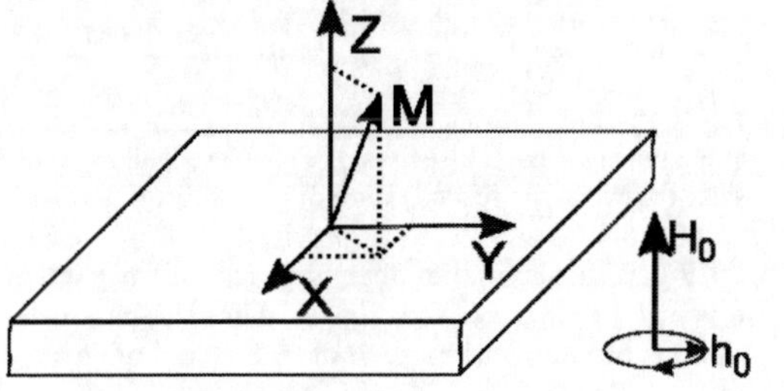

Fig. 1. Geometry of the problem.

Assuming that the total energy density of the plate *U* in the field $\mathbf{H}=\{0;0;\mathbf{H}_0\}$ equals the sum of magnetic, elastic and magnetoelastic energy densities, we obtain:

$$U=-M_0H_0m_z+2\pi M_0^2m_z^2+2c_{44}(u_{xy}^2+u_{yz}^2+u_{zx}^2)+$$

$$+2B_2(m_xm_yu_{xy}+m_ym_zu_{yz}+m_zm_xu_{zx}) \quad (1)$$

where $\mathbf{m}=\mathbf{M}/M_0$ - normalized magnetization vector, M_0 - the saturation magnetization of the magnetic plate, u_{ij} – the components of the strain tensor, c_{44} - the elastic constant and B_2 – the magnetoelastic coupling constant.

The initial equations for the system are the Landau-Lifshitz-Gilbert equation and the equations for the elastic displacement vector components $u_{x,y}$:

$$\frac{\partial \vec{m}}{\partial t}=-\gamma\left[\vec{m}\times\vec{H}_e\right]+\alpha\left[\vec{m}\times\frac{\partial \vec{m}}{\partial t}\right], \quad (2)$$

$$\frac{\partial^2 u_{x,y}}{\partial t^2}=-2\beta\frac{\partial u_{x,y}}{\partial t}+\frac{c_{44}}{\rho}\cdot\frac{\partial^2 u_{x,y}}{\partial z^2}, \quad (3)$$

where γ - the gyromagnetic ratio; α, β - damping constants of the magnetic and elastic subsystems. The effective field is determined by the following expression:

$$\vec{\mathbf{H}}_e=-\frac{1}{M_0}\cdot\frac{\partial U}{\partial \vec{\mathbf{m}}}. \quad (4)$$

The boundary conditions are:

$$c_{44}\frac{\partial u_{x,y}}{\partial z}\Big|_{z=\pm d/2}=-B_2m_{x,y}m_z. \quad (5)$$

* Corresponding author: aff@sfi.komi.com

The system of equations [2] was solved numerically by the Runge-Kutta 7–8 orders method with control of the integration at every step length.

The material parameters used in the calculation are typical for Ni: $\mathbf{H}_0$=10000 Oe, ρ=8.968 g·cm^{-3}, B_2=0.16·10^9 erg·cm^{-3}, M_0=300÷600 G, c_{44}=1.2·10^{11} erg·cm^{-3}, α=0.035.

3 Development of magnetoelastic oscillations in a single-layer sample

Assume that the plate is in the DC magnetic field $\mathbf{H}_0$. The $\mathbf{H}_0$ and $\mathbf{m}$ oriented in the negative direction of the axis Oz in the initial state. The DC field value is greater than the demagnetization field value, i. e. $\mathbf{H}_0<4\pi M_0$. The $\mathbf{H}_0$ changes its direction from negative to positive at the initial time. Accordingly, constant and demagnetization fields become contra-directional. The system is in the unstable equilibrium and a minor fluctuation leads to the magnetization vector tends to rotate towards the field $\mathbf{H}_0$. As a result, the magnetization vector precesses around the axis Oz that accounts for hypersound oscillations excite due to magnetostriction.

The spatial evolution of magnetic oscillations and temporal development of elastic oscillations are shown in Fig. 2 and Fig. 3.

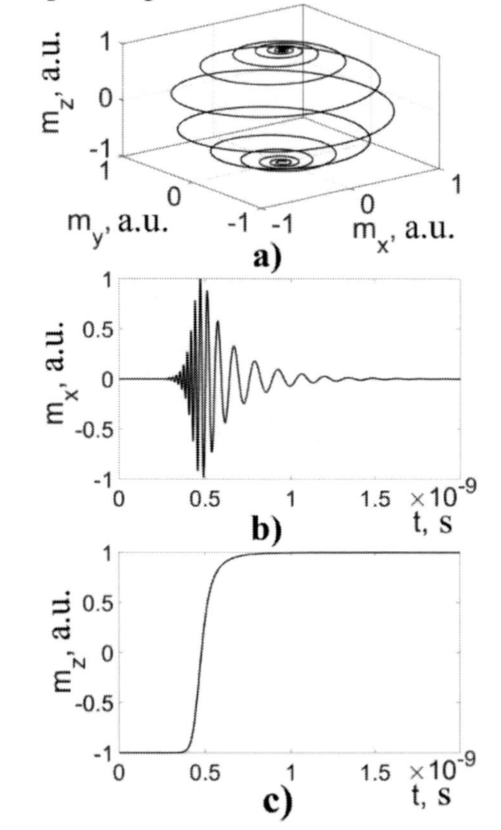

Fig. 2. a) development of magnetization components m$_x$, m$_y$, m$_z$; b) the m$_x$ time dependence; c) the m$_z$ time dependence.

The travelling time of magnetization vector under Oxy-plane is different from the same over Oxy-plane. This is due to correlation of the DC field and the demagnetizing field directions, i.e. at the first stage, when the deviation of the magnetization vector from the field direction increases in time, the force that turns the magnetization vector to the field direction also increases, and at the second stage, when the deviation decreases, the force also decreases gradually to zero.

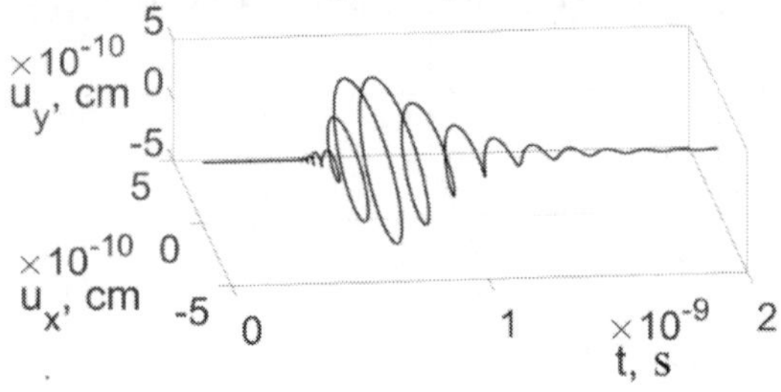

Fig. 3. The temporal development of elastic displacement components u$_x$, u$_y$.

Difference between magnetization precession frequencies is defined by the Kittel formula for normal magnetized infinité thin plate at the first and second stage.

Magnetization oscillations leads to elastic displacement oscillations due to magnetostriction. The elastic oscillations frequency is in accord with frequency of magnetization oscillations. This is apparent from spectra of elastic displacement and magnetization oscillations in Fig. 4. This is because magnetization oscillations relaxation time is much greater than elastic oscillation relaxation time and elastic oscillations follow magnetization oscillations in a quasi-stationary mode. The initiation of the minimum of amplitude of elastic oscillations is described in detail in the paper [3].

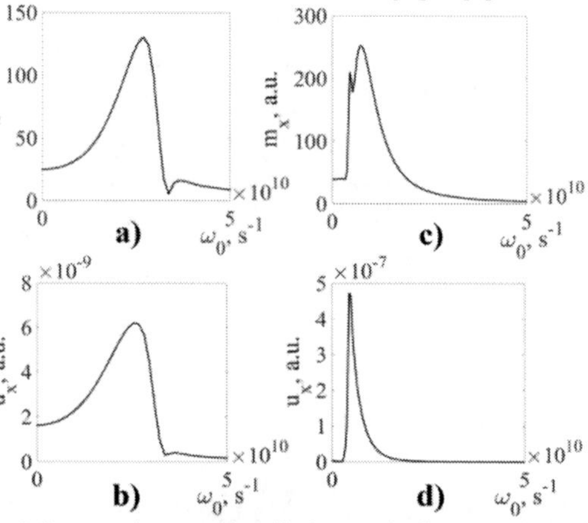

Fig. 4. Spectra of magnetic oscillations at the first stage (a) and at the second stage (c); spectra of elastic oscillations at the first stage (b) and at the second stage (d).

Fig.5 displays the dependencies of magnetization reversal process on the elastic damping parameter. It should be noticed here, that the time of maximum amplitudes development of elastic oscillations is kept in the range of elastic damping parameter values from 10^3 to 10^7 s^{-1} and from 10^8 to 10^{11} s^{-1}. However, a strong change of the maximum amplitudes development time at the second stage is observed between this intervals of elastic damping parameter.

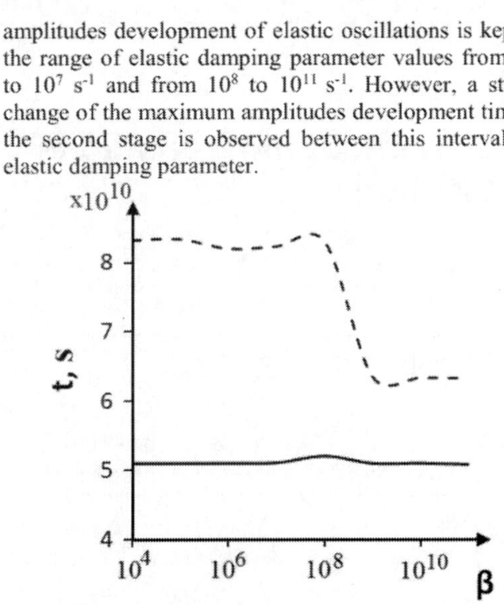

Fig. 5. The maximum amplitudes development time for m_x (solid line) and u_x at the second stage (dash line).

The constancy of the time of maximum amplitudes development stem from the following facts. In case of the small values of elastic damping parameter the elastic oscillation damping is defined by magnetic loss rather than elastic losses. And in case of the larger values of elastic damping parameter the elastic oscillations become nigh aperiodic. In addition to this, the elastic oscillation amplitude follows the magnetization amplitude.

The time of maximum amplitude development for elastic oscillations also depend on the saturation magnetization. The elastic oscillations at the second stage is most strongly influenced through the saturation magnetization. The fastest maximization of elastic oscillations amplitude is possible in the 330-560 G range of the saturation magnetization (Fig. 6).

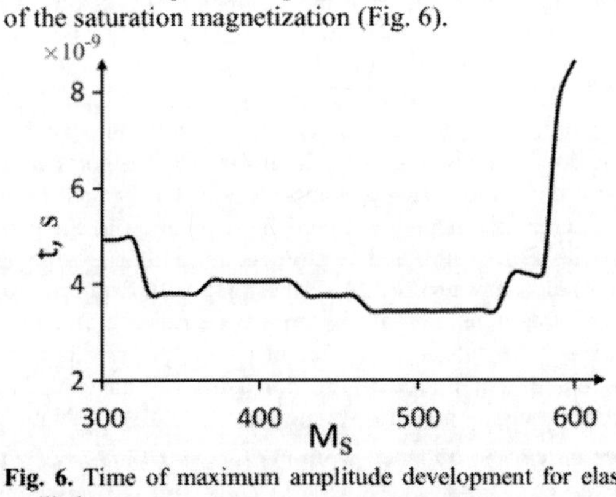

Fig. 6. Time of maximum amplitude development for elastic oscillations.

As is shown in Fig. 7 the amplitude of elastic oscillations exponentially decrease at the first stage of the magnetization reversal and exponentially increases together with increase of the saturation magnetization at the second stage.

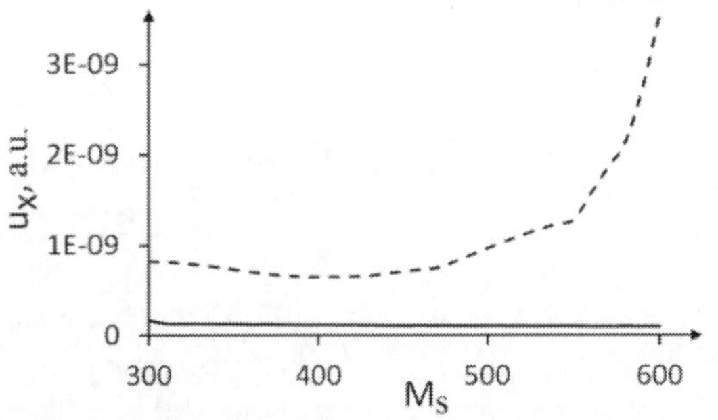

Fig. 7. Dependence of elastic oscillations maximum amplitude on the saturation magnetization.

4 Development of magnetoelastic oscillations in a multilayer sample

The development of magnetization oscillations and elastic displacements in multilayer plates with different layers material parameters of the sample corresponds to the development of oscillations in single-layer samples with equivalent parameters. This is due to weak interaction between layers in the absence of exchange interaction. There are the insignificant distortions of the elastic oscillations in the layers boundaries by this means.

5 Conclusion

Thus the dependencies of the maximum amplitudes development times of the magnetoelastic oscillations on the magnetic damping parameter and the saturation magnetization were defined. The values of the elastic damping parameter β for the most efficient excitation of hypersound oscillations have been determined. The paper reveals the fact that the displacement amplitude of the excited elastic waves in the microwave range can exceed 10^{-9} cm using sample with Ni properties. It is equal to the level of strong nonlinear excitations and increases the level of linear excitation by more than orders of magnitude.

This work was supported by RFBR (grant no. 17-02-01138-a, grant 17-57-150001 НЦНИ-а.).

References

1. V.S. Vlasov, L.N. Kotov, V.G. Shavrov, V.I. Shcheglov, Nonlinear excitation of hypersound in a ferrite plate under the ferromagnetic-resonance conditions, J. Comm. Tech. El. 54 (2009) 821-832.
2. M.S. Kirushev, V.S. Vlasov, D.A. Pleshev, F.F. Asadullin, L.N. Kotov, V.G. Shavrov, V.I. Shcheglov, Solid State Phenomena Vols 233-234 (2015) pp 73-78.
3. V.S. Vlasov, D.A. Pleshev, L.N. Kotov, V.G. Shavrov, V.I. Shcheglov, F.F. Asadullin, S.M. Poleshchikov, Solid State Phenomena Vols 233-234 (2015) pp 480-484.

Rotational remagnetization and orientational transitions in submicron YIG films

Poimanov Vladyslav[1,*], *Nepochatykh* Yurii[2], *Koledov* Victor[3], *Shavrov* Vladimir[3]

[1]Donetsk National University, 83001, Universitetskaya 24, Donetsk, Ukraine
[2]Donetsk institute for physics and engineering named after A.A.Galkin, 83001, R. Luxemburg str. 72, Donetsk, Ukraine
[3]Kotel'nikov institute of radio engineering and electronics of RAS, 125009, Mokhovaya str., Moscow, Russia

Abstract. It is shown that for the quasinormal remagnetization of the submicron YIG films with the normal along the [111] axis, which are in the single-domain state, two types of orientation transitions (OT) due to the influence of easy axes can be distinguished. The nonreciprocity of the FMR spectrum for forward and reverse magnetization due to the influence of easy axes is shown.

1 Introduction

At present a hypothesis exists that submicron ferrite-garnet film (thickness ~0.1 μm), obtained by the two-crucible liquid-phase epitaxy technique from wetting melt-solution on a nonmagnetic substrate [1-2] are in a single-domain state. This opens the prospect of their use as both switching devices and high frequency field generators that do not require the presence of a permanent magnet, which can significantly reduce the weight of the device.

The results of numerical modeling of the remagnetization of these films by a quasinormal field are presented. In the experiment, it is quite difficult to ensure that the external field is applied strictly normal to the film. This also concerns the orientation of the easy axes of cubic symmetry – the growth of the film can occur not strictly along any axis of symmetry. Due to the asymmetry in the plane of the film this leads to a number of interesting effects. In particular, there appears a nonreciprocity of direct and reverse remagnetization both for the angular dependences of the dynamic susceptibility and for the frequency field dependencies.

2 Analysis of expression for density of energy

Taking into account all types of interactions (cubic, magnetodipole, Zeeman and unidirectional), the density of magnetic energy has the form [3]:

$$W = 4\pi M^2 \frac{\cos^2\theta}{2} - H_C M\left(\frac{\sin^4\theta}{4} + \frac{\cos^4\theta}{3} + \frac{\sqrt{2}}{3}\sin^3\theta\cos\theta\cos(3\varphi)\right)$$
$$- HM(\cos\theta\cos\theta_H + \sin\theta\sin\theta_H\cos(\varphi-\varphi_H)) - H_b M\cos\theta \qquad (1)$$

where θ and φ – polar and azimuth angles of orientation of equilibrium orientation of magnetization, θ_H and φ_H – angles of external field orientation.

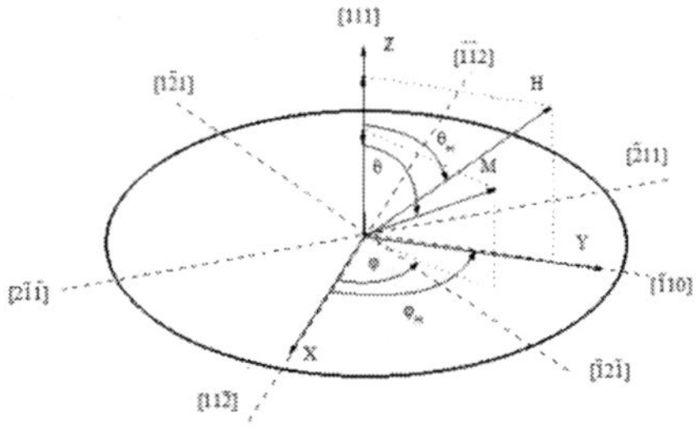

Fig. 1. Geometry of the problem.

Since the cubic anisotropy field H_C=45 Oe is much smaller than the saturation magnetization $4\pi M$ = 1250 Oe, then there are two types of orientational transition (OT). The first one of them (OT1) is observed in external fields of the order of the anisotropy field H_C, when the magnetization M is quasi-planar – the exit angle from the plane in the film is less than 1 deg. Accordingly, the contribution to the total energy from the demagnetizing field of the film ($4\pi M^2$) is comparable with the cubic anisotropy energy ($\sim H_C M$). In such fields, the main role is played by OT with rotation in the plane of the film (rotational remagnetization, RR), while the normal component of M varies weakly. To suppress the normal component and subsequent reorientation to a new metastable state, it is sufficient to apply a small (to the extent of smallness of the constants of the cubic and unidirectional anisotropy) negative field, the value of which can be obtained from (1) $H_1 = 0.47 H_C$=20 Oe. This shows the possibility of controlling a relatively small field along the normal by the magnetization in the plane.

The second type OT (OT2) is associated with remagnetization along the normal and occurs already in

* Corresponding author: Vladislav.Poymanow@yandex.ru

another interval of external fields (order of $4\pi M$). In this case, on the contrary, the orientation of M in the plane varies weakly, to the extent of the smallness of the planar component of the field $H_{\perp}$. The role of cubic anisotropy (CA) in this case is to the fact M is "fixed" in one of the "easy" directions and, with an increase in the field, without hysteresis, it reaches the normal. The planar field $H_{\perp}$ plays the role of a perturbation and leads to a nonreciprocity of the oscillation spectrum with respect to the direction of the field.

3 Results of numerical modeling of the remagnetization process

The results of modeling on the basis of W are given below.

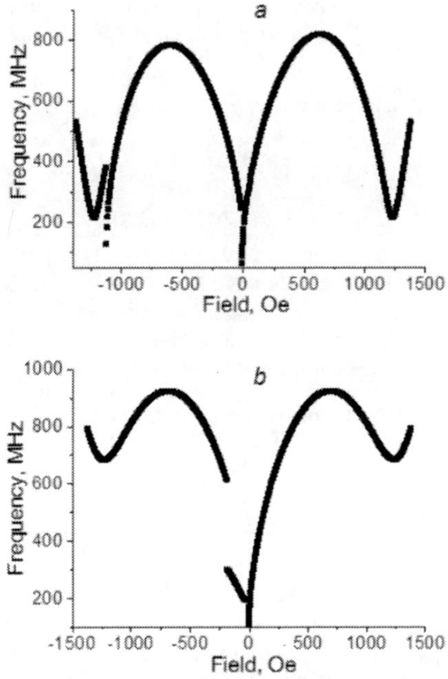

Fig. 2. Frequency-field dependencies for change of the field from –1400 Oe to 1400 Oe in the direction φ_H=30 deg., θ_H=0.5 deg.(a) and θ_H=3 deg.(b)

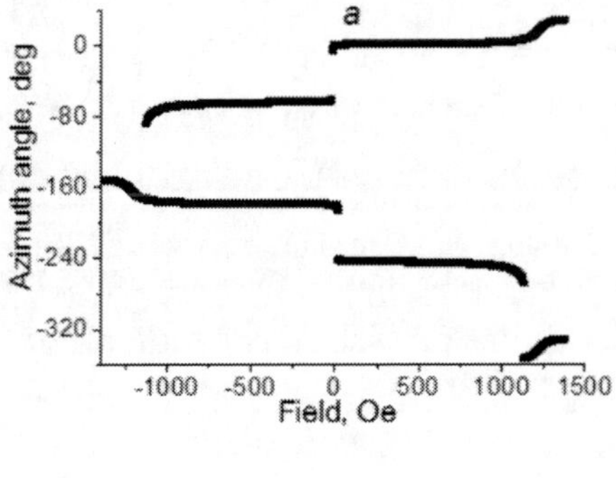

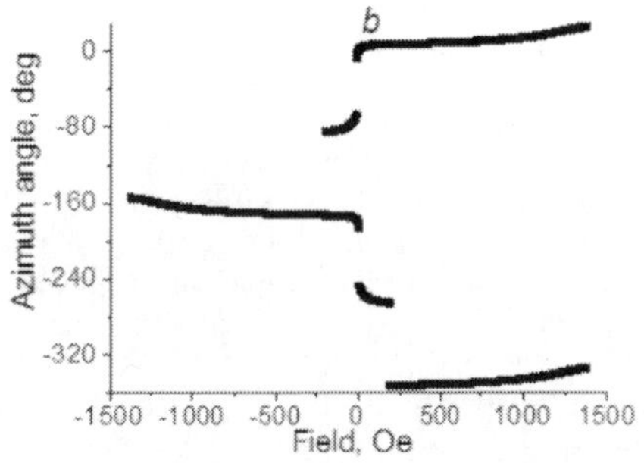

Fig. 3. Rotation of the magnetization in azimuth during one cycle of remagnetization 0 Oe → –1400 Oe → 1400 Oe → 0 Oe in the direction φ_H=30 deg. θ_H=0.5 deg. (a) and θ_H=3 deg. (b)

The gaps on the dependencies caused by orientational transitions. As θ increases, the orientational transition near the saturation field vanishes, and the gap near OT1 increases.

With a full cycle of remagnetization, the magnetization in the plane passes a complete counter-clockwise rotation. This behavior is typical for the azimuthal angles of the application of the field $0 \prec \theta_H \prec \frac{\pi}{2}, \frac{2\pi}{3}n \prec \varphi_H \frac{\pi}{3} + \frac{2\pi}{3}n$. The direction of rotation is reversed at $0 \prec \theta_H \prec \frac{\pi}{2}, \frac{\pi}{3} + \frac{2\pi}{3}n \prec \varphi_H \frac{2\pi}{3} + \frac{2\pi}{3}n$. This effect can be used in the device for switching the direction of rotation. The polar angle is weakly sensitive to the presence of easy axes, and for θ_H>1° the dependence $\theta(H)$ is extrapolated well by the straight line.

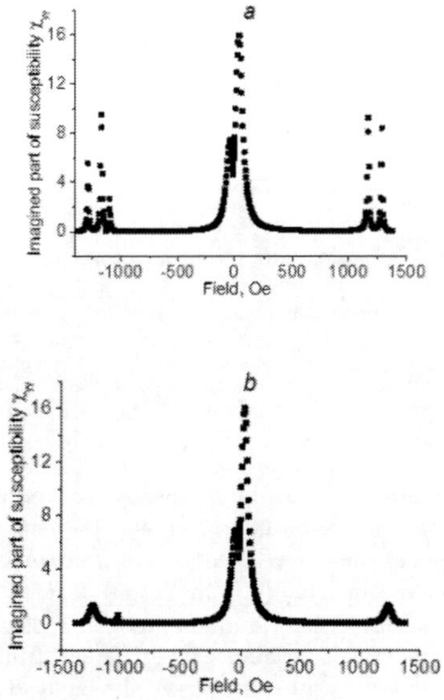

Fig. 4. Dynamic susceptibility χ_{yy} at frequency 300 MHz for φ_H=30 deg., θ_H=0.5 (a) and θ_H=3 deg.(b)

As follows from the shown dependencies if the angle of application of the field deviates from the normal, the value of the field OR2 shifts toward smaller negative fields, which is associated with the increasing role of the planar component. In this case, the gap in the spectrum of the frequency-field dependencies also increases, as evidenced by the disappearance of lateral peaks of susceptibility (fig. 4). For $\theta_H = 0.5°$ there are six distinct peaks arranged symmetrically in the field, and already for $\theta_H=1°$ this symmetry is violated, since OT2 is expressed more clearly.

4 Accounting of unidirectional anisotropy

In the presence of unidirectional anisotropy (UA), in addition to the expected shift of the curve by the value of the field UA, an additional asymmetry also arises. In particular, now the size of the gaps for the positive and negative fields is different. If, without UA, $\Delta\nu$ = 337 MHz, then, if it is present, $\Delta\nu_-$ = 343.4 MHz, $\Delta\nu_+$ = 330.7 MHz.

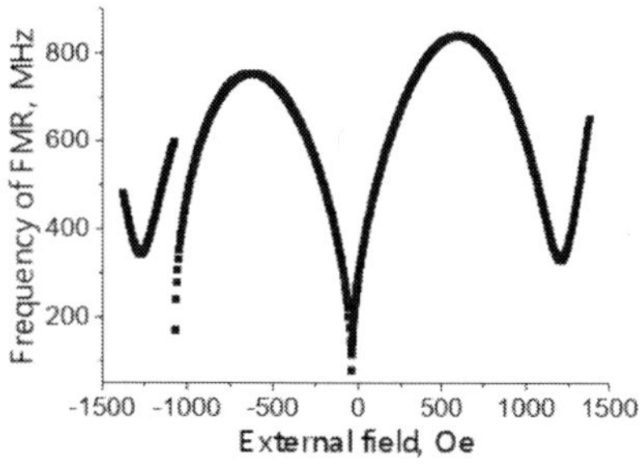

Fig. 5. Frequency-field dependency with presence of unidirectional anisotropy φ_H=30 deg. θ_H=1 deg.

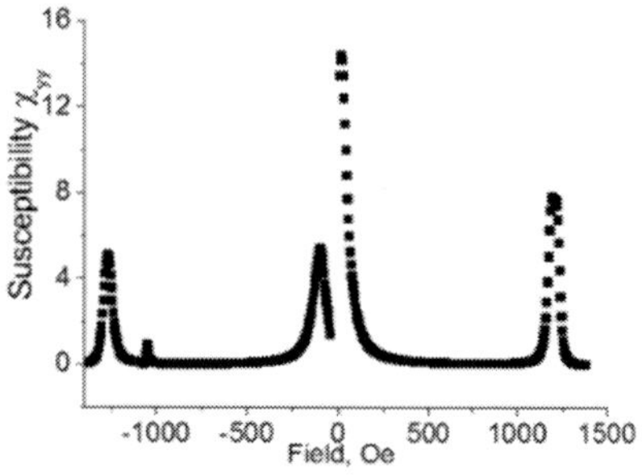

Fig. 6. Asymmetry of the susceptibility peaks for the frequency of 332 MHz, due to different lower FMR frequencies for direct and reverse magnetization.

The simulation data of the frequency-field dependencies for different φ_H show that the lower frequency limit of the frequency-field dependencies is significantly reduced when rotating in azimuth from 0 to 60 degrees. The field corresponding to the minimum frequency shifts toward decreasing with increasing azimuth angle from 0 to 60 degrees, and this shift is the larger, the larger θ_H.

5 Influence of azimuth angle of the applied field φ_H

When the film is rotated along the azimuth, the lower frequency limit of the FMR (gap) changes, reaching the smallest value at φ_H=60°. The effect of unidirectional anisotropy turns to the asymmetry of FMR observation at a given frequency.

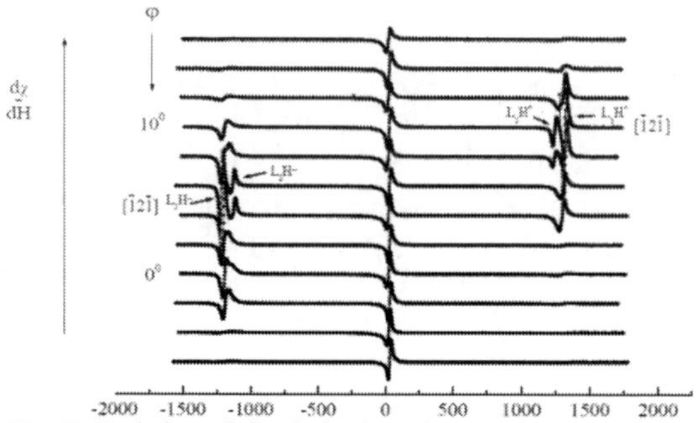

Fig. 7. Variation of the derivative of magnetic susceptibility with quasinormal external (θ_H =1 deg.) field [4] for a range of azimuthal angles (φ_H) separated by an interval of 2 deg. (ν = 300 MHz at room temperature).

Conclusions

1. The presence of an angle of misorientation of the applied field from the normal to the film [111] leads to the appearance of orientational transitions of two types. The first of them occurs in a field of the order of cubic anisotropy, the second – in a field of the order of saturation magnetization.

2. The possibility of controlling the rotation of the magnetization in the plane of the film is shown.

3. With an increase in the polar angle of application of the field to the normal, the value of the gap in the spectrum of the frequency-field dependency increases, and the field OT2 decreases in absolute value.

4. The presence of unidirectional anisotropy, in addition to the shift of the frequency-field dependency, leads to an additional asymmetry in the FMR spectrum – the sizes of the gaps for direct and reverse magnetization turn out to be different.

This work was supported by Russian Science Foundation, grant No. 14–22–00279

References

[1] V.F.Shkar', V.N.Varyukhin, JETP Letters, **88** (2008)

[2] V.F.Shkar', V.N.Varyukhin, JETP Letters **92** (2010)

[3] A.G. Gurevich, Magnetic Resonance in Ferrites and Antiferromagnets [in Russian] (Nauka, Moscow, 1973)

[4] V. Poimanov, V. Shkar, Yu. Nepochatykh al. JMMM, **443** (2017)

Frequency dependence of microwave complex permeability under magnetic bias

Artem O. Shiryaev*, *Stanislav Y.* Bobrovskii, *Aleksey V.* Osipov, *Dmitry A.* Petrov, *Konstantin N.* Rozanov, and *Sergey N.* Starostenko

ITAE RAS, 125412, Moscow, Izhorskaya 13, Russian Federation

Abstract. Measurement of the frequency dependence of the permeability under magnetic bias is proposed as a new method for studying the microwave magnetic properties of magnetic materials. The samples under study are either rolls of thin ferromagnetic films or the composites filled with sendust particles. It is shown that the permeability measured under external magnetic bias depends on sample thickness. The correct interpretation of the measured data is possible only when sample demagnetization is accounted for. The values of the anisotropy field and the saturation magnetization of thin nitrogen doped iron films and flake-shaped sendust particles are calculated considering demagnetization of the washer-shaped sample.

1 Introduction

For the characterization of microwave properties of magnetic materials, ferromagnetic resonance (FMR) measurements with fixed frequency and swept magnetic bias is frequently used [1–3]. This technique is usually performed in a high-quality cavity and has significantly higher signal-to-noise ratio, but the calculated permeability data may differ from that ones obtained under zero bias and the technique is poorly suited for measurements within a range of frequencies. Also, the measurement frequency is typically fixed and not very high, about 10 GHz. The FMR frequency of the measured sample may be above 10 GHz, or FMR may be observed at a low permanent magnetic field and the sample may not be saturated. However, the measurement technique allows easy variation of sample shape, size and orientation.

This technique allows finding the anisotropy field and the saturation magnetization using Kittel's equation:

$$f_{res} = \gamma\sqrt{(H_k + H_{bias})(H_k + H_{bias} + 4\pi M_0)}\,, \qquad (1)$$

where f_{res} is a FMR frequency, H_{bias} is the external magnetic bias, at which FMR is observed, H_k is the anisotropy field, $4\pi M_0$ is the saturation magnetization, and γ is the gyromagnetic ratio. Equation (1) is valid for elongated particles and thin ferromagnetic films [4].

An alternative way of microwave magnetic measurement may be a measurement of frequency dependence of complex permeability μ under external magnetic bias [5]. The permeability is conventionally measured in a coaxial line with high accuracy in a wide frequency range up to 20 GHz. The calibration technique for the coaxial line is well elaborated. However, samples of the coaxial form are nonuniformly magnetized, in contrast to samples of ellipsoidal shape. The response of such samples to the external bias is described by a sum of depolarization factors [6]. Often an effective depolarization factor, which depends not only on the shape and orientation of the sample, but also on its permeability, is used [7]. The problem of demagnetization of a coaxial washer is well studied in the application to the problem of creating a coilgun [8,9].

The demagnetization factor for ring with rectangular cross-section in the direction along axis is described with the equation:

$$N_z(\sigma,\tau) = \frac{1}{\tau(1-\sigma^2)}\int_0^{+\infty}\frac{dq}{q^2}(1-e^{-2q\tau})[J_1(q)-\sigma J_1(\sigma q)]^2\,, \qquad (2)$$

where σ is the ratio of the inner radius to the outer radius, τ is the ratio of the thickness to the outer diameter, and J_1 is the Bessel function.

Because of the sample magnetization, it is necessary to use a field inside the sample, weakened by demagnetization, instead of the external magnetic bias. This field depends on the depolarization factor N and the static permeability μ_{st} of the sample:

$$H_i = H_{bias}/(1+4\pi(\mu_{st}-1)N). \qquad (3)$$

Previously this method without demagnetization account was used only for qualitative estimations of high-frequency properties [4, 10-13]. Accounting for demagnetization allows the magnetic characteristics of the samples to be found quantitatively. Also a microstrip line is often used for measurements under bias [14], but the frequency range is not wide, up to 8 GHz.

The research is aimed at studying the demagnetization effect on the measured permeability of coaxial washer-shaped samples.

* Corresponding author: artemshiryaev@mail.ru

2 Experiment

Two types of coaxial samples are under study: rolls made of thin flexible iron films and composites with paraffin wax as host matrix filled with sendust (Fe-Si-Al) particles.

Iron films of submicrometer thickness are deposited by the RF magnetron sputtering with an iron target onto a flexible mylar substrate [15]. The deposition is performed in $Ar+N_2$ atmosphere at the room temperature and the pressure of 10^{-4} Torr. The fraction of N_2 in the sputtering chamber is 10%. Due to the nitrogen admixture, the iron of the deposited films contains dissolved nitrogen and is further referred to as Fe-N. The thickness of the film is 0.27 μm.

The composite samples are prepared by mixing flake sendust powder with molten paraffin. The mixture is pressed in a shape of 2 mm thick washers fit for a standard 7/3mm coaxial line.

The frequency dependencies of microwave permittivity and permeability of the samples are measured by the Nicolson-Ross-Weir technique in 7/3 coaxial line in the frequency range of 0.1 to 20 GHz. The coaxial cell is placed in a coil producing magnetic bias up to 930 Oe directed along coaxial axis.

3 Thin ferromagnetic films

Because of sample demagnetization, rolls of different thicknesses behave differently under magnetic bias. The measured data is shown in Fig. 1. The black lines correspond to the 5.4 mm roll thickness, the gray lines to 2.5 mm. The strength of the external permanent magnetic bias is shown in the figure. With magnetic bias increase, the static permeability decreases and the FMR frequency shifts toward higher frequencies. The measured permittivity does not depend on bias. Under 190 Oe external bias, the difference between permeabilities of the rolls of different thickness is maximal. The further bias increase leads to the difference decrease due to drop of static permeability.

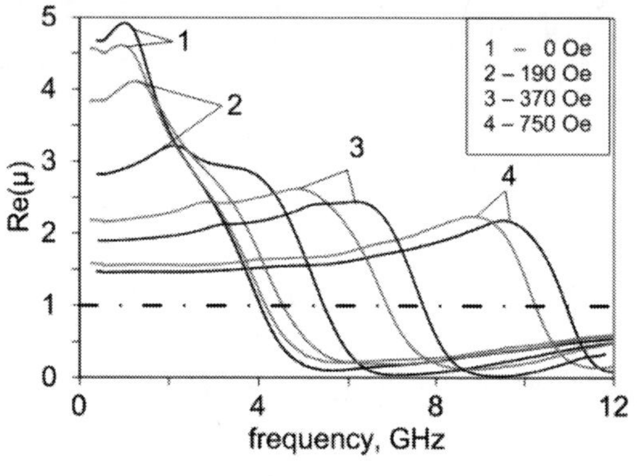

Fig. 1. The measured frequency dependence of the real permeability of coaxial rolls made of thin Fe-N films under magnetic bias (black lines correspond to 5.4 mm sample thickness, gray lines correspond to 2.5 mm, the numbers show the magnetic bias).

The dependence of the FMR frequency on the strength of magnetic bias is shown in Fig. 2. The circles correspond to the dependency of the FMR frequency on the strength of external bias, and the triangles correspond to dependency on the field inside the sample, with the account for demagnetization. The demagnetization factors in the direction along coaxial axis are taken from [8,9]. For the coaxial sample of 7 mm external diameter, 3 mm internal diameter and 2.5 mm thickness, the demagnetization factor is 0.6. For the 5.4 mm sample thickness the demagnetization factor is 0.35. The FMR frequency dependence on the internal field is shown by the same single curve, which can be approximated by Eq. (1). The measurement at 0 Oe doesn't fit the curve, because the sample under zero bias is unsaturated.

The approximation (1) makes it possible to find the saturation magnetization and the anisotropy field values. The obtained values of the anisotropy field is H_k=57 *Oe* and the saturation magnetization is M_0=1480 *G*. The reference book value of saturation magnetization for iron is 1700 G. The difference may be related to the nitrogen doping of the sputtered film.

Vectors of the strength of the permanent external magnetic bias and microwave magnetic field are perpendicular to each other and are parallel to the film plane. The correct interpretation of the measured data is possible for isotropic films only.

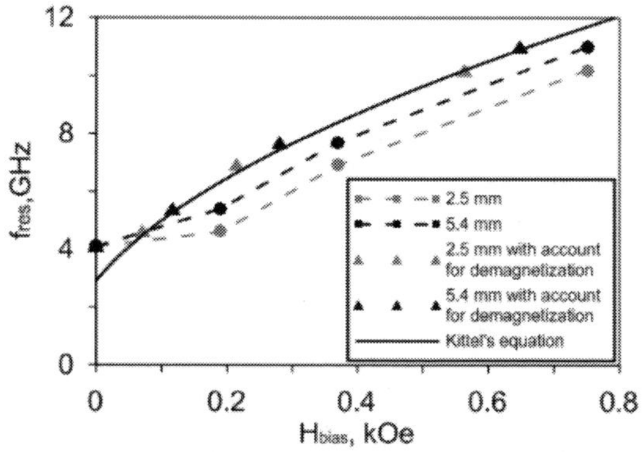

Fig. 2. The measured dependence of the FMR frequency on the strength of magnetic bias: the circles correspond to the measured data without accounting the sample demagnetization, the triangles correspond to the measured data with accounting the sample demagnetization, the black curve corresponds to the approximation by Kittel's equation.

4 Flake-filled composites

The measured frequency dependencies of the permeability of composites filled with different fractions of the sendust particles is shown in Fig. 3. The volume fraction of the inclusions in the samples under study is 2, 4, 7.4, 11.6 and 14.4%.

In the absence of the external magnetic bias, the FMR frequency for all fillings is close to 3 GHz. The effective properties of composites containing inclusions with low demagnetization factors typically obey the Wiener mixing rule [16].

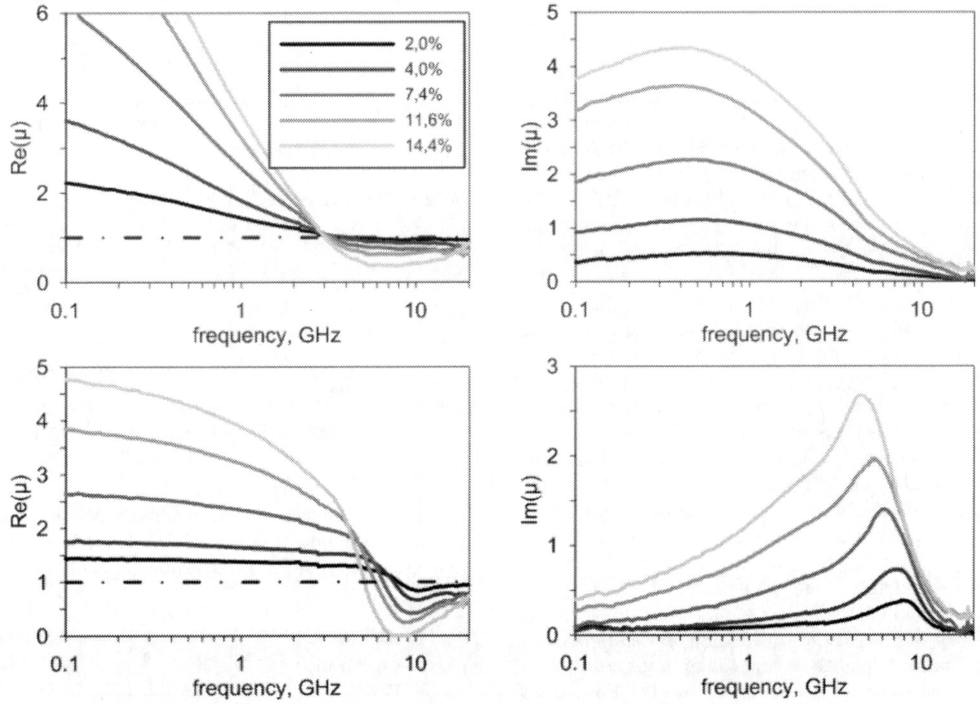

Fig. 3. The frequency dependence of the complex permeability at different volume concentrations of inclusions in composite material: no external magnetic bias (top), external magnetic bias of 930 Oe (bottom), lines of different colors denote different volume fraction of inclusions.

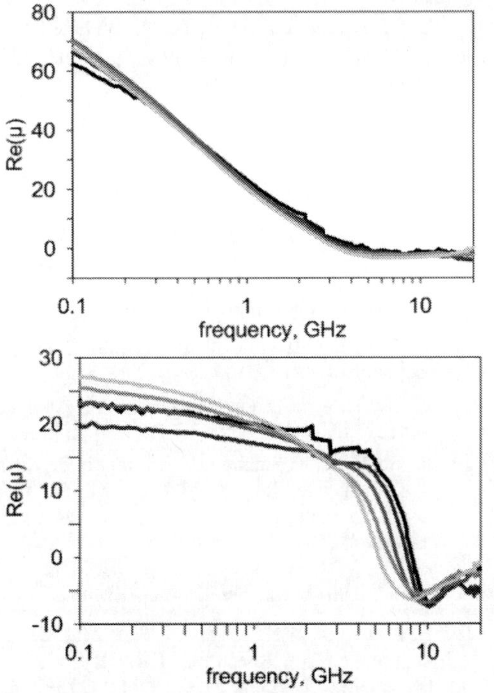

Fig. 4. The frequency dependence of the real intrinsic permeability under 0 magnetic bias (top) and under 930 Oe magnetic bias (bottom), lines of different colors denote different volume fraction of inclusions.

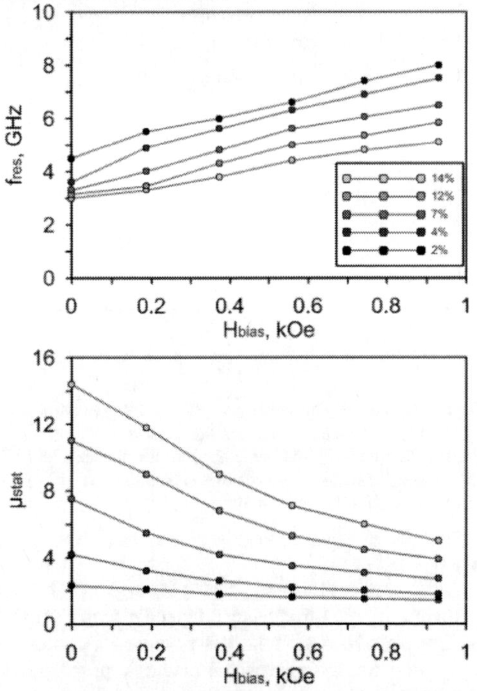

Fig. 5. The measured dependencies of the FMR frequency (top) and static permeability (bottom) of the composites on the external magnetic bias, darker circles denote smaller volume fraction of inclusions.

Therefore the intrinsic susceptibility of sendust particles may be obtained as the ratio of composite susceptibility to the volume concentration. The frequency dependence of intrinsic permeability of sendust particles is shown in Fig. 4. Under zero bias the intrinsic permeability is the same for all fillings.

However, an increase of the external magnetic bias leads to a shift in the resonant frequencies, and the more is the concentration of the sendust in the composites, the larger is the static permeability of the composite and the less is the FMR frequencies shift. Therefore the intrinsic permeability under bias depends on filling, which is inexplicable within the framework of known theories.

Dependencies of the FMR frequency and static permeability of the composites on the strength of the external bias for different fillings are shown in Fig. 5.

The dependence of the composite FMR frequency on bias is reduced to a single curve, when sample demagnetization is accounted for by Eq. (3). The values of the anisotropy field and the saturation magnetization of the sendust particles are found by fitting the parameters of Kittel's equation (1), the fitted results are shown in Fig. 6. The fitted parameters correspond to H_k=160 Oe and M_0=645 G. The handbook value of saturation magnetization for sendust is 780 G.

In case of the composite samples the data scatter is higher than in case of the rolled films. Possible reasons are the contribution of flake orientation in the composite, difference of shape of FMR curve from the Lorentzian type, and the inaccuracy of static permeability determination.

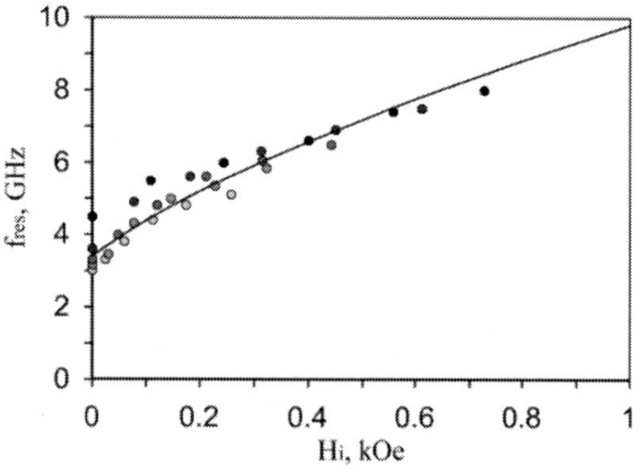

Fig. 6. The measured dependencies of the FMR frequency on the strength of the magnetic bias inside the sample: the circles correspond to the measured data with account for the sample demagnetization, the black curve corresponds to the approximation by Kittel's equation.

5 Conclusion

Measurement of the frequency dependence of the permeability under magnetic bias is proposed as a new method for studying the microwave magnetic properties of magnetic materials. The samples under study are either rolls of thin ferromagnetic films or the composites filled with sendust particles. It is shown that the permeability measured under external magnetic bias depends on sample thickness. The correct interpretation of the measured data is possible only when sample demagnetization is accounted for. The values of the anisotropy field and the saturation magnetization of thin nitrogen doped iron films and flake-shaped sendust particles are calculated considering demagnetization of the washer-shaped sample.

Acknowledgement

The study has been supported by Russian Foundation for Basic Research (RFBR); Project no. 18-08-00491.

References

1. R.M. Bozorth, *Ferromagnetism* (IEEE Press, New York, 1993)
2. I. Neudecker, G. Woltersdorf, B. Heinrich, T. Okuno, G. Gubbiotti, C.H. Back, J. Magn. Magn. Mater. **307**, 148 (2006)
3. S.S. Kalarickal, P. Krivosic, M. Wu, C.E. Patton, M.L. Shnider, P. Kabos, T.J. Silva, J.P. Nibarger, J. Appl. Phys. **99**, 093909 (2006)
4.
5. I.T. Iakubov, O.Yu. Kashurkin, A.N. Lagarkov, S.A. Maklakov, A.V. Osipov, K.N. Rozanov, I.A. Ryzhikov, S. N. Starostenko, J. Magn. Magn. Mater. **324**, 3385–3388 (2012)
6. S.N. Starostenko and K.N. Rozanov, PIER C **8**, 43 (2009)
7. D.J. Bergman, Phys. Rev. B. **19**, 2359 (1979)
8. D.J. Bergman, K.J. Dunn, Phys. Rev. B. **45**, 13262 (1992)
9. M. Beleggia, D. Vokoun, M. De Graef, J. Magn. Magn. Mater. **321**, 1306 (2009)
10. C.P. Bidinosti, Y. Sakamoto, K. Asahi, IEEE Magn. Lett. **5** (2014)
11. I. T. Iakubov, A. N. Lagarkov, S. A. Maklakov, A. V. Osipov, K. N. Rozanov, I. A. Ryzhikov, V. V. Samsonova, and A. O. Sboychakov, J. Magn. Magn. Mater. **321**, 726 (2009).
12. M. Yamaguchi, S. Yabukami, K.I. Arai, Sensors and Actuators **81**, 212–215 (2000)
13. L. Liu, K.N. Rozanov, M. Abshinova, Appl Phys A **110**, 275–279 (2013)
14. K. Chalapat, J. Timonen, M. Huuppola, L. Koponen, C. Johans, R. Ras, O. Ikkala, M. Oksanen, E. Seppala, and G. Paraoanu, Nanotechnology **25**, 485707 (2014)
15. C. Brosseau, S. Mallégol, P. Quéffelec, and J. Youssef, Phys. Rev. B **70**, 092401 (2004)
16. I.T. Iakubov, A.N. Lagarkov, S.A. Maklakov, A.V. Osipov, K.N. Rozanov, I.A. Ryzhikov, S.N. Starostenko, J. Magn. Magn. Mater. **272–276**, 2208 (2004)
17. K.N. Rozanov and M.Y. Koledintseva, J. Appl. Phys **119**, 073901 (2016)

ELECTRICAL PROPERTIES OF FERROMAGNETIC-INSULATOR NANOCOMPOSITES

Kalinin Yu.E.[1], *Makagonov V.A.*[1], *Sitnikov A.V.*[1], *Granovsky A.B.*[2]

[1]Voronezh State Technical University, Voronezh, 394026, Moskovskii prospekt 14, Russia
[2]M.V. Lomonosov Moscow State University, Faculty of Physics, Moscow, 119991, Leninskie Gory, 1-2, Russia

Abstract. Analysis of experimental studying of electrical properties of ferromagnetic-insulator nanocomposites is presented in the paper.

The investigation of composite materials consisting from ferromagnetic metal nanoparticles distributed in insulating matrix is one of promising directions in the condensed state physics. Scientific interest to the such nanocomposites is caused by the fact that presence of the nanometer-sized ferromagnetic particles leads to appearance of unique electrical, galvanomagnetic, optical and magneto-optical properties. As examples of the most significant properties, we can note the tunneling magnetoresistance reaching 12-13% at room temperature [1], the giant Hall effect, which is four orders of magnitude greater than the given effect in pure metals [2], the magnetorefractive effect which is two orders of magnitude greater than the traditional magneto-optical effects [3] and others phenomenon [4-5]. This paper is devoted to analysis of our experimental data on the study of electrical properties of ferromagnetic-insulator nanocomposites and the mechanisms of charge carrier transfer established in them.

Various charge carrier transfer mechanisms are realized in ferromagnetic-insulator nanocomposites in pre-percolation region [6]. The first frequently found mechanism is the hopping conductivity via localized states near the Fermi level with a variable hopping range (Mott's model).

In this case the electrical conductivity can be described by the following equation [7]

$$\sigma = e^2 \cdot R^2 \cdot \nu_{ph} \cdot g(E_F) \cdot \exp\left(-\frac{B}{T}\right)^{1/4}, \quad (1)$$

where

$$B = \frac{16}{a^3 \cdot k \cdot g(E_F)}, \quad (2)$$

e is the electron charge, R is the hopping length, υ_{ph} is the factor of the phonon–phonon interaction spectrum, T is the absolute temperature, $g(E_F)$ is the density of states at the Fermi level, and a is the localization radius of the electron wave function, k is the Boltzmann's constant.

As an example the temperature dependences of electrical conductivity for $(Co_{45}Fe_{45}Zr_{10})_x(Al_2O_3)_{100-x}$ composites with different metallic phase concentrations are shown at the fig. 1 in coordinates $\ln(\sigma/\sigma_0) \propto (1/T)^{1/4}$ at the temperature range of 80 – 180 K. From the data of fig. 1 according eq. 1 it is possible to determine value B for all investigated composites and it allows make estimating the values of effective density of electronic states at the Fermi level. Such estimates for $(Co_{45}Fe_{45}Zr_{10})_x(Al_2O_3)_{100-x}$ and other composites, are shown on fig. 2.

If we extrapolate the dependence of $g(E_F)$ (X) to the concentration of the metal phase corresponding to the percolation threshold, and use the equation

$$g(E_F)_{Xc} = g_0 \cdot X_C, \quad (3)$$

then the g_0 will give the density of states at the Fermi level of metallic granules (fig. 2), where $g(E_F)_{Xc}$ is density of states at the Fermi level, extrapolated to the percolation threshold, X_C – percolation threshold.

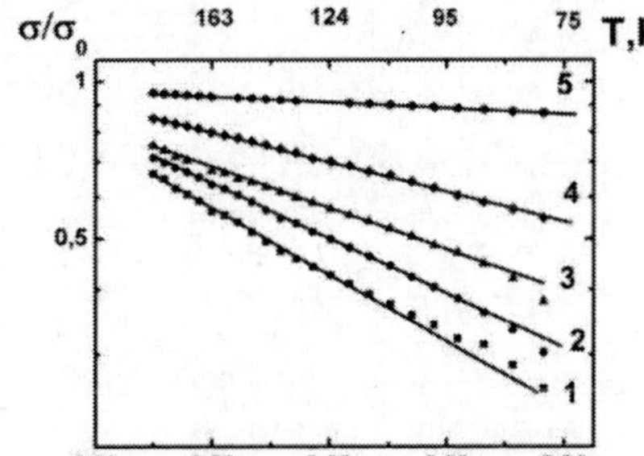

Fig. 1. Temperature dependences of electrical conductivity of $(Co_{45}Fe_{45}Zr_{10})_x(Al_2O_3)_{100-x}$ composites are shown in $\ln(\sigma/\sigma_0) \propto (1/T)^{1/4}$ coordinates, where σ_0 - conductivity at 180 K, X – concentration of metallic phase, at.%: 1-31.7; 2-34; 3- 36.3; 4-37.3; 5 - 38.7

* Corresponding author: vlad_makagonov@mail.ru

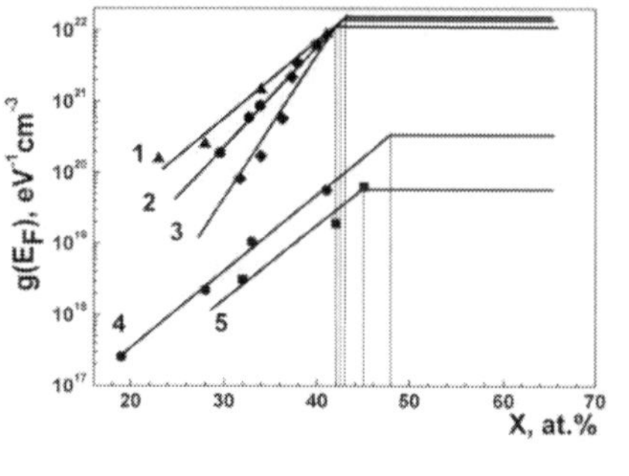

Fig. 2. Dependence of the density of electronic states at the Fermi level from concentration of metallic phase, obtained for nanocomposites $(Fe_{45}Co_{45}Zr_{10})_X(SiO_2)_{100-X}$ (1), $(Co_{45}Fe_{45}Zr_{10})_X[Pb_{0.81}Sr_{0.04}(Na_{0.5}Bi_{0.5})_{0.15}(Zr_{0.575}Ti_{0.425})O_3]_{100-X}$ (2) $(Fe_{45}Co_{45}Zr_{10})_X(Al_2O_3)_{100-X}$ (3), $(Co_{41}Fe_{39}B_{20})_X(SiO_2)_{100-X}$ (4) и $(Co_{84}Nb_{14}Ta_2)_X(SiO_2)_{100-X}$ (5)

The effective density of electron states at the Fermi level and its derivative with respect to energy estimating from the experimental results showed that the electron density of states at the Fermi level ($g(E_F)$) increases in nanocomposites with replacement of material granules according to the following sequence: CoNbTa → CoFeB → CoFeZr → Co (see fig.2).

The next charge transport mechanism observed in ferromagnetic-insulator nanocomposites is the inelastic resonant tunneling mechanism of electrons via insulator's localized states between granules. The changing of charge transfer mechanism for the studied composites is observed in the temperature range of ~ 200 – 300 K. This is expressed in a deviation from the law «1/4» for conductivity. A theoretical model of inelastic tunneling through amorphous dielectric layers was used by the authors [3, 4] to explain the temperature dependence of the conductivity of the granular composites located before of the percolation threshold. According to this model, the observed experimental temperature dependences of conductivity are explained by resonant tunneling through a chain of localized states in an amorphous insulating layer in tunnel junctions, and the temperature dependence of the conductivity in a channel containing n impurities has a power-law form:

$$\sigma_n = P\left(\frac{\Lambda^2}{\rho_0 c^5}\right)^{\frac{(n-1)}{(n+1)}} \frac{\left(ga^2n^2l\right)^n T^{\gamma_n} E^{\beta_n}}{al} \exp\left[-\frac{2l}{a(n+1)}\right], \quad (4)$$

where a is the radius of a localized state, l is the average distance between granules, $\gamma_n = n - 2/(n+1)$; $\beta_n = 2n/(n+1)$, P is a coefficient, Λ - is the deformation potential constant, ρ_0 – density of matrix, c – sound velocity, g – density of localized states, E – depth of occurrence of a localized state in the barrier.

The temperature dependences of the electrical conductivity for $(Co_{45}Fe_{45}Zr_{10})_x(Al_2O_3)_{100-x}$ composites with different concentrations of metallic phase are shown in the coordinates $\ln(\sigma_0/\sigma) \propto f[\ln(T_0/T)]$ (where σ_0 – electrical conductivity at temperature T_0 = 300 K) at fig. 3.

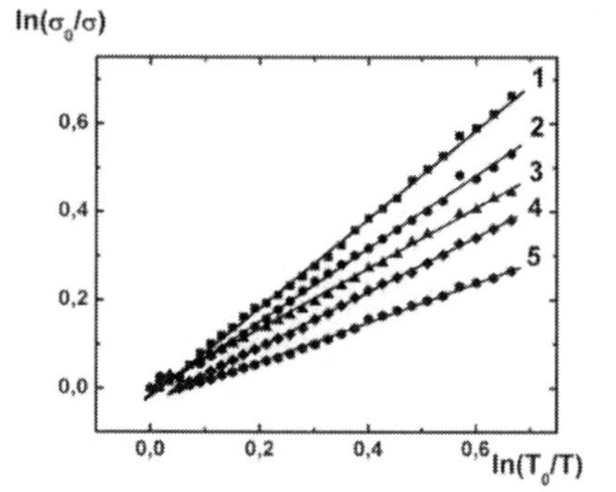

Fig. 3. Temperature dependences of conductivity in model coordinates (eq. 4) for $(Co_{45}Fe_{45}Zr_{10})_x(Al_2O_3)_{100-x}$ composites with different concentrations of metallic phase X, at.%: 1 - 31,7; 2 - 34; 3 - 36,3; 4 - 7,3; 5 - 38,7

The analysis of the obtained experimental results showed, that these dependences obey a power law in wide temperature range (~ 200 - 300 K). This allowed us to estimate index γ in eq. 4 and make it possible to calculate the average number of localized states $\langle n \rangle$ within tunneling channels between isolated conductive clusters according to the equation

$$\langle n \rangle = \frac{1}{2}\left[\gamma - 1 + \left(\gamma^2 + 2\gamma + 9\right)^{1/2}\right]. \quad (5)$$

The results of calculation showed that with increasing of metal phase concentration, the average number of localized states between the granules decreased, and dependence $\langle n \rangle \propto f[X]$ is practically linear (fig. 4).

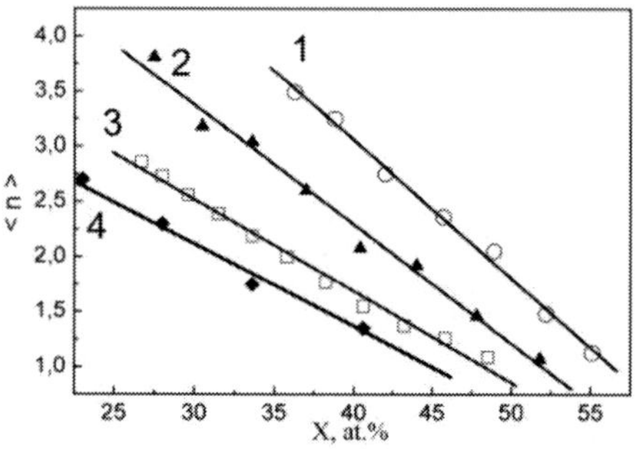

Fig. 4. Dependences of average number of localized states from concentration of metallic phase (X, at.%) for different composites: 1 - $(Co_{41}Fe_{39}B_{20})$ - SiO_n; 2 - $(Co_{86}Ta_{12}Nb_2)$ - SiO_n; 3 - $(Co_{41}Fe_{39}B_{20})$ - Al_2O_n and 4 - $(Co_{45}Fe_{45}Zr_{10})$ - SiO_n.

Estimates of average number of localized states within insulator matrix between metal granules, involved into electron transport, made from the experimental results, indicated that it decreases with increasing of heat treatment temperature. The elastic electron tunneling mechanism between granules through a dielectric barrier have been proposed in works [10, 11].

Also the thermally activated hopping conductivity between nearest neighboring states or inelastic resonant tunneling between granules throughout insulator matrix are obtained in ferromagnetic-insulator nanocomposites. The analysis of the temperature dependences of specific electrical resistance of pre-percolation composites has shown, that there is a deviation from the power law at temperatures above room temperature. At the same time the electrical resistance obey to the exponential law in a wide temperature range, i.e. dependency $\ln(\rho/\rho_o) \propto f(1/T)$ becomes a line, where ρ_o is the specific electrical resistance at T = 333 K. If we use formula for the thermally activated hopping conductivity to describe the temperature dependences

$$\rho = \rho_0 \exp\left(\frac{w}{kT}\right), \tag{6}$$

where w is being an average hopping energy is of the order of localized states band width, we can obtain the values of w from the slope of dependences $\ln(\rho/\rho_o)$ - $f(1/T)$. The values of w obtained for some composites are shown in table. 1.

On the other hand, it is also possible to apply the model of inelastic resonant tunneling for this temperature range. The deviation from the power law can be explained by the inclusion of additional channels of tunneling, which leads to an increase in electrical conductivity. In this case, the temperature dependence of the conductivity can be described by the equation

$$\rho = \rho_0 \exp\left(\frac{E}{2kT}\right), \tag{7}$$

where E is the energy difference between the initial and final states in neighbor granules. The coefficient 2 in the denominator of eq. 7 indicates that energy of the electron can either increase or decrease at inelastic tunneling between the granules.

Table The hopping activation energy (eq. 6) and the energy difference between initial and final states in neighboring granules for inelastic tunneling (eq. 7) for some composites

Composition	w, eV (eq. 6)	E, eV (eq. 7)
$(Co_{45}Fe_{45}Zr_{10})_{23}(Al_2O_3)_{77}$	0,090 ± 0,010	0,180 ± 0,020
$(Co_{45}Fe_{45}Zr_{10})_{28}(Al_2O_3)_{72}$	0,066 ± 0,007	0,130 ± 0,014
$(Co_{45}Fe_{45}Zr_{10})_{34}(Al_2O_3)_{66}$	0,030 ± 0,005	0,060 ± 0,010
$(Co_{45}Fe_{45}Zr_{10})_{29,6}(PZT)_{70,4}^{*}$	0,170 ± 0,005	0,340 ± 0,010
$(Co_{45}Fe_{45}Zr_{10})_{33,9}(PZT)_{66,1}^{*}$	0,080 ± 0,007	0,160 ± 0,014
$(Co_{45}Fe_{45}Zr_{10})_{37,9}(PZT)_{62,1}^{*}$	0,050 ± 0,008	0,100 ± 0,016
$(Co_{45}Fe_{45}Zr_{10})_{39,9}(PZT)_{60,1}^{*}$	0,035 ± 0,009	0,070 ± 0,018

*PZT- $Pb_{0,81}Sr_{0,04}(Na_{0,5}Bi_{0,5})_{0,15}(Zr_{0,575}Ti_{0,425})O_3$

In composites beyond the percolation threshold, the electrical resistivity is determined by labyrinth structure of conducting channels formed from the metal granules and can have temperature dependence with a temperature coefficient of electrical resistance close to zero. For systems near the percolation threshold, the logarithmic law $\sigma(T) = A\,(1 + \alpha \ln T)$ is satisfied for conductivity, where the parameters A and α depend on concentration of conduction phase. This logarithmic dependence, according to the theory of Efetov et al.[12], is associated with features of the Coulomb interaction in nanogranular material during intergranular tunneling in transition range of concentrations from metallic conductivity to the dielectric regime [7].

At conclusion it should be noted that various charge carrier transfer mechanisms has been observed in ferromagnetic-insulator nanocomposites in pre-percolation region.

1. The mechanism of hopping conductivity via localized states near the Fermi level with a variable hopping range according Mott's law.

2. The mechanism of inelastic resonant electron tunneling via insulator's localized states between granules.

3. The elastic electron tunneling mechanism between granules through a dielectric barrier.

4. The thermally activated hopping conductivity between nearest neighboring states or inelastic resonant tunneling between granules throughout insulator matrix.

The electrical resistivity in composites beyond the percolation threshold, is determined by labyrinth structure of conducting channels from metal granules and can have temperature dependence with a temperature coefficient of electrical resistance close to zero.

The work was supported by the Ministry of Education and Science within project part of the state task (project № 3.1867.2017/4.6).

References

1. A.E. Berkowitz, J.R. Mitchell, M.J. Carey, A.P. Young, S. Zhang, F.E. Spada, F.T. Parker, A. Hutten, G. Thomas. Phys. Rev. Lett. **68** (25), 3745 (1992)
2. B.A. Aronzon, A.B. Granovskii, A.B. Davydov, M.E. Dokukin, Yu.E. Kalinin, S.N. Nikolaev, V.V. Rylkov, A.V. Sitnikov, V.V. Tugushev. JETP **103** (1), 110 (2006)
3. A. Granovskii, V. Gushchin, I. Bykov, A. Kozlov, N. Kobayashi, S. Ohnuma, T. Masumoto, M. Inoue. Phys. Solid State **45** (5), 911 (2003)
4. R.S. Iskhakov, E.A. Denisova, S.V. Komogortsev, L.A. Chekanova, Yu.E. Kalinin, A.V. Sitnikov. Phys. Solid State **52** (11), 2263 (2010)
5. D.A. Filippov, T.A. Galkina, V.M. Laletin, G. Srinivasan. Phys. Solid State **53** (9), 1832 (2011)
6. I.S. Beloborodov, A.V. Lopatin, V.M. Vinokur, K.B. Efetov. Rev.Mod. Phys. **79** (2), 469 (2007)
7. N. Mott and E. Davis, *Electronic Processes in Non Crystalline Materials* (Clarendon, Oxford, 1971)
8. L.I. Glazman, K.A. Matveev. Sov. Phys. JETP **67** (6), 1276 (1988)
9. L.I. Glazman, R.I. Shehter. Sov. Phys. JETP **67** (1), 163 (1988)
10. B. Abeles, P. Sheng, M.D. Coutts, Y. Arie. Advances in Physics **24** (3), 407 (1975)
11. P. Sheng, B. Abeles, Y. Arie. Phys. Rev. Lett. **31** (1), 44 (1973)
12. K.B. Efetov, A. Tschersich. Phys. Rev. B **67** (17), 174205 (2003)
13. B.I. Shklovskii, A.L. Efros, *Electronic Properties of Doped Semiconductors* (Springer-Verlag, New York, 1984)

The origin of fine structure in magnetization curve of α-CoV_2O_6

Yuri Kudasov[1,2,3,*] and *Roman* Kozabaranov[1,2]

[1] *Sarov Physics and Technology Institute, National Research Nuclear University "MEPhI", Dukhov str. 6, Sarov, 607186, Russia*
[2] *National Research Nuclear University "MEPhI", Kashirskoe shosse 31, Moscow, 115409, Russia*
[3] *Russian Federal Nuclear Center - VNIIEF, Pr. Myra 37, Sarov, 607188, Russia*

Abstract. Multiple field-induced plateaus in α-CoV_2O_6 at low temperatures were revealed earlier by M. Lenertz et al. [J. Phys. Chem. C 115, 17190 (2011)] and carefully investigated recently by M. Nandi and P. Mandal [J. Appl. Phys. 119, 133904 (2016)]. Four equidistant steps were observed in the magnetization curve. We present a model to describe this phenomenon. A magnetic structure of this substance is formed by highly anisotropic triangular lattice of Ising chains running along the **b** axis. Due to a three-fold degeneracy of three-sublattice magnetic ordering, domain boundaries appear. Their transformation under magnetic field variation leads to two additional steps in the 1/3 magnetization plateau and gives rise to complex magnetic behavior observed experimentally. The domain structure in α-CoV_2O_6 occurs to be strongly anisotropic because a lifetime of the metastable states depends greatly on the configuration orientation. A strong dependence of the magnetization curve on magnetic field sweep time is predicted.

1 Introduction

Ising spin-chain compounds have drawn considerable attention due to unusual magnetic behavior [1–4]. A complicated hierarchy of magnetic interactions determines main features of these systems. The strongest of them act along the chains. That is why in the framework of the rigid-chain model [5] the chains are considered as elements of the magnetic structure with two possible magnetization states. Weak antiferromagnetic (AFM) interchain interactions in the triangular lattice of chains lead to the geometric frustration and complex phase diagram. At high temperatures a partially disordered antiferromagnetic (PDA) phase appears [2]. Very weak next-to-the-nearest interchain interactions or anisotropy partially lift the degeneracy of the ground state and produce various low-temperature magnetic structures: the three-sublattice (3SL) one in $CsCoCl_3$ and related compounds, stripes in $Sr_5Rh_4O_{12}$ [6, 7], and strongly disordered Wannier's state [8] in $Ca_3Co_2O_6$ [3, 9, 10].

One of the most striking features observed in the spin-chain compounds is a multi-step magnetization curve in $Ca_3Co_2O_6$ [3, 9] and Sr_3HoCrO_6 [4]. The plateau 1/3 at high temperatures is related to transition to the 3SL phase. At low temperatures two additional steps appear at the plateau. They are assumed to be a consequence of a three-fold degeneracy of the 3SL configuration which gives rise to a domain structure [1, 11]. This is a long-lived metastable state which determines a slow magnetic dynamics in $Ca_3Co_2O_6$ [1, 9].

In this Brief Report we discuss the multi-step behavior in CoV_2O_6 revealed in Ref. [12] and thoroughly investigated recently [13]. This compound crystallizes in the two allotropic forms [12]: the branneritelike monoclinic structure with the space group C2/m which is schematically shown in Fig. 1 (α-phase) and the triclinic structure with the P-1 space group (γ-phase). They have rather similar magnetic behavior, however in the present article we discuss the α-CoV_2O_6 only. Magnetic Co^{2+} ions in the α phase are placed in the well separated chains made up of edge-sharing distorted CoO_6 octahedra. The chains run along the **b** axis. Pentavalent vanadium ions are nonmagnetic and located in edge-shared VO_5 square-pyramids between the chains being involved in interchain magnetic interactions.

*e-mail: yu_kudasov@yahoo.com

2 Magnetic structure of α-CoV_2O_6

Magnetic measurements of α-CoV_2O_6 revealed an Ising-type anisotropy with the easy axis along **c** axis [14]. It arises from cooperative effects of the crystal electric field of the distorted octahedra and spin-orbit coupling [15]. This combination also leads to unusually high orbital moment of Co^{2+} ions. The magnetic structure is formed by the strong ferromagentic (FM) in-chain coupling between the Co^{2+} ions, and weak AFM interchain interactions. They lead to the AFM state below $T_N = 15$ K. The lattice of the chains can be considered as a distorted 2D triangular lattice with three AFM coupling constants between the nearest neighboring chains J_1, J_2, and J_3 which act along [100], [101], and [001] directions, respectively (see Fig. 1). The values of J_1, J_3 although are different but comparable, J_2 is much weaker than J_1 and J_3. According to Ref. [16] the estimations of the coupling constant are

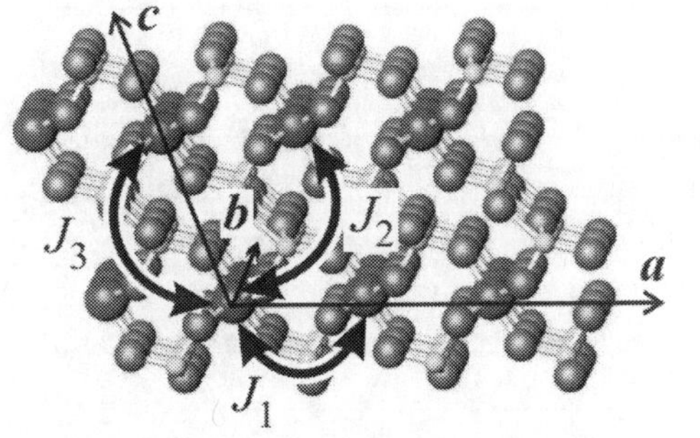

Figure 1. (color online) The crystal structure of α-CoV_2O_6. The large red, smaller blue, and the smallest yellow spheres denote the Co^{2+}, O^{2-}, and V^{5+} ions correspondingly. The arrows show interchain interations.

$J_1 = 0.356$ meV, $J_2 = 0.051$ meV, and $J_3 = 0.463$ meV. As was shown in Ref. [17] interactions between the next to the nearest neighbor are rather weak and can be omitted.

3 Magnetization curves and model of alpha-CoV2O6

Magnetization curves of α-CoV_2O_6 below T_N demonstrated two steps separating plateaus with zero magnetization ($M = 0$) at low magnetic field, one third of the saturation magnetization ($M = 1/3$), and the saturated state ($M = 1$). This structure was observed from 12 K to 5 K. The neutron scattering [16, 18] and Monte Carlo simulation showed that the first plateau corresponds to the stripe magnetic structure, and the 1/3 plateau does to three sublattices (3SL) configurations. The Monte Carlo simulation by the Metropolis algorithm in some cases gave additional features in the 1/3 plateau. However, it should be mentioned that this technique fails to relax correctly a trial state into the equilibrium state on the AFM triangular Ising model and the Wang-Landau algorithm should be applied instead [19]. The Wang-Landau simulation produces the flat 1/3 plateau in case of α-CoV_2O_6 [20].

Two additional steps in the 1/3 plateau were revealed at very low temperatures below 5 K [12]. The four magnetization steps are equidistant on the magnetic field (1.6, 2.2, 2.8, and 3.4 T) [12, 13] similarly to $Ca_3Co_2O_6$ despite the fact the 2D lattice of chains is strongly anisotropic. This phenomenon is worthy of thorough investigation.

The rigid-chain model for α-CoV_2O_6 was formulated by Yao as following [20]

$$\hat{H} = \sum_{<ij>} J_{ij}\sigma_i\sigma_j - hM \quad (1)$$

where $\sigma_i = \pm 1$ is the c-axis projection of the i-th chain magnetic moment, $M = \sum_i \sigma_i$ is the total magnetization, $< \ldots >$ denotes the summation over the nearest neighboring chains only, J_{ij} is a positive constant which assumes values J_1, J_2, or J_3 ($J_1 \approx J_3$, $J_2 < J_1$, $J_2 < J_3$) depending on the bond orientation, h is the magnetic field.

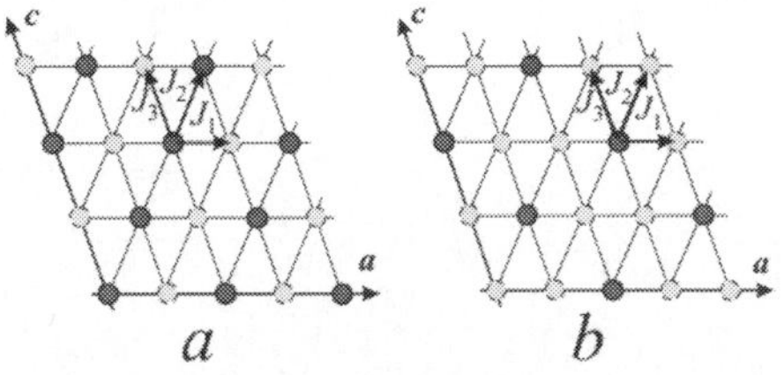

Figure 2. (color online) Magnetic structures of stripe (a) and 3SL phases. The solid (red) and dash (blue) circles denote spin-down and spin-up (along the magnetic field) chain states, repectively.

4 Metastable states and transitions

At the zero magnetic field the equilibrium state is the stripe structure oriented along the [101] direction as shown in the Fig. 2a. While magnetic field increasing a first order transition to the 3SL phase occur at the first critical magnetic field [20]

$$h_{c1} = 2(J_1 + J_3) - 4J_2. \quad (2)$$

The 3SL structure is shown in Fig. 2b. Further increase of the magnetic field leads to the second transition to the saturated FM state at the second critical magnetic field [20]

$$h_{c2} = 2(J_1 + J_3 + J_2). \quad (3)$$

These two critical magnetic fields form the two steps in the equilibrium magnetization curve (Fig. 3) which was observed in the temperature range between 5 and 15 K.

To discuss the multi-step behavior at low temperatures one should take into account frozen metastable states. Necessary conditions of the metastability at the zero temperature have the following form [5, 21]

$$\sigma_i h_i \leq 0 \quad (4)$$

where $h_i = \sum_{j(i)} J_{ij}\sigma_j - h$ is the effective field for the i-th chain. This condition should be satisfied for all the chains.

At finite temperatures the lifetime τ of a metastable state can be estimated by the Glauber theory [22] extended to 2D and 3D lattices [1, 23]. Following to it one assumes that the chains interact not only with the nearest neighbors and external magnetic field but also with a heat reservoir. It was shown that a hard type of the Glauber dynamics describes well the slow dynamics in frustrated spin-chain system [1, 11, 23]. In this case the probability of a spin flip of the i-th chain per time unit can be written down as

$$W_i = \frac{\alpha}{2}\left[1 + \sigma_i \cdot \tanh\left(\frac{h_i}{k_B T}\right)\right] \quad (5)$$

where α is the constant of the interaction of a chain with the heat reservoir, k_B is the Boltzmann constant, T is the

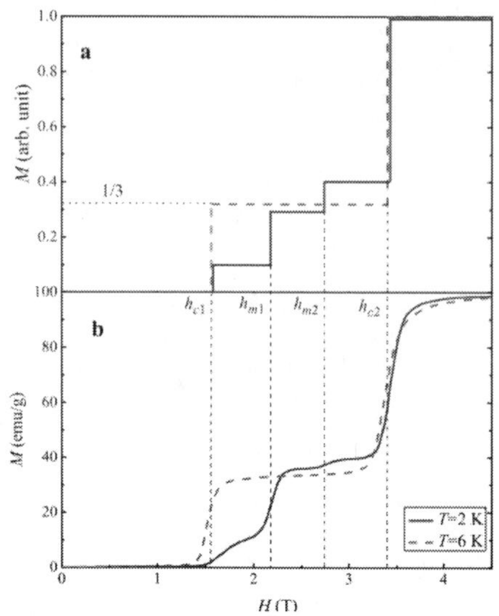

Figure 3. (color online) (a) A schematic view of the magnetization curves of α-CoV_2O_6 in the increasing magnetic field: the dash (red) and the solid (blue) lines correspond to the equilibrium and metastable regimes, (b) experimental magnetization curves at 2 K and 6 K from Ref. [13].

temperature. Then the lifetime of the i-th chain state is $\tau_i = W_i^{-1}$. If $\tau \sim \alpha^{-1}$ the state is unstable, and $\tau >> \alpha^{-1}$ corresponds to a metastable (or stable) state.

There are two basic causes for the metastability just above h_{c1}. Firstly, using the condition Eq. (4) it is easy to show that the stripe phase still exist as a metastable state up to the magnetic field

$$h_{m1} = 2(J_1 + J_3) - 2J_2. \quad (6)$$

The 3SL structure is three-fold degenerate, and when it starts growing from few nucleation centers domain boundaries appear. Let us consider the domain boundaries lying in the (101) planes. The possible configurations are shown in Fig. 4. Since the FM bonds for the spin-down chains in domain boundaries (A and B positions) are along the [101] direction with the weakest AFM coupling constant J_2 this orientation makes the most stable domain boundaries. We discuss this point below. The configurations in Fig. 4a give an excess of spin-down chains as compared with monodomain 3SL structure. That is why just above h_{c1} the magnetization of the metastable state is less than that of the equilibrium one. The chains in the position B become unstable at the magnetic field determined by Eq. (6) that causes the additional step in the magnetization curve (see Fig. 3).

The spin-down chains in the position A in Fig. 4b remain stable up to another critical field

$$h_{m2} = 2(J_1 + J_3). \quad (7)$$

At this point half of the spin-down chains in the A-A pairs (Fig. 3) flips to the spin-up state. This gives rise to an ex-

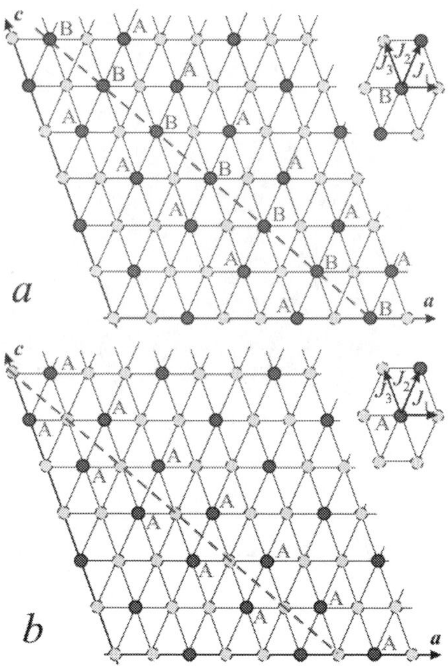

Figure 4. (color online) Two possible structures of the domain boundaries in the 3SL phase in the (101) planes. A and B denote positions of spin-down chains with two different types of the nearest-neighbor environment.

cess of spin-up chains. Therefore, there appears a magnetization step at h_{m2} and the plateau with the magnetization greater than the equilibrium one. In the range between h_{m2} and h_{c2} spin-down chains are stable only in case of all the six nearest neighbors are in the spin-up state.

Until now we discussed the domain boundaries lying in the (101) plane. Let us consider the ones in other planes, e.g. (100), and more generally other types of metastable states. Then we investigate the stability of the chain in the down state similar to the B position but with two nearest neighbors in the down positions along the [100] direction (B′ position see Fig. 5). The corresponding critical magnetic fields can be calculated by a coupling constants permutation. Then the equation Eq. (6) is replaced by $h'_{m1} = 2(J_2 + J_3) - 2J_1$. If $h'_{m1} < h_{c1}$ the B′ states are unstable, that is they exist only if $J_2 > 2/3J_1$.

Let us compare the lifetime for different metastable states. The additional steps at the 1/3 plateau are observed at very low temperatures only when k_BT is much less than any of J_1, J_2, and J_3. Then one can estimate the lifetime from Eq. (5) as $\tau_i \approx \alpha^{-1} \exp(|h_i|/k_BT)$. Therefore the lifetime is large as compared to α^{-1} in all the area of the metastablity except a narrow region in the vicinity of the critical field where $|h_i|$ is close to or less than k_BT. Applying this expression we obtain that the lifetime of a chain in the B position larger than that in the B′ one by a factor of $\exp[2(J_1 - J_2)/(k_BT)] >> 1$.

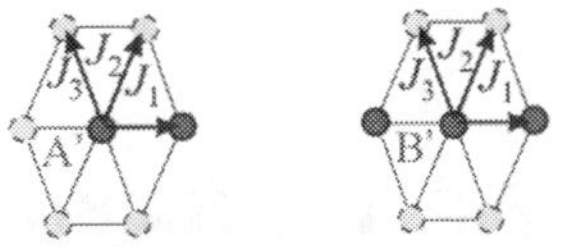

Figure 5. (color online) Examples of possible short-lived metastable configurations.

After a similar reasoning one can show that if a chain in the down state has one or two nearest neighbors in the down states then any configuration, which differs from A or B positions (e.g. A′ or B′ in Fig. 5), has exponentially small lifetime and therefore can be neglected.

5 Conclusion and outlook

It is interesting to mention that the four-step structure in the magnetization curve can appear in the two cases only: perfect triangular lattice ($Ca_3Co_2O_6$) and strongly anisotropic triangular lattice (α-CoV_2O_6). If the anisotropy is weak the critical fields for different orientation of domain boundaries slightly differ from one another. This should smear the additional steps at the plateau 1/3.

It is possible to get a new relation between the coupling constants from the model developed above:

$$\frac{h_{c2} - h_{c1}}{h_{m2}} = \frac{3J_2}{J_1 + J_3}. \quad (8)$$

From here one can see that the value of J_2 obtained in Ref. [16] and mentioned above is underestimated by a factor of 3.4 as compared to J_1 and J_3.

There are very few magnetic systems which demonstrate the four-step magnetization curves (e.g. $Ca_3Co_2O_6$ [3, 9] and Sr_3HoCrO_6 [4]). The α phase of CoV_2O_6 is the only one among them with the first step of the four-step structure shifted from the zero to finite magnetic field. The nature of the additional steps in the plateau 1/3 observed in this compound stems from metastable states induced by domain boundaries in the 3SL phase. In contrast to $Ca_3Co_2O_6$ the lattice of chains in α-CoV_2O_6 is strongly anisotropic. This leads to various possible metastable configurations with particular critical fields. However only few of them with the special orientation have the large lifetime which is sufficient for observation. It is exponentially small for the other configurations and therefore they can be neglected. That is why the domain structure in the 3SL phase should be highly anisotropic, i.e. the domain boundaries of the types shown in Fig. 4 lie in the (101) plane only. It is important to investigate the dependence of the magnetization curve at low temperatures on the magnetic field sweep rate in detail similarly to $Ca_3Co_2O_6$. A simulation of the magnetization dynamics in the framework of Glauber's theory would be also useful for deep insight into the magnetic phenomena in α-CoV_2O_6. We assume that at very slow magnetic field variation the magnetization curve should tend to the two-step structure with the flat plateau 1/3. In the opposite case of large sweep rate the four-step magnetization curve also could be broken by the short-lived metastable states.

References

[1] Y. B. Kudasov, A. S. Korshunov, V. N. Pavlov, and D. A. Maslov, Phys. Usp. **55**, 1169 (2012).
[2] M. Mekata, J. Phys. Soc. Jpn. **42**, 76 (1977).
[3] A. Maignan, V. Hardy, Hébert, M. Drillon, M. R. Lees, O. Petrenko, D. M. K. Paul, and D. Khomskii, J. Mater. Chem. **14**, 1231 (2004).
[4] V. Hardy, C. Martin, G. Martinet, and G. André, Phys. Rev. B **74**, 064413 (2006).
[5] Y. B. Kudasov, Phys. Rev. Lett. **96**, 027212 (2006).
[6] G. Cao, V. Durairaj, S. Chikara, S. Parkin, and P. Schlottmann, Phys. Rev. B **75**, 134402 (2007).
[7] Y. B. Kudasov, JETP Lett. **88**, 586 (2008).
[8] G. H. Wannier, Phys. Rev. **79**, 357 (1950).
[9] V. Hardy, D. Flahaut, M. R. Lees, and O. A. Petrenko, Phys. Rev. B **70**, 214439 (2004).
[10] Y. B. Kudasov, A. S. Korshunov, V. N. Pavlov, and D. A. Maslov, Phys. Rev. B **78**, 132407 (2008).
[11] Y. B. Kudasov, A. S. Korshunov, V. N. Pavlov, and D. A. Maslov, Phys. Rev. B **83**, 092404 (2011).
[12] M. Lenertz, J. Alaria, D. Stoeffer, S. Colis, and A. Dinia, J. Phys. Chem. C **115**, 17190 (2011).
[13] M. Nandi and P. Mandal, J. Appl. Phys. **119**, 133904 (2016).
[14] Z. He and W. Cheng, J. Mag. Mag. Mat. **362**, 27 (2014).
[15] B. Kim, B. H. Kim, K. Kim, H. C. Choi, S.-Y. Park, Y. H. Jeong, and B. I. Min, Phys. Rev. B **85**, 220407 (2012).
[16] M. Lenertz, J. Alaria, D.Stoeffler, S. Colis, A. Dinia, O. Mentré, G. André, F. Porcher, and E. Suard, Phys. Rev. B **86**, 214428 (2012).
[17] A. Saul, D. Vodenicarevic, and G. Radtke, Phys. Rev. B **87**, 024403 (2013).
[18] M. Markkula, A. M. Arevalo-Lopez, and J. P. Attfield, J. Sol. St. Chem. **192**, 390 (2012).
[19] M. H. Qin, K. F. Wang, and J. M. Liu, Phys. Rev. B **79**, 172405 (2009).
[20] X. Yao, J. Phys. Chem. A **116**, 2278 (2012).
[21] E. Kim, B. Kim, and S. J. Lee, Phys. Rev. E **68**, 066127 (2003).
[22] R. Glauber, Jour. Mat. Phys **2**, 294 (1963).
[23] G. M. Buendía, P. A. Rikvold, K. Park, and M. A. Novotny, J. Chem. Phys. **121**, 4193 (2004).

Coexistence of the magnetically ordered and Haldane states in $(Y_{1-x}Nd_x)_2BaNiO_5$

Elena Popova[1,*], *Sergei* Klimin[2], *Marina* Popova [2], *Rüdiger* Klingeler[3], *Bernd* Büchner[4], and *Alexander* Vasiliev[5]

[1]*National Research University Higher School of Economics, 101000 Moscow, Russia*
[2]*Institute of Spectroscopy RAS, 108840 Troitsk, Moscow, Russia*
[3]*Kirchhoff Institute for Physics, University of Heidelberg, INF 227, 69120, Heidelberg, Germany*
[4]*Leibniz-Institute for Solid State and Materials Research IFW Dresden, 01171 Dresden, Germany*
[5]*Low Temperature Physics Department, Moscow State University, 119991 Moscow, Russia*

Abstract. The magnetic properties of $(Y_{1-x}Nd_x)_2BaNiO_5$ with x = 0.01, 0.02, 0.03, 0.04, and 0.05 are investigated by means of specific heat, magnetic susceptibility, and spectroscopic measurements. We estimate a magnetic contribution of the neodymium subsystem into the magnetic susceptibility and specific heat using temperature dependencies of the splitting of the ground Kramers doublet of the Nd^{3+} ion obtained from spectroscopic experiments. We estimate also the contribution caused by nickel chain breaks and by ferrons. Different scenarios of low-temperature magnetic behavior are discussed.

1 Introduction

In the compounds that contain chains of magnetic ions with the spin S = 1, a disordered ground state with a gap in the spectrum of magnetic excitations (the so-called Haldane gap) is realized [1]. The crystal structure [2] of $(Y_{1-x}Nd_x)_2BaNiO_5$ contains NiO_6 octahedra connected by their apical oxygen ions. The chains directed along the a-axis of the crystal are interconnected through the Nd^{3+}, Y^{3+}, and Ba^{2+} ions. The magnetic system of $(Y_{1-x}Nd_x)_2BaNiO_5$ consists of two kinds of magnetic ions: the Ni^{2+} ions with the S= 1 and Nd^{3+} ions with J=9/2.

Y_2BaNiO_5 is a well-known Haldane system. No long-range magnetic order was detected in this system down to at least 0.1 K [3]. The correlation length was found to be seven lattice constants along the a-axis. Neutron scattering and magnetization studies have revealed a singlet ground state and a gap of about 10 meV in the magnetic excitation spectrum of Y_2BaNiO_5 [4-6].

Total or partial substitution of a magnetic rare-earth (RE) ion for the nonmagnetic yttrium leads to an antiferromagnetic (AFM) ordering. Neutron scattering measurements on $(Y_{1-x}Nd_x)_2BaNiO_5$ with $x \geq 0.25$ have shown that the Néel temperature decreases from 49K (x=1) to 17 K (x=0.25), the magnetic moments of both the Nd and Ni ions are reduced, and the magnetic moments of the Nd^{3+} ions are oriented along the *c* axis, while the magnetic moments of the Ni^{2+} ions lie in the (ac) plane [7]. A combined study of thermodynamic and spectroscopic properties has demonstrated a magnetic ordering of the compound with x=0.15 at 13 K [8]. Inelastic neutron scattering experiments have revealed a coexistence of the Haldane-gap excitations and spin waves in the ordered state and an increase of the Haldane gap with decreasing the temperature below T_N [7, 9].

In this situation, it is not clear how does the Ni subsystem behave in the thermodynamic properties. Another question is whether the magnetic ordering occurs down to the lowest concentration of Nd in the compound or there is a critical concentration below which the system is in a disordered state. In this paper we study specific heat, magnetic susceptibility, and spectroscopic properties of $(Y_{1-x}Nd_x)_2BaNiO_5$ compounds with low concentrations of neodymium.

2 Experimental

Polycrystalline samples of $(Y_{1-x}Nd_x)_2BaNiO_5$ with x = 0.05; 0.04; 0.03; 0.02; 0.01 were synthesized by solid state reactions as described in Ref. [10]. The quality and composition of the samples were tested by x-ray diffraction techniques. The temperature dependencies of the magnetic susceptibility in zero-field-cooled (ZFC) and field-cooled (FC) regimes were measured in the magnetic field of 0.01T at temperatures between 1.8 and 300 K using a MPMS-XL5 SQUID magnetometer (Quantum Design). The temperature dependencies of the specific heat in the temperature range 0.4 ÷ 300K were measured using Heat Capacity options of Physical Property Measurement System (Quantum Design). Optical absorption measurements in the spectral range 10000–15000 cm^{-1} were performed using a Fourier spectrometer BOMEM DA3.002, low temperatures were achived using of a helium-bath cryostat with a possibility to pump helium vapor.

* Corresponding author: eapopova@hse.ru

3 Results

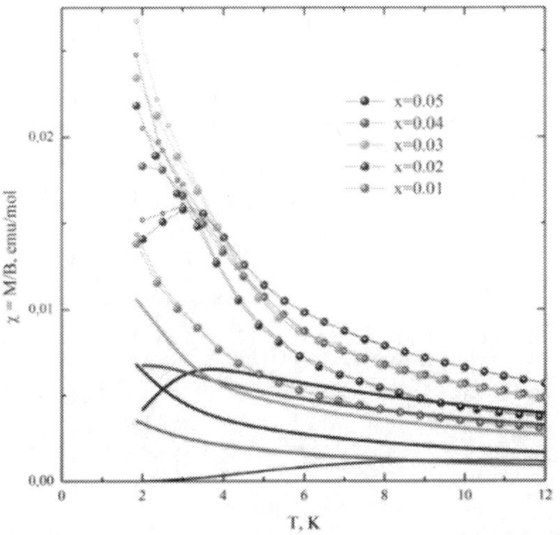

Fig.1. The temperature dependencies of the magnetic susceptibility in a low-temperature range measured in the magnetic field B=0.01 T in ZFC (closed circles) and FC (open circles) regimes for $(Y_{1-x}Nd_x)_2BaNiO_5$. Thick solid lines present the calculated (this work) contribution of the neodimium subsystem. The contribution from chain breaks is represented by thin solid line (only for the compound with x=0.05).

The temperature dependencies of the magnetic susceptibility in the low- and high-temperature ranges are presented in Fig.1 and Fig.2, respectively. The difference in the low-temperature behavior of the magnetic susceptibility measured in ZFC and FC regimes points to a spin-glass state.

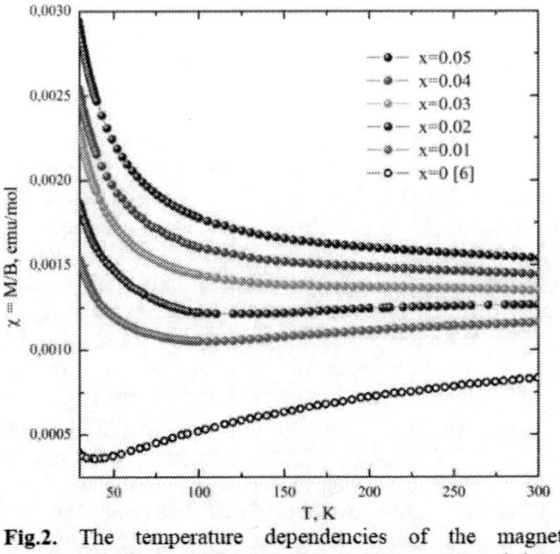

Fig.2. The temperature dependencies of the magnetic susceptibility in a high-temperature range measured in the magnetic field B=0.01 T for $(Y_{1-x}Nd_x)_2BaNiO_5$.

At high temperatures, the behavior of the magnetic susceptibility depends on the Nd concentration in the compound. For compounds with x=0.05, 0.04, and 0.03, the magnetic susceptibility is almost temperature-independent, while at the lowest concentration of Nd, the susceptibility slightly increases with increasing the temperature, similar to the case of the Haldane-gap system Y_2BaNiO_5 [6]. After subtracting the magnetic susceptibility of nickel subsystem taken from the data on Y_2BaNiO_5, we get an estimate for the effective magnetic moment. Its value is close to the free-ion value of Nd in every compound. On these grounds, we argue that the Haldane-gap state of the nickel chains is preserved in the high-temperature range for the studied mixed compounds.

Figure 3 presents the temperature dependencies of the specific heat. A λ-type anomaly observed in the specific heat for compound with x=1, 0.75, 0.50, and 0.25 [11] was not found for compounds with x≤0.05. A maximum observed in specific heat originates mainly from the temperature-driven depopulation of the upper component of the Nd^{3+} ground Kramers doublet (KD) split in an internal magnetic field that appears in the magnetically ordered state.

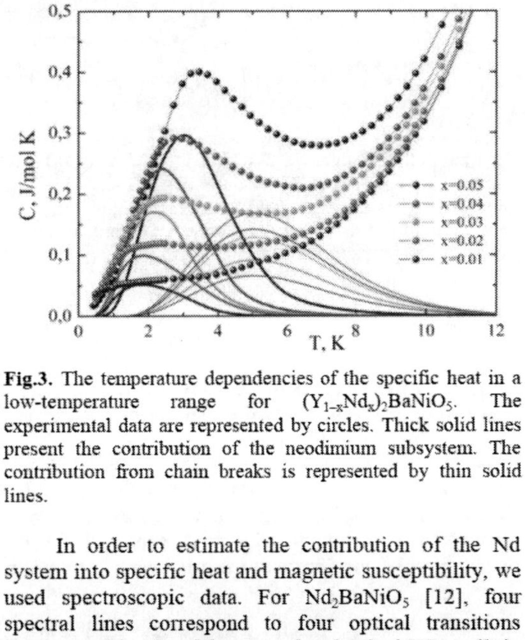

Fig.3. The temperature dependencies of the specific heat in a low-temperature range for $(Y_{1-x}Nd_x)_2BaNiO_5$. The experimental data are represented by circles. Thick solid lines present the contribution of the neodimium subsystem. The contribution from chain breaks is represented by thin solid lines.

In order to estimate the contribution of the Nd system into specific heat and magnetic susceptibility, we used spectroscopic data. For Nd_2BaNiO_5 [12], four spectral lines correspond to four optical transitions between sublevels of the ground and exited KDs split in the internal magnetic field. For mixed compounds, a complex shape of the spectral line is caused by non-equivalent RE centers. Fig. 4a presents the spectral line IA in the $^4I_{9/2}\rightarrow{}^4F_{3/2}$ transition of Nd^{3+} ion in the compound with x=0.03 measured at different temperatures. At temperatures lower that 4 K this line (as well as other spectral lines) starts to shift to higher energies. We argue that such behavior is connected with the splitting of KDs of the Nd^{3+} ion as it is shown on the scheme of Fig. 4b. Indeed, for Nd_2BaNiO_5 (x=1), the splitting Δ of the ground KD is larger than that Δ_A of the terminal level of the considered optical transition [12] and the same should be valid for any x. In this case, with lowering the temperature, two spectral components shown by solid lines in Fig. 4 move to the high-frequency side from the line's center, while two other components marked by dash lines symmetrically move to the opposite side and diminish in intensity, due to thermal depopulation of the upper component of the split ground KD. Small line splittings are hidden in the linewidth but the spectral weight shifts to higher frequencies by Δ/2. A similar behavior was observed in the case of $DyMnO_3$ [13]. Thus, spectroscopic data allowed us to find Δ as a function of temperature. With diminishing x, the shift of spectral lines diminishes and starts at lower temperatures

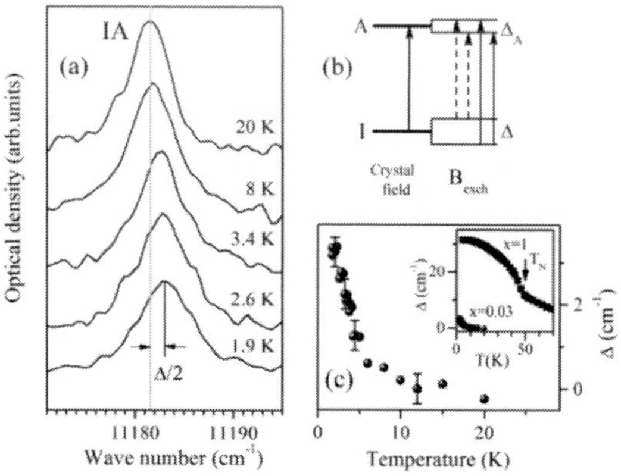

Fig.4. a) The lowest-energy spectral line in the $^4I_{9/2}\to{}^4F_{3/2}$ intermultiplet transition in Nd^{3+} ion in $(Y_{0.97}Nd_{0.03})_2BaNiO_5$. b) An energy level scheme demonstrating splitting of Kramers doublets in an exchange field. c) Δ(T) function for x=0.03 compound. Inset shows a comparison of Δ(T) functions for the cases of x=1 [12] and x=0.03.

4 Discussion

4.1. Contribution of the neodymium subsystem into specific heat and magnetic susceptibility

The smoothed temperature dependencies of the splitting Δ(T) of the ground KD of the Nd^{3+} ion were used for calculation of the Nd contribution into specific heat and magnetic susceptibility. This contribution of the both neodymium sublattices into specific heat can be written as:

$$C = 2xR\left(\frac{\Delta}{kT}\right)^2 \frac{\exp(\Delta/kT)}{(1+\exp(\Delta/kT))^2}, \qquad (1)$$

where R is the gas constant and k is the Boltzmann constant. The factor $2x$ takes into account the concentration of the neodymium ions in the compound.

In order to calculate a contribution of the neodymium subsystem into magnetic susceptibility, we have to take into account the fact that the neodymium sublattices become nonequivalent in an external magnetic field. The sublattice with the magnetic moment components directed along the external magnetic field is magnetized, and the other sublattice with the opposite magnetic moment directions is demagnetized. An effective magnetic field acting on Nd^{3+} ion results from both an external field B (with components B_α, $\alpha = a, b, c$) and an internal field. The latter one is created, mainly, by the nickel subsystem B_{ex}^{Nd} and is due to the Nd-Ni exchange interaction. We take also the Nd-Nd interaction into account. The components of the effective magnetic field $B_{\mathrm{eff}\,1,2\alpha}$ ($\alpha = a, b, c$) at the Nd^{3+} ions belonging to the sublattices 1 and 2 are determined by the following expressions:

$$\begin{cases} B_{\mathrm{eff}\,1\alpha} = B_\alpha + \kappa_{11\alpha} M_{1\alpha}^{Nd} + \kappa_{12\alpha} M_{2\alpha}^{Nd} + B_{ex\ \alpha}^{Nd} \\ B_{\mathrm{eff}\,2\alpha} = B_\alpha + \kappa_{11\alpha} M_{2\alpha}^{Nd} + \kappa_{12} M_{1\alpha}^{Nd} - B_{ex\ \alpha}^{Nd} \end{cases} \qquad (2)$$

Here, M_1^{Nd} and M_2^{Nd} are the magnetic moments of the first (1) and the second (2) neodymium sublattices, respectively. Parameters κ_{11} and κ_{12} determine internal magnetic fields acting on the Nd^{3+} ion with a given magnetic moment, which are induced by Nd^{3+} ions with the same (κ_{11}) and opposite (κ_{12}) directions of their magnetic moments.

For compounds with a high concentration of neodymium, x=1, 0.75, 0.50, and 0.25, the contribution of the Nd subsystem describes well the Schottky anomaly in the C(T) dependence [11]. However, for compounds with low concentration, x≤0.05, the calculated contribution exceeds the experimental data. We assume that only a part of the nominal concentration of the neodymium contributes into the specific heat and magnetic susceptibility, and this part decreases with decreasing x. In according to this assumption, we calculate the contribution of the Nd subsystem into the total magnetic susceptibility and specific heat represented by thick solid lines in Fig.1 and Fig.3, respectively.

4.2 Contribution from chain breaks and ferrons

An additional contribution into specific heat that takes place at temperatures higher than the temperature of the Schottky maximum can be described by chain breaks. Uncontrollable nonmagnetic impurities can result in breaks in nickel chains and the formation of a set of chain segments. According to the valance-bond solid model (VBS) [14], each spin S=1 can be represented in the form of two spins S=½. Unpaired spins S = 1/2, which appear at the ends of chain segments, can interact with each other along the chain. This interaction causes a splitting of the fourfold degenerate level into a singlet (S = 0) and a triplet (S = 1) states. The splitting of the triplet into two sublevels is caused by an anisotropy of the Ni^{2+} ion. For the best agreement with the experimental data, we assume that the splitting reflecting interaction between end spins ½ depends on the temperature and this behavior is similar to the behavior of the Haldane gap in the ordered state. Such behavior has been explained in the framework of a mean-field model that considered Haldane chains in a staggered magnetic field created by the ordered rare-earth magnetic moments. Redistribution of the electrons on the energy sublevels corresponding to chain segments gives a Schottky anomaly at 5 K in the specific heat and a maximum in the magnetic susceptibility at about 10 K. The contribution from chain breaks into the magnetic susceptibility and specific heat is represented by thin solid lines in Fig1 and Fig.3, respectively. A modeling points to the concentration of nonmagnetic impurities of about 3%.

Another additional contribution that cannot be described by the contribution of the Nd subsystem at the lowest temperatures can be explained by a small oxygen nonstoichiometry, resulting in the formation of holes in the 2*p* orbitals of the oxygen ions along a nickel chain [15, 16]. The holes have effective spins ½, as each $2p_z$ orbital containing a hole is occupied by a single electron. The interaction between Ni1 and Ni2 ions which surround a hole is realized by hopping of the hole with a spin S=1/2 from O to Ni1 and from Ni2 to O. This exchange interaction leads to a ferromagnetic coupling between the Ni ions. In the framework of the VBS model, a hole with S=1/2 provides a break in the VBS state of the S = 1 chain. The two Ni-ions, which are the nearest neighbors of the hole, have unpaired spins S=1/2. Thus, a spin bubble, named ferron, consists of three spins S=1/2, the central spin belonging to the oxygen hole. We have to take into account the interaction between two ferrons, because a single ferron does not give anomalies in the specific heat and magnetic susceptibility. Temperature changes in the population of the energy sublevels corresponding to the system of two ferrons give a maximum in the susceptibility at about 3 K and a Schottky anomaly in the specific heat at about 1 K. The modeling points to the concentration of ferrons of about 1%.

4.3 Magnetic ordering

The anomalies observed in the temperature dependencies of specific heat are explained by the contribution from the Ni chain breaks and ferrons, and also from the neodymium subsystem. We had to suppose that only a part of the nominal Nd concentration contributes. Although we checked the concentration of neodymium, we cannot exclude that the real Nd concentration was smaller than the nominal one. Possibly, a more thorough investigation of this point is necessary. Another reason can be the formation of magnetic clusters, inside of which the Nd ions interact with each other through the Ni chain (at the distance of correlation length).

The spectroscopic data and the presence of the Schottky anomaly in specific heat unambiguously point to a splitting of the Nd ground Kramers doublet and, hence, to the presence of an internal magnetic field. Such field could be a result of a magnetic ordering. Meanwhile, the residual splittings of the neodymium ground KD due to strong correlations in Haldane chains exist also in the paramagnetic phase of Nd_2BaNiO_5, giving rise to a "tail" of residual line splittings [12], see Inset to Fig. 4c. The internal field can result also from the short-range order, i.e., ordering inside the magnetic clusters.

Support by the Academic Fund Program at the National Research University Higher School of Economics (HSE) in 2017 (Grant № TZ-89) and by the Russian Academic Excellence Project «5-100» is acknowledged. K.S.A. and P.M.N. acknowledge a support by the Russian Academy of Sciences under the Program "Topical Problems of Low-temperature Physics".

References

1. F.D.M. Haldane, Phys. Rev. Lett. **50**, 1153 (1983).
2. E. García-Matres, J. L. Martínez, and J. Rodríguez-Carvajal, J. Solid State Chem. **103**, 322 (1993).
3. K. Kojima, A. Keren, L. P. Le, G. M. Luke, B. Nachumi, W. D. Wu, Y. J. Uemura, K. Kiyono, S. Miyasaka, H. Takagi, and S. Uchida, Phys. Rev. Lett. **74**, 3471 (1995).
4. J. Darriet, L.P. Regnault, Solid State Commun. **86**, 409 (1993).
5. G.Xu, G.Aeppli, M.E.Bisher, C.Broholm, J.F.DiTusa, C.D.Frost, T.Ito, K.Oka, R.L.Paul, H.Takagi, M.M.J.Treacy, Science **289**, 419 (2000).
6. J.Das, A.V.Mahajan, J.Bobroff, H.Alloul, F.Alet, E.S.Sørensen, Phys. Rev. B **69**, 144404 (2004).
7. T. Yokoo, Z. Zheludev, M. Nakamura, and J. Akimitsu, Phys.Rev. B **55,** 11516 (1997).
8. E.A. Popova, S.A. Klimin, M.N.Popova, R. Klingeler, N.Tristan, B.Büchner, A.N.Vasiliev, J. Magn. Magn. Materials **331**, 133 (2013).
9. A. Zheludev, S. Maslov, T. Yokoo, J. Akimitsu, S. Raymond, S.E. Nagler, and K. Hirota, Phys. Rev. B **61**, 11601 (2000).
10. S. Schiffler, H. Muller-Buschbaum, Z. Anorg, Allg. Chem. **532**, 10 (1986).
11. E. A. Popova, A. N. Vasil'ev, S. A. Klimin, M. V. Narozhnyi, and M. N. Popova JETP **111** (2), 204 (2010).
12. E.A. Popova, S.A.Klimin, M.N.Popova, R. K lingeler, N.Tristan, B.Büchner, A.N. Vasil'ev, JETP **118** (4), 611 (2014).
13. M. A. Kashchenko, S. A. Klimin, A. M. Balbashov, and M. N. Popova, Phys. Status Solidi RRL **10** (6), 462 (2016).
14. E.S. Sorensen, I. Affleck, Phys. Rev. B 51, 16115 (1995).
15. E.Dogotto, J.Riera, A.Sandvik, A. Moreo, Phys.Rev.Lett. **76**, 1731(1996).
16. I.Bose, E. Chattopadhayay, Int. J. Mod. Phys. B **15**, 2535 (2001).

EPJ Web of Conferences **185**, 03004 (2018)
MISM 2017
https://doi.org/10.1051/epjconf/201818503004

Critical phenomena in 1D Ising model with arbitrary spin

Alexey Proshkin[1,*], *Felix* Kassan-Ogly[1,**], *Alexander* Zarubin[1], *Tatyana* Ponomareva[2], and *Ivan* Menshikh[2]

[1] *Institute of Metal Physics of Ural Branch of Russian Academy of Sciences, 620990, Ekaterinburg, Russia*
[2] *Ural Federal University, 620002, Ekaterinburg, Russia*

Abstract. The aim of this work was to study critical phenomena taking place in 1D Ising model with different exchange interactions signs and arbitrary spin values in a magnetic field. Exact analytical formulas for frustration fields, zero temperature magnetization and entropy at these fields are obtained. The general behavior of pair spin correlation function with the accounting of only interactions between nearest neighbors is examined.

1 Introduction

Due to recent developments in synthesizing low-dimensional systems there is now a great interest devoted to unique properties of these materials [1, 2]. Moreover as was shown previously such real systems as lanthanide and actinide monopnictides and monochalcogenides can adequately be described in terms of low-dimensional Ising and Potts models [3–5]. The magnetic and thermodynamic properties of «classical» Ising and Potts models on different lattices were thoroughly investigated in the literature, see, for example, [6–10]. But there is a lack of works in which the accounting of different spin values s is examined. The aim of this work was to study critical phenomena taking place in the Ising model on a linear chain with arbitrary spin values s. The Hamiltonian of the system with the accounting of nearest-neighbors J_1 and next-nearest-neighbors interactions J_2 in a magnetic field H has a standard form:

$$\mathcal{H} = -J_1 \sum_i S_i S_{i+1} - J_2 \sum_i S_i S_{i+2} - H \sum_i S_i, \qquad (1)$$

where S in general takes $2s + 1$ values $-s, -s+1 \ldots s-1, s$.

The investigations were performed by the method of Kramers-Wannier transfer-matrix [11], in which one has to construct transfer-matrix and obtain a maximum eigen value of this matrix. All parameters such as free energy, magnetization and entropy can only be expressed in terms of the maximum eigen value of the corresponding transfer-matrix.

In the first chapter of the article we discuss magnetization and entropy peculiarities of the 1D Ising model with arbitrary spin; in the second chapter the behavior of pair spin correlation function in the cases $s = 1/2$ and $s = 1$ is examined and generalizations for the arbitrary spin values are made.

*e-mail: proshkin_ai@imp.uran.ru
**e-mail: felix.kassan-ogly@imp.uran.ru

2 Magnetization and entropy

2.1 Nearest-neighbors in a magnetic field

It is known that the one-dimensional antiferromagnetic ($J < 0$ in Eq. (1)) Ising model with the only accounting of nearest-neighbors interaction possesses frustrations at certain critical magnetic fields at which at low temperatures the dramatic change of the ground state takes place [4]

$$H_{\text{fr}} = -2sJ. \qquad (2)$$

We were able to derive analytical expressions for magnetization and entropy at these frustration fields as functions of s at $T \to 0$:

$$M_{\text{fr}} = \left(\frac{2s-1}{4}\right) + \left(\frac{2s+1}{4}\right)\frac{1}{\sqrt{1+8s}}, \qquad (3)$$

$$S_{\text{fr}} = \ln\left(\frac{1+\sqrt{1+8s}}{2}\right). \qquad (4)$$

Figure 1 demonstrates the example of magnetization and entropy behavior in the one-dimensional antiferromagnetic Ising model with $s = 3/2$. It is clearly visible that the magnetization at low temperatures at $H = 3$ has a jump from $M = 0$ to $M = 3/2$, magnetization at this field at $T \to 0$ tends to the value $1/2 + 1/\sqrt{13}$ and the zero-temperature entropy tends to non-zero value $\ln\left[(1+\sqrt{13})/2\right]$ in all accordance with the equations (2)–(4). A small change in the magnetic field results in the zero entropy at $T \to 0$. Asymptotic value of the entropy at infinite temperature is equal to the natural logarithm of the number of possible states, in the case $s = 3/2$ it would be $S_\infty = \ln 4$.

2.2 Next-nearest neighbors in the absence of magnetic field

The accounting of the next-nearest-neighbors with antiferromagnetic exchange leads to the two different ground

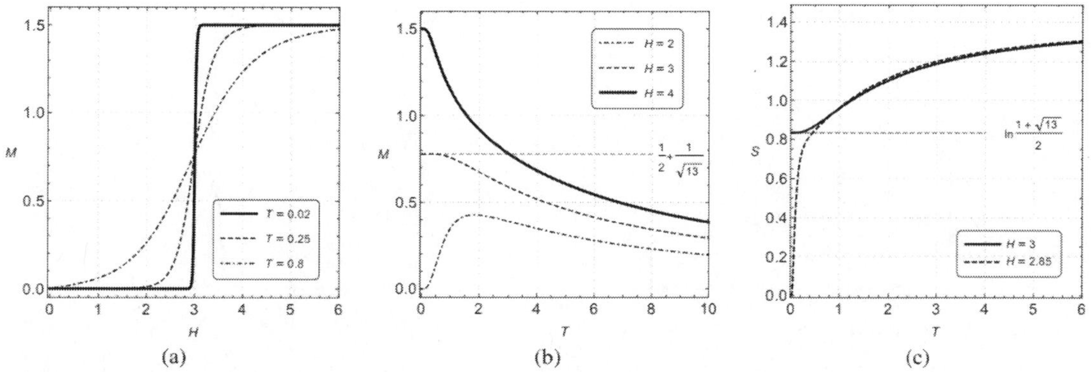

Figure 1. Magnetization as a function of magnetic field (a) and temperature (b); entropy as a function of temperature (c) in 1D antiferromagnetic Ising model with $s = 3/2$.

states at $T \to 0$ depending upon the ratio of $R = J_2/J_1$. In this case frustrations appear at $|R| = 0.5$ for both cases $J_1 < 0, J_2 < 0$ and $J_1 > 0, J_2 < 0$ called competing interactions [12]. It was found that at the frustration point the entropy for both competing interactions at $T \to 0$ has the same value:

$$S_{\rm fr} = \ln\left\{\frac{1}{3}\left[1 + \left(27s - 8 + 3\sqrt{3}\sqrt{27s^2 - 16s}\right) + \left(27s - 8 - 3\sqrt{3}\sqrt{27s^2 - 16s}\right)\right]\right\}. \quad (5)$$

It is interesting to note that the entropy at $s = 1/2$ coincides with the entropy derived for the model with the nearest-neighbor-interactions only (suppose $s = 1/2$ in Eq. (4)) and is equal to natural logarithm of famous golden ratio.

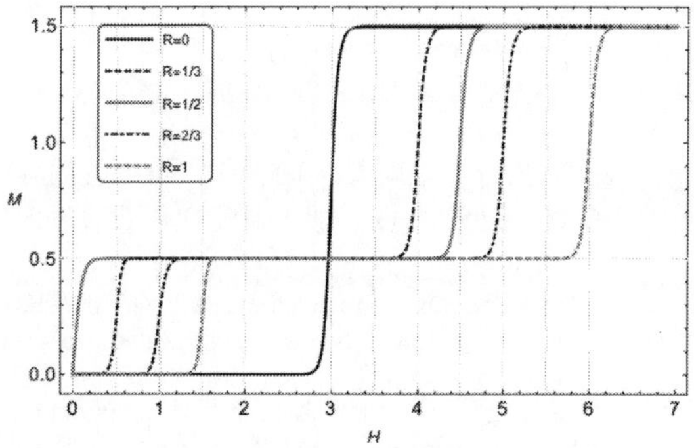

Figure 2. Magnetization of the one-dimensional Ising model with nearest-neighbor and next-nearest-neighbor antiferromagnetic interactions with $s = 3/2$ at $T = 0.05$ for five different values of parameter $R = J_2/J_1$

2.3 Next-nearest neighbors in a magnetic field

General behavior of this model in the «classical» case is represented in [12]. Here we write generalized equations for the case of arbitrary spin. The case when both exchange interaction values are negative (antiferromagnetic) is assumed. The results for the other case will be published elsewhere.

For both antiferromagnetic interactions there exist two different frustration fields depending upon the ratio R:

$$H_1 = \begin{cases} -2sJ_1 + 4sJ_2, & 0 < R \leq 0.5; \\ sJ_1 - 2sJ_2, & R \geq 0.5. \end{cases}$$
$$H_2 = -2sJ_1 - 2sJ_2. \quad (6)$$

In Figure 2 the calculated magnetizations for $s = 3/2$ at $T = 0.05$ for five cases of parameter R are shown. The jumps of magnetizations at critical fields determined by the Eq. (6) are clearly pronounced.

Exact analytical expressions for magnetization and entropy at $T \to 0$ at frustration fields as functions of s are listed in [4].

3 Correlation function

3.1 Correlation function in the absence of magnetic field

This chapter is devoted to the behavior of the Fourier-transform of the pair spin correlation function in the one-dimensional Ising model with the accounting of interactions between nearest-neighbors in the case $s = 1$. Generalizations for the arbitrary spin values s are made at the end of the chapter. The results for the correlation function of the Ising model with allowance of next-nearest-neighbor interactions are to be published.

In the absence of external magnetic field the expression for Fourier-transform of pair spin correlation function with $s = 1$ has the form

$$K(q) = \cos^2\left[\frac{1}{2}\operatorname{arccot}\left(\frac{2\cosh\frac{J}{T} - 1}{2\sqrt{2}}\right)\right]\frac{1 - \eta^2}{1 - 2\eta\cos q + \eta^2}, \quad (7)$$

 https://doi.org/10.1051/epjconf/201818503004

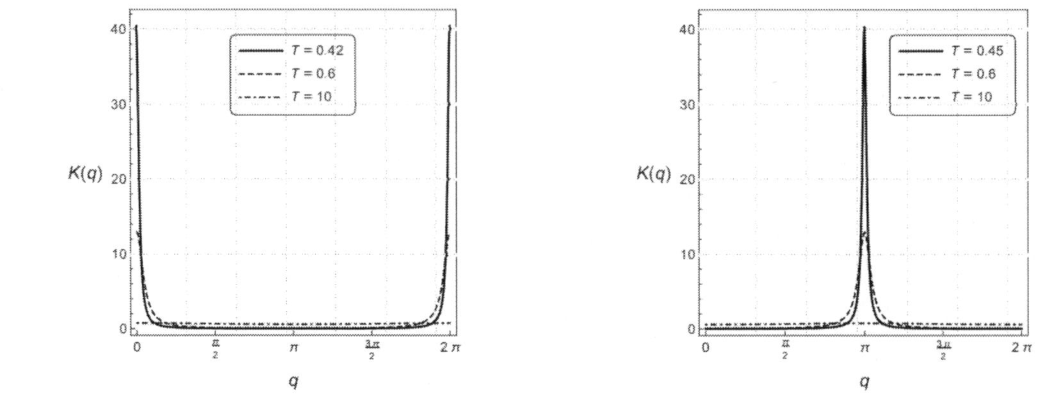

(a) (b)

Figure 3. The dependence of the correlation function upon wave-vector q for the ferromagnetic ($J > 0$) (a) and antiferromagnetic ($J < 0$) interactions (b) in the one-dimensional Ising model with $s = 1$ at zero magnetic field.

where q — wave vector, $\eta = \lambda_2/\lambda_1$:

$$\lambda_1 = \cosh\frac{J}{T} + \frac{1}{2} + \sqrt{\left(\cosh\frac{J}{T} - \frac{1}{2}\right)^2 + 2}, \tag{8}$$

$$\lambda_2 = 2\sinh\frac{J}{T}. \tag{9}$$

In the ferromagnetic case (Fig. 3a) correlation function has peaks exactly at the positions of Bragg reflections $q_n = 2\pi n$, $n \in \mathbb{Z}$ and at $T \to 0$ takes the delta-function form what means that the phase-transition takes place (1D systems undergo a phase-transition at zero-temperature).

Figure 3b demonstrates the calculation of correlation function in the antiferromagnetic case. It is clearly visible that the peaks are located exactly in the middle of Bragg reflections $q_n = \pi(2n+1)$, $n \in \mathbb{Z}$ because the ground state in this case $(+-+-)$ has doubled translation period compared to the lattice. At $T \to 0$ correlation function again takes the form of delta-function.

3.2 Correlation function in a magnetic field

The expression for the correlation function for the Ising model with $s = 1$ in an external magnetic field consists of three terms

$$K(q) = M^2 L(q) + X D_1(q) + Y D_2(q), \tag{10}$$

where M — magnetization, $L(q)$ — Laue function, $D(q)$ — diffuse scattering function. Let us suppose $k = J/T$ and $h = H/T$, that

$$M = \cos^2\beta\cos^2\gamma - (\cos\alpha\sin\beta\cos\gamma + \sin\alpha\sin\gamma)^2, \tag{11}$$

$$\tan\alpha = e^{k+h/2}\frac{\lambda_3 - 2e^{-h}\sinh k}{1 - e^k - \lambda_3}, \tag{12}$$

$$\tan\beta = -e^h\frac{\lambda_3 - 2e^{-h}\sinh k}{\lambda_3 - 2e^h\sinh k}\cos\alpha, \tag{13}$$

$$\cos^2\gamma = \frac{1}{\lambda_1 - \lambda_2}\left(e^{k+h} - \lambda_2 + \tan^2\beta\left(e^{k+h} - \lambda_3\right)\right). \tag{14}$$

The eigenvalues λ ($\lambda_1 > \lambda_2 > \lambda_3$) may easily be found from the equation

$$\lambda^3 - \left(1 + 2e^k\cosh h\right)\lambda^2 - \left(2\cosh h\left(1 - e^k\right) - 2\sinh 2k\right)\lambda + 4\sinh k - 2\sinh 2k = 0, \tag{15}$$

Laue function $L(q)$ defines positions of Bragg reflections while functions $D(q)$ define diffuse scattering.

$$L(q) = \frac{1}{N}\frac{\sin^2(qN/2)}{\sin^2(q/2)},$$
$$D_1(q) = \frac{1 - \eta_2^2}{1 - 2\eta_2\cos q + \eta_2^2}, \tag{16}$$
$$D_2(q) = \frac{1 - \eta_3^2}{1 - 2\eta_3\cos q + \eta_3^2},$$

where $\eta_2 = \lambda_2/\lambda_1$ and $\eta_3 = \lambda_3/\lambda_1$.

$$Y = \cos^2\beta\left[\left(1 + \cos^2\alpha\right)\sin\beta\cos\gamma + \frac{1}{2}\sin 2\alpha\sin\gamma\right] \tag{17}$$

and all coefficients in (10) are connected as follows:

$$X + Y + M^2 = \langle S^2\rangle, \tag{18}$$

where $\langle S^2\rangle$ is the mean squared value of the spin:

$$\langle S^2\rangle = \cos^2\beta\cos^2\gamma + (\cos\alpha\sin\beta\cos\gamma + \sin\alpha\sin\gamma)^2. \tag{19}$$

It is interesting to compare Eq. (11) and Eq. (19).

Laue function defines Bragg scattering, does not depend upon exchange interaction sign and is characterized by lattice parameters. Ferromagnetic system does not possess frustrations and $D(q) = XD_1(q) + YD_2(q)$ in this case is equal to zero at low temperatures. Thus hereinafter we will focus on a behavior of the diffuse scattering part of the correlation function (10) in the antiferromagnetic case.

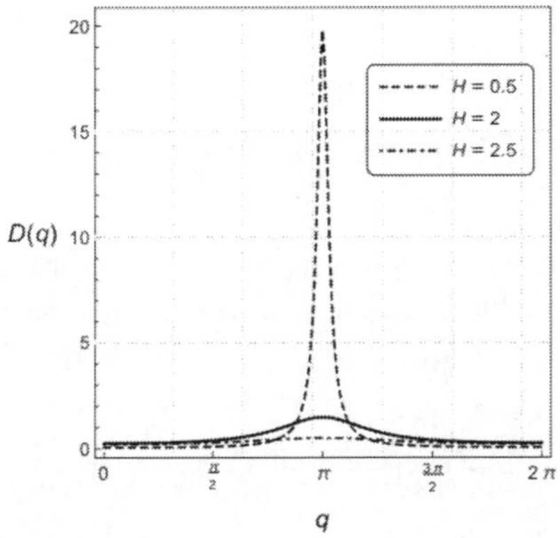

Figure 4. Diffuse scattering in the one-dimensional antiferromagnetic Ising model with $s = 1$ at $T = 0.5$ at three values of magnetic field H.

Figure 4 illustrates the dependence of the diffuse scattering function $D(q)$ upon wave vector q in the one-dimensional antiferromagnetic Ising model with $s = 1$ at $T = 0.5$ for three different values of magnetic field H. It is clearly pronounced that above the frustration field ($H > 2$) diffuse scattering vanishes (both coefficients X and Y in $D(q)$ are equal to zero). Below the frustration field ($H < 2$) the function tends to take the delta-function form that says about the phase transition. Exactly at the frustration field ($H = 2$) the function has a broad maximum and does not tend to take the delta-function form at $T \to 0$. This unusual behavior of the correlation function is the clear evidence that there is no long-range order in the system even at zero temperature. Similar results for the diffuse scattering function were obtained in three- and four-state Potts models [13, 14] and for «classical» Ising model [15]. In the case $s = 1/2$ the analytical expression for the diffuse scattering function has a simple form:

$$D(q) = \left(\frac{1}{4} - M^2\right) \frac{1-\eta^2}{1-2\eta\cos q+\eta^2}, \tag{20}$$

where again $\eta = \lambda_2/\lambda_1$,

$$\lambda_{1,2} = e^{\frac{J}{4T}} \left(\cosh\frac{H}{2T} \pm \sqrt{\sinh^2\frac{H}{2T} + e^{-\frac{J}{T}}}\right), \tag{21}$$

$$M = \frac{\sinh\frac{H}{2T}}{2\sqrt{\sinh^2\frac{H}{2T} + e^{-\frac{J}{T}}}}. \tag{22}$$

Since in the case $s = 1/2$ the asymptotic value of magnetization is equal to $1/2$ it is clearly visible from the Eq. (20) that $D(q) \to 0$ at $M \to 1/2$. The only general difference in the diffuse scattering function behavior between the cases $s = 1/2$ Eq. (20) and $s = 1$ Eq. (10) consists in the different values of frustration field (Eq. (2)). Generally the behavior of the correlation function described above is similar in the one-dimensional antiferromagnetic Ising model for any value of s.

It is evident that frustrated states have peculiar properties. Magnetization at frustration fields at low temperatures has jumps, entropy tends to nonzero values and diffuse scattering function has a broad maximum and does not take the form of delta-function.

4 Summary

In this article the main results obtained for the one-dimensional Ising model are presented. Exact analytical formulas for frustration fields, zero temperature magnetization and entropy at these fields are obtained. The general behavior of pair spin correlation function with the accounting of interactions between nearest neighbors is examined. It is shown that in the antiferromagnetic Ising model a diffuse scattering function at $T \to 0$ does not take the delta-function form that is a clear evidence of the absence of the long-range order in the system.

The reported study was carried out within the state assignment of FASO of Russia (theme «Quantum» n. 01201463332) and was funded by RFBR according to the research project n. 16-32-00032.

References

[1] T. Giamarchi, *Quantum physics in one dimension* (Oxford: Oxford university press, 2003)

[2] S. Roth, D. Carroll, *One-dimensional metals: conjugated polymers, organic crystals, carbon nanotubes and graphene* (Wiley-VCH, 2015)

[3] F.Y. Wu, Rev. Mod. Phys. **54**, 235 (1982)

[4] A. Proshkin, F. Kassan-Ogly, Mater. Sci. Forum **845**, 93 (2016)

[5] F.A. Kassan-Ogly, A.K. Murtazaev, A.K. Zhuravlev, M.K. Ramazanov, A.I. Proshkin, J. Magn. Magn. Mater. **384**, 247 (2015)

[6] L. Onsager, Phys. Rev. **65**, 117 (1944)

[7] A.K. Murtazaev, M.K. Ramazanov, M.K. Badiev, Low Temp. Phys. **37**, 1001 (2011)

[8] F.A. Kassan-Ogly, B.N. Filippov, A.K. Murtazaev, M.K. Ramazanov, M.K. Badiev, J. Magn. Magn. Mater. **324**, 3418 (2012)

[9] A.I. Proshkin, F.A. Kassan-Ogly, J. Magn. Magn. Mater. **383**, 13 (2015)

[10] D.P. Landau, Phys. Rev. B **21**, 1285 (1980)

[11] H.A. Kramers, G.H. Wannier, Phys. Rev. **60**, 252 (1941)

[12] F.A. Kassan-Ogly, Phase Transitions **74**, 353 (2001)

[13] F.A. Kassan-Ogly, B.N. Filippov, Phys. Met. Metallogr. **95**, 518 (2003)

[14] F.A. Kassan-Ogly, B.N. Filippov, Phase Transitions **77**, 261 (2004)

[15] R.J. Baxter, *Exactly solved models in statistical mechanics* (Academic press, 1989)

Magnetization curves and hysteresis loops of nanostructured rare earth titanates

Anatoly B. Rinkevich[1,*], *Alexander* V. Korolev[1], *Mikhail* I. Samoilovich[2], *Sergej* O. Demokritov[1,3], and *Dmitry* V. Perov[1]

[1]M.N: Miheev Institute of Metal Physics, 620108 S. Kovalevskaya st. 18, Ekaterinburg, Russia
[2]Central Research Technological Institute "TECHNOMASH", 121108 I. Franko st. 4, Moscow, Russia
[3]Institute for Applied Physics and Center for Nonlinear Science, University of Muenster, 48149 Corrensstrasse 2-4, Muenster, Germany

Abstract. Magnetic properties of the nanocomposite materials containing particles of rare earth titanates with pyrochlore structure have been investigated. For the nanocomposites with $Gd_2Ti_2O_7$, $La_2Ti_2O_7$ it has been observed that the effective magnetic moment in the nanocomposites differs substantially from that for Gd^{3+} and La^{3+} ions. The hysteresis loops was obtained for the nanocomposites with Dy^{3+}, Gd^{3+}, Yb^{3+}, Er^{+}, Sm^{3+} ions. There is no hysteresis loops for the nanocomposites with $La_2Ti_2O_7$, $Pr_2Ti_2O_7$ and $Nd_2Ti_2O_7$ particles. It has been shown that nanocomposites with $Yb_2Ti_2O_7$, $Dy_2Ti_2O_7$ and $Er_2Ti_2O_7$ particles have one crossing point on the descending branches of hysteresis loop in some temperature range.

1 Introduction

Investigation of low temperature magnetic properties of rare earth titanantes with the pyrochlore structure constitutes a fashionable task. Pyrochlore titanates have the type of $R^{3+}{}_2Ti^{4+}{}_2O_7$ where R^{3+} is a rare-earth ion. They potentially promote to constitute a geometrically frustrated magnet. Short range ordered structures such as spin glasses, spin ices and spin liquids are organized in these systems instead of long range ordered ground states [1]. The problem of "spin ice" established for several titanates attracts the most interest [2].

The aim of this paper is to investigate the magnetization curves and hysteresis loops for several nanocomposite rare earth titanates of $R_2Ti_2O_7$ type. The size of the particles inside the nanocomposites varied from 5 to 60 nm. The particles are composed of La, Dy, Yb, Er, Sm, Nd, Gd and Pr titanates. Major and minor magnetic-hysteresis loops have been measured in temperature range from 2 to 50 K. We observed that for several nanocomposites the descending branches of the hysteresis loops measured at different temperatures cross at one point of the magnetization M – magnetic field H plane.

2 Preparation of samples and their structure

Sintering of monodisperse opal matrices with the submicron SiO_2 spheres is described elsewhere [3]. The nanocomposite materials with embedded titanate nanoparticles are obtained by chemical synthesis method with posterior thermal treatment. Annealing was carried out at temperature 900 °C, whereas impregnations and afterheats were repeated 5 times resulting in gradual infill of the inter-sphere voids. The SEM image of the 3D-nanocomposites structure is shown in Fig.1. The inserted particles have polycrystalline structure and irregular shape. The most part of the particles has the size from 5 to 60 nm. The important feature seen in Fig.1, where structure of the nanocomposite with $Yb_2Ti_2O_7$+TiO_2 particles is shown, that the particles form aggregates, where individual particles are located very close one to another, at the distances of few nanometers.

Fig. 1. Sampling electron microscope image of the nanocomposite contained particles of $Yb_2Ti_2O_7$+TiO_2, magnification x20000.

* Corresponding author: rin@imp.uran.ru

The following scheme of the sample enumeration is accepted. In the sample number No.244/5-900, for example, at first the sequential number of the sample (244) is indicated and then the impregnations number goes (5), and then the temperature of heat treatment (900 °C) is shown. Concluding this section we note that the samples with nanocomposite rare earth titanates have desirable chemical composition and phase content.

3 Magnetization curves and magnetic moments

Magnetic properties of the nanocomposites were studied using the PPMS-9 system. The magnetization curves and hysteresis loops were measured in magnetic fields to 30 kOe at temperatures from 2 to 20 K. The magnetization curves for sample No. 244/5-900 with particles of $Gd_2Ti_2O_7$, which have been measured at different temperatures, are shown in Fig.2a. This figure presents also the appropriate theoretical dependencies calculated from Brillouin formula [4]

$$M = Ng\mu_B JB_J(x)\,, \quad (1)$$

where

$$B_J(x) = \frac{2J+1}{2J}\text{cth}\left(\frac{2J+1}{2J}x\right) - \frac{1}{2J}\text{cth}\frac{x}{2J}\,, \quad (2)$$

M is the magnetic moment, N is the number of magnetic atoms in the sample, μ_B is the Bohr magneton, J is the total angular momentum quantum number, $x = \frac{g\mu_B JH}{k_B T}$, g is the Landé factor, H is external magnetic field, k_B is the Boltzmann constant, T is temperature.

The effective magnetic moment per one rare earth atom in a sample is calculated by the following way

$$\mu = g\mu_B\sqrt{J(J+1)}\,. \quad (3)$$

The approximate curves shown in Fig.2a are corresponding to the quantum number J = 7/2 and effective magnetic moment $\mu = 7.94\mu_B$.

Approximation of the experimental dependences is realized via variation of J at a constant value of g. It is clearly seen from Fig.2a that the magnetization curves measured at different temperatures, differ markedly from each other. These data evidently proves interaction either between Gd^{3+} ions or between the particles. Because of this, the approximation of the experimental magnetization curves measured at different temperatures has been performed, in order to get the temperature variation of effective magnetic moment. The results are summarized in Fig. 2b. It is easy to verify that for $Gd_2Ti_2O_7$ the obtained from approximation values μ varying from 3.92 to 4.81, and also J varying from 1.96 to 2.4 are much close to $^5S_2$ term, than to $^8S_{7/2}$ term which is generally recognized for Gd^{3+} ion. The effective magnetic moment deduced from this approximation for the nanocomposite with $Gd_2Ti_2O_7$ particles $\mu = 4.81\mu_B$ at T = 10 K differs substantially from the effective magnetic moment of ion $\mu = 7.94\mu_B$ [4]. Moreover, it is worthy to mention that the effective magnetic moment in nanocomposite increases with temperature, see Fig.2b.

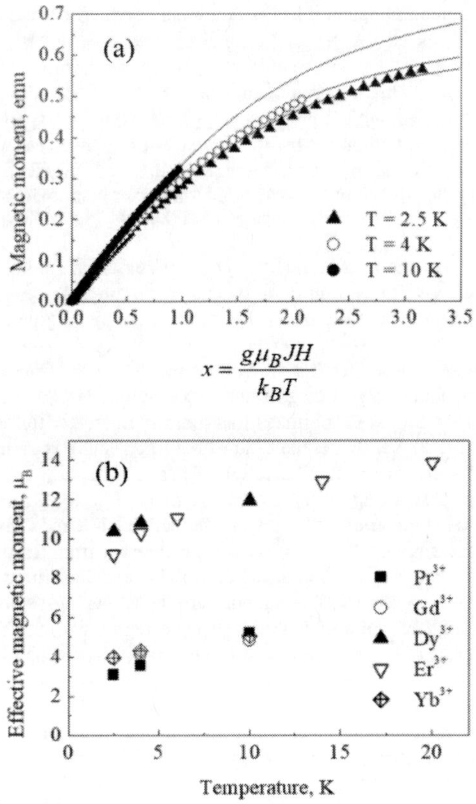

Fig. 2. Magnetization curves for the sample No.244/5-900 with particles of $Gd_2Ti_2O_7$ measured at different temperatures – (a); effective magnetic moment per rare earth ion for several nanocomposite pyrochlore titanates with different rare earth ions – (b).

The hysteresis loops were obtained for the nanocomposites with $Yb_2Ti_2O_7$, $Dy_2Ti_2O_7$, $Er_2Ti_2O_7$ and $Sm_2Ti_2O_7$ particles at cryogenic temperatures. The narrow hysteresis loops for the sample No.228/5-900 with particles of $Yb_2Ti_2O_7 + TiO_2$ measured at different temperatures are shown in Fig.3. The hysteresis loops for the sample No.230/5-900 with particles of $Er_2Ti_2O_7$ measured at two temperatures 2 K and 10 K are shown in Fig.4a. The presence of hysteresis loop itself could be caused by two reasons: 1. Influence of admixture of 3d elements leading to magnetic polarization and magnetic ordering of the nanocomposite. 2. Intrinsic magnetic polarization comes true because of nano-scale size of titanate particles. The hysteresis loop is observed for the samples of several different titanates in which the

https://doi.org/10.1051/epjconf/201818503005

admixture of 3d elements does not found. Because of this circumstance and a fact that the presence of defects in the nanocomposites can stabilize the magnetic structure, the second reason seems to be most probable one. Moreover, confinement for the movement of monopoles, which are quasi-particles in spin ice solids [5], inside a nanoparticle can be a valuable factor for nanocomposite samples to get magnetic ordering. The descending branches of hysteresis loops have been studied in detail, see Fig.4b for example. For several nanocomposites it has been obtained experimentally that discussed descending branches measured at different temperatures intersect each other in one point at *M-H* plane. This fact has been established for nanocomposite titanates with $Yb_2Ti_2O_7$, $Dy_2Ti_2O_7$, $Er_2Ti_2O_7$ particles. For nanocomposite with $Sm_2Ti_2O_7$ particles the area has been defined where these dependences intersect. The hysteresis loop is lacking for the nanocomposites with $La_2Ti_2O_7$, $Pr_2Ti_2O_7$ and $Nd_2Ti_2O_7$ particles.

In the presence of the point or region of intersection of the descending parts of hysteresis loops there should be a temperature interval within which the magnetization measured at the intersection point remains constant or only slow temperature dependent. The temperature dependences of magnetization for sample No.243a/5-900 with $Sm_2Ti_2O_7$ particles measured in magnetic field of H = -100 Oe, that is the field of the intersection region, and the remanent magnetization M_r, are shown in Fig.5. It is clearly seen that these dependences are quite different at low temperatures, Namely, the dependence measured in the field of H = -100 Oe has an interval from 2K to 5K where magnetization varies slowly, and this interval is lacking for M_r. The temperature dependence of coercive field for samples with the point or region of intersection are expected to be unusual. This dependence for sample No.243a/5-900 with $Sm_2Ti_2O_7$ particles is presented in Fig.5b. This dependence contains a low temperature section where the coercive field decreases if temperature diminishes.

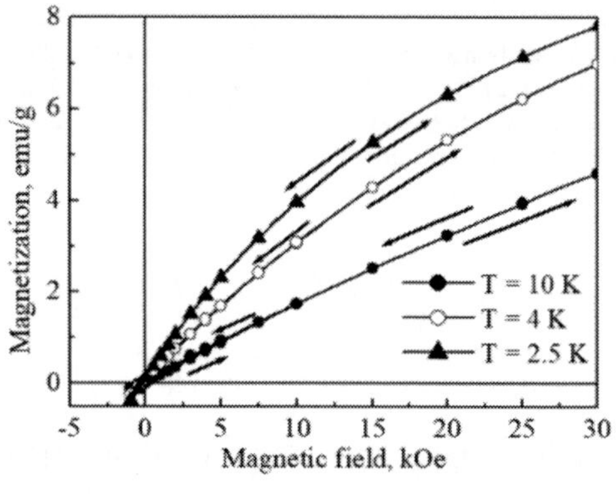

Fig. 3. Hysteresis loops for the sample No.228/5-900 with particles of $Yb_2Ti_2O_7 + TiO_2$ measured at different temperatures.

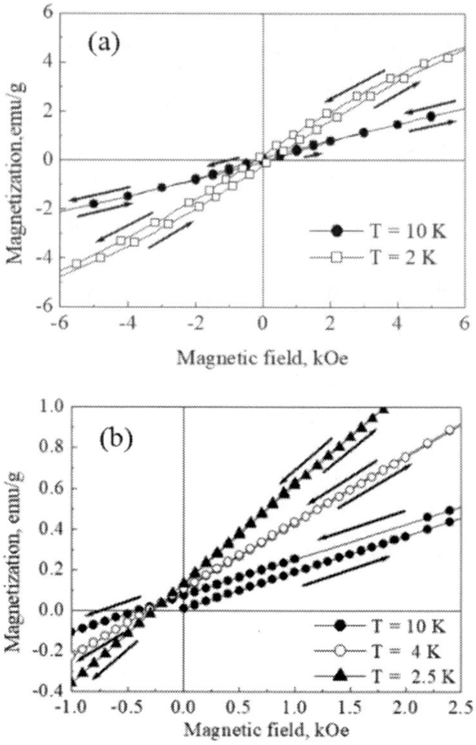

Fig. 4. Hysteresis loops for the sample No.230/5-900 with particles of $Er_2Ti_2O_7$ measured at two temperatures – (a); part of the hysteresis loop for the sample No.228/5-900 with particles of $Yb_2Ti_2O_7 + TiO_2$ showing the crossing point – (b).

4 Conclusions

Magnetic properties of nanocomposite titanates of $R_2Ti_2O_7$ type are investigated where R is a rare earth ion: Nd^{3+}, La^{3+}, Pr^{3+}, Sm^{3+}, Gd^{3+}, Dy^{3+}, Er^{3+}, Yb^{3+}. Magnetization curves of titanates are carefully studied and it is found that the magnetization curve is well approximated by Brillouin function. The effective magnetic moment of most titanates under study is to a certain extent different from the moment of subsequent ion. For nanocomposite titanate $Gd_2Ti_2O_7$, however, obtained quantum number J and effective magnetic moment μ are markedly different from the values for the ion. The effective magnetic moment of all studied titanates increases when temperature increases. A high density of crystallographic defects inevitably presents in the nanoparticles, in particular at their subsurface layers. Consequently, due to this reason a finite-size induced magnetic ordering should be expected.

 https://doi.org/10.1051/epjconf/201818503005

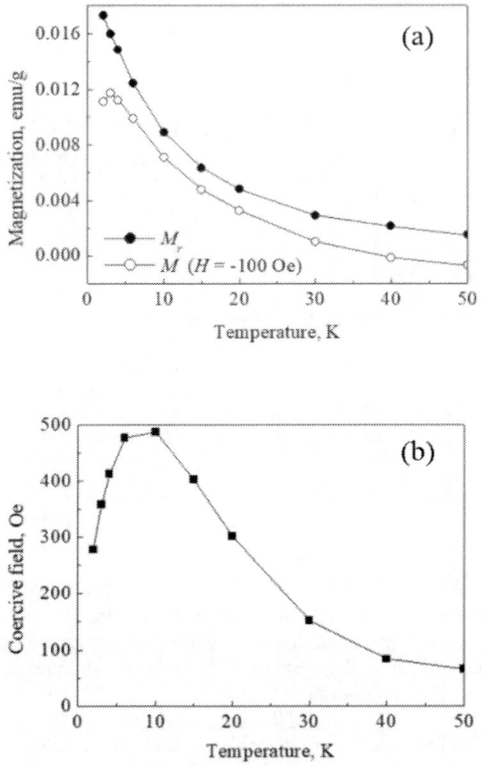

Fig. 5. Temperature dependences of the magnetization for sample No.243a/5-900 with $Sm_2Ti_2O_7$ measured in magnetic field of H = -100 Oe, that is close to the field of the intersection region, and the remanent magnetization M_r – (a); temperature dependence of coercive field for the same sample – (b).

The hysteresis loops were obtained for the nanocomposites with $Yb_2Ti_2O_7$, $Dy_2Ti_2O_7$, $Er_2Ti_2O_7$ and $Sm_2Ti_2O_7$ particles at cryogenic temperatures. The hysteresis loop is lacking for the nanocomposites with $La_2Ti_2O_7$, $Pr_2Ti_2O_7$ and $Nd_2Ti_2O_7$ particles. For several nanocomposites it has been obtained experimentally that discussed descending branches measured at different temperatures intersect each other in one point at M-H plane. In the presence of the point or region of intersection of the descending parts of hysteresis loops there is a temperature interval within which the magnetization measured at the intersection point remains constant or only slow temperature dependent.

The work is carried out within the RAS theme "Spin" No. 01.2.006 01201463330, with partial support of grant from Russian Ministry of Education and Science No. 14.Z50.31.0025 and RFBR grant No. 17-02-00029. Electron microscopic studies and magnetic measurements are carried out in the Testing Center of Institute of Metal Physics.

References

1. J.S. Gardner, M.J.P. Gingras, J.E. Greedan, Reviews of Modern Physics **82**, 53-107 (2010)
2. S.T. Bramwell, M.J.P. Gingras, Science **294**, 1495-1501 (2001)
3. A.B. Rinkevich, A.M. Burkhanov, M.I. Samoilovich, A.F. Belyanin, S.M. Kleshcheva, E.A. Kuznetsov, Russ. J. Gen. Chem. **83**, 2148-2158 (2013)
4. J.S. Smart, *Effective Field Theories of Magnetism* (W.B. Saunders Company, Philadelphia-London, 1966)
5. C. Castelnovo, R. Moessner, S.L. Sondhi, Nat. Lett. **451**, 42-45 (2008)

Appearance of itinerant electrons detected by IR spectroscopy and its correlation with surface magnetism in Co-doped TiO_2 nanopowders

E.V. Mostovshchikova[1*], A.Ye. Yermakov[1,2], M.A. Uimin[1,2], and A.S. Minin[1,2]

[1]M.N. Miheev Institute of Metal Physics of Ural Branch of Russian Academy of Sciences, 620108 Ekaterinburg, Russian Federation
[2]Ural Federal University, 620083, Ekaterinburg , Russian Federation

Abstract. IR absorption and magnetic data for the titanium dioxide nanopowders are studied. Additional absorption in IR range is revealed in nanocrystalline TiO_2 and Co-doped TiO_2 after reducing annealing and explained by appearance of the oxygen vacancies and delocalised charge carriers. Increasing of the absorption in IR range after annealing of Co-doped TiO_2 in vacuum (or decreasing of this contribution after annealing in air) correlates with arising (or vanishing) of ferromagnetic contribution.

1 Introduction

Titanium dioxide in nanostate is one of the intensively studied materials for nanotechnology applications in photocatalysis and photoelectrochemical solar cells, in sensors, spintronics, and medicine [1]. Various reducing treatments, creation of the oxygen vacancies, doping by 3d ions modifies the electronic and magnetic properties of nano-TiO_2 and this is the way for improvement of the properties which are important for practice.

Recently it was shown that doping TiO_2 with Co-ions up to 4% and subsequent annealing in vacuum at 1000 K results in appearance of spontaneous magnetization at room temperature that was associated with the increase in the Co atoms content on the surface of nanoparticles [2]. The results were interpreted within the framework of core−shell model. It was assumed that the ferromagnetic ordering is a result of the interaction of delocalized electrons near the vacancies and the nearest neighbour cobalt atoms due to itinerant type interaction.

One of the methods that can answer the question about the existence of the delocalized states in semiconductor is the infrared spectroscopy and study of the absorption spectra in the "transparency window" between the fundamental and phonon absorption where delocalised charge carriers can manifest themselves. In the present work we try to establish the relation between the evolution of the magnetic properties and IR spectra in Co-doped TiO_2 nanopowders under the reducing treatment as well. To separate influence of the Co-doping and different treatments on the evolution of the IR spectra of the titanium dioxide the absorption data of undoped nanocrystalline TiO_2 were studied also.

2 Experimental details

Nanocrystalline TiO_2 and TiO_2:Co (Co content is 1.5 at%) were prepared by the hydrothermal method [2]. XRD indicates the formation of anatase structure. The size of the nanoparticles determined by TEM is about 60−80 nm in length and 20−25 nm in width.

Before any measurements and annealing the treatment of the synthesized TiO_2 and TiO_2:Co nanopowders in HCl was carried out to remove contamination on the surface, if any. These powders are denoted as the nanopowders in initial state. Then the reducing annealing was carried out in vacuum of 10^{-4} Pa at T~1000 K during 30 minutes. Further, one part of the vacuum treated nanopowders was annealed in air at 670 K during 2 hours and another part was again etched in HCl.

Magnetization of the nanocrystalline TiO_2:Co was measured at room temperature on the Faraday balance in the fields up to 12 kOe.

For the optical measurements nanopowders of TiO_2 or TiO_2:Co were compacted with CsI under the pressure P=5 GPa.To study the evolution of the IR absorption of TiO_2 or TiO_2:Co nanopowders the transmission of the obtained composites in the range 0.9-12.5 μm was measured and optical density was calculated as $D(\lambda)=\ln(1/T(\lambda))$. The transmission spectrum of CsI was taken into account.

3 Results and discussion

Figure 1 demonstrates the magnetization curves of the nanocrystalline TiO_2:Co. It is shown that initial TiO_2:Co nanopowder is paramagnetic. After annealing of the nanopowder in vacuum behaviour of the magnetization curve changes drastically, and the ferromagnetic contribution is clearly observed. Saturation magnetization M_s at T=300 K is about 0.85 emu/g. Subsequent annealing of the TiO_2:Co powder in air

* Corresponding author: mostovsikova@imp.uran.ru

results in decreasing of the magnetization, but M(H) curve still demonstrates ferromagnetic-like character: the saturation of M is observed, the value of M_s is about 0.04 emu/g (the insert in Figure 1).

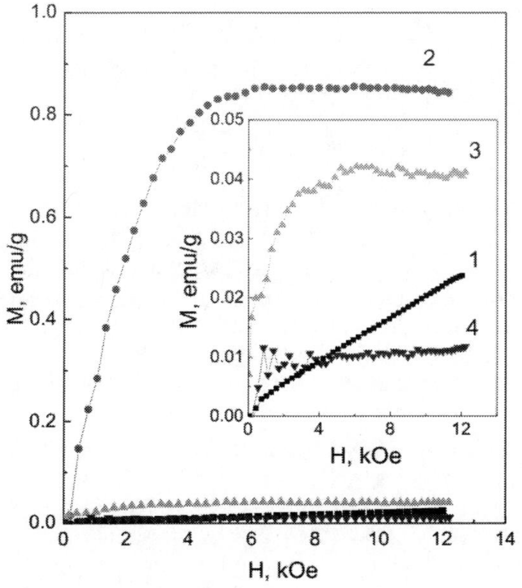

Fig. 1. The magnetization curves for TiO_2:Co nanopowders. 1 – initial state, 2 – annealing in vacuum, 3 – annealing in vacuum and further annealing in air, 4 – annealing in vacuum and etching in HCl.

HCl-etching TiO_2:Co nanopowder annealed in vacuum affects magnetization more significantly then the annealing in air. The curve M(H) becomes close to that of initial powder TiO_2:Co. Disappearing of the magnetization after etching in HCl is in agreement with the conclusion made in [2] that there is the tendency of Co-ions to localise on the surface and subsurface layers resulting in increase in the Co atoms content there as compared with that in the bulk.

The optical density spectra D(λ) of TiO_2:Co nanopowders in comparison with that of undoped TiO_2 are presented in Figure 2. D(λ) curves for the powders with and without Co-ions have the same character. The increase of the absorption in the long-wave range of the spectra is associated with the phonon band with the frequency ν_{LO}=876 cm^{-1} (λ=11.41 μm) [3]. The increase of the absorption in the shortwave range of the spectra is caused by the presence of the band at λ~1 μm (E~1.2 eV) associated with the polaron optical absorption due to self-trapped electrons undergoing hopping between Ti^{4+} and Ti^{3+} [4]. In D(λ) spectra of the powders in the initial state there are the weak absorption bands at λ=7.3 μm, 6.2 μm, and 3.1 μm which correspond to the vibrations of b-CO_3, water, and O-H strongly adsorbed on the nanoparticles surface [5]. Doping with Co does not influence the spectra at λ>4 μm and results in the increase of the absorption at λ<4 μm that is caused by the appearance of the absorption bands attributed to the d-d- transitions in Co-ions [6].

After annealing of nanopowders in vacuum, the following evolution of the optical spectra occurs. The bands at λ=7.3 μm, 6.2 μm, and 3.1 μm are almost disappeared, that can be explained by removal of the corresponding molecules from the nanoparticles surface. The absorption increases both for undoped and Co-doped titanium dioxide (curves 2 in Fig.2). The growth of the absorption in the short-wave range is the result of the appearance additional absorption band at ~0.8 μm attributed to Ti^{3+} ions [8].

The more intriguing change of D(λ) spectra is observed near phonon bands. The increase of the absorption near phonon edge begins at shorter wavelength (at λ~7 μm) and D(λ) curves in the range λ>7 μm is turned up in contrast to the curves of the powders in initial state, that is typical for materials with delocalised charge carriers. It is known that the higher

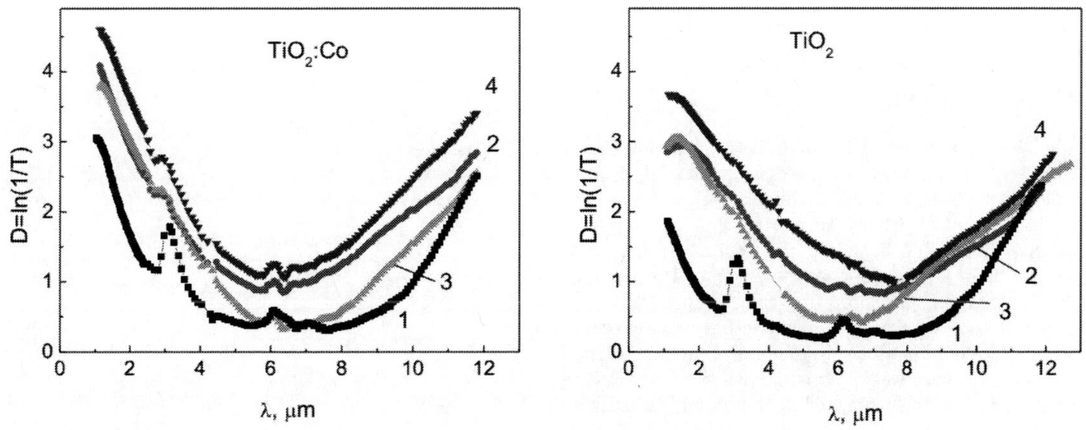

Fig. 2. Optical density spectra of TiO_2:Co (a) and TiO_2 (b) nanopowders. 1 – initial state, 2 – annealing in vacuum, 3 – annealing in vacuum and further annealing in air, 4 – annealing in vacuum and etching in HCl

conductivity of the material corresponds to the higher absorption coefficient in IR range. It is important that delocalised charge carriers appears not only in TiO_2:Co powder but in undoped TiO_2 too. Therefore, the origin of these carriers is not connected with the possible appearance of the droplets of metallic Co on the nanoparticles surface. So, the growth of the absorption and the appearance of Drude-like additional contribution in D(λ) spectra may be explained by appearance of delocalised charge carriers associated with the oxygen vacancies.

Annealing reduced TiO_2 or TiO_2:Co nanopowders in air results in decrease of the absorption (curves 3 in Fig.2), but some residual Drude-like contribution remains. It means that some amount oxygen vacancies still remains. Decrease of the absorption of nanopowders after annealing in air correlates with the decrease of the magnetization (Fig.1) and is in agreement with the hypothesis about itinerant type of magnetism [2].

As one can see, if nanopowder contains Co-ions, change of the absorption after the reducing or oxidizing annealing is somewhat stronger than that for undoped TiO_2. One of the possible reasons is that doping TiO_2 with cobalt is accompanied by a significant increase in the concentration of the oxygen vacancies, as it was shown by study of the electron paramagnetic resonance [2]. Another reason may be so that segregations of Co-ions on the surface of the nanopowder after annealing in vacuum may lead to some segregation of the delocalised charge carriers. As a result, their contribution into the optical spectra is stronger.

Now let us consider the influence of HCl etching TiO_2 or TiO_2:Co nanopowders annealed in vacuum on the optical density spectra. Although the magnetization of TiO_2:Co after HCl-etching decreases more strongly than after annealing in air, the absorption does not decreases (curves 4 in Fig.2) but even slightly increases. It again indicates that delocalised charge carriers do not appear from the metallic Co-droplets, if any. The distinction in the influence of HCl-treatment on the magnetic and optical properties may be explained by the difference of the thickness of surface layer of nanoparticles, in which the charge carriers connected with oxygen vacancies appear or the magnetization is formed.

4 Conclusions

We have reported the IR absorption data (optical density spectra D(λ)) of nanocrystalline TiO_2:Co and TiO_2 after different heat treatment and etching in HCl.

In the optical density spectra of TiO_2 and TiO_2:Co nanopowders after annealing in vacuum an additional absorption of Drude-like character appears. It is revealed that there is the correlation between the appearance and disappearance of the ferromagnetic-like magnetization and Drude-like contribution after reducing or oxidizing annealing, respectively. It is shown that Drude-like contribution is not connected with possible forming of metallic Co-rich droplets. The results support the hypothesis about itinerant type of surface magnetism in Co-doped TiO_2 nanopowders.

This work was supported by the Russian Scientific Foundation, Grant No. 16-12-10004.

References

1. X. Chen, Chin. J. Catal., **30**, 839 (2009).
2. A.Ye. Yermakov, G.S. Zakharova, M.A. Uimin et al., J. Phys. Chem. C **120**, 28857 (2016)
3. R. J. Gonzalez, R. Zallen, and H. Berger, Phys.Rev.B **55**, 7014 (1997)
4. C. A. Triana, C. G. Granqvist, and G. A. Niklasson, J. Appl. Phys. **119**, 015701 (2016)
5. C. Yan, W. Yi, H. Yuan, et al, Environmental Progress & Sustainable Energy **33**, 419 (2014)
6. B. Babu, Ch. Rama Krishna, Ch. Venkata Reddy et al., Spectrochimica Acta Part A: Molecular and Biomolecular Spectroscopy **109**, 90 (2013)
7. V.N. Kuznetsov and N. Serpone, J. Phys. Chem. C, **113**, 15110 (2009)

First-principle calculations of the exchange interaction characteristics for multilayer magnetic structures

Marina Mamonova[1,*] and *Vladimir* Prudnikov[1,**]

[1] *Omsk State University, pr. Mira, 55a, Omsk, 644047 Russia*

Abstract. Magnetic multilayer structures which consist of the ferromagnetic layers separated by nonmagnetic layers with giant magnetoresistance (GMR) are widely used as spintronic devices and read heads of hard disks. In this work magnetic and exchange interaction characteristics of Co/Cu/Co, Ni/Cu/Ni and Fe/Cr/Fe multilayer structures are studied by using VASP software package within the Projector Augmented Wave (PAW) method with generalized-gradient approximation (GGA PBE).The finite temperature magnetism of these structures was studied by Monte-Carlo simulations with the use of the exchange interaction parameters obtained from the ab initio results.

1 Introduction

The ultrathin films on the basis of the magnetic transition metals are used as a components for magnetic multilayer structures with giant magnetoresistance(GMR) which consist of the ferromagnetic layers separated by nonmagnetic layers [1]. These magnetic structures are widely used as read heads of hard disks, memory and spintronic devices [2]. At present, the development of modern computer technology and methods for solving large systems of equations makes it possible to obtain with high accuracy in the calculation of the properties of substances from the first principles of the characteristics of low-dimensional systems that are difficult or expensive to investigate in an experiment. The first-principles calculations of the electronic and crystal structure of magnetic materials are based on the spin theory of the density functional and the numerical solution of the system of Kohn-Sham equations. In this paper atomic structures, magnetic properties and exchange interaction characteristics of ferromagnetic films on Cu and Cr substrates are studied with the use of VASP (Vienna Ab-Initio Simulation Package) software package [3].

2 Model and methods

The layers of ferromagnetic atoms in systems Co/Cu/Co, Ni/Cu/Ni and Fe/Cr/Fe are placed pseudomorphically on each side of the nonmagnetic slab to retain the inversion symmetry. The magnetic moment of atoms is directed collinearly along the z axis. The substrate with 100 orientation surface plane is modeled by the three-layer slab. Supercells for BCC (Fe/Cr)and FCC(Ni/Cu and Co/Cu) systems are given in Fig. 1. We modeled the surface cell using an one atom in monolayer to calculate the energy of film formation and four atoms in a monolayer to calculate the exchange integrals.

We carried out calculations with the use of VASP by means of the Projector Augmented Wave (PAW) method with PBE [4] approximation to describe the exchange-correlation interactions. We calculated the optimized lat-

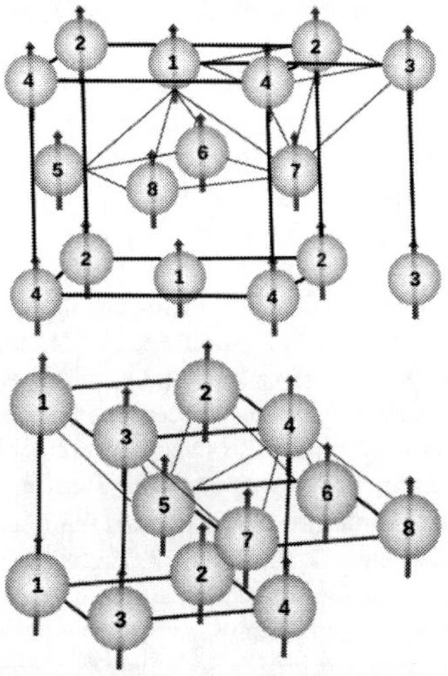

Figure 1. Atoms configurations in FCC and BCC super cells for three monolayers film. The arrows represent magnetic spins directions. The lines show connections the nearest and next-nearest neighbors atoms.

tice constant for bulk Cu through local density approximation (LDA) a = 3.524(1)Åand generalized gradient approximation (GGA) a = 3.6367(5) calculations, that cor-

*e-mail: MamonovaMV@omsu.ru
**e-mail: prudnikv@univer.omsk.su

Table 1. Results of calculation of the total E_{tot} and film formation E_{film} energies and magnetic moment μ_{tot}.

Position	E_{tot},eV	E_{film},eV	μ_{tot}, $\frac{\mu_b}{at}$
Ni/Cu(100)fcc	−27.588	−1.405	0.330
Ni/Cu(111)fcc	−28.118	−1.159	0.430
Ni/Cu(111)bridge	−28.042	−1.120	0.454
Ni/Cu(111)ontop	−27.347	−0.774	0.531

responds to the value a = 3.6369Åobtained by other authors [5]. Also, the application of GGA approach for magnetic calculations is preferable due to more accurate band calculations and further accurate calculations of magnetic moments.

To determine the Ni film formation energy, it is necessary to calculate the difference between total energy of the system and the energy of its components:

$$E_{film} = \frac{1}{2n}(E_{NiCu} - E_{Cu} - 2nE_{Ni}), \tag{1}$$

where E_{Cu} is the total energy of 5-ML Cu slab; E_{NiCu} is the total energy of Ni/Cu/Ni system; E_{Ni} is the total energy of 1-ML Ni slab; n is Ni film thickness in units of monolayers (we used one atom into the monolayer).

We obtained the energy of the film formation for Ni/Cu(100)/Ni system and the magnetic moment atom in 1-ML Ni slab as the functions of convergence parameters We confined ourselves to values of the plane waves cut-off energy E_{max} = $350eV$ and Monkhorst Pack grid size 11 × 11 × 1. Magnetic moments μ=1.02(7)μ_b and μ=0.33 (4)μ_b, which were calculated correspondingly for atom free standing monolayer Ni film and for atom in 1-ML Ni on Cu(001), are in good agreement with results μ=0.95μ_b and μ=0.37μ_b obtained by other authors[6].

Results of calculation of the total and film formation energies, interlayer distance relaxation parametes and magnetic moments are given in table 1. Comparison of the calculated film formation energies for various Ni atom positions (ontop, bridge, fcc) on Cu(111) substrate shows that more energetically favorable position is the position of repeating structure of the substrate. This position of the magnetic atoms are used in further calculations.

We considered the possibility of realization of the Cu/Ni/Cu(001) "sandwich" structure with two additional Cu cap layers, which are placed pseudomorphically on each side of the Ni/Cu(001) slab. The calculations show that the energy of monolayer film system E_{tot} = −27.134 eV is greater than in the case of formation of the sandwich structure E_{tot} = −28.287 eV. Therefore, these results determines that the formation of a sandwich structure is energetically more favorable.

The magnetic moment is the most fundamental quantity in magnetism and yet very difficult to measure experimentally in the case of thin films. The calculated values of the magnetic moments atoms at film layers and 1-st substrate layer for Co/Cu/Co, Ni/Cu/Ni and Fe/Cr/Fe multilayer structures are presented in table 2.

Comparison of the magnetic moments of the atoms in different layers allows to conclude that the atoms most distant from the substrate have greatest magnetic moment.

Table 2. The calculated values of the μ_0 – magnetic moment atom at 1-st substrate layer; μ_i – magnetic moment atom at i film layer; μ_{tot} – average magnetic moment.

	μ_0 μ_B/at	μ_1 μ_B/at	μ_2 μ_B/at	μ_3 μ_B/at	μ_{tot} μ_B/at
Ni/Cu/Ni	0.014	0.489	0.678	0.776	0.622
[7]		0.51	0.71	0.81	
[8]		0.3(1)	0.6(1)	0.8(1)	
Co/Cu/Co	0.024	1.602	1.580	1.825	1.670
Fe/Cr/Fe	-0.318	2.084	2.417	2.932	2.349

This regularity was also established both in experimental [8] and theoretical [7] investigations. Also, we can see in table 2 that the surface atoms of the nonmagnetic substrate are magnetized, and the directions of the magnetic moments for iron and chromium atoms are opposite in the Fe / Cr / Fe system.

3 Methodology and results of calculation exchange interaction parameters

The energy of quantum Heisenberg magnet per unit cell can be written as

$$H = -\frac{1}{2n}\sum_{i,j} J_{ij}S_iS_j, \tag{2}$$

where the summations are over the all sites in the unit cell, S_i are quantum spin operators, n is the number of unit cells in the crystal. The exchange interaction parameters of nearest neighbors and the next-nearest neighbors atoms ($d_1 = (a\sqrt{2})/2$ and $d_2 = a$ for FCC structure as it is shown in Fig. 1) can be obtained from the difference of the total energy for the ferromagnetic and antiferromagnetic configurations of spins in the film:

$$N_{ij}J_j = \Delta E_i = E_i^{AFM} - E^{FM}, \tag{3}$$

where N_{ij} is a number of antiparallel pairs in one unit cell in the i-th AFM state for nearest neighbors (j = 1) and the next-nearest neighbors (j = 2) atoms. The values of parameters in this equation, obtained by our calculations are given in table 3–4. The considered antiferromagnetic configurations of spins are given in table 3. The sign '+' indicates the direction of the spin projection - upwards, the sign '-' the direction down. The order of alternation of signs corresponds to the numbering of atoms, an example of which is shown in Fig 1.

The investigation calculated values of the total energies for the ferromagnetic EF and antiferromagnetic EAF spin configurations of multilayer systems (Table 4) shows that, regardless of the system, a fully ordered ferromagnetic state is energetically more favorable.

4 Results of calculations of the magnetoresistance

On the basis of these values of the exchange interaction integrals, we calculated the temperature dependence of the

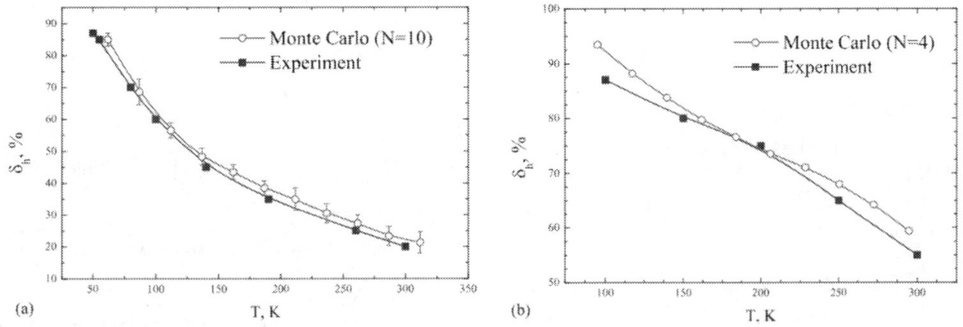

Figure 2. Comparison of the calculated and experimentally measured [1] temperature dependence of the magnetoresistance with the use of the ab initio determined values of the exchange integral for Fe/Cr/Fe (a) and Co/Cu/Co (b) structures.

Table 3. Antiferromagnetic spin configurations, the number of antiparallel pairs N_{ij}, and results of calculation of the i AFM states total energy EAF_i.

	AFM state					N_1	N_2
BCC	1	-	-	+	+	20	24
		+	+	-	-		
		-	-	+	+		
	2	+	-	-	+	24	12
		-	+	+	-		
		+	-	-	+		
FCC	1	+	-	+	-	22	0
		-	+	-	+		
		+	-	+	-		
	2	+	+	-	-	18	6
		-	-	+	+		
		+	+	-	-		

Table 4. The calculated values of the total energies for the ferromagnetic EF and antiferromagnetic EAF spin configurations.

	EF eV	EAF1 eV	EAF2 eV
Ni/Cu/Ni	-167.819	-166.749	-166.691
Co/Cu/Co	-201.463	-197.313	-197.313
Fe/Cr/Fe	-300.036	-295.483	-295.621

Table 5. The calculated values of the exchange integrals of nearest neighbor and the next nearest neighbor interaction.

	$J_1 * 10^{-14}$, erg	$J_2 * 10^{-14}$, erg
Ni/Cu/Ni	0.65	0.56
Co/Cu/Co	2.52	1.68
Fe/Cr/Fe	1.73	0.73

magnetoresistance using the Monte-Carlo simulations for the trilayer structures Fe/Cr(100)/Fe and Co/Cu(100)/Co with different thicknesses N of ferromagnetic films. Magnetoresistance for a multilayer structures is introduced as $\delta_h = (R_{AP} - R_P)/R_P$, where R_{AP} and R_P are the resistance of the structure when the magnetizations of adjacent ferromagnetic layers are aligned antiparallel and parallel orientation of the magnetization of ferromagnetic layers respectively. The procedure of calculation consists of the following steps: the first step is connected with Monte Carlo simulations of a magnetic structure in the equilibrium state at temperature T with determination of the magnetization of ferromagnetic films m_1 and m_2, which gives the possibility of calculating the electron densities $n_{\uparrow\downarrow}$ for film cells; in the second step the average electron velocities V and the current densities $J_{i,\uparrow,\downarrow} = eVn_{\uparrow\downarrow}$ are calculated under relations

$$V = \mu E = \frac{eZa^2}{kT} e^{-\frac{\Delta E_i}{kT}} E \tag{4}$$

$$E_{i,\uparrow,\downarrow} = \mp J_1 \Big[\sum_{j \neq i} S_j^z (n_{j,\uparrow} - n_{j,\downarrow}) + S_i^z (n_{i,\uparrow} - n_{i,\downarrow}) \Big], \tag{5}$$

subject to the spin configuration realized at a given time of the simulation and averaged over Monte Carlo steps at times of the equilibrium state simulation; in the last step the calculation of the magnetoresistance is carried out under relation $\delta_h = \frac{(J\uparrow - J\downarrow)^2}{4J\uparrow J\downarrow}$.

In Monte Carlo calculations the temperature scale is determined through the value of the exchange integral. So for comparison of the calculated and experimentally measured temperature dependence of the magnetoresistance we used the ab initio determined values of the exchange integral.

More detailed description of Monte-Carlo simulations is presented in our papers [9, 10]. The graphs of calculated temperature dependence of the magnetoresistance are given in Fig. 2, which demonstrate a good agreement with experimental data [1].

5 Conclusions

In this paper, magnetic properties of Ni/Cu/Ni Co/Cu/Co and Fe/Cr/Fe trilayer structures was investigated by ab-initio methods. The exchange interaction integrals for these structures were first calculated, that allowed us to obtain the temperature dependence of the magnetoresistance,

https://doi.org/10.1051/epjconf/201818503007

which demonstrate a good agreement with experimental data.

Support by RFBR grant N17-02-00279 is acknowledged.

References

[1] J. Bass and W. P. Pratt, J. Magn. Magn. Mater. **200**, 274–289 (1999).

[2] S. N. Piramanayagam and T. C. Chong, *Developments in data storage : materials perspective* (Wiley-IEEE Press, 2011).

[3] P. G. Kresse and J. Furthmuller, Phys. Rev. B **54**, 11169 (1996).

[4] J. P. Perdew, K. Burke and M. Ernzerhof, Phys. Rev. Lett. **77**, 3865 (1996).

[5] K. Lejaeghere, V. Van Speybroeck, G. VanOost and S. Cottenier, Critical Reviews in Solid State and Materials Sciences, **39**(1), 1–24 (2014).

[6] C. S. Wang and A. J. Freeman, Phys. Rev. B **21**, 4585 (1980).

[7] Z. Yang and V. I. Gavrilenko and R Wu, Surf. Sci. **447**, 212(2000).

[8] A. Ney and A. Scherz and P. Poulopoulos and K. Lenz and H. Wende and K. Baberschke, Phys. Rev. B **65**, 024411 (2001).

[9] V. V. Prudnikov, P. V. Prudnikov, D. E. Romanovskiy, JETP Letters **102**, 668–673 (2015).

[10] V. V. Prudnikov, P. V. Prudnikov, D. E. Romanovskiy, J. Phys. D: Appl. Phys. **49**(23), 235002 (2016).

Monte Carlo simulation of aging phenomena in multilayer magnetic structures

Alexandr N. Purtov[*], *Marina* V. Mamonova, *Vladimir* V. Prudnikov, and *Pavel* V. Prudnikov

Omsk State University, Mira 55A, Omsk, Russia

Abstract. A Monte Carlo simulation of the non-equilibrium behavior of multilayer magnetic structures consisting of alternating magnetic and nonmagnetic nanolayers is realized. The calculated two-time autocorrelation function for the structure during its evolution starting from high-temperature initial state is analyzed. The analysis reveals aging effects characterized by a slowing down of correlation characteristics in the system with increase of the waiting time. The dependence of aging characteristics on thickness of ferromagnetic films is investigated. It is shown that, in contrast to bulk magnetic systems, the aging effects in magnetic superstructures arise not only at the ferromagnetic ordering temperature T_c in the films but also within a wide temperature range at $T \leq T_c$.

1 Introduction

The behavior of systems with extremely slow dynamics stays one of the intriguing problems in theoretical physics [1]. It demonstrates a wide range of interesting phenomena such as aging, and violation of the fluctuation-dissipation theorem. Originally, these features were founded in complex spin glass systems [2] but different studies of these phenomena showed that the main properties of slow dynamics can be observed in simpler systems with second-order phase transition near critical point, because the evolution of such systems is characterized by large relaxation times [3].

The nanoscale periodicity in magnetic multilayer structures gives rise to the mesoscopic effects of the spatial spin correlation with the slow relaxation dynamics of magnetization accompanying the quenching of the system in the non-equilibrium state. In contrast to the bulk magnetic systems, where the slow dynamics and aging effects manifest themselves near the critical point [4], magnetic superstructures with nanoscale periodicity allow increasing the relaxation time owing to the effects related to the larger characteristic spin-spin correlation length. That is why the aging and nonergodicity effects can be experimentally observed in the multilayer Co/Cr magnetic structure [5] within a wider temperature range as compared to that for the bulk magnetic systems.

We have performed in paper [6] a numerical Monte Carlo simulation of the non-equilibrium behavior of the multilayer Co/Cr/Co magnetic structure with the thickness of Co films equals to N=3 ML. Calculations of the two-time correlation functions and the staggered magnetization allow us to reveal the aging effects in the multilayer magnetic structure occurring not only in their non-equilibrium critical behavior at T_c but also within a wide range of temperatures at T<. T_c. The revealed aging effects in the behavior of the magnetization relaxation for our model multilayer structure are in good agreement with the aging effects observed in experiments for the Co/Cr/Co structure [5]. The present work deals with study of dependence of aging characteristics on thickness of ferromagnetic films in the multilayer Co/Cu(100)/Co magnetic structure.

2 Model, methods, and results

We realize in this work a numerical Monte Carlo study of the nonequilibrium behavior of a multilayer magnetic structure (Figure 1) consisting of ferromagnetic films separated by nonmagnetic metal layer. The magnetic films have linear sizes of $L \times L \times N$ with imposed periodic boundary conditions in the film plane.

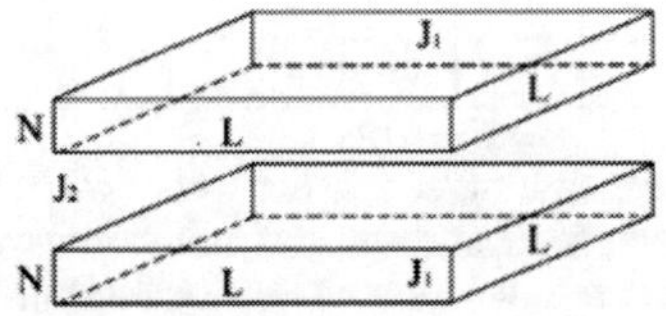

Fig. 1. Model of a multilayer structure consisting of two ferromagnetic films separated by the nonmagnetic metal film; L and N are the linear sizes of the films, J_1 and J_2 are the exchange integrals.

We consider the array consisting of ferromagnetic films with the thicknesses N = 3, 5, 7, 9 in units of monatomic layers. The exchange integral J_1 determining the interaction between the neighboring spins is assumed to be J_1/k_BT=1 (k_B is the Boltzmann constant), whereas that for the interlayer interaction is J_2 = - 0.3 J_1.

[*] Corresponding author: alprtv@gmail.com

The sign of J_2 is negative because the thickness of nonmagnetic spacers in multilayer structures exhibiting the giant magnetoresistance effect is tuned such that both the long-range and oscillating RKKY exchange interactions between the ferromagnetic layers effectively provide antiferromagnetism. Owing to this interaction, the magnetizations of the neighboring ferromagnetic layers have opposite orientations. Thus, the structure under study models artificially created multilayer structures exhibiting the giant magnetoresistance effect [7–9].

The magnetic characteristics of ultrathin films based on Fe, Co, and Ni in contact with the nonmagnetic metal substrate are most adequately described by the anisotropic Heisenberg model [10, 11] with the Hamiltonian

$$H = -\sum_{<i,j>} J_{ij}\{(S_i^x S_j^x + S_i^y S_j^y) + (1-\Delta(N))S_i^z S_j^z\}, \quad (1)$$

where $S_i = (S_i^x, S_i^y, S_i^z)$ is a three-dimensional unit vector at the i-th site. $\Delta(N)$ is the parameter of anisotropy for Co films with the thickness N (N=3,5,7,9) on Cu(100) substrate which values are presented in Table 1. The dependence of the anisotropy parameter $\Delta(N)$ on the film thickness N were chosen proportional to the experimental critical temperatures for Co/Cu(100) system [12]. The form and parameters of Hamiltonian (1) correspond to the Co/Cu(100)/Co multilayer structure, for which the ultrathin Co films exhibit the spontaneous magnetization in the XY plane of the film below the critical temperature.

Table 1. The values of the anisotropy parameter Δ and the critical temperature T_c for films with thicknesses N = 3,5,7,9.

Thickness, N	Anisotropy parameter, Δ	Critical temperature, T_c
3	0.66	2.5413(8)
5	0.476	2.9033(12)
7	0.337	3.0356(6)
9	0.253	3.1014(11)

At the first stage of this work, we calculated the equilibrium characteristics of the multilayer structure with the aim to determine the critical temperatures T_c of the ferromagnetic films with different thicknesses N. For a more accurate determination of the critical temperatures, we considered the structures with films of different linear sizes L = 20, 32, and 40. We calculated such characteristics as the staggered magnetization

$$\mathbf{m}_{stg} = \mathbf{m}_1 - \mathbf{m}_2, \quad (2)$$

where $\mathbf{m}_1$ and $\mathbf{m}_2$ are the magnetizations of the films; staggered susceptibility

$$\chi_{stg} = [<m_{stg}^2> - <m_{stg}>^2]/TN_s, \quad (3)$$

specific heat

$$C_h = [<E^2> - <E>^2]/k_B T^2 N_s, \quad (4)$$

where N_s is the number of spins in the film; the Binder cumulant

$$U_4 = (3 - <m^4>/<m_s^2>^2)/2. \quad (5)$$

The values of the critical temperature $T_c(N)$, presented in Table 1, were obtained by the method of intersection of curves for the Binder cumulant U_4 with different linear sizes L.

At the next stage of our studies, we considered the non-equilibrium behavior of the multilayer structure. We calculated the time dependence of magnetization for a ferromagnetic film located in the structure

$$m(t) = \left\langle \frac{1}{NL^2} \sum_{i=1}^{NL^2} S_i(t) \right\rangle \quad (6)$$

and the two-time autocorrelation function

$$C(t,t_w) = \left\langle \frac{1}{NL^2} \sum_{i=1}^{NL^2} S_i(t) S_i(t_w) \right\rangle - m(t)m(t_w), \quad (7)$$

where brackets < ... > denote a statistical averaging.

The aging effects arise at $t \ll t_{rel}$, where t_{rel} is the relaxation time, and manifest themselves in the dependence of the autocorrelation function on characteristic time variables such as waiting time t_w and time of observation $t - t_w$ with $t > t_w$ not via $t - t_w$ only. The waiting time t_w characterizes the time interval from the preparation of the sample up to starting measurements of its characteristics. It is important that decays for the two-time autocorrelation function as function of time of observation $t - t_w$ are slower for larger waiting times t_w. Initial conditions affect the evolution of the system at $t - t_w$, $t_w << t_{rel}$.

We considered the evolution of the system from the high-temperature initial state created at $T_0 >> T_c$ with normalized staggered magnetization $m^{stg}_0 = 0.05 << 1$. In calculations of the non-equilibrium behavior of the autocorrelation function $C(t,t_w)$, we simulated the structures with the linear size of the film L = 64, which allows us to analyze the nonequilibrium characteristics of the system at times up to 5000 Monte Carlo steps per spin (MCs/s). We used the waiting times t_w = 10, 30, 50, 100 MCs/s in the study of the two-time dependence of $C(t,t_w)$. The averaging of $C(t,t_w)$ was carried out on 500 MC runs for every t_w.

The obtained time dependence of the autocorrelation function $C(t,t_w)$ from observation time $t - t_w$ for different t_w values and thicknesses N of ferromagnetic films are presented in Figures 2 and 3 at the critical temperatures $T_c(N)$ and in the low-temperature phase with temperatures $T_c(N)/2$. The curves of $C(t,t_w)$ demonstrate the aging effects, i.e., the slowing down of time correlations with increasing system age t_w. We must note that the aging effects in the multilayer structures arise not only at $T=T_c$ as in bulk systems, but also at $T<T_c$. Furthermore, the comparison of $C(t,t_w)$ curves in Figure 2 at temperatures $T=T_c$ and $T_c/2$ shows that the aging effects become stronger in the low-temperature phase, i.e. the falling of time correlations at $T_c/2$ with the same waiting times t_w become more slow than at the critical

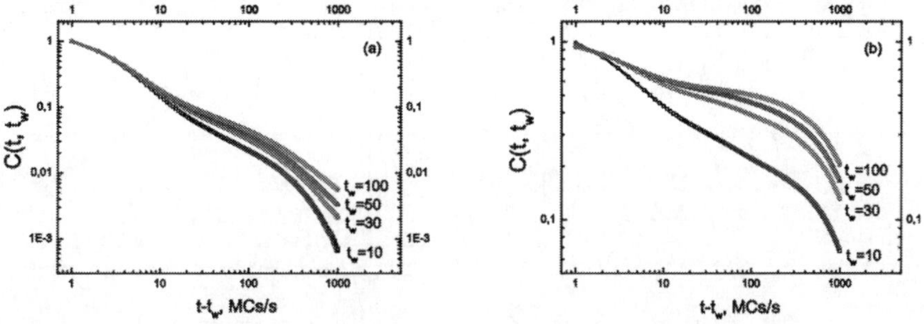

Fig. 2. Time dependence of the autocorrelation function $C(t,t_w)$ as a function $t - t_w$ for different values of t_w for Co/Cu(100)/Co structure with Co film thickness $N = 7$ at temperatures $T_c = 3.0356$ (a) and $T_c/2 = 1.5178$ (b) for evolution from high-temperature initial state.

temperature. Reason for this behavior is connected with *XY*-type anisotropy, which is realized in Co/Cu(100)/Co structures, and with extremely slow dynamics in two-dimensional *XY*-model characterized by aging not only near the temperature of the Berezinskii–Kosterlitz–Thouless phase transition T_{BKT}, but also in the entire range of the existence of the low-temperature phase [13, 14]. These features of non-equilibrium behavior are conserved in ultrathin *XY*-type films.

The dependence of aging characteristics on thickness N of ferromagnetic films is demonstrated in Figure 3 where the curves of the autocorrelation function $C(t,t_w)$ are given for film thicknesses N=3,5 at temperatures T_c and $T_c/2$. The analysis of the $C(t,t_w)$ slowing down with the same waiting times t_w shows that the aging effects in multilayer structure are decreased with increasing thickness of ferromagnetic films N at the critical temperatures $T_c(N)$. This is due to the weakening of the correlation for dimensional transition from quasi-two-dimensional systems to three-dimensional bulk systems. The analysis of the $C(t,t_w)$ curves at $T_c/2$ in Figure 3b indicates the opposite dependence of the autocorrelation function $C(t,t_w)$ with increasing thickness N of ferromagnetic films in considered range of thickness changes.

It is well known that in the aging regime at $t - t_w \sim t_w >> 1$ the two-time autocorrelation function is characterized by the scaling form [4]

$$C(t,t_w) \sim (t_w)^{-b} F_C(t/t_w), \qquad (8)$$

where the exponent b at the critical temperature T_c is expressed in terms of the critical exponents as $b_c = 2\beta/\nu z$. The scaling function $F_c(t/t_w)$ is a homogeneous function of its argument t/t_w and, at the long time stage of its evolution with $t - t_w >> t_w >> 1$, is characterized by the power-law decay

$$Fc(t/t_w) \sim (t/t_w)^{-c_a}, \qquad (9)$$

where the exponent c_a is given at T_c by relations $c_a = d/z - \theta'$ for the evolution from the high-temperature state and by $c_a = 1 + d/z + \beta/\nu z$ for the low-temperature one. Here, β, ν, z and θ' are well known static and dynamic critical exponents, d is a system dimension At the thermal bath temperatures $T_s \neq T_c$, the exponents b and c_a in Eqs. (8) and (9) are not related to the critical exponents.

To check the scaling predictions for $C(t,t_w)$ given by relation (8), we plot the dependences of $t_w^{2\beta/(\nu z)}\, C(t,t_w)$ versus t/t_w in Figure 4 for different Co film thicknesses N at the critical temperatures $T_c(N)$ with the use of exponent $2\beta/\nu z$ chosen to fit the data for different t_w values to a single curve at $t/t_w >> 1$. The exponent $b_c = 2\beta/\nu z$ obtained from this fit is characterized by following values: $2\beta/\nu z$=0.3184(4) for N=3, $2\beta/\nu z$=0.4140(4) for N=5, $2\beta/\nu z$=0.4364(5) for N=7, and $2\beta/\nu z$=0.4562(5) for N=9. These values of exponent $2\beta/\nu z$ demonstrate a dimensional crossover from quasi-two-dimensional to three-dimensional systems.

3 Conclusions

A Monte Carlo simulation of the non-equilibrium behavior of multilayer Co/Cu(100)/Co magnetic

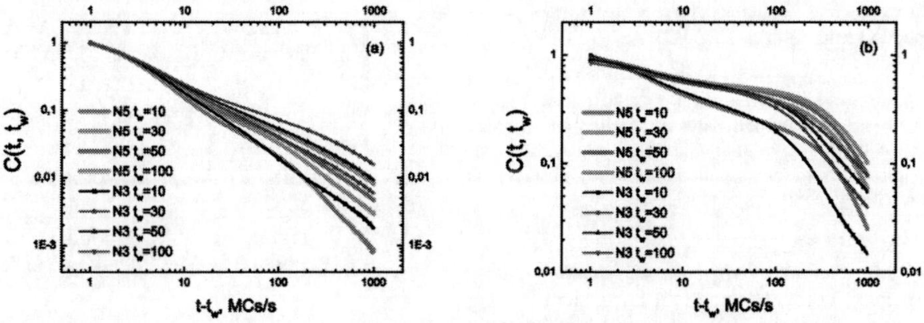

Fig. 3. Time dependence of the autocorrelation function $C(t,t_w)$ as a function $t - t_w$ for different values of t_w for Co/Cu(100)/Co structure with Co film thicknesses $N = 3, 5$ at temperatures T_c (a) and $T_c/2$ (b) for evolution from high-temperature initial state.

https://doi.org/10.1051/epjconf/201818503008

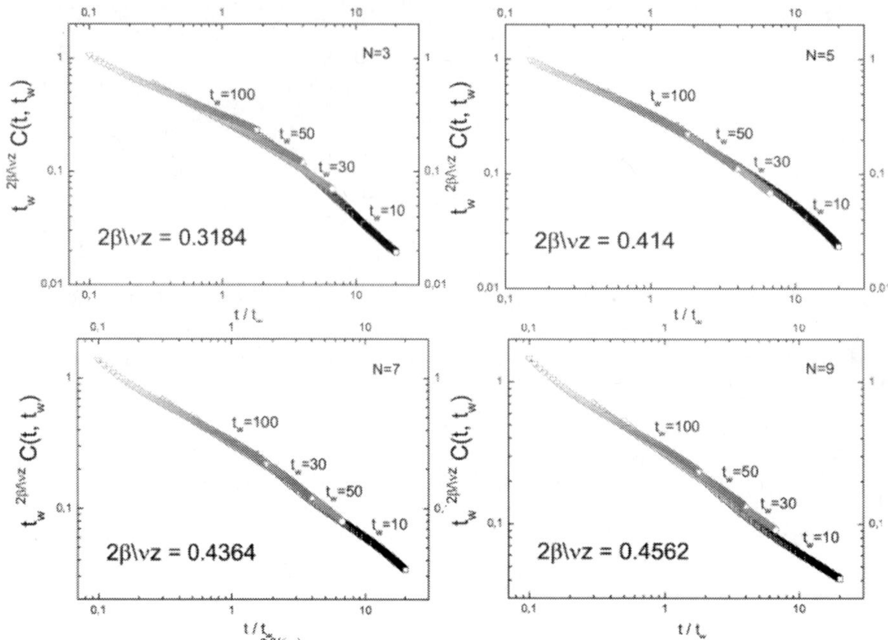

Fig. 4. Scaling functions $F_C(t/t_w) = t_w^{2\beta/(\nu z)}\, C(t,t_w)$ versus t/t_w for different Co film thicknesses $N = 3$ (a), 5 (b), 7 (c), and 9 (d) at the critical temperatures $T_c\,(N)$ for evolution from high-temperature initial state.

structure was realized with use of the anisotropic Heisenberg model for describing magnetic properties of ultrathin Co films. The calculations of the two-time dependent autocorrelation function $C(t,t_w)$ revealed the aging phenomena in multilayer structures in wider temperature range at $T \le T_c$ than in bulk systems by reason of the increase of characteristic spin-spin correlation length in magnetic nanostructures.

The analysis of $C(t,t_w)$ showed that aging effects are characterized by a slowing down of the correlation time in the system with increasing waiting time t_w. It was detected that these effects are increased in the low-temperature phase in comparison with slowing down of $C(t,t_w)$ at the critical temperature T_c. This phenomenon is connected with extremely slow dynamics in Co/Cu(100)/Co magnetic structures with Co films as quasi-two-dimensional *XY* systems.

The study of dependence of aging characteristics on thickness *N* of Co films revealed that the aging effects in multilayer structure are decreased with increasing thickness *N* of ferromagnetic films at the critical temperatures T_c (*N*) but characterized by the opposite dependence at temperatures T_c (*N*)/2.

This work was supported in part by the RFBR through Grant 17-02-00279 and by Grant MD-6024.2016.2 of Russian Federation President. Simulations were carried out on the SKIF-MSU of the Moscow State University and with use of computing resources of the "Applied theoretical physics and parallel computing laboratory" of Omsk State University.

References

1. Henkel M., Pleimling M., *Non Equilibrium Phase Transitions*, Vol. 2: Ageing and Dynamical Scaling far from Equilibrium (Theoretical and Mathematical Physics) (Springer, Heidelberg, 2010)
2. J.P. Bouchaud, E. Vincent, J. Hammann, M. Ocio, L.F Cugliandolo, Lect. Notes Phys. **492**, 184 (1997)
3. P. Calabrese, A. Gambassi, J. Phys. A **38**, R133 (2005)
4. P.V. Prudnikov, V.V. Prudnikov, E.A. Pospelov, P. N. Malyarenko, A.N. Vakilov, Prog. Theor. Exp. Phys. **2015**, 053A01 (2015)
5. T. Mukherjee, M. Pleimling, Ch. Binek, Phys. Rev. B **82**, 134425 (2010).
6. V.V. Prudnikov, P.V. Prudnikov, A.N. Purtov, M.V. Mamonova, JETP Lett. **104**, 776 (2016)
7. V.V. Prudnikov, P.V. Prudnikov, D.E. Romanovskii, JETP Lett. **102**, 668 (2015)
8. J. Bass, W.P. Pratt, J. Magn. Magn. Mater. **200**, 274 (1999)
9. V.V. Prudnikov, P.V. Prudnikov, D. E. Romanovskiy, J. Phys. D: Appl. Phys. **49**, 235002 (2016)
10. P.V. Prudnikov, V.V. Prudnikov, M. A. Medvedeva, JETP Lett. **100**, 446 (2014)
11. P.V. Prudnikov, V.V. Prudnikov, M.A. Menshikova, N.I. Piskunova, J. Magn. Magn. Mater. **387**, 77 (2015)
12. F. Huang, M.T. Kief, G.J. Mankey, R.F. Willis, Phys. Rev. B. **49**, 3962 (1994)
13. L. Berthier, P.C.W. Holdsworth, M. Sellitto, J. Phys. A **34**, 1805 (2001)
14. V.V. Prudnikov, P.V. Prudnikov, S.V. Alekseev, I.S. Popov, Phys. Met. Metallogr. **115**, 1186 (2014)

Structural and magnetic characteristics of the Co/Cu/Co thin-film systems

Elena Shalygina[1], *Anna* Kharlamova[1,*], *Andrey* Makarov[1], *Galina* Kurlyandskaya [2], *Andrey* Svalov[2]

[1]Faculty of Physics, Moscow State University, Moscow, Russia
[2]Ural Federal University, Ekaterinburg, Russia

Abstract. The results on investigation of structural and magnetic characteristics of Co/Cu/Co thin-film systems obtained by magnetron sputtering on glass substrates are presented. The thickness of Co layers in all samples is equal to 5 nm and the Cu layer is varied from 0.5 to 4 nm. It is found that the saturation field, H_S, oscillates in magnitude with increasing Cu layer thickness with the period of the order of 1 nm. The maximum values of H_S are observed for t_{Cu} = 1.4, 2.2 and 3.2 nm. The hysteresis loops measured for these samples in a magnetic field applied along the easy magnetization axis have a two-step form, and for other t_{Cu} – rectangular one. The obtained results are explained by the presence of exchange coupling between the ferromagnetic layers through a Co spacer and its oscillating behavior with changing t_{Cu}.

1 Introduction

Magnetic thin films attracted the attention of researchers since the mid of XX century. The interest to studying these materials is due to the peculiarities of their physical properties. In particular, thin-film systems, in which the size of one of the directions is comparable with the wavelength of an electron, can exhibit quantum-size effects, which manifest themselves in a change of the energy spectrum of the samples along this direction. As a result, the transport and magnetic properties of the samples are changed. For example, giant magnetoresistance (GMR) was discovered for three-layer and multilayer thin-film systems consisting of alternating ferromagnetic (FM) and nonmagnetic (without magnetic ordering, NM) layers of submicron thicknesses. It was found that the GMR value depends on the mutual orientation of the magnetization in the FM layers and may reach several tens of percent in multilayer systems (superlattices) [1]. Furthermore, the oscillatory behavior of the exchange interaction between FM layers with a variation of NM thickness was discovered [2-4]. Due to their unique properties, multilayer thin-film systems are widely used in devices of micro- and nanoelectronics, in particular, in various sensors, including sensors of magnetic fields [5]. Thin-film systems are also used to record and store information in the devices with high density and speed of recording [6].

A large number of experimental works are devoted to the study of the magnetic characteristics of multilayer (including a three-layer) thin-film systems. In particular, thin-film three-layer FM/NM/FM systems have been investigated in [3, 7-13], in which FM – Fe or Co, and NM – Ag, Au, Zr, Ta, Mo, Pt, Pd, Si, Bi, Cu. It was found that the values of saturation field, H_S, oscillate with changing nonmagnetic layer thickness, and the period of these oscillations (except systems with Bi) of the order of 1.0-1.8 nm in the thickness range of FM up to 10 nm and NM up to 5 nm.

Despite the existing large volume of experimental data, the study of the magnetic properties of thin-film magnetic structures still attracts the attention of researchers from both a scientific and a practical point of view. Herewith, the most important problem is the experimental analysis of the influence of the thickness and composition of magnetic and nonmagnetic layers on the magnetic properties of thin-film magnetic systems. It is obvious that for correct solution of this question the study of two- and three-layer samples are required, since the large number of interfaces in the multilayer structures usually complicate the interpretation of the experimental data. From this point of view, three-layer samples with two magnetic cobalt layers and a copper interlayer deserved attention. The aim of this work is the investigation of influence of Cu interlayer on the structure and magnetic properties of the Co/Cu/Co thin-film systems.

2 Sample and Experimental method

The Co/Cu/Co samples were grown by magnetron sputtering using Co and Cu targets and corning 2845 glass substrates. The Ta seed layers of 5 nm thick were deposited on the glass substrates. The background pressure in the vacuum chamber was $4{\cdot}10^{-7}$ mbar and the argon pressure was as high as $3.8{\cdot}10^{-3}$ mbar. A constant magnetic field of 250 Oe was applied parallel to the substrate, H_{SUB}, in order to form an in-plane easy magnetization axis (EMA). The thickness of the Co

Kharlamova* Corresponding author: anna-h-m@mail.ru

layers, t_{Co}, in all samples was equal to 5 nm. The thickness of the Cu layer, t_{Cu}, was varied from 0.5 to 4 nm. The microstructure of the Co/Cu/Co samples and their surface morphology were studied by X-ray diffraction (XRD) (using CuKα radiation) and an atomic force microscope (AFM). The near-surface and the bulk magnetic characteristics of the Co/Cu/Co thin-film systems were measured employing a magneto-optical magnetometer (MOM) by means of the transverse Kerr effect and the vibration sample magnetometer «Lake Shore VSM 7400» with the sensitivity up to 10^{-7} G•cm^3. The hysteresis loops of the studied samples were measured for two directions of the magnetic field, *H*. In particular, the orientation of *H* was parallel or perpendicular to the EMA (denoted, respectively, as D1 and D2 directions).

3 Results and discussion

The results of XRD measurements of Co/Cu/Co samples and «reference» monolayer Co thin film with t_{Co} = 50 nm, deposited directly on a glass substrate, are presented in Fig. 1. As can be seen in Figure 1 (b) X-ray spectra, obtained for the trilayer samples, don't have peaks corresponding to the lines of Ta. The presence of Ta seed layer enhances the texture of the Co film as the most notable remains only a peak corresponding to the line (002) of the hexagonal lattice of Co. The position of this peak for three-layer samples is slightly shifted to the left relative to the corresponding peak position for a single-layer Co film. It is evidence to the fact that at the boundary layers of Ta and Co is a partial mutual diffusion, which leads to a certain increase in the lattice parameter of Co, due to the large size of the Ta atom as compared to the Co atom. The peak on curve 3 in Fig.1 b, corresponding to the (111) line of the fcc lattice of Cu, becomes noticeable. It was found that the Co films have nanocrystalline structure. The size of crystallites, estimated from the half-width of the peaks, is of the order of Co film thickness.

According to AFM investigations, the average value of the surface roughness, *Ra*, of the studied samples does not exceed 0.5 nm that also confirms high quality of the surface. The results of magnetic measurements are presented in Figure 2: MOM data in Fig. 2 a, c, e, g, i and VSM data in Fig. 2 b, d, f, h, j. All measurements were carried out in open air at room temperature. The results of magnetic measurements of hysteresis loops allowed to receive the dependences of saturation field H_S on the t_{Cu} (Fig. 3). As can be seen in Figure 2 the hysteresis loops, measured in the magnetic field, applied along the D1 and D2 directions, differ from each other. It should be noted that the difference of the hysteresis loops was observed at measurements by using both the magneto-optical magnetometer and the vibration one. This fact indicates the formation of induced magnetic anisotropy with EMA parallel to the direction of magnetic field, applied during the deposition process. According to the commonly accepted notions [14], the main mechanism underlying the appearance of induced magnetic anisotropy is the pair ordering of atoms.

The hysteresis loops of all the samples, measured in the magnetic field, applied along the D2 direction, have an inclined shape. Herewith, the values of coercive field, H_C, and remanent magnetization are sufficiently low. In this case, the remagnetization of the samples is carried out mainly by the rotation of the magnetization vector.

The shape of hysteresis loops, measured in the magnetic field parallel to the D1 direction, depends on t_{Cu}. In particular, some samples were characterized almost by rectangular hysteresis loops (Fig. 2 a, b, e, f), while others (Fig. 2 c, d, g, h, i, j) – by more complex two-step loops. The values of H_S of the studied samples oscillate as a function of t_{Cu}. The peaks of H_S are observed at t_{Cu} = 1.4, 2.2 and 3.2 nm, and the hysteresis loops at these t_{Cu} have the two-step shape.

According to calculations, performed by K.A. Zvezdin in [15], and the experimental results [7-13, 16-19] the two-step and almost rectangular hysteresis loops are observed for antiparallel and parallel orientations of magnetization components in Co layers. Thus, the obtained experimental data can be explained by the existence of exchange interaction between the FM layers through the Cu layer and it oscillating behavior with the change of t_{Cu} (transition from the ferromagnetic (F) to the antiferromagnetic (AF) interaction).

The Co/Cu/Co samples with the ferromagnetic coupling between the magnetic layers are characterized by rectangular hysteresis loops. The remagnetization of these samples is similar to the processes of magnetization reversal of thin single- layer films, and can proceed through the irreversible growth of the remagnetization nuclei.

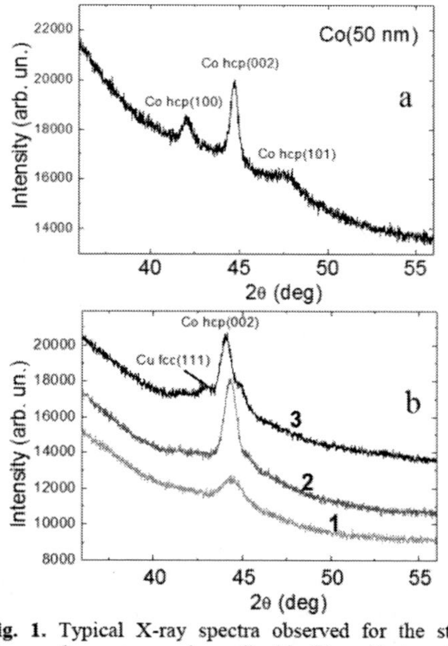

Fig. 1. Typical X-ray spectra observed for the studied samples: «reference» monolayer Co thin film with t_{Co} = 50 nm deposited directly on a glass substrate. (a), and samples based on Ta, Co and Cu (b): curve 1 – Ta/Co(5nm)/Ta, curve 2 – Ta/Co(5nm)/Cu(1nm)/Co(5nm), curve 3 – Ta/Co(5nm)/Cu(4nm)/Co(5nm).

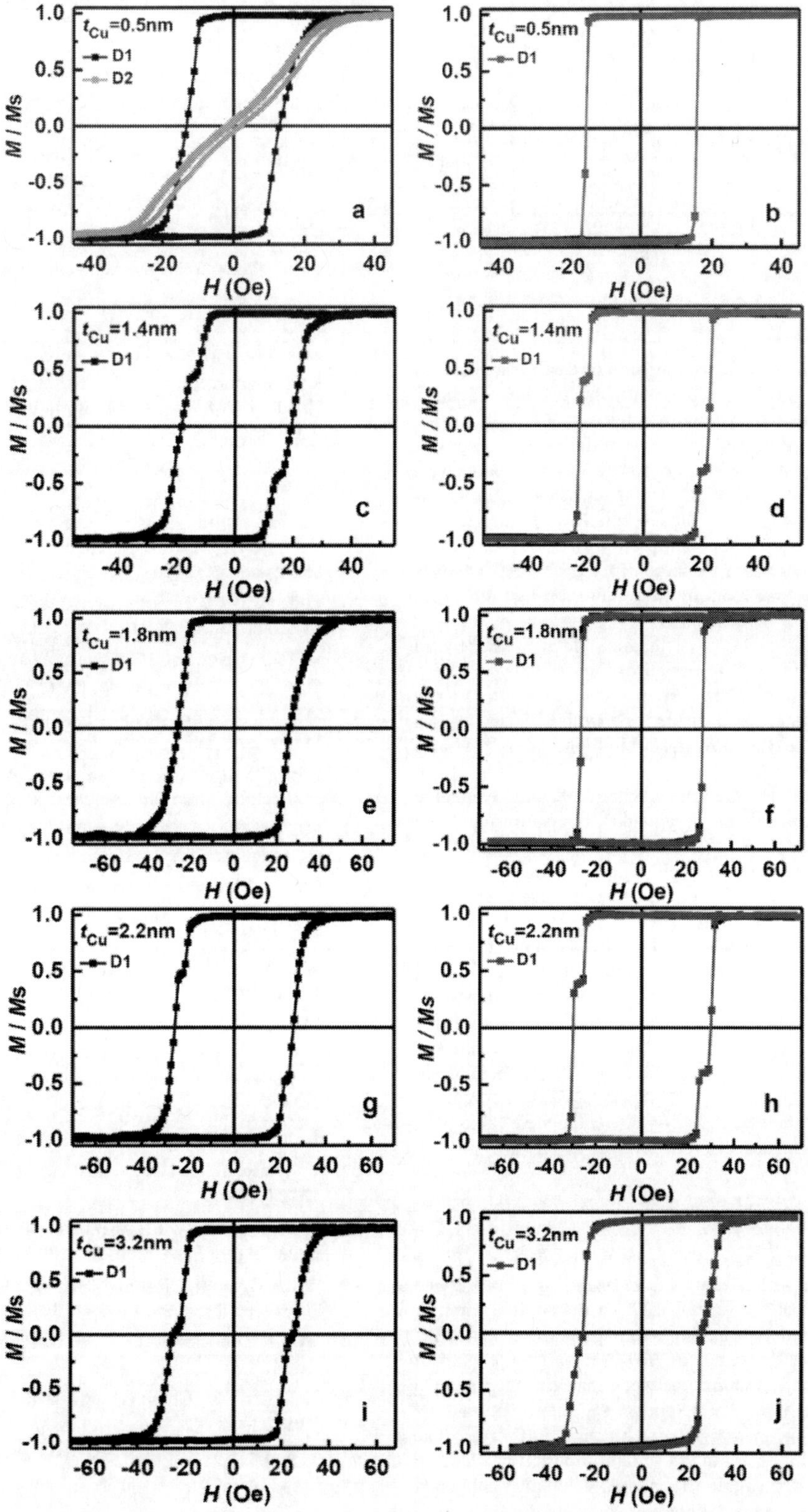

Fig. 2. Hysteresis loops of Co/Cu/Co samples measured employing magneto-optical (a, c, e, g, i) and vibration sample (b, d, f, h, j) magnetometers.

Kharlamova * Corresponding author: anna-h-m@mail.ru

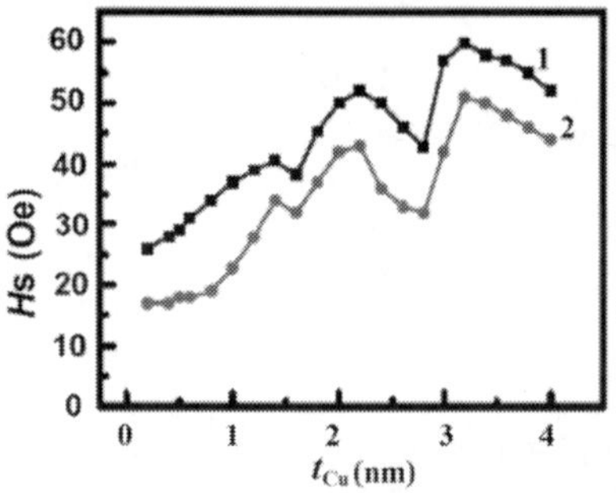

Fig. 3. Dependences of $H_S(t_{Cu})$ obtained by using data of magneto-optical (curve 1) and vibration sample magnetometers (curve 2).

The magnetization components in Co layers, parallel to the EMA, in the samples with the antiferromagnetic coupling have the opposite directions at $H = 0$, i.e., these thin-film structures are similar to uniaxial antiferromagnet. The saturation field of the samples with the antiparallel orientation of magnetization components in the Co layers (AF-exchange coupling) is higher than saturation field of the samples with the F-exchange coupling. It is caused by an additional energy, which is necessary for overcoming the AF-exchange coupling between the magnetic layers.

The analysis of experimental data presented in Figure 3 shows that the period of the oscillations of H_S, Λ, is of the order of 1nm. It should be noted that in the multilayer systems Co/Pt [16], Fe/Cr [17], Fe/Mo [18], Fe/Cu [19] the period of the oscillations of H_S was also of the order of 1 nm.

According to the mechanism of the exchange coupling between the magnetic layers through the nonmagnetic spacer layer via the RKKY interaction the period of the observed oscillations, Λ, must be of the order of π/κ_F [20-22], where κ_F – Fermi wave vector.

For most metals, π/κ_F is of the order of 0.3 - 0.4 nm. The experimentally found value of Λ is more than π/κ_F. The presence of the longer period of the oscillations of H_S in the thin-film systems with metallic nonmagnetic layers was attributed to the influence of the quantum size effects, manifesting in the change of the electronic structure of ultra-thin magnetic layer (the appearance of the so-called Quantum Well States, QWSs) as compared to with bulk samples [23, 24]. The calculated value of Λ with taking into account the QWSs is 1- 1.2 nm.

Conclusions

The investigation of magnetic properties of the Co/Cu/Co thin-film systems have been carried out. It had been found that the values of H_S of the studied samples show oscillatory behavior with increasing Cu layer thickness. The dependences $H_S(t_{Cu})$ have been explained by the exchange coupling between the ferromagnetic Co layers through the nonmagnetic Cu interlayer and its oscillatory behavior as a function of t_{Cu}. The experimentally determined oscillation period of H_S ($\Lambda \sim$ 1 nm) coincided with the calculated value of Λ, obtained in theoretical works taking into account the quantum-size effects, manifesting in the change of electronic structure of ultra-thin films as compared to the bulk samples.

This work was supported by the Russian Foundation of Basic Research, Grants 15-02-02077.

References

1. S.S.P Parkin., Z.G. Li, D.J. Smith, Appl. Phys. Lett. **51**, 2710 (1991).
2. S.S.P. Parkin, N. More, K.P. Roche, Phys. Rev. Lett. **64**, 2304 (1990).
3. S.S.P. Parkin, R. Bhadra, K.P. Roche, Phys. Rev. Lett. **66**, 2152 (1991).
4. R. Coehoorn, Ed. by K. H. J. Buschow. Elsevier, Handbook of Magnetic Materials **15**, 1–197 (2003).
5. A.V. Svalov, I.R. Aseguinolaza, A. Garcia-Arribas, I. Orue et al., IEEE Trans. Magn. **46** №2, 333 (2010).
6. S.I. Kasatkin, N.P. Vasilieva, A.M. Myraviev, Institute of Control Sciences RAS **168**, ISBN 5-85399-068-3 (2005).
7. Z.Q. Qiu, J. Pearson, S.D. Bader, Phys. Rev. B **46** № 13, 8659 (1992-I).
8. M.T. Johnson, R. Coehoorn, J.J. de Vries et al., Phys. Rev. Lett. **69** № 6, 969 (1992).
9. E.E. Shalyguina, Kyung-Ho Shin, J. Magn. Magn. Mater. **220**, 167 (2000).
10. E.E. Shalygina, E.A. Gan'shina, A.M. Kharlamova et al., Solid State Phenomena **233**, 653 (2015).
11. E.E Shalygina, A.V Svalov, A.M. Kharlamova et al., Japanese J. Appl. Phys. **55**, 07MF01 (2016).
12. E.E. Shalygina, A.M. Kharlamova, G.V. Kurlyandskaya, A.V. Svalov, J. Magn. Magn. Mat. **440**, 136 (2017).
13. W.F. Egelhoff, Jr. and M.T. Kief, IEEE Trans. Magn. **28** №5, 2742 (1992).
14. Chikazumi, Physics of Magnetism, New York–London–Sydney: John Willey&Sons, Inc. (2001)
15. A. Zvezdin. Phys. Solid State. **42**, 120 (2000)
16. C.J. Lin, G.L. Gorman, C.H. Lee, R.F.C. Farrow, et al., J. Magn. Magn. Mater. **93**, 194 (1991).
17. Y. Wang, P.M. Levy, J.L. Fry, Phys. Rev. Lett. **65**, 2732 (1990).
18. M.E. Brubaker, J.E. Mattson, C.H. Sowers, S.D. Bader, Appl. Phys. Lett. **58**, 2306 (1991).
19. W.R. Bennett, W. Schwarzacher, W.F. Egelhoff, Phys. Rev. Lett. **65**, 3169 (1990).
20. M.A. Ruderman, C. Kittel, Phys. Rev. **96**, 99 (1954).
21. P. Bruno, C. Chappert, Phys. Rev. Lett. **67**, 1602 (1991).
22. Y. Yafet, Phys. Rev. B **36**, 3948 (1987).
23. M.D. Stiles, Phys. Rev. B **48**, 7238 (1993).
24. P.J. Bruno, Phys. Rev. B **52**, 411 (1995).

Kharlamova* Corresponding author: anna-h-m@mail.ru

Phase Dynamics in Arrays of Coupled Vortex Spin-Torque Nano-Oscillators

Olga Katkova[*], *Ansar* Safin, *Nikolay* Udalov, *Mikhail* Kapranov

National Research University "MPEI", Department of Radio Engineering and Electronics, 111250 Moscow, Russia

Abstract. In this work, the mode analysis technique of complex networks nonlinear self-oscillatory vortex-based spin-torque nano-oscillattors (STNOs) with nonidentity and nonisochrony is developed. We construct adjacency matrices of different type of networks and calculate the normal modes. After the calculation of normal modes we shift to truncated equations for slowly varying amplitudes and phases in the normal coordinates using generalized quasi-Hamiltonian approach. Finally, we present the phase dynamics based on the Kuramoto-approach and compare different networks to the ability of synchronization.

1 Introduction

A great attention has attracted by studies of the phase locking in the arrays of spin transfer nano-oscillators (STNOs), which benefit from the spin-transfer torque phenomenon [1] and giant or tunnel magnetoresistance effects to generate microwave voltage signal. These oscillators in nanoscale can be easily tuned by a dc-current or external magnetic field, and compatible with CMOS-architectures. The enumerated features allow for consideration of STNOs for voltage or current controlled oscillators for telecommunication systems, and elements for building biological inspired computing systems [1]. Bio-inspired computing requires a capability to fabricate dense networks of interacting oscillators and to control the degree of coupling between STNOs. There are several studies of STNOs phase locking achieved by various physical mechanisms: through electrical connection in series of oscillators [2-4], by spin-wave propagation [5,6], by antivortices [7,8], and by dipolar coupling [9,10]. The synchronization and topological excitations in arrays of STNOs through the Kuramoto model described in [11]. Nevertheless, the theoretical description of the synchronization dynamics of STNO is more complicated than traditional limit-cycles oscillators (van der Pole oscillators, Josephson junctions, rotating pendula) which have constant orbit radius, and can be described in single oscillator case by the Adler equation. Due to technologically fabricated nonidentity of STNOs in the network [12] they can be expressed by generalized Adler-like equations.

In real networks there are as many frequencies (we will call them "modes") as the number of oscillators. When more than two oscillators are mutually coupled the finding of each mode structure and determination of their stability (or stability of group of them) becomes the main problem that has to be clarified. When the dissipation parameter is small enough, one can use the linear normal mode formalism for analyzing of the dynamics of the network of oscillators. If this condition is not satisfied, one can use the nonlinear normal mode formalism, which is much harder than the linear one. In general, the majority of these networks has the small dissipation parameter, and therefore it is possible to use the linear normal mode formalism for these systems.

The purpose of this paper is an analyzing of the structure of linear normal mode and mutual synchronization in different types of networks of nonlinear oscillators, from simple to complex cases. We investigate how network topology affects the stability of modes for the vortex-based STNOs. We show the impact of degeneration and synchronization of oscillation modes to the problem of optimal bio-inspired network design and microwave synthesizers. We find that hierarchical networks are characterized by a smaller number of stable nontrivial modes than partially hierarchical, or randomly organized networks. Our analysis gives rise to an approach for specifying of topological transformations of networks that can enhance synchronization.

2 Quasi-Hamiltonian Approach

We need to describe the nonlinear properties of a system of mutually coupled nonidentical and nonisochronous oscillator of complex networks from a common point of view. The quasi-Hamiltonian approach is sufficiently convenient for this description, especially for writing truncated equations (using averaging method). The quasi-Hamiltonian equations with N degrees of freedom are characterized by coordinates $q_1,\ldots,q_N$ and impulses $p_1,\ldots,p_N$. They satisfy the following equations:

[*] Corresponding author: arnellemorte@gmail.com

$$\frac{\partial q_j}{\partial t}=\frac{\partial H}{\partial p_j}+G_{1j};\frac{\partial p_j}{\partial t}=-\frac{\partial H}{\partial q_j}+G_{2j}, \quad (1)$$

where $G_{1;2j}$ are the dissipative functions, and H is the Hamiltonian of the system. If $G_{1;2j}=0$ then (1) gives simple Hamiltonian form of equations. Now lets change over to canonical variables $Q_j=\alpha_j q_j$, $P_j=p_j/\alpha_j$, where α_j are dimensionless constants (Q_j and P_j have the same dimension). After introducing the complex variables:

$$b_j=\left(Q_j+iP_j\right)/\sqrt{2};b_j^*=\left(Q_j-iP_j\right)/\sqrt{2}, \quad (2)$$

and using following expression for Hamiltonian (here $X_j=\{Q_j,P_j\}$):

$$\frac{\partial H}{\partial X_j}=\frac{\partial H}{\partial b_j}\cdot\frac{\partial b_j}{\partial X_j}+\frac{\partial H}{\partial b_j^*}\cdot\frac{\partial b_j^*}{\partial X_j}. \quad (3)$$

We obtain equations of motion in the form

$$\frac{\partial b_j}{\partial t}=-i\frac{\partial H}{\partial b_j^*}+F_j;\frac{\partial b_j^*}{\partial t}=i\frac{\partial H}{\partial b_j}+F_j^*. \quad (4)$$

Here $F_j=\left(\alpha_j G_{1j}+iG_{2j}/\alpha_j\right)/\sqrt{2}$ and the second equation in (4) may be obtained from the first one by complex conjugation. There is a wide range of changes from canonical variables b_j,b_j^* to other variables. The possibility of choosing various canonical variables is an advantage of the Hamiltonian approach.

Assuming b_j,b_j^* to be small, let us expand the function $H\left(b_j,b_j^*\right)$ in a power series in b_j,b_j^*. We are not interested in the zero term $H(0,0)$, and the first order of H is equal to zero. Therefore, the expression of H begins with the second order terms in the form

$$H=H_2+H_3+H_4+... \quad (5)$$

In the general form of network geometry we can write

$$H_2=\sum_{l,k=1}^{N}\left\{A_{lk}b_l b_k^*+0.5\left(B_{lk}^*b_l b_k+B_{lk}b_l^*b_k^*\right)\right\}. \quad (6)$$

The Hamiltonian (6) may be diagonalized by linear uv-transformation

$$c_j=\sum_{m=1}^{N}\left\{u_{jm}b_m+v_{jm}b_m^*\right\}. \quad (7)$$

After special choosing of matrices u_{jm}; v_{jm} in (7) the Hamiltonian (6) takes the form

$$H_2=\sum_{j=1}^{N}\omega_j c_j c_j^*+\sum_{k,m=1;k\neq m}^{N}\Omega_{km}c_k c_m. \quad (8)$$

Here $[\Omega_{km}]$ is the matrix of linear coupling between oscillators. This expression for H_2 we will use for calculating linear modes of oscillator networks. If the oscillator is nonisochronous (frequency is a function of the amplitude), we can obtain H_3 in a simple form

$$H_3=\sum_{j=1}^{N}\mathrm{E}_j c_j c_j^* c_j. \quad (9)$$

After substituting (7)-(9) in (4) equations of motion for c_j we will have the following view

$$\frac{\partial c_j}{\partial t}=-i\omega_j c_j-i\sum_{k=1}^{N}\Omega_{jk}c_k-iN_j\left|c_j\right|^2+\Upsilon_j, \quad (10)$$

where $\Upsilon_j=\sum_{m=1}^{N}\left[u_{jm}F_m+v_{jm}F_m^*\right]$. For several types of oscillators it is possible to write Υ_j in the simple form

$$\Upsilon_j(x)=\Upsilon_{j,-}(x)-\Upsilon_{j,+}(x). \quad (11)$$

Here $x=\left|c_j\right|^2$ - generalized power of complex variable c_j, $\Upsilon_{j,+}$ is a positive damping function (Hilbert damping in ferromagnetic or positive resistance in van der Pole scheme) and $\Upsilon_{j,-}$ is a negative damping function (spin-transfer torque effect in ferromagnetic or negative resistance in the van der Pole scheme). If the power x is small, we can expand the functions $\Upsilon_{-,+}$ in a power series. Now let us shift to the mode analysis of the fully identical nonlinear oscillators.

3 Mode Analysis & Truncated Equations

First, let us calculate normal modes λ_j of the complex networks. An unperturbed equation of (10) for $c_j^{\sim}$ is

$$\frac{\partial c_j^{\sim}}{\partial t}=-i\omega_j c_j^{\sim}-i\sum_{k=1}^{N}\Omega_{jk}c_k^{\sim}, \quad (12)$$

or for $\mathbf{c}=\left[c_1^{\sim},...,c_N^{\sim}\right]^T$ in the vector form

$$\frac{\partial \mathbf{c}}{\partial t}=-i\cdot\Xi\cdot\mathbf{c}, \quad (13)$$

with coupling matrix Ξ. These equations (12) or (13), we can transform into the canonical, using linear matrix transformation $\mathbf{c}=T\cdot\mathbf{a}$, $\mathbf{a}=\left[a_1,...,a_N\right]^T$ into the

$$\frac{\partial \mathbf{a}}{\partial t}=-i\cdot\left(T^{(-1)}\cdot\Xi\cdot T\right)\cdot\mathbf{a}. \quad (14)$$

For eigenmodes λ_j we need

$$T^{(-1)}\cdot\Xi\cdot T=diag\left[\lambda_1,...,\lambda_N\right]. \quad (15)$$

Form (14), (15) we can write

$$c_j^{\sim}=\sum_{m=1}^{N}t_{jm}a_m. \quad (16)$$

Now the unperturbed solution a_j could be expressed as follows (here U_j,ϕ_j are constants)

$$a_j(t)=U_j(t)e^{-i\lambda_j t-i\phi(t)}. \quad (17)$$

For the small coefficients $\Upsilon_{-,+}$, if we regard U_j,ϕ_j as slowly varying functions of time, then the dynamics of $U_j=U_j(t)$ and $\phi_j=\phi_j(t)$ can be represented by averaged equations or truncated equations.

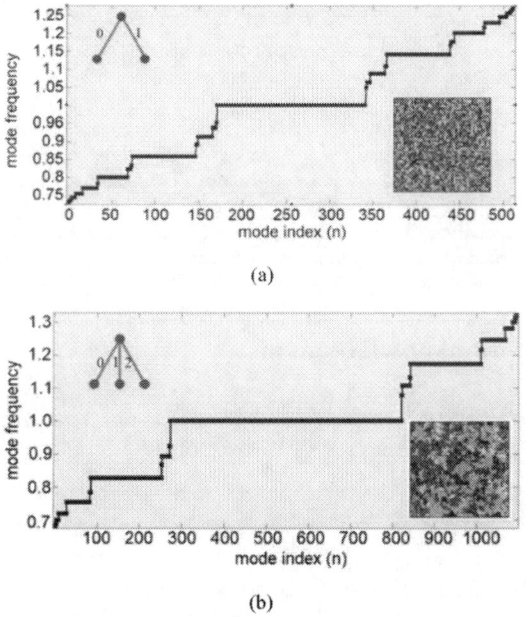

Fig. 1. Mode structures of the 2-adic (a) and 3-adic (b) networks of STNOs and phase clusters.

For the 2-adic and 3-adic tree-like networks (see Fig.1) we have the fractal structure of the normal modes spectrum, which looks like devils staircase fractal. The randomly and hierarchically distributed networks shown in Fig.2. We analyzed the typical examples of the networks with N = 25 number of oscillators.

The structure of linear normal mode of the hierarchical network shown in Fig.2a. Here we can see the nonsymmetrical structure of the dependence and one wide section of the degenerate normal modes, which is equal to 9. Other modes are nondegenerate (nonequal to each other). From the other hand if we move into the hierarchically distributed network (Fig.2b) with the number of modes is equal to the previous one 25, we will see the different picture. Here we have the four blocks of degenerate modes with L=5 for three blocks and L=10 for one block, where L is the number of degenerate modes. Therefore, in general the number of degenerate modes for the hierarchically distributed network is larger than for the randomly organized.

For the task of power summation and mutually phase locking, it is better to use the hierarchically organized network where the number of degenerate modes is high. From the other hand, for the task of multimode communication it is better to use randomly distributed networks, because the number of degenerate modes is low. These effects confirmed by the nonlinear mode analysis of the microwave oscillators with these two topological structures of networks

4 Phase Dynamics & Kuramoto approach

In the particular case, it is possible to use phase-approximation (Kuramoto-approach), when the amplitudes of oscillators signals are established faster than phases. Following this approximation for phases ϕ_j for each vortex STNO is

$$\frac{d\phi_j}{dt} = \omega_j + \sum_{k=1,k\neq j}^{N} \chi_{kj} \cdot \sin\left(\phi_k - \phi_j\right), \qquad (18)$$

where ω_j are eigenfrequencies of oscillator ϕ_j, χ_{jk} are parameters of connection between ϕ_j and ϕ_k, which are decreased with increasing of a distance between oscillators, that is $\chi_{jk} \sim 1/d_{jk}$, where d_{jk} is the distance between vortex STNO.

Analysis of received system of phase equations shows that besides the steady-state regime of equal phases there are also complicated chaotic modes and chimera states, where one part of oscillators is phase-synchronizned and another part is in chaotic regime. It is necessary that non-identify of vortex STNO sizes would be least (units of *nm* when diameter of the sample is hungreds of *nm*) for capacity addition.

In the Fig.3 the phase planes of three Kuramoto-oscillators for completely phase locking, partial phase locking of 1 and 2, of 1 and 3 and asynchronous mode shown. Using same approach we calculated the phase distribution of different networks which are shown in Fig.1,2. It is easy to see that the partially hierarchical network (Fig.2b) is characterized by the lower number of phase clusters.

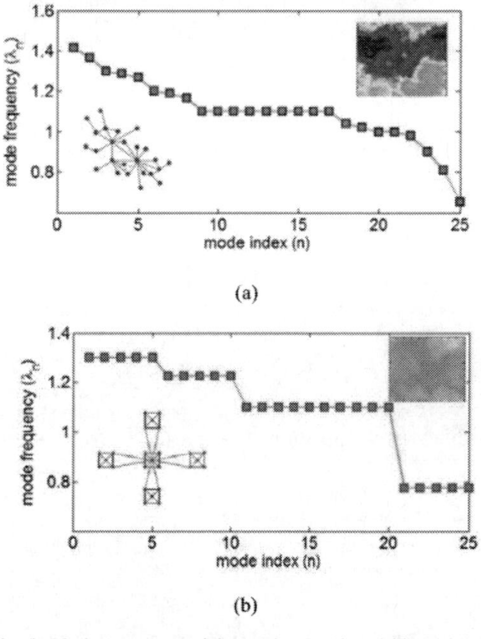

Fig. 2. Mode structures of the randomly (a) and hierarchically (b) distributed networks of STNOs and phase clusters..

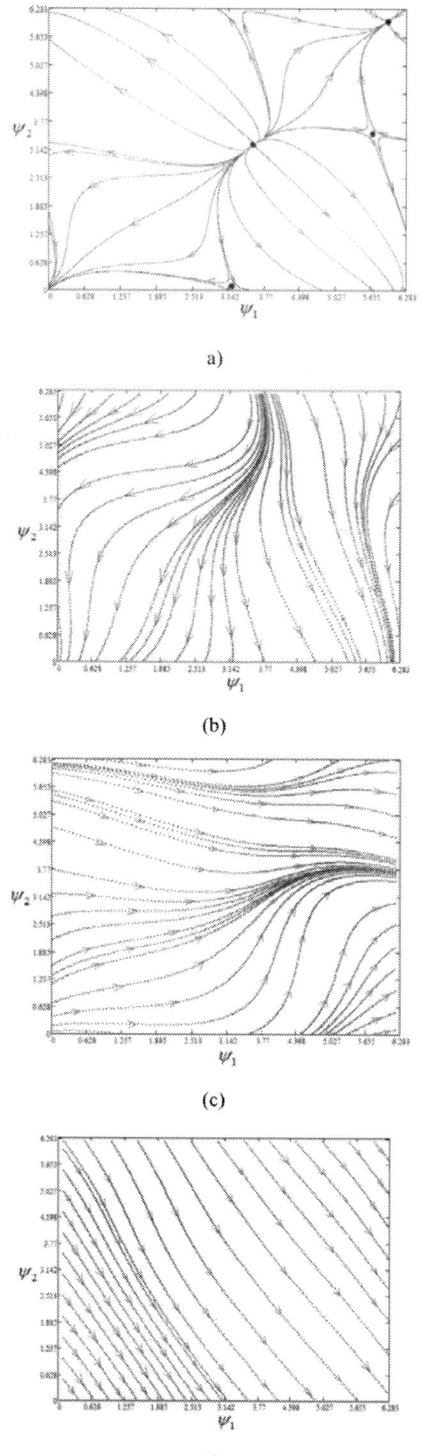

Fig. 3. Phase planes of three Kuramoto-oscillators for completely phase locking (a), partial phase locking of 1 and 2 (b), of 1 and 3 (c) and asynchronous mode (d).

5 Conclusion

In this work we presented mathematical model that allows us to calculate linear normal modes by using quasi-Hamiltonian approach. We demonstrated the fractal structure of the normal modes spectrum and wrote truncated equations for slowly varying amplitudes and phases in the normal coordinates. The stability analysis of the normal modes submitted. The relationship between a structure of a network that consists of Kuramoto-oscillators, and the corresponding number of phase clusters is shown.

6 Acknowledgment

The research was supported by a grant of the President of the Russian Federation for state support of young Russian scientists No.MK-7026.2016.8.

References

1. J. Grollier, D. Querlioz, and M.D. Stiles, *Proc. Of the IEEE* **104**, 10 (2016).
2. J. Grollier, V. Cros, and A. Fert, *Phys. Rev B.* 73, 060409(R) (2006).
3. B. Georges, J. Grollier, M. Darques, V. Cros, C. Deranlot, B. Marcilhac, G. Faini, and A. Fert, *Phys. Rev. Lett.* **101**, 017201 (2008).
4. B. Georges, J. Grollier, V. Cros, and A. Fert, *Appl. Phys. Lett.* **92**, 232504 (2008).
5. X. Chen, and R.H. Victora, *Phys. Rev. B* **79**, 180402 (2009).
6. S. Sani, J. Persson, S.M. Mohseni, Y. Pogoryelov, P.K. Muduli, A. Eklund, G. Malm, M. Kall, A. Dmitriev, and J. Akerman, *Nat. Commun.* **4**, 2731 (2013).
7. C.E. Zaspel, *Appl. Phys. Lett.* **102**, 052403 (2013).
8. A. Ruotolo, V. Cros, B. Georges, A. Dussaux, J. Grollier, C. Deranlot, R. Guillemet, K. Bouzehouane, S. Fusil, and A. Fert, *Nat. Nanotechnol.* **4**, 528 (2009).
9. A.D. Belanovsky, N. Locatelli, P.N. Skirdkov, F. Abreu Araujo, J. Grollier, K.A. Zvezdin, V. Cros, and A.K. Zvezdin, *Phys. Rev. B* **85**, 100409(R) (2012).
10. F. Abreu Araujo, and J. Grollier, *Journ. of Appl. Phys.* **120**, 103903 (2016).
11. V. Flovik, F. Macia, and E. Wahlstrom, *Sci. Rep.* **6**, 32528 (2016).
12. A.R. Safin, N.N. Udalov, M.V. Kapranov, *Eur. Phys. J. Appl. Phys.* **67**, 20601 (2014).

Microwave and magneto-optic properties of ε-Fe_2O_3 nanoparticles arising in borate glasses doped with Fe and Gd

Oxana Ivanova[1,2,*], *Janis* Kliava[3], *Irina* Edelman[1], *Ruslan* Ivantsov[1] and *Eleonora* Petrakovskaja[1]

[1]Kirensky Institute of Physics, FRC KSC RAS, Krasnoyarsk, 660036, Russia
[2]Siberian Federal University, Krasnoyarsk, 660036, Russia
[3] LOMA, UMR 5798 Université Bordeaux 1-CNRS, Talence cedex, 33405, France

Abstract. The ε-Fe_2O_3 nanoparticles in borate glasses co-doped with low concentrations of Fe_2O_3 and Gd_2O_3 were investigated with two experimental techniques – visible magnetic circular dichroism (MCD) and electron spin resonance (ESR). The most prominent features of the MCD spectra are: non-linear dependence of the spectral maxima intensities on the Gd concentration and strong increase with the temperature decrease. The ESR spectra of these glasses exhibit two features with g = 4.3 due to the diluted Fe^{3+} ions and 2.0 associated both with Fe^{3+} ions and with nanoparticles. The integrated spectra intensities do not follow the T^{-1} Curie law suggesting a considerable contribution from the magnetically ordered nanoparticles to ESR. Analysis of the MCD and ESR concentration dependences allow suggestion on the Gd ions incorporation into nanoparticles.

1 Introduction

Iron oxides are among the most used metal oxides with various applications in different scientific and industrial fields. Several stoichiometric phases of iron oxides are under intense study nowadays. The Fe_3O_4, γ-Fe_2O_3, and α-Fe_2O_3 phases have been found in nature as minerals magnetite, maghemite, and hematite, correspondingly. Despite the fact that they are known and serve humanity since ancient times, synthesis of these compounds in the form of nanoparticles, in particular as core-shell structures, allows to discover new properties and develop a lot of new practical applications on their basis. Two other phases β-Fe_2O_3 and ε-Fe_2O_3 are not found in nature and have been synthesized recently [1, 2] as nanoparticles. Structure and physical properties of the nanoparticles were shown to depend critically on the technology which creates difficulties in the identification of the particular iron oxide phase. This is especially important in the case of ε-Fe_2O_3 nanoparticles (e.g., [3-7]).

In the processes of iron oxide transformations under an action of the heat treatment ε-Fe_2O_3 is considered as an intermediate polymorph between γ-Fe_2O_3 and α-Fe_2O_3 [6]. The transformation temperature of ε-Fe_2O_3 to α-Fe_2O_3 is within the range of 500-750 °C in dependence on method of preparation. The particles ε-Fe_2O_3 size prepared by various methods varies within the range of 10-80 nm.

Recently, we have observed the formation of unusual magnetic nanoparticles in borate glasses co-doped with low concentrations of Fe_2O_3 and Gd_2O_3 [8] which were identified as ε-Fe_2O_3 nanoparticles. At that, only part of the observed properties of nanoparticles corresponded well to the properties of ε-Fe_2O_3 nanoparticles presented in literature. Besides, the role of Gd in the formation of nanoparticles and its localization in the glass remained unclear. To elucidate these questions, we have undertaken a study of these glasses with two experimental techniques – visible magnetic circular dichroism (MCD) and electron spin resonance (ESR). The MCD and the ESR are closely related to each other by the underlying physical phenomenon, viz., splitting of electronic energy levels due to breaking the degeneracy in an applied magnetic field. However, the electronic transitions involved in the MCD and the ESR are quite different. MCD, as a rule, is due to electrical dipole transitions between different orbital states, and it usually occurs in the optical wavelength range. ESR is associated with magnetic dipole transitions between different Zeeman sublevels of one and the same orbital state, and it usually occurs in the microwave frequency range. Besides, MCD is associated, mainly, with magnetically ordered phases while ESR includes a contribution of diluted paramagnetic ions, as well. So, these methods are expected to provide consistent and complementary data.

2. Experimental detail

Glasses of the basic composition K_2O–Al_2O_3–B_2O_3 were synthesized using a conventional glass technique. Prior to the synthesis, 1.5 wt.% Fe_2O_3 were loaded in the charge together with Gd_2O_3 in concentrations of 0.1, 0.2, 0.3, 0.4, 0.6, and 1.0 wt.% (samples 1-6,

* Corresponding author: osi@iph.krasn.ru

https://doi.org/10.1051/epjconf/201818503011

correspondingly) with respect to basic glass composition. The mixtures were molten at 1100–1300 °C under oxidizing conditions. The melts were poured onto steel sheets, cooled down in air to 380 °C. Then the glass plates were subjected to additional thermal treatment at 560 °C during 2 h.

The nanoparticles morphology and structure, and the elements distribution in a glass were studied with electron microscope JEM-2200FS, energy dispersive X-ray analysis (EDX), Fast Fourier transform (FFT). The MCD spectral, temperature, and magnetic field dependences were measured using the laboratory made device in the energy range 2-3 eV, at temperatures 90 and 300 K, and in magnetic fields up to 15 kOe. The ESR spectra were recorded in the X band (9.46 GHz) with a Bruker EMX spectrometer equipped with an ER4112HV variable-temperature unit (300 and 120 K).

3. Results and discussion

3.1 TEM

Particle formation in the samples is clearly seen in the TEM images. As an example STEM-HAADF image for sample 6 is shown in Fig. 1. Several meanings of the interplanar distances in the lattice of these particles obtained with the FFT analysis can be related to several modifications of iron oxide. The closest coincidence is observed between the obtained spacing's and the spacing's in disordered ε- Fe_2O_3 [5]. Elemental analysis and elemental mapping show the basic matrix elements (Al, K, and B) distribution to be practically homogeneous through sample with visible "holes" in places of the particles. Fe and Gd ions are detected, mainly, in the particle regions and in the trace amounts in the region free of particles.

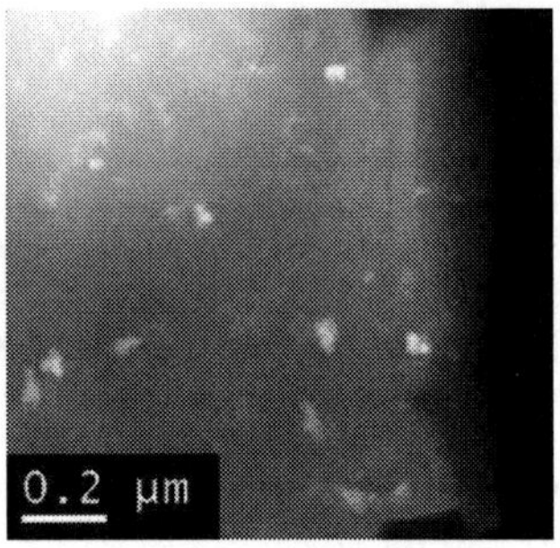

Fig. 1. STEM-HAADF image for samples 6.

Earlier we have already shown that the X-ray diffraction patterns contained several weak reflexes approximately the same for all samples [8]. The positions of the reflexes were closest to those for ε-Fe_2O_3 nanoparticles prepared by the wet chemical pore filling impregnation method [9]. Thus, electron microscopic data suggest that the nanoparticles can be referred to the phase ε-Fe_2O_3.

3.2 MCD

In Fig.2, there are presented the MCD magnetic field dependences for two samples with 0.6 and 1.0 wt.% concentrations of Gd_2O_3 measured at the light wave energy 2.5 eV close to a maximum in the MCD spectral dependence shown in Fig.3. A hysteresis at T=300 K was observed in all samples with a coercive field, H_c, in range 0.05 – 0.5 kOe. Parameters of the hysteresis loop such as H_c and remnant MCD value increase practically monotonic with the increase of the Gd concentration. At the same time, the MCD value in the maximal magnetic field 13 kOe changes non-monotonously with the Gd concentration increase (Fig.4).

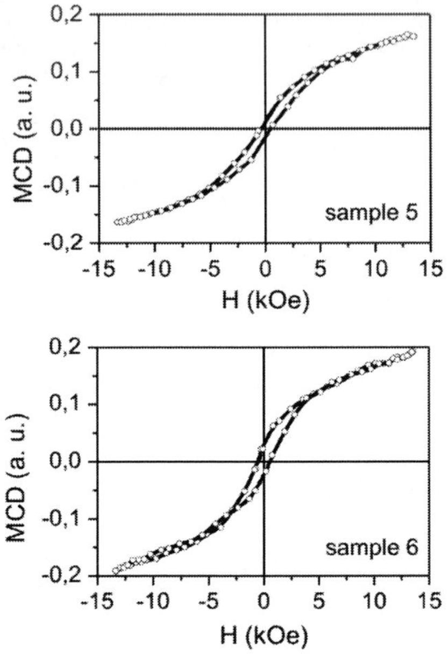

Fig. 2 MCD magnetic field dependences for samples 5 (top) and 6 (bottom) at 300 K (E=2.5 eV).

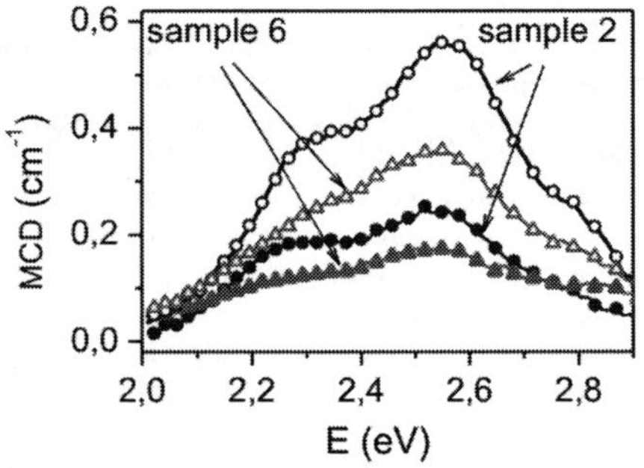

Fig. 3. MCD spectra of samples 2 (circles) and 6 (triangles) at 300 and 90 K (resp., full and empty symbols). H=3.5 kOe.

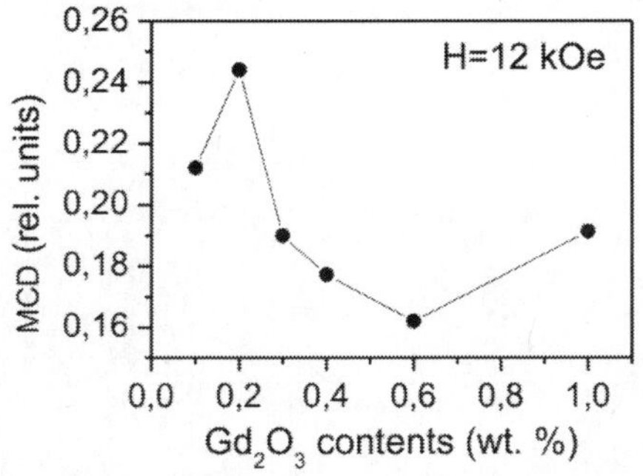

Fig. 4. MCD value in the field 13 kOe for all samples in dependence on the Gd concentration. Here the Gd oxide content in a charge is shown.

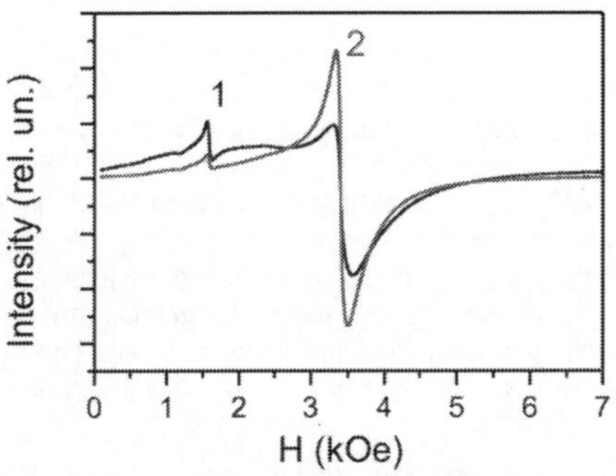

Fig. 5. ESR spectra for sample 3 at 120 (curve 1) and 300 K (curve 2).

The spectral MCD curves are shown in Fig. 3 for samples 2 and 6 at two temperatures. In the region 2.0-3.0 eV, the MCD spectra shape is the same for all samples 1-6 and is close to the shape of the γ-Fe_2O_3 MCD spectrum nanoparticles [10]. However, in the case of ε-Fe_2O_3 nanoparticles maxima are noticeable wider and their gravity centres are shifted to lower energies comparing to those of γ-Fe_2O_3. The most prominent difference between ε-Fe_2O_3 and γ-Fe_2O_3 nanoparticles is the anomalously strong increase of the MCD maximum amplitudes, by a factor of 2.2 – 2.5, with the temperature decrease.

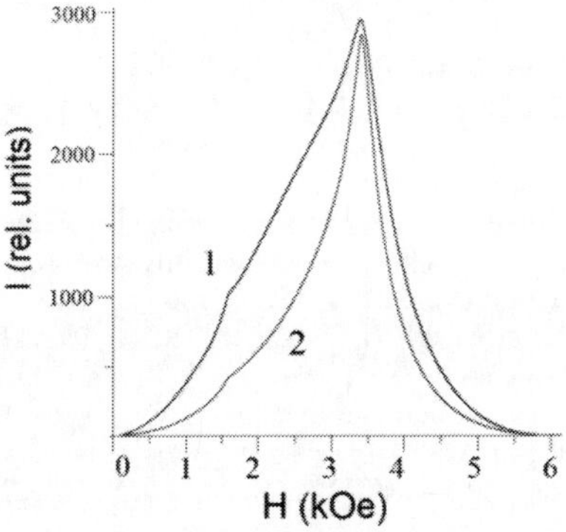

Fig. 6. Integrated experimental ESR spectra for sample 3 at 90 K (curve 1) and 300 (curve 2).

3.3 ESR

Typical differential and integral forms of the ESR spectra for sample 4 are shown in Figs. 5 and 6 at two temperatures. The differential spectra in all glasses exhibit two features with g = 4.3 and 2.0 due to diluted Fe^{3+} ions, and the lowest-field one, with g = 6.0, is due to Gd^{5+} ions. The line with g=2 can be associated also with some contribution of magnetic nanoparticles. Interestingly, with lowering the temperature the g = 2.0 feature is broadened whereas the g = 4.3 one is narrowed. The former can be due to nanoparticles. The integrated spectra intensities do not follow the T^{-1} Curie law suggesting that a considerable contribution to the resonances arises from magnetically ordered nanoparticles.

The double-integrated intensities of the experimental ESR spectra recorded at room temperature (Fig.7) tend to decrease with the Gd concentration increase. The similarity of the Gd concentration dependences of the MCD values and the double-integrated intensities of the ESR spectra allows supposition on the Gd incorporating into particles which is supported by the TEM data: Gd was detected in the regions of nanoparticles together with Fe. If it is so, gadolinium ions replace iron ions reducing the total magnetic moment of the iron sub-lattice and causing distortions in the crystal lattice, because of their much larger radius in comparison with the Fe^{3+} ionic radius.

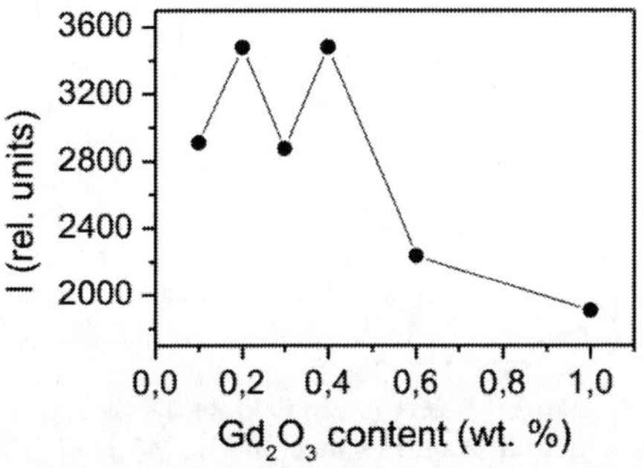

Fig. 7. Double-integrated intensities of the experimental ESR spectra at room temperature.

Distortions of the crystal lattice can lead to an increase in the coercive field and remnant magnetization, which is observed experimentally. The ε-Fe_2O_3 magnetic structure remains a subject of discussion up to now. Various authors attribute it to collinear ferrimagnet or a canted antiferromagnets. Besides, magnetic properties and resonance behavior of ε-Fe_2O_3 were shown to be modified upon partial Fe substitution by Al [11], Ga [12], or In [13-15]. Gd being the magnetic ion can effect in the ε-Fe_2O_3 magnetic properties even in higher extent. The study of the ESR temperature dependences of the ε-Fe_2O_3 containing glasses is in progress now.

4. Conclusions

The ε-Fe_2O_3 nanoparticles in borate glasses co-doped with low concentrations of Fe_2O_3 and Gd_2O_3 were investigated using transmission electron microscopy methods, visible magnetic circular dichroism, and electron spin resonance. Noticeable decreases of the MCD maximum amplitude and ESR intensity have been revealed at the Gd concentration increase. At the same time, the characteristic features of the MCD hysteresis loops, coercivity and remnant magnetization, increase with an increase of the Gd content. These observations evidence that the Gd ions can be incorporated into nanoparticles.

References

1. H. Braun, K.J. Gallagher, Nature, Phys. Sci. 240, 13 (1972)
2. H. Forestier and G. Guiot-Guillain, C. R. Acad. Sci. (Paris) **199** 720 (1934)
3. R. Zboril, M. Mashlan, D. Krausova, In Mossbauer Spectroscopy in Materials Science, Eds.; Kluwer Academic Publishers: Dordrecht, pp 49-56 (1999)
4. J. Kohout, P. Brazda, K. Zaveta, D. Kubaniova, T. Kmjee, L. Kubickova, M. Klementovaá, E. Santava, and A.Laneok, J. App. Phys. **117** 17D505 (2015)
5. A. Dubrovskiy, D. A. Balaev, K. A. Shaykhutdinov, O. A. Bayukov, O. N. Pletnev, S. S. Yakushkin, G. A. Bukhtiyarova, and O. N. Martyanov, J. Appl. Phys. **118** 213901 (2015)
6. Tronc, C.Chaneac, J.P.Jolivet, J. Sol. St. Chem. **139** 93 (1998)
7. S.S. Yakushkin, G.A. Bukhtiyarova, O.N. Martyanov, J. Struct. Chem. **54** 876 (2013)
8. O.S. Ivanova et al., J. Magn. Magn. Mater., **401,** 880 (2016)
9. G.A. Bukhtiyarova, O.N. Martyanov, S S. Yakushkin, M.A. Shuvaeva, O.A. Bayukov, Phys. Sol. St. **52** 826 (2010)
10. I.S. Edelman, O.S. Ivanova, E.A. Petrakovskaja, D.A.Velikanov, I.A. Tarasov, Y.V. Zubavichus, N.N. Trofimova, V.I. Zaikovskii, J. All. Com. **624** (2015) 60
11. A. Namai, S. Sakurai, M. Nakajima, T. Suemoto, K. Matsumoto, M. Goto, S. Sasaki, S. Ohkoshi, J. Am. Chem. Soc. **131** 1170 (2009)
12. S. Ohkoshi, S. Kuroki, S. Sakurai, K. Matsumoto, K. Sato, S. Sasaki, Angew. Chem. Int. Ed. **46** 8392 (2007)
13. M. Yoshikiyo, A. Namai, M. Nakajima, K. Yamaguchi, T. Suemoto, and S. Ohkoshi, J. Appl. Phys. **115** 172613 (2014)
14. A. I. Dmitriev, O. V. Koplak, A. Namai, H. Tokoro, S. Ohkoshi, R. B. Morgunov, Phys. Solid State **55** 2252 (2013)
15. Dmitriev, O. V. Koplak, A. Namai, H. Tokoro, S. Ohkoshi, R. B. Morgunov, Phys. Solid State **56** 1795 (2014).

The reported study was funded by Russian Foundation for Basic Research, Government of Krasnoyarsk Territory, Krasnoyarsk Regional Fund of Science to the research project № 17-42-240080».

On Nature of the Induced Anisotropy in the Chemically Deposited Co–P Films

Anatoly V. Chzhan [1,2,*], *Sergey A.* Podorozhnyak [1], *Mikhail N.* Volochaev [3], *and Gennady S.* Patrin [1,3]

[1]Siberian Federal University, Krasnoyarsk, 660041 Russia
[2]Krasnoyarsk State Agrarian University, Krasnoyarsk, 66049 Russia
[3]Kirensky Institute of Physics, Federal Research Center KSC SB RAS, Krasnoyarsk, 660036 Russia

Abstract. The induced magnetic anisotropy in the Co–P films prepared by chemical deposition from liquid solutions in a magnetic field has been investigated. The pronounced difference between the anisotropy values and relaxation times points out the existence of different physical mechanisms responsible for the occurrence of anisotropy of the Co–P films at low and high pH values. Possible mechanisms of the induced anisotropy in the Co–P films prepared at different pH values are discussed.

Introduction

The magnetic anisotropy is one of the fundamental properties of ferromagnets, since it strongly affects characteristics of these materials. In contrast to the crystallographic anisotropy, the induced anisotropy can be created artificially during fabrication of samples or upon subsequent temperature annealing in an applied magnetic field. The nature of induced anisotropy and its effect on the magnetic properties of samples are often arguable. This is caused by the existence of different physical mechanisms responsible for the occurrence of the induced anisotropy. Therefore, elucidation of the anisotropy origin needs thorough investigations.

The aim of this study was to determine the effect of acidity of working solutions on the value and dynamics of induced anisotropy in the Co–P films prepared by chemical deposition.

1 Experimental

Thin-film samples for investigations were formed by chemical reduction of Co and P from aqueous solutions [1]. The films were deposited onto glass substrates in a dc magnetic field of H = 2.5 kOe, applied to the film's plane. The film thickness was ~30 nm. The solution acidity was changed by adding alkalies in different concentrations [2].

The film thickness and composition were determined by X-ray spectral analysis. The film microstructure was studied on a Hitachi HT-7700 transmission electron microscope (TEM) equipped with a Bruker X-Flash 6T/60 energy-dispersive detector.

The constant of induced anisotropy K_U is determined using a rotational anisometer in magnetic field H = 10 kOe. The anisotropy field H_K is detected through analyzing the magnetic hysteresis loops at the frequency of 50 Hz. To establish time evolution of the induced anisotropy, the films were additionally annealed in a dc magnetic field of H = 3 kOe in the direction perpendicular to the easy magnetization axis.

2 Results

Figure 1 shows the dependence of the induced anisotropy constant K_U and coercivity H_C on the solution acidity.

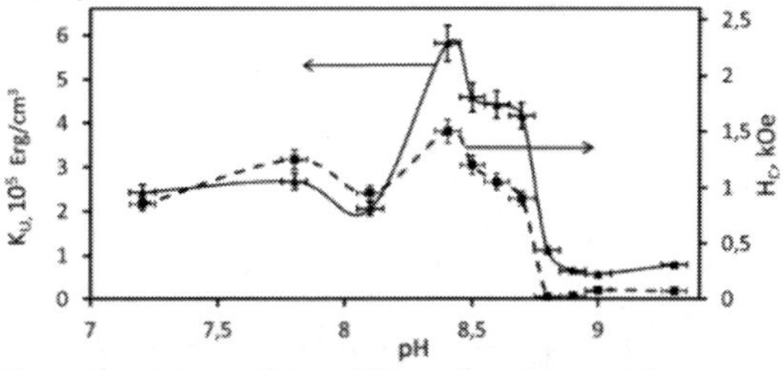

Fig. 1. Dependence of K_U and H_C on the solution acidity.

In the low-pH region, we have $K_U \sim 2.5 \times 10^5$ erg/cm^3. As the pH value increases (pH > 7,2), the anisotropy grows and attains about 6×10^5 erg/cm^3 in the vicinity of pH ~ 8.5. With a further increase in the acidity value, the K_U value abruptly decreases up to 5×10^4 erg/cm^3.

Simultaneously with pH growth up to ~ 8.5, H_C increases from 1 kOe to 1.5 kOe, and then H_C abruptly drops stepwise up to several oersteds with further pH growth.

The saturation field H_S in the films with uniaxial anisotropy being magnetized along the light direction practically coincides with the value of H_C, caused by the high degree of the hysteresis loop's rectangularity.

The observed variation in the induced anisotropy is consistent with the behavior of coercivity. As we

* Corresponding author: avczh@mail.ru

demonstrated in [3], the high-coercivity region corresponds to the submicrocrystalline film structure, while the low-coercivity region corresponds to the nanocrystalline. The region of sharp coercivity variations corresponds to the transitions between these structures.

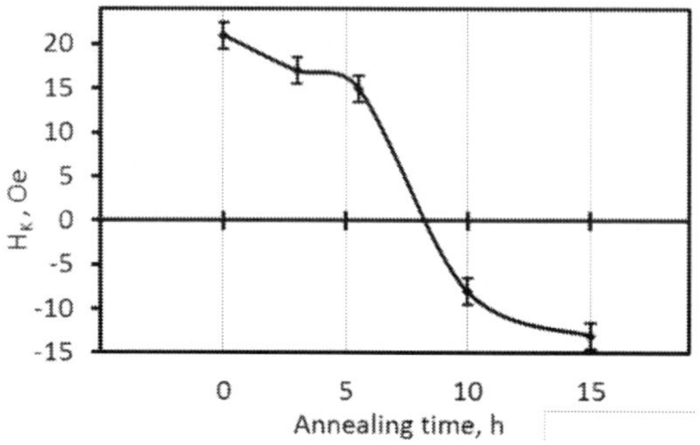

Fig. 2. Dependence of the anisotropy field in the Co–P films on the annealing time. The annealing temperature is 100°C.

The difference in the nature of induced anisotropy in the films formed at the low and high pH values can be followed not only from the anisotropy values, but also from the dynamics of the anisotropy filed upon annealing of the film in a magnetic field. With an increase in the time of annealing of the low-coercivity film, the initial anisotropy decreases to zero and then grows in the applied field direction. Annealing of the high-coercivity film in a magnetic field perpendicular to the initial easy magnetization axis at T = 300°C did no noticeably change the initial anisotropy value.

3 Discussion

The relaxation character of the variation in the anisotropy field shows that the induced anisotropy in the films fabricated at the high pH values originates from the diffusion processes of directional ordering. These processes are most frequently described within the theory of redistribution of pairs of magnetic ions along the preferred direction formed by an applied field. In our case, the diffusion processes are related to the redistribution of axes connecting the dipolar configurations of Co ions upon variation in the magnetization orientation under the action of a magnetic field.

The induced anisotropy relaxation time is determined by the Arrhenius ratio

$$\tau = \tau_0 \cdot \exp(E_A / kT) \qquad (1)$$

where E_A is the activation energy, k is the Boltzmann constant, and τ_0 is the inverse probability of hoppings at high temperatures.

At room temperature, the Co–P film is characterized by $\tau \sim 200$ days [4], while at T = 100°C, by $\tau \sim 3$ h. Then, we have $E_A \sim 1$ eV and $\tau_0 \sim 10^{-9}$.

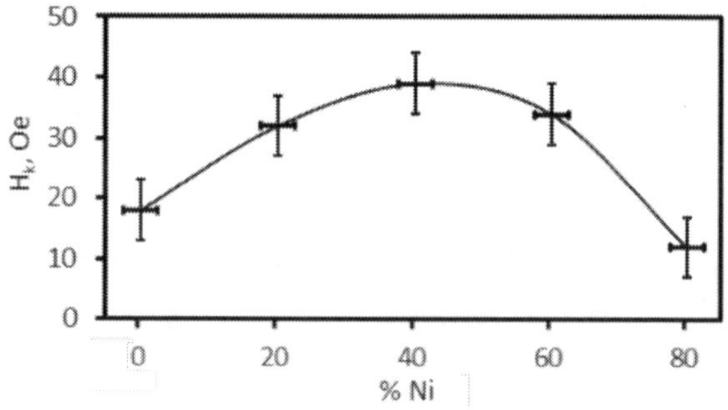

Fig. 3. Dependence of the K_U value on the Ni content.

The pair mechanism is confirmed by the change in the H_K value upon embedding of Ni ions in the Co lattice. As the Ni content increases to ~50%, the H_K value grows and then, with a further increase in the Ni content, drops (Fig. 3). The observed changes in the anisotropy field in the pair ordering model are explained by the fact that the contribution of pairs of different ions to the anisotropy is larger than the contribution of pairs of identical ions [5, 6].

The non-relaxation character and high K_U values in the films obtained at the low pH values (up to 18% of K_U for pure cobalt, which amounts to 3.3×10^6 erg/cm^3 [7]) are indicative of the fact that the induced anisotropy is stimulated, in this case by another mechanism. According to [8], it can be (i) the growth anisotropy, (ii) elastic strains, or (iii) structural changes in the cobalt lattice upon variation in the solution acidity.

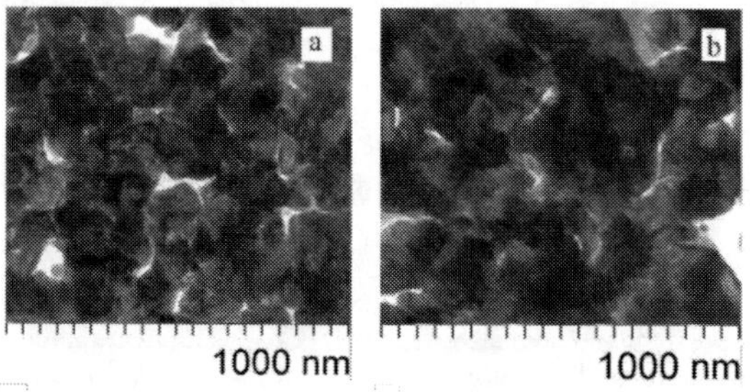

Fig. 4. TEM image of the high-coercivity films deposited in (a) zero and (b) nonzero magnetic field.

The uniaxial magnetic anisotropy caused by the directional film growth in a magnetic field was observed in the Co–P–Ni films prepared using the electrolytic technique [9]. This anisotropy is caused by the Lorentz forces that influence moving ions from the side of magnetic field and ensure the directional growth of cobalt crystallites.

Comparing electron microscopic photos of the CoP films' surfaces (Fig 4a, 4b), manufactured by the method of chemical deposition with or without magnetic field, let us conclude that magnetic field does not influence on shape and size of the crystallites. In both cases crystallites are of irregular shape caused by the island character of film's growing [3].

The established differences between the films grown by the chemical and electrolytic techniques are related to the fundamental difference between the mechanisms of film deposition onto a substrate. During chemical deposition, the ionic current only flows in the solution-substrate contact region and its value is very low.

The stresses induced during the film growth in a magnetic field can be caused by the occurrence of large crystallites and film structure compacting due to disappearance of spacers (Fig. 4). In this case, the stress-induced anisotropy constant is $K_U = 3\lambda\sigma$. The magnetostriction constant λ of hcp Co is $\sim 10^{-5}$ [10]. Then, at $K_U = 6 \times 10^5$ erg/cm^3 the stresses σ should attain $\sim 10^{10}$ dyn/cm^2, which is beyond the film strength limits.

As was mentioned above, the K_U value significantly changes in the region of the transition of the film structure from submicrocrystalline to nanocrystalline. The structural transformations are the probable reason of the sharp drop of the induced anisotropy in the Co–Pfilms with increasing pH value. To fully understand this mechanism, a more detailed structural examination should be performed.

Conclusions

Based on the data obtained, we can make the following conclusions. The observed changes in the induced anisotropy in the chemically deposited Co–P films are related to the structural variations determined by the pH values of the solutions used. The sharp difference between the K_U values and anisotropy relaxation times is indicative of the existence of different physical mechanisms responsible for the occurrence of the induced anisotropy of the Co–P films at low and high pH values.

References

1. A. Brenner and G. Riddell, J. Res. Nat. **37**, 31 (1947)
2. A. V. Chzhan, T. N. Patrusheva, S. A. Podorozhnyak, V. A. Seredkin, and G. N. Bondarenko, Bull Russ. Acad. Sci. Phys. **80**, 11 (2016)
3. A. V. Chzhan, S. A. S.A.Podorozhnyak, G. N. Bondarenko, M. N. Volochaev, and G. S. Patrin, Phys. Sol. St. **59**, 7 (2017)
4. A. V. Chzhan, G. S. Patrin, S. Ya. Kiparisov, and V. A. Seredkin, J. Siberian University. Mathematics and Physics **3**, 1 (2010)
5. S. Taniguchi, Sci. Rep. Res. Inst. Tohoku Univ. **7A**, 269 (1955)
6. L. Neel, J. Appl. Phys. **30**, 4 (1959)
7. S. Chikazumi, Physics of Ferromagnetism, M. (1987), 497 p.
8. A. G. Lesnik, Induced Magnetic Anisotropy, Kiev, Naukova dumka (1976), 163 p.
9. V. Georgescu and M. Daub, Surf. Sci. **600,** 4195 (2006)
10. A. W. Simpson and D. R. Brambley, Phys. Stat. Sol. (b) **43,** 291 (1971)

Magneto-Optical Response Amplification In Multi-Layer Nanocomposite-Semiconductor Structures

Victoria Buravtsova [1, •], *Elena* Gan'shina [1], *Yuri* Kalinin [2] and *Alexander* Sitnikov [2]

[1]Moscow State University, Faculty of Physics, Moscow, 119991, Russia
[2] Voronezh State Technical University, Voronezh, 394026, Russia

Abstract. The spectral and field dependences of the Transversal Kerr Effect (TKE) for $(Co_{40}Fe_{40}B_{20})_Z(SiO_2)_{100-Z}$/Si (or C) multilayer (ML) nanocomposite-semiconductor structures were studied depending on the concentration of the nanocomposite layer ferromagnetic phase (Z), chemical of semiconductor layer and its thickness. It was found that under certain conditions an addition of thin non-magnetic semiconductor layer leads to the increase of a multilayer magneto-optical (MO) response and softening of the magnetic properties of ML structure, up to the appearance of ferromagnetic (FM) ordering. The observed anomalous changes in magnetic and MO properties are well correlated with abrupt changes in the electric properties of ML structures. Decreasing of ferromagnetic phase concentration Z in the nanocomposite layer as well as increasing of the semiconductor layer thickness led to the disappearance of anomalous behavior in the MO properties. The results obtained are discussed taking into account the formation of a new magnetic phase on the FM granule-semiconductor interface.

Introduction

The desire in successful materials creation for spintronics is a combination of semiconducting and magnetic properties in one material. From this perspective there is a promising way of a multilayer systems creation on the basis of ferromagnetic metal and semiconducting layers. Nevertheless, the interpretation of results from such structures is very difficult with the main complexities related to the high influence of interfaces quality and diffusion processes between dissimilar phases on the resulting properties of formed metal–semiconductor compounds [1-6]. A microstructure of the FM/ semiconductor ML film can vary significantly, depending on the alternating layers thickness and their chemical activity; and the variations in the MO response from such ML structures also will be substantially different. As a magnetic layer a granular nanocomposite with concentrations before the percolation threshold can be used. This allows to reduce the influence of granule contacts with the neighbouring layers on macroscopic properties of the whole system and, on the other hand, to reduce the possibility of the diffusive formation of metallic-semiconductor spacers on the composite-semiconductor interface due to the presence of a barrier dielectric layer. Thus, experimental methods that can provide information about the microstructure of such materials and disclose peculiarities of the magnetic interactions inside of them become very actual.

Optical and magneto-optical methods are the most simple, effective and informative for nanostructures studies. Magneto-optical methods have several advantages, unlike optical ones, such as sensitivity to the presence of magnetic inhomogeneities, a particle shape and size changes, their volume distribution, and appearance of new magnetic phases in a nanostructure. Previous studies of nanocomposite-silicon multilayers $[(Co_{45}Fe_{45}Zr_{10})_Z(Al_2O_3)_{100-Z}(X)/Si(Y)]_n$ [7-9] with composite concentrations near and under the percolation threshold have shown that the addition of silicon layers leads to anomalous behaviour of electrical, magnetic and magneto-optical properties of ML with small thickness of Si (< 2nm). In [9] it was assumed that TKE spectra and anomalies in field dependences were related to competition of contributions from several magnetic phases.

In the present work we focus on MO properties of nanocomposite-semiconductor multilayers $(CoFeB)_z(SiO_2)_{100-z}$/Si and $(CoFeB)_z(SiO_2)_{100-z}$/C with Z = 33.9 and 40 at.%[a] in order to trace the influence of the chemical composition and the thickness of ML layers on their MO behavior.

Samples and experimental details

Nanomultilayer systems nanocomposite-silicon and nanocomposite-carbon were prepared by ion-beam deposition with different combination of granular and semiconductor layer thicknesses (see Table 1). A sketch of a model structure is shown in Fig. 1.

• Corresponding author: v.e.buravtsova@gmail.com

[a] The percolation threshold in the bulk nanocomposite $(CoFeB)_z(SiO_2)_{100-z}$ takes place at 43 at.%. [10].

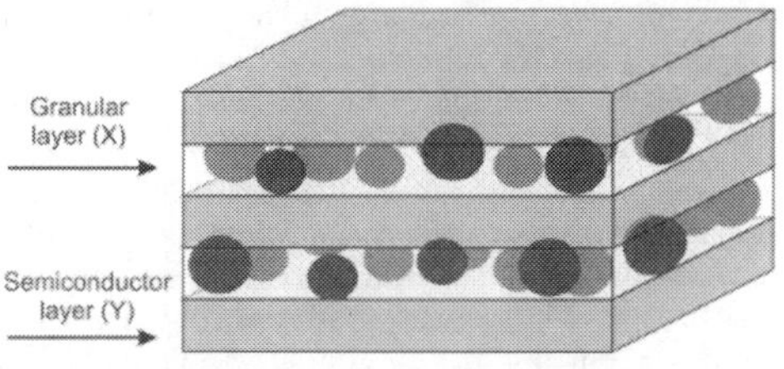

Fig. 1. Model of a nanocomposite-semiconductor multilayer structure.

By a variation of the forming layers thickness, their chemical composition and the concentration of the FM phase in the granular layers, it is possible to control the contributions from interfaces to the resulting magneto-optical response.

Magneto-optical properties of the samples have been measured in the Transversal Kerr Effect (TKE) geometry.

The value $\delta = \dfrac{I(H) - I(-H)}{2I(0)}$, where *I(H)* and *I(0)* are the reflected p-polarized light intensities in the presence and absence of a magnetic field, respectively, was directly measured in the experiment. The TKE spectra, δ(hν), were obtained in the energy range of hν = 0.5 - 4.0 eV at the light incidence angle of 68° and in the magnetic fields up to 3.0 kOe. The TKE value depends on magnetization linearly and we have measured TKE dependences on magnetic field, δ(H), at some fixed energies to characterize magnetic state of the ML structures.

Table 1. Number of series, chemical composition, structure formula and thickness of based layers.

№	Multilayer structures	X, nm	Y, nm	Z, at.%
1	$(Co_{40}Fe_{40}B_{20})_z(SiO_2)_{100-z}(X)/Si(Y)$	2.8 – 4.8	0.2 – 3.5	40
2	$(Co_{40}Fe_{40}B_{20})_z(SiO_2)_{100-z}(X)/Si(Y)$	2.0 – 4.9	0.2 – 4.5	33.9
3	$(Co_{40}Fe_{40}B_{20})_z(SiO_2)_{100-z}(X)/C(Y)$	2.8 – 7.9	0.2 – 1.9	33.9

Results

Field, spectral and isoenergetic TKE dependences for multilayer structures from the first series (with the concentration of the magnetic phase Z in the granular layer near the percolation threshold) are shown in Figure 2 and 3. It was found that an addition of a thin non-magnetic semiconductor interlayer leads to anomalous behaviour of resulting magnetic and magneto-optical properties of the multilayer structure of this series. It can be seen that with silicon thickness increasing the shape of the magnetization curve evolves from superparamagnetic (Y = 0.19 nm) to ferromagnetic one (Y = 0.92, 2.47 nm) with the saturation field of less than 1 kOe (Fig. 2A). Meanwhile, the negative effect amplitude grows (Fig. 2B) and reaches its maximum at Si thickness of approximately 0.6 nm. This is also clearly seen at the isoenergetic dependences of the effect on Si thickness (Fig. 3A). A further increase of the interlayer thickness leads to a weakening of the effect and a reduction of the magnetization curve back to the superparamagnetic type. The spectrum shape changes as well: the effect becomes positive over the entire wavelength range, the spectrum shape becomes similar to the bulk nanocomposite with CoFeB granules in a complex SiO_2+Si matrix or/and silicides like in [11].

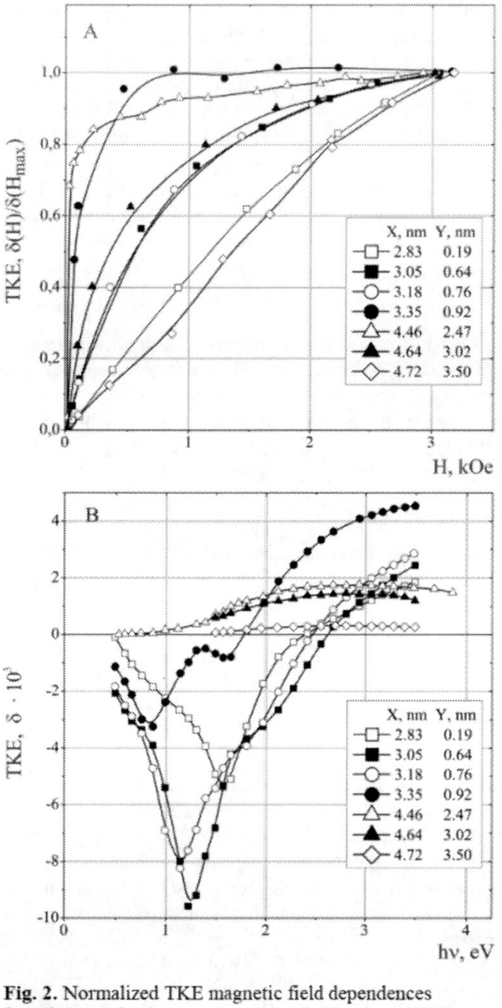

Fig. 2. Normalized TKE magnetic field dependences $\delta(H)/\delta(H_{max})$ for the wavelength hν=2 eV (A), and TKE spectra (B) for series 1.

Such behavior of the first ML series can be explained by a fact that a new nanocomposite (CoFeB) in a complex matrix (SiO_2+Si) is formed on the interface of ferromagnetic granule-silicon, as in the case considered in [11]. The percolation threshold in this new nanocomposite is reached at approx. 0.6 nm silicon thickness. For the same silicon thickness one can see that TKE in the near-IR range reaches its maximum value and there is a good correlation with the beginning of the electrical percolation (Fig.3 B).

The formation of Si layer occurs over the CoFeB granules due to the surface energy difference between metal granules and the composite layer matrixes [7]. The first atomic layer of the semiconductor can cooperate with metal in the form of silicide with a high probability, and then can grow on a surface in the form of island structures. At the early stage of semiconductor island growth, when the surface area occupied by the islands is smaller, than that occupied by metallic granules emerging at the interface two channels of electrotransport are possible. The first is a granule–insulator–granule way, as in a common bulk granular nanocomposite. The second (new) way of electrotransport is a granule– semiconductor–insulator–granule channel. The resulting island structure of the semiconductor layer will have a minor impact on the resistivity of the whole ML up to a thickness at which the semiconductor layer becomes continuous over a distance larger than the intergranular spacing. Under such conditions, granule–spacer–granule electrotransport channels arise and the resistivity sharply drops (Fig3. model II). So the percolation threshold comes in the granule-semiconductor-granule and/or granule-(silicide+semiconductor)-granule systems. Such percolation transition is responsible for the increase of the effective magnetic interaction between ferromagnetic granules through the semiconductor layer (or a silicide) what explains the observed features in the magnetic and MO properties of the samples with small thickness of Si layer. Obtained results allow to conclude that MO response in the studied ML are formed both by magnetic granular layer and new magnetic phase formed on the magnetic granule – semiconductor interface.

Further growth of semiconductor spacers causes a separation of the nanocomposite layers and their interaction through the semiconductor becomes weaker and amplification of the MO response disappears. The insignificant growth of the ML electrical resistivity is determined mainly by the silicon conductivity in this range of the semiconductor thickness (Fig. 3. ML model III).

In the second series of ML the magnetic phase concentration Z within the composite layers was slightly less than in the first one (farther away from the percolation), but it leads to abrupt changes in the MO properties. TKE value for series 2 (Fig.4) not only did not grow, but fell more than an order of magnitude compared to the nanocomposite $(Co_{40}Fe_{40}B_{20})_{33,9}(SiO_2)_{66,1}$ (witness). Maximal magneto-optical response for series 2 structures coincides with a beginning of the electrical percolation at the region near 0.5 nm of silicon thickness. Nevertheless, the effect is almost negligible, so it was possible to obtain TKE field dependencies only for several samples. All these dependences δ(H) show a paramagnetic type of magnetization for ML structure.

Such MO behavior testifies that a decrease of the ferromagnetic phase concentration in the nanocomposite layer leads to an increase of distances between the granules both inside of the nanocomposite and on the interface. In this case, the exchange interaction between granules of the adjacent layers through the semiconductor spacer weakens and it causes a decline of contribution from a new FM phase formed at the boundary of the FM granule-semiconductor in resulting MO response. It leads to intergranular exchange interaction attenuation, thus, MO response amplification vanishes. Moreover, the formation of non-magnetic silicides on the FM granule - silicon interface causes further TKE reduction [6].

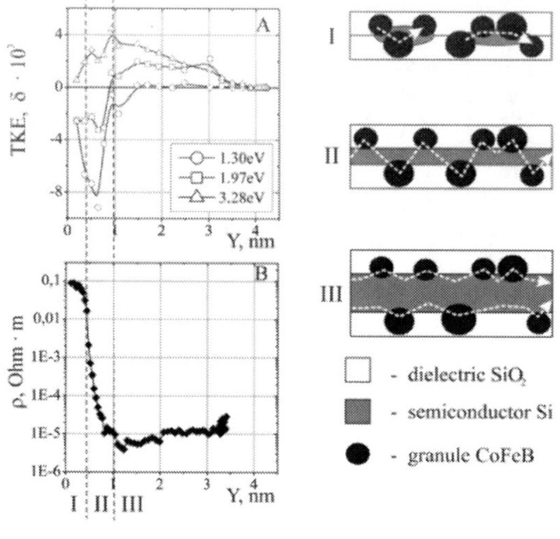

Fig. 3. Isoenergetic Si thickness dependences of TKE (A), electric resistance (B) for series 1, and ML models for three specific ranges of the thickness.

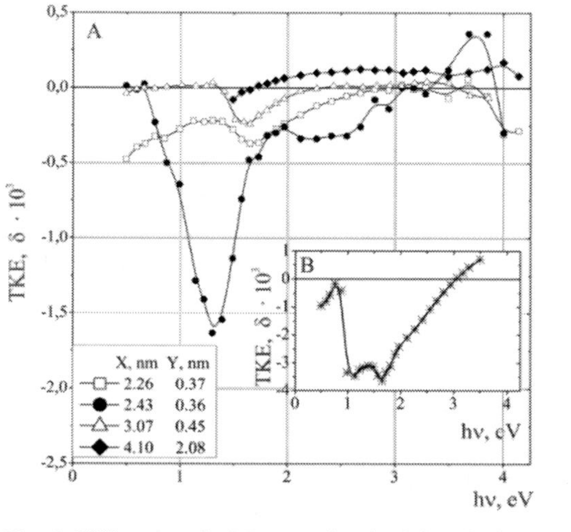

Fig. 4. TKE spectra for ML second series (A) and witness – deposited layer-by-layer composite $(Co_{40}Fe_{40}B_{20})_{33,9}(SiO_2)_{66,1}$ with based layer thickness X = 3.2 nm (B).

For the third ML series, where silicon was replaced by carbon, the studies of TKE spectra revealed that the shape of δ(hν), curves were generally very similar to TKE spectra shape for witnesses (granular nanocomposite of the same chemical composition). The magnitude of MO signal was increasing with increasing of the granular composite layer thickness (X) and decreasing of carbon layer thickness (Y). It was found that adding of a thin non-magnetic semiconductor layer leads to the multilayer MO response decrease as compared to the nanocomposite witness (Fig. 5).

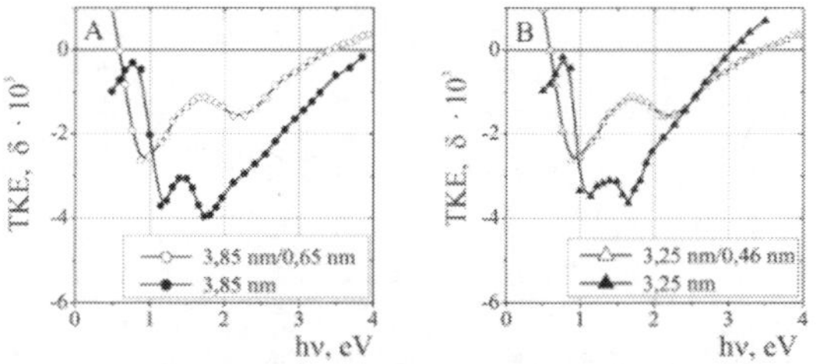

Fig. 5. TKE spectra for nanocomposite witness $(Co_{40}Fe_{40}B_{20})_{33.9}(SiO_2)_{66.1}$ without carbon interlayer (black marker) and for series 3 ML with the same thicknesses of granular layers (X).

TKE field dependencies exhibit a superparamagnetic kind of magnetization and practically matched. This fact demonstrates that $(Co_{40}Fe_{40}B_{20})_{33.9}(SiO_2)_{66.1}$ composite in the ML structure of the third series is superparamagnetic, and ferromagnetic granules do not interact. Electric resistance of the third ML series with carbon layers thickness ranging from 1 to 1.6 nm abruptly decreases. Apparently, the electrical percolation transition is observed in the ML.

With the increase of carbon spacer thickness carbon bridges begin to be formed between the ferromagnetic granules, protruding from the composite layers. These carbon bridges merge into a continuous layer at the carbon thickness about 1.7 nm. The formation of the latter layer leads to the electrical resistivity percolation, whereas no sharp TKE magnitude changes occur at this thickness.

The ML MO response considerably weakens as C is being replaced with Si. The decrease of magneto-optical response could be associated with the higher chemical activity of Si compared with C as well as with the mutual compensation between TKE of the base composite layer and TKE of a new composite formed at the "FM granule-semiconductor" interface with Si, which may have opposite signs [11]. The assumption of mutual compensations is confirmed by positive sign of TKE signal in whole investigated wavelength range from samples with the thickest silicon layers (Fig.4).

Conclusion

The spectral and field dependences of the Transversal Kerr Effect for $(Co_{40}Fe_{40}B_{20})_Z(SiO_2)_{100-Z}$/Si (or C) multilayer nanocomposite-semiconductor structures with different thicknesses of nanocomposite and semiconductor layers as well as magnetic phase concentrations Z have been studied. It has been found that MO response from studied ML is formed by magnetic composite layer and a new magnetic phase formed on the magnetic granule – semiconductor interface. The contribution from the new magnetic phase to the resulting MO response depends on thicknesses of nanocomposite and semiconductor layers and magnetic phase concentration Z within the nanocomposite layer.

Increase of a semiconductor layer thickness leads to the continuous granule-semiconductor-granule and/or granule-(silicide+semiconductor)-granule chain formation and causes the appearance of transport percolation transition in nanocomposite-semiconductor structures. Such a percolation transition is induced by increasing of the effective magnetic interaction between (CoFeB) granules through the semiconductor (or a silicide) layer and explains anomalous behavior of observed magnetic and MO properties of structures with thin Si layers and, at the same time, short distances between FM granules within the composite layer (where Z is near percolation). At lower FM phase concentrations Z as well as an increase of a layer thickness of the semiconductor, the distance between the granules become larger, so it leads to the exchange interaction weakening between granules. This explains TKE reduction and anomalous behavior of the MO properties of ML disappearance.

The contribution from the new phase at the interfaces is considerably reduced as Si is being replaced with C, and can be associated with the higher chemical activity of Si compared with C.

This research was supported by the Russian Foundation for Basic Researches №15-02-02077

References

1. V. O. Vas'kovskiya, G. S. Patrin, et.al. Physics of the Solid State **49,** 302 (2007)
2. S.M. Valvidares, C. Quiros, et.al. Phys. Rev. Lett. B **78,** 064406 (2008)
3. I. I. Pronin, M. V. Gomoyunova, et. al. Physics of the Solid State **53**, 606 (2011)
4. I. V. Belousov, G. V. Kuznetsov, O. P. Pchelyakov. Semiconductors **40**, 881 (2006)
5. M. V. Gomoyunova, G. S. Grebenyuk, et.al. Physics of the Solid State **55**, 437 (2013)
6. S. R. Naik, S. Rai, et. al. J. Phys. D: Appl. Phys. **41**, 115307 (2008)
7. E. A. Gan'shina, N. S. Perov, et. al., Bulletin of RAS: Physics. **72**, 1379 (2008)
8. V. Buravtsova, E. Gan'shina, E. Lebedeva, et. al. SSP **168**, 533(2011)
9. E. Gan'shina, V. Buravtsova, A. Novikov, et .al. SSP **190**, 361 (2012)
10. E.A.Gan'shina, M.V.Vashuk, A.N.Vinogradov et al. JETP **98**, 1027 (2004)
11. V. Buravtsova, E. Gan'shina, Yu. Kalinin, A. Sitnikov, D. Zubakin. SSP **233**, 603(2015)

Effect of chemical ordering on optical properties of Fe_3Si epitaxial films

Ivan Tarasov[1,*], *Zakhar* Popov[2], *Maxim* Visotin[1,3], *Ivan* Yakovlev[1] and *Sergey* Varnakov[1]

[1]Federal Research Center KSC Siberian Branch Russian Academy of Sciences, 660036 Krasnoyarsk, Russia
[2]The National University of Science and Technology MISiS, 11999, Moscow, Russia
[3]Siberian Federal University, 660041 Krasnoyarsk, Russia

Abstract. Optical characteristics (electron energy loss function, optical conductivity σ, permittivity ε, refractive index n, extinction coefficient k, and absorption coefficient α) of a 30 nm thick epitaxial Fe_3Si iron silicide films grown at different silicon substrate temperature (26, 100, 200, 300 °C) were determined within E = 0.74–6.46 eV photon energy range using spectroscopic ellipsometry technique. The experimental data are compared to the optical characteristics calculated in the framework of the density functional theory using the GGA–PBE approximation. Variations of the optical characteristics spectra are discussed from the point of view of chemical ordering of DO3 type crystal structure. It is asserted that the electron energy-loss function, optical conductivity and extinction coefficient of the Fe_3Si iron silicide films undergo noticeable changes in different spectral ranges over the whole spectrum between 0.74 and 6.46 eV due to variation in the chemical order. Information on the effect of chemical ordering on the optical properties obtained here allows one to carry out quick qualitative analysis of Fe_3Si film crystal quality during the synthesis procedures by ellipsometry method in situ.

1 Introduction

Epitaxial thin films of iron silicide Fe_3Si grown on semiconducting substrates, along with other Heusler compounds [1] attract a lot of attention on account of their possible application in spintronics devices where effective injection of spin-polarised current into a semiconductor and its detection is essential. Except for their potential application as spin injectors, these materials are interesting in the other field of applied physics. Since they possess excellent soft magnetic properties, small values of the width of the ferromagnetic resonance line and the coercive field, high permeability and saturation magnetisation, while having a higher resistivity for lower eddy losses, they might be used in different electrically controlled devices of the microwave range. Larger magnetooptical Kerr effect (MOKE), and higher spin polarisation degree in comparison with pure bcc-Fe make this Heusler compound a prominent candidate for fabrication of ferromagnetic periodic structures such as magnetophotonic crystals. However, the optical properties of such films were poorly investigated whereas this information could make easier understanding of the situation with enhancement of magnetooptical response of iron-rich Fe-Si alloy thin film. From another point of view, deeper understanding interrelationship between optical and structural, electronic properties and defect formation of epitaxial Fe_3Si alloys thin films would bring us a quick tool for estimating the quality of as-grown or being grown thin films by in situ or ex situ optical methods, such as ellipsometry.

2 Experiment and calculations

2.1 Sample preparation

Five epitaxial $Fe_3Si(111)$ thin films (denoted FS1-4 and FSC) on silicon substrate Si(111) were prepared according to the following technological procedures. The experiment was carried out with ultrahigh vacuum molecular beam epitaxy "Angara" set-up equipped with a system of reflection high-energy electron diffraction (RHEED) and a one-wave ellipsometer. The base pressure in the growth chamber was 6.5×10^{-8} Pa. Silicon substrates with 1.0 ° Si(111) plane miscut were used. The substrates were prepared by the special chemical treatment in air and a thermal flashing at 900 °C in vacuum.

The Fe_3Si films were prepared by molecular-beam epitaxy (MBE) technique with simultaneous deposition of Si and Fe on Si(111) 7×7 substrate at room temperature for the FS1 sample and heated at 100, 200, and 300 °C for the FS2, FS3 and FS4 samples. The substrate temperature in the case of the FSC sample was 200 °C. Deposition rates were Fe – 1.24 nm/min, Si – 0.88 nm/min. The component materials were evaporated from Knudsen effusion cells. The film thickness estimated by laser ellipsometry technique was about 30 nm. FSC sample was obtained to carry out structural characterisation by X-ray diffraction analysis (XRD). In order to obtain symmetrical scan with well distinguishable reflections, the thickness of the FSC sample was increased up to 70 nm. The films were deposited at the oblique incidence angle, so-called the oblique sputtering.

* Corresponding author: tia@iph.krasn.ru

The angle between the Fe flow direction and the normal to the substrate surface was 14°. XRD analysis was performed on a PANalytical X'Pert PRO diffractometer equipped with a solid state detector PIXcel on Cu and Co Kα radiation. The ellipsometric spectra were measured with help of VASE J.A. Woollam ellipsometer in unique scientific installation (USI) "Physics, Chemistry, and Mechanics of Crystals and Thin Films" in Institute of problems of mechanical engineering Russian Academy of Sciences (IPME RAS), St. Petersburg, Russia.

2.2 DFT calculation of optical properties

The electronic and geometric structures of DO3 type Fe_3Si silicide were calculated by quantum chemistry simulations based on the licensed program package VASP 5.3 [2] in the framework of the density functional theory (DFT) using the plane wave basis and the projector augmented wave (PAW) formalism. To describe the exchange-correlation functional, we use the generalised gradient approximation (GGA) involving also the Perdew–Burke–Ernzerhof (PBE) approximation. For optimisation of the unit cell geometry of Fe_3Si (Fm3m), the first Brillouin zone in the reciprocal space is automatically divided into a 6 × 6 × 6 mesh chosen according to the Monkhorst–Pack scheme. In the calculations, the cutoff energy E_{cutoff} is equal to 293 eV. In the modelling of the structure under study, the optimisation of the geometry is performed up to maximum values of the forces acting on atoms and equal to 0.01 eV/A. After calculation of the ground state by the VASP program, we calculate the frequency dependent dielectric matrix. The details of this method are described in [3]. In order to illuminate how the chemical disorder in ideal DO3 crystal structures of the Fe_3Si compound can affect the optical characteristics the ideal DO3 (Fig.1a) and with a particular chemical disorder (Fig1b) Fe_3Si unit cells were used for calculations.

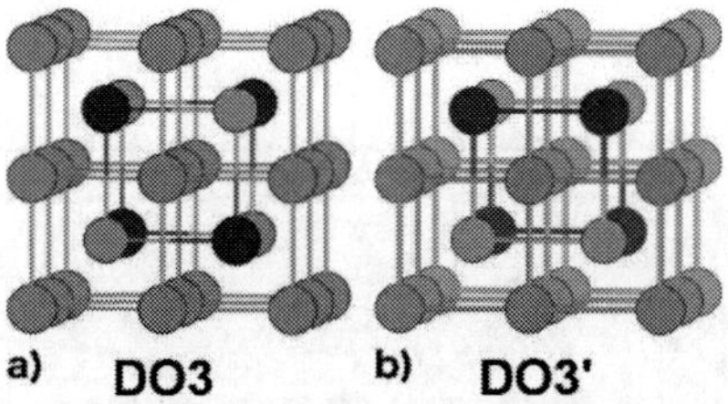

Fig. 1. (a) Ideal DO3 crystal structure of Fe_3Si (a) compared to a disordered DO3 (b). Red, blue spheres represent Fe, green – Si.

3 Results and discussion

3.1 XRD analysis

The in-plane epitaxial orientation was analysed using the asymmetrical φ-scans of reflections {224} from the Fe_3Si iron silicide film and Si substrate. Typical φ-scan for the FSC sample (Fig. 2b) reveals the following orientation relationship: Fe_3Si(111)[1-21] || Si(111)[11-2]. The lattice parameter of ordered DO3 (Fm3m) and disordered B2 (Pm3m) and A2 (Im3m) type crystal structures is strongly dependent on silicon content. For the crystal structure analysis, a series of characteristic reflections 111, 222, 333 and 444 from the Fe_3Si film (FSC sample) were measured (Fig. 2(a)). To reduce the 111 reflection from the Si substrate, an additional tilt of 1° was applied. The cubic (Fm3m, DO3 type) lattice parameter $a = 5.666(1)$ Å was determined from the XRD pattern by the derivative difference minimization method [4] and corresponds to stoichiometric Fe_3Si compound, with chemical composition Fe (75 at. %) and Si (25 at. %) [5]. The silicon content of the $Fe_{1-x}Si_x$ alloy films was calculated by estimation of specific volume for one silicon atom in the $Fe_{1-x}Si_x$ unit cell.

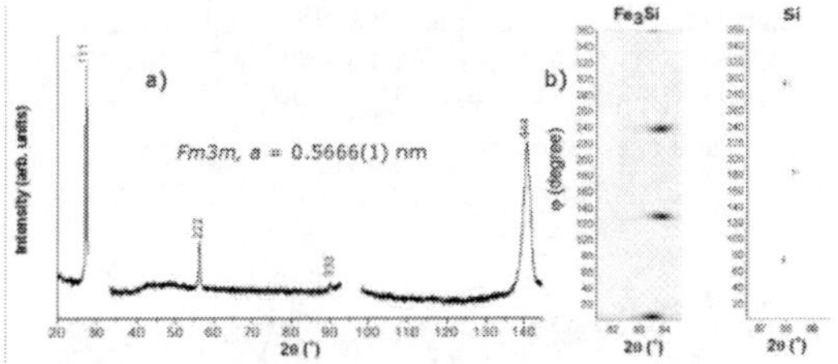

Fig. 2. Observed XRD patterns for 4 characteristic diffraction peaks of Fe_3Si (a). 2D-images for φ-scans of reflections {224} from Fe_3Si film and Si-substrate (b).

3.2 Optical characteristics

One could expect that in metallic compounds such as stoichiometric Fe_3Si iron silicide effect of chemical disorder on optical properties would be negligible. Nevertheless, that scarce data on optical properties of Fe-rich Fe-Si alloys, which exist in the literature available [6,7] suggests that changes in optical characteristics caused by thermal annealing can result from DO3-B2-A2 transformations. It is known that optical conductivity is most sensitive and informative spectral parameter characterising the frequency dependence and the intensity of the optical response of the medium [8]. However, a survey of other optical parameters characterising the optical response of the Fe_3Si compound seems to be essential to define which optical parameter could be easily used for qualitative controlling chemical ordering by optical methods. In the section information about real and imaginary parts of permittivity, refractive index n and extinction coefficient k, electron energy loss function, absorption coefficient α, optical conductivity σ and reflectivity R is given. All optical characteristics were calculated using real and imaginary parts of permittivity obtained by converting ellipsometric parameters with the semi-infinitive medium model. Reflectance from the interface Fe_3Si/Si is not considered and expected to give a small contribution due to high absorbance of the Fe_3Si compound. Nevertheless, more complex optical models will be considered in the nearest future.

3.2.1 Permittivity

Overlook of spectral dependence of experimental and calculated permittivity (Fig.3) demonstrates their qualitative accordance. Real part of permittivity has values in the range between 2-6.46 eV typical for metals and is caused by both commensurable impact of free-carrier absorption and bound electrons. At lower photon energies, < 2 eV, impact of bounded *d*-electrons increases and becomes prevailing. Further decrease of Real(ε) with lowering photon energy is expected due to intraband acceleration of free-carries by electromagnetic field. Comparing to metals with high concentration of free electrons Fe_3Si compound demonstrates high impact of interband transition at energies about 1 eV, which mainly caused by transitions from flat bands formed by d_{xy}, d_{xz}, and d_{yz} (T_{2g}) to those ones created by d_{z2} and $d_{x2\text{-}y2}$ states (E_g) [9]. Effect of synthesis temperature is negligible over the whole spectral range and mainly can be easily noticed in imaginary part of permittivity at photon energies about 1 eV due change of Im(ε) value from 45 up to 55 and slight shift of the peak position. Consideration of the specific chemical disorder in Fe_3Si compound (Fig. 1) reveals a possible reason for shifting of low energy prominent absorption peak, from ~1.8 to ~1.3 eV.

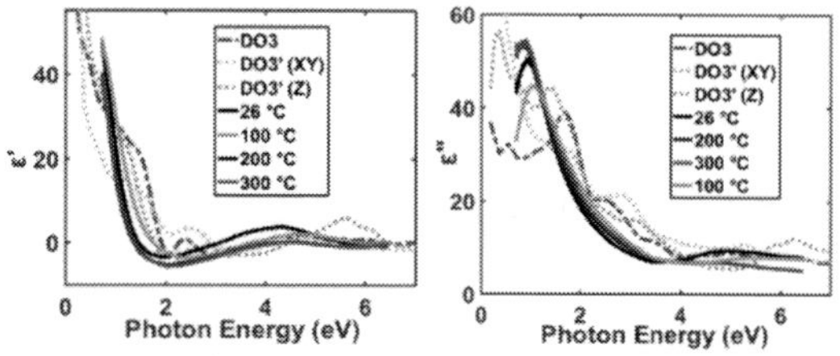

Fig. 3. Real (left) and imaginary (right) part of permittivity of Fe_3Si thin films obtained at different temperatures of Si substrate (solid lines) along with ab initio calculation (dashed lines)

3.2.2 Refractive index and Extinction coefficient

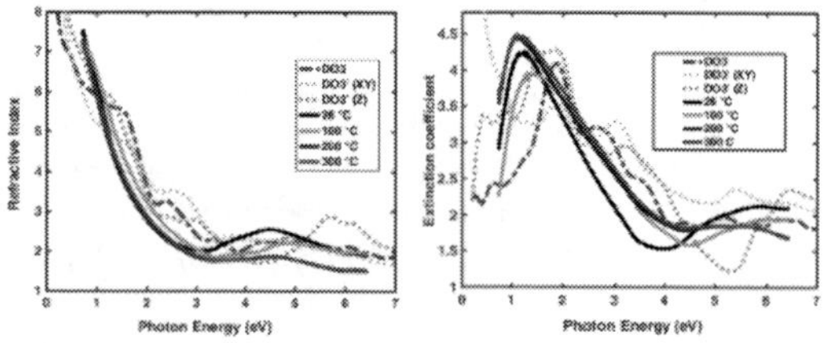

Fig. 4. Refractive index (left) and extinction coefficient (right) of Fe_3Si thin films obtained at different temperatures of Si substrate (solid lines) along with ab initio calculation (dashed lines)

In turn, refractive index and extinction coefficient spectra (Fig. 4) are more sensitive to the synthesis temperature and chemical disordering than permittivity spectra. One can readily note prominent difference between spectra corresponded to Fe_3Si films obtained at room temperature (RT), 100 °C and 200, 300 °C. The spectra referred to the 200, 300 °C synthesis temperatures are in almost full coincidence, which could be evidence of termination atom interdiffusion between different Wyckoff position in Fe_3Si unit cell. Another noticeable feature of the extinction spectrum is a profound minimum close to 4 eV for RT-grown sample and 4.5 eV for 100 °C. As theoretical curve for disordered Fe_3Si unit cell shows the chemical disordering can result in such shifts. The photon energy value of the minimum location is about 5.3 eV.

3.2.3 Electron energy loss function (ELF)

The electron energy-loss function (ELF) describes the interaction by which energy is lost by a fast moving electron travelling throughout the material. Interactions may include phonon excitation, interband and intraband transitions, plasmon excitations, inner shell ionisations and Cerenkov radiations [10]. As it can be seen, photon energy dependence of ELF has one distinctive feature, a broad peak around 3.5 eV, which is sensitive to chemical order according to the theoretical curves (Fig. 5).

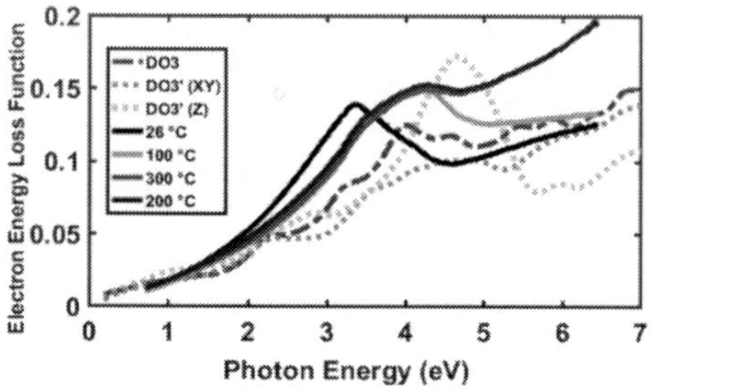

Fig. 5. Electron energy-loss function of Fe_3Si thin films obtained at different temperatures of Si substrate (solid lines) along with ab initio calculation (dashed lines)

The synthesis temperature strongly affects ELF. Thus, RT results in formation an abrupt prominent peak with position at ~3.2 eV, whereas increasing temperature up to 100 ° C causes its shift to higher energy (~4 eV) along with spreading. This feature is in good correspondence with theoretical calculation for the ideal DO3 structure. Further increase of the synthesis temperature does not influence on the peak position. However, it activates subsequent peak broadening with strong increase of ELF value in UV region.

3.2.4 Optical conductivity

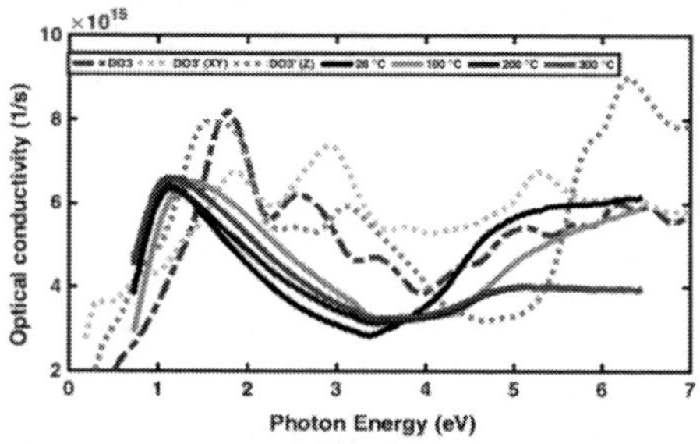

Fig. 6. Optical conductivity of Fe_3Si thin films obtained at different temperatures of Si substrate (solid lines) along with ab initio calculation (dashed lines).

Optical conductivity is mainly formed [10] by free-carriers and links the current density to the electric field. Most prominent peak in experimental curves (Fig. 6) located around 1 eV, which corresponds to high density of electrons (T_{2g} for Fe2 atoms) slightly below Fermi level [9]. Although the carrier densities are low for high energies (5-6.4 eV) and mainly formed by more localised electrons (T_{2g} for Fe1 atoms), the high carrier mobilities of T_{2g} electrons of Fe2 atoms [9] nevertheless guarantee a large contribution of the free carriers. Thus, this energy region is most sensitive to chemical ordering as it follows from comparison with theoretical curves (Fig. 6).

3.2.5 Absorption coefficient

Absorption coefficient characterises the value of penetration of electromagnetic wave in material and is formed by the excitations due to the interaction of photons and electrons, interband and intraband transitions. Low values of α can cause some inaccuracy in description of the ~30 nm thick Fe_3Si film optical properties with semi-infinite medium model. The low photon energy peak positions (~1 eV) in σ, k and Im(ε) parameters are expected to be the most sensitive to this discrepancy. Thus, this issue requires further detailed examination.

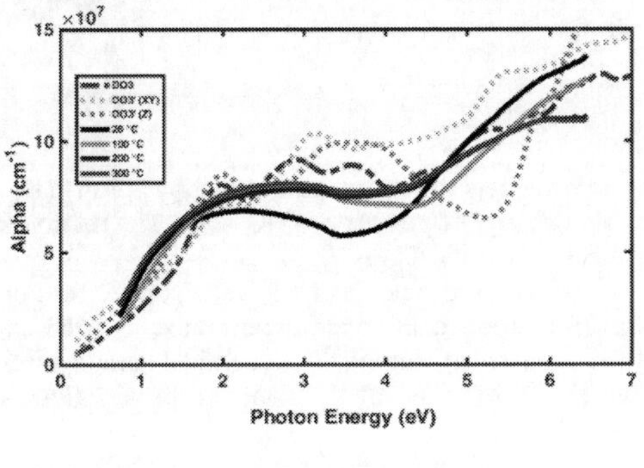

Fig. 7. Absorption coefficient of Fe_3Si thin films obtained at different temperatures of Si substrate (solid lines) along with ab initio calculation (dashed lines).

In general, consideration of α parameter for detection of changes in chemical order is not informative as optical conductivity and ELF, although undergo some distinctive alterations between 3.5 and 4.5 eV.

3.2.6 Reflectivity

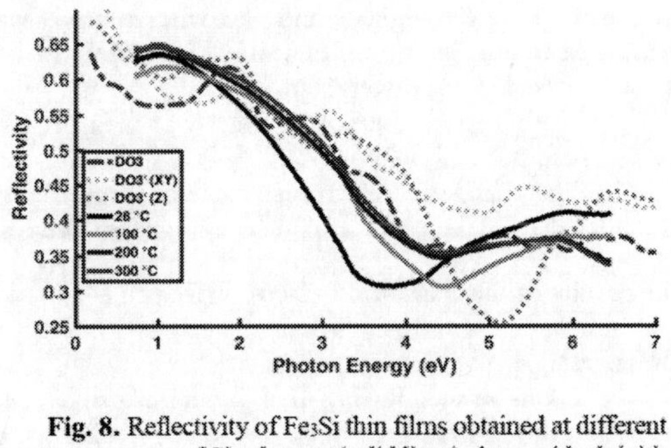

Fig. 8. Reflectivity of Fe_3Si thin films obtained at different temperatures of Si substrate (solid lines) along with ab initio calculation (dashed lines).

Another important optical parameter is reflectivity, which can be measured directly without utilisation of polarimetric techniques. Thus, effect of chemical ordering on reflectivity is found to be essential. Figure 8 reveals that the most sensitive region to chemical ordering lies in higher photon energies (3.2 – 6.4 eV).

4 Conclusions

This work is intended to develop an approach, which would allow us to carry out simple optimisation of growth and annealing conditions of Fe_3Si thin films, by spectroscopic techniques in situ. A simple consideration of the effect of the chemical ordering in Fe_3Si compound on optical parameters by theoretical and experimental methods reveal that electron energy-loss function, optical conductivity and extinction coefficient are sensitive to this phenomenon in different photon energy regions and thus make the whole range (0.74–6.46 eV) useful for the estimation. It is clear, that more deep insight to the problem with consideration of different optical models and various chemical ordering along with involvement of magnetic properties investigation is required and will be discussed soon.

5 Acknowledgements

The reported study was funded by Russian Foundation for Basic Research, Government of Krasnoyarsk Territory, Krasnoyarsk Region Science and Technology Support Fund to the research projects No. 16-42-243060, 16-42-243035 and the President of the Russia Federation Program (SP-3382.2016.3), RFBR (Grants No. 16-32-00291 and 17-42-190308). We also thank Solovyov L.A. for help in XRD analysis.

References

[1] J. Karel, J. Juraszek, J. Minar, et al., Phys. Rev. B **91**, 144402 (2015).
[2] G. Kresse and J. Furthmüller, Phys. Rev. B **54**, 11169 (1996).
[3] M. Gajdoš, K. Hummer, G. Kresse, et. al, Phys. Rev. B - Condens. Matter Mater. Phys. **73**, 1 (2006).
[4] L. A. Solovyov, J. Appl. Crystallogr. **37**, 743 (2004).
[5] in *Powder Diffr. File (PDF 4+) Inorg. Phases* (Swarthmore (PA, USA): International Center for Diffraction Data, 2012), pp. 1–1.
[6] Y. V. Kudryavtsev, V. V. Nemoshkalenko, Y. P. Lee, et al., J. Appl. Phys. **90**, 2903 (2001).
[7] Y. V. Kudryavtsev, Y. P. Lee, J. Dubowik, and J. Y. Rhee, Phys. Rev. B **68**, 134416 (2003).
[8] Y. V. Knyazev, A. V. Lukoyanov, and Y. I. Kuz'min, Phys. Solid State **57**, 1705 (2015).
[9] I. Sandalov, N. Zamkova, V. Zhandun, et al., Phys. Rev. B **92**, 205129 (2015).
[10] S. Naderizadeh, S. M. Elahi, M. R. Abolhassani, et al., Eur. Phys. J. B **85**, 144 (2012).

Atomic ordering and coercive force of nanocrystalline Fe-Co alloy films

Nataliya Repina

Krasnoyarsk Institute of Railway Engineering, 89, L.Ketskhoveli St, Krasnoyarsk, 660028, Russia

Abstract. In single crystal films of the $Fe_{0.5}Co_{0.5}$ alloys grown by the method of vacuum condensation, a metastable condition may be received that is similar to the completely disordering condition. Approximately during ten days after the condensation in the film at the room temperature the short range and then long range atomic ordering is developed. Simultaneously the changes of magnetic anisotropy, electroresistance and coercive force of the films were investigated. This investigations show that the appearance of the shot range atomic ordering increases the electroresistance of films and the long range atomic ordering reduces it. The dependence of coercive force from time to time at room temperature atomic ordering films $Fe_{0.5}Co_{0.5}$ alloys, and $Fe_{0.75}Co_{0.25}$. Found that the biggest change Hc (double) is observed in single-crystal films when changing mechanism of magnetization reverse. Most is a consequence of the changes to the crystallographic anisotropy.

Iron–cobalt alloys, possessing interesting magnetic properties have been long attracting the researcher's attention.

It is known that the processes of atomic ordering in ferromagnetic alloys with the high Curie temperature, presenting the phase transition of the second kind, proceeds comparatively fast.

The speed of temper necessary for the suppress of the ordering condition in stechiometrical composition Fe–Co alloy is 6000°/s. Therefore it is very difficult to receive a disordering condition of alloy $Fe_{0.5}Co_{0.5}$ in massive specimens.

In the given work the processes taking place in epitaxial magnetic films at the room temperature were investigated. The changes of magnetic anisotropy, electroresistance and coercive force at natural aging were observed.

Single (001) films $Fe_{0.5}Co_{0.5}$ and $Fe_{0.75}Co_{0.25}$ were grown by the method of vacuum condensation on the surfaces of the splits of LiF and MgO crystals heating to 150° C. In the just condensed films, having the body–centered cubic (bcc) structure, showed the absence of the super–structure phase. If you store the tapes at room temperature, i.e. as a result of natural aging.

The thickness of research films was laid in the interval 30-50 nm.

Polycrystalline iron-cobalt alloy films received on glass substrates.

Magnetic anisotropy was measured by the method of the revolving moments. Easy axis of magnetization these films depending on alloy composition are guided either along the underlying sample plane crystallographic directions [100] ($K_1 > 0$) or along the [110] directions ($K_1 < 0$). The relative error of measurement of K_1 is 7.5%.

The character and the value of the tensions in the film specimens were determined by the X–ray diffractometrical method. The chemical composition was studied with the aid of the rentgenospectral fluorescent method.

According to these observations and the structure's investigations it was shown that the alloy $Fe_{0.5}Co_{0.5}$ may be received in the disordering condition comparatively easily in the thin magnetic films. The process of atomic ordering in these films may take place at room temperature and even at much lower temperatures.

As reported in the papers [1, 2], Fe-Co alloy films, under the process conditions are disordering and ordered then at room temperature.

To study the ordering in the alloys it was used «the method of the small makeweights», theoretic proposed by Krivoglas and experimentally was confirmed by us [3]. Au up to 1 at. % and Pd up to 1–3 at. % was used as makeweights.

The electroresistance of films was measured by the potenciometrical method.

The changes of elecroresistance of single crystal films $Fe_{0.5}Co_{0.5}$ take place at natural aging (for 10 days). The results of the measuring electroresistance ($\frac{\Delta\rho}{\rho}$) are shown in fig.1.

It can be seen in fig.1 that the most changes take place during the first three days. The electroresistance of films continue to increase until the sixth day. Then it begins to decrease slowly to the tenth day. After that the quantity of electroresistance is stabilized.

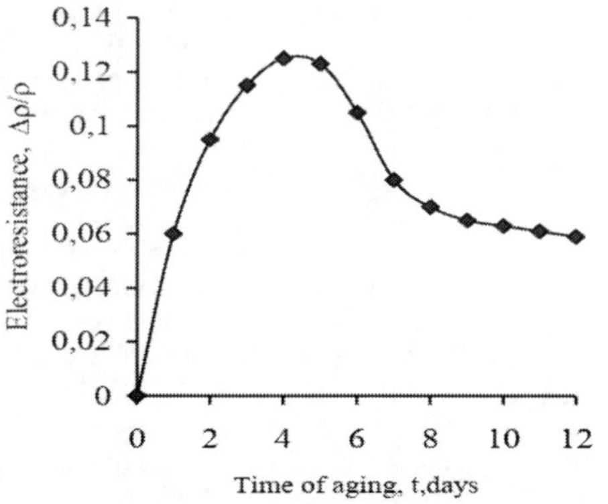

Fig.1. The changes electroresistance $\Delta\rho/\rho$ of the nanocrystalline films $Fe_{0,5}Co_{0,5}$ during the natural aging

Because of the coercive force H_c of ferromagnetic being a structurally sensitive characteristic it aroused our interest to trace the changes H_C of single crystal films Fe–Co alloys during the time at the room temperature. The coercive force of films was defined from the magnetic hysteresis loops, with the help of magnetooptical Kerr effect. In fig.2 the dependence of H_C the films on the aging time of equiatomic composition Fe–Co alloy is depicted.

The coercive force was measured under the magnetization reversal of films along one of the easy magnetic axis. The relative error of measurement of H_C is 5%.

The figures show that the changes of H_C were absent during the first three days. After three days of aging the coercive force of film begins to decrease slowly. Between the seventh and tenth days of aging the sharp (approximately 2 times) decrease of H_C was observed.

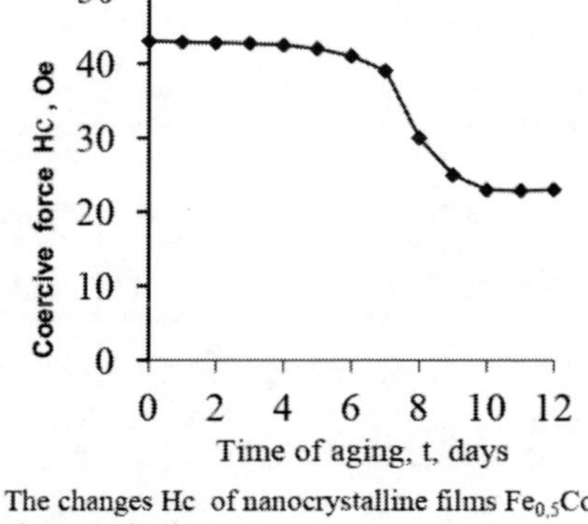

Fig.2. The changes Hc of nanocrystalline films $Fe_{0,5}Co_{0,5}$ during the natural aging

In the course of the investigation of magnetocrystalline anisotropy stechiometrical compositions the epitaxial films it was discovered that the first constant of this anisotropy changes its value from -5×10^5 to $-0,5\times10^5$ erg/sm^3. The changes of K_1 are insignificant during three days. The principal changes of K_1 films begin for three days and they finish for nine days. The energy of activation process of the change K_1 process during the natural aging (E=0,25 eV) was measured. The change of K_1 films at natural aging are similar to the change of K_1 at ordering annealing of massive specimens.

The electronographical investigations of the films were conducted with the application of «the method of the small makeweights». 0, 4 % Au were used as makeweights. In the just condensed films, having the body–centered cubic (bcc) structure, showed the absence of the super–structure phase (fig.3 a). The diffuse background on the electronograms of films appears after two day, (fig.3 b), which is intensified during the time (fig.3 c), the electronogram took off six day), which is intensified during the time.

On the eighth day the weak reflexes type {100} appears (fig.3 d), and a diffuse background on the electronograms of films begins to weaken, however it does not disappear completely.

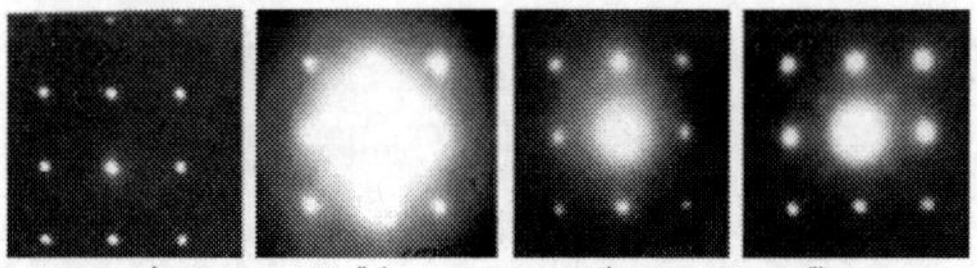

Fig.3. Electronograms of single magnetic films . $Fe_{0,5}Co_{0,5}$:

a) at once after the condensation of film;
b) at three days of the natural aging;
c) at six days of the natural aging;
d) at ten days of the natural aging

The electronomicroscopical investigations of films show that just after condensation they are homogeneous in structure. During the time the inhomogeneous structure appears along the edges of the grains which then turn into pores, disorderly distributed on the film, the number and size of which increases during the time.

In the electronomicroscopical photographs of films vacancy's loops appear along the edges of the grains.

The formation of the film in greatly unequilibrium conditions of the vacuum condensation leads to the saturation defects of the «vacancy» type. At the natural aging of the films the transfer of the vacancies, take place the formation of the accumulations, to slam of vacancy discs and the formation of pores take place too. It is confirmed accumulations, to slam of vacancy discs and accumulations, to slam of vacancy discs and the formation of pores take place too. It is confirmed well with the electronografical and electronomicroscopic investigations.

The appearance of the strong diffuse background on the electronograms of the films during the first days of aging of the films can be explained by two reasons: the formation of vacancy accumulations and the formation of the short–range ordering.

The presence of reflexes {100} is connected with the formation of the super–structure type B2.

It is obvious that the presence of the great quantity of vacancies facilitates the process of atomic ordering. The quantity of energy activation of change K_1 (E=0, 25 eV) corresponds to energy the movement of vacancies.

The above described changes of electroreistance films at natural aging can be explained by the appearance of the short–range and long–range atomic ordering. The formation of the short–order in film and the appearance such inhomogeneous structure of crystal construction such as the vacancies loops and pores lead to the increase the electroresistance. ($\frac{\Delta\rho}{\rho}$) increases to the appearance of superstructure reflexes. Since this moment the electroresistance begin to decrease, that can be explained by the increase of the degree of the long range atomic ordering. The measure of ($\frac{\Delta\rho}{\rho}$) after ten days of aging of the film does not fall under the original measure of ($\frac{\Delta\rho}{\rho}$) in the film that is connected with the presence comparatively great defects (pores, vacancy loops and disks) [4].

The investigations show that process of atomic ordering influences on the coercive force of single crystalline films weakly. The observed sharp decrease of H_c on the 7–8 days of aging can be explained by the change of the process magnetization reversal of the film, connected with the decrease of the magnetocrystalline anisotropy. With the high measure K_1 of the film was reversely magnetized by the formation of the wedge manner domains of 90°– neighbouring, orientated along hard magnetic axis and by the removal of domain walls.

This magnetization reversal is typical for two axis films. When the measure $|K_1|$ becomes little, the film is reversed as uniaxial, that is by the formation and growth of the wedges of 180°–neighbouring.

According to the theory of Kondorsky [5] the field sufficient for the domain–wall movement is determined with the expression:

$$H_C = \frac{P}{I_S}(Cos\vartheta_1 - Cos\vartheta_2)^{-1}, \qquad (1)$$

where P–constant; determined by the structural inhomogeneous and the strains of the material; I_s – saturation magnetization of material; θ_1 and θ_2 – angles between the direction a field and the magnetization vectors on both sides of the domain wall.

In case of the magnetization reversal with 180° domain–wall movement according to this formula the coercive force must be 2 times less that under the magnetization reversal 90°domain–wall movement.

For polycrystalline films alloys $Fe_{0.5}Co_{0.5}$ и and $Fe_{0.75}Co_{0.25}$ studied the dependence of coercive force from the time of natural aging. Growth of coercive force alloy films $Fe_{0.75}Co_{0.25}$ due to an increase in anisotropy of individual crystallites films, and, therefore, the variance of the anisotropy of the film as a whole. For polycrystalline films alloy $Fe_{0.5}Co_{0.5}$ increase driving order, on the contrary, by lower anisotropy of individual crystallites.

Obtained in this study results indicate great opportunities offered by the practice of magnetic phenomena research samples of alloys, Atomic ordering in which occurs at room temperature. In particular, on the basis of the observed changes when you change the coercive force of the mechanism of magnetization reverse films, it can be concluded that the change in the mechanism of magnetization reverse sample can substantially alter the loss for hysteresis (in duplicate).

The results of above described the conclusion that in film specimens Fe–Co may be received in the disordering condition comparatively easily. This condition of alloy is metastable and at the room temperature it takes place gradually in the atomic ordering condition. These data facilitate the study of atomic ordering influence on the physical properties of alloy.

References

1. N.M. Repina, Izvestiya RAN. Seriya fizicheskaya, **77**, 3 .(2013)
2. N.M. Repina, Bull. R: Phys. Acad. Sci. Physics, **77**, 3 (2013)
3. A.S. Komalov, N.M. Repina: Fiz. Met. Met. **.37** (1974)
4. N. Hayashi, F. Stromberg and W. Keune, T. Toriyama and M. Yamashiro, I. Sakamoto: J. Appl. Phys. **101** (2007).
5. E.I. Kondorsky, J.E.T.P, **7** (1937)

Heat Treatment Effect on Magnetic Microstructure of $Fe_{73.9}Cu_1Nb_3Si_{13.2}B_{8.9}$ Thin Films

Evgeniya Mikhalitsyna[1,*], *Ivan* Zakharchuk[2], *Ekaterina* Soboleva[2], *Pavel* Geydt[2], *Vasiliy* Kataev[1], *Vladimir* Lepalovskij[1], and *Erkki* Lähderanta[2]

[1]Ural Federal University, Department of Solid State Magnetism, 620002 Ekaterinburg, Russia
[2]Lappeenranta University of Technology, School of Engineering Science, 53851 Lappeenranta, Finland

Abstract. $Fe_{73.9}Cu_1Nb_3Si_{13.2}B_{8.9}$ (Finemet) thin films were deposited on the glass substrates by means of radio frequency sputtering. The films thickness was varied from 10 to 200 nm. Heat treatment at temperatures of 350, 400 and 450 °C were performed for 30 minutes in order to control thin film structural state. The X-ray powder diffractometry revealed that the crystallization of α-FeSi nanograins took place only at 450 °C whilst the other samples stayed in the amorphous state. Relation between the structure and magnetic properties of the films was discussed in the framework of random magnetic anisotropy model and the concept of stochastic magnetic domains. The latter was investigated using magnetic force microscopy (MFM). MFM data showed formation of such magnetic domains only in samples thermally treated at 450 °C. There was a tendency of the magnetic domain size reduction with the thickness decrease.

1 Introduction

Nanocrystalline $Fe_{73.9}Cu_1Nb_3Si_{13.2}B_{8.9}$ alloy that is well known as Finemet retains a high scientific and technological interest due to its excellent magnetic properties among the soft magnetic materials [1-3]. Low coercivity and both high saturation magnetization and permeability determines a wide range of its applications as functional materials in power transformers, microwave devices, magnetic sensors and other [4-6]. These properties are caused by special microstructure of the alloy, which is characterized by randomly oriented ultrafine bcc α-FeSi grains uniformly dispersed in a residual Fe–Nb–B amorphous matrix. Macroscopic properties of Finemet alloy are closely related to the microscopic structural and magnetic parameters of the material such as grain size, saturation magnetization, magnetic anisotropy constant and exchange parameter. This relationship has been successfully described within the framework of random magnetic anisotropy model (RAM) [7].

In this report, we present a study of magnetic microstructure of $Fe_{73.9}Cu_1Nb_3Si_{13.2}B_{8.9}$ thin films depending on temperature of heat treatments and film thicknesses.

2 Experiment

2.1 Preparation of samples

Thin films were prepared by means of radio frequency ion-plasma sputtering of $Fe_{73.9}Cu_1Nb_3Si_{13.2}B_{8.9}$ target onto a Wilmad glass substrates. The deposition was carried out in the presence of technical magnetic field of 100 Oe oriented parallel to the substrate. For all samples the achieved base pressure was of 10^{-6} Torr and the 99.987% Ar process pressure was of 10^{-3} Torr. The substrate temperature did not exceed 40-50 °C during the deposition. The structural state was controlled by a heat treatment. The samples were annealed at the pressure not worse than 10^{-3} Torr and at temperatures of 350, 400 and 450 °C for 30 min. Magnetic field was also applied during the heat treatments in the same direction as at depositions.

The thin films thicknesses of 10, 50, 100 and 200 nm were verified by means of Dektak 150 Stylus Profilometer.

2.2 Study techniques

Microstructure of the samples were identified using the Bruker D8 Advance X-Ray diffractometer operating at 40 kV and 40 mA, in theta-2theta configuration, and secondary monochromator with Cu-K_α radiation (λ = 1.5418 Å).

The atomic force (AFM) and magnetic force (MFM) microscopies were obtained using the Scanning Probe Microscope Bruker Multimode 8. The MFM mode of measurements is based on two-pass scanning technique. The sample topology was recording in the tapping mode during the first pass. During the second one, magnetic phase contrast was measuring in the lift mode that means scanning with a constant distance between the probe and the sample surface (typically of about 80-100 nm). The scans were performed in the presence of external

* Corresponding author: evgenia.mihalitsyna@urfu.ru

magnetic field ($H \sim 180$ Oe) directed out-of-plane to the sample surface. The scan area was 10 x 10 μm^2 (resolution 512 x 512 pixels) and the scan rate was 0.150 Hz. Tip with approximate radius of 40 nm, magnetic CoCr coating, and spring constant of 5 N/m was used in experiments.

The in-plane hysteresis loops in both easy magnetization axis (EA) and hard magnetization one (HA) were measured at room temperature by Evico magnetics GmbH magneto-optical Kerr effect (MOKE) microscope using an overview mode.

3 Results and discussion

Room temperature X-Ray diffraction (XRD) patterns of the as-prepared and annealed samples were measured in the angle range of 10° ≤ 2θ ≤ 120°. Large halo in the small angles range was corresponded to the signal from the substrate. As-prepared and annealed at 350 and 400 °C films were mainly in roentgen-amorphous state (Fig. 1, top picture). The presence of crystalline phase was identified in samples annealed at 450 °C. The broadened peak for an angle of order 2θ=45° in Fig. 1 (bottom picture) indicates that the phase α-FeSi is formed in the films [1,2]. The average size of nanocrystalline α-FeSi grains was estimated of 16±6 nm using the Scherrer's formula for cubic particles.

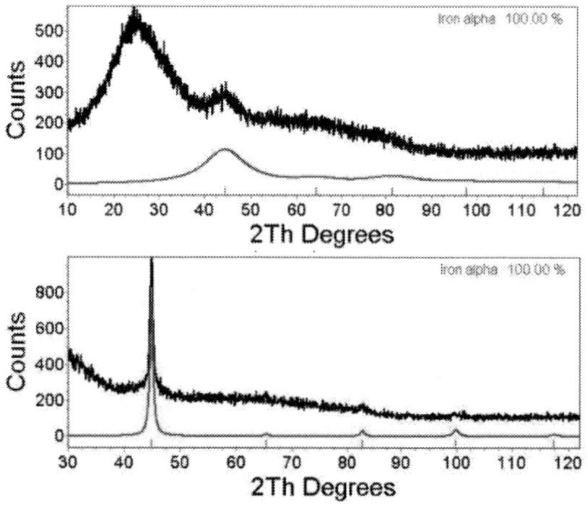

Fig. 1. XRD patterns of 200 nm films in the as-prepared state (top) and after the heat treatment at the temperature 450 °C for 30 min (bottom).

As can be seen from the joint consideration of the XRD data and hysteresis loops in Figs. 2a-2c the thin films before crystallization are characterized by a soft magnetic behavior with in-plane magnetic domains. Induced magnetic anisotropy is observed for the films in as-prepared state. Increase of the annealing temperature leads to decrease of the induced anisotropy field and its total disappearance for the film annealed at 450 °C (Fig. 2d). Assuming magnetoelastic nature of the induced anisotropy, decreasing of anisotropy filed can be explained by the internal stresses relaxation.

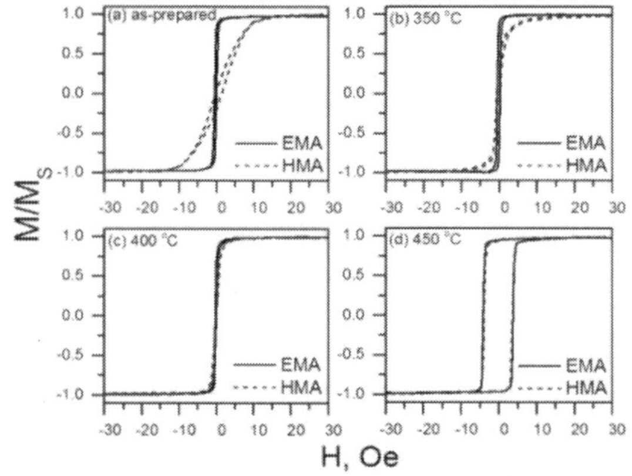

Fig. 2. MOKE hysteresis loops measured along easy and hard magnetization axes of the 200 nm films in the as-prepared state (a) and after the heat treatment at 350 (b), 400 (c) and 450 °C (d).

A surface topography of the 200 nm films annealed at 400 and 450 °C are presented in Figs. 3a and 3c. As can be seen in Fig. 1 and Fig. 3c heat treatment at the temperature of 450 °C leads to the crystallization and formation of the grain-like structure on the film's surface. The grains size is about 60-90 nm in diameter and 7-10 nm in height. The result of AFM study shows that root mean square roughness increases approximately from 1.6 nm to 4.1 nm when the annealing temperature goes up.

The MFM image of the 200 nm thick film annealed at 450 °C shows appearance of magnetic contrast on the surface. A size of the uniform contrast regions is about of 300-600 nm in width and 10 μm in length (Fig. 3d). However, there is no visible magnetic contrast on the surface of the same thickness thin film annealed at 400 °C as shown in Fig. 3b.

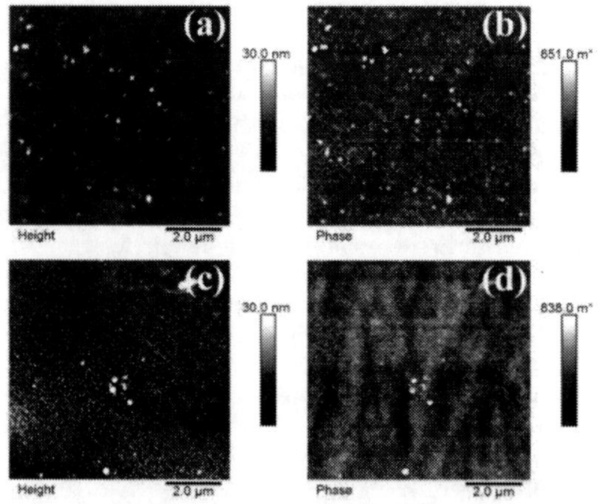

Fig. 3. AFM (a, c) and MFM (b, d) patterns of the 200 nm films after the annealing at 400 °C (a, b) and 450 °C (c, d).

The absence of magnetic contrast on the surface of 400 °C annealed film could be connected with a very low coercivity of the film. So, a probe magnetic field could affect on a sample magnetic structure.

The areas of uniform contrast on the MFM images in Fig. 3d can be correlated with the extension of the magnetic correlation region. And the image as a whole can represent a stochastic magnetic domain structure. The change in color contrast will correspond to an inhomogeneous distribution of the magnetization.

As mentioned earlier, the heat treatment at 450 °C leads to crystallization. According to the assumption of random magnetic anisotropy model, easy magnetization axes of grains are randomly oriented. Due to strong exchange coupling (that needs small enough grain sizes $D << \sqrt{A/K}$, where A is an exchange stiffness constant and K is a local anisotropy constant) between grains, correlated magnetic microstructures are formed. Within these correlated regions averaging of local magnetic properties provides a decreased efficient anisotropy constant and, subsequently, low coercivity. It should be noted that the sample annealed at 450 °C has a markedly greater value of coercivity (Fig. 3d). The possible reason of it is a sufficiently large grain size for the effective averaging within the RAM framework, but it does not exclude the existence of stochastic domains.

Stochastic magnetic structure appeares for all samples annealed at 450 °C. Consequently, a next step is to investigate the evolution of the domain structure with the film thickness. The surface topology and magnetic contrast images of the samples with the thicknesses of 10, 50, 100 and 200 nm are presented in Fig. 4. All samples demonstrate the presence of the grains on the film's surface as the result of crystallization process [8]. Here the size of the grains increases from 60-90 up to 300 nm with the thickness decreasing. Topography of very thin films can be affected the glass substrate characterized by prolonged irregularities. Also, thickness decreasing can restrict structural transformation in the direction perpendicular to the film plane.

It is clearly seen in MFM images in Fig. 4 the domains become thinner and almost disappear in the 10 nm film with the film thickness decreasing. Similar results were observed in Refs. [9-10] for FeSiB thin films.

Parameters of stochastic magnetic domains are defined by microscopic parameters such as exchange constant, saturation magnetization, local magnetic anisotropy constant and grain size. The influence of each parameter should be studied in details. One can only conclude that the decrease of thickness has an effect on the size of heterogeneous magnetic contrast regions. A dimensionality of magnetic correlations can decrease when film becomes thinner. As a result, efficiency of local anisotropy axes averaging deteriorates, that is directly connected with the stochastic domain size [11].

Moreover anisotropic form of the magnetic domains in MFM images of the film with thickness of 50, 100 and 200 nm should be mentioned despite the samples are completely isotropic after the heat treatment at 450 °C as shown in Fig. 2d.

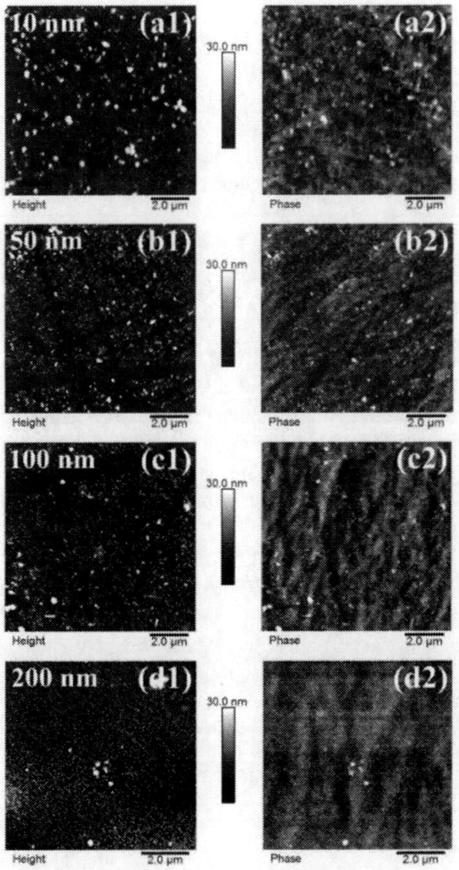

Fig. 4. AFM (a1-e1) and MFM (a2-e2) images of the films with thicknesses of 10 (a), 30 (b), 50 (c), 100 (d) and 200 nm (e) after the heat treatment at 450 °C.

4 Conclusion

To conclude, magnetic and morphology microstructure of $Fe_{73.9}Cu_1Nb_3Si_{13.2}B_{8.9}$ thin films of various thickness were investigated using XRD, AFM, MFM, MOKE microscope, and SQUID magnetometer. It was found that the heat treatment leads to increase of surface roughness and size of grains on the surface. The MFM detected presence of magnetic contrast on the surface of the films annealed at 450 °C that could be interpreted as stochastic magnetic domains. The domain size decrease with the film thickness decreasing.

The research was supported by the Ministry of Education and Science of the Russian Federation Agreement no. 02.A03.21.0006 and project no. 3.6121.2017.

References

1. Y. Yoshizawa, S. Oguma, K. Yamaguchi, J. Appl. Phys. **64**, 6044 (1988)

2. Y. Yoshizawa, K. Yamaguchi, Mater. Trans. JIM **31**, 307 (1990)
3. K. Hono, A. Inoue, T. Sakurai, Appl. Phys. Lett. **58**, 2180 (1991)
4. J. Moulin, I. Shahosseini, F. Alves, F. Mazaleyrat, J. Micromech. Microeng. **21**, 074010 (2011)
5. V. Prida, P. Gorrıa, G. Kurlyandskaya, M. Sanchez, B. Hernando, M. Tejedor, Nanotechnology **14**, 231 (2003)
6. M. Vazquez, G. Kurlyandskaya, J. Garcia-Beneytez, J. Sinnecker, IEEE Trans. Mag. **35**, 3358 (1999)
7. G. Herzer, Acta Mater. **61**, 718 (2013)
8. E. Mikhalitsyna, V. Kataev, P. Geydt, V. Lepalovsky, E. Lähderanta, Solid State Phenom. **233–234**, 699 (2015)
9. M. Coïsson, F. Vinai, P. Tiberto, F. Celegato, J. Magn. Magn. Mater. **321**, 806 (2009)
10. M. Coïsson, F. Celegato, E. Olivetti, P. Tiberto, F. Vinai, M. Baricco, J. Appl. Phys. **104**, 033902 (2008)
11. R. S. Iskhakov, S. V. Komogortsev, Phys. Met. Metallogr. **112**, 666 (2011)

MAGNETO-OPTICAL AND MAGNETIC RESONANCE PROPERTIES OF NANO-SCALED GRANULAR FILMS $(CoFeB)_x(SiO_2)_{100-x}$ and $(CoFeB)_xC_{100-x}$

Sergey Vyzulin[1], *Elena* Gan'shina[2], *Vladimir* Garshin[2], *Natalia* Perova[2,3], *Nikolaj* Syr'ev[2]

[1]Army General S.M Shtemenko Higher Military School, 350035, Krasnodar, Russia
[2]M.V. Lomonosov Moscow state University, Physics faculty, 119991, Moscow, Russia
[3]Immanuel Kant Baltic Federal University, Kaliningrad, 236041, Russia

Abstract. The results of experimental studies of magneto-optical (MO) and magnetic resonance (MR) properties of nanogranular film structures $(CoFeB)_x(SiO_2)_{100-x}$ and $(CoFeB)_xC_{100-x}$ obtained by the ion-beam sputtering method are presented. Magnetic spectra were registered by the method of ferromagnetic resonance (FMR), and magneto-optical spectra were registered by observation of Transversal Kerr Effect (TKE). A comparative analysis of the experimental MO spectra and FMR spectra for systems $(CoFeB)_x(SiO_2)_{100-x}$ and $(CoFeB)_xC_{100-x}$ indicates that a multi-phase structure realizes in the $(CoFeB)_xC_{100-x}$ system. The existence of a multiphase structure is associated with possible plasma-chemical processes that can occur during the synthesis of nanosystems with carbon.

1 Introduction

Nanoheterogeneous structures are promising functional materials for various practical applications (for example, in spintronics, one-electron electronics, etc.). One of the main provisions that technologists usually follow when synthesizing such structures is the impossibility of forming chemical compounds between the phases of a heterogeneous system. The relevance of this restriction can be confirmed by the following example. In the synthesis of film nanostructures of the CoFeB- Si type, silicon enters into chemical reactions with metals to form magnetic and nonmagnetic complexes. As a result, the introduction of even a small amount of silicon led to a significant change in the magnetic and magneto-optical properties up to disappear [1- 3]. Unfortunately, the number of combinations of compounds that by means of self-organization processes can form a multiphase nanoscale structure and, at the same time, are mutually inert, is limited. In synthesis of ordered nanostructures the requirement of mutual inertness for the starting materials may in some cases be redundant. It is shown experimentally [4] that the process of self-organization with formation of a heterogeneous structure is possible also when using starting materials which themselves and / or their individual elements can engage each other in a chemical reactions to form a new (previously absent) compounds. For this reason, it seems very promising to study the cases for which the process of self-organization with formation of a heterogeneous structure is possible against the background of plasma-chemical reactions in the synthesis by ion-plasma sputtering. An example of such actually realized systems can be thin nanogranular film structures of the type a ferromagnet- carbon matrix. How will the plasma-chemical synthesis processes affect physical properties of the new heterogeneous materials? The answer to this question is currently lacking.

The aim of the present investigation is to compare the magneto-optical and magnetic resonance characteristics of nanogranular magnetic films synthesized by the ion-plasma sputtering from substances for which plasma-chemical processes are possible, with the similar characteristics of the systems synthesized from mutually chemically neutral materials.

2 Object of research

We have studied nanogranular film structures with magnetic granules of amorphous CoFeB embedded in the non-magnetic SiO_2 matrix (system 1) and in the carbon matrix (system 2). Nano-systems were obtained by ion-beam sputtering. Thin films with thickness h and magnetic phase concentration x both varying along the length of the samples have been prepared using alternate deposition without interlayers. For the system 1 parameter h varied from 210 to 420 nm and x from 10 to 40 at% (thickness of one deposited layer varied from 4,6 to 8 nm). For the system 2 the intervals were 50 nm $< h < 250$ nm and 17 at% $< x <$ 56 at% (thickness of one layer varied from 1.0 to 4.9 nm). In the direction transverse to the length of the structure, the thickness of

*Corresponding author: vyzulin@mail.ru

 https://doi.org/10.1051/epjconf/201818504002

the sputtered layer and the magnetic phase concentration did not change. The direction in the plane of the sample, coinciding with the longitudinal direction of the original nanostructure, is further denoted by the vector **i**, and the direction in the plane of the film transverse to it is vector j.

It is known that carbon can interact both with transition metals and alloys on their basis, and with boron [4]. In our case, along with nano-sized CoFeB granules, it is possible to form nanoparticles from metastable (at room temperature) compounds of the Co_3C, Co_2C, Fe_3C, and B_4C type. Since nanoparticles of carbides with FM metals exhibit magnetic properties [5,6], their presence in the studied system $(CoFeB)_xC_{100-x}$ should manifest itself both in magnetic resonance and transversal Kerr effect (TKE).

3 Experimental methods

A complex approach to the study of nanocomposites was used based on the study of magnetic and magneto-optical spectra. Magnetic resonance properties were studied by FMR, and magneto-optical properties were investigated by observing TKE spectra.

The FMR spectra were measured with highly sensitive EPR spectrometer JEOL FA-300 at a frequency of 9.44 GHz at room temperature. The spectra for each sample were measured for different orientations of the magnetization field **H** with respect to the film plane, from $\alpha = 0$ (**H** || **n**) to $\alpha = 90°$ (**H** $\perp$ **n**), where **n** is the normal to the film plane. Two series of experiments were performed. In one, the external magnetic field is **H**$\perp$**i**, in the other – **H**$\perp$**j** (Fig.1) The spectra were used to determine the resonance fields H_r, the absorption line width H and the intensity of the first derivative of the absorption signal J.

Magneto -optical properties were studied in the energy range of 0.5 - 4.2 eV in TKE geometry. An alternating magnetic field up to 3 kOe was aligned parallel to the sample surface and perpendicular to the light incidence plane.

4 Experimental results

4.1 Magneto-optical properties

Fig.2a shows TKE spectra for the system 1 samples with various concentrations of the ferromagnetic phase, as well as spectra of the reference specimen – alloy $Co_{40}Fe_{40}B_{20}$. The TKE spectra of the amorphous nanocomposite $(Co_{40}Fe_{40}B_{20})_x(SiO_2)_{100-x}$ essentially differ from the corresponding spectra of the amorphous alloy $Co_{40}Fe_{40}B_{20}$. Particularly pronounced changes are seen in the energy range $E < 1.5$ eV. While in the alloy $Co_{40}Fe_{40}B_{20}$ the decrease of the photon energy leads to diminishing of the TKE to zero (TKE ≈ 0 at $E \approx 0.5$ eV), in the nanocomposites the TKE changes its sign and reaches extremal negative magnitude in the energy range 1–1.4 eV. In this interval the TKE magnitude of the nanocomposites by its absolute value is approximately 5 times as much than the one of the amorphous alloy $Co_{40}Fe_{40}B_{20}$.

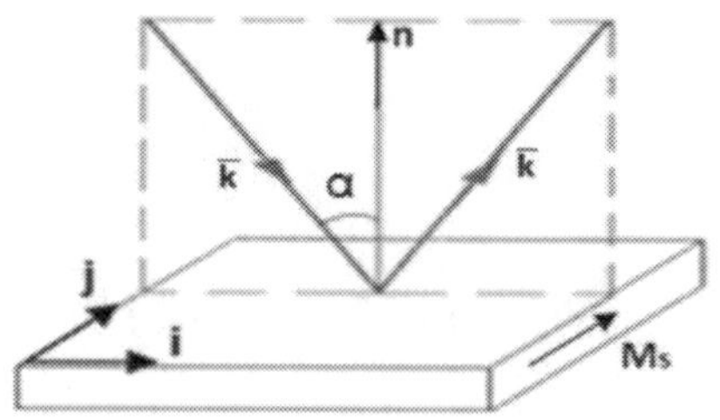

Fig.1 Schematic representation of the light wave vector **k** and the magnetization M at the TKE measurements, and the orientation of the magnetic field at FMR.

The general form of the TKE curves corresponds to the curves obtained in [7] for bulk composites of the identical composition deposited onto a fixed substrate. Fig.2b demonstrates the TKE dependence on the magnetizing field magnitude. This data was obtained at the fixed energy of the light quanta $E = 1.97$ eV. It is clearly seen that the samples with low content of the ferromagnetic phase (15-30%) exhibit a linear dependence of the TKE on the magnetic field magnitude. Such behavior of the curves indicates a superparamagnetic character of magnetization process, which is specific for the samples in the state below the percolation threshold [7,8]. Then there is the transitional sample with magnetic phase content $x = 35\%$, whose magnetic field dependence acquires the shape typical for ferromagnetics. With the further increase of concentration to 40% the contribution from the ferromagnetic component still grows. In the percolation region, the ferromagnetic component of the sample is magnetized as a single big cluster, yet it seems there still remain separate small magnetic clusters, which do not interact with each other and produce a superparamagnetic contribution to the magneto-optic response of the nanostructure. For this system the percolation threshold is in the region 35-40.7%, what is a little less than obtained in [7].

Fig. 2c shows TKE spectra of the $(Co_{40}Fe_{40}B_{20})_x(C)_{100-x}$ composites. The TKE spectra of the system 2 substantially differ from the spectra of the system 1. However, by their shape they are identical to the corresponding dependence of the spectrum of the $Co_{40}Fe_{40}B_{20}$ amorphous alloy. For the nanocomposites $(Co_{40}Fe_{40}B_{20})_x(C)_{100-x}$ in the spectral interval with the maximal MO response the TKE absolute magnitude is roughly 2.5-3 times as less than in the alloy $Co_{40}Fe_{40}B_{20}$. The major difference between these two systems is a virtually complete absence of the negative TKE in the metal-carbon system.

In fig.2d are given the results obtained for the TKE magnetic field dependence in the system with carbon content. Comparing this dependence with the similar one for the metal-dielectric system, we can say that already at low concentrations of the ferromagnetic phase the samples demonstrate non-linear magnetic field dependence, testifying the ferromagnetic behavior of the

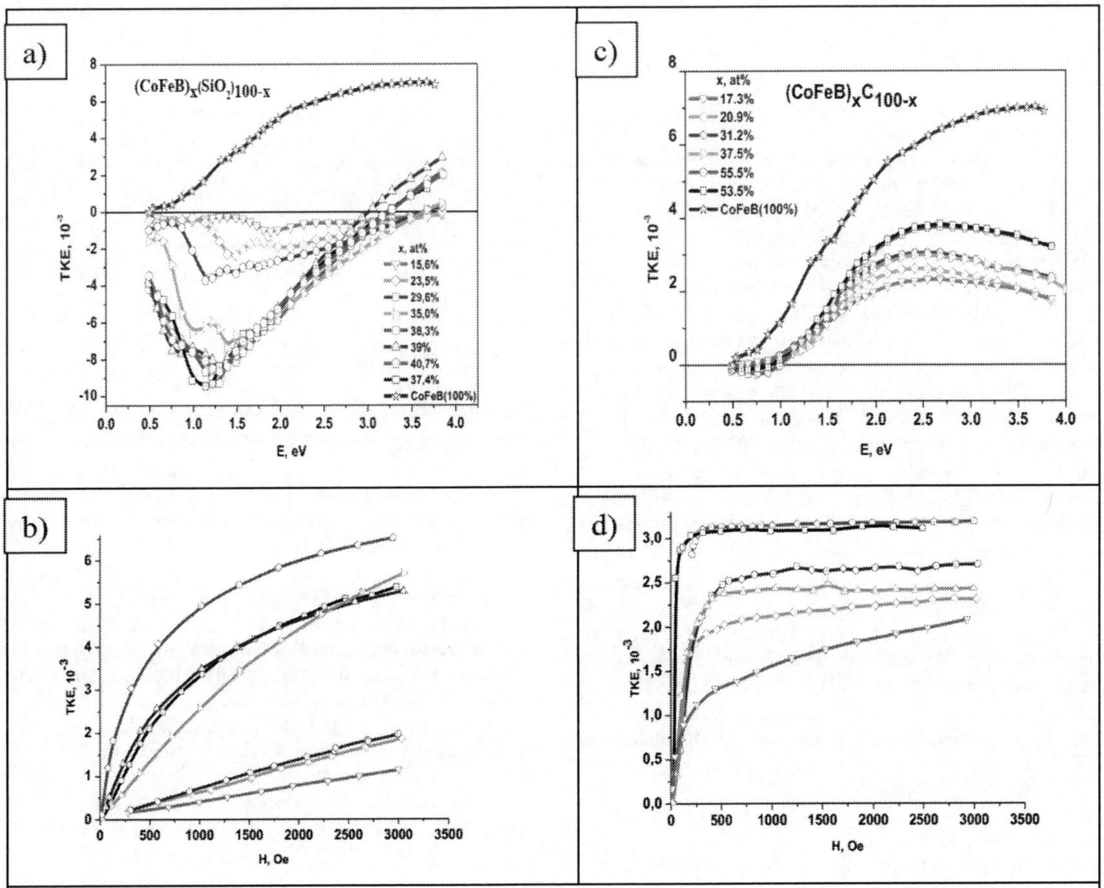

Fig. 2. Spectral and field dependences of TKE for system 1 (a, b) and system 2 (c, d)

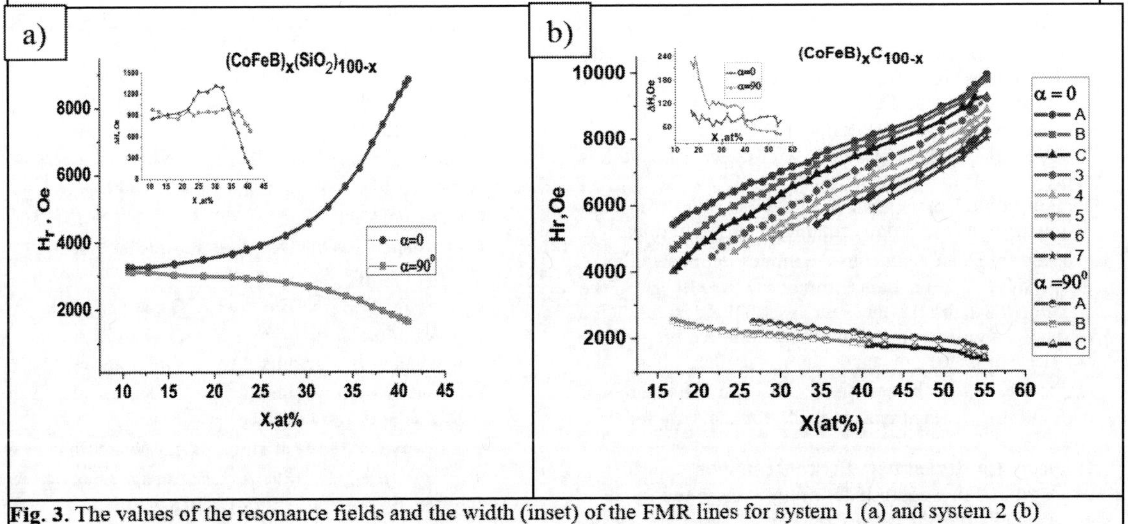

Fig. 3. The values of the resonance fields and the width (inset) of the FMR lines for system 1 (a) and system 2 (b)

system 2 in all investigated concentration range.

4.2 FMR properties

In the samples of the system 1 for $\alpha \in [0, 90°]$ only one mode was observed in FMR spectra. In the spectra of the system 2 samples as a rule, several absorption lines were seen at all orientations of **H**. With x decreasing from 56 to 17 at.% the number of modes in the spectra also decreased, namely, from 8 to 3 for $\alpha = 0$, and from 3 to 1 for $\alpha = 90°$. The modes observed at normal magnetization (α=0) in the spectra of all samples of system 2 with concentrations $17 \leq x \leq 56$ at.% will be further designated as A, B and C (in decreasing order by H_r). For tangential magnetization (α=90°) the mode A was observed in samples with $x \geq 23$ at.%, the mode B with $x \geq 17$ at.%, and the C mode - with $x \geq 40$ at.%.In approximation of weakly coupled oscillation circuits these modes can be interpreted as homogeneous oscillation modes of magnetization vectors of different magnetic sub-systems, each formed by nanoparticles of one particular composition (carbides of Co and Fe and/or CoFeB).

In the film plane an anisotropy of H_r was observed. At tangential magnetization the difference between H_r magnitudes for the cases $\boldsymbol{H} \perp \boldsymbol{i}$ and $\boldsymbol{H} \perp \boldsymbol{j}$ may reach magnitudes near 30, 100 and 200 Oe for the modes *A*, *B* and *C* correspondingly. The observed variations of H_r can be explained (in the framework of the theory of FMR in granular nanostrucures [9]) taking into account the following assumptions:

– the FMR properties are mainly influenced by shapes of magnetic granules;

– magnetic granules are ellipsoids with one of the main axes oriented normally to the film plane.

With α varying from 0 to 90° H_r monotonically decreases. The curves $H_r(\alpha)$ are almost symmetrical to the axis $\alpha = 0$.

The concentration dependence of resonance fields H_r for the normal and in-plane magnetization for the systems 1 and 2 are shown in fig.3a and fig.3b, correspondingly (the curves $H_r(x)$ are shown only for the case $\boldsymbol{H} \perp \boldsymbol{i}$).

In the samples of the system 1 with concentrations less than $x \approx 33$ at.% the resonance fields at normal magnetization are smaller, and at the in-plane magnetization, on the contrary, bigger, than the magnitudes H_r of the resonance modes in the samples of the system 2. For concentrations $x > 39$ at.% the situation changes. In the samples of the system 2 the resonance fields at normal magnetization are bigger, and at the in-plane magnetization smaller, than the magnitudes H_r of the resonance modes in the samples of the system 1 (comparison is made for the samples with equal x).

Fig.3a,b also demonstrate the concentration dependence of absorption line width H for the systems 1 and 2, correspondingly (the curves $H(x)$ are shown only for the case $\boldsymbol{H} \perp \boldsymbol{i}$).

Absorption lines of the *A*, *B* and *C* modes in the samples of the system 2, as a rule, have the widths around 50 – 200 Oe both for in-plane and normal magnetizations.

The width of a singe resonance mode in the samples of the system 1 with c $x < 32$ at.% is close to 1000 Oe. It is substantially larger (up to 10 times), than the widths of *A*, *B* and *C* absorption lines in the samples of the system 2 with the same x.

The observed abrupt decrease of H in the samples of the system 1 with increase of x from 32 to 41 at.% is explained by approaching to the percolation threshold. For the composite $(CoFeB)_x$-$(SiO_2)_{100-x}$ the percolation threshold – i.e. region of the structural transformation from superparamagnetic to ferromagnetic state, was observed at the concentration of magnetic granules $x \approx 43$ at.% [7].

5 Conclusion

A comparative analysis of experimental MO and FMR spectra of the systems $(CoFeB)_x(SiO_2)_{100-x}$ and $(CoFeB)_xC_{100-x}$ testifies that in the synthesis of the system 2 some plasma-chemical processes probably take place – chemical reactions of carbon with metals and boron. As a result there appear new (previously absent) compounds (presumably, Co_3C, Co_2C, Fe_3C, B_4C). Some formed nanoparticles exhibit magnetic properties. The multimode type of the FMR spectra of the system 2 can be explained by different compositions of the particles. The shape of the forming particles depends on their composition. With increase of the metallic phase content x the shape of the granules (independently of composition) come nearer to the spherical one.

This research was supported by the Russian Foundation for Basic Researches №15-02-02077

1. Vysulin S.A., Gorobinsky A.V., Lebedeva E.V., et.al., *Bulletin of the Russian Academy of Sciences: Physics*, 75 (2011) 1027-1030.
2. Buravtsova V., Gan'shina E., Lebedeva E., et.al. Solid State Phenomena Vol.168 (2011) p. 53
3. Chekrygina Ju., Devizenko A., Kalinin Yu., et.al., *Solid State Phenomena*, 2012.V. 190. P. 605-608.A.
4. Aleshnikov AA, Kalinin Yu.E., Kryachko AV, Sitnikov A.V., Bulletin of VSTU, 8 (2012) 77-82
5. Carroll K. J. and all., Applied physics letters, 2012. V. 101. 012409. P. 1-5.
6. Komogortsev SV, Iskhakov RS, Balaev AD et.al., *Physics of the Solid State*, 49(2007) 700-703.
7. Gan'shina E.A., Vashuk M.V., Vinogradov A.N. et al. Journal of Experimental and Theoretical Physic v. 98, № 5, c. 1027 (2004).
8. Buravtsova V.E., Gan'shina E.A., GustchinV.S. et al. *Bulletin of the Russian Academy of Sciences: Physics*, 67, № 7, 67 (2003) c. 918-920.
9. Vyzulin S.A., Lebedeva E.V., Lysak D.A., Syr'ev N.E., *Bulletin of the Russian Academy of Sciences: Physics*, 74 (2010) 1687–1689.

Influence of hydrogenation on magnetoresonance characteristics of nanocomposite $(CoFeB)_mC_{100-m}$ films

Sergei Vyzulin[1], *Alexandr* Kevraletin[1,*], and *Nikolaj* Syr'ev[2]

[1]Higher Military School named after General of the Army S.M. Shtemenko, 350063 Krasnodar, Russia
[2]Lomonosov Moscow State University, Faculty of Physics, 119991 Moscow, Russia

Abstract. The influence of hydrogenation on the magnetoresonance characteristics of nanocomposite films during synthesis is studied by the method of ferromagnetic resonance (FMR). It is shown that introduction of hydrogen into the working chamber during the synthesis of film nanogranular structures $(CoFeB)_mC_{100-m}$ leads to a change in a resonant field, to increase of the absorption line width, to reduction of anisotropy field in the film plane.

Introduction

Granular nanocomposite films, comprising magnetic and nonmagnetic phase, are promising functional materials for such varied practical applications as spintronic, single-electron electronics and so on [1].

Properties of nanocomposite films are under intensive research. The interest of researchers is associated with metal nanoparticles having the unique magnetic properties. The magnetic properties of nanoparticles are determined by many factors. One of these factors is the size and shape of nanoparticles. Changing the size and shape of nanoparticles, it is possible, within certain limits, to manage the magnetic properties of materials, based on them [2-3]. It is possible to change the size and shape of nanoparticles in the composites synthesis process. The effect of hydrogen during the synthesis to form carbon nanostructures was shown [4].

The aim of this work – to study the influence of hydrogenation on the magnetoresonance properties of nanocomposite $(CoFeB)_mC_{100-m}$ film structures by the basic characteristics of the ferromagnetic resonance spectra.

Object of investigation

Nanogranule ferromagnetic – nonmagnetic matrix film structures were studied. Samples are synthesized via the ion-beam sputtering of amorphous CoFeB with samples of C on to sitall substrate.

Two series of samples are investigated. The samples of 1 series were sputtered in an argon atmosphere, 2 series – in an atmosphere of argon with addition of hydrogen. Magnetic phase concentration m and thickness h were varied along the structure for both series in the same way ($35,3 \leq m \leq 65,0$ at. %, $1 \leq h \leq 2$ μm). In the transverse direction of the structure the thickness and concentration of magnetic phase were not changed. Direction in the sample plane coinciding with the longitudinal direction of the reference nanostructure will be denoted as vector **i**, and transverse direction as vector **j**.

Method of analysis

The FMR spectra were measured using conventional modulation equipment with resonator at a frequency of 9,44 GHz at a room temperature.

The spectra were measured for each sample under differently oriented magnetization field **H** relative to the film plane, from $\alpha=0$ (normal magnetization –$\mathbf{H}\|\mathbf{n}$) to $\alpha=90°$ (tangential magnetization – $\mathbf{H}\perp\mathbf{n}$), where **n** – is the vector normal to the film plane.

Two series of experiments were performed. In one the external magnetic field was applied as $\mathbf{H}\perp\mathbf{i}$, in another – $\mathbf{H}\perp\mathbf{j}$. The acquired spectra enabled us to determine resonance field H_r and absorption line width ΔH.

Experimental results

The form of the observed FMR spectra, the quantitative composition of the absorption lines and value of H_r are significantly dependent on the magnetic phase concentration m and the orientation of the external magnetic field relative the normal to the film plane.

With increasing m, with $\alpha=0$, number of absorption lines for samples of 1 series are varied from 1 to 3, for 2 series samples the quantity of modes in the spectrum reached 11 (fig. 1a, b).

In case $\alpha=90°$ for samples of 1 series there was mainly a single resonance mode. With increasing m the maximum number of mode reached 3. For samples 2 series, regardless of the values of m, the number of mode in the spectrum does not exceed 1 (fig. 1c, d).

* Corresponding author: kevr@list.ru

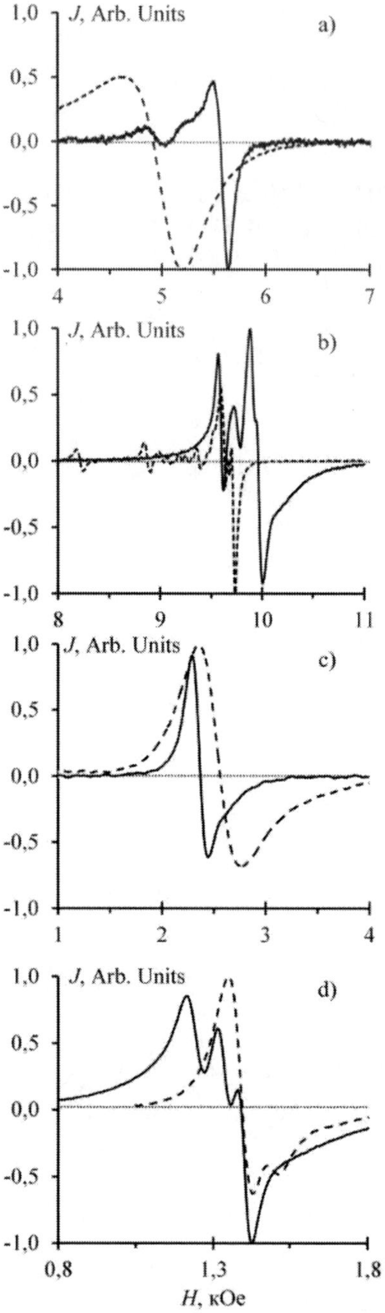

Fig. 1. FMR spectra (the solid line – for series 1, the dotted line – series 2):
a – m=38,1 at. %, b – m=65,3 at. % (α=0);
c – m=38,1 at. %, d – m=65,3 at. % (α=90°).

For all films the value of H_r at α=90° is smaller than at α=0. For samples of series 2 H_r with tangential magnetization is more, and in normal – less than that of the film structures of series 1 (fig. 2).

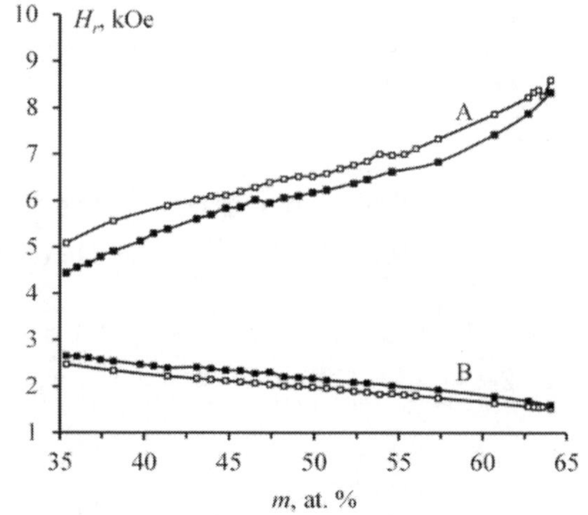

Fig. 2. Dependences H_r of m: □ – series 1, ■ – series 2; A - α=0; B - α=90°.

As m is increasing the value of H_r decreases when α=90°, and increases at α=0.

At α=90° the resonance field, observed in the same experimental conditions for the cases **H⊥i** and **H⊥j** differed for samples of both series. The differences registered in values of H_r are quantitatively characterized by the value of δH_r, that is determined by the ratio

$$\delta H_r = |H_r(\mathbf{H}\perp\mathbf{i}) - H_r(\mathbf{H}\perp\mathbf{j})|. \qquad (1)$$

Values of δH_r with the tangential magnetization for the samples of series 1 were up to 215 Oe, for series 2 - 119 Oe (Fig. 3). The solid line represents the dependence δH_r of m for the samples of series 1, the dotted – for samples of series 2. Experimental points are selected with markers.

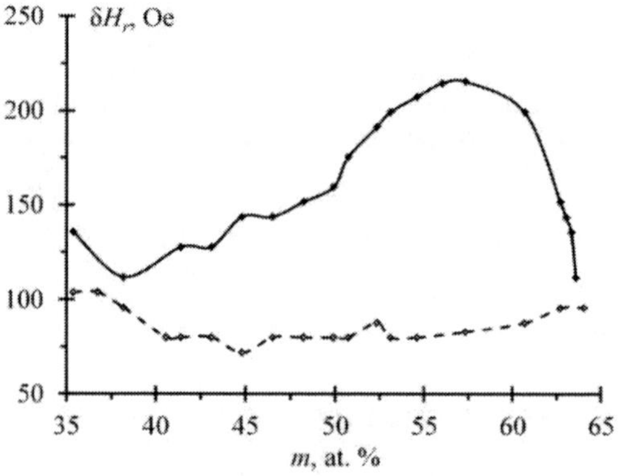

Fig. 3. Concentration dependences δH_r, α=90°: the solid line – series 1, the dotted line – series 2.

The presence of anisotropy H_r can be explained by the asymmetric shape of the magnetic granules with respect to the selected orientation axes **i** and **j** of the original nanostructures and **n**.

The width of the absorption line regardless of the orientation of the magnetization field for the samples of series 1 at a variation of m does not change significantly and has a magnitude of 100-200 Oe, for samples of series 2, with m>48 at. % the values of $\Delta H \approx 150$-200 Oe.

The Dependences ΔH from m for the films of series 1 and 2, with the tangential and normal magnetization are shown in figures 4a and 4b respectively.

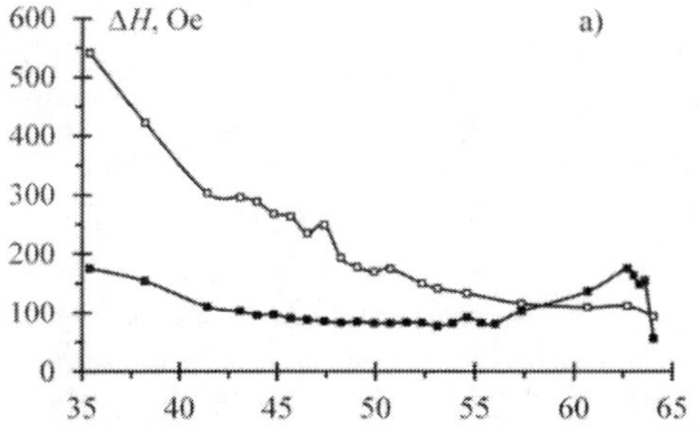

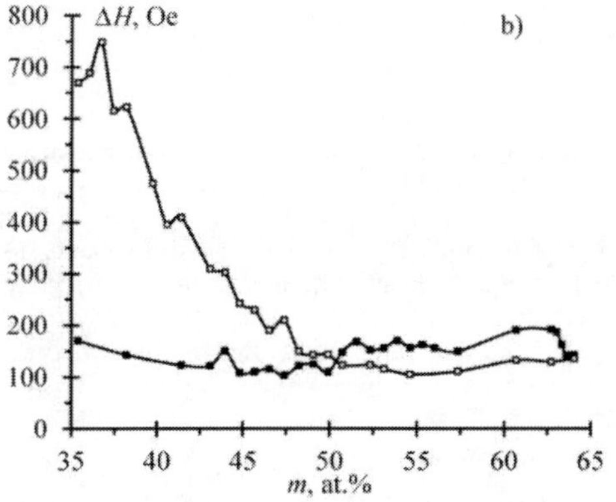

Fig. 4. Concentration dependences ΔH (□ – series 1; ■ – series 2): a – α=90°; b – α=0.

It is seen that for samples of series 2 with decreasing of m there is a sharp growth of ΔH up to the values of 550 Oe (with α=90°) and 750 Oe (with α=0). The latter indicates the possibility of transition of the magnetic system at 35<m<45 at. % from ferromagnetic to superparamagnetic state. For series 1, the percolation threshold wasn't observed. For a system similar to (composition and synthesis conditions) series 1, 48≤m≤72 at. %, the percolation threshold was observed at $m \approx 62$ at.% [5]. A possible explanation for the observed discrepancies is the difference in performance, by which the phenomenon of percolation is recorded. In [5] it is performed by electrical resistivity along the sample, in our work – on ΔH.

Evaluation of the shape of magnetic granules is made in the framework of the theory of FMR in granular nanostructures [6] by calculating N_x, N_y and N_z demagnetizing factors of magnetic granules in longitudinal and transverse directions of the film plane, respectively. The saturation magnetization of the samples was taken equal to 1050 G. Figure 5 shows dependences of the aspect ratio $\delta = a/c$ from m, a – is the magnetic granules semiaxes length along the normal to the film plane, c – in plane of the film. The values of δ are calculated on experimental values of H_r in the approximation of spheroidal form of the granules (the

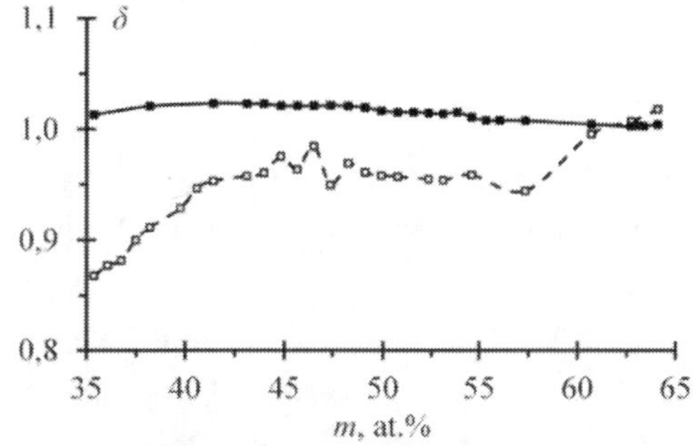

Fig. 5. Dependences δ of m: ■ – series 1; □ – series 2.

calculation method described in [6]).

It is seen that the magnetic granules of the samples of series 1 are stretched, and of series of 2 – are flattened along **n**. With the exception of nanocomposite film series 2 with m>60 at. %.

The absence of percolation threshold for series 1 is explained by the distance between the magnetic granules of the samples. The distance between the granules, elongated in the plane of the film is smaller than that between the granules, extended along the normal direction. Magnetic granules of the samples of series 1 interact with each other at lower concentrations than for series 2.

The observed differences in the resonance field for systems 1 and 2 with tangential magnetization in cases **H⊥i** and **H⊥j** are explained by the different form of magnetic granules. The shape of the magnetic granules varies as a result of hydrogenation in the synthesis of composite films.

Conclusion

It was established experimentally that the introduction of hydrogen into the working chamber during the synthesis of nanostructures film $(CoFeB)_mC_{100-m}$ leads to changes of values of the resonant fields: an increase in tangential and a decrease in the normal magnetization; to 1,5-2 times increase of the absorption line width. The value of anisotropy field in the plane of the samples subjected to hydrogenation during synthesis, decreases with the tangential magnetization in 1,5-2 times. The differences between magnetoresonance characteristics are explained by the shape of magnetic nanoparticles.

References

1. S. A. Vyzulin, A. L. Kevraletin, N. E. Syr'ev, Bull. Russ. Acad. Sci.: Phus., **81**, 3 (2017).
2. G. Yu. Yurkov, Russ. Chem. Reviews, **74**, 6 (2005).
3. D. S. Schmool, et. al., J. Appl. Phys. **101**, 103907 (2007).
4. R. K. Sahoo, H. Mamgain, C. Jacob, Bull. Mater. Sci., **37**, 6 (2014).
5. A. A. Abryckin, et. al., Bull. VSTU, **8** (2012).
6. S. A. Vyzulin et al. Bull. Russ. Acad. Sci.: Phys., **74** (2010).

Ferromagnetic resonance and random magnetic anisotropy in nanocrystalline Fe-Zr-N films

Elena Sheftel[1], *Eugene* Harin[1,*], *Valentin* Tedzhetov[1], and *Yuri* Koksharov[2]

[1]A.A. Baikov Institute of Metallurgy and Materials Science RAS, Moscow, Russia
[2]M.V. Lomonosov Moscow State University, Moscow, Russia

Abstract. The effective magnetic anisotropy of the ferromagnetic nanocrystalline Fe-Zr-N films prepared by magnetron sputtering under different conditions has been determined by both correlation magnetometry and ferromagnetic resonance methods. The effective magnetic anisotropy values obtained by the two methods are compared and the origin of the values is discussed.

1 Introduction

At present, efforts are being made to design soft magnetic nanocrystalline films intended for devices operating at high frequencies [1,2]. Thus, it is of interest to find correlations between parameters describing static magnetic properties (magnetic anisotropy and exchange interaction) and high frequency properties.

The random magnetic anisotropy model [3] is widely used to describe the hysteresis for nanocrystalline ferromagnets. In accordance with the model, the magnetization of individual grains is randomly oriented and can be described by an autocorrelation function for the magnetization. The magnetic autocorrelation length (R_L) is an important parameter of the function and is a typical magnetic microstructure radius corresponding to relatively uniform magnetization (a stochastic magnetic domain). The model is valid when $R_L > R_c$, where R_c is a grain radius, and grains are exchange coupled. In this case, magnetic hysteresis is determined by magnetic structure parameters, such as R_L, the effective magnetic anisotropy $<K_{eff}>$ at R_L scale, and the effective magnetic anisotropy K_{eff} at R_c scale. The correlation magnetometry method [4] allows one to determine these magnitudes using a magnetization curve.

In this case, not only the correlation magnetometry, but also ferromagnetic resonance (FMR) method allows one to determine the effective magnetic anisotropy.

The correspondence of two effective magnetic anisotropy values determined by the two methods, namely, by correlation magnetometry (static magnetic properties) and FMR methods, will allow us to estimate the high frequency properties using the data on measured static magnetic properties. This is of substantial interest since the measurement of high frequency magnetic permeability is difficult because of strict requirements for the dimensions and shape of films, kind of substrate and its thickness, and other experimental difficulties.

Among soft magnetic films studied in the recent years, the nanocrystalline Fe-Zr-N films are of substantial interest for investigators. This is related, on the one hand, to the possibility to prepare such films in amorphous and nanocrystalline states, in particular by magnetron sputtering, and, on the other hand, to the prospect to prepare such films with the unique static and high frequency soft magnetic properties [5,6].

The aim of the present study is to measure the effective magnetic anisotropy values for the Fe-Zr-N films using correlation magnetometry and FMR and to analyze the origin of the obtained values.

2 Experimental

The ferromagnetic nanocrystalline Fe-Zr-N films of 0.3-1.7 μm in thickness were prepared by magnetron sputtering of the Fe and Fe-Zr targets containing 2.4 and 5.3 wt.% Zr. The films were deposited on the glass substrates. The N_2 content in the $Ar+N_2$ gas mixture was 0, 5, and 15%; the gas mixture pressure was 0.13-0.2 Pa. The preparation conditions are described in detail in [6,7].

The structure of the films (phase composition, lattice parameter, and R_c grain radius) was studied by X-ray diffraction analysis [6]. The chemical composition of the as-sputtered films was determined by glow-discharge optical emission spectroscopy [8].

The saturation magnetization (M_s) was determined in magnetic fields up to 16 kOe using a vibrating sample magnetometer and Akulov's law of approach to saturation magnetization; depending on the preparation conditions of the films, their saturation magnetization was ranged within 1300-1700 G (1.6-2.1 T). The magnetic structure parameters (K_{eff}, $<K_{eff}>$, and R_L) were determined by correlation magnetometry [4].

The dependencies of FMR spectrum on applied dc magnetic fields up to 6 kOe were measured at the frequency f = 9.54 GHz using a Varian E-4 spectrometer.

* Corresponding author: harin-eugene@ya.ru

All measurements were performed at room temperature.

3 Results and discussion

3.1 Phase composition and structure

According to X-ray diffraction data, the films under study contain α-Fe, Fe_4N, Fe_3N, Fe_2N phases coexisting in various proportions; depending on the composition and the preparation conditions, the grain size of the phases is 2-50 nm (Table 1).

Table 1. Chemical composition and structure of the films under study

Film No.	Chemical composition, at.%	Phase composition, vol.%	α-Fe grain size ($2R_c$), nm
1	Fe	α-Fe	39±11
2	$Fe_{90}N_{10}$	α-Fe(N) + 5% Fe_2N	15
3	$Fe_{95}Zr_5$	α-Fe(Zr)	34±3
4	$Fe_{85}Zr_5N_{10}$	α-Fe(N,Zr) + 7% Fe_2N	11±1
5	$Fe_{86}Zr_{4.3}N_{6.7}O_{2.9}$	α-Fe + 42% Fe_4N	4±2
6	$Fe_{87.3}Zr_{2.4}N_{7.4}O_{2.9}$	α-Fe + 64% Fe_4N + 5% Fe_3N	7±3
7	$Fe_{90.4}Zr_{2.9}N_{4.7}O_2$	α-Fe	17±3
8	$Fe_{75.8}Zr_{8.7}N_{6.8}O_{8.7}$	α-Fe	5±1
9	$Fe_{82.5}Zr_{3.1}N_{6.4}O_8$	α-Fe + 42% Fe_4N	10±2
10	Fe	α-Fe	46

3.2 FMR study

Differential microwave absorption spectra for the films under study were measured in dc magnetic fields applied in the film plane (Fig. 1). The in-plane FMR condition is determined by Kittel's law [9]

$$(f/\gamma)^2 = (H_{res} + 4\pi M_s)(H_{res} + 2K/M_s), \quad (1)$$

where $\gamma = 2.9$ MHz/Oe is the gyromagnetic ratio, K is the FMR-driving magnetic anisotropy, and H_{res} is the resonant field. The H_{res} values were determined as the applied dc field values corresponding to the midpoint between differential peaks using the in-plane FMR spectra ("peak-to-peak" method, Fig. 1, Table 2). After that, the K values were determined by Eq. (1).

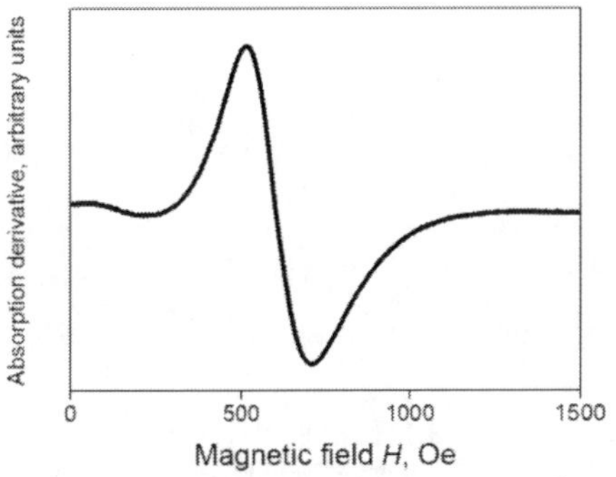

Fig. 1. In-the-film-plane FMR spectrum of the film No. 2.

3.3 Correlation magnetometry study

In terms of correlation magnetometry theory, the magnetization curves measured in high magnetic fields were recalculated for the magnetization dispersion $d_m = (M_s - M)/M_s$ and plotted on log-log scale. This allowed us to find two laws of approach to saturation magnetization (Fig. 2): $d_m = (1/2)(DH_a)^2/(H_R^{3/2}H^{1/2})$ in fields $H < H_R$ and $d_m = (1/2)(DH_a/H)^2$ in fields $H > H_R$ ($D = 1/15^{1/2}$ for the uniaxial anisotropy), which were used to determine the local magnetic anisotropy ($K_{eff} = H_aM_s/2$) and the exchange field (H_R, Table 2).

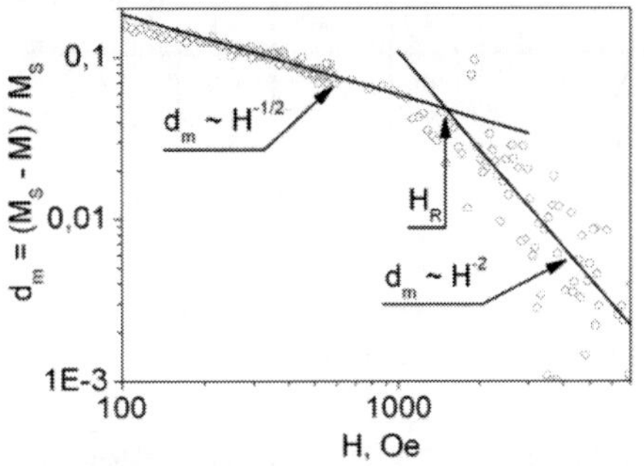

Fig. 2. The magnetization dispersion of the film No. 2. Solid lines are described in the text.

When such a material is under magnetization process, three phenomena occur simultaneously, which, as the applied magnetic field increases, become dominant in the following sequence: (1) stochastic domain magnetization rotation within $2R_L$, (2) progressive decrease in the stochastic domain size down to $2R_c$ grain scale, and (3) in fields $H > H_R$, magnetization rotation within individual grains (K_{eff} within $2R_c$). The magnetic structure of the films under study was discussed in detail in our previous works [6,7].

According to [4], the dependence of effective magnetic anisotropy of nanocrystalline ferromagnets on the applied dc magnetic field can be described by several different functions. In particular, at $H > H_R$, the effective magnetic anisotropy [4] is constant and equal to the K_{eff}

magnitude within a ferromagnetic grain. At $H < H_R$, the effective magnetic anisotropy equals

$$K_H = K_{eff}(R_c/R_H)^{3/2}, \quad (2)$$

where $R_H = (2A/M_sH)^{1/2}$ is the magnetic autocorrelation radius in H field and A is the exchange stiffness.

Table 2. Magnetic properties of the films under study

Film No.	Chemical composition, at.%	M_s, G	H_R, Oe	H_{res}, Oe	K_{eff}, 10^5 erg/cm^3	K, 10^5 erg/cm^3
1	Fe	1683 ±8	1700 ±700	1690 ±30	10.1 ±3.4	10.2 ±0.4
2	$Fe_{90}N_{10}$	1332 ±20	1500 ±400	830 ±30	12 ±2.9	1.4 ±0.3
3	$Fe_{95}Zr_5$	1598 ±9	280 ±40	1840 ±30	6.4 ±0.8	10.8 ±0.3
4	$Fe_{85}Zr_5N_{10}$	1310 ±7	800 ±50	1920 ±30	10.5 ±0.7	8.7 ±0.3
5	$Fe_{86}Zr_{4.3}N_{6.7}O_{2.9}$	1439 ±5	21389	643	7.2	0.47 ±0.01
6	$Fe_{87.3}Zr_{2.4}N_{7.4}O_{2.9}$	1318 ±0.5	13000 ±700	596 ±1	4.1 ±0.1	0.23 ±0.01
7	$Fe_{90.4}Zr_{2.9}N_{4.7}O_2$	1720 ±12	25022	563 ±1	8.7	0.64 ±0.05
8	$Fe_{75.8}Zr_{8.7}N_{6.8}O_{8.7}$	1371 ±1	16000 ±2600	695 ±1	4.3 ±0.5	0.62 ±0.01
9	$Fe_{82.5}Zr_{3.1}N_{6.4}O_8$	1404 ±9	25120	627 ±1	7.2 ±0.1	0.24 ±0.03
10	Fe	1414	22000 ±1300	275 ±15	6.7 ±0.7	2.3 ±0.11

3.4 Comparison of magnetic anisotropy from FMR and correlation magnetometry

In accordance with the resonant field value (H_{res}) from Eq. (1), the films under study can be divided into two groups: $H_{res} \geq H_R$ (group I, 3 films in Fig. 3) and $H_{res} < H_R$ (group II, 7 films in Fig. 3).

In accordance with the aforementioned considerations, the magnetic anisotropy determined from the resonant condition is $K \approx K_{eff}$ for group I (3 films in Fig. 4) and $K < K_{eff}$ for group II (7 films in Fig. 4).

In the case of group II, the K_H and K magnitudes correlate adequately. In particular, for film No. 2 (Figs. 1 and 2, Table 2), the resonant field is $H < H_R$. In accordance with the aforementioned considerations, for this film, we obtain $K < K_{eff}$. For this film, Eq. (2) gives $K_H = (7.4\pm4.3)\cdot10^5$ erg/cm^3, which agrees adequately with $K = (1.4\pm0.3)\cdot10^5$ erg/cm^3.

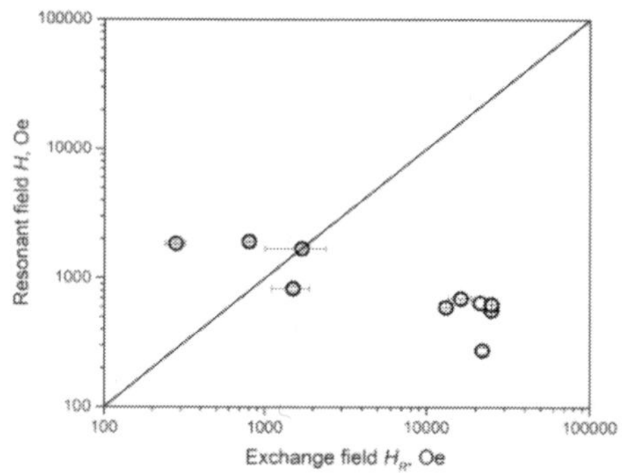

Fig. 3. Comparison of the resonant (H_{res}) and exchange (H_R) fields for the films under study. Solid line corresponds to $H = H_R$.

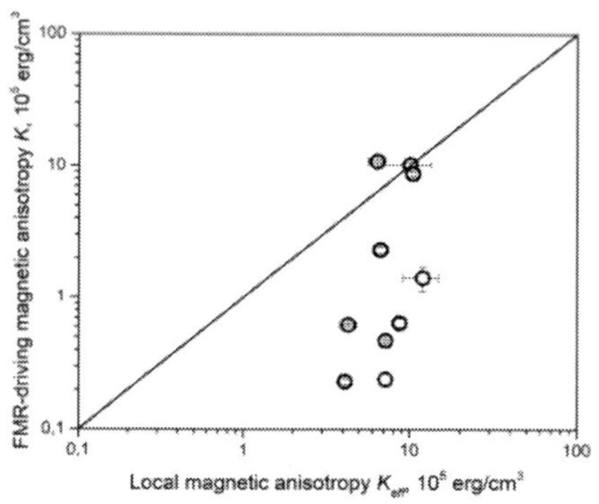

Fig. 4. Comparison of the FMR-driving magnetic anisotropy (K) and the local magnetic anisotropy (K_{eff}) for the films under study. Solid line corresponds to $K = K_{eff}$.

4 Conclusions

The effective magnetic anisotropy of the ferromagnetic nanocrystalline Fe-Zr-N films prepared by magnetron sputtering under different conditions has been measured by correlation magnetometry and FMR methods.

For some films, the resonant microwave power absorption takes place when the applied dc magnetic field is lower than the exchange field ($H_{res} < H_R$). It means, in this external magnetic field ($H_{res} < H_R$), the resonance condition is determined by the magnetic anisotropy within a stochastic domain, which exists in this applied dc magnetic field, but not the magnetic anisotropy K_{eff} within a grain.

Future investigations will be related to the estimation of the dependence of the resonant field value on the frequency of the ferromagnetic resonance, which will allow us to obtain the dependence of effective magnetic anisotropy of stochastic domain on the applied magnetic field for a material with the stochastic domain structure.

This study was supported by RFBR (project No. 15-08-02831a).

References

1. S.S. Kalarickal, P. Krivosik, M. Wu, C.E. Patton, M.L. Schneider *et al.* J. Appl. Phys. **99**, 093909 (2006)
2. H.-S. Song, K.-D. Lee, Ch.-Y. You, B.-G. Park, J.-I. Hong, J. Magn. Magn. Mat. **406**, 129 (2016)
3. G. Herzer, Acta Materialia **61**, 718 (2013)
4. R.S. Iskhakov, S.V. Komogortsev, Phys. of Metals and Metallography **112**, 666 (2011)
5. R.S. Iskhakov, M.M. Brushtunov, A.G. Narmonev, *et al.* Phys. Metals and Metallography **79**, Iss. 5, 122 (1995)
6. E.N. Sheftel, E.V. Harin, V.A. Tedzhetov, *et al.* Physica B **494**, 13 (2016)
7. E.V. Harin, E.N. Sheftel, Phys. Metals and Metallography **116**, 753 (2015)
8. Ph.V. Kiryukhantsev-Korneev, Prot. Met. Phys. Chem. **48**, 585 (2012)
9. F. Gerhardter, Y. Li, K. Baberschke, Phys. Rev. B **47**, 11204 (1993)

Magnetic properties of poly(*p*-xylylene)-Ni nanocomposites

Artem Vdovichenko[1,2,*], *Dmitry* Streltsov[1,2], *Anton* Orekhov[1], *Alexander* Vasiliev[1], *Yan* Zubavichus[1], *Evgenii* Grigoriev[1], *Sergei* Zavyalov[1], *Leonid* Oveshnikov[1,3], and *Sergei* Chvalun[1,2]

[1]*National Research Center "Kurchatov Institute", 123182 Moscow, Russia*
[2]*N.S. Enikolopov Institute of Synthetic Polymer Materials RAS, 117393 Moscow, Russia*
[3]*P.N. Lebedev Physical Institute RAS, 119991 Moscow, Russia*

Abstract. The dependence of electrical and magnetic properties on structure of thin poly(*p*-xylylene) - nickel nanocomposite films with Ni concentrations from 5 to 30 vol.% was studied. It was found that metal concentration strongly affects size and oxidation state of the nanoparticles. Effect of the filler content on electrical and magnetic properties of the nanocomposites was revealed. These properties are determined by percolation phenomenon, with the percolation threshold value of about 10 vol.%. The well pronounced magnetic hysteresis as well as ferromagnetic ordering were observed at Ni content above the percolation threshold. The diagrams of magnetic properties of these composites as a function of composition and temperature were elaborated.

1 Introduction

Magnetic polymer nanocomposites are perspective materials due to possibility to combine several components with different properties in a single material. Organic-inorganic synergies add new properties that cannot be achieved in just organic or inorganic materials. Furthermore, when magnetic particles are of nano-size its magnetic properties are different qualitatively from the bulk material, they vary with the particle size, and can be affected by the polymer interphase. A lot of papers on magnetic, magnetoresistive properties of the polymer nanocomposites and spintronic devices have been published revealing recent progress in spintronics research [1–6].

Spin-valve structures are the subject of special interest in these studies [7]. The devices employing the effect of the giant (or tunneling) magnetoresistance are generally three layer systems. A non-magnetic layer is sandwiched between hard- and soft-ferromagnetic ones. On the basis of such structures magnetoresistive memory cells as well as scanning magnetic heads for high-density HDDs can be produced. It should be noted that polymer materials are often used as the non-magnetic interlayer only [4, 5, 8]. However, the use of single nanocomposite material with embedded magnetic particles in spin-valve applications seems to be technologically promising.

2 Experimental

The poly(*p*-xylylene) (PPX) - Ni nanocomposite films have been prepared by low-temperature vapor deposition polymerization (VDP) technique [9], which is a powerful

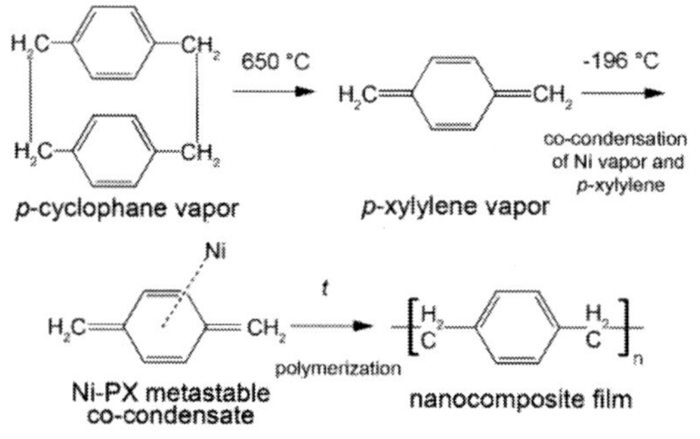

Figure 1. Scheme of synthesis of Ni-PPX nanocomposite films.

vacuum deposition technique enabling preparation of thin hybrid nanocomposite films with a high content of inorganic filler (up to several tens of volume percents) [9, 10]. The process is carried out without solvent or catalyst and allows to form nanocomposite films of controllable thickness in the range from several nanometers to several hundred micrometers with sizes of inorganic nanoparticles down to 2 nm and adjustable concentration of the filler.

The process consists of two stages (see Fig. 1). The first stage is a co-condensation of Ni vapor and *p*-xylylene (PX) monomer on the substrate cooled down to -196 °C. During the deposition, a vacuum better than 10^{-5} Torr was maintained in the VDP chamber. Upon the second stage of the process, during slow heating of the co-condensate up to room temperature, the polymerization of PX into PPX and aggregation of the nickel clusters and atoms into nanoparticles occur. The PX monomer was prepared by pyrolysis of cyclic dimer of PX ([2.2]*p*-cyclophane) (Daisan Kasei, Japan) at 650 °C using classical Gorham's method [11].

*e-mail: vdartem@ya.ru

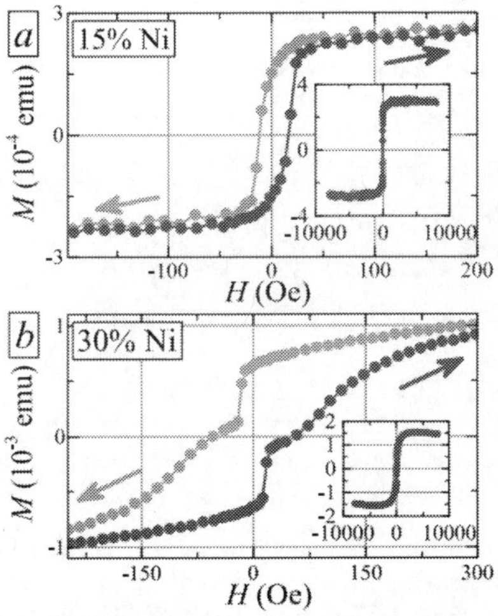

Figure 2. (color online) Magnetic field dependences of magnetic moments for the samples a) 15 vol.% of Ni and b) 30 vol.% of Ni at room temperature (300K). Insets show the same dependences in the wider range of magnetic fields up to 10000 Oe. These curves clearly demonstrate the saturation of magnetization in fields of approximately 200 Oe for the sample 15 vol.% and 4000 Oe for the sample 30 vol.%.

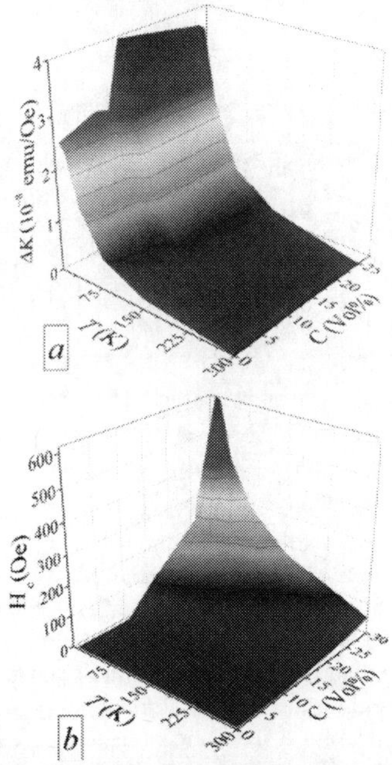

Figure 3. (color online) Temperature and composition dependences of a) ΔK and b) coercive force H_c.

Ni vapor was produced by thermal evaporation. The PPX-Ni samples with Ni concentration in the range from 5 to 30 vol.% were formed. The film thickness was about 300 nm for X-ray diffraction, magnetic and electrical measurements, while for transmission electron microscopy (TEM) studies it was less than 100 nm.

Magnetic properties measurements of the synthesized nanocomposites were carried out with a magnetometer LakeShore 7407 in magnetic fields up to 16000 Oe and with a SQUID magnetometer (Quantum Design, USA) in magnetic fields up to 70000 Oe at temperatures in the range from 4.2 to 300 K. The magnetic field was parallel to the film surface.

The microstuctural analysis was performed in a transmission/scanning electron microscope Titan 80-300 TEM/STEM (FEI, USA). The crystalline structure of the metal/polymer nanocomposites was studied using Rigaku SmartLab (Japan).

3 Results

3.1 Magnetic properties

The synthesized nanocomposite films reveal metal type of conductivity for both 15 and 30 vol.% samples with longitudinal resistance R_{xx} increasing linearly with temperature. The resistivity of the 30 vol.% sample is about $2 \cdot 10^{-3}$ Ohm·cm and the temperature coefficient of resistance $\alpha = 0.015$ K^{-1} is just twice as large as α for bulk nickel.

Fig. 2 presents the dependence of magnetic moment on external magnetic field for two samples with Ni content above the percolation threshold (15 vol.% and 30 vol.%, correspondingly) measured at room temperature. Both samples demonstrate the ferromagnetic behavior showing well pronounced magnetic hysteresis. These plots reveal the establishment of long-range ordering and magnetic percolation even in the sample with 15 vol.% of Ni. These observations confirm that the magnetic percolation threshold is less than 15 vol.%.

The magnetization saturates at room temperature at fields of approximately 200 and 4000 Oe for the samples with 15 and 30 vol.% of Ni respectively. The value of the coercive force at room temperature is found to be 13 Oe for the sample with 15 vol.% of Ni and 50 Oe for the sample with 30 vol.% of Ni. The saturation magnetic moment of the sample with 30 vol.% of Ni is more than 5 times higher than that of the sample with 15 vol.% of Ni (see Fig. 2). The difference in magnetic properties of these samples is in agreement with the difference in Ni content taking into account the percolation character of exchange interaction in these type of systems. In particular, such behavior is related to the nonlinear increase of percolation cluster volume at such C values.

The magnetic hysteresis curve of the sample with 30 vol.% of Ni shows narrowing in the vicinity of zero magnetic field, unlike that of the sample with 15 vol.% of Ni (Fig. 2). Such narrowing is typical for ferromagnetic material with wide particle size distribution or at least two fractions of particles of different size because of their different contribution to the sample magnetization. One can explain this observation by wider particle size distribution in the PPX-Ni nanocomposite films with nickel concentration significantly above the percolation threshold, where some metal particles aggregates into bigger clusters while others stay individual. However, if the metal concentration is close to the percolation threshold, fraction of aggregated particles is low compared to individual ones.

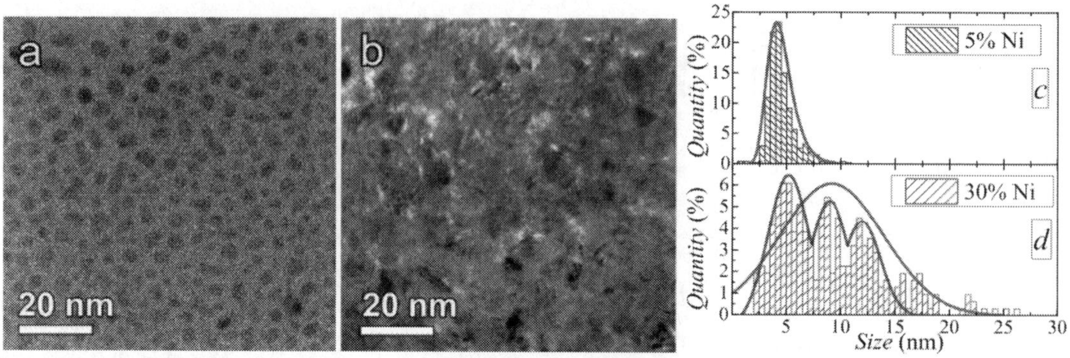

Figure 4. (color online) Bright-field TEM images of the nanocomposite films a) with $C = 5$ vol.% of Ni and b) 30 vol.% of Ni. c) and d) - corresponding nanoparticle size distributions. Solid curves - normal distribution functions; red - single-modal, blue - multi-modal (see text).

Fig. 3 shows the summarized results of magnetic measurements for the PPX-Ni composite films with various concentrations of nickel. All studied samples including pure PPX film demonstrate linear $M(H)$ dependence for $H > 40000$ Oe in the temperature range from 150 K down to 10 K. The diamagnetic contribution of substrate was subtracted. Plotting the change of linear slope ΔK as a function of temperature and concentration and using data for pure PPX sample as a reference one can evaluate paramagnetic contribution related to the Ni filler. One can see that an increase of Ni concentration leads to enhancement of paramagnetic response (Fig. 3a). Its magnitude has non-linear dependence on the filler concentration C, which can be attributed to the presence of oxidized Ni particles (e.g. NiO particles).

$M(H)$ dependences for samples with $C > 15$ vol.% at low magnetic fields reveal well-pronounced hysteresis loops (see Fig. 2), that have clear temperature dependences. Using the coercive force H_c as a marker of ferromagnetic properties, one can plot corresponding temperature-composition diagram (Fig. 3b). A growth of the nickel concentration C from 15 to 30 vol.% results in an increase of the coercive force H_c at temperatures below 40 K by a factor of 6. Thus, introduction of Ni into PPX matrix leads to an increase of paramagnetic contribution and at $C > 15$ vol.% ferromagnetic phase is formed, even at room temperature.

3.2 Structure

The magnetic and electrical properties of the PPX-Ni nanocomposite films described above should be caused by structural changes with the increase of Ni content.

The structure of the PPX-Ni nanocomposite films was studied by transmission electronic microscopy (TEM), high resolution TEM (HRTEM), and X-ray diffraction patterns (XRD).

Bright-field (BF) TEM images of the PPX-Ni films with 5 and 30 vol.% of Ni are shown in Fig. 4a and 4b, respectively. In the micrographs, the Ni particles can be identified by a darker appearance. As one can see in Fig. 4a, in the sample with low nickel concentration the metal particles are isolated and homogeneously distributed in the polymer matrix. The nanoparticles are spherical or slightly elongated with the aspect ratio of up to 2. Their average size is about 4-5 nm (see Fig. 4c). The surface density (the number of particles per unit area) is about $3 \cdot 10^{11} \mathrm{cm}^{-2}$. At high filler concentration the inorganic particles are distributed quite uniformly (see Fig. 4b). Their average size is about 8-10 nm, if we assume that their size distribution is also single-modal (see Fig. 4d - red line). Thus, the nanoparticles in studied films with high filler concentration are about twice as big as the ones in the sample with low filler content. However, their density is more or less the same for the samples with 5 and 30 vol.% of Ni. It should be noted that particle size distribution for the sample with 30 vol.% of Ni can be also interpreted as a multi-modal while for sample with 5 vol.% of Ni it is clearly single-modal. As an example we provide three-peak line as a guide for an eye, which is just a superposition of three normal distribution curves (see Fig. 4d). One can see that multi-modal curve gives a better fit for present data than single-modal one. The actual number of peaks in multi-modal distribution cannot be retrieved from our data, but even a single-modal distribution (being wide enough, as it is in present case) can explain the complex form of hysteresis loop for this sample (see Fig. 2b).

Chemical nature of nanoparticles is of key importance for rationalization of their electrical and magnetic properties. The difference in structure of the PPX-Ni nanocomposites with 5 and 30 vol.% of nickel is mainly related to size of the metal nanoparticles. A two-fold increase in size of the nanoparticles results in a decrease in average distance between adjacent nanoparticles and as a consequence leads to dramatic difference in their electric and magnetic properties.

The X-ray diffraction (XRD) patterns of the samples with 5, 9, and 30 vol.% of Ni are shown in the Fig. 5. The patterns corresponds to those of bulk Ni and NiO crystals. Observed peaks are considerably broadened compared to the single crystal case, which signifies the finite size of nanoparticles.

One can see that sample with Ni concentration of 30 vol.% have peaks at $d = 0.203$ and 0.178 nm. These peaks are attributed to the planes (111) and (200) respectively of the fcc-lattice of metallic nickel.

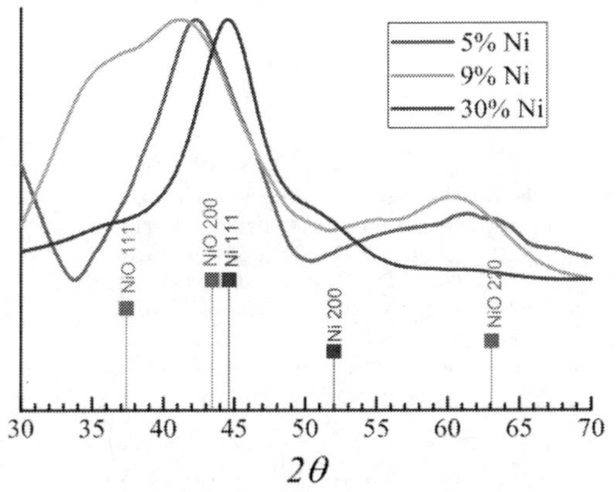

Figure 5. (color online) XRD patterns of three samples with 5 (red), 9 (green), and 30 (blue) vol. % Ni. $\lambda = 1.5406$Å.

At concentration 9 and 5 vol.% of Ni the NiO-related peaks in obtained patterns are observed. We can identify these broadened peaks as (111), (200) and (220) planes of NiO.

Therefore, in addition to structural changes observed by TEM, the growth of Ni concentration in the PPX-Ni nanocomposite films also affects the oxidation state of the inorganic nanoparticles.

4 Conclusions

The growth of Ni concentration in the PPX-Ni nanocomposite films affects the oxidation state of the inorganic nanoparticles. Most of the particles in the composite films with low filler content (C = 5 vol.% of Ni) are oxidized (NiO), whereas the particles in the PPX-Ni composite with 30 vol.% of Ni are mainly metallic.

The magnetic and electrical properties of the PPX-Ni nanocomposite films are determined by their structure being of the percolation type. The well pronounced magnetic hysteresis as well as ferromagnetic ordering were observed at Ni content above the percolation threshold. On the base of obtained data the diagrams of magnetic properties of the PPX-Ni nanocomposite as a function of composition and temperature including both paramagnetic and ferromagnetic contributions were elaborated. These graphs demonstrate that variation in composition of the nanocomposite films results in significant changes of their magnetic properties. In particular, pronounced ferromagnetic phase appears at room temperature in samples with $C > 15$ vol.%, although introduction of Ni into PPX matrix leads to an increase of paramagnetic contribution for all studied C values (6-30 vol.%).

This work was partially supported by the Russian Foundation for Basic Research (grant #15-29-01267). The study was carried out using the equipment of the Resource Centers of Kurchatov's Complex of NBICS-Technologies.

References

[1] S. Majumdar, H. S. Majumdar, H. Aarnio, and R. Osterbacka, Phys. Status Solidi - Rapid Res. Lett. **3**, 242 (2009).

[2] E. I. Grigoriev, S. A. Zavyalov, and S. N. Chvalun, Polym. Sci. Ser. A **45**, 1308 (2003).

[3] P. K. Choudhury, S. Ramaprabhu, K. P. Ramesh, and R. Menon, J. Phys. Condens. Matter **23**, 265303 (2011).

[4] Y. Liu, S. M. Watson, T. Lee, J. M. Gorham, H. E. Katz, J. A. Borchers, H. D. Fairbrother, and D. H. Reich, Phys. Rev. B **79**, 75312 (2009).

[5] M. Grunewald, J. Kleinlein, F. Syrowatka, F. Wurthner, L. W. Molenkamp, and G. Schmidt, Org. Electron. **14**, 2082 (2013).

[6] K. Akamatsu, S. Adachi, T. Tsuruoka, S. Ikeda, S. Tomita, and H. Nawafune, Chem. Mater. **20**, 3042 (2008).

[7] E. Hirota, H. Sakakima, and K. Inomata, *Giant Magneto Resistance Devices* (Springer Science and Business Media, 2002).

[8] A. Riminucci, I. Bergenti, L. E. Hueso, M. Murgia, C. Taliani, Y. Zhan, F. Casoli, M. P. de Jong, and V. Dediu, arXiv:cond-mat/0701603 1 (2007).

[9] H. Hopf, G. N. Gerasimov, S. N. Chavalun, V. I. Rozenberg, E. L. Popova, E. V. Nikolaeva, E. I. Grogoriev, S. A. Zavjalov, and L. I. Trakhtenberg, Chem. Vap. Depos. **3**, 197 (1997).

[10] G. N. Gerasimov, V. A. Sochilin, S. N. Chvalun, L. V. Volkova, and I. Y. Kardash, Macromol. Chem. Phys. **197**, 1387 (1996).

[11] W. F. Gorham, J. Polym. Sci. Part A-1 Polym. Chem. **4**, 3027 (1966).

Interaction of magnetic elements in metallic ferromagnetic nanofilms

Panaetov V. P[1].

[1] *State establishment "High mountainous geophysical institute", Kabardino-Balkar Republic. Nalchik, Lenin Street 360030. 2. Rossiya. _email: Valen_pvs@rambler.ru*

Abstract. In this paper, a comparative analysis of the distances between magnetic elements is made, under which a noticeable effect arises when they interact. The sides of the domain (domain walls), the terminal part of the domain (domain apex), we called magnetic elements. During magnetization, the domain structure is restructured. The distances between the domain walls, between the domain wall and domain apex, between the apexes of the domains change. A metallic ferromagnetic nanofilm can be a medium for recording information with the help of magnetic moments of electrons. Interaction of domain walls, the interaction of the domain wall and the domain apex, the interaction of the domain apexes in the design of storage devices must be taken into account.

1 The interaction of homogeneous domain walls

Micro magnetization processes were observed at high magnification in a transmission electron microscope UMV-100K by the method of Lorentz microscopy. The films were obtained by the method described in [1]. In [1] it was shown theoretically and experimentally in which cases the existence of double domain walls in nanofilms with definite magnetic parameters is energetically more advantageous The width of the domain walls changes as the distance between them changes. In [2], double walls with the same and opposite direction of rotation of the magnetization vectors in nanofilms were experimentally proved. It follows from theoretical calculations that for the same direction of rotation of the magnetization vectors in the walls, the width of the walls should decrease with decreasing distance between the walls and increase for walls with the opposite direction of rotation of the magnetization vectors in the walls. By photometry of domain walls at different distances between them, the following graphs of the dependence of the width of the domain walls on the distance between them were obtained. With the same direction of rotation of the magnetization vectors in the walls, the walls are repelled from each other. With an opposite direction of rotation of the magnetization vectors, attraction arises in the walls. These studies are presented in [2-3].

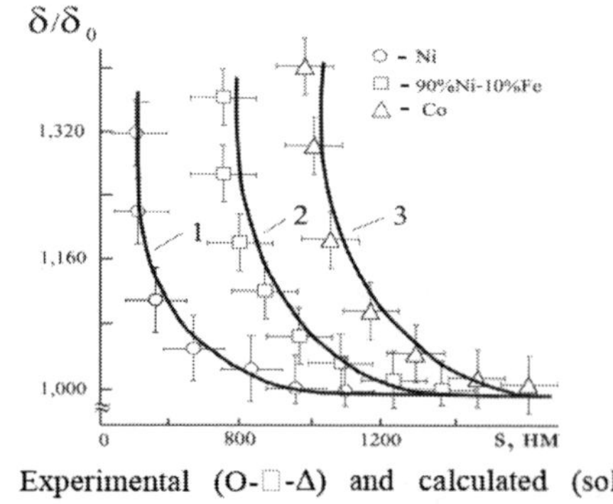

Fig.1. Experimental (O-□-Δ) and calculated (solid) curves for the dependence the width domain walls on the distances between them. δ – The width of the interacting wall. δ_0 – The width no interacting wall.

The experimental dependences of δ / δ_0 on the distance s for nickel films (I_s = 0,48 A / m), cobalt (I_s = 1,40 A / m) and $Ni_{0,90}$-$Fe_{0,1}$ (I_s = 0,65 A /m) are shown in Fig. 1. I_s - saturation magnetization. The distance between the walls increases with increasing saturation magnetization.

2 Interaction of periodic domain walls

Periodic domain walls do not interact with each other in Fig. 2a. This is manifested in the fact that the transverse bonds on the domain walls smoothly transform into the ripple of the magnetization, there are no curvatures of the transverse bonds, there are no bends of the domain walls. If an external magnetic field is applied along the axis of easy magnetization (AEM), it is possible to reduce the distance between the walls so that the transverse bonds of one wall are closed onto the circular lines of the other wall. Such an interaction is shown in Fig. 2b.

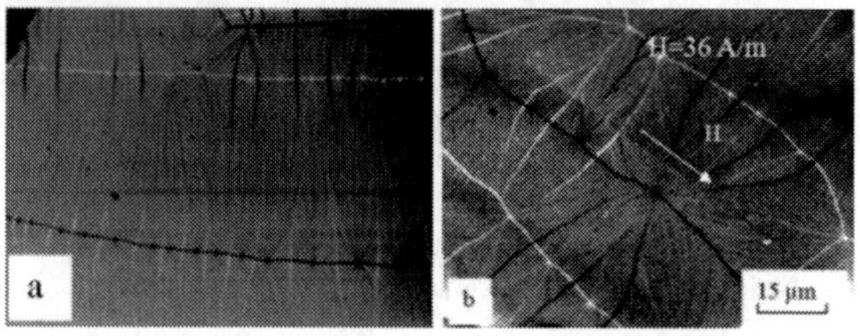

Fig.2. a – Non interacting periodic domain walls. b - Interactive periodic domain walls. Composition $Fe_{0,25}$-$Ni_{0,75}$, thickness of film 40 nm. Coercive force H_c=28 A/m, uniaxial anisotropy, H_k=160 A/m.

3. Interaction of the domain wall with of the domain apex

On the walls and of domains apexes there is a violation of the continuity conditions for the magnetization vectors I_s (div I_s = 0), then the walls and apices are sources of scattering fields. Therefore, they must interact with each other. When the external magnetic field applied along the AEM is changed, the domain increases or decreases depending on the direction of the field, and the distance between of the domain apex and the domain wall changes. The films were demagnetized before the observation by an alternating magnetic field. Then a DC magnetic field was applied along the axis of easy magnetization (AEM) or the axis of difficult magnetization (ADM) of the film, and either a domain wall displacement process or a magnetization reversal was observed, respectively. Several equilibrium states of a demagnetized film with homogeneous walls 25 nm in thickness are $Fe_{0.20}$-$Ni_{0.80}$ in Fig. 3, a, b, c. The arrows in the figure indicate areas in which the upper part of the domain and part of the domain wall interact with each other. In the initial state, the domain apex has a weak effect on the wall (Fig. 3a). When a magnetic field H <H_c is applied along the AEM, the domain structure as a whole remains unchanged with the exception of the section of the wall located near the apex.

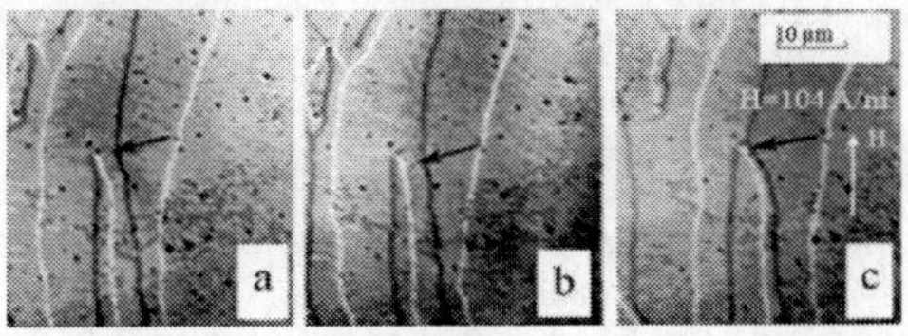

Fig.3. Interaction of domain apex with a domain wall. H_c=100 A/m, H_k=360 A/m.

The distance between of the domain apexes and the domain wall is reduced. On this section the wall weakly bends toward the apex (Fig. 3, b). When the magnetic field H is increased to the value of the field H ~ equal to the field H_c, (H_c is the coercive force), the bending of the wall increases noticeably (Fig. 3c), which leads to an increase length of wall and, consequently, its energy.

4 Interaction of domain apexes in films

Figure 4a shows a photograph of the magnetic structure of the two apexes from different domains located at a distance ~ 150 μm from each other. The structure of the domain apexes is symmetrical about the axis of the easy magnetization. Domain apexes are located not strictly along the line passing through the apexes under consideration, but there is a displacement along the axis of the difficult magnetization. Under the action of an external magnetic field, applied along the axis of AEM, whose direction coincides with the direction of the magnetization vectors in the domains, the domains apexes begin to approach each other. At a certain distance between domains, a magnetostatic interaction arises. The interaction manifests itself in the fact that the distribution of the magnetization vectors at of the apexes of domains varies, this is noticeable from a change in the ripple of magnetization (or the dispersion of the magnetic anisotropy). In particular, near the apexes of both domains appear domains formed from transverse bonds, in the form of triangles, the magnetization in which is directed along the ADM. Between the apexes there is a domain through which the magnetic flux closes between domains.

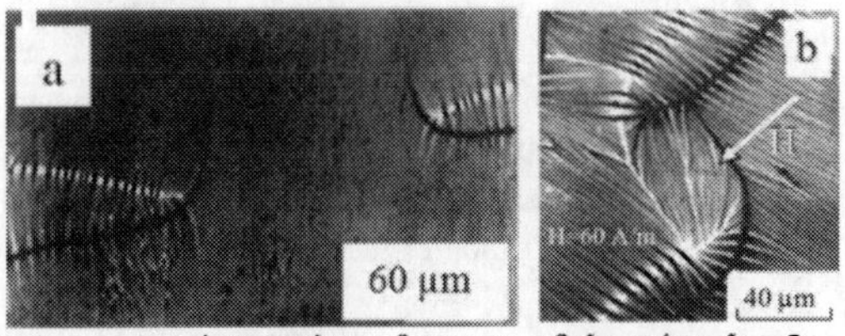

Fig.4. a - non interacting of apexes of domains. b - In-interacting apexes; thickness 50 nm, composition $Fe_{0.20}$-$Ni_{0.80}$. H_c=40 A/m, H_k=250 A/m.

The resulting distribution of the magnetization vectors leads to a decrease in the energy of the scattering fields between the apexes of the domains. The distance between the interacting of the domain apexes is 80 μm. Further convergence of domain apexes causes their annihilation.

5 Conclusions

Many experimental results of the interaction of magnetic elements could not be placed within the scope of this article. It follows from the above results that the greatest distance at which the there is interaction ~ 100 μm is the interaction between of apexes of domains. Between the periodic walls, the interaction manifests itself at distances between them of ~ 20 -30 μm. Between the homogeneous walls, the interaction is manifested at distances of ~ 1 - 2 μm. All these results depend on the specific magnetic parameters of the film. Such parameters are saturation magnetization, perpendicular anisotropy constant, film thickness, etc.

References

1. V. P. Panaetov. JNTM Vol. 05, - N°02 (2015)
2. V. P. Panaetov. Physics of the Solid State, , Vol. 51, No. 10 (2009)
3 .V.P. Panaetov. Bulletin of Higher Educational Institutions Vol. 60 №2 (2017)

https://doi.org/10.1051/epjconf/201818504007

MAGNETIC PROPERTIES OF THE NATURAL AND ISOTOPE-MODIFIED DIAMOND AND SILICON CARBIDE

Taldenkov A.N. [1,*], *Inyushkin A.V.* [1], *Chistotina E.A.* [1], *Ralchenko V.G.* [2], *Bolshakov A.P.* [2], *Mokhov E.N.* [3]

[1]NRC "Kurchatov Institute", Moscow, 123182, Russia
[2]A.M. Prokhorov General Physics Institute RAS, Moscow, Russia
[3]Ioffe Physico-Technical Institute RAS, St. Petersburg, Russia

Abstract. The magnetic properties of single crystals of synthetic diamond and crystals of silicon carbide were studied. High-purity samples of diamonds synthesized with HPHT and CVD technologies were used. The crystals of silicon carbide were grown by sublimation and industrial technology. Along with samples with a natural isotopic composition, monoisotopic crystals of diamond (99.96% ^{12}C and 99.96% ^{13}C) and silicon carbide (99.993% of ^{28}Si) were studied. On the basis of the data obtained, the diamagnetic susceptibility was determined and the concentration of paramagnetic centers and the content of the ferromagnetic component were evaluated. The results are discussed.

1 Introduction

One of the trends in solid-state electronics is the increasing research activity in the field of wide-gap semiconductors – silicon carbide and diamond. The transport of spin-polarized electrons attracts special attention of theorists and experimenters.

The main channel for the relaxation of the electron spin is the scattering on nuclear magnetic moments of the odd isotopes of carbon ^{13}C and silicon ^{29}Si. For this reason, the investigation of isotopically modified crystals is an urgent task.

Diamond is regarded as an ideal material for spin transport and promising "quantum computations". Due to the weak orbital interaction in comparison with other materials, a very long spin diffusion length is expected in it. In isotopically pure diamond, where the content of ^{13}C isotope is reduced, the spin relaxation time may exceed 1 ms at room temperature. Despite the great progress in the production of synthetic diamonds, their magnetic properties remain practically unstudied. Nothing like a study of the magnetic properties of natural diamonds [1] was made for synthetic diamonds. Magnetic properties of synthetic diamonds with modified isotopic composition have been also not studied yet.

Investigation of the magnetic properties of SiC became very important after the detection of ferromagnetism (FM) in SiC crystals [2]. FM was induced by neutron irradiation, neon and xenon ions, protons, and also by doping with aluminium. The combination of ferromagnetism and isotopic modification in one crystal can be promising for the creation of effective spin transport.

2 Experimental details

The magnetic moments of the crystals were measured with MPMS XL-7 SQUID magnetometer (Quantum Design) using the reciprocating sample option (RSO) in temperature range temperature 2 K < T < 300 K and in magnetic field H < 7 T.

SiC single crystals under investigation were produced with various techniques (Table 1). Samples #2 is the crystal ^{28}SiC with 99.993% of ^{28}Si and sample #1 is its counterpart with natural isotopic abundance. Both were synthesized with low temperature sublimation technique. Sample #3 was grown with high temperature Lely technique. Samples #4 – #13 were synthesized with industrial technique: #5 and #6 are 6H and 4H substrates made by CREE Inc. ($n = 10^{18}$ cm^{-3}); 6H crystals #8 and #9 were irradiated with neutron dose about 2.2×10^{18} and crystal #7 is its unirradiated counterpart; samples #10 and #11 are 6H crystals with free carrier concentration of $n \sim 10^{16}$ cm^{-3} and $n \sim 10^{18}$ cm^{-3}; samples #12 and #13 are 6H SiC crystals doped with Fe.

The samples of synthetic diamond have been grown by high-pressure high-temperature process (HPHT) #1 and CVD method (samples #3 - #7) (Table 2). High-purity CVD single crystals, #3 from Element Six Corp., #1 from New Diamond Technology (Russia), and #4 synthesised in General Physics Institute, RAS (Moscow), were investigated. Also, the unique diamond crystals #5 with isotope ^{13}C concentration 99.96% and #6 enriched with ^{12}C to 99.96% obtained by the CVD method also were studied. The CVD diamond #7 had unique light pink colour. The pure natural diamond #2 was investigated for comparison.

* Corresponding author: box-n3@bk.ru

https://doi.org/10.1051/epjconf/201818504007

3 Results and discussion

The characteristic magnetization dependencies measured at permanent temperature are shown in Fig. 1a. Three main contributions can easily be distinguished: diamagnetic, paramagnetic, and ferromagnetic. The diamagnetic term can be seen as linear in field

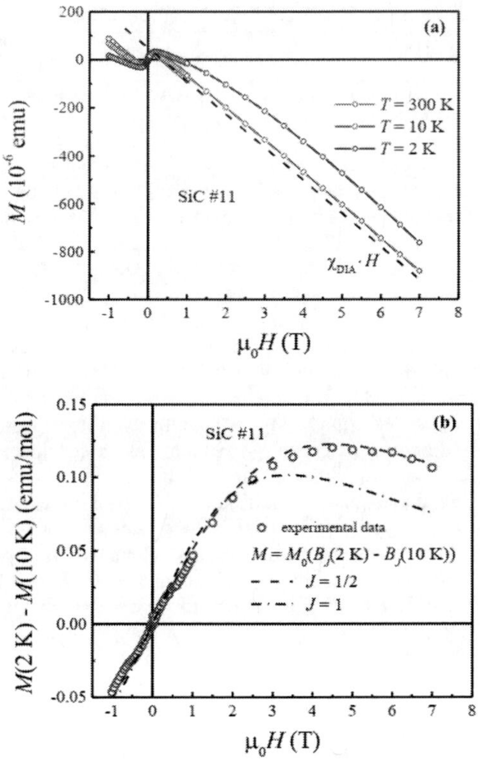

Fig. 1. The isothermal magnetization curves at different temperatures (a), dash line demonstrates diamagnetic contribution at T = 300. The field dependence of difference $\Delta M(H) = M(2\text{ K}) - M(10\text{ K})$ (b). Lines were computed with difference of the Brillouin functions $M_0B_J(x)$ with saturation value M_0.

contribution $\chi_{DIA2} \cdot H$ at high temperature. The FM contribution with saturated value M_{FM} is manifested in the form of singularities at $H < 1$ T in the entire temperature range. The paramagnetic contribution can be identified by saturation in high fields at the lowest temperature. It can be assumed that terms $\chi_{DIA2} \cdot H$ and M_{FM} do not depend on temperature at $T < 20$ K. The field dependence of the difference $\Delta M(T_1,T_2) = M(T_1,H) - M(T_2,H)$ contains only the paramagnetic contribution, and can be described using Brillouin function $B_J(x)$, where $x = gJ\mu_B B/k_B T$. In this case, the quantum number J and the concentration of the paramagnetic centres N_{P2} are free fitting parameters. An example of such data treatment is shown in Fig. 2b. For most crystals, the value of J was close to 1/2 and never exceeded unity.

The temperature dependencies of molar static susceptibility $\chi_M = (M(T,H) - M_{FM})/H$ in applied field $\mu_0 H = 1$ T are shown on Fig. 2. Susceptibility was fitted with Curie-Weiss function $\chi_M(T) = \chi_{DIA1} + C/(T - \Theta)$, with χ_{DIA1}, C and Θ as the free parameters. The value of Θ was less than 2 K. Using the Curie-Weiss constant C one can estimate the concentration of paramagnetic

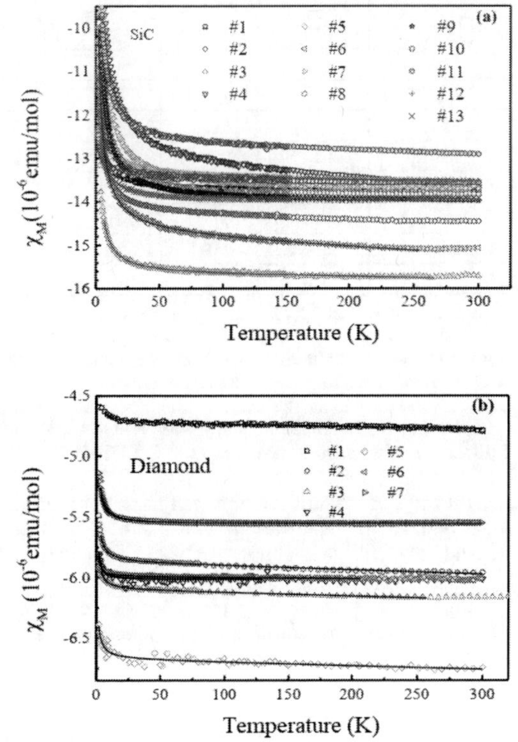

Fig. 2. Temperature dependencies of the static susceptibility of silicon carbide (a) and diamond (b). Lines are approximation by the Curie-Weiss law.

centers N_{P1} with spin J. The results of the processing of all the data are presented in Tables 1 and 2.

The diamagnetic susceptibility χ_{DIA} is about -13.5 emu/mol for most of the crystals within 10% range. It is close to the value published in Ref. [2] for unirradited 6H SiC sample synthesised by industrial technique. Significant change in χ_{DIA}, exceeding the experimental error, is observed for crystal #3. This sample was grown by the high-temperature method, and it can be assumed that it has the lowest concentration of vacancies.

There is a positive correlation between the measured values of paramagnetic centers N_P and the free carrier concentrations (#10 and #11). Thus, the measurement of low-temperature paramagnetic contribution to the static magnetization demonstrates the possibility to determine the concentration of electrically active impurities. The

Table 1. Parameters of the single crystals of SiC

#	$-\chi_{DIA1}$ (emu/mol)	$-\chi_{DIA2}$ (emu/mol)	M_{FM} (emu/g)	N_{P1} (cm^{-3})	N_{P2} (cm^{-3})
1	$12.8\cdot10^{-6}$	$13.7\cdot10^{-6}$	$2.0\cdot10^{-4}$	$7.9\cdot10^{17}$	$7\cdot10^{17}$
2	$14.3\cdot10^{-6}$	$14.3\cdot10^{-6}$	$6.8\cdot10^{-4}$	$1.2\cdot10^{18}$	$1.0\cdot10^{18}$
3	$15.5\cdot10^{-6}$	$15.9\cdot10^{-6}$	$3.3\cdot10^{-4}$	$1.1\cdot10^{18}$	$9\cdot10^{17}$
4	$12.9\cdot10^{-6}$	$13.2\cdot10^{-6}$	$2.13\cdot10^{-3}$	$1.5\cdot10^{18}$	$7\cdot10^{17}$
5	$13.4\cdot10^{-6}$	$13.4\cdot10^{-6}$	$3.9\cdot10^{-4}$	$1.1\cdot10^{18}$	$1.0\cdot10^{18}$
6	$15.0\cdot10^{-6}$	$14.8\cdot10^{-6}$	$5\cdot10^{-5}$	$2.0\cdot10^{18}$	$1.3\cdot10^{18}$
7	$13.7\cdot10^{-6}$	$13.8\cdot10^{-6}$	$4\cdot10^{-5}$	$3.2\cdot10^{18}$	$2.5\cdot10^{18}$
8	$13.8\cdot10^{-6}$	$14.0\cdot10^{-6}$	$1.2\cdot10^{-4}$	$2.6\cdot10^{18}$	$1.6\cdot10^{18}$
9	$13.8\cdot10^{-6}$	$14.0\cdot10^{-6}$	$3.3\cdot10^{-4}$	$2.4\cdot10^{18}$	$1.9\cdot10^{18}$
10	$13.6\cdot10^{-6}$	$13.4\cdot10^{-6}$	$4\cdot10^{-5}$	$1.3\cdot10^{17}$	$1.2\cdot10^{17}$
11	$12.9\cdot10^{-6}$	$12.7\cdot10^{-6}$	$1.26\cdot10^{-3}$	$2.0\cdot10^{18}$	$3.5\cdot10^{18}$
12	$13.8\cdot10^{-6}$	$13.9\cdot10^{-6}$	$2.1\cdot10^{-4}$	$1.0\cdot10^{18}$	$1.5\cdot10^{18}$
13	$13.5\cdot10^{-6}$	$13.5\cdot10^{-6}$	$2\cdot10^{-5}$	$7\cdot10^{17}$	$7\cdot10^{17}$

χ_{DIA1} is calculated from $\chi(T)$. χ_{DIA2} is calculated from $M(H)$ at $T > 150$ K. M_{FM} is the ferromagnetic contribution evaluated from $M(H)$ at $T > 150$ K. N_{P1} is the concentration of the paramagnetic centers evaluated from the Curie-Weiss law at 2 K < T <300 K. N_{P2} is the concentration of paramagnetic centers evaluated from the fitting of the $M(H)$ at $T = 2$ K with Brillouin functions. One mole of natural SiC is 40.0962 g.

SiC crystal #2 enriched with ^{28}Si has a concentration of paramagnetic impurities close to that observed in the crystal #1 with the natural isotopic composition. In general, crystals obtained by the low-temperature sublimation technique in their parameters are not inferior to crystals grown by industrial high-temperature method.

Single crystals #8 and #9 of silicon carbide irradiated with a neutron flux have an increased concentration of the ferromagnetic component compared to the pristine crystal #7. The rest parameters of these samples are similar. It can be assumed that the main result of irradiation is the formation of $V_{Si}V_C$ divacancies, as suggested in [2]. The maximal value of M_{FM} is nearly the same as that published in [2] for irradiated crystals. For the samples #12 and #13 doped with iron, we have not observed any significant effect of the doping on their magnetic properties.

For diamonds the diamagnetic susceptibility χ_{DIA} of four of our single crystals is about $-6\cdot10^{-6}$emu/mol. Magnetic susceptibility in natural diamonds ranging from $-5.5\cdot10^{-6}$emu/mol to $-4.9\cdot10^{-6}$ emu/mol was obtained in Ref. [1] and values $-5.9\cdot10^{-6}$emu/mol and $-5.4\cdot10^{-6}$emu/mol was published in Refs. [3] and [4] correspondingly. Despite the fact that our data are systematically lower, these results are considered to be relevant to each other. The values of χ_{DIA} observed in crystals #1 and #5 are considerably knocked out from the range under consideration. We cannot follow assumption in Ref. [1] that the larger value of χ_{DIA} in sample #1 is due to the significant content of the FM phase, since measured M_{FM} was not large, and was subtracted when determining χ_{DIA}. Perhaps this result relates to the processes occurring during HPHT synthesis. A significantly lower value for sample #5 may be caused by isotope effect due to large change in atomic mass in ^{13}C diamond. But we have to note, that a systematic error for sample #5 may by enhanced due to small sample mass (6.32 mg).

Table 2. Parameters of the single crystals of diamonds

#	$-\chi_{DIA1}$ (emu/mol)	$-\chi_{DIA2}$ (emu/mol)	M_{FM} (emu/g)	N_{P1} (cm^{-3})	N_{P2} (cm^{-3})
1	$4.7\cdot10^{-6}$	$4.8\cdot10^{-6}$	$2.6\cdot10^{-5}$	$2.7\cdot10^{17}$	$1.6\cdot10^{17}$
2	$5.7\cdot10^{-6}$	$5.9\cdot10^{-6}$	$3.2\cdot10^{-4}$	$4.1\cdot10^{17}$	$3.1\cdot10^{17}$
3	$6.1\cdot10^{-6}$	$6.1\cdot10^{-6}$	$5.0\cdot10^{-4}$	$2.6\cdot10^{17}$	$2.2\cdot10^{17}$
4	$6.0\cdot10^{-6}$	$6.0\cdot10^{-6}$	$8.0\cdot10^{-5}$	$1.8\cdot10^{17}$	$2.2\cdot10^{17}$
5	$6.7\cdot10^{-6}$	$6.7\cdot10^{-6}$	$6.3\cdot10^{-5}$	$3.2\cdot10^{17}$	$1.3\cdot10^{17}$
6	$6.0\cdot10^{-6}$	$6.0\cdot10^{-6}$	$7.2\cdot10^{-5}$	$9\cdot10^{16}$	$4\cdot10^{16}$
7	$5.5\cdot10^{-6}$	$5.6\cdot10^{-6}$	$<5\cdot10^{-6}$	$4.6\cdot10^{17}$	$4.3\cdot10^{17}$

Notations are the same as in **Table** 1. One mole of natural diamond is 12.01 g.

The isotopically engineered crystals have concentration of paramagnetic impurities and ferromagnetic inclusions close to that in crystal commercially available in the market. The parameters of crystal ^{12}C(99.96%) #6 slightly differ from the parameters of crystal #4, which can be considered as a counterpart with a natural isotopic content. Moreover, the paramagnetic contribution received in isotopically pure crystal #6 is the lowest among all measured samples, which indicates its chemical purity. The paramagnetic contribution estimated in CVD and HPHT crystals are generally close to each other and the best samples of synthetic diamonds are superior in quality to natural.

The work is supported by RFBR grants 16-07-00979 and 16-0701188, and by the RSF grants 16-42-01098 and 14-12-01403.

References

1. A. Sigamony, Indian Acad. of Science, Proc.A, **19** (1944), p.310-314.
2. Yutian Wang et al., Scientific Reports **5** (2015), 8999.
3. S. Hudgens, M. Kastner, and H. Fritzsche, Phys. Rev. Lett. **33** (1974), 1552.
4. J. Heremans, C.H. Olk, and D.T. Morelli, Phys. Rev. **49** (1994), 15122.

Effects of spin polarization on resonant photoemission from d-f states in $TbNi_2Mn_x$ compounds

Vladimir I. Grebennikov[1,2,*], *Tatyana V.* Kuznetsova[1,2], *Evgenii.G.* Gerasimov[1], and *Nikolai. V.* Mushnikov[1]

[1]M.N. Miheev Institute of Metal Physics UD RAS, 620990 Yekaterinburg, Russia
[2]Ural Federal University, 620002 Yekaterinburg, Russia

Abstract. Resonant photoemission in narrow-band materials is described by the sum of first- and second-order transitions, their quantum-mechanical interference leads to an increase in the spectrum from the valence bands and the appearance of an asymmetric dependence on the photon energy. These effects are studied theoretically and experimentally using the example of three-component intermetallic compounds $TbNi_2Mn_x$. The competition between the elastic and inelastic photoemission channels leads to a different dependence of photoemission spectra from nickel and manganese on photon energy. The elastic channel is realized on atoms with large magnetic moments, the inelastic Auger decay occurs on atoms with small moments.

1 Introduction

Effects of superposition and interference of quantum states can be observed in resonance X-ray spectroscopy of solids. For the first time such effects were investigated in detail by Hugo Fano [1] in 1961 in connection with experiments on inelastic scattering of electrons by helium atoms. He developed a theory of a resonance with an asymmetric profile, called the Fano resonance, which arises as a result of interference of two wave processes. The nature of the interfering processes can be very different, in particular, the resonance was observed in the transmission and reflection spectra of photonic crystals. Let us turn to the process of photoemission of electrons from a solid with a narrow (3*d*- or 4*f*-) band.

Left Fig. 1 shows the direct production of a photoelectron when a photon is absorbed in the valence band (VB). An additional channel is opened if the photon energy reaches an excitation edge for a core level. The photon first transfers the electron from the core level to the unoccupied state of the valence band (VB), and then the Coulomb interaction by

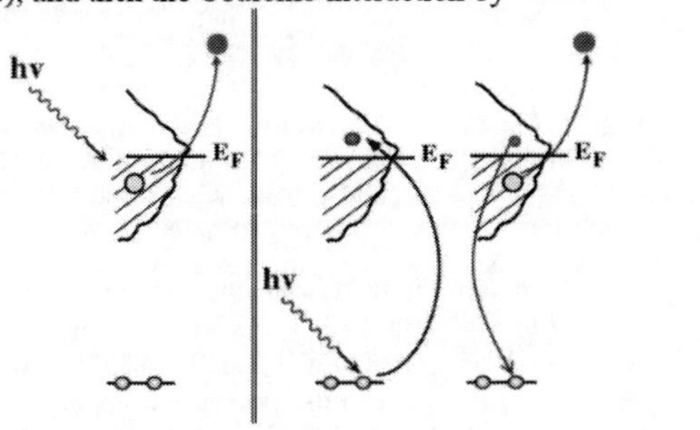

Fig. 1. Direct (left) and resonant (right) photoemission of electrons from a valence band.

autoionization returns the excited electron back to the core level with the ejection of another VB electron to the free state. The same final state (photoelectron plus a hole in the valence band) is obtained as a result of two types of transitions.

2 A theoretical background

According to the principles of quantum mechanics, it is necessary to add the amplitudes of all such transitions and take the square of the modulus. As a result, the intensity of photoemission will be determined by the equality

$$I(hf, E_B) = \sum_{k_1} \left| A + \sum_{k_2} \frac{M_1 M_2}{hf - (E_2 - E_c) - ig} \right|^2 \delta(E_B - E_1) \quad (1)$$

Here $A = \langle q|d|k_1\rangle$ is a matrix element (amplitude) of the electron dipole transition from the state k_1 with energy E_1 to the free state q with the energy E_{kin} when the photon with energy hf is absorbed. The same final state is the result of the second-order process whose amplitude is described by a fraction in (1): $M_1 = \langle k_2|d|c\rangle$ is the dipole matrix element of the photon absorption, accompanied by the electron excitation from the core level c to the free VB states k_2, the denominator is the difference of the corresponding energies, the imaginary additive g stands for the half-width (the inverse lifetime) of the intermediate state. Then an autoionization Auger transition occurs due to the Coulomb interaction $M_2 = \langle cq|Q|k_2k_1\rangle$, the electron k_2 returns to the core level c with the emission of the valence electron k_1 into

* Corresponding author: vgrebennikov@list.ru

https://doi.org/10.1051/epjconf/201818504008

the final (registered) state q. The process amplitude contains summation over all possible intermediate states k_2, and its squared modulus is summed over all finite states k_1. We note that there is also a contribution from the exchange matrix element $M_2' = \langle cq|Q|k_1k_2\rangle$. The delta function takes into account the energy conservation. The kinetic energy of photoelectrons and the energy of photons are measured experimentally, their difference has the meaning of the binding energy E_B of the states k_1 from which the photoelectrons exit. Passing from summation over states to integration over energy (in the approximation of the weak energy dependence of the matrix elements), we write equation (1) in the form

$$I(hf, E_B) = D(E_B)\left|A + M_1M_2G(hf + E_c - ig)\right|^2. \quad (2)$$

Here $D(E)$ is the density of states (DOS), the integral $G(z) = \int dE_1 D(E_1)/(z - E_1)$ is called the single-site Green function. We note that for a Lorentz DOS (width Γ with center at E_0) the function has a simple form $G(z) = (z - E_0 - i\Gamma)^{-1}$. For rectangular DOS (width 2Δ) $G(z) = (\ln(z - E_0 + \Delta) - \ln(z - E_0 - \Delta))/2\Delta$. The photoemission intensity as a function of the photon energy hf and the binding energy of electrons E_B

$$I(hf, E_B) = D(E_B)\left|A + B/(e - i)\right|^2,\ e = (hf - E_{edge})/\Gamma \quad (3)$$

is determined by the amplitudes of the first-order transition A and the second order B ($= M_1M_2/\Gamma$), the latter depending on the energy e: its position is determined by the energy difference $E_{edge} = E_0 - E_c$ of the unoccupied VB part and the core-level energy. The total width of 2Γ includes the width of the core level and the width of the unoccupied part of the VB. In the approximation of the zero mean phase difference for the matrix elements of the first and second order, we obtain the equation

$$I(e, E_B) = D(E_B)\left(A^2 + (B^2 + 2tABe)/(e^2 + 1)\right), \quad (4)$$

which includes modules A and B of the matrix elements, and the attenuation coefficient t taking into account their phase fluctuations. Formula (4) describes the resonance enhancement of photoemission at the core-level excitation edge (e = 0). The interference of two transitions (the crossed term proportional to the energy e) generates an asymmetry of the profile with respect to photon energy. After normalization to the the first order intensity (or to the asymptotic value as e $\to\infty$), the equation (4) can be written as

$$f(e) = 1 + (q^2 + 2tqe)/(e^2 + 1) \quad (5)$$

where the parameter $q = B/A$ stands for the ratio of the second to first order matrix elements. It depends on the binding energy E_B of the valence states and does not depend on the photon energy e. In the following we consider the profile of the function (5) with respect to the variable e for fixed values of the photoelectron binding energy. The function $f(e)$ has a maximum and a minimum $f_\pm = 1 + (q^2/2)\left(1 \pm \sqrt{1 + (2t/q)^2}\right)$ at the points $e_\pm = -(q/2t) \pm \sqrt{(q/2t)^2 + 1}$. The full width at half maximum of the function (the middle between maximum and minimum) is $e_> - e_< = 2\sqrt{(q/2t)^2 + 1}$. The asymmetry index p is determined by the difference between the right and left half-widths of the spectrum divided by their sum (the full width)

$$p = \frac{(e_> - e_+) - (e_+ - e_<)}{e_> - e_<} = \frac{t}{q}\left(\frac{2}{1 + \sqrt{1 + (2t/q)^2}}\right) \approx \frac{t}{q} \quad (6)$$

Here e_+ is the energy of the maximum, $e_>$ and $e_<$ the right and left energy values at the half maximum. The asymmetry index (6) is reduced by a factor of t compared with the Fano index $1/q$ [1].

3 Experimental analyses

Let us now analyze the experimental spectra. Fig. 2 shows the contour plot of the intensity of photoemission in the space of the electron binding energy and photon energy.

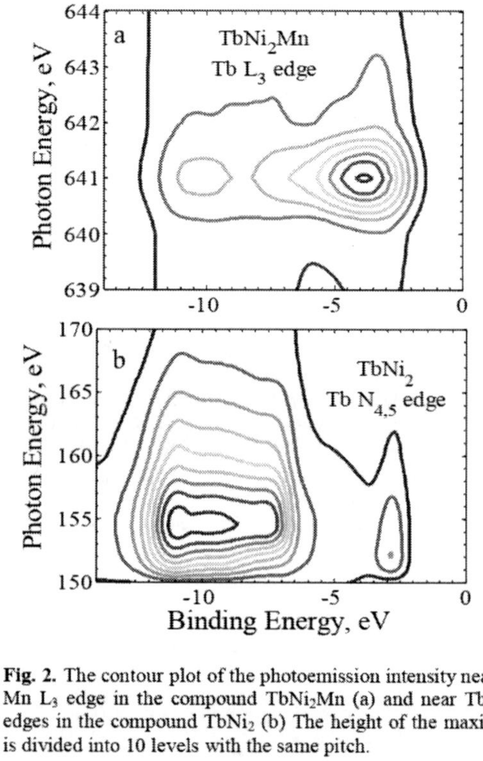

Fig. 2. The contour plot of the photoemission intensity near the Mn L_3 edge in the compound $TbNi_2Mn$ (a) and near Tb $N_{4.5}$ edges in the compound $TbNi_2$ (b) The height of the maximum is divided into 10 levels with the same pitch.

A tenfold increase in the photoemission yield is observed at the Mn L_3 edge ($2p \to 3d$ transition) in the intermetallic compound $TbNi_2Mn$. The width Mn of the spectrum with respect to the photon energy is ~ 1.0 eV. The profile width at Tb $N_{4.5}$ edge (the transition $4d \to \varepsilon f$,

https://doi.org/10.1051/epjconf/201818504008

the so-called giant resonance, or shape resonance) is much larger ~ 7.7 eV. The regions of localization of manganese state 2 – 6 eV and terbium states 7 – 12 eV are determined from the resonance enhancement. Fig. 3 shows the Mn $L_{2,3}$ and Tb $N_{4,5}$. x-ray absorption near edge structures (XANES). One can clearly see the strong multiplet splitting, which is very characteristic for transition and rare-earth elements compounds [2, 3].

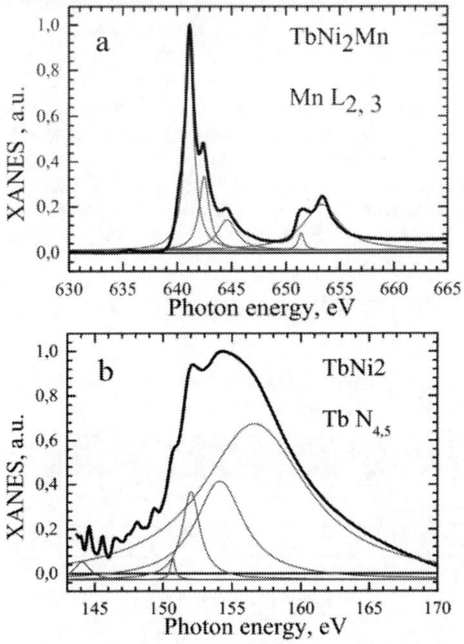

Fig. 3. X-ray absorption spectra at the Mn L_3 edge in $TbNi_2Mn$ (a}and on the Tb $N_{4,5}$ edge in $TbNi_2$ (b). The decomposition of the curves into Lorentz peaks is shown.

Let us now consider the effects of the interference of the first and second order transitions as the photon energy changes. Formula (5) can be used for any binding energy E_B, but it is convenient to consider integrally two VB regions with binding energy E_B 1 – 6 eV, formed predominantly to manganese (and nickel) states and a region of 7 – 12 eV in which terbium states predominate. The experimental photoemission intensity on the Mn L_3 edge in $TbNi_2Mn$ (a) and on Tb $N_{4,5}$ edge in $TbNi_2$ (b) averaged over these intervals is shown by the dots in Fig. 4 as a function of the energy of the exciting photon energy *hf*. The solid lines show the results of fitting according to the formula (5) with three fitting parameters Γ (0.48 for $TbNi_2Mn$ (panel *a*) and 3.60 eV for $TbNi_2$ (panel *b*)), *q* (2.78 for 1-6 eV and 1.07 for 7-12 eV, curves *a* and 1.82 for 1-6 eV and 3.59 for 7-12 eV, curves *b*) and *t* (see the values below). Our simple model gives a good description of profiles on the excitation edges of manganese and terbium.

Note that the resonance on manganese not only leads to an increase in the electron yield from the manganese bands, but also more than doubles the photoemission from the terbium states of 7 – 12 eV (which seem to be on the side). In turn, the resonance in terbium leads to a threefold increase in the photoemission yield from the nickel band (1 – 6 eV). There is a strong interaction and competition between transitional and rare-earth components.

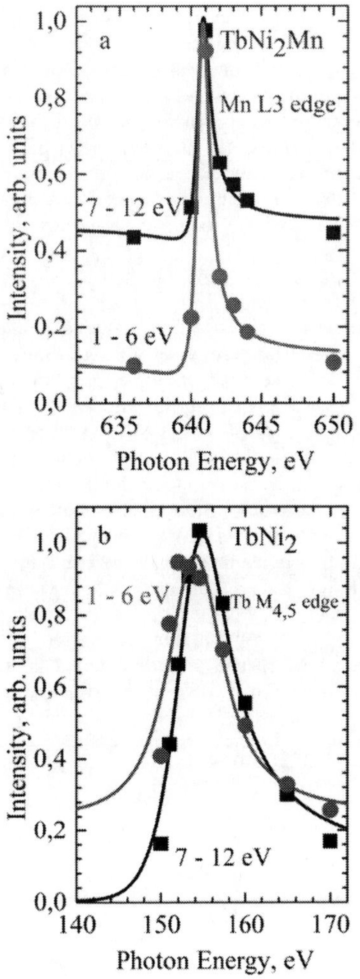

Fig. 4. The experimental photoemission intensities in the binding energy range 7 – 12 eV (black squares) and 1 – 6 eV (red dots) on the Mn L_3 edge in $TbNi_2Mn$ (a) and on Tb $N_{4,5}$ edge in $TbNi_2$ (b). The lines show a fit according to Eq. (5).

The second-order photoemission process (with allowance for the resonant denominator) is much more efficient than the direct photoelectric effect from the valence band at both edges. The point is that the photoionization cross section (the Fourier transform of the wave function) is small for valence states weakly localized in space. It is much larger for strongly localized wave functions of the core levels. The Coulomb interaction acting in the second stage is also stronger than the interaction with radiation. The evidence of interference is the asymmetry of the profiles, which is seen directly in Fig. 4. Fig. 5 shows the values of the asymmetry index *p*. They are large in terbium and manganese (*a*) and much less on nickel (1 – 6 eV band, panel *b*).

The correlation of the two transitions *t* is given below

Mn L_3(Tb)	Mn L_3 (Mn+Ni)	Tb $N_{4,5}$ (Tb)	Tb $N_{4,5}$ (Ni)
0.29	0.58	1	0.09

It is equal to 1 only for terbium resonance in the range of terbium states. In other cases it is much less. The reason here is not only in the probable accidental phase failure of the intermediate state wave function, but also the states obtained as a result of the transitions of the first and second order can be somewhat different. Auger autoionization involves overlapping the wave functions of the valence and core states, but direct photoemission, of course, does not. This difference leads only to partial interference.

The resonance emission not always follows the elastic scenario described in Fig. 1. Now we consider what processes occur in the same compound on the nickel atom whose resonant XPS is shown in Fig. 6. Again, the resonance multiple amplification of the photoemission yield from the valence states is observed at the Ni L_3 edge (hf = 854 eV). At the same time, the spectrum shape changes greatly above the edge. The maximum shifts evenly along the binding energy scale (and it is stationary in the kinetic photoelectron energy scale). This is a typical LVV Auger line. The final state contains two holes in the valence band and an excited electron above the Fermi level, which carries away the excess photon energy. It should be noted that in the region of the Auger line generation the resonant photoemission spectra provide detailed information for the investigation of the two-hole correlation effects [4].

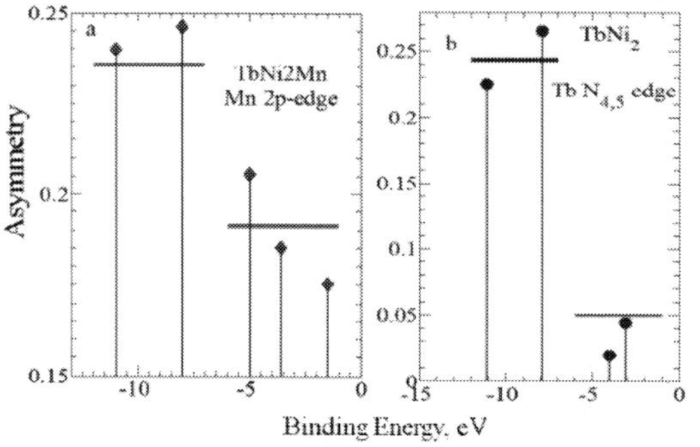

Fig. 5. The asymmetry index of profiles at different binding energies (points). The horizontal lines show mean values in the intervals 1 – 6 and 7 – 12 eV at the excitation edges of manganese (a) and terbium (b).

Thus, there is a competition between the elastic channel of virtual excitation of the core level, as described above, and the real (energy conservation in the intermediate state) Auger process in which the 2*p* hole is occupied by one valence electron, whereas another valence electron leaves the sample and it is recorded by the detector. The first process is dominant in case of the minganise atoms, and the second process prevails for the nickel atoms. This is closely related to the spin magnetic state of the atoms [5, 6], as it is easily seen from a simple model. Let the manganise atom have five *d* electrons. According to Hund's rule, they all have the same spin projections ↑, and they completely fill the corresponding spin subband. As a result, only the electron with the spin ↓ takes part in the 2p →3d transition due to spin conservation. Consequently, only one electron from the six *d* electrons can occupy the 2*p* hole (with the spin ↓), and this is the electron occupied the *d* state due to the photon absorption, i.e. in this case, the elastic resonant photoemission channel is implemented. If now the ground state has 8 electrons in the (5↑3↓), configuration, then again, a hole with the spin down arises, and the L↓(5↑3↓1*↓) excited state is obtained. Now, 4 VB electrons ↓ can occupy the L↓ photohole. Hence, with a decrease in the spin polarization (occupation of the 3*d* band), the probability of the elastic channel decreases, and the contribution of the inelastic Auger channel of electron emission increases.

Thus, resonant photoemission spectroscopy has a selective sensitivity to the components of compounds and makes it possible to measure the fine coherence effects in materials.

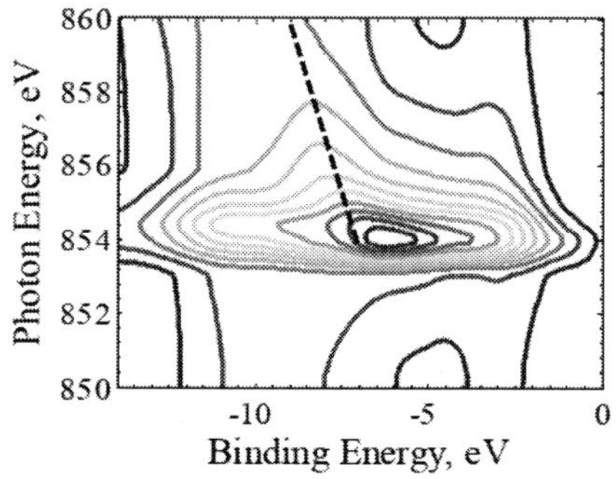

Fig. 6 The contour plot of the photoemission intensity near the Ni L_3 edge in $TbNi_2Mn$.

The research was carried out within the state assignment of FASO of Russia (theme "Quant" No. 01201463332), supported in part by RFBR (No. 18-02-00060) and UD RAS (No.15-8-2-10). The authors express their gratitude to A. Preobrazhenskii and N. Vinogradov (synchrotron MAX-Lab, Lund, Sweden) for their assistance in carrying out experiments.

References

1. U. Fano, Phys.Rev.. **124**, 1866 (1961)
2. T. V. Kuznetsova, V. I. Grebennikov, N. V. Mushnikov, E. G. Gerasimov, A. Buling, C. Derks, M. Neumann, Bull. RAS: Physics, **77**, 226 (2013)
3. V.I. Grebennikov, T.V. Kuznetsova, A.G. Kuchin, Bull. RAS: Physics. **79**, 134 (2015)
4. V. I. Grebennikov, T. V. Kuznetsova, and M. V. Yakushev, Bull. Russ. Acad. Sci.: Phys., **77**, 1123 (2013)
5. M. V. Yablonskikh, Yu. M. Yarmoshenko, V. I. Grebennikov, E. Z. Kurmaev, S. M. Butorin, L.-C. Duda, J. Nordgren, S. Plogmann, and M. Neumann, Phys. Rev. B, **63**, 235117 (2001)
6. V. I. Grebennikov, Poverkhnost, **11**, 41 (2002)

Magnetic structures of $(Gd_{1-x}Y_x)Mn_2Ge_2$ and $(Tb_{1-x}Y_x)Mn_2Ge_2$ studied by neutron powder diffraction

Sergey Granovsky[1,2,*], *Irina* Gaidukova[1] and *Clemens* Ritter[3]

[1]Faculty of Physics, M.V.Lomonossov Moscow State University, 119991 GSP-1 Moscow, Russia
[2]IFMP, TU Dresden, D-01062, Dresden, Germany
[3]Institut Laue-Langevin, F-38042 Grenoble Cedex 9, France

Abstract. Magnetic structures and magnetic phase transitions in "naturaly-layered" $(Gd_{1-x}Y_x)Mn_2Ge_2$ and $(Tb_{1-x}Y_x)Mn_2Ge_2$ compounds have been studied by electrical resistivity and neutron powder diffraction experiments. Hot neutrons were used to probe the magnetic structures of Gd-containing materials. The variable R-Mn exchange forces Gd moments to realign from the tetragonal axis into the plane and back to this axis, whereas canted and tripled structures were detected in the series of compounds containing the highly anisotropic Tb-ion.

1 Introduction

Recent achievements in spintronics require better understanding and classification of antiferro- and ferrimagnetic materials. In the range of hundreds of such materials, tetragonal RT_2X_2 (R- rare earth or Y, T – transition metal, X- Ge or Si, $ThCr_2Si_2$-type structure, space group *I4/mmm*) receives a special interest: superconductivity, mixed valence, Kondo effects have been investigated during last decades. The long-range magnetic order in 3*d*-sublattice is established only in Mn-containing compounds while in those with Fe, Co and Ni only the rare earth sublattice is magnetic. The magnetic transition in Mn sublattice occurs in RMn_2T_2 at 350-500K; all compounds with heavy rare earths exhibit the separate transition to ferrimagnetic phase at T less than 100K while for light rare earths the "separate ordering" takes place in a number of cases too, with transition temperatures ranging from 100K to 250K.

The manganese atoms form a primitive tetragonal sublattice in the $ThCr_2Si_2$-type structure; the distance between Mn atoms in the layer is approximately half of the distance between the adjacent Mn-layers. This feature leads to a significant spread of the 3*d* electrons within the layers and this structure could be considered as "naturally layered" in that sense. Consequently, the magnetic structure of such a compound is determined by the interplay of four types of magnetic exchange: Mn-Mn exchange in one particular layer J^{L}_{Mn-Mn}, interlayer exchange J^{IL}_{Mn-Mn}, R-Mn exchange J_{R-Mn} and R-R interaction J_{R-R}.

Magnetic structures of $GdMn_2Ge_2$ and $TbMn_2Ge_2$ were widely studied by macroscopic as well as microscopic methods. In both compounds the interplanar Mn-Mn exchange has approximately the same value as the R-Mn interaction and the competition between these two types of couplings determines the low temperature properties of these compounds. The Mn-sublattice orders collinearly and antiferromagnetically (*AFil* structure, here and below we follow the classification given in Ref. [1]) at $T_N = 363$ K in the Gd-based compound and at $T_N = 414$ K in $TbMn_2Ge_2$. Another transition from *AFil* to collinear ferrimagnetic *F* structure where the R-sublattice is now aligned antiparallel to the Mn-sublattice occurs in both compounds at $T_C = 95$K. This transition is of the first-order type and is accompanied by a significant change of the lattice constants and temperature-dependent resistivity.

The substitution of magnetic rare earth element to non-magnetic Y, La or Lu should, generally, decrease the R-Mn interaction and also the magnetic anisotropy of the R ion. As in- and interlayer interactions between Mn ions depend mostly on Mn-Mn separations, these will be not affected by substitution when the ionic radii of substitutional and replaceable elements are close to each other. Thus, pseudo-ternary $(Gd_{1-x}Y_x)Mn_2Ge_2$ and $(Tb_{1-x}Y_x)Mn_2Ge_2$ are good objects to investigate magnetic structures and phase transitions caused by the interplay of R-Mn and interlayer Mn-Mn interactions.

Macroscopic properties of $(Gd_{1-x}Y_x)Mn_2Ge_2$ were studied in [2, 3]. Following [2], the collinear low-temperature phase should be replaced by a canted *Fmc* structure with increasing of Y-content while the sample with composition $(Gd_{0.6}Y_{0.4})Mn_2Ge_2$ undergoes two transitions *AFil-Fmc-F*. According to calculations and experimental data in the ref. [3], the specific canted ferrimagnetic structure with Gd-moment in the *c*-plane is expected to be a low-temperature phase for compounds with $x \sim 0.4$.

The $(Tb_{1-x}Y_x)Mn_2Ge_2$ system is more complicated with regard to the anisotropy of Tb ions. The magnetisation and X-ray thermal expansion experiments

* Corresponding author: ser@plms.ru

 https://doi.org/10.1051/epjconf/201818504009

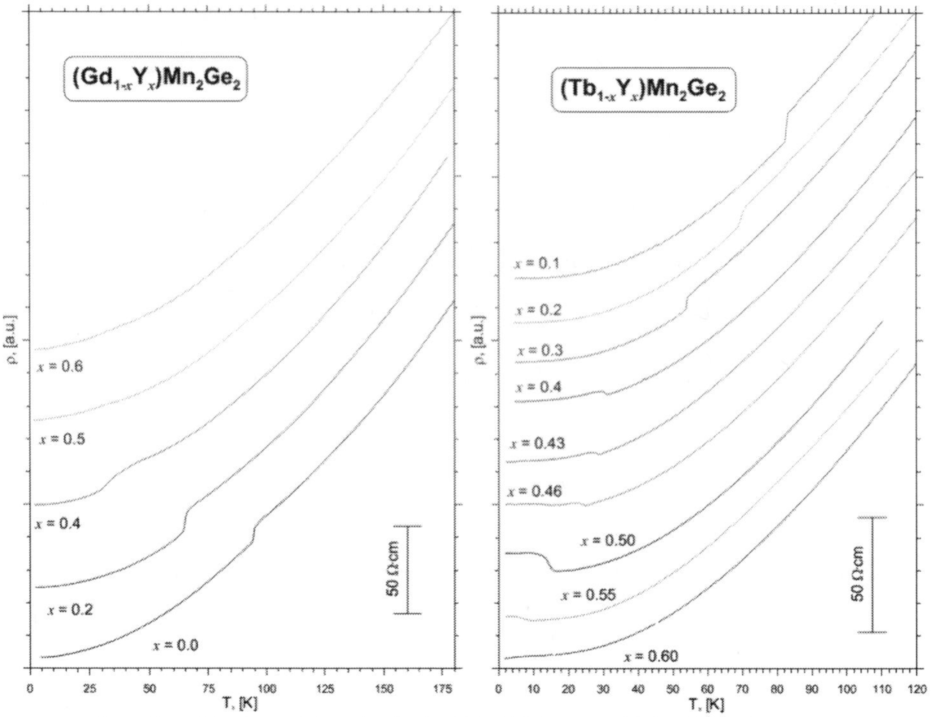

Fig. 1. Resistivity thermal dependencies for $(Gd_{1-x}Y_x)Mn_2Ge_2$ and $(Tb_{1-x}Y_x)Mn_2Ge_2$ series.

are presented in [4]. Following this work, the magnetic structure changes in the consequence *AFil-F-Fmc* in all diluted compounds. At the same time, the volume effect, characteristic of the first-order type *AFil-F* transition was seen only in $TbMn_2Ge_2$ and $(Tb_{0.9}Y_{0.1})Mn_2Ge_2$.

The aim of the current work was to probe directly the magnetic structures in pseudo-ternary systems $(Gd_{1-x}Y_x)Mn_2Ge_2$ and $(Tb_{1-x}Y_x)Mn_2Ge_2$.

2 Experiments

Large 7-8 g polycrystalline samples $(Gd_{1-x}Y_x)Mn_2Ge_2$ $x = 0{,}2$, 0,4, 0.5, 0.6 and $(Tb_{1-x}Y_x)Mn_2Ge_2$ $x = 0.1$, 0.2, 0.3, 0.4, 0.43, 0.46, 0.50, 0.55, 0.6 were synthesised using the induction melting technique in a quasi-levitation cooper crucible of a special design which provided the minimum contact of the melted elements with the crucible. Preferential Mn evaporation was taken into account; therefore, weights of the components were calculated in accordance to the results of a number of test meltings. After the melting procedure, samples were annealed in dynamic or static vacuum for 7 days.

Magnetisation, ac-susceptibility and resistivity temperature dependencies were measured on synthesized samples for their characterisation.

The neutron diffraction experiments were performed at ILL, Grenoble. Diffractograms for Gd-contained samples were collected at the two-axis diffractometer D4 which is situated at a hot neutron source. The neutron wavelength was set to $\lambda = 0.4972$ Å using the Ge monochromator. Diffraction patterns were collected at different temperatures with a counting time of 2 hours per pattern. The background scattering was measured with an empty sample container and the same container filled with highly absorbing enriched ^{10}B powder. Then the linear combination of both measured backgrounds was subtracted from the measured scattering intensity. Thermodiffractograms for the Tb-containing series were collected with the high intensity powder diffractometer D1B (dynamic temperature sweep: between 5 - 130 K, 2θ range: 10 - 90 deg., total counting time: approx. 10 hours per sample).

3 Results and discussion

Figure 1 shows the thermal dependence of the electrical resistivity $\rho(T)$ for synthesized samples of both series. As can be seen from these dependencies, the low temperature behaviour of studied samples is significantly different. One first order transition likely from *AFil* to *F* structures accompanied by jump-like decrease in resistivity could be seen in $Gd_{0.8}Y_{0.2}Mn_2Ge_2$ and in Tb-series with $x \leq 0.3$. $Gd_{0.6}Y_{0.4}Mn_2Ge_2$ sample shows two consequent magnetic transitions, which is in agreement with the data from the Ref.[2]. Only one phase transition is detected for Tb-based compounds, however, the $\rho(T)$ dependence for $0.4 \leq x \leq 0.46$ is different from those for $0.5 \leq x \leq 0.55$. Samples with yttrium content of 60% show no transitions and should stay in *AFil* phase down to lowest temperatures.

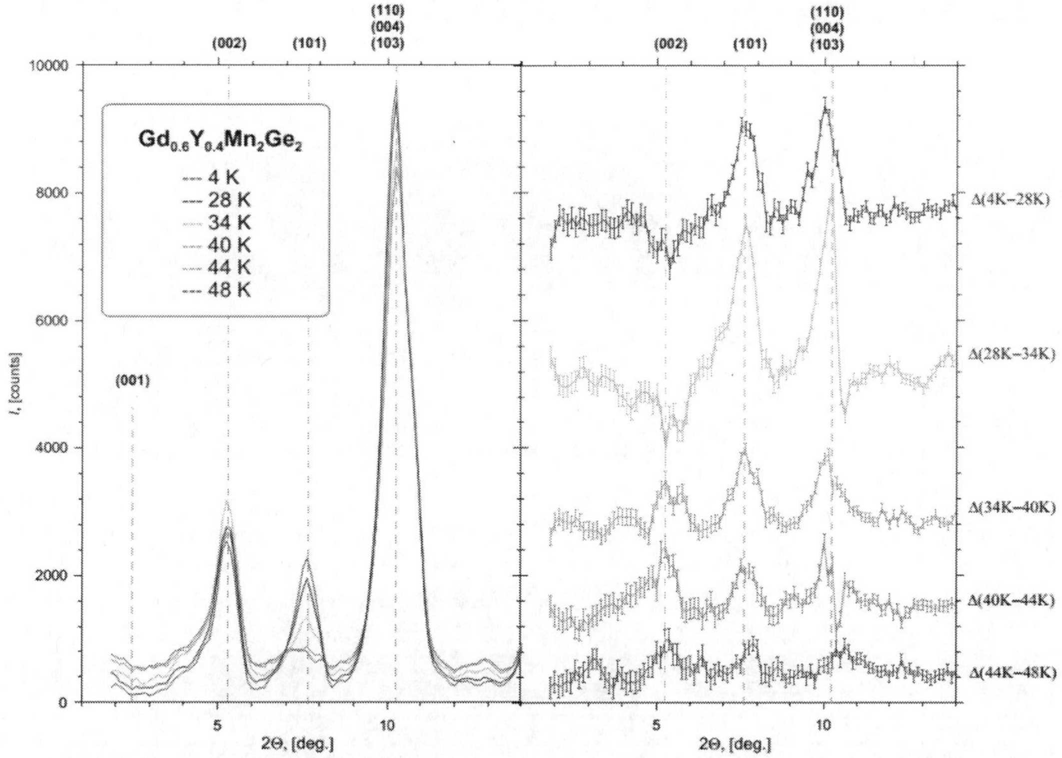

Fig. 2. Hot neutron diffraction patterns for $Gd_{0.6}Y_{0.4}Mn_2Ge_2$ for different temperatures (left) and their difference plots (right). Bragg indexes are indicated above.

The interpretation of hot-source diffractograms obtained for $(Gd_{1-x}Y_x)Mn_2Ge_2$ series is complicated as the short wavelength of the hot neutrons leads to overlapping of reflections. Nevertheless, several conclusions about the magnetic structure could be drawn from the analysis of the first diffraction peaks and the differences of diffraction patterns for different temperatures. The diffraction patterns for $Gd_{0.2}Y_{0.8}Mn_2Ge_2$ are similar to those of non-diluted $GdMn_2Ge_2$ where collinear *F* low temperature phase was confirmed [5].

Two magnetic transitions are expected for $Gd_{0.6}Y_{0.4}Mn_2Ge_2$ [2,3]. Neglecting the thermal expansion in this temperature and angle range, the differences between diffraction patterns are caused only by different magnetic contributions. The difference plots are shown in the Fig.2. One can see that the (001) magnetic reflection, characteristic of the *Fmc* phase is not observed at any temperature. At the same time, the magnetic contribution to the (002) peak related to the in-plane magnetic component is strongly temperature dependent: it increases down to 34 K before it decreases again when cooling to the base temperature of 4 K. Then, comparing the high-temperature *AFil* phase and the low-temperature phase, there is no change in (002) intensity. Thus, our hot source diffraction experiments give the following scenario, which is different from both [2] and [3]. Gd ions order within the tetragonal basis plane first at $T \sim 47$ K. The negative R-Mn interaction force the Mn moments to rearrange from antiparallel *AFil* orientation into the tetragonal basis plane too (*T'* phase in the ref. [3]). The second magnetic transition directs Gd and Mn moments back to the *c*-axis and the resulting structure is collinear *F*.

The Neutron data for the sample with composition $Gd_{0.5}Y_{0.5}Mn_2Ge_2$ show only one magnetic transition from *AFil* to *T'* magnetic phases.

Figure 3 shows diffraction angle-temperature-intensity 3D-plots for selected compounds of $(Tb_{1-x}Y_x)Mn_2Ge_2$. The collinear ferrimagnetic *F* low temperature phase is confirmed for compounds with high terbium content. The low temperature phase for samples with Tb-content $0.4 \leq x \leq 0.46$ must be different from *F* following our $\rho(T)$ experiments. Diffraction data show the presence of intense $(0,0,2n+1)$ peaks, which are the indication of antiferromagnetic component in *a*-direction, below the Curie temperature. The Rietveld analyse clearly proves the canted *Fmc* low temperature phase for these concentrations.

As Tb content decreases the diffraction patterns change again. The low temperature plot for $Tb_{0.5}Y_{0.5}Mn_2Ge_2$ and $Tb_{0.55}Y_{0.45}Mn_2Ge_2$ reveals magnetic peaks of $(h,k,l/3)$-type together with $(0,0,2n+1)$ reflections, characteristic of *Fmc* structure. The magnetic state is then a mixture of the *Fmc* phase and the "triplet" collinear phase, whose magnetic cell is three time larger in the tetragonal direction. One third of Mn ions ordered antiparallel to another 2/3 in this cell, then one third of Tb ions, located between different Mn orientations, are non-magnetic. This magnetic structure has been previously observed in $DyMn_2Ge_2$ and $TbMn_2Si_2$ and could be interpreted using additional interlayer J^{IL-3}_{Mn-Mn}

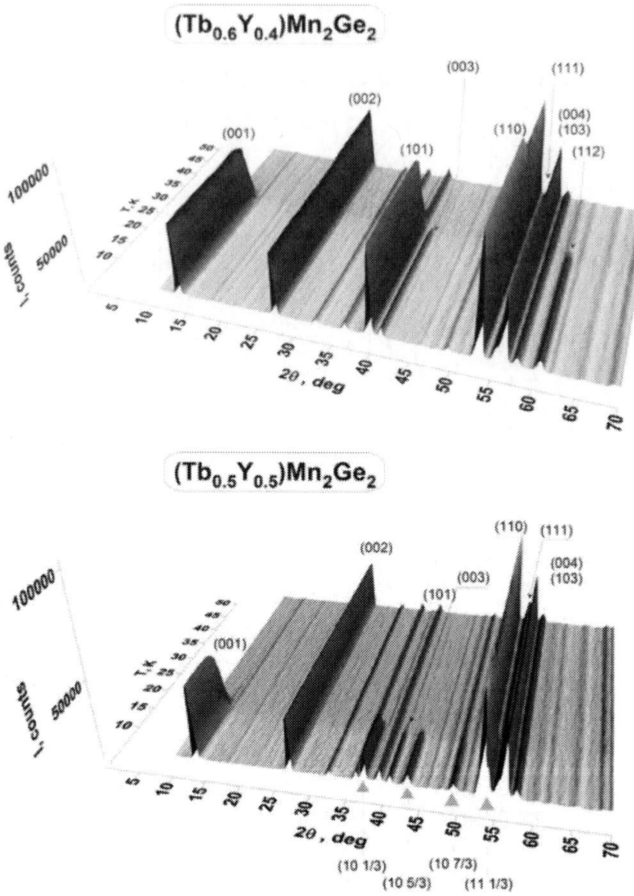

Fig. 3. Neutron powder diffraction patterns for $Tb_{0.6}Y_{0.4}Mn_2Ge_2$ and $Tb_{0.6}Y_{0.4}Mn_2Ge_2$. Bragg indexes are indicated.

term [6,7]. This long-periodic magnetic structure creates a band gap in the first Brillouin zone in tetragonal direction, explaining the significant increase of electrical resistivity in comparison to antiferromagnetic *AFil* phase.

As a summary, magnetic structures, realized in RMn_2Ge_2 with tuneable R-Mn exchange were directly probed in this work. Results of neutron diffraction experiments found to be in good agreement with bulk measurements and partly in agreement with theory expectations. Although the hot neutron diffraction data could not be used for quantitative refinements, it was possible to use the information to select between relatively simple magnetic structures. Thus, we demonstrated that variable R-Mn exchange leads Gd moments to realign from the tetragonal axis into the plane and back to this axis in low-anisotropic Gd-series. In contrast, high-anisotropic Tb holds itself to the *c*-axis and forces reorientation of Mn moments. *Ab-initio* calculations with known CEF-parameters for rare earth ions could provide the quantitative data on magnetic structures and phase transitions in these series.

References

1. G.Venturini, B. Malaman, E.Ressouche, J.Alloys Comp. **241** (1996).
2. H.Wada, Y.Kawaguchi, J.Alloys Comp. **379** (2004).
3. A.Yu.Sokolov, G.Guanghua, S.A.Granovskii, R.Z.Levitin, H.Wada, M.Shiga, T.Goto, JETP **89** (1999).
4. I.S.Dubenko, I.Yu.Gaidukova, S.A.Granovsky, K.Inoue, A.S.Markosyan, S.Roy, N.Ali, J.Appl.Phys **93** (2003).
5. S.A.Granovsky, A.Kreyssig, M.Doerr, C.Ritter, E.Dudzik, R.Feyerherm, P.C.Canfield, M.Loewenhaupt, J.Phys: Cond.Matter **22** (2010).
6. S.A.Granovsky, I.Yu.Gaidukova, M.Doerr, M.Loewenhaupt, A.S.Markosyan, C.Ritter, Physica B **391** (2007).
7. A.S.Lagutin, R.Z.Levitin, J.Vanacken, Y.Buynseraede, J.Magn. Magn. Matter 241 (2002).

Structural and Magnetic Properties of Copper Substituted Mg-Ferrites

Tatiana Kiseleva[1], *Vladislav Kabanov*[1], *Alexander Ilyushin*[1,2], *Gennadiy Markov*[3], *Deleg Sanga*[4] *and H. Hirazawa*[5]

[1]Moscow M.V.Lomonosov State University, Physics Faculty, Mocow, Russia
[2]The Ibragimov Complex Institute, RAS, Grozniy, Russia
[3]The Schmidt Institute of Physics of the Earth RAS, Moscow, Russia
[4]Institute of Physics and Technology, Mongolian Academy of Science, Ulan Bator, Mongolia
[5]National Technological Institute, Ehime, Japan

Abstract. Polycrystalline ferrite powders of $Mg_{1-x}Cu_xFe_2O_4$ (x = 0.2, 0.4, 0.6, 0.8, 1) system synthesized by ceramic technology have been investigated. Samples showed the non-monotonic dependency of heat generation effect in AC magnetic field with increasing concentration of copper. To reveal peculiarities of the structural and magnetic state of the samples and their influence on the heat generation ability we performed a complex study, including X-ray diffractometry, Mössbauer spectroscopy, Scanning electron microscopy, measurements of temperature dependencies of susceptibility and saturation magnetization, hysteresis parameters and FORC.

Typical ferrimagnetic character with small coercivity and saturation magnetization was found. We carried out that anomalous influence of Cu^{2+} ion substitution respectively to the $Mg_{1-x}Cu_xFe_2O_4$ ferrite powder manifested in heat generation ability rise up to x=0.6. The subsequent sharp reducing of this characteristic were accompanied by the main phase crystal structure distortion followed by phase separation to cubic and tetragonal structure. This was matched by in an increase of ferrite particles crystallite size and size distribution appearance. The saturation magnetization and Curie temperature dependencies observed for powders via Cu substitution was explained by phase composition, the cations distributions between ferrite sublattices, modulation of exchange interaction.

1 Introduction

The new opening opportunities for ferrite's systems applications in different fields of modern industry (catalysis, adsorption, humidity sensors, ferrofluids and ferroelastomers technology, components of health monitoring systems and medicine) require the involvement of modern and powerful experimental techniques. The properties-structure-sizes interdependence study allows to manage ferrite's functionality [1-3]. Among a number of industrial important ferrite compositions magnesium ferrite ($MgFe_2O_4$) particles have favorable magnetic properties and biological compatibility. It is worth mentioning that $MgFe_2O_4$ particles of micrometer-range sizes are reported to exhibit greater magnetic heating than do other ferrites [4-5], but a lot of afforts have veen made to synthesize and research these compounds in nanoparticulate form. Despite a lot of efforts to modulate specific magnetic properties of this ferrite as by types of substituting metals and their combinations, included in composition, the cationic distribution and interaction between them, as particles' sizes distributions, the issue of the best composition of ferrite particles to obtain effective heating abilities under application of alternating magnetic field remains controversial [6-7]. The investigation of the different effects influence on the properties of ferrites, which are used as components of complex composite systems for the purposes of therapeutic implants, seems to be important.

It is known, that ferrites have the general formula $(M_{1-z}Fe_z)[M_zFe_{2-z}]O_4$. The divalent metal atoms M (Mg, Fe, Cu or mixture of them) can occupy either eight tetrahedral (A) or sixteen octahedral [B] sites of a cubic mineral spinel structure as depicted by the parentheses or brackets, respectively, z represents the degree of inversion. The complexity of the above formula and the nature of preparation techniques strongly suggest a variation in unit cell composition and possibly a variation in the arrangement of Mg^{2+} and Fe^{3+} ions over the available tetrahedral and octahedral sites. Depending on the cations distribution in (A) and [B] sites, two extreme states—normal (z = 0) and inverse (z = 1) or an intermediate mixed state may be realized. Ultimate magnesium ferrite ($MgFe_2O_4$) is an inverse spinel taken to be collinear ferrimagnetic [8,9], whose degree of inversion depends on the synthesis temperature and cooling rate [8,10]. The ultimate copper ferrite ($CuFe_2O_4$) exists in tetragonal and cubic structures [10]. The distortions from one structure to another is directly related to the magnetic properties. The cubic structure posesses a larger magnetic moment than that of the tetragonal one, because there are more cupric ions (Cu^{2+}) at tetrahedal sites in cubic structure as compared to that in the case of tetragonal structure. As the ionic radii of the Fe, Mg and Cu cations (the ionic radius of Cu^{2+}(0.85

* Corresponding author: Kiseleva.TYu@gmail.com

Å) Mg^{2+}(0.78 Å) (Fe^{2+}(0.83 Å) Fe^{3+}(0.77 Å) Mg^{2+}(0.78 Å) Fe^{2+}(0.83 Å) Fe^{3+}(0.77 Å) are not quite different, random occupancy of cations in any site is possible and will also leads to produce several structural and hyperfine effects that may reveal modulation of ferrite magnetic properties.

The aim of this works was to investigate the practically important structure and magnetic properties correlations in $Mg_{1-x}Cu_xFe_2O_4$ samples series for varying Cu-substitution (0<x<1).

2 Experimental

Polycrystalline ferrite powders of $Mg_{1-x}Cu_xFe_2O_4$ (x = 0.2, 0.4, 0.6, 0.8, 1) system were synthesized by ceramic technology at 1473 K [6].

X-ray diffraction (XRD) has been performed on Empyrean (PAnalytical) diffractometer equipped with a PIXcel3D detector (Bragg-Brentano geometry, CuK_α radiation, Ni filter). ISCD database was used for phase identification. Quantitative phase analysis was carried out using the Rietveld refinement method with derivative difference minimization applying HighScore software.

Scanning electron microscopy (SEM) images of samples have been obtained using Quanta 3D FEG (FEI) microscope.

Fe^{57} Mössbauer study of structure and magnetic state of samples was carried out at 80 and 300 K with ^{57}Co(Rh) source. Analysis of the spectra have been performed with UnivemMS software. All spectra were referenced to α-Fe at room temperature.

Temperature generation ability of ferrites in AC (370kHz, 1.50kA/m) magnetic field was measured on laboratory equipment with IR-thermometer [7].

Hysteresis loops and FORC (first-order-reversal-curves) diagrams were measured at room temperature in applied magnetic fields up to 10 kOe using VSM 3900 magnetometer (LakeShore), temperature dependencies of saturation magnetization and susceptibilities were measured using Curie balance (Ltd "Orion") in temperature range 293-1073 K in applied magnetic field of 4,5 kOe, and Cappabridge MFK-1FA (AGICO) in magnetic field 200 A/m and frequency ~1000 Hz, correspondingly.

3 Results and discussion

Plot of the temperature enhancement from room temperature (ΔT) for powders in the AC(370kHz, 1.50kA/m) magnetic field (Fig.1, a) shows that heat generation ability was improved with increasing the Cu substitution, the highest heat generation ability ($\Delta T \approx 37°$) was obtained at x=0.6. Then, the heat generation ability abruptly decreases and lost over x=1.0. Similar nonmonotoneous dependencies were observed for other ferrite compositions synthesized by the same technology and had different degree of heat generation. In order to gain more information regarding the structure and properties correlations in our synthesized $Mg_{1-x}Cu_xFe_2O_4$ powders we have undertaken X-ray

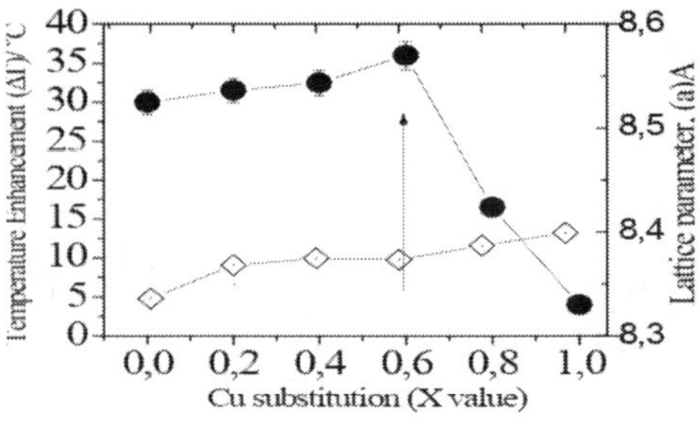

Fig.1. Temperature generation ability of ferrites in AC (370kHz, 1.50kA/m) magnetic field (circles) and lattice parameter (rhombuses) via Cu substitution in $Mg_{1-x}Cu_xFe_2O_4$.

diffraction, Moössbauer spectroscopy and complex magnetic properties study.

The X-ray diffraction patterns of $Mg_{1-x}Cu_xFe_2O_4$ ferrite system for x=0-1.0 (samples 1-6) analysis revealed the polycrystalline structure with μm-sized crystallites. The main phase in all samples is *fcc* (cubic) spinel. Several additional impurities of crystalline phases (MgO, Fe_2O_3, $CuFeO_3$) were identified as a reaction by-products. For the ultimate Cu-substitution (x=1) the appearing of tetragonal ferrite structure in a very small amount have been determined. Lattice parameter of the cubic phase increases with Cu^{2+} content with small distortion from the linear dependence at x=0.4-0.6 (Fig.1, a). The dependence of the integrated intensity ratio $I(220)/I(222)$ on the Cu content revealed that there is a cations redistribution between (A) and $[B]$ sites.

Fig.2 shows the concentration variation of Mössbauer spectra, measured at 300K. The spectra indicate a magnetic ordering with spectral lines broadening via Cu-substitution. Mössbauer spectra analysis allowed to resolve at first exact phase composition: Fe_2O_3 ,$CuFeO_3$, Cu(Fe) phases additionally to the main spinel phase have been determined for x=0 and x=1, correspondingly. Then, spectra profile fitting in a simple model resolving (A) and $[B]$ subspectra due to iron atoms in the two sublattices was applied. Mossbauer spectrum of $MgFe_2O_4$ is known to have an asymmetric broadening of the $[B]$-site lines indicating the presence of several subpatterns arising from the different possible nearest-neighbor (A)-site configurations via supertransfer mechanism [9]. According to [11] the more covalent character of the Fe(A) –O bond compared to the Fe$[B]$ –O bond explains qualitatively why the spin-density transfer from (A) to $[B]$ in the spinel structure is more effective than vice versa. As a consequence, the (A)-site lines usually do not show any structure that indicates the presence of a narrower hyperfine field's distribution. Mössbauer spectra parameters resolved from fitting allowed to determine the concentration dependencies of hyperfine magnetic fields on Fe^{57} atoms in different sublattices so as an inversion degree (z) changes.

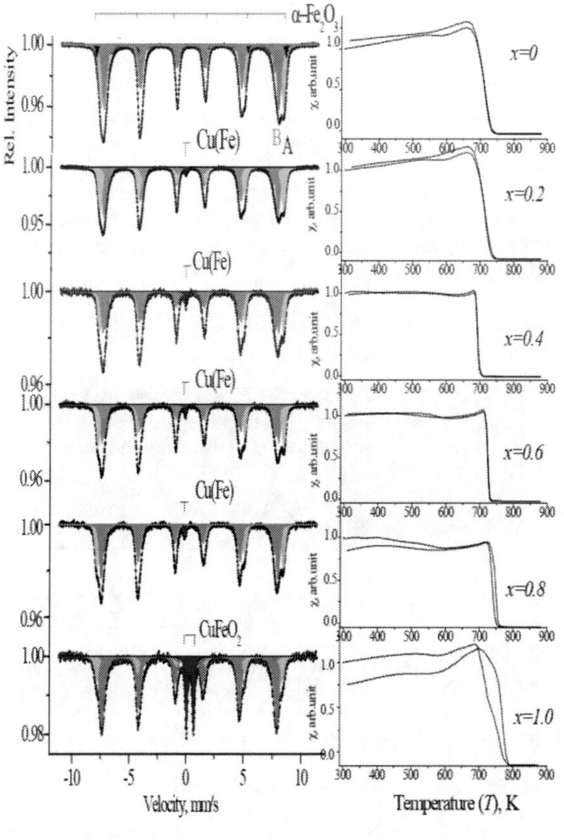

Fig.2. Mossbauer spectra at 300 K (left panel) and temperature dependence of susceptibility (right panel) in dependence on Cu substitution correspondingly)

The degree of inversion was calculated from the subspectral areas (I(A)/I(B)=f(A)/f(B)*z/(2-z)), assuming that the ratio of the recoilless fractions is f(B)/f(A)=0.94 at room temperature. The resulting composition dependence of inversion degree plotted on Fig.3 demonstrates the changes at x=0.4-0.6.

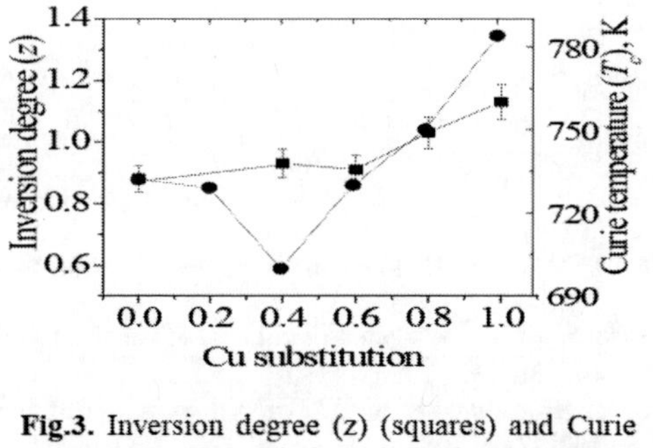

Fig.3. Inversion degree (z) (squares) and Curie temperature (Tc) (circles) dependencies vs Cu substitution

Fig.2 (right panel) shows the temperature dependence of susceptibility χ(T) for samples. The results exhibit normal ferrimagnetic behavior. The pure sample corresponds to x=0.2-0.8 and in accordance with XRD and Mössbauer phase analysis data. The ultimate ferrite sample with x=1 demonstrated multiphase behavior. The Curie temperature (Tc) obtained from susceptibility data is shown in Fig.3.

The Curie temperature (T_c) of the main phase in the ferrite samples measured from χ(T) curves has nonmonotonic dependence via Cu substitution: gradual decrease of the Tc value to the sample with x = 0.4 and followed by an Tc increase. It is known that Tc is determined by the overall strength of the intersublattice AB interactions, but sometimes the intrasublattice AA and BB interactions may become important. The decrease in Tc with increasing concentration of Cu may be explained by the modification of A-B exchange interaction strength due to the change of cations distribution between A and B sites.

Hysteresis loops J(H) for $Mg_{1-x}Cu_xFe_2O_4$ ferrite samples (Fig.4) showed nonmonotonic dependence: it is divided into three groups: I (x=0-0.2, curves 1,2); II (x=0.8-1, curves 5,6); III (x=0.4-0.6, curves 3,4). Similar non-monotonic behavior of hysteresis loops has been reported in several works subjected to study of substitution influence [12]. Magnetic parameters determined from the curves were analysed and plotted in as a Day's diagram [13-14] (J_{rs}/J_s vs H_{cr}/H_c) (Fig.5, a). This diagram clearly demonstrated the three ranges that correspond to different sizes or domain state in the particles: I - the smallest particles are probably a mixture of SD (single domain) and PSD (pseudo single domain) particles, II - larger particles are mixture PSD and MD (multi domain) particles, III - the largest particles (multi-domain state of ferrite powders).

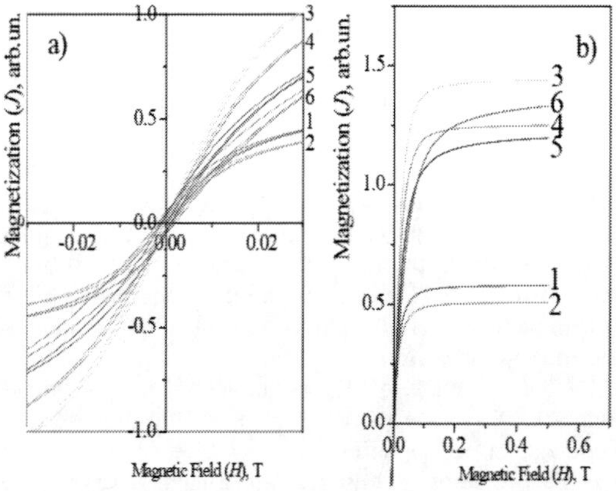

Fig.4. Hysteresis loops vs Cu substitution in small fields (a) and in high fields (b) for $Mg_{1-x}Cu_xFe_2O_4$ (x=0-1, marked 1-6, correspondingly)

Much more detailed information about magnetic assemblages than standard hysteresis curves was derived from the measured FORC (first-order-reversal-curves) diagrams. Characteristic FORC diagram for the samples with concentrations that fall within the determined ranges are shown on (Fig 5 b, c, d). FORC was able to detect the coercive force distribution as well as the magnetic interactions within particles assemblage [15].

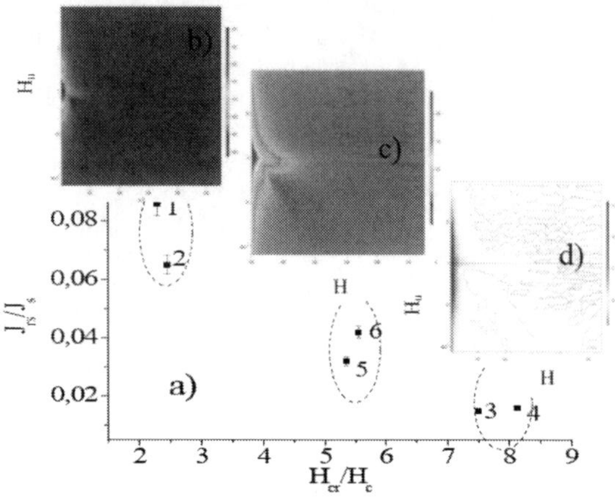

Fig.5. Day's diagram (a), typical FORC diagram for corresponding samples groups (b,c,d)

FORC diagrams were measured by saturating a sample in VSM magnetometer in a field H_{sat}, decreasing the field to a reversal field H_a, then sweeping the field back to H_{sat} in a series of regular field steps H_b as illustrated in color on Fig.5 (b,c,d). This process is repeated for many values of H_a yielding a series of FORCs, and the measured magnetization at each step as the function of H_a and H_b gives $J(H_a, H_b)$. The FORC distribution $\rho(H_a, H_b)$ is defined as the mixed second derivative of the interaction surface.

For each sample we determined the main FORC peak, which is the coercivity field corresponding to the maximum of the FORC distribution (plotted on Fig.5 as $H_u=(H_a+H_b)/2$, $H=(H_b-H_a)/2$). The interaction field is quantified with the full width at half-maximum of the main peak of the FORC distribution parallel to the abscissa axis through the maxima peak H. It was revealed from FORC diagram that for x=0.4-0.6 large multidomain particles were synthesized; the smallest particles sizes corresponds to x=0-0.2 ; intermediate for x=0.8-1. This fact was confirmed by SEM imaging (Fig.6, a,b,c) of particles with restoration of particles sizes distribution (Fig.6, d) where they measured from a few to several tens of microns. Cu substitution tends to increase particles sizes.

The decrease of H_c is in accordance with the increase of average particle's size and the immoderate nonmagnetic ions entering into the lattice that may result in the energy reduction of magneto-crystalline anisotropy that leads to the resulting heat generation ability.

Summary

Synthesized by ceramic technology at 1473 K polycrystalline $Mg_{1-x}Cu_xFe_2O_4$ ferrite powders were investigated by structural and magnetic methods. The effect of Cu-substitution (x=0, 0.2, 0.4, 0.6, 0.8, 1) was investigated. To sum up the results we found that:

1)Synthesized ferrites demonstrated anomalous heating ability via Cu substitution that is connected with their structural and magnetic properties variation.

2) All samples demonstrated soft magnetic behaviour with coercivity depending on particles sizes and domain state.

3) Magnetic hyperfine interaction measured by Mossbauer spectrometry showed that Cu^{2+} has an appreciable effect on the magnetic field of the octahedral sites.

4) It was determined that magnetic properties of synthesized ferrites are in accordance with particles sizes and cations distribution.

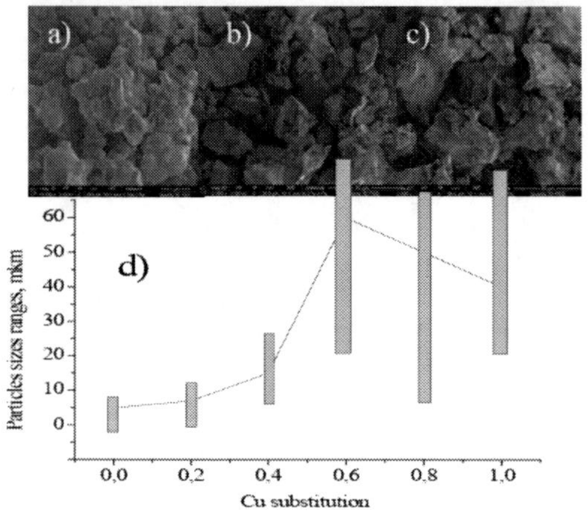

Fig.6. SEM images for x=0 (a), 0.4 (b) and 0.8 (c) and particles sizes distributions (d) from SEM images analysis via Cu substitution

Acknowledgments

Authors thank Moscow University Program of Development and the Ministry of education and science of the Russian Federation (grant 14.Z50.31.0017) for technical and financial support.

References

1. S. Kumar, A. Daverey, et.al, J. Mater. Chem. **B**(1) 3652 (2013)
2. E.A. Schultz-Sikma, H.M. Joshi, et al. , Chem. Mater. **23** (10) 2657–2664 (2011)
3. A.Yu. Zubarev, A.S. Elkady, Physica **A** 413400–408 (2014)
4. T. Maehara, K. Konishi, et al., Jpn. J. Appl. Phys. **41** 1620–1621 (2002)
5. T.Maehara, K.Konishi et al. , J. of Materials Science **40**, 135-138 (2005)
6. H.Hirazawa et.al. , J. of Alloys and Comp., **461**, 467-471 (2008)
7. Hirazawa H., Ito Y., Sangaa D., AIP Conference Proceed. **1763**, 020009 (2016)

8. M.R. Barati, C.Selomulya, J. of Appl.Phys. **115** 17B522 (2014)
9. G. A. Sawatsky, F. Van Der Woude, *Phys.Rev.* **187** 2, 747-757 (1969)
10. J. Smit and H. P. J. Wijn, "*Ferrites*," John Wiley & Sons Inc. New York, (1959)
11. F. Van Der Woude and G. A. Sawatzky, Phys. Rev. B **4**, 3159 (1971)
12. P.K. Roy and J.Bera, J.Magn.Magn.Mater **298** 38 (2006)
13. R.Day, M.Fuller, Phys. Earth Planet. Inter., **13**, 260—267 (1977)
14. D.J.Dunlop J. Geophys. Res. S Solid Earth, 107, B3, 2046—2067 (2002)
15. A.R.Muxworthy, A.P. Roberts First-order reversal curve (FORC) diagrams, in *Encyclopedia of Geomagnetism and Paleomagnetism*, Springer, NY (2006)

THE INVESTIGATION OF CRYSTAL AND MAGNETIC STRUCTURES OF SOLID SOLUTIONS $BaFe_{12-x}D_xO_{19}$ (D= In AND Ga; x= 0.1 – 1.2)

Vitalii Turchenko[1,2,★], *Alex V.* Trukhanov[3,4], *Sergei V.* Trukhanov[3,5], and *I.S.* Kazakevich[3]

[1] *Frank Laboratory of Neutron Physics, Joint Institute for Nuclear Research Joliot-Curie str., 6, 141980 Dubna, Russia*
[2] *Donetsk Institute of Physics and Technology named after O.O. Galkin of the NASU, 46 Nauki Ave, 03680, Kiev, Ukraine*
[3] *SSPA "Scientific and practical materials research centre of NAS of Belarus", 19 P. Brovki str., 220072 Minsk, Beloruss*
[4] *South Ural State University, Lenin's prospect 76, 454080, Chelyabinsk, Russia*
[5] *National University of Science and Technology MISiS, Leninsky Prospekt, 4, 119049, Moscow, Russia*

Abstract. The influence of partially substituted of barium ferrites by diamagnetic In and Ga ions with neutron diffraction method was studied. The substitution of Fe by In ions leads to increase of volume of unit cell while replacing with Ga ions insignificantly change the lattice parameters. The concentration dependence of the T_c Curie temperature as well as the magnetization is constructed. The refinement of the magnetic structure was carried out within the framework of Gorter's model. The microstructural parameters of solid solutions $BaFe_{12-x}D_xO_{19}$ (D= In and Ga; x = 0.1–1.2) were determined by the high-resolution neutron diffraction method.

1 Introduction

The crystalline structure of barium ferrites and their solid solutions has a structure of magnetoplumbit $PbFe_{12}O_{19}$ which was first studied by Adelskold [1] in 1938. As a rule, it is well described in the framework of the space group $P6_3/mmc$ (No. 194) with a hexagonal unit cell a = b ≈ 5.90 Å, c ≈ 23.30 Å containing two formula units. Iron ions are localized in five nonequivalent crystallographic positions: 2a, $4f_{VI}$ and 12k - with octahedral, 2b - with trigonally bipyramidal and $4f_{IV}$ - with a tetrahedral oxygen environment.

In order to improve functional properties of barium ferrites for their application use the substitution of iron ions with diamagnetic ions Al, Sc, Ga, etc. [2, 3], which change the number of bonds between magnetoactive ions. The replacement of iron ions in certain crystallographic positions influences on valence angles and interatomic distances in Fe–O–Fe [4] bonds, and consequently, affects to the pair exchange integral and the magnitude of the sublattice exchange [5].

In this paper, the effect of partial replacement of magnetoactive Fe ions by diamagnetic In and Ga ions on the crystal and magnetic structures of $BaFe_{12-x}D_xO_{19}$ in a wide range of concentrations (D= In and Ga; x = 0.1–1.2) was studied by neutron diffraction method. The substitution of solid solutions $BaFe_{12-x}D_xO_{19}$ by diamagnetic ions D= In and Ga allows research the influence of appearing chemical pressure on distortions of unit cell parameters and on the value of bond length Fe/In/Ga-O and determine the influence to magnetic properties (temperature of magnetic phase transition, value of magnetic moments, etc). Concentration range of diamagnetic substitution (x= 0.1–1.2) was determined due to the preservation of optimal magnetic characteristics in doped Ba-hexaferrites. For concentration more than 10 at.% is high risk of observation some magnetic anomalious like frustration of magnetic structure [6] instead of collinear ferrimagnetic structure. Another feature of the work is the use of a high-resolution diffractometer for the measurement of neutron diffraction patterns $BaFe_{12-x}D_xO_{19}$ (D= In and Ga; x = 0.1–1.2), which made it possible to obtain information about both the crystalline and magnetic structures and the microstructure of the samples studied.

2 Experimental Details

Solid solutions of $BaFe_{12-x}D_xO_{19}$ (D= In and Ga; x = 0.1–1.2) have been prepared from high purity Fe_2O_3 and In_2O_3 and Ga_2O_3 oxides and carbonate $BaCO_3$ using 'two-steps' topotactic reactions (conventional solid reaction method). Oxides and carbonate have been mixed with design ratio and synthesized at 1200^oC (6 h) and at 1300^oC (6 h) in air. After synthesis the sample has been slowly cooled ($100^oC/h$).

$$BaCO_3 + (\frac{1-x}{2}) \cdot Fe_2O_3 + (\frac{x}{2}) \cdot (In/Ga)_2O_3 \rightarrow \rightarrow BaFe_{12-x}(In/Ga)_xO_{19} + CO_2 \uparrow \quad (1)$$

The powder neutron investigations of sample was performed by neutron time–of–flight method at High Resolution Fourier Diffractometer (HRFD, Dubna) at room temperature. The refining of crystal and magnetic structures

★e-mail: turchenko@jinr.ru

was performed by Rietveld analysis [7], using the FullProf [8] software program. The resolution of diffractometer is $\Delta d/d_0 \approx 0.001\%$ and lattice parameters were defined more exactly at standard Al_2O_3 powder (standard SRM-676 of NIST, USA).

Field dependences of specific magnetization were measured at 10–730 K by Liquid Helium Free High Field Measurement System (VSM) [9]. The "ferrimagnet-paramagnet" phase transition temperature for the $BaFe_{12-x}D_xO_{19}$ (D= In and Ga; x = 0.1–1.2) has been defined as the inflection point in the temperature dependence of the specific magnetization.

3 Results and discussion

3.1 Magnetic properties

Fig. 1 demonstrates temperature dependences of the specific magnetization for the $BaFe_{12-x}D_xO_{19}$ (D= In and Ga; x = 0.1-1.2) obtained by VSM. The concentration dependence of the Curie temperature determined as the inflection point in the temperature dependence of the specific magnetization are shown on the inset in Fig. 1. It necessary to note that doping of diamagnetic ions In and Ga leads to reducing of the number of neighbors of magnetic iron cations and so that the magnetic order is destroyed at lower temperatures [10]. Therefore the Curie temperature decreases as the doping of In or Ga ions is increased whereas for the un-doped $BaFe_{12}O_{19}$ the Curie temperature is equal 740 K [11].

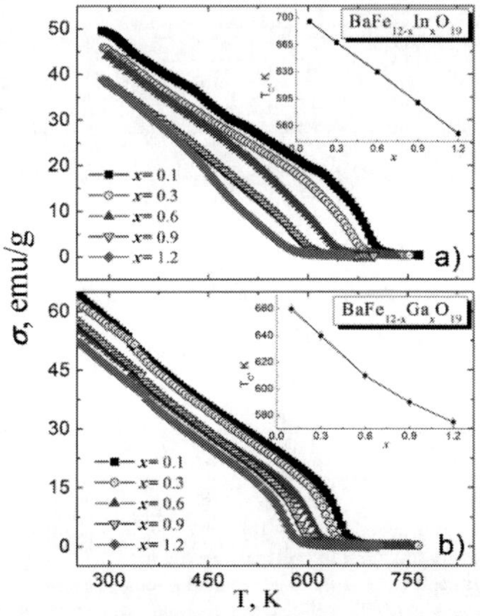

Figure 1. Temperature dependencies of the specific magnetization of $BaFe_{12-x}D_xO_{19}$ (D= In – (a) and Ga – (b); x= 0.1–1.2). In the inset, the concentration dependencies of the Curie temperature are shown.

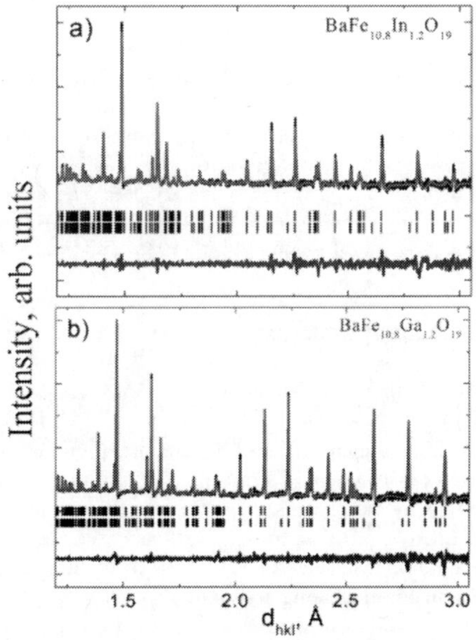

Figure 2. Neutron diffraction patterns of $BaFe_{10.8}D_{1.2}O_{19}$ (D= In – (a) and Ga – (b) obtained with the HRFD at room temperature and processed via the Rietveld method. Experimental points, calculated curves and difference functions normalized to steady state error are shown. For the crystalline and magnetic structures of barium hexaferrite vertical bars designate the calculated diffraction peak positions.

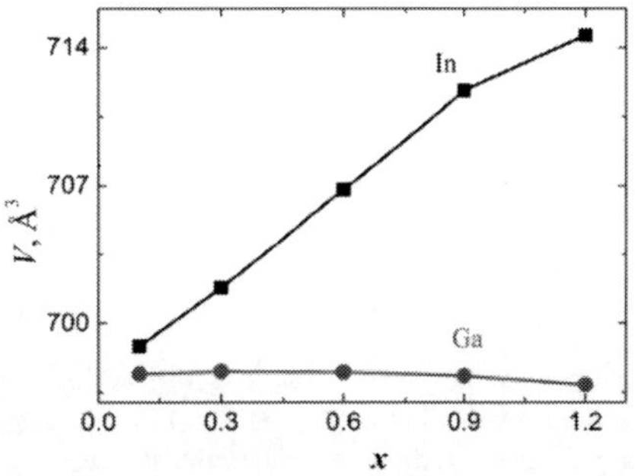

Figure 3. The concentration dependence of volume of unit cell of barium ferrites $BaFe_{12-x}D_xO_{19}$ (D= In and Ga; x = 0.1–1.2).

3.2 Crystal structure

According to neutron data all investigated polycrystalline samples possesses a hexagonal structure with space group ($P6_3/mmc$) with two formula units (Z = 2). The powder neutron diffraction patterns for the $BaFe_{10.8}D_{1.2}O_{19}$ (D= In and Ga) are presented in Fig. 2.

The high resolution of diffractometer and, correspondingly, the large number of well-separated peaks provided the good convergence of the minimization process. Figures of merit (R_{wp}, R_{exp}, χ^2) that quantify the quality of the refinement using the Rietveld method are in range: R_{wp} - from 14.8% to 16.9%, R_{exp} - from 10.0% to 11.9%, χ^2 - from 2.2 to 2.0. The dependencies of lattice parameters of unit cells versus temperature are shown in Fig. 3. Volumes of unit cells of solid solution of barium hexaferrite increase as the concentration of In ions is increased because of significant difference of ionic radii between In^{3+} (r = 0.94 $\AA$) and Fe^{3+} (r = 0.645 $\AA$) [12]. The increase of concentration of Ga^{3+} (r= 0.62 $\AA$) ions leads to insignificant decreasing of volume of unit cell.

3.3 Magnetic structure

In hexaferrites the magnetic Fe^{3+} ions are located in positions which have octahedral (Fe1-2a, Fe4-4f_{VI} and Fe5-12k), tetrahedral (Fe3-4f_{IV}) and bipyramidal (Fe2-2b) oxygen environment. Therefore, partially replacement of iron ions by diamagnetic In or Ga ions that are distributed statistically equivalent for all positions of the magnetic lattice are able to change in the values of the magnetic moments in corresponding positions.

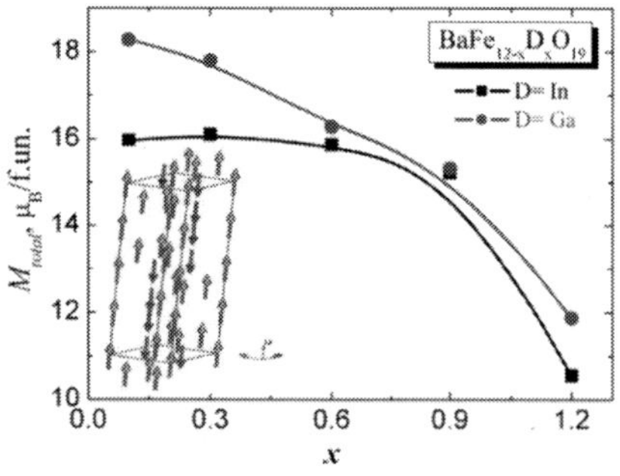

Figure 4. The dependence of the total magnetic moment of ferrite $BaFe_{12-x}D_xO_{19}$ (D= In and Ga; x = 0.1–1.2) versus concentration of diamagnetic ions.

The absence at powder neutron diffraction patterns Fig.2 of the additional magnetic peaks allows to determine the wave vector of the ferromagnetic structure as **k** = [0,0,0]. The magnetic structure for all compositions fully satisfies the model proposed by Gorter [13], according to which, all magnetic moments of the Fe^{3+} cations are oriented along the easy magnetization axis which coincides with the hexagonal c axis. The schema of magnetic structure of $BaFe_{12-x}D_xO_{19}$ (D= In and Ga; x = 0.1–1.2) is shown on the inset in Fig.4. The substitution of iron by diamagnetic cation breaks the exchange interactions between the magnetic positions and sublattices that leads to a decrease in the value of their magnetic moments, as it was previously shown in Ref. [14]. The total magnetic moment M_{total} per formula unit for the barium hexaferrite ($BaFe_{12}O_{19}$) at T temperature can be calculated according to the formula [15]:

$$M_{total}(T) = 1[m_{2a}(T)] + 1[m_{2b}(T)] - \\ -2[m_{4f_{IV}}(T)] - 2[m_{4f_{VI}}(T)] + 6[m_{12k}(T)] \quad (2)$$

where m_i is the magnetic moment of Fe^{3+} ion in the i-th sublattice. If the magnetic moment of Fe^{3+} ion at 0 K is equal $5\mu_B$ then magnetic moment of pure $BaFe_{12}O_{19}$ ferrite will be equal to $20\mu_B$ per formula unit. Less values of magnetic moments are explained by influence of diamagnetic ions and thermal factor causing the disorientation of the magnetic moments in space due to the increase of the thermal fluctuations of the ions forming the crystal lattice. The dependencies of total magnetic moment per formula unit for the barium hexaferrite versus temperature are shown in Fig. 4.

3.4 Microstructure

The high resolution of Fourier Diffractometer allowed making analysis of influence the concentration of diamagnetic ions to microstructural parameters, which were determined from broadening of powder neutron peaks.

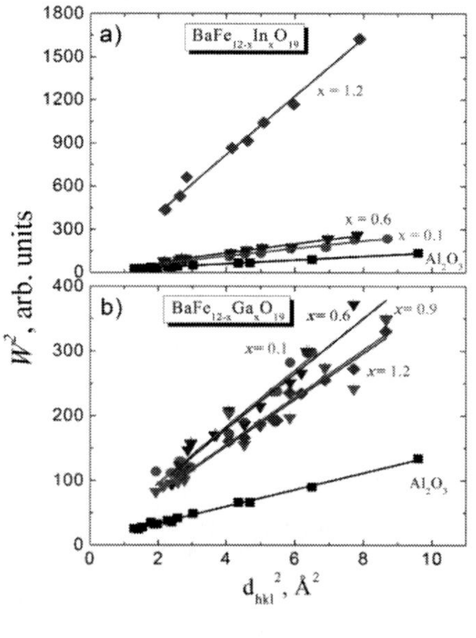

Figure 5. Dependencies of the square width of diffraction peak versus the square of interplanar spacing for $BaFe_{12-x}D_xO_{19}$ (D= In – (a) and Ga – (b)). Experimental points lie on linear curves, i.e., there is no size effect (the coherent scattering region size is large). The lowest dependence corresponds to the diffractometeres solution function determined using the standard sample Al_2O_3.

The presence of microstresses and variation of average size of coherent blocks have different effect on the width of the diffraction peaks, which makes it possible to split their contributions. The widths of the diffraction peaks W_{hkl} of the tested samples have the following dependence from d_{hkl} [16]:

$$W^2 = C_1 + (C_2 + C_3) \cdot d^2 + C_4 \cdot d^4 \quad (3)$$

https://doi.org/10.1051/epjconf/201818504011

where C_1, C_2, C_3, C_4 – refining constants; W – full width at half maximum of diffraction peak; $(C_1 + C_2 \cdot d^2)$ – resolution function of HRFD obtained from standard Al_2O_3 (SRM-676 of NIST, USA); $C_3 \cdot d^2$ – contribution determined by size effect; $C_4 \cdot d^4$ – contribution determined by microstraine effect.

The contribution of true physical line broadening of powder neutron peaks for the $BaFe_{12-x}D_xO_{19}$ (D= In and Ga; x = 0.1–1.2) was determined by FullProf software calculated as difference of widths among experimental samples and standard Al_2O_3. In our case, the line broadening is connected with only microstresses factor. In ferrimagnetic crystals, the separate sublattices give different contributions to general deformation. Besides, the local environment symmetry of magnetic ions in ferrimagnetic crystals differs from macroscopic symmetry. This leads to increase of the microscopic parameters number in comparison with macroscopic. The slope of approximated function increases with temperature decreasing. This behavior indicates that the microstress increases in crystallites. It is shown in Fig. 5 the experimental points correspond to different combinations of Miller indices, which indicates the absence of explicit anisotropic effects in the broadening of diffraction peaks.

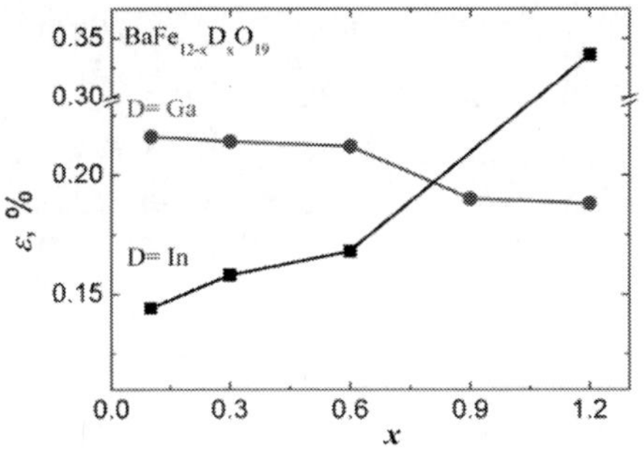

Figure 6. The dependencies of microstresses versus temperature in crystallites of $BaFe_{12-x}D_xO_{19}$ (D = In and Ga).

The dependencies of microstress values versus temperature for the $BaFe_{12-x}D_xO_{19}$ (D= In and Ga; x = 0.1–1.2) are shown in Fig. 6. The calculation of microstresses has been carried out for isotropic approximation i.e. (L > 3000 Å) size effect is absent. The increase in the microscopic microstress with the concentration of diamagnetic ions increasing is associated with rising of the system disorder, as a result of the statistical distribution of indium or gallium in magnetic sublattices, which can make different contributions to the total deformation.

4 Conclusions

The behavior of crystal and magnetic structures of barium ferrites doped by diamagnetic ions was investigated by neutron diffraction method with high resolution. The refinement of crystal structure of solid solutions of $BaFe_{12-x}D_xO_{19}$ (D= In and Ga; x= 0.1 – 1.2) has been performed in frames of space group $P6_3/mmc$ (No. 194). Larger volume of unit cell of barium ferrites doped with In ions is explained by the larger value of ionic radii of In^{3+} unlike of Ga^{3+} ions. The magnetic structure of $BaFe_{12-x}D_xO_{19}$ (D= In and Ga; x= 0.1 – 1.2) is well described by the Gorter model: according to which the magnetic moments of the Fe^{3+} ions are oriented along the hexagonal axis that is the axis of easy magnetization. Microstresses in crystallites increase as the concentration of In ions is increased whereas the rising of concentration of Ga ions decreases microstresses. This behavior was explained by different distortion of crystal sublattices due to diference of ionic radii of indium and gallium ions at substitution of iron ions.

The work was supported by JINR (Grant 04-4-1121- 2015/2017) and BRF (Grant no. F17D- 003); Ministry of Education and Science of the Russian Federation (Grants of NUST «MISiS» K4-2017-041and K3-2017-059; in SUSU - 4.1346.2017/4.6 by Act 211 of the Government of the Russian Federation No. 02.A03.21.0011).

References

[1] V. Adelskold, Avk. Miner. A. **12**, 1 (1938).
[2] D.A. Vinnik, D.A. Zherebtsov, L.S. Mashkovtseva, S. Nemrava, M. Bischoff, N.S. Perov, A.S. Semisalova, I.V. Krivtsov, L.I. Isaenko, G.G. Mikhailov, R. Niewa, J. Alloys Compd., **615**, 1043 (2014).
[3] 4) V. Kostishyn, V. Korovushkin, I. Isaev, A. Trukhanov, East.-Europ. J. Ent. Techn., **1**(85), 10 (2017)
[4] S. V. Trukhanov, A. V. Trukhanov, V. G. Kostishin, L. V. Panina, I. S. Kazakevich, V. A. Turchenko, V. V. Kochervinskii, JETP Letters, **103**, 100 (2016).
[5] John B. Goodenough, *Magnetism and the chemical bond* (Interscience Publishers, New York-London, 1963) 393.
[6] D.A. Vinnik, A.B. Ustinov, D.A. Zherebtsov, V.V. Vitko, S.A. Gudkova, I. Zakharchuk, E. Lähderanta, R. Niewa, Ceram. Intern., **41**, 12728 (2015).
[7] H. M. Rietveld, J. Appl. Crystallogr., **2**, 65 (1969).
[8] J. Rodríguez-Carvajal, Physica B **192** 55 (1993).
[9] S.V. Trukhanov, A.V. Trukhanov, A.N. Vasiliev, A.M. Balagurov, H. Szymczak, JETP, **113**, 820 (2011).
[10] S.V. Trukhanov, JETP, **100**, 95 (2005).
[11] T. Tsutaoka, N. Koga, J. Magn. Magn. Mater., **325**, 36 (2013).
[12] R.D. Shannon, Acta Cryst. A. **32**, 751 (1976).
[13] E.W. Gorter, Proc. IEEE Suppl., **104B**, 225 (1957).
[14] S.V. Trukhanov, A.V. Trukhanov, V.G. Kostishyn, L.V. Panina, V.A. Turchenko, I.S. Kazakevich, An.V. Trukhanov, E.L. Trukhanova, V.O. Natarov, A.M. Balagurov, J. Magn. Magn. Mater., **426**, 554 (2017).
[15] J. Smit, H. P. J. Wijn *Ferrites*, (Hume Press Ltd., Cleaver, 1959) 142.
[16] M.V. Cabanas, P. Germi, J.M. Gonzalez-Calbet, M. Pernet, M. Vallet-Regi, J. Solid State Chem., **114**, 534 (1995).

Magnetism of purified amorphous carbon

Anton Komlev[1,2*], *Erkki* Lähderanta[1], *Evgeniy* Shevchenko [3] and *Nikolay* Vorob'ev-Desyatovskii[4]

[1] Lappeenranta University of Technology, School of Engineering Science, 53850 Lappeenranta, Finland
[2] Saint-Petersburg State Electrotechnical University "LETI", 197376 Saint-Petersburg, Russia
[3] Saint-Petersburg State University, Research Park, 199034 Saint-Petersburg, Russia
[4] JSC Polymetal Engineering, Hydrometallurgical Department, 198216 Saint-Petersburg, Russia

Abstract. In this work, magnetic properties of activated carbon coal, which is also named amorphous carbon, were investigated. Composition and structure analysis was performed by XRD, SEM and EDS techniques. Position and width of peaks on XRD pattern confirmed amorphous structure of coals with small degree of graphitization. Magnetic measurements were performed by SQUID magnetometer. According to thermomagnetic measurements, in part of the samples, antiferromagnetic transition was observed near 120 K. Nature of such phenomena, as well as effect of oxygen and magnetic contaminations on thermomagnetic measurements are discussed referring to similar results from previous researches and data from literature.

1 Introduction

From ancient times carbon materials are widely used for many applications. After recent discoveries of fullerenes, carbon nanotubes and graphene, carbon-based materials became one of the most promising and universal materials. Among a variety of their outstanding mechanical, conductive and many other characteristics [1], also exists a number of controversial topics. One of such problematic areas is possibility of making carbon materials magnetic.

Despite the decades of investigations, magnetic properties of carbon allotropes is still a debatable topic. While the defect-induced paramagnetic behaviour is acknowledged, the problem of possibility of magnetic ordering in carbon materials is not solved yet. Many researches have demonstrated impurity-free carbon materials, where spontaneous magnetization was observed (e.g. Ref. [2, 3]). Such possibility for carbon materials is also predicted by many theoretical works (e.g. Ref [4]), meaning that magnetic carbon does not confront to general principles of magnetism. However, the question of intrinsic or extrinsic character and problem of the reproducibility of such results exists thus far.

In this work, thermomagnetic behavior of activated carbon coal was investigated. This material is also named amorphous carbon and is widely used on practice for extraction of noble metals from ores and for filtration processes. Amorphous carbon can be considered as many disorder graphite planes of different size and with big amount of defects. Thus, if carbon-based magnetism is promoted by defects, there should be a possibility to observe it in such structures. Investigation of magnetic properties of coals is important from both theoretical and practical points of view: first of all, it will give a contribution to the discussions about carbon-based magnetism; secondly, understanding of magnetic behavior of this material is essential since some efforts are undertaken to produce magnetic filtering material based on coals [5, 6].

2 Materials and Methods

Commercially available coal Goldcarb WSC208C-GR produced from coconut shells and provided by Chemviron Carbon company was used as a raw material. Activation of the raw material was done by ball-milling. To remove the acid-soluble ash content, coal was washed with 1 M solution of $HClO_4$ for one day, after what it was washed with distilled water for three weeks in Soxhlet extractor. Water in the extractor was changed every day and the conductivity of the water was checked to detect any contaminations, removed from coal. Part of the samples was saturated by copper complexes from an aqueous solution of ($Na[Cu(CN)_2]$). Motivation for such modification was a prior research related to absorption of a copper complexes during extracting of gold from cyanide solutions, which will not be discussed here. Part of both clean and modified samples was additionally dehydrated at 80°C temperature and 10 mbar vacuum before magnetic measurements.

Structure and composition of the samples were analyzed by means of scanning electron microscope (SEM) Hitachi SU3500 with built-in energy-dispersive X-ray spectroscopy (EDS) radiation and X-ray diffraction (XRD) device Bruker D8 Advance using Cu Kα (λ=1.5418 Å). Magnetic characterization was performed on a superconducting quantum interference device (SQUID) Quantum Design MPMS-XL-1 SQUID magnetometer with an up to 1 T magnet and working temperatures range of 1.8 – 400 K.

* Corresponding author: KomlevAnton@hotmail.com

https://doi.org/10.1051/epjconf/201818504012

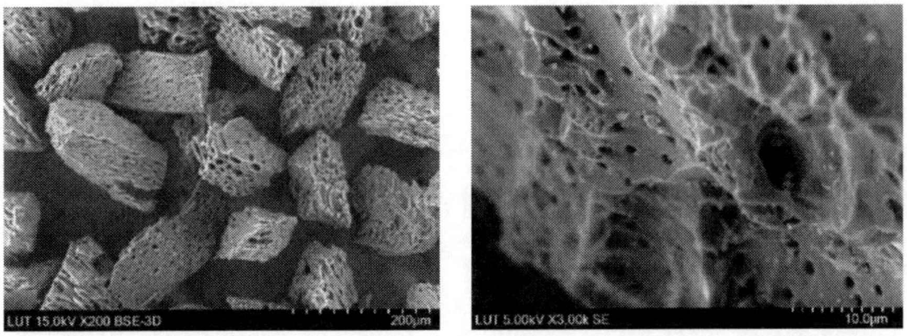

Fig. 1. SEM images of carbon coal, demonstrating high porosity of the samples: ×200 magnification on the left and ×3000 on the right pictures.

3 Results and discussion

According to SEM images on Fig. 1, after activation and purification, the material consists of particles of ~100÷200 μm size. These particles demonstrate highly developed surface and structure of the pores. Theirs size varies from ~15 μm to less than 100 nm, what is a characteristic of typical activated carbon. Big surface area and, thus, big amount of edge states is essential and desirable for both adsorption processes and possible magnetic ordering due to edge-states magnetism.

Structure of the coal was also examined by XRD technique (Fig. 2). The XRD pattern demonstrated two broad peaks at $2\theta \approx 25°$ and $2\theta \approx 45°$, which correspond to reflection from graphite (002) and (101) planes respectively. High ratio of (101) peak intensity to (002) can be an indicator of very high porosity of the sample [7]. Overall, such pattern is typical for disordered carbon structures with small degree of graphitization, meaning that in this material exists some disordered graphite planes, but it is not completely amorphous carbon [8].

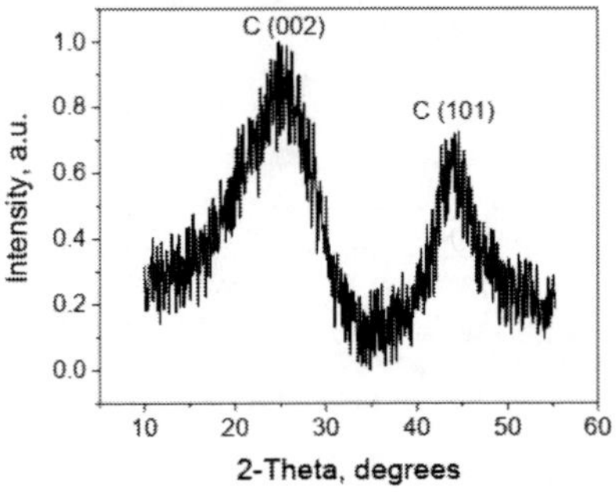

Fig. 2. XRD pattern of purified activated carbon coal.

Composition analysis of clean coal, performed by EDS with 20 kV accelerating voltage, showed 300 ppm of both S and Al. For modified with copper samples, in addition to the same amount of S and Al, 300 ppm of Cl and 3000 ppm of Cu were detected. Precise numbers are not crucial here, because the most important result is absence of any magnetic impurities (e.g. Fe, Co, Ni). However, it has to be mentioned, that in the similarly prepared batch of samples, EDS analysis performed on EDX-8000 (Shimadzu) in other laboratory detected additional contaminations: 350 ppm of Si, 50 ppm of Ca, traces of Br and what is the most essential – 20 ppm of Fe and Co. Such result can be explained by some contamination during the preparation before EDS measurements. However, some possible inhomogeneity of the samples can not be completely excluded and should be kept in mind during the interpretation of the magnetic measurements.

3.1 Magnetic measurements of clean and copper–modified coals

Thermomagnetic SQUID measurements of clean and copper–modified carbon coals are given in Fig. 3. The magnetization was measured in two ways: when the samples were cooled without magnetic field (Zero Field Cooled – ZFC), after what the field was turned on and magnetization was measured during heating; and FC – Field Cooling, when the magnetization was measured during cooling down with the magnetic field turned on. Fig. 3a represents the magnetic susceptibility of simple purified coal. There are several key features:

- Sharp decrease of magnetic susceptibility near 120 K on both FC and ZFC curves;
- Paramagnetism at room temperatures;
- Strong paramagnetic contribution increasing at low temperatures.

Starting from the end, strong paramagnetic contribution increasing at low temperatures is typical for carbon materials. It follows the Curie-Weiss law and is obviously connected with number of paramagnetic spins. This number is also proportional to the amount of defects in graphite or graphene–based materials [9] and can be calculated by approximation of the curve by Curie-Weiss law, where Curie constant will be proportional to the amount of paramagnetic spins. In this case, the amount was calculated to be 7.2×10^{18} spins/g. This means that there is one paramagnetic spin per ~7000 carbon atoms, if only carbon atoms are considered as the source of this signal.

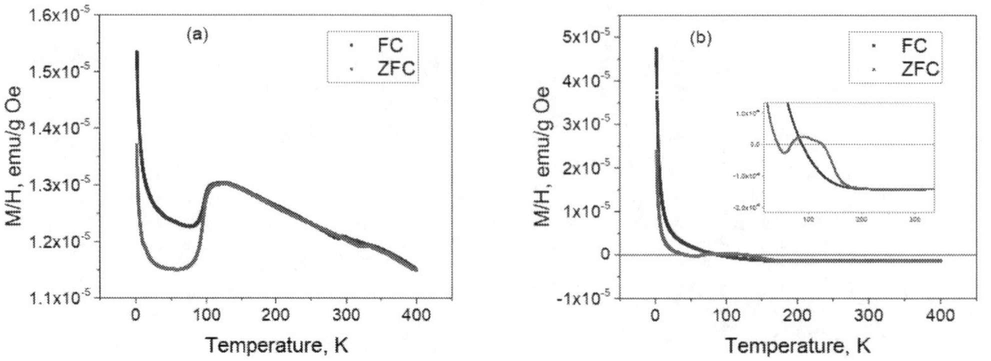

Fig. 3. SQUID FC and ZFC thermomagnetic measurements at 3300 Oe: (a) clean coal; (b) carbon coal modified by copper.

Two other features of magnetic susceptibility of this coal are not typical for carbon materials. While carbon materials usually demonstrate diamagnetic behavior without any sharp changes [9, 10]. Jump of magnetic susceptibility near 120 K can be interpreted as antiferromagnetic transition. Nature of this effect is debatable. It can be easily explained as a signal from some iron contamination, since iron oxide has similar Verwey transition in the same temperature region. Such explanation is also supported by the shift to paramagnetic region at room temperatures, which is not expected to be a feature of carbon. This can be also a trace of ferromagnetism. However, in our previous work [10] the same jump near 120 K, but without shift to paramagnetic region, was observed for functionalized graphene samples. In that work, carbon origin of this effect was approved by ESR measurements. Thus, the effect can originate from either magnetic contaminations or is connected with carbon magnetic behavior.

In case of modified by copper coal (Fig. 3b), no transitions near 120 K are observed and the sample is diamagnetic at room temperatures. Nevertheless, increase of paramagnetic signal at 50–170 K is present. The same phenomena is sometimes observed for different carbon samples [11-13]. Under more precise examination of Fig. 3b, one can see that this increase is present on ZFC curve and vanishing during cooling on FC curve. This effect can be easily explained by absorbed oxygen, which has antiferromagnetic transition near 50 K. If there is a leakage in SQUID system, or the sample was not completely purged, some oxygen can appear inside measuring area. After oxygen spins are unfrozen at 50 K, they go from antiferromagnetically ordered to paramagnetic state, resulting in increase of paramagnetic contribution. After this, with increasing temperature, oxygen starts to boil and evaporates, resulting in gradual decrease of this effect. If the oxygen has appeared only because of bad purging before measurement, it will be removed by nitrogen flow. The effect of oxygen on magnetic susceptibility was well examined for carbon samples in work [13], using Faraday method. However, usually such effect not discussed at all [11], while in others the similar behavior is called a ferromagnetic-like [12] and possible role of oxygen is not mentioned.

Approximation of low-temperature paramagnetic contribution by Curie-Weiss law gives 2.8×10^{20} spins/g, what is 40 times higher than for the pure coal and probably is explained by copper content, which can be paramagnetic.

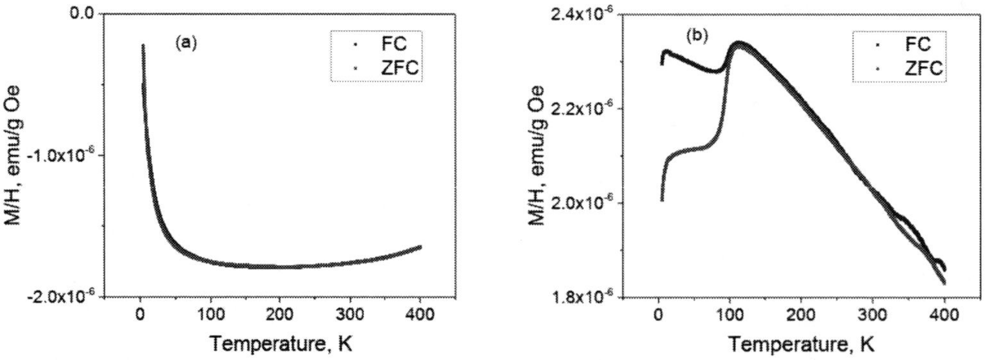

Fig. 4. SQUID FC and ZFC thermomagnetic measurements of dehydrated coal samples at 3300 Oe: (a) clean dehydrated coal; (b) dehydrated carbon coal modified by copper.

3.2 Magnetic measurements of dehydrated coals

In case of dehydrated coals (Fig. 4), in contrast with previous results, antiferromagnetic transition is observed in copper-modified sample, while clear coal demonstrates classical diamagnetic behavior with weaker paramagnetic contributions at low temperatures. No peaks at 50-170 K is observed, indicating that oxygen in previous experiment appeared because of bad purging.

Comparison of the clean non-dehydrated coal and copper-modified dehydrated sample can not be fully adequate, because in copped-modified samples magnetic effects can be dominantly going from copper instead of carbon. However, this can be an argument for the version, that antiferromagnetic transition near 120 K is connected with magnetic contaminations, which are distributed irregularly and because of this were not detected in fraction, measured by EDS.

One more extraordinary effect is observed in dehydrated carbon coal modified by copper (Fig. 4b). At low temperatures, no paramagnetic contribution is observed. Instead of it, rapid decrease of paramagnetic spins or increase of diamagnetic susceptibility is present. This can be explained as antiferromagnetic ordering of some copper compounds or even strong increase of diamagnetism because of superconductivity.

4 Conclusion

A number of amorphous carbon-based materials were investigated by means of SQUID magnetometry. The effect of oxygen on thermomagnetic characteristics was shown. Antiferromagnetic transition was observed for both clean and copper-modified samples. Unfortunately, on the current stage it is not possible to unambiguously connect this effect to either magnetic contaminations or carbon-based magnetism, because there are arguments for both possibilities. According to compositions analysis, up to 20 ppm of Fe are possible in such material. Rough estimation will give approximately the same value of change of magnetic susceptibility due to antiferromagnetic transition, as going from 20 ppm of Fe_2O_3 [12]. However, the similar phenomenon was observed in other carbon samples by ESR [10], indicating that this effect is going from defect-induced carbon paramagnetic spins. Furthermore, observed shift of magnetic susceptibility of carbon coal to paramagnetic region at room temperatures, already needs hundreds ppm of Fe. Thus, this effect should be examined in more details in any case, either if it is going from contaminations or carbon, to avoid further missteps.

References

1. Geim, Andre K., and Konstantin S. Novoselov. "The rise of graphene." Nat. mat. **6.3** (2007): 183-191.
2. Červenka, J., M. I. Katsnelson, and C. F. J. Flipse. "Room-temperature ferromagnetism in graphite driven by two-dimensional networks of point defects." Nat. Phys. **5.11** (2009): 840-844.
3. Kiguchi, Manabu, et al. "Magnetic edge state and dangling bond state of nanographene in activated carbon fibers." Phys. Rev. B **84.4** (2011): 045421.
4. Yazyev, Oleg V. "Magnetism in disordered graphene and irradiated graphite." Phys. rev. let. **101.3** (2008): 037203.
5. Schwickardi, Manfred, et al. "Scalable synthesis of activated carbon with superparamagnetic properties." Chem. Comm. **38** (2006): 3987-3989.
6. Faulconer, Emily K., Natalia V. Hoogesteijn von Reitzenstein, and David W. Mazyck. "Optimization of magnetic powdered activated carbon for aqueous Hg (II) removal and magnetic recovery." J. of Haz. Mat. **199** (2012): 9-14.
7. Arsyad, Nor Amiera Syahida, et al. "Effect of Chemical Treatment on Production of Activated Carbon from Cocos nucifera L.(Coconut) Shell by Microwave Irradiation Method." J. Trop. Resour. Sustain. Sci **4** (2016): 112-116.
8. Babu, V. Suresh, and M. S. Seehra. "Modeling of disorder and X-ray diffraction in coal-based graphitic carbons." Carbon **34.10** (1996): 1259-1265.
9. Sepioni, M., et al. "Limits on intrinsic magnetism in graphene." Phys. Rev. let. **105.20** (2010): 207205.
10. Komlev, A. A., et al. "Magnetism of aniline modified graphene-based materials." J. of Mag. and Mag. Mat. **415** (2016): 45-50.
11. Osipov, V. Yu, et al. "Magnetic and high resolution TEM studies of nanographite derived from nanodiamond." Carbon **44.7** (2006): 1225-1234.
12. Raj, K. Govind, and P. A. Joy. "Magnetism in disordered carbon as a function of the extent of graphitization." S. S. Com. **177** (2014): 89-94.
13. Kawamura, K., Makishima. Y. and Ochiai, Y. "Magnetic susceptibility of oxygen adsorbed on the surface of spherical and fibrous activated carbon." Carbon: Science and Technology. **2(1)** (2009): 73-77.

Ab initio study of magnetic and structural properties of Fe-Ga alloys

Mariya Matyunina[1,*], *Mikhail* Zagrebin [1,2], *Vladimir* Sokolovskiy [1,3] and *Vasiliy* Buchelnikov[1]
[1] Chelyabinsk State University, 454 001 Chelyabinsk, Russia
[2] National Research South Ural State University, 454001 Chelyabinsk, Russia
[3] National University of Science and Technology 'MIS&S', 119991 Moscow, Russia

Abstract. The structural and magnetic properties for a series of $Fe_{100-x}Ga_x$ alloys (x = 18 – 30 at.%) are studied in the framework of first-principles calculations and Monte Carlo simulations. The both, general gradient approximation and local density approximation are considered for the exchange-correlation functional. The ground state *ab initio* calculations are performed for both $D0_3$ and $L1_2$ crystal structures. It is shown that for general gradient approximation, the optimized lattice parameters and total magnetic moments are found in the better agreement with experimental ones. Using the calculated exchange coupling constants for studied compositions, Curie temperatures are estimated by means of Monte Carlo simulations of Heisenberg Hamiltonian.

1 Introduction

The Fe-Ga alloys are successful magnetostrictive materials which are very attractive to use in magnetostrictive actuators, sensor and spintronic devices [1-8]. Doping Ga up to 30 at% into the bcc Fe enhances greatly both the value of tetragonal magnetostriction λ_{001} in the [001] direction and ductility [1-3]. Experimental investigations of $Fe_{100-x}Ga_x$ alloys have shown that the value of tetragonal magnetostriction λ_{001} has two peaks (395×10^{-6} and 350×10^{-6}) for composition with 19 and 27 at% Ga concentration at room temperature, respectively [2].

In the past decade, Fe-Ga alloys became the objects of intensive experimental and theoretical studies. For instance, the experimental investigations of magnetic and structural properties by means of X-ray diffraction, mechanical spectroscopy techniques, dynamical mechanical analyzer, neutron diffraction, Mössbauer spectroscopy can be found in Refs. [2-8]. The theoretical studies of Fe-Ga and related alloys using first principles approach are presented in Refs. [9-16]. Wang et al. [12-13] and Zhang et al. [14-16] calculated the magnetostrictive coefficient λ_{001}, strain-dependent total energies and magnetocrystalline anisotropy (MAE) of the $Fe_{100-x}Ga_x$ ($5 < x \leq 25$). The results were found in a good agreement with experimental data. Wu [9] considered an enhancement in magnetostriction for ordered $D0_3$, $L1_2$, and B2-like structures in $Fe_{75}Ga_{25}$ due to a splitting of Fe 3*d*-bands in minority spin channel around Fermi level. Khmelevska et al. [10] proposed the more complex origin of giant magnetostriction in $Fe_{100-x}Ga_x$ ($0 < x < 25$) due to disorder effects in the framework of coherent potential approximation (CPA).

The aim of this paper is a complex study of magnetic and structural properties of $Fe_{100-x}Ga_x$ (x = 18 – 30 at%) alloys by the density of functional theory with the different approximations for the exchange-correlation energy and finite-temperature Monte Carlo (MC) simulations.

2 Computation details

In the first step of our calculations, we have done the geometric optimization of crystal structures by using the SPR-KKR (spin polarized relativistic Korringa-Kohn-Rostoker) software package based on the Korringa-Kohn-Rostoker Green's function [17]. For all *ab initio* calculations, the exchange-correlation energy was treated by the generalized gradient approximation in the Perdew-Burke-Ernzerhof (GGA-PBE) formulation [18] and the local density approximation in the form of Vosko-Wilk-Nusair (LDA-VWN) [19]. For self-consistent cycles (SCF) calculations, 6348 $\boldsymbol{k}$ points were generated by a $\boldsymbol{k}$-mesh grid of 45^3. The total energy for all calculations converged to 0.01 mRy. To construct off-stoichiometric compositions, the CPA was considered.

To perform the crystal structure optimization, we used two face-centered cubic $L1_2$ ($Pm\bar{3}m$, #221) and body-centered cubic $D0_3$ ($Fm\bar{3}m$, #225) structures with their associated unit cells, which contain 4 atoms. Note that the stoichiometric $Fe_{75}Ga_{25}$ (Fe_3Ga) can crystallize in the $D0_3$ structure [6]. There are three Fe atoms per unit cell belonging to two different sublattices. For the formation of non-stoichiometric compositions of $Fe_{100-x}Ga_x$, Ga atoms taken in the required percentage were fixed at next sites as listed in Table 1.

In the second step, using the optimized lattice parameters obtained within GGA-PBE and LDA-WVN

* Corresponding author: matunins.fam@mail.ru (M. Matyunina)

approximations, the calculations of exchange interaction parameters J_{ij} for both structures were performed. The Heisenberg exchange coupling constants were calculated using the expression proposed by Liechtenstein et al. [20]. All calculations converged to 0.01 mRy of total energy.

Table 1. The atomic distribution for off-stoichiometric $Fe_{100-x}Ga_x$ alloys with the $L1_2$ and $D0_3$ structures

Space group	Wyckoff positions	x<25%	x=25%	x>25%
		Atoms	Atoms	Atoms
$Pm\bar{3}m$	1*a*: 0, 0, 0	Ga, Fe	Ga	Ga
(#221)	3*c*: 0, 0.5, 0.5	Fe	Fe	Ga, Fe
$Fm\bar{3}m$	4*a*: 0, 0, 0	Ga, Fe	Ga	Ga
	4*b*: 0.5, 0.5, 0.5	Fe	Fe	Ga, Fe
(#225)	8*c*: 0.25, 0.25, 0.25	Fe	Fe	Fe

The final step is the Curie temperature estimation with the help of the MC simulations of the three-dimensional Heisenberg model ($H_{mag} = -\sum_{\langle ij \rangle} J_{ij}\mathbf{S}_i\mathbf{S}_j$). Here $\mathbf{S}_i = (S_i^x, S_i^y, S_i^z)$ is a classical magnetic degree of freedom $|\mathbf{S}_i| = 1$. The exchange parameters J_{ij} and partial magnetic moments μ were taken as input parameters from *ab initio* calculations. The model lattice with periodic boundary conditions for $Fe_{75}Ga_{25}$ alloy contains 2826 (3300) Fe and 1099 (1331) Ga atoms for $D0_3$ ($L1_2$) phase, respectively. The MC simulations were carried out using the Metropolis algorithm [21]. As time unit, we used on MC step consisting of N attempts to change the spin variables. The total number of MC steps per the temperature step was 10^6. The magnetic order parameter m and total magnetization M are defined by the following way

$$m^{Fe} = \frac{1}{N^{Fe}}\sum_i \sqrt{(S_i^{Fe,x})^2 + (S_i^{Fe,y})^2 + (S_i^{Fe,z})^2}, \quad M = 3\mu_{Fe}m^{Fe},$$

here μ_{Fe} is the magnetic moment of Fe taken from first principals calculations.

To estimate the Curie temperature from $M(T)$ curves, we plotted the $M^{1/\beta}(T)$ function that decreases almost linearly with increasing temperature. The Curie temperature can be evaluated at the intersection of a $M^{1/\beta}$ curve with the T axis. Here, β is the critical index and it is equal to 0.3646 for the three-dimensional Heisenberg model.

3 Results and discussions

3.1. Lattice parameters and magnetic moments

In this subsection, we present the results of calculations of the optimized lattice parameters and total magnetic moments for studied compositions with $D0_3$ and $L1_2$ structures.

Fig. 1 shows the variation of equilibrium lattice parameter a_0 (which corresponds to a minimum value of energy E_0) as a function of Ga concentration for both $D0_3$ and $L1_2$ structures of $Fe_{100-x}Ga_x$ alloys in comparison with available experimental data taken from Ref. [22]. Here we present results obtained by using GGA-PBE and LDA-WVN approximations for the exchange-correlation functional.

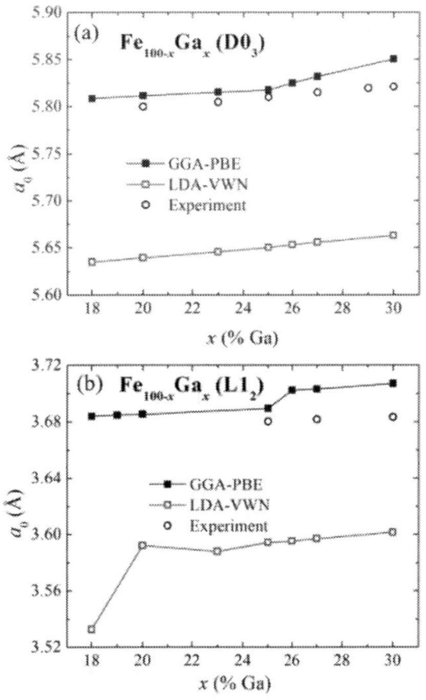

Fig. 1. The calculated equilibrium lattice parameter (a_0) as a function of Ga concentration (x) for the (a) $D0_3$ and (b) $L1_2$ structures of $Fe_{100-x}Ga_x$ alloys in comparison with experimental data. The experimental data were taken from [22].

Our calculation have shown that for both $D0_3$ and $L1_2$ structures, lattice constant increases with increasing Ga concentration. For both bcc and fcc structures, the results obtained within the GGA-PBE approximation are in the better agreement with the experimental data than those obtained for the LDA-VWN approximation.

In Fig. 2, we present the calculated total magnetic moments for the studied compositions $Fe_{100-x}Ga_x$ (x = 18 – 30 at%) with $D0_3$ and $L1_2$ structures. It is clearly seen that for both structures, the total magnetic moment decreases with increasing Ga content. Concerning the effect of choice for the exchange-correlation potential, we can see the similar trend as in Fig. 1. Namely, for both $D0_3$ and $L1_2$ structures, the values of magnetization obtained within the GGA-PBE approximation are closer to the experimental data taken from [8].

Generally, according to Ref. [23], exchange-correlation functional GGA reproduces the equilibrium volume of 3d-metals better than the LDA. Believe that the LDA fails to correctly reproduce the ground state bcc structure of Fe because the magnetic contribution to the stabilization of the bcc structure is weakened by this exchange-correlation functional while at the same time the GGA corrects the volume and thereby the magnetic contribution. We could equally well say that for the ground state calculations of Fe-Ga alloys with different phases the method of Green's function within GGA-PBE gives better results than LDA-VWN.

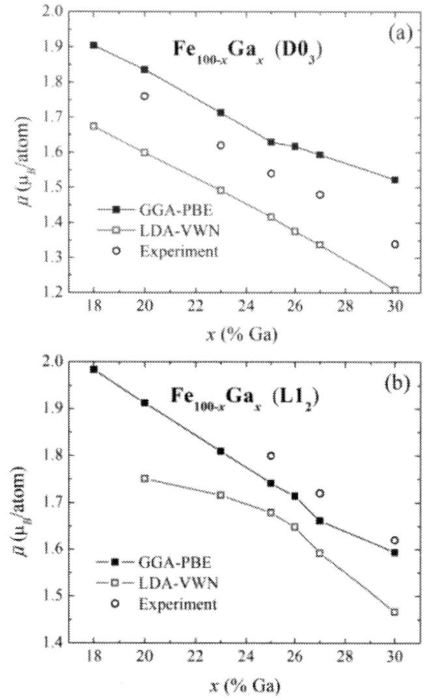

Fig. 2. The calculated total magnetic moment $\bar{\mu}$ per atom (in μ_B/ atom) as a function of Ga concentration (x) for the (a) (a) $D0_3$ and (b) $L1_2$ crystal structures of $Fe_{100-x}Ga_x$ alloys in comparison with experimental data. The experimental data were taken from [8].

3.2. Magnetic exchange interaction and Curie temperature

The knowledge of optimized lattice parameters allows us to calculate further the Heisenberg exchange interaction parameters J_{ij} for both bcc and fcc structures of $Fe_{100-x}Ga_x$ alloys. Since the results of lattice parameter calculations obtained within the LDA-VWN differ significantly from the experimental ones, we further used only the set of a_0 calculated within the GGA-PBE. Nevertheless, the calculations of J_{ij} parameters were calculated using both approaches: GGA-PBE and LDA-WVN. As a result, we will use two new denotations: GGA-GGA and GGA-LDA. The first (second) one denotes the J_{ij} calculations within the GGA-PBE (LDA-WVN) using the set of a_0 calculated within the GGA-PBE, respectively.

Fig. 3 displays the magnetic exchange parameters for $D0_3$ and $L1_2$ phases of the $Fe_{73}Ga_{27}$ alloy as a function of the distance between atoms. The $Fe_{73}Ga_{27}$ composition is of interest due to the presence of structural phase transition between $D0_3$ and $L1_2$ phases [4]. For both phases, the oscillating damped behavior of J_{ij} can be observed. We can see that for $D0_3$ structure, the strongest ferromagnetic interaction ($J_{ij} > 0$) is found between nearest-neighbors Fe_1-Fe_2 (which are located at 4*b* and 8*c* Wyckoff positions, respectively). In a case of $L1_2$ structure, the ferromagnetic contribution to the total exchange energy between nearest neighbors is found to be slightly smaller. Moreover, it is seen that in the second coordination shell ($d/a = 1$) for $L1_2$ structure, the Fe-Fe exchange parameters split into two FM contributions. In this case, each Fe at the distance of a_0 interacts ferromagnetically with six nearest neighbors: two Fe atoms provide $J_{ij} = 7.03$ (7.14) meV (GGA (LDA)) and four Fe atoms locating in (x, y) plane provide J_{ij} =2.59 (2.65) meV (GGA(LDA)), respectively. As can be seen, the difference between values J_{ij} obtained with two approximations is rather small, especially for $L1_2$ phase (see the inset in Fig. 3b).

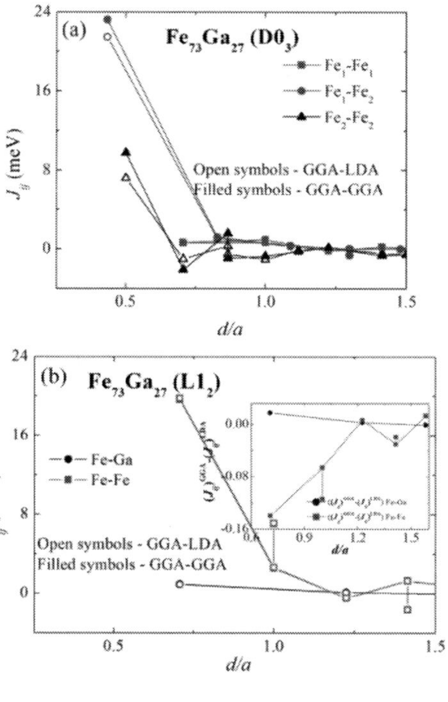

Fig. 3. The calculated exchange interaction parameters J_{ij} as a function of distance (d/a) between atoms i and j for the crystal structures (a) $D0_3$ and (b) $L1_2$ of $Fe_{73}Ga_{27}$ alloy. The inset shows the difference $\Delta J_{ij} = (J_{ij})^{GGA}$-$(J_{ij})^{LDA}$ between values of J_{ij} obtained with GGA and LDA approximations, respectively.

The knowledge of the constants of magnetic exchange interactions and magnetic moments allows us to simulate the temperature dependences of magnetization and to estimate the Curie temperatures for both investigation structures by means of Monte Carlo simulations of the classical Heisenberg Hamiltonian without a magnetic anisotropy term.

Fig. 4 shows the results of evaluations of the Curie temperature for the crystal structures of $D0_3$ and $L1_2$ of

$Fe_{100-x}Ga_x$ alloys. In general, theoretical results of Curie temperature estimations are in a good agreement with experimental data [4, 22]. It should be noted that the using of the magnetic exchange interactions obtained for GGA-LDA approach gives the results, which are the closer to the experimental data.

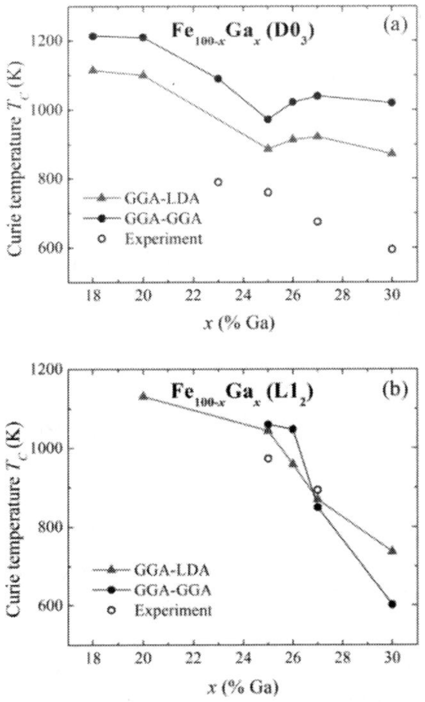

Fig. 4. The Curie temperature for the crystal structures (a) $D0_3$ and (b) $L1_2$ of $Fe_{100-x}Ga_x$ alloys in comparison with experimental data, which were taken from [4, 22].

4 Conclusion

In this work, we have introduced ab initio calculations and Monte Carlo simulations to study the structural and magnetic properties of $D0_3$ and $L1_2$ phases of $Fe_{100-x}Ga_x$ ($x = 18 - 30\%$). The geometric optimization of $D0_3$ and $L1_2$ structures and calculations of exchange magnetic interactions have been performed by using the SPR-KKR package with treated of exchange-correlation energy in different approximations. It is shown that for the ground state calculations of both $D0_3$ and $L1_2$ structures of Fe-Ga alloys, the using of the GGA-PBE functional gives better results than LDA-VWN. However, The LDA-VWN approximation is favored for calculations of the exchange parameters J_{ij} and estimation of Curie temperatures by using the MC simulations of Heisenberg model. For the latter case, the theoretical values of Curie temperature of studied compositions for both phases are found in the better agreement with experimental ones.

Acknowledgments

This work is supported by Russian Science Foundation No. 17-72-20022 (LDA calculations) and No. 18-12-00283 (GGA calculations).

References

1. Q. Xing, Y. Du, R.J. McQueeney, T.A. Lograsso, Acta Mater, **56** 4536 (2008)
2. A.E. Clark, K.B. Hathaway, M. Wun-Fogle, J.B. Restorff, T.A. Lograsso et al., Jpn J Appl Phys, **93** 8621 (2003)
3. A.E Clark., M. Wun-Fogle, J.B. Restorff, T.A Lograsso., J.R. Cullen, IEEE Trans Magn, **37** 2678 (2001)
4. I.S. Golovin, A.M. Balagurov, I.A. Bobrikov, V.V. Palacheva, J. Cifre, J Alloy Compd, **675** 353 (2016)
5. T.A. Lograsso, A.R. Ross, D.L. Schlagel, A.E. Clark, J Alloys Compd, **350** 95 (2003)
6. I.S. Golovin, V.V. Palacheva, A.I. Bazlov, J. Cifre, J. Pons, J Alloy Compd, **644** 959 (2015)
7. I.S. Golovin, A.M. Balagurov, V.V. Palacheva, I.A. Bobrikov, V.B. Zlokazov, Mater. Des. **98** 113 (2016)
8. N. Kawamiya, K. Adashi, Y. Nakamura, J. Phys. Soc. Japan, **33** 1318 (1972)
9. R. Wu, Jpn J Appl Phys, **91** 7358 (2002)
10. T. Khmelevska, S. Khmelevskyi, P. Mohn, J. Appl. Phys. **103** 073911 (2008).
11. C. Paduani, C. Bormio-Nunes, J. Appl. Phys. **109** 033705 (2011)
12. H. Wang, Y.N. Zhang, T. Yang, Z.D. Zhang, L.Z. Sun, R.Q. Wu, Appl. Phys. Lett. **97** 262505 (2010).
13. H. Wang, Y.N. Zhang, R.Q. Wu, L.Z. Sun, D.S. Xu, Z.D. Zhang, Sci. Rep. **3** 3521 (2013)
14. Y. N. Zhang, R. Q. Wu, Phys. Rev. B **82** 224415 (2010).
15. Y. Zhang, R. Wu, IEEE Trans. Magn. **47** 4044 (2011)
16. Y. Zhang, H. Wang, R. Wu, Phys. Rev. B **86** 224410 (2012)
17. H. Ebert, D. Koedderitzsch, J. Minar, Rep. Prog. Phys., **74** 96501 (2011)
18. J. P. Perdew, K. Burke, M. Ernzerhof, Phys. Rev. Let., **77**, 18 3865 (1996)
19. S.H. Vosko, L.Wilk, M. Nusair, Canadian J. of Phys. **58** 1200 (1980)
20. A.I. Liechtenstein et al., J. Magn. Magn. Mater., **67** 65 (1987)
21. D.P. Landau, K. Binder, A. Guide to Monte Carlo Simulations in Statistical Physics, Cambridge, 2009
22. H. Okamoto, Bulletin of Alloy Phase Diagrams, **11** 576 (1990)
23. A.V. Ruban, I.A. Abrikosov, Rep. Prog. Phys. **71** 046501 (2008)

Optical and magnetic properties of Al/NiFe and Al/Ge/NiFe nanosized films

H. Hashim[1,*], *S.P. Singh*[1], *L.V. Panina*[1], *F.A. Pudonin*[2], *I.A. Sherstnev*[2], *S.V. Podgornaya*[1], *and I. Shpetnyy*[3]

[1]National University of Science and Technology (MISIS) Leninsky pr., 4, 119049, Moscow, Russia.
[2]Lebedev Physical Institute (RAS), Leninsky Prospect, 53, 119991, Moscow, Russia.
[3]Sumy State University, 2, Rimsky Korsakov St. 40007 Sumy, Ukraine.

Abstract. Nanosized films with ferromagnetic layers are widely used in nanoelectronics, sensor systems and telecommunication. The physical and magnetic properties of nanolayers may significantly differ from those known for bulk materials due to fine crystalline structure, influence of interfaces, roughness, and diffusion. In this work, we are employing a spectral ellipsometry method, magneto-optical Kerr magnetometry and VSM to investigate the impact of layer thickness on the optical constants and magnetization processes for two and three layer films of the type Al/NiFe/sitall, Al/Ge/NiFe/sitall on sitall substrate for different thickness of the upper Al layers. The refractive indexes of two layer films are well resolved by spectral ellipsometry demonstrating their good quality. Modelling data for three-layer films show considerable discrepancy with the experiment, which can be related to a stronger influence of interfaces. The magnetization processes of two-layer films weakly depend on the type and thickness of the upper non-ferromagnetic layers. However, the coercivity of three layer films may significantly change with the thickness of the upper layer: more than twice when the thickness of Al layer increases from 4 to 20 nm.

1 Introduction

Thin films with magnetic layers are an important class of nanostructured materials for applications in nanoelectronics and spintronics [1-2]. Considerable interest in such systems is also related to a possibility of strengthening and modifying the magneto-optical effects, for example, by excitation of long-living modes of surface plasmon polaritons [3], or by spatial diffusion of spins in the ferromagnetic layers [4-6].

The technological production methods of thin films allow controlling the layer thickness with a precision of a nanometer. However, the properties of thin films may differ significantly from those of bulk materials [7-9]. This is due to the structural parameters such as the size of crystallites, the quality of the interfaces of intermediate layers, and diffusion. The structure of multilayer films can be complicated by the need to protect the functional layers, with the physical properties of interest. Therefore, the task is to control the physical properties of the individual layers along with geometry and overall magnetic response.

Ellipsometric methods are promising for such monitoring, since they permit the measurement of optical parameters of individual nanolayers (and growing layers) with high accuracy and selectivity in the process of production [10-15]. The determination of optical and geometrical parameters is based on the minimization of a functional involving a quadratic form of the difference between the experimental and modelling data. In the case of magnetic films it is of interest to combine the two approaches: traditional ellipsometry and magneto-optical (MO) magnetometry [16]. In particular case of the equatorial MO Kerr effect, a general approach within the concept of matrix optics can be used for the model calculations.

In this work, the methods of magneto-ellipsometry are used to study the optical parameters and magnetization processes of two-layer and three-layer films of Al/NiFe, Al/Ge/NiFe on sitall substrate for varying thicknesses of upper Al layer. The behaviour of the refractive index of individual layers depending on the film structure may disclose the role of interfaces on magnetic properties. We have found that the hysteresis curves of three-layer films undergo significant changes depending on the thickness of the upper nonmagnetic layer.

1.1 Magneto-ellipsometry analysis of multilayered films

The ellipsometric parameters are determined from the measurement of the ratio of complex reflection coefficients for the two polarizations of the light wave: (p) and (s), with the electric field parallel and perpendicular to the plane of incidence, respectively. This complex ratio is usually expressed in the polar form introducing the ellipsometric parameters ψ and Δ, which characterize the relative change in amplitudes and phase shift for p- and s-polarizations, respectively:

$$\frac{r_p}{r_s} = \tan\psi \exp i\Delta \qquad (1)$$

* Corresponding author: *hh@science.tanta.edu.eg , drlpanina@gmail.com

In (1), r_p, r_s are the complex reflection coefficients of p- and s-polarized waves, respectively. Ellipsometric measurements are not direct. In general it is necessary to solve the inverse problem, which may pose considerable difficulties, then, it is very important to build initially an adequate model, which is already using the regression techniques to fit the experimental results.

The magneto-ellipsometry analysis of multilayered films is based on the determination of the characteristic matrices of individual layers, which can have a magnetization in the film plane and perpendicular to the plane of incidence (along the x-axis). This corresponds to the geometry of the equatorial Kerr effect, and allows simultaneous determination of magneto-optical and ellipsometric parameters, since the electromagnetic field components with the polarization s and p remain the eigenfunctions of the system. The influence of the change in magnetization of a layer within multilayered system with a given geometry is the change in the intensity and phase shift of the p-wave, whereas the scattering parameters of s –waves do not depend on the magnetization. This approach is more convenient for numerical calculations than using the Mueller matrix or Jones matrix.

In the model, the equatorial MO effect can be taken into account by the introduction of the antisymmetric tensor of the second rank for the matrix of the dielectric constant [17,18]:

$$\hat{\varepsilon}_j = \varepsilon_j \begin{pmatrix} 1 & 0 & 0 \\ 0 & 1 & iQ_j m_{xj} \\ 0 & -iQ_j m_{xj} & 1 \end{pmatrix} \quad (2)$$

In equation (2) the dielectric constant is written for a ferromagnetic layer (distinguished by index j in the multilayer system). It is assumed that the plane of incidence is the y-z plane, and the surface of the sample is y-x plane. $\varepsilon_j = (\tilde{n}_j + i\kappa_j)^2$, m_{xj} and Q_j are the complex permittivity, the projection of the normalized magnetization on the x-axis (perpendicular to the plane of incidence), and the magneto-optical constant of the j-th layer, respectively. For waves with p-polarization the reflection r_p and transmission t_p coefficients for a multilayer system are found from the matrix equation

$$\begin{pmatrix} 1 + r_p \\ (-1 + r_p)p_{p1} \end{pmatrix} = \hat{M}_p t_p \begin{pmatrix} 1 \\ -p_{pN} \end{pmatrix} \quad (3)$$

$$p_{p1} = \frac{\cos\theta}{n_1}, \qquad p_{pN} = \frac{\cos\theta_N}{n_N} \quad (4)$$

In (4), $\hat{M}_p$ is the characteristic matrix of the system, θ is the angle of incidence from the outside medium with the refractive index n_1, θ_N is the angle of refraction in the last medium (substrate) with a refractive index n_N. The matrix $\hat{M}_p$ is determined by the product of the corresponding matrices of the individual layers with the layer parameters: $n_j = \sqrt{\varepsilon_j}$ is the complex refractive index and h_j is the layer thickness:

$$\hat{M}_p = \prod_{j=2}^{N-1} \hat{M}_{pj} \quad (5)$$

$$\hat{M}_{pj} = \begin{pmatrix} \cos(\beta_j h_j) - i\,\delta_j \sin(\beta_j z) & \frac{i}{p_{pj}} \sin(\beta_j h_j) \\ i p_{pj} \sin(\beta_j h_j) & \cos(\beta_j h_j) + i\,\delta_j \sin(\beta_j z) \end{pmatrix} \quad (6)$$

$$\beta_j = k_0 n_j \cos\theta_j\,, p_{jp} = \frac{\cos\theta_j}{n_j}, \delta_j = iQ_j m_{xj} \tan\theta_j \quad (7)$$

Here λ is the wavelength, $k_0 = 2\pi/\lambda$. The angle θ_j is defined as follows:

$$n_j \cos\theta_j = \sqrt{n_j^2 - n_1^2 \sin^2\theta}$$

Thus, in the developing approach the magnetic layers are treated within the same characteristic matrix formalism using the off-diagonal magneto-optical parameter Q_j.

For s-polarization the intensity of the reflected light does not depend on the magnetization. The basic equations remain the same with the replacement of p_{pj} to $p_{sj} = n_j \cos\theta_j$ and putting $\delta_j = 0$.

The characteristic matrices depend on the refractive indexes which have to be determined by comparing the experimental and modelling results. Since the ellipsometric method is not direct, certain assumptions on the permittivity spectra should be made. For example, in a certain range of wavelengths for layers with attenuation simple approximations (Cauchy-Urbach) are possible for complex refractive index [10]:

$$\tilde{n}_j(\lambda) = A_{nj} + \frac{B_{nj}}{\lambda^2} + \frac{C_{nj}}{\lambda^4} \quad (8)$$

$$k_j(\lambda) = A_{kj} \exp\left(\frac{B_{kj}}{\lambda} - C_{kj}\right) \quad (9)$$

With the help of this model it is possible to achieve a good match with the experimental data on the spectral characteristics of ψ and Δ to determine the optical parameters of the layers.

2 Materials and methods measurements

The investigated multilayered systems of Al/NiFe/ and Al/Ge/NiFe/ on a sitall substrate were fabricated by vacuum thermal evaporation method utilizing VUP-5M (vacuum universal post) using layer-by-layer condensation. The sample holder was rotating to insure good uniformity. In NiFe permalloy alloy the concentration of Ni was approximately 50%. After deposition of the bottom layer of permalloy with constant thickness of 20 nm on the substrate the samples were annealed in vacuum at T=550 K to refine the structural properties. The film was polycrystalline and isotropic in the plane. The thickness of the upper nonmagnetic layer of Al was varied whilst the thickness of the Ge layer kept fix at 2 nm. The quartz resonator method was used to measure the film thickness.

The ellipsometric parameters were measured using spectral ellipsometry (Ellipse 1891) based on the static measuring system with fixed incidence angle of 70° in the wavelength range of 350-1050 nm with a step of 2 nm. The fitting model with available literature data was applied in the range of 400-800 nm. At this stage, the magneto-optical (MO) measurements in the equatorial configuration were carried out separately by measuring

https://doi.org/10.1051/epjconf/201818504014

the change in the intensity of p-polarized light at the wavelength of 632 nm. The magnetic hysteresis curves were also measured with a vibrating magnetometer BM-07 (VSM) at the magnetic field strength up to 200 mT. The sample position during the magnetic measurements varied to investigate the existence of in-plane magnetic anisotropy.

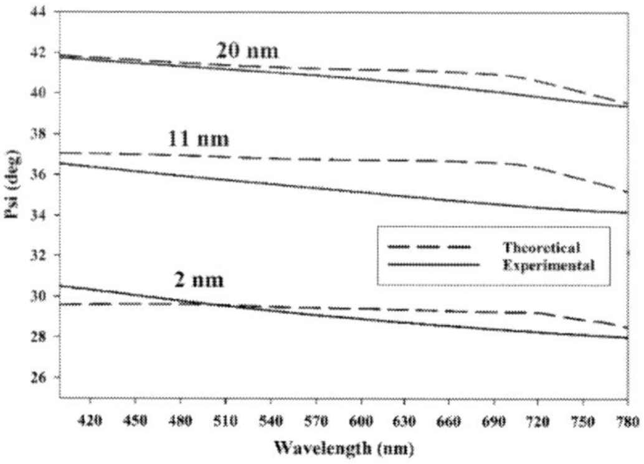

Fig. 1. Spectra of ellipsometric parameter ψ for Al/NiFe/sitall film system for different values of the thickness of Al layer. Solid lines – experiment, dotted lines- two-layer optical model.

3 Results and discussion

Fig. 1 shows the experimental and model spectra of the ellipsometric parameter ψ for the films Al/NiFe/sitall for different values of Al-layer thickness. The refractive indices n_j of the layers were approximated by formulas (7)-(8). The value of n (and thickness h) of individual layers were determined by achieving a good matching between the experimental and modelling data by minimizing a quadratic form which defines the difference between them. Thus, the coefficients in expansion (7)-(8) were determined. The process was initially completed for the Al layer of 20 nm thick. If the same optical parameters (n_j) are used to describe the ellipsometric data for thinner Al- layers, some discrepancies with the experiment could be seen, which indicates the dependence of n on the layer thickness (in the range of 10-15%) for $h < 20$ nm. The values of the refractive indexes of Al and NiFe layers determined from the ellipsometric spectra of Fig. 1 for Al-layer thickness h =20 nm are shown in Fig. 2. The obtained results quite well (less than 5%) coincide with the published data for bulk materials [19].

The magnetic hysteresis loops obtained by MO and VSM methods for the bi-layer system are shown in Fig. 3 for different values of h_{Al}. The hysteresis loops did not depend on the sample alignment proving the absence of in-plane anisotropy. The coersivity and the hysteresis shape were slightly dependent on the Al layer thickness. VSM measurements indicate the existence of rather large anisotropy since the loops does not go to saturation even in fields of 2 kOe.

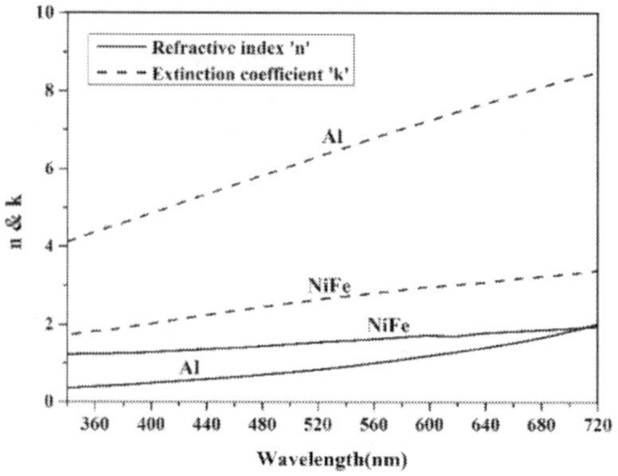

Fig. 2 .Complex refractive index $n = \tilde{n}_j + i\kappa_j$ of Al and permalloy layers determined from the experimental ellipsometric spectra of Fig. 1 for $h_{Al} = h_{NiFe} = 20$ nm. Solid lines represent the real part $\tilde{n}$, the dotted lines show the imaginary part k.

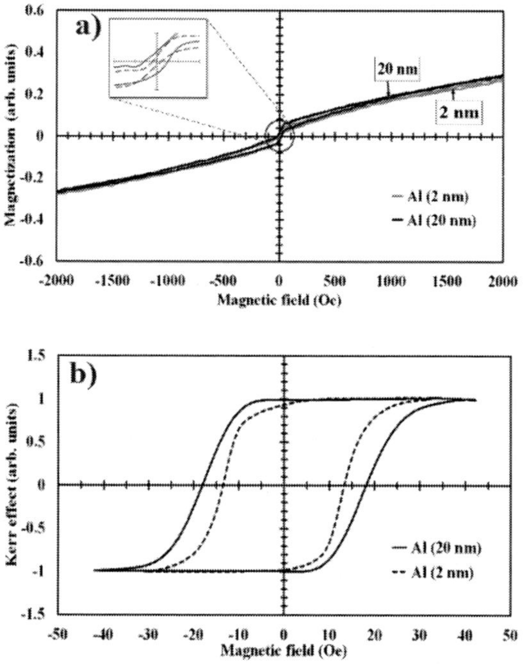

Fig. 3. Hysteresis curves for the films Al/NiFe/sitall for different thickness of Al layer. a) VSM method. b) MO method.

The greatest differences between the experimental and model data are obtained for three-layer films with an intermediate Ge layer when using the optical parameters determined for bilayer films, as can be seen from Fig. 4. The variation in n_j within a reasonable range did not produce matching either. This means that a simple three-layer model does not correspond to the real samples. Very probably, there is a mutual diffusion or roughness, and it is necessary to introduce additional intermediate layers or optical parameter gradients.

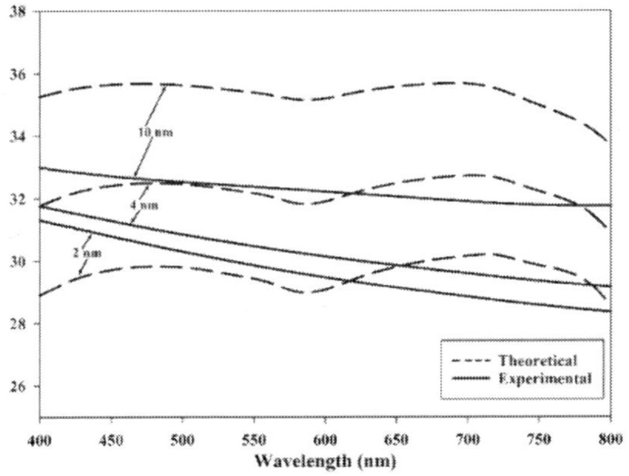

Fig. 4. Spectra of ellipsometric parameter ψ for the film Al/Ge/NiFe/sitall for different values of the thickness of the layer Al. Solid line – experiment, dashed lines– three-layer optical model.

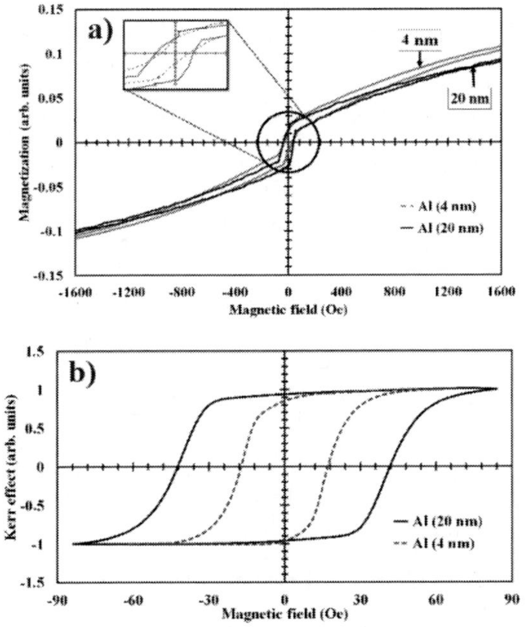

Fig. 5. Hysteresis curves for the films Al/Ge/NiFe/Sitall for different thickness of Al layer. a) VSM method. b) MO method.

In addition, the three-layer films demonstrated unusual strong dependence of coercivity on the thickness of the upper Al layer as shown in Fig.5. MO loops show greater coercivity dependence on h_{Al}, indicating a larger contribution of NiFe surface region in the coercivity increase. It is known that the coercivity of permalloy films is very sensitive to internal stress/strains. The deposition of Al layers did not result in noticeable changes in hysteresis loops (compare with Fig. 3). Therefore we attribute the coercivity increase with increasing h_{Al} to the structural changes produced by Ge layer at NiFe and Al interfaces.

4 Conclusion

Optical and structural parameters of the multilayer films of the type Al/NiFe/, Al/Ge/NiFe/sitall were investigated by ellipsometric and magneto-optical methods. For double-layer systems without the inner Ge layer there is a good agreement between the experimental and two-layer optical model data. Three-layer optical model for Al/Ge/NiFe/sitall does not fit the ellipsometric spectra, which indicates a more complex structure with possible intermediate layers. The measurement of the magnetic hysteresis loops for the three-layer films by MO and VSM techniques shows a strong dependence of coercivity on the thickness of the top layer, which also indicates the structural modification.

Acknologement. L. V. Panina acknowledges the support for this work under the Russian Federation State Contract for organising a scientific work (grant No 3.8022.2017).

References

[1] S.A. Wolf. Science **294**, 1488 (2001).
[2] P.P. Freitas, R. Ferreira, S. Cardoso, F. Cardoso. J. Phys.: Condens. Matter **19**, 165221 (2007).
[3] G. Armelles, A. Cebollada , A. García-Martín, M. U. González, Advanced Optical Materials **1**, 10 (2013).
[4] S. Demokritov, J.A. Wolf, P. Grünberg, Europhys. Lett. **15,** 881 (1991).
[5] V.A. Skidanov. Book of Abstracts, Mo.H-P45. 20th ICM. Barcelona, (5–10 Jul 2015).
[6] V.A. Skidanov. International Symposium "Spin Waves". Saint-Petersburg, 158 (7–13 June 2015).
[7] G. Neuber, R. Rauer, J. Kunze, J. Backstrom, M. Rubhausen. Thin Solid Films **455**, 39 (2004).
[8] S. Visnovsky, R. Lopusnik, M. Bauer, J. Bok, J. Fassbender, B. Hillebrands. Optics express **9**, 121 (2001).
[9] E. A. Ganshina, V.S. Gyschin, C.I. Kasatkin, M.A. Murav'ev, Physics of the Solid State **46**, 864 (2004).
[10] H. Fujiwara. Spectroscopic Ellipsometry: Principles and Applications. John Wiley & Sons Ltd, 170 (2007).
[11] L. Pereira, H. Águas, E. Fortunato, R. Martins. App. Surf. Sci. **253**, 339 (2006).
[12] E. Langereis, S. B. S. Heil, H. C. M. Knoops, W. Keuning, M. C. M. van de Sanden, and W. M. M. Kessels. J. Phys. D: Appl. Phys. **42**, 073001 (2009).
[13] G. E. Jellison, Generalized ellipsometry for materials characterization, Thin Solid Films 450 (2004) 42–50.
[14] Hilfiker J.N. *In situ* spectroscopic ellipsometry (SE) for characterization of thin film growth. – Woodhead Publishing Limited (2011).
[15] K. Postava, D. Hrabovsky, J. Hamrlov, J. Pistora, A. Wawro, L.T. Baczewski, Sveklo I., A. Maziewski. Thin Solid Films **519**, 2627 (2011).
[16] G. Neuber, R. Rauer, J. Kunze, J. Backstrom, M. Rubhausen. Thin Solid Films, **455 – 456**, 39 (2004).
[17] R. P. Hunt, J. Appl. Phys. **38**, 1652 (1967).
[18] Y. J. Yang and M. R. Scheinfein, J. Appl. Phys. **74**, 6810 (1993).
[19] Palik E.D. Handbook of Optical Constants of Solids III. – San Diego, CA: Academic Press, 68 (1998).

Giant forced magnetostriction in $Tb_{0.2}Gd_{0.8}$ single crystal

Sergey Nikitin[1], *Tatiana* Ivanova[1], *Alexey* Zvonov[1,*], *Krzysztof* Rogacki[2,3], *Yurii* Koshkid'ko[2,3] and *Jacek* Ćwik[2,3]

[1]M.V. Lomonosov Moscow State University, Moscow, Russia
[2]W. Trzebiatowski Institute of Low Temperature and Structural Research, Wroclaw, Poland
[3]International Laboratory of High Magnetic Fields and Low Temperatures, Wroclaw, Poland

Abstract. The results of investigation of the $Tb_{0.2}Gd_{0.8}$ single crystal forced magnetostriction are presented. Temperature dependences of magnetostriction have been measured from 4 to 300°K in applied magnetic fields up to 14 T. The giant field induced magnetostriction ~ $1.5 \cdot 10^{-3}$ was discovered in the room temperature region in magnetic fields up to 14 T.

1 Introduction

The magnetostriction is the mechanical deformation of magnetic samples due to the changes in the magnetization degree or in the magnetization direction [1]. The rare earth metals (REM) and their alloys form an important class of materials with giant magnetostriction [1-3]. The theory of magnetostriction of REM was developed and formulated by Callen [4] and Clark [5]. Previously, the study of forced magnetostriction was carried out on the polycrystalline and single-crystal samples of REM [6-9] and their alloys [1,2,10-12]. Tb-Gd single crystal alloys are widely used in prototypes of magnetic refrigerators due to giant reversible magnetocaloric effect observed in them near room temperature [13-16]. The volume magnetostriction values are crucial for determination of exchange interactions dependencies on the unit cell volume and on the interatomic distances. Moreower in the modern world, magnetostrictive materials are widely used in wide range of technical devices such as, for example, high power transducers and actuators [1,3,17].

In this work the magnetostriction of $Tb_{0.2}Gd_{0.8}$ single crystal in strong magnetic fields up to 14 T was investigated. This alloy has a relatively high Curie temperature due to high concentration of Gd and giant values of magnetostriction caused by Tb content, which has high values of single-ion magnetostriction. The combination of strong exchange interactions and high magnetostriction in Tb_xGd_{1-x} alloys makes them promising objects in terms of theoretical research and practical applications. For these applications it is essential to know about field and temperature dependences of magnetostriction.

2 Experimental details

We have performed an experimental study of magnetostriction of $Tb_{0.2}Gd_{0.8}$ single crystal in magnetic fields up to 14 T and in temperature range 4.2 K - 300 K.

For a comprehensive study of magnetic and magnetostrictive properties of rare-earth metals and their alloys large single crystals with a perfect crystal structure and a uniform distribution of alloy components on the length and section of the crystal are required. The most suitable for this purpose is the method of pulling crystals from the melt (Czochralski method). For the growth of the single crystal terbium and gadolinium of high purity (99.99 at.%) were used. Terbium and gadolinium form a continuous series of substitutional solid solutions when alloying, which makes distribution of alloy components uniform all over the sample.

Single crystal pulling was carried out at a speed of 0.3 mm per minute and the rotation rate of 20 rpm in a vacuum of 10^{-6} Torr. The grown single crystal has a shape of cylindrical bar of a 6-10 mm diameter and length of 40-60 mm.

For magnetostriction measurements parallelepiped 6*4*4 mm samples were used. Each sample was cut to make surfaces perpendicular to one of the main crystallographic directions. Orientation of the samples was carried out by X-ray diffraction method.

To remove surface-deformed layer after cutting the samples were chemically etched by 30% HNO_3 ethanol solution for depth of approximately 0.2-0.5 mm on each cut side. Then the samples were annealed for 10 hours at a temperature of 800°C in a pure argon atmosphere.

The $Tb_{0.2}Gd_{0.8}$ single crystal has a hexagonal closed packed crystal structure with $P6_3/mmc$ space group.

The measurement of magnetostriction was carried out using the SK-06-030 TY-350 type strain gauge. This type of strain gauge allows to measure the deformation of the sample caused by the simultaneous magnetostrictions in two mutually perpendicular directions. The gauge was glued to the surface of the sample containing c and b crystallographic directions. Magnetostriction measurements were carried out on the PPMS magnetotransport equipment. The measurements were carried out on the $Tb_{0.2}Gd_{0.8}$ single crystal in magnetic fields directed along the ***c*** axis - λ_{cc}, λ_{cb} and

* Corresponding author: zvonov@physics.msu.ru

directed along the ***b*** axis - λ_{bb}, λ_{bc} (the first index indicates the direction of the magnetic field, the second – the direction of magnetostriction measurement).

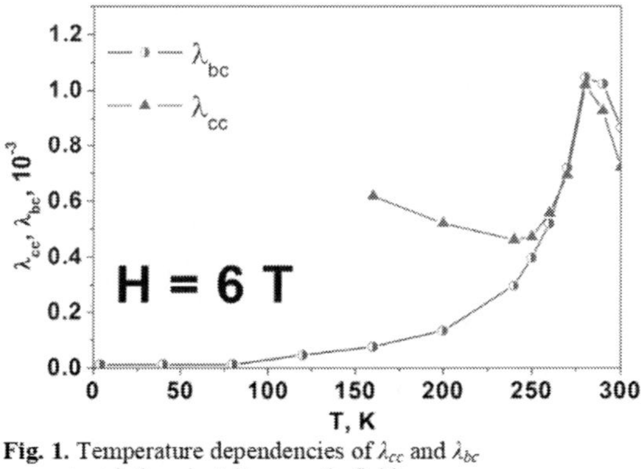

Fig. 1. Temperature dependencies of λ_{cc} and λ_{bc} magnetostrictions in 6 T magnetic field.

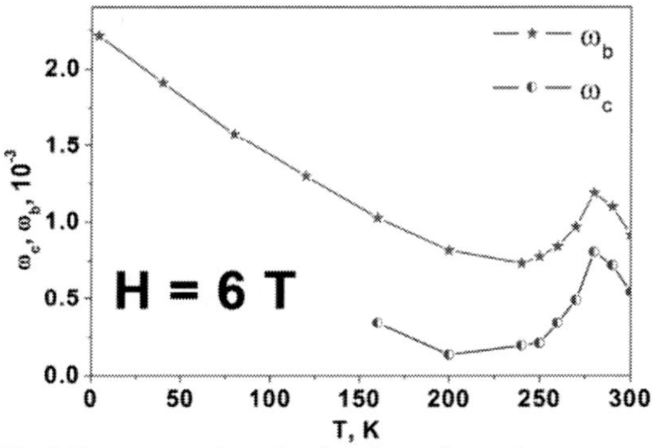

Fig. 2. Temperature dependencies of ω_c and ω_b volume magnetostrictions in 6 T magnetic field.

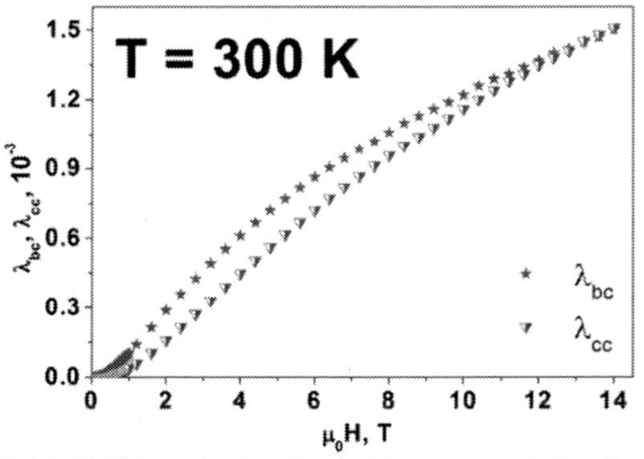

Fig. 3. Field dependencies of λ_{cc} and λ_{bc} magnetostrictions in magnetic field up to 14 T at $T = 300$ K.

3 Results and discussion

The temperature dependencies of λ_{cc} and λ_{bc} magnetostrictions in 6 T magnetic field are shown in fig. 1. These curves have been plotted using experimental data of magnetostriction field dependences measured at constant temperatures. It should be noted that due to strong anisotropy of $Tb_{0.2}Gd_{0.8}$ single crystal (basal plane ***b*** axis is the axis of easy magnetization) λ_{cc} increase with decreasing temperature (fig. 1). The value of λ_{cc} magnetostriction significantly exceeds the value of λ_{bc} in the low temperature region. This effect was also observed before [2]. Temperature dependencies of λ_{cc} and λ_{bc} magnetostrictions in various magnetic fields exhibit maximum near $T_C = 280$ K, the magnetic phase transition temperature of $Tb_{0.2}Gd_{0.8}$ alloy (Curie point). The magnetostriction of $Tb_{0.2}Gd_{0.8}$ single crystal reaches the giant positive values: $\lambda_{bc} \sim 1.5 \cdot 10^{-3}$ in 14 T magnetic field.

We have calculated ω_c and ω_b using experimental values of λ_{cc}, λ_{cb}, λ_{bb} and λ_{bc} and well known equations [11,12,18]:

$$\omega_c = \lambda_{cc} + 2\lambda_{cb}, \quad (1)$$

$$\omega_b = \lambda_{bc} + 2\lambda_{bb}. \quad (2)$$

The temperature dependences of volume magnetostrictions ω_c and ω_b are presented on fig. 2. Temperature dependencies of volume magnetostrictions ω_c and ω_b also exhibit maximum near the magnetic phase transition temperature of $Tb_{0.2}Gd_{0.8}$.

The experimental magnetic field dependences of λ_{cc} and λ_{bc} magnetostriction show an abrupt increase in the area of the magnetic phase transition without saturation up to 14 T magnetic field (fig. 3). It is supposed to calculate the magnetostriction constants for the $Tb_{0.2}Gd_{0.8}$ single crystal using obtained experimental data in the near time. These results suggest that we observe a giant forced magnetostriction. The results obtained in this work indicate that $Tb_{0.2}Gd_{0.8}$ alloy is a promising magnetostrictive material for technical applications.

4 Conclusion

The giant field induced volume magnetostriction exists in the $Tb_{0.2}Gd_{0.8}$ single crystal in the temperature range of ferromagnetic-paramagnetic phase transition (Curie point). However 14 T magnetic field is not enough for saturation of magnetostriction in this single crystal. Therefore the $Tb_{0.2}Gd_{0.8}$ single crystal may be used as magnetostrictive material in various technical devices.

This work is supported by: RFBR grants #16-02-00472 and #16-52-00223; NCN grant, SONATA program (2016/21/D/ST3/03435); ERA.Net RUS Plus project #146 MAGNES.

References

1. K.P. Belov, Magnetostriction Phenomena and Their Technical Applications (Nauka, Moscow, 1987, in Russian).
2. S.A. Nikitin, Magnetic Properties of Rare Earth Metals and Alloys (Moscow University, Moscow, 1989, in Russian).

3. K.P. Belov, G.I. Katayev, R.Z. Levitin, S.A. Nikitin, and V.I. Sokolov. Uspekhi Fizicheskikh Nauk, 140 (2) 271–313, 1983.
4. E. Callen and H. Callene, Phys. Rev., 139 (1965) A 455.
5. A.E. Clark, B.F. De Savage, and R. Bozorth, Phys. Rev., 138 (1965) A 216.
6. S. Khmelevskyi, I. Turek, and P. Mohn. Phys. Rev. B, 70, 132401 (2004).
7. U. Nitzsche, M. Richter, I. Chaplygin, K. Koepernik, I. Opahle, H. Eschrig. JMMM, 272-276(6), E249-E250.
8. G.S. Burkhanov, V.B. Chzhan, G.A. Politova, J. Cwik, N.B. Kolchugina, I.S. Tereshina. Doklady Physics, (2016) 61, 4, 168-171.
9. M. Feuerbacher, M. Heidelmann and C. Thomas. Mater. Res. Lett., 2015 Vol. 3, No. 1, 1–6.
10. K.P. Belov, R.Z. Levitin, S.A. Nikitin, and A.V. Ped'ko, Zh. Eksperim. i Teor. Fiz., 40 (1961) 1562 (in Russian).
11. S.A. Nikitin, Zh. Eksperim. i Teor. Fiz., 43 (1962) 31 (in Russian).
12. J.J. Rhyne, S. Legvold, Phys. Rev., 138 (1965) A 507.
13. A.M. Tishin and Y.I. Spichkin, The Magnetocaloric Effect and Its Applications, Series in Condensed Matter Physics (Institute of Physics, Bristol, 2003).
14. K.A. Gschneidner, V.K. Pecharsky. International Journal of Refrigeration, 31 6, 2008, 945-961.
15. K.A. Gschneidner, V.K. Pecharsky A.O. Tsokol. Reports on Progress in Physics, 68(6), 2005, 1479-1539.
16. J. Kaštil, P. Javorský, J. Kamarád, E. Šantavá. Appl. Phys. A. 104. 2011. 205-209.
17. G. Engdahl and C.B. Bright, "Device application examples" in Handbook of Giant Magnetostrictive Materials (Academic Press, San Diego, 2000), Chap. 5, pp. 287-322.
18. S.A. Nikitin, N.Yu. Pankratov, A.I. Smarzhevskaya, G.A. Politova, Yu.G. Pastushenkov, K.P. Skokov, and A. Del Moral. J. Appl. Phys., 117, 193908 (2015).

Effect of spin-orbital interaction on continuous and jumpwise reversal magnetization in $[Mn(II)(HL)(H_2O)][Mn(III)(CN)_6]\cdot 2H_2O$ molecular ferrimagnet

Marina Kirman* and *Roman* Morgunov

Institute of Problems of Chemical Physics of Russian Academy of Sciences, 142432, Chernogolovka, Russia

Abstract. Magnetic jumps were revealed during the magnetization reversal of $[Mn(II)(HL)(H_2O)]$ $[Mn(III)(CN)_6]\cdot 2H_2O$ molecular ferrimagnet. Amplitudes of the jumps were 0.01 – 0.1 % of the saturation magnetization. Fourier transform of the time series of jumps magnetization indicates that frequency spectrum is close to the white noise. Statistic distribution of the jumps versus time reveals the appearance of the most jumps at the beginning of the demagnetization. Effect of spin-orbit coupling on statistical distribution of magnetization jumps was considered by comparison of two compounds with different single ion anisotropies. The increase of spin-orbit interaction leads to the decrease of power of spectral density of magnetization jumps.

1 Introduction

One of the most interesting new classes of materials is molecular magnets. The exchange interaction in molecular magnets occurs between the metal atoms through organic C-O, C-N bridging and other organic ligands. The atomic structures of these chains and their lengths, as well as the chemical compositions, are variable that attracts researchers of molecular magnet chemistry. However, the interest of physicists is searching of new magnetic effects in these artificial materials. In addition, the ratio between their magnetic characteristic parameters (the domain wall width, the exchange radius, etc.), and geometry of the crystal lattice (lattice parameters) are different from the conventional inorganic magnets. Mentioned above feature of the metal organic magnets lead to the unusual effects, such as a postponed emission of magnetic noise detected in $K_{0.4}[Cr(CN)_6][Mn(R/S)\text{-pn}](R/S)\text{-pnH}_{0.6}$ chiral ferrimagnet (Yellow needle, (YN)) [1, 2]. The novelty of this phenomenon is that these magnetization jumps occur in a constant magnetic field, while Barkhausen jumps occur in a sweeping magnetic field. This fact allows us to consider magnetization jumps as spontaneous nonequilibrium events manifesting non-linear effects in the motion of the domain wall or other spin ensembles (spin soliton and skyrmions in YN were discussed in [3, 4]). The physical nature of jumps in the magnetic reversal constant field is not clear.

In this paper we were focused on searching of similar phenomena (stepwise demagnetization) in other types of molecular magnets, which structure is similar to early studied YN, but driving parameters (exchange interaction, spin-orbital coupling etc) can be varied. Chemical design allows one to create a series of such analogues possessing similar crystal structure and magnetic properties. Thus, metal atoms in studied compound were replaced by other atoms as well as exchange bridges were restructured. The main difference of $[Mn(II)(HL)(H_2O)][Mn(III)(CN)_6]\cdot 2H_2O$ (Brown Needle - BN) compound studied in this paper with those studied in [1, 2] is strong spin-orbit interaction in Mn^{3+} ions (single-ion anisotropy $D \sim 10$ cm^{-1} [5]). In the previously studied YN magnets, weak spin-orbit coupling of Mn^{2+} and Cr^{3+} ions took place (single-ion anisotropy of Mn^{2+} is $D \sim 0.004$ cm^{-1} [6]). Thus, the aim of our work is to establish the laws of the continuous and jumpwise parts of the relaxation in the magnetization reversal $[Mn(II)(HL)(H_2O)]$ $[Mn(III)(CN)_6]\cdot 2H_2O$ molecular magnets.

2 Experiment

Method of synthesizing $[Mn(II)(HL)(H_2O)]$ $[Mn(III)(CN)_6]\cdot 2H_2O$ (BN) microcrystals is described in detail in [7]. The compound (C9H17N8O3Mn2) with the formula weight = 395.16 g/mol was synthesized as dark-red plates (0.28×0.12×0.05 mm), orthorhombic, a = 7.5145(5) Å, b = 14.261(1) Å, c = 14.8844(1) Å, V = 1590.7(2) Å^3, Z = 4.

The crystal structure of BN is quasi-two-dimensional layers of alternating Mn^{2+} ions and Mn^{3+} linked by -CN-

* Corresponding author: kirmanm@yandex.ru

https://doi.org/10.1051/epjconf/201818504016

cyanide bridges (Fig. 1). The individual layers are connected by van der Waals forces. The chiral ligand L = R-pn (1,2-diaminopropane) is in the interlayer space. Space group of BN crystals is chiral ($P2_12_12_1$). Curie temperature is T_C = 21.2 K, saturation magnetization is M_S = 2.79 emu/g at T = 20 K, M_S = 3.94 emu/g at T = 2 K. The coercive force is H_C = 400 Oe at T = 2 K.

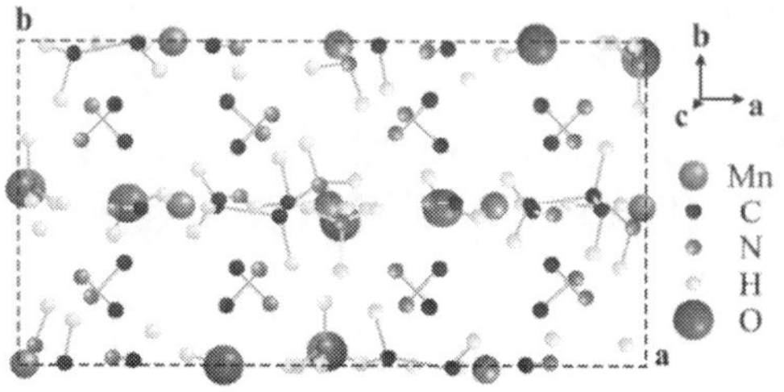

Fig. 1. Atomic structure of [Mn(II)(HL)(H_2O)][Mn(III)$(CN)_6$] molecular ferrimagnet [7].

Dependences of magnetic moment on time, field and temperature were recorded by a MPMS 5XL Quantum Design SQUID magnetometer. Measurements were performed in the magnetic field range from + 50 kOe to - 50 kOe at temperature T = 2 – 300 K. The temperature was stabilized with an accuracy of 0.1 K during the measurement of the sample. The experimental method of studying the dynamics of magnetization reversal is schematically shown in Fig. 2.

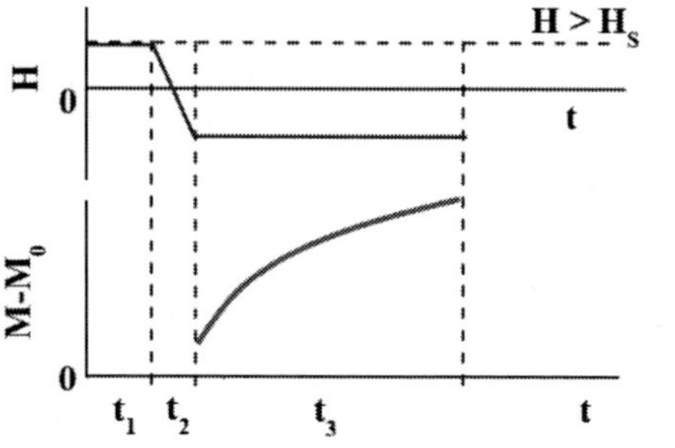

Fig. 2. The scheme of switching of the magnetic field H (upper panel) and the corresponding response of the magnetization of sample M (bottom panel).

Before measurements the sample was cooled down in zero magnetic field. Then the sample is magnetized during t_1 = 300 s in saturation field H = 5 kOe much larger saturation field of sample H_S ~ 1 kOe. After saturation the magnetic field of magnetometer was switched on against the vector of initial magnetization of the sample. "Dead time" necessary for field switching was $t_2 \approx$ 50 s. After magnetic field stabilization in "no overshoot" mode, the time dependence of the magnetic moment was measured during time t_3. On the bottom panel of Fig. 2 shows a time dependence of the magnetic moment (M-M_0), where M_0 - the magnetic moment at the beginning of the measurement $M(t)$.

3 Experimental results and discussion

3.1 Continuous component of magnetic relaxation

Fig. 3 shows the time dependence of the magnetic moment (M-M_0)(t). The magnetic field is switched in accordance with the scheme (Fig. 2) and directed against the vector magnetization of the sample. Switch on the field results in the appearance of the magnetic aftereffect, i.e. the change in the magnetization of the sample with time. If the relaxation of the magnetization associates with a single-barrier process, the magnetic moment is exponentially dependent on time. If there are barriers of different height, the time dependence of magnetization is logarithmic [8].

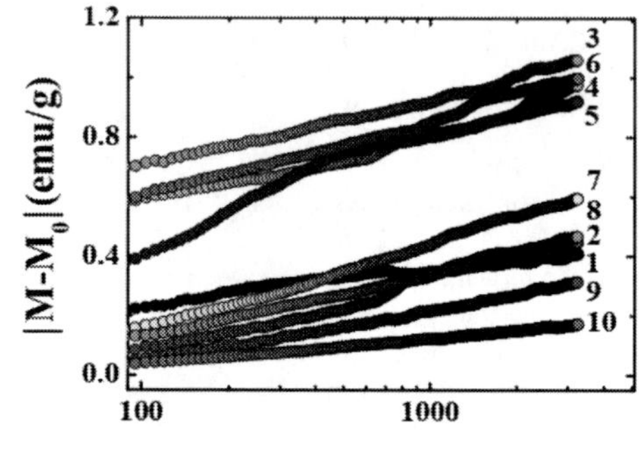

Fig. 3. Time dependences of the absolute value of the change in the magnetic moment $|M$-$M_0|$ at T = 10 K in different magnetic fields: +10 Oe (1), 0 Oe (2), – 5 Oe (3), – 30 Oe (4), – 35 Oe (5), – 40 Oe (6), – 100 Oe (7), –200 Oe (8), – 400 Oe (9), – 1000 Oe (10).

Dependencies $M(t)$ shown on Fig. 3 were approximated by the formula:

$$M = M_0 - S \cdot \ln(t)\,, \qquad (1)$$

where S is magnetic viscosity. The resulting values of S are calculated into a normalized magnetic viscosity S_V = S/χ_{irr} (Fig. 4), where χ_{irr} - irreversible magnetic susceptibility of the sample [9].

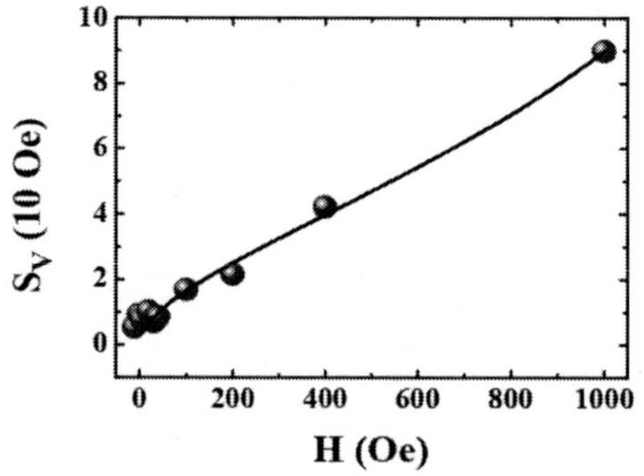

Fig. 4. Dependence of the normalized magnetic viscosity S_V on the magnetic field H at a temperature T = 10 K. The solid line shows the approximation by the expression (3).

The value of S depends strongly on the field, dependence of $S(H)$ has a maximum near the coercive force of the sample H_C = 40 Oe at T = 10 K, which can be attributed to strong pinning of the domain wall by casual obstacles (structural defects) [10].

In the strong pinning limit, the activation energy of the relaxation given by equation [11]:

$$E = (4/3)fb[1-(H/H_0)^{0.5}]^{1.5}, \qquad (2)$$

where f is the force required for a single depinning, $4b$ is the thickness of the domain wall, H_0 is the threshold value of the magnetic field necessary for separation of the domain wall from a defect in the absence of thermal activation.

Dependence $S_V(H)$ (Fig. 4) was approximated by the expression for the magnetic viscosity [12]:

$$S_V = (k_BT/fb)(HH_0)^{0.5}/[1-(H/H_0)^{0.5}]^{0.5} \qquad (3)$$

The value fb = 4 · 10^{-18} J and the value of the threshold magnetic field H_0 = 1823 ± 312 Oe were determined from the approximation of the dependence $S_V(H)$. The value fb is corresponding to the work hardening of a single depinning of the domain wall from the obstacle. The value of the activation energy E = 5.33 · 10^{-18} J was found in the case of strong pinning by the equation (2).

3.2 Magnetization reversal jumps and statistical analysis

The dependence $M(t)$ (Fig. 5) manifests the stochastic magnetization jumps. Statistical time distribution of the jumps probability P and distribution of jump amplitudes ΔM are shown in the inset on Fig. 5 and on Fig. 6, respectively. The initial magnetic moment and the amplitude of the jumps were normalized on saturation magnetization M_s.

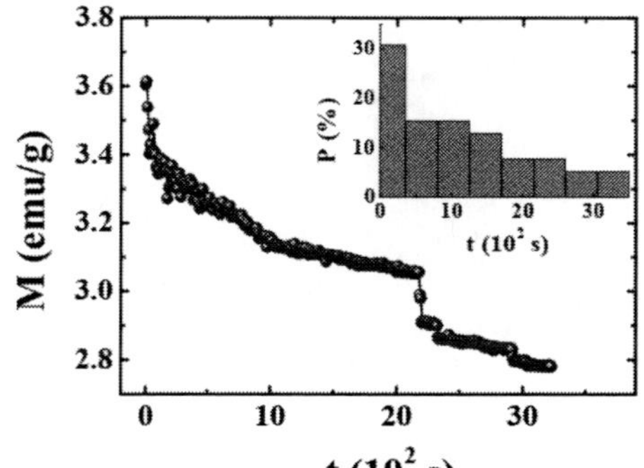

Fig. 5. The time dependence of the magnetic moment M and the jumps probability P (inset) at H = -50 Oe, T = 10 K.

The value of the correlation coefficient $r_{t\Delta M}$ = 0.23 between the time and the jump amplitude ΔM was found by the formula:

$$r_{t\Delta M} = \mathrm{cov}\,(t,\Delta M)\,/(\delta_t\delta_{\Delta M})\ , \qquad (4)$$

cov(t, ΔM) is the covariance, σ_t и $\sigma_{\Delta M}$ are the standard deviations of the t and ΔM, respectively. The correlation coefficient $r_{t\Delta M}$ value is found between 0 and 3 accordingly with Student's scale, which indicates absence of correlation between the time and the amplitude of the jump.

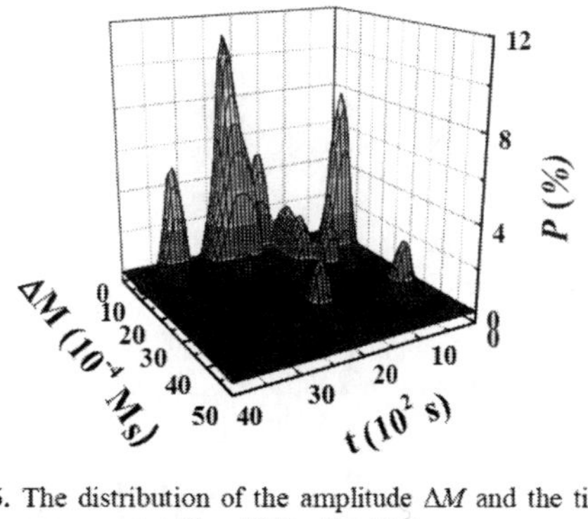

Fig. 6. The distribution of the amplitude ΔM and the time of jump appearance t at H = -50 Oe, T = 10 K.

The distribution of the jumps amplitude ΔM (Fig. 6) show that most jumps occur at the beginning of the measurement.

The atomic structure of the samples is chiral due to Dzyaloshinsky-Moriya interaction. Noncollinear spin states can be formed in chiral molecular magnets. The magnetization of these states is smaller than the magnetization of the commensurate magnetically ordered state. In addition, the energies of noncollinear spin states are characterized by a discrete spectrum. Thus, the transition from a magnetically ordered to a magnetochiral state can be explained by the observed demagnetization jumps.

3.3 Fourier analysis of demagnetization jumps

The time sequence of jump amplitudes (t_n, ΔM_n) is shown in Fig. 7, where n = 1, 2, ..., N is the number of the time interval $t_n = n \cdot \Delta t$, ΔM_n is the jump amplitude occurring in the interval (t_n – $\Delta t/2$; t_n + $\Delta t/2$). Correlogram $r(k)$ of the time series (t_n, ΔM_n) was obtained by the described technique in [1].

The spectral density of the jumps $S(f)$ (inset on Fig. 7) was obtained by Fourier transform of the correlogram $r(k)$ in accordance with the Wiener-Khinchin theorem [13]:

$$S(f_n) = \sum_{k=1}^{N} r(k)\cdot \exp(-(2\pi\ i/N)k\,f_n\Delta t) \qquad (5)$$

where $f_n = n/(N\Delta t)$ corresponds to the set of frequencies of the discrete Fourier transform. Frequency dependence of the spectral density $S(f)$ corresponds to white noise, which is typical for random processes [14, 15].

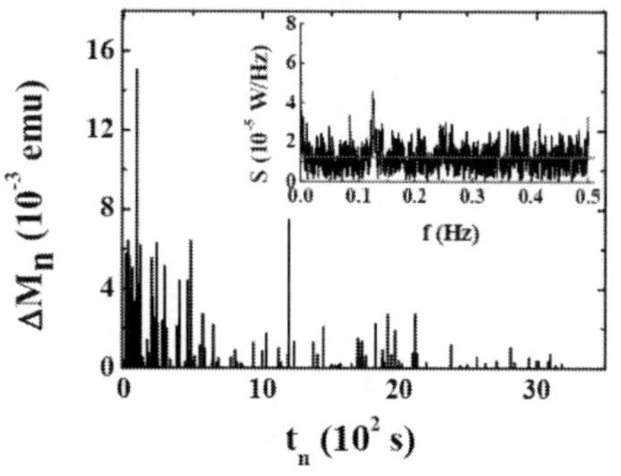

Fig. 7. The time sequence of jump amplitudes collected from 14 independent measurements of $M(t)$ at H = -100 Oe, T = 10 K. Inset: the spectral density S of magnetization jumps in stationary magnetic field at 10 K. The solid line shows the approximation by $(1/f)^{\alpha}$ function, where $\alpha = 0 \pm 0.1$.

White noise was detected as well in YN chiral ferrimagnet in stationary reverse magnetic field [1]. However spectral density of the stochastic magnetization jumps in YN was higher than spectral density in BN at similar conditions. Reduction of amplitude of magnetization jumps can be due to increase of spin-orbit interaction.

4 Conclusions

There are obtained data on the continuous and jumpwise reversal magnetization of [Mn(II)(HL)(H_2O)] [Mn(III)$(CN)_6$]·$2H_2O$ molecular magnet. The values of the magnetic viscosity as well as the activation energy corresponding to the domain wall switching regime were determined for a continuous demagnetization component. It was establish that the frequency spectrum of the jumps corresponds to white noise. The reversal of the magnetization reversal can be caused by the transition of the sample to a state with a magnetochiral structure as a result of competition between the Heisenberg symmetric exchange interaction and the Dzyaloshinsky-Moriya antisymmetric interaction. Spin-orbit interaction effects power of spectral density of magnetization jumps and the energy barriers of domain movements.
This work was supported by RFBR No. 15-02-05149 a.

References

1. R.B. Morgunov, A.D. Talantsev, Phys. Rev. B **94**, 144421 (2016)
2. A. Talantsev, M. Kirman, R. Morgunov, Phys. Status Solidi B **253**, 1222 (2016)
3. R.B. Morgunov, M.V. Kirman, K. Inoue, Y. Tanimoto, J.I. Kishine, A.S. Ovchinnikov, and O. Kazakova, Phys. Rev. B **77**, 184419 (2008)
4. M.V. Kirman, A.D. Talantsev, O.V. Koplak, and R.B. Morgunov, JETP Lett. **101**, 398 (2015)
5. L.J. Challis, A.A. Chazi, K.J. Maxwell, J. Phys. C: Solid State Phys. **12**, 303 (1979)
6. A. Edgar, E. Siegel and W. Urban, J. Phys. C: Solid St. Phys. **13**, 6649 (1980)
7. W. Kaneko, S. Kitagawa, M. Ohba, JACS **129**, 248 (2007)
8. D.K. Lottis, E.D. Dahlberg, J.A. Christner, J.I. Lee, R.L. Peterson, and R.M. White, J. Appl. Phys. **63**, 2920 (1988)
9. D. Givord, Q. Lu, M.F. Rossignol, P. Tenaud, T. Viadieu, JMMM **83**, 183 (1990)
10. P. Gaunt, J. Appl. Phys. **59**, 4129 (1986)
11. P. Gaunt, Philos. Mag. B **48**, 261 (1983)
12. A.I. Dmitriev, V.V. Kucheryaev, E.I. Kunitsyna, R.A. Valeev, R.B. Morgunov, V.P. Piskorskii, O.G. Ospennikova, E.N. Kablov, Physics of the Solid State **58**, 1582 (2016)
13. G.M. Jenkins, D.G. Watts, *Spectral analysis and its applications* (Holden-Day, San Francisco, 1968)
14. H.D. I. Abarbane, R. Brown, J.J. Sidorowich, and L.Sh. Tsimring, Rev. Mod. Phys. **65**, 1331 (1993)
15. N.J. Kasdin, Proceedings of the IEEE **83**, 802 (1995)

Monte Carlo study of magnetic structures in rare-earth amorphous alloys

Alexey Bondarev* and *Igor* Bataronov

Voronezh State Technical University, 14 Moskovski prospekt, 394026 Voronezh, Russian Federation

Abstract. Using the Monte Carlo method, we studied magnetic properties of the models of Re-Tb and Re-Gd amorphous alloys as well as pure amorphous Tb and Gd. The magnetic phase diagrams for the models of amorphous Tb and Gd were constructed. We determined the values of D/J_0 (for Tb) and $J_1/|J_2|$ (for Gd) at which the transition into spin-glass-like state takes place. The local magnetic structure was studied by the spin correlation functions and the angular spin correlation functions. The difference in magnetic structures in amorphous alloys with the random anisotropy and with fluctuations of exchange interaction was revealed.

1 Introduction

Spin glasses are of great interest as a new class of magnetic materials having the unique physical properties [1, 2].

Binary amorphous alloys of heavy rare-earth metals (REM) with nonmagnetic transition metals are very little studied, although in this case the transition to the spin-glass state takes place. In amorphous alloys of the rhenium-terbium (Re-Tb) and rhenium-gadolinium (Re-Gd) systems the typical for spin glasses maximum on the temperature dependence of dynamic magnetic susceptibility $\chi(T)$ and irreversibility of magnetization $M(T)$ were experimentally revealed [3].

However, the nature of the spin-glass state on the microscopic level is studied insufficiently. This encourages us to construct and analyze computer models of atomic structure and magnetic properties of these materials.

As an object of our study, we chose the Re-Tb system with the strong random anisotropy and the Re-Gd system with fluctuations of exchange interactions. The choice of these alloys is caused by the prospects of their application as memory elements.

2 Simulation Technique

Using the molecular dynamics method, we constructed the models of atomic structure of the Re-Tb and Re-Gd amorphous alloys and of pure amorphous Tb and Gd in the wide compositional region. Each model consisted of 100 000 atoms. Interatomic interaction was described by an empirical polynomial potential [4]. The radial distribution functions (RDFs) calculated for the models are in good agreement with the results of the X-ray diffraction experiment [5].

Using the Monte Carlo method in the frame of the Heisenberg model, we carried out simulation of magnetic properties of Re-Tb amorphous alloys and of pure amorphous Tb. The standard Metropolis algorithm of the Monte Carlo method was used [6]. Exchange interaction between the neighbouring magnetic ions as well as the random magnetic anisotropy play the most important role in forming of the magnetically ordered structures in amorphous alloys containing REM. Therefore, we used the following model Hamiltonian [7]:

$$H = -\frac{1}{2}\sum_{i,j} J_{ij}\left(\vec{S}_i \cdot \vec{S}_j\right) - D\sum_i \left(\vec{n}_i \cdot \vec{S}_i\right)^2, \quad (1)$$

where J_{ij} is exchange integral between the spins i and j; D is anisotropy constant; S_i is classical Heisenberg spin; n_i is unit vector determining the direction of the local anisotropy axes.

Dependence of the exchange integral on the interatomic distance r was chosen as a linear function:

$$J(r) = \begin{cases} J_0 \dfrac{r_{min} - r}{r_{min} - r_1}, & r \le r_{min} \\ 0, & r > r_{min} \end{cases}, \quad (2)$$

where $J_0 = 19,26\,\mathrm{K}$ is average value of the exchange integral which is equal to the corresponding value for crystalline Tb; r_1 is position of the first peak on the partial pair RDF $g_{Tb\text{-}Tb}(r)$; r_{min} is position of the first minimum of the $g_{Tb\text{-}Tb}(r)$ function.

Directions of local anisotropy axes were chosen randomly at each site. The value of anisotropy constant in this model varied in a wide region ($D/J_0 = 0-20$).

* Corresponding author: bondarev_a_v@mail.ru

We chose the random anisotropy constant for the Re-Tb alloys linearly dependent on the concentration of Tb atoms:

$$D/J_0 = 6,6 \cdot x/100, \quad (3)$$

where x is concentration of Tb atoms, %. For pure terbium (x = 100 at. %) this formula yields $D/J_0 = 6,6$.

For amorphous Gd and Re-Gd amorphous alloys the Hamiltonian was chosen as follows [8]:

$$H = -\frac{1}{2}\sum_{i,j} J_1\left(\vec{S}_i \cdot \vec{S}_j\right) - \frac{1}{2}\sum_{i,k} J_2\left(\vec{S}_i \cdot \vec{S}_k\right), \quad (4)$$

where $J_1 > 0$ is exchange integral among spins which are in the distances not exceeding $r_1^{min} = 0,455$ nm – the position of the first minimum of the pair RDF $g(r)$; $J_2 < 0$ is exchange integral among spins which are in the distances in the interval from r_1^{min} to $r_2^{min} = 0,770$ nm – the first and second minima of the pair RDF $g(r)$.

The value of the exchange integral in the second coordination sphere J_2 varied in a wide range ($J_1/|J_2| = 2-16$).

3 Magnetic phase diagram for amorphous Tb

For the model of pure amorphous Tb we calculated the temperature dependencies of spontaneous magnetization (Fig. 1) at different values of ratio of the anisotropy constant to the average value of the exchange integral D/J_0. Analyzing the *M(T)* curves, we can conclude that at $D/J_0 = 0$ in the system the ferromagnetic ordering takes place ($M/M_S = 1$ where M_S is saturation magnetization), at $D/J_0 > 0$ – the asperomagnetic ordering ($M/M_S < 1$) which is typical for spin glasses with high concentrations of the REM atoms.

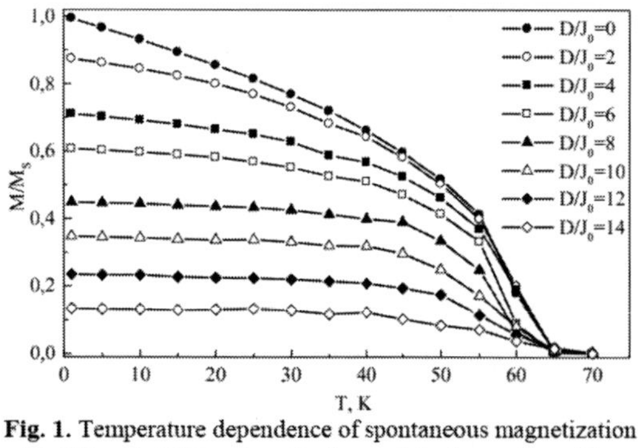

Fig. 1. Temperature dependence of spontaneous magnetization with different D/J_0 values for the model of amorphous Tb

The temperatures of magnetic phase transitions were determined as the positions of peaks on the temperature dependencies of magnetic susceptibility $\chi(T)$. The values of susceptibility were obtained during cooling of the model, and at each temperature the values were averaged over 10 cycles of 10^3 MC-steps/spin.

As a result, we obtained the magnetic phase diagram for amorphous Tb in the $D/J_0 - T$ coordinates (Fig. 2). It allows one to determine the phase state of the system depending on the emperature and the value of anisotropy constant.

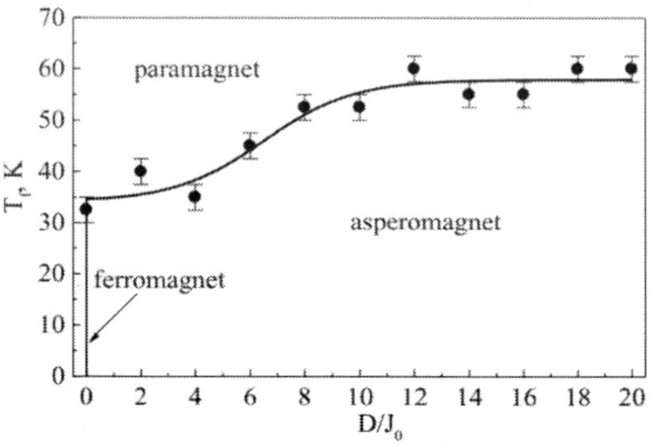

Fig. 2. Magnetic phase diagram $T_f(D/J_0)$ for the model of amorphous Tb

4 Magnetic phase diagram for Re-Tb amorphous alloys

We also studied magnetic properties of the models of $Re_{100-x}Tb_x$ (x = 5, 10, 13, 15, 20, 29, 49, 59 and 91 at. %) amorphous alloys.

In all the models the transition to the spin-glass-like state was observed (Fig. 3), except of $Re_{95}Tb_5$ and $Re_{90}Tb_{10}$ where the paramagnetic phase exists in the entire temperature region up to 1 K. Magnetization at low temperatures for all compositions did not exceed $0,1M_S$. With increasing concentration of Tb atoms the transition temperature linearly increases. This linear dependence qualitatively agrees with the experimental results [3].

The spin-glass transition takes place only if the concentration of Tb atoms $x \geq 13$ at. %, i.e. above the percolation threshold in this system.

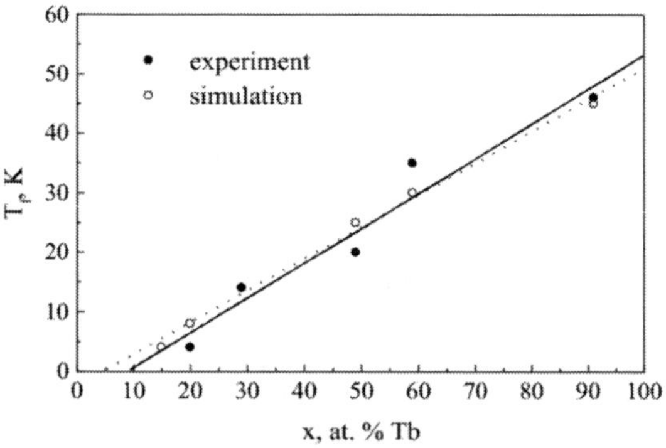

Fig. 3. Compositional dependence of the spin-glass transition temperature for the models of Re-Tb amorphous alloys

5 Spin correlation functions of amorphous Tb

For the models of pure amorphous Tb, the spin correlation functions were calculated (Fig. 4):

$$K(r)=\langle \vec{S}(r)\cdot\vec{S}(0)\rangle . \tag{5}$$

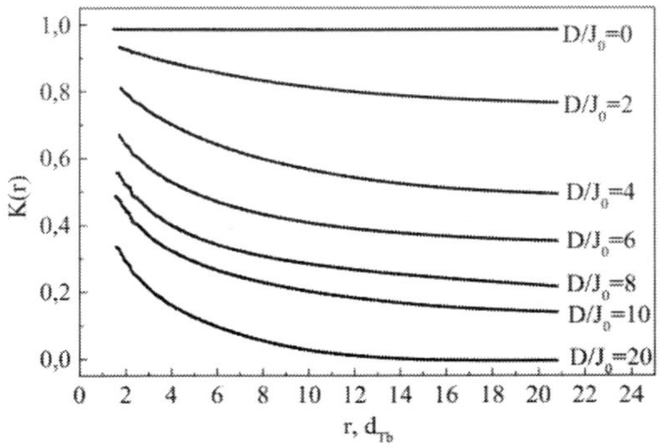

Fig. 4. Spin correlation functions for the model of amorphous Tb (T = 1 K) at various values of the D/J_0 ratio

At all the values of D/J_0, the $K(r)$ functions exponentially decrease with distance:

$$K(r)=K_0+Ae^{-\frac{r-r_0}{\xi}}, \tag{6}$$

where K_0, A, r_0 are constants; ξ is correlation length.

We calculated the dependence of the correlation length ξ from the approximation of the $K(r)$ function by the eq. (6) on D/J_0 for amorphous Tb at T = 1 K (Fig. 5). In Fig. 5 d_{Tb} is the diameter of the terbium atom (d_{Tb} = 0.354 nm).

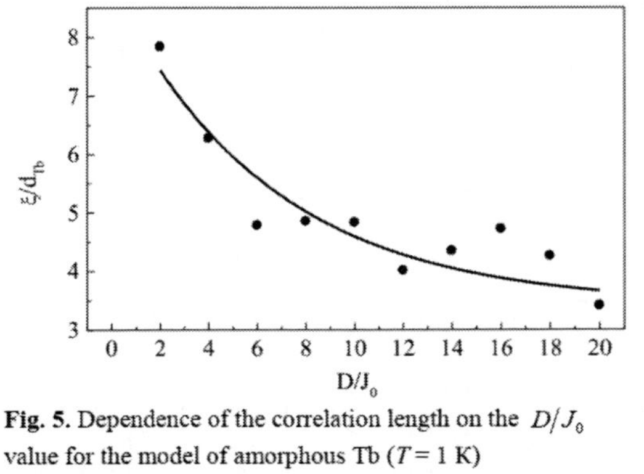

Fig. 5. Dependence of the correlation length on the D/J_0 value for the model of amorphous Tb (T = 1 K)

The temperature dependence of the correlation length for the model of amorphous Tb at $D/J_0=6,6$ is shown in Fig. 6. Near the phase transition temperature T_f the correlation length sharply decreases, and in the paramagnetic phase it is near one diameter of Tb atom.

For these models we also calculated the angular spin correlation functions $P(\theta)$ which are the distribution functions of the angles between directions of spins arranged within the first coordination sphere (Fig. 7). The $P(\theta)$ functions characterize the local magnetic ordering in this system. All these functions have one distinct maximum, its position θ_{max} monotonically increase with increasing the D/J_0 value.

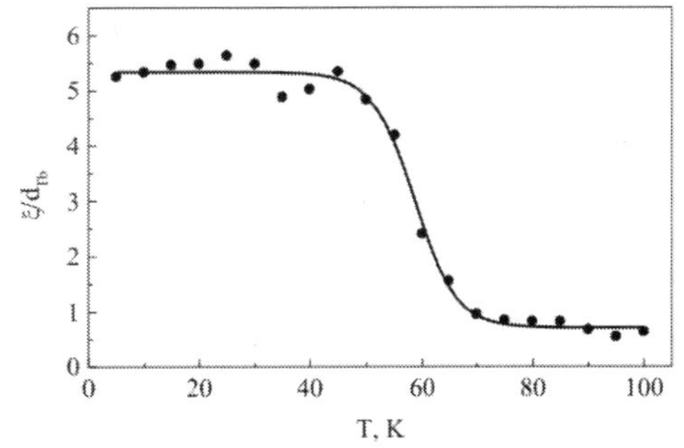

Fig. 6. Temperature dependence of the correlation length for the model of amorphous Tb at $D/J_0=6,6$

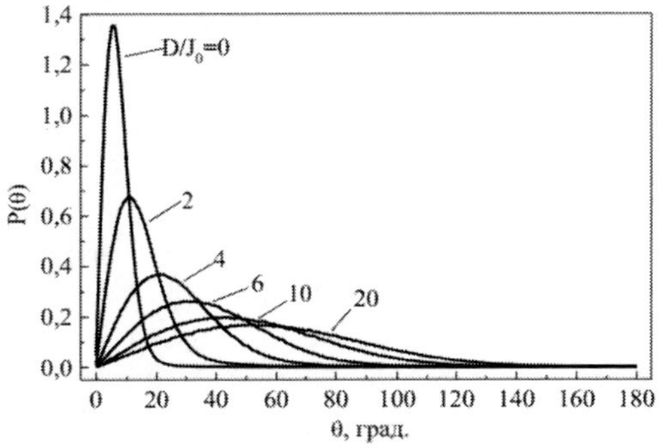

Fig. 7. Angular spin correlation functions for the model of amorphous Tb at $T=1$ K

6 Magnetic phase diagram for amorphous Gd

For the model of pure amorphous Gd, we calculated the temperature dependencies of spontaneous magnetization M at variuos values of the $J_1/|J_2|$ ratio ($J_1/|J_2|=10$, 11, 12, 13 и 14). In the Fig. 8 we present the dependence maximum spontaneous magnetization (at T = 1 K) on the $J_1/|J_2|$ ratio for the model of amorphous Gd.

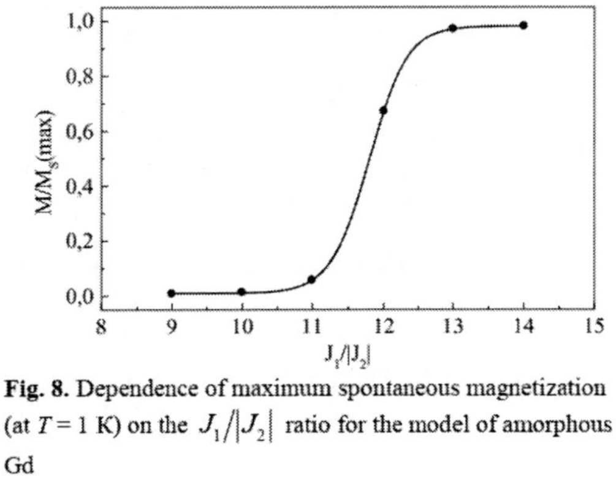

Fig. 8. Dependence of maximum spontaneous magnetization (at T = 1 K) on the $J_1/|J_2|$ ratio for the model of amorphous Gd

At the low temperatures at $J_1/|J_2| \geq 13$ the ferromagnetic state takes place, at $11 < J_1/|J_2| < 13$ – the asperomagnetic state, and at $J_1/|J_2| \leq 11$ the system transits into the spin-glass state.

Thus, we obtained the magnetic phase diagram for amorphous Gd in the $J_1/|J_2|$ – T coordinates (Fig. 9). It allows one to determine the phase state of the system depending on the temperature and the exchange integral within the second coordination sphere J_2.

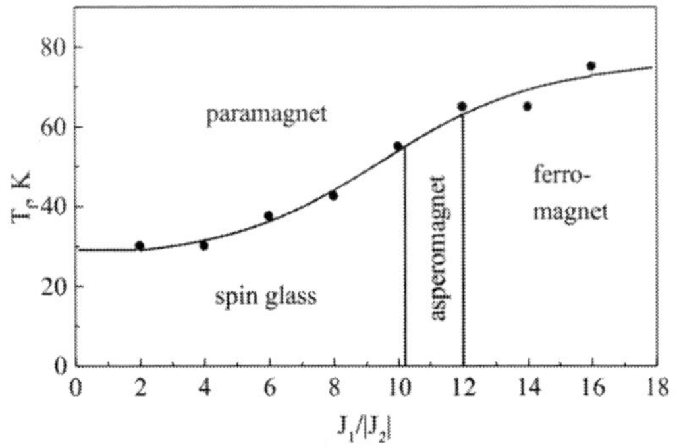

Fig. 9. Magnetic phase diagram $T_f(J_1/|J_2|)$ for the model of amorphous Gd

For the models of Re-Gd amorphous alloys with $J_1/|J_2| = 10$ the transition into spin-glass state is observed. The temperature dependence of the transition temperature is linear and agrees well with the experimental results [3].

7 Spin correlation functions of amorphous Gd

In Fig. 10 we present the spin correlation functions $K(r)$ for the model of amorphous Gd calculated at the values $J_1/|J_2| = 9$, 10, 11, 12, 13 and the temperature $T = 1$ K. The values $J_1/|J_2| = 9$, 10, 11 correspond to the spin-glass state, $J_1/|J_2| = 12$ – asperomagnet, and $J_1/|J_2| = 13$ – ferromagnetic state. In the spin-glass phase, the $K(r)$ functions oscillate in the large distances. This indicates of positive as well as negative correlations in arrangement of magnetic moments of the Gd atoms.

The difference between spin correlation functions with the random anisotropy and with fluctuations of exchange interaction for Tb-based and Gd-based amorphous alloys indicates of the difference in magnetic structures in amorphous alloys with the random anisotropy and with fluctuations of exchange interaction.

In Fig. 11 we present the angular spin correlation functions $P(\theta)$ for the model of amorphous Gd calculated at $J_1/|J_2| = 6$, 8, 10, 12 and the temperature $T = 1$ K.

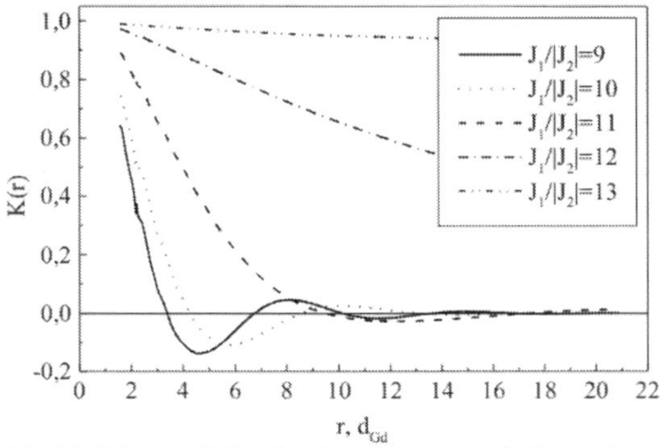

Fig. 10. Spin correlation functions for the model of amorphous Gd at $T = 1$ K at different values of $J_1/|J_2|$ ratio

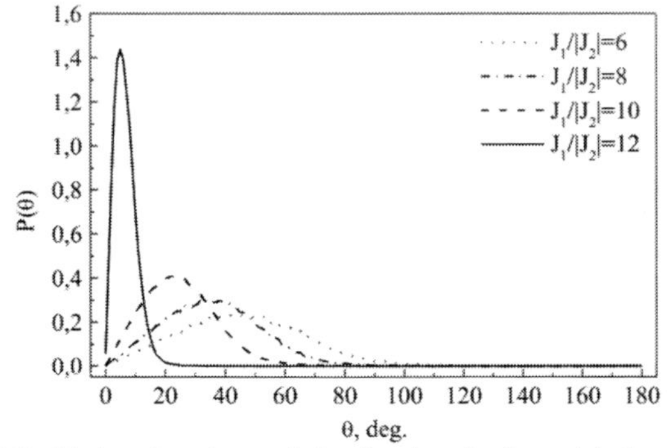

Fig. 11. Angular spin correlation functions for the model of amorphous Gd at $T = 1$ K

The $P(\theta)$ functions have one distinct maximum, its position monotonically decreases with increasing the $J_1/|J_2|$ ratio.

With increasing the temperature, the maximum on the $P(\theta)$ curve becomes lower and wider and shifts to the large angles. The position of the maximum on the angle spin correlation function linearly depends on the temperature.

References

1. K.H. Fischer, Phys. Stat. Sol. (b) **116**, 57 (1983)
2. K. Binder, A.P. Young, Rev. Mod. Phys. **58**, 801 (1986)
3. Yu.V. Barmin, S.Yu. Balalaev, A.V. Bondarev et al., Bull. Russ. Acad. Sci. Physics. **70**, 1308 (2006)
4. I.L. Bataronov, A.V. Bondarev, Yu.V. Barmin, Bull. Russ. Acad. Sci. Physics. **64**, 1329 (2000)
5. A.V. Bondarev, I.L. Bataronov, Yu.V. Barmin, Vestnik VGTU. Ser. Materialovedenie. **1.15**, 39 (2004) (*in Russian*)
6. K. Binder, Rep. Prog. Phys. **60**, 487 (1997)
7. R. Harris, M. Plischke, M.J. Zuckermann, Phys. Rev. Lett. **31**, **160** (1973)
8. A.V. Bondarev, V.V. Ozherelyev, I.L. Bataronov et al., Bull. Russ. Acad. Sci. Physics. **74**, 1474 (2010).

Magnetic properties of Tm_2Fe_{16} under pressure

Anatoly Kuchin[1,*], *Zdenek* Arnold[2], *Jiri* Kamarád[2], *Sergey* Platonov[1]

[1]M.N. Miheev Institute of Metal Physics UD RAS, Ekaterinburg, Russia
[2]Institute of Physics AS CR v.v.i., Prague 8, Czech Republic

Abstract. The magnetic properties of the non-stoichiometric Tm_2Fe_{16} compound under hydrostatic pressures up to 1 GPa were studied. We have revealed that the high-temperature ferrimagnetic state easily magnetized in the basal plane is very sensitive to the volume changes and even moderate pressure is sufficient to its complete suppression and transformation to a helimagnetic state. At the same time, the low-temperature ferrimagnetic state easily magnetized along the hexagonal axis does not disappear under pressure and the temperature of its transition to the high-temperature magnetic states increases under pressure. The remarkable stability of the ground ferrimagnetic state under external pressure can be attributed to the strengthening of the uniaxial magnetic anisotropy and to the mutual perpendicular orientation of the magnetic moments in the ground and the high-temperature magnetic states.

1 Introduction

The majority of R_2Fe_{17} intermetallics (R = rare earth or Y) exhibit a collinear ferro- or ferrimagnetic arrangement of magnetic moments depending on the presence of the light- or heavy-R element, respectively. The R_2Fe_{17} intermetallics with small R-ions are the exceptions; incommensurate helimagnetic structure has been described in the Ce_2Fe_{17}, Lu_2Fe_{17} and Tm_2Fe_{17} intermetallics below the Néel temperature T_N [1-3]. The spontaneous magnetization was observed at temperatures below a critical temperature $\Theta_T < T_N$. Between Θ_T and T_N, the magnetic moments of Fe ions are parallel to each other and antiparallel to those of Tm ions (in Tm_2Fe_{17}) in the atomic layers perpendicular to the crystalline *c*-axis, and they rotate by a certain angle in the consecutive layers. Tm_2Fe_{17} is a collinear ferrimagnet below the transition temperature Θ_T [1, 4]. Unlike Ce_2Fe_{17} and Lu_2Fe_{17}, Tm_2Fe_{17} is characterized by a spontaneous spin-reorientation transition at T_{sr} due to the competition between the uniaxial anisotropy of the Tm-subsystem and the planar-type anisotropy of the Fe-subsystem [4]. Both the Tm and Fe magnetic moments in Tm_2Fe_{17} orient perpendicular to the hexagonal *c*-axis between the magnetic ordering temperature T_N and T_{sr}, and they are parallel to this axis below T_{sr}.

The ferromagnetic state in Ce_2Fe_{17} and Lu_2Fe_{17} is rather unstable and transforms to a helimagnetic state under moderate hydrostatic pressure [2, 3]. This transformation is explained based on the hypothesis of a strong competition of the positive and negative Fe-Fe exchange interactions due to the critical dependence of these interactions on the interatomic Fe-Fe distance [1]. In the present paper, variations of the temperatures of the magnetic phase transitions in the non-stoichiometric Tm_2Fe_{16} compound [4] under a hydrostatic pressure of up to 1 GPa is studied. Tm_2Fe_{16} is a ferrimagnet [4] unlike Tm_2Fe_{17} and non-stoichiometric Tm_2Fe_{18} and Tm_2Fe_{19} compounds having the same magnetic states [4]. The non-stoichiometric Tm_2Fe_{16} – Tm_2Fe_{19} compounds crystallize in the disordered variant of the hexagonal Th_2Ni_{17}-type structure due to the existence of the new defected atomic positions as it was discussed in Ref. [4].

2 Experimental details

The Tm_2Fe_{16} alloy was prepared by induction melting in Ref. [4]. The ingot was homogenized at 1450 K for 10 h and then quenched in water. The phase composition of the sample was investigated by microprobe analysis, X-ray diffraction using Cr radiation, neutron diffraction and magnetometric analysis (see details in [4]). X-ray and neutron diffraction analysis indicated that the Tm_2Fe_{16} compound crystallizes in the disordered variant of the hexagonal Th_2Ni_{17}-type structure [4]. The fraction of free α-Fe was found to be not larger than 1 wt.%. The magnetization measurements at a high hydrostatic pressure of up to 1 GPa (at 5 K) were performed in a SQUID magnetometer (Quantum Design Co.) using a miniature piston-cylinder CuBe pressure cell with a mixture of mineral oils as a pressure transmitting medium [5]. The pressure was determined at low temperatures using the known pressure dependence of the critical temperature of the superconducting state of the pure Pb (5N) sample. The high-pressure studies were performed in magnetic fields up to 7 T in the temperature range from 5 to 320 K. A polycrystalline sample was used for the magnetic measurements. The evolutions of the transition temperature Θ_T, the Néel temperature T_N and the spin-reorientation temperature T_{sr} under different pressures were determined from the

* Corresponding author: kuchin@imp.uran.ru

temperature dependences of the magnetization *M(T)* measured in a magnetic field of 0.01 T with the heating rate of 1 K per minute. T_C or Θ_T were defined as a minimum of derivative of *M(T)* or (at high pressures, see text) as a peak in *M(T)* in the range ~ (130-270) K. T_N was obtained as a local maximum in *M(T)* in the range ~ (230-250) K. T_{sr} was defined as a peak in *M(T)* in the range ~ (90-130) K.

3 Experimental results

The temperature dependences of the magnetization *M(T)* of Tm_2Fe_{16} measured at 0.01 T, 0.1 T and 1 T under ambient and high pressures are presented in Fig. 1.

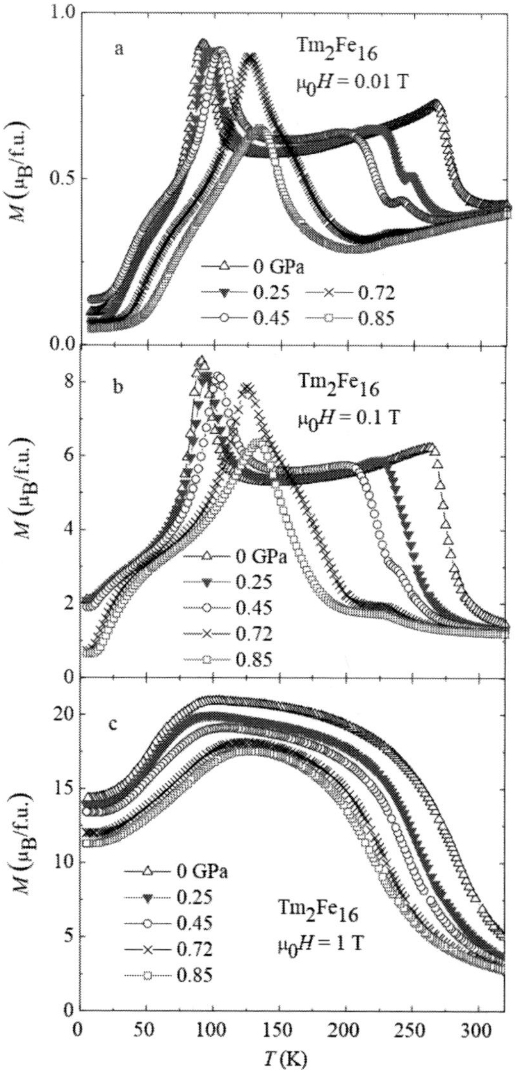

Fig. 1. Temperature dependences of magnetization of the Tm_2Fe_{16} polycrystal measured in a field of 0.01 T (a), 0.1 T (b) and 1 T (c) under different pressures: 0, 0.25, 0.45, 0.72, 0.85 GPa (right-to-left).

The drastic rise of *M(T)* at T ~ 90 K is a result of the well-known spontaneous spin-reorientation magnetic phase transition from the hexagonal axis to the basal plane. The large peak at T ~ 90 K and the bump at $T < T_{sr}$ in the *M(T)* curves in Figs. 1a,b are the contributions from the magnetizations of the microcrystals whose *c*-axis is oriented along the external magnetic field [4].

Tm_2Fe_{16} is a ferrimagnet under ambient pressure [4]. However, a small peak in the *M(T)* curve at the Néel temperature is appeared under the pressure 0.25 GPa in a field of 0.01 T. Starting from this pressure, the types of magnetic states in Tm_2Fe_{16} in a field of 0.01 T become the same as in Tm_2Fe_{17} under ambient pressure (see Introduction). A field of 0.1 or 1 T destroys this induced helimagnetic state (Figs. 1b,c). The higher is the pressure, the stronger is the field to destroy the helimagnet. The increase in the magnetization under pressure at the temperatures below T_N is associated with the transition from the pressure-induced helimagnet to the ferrimagnetic state at Θ_T. In the range P = 0.25-0.72 GPa, Θ_T characterizes the transition between the ferrimagnet and helimagnet, both easily magnetized in the basal plane. Under pressure, T_{sr} moves to the right, and Θ_T and T_N move to the left along the *T*-axis of the plot in Fig. 1a. The lower is the temperature, the higher is the pressure to disappear the ferrimagnetic state easily magnetized in the basal plane as is followed from Fig. 1a. This regularity is supported by the magnetization curves measured at T = 150 K, 200 K or 250 K under P = 0.85 GPa, 0.72 GPa or 0.25 GPa and lower pressures in Figs. 2b,c,d. All of them have apparently different slopes of the initial magnetization portions. The ground ferrimagnetic state easily magnetized along the *c*-axis and the high-temperature helimagnetic state exist under all pressures applied. The existence of the low-temperature spontaneous ferrimagnetism is supported by the same slope of the initial magnetization portions under all pressures applied at T = 50 K in Fig. 2a.

An intense peak in *M(T)* is retained at T ~ 135 K under P = 0.85 GPa in Fig. 1a. It is obvious that this peak characterizes both the T_{sr} and Θ_T values. T_{sr} grows as the pressure increases with the rate dT_{sr}/dP = 49.4 K/GPa. The pressure dependence of Θ_T seems to be non-linear, as is presented in Fig. 3. The linear fit of this dependence gives the average slope of $d\Theta_T/dP$ = -122.6 K/GPa (Fig. 3). The pressure-induced Néel temperature decreases linearly as the pressure increases with a slope of dT_N/dP = -22.3 K/GPa, as is presented in Fig. 3.

The samples are polycrystalline with a magnetic anisotropy field of about 6-7 T [4]. This is why we are not able to saturate the samples and, hence, it is not easy to determine the spontaneous magnetization. For example, *M(H)* curves for Tm_2Fe_{16} at T = 50 - 250 K under different pressures are presented in Fig. 2. Therefore, we estimate the values of the pressure-induced decrease in the magnetization from the values observed in a field of 5 T. The results are presented in Fig. 4. The value $d(\ln M)/dP = -2.10^{-3}$ GPa^{-1} at 5 K for Tm_2Fe_{16} is the quite a reasonable value for the R_2Fe_{17} compounds, compared with $d(\ln M)/dP = -8.2.10^{-4}$ GPa^{-1} for Y_2Fe_{17} in Ref. [6].

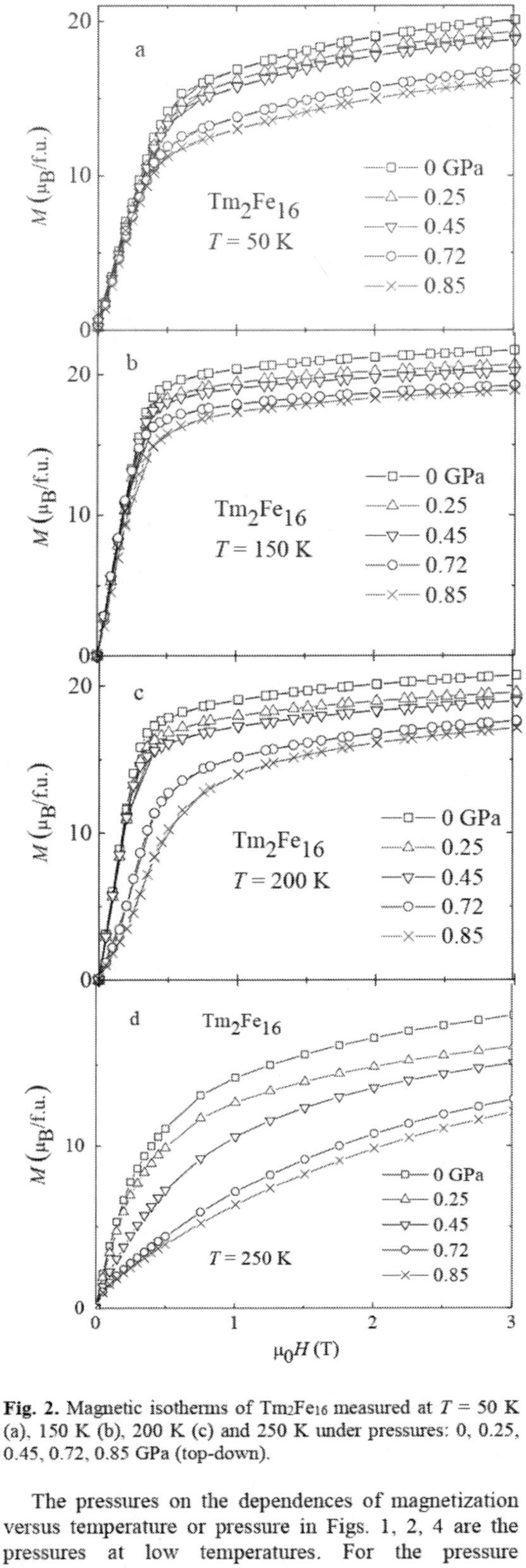

Fig. 2. Magnetic isotherms of Tm_2Fe_{16} measured at T = 50 K (a), 150 K (b), 200 K (c) and 250 K under pressures: 0, 0.25, 0.45, 0.72, 0.85 GPa (top-down).

The pressures on the dependences of magnetization versus temperature or pressure in Figs. 1, 2, 4 are the pressures at low temperatures. For the pressure dependences of the transition temperatures in Fig. 3 we use the corrected pressures at the temperatures of the transitions.

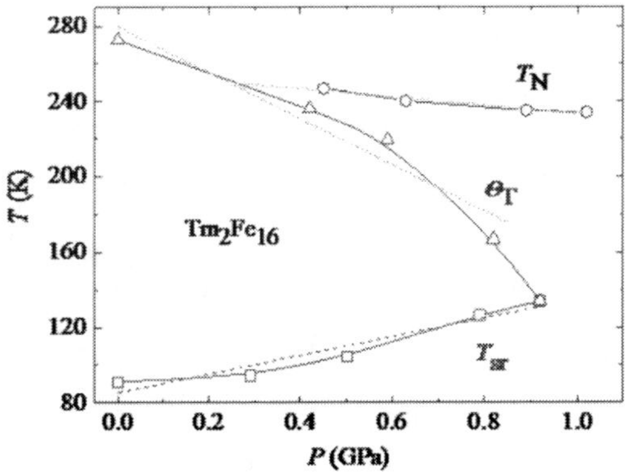

Fig. 3. Pressure dependences of T_N (o), Θ_T (Δ) and T_{sr} ($\square$) of Tm_2Fe_{16} measured at 0.01 T. T_C (Δ) under P = 0 GPa. The dashed lines are the linear fits of the dependences.

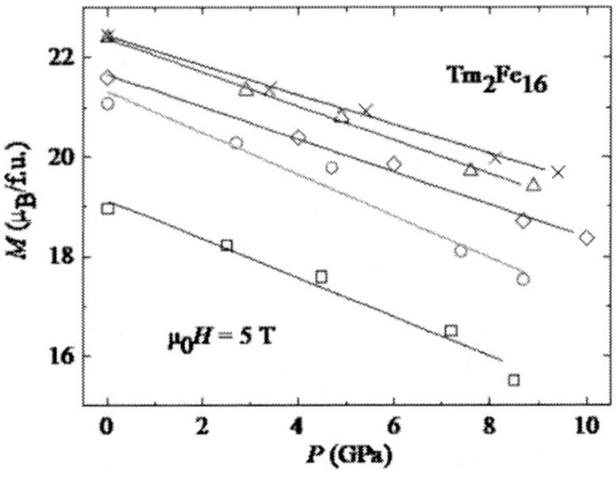

Fig. 4. Pressure dependences of the magnetization of Tm_2Fe_{16} measured in a field of 5 T at T = 5 K ($\square$), 50 K (o), 100 K (Δ), 150 K ($\times$), 200 K ($\Diamond$). The lines are the linear fits of the dependences.

4 Discussion

The increase in the negative exchange interactions in Tm_2Fe_{16} under pressure (Figs. 1a,b) is analogous to the increase in the negative exchange interactions in the system Tm_2Fe_{16}, Tm_2Fe_{17}, Tm_2Fe_{18} and Tm_2Fe_{19}, as the Fe content increases in Ref. [4]. The behavior of Tm_2Fe_{16} under pressure is rather similar to the cases of Ce_2Fe_{17} [2] and Lu_2Fe_{17} [3]. This is because of the strong competition of the positive and negative Fe-Fe exchange interactions due to the critical dependence of these interactions on the interatomic Fe-Fe distance [1]. The similar values of the pressure dependences of the Néel temperature dT_N/dP = -22.3, -19 and -17 K/GPa were obtained for Tm_2Fe_{16} (Fig. 3), Lu_2Fe_{17} and Ce_2Fe_{17} [2, 3], respectively. The closeness of these values of dT_N/dP let us suggest that the high-temperature pressure-induced state in Tm_2Fe_{16} is a helimagnet, as is in the other alloys under ambient pressure. The obtained pressure derivative of the temperature of the transition from the high-temperature ferrimagnet to the helimagnetic state $d\Theta_T/dP$ for Tm_2Fe_{16} (-122.6 K/GPa) is of the same order as for Ce_2Fe_{17} (-380 K/GPa [2]) and Lu_2Fe_{17} (-195 K/GPa [3]). They are also by one order higher than the pressure derivatives of the Néel temperature. Feasibly, this similar correlation between dT_N/dP and $d\Theta_T/dP$ for Tm_2Fe_{16}, Lu_2Fe_{17} and Ce_2Fe_{17} can be explained by the affinity of the magnetic states realized in these compounds.

As a rule, a hydrostatic pressure leads to the decrease in the temperature of the ferro- or ferrimagnetic ordering in R_2Fe_{17} [2, 3, 6]. The low-temperature ferromagnetic phase in the Lu_2Fe_{17} and Ce_2Fe_{17} compounds disappears completely above 0.5 and 0.25 GPa, respectively [2, 3]. As for Tm_2Fe_{16}, the temperature T_{sr} of the transition from the ground ferrimagnetic state to the high-temperature magnetic states increases under pressure. Therefore, the low-temperature ferrimagnetic state is quite stable in this compound. The increase of T_{sr} in Tm_2Fe_{16} under pressure is a result of the weakening of the planar-type magnetic anisotropy in the Fe-subsystem. The weakening of the easy-plane type magnetic anisotropy under pressure was obtained for Y_2Fe_{17} in Ref. [6].

Two lines $\Theta_T(P)$ and $T_{sr}(P)$ in the magnetic phase diagram intersect each other at the critical point (0.92 GPa, 133.7 K) for Tm_2Fe_{16} (Fig. 3). The ground collinear ferrimagnet easily magnetized along the hexagonal axis transforms entirely to the high-temperature helimagnet easily magnetized in the basal plane at this critical point.

5 Conclusions

The influence of the hydrostatic pressure on the Θ_T, T_N and T_{sr} temperatures of the Tm_2Fe_{16} compound has been studied: $d\Theta_T/dP$ = -122.6 K/GPa, dT_N/dP = -22.3, dT_{sr}/dP = 49.4 K/GPa. The values of the pressure derivatives of Θ_T and T_N for Tm_2Fe_{16} are of the same order as for Lu_2Fe_{17} and Ce_2Fe_{17} [2, 3], apparently, due to the similar magnetic structures in these compounds [1]. It was revealed that the high-temperature ferrimagnetic state easily magnetized in the basal plane transforms entirely to the helimagnetic state under P = 0.85 GPa in Tm_2Fe_{16}. At the same time, the ground ferrimagnetic state easily magnetized along the hexagonal axis is retained under all pressures applied, and the temperature T_{sr} of its appearance increases under pressure. This remarkable stability of the ground ferrimagnetic state under external pressure can be attributed to the strengthening of the uniaxial magnetic anisotropy and to the mutual perpendicular orientation of the magnetic moments in the ground and high-temperature magnetic states. The magnetic phase diagram in the plane (T–P) for the Tm_2Fe_{16} compound is drawn. Two lines $\Theta_T(P)$ and $T_{sr}(P)$ in the magnetic phase diagram intersect each other at the critical point (0.92 GPa, 133.7 K).

Acknowledgements

The work was done within RAS Program "Magnet" (project № AAAA-A18-118020290129-5), with support of Grant Agency of CR (Project No. 15-03777S).

References

1. D. Givord, R. Lemaire, IEEE Trans. Magn. **Mag-10**, 109 (1974)
2. I.V. Medvedeva, Z. Arnold, A.G. Kuchin, J. Kamarád, J. Appl. Phys. **86**, 6295 (1999)
3. J. Kamarád, Z.Arnold, I.V. Medvedeva, A.G. Kuchin, J. Magn. Magn. Mater. **242-245**, 876 (2002)
4. A.G. Kuchin, S.P. Platonov, A.V. Korolyov, A.S. Volegov, V.I. Voronin, I.F. Berger, L.V. Elokhina, G.M. Makarova, E.V. Belozerov, J. Alloys Compd. **599**, 26 (2014)
5. J. Kamarád, Z. Machatova, Z. Arnold, Rev. Sci. Instrum. **75**, 5022 (2004)
6. J. Kamarád, O. Mikulina, Z. Arnold, B. Garcia-Landa, M. R. Ibarra, J. Appl. Phys. **85**, 4874 (1999)

Magnetic investigation of low temperature phase transition in iron selenides

Baluyan Tigran[1,*], *Novakova Alla*[1], *Khairullin Marat*[1]

[1]Moscow M.V. Lomonosov State University, Department of physics, Russia

Abstract. 3c- and 4c- Fe_7Se_8 contain large amount of ordered vacancies that define its magnetic structure. Combined comparative magnetic and Mossbauer investigation were performed to find out the peculiarities of the temperature magnetic transitions in these compounds. It was found that adding an additional layer with ordered vacancies in 4c superstructure increased the temperature of spin rotation.

1 Introduction

Vacancies can strongly change material properties, which is important for producing lithium-ion batteries, solar cell units, photocatalytical agents etc[1,2]. As it was previously found out in our work [3], Fe_7Se_8 with 3c superstructure shows a special behaviour in electrical resistivity in a temperature range from 30 to 100K, which leads to a metal-insulator phase transition. It occurs because vacancies are ordered in the selenide structure. It was also shown that this phase transition is preceded by a magnetic phase transition that occurs at 120K. This magnetic transition is accompanied by the spin rotation. 3c-Fe7Se8 is ferrimagnetic above 120K [4,5], but lowering the temperature causes spin rotation towards C axis. As vacancies are strongly ordered in this structure and define its magnetic properties, it was interesting to make a comparison of the observed effect in the same structure with additional atomic layers with vacancies – 4c-Fe7Se8 (Fig.1).

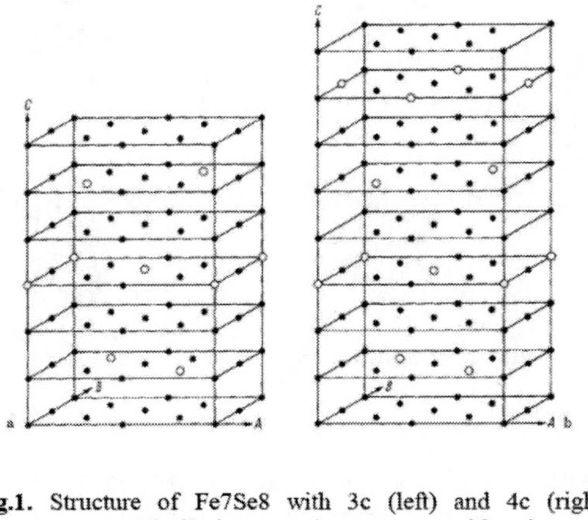

Fig.1. Structure of Fe7Se8 with 3c (left) and 4c (right) superstructure. Black dots are iron atoms, white dots are vacancies. Selenium atoms are omitted for clarity.

Vacancies added in a 4c structure are placed between the layers with no vacancies and next to the diagonal vacancies layer [6]. It was interesting to define the magnetic properties of 4c structure and apply Mossbauer and magnetic methods to define the magnetic peculiarities of this compound

2 Experimental

High purity 3c- and 4c- Fe_7Se_8 samples were investigated in this work. Pure powders (99.999% purity) of Iron and Selenium were sealed in the quartz tube at heated to 1200°C and kept at this temperature for an hour. Then the temperature was lowered to 800°C and held for another hour. Then it was slowly quenched in order to obtain 3c superstructure or quickly cooled to room temperature to obtain 4c superstructure.

Comparative magnetic measurement and Mossbauer investigation was performed for 3c and 4c structures to find out the difference of magnetic phase transitions.

Magnetic investigation was performed on the VSM LakeShore 7407 magnetometer in a temperature range from 80K to 320K. The external field applied was 1 kOe. We also obtained the hysteresis loops for both structures at a temperature range from 80K to 300K to find out the temperature of magnet phase transition completion.

Mossbauer investigation was performed on a MS1104Em Mossbauer spectrometer in a temperature range from 80K to 300K. The obtained spectra were analysed using the UnivemMS tool, which allows spectra fitting with lorentzians and calculates Mossbauer parameters e.g. effective magnetic fields.

3 Results and discussion

3.1 Magnetic measurements

Magnetic moment measurements for 3c- and 4c-Fe_7Se_8 are presented on figures 2 and 3 respectively. Both of them show the similar behaviour – abrupt drop at a certain temperature range : 100-120K for 3c and 100-135K for 4c. The magnitude of magnetic moment is five time higher for 4c structure (0.25 compared to 0.038 emu). Magnetic moment of 4c structure

* Corresponding author tarinax@yandex.ru

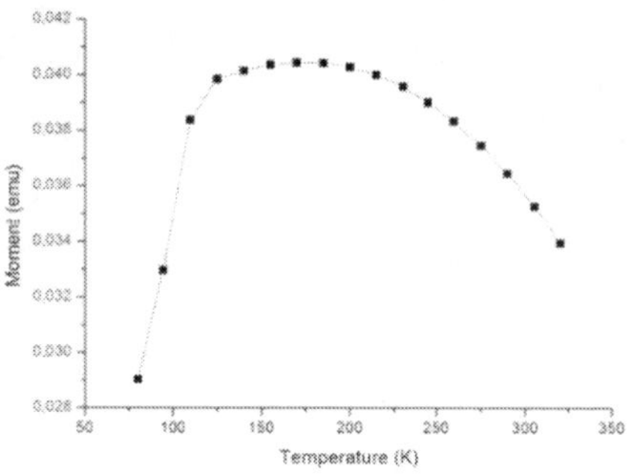

Fig.2. Magnetic moment measurements for 3c-Fe_7Se_8

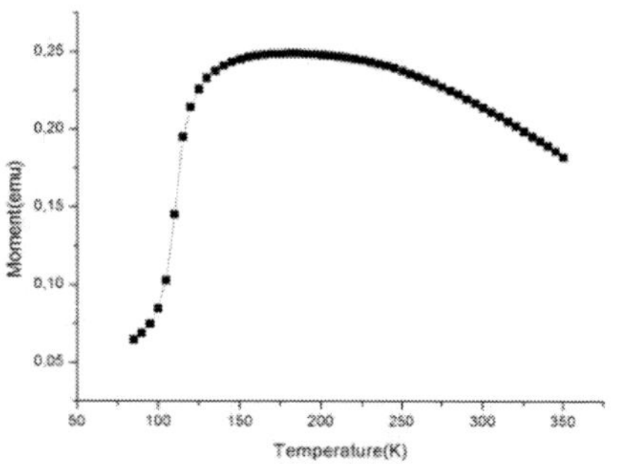

Fig.3. Magnetic moment measurements for 4c-Fe_7Se_8

is greater than 3c because additional atomic layers increase the spin influence of the corresponding ferrimagnetic sublattices[7]. The temperature drop indicates the magnetic phase transition beginning, which is caused by spin rotation toward C axis in both superstructures. This phase transition leads to anisotropy increasing [8], which allows system to reduce it's electric resistivity with further temperature lowering.

This spin rotation, however is not instant, it begins at the drop temperature (120K for 3c and 135K for 4c) and finishes at temperatures around 85K. It is known because all the hysteresis loops for both 3c and 4c superstructures remained classically shaped for all temperatures until 85K. Corresponding hysteresis loops are strongly distorted. This distortion shows the completion of magnetic phase transition in the whole structure[9].

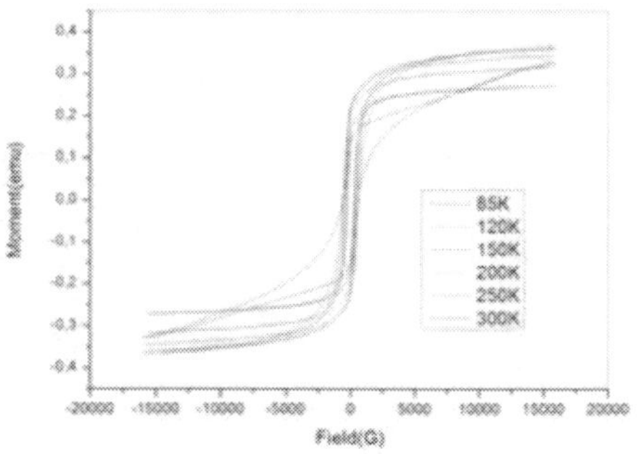

Fig.4. Hysteresis loops for 4c-Fe_7Se_8

3.2 Mossbauer investigation

Mossbauer spectra for 3c and 4c superstructures are different because the ordered vacancies' position defines the iron atom states and 4c structure has additional atomic layer with vacancies. 3c spectra consist of three sextets, each corresponding to specific iron sublattice – A, B and C. A sublattice is bound to the iron atoms in the layers with no vacancies, B lattice is bound to the iron atoms in the layers with diagonal vacancies and C lattice indicates the iron atoms in the layers with other vacancies. In 4c structure, A sublattice is split into A1 and A2. Sublattice A1 is the same as in 3c and A2 represents iron atoms in the layers next to the additional vacancies' layer in 4c structure. B and C sublattices are the same as in 3c.

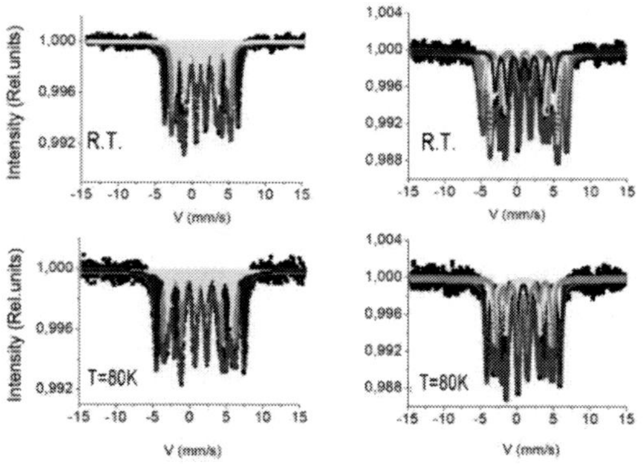

Fig.5. Mossbauer spectra obtained at room temperature and at 80K for 3c-Fe_7Se_8 (right) and 4c-Fe_7Se_8 (left)

All the spectra for both compounds were obtained in a wide range of temperatures from 80K to the room temperature (Fig.5). They were mathematically analysed and the temperature dependence of all sublattices' effective magnetic field for both compounds were calculated. They are shown on Fig. 6 and Fig.7.

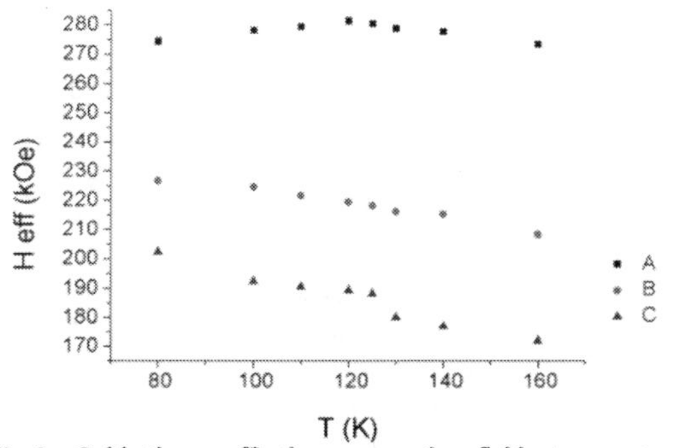

Fig.6. Sublattices effective magnetic field temperature dependence for 3c-Fe_7Se_8

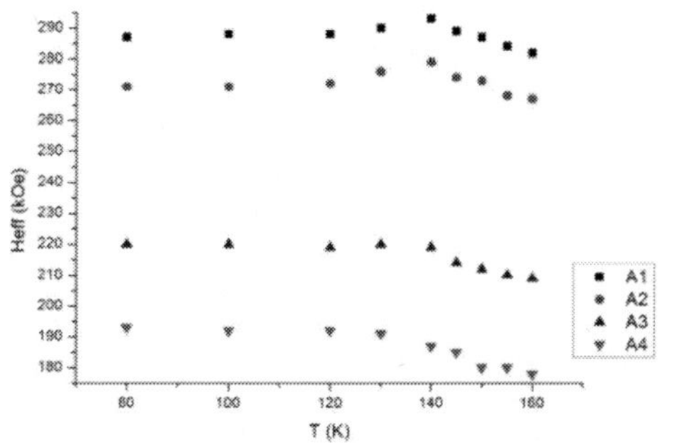

Fig.7. Sublattices effective magnetic field temperature dependence for 4c-Fe_7Se_8

Such behaviour of A sublattices for both compounds is abnormal. There are peaks in temperature dependence of A sites, which indicates the beginning of magnetic disordering with the temperature lowering, that leads to a magnetic phase transition. It occurs at 120K for 3c-Fe_7Se_8 and at 135K for 4c-Fe_7Se_8. What is more, effective magnetic field for 4c A sites increased by 10 kOe comparative to 3c, which also shows the A site spin influence increase because of vacancies adding.

4 Conclusion

In this work we performed the complex comparative Mossbauer and magnetic investigation of 3c- an 4c-Fe_7Se_8 . We found out that adding a cell unit with the ordered vacancies makes influence on the existing ordered structure and leads to the strong increment in the magnetic moment of the whole structure. It also increases the temperature of magnetic phase transition by 15K. It occurs because adding the additional magnetic unit cell with four vacancies splits the A sublattice into two similar sublattices A1 and A2. It makes magnetic subsystem more flexible and increases the whole system entropy, which provides more ways for the whole 4c selenide to minimize it's energy without changing a ferrimagnetic order in the structure.

References

1. 14. Pan, X.; Yang, M. Q.; Fu, X et al Nanoscale 2013, 5,(9), 3601-3614
2. 15. Battaglia, C; Yin, X.; Zheng, M. et al, Nano Lett 2014,14,(2) 967-71
3. G. Li, B. Zhang, T. Baluyan et al *Inorg. Chem.*, 2016, 55 (24), pp 12912–12922
4. A. F. Andresen and J. Leciejewicz, *Journal de Physique*, 1964, **25**, 574-578.
5. M. Kawaminami and A. Okazaki, *J. Phys. Soc. Jpn.*, 1970, **29**, 649-655.
6. C. Boumford, and A. H. Morrish, *Phys. Stat. Sol.*, 1974, **22**, 435-444
7. I. C. Nlebedim, M. Vinitha, P. J. Praveen, D. Das and D. C. Jiles, *Journal of Applied Physics*, 2013, **113**, 193904.
8. U. Kobler, A. Hoser, J. Englich, A. Snezhko, M. Kawakami, M. Beyss and K. Fischer, *J Phys Soc Jpn*, 2001, **70**, 3089-3097.
9. Y. J. Tang, B. Roos, T. Mewes, S. O. Demokritov, B. Hillebrands and Y. J. Wang, *Applied Physics Letters*, 1999, **75**, 707.

Effect of aging on the magnetic properties of rapidly quenched Nd-rich Nd-Fe alloys

Dennis Wiafe-Akenteng, *Vladimir* Menushenkov, *Igor* Shchetinin, *Alexander* Savchenko

National University of Science and Technology «MISiS», Leninskiy pr. 4, Moscow, Russia

Abstract. The Fe_xNd_{100-x} (x = 14, 28) alloys were successfully prepared by melt-spinning. Hysteretic and thermomagnetic properties of the as-melt-spun alloys and short-aged ribbons were investigated. Both ribbons exhibit a soft magnetic behavior at room temperature but at 5 K, the coercive force (H_{ci}) proliferated to 8.6 kOe and 11.7 kOe, respectively. Thermomagnetic measurements, Zero Field Cooling (ZFC) and Field Cooling (FC), demonstrated that the rise in magnetization of the ribbons at temperatures lower 34 K is associated with ferromagnetism of *fcc* Nd-Fe solid solution clusters. Positive effect of aging on the coercivity of the $Fe_{14}Nd_{86}$ and $Fe_{28}Nd_{72}$ ribbons was observed. At 300 K, H_{ci} of the ribbons upsurge to 2 - 2.8 kOe after aging at 400 - 450°C for 15 minutes but sharply diminished after aging at temperatures higher than 500°C. It was suggested that the increase in RT coercivity during aging was as a result of the decomposition of the amorphous phase and formation of the hard magnetic Fe-rich clusters. For both ribbons aged at 500°C, the magnetic transition at 340 K was observed to be in relation to the T_C of the Nd_2Fe_{17} compound. The appearance of this soft magnetic phase in the microstructure of the aged ribbons was linked with the drastic diminution of the coercivity at that temperature.

1 Introduction

The interest in the study of the Nd-Fe alloys was renewed during the last decades owing to the significant role Nd-rich intergranular phase play in the development of high coercivity of sintered Nd-Fe-B-based magnets. Usually, a high coercivity in the sintered magnets is obtained after post sintering aging to produce the optimal microstructure. The effect of the intergranular Nd-rich phase on coercivity was divided into two contributions. The first is the effect of the thin layers between the $Nd_2Fe_{14}B$ grains which consists of amorphous Nd-Fe phase. These layers act to reduce or remove defects at the $Nd_2Fe_{14}B$ grain surfaces. Secondly, they tend to decouple the $Nd_2Fe_{14}B$ grains with respect to magnetic exchange interactions [1].
The motivation for the study of binary Nd-Fe alloys is the large coercivity of 4.5 kOe of as-cast Nd-rich Nd-Fe alloys without any additional processing step. It is considered that the high coercivity is attributed to the Nd-Fe phase with high magnetocrystalline anisotropy [2, 3] which forms during solidification of Nd-rich alloys. The as-cast alloys are characterized by the metastable eutectic structure consisting of the mixture of Nd grains and the amorphous-like Fe-Nd regions denoted in literature as "A_1" phase. The HRTEM imaging from the Fe-Nd regions reveals that A_1 phase is composed of nanocrystals embedded in an amorphous matrix [4-7]. Unfortunately, the components of A_1 regions cannot be identified because of its very fine size.

The coercivity of as-melt-spun Nd-rich alloys is low at room temperature (RT) but increases with decreasing temperature. The microstructure of as-melt-spun Nd-rich alloys consists of an amorphous-like region composed of nanocrystals of unknown phase embedded in an amorphous matrix [6,7]. It is, therefore, expected that aging may lead to additional decomposition of the amorphous phase and formation of the microstructure similar to the one responsible for the high coercivity in as-cast Nd-rich alloys.
In this work, the effect of aging on the magnetic properties - hysteresis loops, thermomagnetic ZFC-FC dependences of Fe_xNd_{100-x} (x = 14, 28) as-melt-spun alloys and polycrystalline Nd - in a temperature range of 5-300 K were investigated.

2 Experimental

Two batches of samples – $Fe_{14}Nd_{86}$ and $Fe_{28}Nd_{72}$ were prepared from starting materials of high purity elements, iron (99.95%) and neodymium (99.9%). The samples were induction melted and casted under argon atmosphere. The ingots were re-melted and quenched by melt-spinning with a copper-wheel speed of approximately 30 ms^{-1}. The aging of the ribbons was performed in vacuum furnace at temperatures ranging from 300–650°C for 15 minutes. Magnetic measurements at temperatures 5 K and 300 K were performed using a Quantum Design PPMS EverCool-II

* Corresponding author: nanaortchere1@gmail.com

magnetometer in the magnetizing field of 90 kOe. The temperature dependence of magnetization (M) on ZFC and FC for the aged ribbons were measured in an operating field of H = 0.5 T over a wide temperature range (5 - 300 K). The specimens were prepared in the form of thin disc from the mixture of the ribbon fragments with epoxy. The error bar of H_{ci} values at repeated measurements was less than 3%.

3. Results and discussion

Hysteresis loops of the Fe_xNd_{100-x} (x = 14 and 28) as-melt-spun ribbons measured at 300 K and 5 K are shown in Figure 1. At 300 K, the hysteresis loops of the ribbons x = 14, 28 showed a soft magnetic behavior with low magnetization and H_{ci} - 75 and 15 Oe respectively. However, both samples displayed a hard magnetic behavior at 5 K with an upsurge in coercivity, 8.6 kOe and 11.7 kOe respectively. Maximum value of magnetization in the field of 90 kOe was recorded at 5 K, 4 times as high as the value measured at 300 K. It should be noted that saturation of the hysteresis loops was not achieved even in magnetic field of 90 kOe.

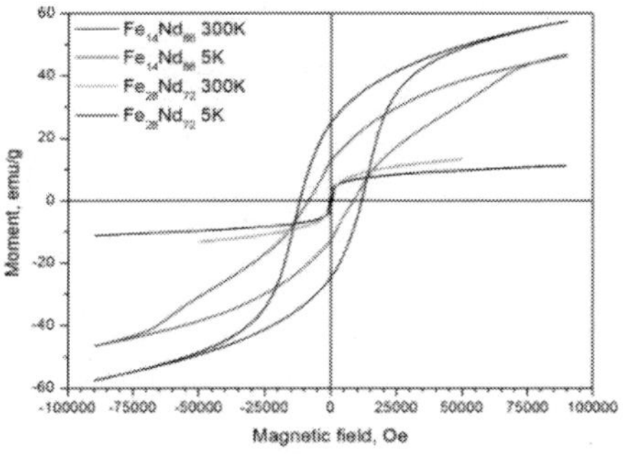

Figure 1. Hysteresis loops of the $Fe_{14}Nd_{86}$ and $Fe_{28}Nd_{72}$ as-melt-spun ribbons measured at 5 K and 300 K

Figure 2a shows a rise in coercivity of $Fe_{14}Nd_{86}$ and $Fe_{28}Nd_{72}$ ribbons when temperature decreases from 300 K to 5 K. The temperature dependence of coercivity show a linear relationship between $H_c^{1/2}$ and $T^{2/3}$ in the temperature range 25 – 200 K (Figure 2b).

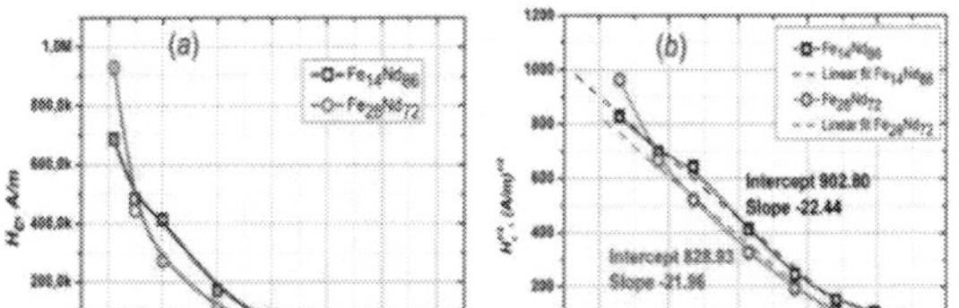

Figure 2. Temperature dependences of $H_c(T)$ (a) and dependences of $Hc^{1/2}(T^{2/3})$ (b) for $Fe_{14}Nd_{86}$ and $Fe_{28}Nd_{72}$ ribbons over the range 5–300 K

These data demonstrate an agreement with strong pinning model of domain walls, proposed by Gaunt [8]. According to [6,7], the coercivity of the as-melt-spun $Fe_{14}Nd_{86}$ and $Fe_{28}Nd_{72}$ ribbons at lower temperatures is associated with a domain wall pinning on the Nd-rich nanocrystals embedded in an amorphous-like matrix phase which are observed in the ribbons microstructure.

For the interpretation of the temperature dependence of coercivity, temperature dependences of magnetization in the range 5–300 K were made to reveal the possible magnetic transitions. ZFC and FC curves (Figure 3) displays the magnetization behavior of the as-melt-spun $Fe_{14}Nd_{86}$ ribbon and polycrystalline Nd in operating field of H = 0.5 kOe over a temperature range 5-300 K. Comparing the ZFC and FC curves for Nd and $Fe_{14}Nd_{86}$ ribbon at temperatures lower than 50 K shows the magnetic transitions at ≈ 9 K, ≈ 17 K and ≈ 34 K in both Nd and ribbon. The observation of these magnetic transitions suggests the presence of several magnetic phases in microstructure of both the crystalline Nd and $Fe_{14}Nd_{86}$ ribbon. According to [9], the first two transitions correspond to antiferromagnetic ordering temperature of the *dhcp* Nd (the onset of ordering on cubic sites). This suggests that the ferromagnetic transition at 34 K is related to the metastable *fcc* allotrope of Nd or Nd-Fe solid solution (Nd-based nanocrystals) [6,7]. Hence, the possible reason for an upsurge in the magnetization at high fields at temperatures lower than 34 K is the presence of *fcc* Nd and Nd-based phases.

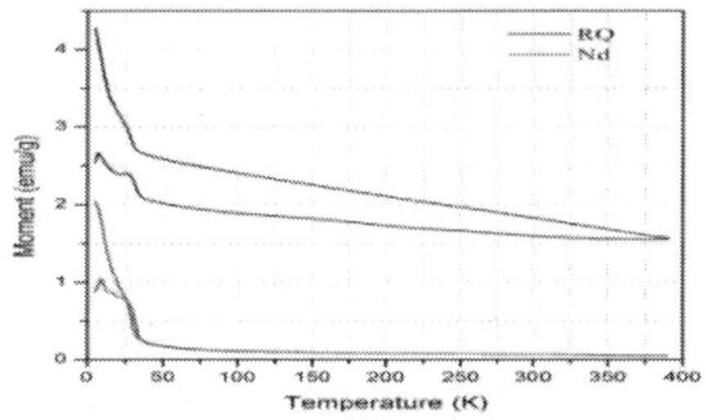

Figure 3. ZFC and FC thermomagnetic curves of the $Fe_{14}Nd_{86}$ ribbon (RQ) and polycrystalline Nd

The paramagnetic susceptibility of the *fcc* Nd and Nd-Fe solid solution resulted in unsaturated magnetization curves for the $Fe_{14}Nd_{86}$ ribbon at temperatures above 34 K. But the fact that saturation of the hysteresis loops is not achieved at 5 K shows the presence of high magnetocrystalline anisotropy of Nd which connotes magnetic moments remain noncollinear down to 5 K. For this reason, the random-anisotropy model [10] of the Fe-contained clusters may be used to explain the magnetic properties of as-melt-spun ribbons. In the zero field, the magnetization moments of the Nd-Fe clusters are directed along the easy axes which have random orientation. An increasing magnetic field overcomes the anisotropy energy and aligns the magnetization moments of the clusters towards the field direction.

It was observed that a short aging of the $Fe_{14}Nd_{86}$ and $Fe_{28}Nd_{72}$ ribbons at temperatures lower than the eutectic temperature in Fe-Nd system (685°C) increased the coercivity. It might be as a result of transforming their microstructure which was composed of Nd-based nanocrystals and residual amorphous phase. It could also be caused by decomposition of the amorphous phase and precipitation of the Fe-rich clusters with high magnetocrystalline anisotropy which increased the coercivity. On the other hand, the precipitation of the soft magnetic Fe_2Nd_{17} clusters resulted in lessening the coercivity. Secondly, the increase of the size of Nd-rich nanocrystals during aging may have resulted in the nanocrystals losing the role as the domain walls pinning centers.

The hysteresis loops of the $Fe_{14}Nd_{86}$ and $Fe_{28}Nd_{72}$ ribbons measured at 300 K and 5 K after aging at 300 – 600°C for 15 minutes are shown in Figures 4 and 5 respectively. The temperature dependences of coercivity for the $Fe_{14}Nd_{86}$ and $Fe_{28}Nd_{72}$ ribbons over the aging temperature range 300–600°C are shown in Figure 6. Maximum values of coercivity, 3.1 kOe and 2.6 kOe, were received after aging at 400 - 450°C for the ribbons x = 14, 28 respectively measured at 300 K. These values of coercivity is only slightly less than the coercivity of as-cast Nd-rich Fe-Nd alloys, H_{ci} = 4-4.5 kOe [2,3]. Aging at temperatures higher than 450°C recorded a reduction in coercivity for the $Fe_{14}Nd_{86}$ and $Fe_{28}Nd_{72}$ ribbons.

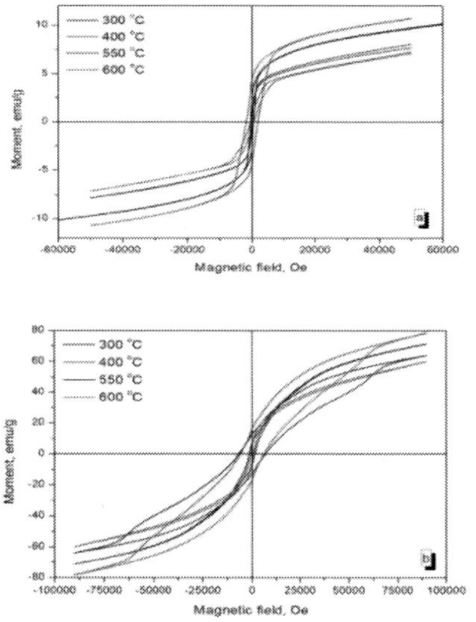

Figure 4. Hysteresis loops of the $Fe_{14}Nd_{86}$ ribbons after aging at 300–600°C for 15 min measured at 5 K (a) and 300 K (b)

As seen in Figure 6a, at 5 K, the coercivity of as-melt-spun $Fe_{14}Nd_{86}$ and $Fe_{28}Nd_{72}$ ribbons (8.6 kOe and 11.7 kOe, respectively), slightly reduced after aging in the temperature range of 300 – 450°C and decreased drastically after aging at 500°C and higher temperatures. At 300 K, the coercivity of as-melt-spun ribbons $Fe_{14}Nd_{86}$ and $Fe_{28}Nd_{72}$ (75 and 15 Oe, respectively) rose to 2 - 2.8 kOe after aging at 400 - 450°C. A sharp drop was observed after aging at temperatures higher than 500°C.

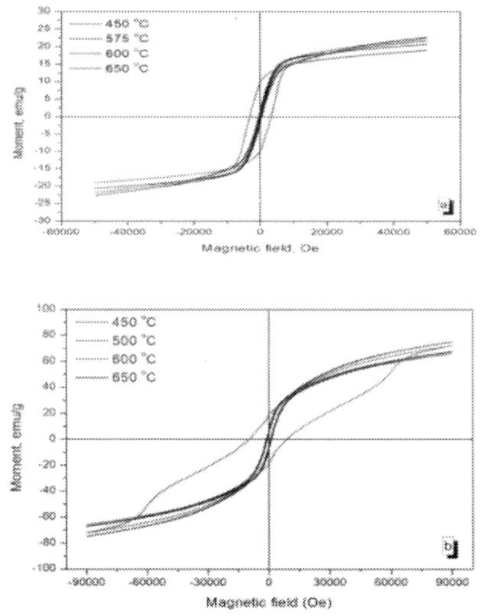

Figure 5. Hysteresis loops of the $Fe_{28}Nd_{72}$ ribbons after aging at 300–600 C for 15 min measured at 5 K (a) and 300 K (b)

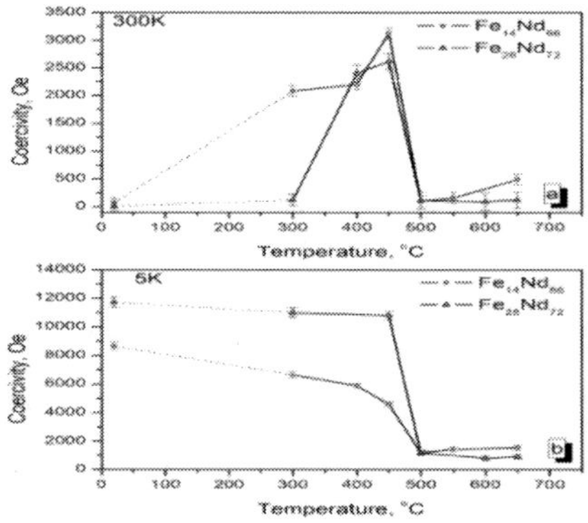

Figure 6. Dependences of H_c versus the aging temperature for the ribbon $Fe_{14}Nd_{86}$ (a) and $Fe_{28}Nd_{72}$ (b) measured at 5 K and 300 K after aging at 300 – 650 C for 15 min

For the interpretation of this data, it was suggested that the main reason for an increase in RT coercivity after short aging at 400 - 450°C is decomposition of the amorphous phase and formation of the Fe-rich clusters of the hard magnetic phase like that in as-cast Nd-Fe alloy. The low level of coercivity of the aged ribbons at 300 K (1.8 - 2 kOe) in comparison with H_{ci} of the as-cast alloy (4.5 kOe) can be explained as due to the non-optimal composition or size of the Fe-rich clusters and a coarsening of the Nd-rich nanocrystals during aging hence decreasing the domain wall pinning.

Figure 7 shows the ZFC and FC thermomagnetic curves of the ribbons $Fe_{14}Nd_{86}$ (a) and $Fe_{28}Nd_{72}$ (b) after aging at 300 – 650 C for 15 minutes. The comparison of the ZFC and FC curves of $Fe_{14}Nd_{86}$ and $Fe_{28}Nd_{72}$ ribbons after melt spinning and after aging in Figures 3 and 7 shows that aging has no effect on the temperature of the bends on the M(T) curves at temperatures lower than 50 K. Hence, the composition of Nd or Nd-based solid solution in these ribbons remains unchanged.

As demonstrated in Figure 7, the temperature dependences of magnetization (FC) for the $Fe_{14}Nd_{86}$ and $Fe_{28}Nd_{72}$ ribbons aged at temperatures 300 - 450°C shows that magnetic transition (T_C) takes place at temperatures higher than 500 K. But for both ribbons aged at 500°C, the magnetic transition was observed at 330-340 K which is close to the T_C of the Nd_2Fe_{17} compound (335 K). The appearance of this soft magnetic phase in the microstructure of the aged ribbons resulted in drastic decrease in the coercivity.

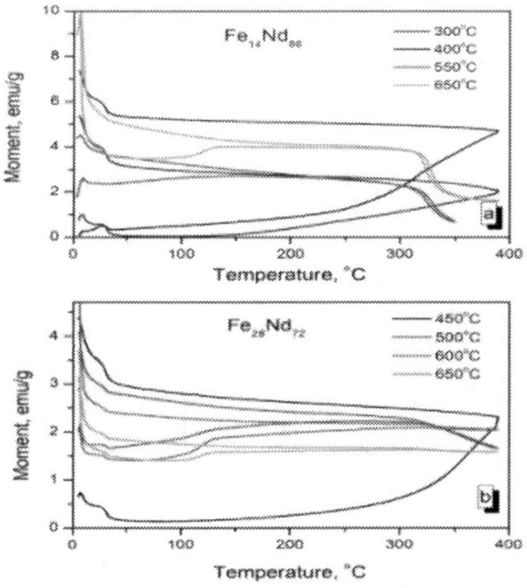

Figure 7. Zero-field cooling (ZFC) and field cooling (FC) thermomagnetic curves of the ribbon $Fe_{14}Nd_{86}$ (a) and $Fe_{28}Nd_{72}$ (b) after aging at 450 – 650 C for 15 min

Based on the obtained data, it can be discussed that the possible mechanism of the intergranular phase has an influence on the H_{ci} of sintered magnets during post-sintering aging. Usually, in sintered magnets the Nd-rich intergranular phase is an amorphous region similar to the amorphous-like phase in as-melt-spun Nd-rich ribbons. It may be assumed that the post-sintering aging of the ribbons led to the decomposition of the intergranular amorphous phase and appearance of the Fe-Nd clusters, similar to the clusters in as-melt-spun ribbons that resulted in a rise in the H_{ci} of aged magnets.

Conclusion

Hysteretic and thermomagnetic properties of the $Fe_{14}Nd_{86}$ and $Fe_{28}Nd_{72}$ as-melt-spun alloys and short aged ribbons were investigated. The present study showed that as-melt-spun ribbons exhibit a soft magnetic behavior at room temperature but with decreasing temperature from 300 to 5 K, H_{ci} proliferated up to 8.6 kOe and 11.7 kOe, respectively. The increase in H_{ci} can be associated with a domain wall pinning on the Nd-rich nanocrystals embedded in an amorphous-like matrix which were observed in the microstructure of the ribbons.

Thermomagnetic measurements (ZFC and FC) for the as-melt-spun $Fe_{14}Nd_{86}$ and $Fe_{28}Nd_{72}$ ribbons show magnetic transitions at 9, 17 and ≈ 34 K. The first two transitions, 9 and 17K, originated from *dhcp* Nd and the third transition is associated with ferromagnetism of *fcc* Nd-Fe solid solution which resulted in an increase of the magnetization of the ribbons at temperature lower than 34 K.

The positive effect of aging on the H_{ci} of the as-melt-spun $Fe_{14}Nd_{86}$ and $Fe_{28}Nd_{72}$ ribbons was observed. At 300 K, the H_{ci} of the as-melt-spun ribbons (75 and 15 Oe, respectively), escalated to 2 - 2.8 kOe after aging at 400 - 450°C but sharply declined after aging at temperatures higher than 500°C. It was suggested that the main reason for the rise in RT H_{ci} during aging is the decomposition of the amorphous phase and formation of the hard magnetic Fe-rich clusters. For both ribbons aged at 500°C, magnetic transition was observed at 340 K, closer to the T_C of the Nd_2Fe_{17} compound (T_C = 335 K). The appearance of this soft magnetic phase in the microstructure of the aged ribbons resulted in H_{ci} reducing drastically.

Further studies must be mainly directed to structural analysis of the aged samples to help clarify the effect of aging on its magnetic properties.

Acknowledgement. This work was supported by the Ministry of Education and Science of the Russian Federation (agreement no. 14.587.21.0028, a unique no. RFMEFI58716X0028).

References

[1] T.G. Woodcock, Y. Zhang, G. Hrkac, G. Ciuta, N.M. Dempsey, T. Schrefl, O. Gutfleisch, D. Givord. Scripta Materialia 67 (2012) 536.
[2] Drozzina V, Janus R. Nature, 1935, 135, 36.
[3] J.J. Croat, Appl. Phys. Lett., 1981, 39, 357.
[4] V. P. Menushenkov, A. S. Lileev, M. A. Oreshkin, S. A. Zhuravlev, J. Magn. Magn. Mater., 1999, 203, 149.
[5] Kumar, J. Eckert, S. Roth, W. Loser, L. Schultz, S. Ram. *Acta Materialia*, 51 (2003) 229.
[6] Menushenkov, V. P., Shchetinin, I. V., Gorshenkov, M. V., Savchenko, A. G., & Ketov, S. V. (2016). *IEEE Magnetics Letters*, *7*, 1-4.
[7] V.P. Menushenkov, I.V. Shchetinin, M.V. Gorshenkov, A.G. Savchenko, S.V. Ketov, Proc. of the 24-th Int. Workshop on Rare-Earth and Future Permanent Magnets and their Applications (REPM 2016), 2016, Darmstadt, Germany.
[8] P. Gaunt. Philos. Mag. B, 1983, 48, 261.
[9] E. Bucher, C. W. Chu, J. P. Maita, K. Andres, A. S. Cooper, E. Buehler, and K. Nassau, Phys. Rev. Lett., 1969, 22, 1260.
[10] R. Ribas, B. Dieny, B. Barbara, A. Labrata, J. Phys. Condens. Matter, 1995, **7**, 3301.

Magnetic properties of the non-stoichiometric $TbCo_2Ni_x$ alloys

Alexander Inishev*, *Evgeny* Gerasimov, *Nikolay* Mushnikov, *Pavel* Terentev, and *Vasily* Gaviko

M.N. Miheev Institute of Metal Physics UB RAS, Ekaterinburg, Russia
Institute of Natural Sciences and Mathematics, Ural Federal University, Ekaterinburg, Russia

Abstract. The magnetic and magnetothermal properties of the non-stoichiometric $TbCo_2Ni_x$ ($0 \leq x \leq 0.2$) alloys were studied. It was found that the concentration dependence of the Curie temperature and magnetic moment of the 3d-sublattice have a maximum at $x = 0.025$. The obtained experimental magnetic properties of the $TbCo_2Ni_x$ alloys were discussed under assumption that the Co magnetic moment in the compounds changes with increasing x. The magnetic entropy change was determined using the temperature dependences of the magnetization and Maxwell's thermodynamic relation. The obtained results for $TbCo_2Ni_x$ were compared with those for the $ErCo_2Mn_x$ alloys.

1 Introduction

It was found recently that ternary RM_2Mn (R – rare-earth metal, M – 3d transition metal Ni, Co) compounds crystallize in the Laves-phase $MgCu_2$-type cubic structure in spite of the ratio of the R and M components is 1 to 3. The Mn atoms in Laves-phase structure of the RM_2Mn compounds partially occupy both the M (16d) and rare earth (8a) sites [1-3]. The Curie temperatures T_C are considerably higher for the RNi_2Mn alloys in comparison with T_C for corresponding binary RNi_2 and RMn_2. Later, the structure and magnetic properties of the non-stoichiometric RM_2Mn_x ($0 \leq x \leq 1.5$) compounds with R = Tb, Dy, Gd were studied [4-7]. It was shown that the $MgCu_2$-type structure exists up to the manganese content x = 0.8 for $ErCo_2Mn_x$ and up to x = 1.5 for $ErNi_2Mn_x$. It was established that T_C is a non-monotonous function of Mn concentration, which has a maximum at $x < 1$. It was found also that the maximum isothermal entropy change and the adiabatic temperature change remain weakly changed in a wide temperature range in $ErCo_2Mn_x$. Thus, there is new opportunity to modify magnetic properties of the Laves phase compounds by adding Mn and forming non-stoichiometric RM_2Mn_x compounds.

Up to now the properties of $RM_2M'_x$ – type compounds with M'=Mn were studied only. However, we expect that non-stoichiometric $RM_2M'_x$ compounds with another 3d transition metals can exist. In particularly, such type compounds can forms in case M'=Ni.

In the present paper, we report the results of studies of the structure, magnetic and magnetothermal properties of the non-stoichiometric $TbCo_2Ni_x$. The obtained results are compared with those for $ErCo_2Mn_x$ which we previously studied.

2 Experimental details

The ingots of the $TbCo_2Ni_x$ alloys were prepared by induction melting of the constituents in alumina crucibles in argon atmosphere. In order to obtain the equilibrium phase state, the ingots were annealed at 800°C for 7 days. Structural and magnetic studies were performed at the Centre of Collective Use of the Institute of Metal Physics UB RAS. Room temperature X-ray diffraction patterns were measured for powdered samples with the average particle size 30–50 μm using a DRON-type diffractometer with Cr Kα radiation. The x-ray diffraction patterns were analysed with the PowderCell 2.4 program. To determine the Curie temperature of the compounds, we measured temperature dependences of ac magnetic susceptibility χ_{ac} and used the position of minimum in the temperature dependences of derivative $d\chi_{ac}(T)/dT$, which corresponds to the more abrupt decrease in the susceptibility with increasing temperature. The spontaneous magnetization μ_s of compounds was determined from the demagnetization curves as magnetization extrapolated to zero internal field.

3 Results and discussion

Fig. 1 shows the room temperature X-ray diffraction patterns of $TbCo_2Ni_x$ alloys. All main reflections in the X-ray diffraction patterns are adequately described in terms of the $MgCu_2$-type structure ($Fd3m$); the volume fraction of impurity phases with the $PuNi_3$-type structure ($R3m$) does not exceed 8 vol. %. According to X-ray diffraction studies, the single phase non-stoichiometric $TbCo_2Ni_x$ compounds with the cubic Laves-phase structure exist up to critical nickel concentration x = 0.1 [8]. This value is much lower than x = 0.8-1.5 which

* Corresponding author: inishev@imp.uran.ru

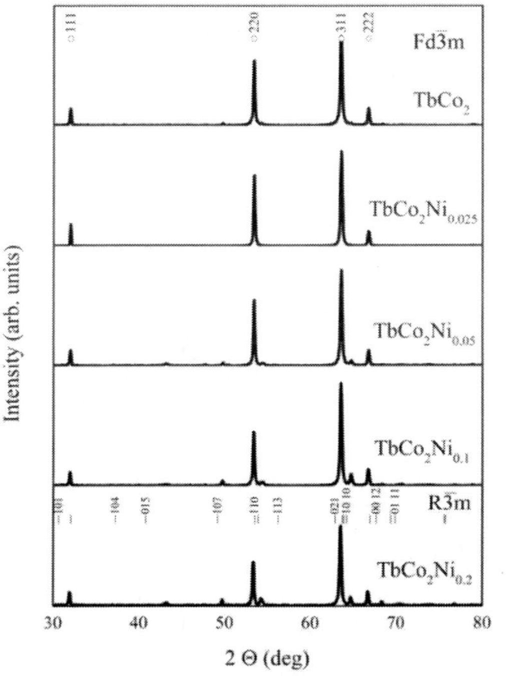

Fig. 1. X-ray diffraction patterns of $TbCo_2Ni_x$ alloys with different x. Bragg peak positions are indicated by the markers for $MgCu_2$-type structure (open circles). $PuNi_3$-type structures (line bars).

were observed early for non-stoichiometric RM_2Mn_x compounds. It is known that the stability of the Laves phases is determined by factors such as the relationship of atomic radii of R and M elements and the average number of valence electrons per atom. The chemical bonds in Laves phases are assumed to be mixed metallic–covalence–ionic. Because of this, to analyse the stability of the phases, atomic radii of elements are sometimes used. It is likely that the same factors must affect the concentration boundaries of the existence of the nonstoichiometric compounds. This may be due to that the atomic and ionic radii of Ni are less than those of Mn. And, correspondingly, large difference between Ni and R does not allows to occupy enough amount of R (8a) sites by Ni atoms. Thus, the non-stoichiometric RCo_2Ni_x alloys with cubic $MgCu_2$-type Laves phase structure can be obtained in more narrow concentration range than in case of alloying with Mn.

The lattice parameter of the Ni-containing compounds are virtually independent of x (Fig. 2). Such behavior caused by two opposite factors. It is known that the atoms of the alloying element (Ni or Mn) occupy both the R (8a) and the M (16d) sites. The partial substitution of the R atoms by Ni or Mn at the 8a sites leads to decrease the lattice parameter. On the other hand, the substitution of the Co atoms at the 16d sites by a 3d element with a larger ionic or metallic radius leads to increasing of the lattice parameter. Two opposite tendencies lead to a weak concentration change of the lattice parameter for $RM_2M'_x$.

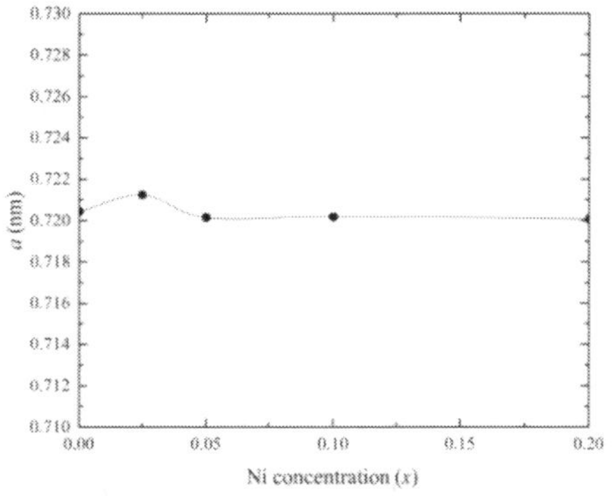

Fig. 2. Concentration dependences of the lattice parameter for $TbCo_2Ni_x$ compounds.

Fig. 3 shows the concentration dependences of the Curie temperature for $ErCo_2Mn_x$ and $TbCo_2Ni_x$ compounds in range of concentration from $x = 0$ up to maximal concentration for which exist single phase non-stoichiometric $TbCo_2Ni_x$ compounds with $MgCu_2$-type structure.

As can be seen in Fig. 3, concentration dependence of the Curie temperature increases sharply at small Ni concentrations and has maximum at $T = 233$ K. With further increasing the Ni content, the concentration dependence of the T_C decreases. The Curie temperature of $ErCo_2Mn_x$ compounds monotonically increases and reaches maximum value 212 K at manganese concentration $x = 0.6$ [6].

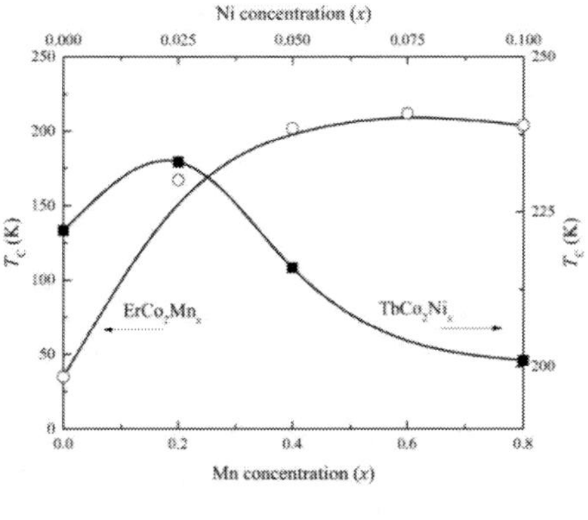

Fig. 3. Concentration dependences of the Curie temperature for $ErCo_2Mn_x$ (○) and $TbCo_2Ni_x$ (■) compounds.

The spontaneous magnetic moments μ_s decreases with increasing x both for $TbCo_2Ni_x$ and for $ErCo_2Mn_x$ as result of antiparallel ordering of magnetic moments of heavy rare-earth ions and 3d-transition atoms (Fig. 4).

Fig. 5 shows the concentration dependence of the magnetic moment of 3d-sublattice of $ErCo_2Mn_x$ and $TbCo_2Ni_x$ compounds. We calculated magnetic moment of 3d-sublattice μ_{3d} in assumption that the R and 3d magnetic moments have collinear antiparallel alignment:

$$\mu_{3d} = (\mu_R - \mu_s)/(2+x), \quad (1)$$

where μ_s - spontaneous magnetic moment of the compounds, μ_R - magnetic moment of the rare-earth ion. The maximum μ_{3d} value is 1.38 μ_B for $TbCo_2Ni_{0.025}$ compound and 1.5 μ_B for $ErCo_2Mn_{0.6}$ compound. It is seen that character of the $\mu_{3d}(x)$ concentration dependencies both for $ErCo_2Mn_x$ and for $TbCo_2Ni_x$ is very similar to those for Curie temperatures.

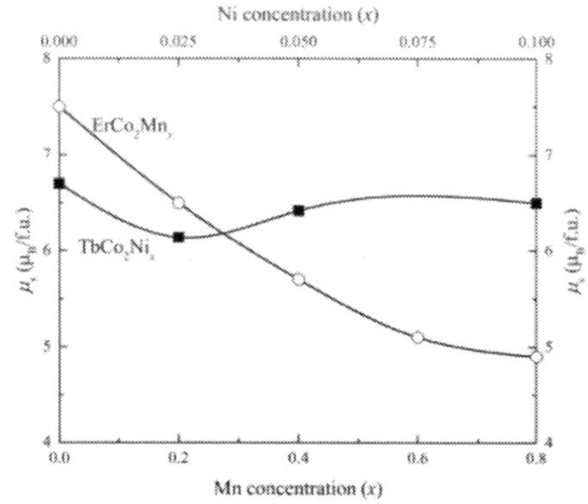

Fig. 4. Concentration dependences of the spontaneous magnetic moment at 4 K for $ErCo_2Mn_x$ (○) and $TbCo_2Ni_x$ (■) compounds.

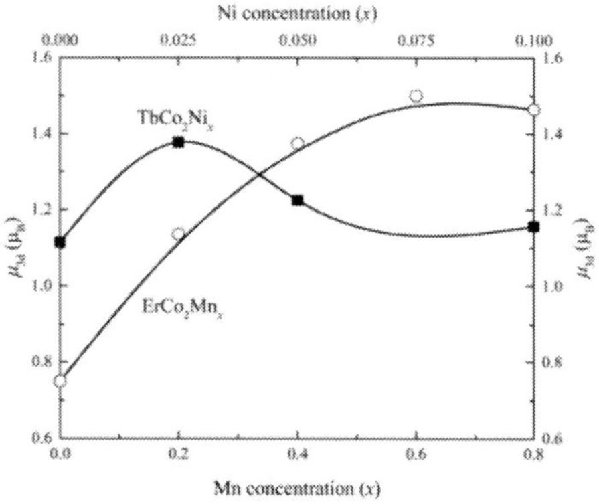

Fig. 5. Concentration dependences of the magnetic moment of 3*d*-sublattice for $ErCo_2Mn_x$ (○) and $TbCo_2Ni_x$ (■) compounds.

The $\mu_{3d}(x)$ dependence has maximum for $TbCo_2Ni_x$ and is monotonous for $ErCo_2Mn_x$. Our analysis of magnetic moments shows that the increase in the Curie temperature of the compounds is mainly due to changes in the band structure. It is possible, that alloying with Mn and Ni increase the Co magnetic moment due to the additional splitting of the 3d band. This, in turn, leads to an increase in the *R*-Ni (Mn), Co-Ni (Mn) exchange interaction.

We measured temperature dependences of magnetization *M*(*T*) in magnetic field H_1 = 1 T. It allows us to determine temperature variation of the magnetic entropy change $\Delta S_m(T)$ in $TbCo_2Ni_x$ compounds for magnetic field change $\mu_0\Delta H = H_1 - 0$ = 1 T using well known Maxwell's thermodynamic relation:

$$\Delta S_m(T,\Delta H) = \int_0^{H_1} \left(\frac{\partial M(T,H)}{\partial T}\right)_H dH \approx \frac{\partial M(T,H_1)}{\partial T} H_1. \quad (2)$$

Fig. 6 shows the temperature dependences of entropy change for $ErCo_2Mn_x$ and $TbCo_2Ni_x$ compounds. Maximum isothermal entropy change is -1.1 J kg^{-1}K^{-1} for $TbCo_2Ni_{0.025}$ and -1.15 J kg^{-1}K^{-1} for $ErCo_2Mn_{0.2}$ for the magnetic field change 1 T. These values are close to those observed, for the binary $TbCo_2$-based compounds.

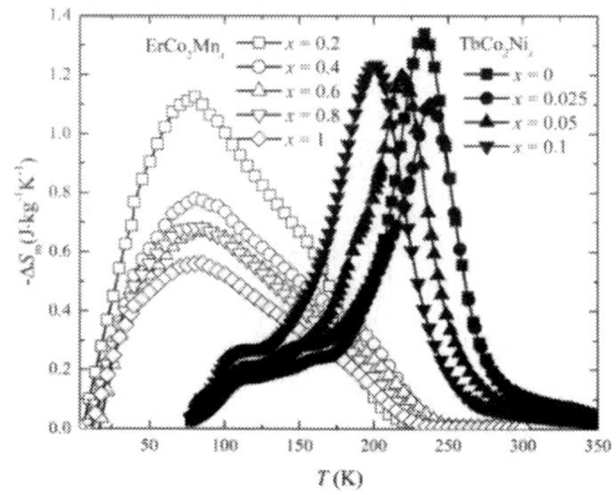

Fig. 6. Temperature dependences of entropy change of $ErCo_2Mn_x$ and $TbCo_2Ni_x$ compounds for magnetic field change $\mu_0\Delta H$ = 1 T.

The magnitudes of $\Delta S_m(T)$ decrease with increasing *M*' concentration because of decreasing the spontaneous magnetic moment of compounds. The behaviour of the $\Delta S_m(T)$ curves strongly differs. For $TbCo_2Ni_x$, $\Delta S_m(T)$ curves have maximum near T_C whereas, for $ErCo_2Mn_x$ maxima are observed at temperatures which are much lower than T_C. Such difference arises because $\Delta S_m(T)$ shape depends on shape of temperature dependence of resultant magnetization which reflects a non-Brillouin behavior of the resultant magnetization of the ferrimagnetically ordered *R* and *M*(*M*') magnetic sublattices.

4 Conclusion

We studied the effect of Ni alloying on the structure, magnetic and magnetothermal properties of $TbCo_2Ni_x$ alloys. It was established that single-phase non-stoichiometric compounds with the cubic $MgCu_2$-type structure are formed at the Ni content $x \leq 0.1$.

We found that concentration dependences of the Curie temperature and magnetic moment of 3d-sublattice have maximum at Ni content x = 0.025. Our analysis shows that all these changes are associated with changes in the Co magnetic moment which can arise from changes of compounds electron band structure.

Using magnetization measurements, we estimated the magnetocaloric effect in the $ErCo_2Mn_x$ and $TbCo_2Ni_x$ compounds. The maximum values of the ΔS_m of $TbCo_2Ni_x$ compounds are sufficiently close for all nickel concentrations.

The research was carried out within the state assignment of FASO of Russia (theme "Magnet" No. AAAA-A18-118020290129-5) and supported by project of UB RAS (No. 18-10-2-5).

References

1. J.L. Wang, C. Marquina, M.R. Ibarra, G.H. Wu, Phys. Rev. B **73**, 094436 (2006)
2. M. Bibekananda, K.G. Suresh, A.K. Nigam, J. Magn. Magn. Mater. **322**, 2415 – 2418 (2010)
3. J.L. Wang, S.J. Campbell, M. Hofmann, M. Hoelzel, R. Zeng, S.X. Dou, S.J. Kennedy, J. Appl. Phys. **111**, 07E334 (2012)
4. N.V. Mushnikov, V.S. Gaviko, E.G. Gerasimov, P.B. Terent'ev, I.A. Tkach, Phys. Met. Metallogr.**110**, 210 – 217 (2010)
5. E.G. Gerasimov, N.V. Mushnikov, P.B. Terentev, V.S. Gaviko, A.A. Inishev, J. Alloys and Compd. **571**, 132 – 137 (2013)
6. E.G. Gerasimov, N.V. Mushnikov, A.A. Inishev, P.B. Terentev, V.S. Gaviko, J. Alloys and Compd. **680**, 359 – 365 (2016)
7. A. Aryal, A. Quetz, S. Pandey, T. Samanta, I. Dubenko, D. Mazumdar, S. Stadler, N. Ali, J. Appl. Phys. **119**, 043905 (2016)
8. A.A. Inishev, E.G. Gerasimov, N.V. Mushnikov, P.B. Terentev, V.S. Gaviko, Phys. Met. Metallogr. **118**, 1059 (2017)

Structural and magnetic anisotropy of directionally-crystallized ferromagnetic microwires

Svetlana Evstigneeva[1], *Alexander* Morchenko[1], *Alexey* Trukhanov[1,2], *Larissa* Panina[1,3], *Vladimir* Larin[4], *Natalia* Volodina[1], *Nikolay* Yudanov[1], *Makhsudsho* Nematov[1], *Hisham* Hashim[1], and *Haseeb* Ahmad[1]

[1]NUST "MISIS", 119049 Moscow, Russia
[2]Scientific Practical Materials Research Centre, NAS of Belarus, 220072 Minsk, Belarus
[3]Institute for Design Problems in Microelectronics RAS, Moscow 124681, Russia
[4]MFTI Ltd, MD2021 Kishinev, R. Moldova

Abstract. Amorphous ferromagnetic microwires have drawn attention primarily due to their excellent soft magnetic properties as the elements of sensors. In this work, semi-hard magnetic microwires of composition $Fe_{4.3}Co_{67.7}Si_{11}B_{14}Cr_3$ were obtained after a process of directional crystallization from amorphous state. The XRD analysis of crushed and whole wires identified the formation of the face-centered Co-modification as the main phase of directionally crystallized core alloy, which explains a large increase in anisotropy and coercivity. The application of a magnetic field during crystallization may orient the easy anisotropy axis of crystallites along the wire. This is confirmed by the investigation of crystallite orientation with respect to the wire axis and X-rays direction in diffractometer. The obtained results revealed that the formation of crystallites in amorphous Co-rich microwires occurs with the predominant orientation of the crystallographic direction [111] along the wire axis and the direction of a magnetic field during the directional crystallization process.

1 Introduction

Amorphous soft ferromagnetic wires exhibiting fast magnetization reversal and giant magnetoimpedance are promising materials for developing sensitive sensors of weak magnetic field, current, mechanical stress for various applications including structural health monitoring. In order to tailoring the operational magnetic parameters, the wires are subjected to different thermal treatments [1]: annealing in a magnetic field and/or under stress, current annealing, etc. After annealing treatment at temperatures close to the crystallization temperature, initially amorphous microwires change their structural and magnetic properties. Thus, the coercivity of Co-rich glass-coated microwires was significantly increased as a result of directional crystallization of the amorphous core [2-3].

The transformation from soft to hard magnetic properties was explained by the formation of crystallites of a hexagonal modification of cobalt [2] and their alignment along the wire during the crystallization process. A more detailed XRD analysis performed in [3] revealed the presence of face-centered modification of cobalt after the crystallization procedure, which could have cubic crystalline anisotropy. The coexistence of face-centered and hexagonal phases of nanocrystalline cobalt in CoSiB amorphous matrix was found in microwires, prepared by a special casting method [4].

In the present work, the mechanisms of coercivity and anisotropy increase in microwires after directional crystallization were investigated. A number of experiments with different orientation of the wire samples with respect to the measuring directions was carried out with the help of diffractometric and magnetometric methods.

2 Experimental procedures

Glass-coated amorphous microwires of the composition $Fe_{4.3}Co_{67.7}Si_{11}B_{14}Cr_3$ were used as initial materials. The metal core diameter and glass shell thickness were about 30 and 3.5 μm, respectively.

In order to form a nano- and/or microcrystalline structure, the wires were subjected to different annealing treatments as described in [2]. We consider that a directed crystallization may occur in the process of annealing at temperatures slightly lower than the primary crystallization temperature T_{cr} when one end of microwires is overheated (by ~60 °C, named as "ignition") due to release of latent heat and propagation of the crystallization front. In accordance with the annealing procedure and resulting coercivity (estimated by the induction method), the wire samples were conditionally classified as explained in Table 1.

Calorimeter DSC 204 F1 Netzsch was used to study the degree of crystallization of the obtained microwires. The sample heating was at the rate of 10 K/min in an

* Corresponding author: dratm@mail.ru

https://doi.org/10.1051/epjconf/201818504022

inert atmosphere. The values of T_{cr} and the Curie temperature T_c of the amorphous state estimated from DSC curves were 490 °C and 360 °C, respectively. The value of T_{cr} was used to set the annealing temperature T_{an}.

X-ray Diffractometer (DRON-3, Co Kα- source) was used to obtain the diffraction patterns. The diffractogramms were detected with different orientation of the wire-samples relative to the X-ray direction. X-ray phase analysis was performed. In addition, Mössbauer spectra were used to study the magnetic texture in microwires.

Table 1. Annealing regimes and sample classification.

Type	The processing method and the state of the wire after processing	Coercive force H_c, Oe (induction test)	Coercive force H_c, Oe (VSM curves)
1	Initial state (as-cast wires)	0.2-0.6	0.5-2.5
2	Crystallized by ordinary annealing at T_{an} =560 °C >T_{cr} for 10 min	~12.5	10-15
3	Not crystallized after ordinary annealing at T_{an}~485 °C <T_{cr} .	<12.5	~3
4	Not crystallized after annealing at T_{an}~485 °C <$T_{cr}$with ignition	<12.5	~5
5	Directionally crystallized (T_{an} ~485 °C <T_{cr} with ignition)	>>12.5	20-90
6	Directionally crystallized (T_{an}~485 °C <T_{cr} with ignition in a magnetic field of 1 kOe)	>>12.5	20-150
7	Partially crystallized (T_{an}~485 °C <T_{cr} with ignition in a magnetic field of 1 kOe)	<12.5	10-20

The magnetization loops were measured with the help of vibrating sample magnetometer (modernized MV-07 and Liquid Helium "CryogenicLTD") for different wire orientation (along and perpendicular to the magnetizing field direction).

3 Result and discussion

Figure 1 presents the DSC curve for as-prepared microwires (type 1) in comparison with those for wires passed the directional crystallization (types 5 and 6) and partial directional crystallization (type 7). The DSC curve for as-prepared wire has two sharp peaks, indicating that the crystallization process from amorphous state proceeds in two stages. The wires after directional crystallization (type 5 and 6) are fully crystalline, because there are no sharp peaks in their DSC curves. The wires of type 7 passed only partial crystallization: there is only one peak in the DSC curve corresponding to the second crystallization stage. The DSC data for samples of types 2-4 are not presented as they are fully consistent with the crystalline state (flat curve) and quasi-amorphous state (broad peak).

Previosly [2, 3], the crystal structure of microwires was studied by X-ray diffraction methods. It was shown that cobalt is the main crystalline microphase in all samples. However, in contrast to [2], we found that this cobalt modification is the face-centered (fcc-Co) which was discussed in detail in [3].

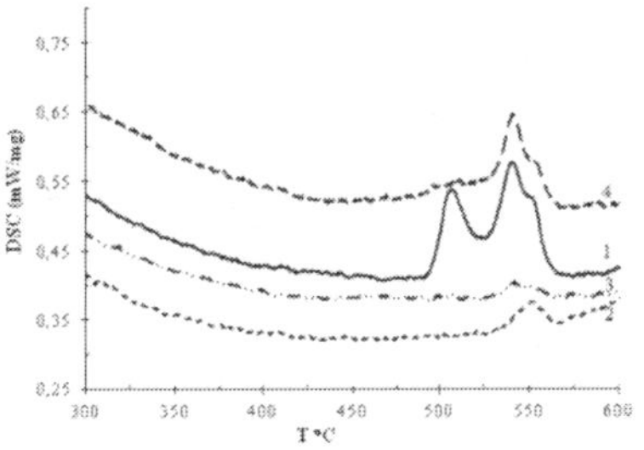

Fig. 1. DSC of glass coated microwires in various state: as-cast (1) and after directional crystallization: type 5 (2), type 6 (3), type 7 (4).

The XRD patterns for microwires after various annealing treatments are presented in Figs. 2-5. XRD-spectra of as-prepared microwires are not shown as they display only the diffuse peak of the Co phase. The annealed microwires reveal the presence of Co, Fe-Co, $CoSi_2$, Fe-Si and Co-B nanosized phases. In Fig. 2, a sharp peak (associated with Co phase) is seen for microwires subjected to standard crystallization (type 2). In the case of wires of type 4 which failed the directional crystallization (Fig.3), the intensity of the peaks decreases and the estimated average size of Co-phase crystallites is in the range of 1.7-1.8 nm, which can be regarded as existence of a quasi-amorphous state. For the wires undergone directional crystallization (types 5), the average grain size of Co increases to 5.0-5.1 nm. In samples additionally subjected to a magnetic field during the directional crystallization, the crystallite size is slightly increased to 5.9-6.1 nm.

The experiments with different arrangement of wire-pieces in the X-ray apparatus relative to the axis of rotation (rocking) revealed the dependence of the XRD patterns on the wire position in the cell (Figs. 2-5).

It is seen that when scanning is performed with the location of the wire bundle along the beam of X-ray, additional peaks arise in the spectra of all the crystalline wires (for example, first reflex in Figs. 2-4 at the same angle 2θ position).

The spectra descend from top to bottom without deviations when angle is increasing, which indicates that the clusters are elongated rather than rounded, and there is a preferable orientation in the wire. The formation of crystallites of the dendrite type was also reported in [2]. The most obvious difference of the diffraction patterns obtained at various positions of the wires relative to the direction of the X-radiation is observed in samples of type 6 (Fig. 4) which were directionally crystallized in the presence of a large magnetic field, indicating the orientational influence of the field.

Additional weak peaks appearing at small reflection angles in the samples undergone directional crystallization (types 5-7), can presumably refer to SiO_2 phases occurring in the glass shell and to the formation of superstructure with doubled lattice parameters (possibly, Fe-B, Co-B, Co-Si and Co-Si-B [4, 5]).

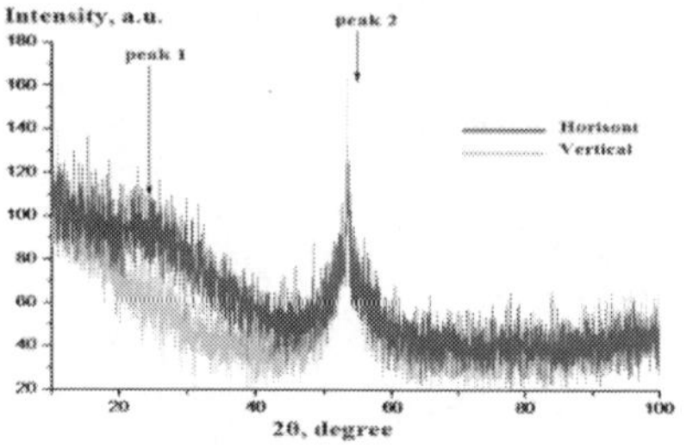

Fig. 2. XRD patterns for microwires of type 2 crystallised by standard annealing treatment (T_{an} =560 °C >T_{cr}).

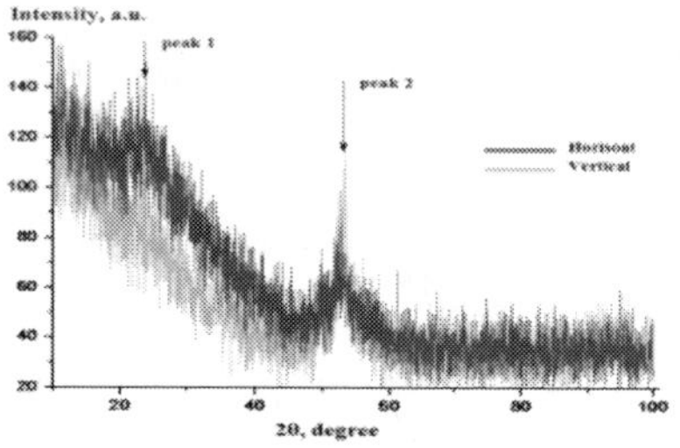

Fig. 3. XRD patterns for microwires of type 4 which failed the directional crystallization.

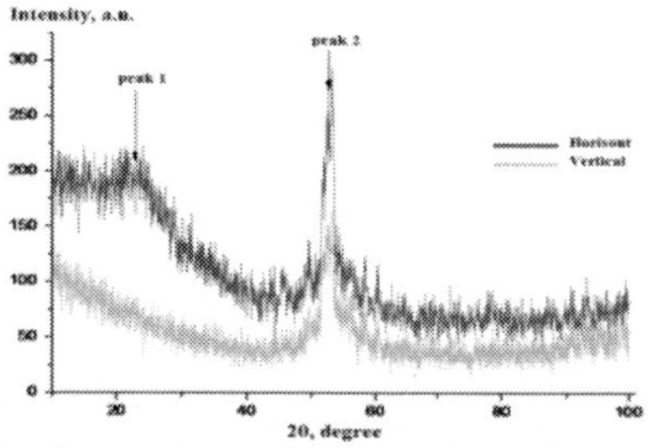

Fig. 4. XRD patterns for microwires after directional crystallization assisted by a dc magnetic field (type 6).

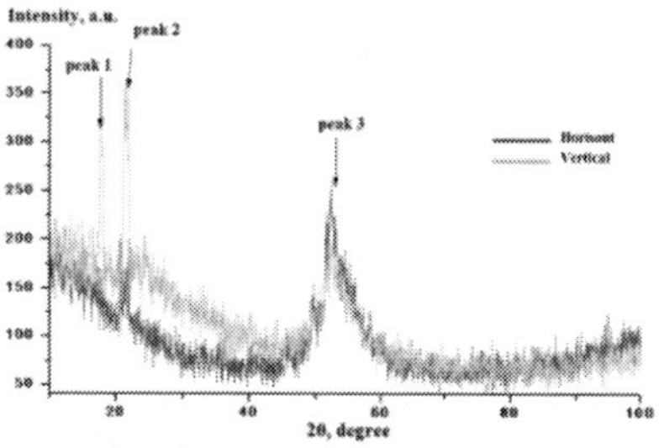

Fig. 5. XRD patterns for microwires after partial directional crystallization assisted by a dc magnetic field (type 7).

The factors that cause the preferential orientation of crystallites can be related to the magnetic field applied in the process of thermal treatment and the demagnetizing fields at temperatures lower than the Curie temperature T_c (which considerably increases in the crystalline state). The other mechanism of oriented crystallization may be related with a change in elastic and magnetoelastic energies. In the process of directional crystallization, the clusters of the nascent crystalline Co phase under the action of a magnetic field are oriented along the wire axis. For these conditions, the total energy of the system will be lowered also by decreasing the magnetostatic energy. Since for cubic crystals with a negative constant of magnetocrystalline anisotropy (K_1), the easy directions of magnetization are crystallographic axes of type <111>, the growing crystallites align their easy axes in the direction of the magnetic field vector, that is along the wire. Simultaneously, the other three axes of the same type will line up at the angle of 70.5° to the wire axis (polar axis), and will be oriented with equal probability in different directions relative to the azimuth angle. Thus, a texture must be formed in the wire, which manifests itself in the appearance of additional complex anisotropy of a 'light cone' type. This assumption is consistent with the analysis of XRD patterns measured for different wire orientation relative to the X-rays direction.

The existence of magnetic texture is also supported by VSM measurements which were made for different wire positions with respect to the magnetizing field. Figure 6 presents the hysteresis loops of the as-prepared microwires for the two orientations. When the magnetizing field is parallel to the wire, the hysteresis loop is almost rectangular with the remanence close to saturation and very small coercivity of 0.5-2.5 Oe. This confirms the existence of the axial anisotropy almost in the entire wire.

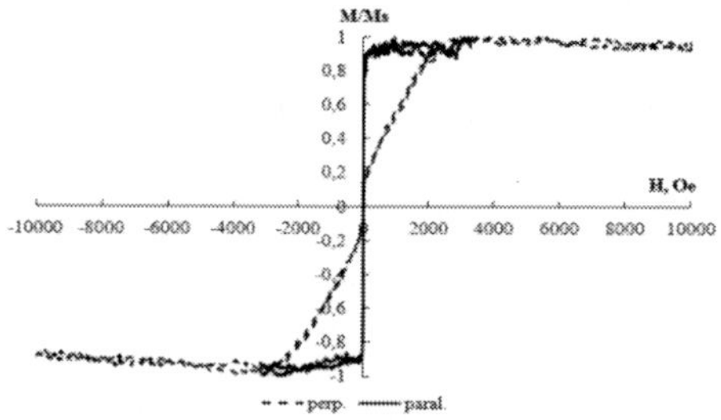

Fig. 6. Hysteresis loops of ferromagnetic microwire of type 1 with Rayleigh's correction.

In amorphous glass-coated microwires with positive magnetostriction, the axial magnetic anisotropy is formed due to a predominant axial tensile stress. In the surface layer, a radial closure domain structure may exist, but its effect on the total axial hysteresis is small. The remagnetization in the perpendicular field is described by an inclined curve with small hysteresis which could be due to transverse domain structure. Much higher saturation fields for perpendicular magnetization are caused by the demagnetization effect.

The effect of the perpendicular demagnetizing field on the magnetization can be eliminated applying the Rayleigh parabolic model of hysteresis [6]. The curve is shifted along the H-axis by the amount of the demagnetizing field. Then, the value of the average axial anisotropy constant K_{an} can be estimated as the area between the hysteresis loop for positive values of the magnetizing field H and y-axis [7]. The corrected magnetization curve for the wire sample of type 5 (directionally crystallized without a magnetic field) is given in Fig. 7. This curve is used to estimate the average axial anisotropy field which is found to be within 500-600 Oe. The wires directionally crystallized in the presence of the external field have larger anisotropy field in the range of 0.8-1 kOe as found from corrected curve of Fig. 8. If the wires have only partial crystallization (type 7), the average anisotropy field drops below 100 Oe (see Fig. 9). The insert in Fig. 9 demonstrates the hysteresis shape for parallel magnetization which shows the presence of two magnetic phases with low (<15 Oe) and high (~200 Oe) coercivities.

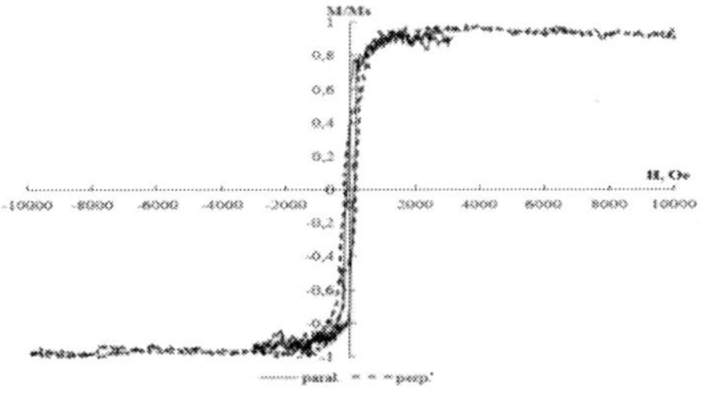

Fig. 7. Hysteresis loops of ferromagnetic microwire of type 5 with Rayleigh's correction.

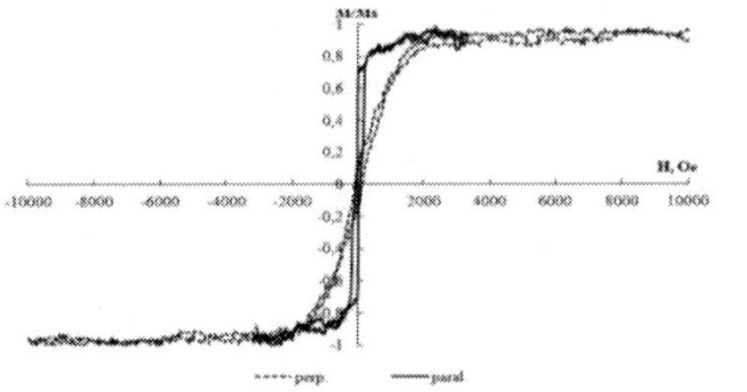

Fig. 8. Hysteresis loops of ferromagnetic microwire of type 6 with Rayleigh's correction.

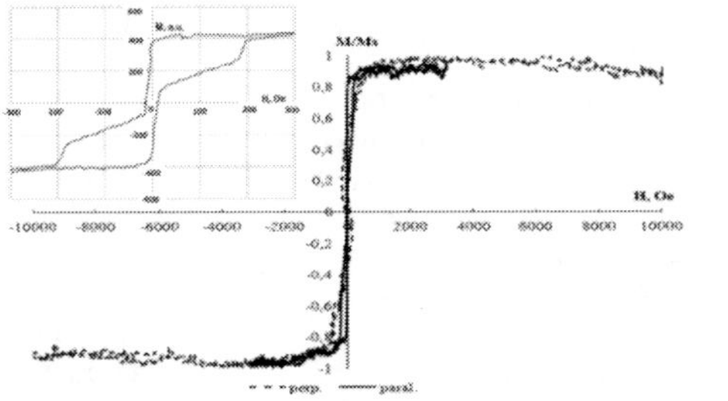

Fig. 9. Hysteresis loops of ferromagnetic microwire of type 7 with Rayleigh's correction. The inset shows an enlarged image of the hysteresis loop for parallel magnetization.

The estimated values of the anisotropy field H_K can be compared with those due to the magnetocrystalline anisotropy of <111> type (fcc-Co): $H_K = |4K_1|/(3M_s)$. Taking the magnetic parameters for crystalline Co: $K_1 =$ –700 kerg/cm^3 and $M_s =$ 1.5 kGs [8] ($T \ll T_c = 1121°C$), the anisotropy field is about 620 Oe at room temperature (and ~430 Oe at T_{cr}). This value is comparable with those estimated from Figs. 7, 8 and confirms the formation of predominately fcc-Co phase. A larger value of the anisotropy found for wires of type 6 may be due to additional contribution from induced strains.

Thus, we can estimate the average anisotropy field and the anisotropy constant of the crystallite matter. Comparing the obtained values with the literature data it is possible indirectly confirm the formation of a particular phase in the process of directional crystallization of ferromagnetic microwires.

4 Conclusion

It is established that the phase composition of initially amorphous ferromagnetic Co-enriched microwires changes as a result of their annealing treatment at the crystallization conditions. The main phase of the crystallized samples is the face-centered cobalt. Also it was found that the crystallites have an elongated shape and there exists a preferential orientation along the wire. This tendency is stronger in wires undergone a directional crystallization in the presence of a magnetic field.

The easy magnetization axes of crystallites correspond to the crystallographic directions of the <111> type. Due to the effect of orientating action of the magnetic field, the crystallites gross direction coincides with one of the crystallographic direction [111], i.e. along the wire.

The magnetometry of the samples carried out for different orientations relative to the field direction of the electromagnet, made it possible to detect a noticeable anisotropy, which confirms the conclusion about the formation of a texture in the directionally crystallized wires and explains the appearance of hard magnetic state.

L. V. Panina acknowledges the support for this work under the Russian Federation State Contract for organising a scientific work (grant No 3.8022.2017).

References

1. Rodionova V., Chichay K., Zhukova V., et al., J. Supercond. Nov. Magn. **28**, 977 (2015).
2. Larin V.S., Panina L.V., Patroi E. A., et al., Phys. Status Solidi (a) **213**, 372 (2016).
3. Morchenko A. T., Panina L. V., Churyukanova M. N., et al., J. Alloys Compd. **698**, 685 (2017).
4. Rodionova V., Baraban I., Chichay K., et al., *Moscow International Symposium on Magnetism (MISM). Book of Abstracts*, 466 (2017).
5. Životský O., Titov A., Jirásková Y., et al., J. Alloys Compd. **681**, 402 (2016).
6. Bertotti G., *Hysteresis in Magnetism* (Academic Press, San Diego, 1998).
7. Herzer G. Acta Mater. **61**, 718 (2013).
8. Takahashi M., Kim C.O., Koshimura M., Suzuki T., Jpn. J. Appl. Phys. **17**, 1911 (1978).

Competing anisotropy in the $(Tm_xPr_{1-x})_2Fe_{17}$ system

Sergey Platonov [1,*], *Anatoly* Kuchin [1], *Alexey* Volegov [2], *Alexander* Korolev [1], *Dmitriy* Neznakhin [2], *Vladimir* Voronin [1], *Natal'ya* Proskurnina [1], and *Denis* Kolodkin [1]

[1]M.N. Miheev Institute of Metal Physics, Ural Division of RAS, 620219, 18 S. Kovalevskaya Str., Ekaterinburg, Russia
[2]Ural Federal University named after First President of Russia B.N. Yeltsin, 620002, 19 Mira Str., Ekaterinburg, Russia

Abstract. The magnetization curves of magnetically aligned finely powdered samples of the $(Tm_xPr_{1-x})_2Fe_{17}$ compounds have been measured at 4 K. The easy magnetization axis is oriented in the basal plane or along the hexagonal axis for the compounds with x = 0-0.3 and 0.7-1, respectively. This is because of the absence of magnetic ordering in the Tm and Pr subsystems in these ranges, respectively, and because of competing anisotropy of the subsystems. For the compositions with x = 0.4-0.6, both rare-earth subsystems are magnetically ordered and the easy magnetization axis is oriented between the basal plane and the hexagonal axis. The critical fields of FOMPs decrease quickly as the Pr or Tm content decreases in the ranges 0-0.3 and 0.7-1, respectively. The magnetization anisotropy also diminishes as the Tm content becomes smaller than x = 0.7. No influence of the intrinsic microdeformations on the magnetization of the compounds was detected.

1 Introduction

The Pr_2Fe_{17} and Tm_2Fe_{17} compounds exhibit different crystal structures of the Th_2Zn_{17}- and Th_2Ni_{17}-type and parallel or antiparallel coupling between the moments of the Pr-, Tm- and Fe-sublattices, respectively [1,2]. Pr_2Fe_{17} is a ferromagnet easily magnetized in the basal plane. Tm_2Fe_{17} is unique among the R_2Fe_{17} compounds because it is an easy-axis ferrimagnet below 77 K; at higher temperatures, the magnetization vector lies in the basal plane, and at 234 K, the ferrimagnet transforms to a helimagnet. The temperatures of magnetic ordering equals 284 K for Pr_2Fe_{17} and 274 K for Tm_2Fe_{17}. Thus, a competing anisotropy exists and a concentration spin-reorientation process has to take place in the $(Tm_xPr_{1-x})_2Fe_{17}$ system at low temperatures. Both the Pr_2Fe_{17} and Tm_2Fe_{17} binaries are characterized by FOMP (first order magnetization process) along the c-axis and in the basal plane, respectively. The magnetization anisotropy in Tm_2Fe_{17} takes place at T = 4 K. Microdeformations in the $(Tm_xPr_{1-x})_2Fe_{17}$ alloys with x = 0.6-0.9 have been earlier detected by means of neutron diffraction at room temperature [3]. Feasibly, these microdeformations stipulate a nonmonotonous variation of the temperatures of magnetic phase transitions in the $(Tm_xPr_{1-x})_2Fe_{17}$ system. It is not clear whether these microdeformations influence the magnetization of the compounds or not. Thus, the properties of the Pr_2Fe_{17} and Tm_2Fe_{17} binary compounds are quite different. In the present paper, the magnetization curves of the $(Tm_xPr_{1-x})_2Fe_{17}$ compounds measured along the hexagonal axis and in the basal plane are studied.

2 Experimental details

Polycrystalline ingots of the $(Tm_xPr_{1-x})_2Fe_{17}$ compounds with x = 0, 0.1, 0.2, 0.3, 0.4, 0.5, 0.6, 0.7, 0.75, 0.8, 0.85, 0.9, 1 were prepared by induction melting [3]. The ingots were homogenized at 1293 K for two weeks and then quenched in water. The phase composition of the samples was investigated by the methods of neutron and X-ray diffraction. The compounds investigated crystallize into the hexagonal Th_2Ni_{17}–type and rhombohedral Th_2Zn_{17}-type structures, the fraction of free α-Fe being no larger than 1-3 wt.%. The MPMS technique was used for magnetic measurements in magnetic fields up to 7 T. The magnetization parallel or perpendicular to the c-axis of the magnetically aligned samples was measured at 4 K. To obtain optimal magnetic alignment, the finely powdered samples were magnetically aligned by means of the sample-rotation magnetic alignment technique, as in Ref. [4]. This alignment was performed at the temperatures T=270-272 K which are somewhat lower than the temperatures of the magnetic ordering of all the alloys except for the alloys with x=0.7-0.8 [3]. The X-ray diffraction patterns indicate a good magnetic alignment in the case of the single-phase compounds with x=0-0.3 (rhombohedral structure) and x=0.8-1 (hexagonal structure). A working magnetic alignment was realized in the two-phase compounds with x=0.7, 0.75. The magnetic alignment in the compounds with x = 0.4-0.6 was not successful, as it follows from the X-ray diffraction data.

* Corresponding author: platonov@imp.uran.ru

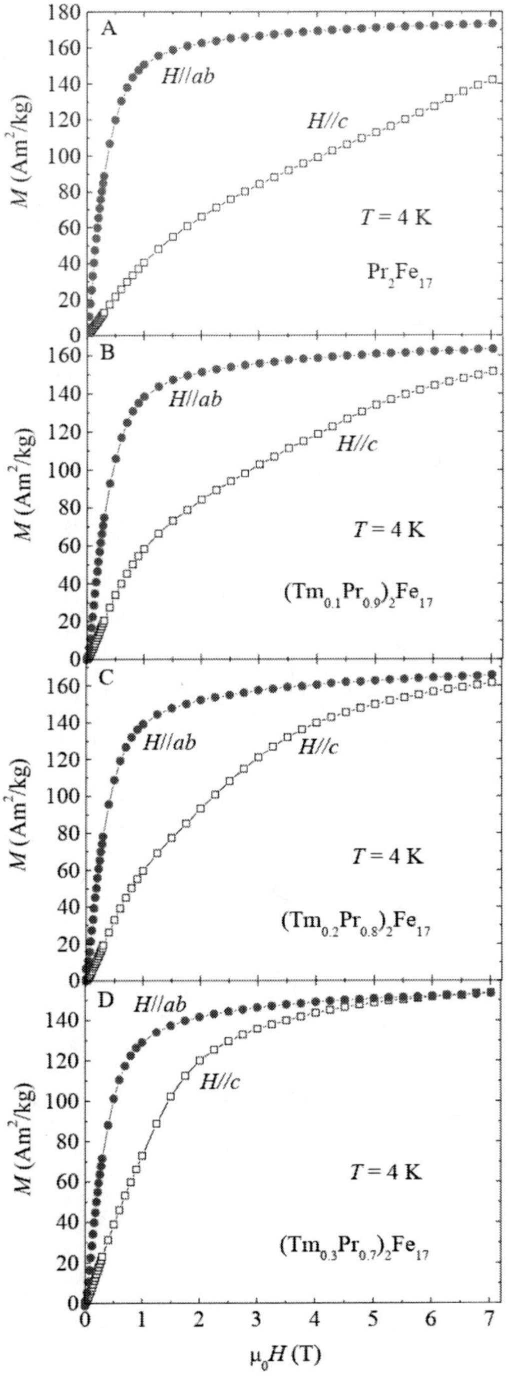

Fig. 1. Magnetization measured at 4 K on magnetically aligned $(Tm_xPr_{1-x})_2Fe_{17}$ in external field applied along the *ab* plane (●) and the *c*-axis (□), *x*=0 (A), 0.1 (B), 0.2 (C), 0.3 (D).

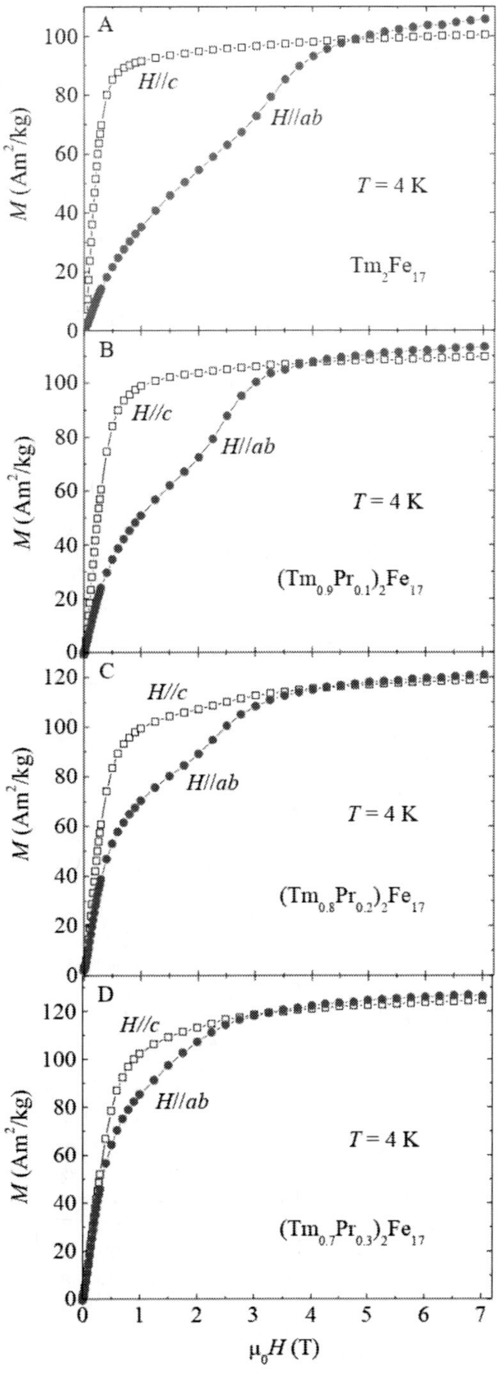

Fig. 2. Magnetization measured at 4 K on magnetically aligned $(Tm_xPr_{1-x})_2Fe_{17}$ in external field applied along the *ab* plane (●) and the *c*-axis (□), *x*=1 (A), 0.9 (B), 0.8 (C), 0.7 (D).

3 Experimental results and discussion

The magnetization curves of the compositions with x=0-0.3 and 0.7-1 are shown in Figs. 1 and 2, respectively. The magnetization curves for the binary compounds are about the same as in Ref. [4]. One can suppose from the magnetization data (Figs. 1, 2) that the magnetic anisotropy in the compositions with x=0-0.3 and 0.7-1 is an easy-plane type or an easy-axis type, respectively, or close to them. Such different orientations of the easy-magnetization axis in the $(Tm_xPr_{1-x})_2Fe_{17}$ system is due to the competing anisotropy of the Tm and Pr subsystems. The FOMP with lower critical fields as compared with the binary compounds is retained for the compositions x=0.1, 0.2, and 0.7-0.9. The magnetization anisotropy of Tm_2Fe_{17} decreases when Pr is substituted for Tm. This result confirms the supposition from Ref. [4] that the magnetization anisotropy of Tm_2Fe_{17} is due, in particular, to canting of the magnetic moments in the Tm-subsystem under the external field applied along the basal plane.

The magnetization curves of the magnetically aligned compositions with x = 0.4-0.6 are about the same along different directions, i.e. these samples are truly polycrystalline. This is in agreement with the data of X-ray diffraction testifying to the absence of magnetic alignment in the samples with x = 0.4-0.6. The non-single-phase state of the compounds with x = 0.5, 0.6 and a harmful effect of magnetic moments of Tm at x = 0.4 (see below for more details) can be the reasons for this unsuccessful alignment. The orientation of the easy-magnetization axis of the compounds with x=0.4-0.6 was not obtained from our magnetization data. Apparently, the easy-axis orientation in the $(Tm_xPr_{1-x})_2Fe_{17}$ compounds with x = 0.4-0.6 changes gradually with the composition variation in accordance with the content of the Th_2Zn_{17}- and Th_2Ni_{17}-type structures. Feasibly, the magnetic neutron diffraction technique will be useful to study the easy-axis orientation in the compositions with x = 0.4-0.6.

Fig. 3 shows the concentration dependences of the magnetization M of the $(Tm_xPr_{1-x})_2Fe_{17}$ compounds measured along the c-axis and in the basal plane in a field of 7 T at T = 4 K. The M values along these two different directions are close to each other for x = 0.2-1. The noticeable difference appears only for the compositions with x=0 and 0.1 because of their large magnetic anisotropy (Fig. 1), as is for the Pr_2Fe_{17} compound in Ref. [4]. As a result, the magnetization along the basal plane decreases monotonically with increasing x, whereas the magnetization along the c axis does not change monotonically in the range x = 0-0.3. We suppose that the spontaneous magnetic moments of the Pr and Fe subsystems lie in the basal plane for the compounds with x = 0-0.3. The Tm subsystem is not magnetically ordered in the compounds with x = 0.1-0.3, as one can suppose from the data on Fig. 1. The Tm magnetic moment has to be oriented against the magnetic moments of the Fe and Pr ions in the basal plane, whereas the easy-axis type magnetic anisotropy of the Tm ions causes the orientation of the Tm moments along the c axis. Apparently, the competition of these two interactions causes the spontaneous orientation of the Tm moments at an angle to the c-axis for x = 0.1-0.3, and the Tm moments create a fan around the c-axis. The magnetic field applied along the c-axis deflects the moments of the Pr and Fe ions from the basal plane and change the angle of the fan of the Tm moments. This picture allows us to explain the non-monotonic variation of the magnetization along the c-axis in a field of 7 T (Fig. 3). The decrease in the contribution from the Pr subsystem to the total magnetization is compensated for or even exceeded by the increase in the contribution from the Tm subsystem for the compositions with x = 0.1-0.3.

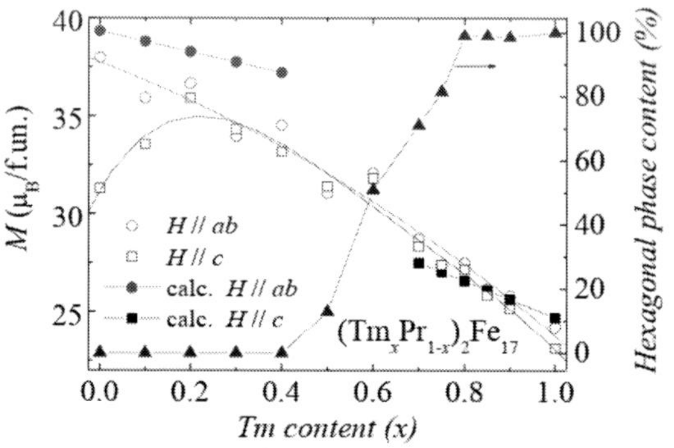

Fig. 3. Concentration dependences of the magnetization M of the $(Tm_xPr_{1-x})_2Fe_{17}$ compounds measured along the ab plane (o) and the c-axis (□) at T = 4 K in a field of 7 T. The M values for x = 0.4-0.6 were measured for the polycrystalline samples (see text). The estimated M values along the ab plane (●) and the c-axis (■) are presented, see text for details. The percentage content of the Th_2Ni_{17}-type hexagonal phase is shown by (▲) [3].

Let us estimate the concentration variation of the magnetization of the $(Tm_xPr_{1-x})_2Fe_{17}$ compounds measured in a field of 7 T applied along the c axis and in the basal plane. We suppose that the spontaneous magnetization lies in the basal plane or along the c axis (or close to these crystallographic directions) for the compounds with x=0-0.3 and 0.7-1, respectively. Such supposition can be made from the $M(H)$ curves in Figs. 1 and 2. We take into account a real composition, $Tm_2Fe_{19.35}$, for the compounds with x=0.7-1, as it was obtained by means of neutron diffraction technique in Ref. [3]. The magnetic moments of the Pr, Tm and Fe ions were taken as 2.66 μ_B [1], 7 μ_B [2] and 2 μ_B, respectively. As can be seen from Fig. 3, there is a good fit of the estimated data to the experimental ones for the compounds with x = 0.7-1 measured along the c axis (along the basal plane in the case of x=1 because of the strong magnetization anisotropy). The estimated data of the magnetization measured along the basal plane in the compounds with x = 0-0.3 are larger than the experimental ones, perhaps, because of the neglect of the magnetic anisotropy of the Pr ions in the basal plane and contribution from the Tm sublattice. The large difference between the calculated and the experimental data for x=0.4 can be explained by the orientation of the easy magnetization axis of the compound at an angle to the basal plane. This is an interesting result because the composition with x=0.4 crystallizes in the rhombohedral phase with the intrinsic easy-plane type magnetic anisotropy. The result can be explained by the contribution of the easy-axis type anisotropy of the Tm ions, which substitute in part for the Pr ions in the rhombohedral structure and have a much larger magnetic moment.

Microdeformations do not influence the magnetization of these alloys (as one can see from Fig. 3) unlike the early obtained strong decrease of the temperatures of the magnetic phase transitions for the compositions with x = 0.6-0.9 [3]. This result confirms the supposition from Ref. [3] that an inhomogeneous distribution of Pr and Tm atoms in the defected hexagonal lattice and the large difference of their atomic radiuses can be the reasons for the nonmonotonic variation of the temperatures of the magnetic phase transitions in the system. It is well known that the temperatures of the magnetic phase transitions in the R_2Fe_{17} compounds strongly depend on the interatomic Fe-Fe distances [1-3].

4 Conclusions

It has been established that the easy magnetization axis in the $(Tm_xPr_{1-x})_2Fe_{17}$ compounds is oriented in the basal plane for x = 0-0.3 or along the hexagonal axis for x = 0.7-1 at T = 4 K. This is because of the absence of magnetic ordering in the Tm and Pr subsystems in these ranges, respectively, and because of competing anisotropy of the subsystems. For the compositions with x = 0.4-0.6, both these rare-earth subsystems are magnetically ordered and the easy magnetization axis has to be oriented between the basal plane and the hexagonal axis. To study the easy axis orientation in the compositions with x = 0.4-0.6 it is necessary to use the magnetic neutron diffraction technique. The critical fields of FOMPs decrease quickly as the Pr or Tm content decreases in the ranges 0-0.3 and 0.7-1, respectively. The magnetization anisotropy also diminishes as the Tm content becomes smaller x = 0.7. No influence of the intrinsic microdeformations on the magnetization of the compounds was detected unlike the early obtained strong decrease of the temperatures of the magnetic phase transitions for the compositions with x = 0.6-0.9 under the action of these microdeformations [3]. This result confirms the supposition from Ref. [3] that an inhomogeneous distribution of Pr and Tm atoms in the defected hexagonal lattice and the large difference of their atomic radiuses can be the reasons for the nonmonotonic variation of the temperatures of the magnetic phase transitions in the system.

The work was done within RAS Program "Magnet" (project № AAAA-A18-118020290129-5).

References

1. O. Isnard, S. Miraglia, J.L. Soubetoux, D. Fruchart, J. Pannetier, Phys. Rev. B **45**, 2910 (1992-II)
2. O. Isnard, A.V. Andreev, M.D. Kuz'min, Y. Skourski, D.I. Gorbunov, J. Wosnitza, N.V. Kudrevatykh, A. Iwasa, A. Kondo, A. Matsuo, K. Kindo, Phys. Rev. B **88**, 174406 (1992-II)
3. A.G. Kuchin, S.P. Platonov, V.I. Voronin, W. Iwasieczko, A.V. Korolev, A.S. Volegov, D.S. Neznakhin, A.V. Protasov, I.F. Berger, D.A. Kolodkin, N.V. Proskurnina, J. Alloys Compd., **726**, 330 (2017).
4. X.C. Kou, F.R. de Boer, R. Grössinger, G. Wiesinger, H. Suzuki, H. Kitazawa, T. Takamasu, G. Kido, J. Magn. Magn. Mater., **177-181**, 1002 (1998)

Static and high-frequency magnetic properties of nanocrystalline Fe-Zr-N films

Elena Sheftel[1], *Eugene* Harin[1], *Valentin* Tedzhetov[1,*], *Konstantin* Rozanov[2], *Polina* Zezyulina[2], *Stanislav* Bobrovskii[2], and *Philipp* Kiryukhantsev-Korneev[3]

[1]A.A. Baikov Institute of Metallurgy and Materials Science RAS, Moscow, Russia
[2]Institute for Theoretical and Applied Electromagnetics RAS, Moscow, Russia
[3]National University of Science and Technology "MISIS", Moscow, Russia

Abstract. The phase and structural states of nanocrystalline Fe-based films alloyed with Zr and N, which were prepared by reactive magnetron sputtering under different conditions, magnetic-structure parameters of the films and their static and high-frequency magnetic properties have been studied. The interrelation of static magnetic properties and magnetic-structure parameters with the value of real effective magnetic permeability μ' and frequency range, for which the value is unchanged, has been studied.

1 Introduction

The designing of miniature high-frequency induction electronic devices used for radiocommunication, data storage, etc., which operate in the high-frequency (up to gigahertz range) ac magnetic fields, is the today's trend of the modern microelectronics. Film amorphous and nanocrystalline soft magnetic materials intended for these devices should exhibit the high saturation magnetization M_s, high magnetic permeability μ under static conditions and in high-frequency ac magnetic fields, and high frequency of natural ferromagnetic resonance (nFMR) f_r, which should exceed operating frequencies of electronic device.

The ferromagnetic resonance effect in a film ferromagnet characterized by both high saturation magnetization M_s and effective magnetic anisotropy field H_{eff} can be shifted to the high-frequency (HF) range; in this case, the films should be characterized by high magnetic permeability $\mu = M_s/H_{eff}$ in the HF spectral range $f_r = \gamma(4\pi M_s H_{eff})^{1/2} = \gamma H_{eff} (4\pi\mu)^{1/2}$, where $\gamma = 2.9$ MHz/Oe [1-4]. Hence, one can see that these requirements for a material are conflicting, namely, to increase f_r, the H_{eff} should be increased; in this case, a decrease in $\mu = M_s/H_{eff}$ takes place.

It is known that H_{eff} is determined by the magnetic structure of a film ferromagnet [5, 6], which in its turn, is determined by preparation and heat-treatment conditions. Investigation of a correlation between the magnetic structure and both static and HF properties, in particular magnetic permeability μ, as well as the film ferromagnet preparation conditions pays the way to optimization of the magnetic anisotropy field H_{eff}.

Among soft-magnetic film materials studied in recent years, nanocrystalline Fe-Zr-N system films are of substantial interest for investigators. This is related, on the one hand, to the possibility to prepare such films in amorphous and nanocrystalline states, in particular, by magnetron sputtering, and, on the other hand, to the prospect to prepare such films with unique static and HF soft-magnetic properties [7-11].

The aim of the present work is to study the correlation between static magnetic properties, magnetic-structure parameters and a value of real effective magnetic permeability μ' in frequency range, in which μ' is unchanged (nFMR), of Fe-Zr-N films prepared by magnetron sputtering under different conditions.

2 Experimental

The Fe-Zr-N films were prepared by magnetron sputtering of the Fe-2.4 wt. % Zr target being a Fe disc with evenly distributed Zr chips. In the magnetron chamber under exposure to plasma flame there occurred surface melting of Fe and Zr forming different alloys of Fe-Zr system in the erosion zone. Before each sputtering, the target was heated above Curie temperature using plasma at which sputtering process was held. Deposition took place in gas atmosphere of Ar + 5 [%] N_2 or Ar + 15 [%] N_2 under total pressure of p_{Ar+N2} 0.3 Pa. As substrates, we used glass plates with the thickness of $d_s = 0.4$ mm. The thickness of prepared films was estimated by glow-discharge optical emission spectroscopy [12], which was performed on a Horiba Jobin Yvon Profiler-2 instrument; the thickness was $d_f = 1.2\pm0.2$ μm.

The phase composition and fine structure (the coherent domain size corresponding to the grain size $2R_c$ and the microstrain ε at grain scale) were studied by X-ray diffraction (XRD) analysis using a Rigaku Ultima IV diffractometer equipped with a graphite monochromator, CuK_α radiation, Bragg-Brentano geometry, and original

* Corresponding author: velmurad@ya.ru

software for full-profile Rietveld analysis [13].

Magnetic hysteresis loops were measured in magnetic fields up to 16 kOe using a LakeShore 7407 vibrating-sample magnetometer. The error of measurement of saturation induction, which is related to the difference in the shape and dimensions of Ni standard sample (3 mm in diameter) and used samples, did not exceed ~10%. Magnetic structure parameters were estimated by correlation magnetometry [6].

The microwave permeability of the films is measured by the stripline technique [14] using square-shaped plate samples 23 mm in the plate dimension. The technique involves calibration measurements of two samples with a known permeability to eliminate the measurement uncertainty appeared because of intrinsic non-uniformities in the measuring setup. The measurements are made in the frequency range of 0.1 to 2 GHz in two perpendicular directions (x and y). From the measured effective permeability of the sample, $\mu_{measured}$, the intrinsic permeability of the film, $\mu_{intrinsic}$, is found by equation $\mu_{measured} = 1 + p(\mu_{intrinsic} - 1)$, where $p = d_f/d_s \approx 0.003$ is the content of the ferromagnetic constituent in the sample.

3 Results and discussion

3.1 Phase composition and structure

According to XRD data, the films under study are either amorphous (in terms of X-ray diffraction) or nanocrystalline (Fig. 1), depending on the nitrogen content in the magnetron chamber. At the nitrogen content of 5%, the phase composition of prepared film consists of two nitride nanocrystalline phases, such as Fe_4N and Fe_3N with a grain size of 3-6 nm. As the nitrogen content in the magnetron chamber increases from 5 to 15%, the amorphous structure enriched in Fe is formed in the films (Table 1, Fig. 1a). According to our previous magnetic studies [15], the Fe-Zr-N films with the amorphous structure, which gives a wide reflection in X-ray diffraction patterns, consist of α-Fe-based ferromagnetic grains less than 2 nm in size.

Table 1. Phase composition and fine structure of studied films.

N_2 during deposition	Phase composition, vol. %	Lattice parameter, Å	Grain size $2R_c$, nm	Micro-strain ε, %
15%	XRD amorphous	–	–	–
5%	Fe_4N 12.0±0.4	3.797±0.004	5.5±0.8	0.3±0.8
	Fe_3N 88.0±0.4	4.711±0.005 4.291±0.005	3.1±0.1	0.3±0.2

3.2 Static magnetic properties and magnetic structure

As is known, the origin of static coercivity and magnetic permeability in nanocrystalline ferromagnets is explained by a random magnetic anisotropy model [5,6]. According the model the local effective magnetic anisotropy at the grain scale ($D^{1/2}H_a$) is suppressed when the grain size $2R_c$ is less than the exchange interaction length R_L. In this case, uniform magnetization domains $2R_L$ (the stochastic domains) determined by exchange interaction are formed in the nanocrystalline ferromagnet and its magnetic hysteresis in low magnetic fields is determined by the magnetic anisotropy field at the stochastic domain scale ($D^{1/2}\langle H_a\rangle$). Note, that several the stochastic domain with different $D^{1/2}\langle H_a\rangle$ magnitude can be formed in the nanocrystalline ferromagnet. Here, we report results of determination of $D^{1/2}H_a$ and $D^{1/2}\langle H_a\rangle$ magnitudes and static magnetic properties of studied films.

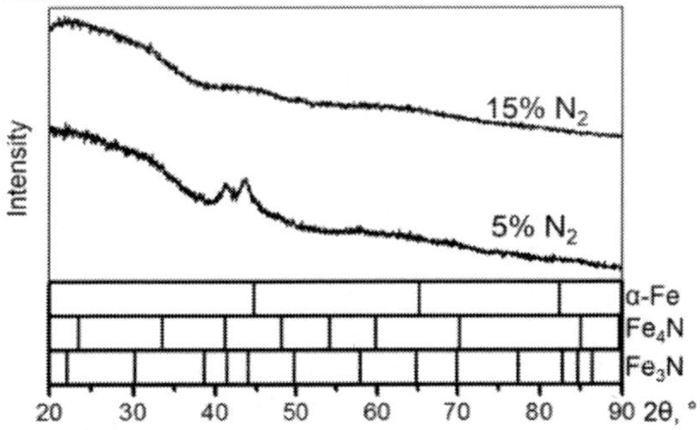

Fig. 1. XRD patterns of studied films deposited at the nitrogen content in the magnetron chamber of 15 and 5% N_2.

The shapes of hysteresis loops (Fig. 2) allow us to find the simultaneous presence of, at least, two main magnetic anisotropies differing in the value and symmetry. This is indicated by the change in the magnetization curve slope inwards the hysteresis loop in the first and third quadrants (Figs. 2a and 2b) and extended hysteresis-free portion of magnetization curve, which corresponds to the magnetization of more than half of film volume (Fig. 2a).

Magnetization curves in high magnetic fields are described by law of approach to magnetic saturation in terms of correlation magnetometry method [16]

$$M(H) = M_s[1 - (1/2)(D^{1/2}H_a)^2 / (H^2 + H^{1/2} H_R^{3/2})]. \quad (1)$$

Using Eq. (1) for the films under study, the saturation magnetization (M_s), the root-mean-square fluctuation of effective local (at the grain scale $2R_c$) anisotropy field ($D^{1/2}H_a$) and exchange field (H_R) were determined (Table 2). These magnitudes were used to estimate the first (of two anisotropy fields found from the hysteresis loop shape) anisotropy field of stochastic domain (Table 2): $D^{1/2}\langle H_a\rangle_1 = (D^{1/2}H_a)^4/H_R^3$.

Magnetization dispersion curves $1-M(H)/M_s$ plotted on log-log coordinates (insets on Fig. 3) indicate three asymptotes: (*I*) $1 - M(H) / M_s = (1 / 2) (D^{1/2}\langle H_a\rangle_2 / H)^2$, (*II*) $1 - M(H) / M_s = (1 / 2) (D^{1/2}H_a)^2 / (H^{1/2}H_R^{3/2})$ and (*III*) $1 - M(H) / M_s = (1 / 2) (D^{1/2}H_a / H)^2$. The curve connected the second and third asymptotes (*IV*) is described by Eq. (1). The second root-mean-square fluctuation of effective anisotropy field of stochastic domain $D^{1/2}\langle H_a\rangle_2$ (Table 2) was estimated using asymptote I.

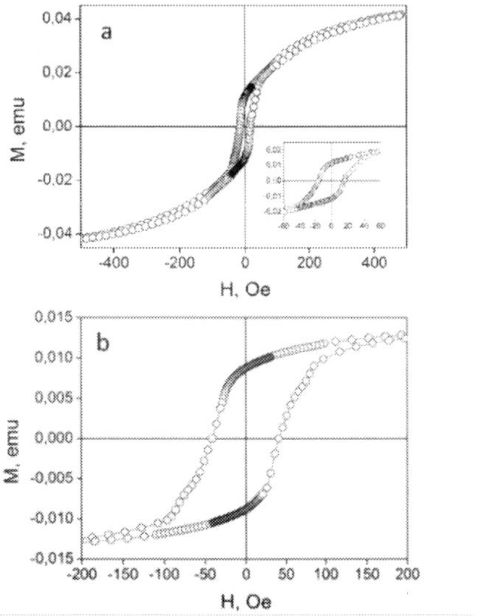

Fig. 2. The hysteresis loops of the studied films prepared with (a) 15 and (b) 5% N_2 in the magnetron chamber atmosphere.

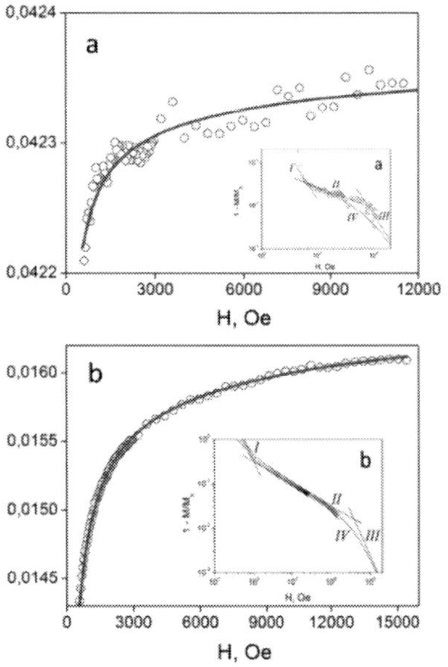

Fig. 3. High-field magnetization curves for the studied films prepared at a nitrogen content of (a) 15 and (b) 5% in the magnetron-chamber atmosphere. Solid lines are fitting by Eq. (1). Insets: magnetization dispersion curves ($1-M/M_s$) for the studied films prepared at a nitrogen content of (a) 15 and (b) 5% in the magnetron-chamber atmosphere. Solid lines are described in the text.

This means that the magnetic hysteresis in the studied films is induced by rotation of stochastic domains with different anisotropy filed magnitudes ($D^{1/2}<H_a>_1$ and $D^{1/2}<H_a>_2$, Table 2). In this case, in accordance with theoretical estimations [6], the coercive field of the film should be equal (to a certain approximation) to the anisotropy field of stochastic domains abundantly present in the magnetic structure.

Table 2. Parameters of magnetic structure of studied films.

N_2 content during depo-sition	M_s, G (T)	H_c, Oe	$D^{1/2}$ $<H_a>_1$, Oe	$D^{1/2}$ $<H_a>_2$, Oe	$D^{1/2}$ H_a, Oe	H_R, Oe
15%	1060 ±270 1.3 ±0.3	16 ±1	0.03 ±0.01	90 ±70	350 ±150	8000 ±4500
5%	360 ±30 0.46 ±0.04	41 ±3	39 ±14	80 ±10	6900 ±400	38600 ±2700

3.3 High-frequency magnetic properties

The microwave permeability of the sample deposited in the atmosphere containing 5% N_2 is negligible as compared to the measurement error and is not discussed below. The measured microwave permeability of the sample produced at 15% nitrogen content is shown in Fig. 4 for two mutually perpendicular orientation of the film. The magnetic properties of the film are slightly anisotropic in the film plane. The magnetic loss peak observed at frequencies of about 700 MHz is due to the nFMR arising at the effective anisotropy field H_{eff} [8]. For thin films, the relation between nFMR frequency f_r and the effective anisotropy field is given by Kittel's equation,

$$f_r = \gamma(4\pi M_s H_{eff})^{1/2} \quad (2)$$

where γ is gyromagnetic ratio. The values of H_{eff} estimated by Eq. (2) from the measured location of the magnetic loss peak are given in Table 3.

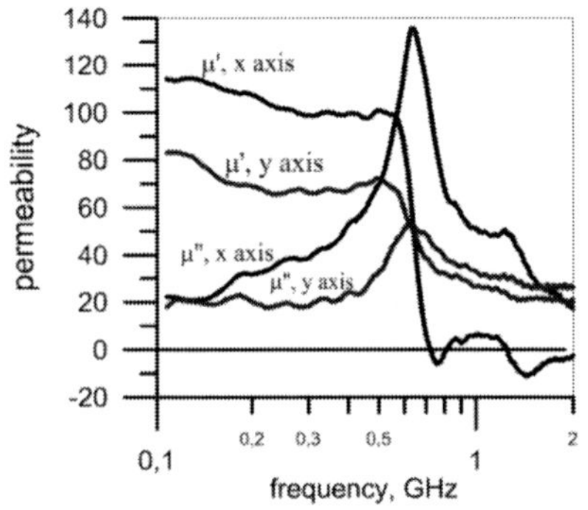

Fig. 4. The real (μ') and imaginary (μ'') permeability of the films, obtained in atmosphere with 15% N_2, measured along two mutually perpendicular directions, *x* and *y*.

It is seen from Tables 2 and 3 that, as the deposition conditions changes (5 and 15 %N_2 in magnetron-chamber atmosphere), the H_{eff} and H_c magnitudes demonstrate the similar character of variations. This agrees with data of [17], which showed that the H_{eff} magnitude should be proportional to the H_c coercive force magnitude. This indicates the fact that the static magnetic structure affects the dynamic magnetic properties. It should be noted that the measured magnitudes (Table 3) depend on the measurement direction (x and y) and the fact that the unsmoothed magnetization permeability curves (Fig. 4) are likely to indicate the magnetic-structure inhomogeneity.

Table 3. HF magnetic permeability of the studied films.

N_2 content during deposition	Measure-ment direction	f_r, GHz	H_{eff}, Oe	μ'	f, GHz
15%*	x	0.7	4.3±0.8	80	0.58
	y	1.17	12.2±4	60	0.6
15%*	x	0.57	2.9±0.6	160	0.1
	y	0.64	3.7±0.8	130	0.1
5%	x	0.82	17.4±1.4	12	0.45
	y	0.72	13.5±1.1	12	0.3

* Two samples were deposited simultaneously.

The difference between field magnitudes H_{eff}, H_c and $D^{1/2}\langle H_a\rangle$ can be explained by decreasing scales at which they are realized, i.e., from macroscopic scale, H_{eff} for whole film, to the maximum fraction with a some magnetic anisotropy close to the H_c and, finally, to microscopic scale, for stochastic domain $D^{1/2}\langle H_a\rangle$. At a frequency of 1 GHz (the wavelength is 30 cm), the magnetic moment of whole film 2.3 cm in length is in resonance oscillations; this determines the H_{eff} magnitude. As it was noted above, the analysis of hysteresis loop shape allows us to assume the presence of, at least, two volumes differing in the magnetic anisotropy in the studied films, whereas the H_c field of the film is determined by the magnetic anisotropy of biggest volume. The $D^{1/2}\langle H_a\rangle$ field magnitude is determined from the high-field portion of magnetization curve (Fig. 3) and corresponds to the anisotropy field of stochastic micron-sized domain.

The real effective magnetic permeability μ' and the frequency range of its "operation" are the applied characteristics for high-frequency operation. Table 3 shows the μ' magnitudes for the studied films and the maximum f frequency for which the μ' value remains unchanged. Taking into account the measurement accuracy of the film thickness (1.2±0.2 μm), the error of measurement of μ' is ±17%. The measured μ' values qualitatively obey the relationship $\mu' = M_s/H_{eff}$. Note that the Fe-Zr-N films, in particular, $Fe_{89-88}Zr_{10}N_{11-12}$ [11,15], have the substantially higher M_s values as compared to those of the films studied in the present work, which reach 1.8-1.9 T. For the films with M_s =1.8-1.9 T, the higher f_r value and the higher permeability $\mu' = M_s/H_{eff}$ can be expected.

4 Conclusions

Nanocrystalline (with a grain size of 3-6 nm or < 2 nm) soft magnetic FeZrN films prepared by dc magnetron sputtering of the Fe – 2.4 wt% Zr target in the (Ar + N_2) atmosphere containing 5 and 15% N_2 have been studied. Parameters of magnetic structure ($D^{1/2}\langle H_a\rangle_1$, $D^{1/2}\langle H_a\rangle_2$, $D^{1/2}H_a$, H_R), saturation magnetization M_s (0.5-1.34 T), and magnetic permeability μ' (80-100) remained unchanged to frequencies of 500 GHz were determined. The dependence of the high-frequency magnetic permeability μ' and the ac magnetic-field frequency, at which the μ' magnitude remains unchanged, on the magnetic-structure parameters of the films was considered. It was shown that the static magnetic structure determines the dynamic magnetic properties.

This study was supported in part by the Russian Foundation for Basic Research (projects nos. 18-03-00502 and 18-08-00491).

References

1. S.X. Wang, N.X. Sun, M. Yamaguchi, S. Yabukami, Nature **407**, №6802, 150 (2000)
2. I. Kim, J. Kim, K.H. Kim, M. Yamaguchi, IEEE Trans. Magn. **40**, 2706 (2004)
3. S. Kong, T. Okamoto, Sh. Nakagawa, J. Appl. Phys. **93**, 6778 (2003)
4. I.T. Iakubov, A.N. Lagarkov, S.A. Maklakov, *et al.* J. Magn. Magn. Mat. **321**, 726 (2009)
5. G. Herzer, Acta Materialia **61**, 718 (2013)
6. R.S. Iskhakov, S.V. Komogortsev, Phys. of Metals and Metallography **112**, 666 (2011)
7. R.S. Iskhakov, M.M. Brushtunov, A.G. Narmonev, *et al.* Phys. Met. Metallogr. **79**, Iss. 5, 122 (1995)
8. C. Kittel, Phys. Rev. 73, 155 (1948)
9. O.M. Zhigalina, D.N. Khmelenin, A.L. Vasil'ev, E.N. Sheftel', *et al.* Cryst. Reports 58, 344 (2013)
10. E.N. Sheftel', E.V. Kharin, V.A. Tedzhetov, *et al.* Russian Metallurgy (Metally) **2016**, 826 (2016)
11. E.N. Sheftel, E.V. Harin, V.A. Tedzhetov, *et al.* Physica B **494**, 13 (2016)
12. Ph.V. Kiryukhantsev-Korneev, Protection of Metals and Physical Chemistry of Surfaces **48**, 585 (2012)
13. E.V. Shelekhov, T.A. Sviridova, Metal Science and Heat Treatment **42**, 309 (2000)
14. S.N. Starostenko, K.N.Rozanov, A.V. Osipov, J. Appl. Phys. **103**, 07E914 (2008)
15. E.N. Sheftel, V.A. Tedzhetov, E.V. Harin, *et al.* Phys. Stat. Sol. C **13**, 965 (2016)
16. S.V. Komogortsev, R.S. Iskhakov. J. Magn. Magn. Mat. **440**, 213 (2017)
17. S.S. Maklakov, S.A. Maklakov, I.A. Ryzhikov, *et al.* J. Magn. Magn. Mat. **324**, 2108 (2012)

https://doi.org/10.1051/epjconf/201818504025

Electomagnetic absorption of composites based on epoxy resin and metallic iron nanoparticles

Roman V. Krekhno*[1] *Alexander* P. Safronov[1,2], *Anna* I. Beketova[1,2], *Igor* V. Beketov[1,2]

[1] Ural Federal University, Ekaterinburg, Russian Federation
[2] Institute of Electrophysics UD RAS, Ekaterinburg, Russian Federation

Abstract. The heat losses originated from the electro-magnetic absorption in magnetic epoxy-based composites with embedded metallic iron nanoparicles were studied by Calvet microcalorimetry. Iron magnetic nanoparticles (MNPs) were synthesized by electrical explosion of wire (EEW) method; they were non-agglomerated, spherical in shape and had a weight average diameter 85 nm. Composites based on the cured epoxy-dian resin contained MNPs in weight content varying from 10% up to 70% . To study the heat loss in alterating magnetic field commercial Calvet microcalorimeter was equiped by two coils in the serial connection placed in the calorimeter cells ; one of the coils contained a sample of composite the other was a reference. The electromagnetic adsorption was studied in the alternating magnetic field up to 1700 A /m in 67 – 214 kHz frequency range. The measured values of the specific power losses revealed linear dependence on iron MNPs content in composite and non-linear increasing function of the field frequency.

1 Introduction

Modern polymer composites are multicomponent systems comprising a polymer matrix and particulate filler material that impart various functional properties. The composite materials based on epoxy resins are of particular interest, since this class of composites has a mechanical strength, chemical resistance, high dielectric properties after curing, good adhesion to metals. Epoxy composites are promising materials for use in aviation and space industry. Magnetically filled polymer composites containing magnetically soft particles are widely used to produce magnetic screens for the absorption of electromagnetic radiation of various frequencies and coatings to protect instruments and sensors sensitive to electromagnetic radiation. At the present time there is a number of works devoted to the electro-magnetic absorption of polymer composites based on thermoplastic polymers with embedded magnetic particles [1-9].

The absorption of electromagnetic emission often means conversion of electro-magnetic energy to heat. Consequently, the direct measurement of the thermal power emitted by magnetic filled composites in the absorption of electromagnetic fields can be a precise quantitative assessment of the magnitude of the absorbed emission. Calorimetry is a direct experimental method for direct measurement of the heat evolution in various chemical and physico-chemical processes. Recently the usage of the modified Calvet microcalorimeter was reported for the measurement of specific power losses in the hyperthermia studies of ferrofluids [10, 11].

In the present work we intended to elaborate the direct microcalorimetry measurenets to evaluate heat losses of the epoxy-based solid polymeric composites with embedded iron magnetic nanoparticles (MNPs).

2 Experimental

The dual-component system of ED-20 (epoxy-diane resin based on diphenylolpropane) and a curing agent polyethylenepolyamine (PEPA) was used as a matrix for composite materials.

The metallic iron MNPs (Ssp = 9.0 m^2/g, weight average particle size 85 nm) were synthesized by the electrical explosion of wire (EEW). The details of EEW method were given elsewhere [12-14]

For the preparation of a composite the iron powder was gradually filled in the resin ED-20 under continuous stirring. Epoxy resin was solidified at room temperature. The hardener PEPA was added to placing the composite in a mold and the resin began curing immediately thereafter. The samples were obtained in the form of cylinders 4 mm in diameter and 30 mm in length. Composites with increasing weight content of iron MNPs 10, 20, 30, 40, 50, 60, 70% were prepared.

The powder X-ray diffraction (XRD) patterns were recorded using Bruker D8 Discover with Cu K_α radiation (λ = 1.542 Å) with graphite monochromator. The Rietveld refinement of XRD patterns were performed using Topas-3 software. The morphology of MNPs was examined using JEOL JEM2100 transmission electron microscope (TEM) operating at 200 kV. The specific surface area of MNPs was measured by the low-temperature adsorption of nitrogen (Brunauer-Emett-Teller (BET) approach) using Micromeritics TriStar3000 analyzer. The magnetic measurements were carried out using (Cryogenics Ltd. VSM) vibrating sample magnetometer (VSM) at room

* Corresponding author: lipka-lipka-lipka@yandex.ru

temperature for powder samples placed in a gelatine capsule. The magnetization values in a field of 1.8 T were designated as the saturation magnetization values (M_s).

The Calvet microcalorimetry method was used to study the specific power losses of magnetically filled polymer composites in altering homogeneous magnetic fields. The commercial microcalorimeter was modified using solenoids with a number of turns of 12 000 per meter, which were placed in calorimeter cells and serially connected to an alternating current source up to 150 mA, as shown in Figure 1. The field intensity was 1700 A/m. The frequency range of sound-wave generator was 66-214 KHz. The sensitivity of calorimetric cells was 10^{-6} Joule/sec.

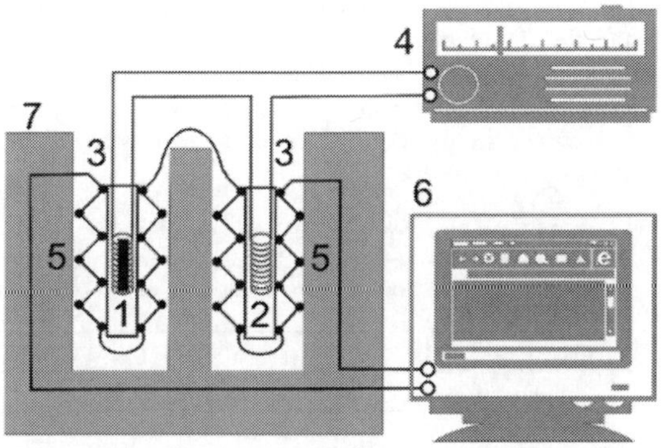

Fig. 1. Schematic diagram of the experimental setup for the microcalorimetry measurement of the heat losses of epoxy composites with iron MNPs in the alternating magnetic field. 1 – Calorimetric cell with the sample of composite. 2 – Reference cell. 3 – Identical serially connected solenoids mounted in cells. 4 – Sound-wave generator. 5 – Identical differentially connected batteries of thermocouples (450 junctions in each). 6 – Computerized control unit for the measuring of calorimetry signal. 7 – Thermostat.

3 Results and discussion

Figure 2 presents TEM image of EEW iron MNPs used for the preparation of epoxy composites. It can be seen MNPs have a spherical shape. XRD plots are given in the inset in Figure 2. The iron MNPs had a phase composition of 98% αFe and 2% γFe. αFe had a lattice period a = 2.866 Å, a coherent scattering region of 80 nm. The dimensions of the coherent scattering regions are consistent with the particle diameters obtained in the analysis of micrographs.

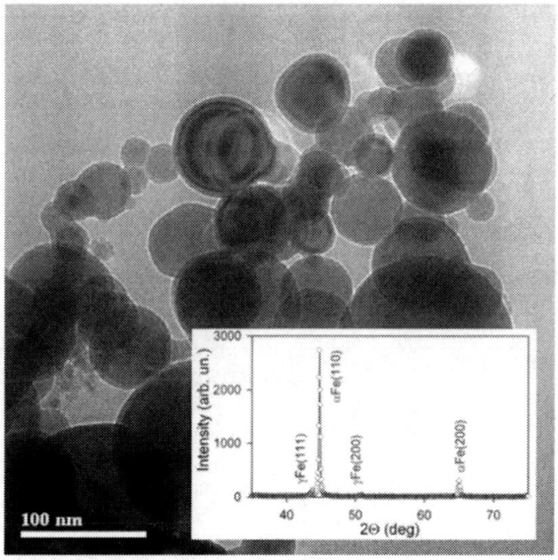

Fig. 2. TEM image of iron MNPs used for the preparation of epoxy composites. Inset – XRD plot of iron MNPs

Magnetization loops of ED-20 + Fe composites, recalculated per gram of composite, in comparison with Fe powder magnetization loop are presented in Figure 3.

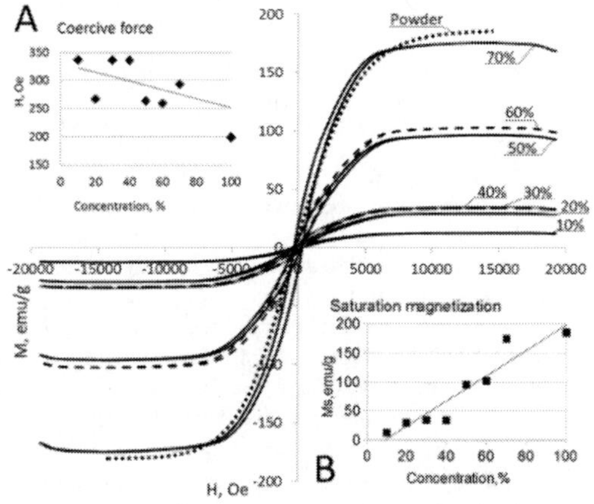

Fig. 3. Magnetization loops for epoxy composites with embedded iron MNPs. Inserts show the dependence of coercivity(A) and saturation magnetization(B) on Fe content in composite.

The comprehensive breakdown of this figure is summarized in the Table 1. The saturation magnetization increases linearly with the increase in the weight content of iron in the polymer composite. The coercive force alters in small limits , but generally has a tendency to decrease with increasing iron content, approaching the coercive force of the iron powder.

Table 1. Magnetic parameters of epoxy composites with iron MNPs

	Saturation magnetization	Residual magnetization	Coercivity
% Fe	Ms, emu/g	Mr, emu/g	Hc, Oe
10	12,4	1,27	337,2
20	29,3	1,95	266,5
30	34,6	3,69	337
40	33,9	3,63	335,5
50	95,8	6,76	262,9
60	102,2	7,21	259
70	174,3	18,8	293,5
100	185	30	199,8

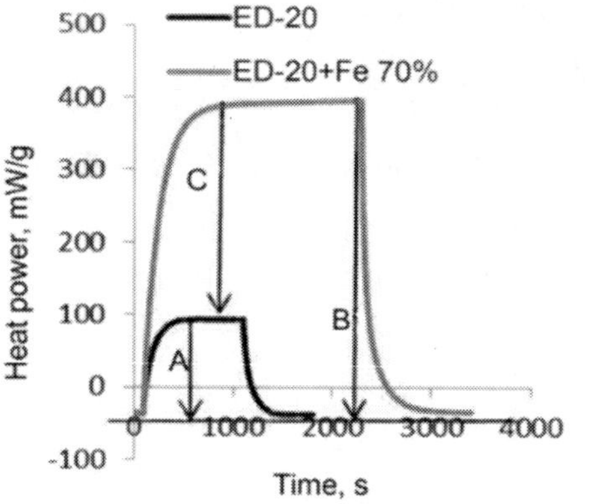

Fig. 4. Heat evolution curves of pure epoxy resin ED-20 without filler and a composite ED-20 + 70% Fe in an alternating magnetic field. A – thermal effect detected in pure epoxy resin. B – thermal effect detected in composite with 70% Fe particles content. C – substraction results, which shows the remagnetization power loss in composite.

Figure 4 shows the examples of heat evolution curves registered in the measurement of specific power losses by calorimetry. The plots correspond to the power losses from ED-20 epoxy resin and to the power losses of a composite. The curves have the same shape due to the steps of measurement process. Initially they have the same baseline during the current generator offline time. Then, as the wave generator was turned on, the calorimetric signal increased due to the heat evolution in the cell. During this phase the signal eventually came to saturation. It meant that the thermal balance was established between the heating power in the cell and the dissipation of the heat from the cell due to the thermal conductivity of thermocouples. According to the theory of Calvet microcalorimetry the saturation value is proportional to the heating power evolved in the cell [15]. The proportionality constant was determined in the separately done calibration of the cell by the Joule heating. After the saturation of the calorimetric signal was established the current generator was turned off and the thermal emf of the calorimetric cell exponentially returned to the baseline.

The thermal response of the epoxy resin ED-20 without magnetic nanoparticles appears due to the Joule heat evolved in solenoids when an alternating voltage was applied to them. The heat evolution curve corresponding to a composite system, ED-20 + 70% Fe is substantially higher than the curve corresponding to the pure epoxy resin ED-20. The difference in saturation values correspond to the evolution of heat due to the remagnetization of embedded MNPs in alternating magnetic field.

To calculate the specific power losses of the remagnetization of MNPs one have to calculate the net heat of pure ED-20 for use as a reference first. To do this, the value of the thermal effect, which was released on the saturation value was corrected by the baseline (Figure 4, arrow A). Further, in the same way the value of the thermal effect from the sample was calculated (Figure 4, arrow B). Finally, the heat power of epoxy ED-20 was subtracted from heat power of each sample and related to the weight of iron MNPs in the composite. The result of this subtraction can be visualized as an arrow C in Figure 4. This way the power losses for 7 samples of epoxy composites of different concentrations at different frequency of alternating field were obtained and given in Table 2.

Table 2. Specific power losses during remagnetization of composites ED-20 + Fe with different concentrations at different frequencies, of an uniform magnetic field 1700 A / m.

Frequency, kHz	67	100	167	214
concentration, %	heat power, mW/g			
10	70,8	78,7	214,8	222,1
20	63,2	78	217,2	241,7
30	67,8	84,6	242,7	239,1
40	65,5	83,6	245,9	276,3
50	60,2	72,6	246,9	
60	78	96,9	303,6	330,1
70	72,6	98,3	313,1	397,7

Dependencies of the specific power losses on the content of iron MNPs in composite and on the frequency of the applied field are presented in Figures 5 and 6 respectively

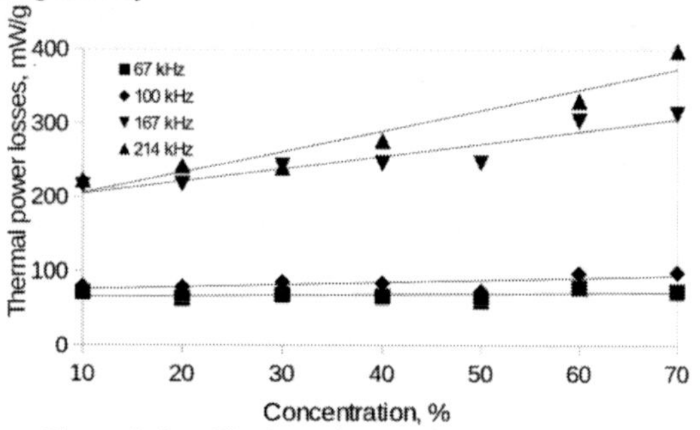

Figure 5. Specific thermal power losses per 1 g of Fe in remagnetization of magnetic filled samples in the alternating field 1700 A/m at different frequencies, depending on the concentration of filler in the composite.

Figure 5 shows that power losses at frequencies of 67 kHz and 100 kHz is much lower than the power losses at 167 kHz and 214 kHz. At 67 kHz and 100 kHz power losses very slightly depend on MNPs content in the composite, while at 167 kHz and 214 kHz they substantially increase with increasing concentration of iron MNPs.

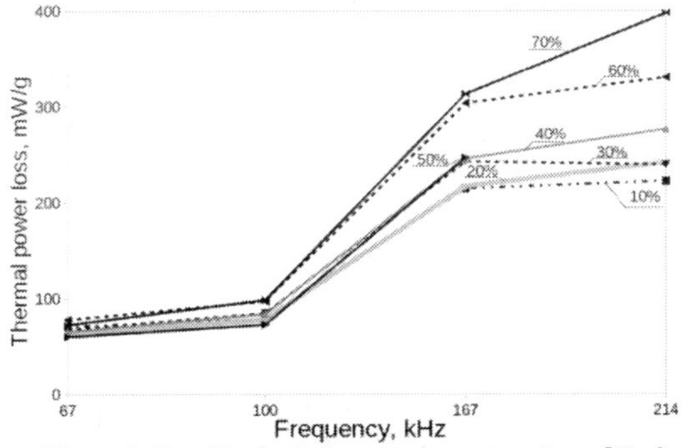

Figure 6. Specific thermal power losses per 1 g of Fe in remagnetization of magnetic filled samples of various concentrations, depending on the frequency of the uniform field 1700 A / m.

Figure 6 shows the thermal power losses in the remagnetization of magnetic filled composite of different concentrations, depending on the frequency. The data reveals that in general the thermal power loss increases with frequency. At the same time, the largest increase of heat occurs between 100 kHz and 167 kHz. Moreover, at the frequency above 167 kHz, the significant discrepancy between the curves appear. At a frequency of 67 kHz the difference in values is less than 10 mW, while at a frequency of 214 kHz the difference reaches 200mW.

4 Conclusions

1. Vibrational magnimetry method was used to obtain magnetization curves of epoxy resin composite filled with iron nanoparticles. It was found that the saturation magnetization increases linearly with the increase in the weight content of iron in the polymer composite. The coercive force decreases with increasing iron content, approaching the coercive force of the iron powder.

2. Calvet microcalorimetry method was used to study remagnetization power loss in magnetic filled polymer composites ED-20 + Fe MNPs with various concentrations and mass in an alternating magnetic field 1700 A/m. The heat dissipation power varied from 60 to 400 mW/g , reaching a maximum value of 400 mW/g at a maximum frequency of 214 kHz and with a maximum iron particle concentration of 70%. A linear dependence of the heat power with the concentration of iron and nonlinear, monotonically increasing frequency dependence was found.

Acknowledgements

This work is supported in part by the project for basic research of UD RAS. The assistance of Shchipanova T. A. in the measurements of hysteresis loops for composites is appreciated.

References

1. K. Sonoda, M. Teirikangas, J. Juuti, Y. Moriya, H. Jantunen, JMMM, **323,** 2281–2286 (2011)
2. I. Kong, S. Hj Ahmad, M. Hj Abdullah, D. Hui, A. Nazlim Yusoff , D. Puryanti ,JMMM **322, 3401–3409** (2010)
3. M.T. Sebastian, H. Jantunen, review, Int. J. Appl. Ceram. Technol. **7,** 415–434 (2010)
4. T. Hu, J. Juuti, H. Jantunen, T. Vilkman, J. Eur. Ceram. Soc. **27** 3997–4004 (2007)
5. J. Juuti, M. Teirikangas, K. Sonoda, H. Jantunen, T, Int. J. Appl. Ceram. Technol. **7,** 452–460 (2010)
6. A.C. Barazs, T. Emrick, T.P. Russell , Science **314,** 1107–1110 (2006)
7. E. Muhammad Abdul Jamal, P.A. Joy, P.Kurian, M.R. Anantharaman, MSE, **B 156** 24–31 (2009).
8. Ta-I Yang, Rene N.C. Brown, Leo C. Kempel, P. Kofinas , JMMM **320,** 2714– 2720 (2008)
9. T. Wang, Z. Liu, M. Lu, B. Wen, Q. Ouyang, Y. Chen, C. Zhu, P. Gao, C. Li, M. Cao, L. Qi, JAP **113**, 024314 (2013)
10. Iu. P. Novoselova, A. P. Safronov, O. M. Samatov, G. V. Kurlyandskaya, – AIP Conference Proceedings **1767**, 020016 (2016)
11. A.P. Safronov, O.M. Samatov , I.S. Tyukova , E.A. Mikhnevich , I.V. Beketov, JMMM **415,** 24–29 (2016)
12. Yu.A. Kotov, Nanoparticle Research, 539–550 (2003)
13. G.V. Kurlyandskaya , S.M. Bhagat , A.P. Safronov , I.V. Beketov , A. Larrañaga , *AIP Advances* **1,** 042122, (2011)
14. A.P. Safronov, G.V. Kurlyandskaya, A.A. Chlenova, M.V. Kuznetsov, D.N. Bazhin, I.V. Beketov, M.B. Sanchez-Ilarduya, A. Martinez-Amesti, Langmuir, **30**, 3243 (2014)
15. Calvet, E., Prat, H. Applications physico-chimiques et biologiques. Paris. Masson and Cie (1956)

Effect of alloying elements (Zr, Hf, Co), heat and mechanical treatment conditions on the phase composition and magnetic properties of $SmFe_{11}Ti$ compounds with $ThMn_{12}$ structure

Andrey Urzhumtsev[1,*], *Maksim* Anikin[1], *Evgeniy* Tarasov[1], *Mikhail* Semkin[1,2], *Maksim* Cherepkov[1], *Nikolay* Kudrevatykh[1], *Aleksandr* Zinin[1], and *Vladimir* Moskalev[3]

[1]Ural Federal University, Ekaterinburg, Russia
[2]M.N. Mikheev Institute of Metal Physics of Ural Division of RAS, Ekaterinburg, Russia
[3]POZ - Progress Co, Verkhnaya Pushma, Sverdlovsk region, Russia

Abstract. The results of thermomagnetic, metallographic and X-ray diffraction phase analysis as well as the measurements of specific magnetization (σ_s), Curie temperature (T_C), coercive force (H_C) of $(Sm,M)(Fe,M)_{12-x}Ti_x$ alloys samples, where M = Zr, Hf, Co with the $ThMn_{12}$ main phase structure (1-12) are presented. The effect of the annealing temperature and the cooling rate on the formation of 1-12 phase and its magnetic properties, including the effect of high-energy milling on the magnetic hysteresis properties and alloys structure are described. It was found that the highest magnetic characteristics such as $\sigma_s = 112.6$ emu/g and $T_C = 600$ °C are attained in the $(Sm_{0.8}Zr_{0.2})(Fe_{0.75}Co_{0.25})_{11.4}Ti_{0.6}$ alloy after its annealing at 1050 °C and rapid cooling. It is noted that a mechanical milling of the alloy leads to 1-12 phase amorphization which accompanied by an α-(Fe) or metal Co phases impurity formation.

1 Introduction

Rare Earth – 3d-metal intermetallic compounds with the $ThMn_{12}$ crystal structure are considered as a promising material for high magnetic energy product permanent magnets due to the favorable potential magnetic parameters for several of them [1]. However, the Fe-containing 1-12 phase is formed only with some amount of another 3d-transition metals (Cr, V, Ti – stabilizating elements (SE)), which prevents to attain the high magnetization value and Curie temperature T_C of this phase [2]. The average magnetic moment per 3d-ion is inversely proportional to the concentration of SE and monotonously decreases [3]. Neutron diffraction and Mossbauer spectroscopy studies show that the SE in the $ThMn_{12}$ type structure of $RFe_{(12-x)}Ti_x$ alloys occupies the 8*i* positions without forming its own sublattice, since the values of x in the homogeneity region are smaller than the fraction of 3d-metal atoms in each of the sublattices [4]. In connection with this, the main tasks in the development of hard magnetic materials based on this class of compounds are the stabilizing element amount reduction in the alloy and technological methods finding to keep in the alloy the 1-12 phase presence in a highest volume [5].

The dependence of T_C value on the R – metal type in 1-12 phases is common for R – 3d intermetallics with a large 3d-atoms molar portion. The difference between the maximal and minimal T_C values in the R-series does not exceed 25 %, indicating that the Fe – Fe exchange interaction is dominating. The contribution from the R – Fe interaction is proportional to the spin of the R-ion in the first approximation [4].

Despite the high values of σ_s, T_C and the anisotropy field (H_a) of the $SmFe_{11}Ti$ type compounds, the value of their coercive force (H_C) is still very far from the theoretical limit (~ 50 kOe). For sintered magnets the highest H_C value of 1.8 kOe is reported, for samples obtained by mechanical alloying – 4.4 kOe, for thin films is 3.9 kOe and 5.8 kOe – for rapidly quenched materials [3, 6]. The reasons of the large discrepancy between the theoretical and experimentally obtained H_C values presumably lie not only in the presence of secondary phases in these type alloys (besides 1-12), but also in the fact that their coercivity nature is not fully understood and requires the further studies [5, 7]. Thus, the task of this work was to investigate the effect of alloying elements and a heat treatment conditions on the structure and magnetic properties of $(Sm,M)(Fe,Co)_{12-x}Ti_x$ type alloys with $ThMn_{12}$ main phase as in the bulk and powdered states prepared by its mechanical milling.

2 Experimental details

Samples of the following series: No. 1 – $SmFe_{(12-x)}Ti_x$, x = (0.5, 1.0); No. 2 – $(Sm_{0.8}M_{0.2})(Fe_{0.75}Co_{0.25})_{11.5}Ti_{0.5}$, where M = (Sm, Zr, Hf); No. 3 – $(Sm_{(1-x)}Zr_x)(Fe_{0.75}Co_{0.25})_{(11.2+x)}Ti_{(0.8-x)}$, x = (0.05 – 0.25) were melted in an electric arc furnace under a pure helium gas protection. An excess of rare earths metal

* Corresponding author: andrej707@bk.ru

(~ 20 wt %) was added to the starting compositions for the formation preventing of α-(Fe,Co) phases. After the melting, each sample was sealed in to a quartz ampoules, which pumped to the pressure $P = 10^{-3}$ torr and subjected to the further annealing. For alloys samples of No. 1 and No. 2 series it was made in a furnace at temperature (T) 850 °C during 5 hours. The samples of No. 3 system were split into two parts, each of them was also placed in an ampoule with the following pumping and sealing. Annealing of these samples was carried out at T = 1050 °C for 8 hours, after that a one ampoule with the sample was rapidly cooled (RC) by dropping into a water, and the other one was slowly cooled (SC) to a room temperature in the furnace.

Magnetic susceptibility temperature dependence measurements χ(T) of the samples were carried out in an alternating magnetic field of 50 Oe amplitude in the temperature range from 25 up to 800 °C.

The samples crystalline structure was determined using X-ray diffraction technique (D8 Advance, Bruker with Cu K_α radiation source). The morphological structure of the alloys was revealed with a help of MIM-7 optical microscope. For that aim prepared by mechanical polishing a flat surface of the studied sample was treated with 3 % nitric acid for 10 seconds.

Magnetization was measured using the vibrating sample magnetometer of VM-114 type at room temperature in a magnetic field up to 12 kOe.

Mechanical milling of the alloys was carried out in a ball mill of the attritor type in ethyl alcohol media at the ratio of the milling balls to the material sample masses of 500/8, and rotational speed of the impeller blade – 4000 rpm.

3 Results

The χ(T) dependences plotting of the No. 1 and No. 2 series (Fig. 1) shown to the presence of a several specific places which correspond to the temperatures of magnetic phase transition close to T_C values. As it is seen, the homogenized samples of $SmFe_{11}Ti$, $SmFe_{11.5}Ti_{0.5}$ and $Sm(Fe_{0.75}Co_{0.25})_{11.5}Ti_{0.5}$ alloys contain at least two magnetic phases.

The lowest temperature Curie, corresponding to the $ThMn_{12}$ type phase is observed for $SmFe_{11.5}Ti_{0.5}$ sample. Its T_C value = (210 ± 5) °C. For alloys where a 25 % of Fe is replaced by Co, the T_C of the main 1-12 phase is increased up to (600 ± 5) °C. A sharp kink in the χ(T) dependences at T = (780 ± 5) °C corresponds to a ferro – paramagnetic phase transition for the second phase which can be identified as α-Fe. Unlike of the above event, the samples of $(Sm_{0.8}M_{0.2})(Fe_{0.75}Co_{0.25})_{11.5}Ti_{0.5}$ alloys, where M = (Zr, Hf), do not contain the α-Fe phase. Therefore, the substitution of 20 % Sm for Zr or Hf prevents this phase formation.

The results of the thermomagnetic phase analysis (TMA) of the $(Sm_{(1-x)}Zr_x)(Fe_{0.75}Co_{0.25})_{(11.2+x)}Ti_{(0.8-x)}$ homogenized samples showed that in the entire series there is no contribution from the α-Fe phase in the χ(T) dependence. Nevertheless on the χ(T) curves of slowly cooled samples (Fig. 2), the contributions from another magnetic phases with T_C close to the main 1-12 phase are observed. As it seen, the rapid alloy cooling treatment prevents their formation.

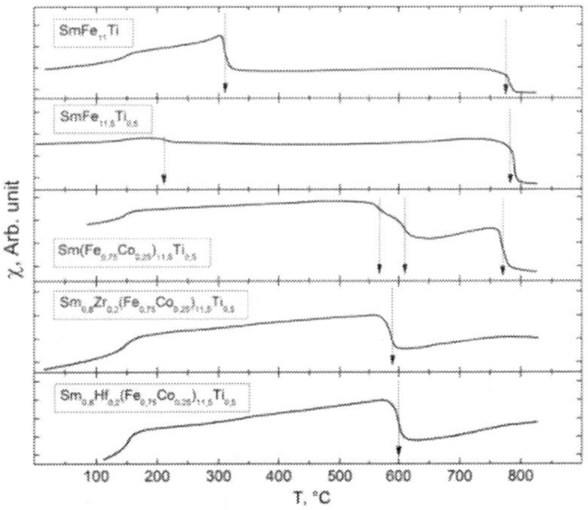

Fig 1. Magnetic susceptibility temperature dependences χ(T) of No. 1 and No. 2 series homogenized samples

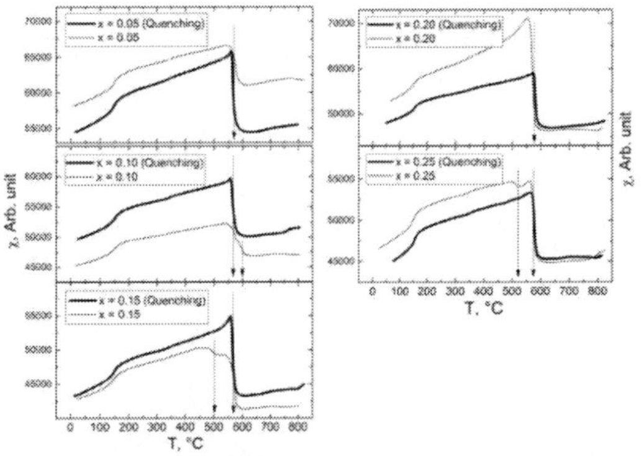

Fig. 2. Magnetic susceptibility temperature dependences – χ(T) of No. 3 series homogenized samples. A wide line corresponds to the rapidly cooled samples, and a thin one for slowly cooled.

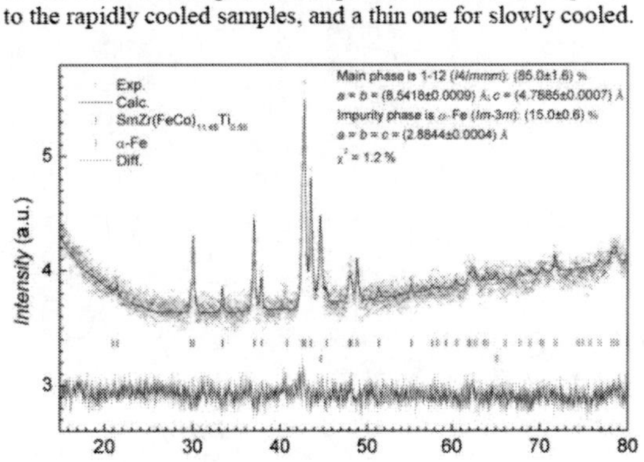

Fig. 3. The X-ray diffraction pattern of $(Sm_{0.8}Zr_{0.2})(Fe_{0.75}Co_{0.25})_{11.4}Ti_{0.6}$ after annealing and rapid cooling.

The X-ray diffraction analysis of the $(Sm_{0.8}Zr_{0.2})(Fe_{0.75}Co_{0.25})_{11.4}Ti_{0.6}$ RC alloy sample (pattern in Fig. 3) shows, that in addition to the main 1-12 phase

($a = b = 8.5418(9)$ Å, $c = 4.7885(7)$ Å) there are peaks from the second α-Fe phase ($a = b = c = 2.8844(4)$ Å), whose fraction is 15 at%, but contribution from it does not observed in χ(T) - dependence (see, Fig. 2).

Comparing the samples specific magnetization values for the No. 1 and No. 2 series alloys before and after the homogenizing annealing (Fig. 4A) we could see that annealing increase one for $Sm(Fe_{0.75}Co_{0.25})_{11.5}Ti_{0.5}$ and $(Sm_{0.8}Hf_{0.2})(Fe_{0.75}Co_{0.25})_{11.5}Ti_{0.5}$ alloys. Presumably, in these alloys in addition to α-Fe the Co-metal impurity are formed during the homogenization, as a result of the peretectic reaction. The particular substitution of Sm for Hf prevents the α-Fe phase precipitation and reduces the amount of Co- impurity. A similar substitution of Sm for Zr practically prevents the precipitation of Co-metal also.

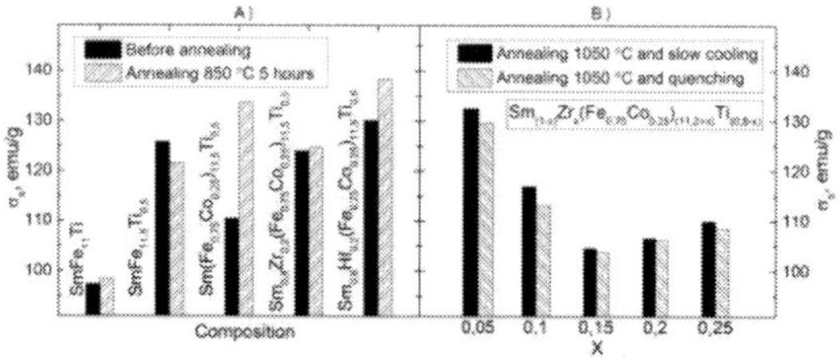

Fig. 4. A) Specific magnetization σ_s values in a magnetic field H = 12 kOe for No. 1 and No. 2 series samples before and after homogenizing annealing. B) The same as in (A) but for No. 3 series as the dependence on the cooling rate.

Comparison of the specific magnetization values for No. 3 series homogenized alloys samples after SC and RC treatments is shown in Fig. 4B. The slowly cooled samples magnetization is higher than that of RC samples, which indirectly indicates to the presence of additional phases with a large magnetization, probubly α-(Fe,Co).

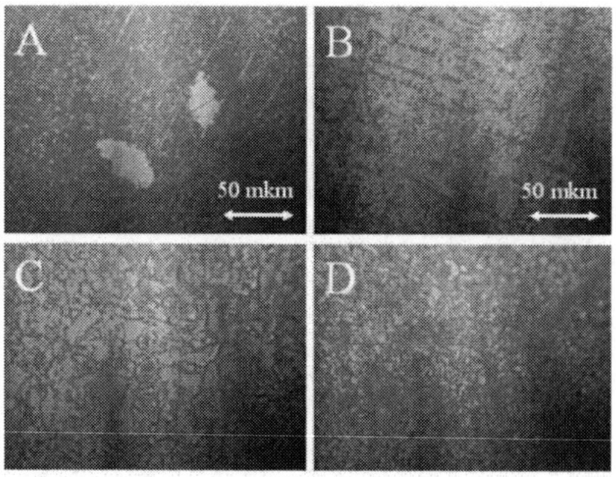

Fig. 5. Micrographs of homogenized samples of the $(Sm_{0.8}Zr_{0.2})(Fe_{0.75}Co_{0.25})_{11.4}Ti_{0.6}$ alloy. A and C – material is after a slow cooling. B and D – after rapid cooling in a water.

Figure 5 shows the results of the metallographic analysis of the $(Sm_{0.8}Zr_{0.2})(Fe_{0.75}Co_{0.25})_{11.4}Ti_{0.6}$ alloy samples after the homogenizing annealing with slow final cooling to RT (Fig. 5A and Fig. 5C) and after RC (Fig. 5B and Fig. 5D). At the micrograph of Fig. 5A there are two large white spots which present the inclusions of additional phase, while in Fig. 5B there are no them. Figures 5C and 5D show a matrix of dark fine lines presumably being an intergrain boundaries.

Figures 6, 7 and 8 show the coercive force changes of the studied alloys powders on the milling time. As a result for No. 1 sample (x = 1) the coercive force is monotonously reduced from 450 Oe to 50 Oe during the first 10 minutes of milling (Fig. 6) and remains unchanged up to the 60 minutes treatment.

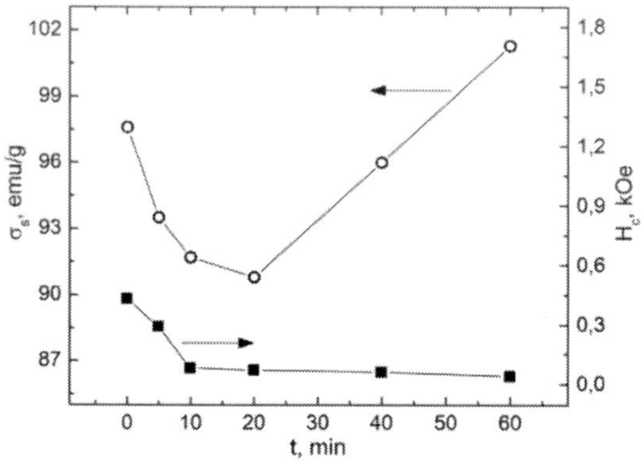

Fig. 6. σ_s (H = 12 kOe) and H_C dependences of the No. 1 series sample (x = 1) on the milling time in the attritor.

For No. 2 and No. 3 series of quenched alloys samples, a slightly different picture is observed. For the $(Sm_{0.85}Zr_{0.15})(Fe_{0.75}Co_{0.25})_{11.35}Ti_{0.65}$ alloy H_C value increases from 30 Oe up to 500 Oe during the first 8 minutes of the process (Fig. 7). Then in the interval from 8 up to 20 minutes it oscillates in the interval of (300 ± 80) Oe, depending on the milling time. What is explained by the way of powders probe taking and testing. It is noted that the σ_s and H_C curves correlate well.

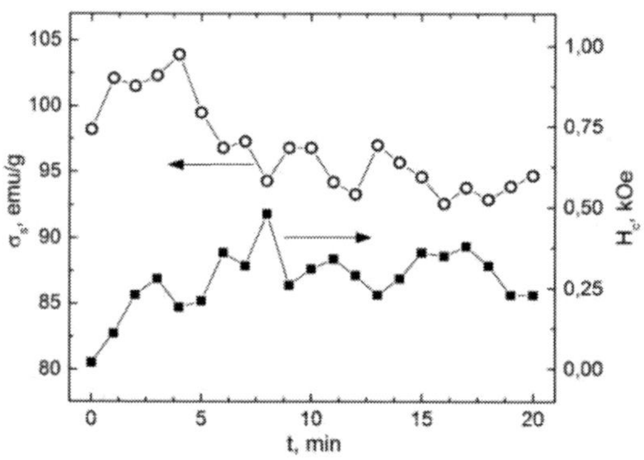

Fig. 7. σ_s (H = 12 kOe) and H_C dependences of series No. 3 sample (x = 0.15) on the milling time in the attritor.

For alloy $(Sm_{0.9}Zr_{0.1})(Fe_{0.75}Co_{0.25})_{11.3}Ti_{0.7}$, the increment of H_C value from 100 Oe up to 330 Oe in the first 20 minutes milling and a smooth drop from 250 Oe up to 50 Oe in the interval (90 – 120) minutes are found (Fig. 8).

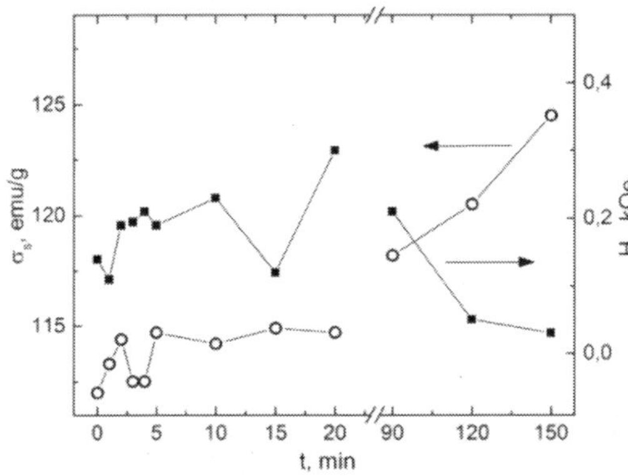

Fig. 8. σ_s (H = 12 kOe) and H_C dependences of the series No. 3 sample (x = 0.10) on the milling time in the attritor.

Figure 9 shows the X-ray diffraction intensity-normalized pattern at room temperatures for $(Sm_{0.8}Zr_{0.2})(Fe_{0.75}Co_{0.25})_{11.4}Ti_{0.6}$ alloy before and after its 150 minutes mechanical milling. A significant broadening of the peaks is observed on the X-ray pattern of the sample, which indicates to a main 1-12 phase transformation to the amorphous state. It accompanies by σ_s increase, which is explained by the bcc-Fe phase precipitates formation having the higher magnetization.

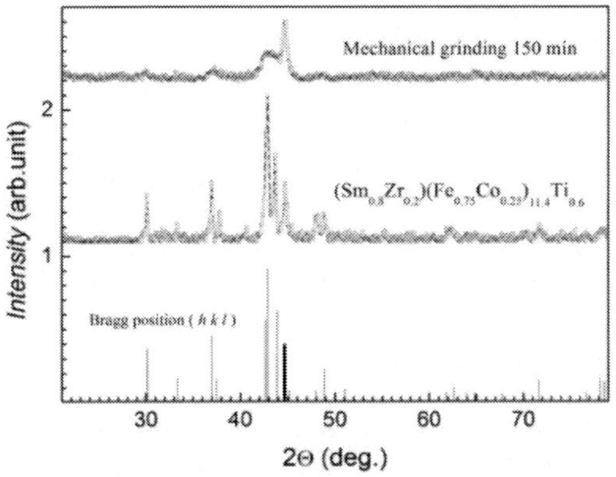

Fig. 9. The X-ray diffraction intensity – normalized pattern of $(Sm_{0.8}Zr_{0.2})(Fe_{0.75}Co_{0.25})_{11.4}Ti_{0.6}$ before and after high-energy 150 minutes milling in the attritor.

4 Conclusions

Based on the results of the work, the effect of the composition and processing conditions on the crystal structure and magnetic properties of $SmFe_{11}Ti$ type alloys, with the $ThMn_{12}$ (1-12) main phase structure were studied.

Substitution of the 25 % Fe atomic fraction by Co raises the Curie temperature of 1-12 phase from 300 up to 600 °C. The substitution of 20 % Sm for Hf prevents the α-Fe phase precipitation from the melt, and the substitution of 20 % Sm for Zr suppress the precipitation both α-Fe and Co-metal phases.

The alloys rapid cooling after annealing significantly reduces the content of the secondary phases giving raise the main 1-12 phase volume. The specific saturation magnetization of the most homogeneous polycrystalline $(Sm_{0.8}Zr_{0.2})(Fe,Co)_{11.4}Ti_{0.6}$ sample is found to be 112.6 emu/g in a magnetic field of 12 kOe, which confirms its perspective for a high $(BH)_{max}$ permanent magnets manufacturing from them.

Mechanical milling amorphizes the crystal structure of 1-12 compounds and does not allow to achieve the higher coercive force values.

Acknowledge

The work was supported by the State contracts No. 3.6121.2017/8.9 between UrFU and the Ministry of Education and Science of Russian Federation and by the Fund of assistance to development of small enterprises in scientific-technical sphere No. 11996GU/2017.

References

1. T. Kuno, S. Suzuki, K. Urushibata, K. Kobayashi, N. Sakuma, M. Yano, A. Kato, A. Manabe, J. AIP Adv. **6**, 025221 (2016)
2. T. Saito, H. Miyoshi, D. Nishio-Hamane, J. Alloys and Comp. **519**, 144 (2012)
3. B. Reinsch, B. Grieb, E.-Th. Henig, G. Petzow, IEEE Trans. on Magn. **28**, 2832 (1992)
4. A. Andreev, A. Bogatkin, N. Kudrevatykh, S. Sigaev, E. Tarasov, J. FMM. **1**, 70 (1989)
5. A. Hutten, G. Thomas, J. Appl. Phys. **70**, 6113 (1991)
6. R. Verhoef, F. De Boer, Z. Zhang, J. Magn. Magn. Mat. **75**, 319 (1988)
7. K. Kobayashi, S. Suzuki, T. Kuno, K. Urushibata, N. Sakuma, M. Yano, A. Kato, A Manabe, J. Alloys and Comp. **694**, 39145 (2016)

Stress-induced magnetic domain structure in $DyFe_{11}Ti$ compound

Elena Semenova[1,*], *Marina* Lyakhova[1], *Dmitriy* Karpenkov[2], *Yulia* Kuznetsova[1], *Alexey* Karpenkov[1], and *Konstantin* Skokov[3]

[1]Tver State University, Faculty of Physics, 170002, Tver, Russia
[2]Chelyabinsk State University, Faculty of Physics, 454001, Chelyabinsk, Russia
[3]Technische Universität Darmstadt, 64287, Darmstadt, Germany

Abstract. In this work, we present the results of the $DyFe_{11}Ti$ compound magnetic domain structure (DS) exploration. It was found that mechanical polishing leads to a change of the anisotropy of the surface layer, which affects the magnetic DS configuration. It is established that the equilibrium structure of the surface domains is restored through time under unchanged external conditions. It is suggested that the mechanism of DS relaxation has thermal fluctuation nature.

1 Introduction

The analysis of magnetic domain structure (DS) allows to estimate some intrinsic parameters of the obtained phases [1, 2]. Among a large number of the experimental techniques for DS investigation, magneto-optical Kerr microscopy is the most appropriate, because it can provide the researchers with a high-throughput characterization data.

Nevertheless, there are many reasons why the quantitative analysis of magnetic properties from domain structure can be erroneous. It is known that the magnetoelastic contribution can change the magnetic anisotropy type of the sample surface layer in comparison with the anisotropy of the sample volume [3, 4]. In this work the results of experimental study of the magnetic domain structure of the $DyFe_{11}Ti$ compound, as well the analysis of the mechanical stress influence on the surface DS of bulk crystals are presented.

The $RFe_{11}Ti$ intermetallic compounds are now reconsidered as promising materials for permanent magnets with low contents of rare earth materials. This compound has $ThMn_{12}$-type tetragonal crystal structure (space group I4/mmm) and is characterized by reasonably high values of the Curie temperature and of the spontaneous magnetic moment [5-7]. $DyFe_{11}Ti$ compound has uniaxial magnetic anisotropy at room temperature. This compound is characterized by the two spin-reorientation phase transitions (SRT). According to the literature data, SRT temperatures significantly vary: the SRT1 is observed in the temperature range of 98–120 K, SRT2 – 214–250 K [5-10]. Besides, $DyFe_{11}Ti$ compound possess the large magnetostriction [11]. Temperature study of the magnetic DS of $DyFe_{11}Ti$ single crystals was carried out in [3,4], where it was shown by means of magnetic measurements that two SRT occurred in the sample (T_{srt1}=275 K and T_{srt2}=125 K), and at the same time two types of areas with different SRT temperatures were observed by means of DS study. The presence of impurities of the second phase can be excluded since the studies were carried out on single crystals.

2 Experimental

The $DyFe_{11}Ti$ compound was synthesized by induction melting of stoichiometric composition under argon atmosphere. The resulting ingot was sealed in an evacuated quartz tube and annealed in a resistance furnace. Several large grains were extracted and checked for single crystallinity. According to metallographic, thermomagnetic and X-ray analyses the samples were single phased and had $ThMn_{12}$ structure.

The preparation of the sections was carried out by using the set of diamond pastes. After polishing the sample was cleaned with ethanol. The time between the final stage of preparation procedure of the sample surface and the beginning of observations did not exceed 1 minute. To verify the effect of mechanical stress on surface of the section this procedure was repeated several times on the samples with different crystallographic orientation of the observation plane. It was established that even after a brief polishing the visible transformation of the surface DS began only 30 minutes after starting of the observations.

The magnetic domain structure of the samples was investigated by means of the polar magneto-optical Kerr effect on Axiovert 200MAT (Zeiss) in the temperature range from 273 K up to 423 K.

MFM measurements were carried out at room temperature on Scanning Probe Microscope Solver Next (NT-MDT (Russia)) using magnetic cantilevers. The measurements were executed by two-pass method.

* Corresponding author: Semenova_E_M@mail.ru

During the first pass, the topography is studied by standard tapping AFM mode. In the second pass the cantilever is lifted to a certain height (150 nm) for each scan line and performs the scanning, subtracting the stored topography.

3 Results and discussion

The surface domain configuration of a uniaxial magnetic material is determined by the quality factor $Q = K_1/(2\pi I_S^2)$ (K_1 – magnetocrystalline anisotropy constant, I_S – magnetization saturation value). $DyFe_{11}Ti$ compound reveals $Q \sim 2.98 > 1$, thus it is a highly anisotropic magnetic material, in which the surface domain configuration is determined by the energy of the stray fields of the sample. At room temperature, the equilibrium surface DS on the basal plane of the samples with uniaxial anisotropy demonstrate "stars-like" domain. The impact of abrasive materials on the sample surface leads to the formation of stressful surface layer. This effect can be attributed to tensile stresses, which affect the domain configuration. In particular, the magnetic domains on the basal plane of the $DyFe_{11}Ti$ single crystal are not immediately detected by the polar Kerr effect after mechanical polishing of the surface. However, the labyrinth domain structure is detected by the Bitter method and by using ferrite-garnet indicator tape. This is because the surface layer is thin enough for the Bitter method and indicator tape to depict the stray field of the main domains. The main feature of the presented experimental DS observation was the fact that the surface domain structure returns to its equilibrium state through time without impact of an external magnetic field and at constant temperature. Fig.1a presents the labyrinth domain structure on the $DyFe_{11}Ti$ single crystal 1 hour after a short polishing. After 3 hours, the contrast of domains significantly increased (fig.1b), while the boundaries of the labyrinth domains began to take on sharp outlines, in which one can see the emerging "star" domains. The typical surface domain structure formed on the basal plane after 10 hours under constant external conditions (fig.1c). The reason that the method of the polar Kerr effect does not depict the domains in the basal plane immediately after mechanical polishing is that the magnetic moment is oriented in the observation plane due to stresses. Through time, the magnetic moment begins to deviate from this direction and DS is visualized in polarized light. The necessary relaxation time depends on the duration and severity of

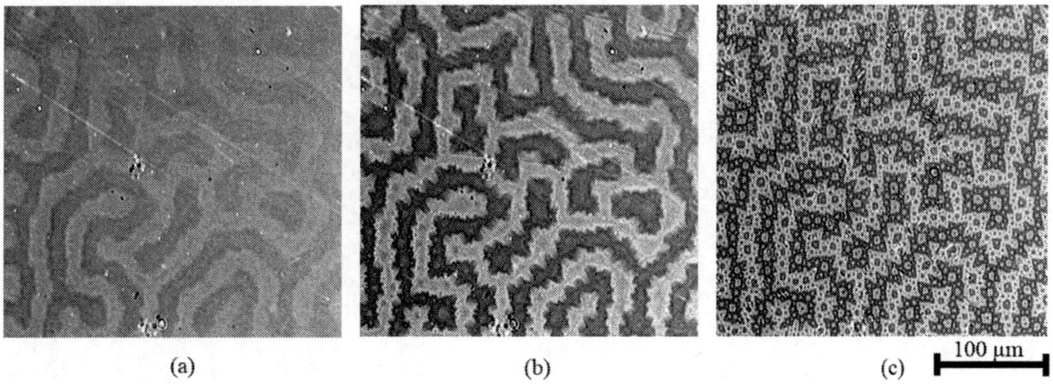

Fig. 1. The magnetic domain structure transformation on the basal plane of $DyFe_{11}Ti$ single crystal detected by Kerr magneto-optical microscopy: a) after 60 min, b) after 3 h, c) after 10 h.

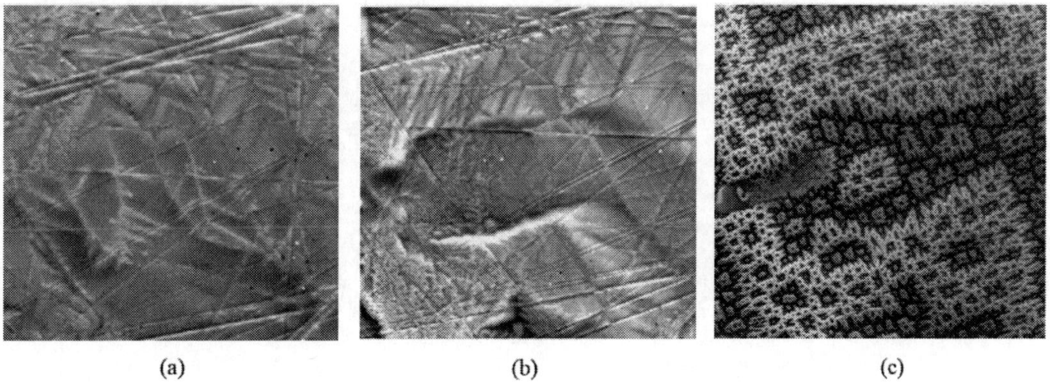

Fig. 2. MFM images of the magnetic domain structure relaxation process on the basal plane (001) of $DyFe_{11}Ti$ single crystal at room temperature: a) after mechanical polishing immediately, b) after 45 h, d) after 300 h (100x100 μm)

mechanical polishing and temperature of the sample.

It should be noted that the heating of the single crystalline specimen up to 423 K after mechanical polishing the relaxation time reduced down to 1-2 minutes. Moreover, the stages of the DS relaxation transformation remain exactly the same as at the room temperature. The restoration of the equilibrium structure always begins at the edge of a single crystal or a grain.

An MFM study allowed to identify the DS on the (001) surface of the sample just after polishing. Registration of magnetic stray fields was carried out at the distance of 150 nm above the surface. This allows to visualize not only the surface defects, but also the domain boundaries.

In contract with Fig.1, where the relaxation of DS after a short-term effect of the polishing on surface of the section was shown, Fig. 2 depicts DS images obtained after a complete surface treatment of the same single crystal, including the grinding and polishing steps. The degree and duration of the impact on the surface significantly affect the speed of the relaxation process.

As it is shown in Fig.2a, the $DyFe_{11}Ti$ sample demonstrates typical multi-axis magnetic DS. After 45 hours in the same region of the surface domain structure has changed: the resulting image corresponds to the stray fields of the labyrinth structure. After 300 hours the equilibrium typical branched (closure-type or star-like) domains structure has formed on the surface. Thus, the rate of restoration of the equilibrium domain structure has increased by a factor of 30.

Moreover, the relaxation process study of the polycrystalline samples has shown that the relaxation rate of the grains with various crystallographic orientations differs: restoration of the surface DS of the grains with an orientation close to the basal plane was faster than for grains with the prismatic orientation.

Analysis of the surface domain parameters allowed us to calculate values of the surface density of energy (γ) and width (δ) of domain walls. To determine the values of γ and δ theoretical and experimental approaches were used.

For the theoretical calculation the expression obtained for the 180° domain walls was applied:

$$\gamma = 4\sqrt{A \cdot K_1} \quad (1)$$

$$\delta = \pi\sqrt{A/K_1} \quad (2)$$

where A=159,12×10^{-13} J/m is the exchange parameter, K_1= 8,31×10^5 J/m^3 is magnetocrystalline anisotropy constant. The set of obtained parameters values is summarized in Table 1.

The second Bogenberger-Hubert method based on the analysis of experimental images of the magnetic domain structure gives the following equation:

$$\gamma = \frac{L I_S^2}{4\pi\beta} \quad (3)$$

where L is the average distance between domain boundaries, I_S = 665 Gs – saturation magnetization value, β=0,31±0,02.

Table 1. The surface density of the energy (γ) and width (δ) of domain walls

δ, Å	γ, J/m³	
	calc.	exp.
137	145,0	108,0

On arbitrarily oriented plane, the structure of closure-type domains of the sample and the DS of surface stressed layer is identified by the polar effect Kerr technique at the same time. Fig.3a shows images of the nonequilibrium DS obtained in the grains of polycrystalline sample, which orientations are close to the prismatic plane. It is seen that the walls of surface layer domains are oriented perpendicular to the walls of the main domains.

Analysis of all obtained data suggests that the mechanism of relaxation of the surface DS has thermal fluctuation character.

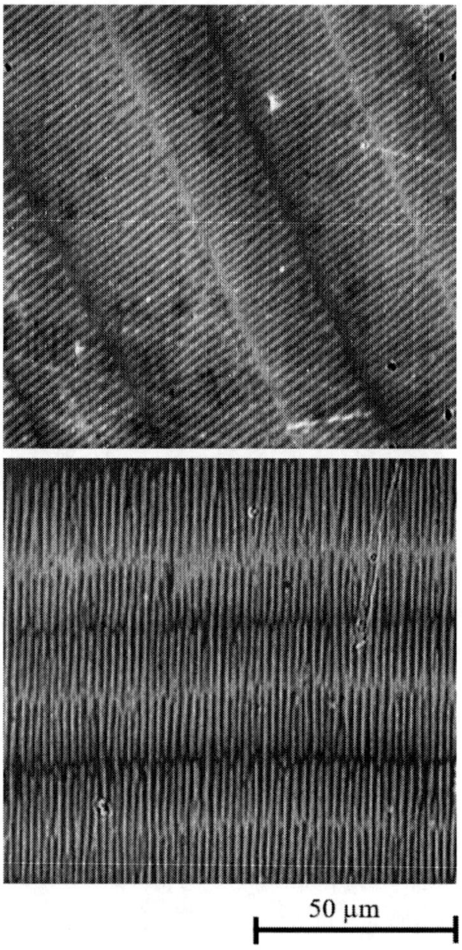

Fig. 3. Magnetic domain structure on the arbitrary plane of $DyFe_{11}Ti$ polycrystalline sample detected by Kerr magneto-optical microscopy at room temperature after mechanical polishing.

Conclusions

It was established that the mechanical impact induced during polishing in the single crystal and polycrystalline $DyFe_{11}Ti$ compound leads to configuration change of the surface domain structure.

Through time in the absence of external magnetic fields stress-induced surface domain structure relaxes into an equilibrium configuration that is typical for a magnetic material with the uniaxial anisotropy.

The relaxation time of the surface domain structure depends on the duration and severity of exposure to the surface, and the sample temperature. Heating the sample enhances the recovery process of the surface domain structure, which indicates the thermal fluctuation nature of the relaxation process.

Summarizing the obtained results, the observation of domain structure by means of magneto-optical Kerr microscopy can be used as high-throughput analysis technique for characterization of new uniaxial phases within combinatorial analysis. Nevertheless, it should be mentioned that detection and quantitative analysis of uniaxial phases can be erroneous due to the impact of external stresses during polishing. To eliminate the possible errors one need to use some chemical etching or thermal or magnetic treatments.

This work was supported by grant of Ministry of Education and Science of the Russian Federation № 3.7849.2017/БЧ and Russian Science Foundation Grant № 15-12-10008.

References

1. A. Hubert, R. Schäfer, Springer Berlin Heidelberg NewYork, 707 (2009).
2. Yu.G. Pastushenkov, Zeitschrift für Metallkunde 93(10) 991 (2002).
3. Yu.G. Pastushenkov, J. Bartolome, A. Larrea, K.P. Skokov, T.I. Ivanova, L.V. Lebedeva, A.G. Grushichev, J. Magn. Magn. Mater. **300**, e514 (2006)
4. Yu. Pastushenkov, J. Bartolome, N. Suponev, K. Skokov, T. Ivanova, A. Larrea, M. Lyakhova, E. Semenova, S. Smirnov, J. Alloys Comp. **451**, 488 (2008)
5. B.-P. Hu, H.-S. Li, J.P. Gavigan, J.M.D. Coey, J. Phys.: Condens. Matter **1** , 755 (1989)
6. X.C. Kou, T.S. Zhao, R. Grössinger, H.R. Kirchmayr, X. Li, F.R. de Boer, Phys. Rev. B **47**, 3231 (1993)
7. W. Suski, Handbook on the Physics and Chemistry of Rare Earths **22**, 143 (1996)
8. K.Yu. Guslienko, X.C. Kou, R. Grössinger, J. Magn. Magn. Mater. **150**, 383 (1995)
9. L.M. García, J. Bartolomé, P.A. Algarabel, M.R. Ibarra, M.D. Kuz'min, J. Appl. Phys. **73**, 5908 (1993)
10. S.A. Nikitin, I.S. Tereshina, N.Yu. Pankratov., Phys. Solid State **41**, 1508 (1999)
11. A.V. Andreev, M.I. Bartashevich, N.V. Kudreva-tykh, E.N. Tarasov, Physica B: Condens. Matter **167**, 139 (1990)

Spontaneous magnetization reversal caused by magnetic noise in ε-$In_{0.24}Fe_{1.76}O_3$ nanoparticles

Alexei Dmitriev*

Institute of Problems of Chemical Physics, Russian Academy of Sciences, 142432 Chernogolovka, Russia

Abstract. Kinetics of magnetization relaxation of the exotic ε-$In0_{.24}Fe_{1.76}O_3$ nanoparticles under applied magnetic field has been studied. The fluctuation field and the activation volume have been calculated from the magnetic viscosity data. The relation between magnetic viscosity and magnetic noise caused by the random thermally activated magnetization reversal of a single nanoparticle has been established. Stepped sweeping of magnetic field expands the windows of experimentally detectable magnetic fluctuations. The changes in the reversal magnetic field provide ε-$In_{0.24}Fe_{1.76}O_3$ nanoparticles scanning and sorting them by magnetic noise frequency.

1 Introduction

Recently it has become possible to create new nanomagnets based on the epsilon phase of iron (III) oxide ε-Fe_2O_3 with a giant (up to 2.34 T) coercive force [1]. This exceeds all known values revealed by such magnets as $Fe_{14}Nd_2B$ or $SmCo_5$. Permanent magnets based on ε-Fe_2O_3 nanoparticles are much cheaper than rare-earth magnets such as $Nd_{14}Fe_2B$ and $SmCo_5$. Static magnetic properties of ε-Fe_2O_3 nanoparticle ensembles were studied in detail [2]. Time stability of parameters of permanent magnets is one of the main criteria of applicability of materials in various devices. Therefore, in this work, slow relaxation of magnetization of the epsilon phase of iron oxide nanoparticle ensembles has been analyzed.

Slow magnetization reversal (magnetic viscosity) in different magnets is caused by magnetic noise [3]. Despite intensive investigations of ε-Fe_2O_3 nanomagnets, the problems of the spectrum of magnetic fluctuations, spontaneous demagnetization mechanisms and time stability are still open questions. The purpose of this work is to find a relation between magnetic viscosity and spectral characteristics of magnetic fluctuations in epsilon phase of iron oxide nanoparticle ensembles.

2 Experimental

The epsilon phase of iron (III) oxide ε-$In_{0.24}Fe_{1.76}O_3$ nanoparticles with a length of 80 nm and a diameter of 35 nm (Fig. 1) were fabricated by two methods, namely, synthesis in inverse micelles and sol-gel method. The nanoparticles have orthorhombic crystal structure with four non-equivalent cationic positions of Fe^{3+} ions. One of them has tetrahedral and three others octahedral environment. Indium ions are substitution impurities in the octahedral positions of iron ions. Synthesis and characterization of the nanowires were described in detail in [4, 5].

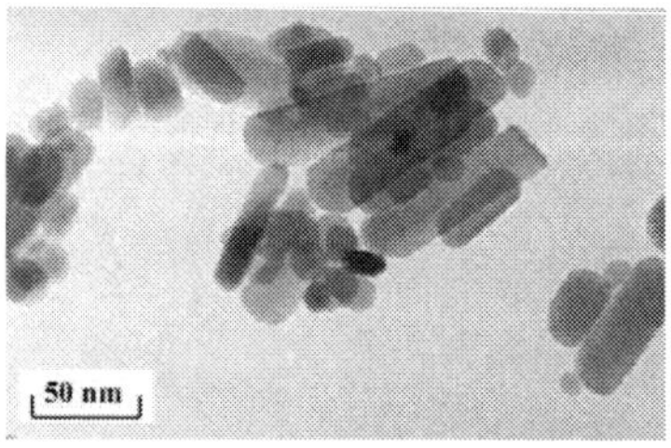

Fig. 1. TEM image of ε-$In_{0.24}Fe_{1.76}O_3$ nanoparticles.

A Quantum Design MPMS 5XL SQUID magnetometer was used in standard measurements of the field dependence *M*(*H*) of sample magnetic moment. The SQUID magnetometer was also used to obtain time *M*(*t*) dependences of magnetic moment for magnetic fields in the range $H = 1 - 13$ kOe at $T = 300$ K. Fig. 2(a) shows the scheme of applied magnetic field switching in measurements of the *M*(*t*) dependences.

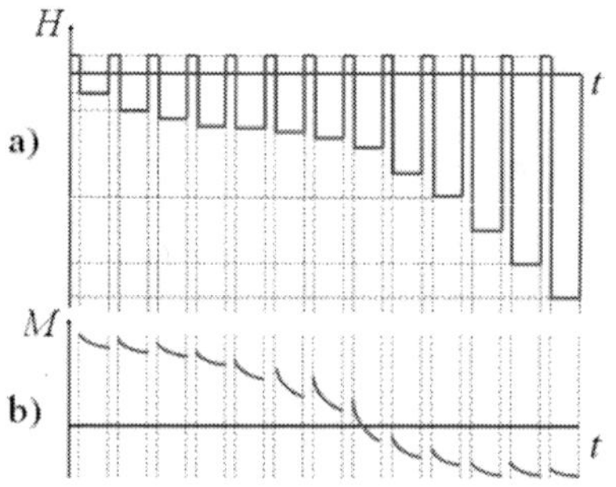

Fig. 2. Scheme of applied magnetic field switching in the experiments.

* Corresponding author: aid@icp.ac.ru

A sample was first magnetized to saturation in magnetic field H = 50 kOe, which is higher than the saturation field. Then the field directed opposite to the magnetization of the sample was turned on. After negative magnetic field became steady, the time dependence of magnetization of a sample was recorded. Fig. 2(b) shows the schematic diagram of relaxation of the magnetic moment of a sample.

3 Results and discussion

3.1 Magnetic viscosity

The time dependence of magnetic moment increment $\Delta M(t)$ recorded in the reverse magnetic field H = 9 kOe at T = 300 K is shown in normal and semi-logarithmic coordinates in Fig. 3.

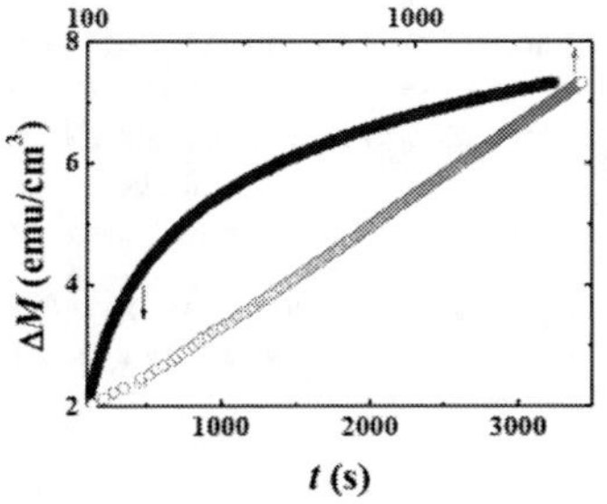

Fig. 3. Time dependence of magnetic moment $\Delta M(t)$ at T = 300 K in external magnetic field H = 9 kOe in normal and semi-logarithmic coordinates.

The dependence $\Delta M(t)$ follows the logarithmic law:

$$\Delta M \sim S \ln t \quad (1)$$

Tangent of the slope of $\Delta M(\ln t)$ curve corresponds to magnetic viscosity S:

$$S = k_B T/\Delta E \quad (2)$$

where k_B is the Boltzmann constant, T – temperature, ΔE – range of activation energy of magnetic fluctuations.

Time dependences $\Delta M(t)$ were recorded in different magnetic fields at T = 300K (Fig. 4).

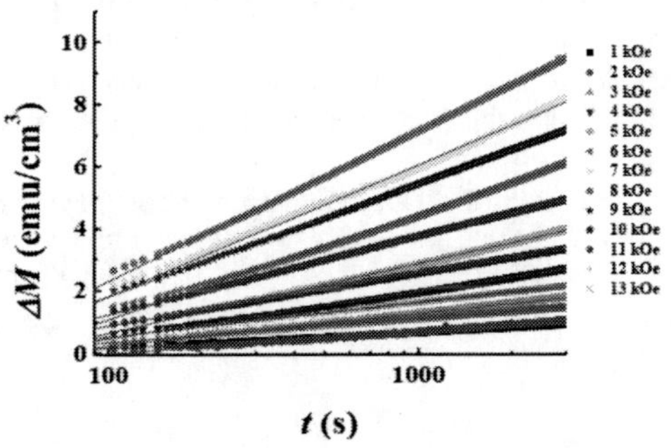

Fig. 4. Time dependences of magnetic moment $\Delta M(t)$ at T = 300 K in different magnetic fields. Solid lines show approximations.

Slopes of these curves and correspondent magnetic viscosities are non-monotonic functions of magnetic field (Fig. 5).

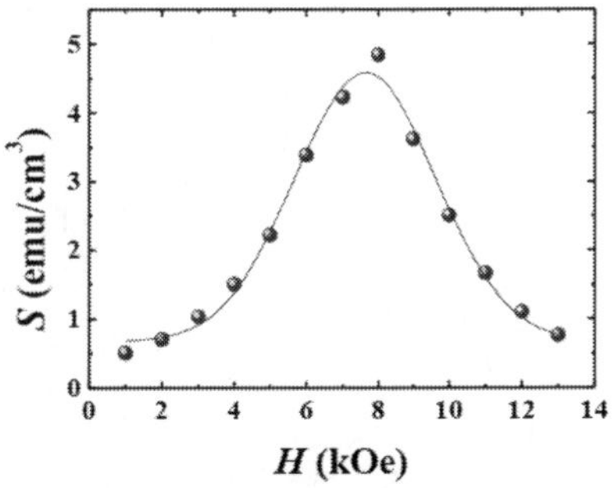

Fig. 5. Field dependence of viscosity S at T = 300 K. Solid line is a guide for the eye.

Field dependences $S(H)$ at T = 300K show maximum corresponding to 8 kOe. The shape of the $S(H)$ depends on volume distribution of nanoparticles $f(V_C)$. Those nanoparticles contribute to magnetic relaxation only, which volume is equal to critical nanoparticle volume $V_C = 25k_BT/(K(1-(H/H_a)^2))$. Here, K is constant of magnetic anisotropy, k_B is Boltzmann constant, H_a is magnetic anisotropy field. Relaxation of nanoparticles of smaller volume $V < V_C$ is finished when new value of magnetic field is attained, i.e. before time dependence recording. The nanoparticles of larger volume $V > V_C$ do not contribute to magnetic relaxation because their energy barrier separating opposite directions of magnetization is too high in comparison with the energy of thermal fluctuations. In case of lognormal distribution $f(V_C)$ shape of the $S(H)$ dependence is close to the lognormal distribution function. The peak position of $S(H)$ dependence corresponds to the magnetization reversal of the nanoparticles coinciding to the mode of distribution $f(V_C)$.

3.2. Fluctuation field and activation volume

Fluctuation field H_f is another fundamental parameter additionally to magnetic viscosity S [6]. The fluctuation field was used to compare magnetic fluctuations intensity with typical parameters of the nanoparticles expressed in the same units (anisotropy field, coercive field, etc). Fuctuation field is defined by equation [7]:

$$H_f = S/\chi_{irr} \quad (3)$$

where χ_{irr} is irreversible susceptibility expressed as

$$\chi_{irr} = \chi/(1+N\chi) \quad (4)$$

where $N = 2\pi$ is the demagnetizing factor of cylinder and $\chi = dM/dH$ is the slope of tangent to demagnetizing curve of magnetic hysteresis loop [8, 9] (Fig. 6). Field dependence $\chi_{irr}(H)$ at T = 300 K show maximum corresponding to 7 kOe (fig. 7). Positions of the $S(H)$ and $\chi_{irr}(H)$ maxima are close to coercive field H_c = 6 kOe determined from the hysteresis loop [8, 9] (Fig. 6).

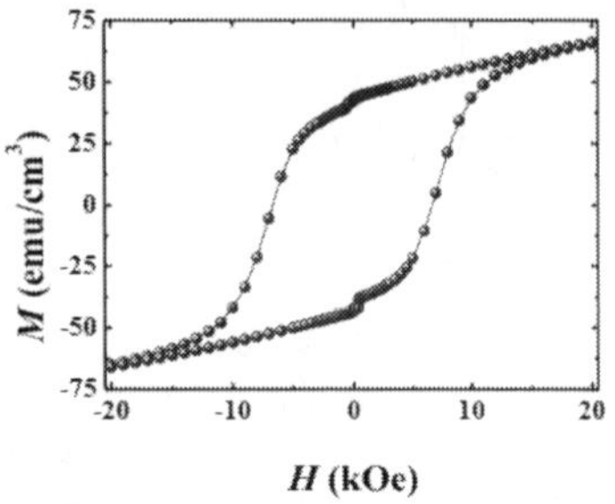

Fig. 6. Magnetic hysteresis loop in ε-$In_{0.24}Fe_{1.76}O_3$ nanoparticles at $T = 300$ K. Solid line is a guide for the eye.

A small discrepancy is caused by measurement uncertainty of magnetic field, defined as half of the step of the magnetic field sweep, which is 1 kOe.

The average fluctuation field $H_f = 315$ Oe at $T = 300$ K was estimated from Eqs. (3) and (4). The H_f value is close to typical values in hard magnetic materials such as $Nd_2Fe_{14}B$ [10].

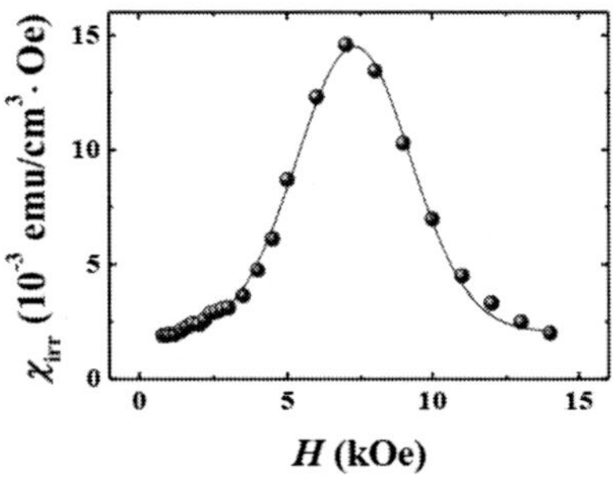

Fig. 7. Field dependence of irreversible magnetic susceptibility χ_{irr} at $T = 300$ K. Solid line is a guide for the eye.

The H_f value can be used to calculate activation volume [8]:

$$V_a = k_B T / M_s H_f \quad (5)$$

$M_s = 84.4$ emu/cm^3 at $T = 300$ K [7]. The activation volume corresponds to the average volume of thermal activation. Average activation volume $V_a = 1.6 \cdot 10^{-18}$ cm^3 at $T = 300$ K was estimated from Eq. (5). The V_a value is noticeably smaller than the volume of the single ε-$In_{0.24}Fe_{1.76}O_3$ nanoparticle. This implies that only a small part of its volume is involved in the elementary act of magnetization reversal.

3.3. Spectral density of magnetic fluctuations

On the assumption of a relaxation time distribution $F(\tau)$, spectral density of magnetic fluctuations $W(f)$ is defined by the following equation [3]:

$$W = k_B T / 2f\Delta E \quad (6)$$

When comparing Eqs. (2) and (6), we can find a relation between magnetic viscosity and spectral density of thermal fluctuations:

$$W(f) = S/2f \quad (7)$$

Equation (7) is a "bridge" connecting the macroscopic phenomenon of magnetic viscosity and the microscopic fluctuation phenomena of magnetization reversal of a single nanoparticle. Thus, when controlling magnetic viscosity by changing magnetic field, we change the width of "window" for detecting a spectrum of magnetic noise.

4 Conclusion

Time dependences of magnetization reversal of ε-$In_{0.24}Fe_{1.76}O_3$ nanoparticles in the reverse magnetic field were measured and analyzed in a wide range of magnetic field strength. Logarithmic kinetics of magnetization reversal allowed field dependences of magnetic viscosity to be extracted. The logarithmic kinetics of magnetic viscosity is determined by a superposition of thermally activated magnetization reversal of a single nanoparticle with a wide relaxation time distribution. A relation between magnetic viscosity and spectral characteristics of magnetic fluctuations in nanoparticle ensembles was found. The fluctuation field and the activation volume were calculated from the magnetic viscosity data.

Supported by RFBR, project no. 16-07-00863 a. Author is grateful to Prof. R.B. Morgunov for samples providing, measurements support and discussion.

References

1. J. Jin, S. Ohkoshi, K. Hashimoto, Adv. Mater. **16**, 48 (2004).
2. J. Tucek, R. Zboril, A. Namai, S. Ohkoshi, Chem. Mater. **22**, 6483 (2010).
3. A.I. Dmitriev, A.D. Talantsev, E.I. Kunitsyna, R.B. Morgunov, V.P. Piskorskii, O.G. Ospennikova, E.N. Kablov, J. Exp. Theor. Phys. **123**, 303 (2016).
4. S. Sakurai, S. Kuroki, H. Tokoro, K. Hashimoto, S. Ohkoshi, Adv. Funct. Mater. **17**, 2278 (2007).
5. K. Yamada, H. Tokoro, M. Yoshikiyo, T. Yorinaga, A. Namai, S. Ohkoshi. J. Appl. Phys. **111**, 07B506 (2012).
6. L. Neel, J. Phys. Radium **11**, 49 (1950).
7. E.P. Wohlfarth, J. Phys. F: Met. Phys. **14**, L155 (1984).
8. A.I. Dmitriev, O.V. Koplak, A. Namai, H. Tokoro, S. Ohkoshi, R.B. Morgunov, Phys. Solid State **55**, 2252 (2013).
9. A.I. Dmitriev, O.V. Kolpak, A. Namai, H. Tokoro, S. Ohkoshi, R.B. Morgunov, Phys. Solid State **56**, 1795 (2014).
10. J.C. Martinez, F.P. Missell, J. Appl. Phys. **64**, 5726 (1988).

STUDY OF REVERSIBILITY OF COERCIVITY UNDER HEAT TREATMENT IN PERMANENT MAGNETS

Alexey Lileev, *Victoria* Pinkas, *Kristina* Voronchihina, and *Anton* Gunbin

NUST "MISIS", Moscow, Russia

Abstract. This article is aimed at observation all discussion and results that have ever been found about the phenomenon of reversibility of coercivity under heat treatments in permanent magnets including the latest researches. Despite the fundamental differences in the nature of coercivity as sensitive to structure property and magnetization reversal processes all works illustrate no changes in structure under heat treatment that causes doubts about this phenomenon.

Introduction

The phenomenon of reversibility of coercivity under heat treatments in permanent magnets have known for a long time. Short time heating of the magnets at a high temperature after the complete technological cycle leads to a sharp reduction of magnetic properties, especially the coercivity. Low-temperature aging reverses alloys to initial magnetic properties. Analysis of previous investigations and additional experiments show that reversibility of coercivity is characteristic of almost all magnets. This phenomenon occurs in almost all alloys for permanent magnets, despite the fundamental differences in the nature of coercivity as sensitive to structure property and magnetization reversal processes. Present work based on studies of changes and the reversibility of magnetic properties under different heat treatments attempted to explain this phenomenon in different magnetic alloys.

Comparative study of reversibility of coercivity

Reversibility of coercivity of the permanent magnets under heat treatments discovered in current research and previous works remains unclear and ambiguous because of many issues of this phenomenon at present. In current research, the results of studies and literature data describe the samples of alloys obtained from industrial enterprises using a standard, common technology, and the measurements were performed with standard equipment.

Table 1 Reversibility of coercivity under heat treatments in permanent magnets

Year	Material	Hc optimal кA/m(kOe)	Hc spoiling кA/m (kOe)	Hc reconditioning кA/m (kOe)	Source
1964	Alnico	48 (0.6)	32 (0.4)	48 (0.6)	[1]
1973	Ticonal	116 (1.45)	36 (0.45)	116 (1.45)	[2]
1975	$SmCo_5$	1840 (23.0)	1240 (15.0)	1840 (23.0)	[3]
1979	Sm-Co-Cu-Fe-Zr	1120 (14.0)	720 (9.0)	1120 (14.0)	[4]
2014	$Sm(Co_{0.65}Fe_{0.26}Cu_{0.07}Zr_{0.02})_7$	2360 (29.5)	96 (1.2)	2360 (29.5)	[5]
2014	$Nd_2Fe_{14}B$ (1050°C)	1520 (19.0)	1190 (14.9)	1520 (19.0)	[6]
2016	$Nd_2Fe_{14}B$ (1000°C)	1740 (21.8)	1247 (15.7)	1662 (20.9)	present research
2015	$Nd_2Fe_{14}B$ (900 °C)	1740 (21.8)	1340 (16.9)	1740 (21.8)	[7]
2017	$Nd_2Fe_{14}B$ (800 °C)	1740 (21.8)	1367 (17.2)	1740 (21.8)	present research
2017	$Nd_2Fe_{14}B$ (700 °C)	1740 (21.8)	1332 (16.7)	1740 (21.8)	present research
2017	FeCrCo	44 (0.55)	2.0 (0.025)	44 (0.55)	present research

Table summarizes data for a reversible change of the coercivity of the magnets under heat treatments obtained on the various alloys with reference to authors (the coercivity in the initial condition – "Hc optimal", after high-temperature heating of the initial state – "Hc spoiling" and after the low temperature aging – "Hc reconditioning").

This table shows only values of the coercivity on the magnetization as this property is sensitive to structure and determined by the mechanism of magnetization reversal. High temperature range that leads to reduction of properties ('spoiling') and low temperature aging that reverses initial properties ('reconditioning') are described in details later. The essential point is the reversible changes of the coercivity, as the structure of the alloy is not changed reversibly by heat treatment. The aim of the present work is to analyse studies of magnetization reversal processes in real permanent magnets for the possibility of further improvement of technology.

The analysis starts with the system Fe-Co-Ni-Al-Ti (Alnico and ticonal), in which the coercivity is determined by the rotation process of the magnetization vector in the elongated single-domain particles of the ferromagnetic phase. Heating of the samples with coercivity 116 kA/m (1.45 kOe) at a temperature of 800 °C for 1-3 minutes causes it to reduce to 36 kA/m (0.45 kOe). Aging at 650 °C within 5 hours and 560 °C in 24 hours reverse of coercivity to the initial value.

In this case, the reversible changes $_IH_C$ were about 70 % connected with reversible local change in the composition of α_1 and α_2 – phases and possible atomic ordering in one of the phases [2], so the quantitative changes are not explained only by the composition changes. However, in these studies it is ignored the fact of changes in magnetostatic interaction between particles that is related to changes of the magnetic moment connected with changes in the composition of the phases.

In the magnets of system Fe-Co-Cr coercivity is also associated with the rotation process of the magnetization vector in single-domain particles. The coercivity after the full treatment cycle – 620 °C 45 minutes TMT (Thermomagnetic Tempering) and step aging 610 °C in 2 hours, 580 °C in 3 hours, 560 °C in 4 hours and 540 °C in 6 hours) is 44.9 kA/m. The "spoiling" (620 °C for 15 minutes) reduces the coercivity to 0.6 kA/m. Step aging reverse the coercivity to 43.7 kA/m. Reversibility of 95 % of coercivity is not able to be explained by local changes of structure. Even accounting for the extended partition of particles, which is observed in these alloys does not eliminate the problem of understanding the phenomenon. These results were obtained for the first time and there was no a detailed study of the influence of temperature and duration of the "spoiling", but the result deserves attention.

In sintered magnets based on $SmCo_5$ and Nd-Fe-B coercivity is determined by the difficulty of nucleation in highly anisotropic phase. In [3] a sintered permanent magnet $SmCo_5$ with the coercivity 1840 kA/m (23.0 kOe) was heated to a temperature 700 °C for 1 minute that reduced the coercivity to 1240 kA/m (15.0 kOe). Aging at 400 °C for 60 minutes reverses the value of the coercivity. Reversible changes was about 30 %.

In [6] "spoiling" sintered magnets Nd-Fe-B of different composition at high temperature 1050 °C within 30 minutes was performed, and then spent the cooling at 500 °C for 1 hour. Reversible changes of the coercivity were about 30 %.

In [7] and in subsequent studies, experiments of 'spoiling' were performed at temperatures of 1000, 900, 800 and 700 °C. Aging at 500 °C for 1 hour reversed the value of the coercivity to the initial. Reversible changes of the coercivity were about 30%.

It is difficult to explain the reversibility of changes of structural properties in single-phase materials. In several papers it is assumed that the process of aging at 500 °C makes grain boundaries smooth which leads to increase of coercivity. However, it is difficult to suppose explanation of the phenomenon of reversibility. There is hypothesis that excess phases occur transformation at the boundary of cell phase [8].

The most interesting results and discussions are related to the reversibility of coercivity that exists in system of Sm-Co-Fe-Cu-Zr, in which the coercivity is determined by pinning of domain walls at the boundaries of cells phase 2:17. The boundary cells similar to phase 1:5 [9]. In [5] a study of the "spoiling" at a temperature of 800 °C within 16 hours at which the structure of the alloy is formed, followed by controlled cooling from 400 °C to 800 °C for 4 hours. Heating of the samples with coercivity of 2360 kA/m (29.5 kOe) for 5 minutes reduces the coercivity to 96 kA/m (1.2 kOe). Controlled cooling reverses coercivity and a reversible change is 95 % or almost 30-fold change compared to the state with zero coercivity. It is important to mention that there are no changes in value of reversal coercivity with increasing aging time at 800 °C up to 22 hours.

In [9-11] detailed chemical analysis was performed with using 3D atom probe of the boundary-phase 1:5 in the state after rapid cooling from 800 °C and after controlled cooling. It is shown that on the border there is a redistribution of Cu and Zr, and changes the nature of the distribution of the components. In [11] the calculation of the distribution constants of anisotropy in the phase boundary takes into account the changing nature of the distribution of the components. However, the question how the changing nature of distribution of the constants of anisotropy can affect multiple changes in the coercivity is not solved.

As can be seen from the listed examples of reversible changes of the coercivity under heat treatments, many questions are related not only to the formation of the alloy structure, but also to the nature of coercivity and magnetization reversal processes.

References

[1] A.S. Ermolenko. Diss. candidate of ph. and math. sciences, Institute of metal physics, Sverdlovsk USSR (1964)

[2] A.S.Lileev, V.P.Menushenkov, V.I. Sumin, *PhMM*, (1973) 183-186.

[3] B.G. Livshitz, A.S. Lileev, V.P. Menushenkov, *IEEE Trans. Magn.*, (1975) 1673 – 1675.

[4] V.I. Habarov, T.Z. Puzanova, Ya.S. Shur, *PhMM*, (1979) 921-926.
[5] A.S. Lileev, O.A. Arinicheva, M. Reissner, F. Kubel, V. A. Sein, *J. MSHT*, **56**, (2016) 591-594.
[6] T.G. Woodcock, F. Bittner, T. Mix, *JMMM*, **360** (2014) 157-164..
[7] *Labour Digest VI INTERNATIONAL CONFERENCE Functional nanomaterials and high-purity substances*. Suzdal, Russia (2016).
[8] V.P. Menushenkov, A.G. Savchenko, *JMMM*, **272** (2004) 1891-1893.
[9] X.Y. Xiong, T. Ohkubo, T. Koyama, K. Ohashi, Y. Tawara, K. Hono, *Acta Mater*, **52**, (2004) 737–748.
[10] O.Gutfleisch, K.-H. Müller, K. Khlopkov, M. Wolf, A. Yan, R. Schäfer, T. Gemming, L. Schultz, *Acta Mater*, **54**, (2006) 997–1008.
[11] H. Sepehri-Amin, J. Thielsch, J. Fischbacher, *Acta Mater*, **126**, (2017) 1-10.

Stress Effects on Magnetic Properties of Amorphous Microwires Subjected to Current Annealing

Nematov Makhsudsho G.[1,3,*], *Yudanov Nicolay A.*[1], *Yakubtsev Rodion A.*[1], *Panina Larissa V.*[1,2], *Beklemisheva Anna V.*[1], *Morchenko Alexander T.*[1], *Thakur Atul*[4]

[1]National University of Science and Technology, MISIS, Moscow 119991, Russia
[2]Institute for Design Problems in Microelectronics RAS, Moscow 124681, Russia
[3]Tajik Technical University named after ac. M.S. Osimi, Dushanbe, 734042, Tajikistan
[4]Centre of Nanotechnology, Amity, University Haryana, Manesar-122413, India

Abstract. Effects of current annealing on magnetic hysteresis properties and magnetoimpedance (MI) of glass-coated amorphous microwires with nominal composition of $Co_{71}Fe_5B_{11}Si_{10}Cr_3$ and small positive magnetostriction were investigated with the purpose to control the magnetic anisotropy. We have demonstrated that current annealing can produce a controllable change in the easy anisotropy direction from almost axial to circular, depending on the annealing time. The induced magnetization configuration is very sensitive to the applied tensile stress. The combination of positive magnetostriction and helical anisotropy makes it possible to realise large stress-magnetoimpedance (S-MI) effect without use of any dc bias fields owing to the directional change in magnetization. This is especially important for microwave frequency S-MI effect and can be very interesting for developing stress-sensors operating at the high frequency region.

1 Introduction

In recent years, a considerable interest has been devoted to studying the magnetization processes and magnetoimpedance (MI) in Co-Fe based amorphous glass-coated microwires (diameter is between 1 to 100 μm), which is related to their technical applications in miniature sensor devices [1-4]. Owing to the absence of a crystalline structure the magnetoelastic interactions play the main role in determining the magnetic anisotropy in amorphous microwires. The easy anisotropy axes are therefore affected by the value of the saturation magnetostriction and the distribution and values of internal stresses both of which may strongly depend on temperature. This could be useful to modify the magnetic anisotropy by various heat treatments partially releasing the internal stress occurring during fast solidification [5-7]. Among thermal treatments annealing in the presence of magnetic field has a special importance since it is possible to form easy anisotropy direction along the field due to short-range pair ordering [8-9].

Co-Fe alloys on Co-rich side possess a near-zero magnetostriction and excellent soft magnetic properties. In these systems, it is possible to control the magnetic anisotropy by current annealing which combines the Joule heating and circular magnetic field [10-12]. This technique also helps to avoid the deviations in the easy anisotropy due to inhomogeneous distribution of internal stresses. In microwires with a small negative magnetostriction where the magnetoelastic anisotropy is predominantly circumferential the current annealing helps to improve this anisotropy and to enhance the relative change in impedance up to 600% [10].

In the present work, we report on the influence of dc current annealing on the magnetic anisotropy in Co-rich amorphous microwires having a small positive magnetostriction. We have demonstrated that the current annealing produces a controllable change in the easy anisotropy direction from almost axial to helical accompanied by corresponding changes in the magnetic hysteresis and MI. The induced helical anisotropy not only increases the relative change in impedance (MI ratio) up to 200% (from 10% in as-prepared state) in the low-field region, but makes it possible to realize highly stress-sensitive MI. This has a potential for applications in miniature stress sensors, in particular, for embedded sensing techniques [13-15].

2 Analysis of induced anisotropies

In amorphous alloys, the effective magnetocrystalline anisotropy is low due to the averaging effect of exchange interaction. The additional contribution to the anisotropy before annealing comes from magnetoelastic interactions. In glass-coated amorphous wires the internal stress σ_i originated from quenching and drawing processes is predominantly tensile and directed along the wire [15-18].

The combination of a negative magnetostriction and tensile stress produces a circumferential easy anisotropy. Then, the external tensile stress strengthens this anisotropy, but the character of magnetization loops does not change. This could present a problem when realizing

* Corresponding author: nematovmaqsud@misis.ru

stress-sensitive MI, since the highest sensitivity of the impedance change is caused by the directional change in magnetization [19-20].

For positive magnetostriction the tensile stress sets the axial easy anisotropy. Heat treatments produce easy anisotropies along the direction of local magnetization. The presence of a magnetic field during annealing aligns the magnetization and induces a uniform anisotropy with the easy axis along the field, that is, along circumference. In this case, the applied tensile stress changes the direction of the easy axis which is needed for large changes in impedance, especially at microwave frequencies [13].

To analyze the change in the magnetic anisotropy due to current annealing and application of external stresses σ_{ex} it is assumed that the wire has also a frozen-in torsion stress σ_t, that corresponds to the effects of tension and compression $\pm\sigma_t$ directed at 90° to each other and at 45° with respect to the wire axis. The presence of torsion will allow us to introduce a generalized helical magnetic anisotropy and to describe a gradual transition from axial to circular anisotropy due to annealing and from circular to axial anisotropy after annealing upon the application of an external tensile stress. The current annealing induces a circumferential easy anisotropy the value of which K_a depends on the annealing conditions. The magnetic energy U_m can be then written in the form:

$$U_m = -(K - K_a)\cos^2\theta - \frac{3}{2}\lambda(\sigma_i + \sigma_{ex})\cos^2\theta - \frac{3}{2}\lambda\sigma_t\left[\cos^2\left(\theta - \frac{\pi}{4}\right) - \cos^2\left(\theta + \frac{\pi}{4}\right)\right] \quad (1)$$

Here K is the uniaxial anisotropy constant, which is not related to the magnetostriction interactions, θ is the angle between the magnetization and the wire axis, and $\lambda > 0$ is the magnetostriction coefficient. Equation (1) can be converted into the form typical of a uniaxial anisotropy with the easy anisotropy axis defined by angle α with respect to the wire axis:

$$U_m = -|\tilde{K}|\cos^2(\alpha - \theta) \quad (2)$$

The equivalent constant of uniaxial anisotropy $\tilde{K}$ is defined as :

$$\tilde{K} = [K - K_a + \left(\frac{3}{2}\right)\lambda(\sigma_i + \sigma_{ex})]/\cos 2\tilde{\alpha} \quad (3)$$

$$\tilde{\alpha} = \frac{1}{2}\tan^{-1}\frac{3|\lambda\sigma_t|}{|K - K_a + (3/2)\lambda(\sigma_i + \sigma_{ex})|} \quad (4)$$

The easy angle α in equation (2) depends on the sign of $\tilde{K}$: $\alpha = \tilde{\alpha}$ for $\tilde{K} \geq 0$ and $\alpha = 90° - \tilde{\alpha}$ for $\tilde{K} < 0$.

In the case of $K > 0$ and a positive magnetostriction $\lambda > 0$, before annealing the angle $\alpha < 45°$ and the easy anisotropy axis is close to the wire axis, if $\sigma_i \gg \sigma_t$. Upon annealing, there is an increasing contribution of K_a so the sign of $\tilde{K}$ becomes negative which allignes the anisotropy closer to circumference ($\alpha > 45°$). This will change the character of the magnetization curves: the rectangular loop is transformed into an inclined flat loop, and the hysteresis region decreases significantly. Upon the application of external tensile stresses, the effective anisotropy constant $\tilde{K}$ becomes positive and the anisotropy direction tends towards the wire axis ($\alpha <$ 45°).

3 Experimental details

Amorphous glass-coated microwires with the nominal composition of $Co_{71}Fe_5B_{11}Si_{10}Cr_3$ produced by the Taylor–Ulitovski technique [21] were kindly supplied by MFTI Ltd. (Republic of Moldova). The total diameter was 29.5 µm and the metal core diameter was about 23.9 µm. The insulating glass was of Pyrex type. The wires have a small positive magnetostriction in the order of 10^{-7}.

The Curie temperature (T_c = 637 K) and the crystallization temperature (T_{cr} = 736 K) were estimated from DSC curves (by using the DSC 204 F1 Netzsch instrumentation) with the help of standard IT application. The DSC curves were also measured after annealing confirming that the amorphous structure did not undergo noticeable changes.

The wire samples of 15 cm long were annealed by passing a dc current (I) of 50 mA during 1 - 3 hours in air. All current treatments were performed in the same ambient conditions. The current magnitude was chosen in order to realize a moderate heating effect in the range of temperatures 470 - 530 K, which are considerably lower than T_c and T_{cr}. The electrical contacts were made by standard soldering technique removing the glass coat at the ends. The current (I) produces a circular magnetic field H_a the value of which at the sutface $H_a = I/2\pi a$ was about 400 A/m which is sufficient to align the local magnetization.

The hysteresis loops under the effect of applied load up to 340 MPa were obtained by using induction method with two differential detection/magnetization coils having an inner diameter of 3 mm. The magnetization coil was driven by a current of a frequency of 500 Hz, which produced a magnetizing field with the amplitude of up to 1000 A/m. Such magnitude is sufficient for re-magnetizing the wires. The load was hang at one end of the microwires whilst the other was kept fixed. The magneto-impedance was measured using a Hewlett-Packard 8753E Vector Network Analyzer in the frequency range between 1 and 100 MHz at room temperature. Two-port measuring scheme (S_{21}) was used. The sample was placed inside of a solenoid coil that produced a slowly varying magnetic field up to 600 A/m. In order to introduce a tensile stress during S_{21} measurement, stripes of paper of equal thickness were inserted under the middle part of the wire. It is difficult to evaluate an exact value of the applied stress but it is possible to continuously increase its magnitude by increasing the number of stripes.

4 Results and discussion

The longitudinal hysteresis loops for as-cast and current annealed wires are shown in Figure 1. Obviously, the current annealing produced a change in the direction of the easy anisotropy as discussed above. For sufficiently long annealing times (about 2 hours or longer) the current induced anisotropy dominates over the residual contributions from the magnetocrystalline and magneto-elastic anisotropies, so the analysis based on equations (1) - (4) can be applied. Accordingly, a near rectangular loop transforms into a flat-type loop and for the annealing time of 2 hours the remanence-to-saturation ratio reduces to about 0.15. This corresponds to almost a circular easy anisotropy. For longer annealing times the type of magnetic hysteresis does not change but the effective anisotropy continuous to grow. This further increase in the anisotropy field is unfavorable for various sensing applications and an optimal annealing time should be used.

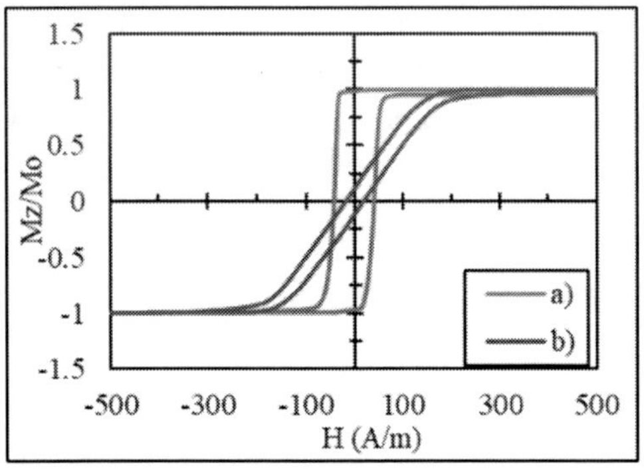

Fig. 1. Hysteresis loops of amorphous $Co_{71}Fe_5B_{11}Si_{10}Cr_3$ microwires: a) as-prepared and b) after annealing by current of 50 mA for 2 hours.

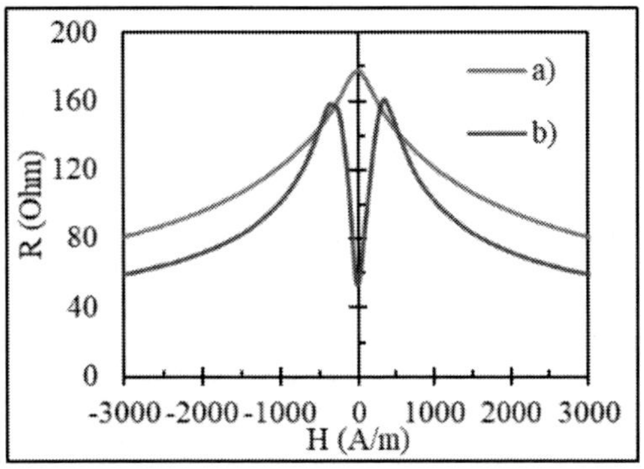

Fig. 2. Real part of the impedance vs. magnetic field at 50 MHz for amorphous $Co_{71}Fe_5B_{11}Si_{10}Cr_3$ microwires: a) as-prepared and b) after annealing by current of 50 mA for 2 hours.

The coercivity value continuously decreases with increasing annealing time. For as-prepared wires, H_c is about 25 A/m whereas it decreases down to 13 A/m for annealing time of 2 hours. The coercivity decrease after annealing is associated firstly with a different magnetization configuration: the driving field becomes almost orthogonal with respect to the easy anisotropy axis.

We further investigated the magnetoimpedance characteristics in the frequency range of 1 - 100 MHz. It is interesting to compare the MI behavior in microwires before and after current annealing, which confirms a drastic change in the anisotropy by current annealing. The MI plots are presented in Figure 2. For the as-prepared wires, the impedance has a maximum in the zero field, which is typical of the systems with the easy anisotropy axis parallel to the high-frequency current and dc external magnetic field. After current annealing, the impedance behavior has changed showing two symmetrical peaks, which is consistent with the transverse anisotropy with respect to the high frequency current and dc magnetic field [19-20].

It is also seen that the MI ratio after current annealing increased: in the low field region (<400 A/m), it is about 200%. For as prepared wires, the sensitivity is much lower, about 10% in the same field region. This indicates that the induced anisotropy constant K_a can be tailored to remain small enough and the anisotropy axis can be uniformly oriented in order to achieve high permeability and high sensitivity of MI [19, 22]. The value of the effective circular anisotropy after current annealing estimated from the impedance plot is about 400 A/m which is about twice as large comparing to that found from the hysteresis loops. This suggests that the induced anisotropy may depend on the wire radius: MI reflects the anisotropy in the surface layer whereas the hysteresis is associated with the averaged anisotropy.

The induced anisotropy may be referred to as a reverse anisotropy with respect to the sign of magnetostriction. Customarily, amorphous wires have an axial or circular anisotropy for positive and negative magnetostriction, respectively. In those cases, in order to observe the tensile stress effect on either the magnetization or MI it is necessary to apply a bias current or a bias field, which is not desirable for many applications. Inducing a circular anisotropy in positive magnetostriction wires makes it possible to realize strong influence of the applied tensile stress σ_{ex} on both the magnetization behavior and MI. Figure 3 shows the transformation of the hysteresis loop under the effect of σ_{ex}: for sufficiently large σ_{ex} a perfect rectangular loop is observed. Therefore, the effective anisotropy has the axial easy axis as a result of the magnetoelastic contribution which is consistent with equations (2) - (4).

The MI behavior in the presence of σ_{ex} shown in Figure 4 follows the change in the magnetization. As σ_{ex} is increasing the two MI peaks gradually merge and finally the MI plot with a single peak at zero field is observed. This happens since the external stress is capable of changing the direction of the static magnetization rotating it towards the axis. The change in impedance by σ_{ex} at zero external field is more than 250% for the highest stress which is estimated to be in the range of 350 MPa. The observed effect can be referred to as a giant stress-magnetoimpedance effect (S-MI).

Conclusion

The obtained experimental results on Co-based amorphous glass-coated microwires subjected to current annealing and tensile stress indicate the possibility to control the magnetic anisotropy, that is, to establish well-defined easy anisotropy axes in a desirable way whilst keeping the anisotropy value low, so the wires preserve soft magnetic properties. This makes it possible to tailor the magnetic hysteresis and magnetoimpedance by small external stimuli. Thus, inducing a circumferential easy anisotropy in positive magnetostriction wires creates the condition for very large impedance change by applying a tensile stress which could be named as giant stress-magnetoimpedance effect (S-MI). The latter has a potential for developing stress-sensitive elements especially for wireless operation at microwave frequencies.

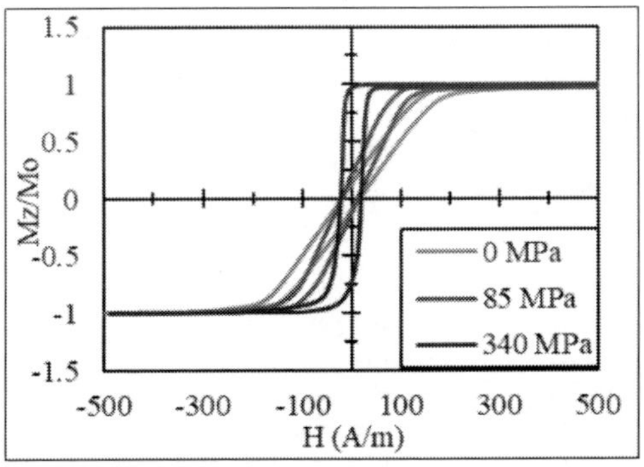

Fig. 3. Effect of tensile stress on the axial hysteresis loops of amorphous $Co_{71}Fe_5B_{11}Si_{10}Cr_3$ microwires after current annealing (50 mA, 2 hours). The load was attached at the free end of the wire.

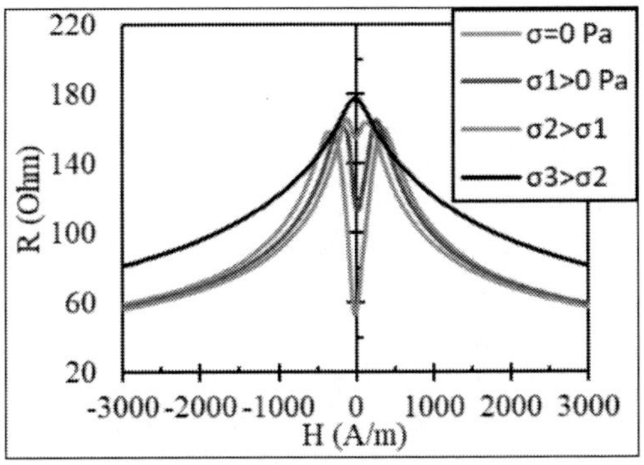

Fig. 4. Effect of tensile stress on the real part of the impedance vs. magnetic field at 50 MHz in amorphous $Co_{71}Fe_5B_{11}Si_{10}Cr_3$ microwires after current annealing (50 mA, 2 hours). The stress is applied by inserting paper-stripes underneath the microwire the both ends of which are soldered at the microstrip cell.

Acknowledgement

L. V. Panina acknowledges the support for this work under the Russian Federation State Contract for organising a scientific work (grant No 3.8022.2017). The work was also partially supported by the grant from the Ministry of Education and Science of the Russian Federation in the framework of increase Competitiveness Program of NUST "MISIS", implemented by a governmental decree dated 16th of March 2013, No 211.

References

1. M. Vázquez, H. Chiriac, A. Zhukov, L. Panina and T. Uchiyama, Phys. Status Solidi A, **208**, 493 (2011).
2. A. Zhukov and V. Zhukova, Barcelona: International Frequency Sensor Association Publishing, **164**, 156 (2013).
3. K. Mohri, T. Uchiyama, L.V. Panina, M. Yamamoto, K. Bushida, J. of Sensors, **2015,** 718069 (2015).
4. V.M. García-Chocano, H.G. Miquel, J. Magn. Magn. Mat., **378**, 485 (2015).
5. Y. Zhao, H. Hao, Y. Zhang, Intermetallics, **42**, 62 (2013).
6. A. Talaat, V. Zhukova, M. Ipatov, J.M. Blanco, L. Gonzalez-Legarreta, B. Hernando, J.J. del Val, J. Gonzalez, A. Zhukov, J. Appl. Phys., **115,** 17A313 (2014).
7. N.A. Yudanov, S.A. Evstigneeva, L.V. Panina, A.T. Morchenko and X.H. Peng, Phys. Status Solidi A, **213**, 372 (2016).
8. C. Gomez-Polo and M. Vázquez, J. Magn. Magn. Mat., **118**, 86 (1993).
9. D. Mishra, P. Saravanan, A. Perumal and A. Shrinivasan, J. Appl. Phys., **109**, 07A306 (2011).
10. L. Kraus, Z. Frait, K.R. Pirota, H. Chiriac, J. Magn. Magn. Mat., **255**, 399 (2003).
11. A. Amirabadizadeh, R. Mardani, M. Ghanaatshoar, J. of Alloys and Comp., **661**, 501 (2015).
12. B. Tian, L.H. Wang, L.Y. Zhou, Advanced Materials Research, **79-82**, 1407-1410 (2009).
13. D.P. Makhnovskiy, L.V. Panina, C. Garcia, A. Zhukov, J. Gonzalez, Phys. Rev B, **74,** 064205 (2006).
14. Y. Luo, F.X. Qin, F. Scarpa, J. Carbonell, M. Ipatov, V. Zhukova, A. Zhukov, J. Gonzalez, L.V. Panina, H.X. Peng, J. Mag. and Mag. Mat., **416**, 299 (2016).
15. D. Makhnovskiy, V. Zamorovskii, J. Summerscales, Composites A., **61**, 216 (2014).
16. H. Chiriac, T-A. Ovari, A. Zhukov, J. Magn. Magn. Mat., **254–255**, 469 (2003).
17. E.E. Shalygina, V.V. Molokanov, M.A. Komarova, J. Experimental and Theoretical Physics, **95**, 511 (2002).
18. A.S. Antonov, V.T. Borisov, O.V. Borisov, A.F. Prokoshin and N.A. Usov, J. Phys. D, **33**, 1161 (2000).
19. D.P. Makhnovskiy, L.V. Panina, and D.J. Mapps, Phys. Rev. B, **63,** 144424 (2001).
20. L.V. Panina, S. I. Sandacci, D.P. Makhnovskiy, J. Appl. Phys., **97**, 013701 (2005).
21. A. Zhukov, J. Gonzalez, J.M. Blanco, M. Vázquez and V. Larin, J. Mater. Res. **15**, 10 (2000).
22. A.T. Le, M.H. Phan, C.O. Kim, M. Vázquez, H. Lee, N.Q. Hoa, and S.C. Yu, J. of Phys. D. Applied Physics, **40**, 4582 (2007).

Influence of additional compounds on coercivity of sintered Nd-Fe-B magnets

Thanh Pham[1,2], *Duong* Nguyen[2,3], *Yen* Nguyen[1,2], *Lam* Nguyen[3], *Quang* Dinh[4], *Ngoc* Nguyen[4], *Koledov* Victor[5], *Kamantsev* Alexander[5], *Mashirov* Alexey[5], and *Dan* Nguyen[1,2]

[1] Institute of Materials Science, Vietnam Academy of Science and Technology, 18 Hoang Quoc Viet, Hanoi, Viet Nam
[2] Graduate University of Science and Technology, Vietnam Academy of Science and Technology, 18 Hoang Quoc Viet, Hanoi, Viet Nam
[3] Faculty of Physics, Hanoi Pedagogical University 2, 32 Nguyen Van Linh, Vinh Phuc, Viet Nam
[4] Hong Duc University, 565 Quang Trung, Dong Ve, Thanh Hoa, Viet Nam
[5] Kotelnikov Institute of Radio-engineering and Electronics of RAS, Moscow, Russia

Abstract. In this paper, we investigated the influence of additional compounds of Dy-Nb-Al, Nd-Cu-Al… on the coercivity of the sintered $Nd_{16.5}Fe_{77}B_{6.5}$ magnets. The additional compounds were first prepared by arc-melting method and then ground into particles with size in the range of 40 - 80 nm using a high energy ball milling method. After that, the additional powder were mixed with micrometer Nd-Fe-B powder before magnetic anisotropic pressing, vacuum sintering and annealing. The structure of the magnets was thoroughly analyzed using X-ray diffraction and electron microscopy techniques. The magnetic properties of the magnets were investigated on a pulsed field magnetometer. The results show that the coercivity of the sintered Nd-Fe-B magnets can be improved by introducing additional nanoparticles to their grain boundaries. The improvement of the coercivity of the magnets is clearly dependent on composition and fraction of the additional compounds. The coercivity has been enhanced 40% for the magnets by adding 3 wt% of the Dy-free compound of $Nd_{40}Cu_{30}Al_{30}$.

1. Introduction

Due to excellent magnetic properties such as the high saturation magnetization (M_s = 16 kOe) and large magnetocrystalline anisotropy (H_A = 75 kOe) of the $Nd_2Fe_{14}B$ phase, the sintered Nd-Fe-B magnets are widely applied in practice [1, 2]. However, the rapid decrease of the coercivity H_c by the thermal demagnetization limits applications of the magnets in the devices with high operating temperature (~ 200°C) such as motors, generators… [3]. Because the decrease of the coercivity leads to the reduction of power and efficiency of the devices, a requirement of sintered Nd-Fe-B magnets for these applications is high coercivity. Previous investigations have shown that the coercivity of this type of Nd-Fe-B magnets can be enhanced by new technology measures such as multi-stage heat treatment, grain boundary diffusion, element addition... [4-10]. Especially, the addition of compound containing Dy, Al, Cu, Nb… to grain boundary is more and more concerned to study [11-18]. Each element in additional compounds differently affects on the magnetic properties of the magnets. With addition of compound containing Dy, the coercivity is considerably increased by the formation $(Nd,Dy)_2Fe_{14}B$ phase with high magnetocrystalline anisotropy. Meanwhile, the remanence B_r of the magnets is little decreased due to the anti-ferromagnetic coupling between Fe and Dy. It is known that the magnetic properties of sintered Nd-Fe-B magnets are dependent not only on the intrinsic properties of the hard magnetic phase $Nd_2Fe_{14}B$, but also on the microstructure of the material. Other research results have been confirmed that both of Al and Cu do not enter the $Nd_2Fe_{14}B$ phase, hence the intrinsic properties (M_s and H_A) of the $Nd_2Fe_{14}B$ grains remain unchanged. These elements of Al and Cu mainly help to improve microstructure such as uniform distribution of grain boundaries, controlling of the particle size… As for Nb, it has the effect to destroy the disadvantage crystalline Fe-rich areas to improve the corrosion resistance ability and increase the coercivity of the magnets. In this work, the influence of the additional compounds of Dy-Nb-Al, Nd-Cu-Al… on the coercivity of the sintered Nd-Fe-B magnets was investigated.

2. Experimental

The pre-alloys with nominal composition of $Nd_{16.5}Fe_{77}B_{6.5}$ were prepared from Nd, Fe and FeB (18% B) on an induction furnace for about 30 minutes under Ar gas to avoid oxidation. The resulted ingots was crashed into small pieces with size smaller 2 centimeters, coarsely milled and was then fine milled in 8 h by using ball milling method. Industrial white gasoline was used as a solvent during milling process to avoid oxygenation.

* Corresponding author: thanhpt@ims.vast.ac.vn

The additional alloys with compositions of $Nd_{40}Cu_{30}Al_{30}$ (P1), $Dy_{40}Nb_{30}Al_{30}$ (P2), $Dy_{40}Zr_{30}Al_{30}$ (P3) and $Nb_{40}Cu_{30}Al_{30}$ (P4) were first prepared by an arc-melting furnace. These additional alloys were then pulverized by high energy ball milling method with milling time of 4 h to create nanoparticles with size smaller than 100 nm. The solvent/material and ball/powder ratios are 1/1 and 4/1, respectively. The Nd-Fe-B powder were mixed with the powder of additional compounds with the various weight fractions. The mixed powder was compressed with a pressure of 15 MPa, into rectangular parallelepiped tablets (6.5x5.5x3.5 cm^3) in an oriented magnetic field, yielded by an electromagnet, of about 20 kOe. The pressed magnet tablets were then sintered at 1080°C for 1 h in vacuum. After that, these magnets were heat-treated to improve their coercive force. A two-stage heat treatment process was chosen and carried out using a vacuum furnace. At the first stage, the magnets were heated at T_{a1} = 820°C for 1 h and then fast cooled down to room temperature. For the second stage, magnets were heated at temperatures, T_{a2}, of 540°C for 1 h and quenched rapidly by Ar gas. For both the stages, the samples were heated with the rate of 30°C/min and quenched rapidly with the rate of about 50°C/min. The structure of the materials was thoroughly analyzed by using X-ray diffraction (XRD), scanning electron microscope (SEM). In order to investigate magnetic properties of the sintered magnets on a pulsed high field magnetometer, the specimens of cylinders with 3 mm diameter and 3 mm height were cut from the magnets. To calculate the maximum energy product $(BH)_{max}$ demagnetization factor was estimated based on a semi-experimental data sheet.

3. Results and discussion

Fig. 1 shows SEM images of additional compounds powder with milling time of 4 h. We can see that with the P2 sample, the grains is coalesced so it is difficult to observe individual grains.

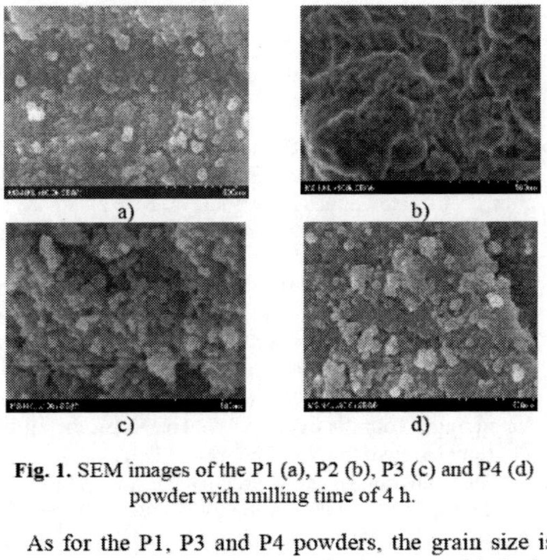

Fig. 1. SEM images of the P1 (a), P2 (b), P3 (c) and P4 (d) powder with milling time of 4 h.

As for the P1, P3 and P4 powders, the grain size is relatively uniform, in the range of 40-80 nm. This is the desired size to mix with the master Nd-Fe-B alloy powder. X.G. Cui et al. suggested that nanopowders are probably distributed on the surface of $Nd_2Fe_{14}B$ grains more evenly than micro-scale powders [19]. With the increased activity due to the higher surface energy, nanopowders would be melted first and then dissolved in the liquid phase during sintering, leading to a homogenous distribution of intergranular phase.

The structure of addition powders was further investigated through the XRD patterns (Fig. 2). However, the use of X-ray diffraction patterns is not to determine the crystalline phases in the samples, but only to confirm nanostructure of the obtained powders with above milling time. The results show that with the 4 h milling time, compounds of P1 and P4 still have relatively high crystalline diffraction peaks, but the fraction of these peaks is negligible. Meanwhile, with the other additional compounds, expanding, low intensity and non-sharp of diffraction peaks characterize for crystalline phase in nanometer size. It is agreed with obtained results from the SEM images.

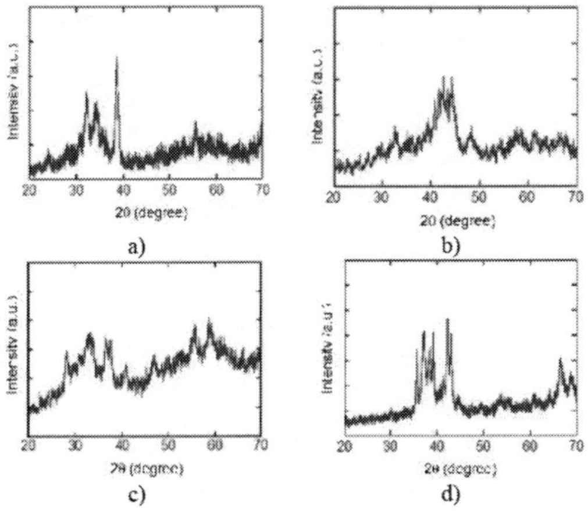

Fig. 2. XRD patterns of P1 (a), P2 (b), P3 (c) and P4 (d) powder with milling time of 4 h.

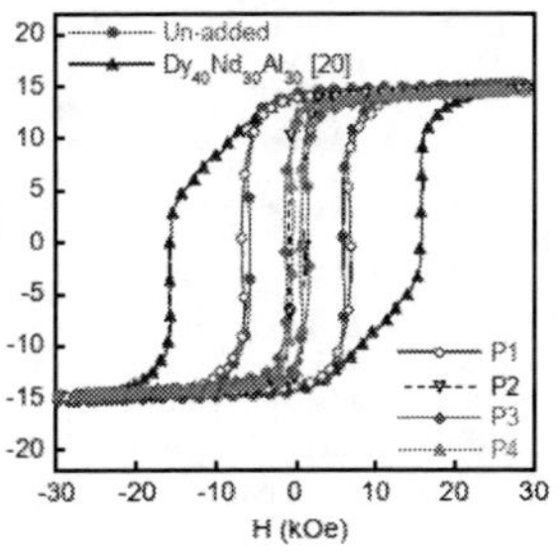

Fig. 3. Hysteresis loops of the magnets added with 2 wt% of additional compounds after sintering at 1080°C for 1 h.

In our previous study, the influence of $Dy_{40}Nd_{30}Al_{30}$ nanopowders on magnetic properties of the sintered Nd-Fe-B magnets was investigated. With 2 wt% of $Dy_{40}Nd_{30}Al_{30}$ nanoparticles, the coercivity is significantly enhanced (above 20 kOe) [20]. For this study, 2 wt% of nanoparticles of the additional compounds were mixed with Nd-Fe-B master alloy powders having size of about 2-5 μm. Hysteresis loops of un-added and added (without and with additional compounds) magnets after sintering at 1080°C for 1 h (as-sintered) are presented in Fig. 3. We can see that the hard magnetic properties of the added magnets are degraded with coercivity (H_c < 2 kOe) lower than that of the un-added magnets (H_c = 5.5 kOe) except the one added with P1 compound (H_c = 6.5 kOe). We also realize that, the saturation magnetization of all the added magnets is decreased. This can be explained that, all of the addition compounds are non-ferromagnetic which reduces the volume fraction of ferromagnetic phase. On the other hand, the additional compounds can effect the ferromagnetic interaction of the materials, leading to the decrease of the saturation magnetization. However, the squareness of the hysteresis loops of the magnets is better. The reduction of coercivity probably is due to different optimal fabrication conditions for each addition compound, which contains different elements with different effects on sintering temperature and time. The $Dy_{40}Nd_{30}Al_{30}$ additional compound of our previous study has the largest concentration of rare earth (70%), which does not affect on sintering regime and creating high coercivity for the magnets.

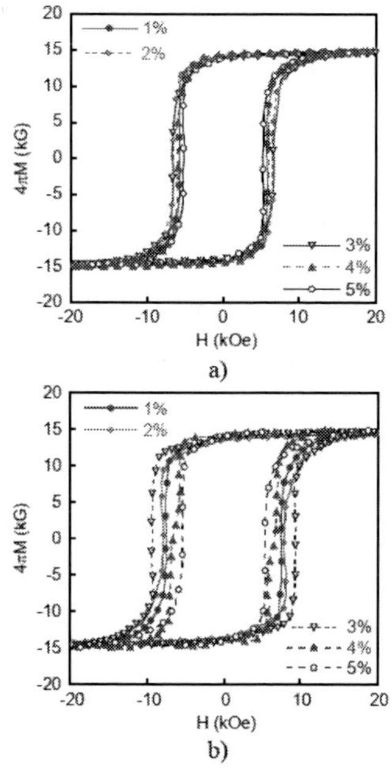

Fig. 4. Hysteresis loops of the magnets added with various fractions of $Nd_{40}Cu_{30}Al_{30}$ before (a) and after (b) heat treatment.

As well-known that, the Dy adding is an effective way to produce sintered Nd-Fe-B magnets with the high coercivity, which meets the application requirements for the motors and generators. However, the natural abundance of Dy is about 10 times less than that of Nd and its cost is much more expensive than that of Nd (about 4 times). Thus, investigations to enhance the coercivity without using Dy are important. In the fabricated additional compounds, the $Nd_{40}Cu_{30}Al_{30}$ alloy in form of nanoparticles was chosen to investigate the influence of its weight fraction on magnetic properties of the magnets. Fig. 4 shows hysteresis loops of the magnets added with various fractions of $Nd_{40}Cu_{30}Al_{30}$ before and after heat treatment. It can be realized that, the change of the coercivity on various fractions before heat treatment is not significantly. However, after heat treatment the coercivity was considerably enhanced and relatively changed with the different additional fractions. Nevertheless, the squareness of the hysteresis loops of the magnets is worse. In the second quadrant demagnetization curves of the heat-treated magnets, a small dip appears. These dips can be caused by physical or chemical modifications of the grain boundaries, which may reduce the local nucleation field. For examples, the formation of the soft magnetic α-Fe phase would create nucleation centres for reversed magnetic domains resulting in the dips in the demagnetization curves of the added magnets.

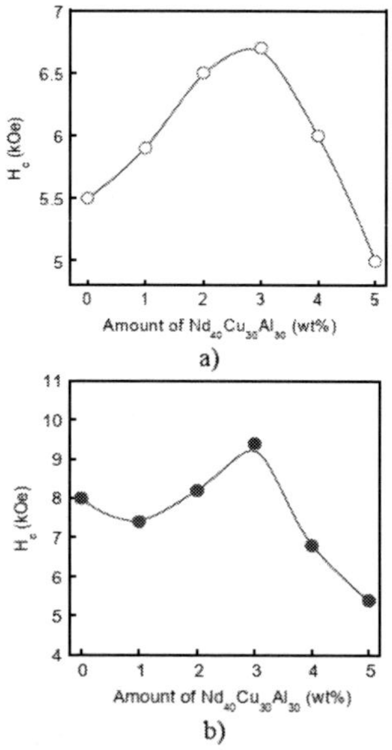

Fig. 5. The dependences of coercivity H_c of the added magnets on various fractions of $Nd_{40}Cu_{30}Al_{30}$ before (a) and after (b) heat treatment.

The dependence of coercivity H_c of the added magnets on various fractions of $Nd_{40}Cu_{30}Al_{30}$ before and after heat treatment are shown in Fig. 5. We can see that the H_c slightly increases as the $Nd_{40}Cu_{30}Al_{30}$ amount increases from 1 to 2 wt% (from 5.8 to 6.5 kOe for magnet before heat treatment, and from 7.5 to 8 kOe for magnet after heat treatment). The highest coercivity for added magnets before and after heat treatment was obtained to be 6.7 kOe and 9.5 kOe, respectively, with fraction of 3 wt%. The H_c decrease rather rapidly when the adding amount is over 3 wt%. Therefore, the optimal additional amount of $Nd_{40}Cu_{30}Al_{30}$ is 3 wt%.

The reduction of the coercivity with large additional fractions (4 and 5 wt%) probably is due to formation of disadvantage phases at grain boundaries for hard magnetic properties of the magnets. At the same time, heat treatment process might not create the desirous microstructure. Thus, the further enhancement of the coercivity of Dy-free sintered Nd-Fe-B magnets needs more carefully and systematically researches.

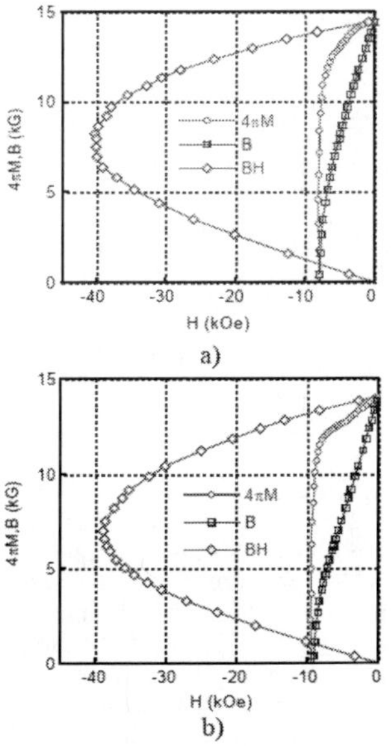

Fig. 6. Magnetic characteristic curves of un-added (a) and 3% $Nd_{40}Cu_{30}Al_{30}$ added magnets (b) after heat treatment.

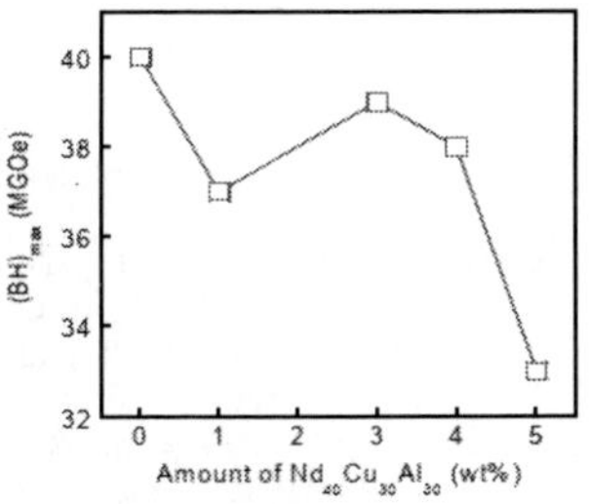

Fig. 7. The dependences of maximum energy product, $(BH)_{max}$ of the added magnets on various fractions of $Nd_{40}Cu_{30}Al_{30}$ after heat treatment.

Maximum energy product, $(BH)_{max}$, of the un-added and 3 wt% $Nd_{40}Cu_{30}Al_{30}$ added magnets after heat treatment was calculated to be ~ 40 MGOe and ~ 38 MGOe, respectively (Fig. 6). The lower $(BH)_{max}$ of the added magnets with various fractions of $Nd_{40}Cu_{30}Al_{30}$, as shown in Fig. 7, probably is due to a reduction of saturation magnetization by these non-ferromagnetic additional nanoparticles. These values of $(BH)_{max}$ might be improved by an optimal heat treatment (further investigation of heat treatment is needed).

4. Conclusion

The influence of additional compounds on coercivity of the sintered $Nd_{16.5}Fe_{77}B_{6.5}$ magnets has been investigated. The highest coercivity, H_c, and the largest maximum energy product, $(BH)_{max}$, have been obtained to be 9.5 kOe and 38 MGOe, respectively, with 3% $Nd_{40}Cu_{30}Al_{30}$ additional nanopowders and suitable technological conditions. The coercivity of the sintered Nd-Fe-B magnets can be increased 40% by addition of $Nd_{40}Cu_{30}Al_{30}$ nanoparticles.

This work was supported by Vietnam Academy of Science and Technology under Grant number of VAST03.05/16-17. Some experiments were done in the Key Laboratory for Electronic Materials and Devices, Institute of Materials Science, Vietnam Academy of Science and Technology, Viet Nam.

References

1. Y. Matsuura, J. Magn. Magn. Mater. **303**, 344 (2006)
2. J.M.D. Coey, *Rare-earth iron permanent magnets* (Clarendon Press, Oxford, 1996)
3. B.C. Chen, X.M. Liu, R.J. Chen, S. Guo, C.J. Yan, D. Lee, A.R. Yan, J. Appl. Phys. **111**, 07A710-1 (2012)
4. H. Nakamura, K. Hirota, T. Ohashi, T. Minowa, J. Phys. D: Appl. Phys. **44**, 064003-1 (2011)
5. Y. Liu, S. Guo, R. Chen, D. Lee, A. Yan, IEEE Trans. Magn. **47**, 3270 (2011)
6. H. Sepehri-Amin, Y. Une, T. Ohkubo, K. Hono, M. Sagawa, Scri. Mater. **65**, 396 (2011)
7. T. Kim, S. Lee, J. Kim, Y. Kim, H. Kim, M. Lee, T. Jang, J. Appl. Phys. **115**, 17A770-1 (2014)
8. K. Kobayashi, K. Urushibata, Y. Une, M. Sagawa, J. Appl. Phys. **113**, 163910-1 (2013)
9. N.Y. Yu, X.M. Pan, Y.P. Zhang, L.H. Ge, J. Magn. **18**, 235 (2013)
10. X. Liu, T. Ma, X. Wang, M. Yan, J. Magn. Magn. Mater. **382**, 26 (2015)
11. S.E. Park, T. Kim, S.R. Lee, S. Namkung, T.S. Jang, J. Appl. Phys. **111**, 07A707-1 (2012)
12. M.W. Lee , D.R. Dhakal, T.H. Kim, S.R. Lee, H.J. Kim, T.S. Jang, Arch. Metal. Mater. **60**, 1407 (2015)
13. L. Liang, T. Ma, P. Zhang, J. Jin, M. Yan, J. Magn. Magn. Mater. **355**, 131 (2014)
14. Y. Zhang, T. Ma, X. Liu, P. Liu, J. Jin, J. Zou, M. Yan, J. Magn. Magn. Mater. **399**, 159 (2016)
15. B. Zhou, X. Li, X. Cao, G. Yan, A. Yan, Chin. Phys. B **25**, 117504-1 (2016)
16. M. Pan, P. Zhang, Q. Wu, H. Ge, Int. J. Electrochem. Sci. **11**, 2659 (2016)
17. J.J. Ni, T.Y. Ma, Y.R. Wu, M. Yan, J. Magn. Magn. Mater. **322**, 3710 (2010)
18. S. Lee, J. Kwon, H. Cha, K.M Kim, H. Kwon, J. Lee, D. Lee, Met. Mater. Int. **22**, 340 (2016)
19. X.G. Cui, M. Yan, T.Y. Ma, L.Q. Yu, Phys. B **403**, 4182 (2008)
20. N.H. Dan, P.T. Thanh, N.H. Yen, L.T. Hung, IEEE Trans. Magn. **50**, 210204-1 (2014)

Investigation of magnetic phase transition and magnetocaloric effect of (Ni,Co)-Mn-Al melt-spun ribbons

Yen Nguyen[1,*], *Mai* Nguyen [4], *Quang* Vu[2,3], *Thanh* Pham[1,2], *Victor* Koledov[5], *Alexander* Kamantsev[5], *Alexey* Mashirov[5], *Thanh* Tran[1,6], *Hau* Kieu[6], *Yu* Seong[6], and *Dan* Nguyen[1]

[1] Institute of Materials Science, Vietnam Academy of Science and Technology, 18 Hoang Quoc Viet, Ha Noi, Viet Nam
[2] Graduate University of Science and Technology, Vietnam Academy of Science and Technology, 18 Hoang Quoc Viet, Ha Noi, Viet Nam
[3] Faculty of Physics, Hanoi Pedagogical University 2, 32 Nguyen Van Linh, Vinh Phuc, Vietnam
[4] Faculty of Physics, VNU University of Science, 334 Nguyen Trai, Ha Noi, Viet Nam
[5] Kotelnikov Institute of Radio-engineering and Electronics of RAS, Moscow, Russia
[6] Department of Physics, Chungbuk National University, Cheongju 361 - 763, South Korea

Abstract. Magnetic phase transition, magnetocaloric effect and critical parameters of $Ni_{50-x}Co_xMn_{50-y}Al_y$ (x = 5 and 10; y = 17, 18 and 19) rapidly quenched ribbons have been studied. X-ray diffraction patterns exhibit a coexistence of the $L2_1$ and $10M$ crystalline phases of the ribbons. Magnetization measurements show that all the samples behave as soft magnetic materials with a low coercive force less than 60 Oe. The shape of thermomagnetization curves considerably depends on Co and Al concentrations. The Curie temperature (T_C) of the alloy ribbons strongly increases with increasing the Co concentration and slightly decreases with increasing the Al concentration. The $Ni_{45}Co_5Mn_{31}Al_{19}$ and $Ni_{40}Co_{10}Mn_{33}Al_{17}$ ribbons reveal both the positive and negative magnetocaloric effects. Under magnetic field change (ΔH) of 13.5 kOe, the maximum magnetic entropy change ($|\Delta S_m|_{max}$) of the $Ni_{45}Co_5Mn_{31}Al_{19}$ ribbon is about 2 and -1 $J{\cdot}kg^{-1}{\cdot}K^{-1}$ for negative and positive magnetocaloric effects, respectively. Basing on Arrott - Noakes and Kouvel - Fisher methods, critical parameters of the $Ni_{45}Co_5Mn_{31}Al_{19}$ ribbon were determined to be $T_C \approx 290$ K, $\beta \approx 0.58$, $\gamma \approx 0.92$ and $\delta \approx 2.59$. The obtained values of the critical exponents indicate that the magnetic order of the alloy ribbon is close to the mean-field model.

1. Introduction

The magnetocaloric effect (MCE) is an intrinsic property of the magnetic material. Basically, magnetocaloric effect is the conversion of magnetic energy into heat energy when the material is exposed in changing magnetic field. This effect appears at all materials with different intensities and is assessed by the magnetic entropy change (ΔS_m) and adiabatic temperature variation (ΔT_{ad}). MCE has been widely utilized in magnetic materials to reach low temperatures. Recently, the giant magnetocaloric effect (GMCE) at room temperature has attracted the attention of researchers by application potential for magnetic refrigeration [1 - 5].

Among magnetocaloric materials, an interesting class is Ni–Mn based Heusler alloys. The Ni-Mn based alloys have been shown to exhibit large magnetocaloric effects, such as providing both the negative and positive magnetocaloric effects. Besides that, the shape memory effect and other interesting properties were also observed [6 - 9]. In the Ni-Mn-Al Heusler alloys, the Neel temperature $T_N \approx 300$ K was found to be virtually independent of composition [10]. The Ni-Mn-Al alloys with theirs relatively low cost and high ductility is a potentially attractive candidate material as a magnetic refrigerant.

Recently, it was reported that Co doping had a strong effect on the magnetic phase transformation of materials. Partial substitution of Co for Ni in the alloys leads to the significant increase of the magnetization change at the martensitic transformation (ΔM) which greatly enhanced the MCE [11-13]. The martensite-austenite temperature decreases gradually, while the Curie temperature increases greatly with increasing the Co concentration. Substitution of Co for Ni will reinforce the magnetic exchange interactions and the ferromagnetic phase of alloys [14-17]. However, the magnetic phase transitions and MCEs of the (Ni,Co)-Mn-Al bulk alloys are very sensitive to crystallize structure and it normally takes a long time (several days) to make structure stable [18-20]. Comparing the ribbon samples with bulk samples of the same composition, they have different structure and magnetic properties. Specifically, martensitic transformation temperature in $Ni_{50-x}Co_xMn_{31+y}Al_{19-y}$ (x = 5, 10) ribbon alloys is significantly lower bulk alloys. As compared to a weakly magnetic of the $Ni_{50}Mn_{31}Al_{19}$ ribbon, the $Ni_{50-x}Co_xMn_{31+y}Al_{19-y}$ (x = 5, 10) ribbons demonstrate ferromagnetic properties. The $Ni_{45}Co_5Mn_{32}Al_{18}$ ribbon was found to exhibit a well-defined martensitic-austenitic phase transition [20-23].

In this work, we investigated magnetic, magnetocaloric and critical properties of $Ni_{50-x}Co_xMn_{50-y}Al_y$ (x = 5 and 10; y = 17, 18 and 19) rapidly quenched ribbons.

2. Experiment

Six alloy ingots with nominal compositions of $Ni_{50-x}Co_xMn_{50-y}Al_y$ (x = 5 and 10; y = 17, 18 and 19) were prepared from pure elements (99.9%) of Ni, Co, Mn and Al

* Corresponding author: yennh@ims.vast.ac.vn

by using arc-melting method in Argon gas. Melt-spinning method was then used to fabricate the alloy ribbons with tangential velocity of copper wheel of 40 m/s. The thickness and width of the ribbons are about 20 μm and 1.5 mm, respectively. The structure of the alloys was investigated by powder X-ray diffraction (XRD) method. The magnetic properties and magnetocaloric effect of the alloys were characterized on a vibrating sample magnetometer.

3. Result and discussion

XRD patterns of the $Ni_{50-x}Co_xMn_{50-y}Al_y$ (x = 5 and 10; y = 17, 18 and 19) ribbons measured at room temperature are shown in figure 1. Structural analyses revealed that besides the main phase associated with an austenitic $L2_1$ structure (space group: Fm3m), some XRD peaks with low intensity corresponding to martensitic 10*M* (space group: Pmma) phase is also present. With high concentration of Al, the 10*M* phase develops strongly in the alloy ribbons. It should be noted that, a small change of structure in this kind of materials can greatly affect on magnetic properties of the ribbons as presented below.

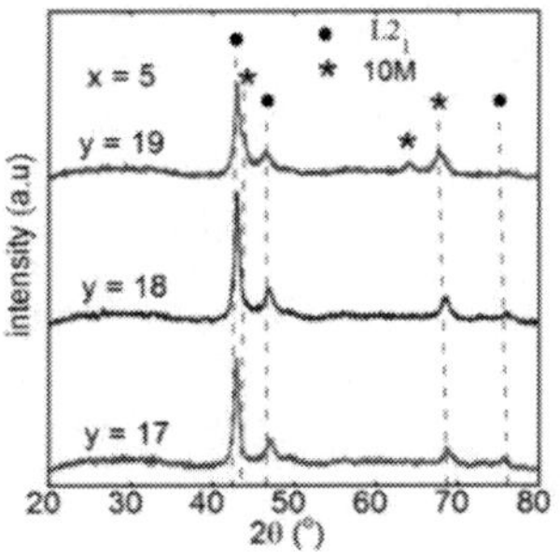

(a)

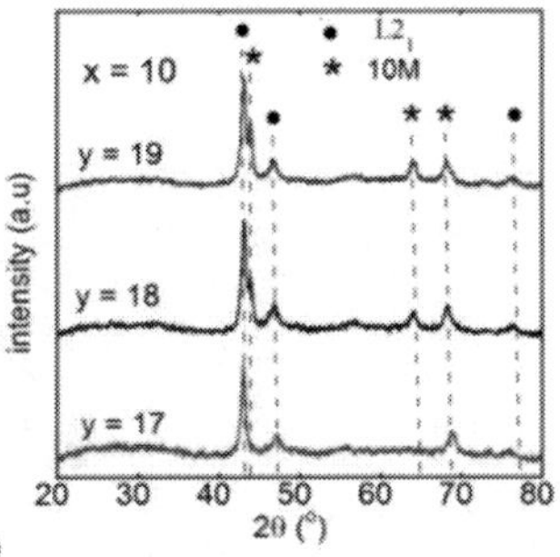

(b)

Fig. 1. *XRD patterns of $Ni_{50-x}Co_xMn_{50-y}Al_y$ ribbons with x = 5 (a) and x = 10 (b).*

The magnetic properties of the samples were characterized by the magnetization measurements (figure 2). The results show that all the samples with Co-concentration of 10 at% behave as the soft magnetic material with a low coercivity less than 20 Oe (figure 2a). The temperature dependence of magnetization, *M*(*T*), for the ribbons in the field 100 Oe are showed in figure 2 (b). One can see that the shape of thermomagnetization curves considerably depend on both the concentrations of Co and Al. As for examples, the magnetization of the samples with x = 5 is increased from ~ 0.1 emu/g (for y = 17) to ~ 4.7 emu/g (for y = 19). In addition, the $Ni_{45}Co_5Mn_{31}Al_{19}$ and $Ni_{40}Co_{10}Mn_{33}Al_{17}$ ribbons reveal both the first-order (FOPT) and second-order (SOPT) magnetic phase transition. While the FOPT is strongly influenced by Al-concentration, the SOPT (FM-PM) is almost unchanged. By increasing the Co-concentration from 5 at% to 10 at%, the Curie temperature (T_C) is strongly increased from ~ 290 K (for x = 5 and y = 19) to ~ 393 K (for x = 10 and y = 19). The magnetization of the alloy ribbons is also increased considerably with the Co-concentration.

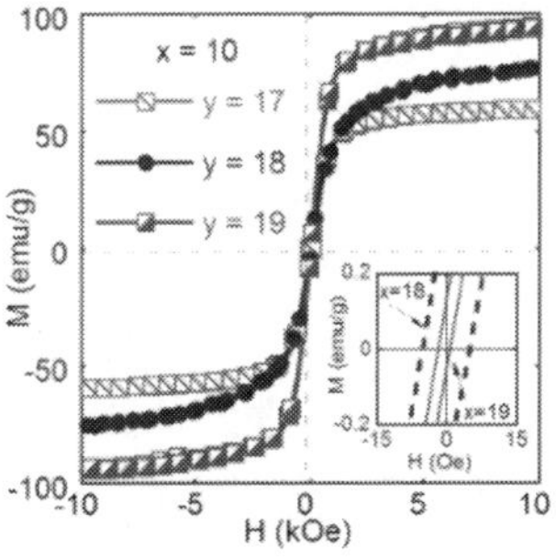

(a)

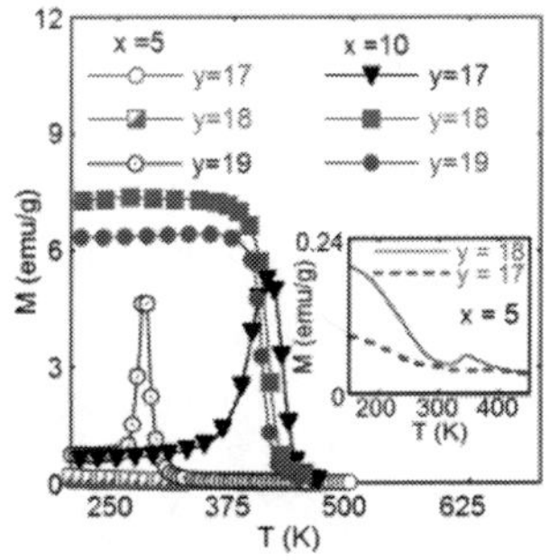

(b)

Fig. 2. *M(H) loops at room temperature (the inset enlarges the typical loops at low magnetic field) (a), M(T) curves measured in magnetic field of 100 Oe (the inset enlarges M(T) curves at low magnetization) (b) of $Ni_{50-x}Co_xMn_{50-y}Al_y$ (x = 5 and 10; y = 17, 18 and 19) ribbons.*

The different influence of Co and Al on T_C can be explained by the dependence of T_C on the exchange interaction in the materials. The ferromagnetic exchange interaction between Ni and Mn atoms in the (Ni,Co)-Mn-Al alloys essentially decides value of their T_C [24, 25]. Partial substitution of Mn by Al (paramagnetic material) hardly affects the exchange interaction of Ni and Mn atoms. But the substitution of Ni by Co (ferromagnetic material) enhances the exchange interaction of the alloy resulting in the large change of T_C. As compared to the $Ni_{45}Co_5Mn_{31}Al_{19}$ bulk alloy with annealing at 1370 K for 24 h, martensite transformation temperature in ribbon is significantly higher, while Curie temperature is almost unchanged [20].

Among this series, the $Ni_{45}Co_5Mn_{31}Al_{19}$ sample shows two strong magnetic phase transitions near room temperature region. Therefore, it was chosen as a representative one for analyzing magnetic properties. For further understanding the type of the phase transition and MCE of the alloy ribbons, the isothermal magnetization curves, *M*(*H*), with magnetic field up to 13.5 kOe were performed. Figure 3a shows *M*(*H*) curves at various temperatures of the $Ni_{45}Co_5Mn_{31}Al_{19}$ sample.

The magnetic orders in a second-order magnetic phase transition of the material can be additionally understood by determining the critical parameters upon using Arrott plots [26]. The Arrott plots, *H*/*M* versus M^2 (figure 3b), were constructed from *M*(*H*) data. The $M^2 - H/M$ curves are non-linear at low magnetic field and linear at high magnetic field. Values of the spontaneous magnetization (M_S) and the inverse initial susceptibility (χ_0^{-1}) at different temperatures were derived from Arrott plots.

https://doi.org/10.1051/epjconf/201818505001

The critical parameters of β, γ and δ relate to the two above quantities by following equations [26, 27]:

$$M_S(T) = M_0(-\varepsilon)^{\beta} \qquad \varepsilon < 0 \qquad (1)$$

$$H = DM^{\delta} \qquad \varepsilon = 0 \qquad (2)$$

$$\chi_0^{-1}(T) = \frac{H_0}{M_0}\varepsilon^{\gamma} \qquad \varepsilon > 0 \qquad (3)$$

$$\delta = 1 + \gamma / \beta \qquad (4)$$

where M_0, H_0 and D are the critical amplitudes and $\varepsilon = (T - T_C)/T_C$ is the reduced temperature.

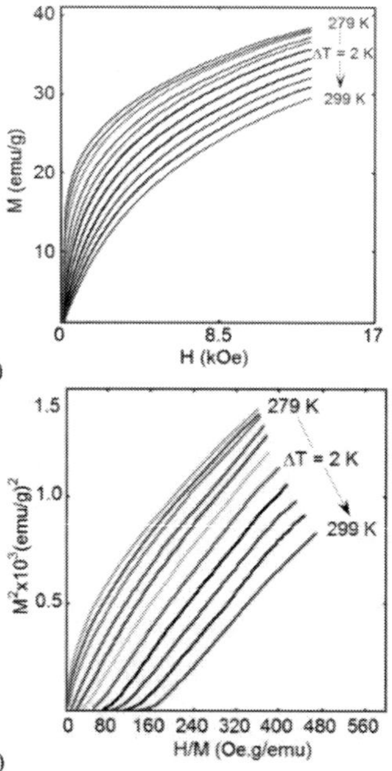

Fig. 3. *Isothermal magnetization curves around T_C (a) and Arrott plots (b) for $Ni_{45}Co_5Mn_{31}Al_{19}$ ribbon.*

The linear extrapolation from high field to the intercepts with the M^2 and H/M axes gives the values of $M_S(T)$ and $\chi_0^{-1}(T)$, respectively. The critical parameter T_C, β and γ obtained from fitting $M_S(T)$ and $\chi_0^{-1}(T)$ data (figure 4a) by using the according formulas (1) and (3), while δ was calculated by using the Widom scaling relation, equation (4) [27, 28]. As resulted, the $Ni_{45}Co_5Mn_{31}Al_{19}$ ribbon possesses $T_C \approx 290$ K, $\beta \approx 0.59$, $\gamma \approx 0.92$ and $\delta \approx 2.56$. The critical parameters are quite close to mean-field theory ($\beta = 0.5$, $\gamma = 1$ and $\delta = 3$) [28].

Alternatively, the critical parameters can be obtained more accurately by the Kouvel - Fisher method [29]. Similarity with Arrott – Noakes method, $M_S(T)$ and $\chi_0^{-1}(T)$ are determined by plotting $M^{1/\beta}$ versus $(H/M)^{1/\gamma}$ curves. The critical parameters of β and γ relate to the two above quantities by these equations:

$$M_S(T)\left[dM_S / dT\right]^{-1} = (T - T_C)/\beta \qquad (5)$$

$$\chi_0^{-1}(T)[d\chi_0^{-1}(T)/dT]^{-1} = (T - T_C)/\gamma \qquad (6)$$

Again, the critical parameters T_C, β and γ obtained from fitting $M_S(T)$ and $\chi_0^{-1}(T)$ data by using the according formulas (5) and (6). Figure 4b shows Kouvel-Fisher curves of the $Ni_{45}Co_5Mn_{31}Al_{19}$ ribbon with fitting results of $T_C \approx 291$ K, $\beta \approx 0.58$ and $\gamma \approx 0.92$. Basing on the formula (4), the δ value was calculated to be 2.59. One can see that the critical parameter values determined from the Kouvel-Fisher method are in good agreement with those obtained by the Arrott-Noakes fittings.

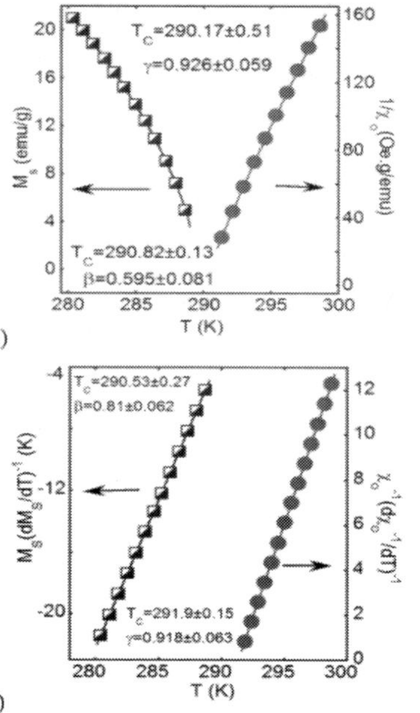

Fig. 4. *$M_S(T)$ and $\chi_0^{-1}(T)$ along with fittings to Arrott-Noakes relations (a) and Kouvel-Fisher plots (b) for $Ni_{45}Co_5Mn_{31}Al_{19}$ ribbon.*

Assessing the reliability of the critical parameters can be carried out by using the static-scaling theory [28, 30]. The isothermal magnetization is determined by the formula:

$$M(H,\varepsilon) = \varepsilon^{\beta} f_{\pm}(H|\varepsilon|^{-(\beta+\gamma)}) \qquad (7)$$

Where f_+ and f_- are regular functions for temperatures $T > T_C$ and $T < T_C$, respectively.

Figure 5 shows the static-scaling plots of $M|\varepsilon|^{-\beta}$ versus $H|\varepsilon|^{-(\beta+\gamma)}$ in the log scale with the obtained critical parameters. We can see that points of the data fall on two universal branches with $T > T_C$ and $T < T_C$. Consequently, the critical values in our study are reliable results.

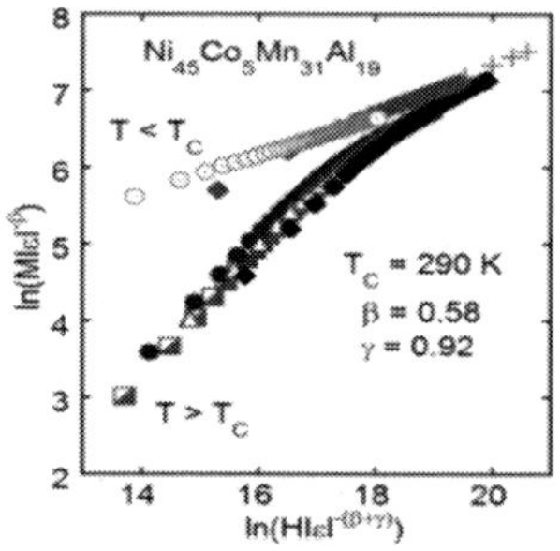

Fig. 5. *The static-scaling performances for $Ni_{45}Co_5Mn_{31}Al_{19}$ ribbon.*

Magnetocaloric effect in the alloys was assessed by the magnetic entropy change (ΔS_m). This value was calculated basing on the isothermal magnetizations (figure 3a). Fig 6 shows $\Delta S_m(T)$ curves of the $Ni_{45}Co_5Mn_{31}Al_{19}$ ribbon under several magnetic field change (ΔH = 2, 6, 10 and 13.5 kOe). With ΔH = 13.5 kOe, the maximum magnetic entropy change are about 2 and -1 J·kg^{-1}·K^{-1} for negative and positive magnetocaloric effects, respectively. These results are similar to reports by Y. Kim et al [18, 19]. In addition, the $|\Delta S_m|_{max}$ value increases with increasing the magnetic field (the above inset of figure 6).

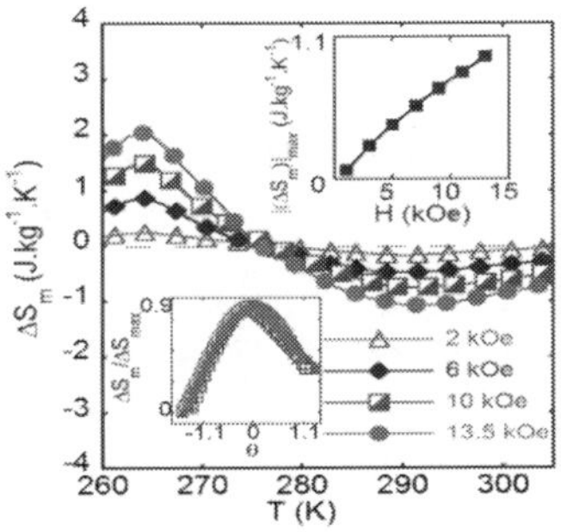

Fig. 6. *Magnetic entropy change (ΔH = 2, 6, 10 and 13.5 kOe) versus temperature for* $Ni_{45}Co_5Mn_{31}Al_{19}$ *ribbon. The above inset shows field dependences of* $|\Delta S_m|_{max}$ *at* $T \approx T_C$. *The below inset shows universal master curves of* $\Delta S_m / \Delta S_{max}$ *versus θ.*

The temperature dependence of ΔS_m under different applied field for the second-order phase transition of materials can be described by the universal master curves [31, 32]. Basing on the $\Delta S_m(T)$ curve, the $\Delta S_m / \Delta S_{max}$ versus θ plots are constructed. White θ value is determined by the formula:

$$\theta = \begin{cases} -(T - T_C)/(T_{r1} - T_C), & T \le T_C \\ (T - T_C)/(T_{r2} - T_C), & T > T_C \end{cases} \qquad (8)$$

Where T_{r1} and T_{r2} are the temperature of the two reference points. For the present study, they are selected as those corresponding to $\Delta S_m(T_{r1,2}) = k.\Delta S_{max}$ (k = 0.5). This choice of k does not affect the actual construction of the universal curve, as it implies only proportionality constant. The below inset of figure 6 shows the universal master curve of the $Ni_{45}Co_5Mn_{31}Al_{19}$ ribbons. All $\Delta S_m(T)$ data under different applied fields are collapsed into a universal master curve. This is an interesting property of the second-order phase transition of materials.

4. Conclusion

Structure of the $Ni_{50-x}Co_xMn_{50-y}Al_y$ (x = 5 and 10; y = 17, 18 and 19) ribbons exhibits multi-crystalline phases of the $L2_1$ and $10M$ types. All the alloy ribbons behave as the soft magnetic materials and their Curie temperature strongly increases with increasing the Co-concentration and hardly changes with increasing the Al-concentration. The maximum magnetic entropy change with variation of magnetic field of 13.5 kOe for the $Ni_{45}Co_5Mn_{31}Al_{19}$ ribbon is about 2 and -1 J·kg^{-1}·K^{-1} for negative and positive magnetocaloric effects, respectively. Critical parameters of the alloy are close to those of the mean-field model for long-range ferromagnetic orders.

This work was supported by Vietnam Academy of Science and Technology under grant No. VAST.HTQT.NGA.05/17-18 and Russian Foundation for Basic Research under grant № 17-58-540002. A part of the work was done in Key Laboratory for Electronic Materials and Devices and Laboratory of Magnetism and Superconductivity, Institute of Materials Science, Vietnam.

References

1. O. Tegus, E. Brück, K.H. Buschow, F.R. Boer, Nature. **415,** 6868 (2002).
2. X. Zhang, B. Zhang, S. Yu, Z. Liu, W. Xu, G. Liu, Phys. Rev. B **76,** 132403 (2007).
3. E. Bruck, J. Appl. Phys **38**, 381 (2005).
4. Y.B. Yang, X.B. Ma, X.G. Chen, J.Z. Wei, R. Wu, J.Z. Han, H.L. Du, C.S. Wang, S.Q. Liu, Y.C. Yang, Y. Zhang, J.B. Yang, J. Appl. Phys. **111**, 07A916 (2012).
5. K.A. Gschneidner, V.K. Pecharsky, Int. J. Ref. **31**, 945 (2008).
6. V.K. Pecharsky, K.A. Gschneidner, A.O. Pecharsky, A.M. Tishin, Phys. Rev. B **64**, 144406 (2001).
7. T. Krenke, E. Duman, M. Acet, E.F. Wassermann, X. Moya, L. Mañosa, Nat. Mater. **4**, 450 (2005).
8. A. Planes, L. Mañosa, M. Acet, J. Phys. **21**, 233201 (2009).
9. R. Kainuma, W. Ito, R.Y. Umetsu, K. Oikawa, K. Ishida, Appl. Phys. Lett. **93**, 091906 (2008).
10. S. Morito, T. Kakeshita, K. Hirata, K. Otsuka, Acta Mater. **46**, 5377 (1998).
11. P. Felcher, J. Cable, K. Wilkinsonm, J. Phys. Chem. Solids **24**, 1663 (1963).
12. F. Gejima, Y. Sutou, R. Kainuma, K. Ishida, Metall. Mater. Trans. A **30**, 2721 (1999).
13. H.C. Xuan, F.H. Chen, P.D. Han, D.H. Wang, Y.W. Du, Intermetallics **47**, 31-35 (2014).
14. M.V. Lyange, E.S. Barmina, V.V. Khovaylo, Mater. Sci. Forum **81**, 232-242 (2015).
15. C. Liu, W. Zhang, Z. Qian, Z. Hu, Q. Zhao, Y. Sui, W. Su, M. Zhang, Z. Liu, G. Liu, G. Wu, J. Alloys Comp. **433**, 37 (2007).
16. A.C. Paduani, A. Migliavacca, M.L. Sebben, J.D. Ardisson, M.I. Yoshida, S. Soriano, M. Kalisz, Solid State Commun. **141**, 145 (2007).
17. T. Busgen, J. Feydt, R. Hassdorf, S. Thienhaus, M. Moske, Phys. Rev. B **70**, 014111 (2004).
18. Y. Kim, W.B. Han, H.S. Kim, H.H. An, C.S. Yoon, J. Alloys Compd. **557**, 265 (2013).
19. Y. Kim, S.J. Lee, W.B. Han, H.-S. Kim, H.H. An, C.S. Yoon, J. Appl. Phys. **113**, 17B102 (2013).
20. X. Xu, W. Ito, M. Tokunaga, R.Y. Umetsu, R. Kainuma, K. Ishida, Mater. Trans. **51**, 1357 (2010).
21. M.V. Lyange, V. Khovaylo, R. Singh, S.K. Srivastava, R. Chatterjee, L.K. Varga, J. Alloys Compd. **586**, 218 (2014).
22. V. Khovaylo, M. Lyange, K. Skokov, O. Gutfleisch, R. Chatterjee, X. Xu, R. Kainuma, Mater. Sci. Forum **738**, 446 (2013).
23. X. Xu, W. Ito, T. Kanomata, R. Kainuma, Entropy **16**, 1808 (2014).
24. J. Rusz, L. Bergqvist, J. Kudrnovský and I. Turek, Phys. Rev. B **73**, 214412 (2006).
25. D.T. Huu, N.H. Yen, P.T. Thanh, N.T. Mai, T.D. Thanh, T.L. Phan, S.C. Yu, N.H. Dan, J. Alloys Compd. **622**, 535 (2015).
26. A. Arrott, J.E. Noakes, Phys. Rev. Lett. **19**, 786 (1967).
27. B. Widom, J. Chem. Phys. **41**, 1633 (1964).
28. H.E. Stanley, *Introduction to phase transitions and critical phenomena* (Oxford University Press, 1971).
29. J.S. Kouvel, M.E. Fisher, Phys. Rev. **136**, A1626 (1964).
30. S.K. Banerjee, Phys. Lett. **12**, 16 (1964).
31. V. Franco, J.S. Blazquez, A. Conde, Appl. Phys. Lett. **89**, 2 (2006).
32. V. Franco, A. Conde, Int. J. Ref. **33**, 3 (2010).

Phase formation and magnetocaloric effect in (Pr,Nd)-Fe alloys prepared by rapidly quenched method

Dan Nguyen[1,2,*], *Ha* Nguyen[2,3], *An* Nguyen[3], *Yen* Nguyen[1,2], *Thanh* Pham[1,2], *Victor* Koledov[4], *Alexander* Kamantsev[4], *Alexey* Mashirov[4], *Thanh* Tran[1,5], *Hau* Kieu[5] and *Seong* Yu[5]

[1] Institute of Materials Science, Vietnam Academy of Science and Technology, 18 Hoang Quoc Viet, Ha Noi, Viet Nam
[2] Graduate University of Science and Technology, Vietnam Academy of Science and Technology, 18 Hoang Quoc Viet, Ha Noi, Viet Nam
[3] Hong Duc University, 565 Quang Trung, Dong Ve, Thanh Hoa, Viet Nam
[4] Kotelnikov Institute of Radio-engineering and Electronics of RAS, Moscow, Russia
[5] Department of Physics, Chungbuk National University, Cheongju 361 - 763, South Korea

Abstract. In this work, $Pr_{2-x}Nd_xFe_{17}$ (x = 0 - 2) ribbons with thickness of about 15 μm were prepared by melt-spinning method. The alloy ribbons were then annealed at different temperatures (900 - 1100°C) for various time (0.25 - 2 h). The formation of the $(Pr,Nd)_2Fe_{17}$ (2:17) crystalline phase in the alloys strongly depends on the Pr/Nd ratio and annealing conditions. Annealing time for the completed formation of the 2:17 phase in the rapidly quenched ribbons is greatly reduced in comparison with that of bulk alloys. Curie temperature, T_C, of the alloys can be controlled in room temperature region by changing Pr/Nd ratio. Maximum magnetic entropy change ($|\Delta S_m|_{max}$) and full width at haft the maximum peak (FWHM) of the magnetic entropy change of the alloys were respectively found to be larger than 1.5 $J.kg^{-1}K^{-1}$ and 40 K in room temperature region with magetic field change ΔH = 12 kOe.

1. Introduction

Magnetocaloric effect (MCE) is defined as the adiabatic temperature change of a magnetic material upon the application of a magnetic field. The MCE of the material is of interest in research by virtue of its potential application in the field of magnetic refrigeration. The magnetic refrigeration bases on the principle of magnetic entropy change of the material. To have large magnetic cooling efficient, the MCE of the material should be large (magnetic entropy change - ΔS_m, adiabatic temperature variation - ΔT_{ad} and refrigerant capacity – RC should be large) with low field change. In later years, some materials have large magnetocaloric effect (so-called giant magnetocaloric effect - GMCE) were discovered, such as: rare earth-containing alloys, As-containing alloys, La-containing alloys, Heusler alloys, Fe and Mn based rapidly quenched alloys, the ferromagnetic perovskite maganites… [1-8]. Among them, alloys containing rare earth elements (RE) such as Gd, Pr, Nd… are considered as very attractive systems [9-15]. In particular, Fe-rich RE_2Fe_{17} alloys exhibit ferromagnetic order with high values for the Fe magnetic moment, and the Curie temperature, T_C, around room temperature in the case of Pr and Nd [16, 17]. Moreover, these alloys show large MCE with the maximum magnetic entropy change ($|\Delta S_m|_{max}$) of 6 $J.kg^{-1}K^{-1}$ under 50 kOe magnetic field at room temperature (for RE = Nd, Pr) [18, 19]. Although ΔS_m of $(Nd,Pr)_2Fe_{17}$ alloys is lower than that of Gd-Si-Ge alloys [20, 21] and La-Fe-Si alloys [22], the Pr_2Fe_{17} and Nd_2Fe_{17} alloys have GMCE around room temperature, broad ΔS_m peak around the Curie temperature, low cost and no toxicity, making them become potential candidates for magnetic refrigeration [23-25]. However, there are still some problems in fabricating $(Nd,Pr)_2Fe_{17}$ single phase and regulating magnetic phase transition for these alloys. Thus, in this paper, we present the results of study on dependence of structure, magnetic properties and magnetocaloric effect on heat treatment and Nd/Pr ratio of $Pr_{2-x}Nd_xFe_{17}$ (with x = 0, 1 and 2) alloy ribbons.

2. Experiment

Ingots with nominal compositions of $Pr_{2-x}Nd_xFe_{17}$ (x = 0, 1 and 2) were prepared from pure components of Pr, Nd and Fe on an arc-melting furnace to ensure their homogeneity. The ribbons were then fabricated on a single wheel melt-spinning system. The quenching rate of the ribbons could be changed by changing tangential velocity, v, of the copper wheel. In this study, the ribbons were prepared with v = 40 m/s. The ribbons samples with the ~ 3 mm width and ~ 15 μm thickness were obtained. Subsequently, the ribbons were annealed from 900°C to 1000°C for various times. All of the arc-melting, melt-spinning and annealing process were performed under Ar atmosphere to avoid oxygenation. Structure of the ribbons was analyzed by X-ray diffraction (XRD). Magnetization measurements in the temperature range of 77 - 600 K were performed on a

* Corresponding author: danh@ims.vast.ac.vn

vibrating sample magnetometer (VSM). The values of magnetic entropy change ΔS_m which caused by a variation of applied magnetic field was calculated via:

$$\Delta S_m = -\int_0^H \left(\frac{\partial M}{\partial T} \right) dH \qquad (1)$$

3. Result and discussion

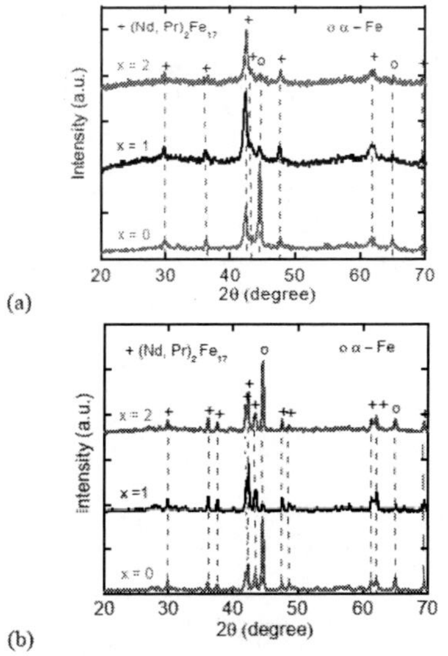

Fig. 1. XRD patterns of $Pr_{2-x}Nd_xFe_{17}$ (x = 0, 1 and 2) ribbons before (a) and after annealing (b).

Figure 1 is the XRD patterns of the $Pr_{2-x}Nd_xFe_{17}$ (x = 0, 1 and 2) alloy ribbons. The results show that the structure of the ribbons clearly depends on Nd-concentration. There are two diffraction peaks of the XRD patterns corresponding to α-Fe phase. Other peaks are of $(Nd,Pr)_2Fe_{17}$ (2:17) phase. Intensity of diffraction peaks of α-Fe phase is reduced with increasing Nd-concentration. This is important in controlling the formation of undesirable phases of the material. Annealing process has also clearly influenced on the structure of the alloy. After annealing, more diffraction peaks characterizing $(Nd,Pr)_2Fe_{17}$ phase are formed. Intensity of these peaks is higher and sharper, especially, in the sample with x = 1. It should be noted that intensity of diffraction peaks of α-Fe is weakened with Nd/Pr ratio of 1. Thus, the sample is relatively single phase with appropriate Nd-concentration. This will affect the magnetic properties and MCE of the alloy as presented below the alloy.

To determine the coercivity, H_c, and the saturation magnetization, M_s, of the samples, we measured hysteresis loops of $Pr_{2-x}Nd_xFe_{17}$ (x = 0, 1 and 2) ribbons at room temperature (figure 2). The results show that all the samples manifest the soft magnetic behavior. The coercivity of the samples is smaller than 160 Oe. Figure 3 presents the dependence of the saturation magnetization on Nd-concentration. From this figure, we see that the M_s of the samples is quite high, above 50 emu/g. Besides, the saturation magnetization is reduced with increasing Nd-concentration (or reducing Pr-concentration). In detail, the M_s value of the samples with x = 0, 1 and 2 is 105, 101 and 54 emu/g, respectively.

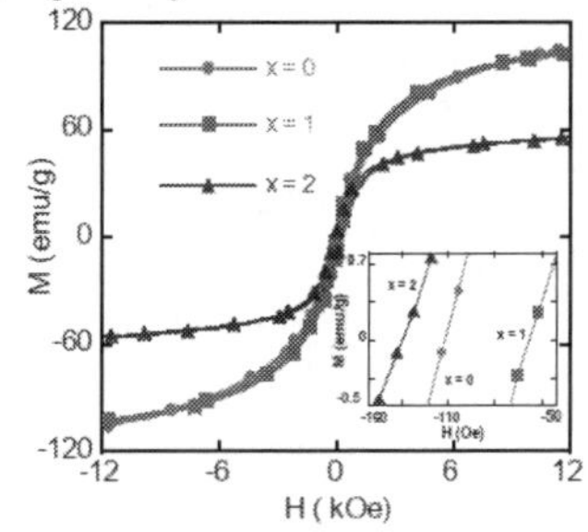

Fig. 2. M(H) loops at room temperature (the inset enlarges the loops at low magnetic field) of $Pr_{2-x}Nd_xFe_{17}$ (x = 0, 1 and 2) ribbons.

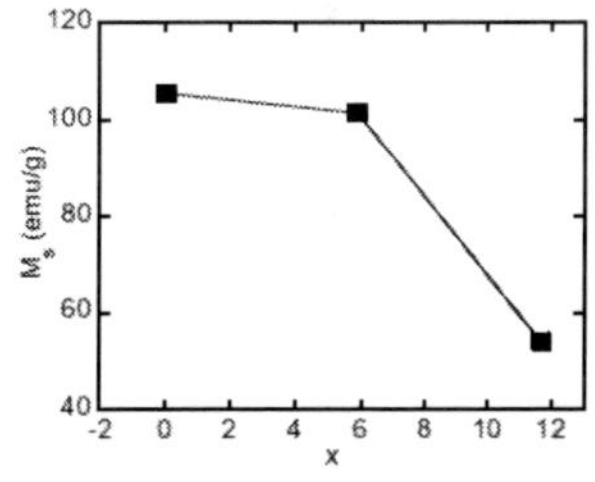

Fig. 3. The dependence of the saturation magnetization on Nd concentration (x) of $Pr_{2-x}Nd_xFe_{17}$ (x = 0, 1 and 2) at room temperature.

In order to investigate influence of annealing process on magnetic properties of the alloy, we measured thermomagnetization curves of the samples before and after annealing. Figure 4 reveals the dependence of reduced magnetization on temperature of the as-quenched ribbons in an applied magnetic field of 100 Oe. In as-quenched state, the samples manifest the multi-phase behavior. These ribbons have magnetic phase transitions in the temperature range of 250 - 450 K. After the first transition, the magnetization of these ribbons does not decrease to zero, but maintains values characterizing the other crystalline phase. The x = 0 sample has a magnetic phase transition at about 300 K. The x = 1 sample owns the first phase transition at 380 K and the second phase transition at 425 K. Similarly, the x = 2 sample has a phase transition at 370 K and the next phase transition at 440 K. With purpose to create single phase for the alloy, heat treatment was carried out. Figure 5 is the reduced thermomagnetization curves in an applied field of 100 Oe of $Pr_{2-x}Nd_xFe_{17}$ (x = 0, 1 and 2) alloy ribbons after annealing at 900°C for 2 h. The results show that the magnetic properties are considerably improved after annealing. Specially, the ferromagnetic-paramagnetic (FM-PM) transition of the x = 1 sample is quite sharp. According to XRD patterns of the samples (figure 7), the sample with x = 0 is almost single phase with the 2:17 structure after annealing. As for the samples with x = 0 and x = 2, diffraction peaks of the α-Fe crystalline phase are comparable with those of the 2:17 phase. That is reason for the single magnetic phase feature of the sample with x = 1 and the multi-magnetic phase behavior of the samples with x = 0 and x = 2. Curie temperature of the α-Fe phase is higher than that of the 2:17 phase leading to the maintenance of magnetization in high temperature region after the transition of the 2:17 magnetic phase near room temperature.

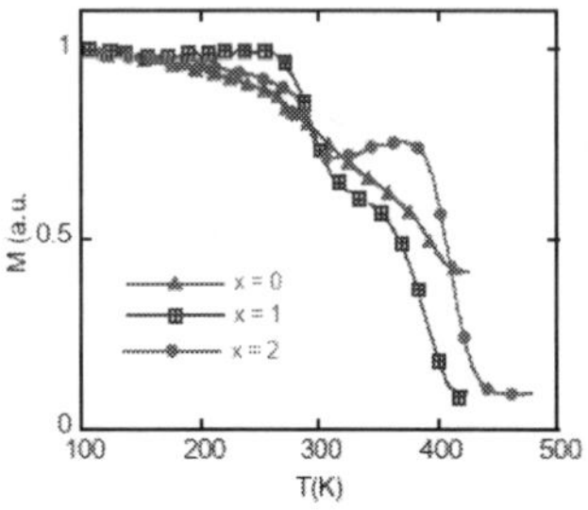

Fig. 4. Reduced thermomagnetization curves in an applied magnetic field of 100 Oe of $Pr_{2-x}Nd_xFe_{17}$ (x = 0, 1 and 2) as-quenched ribbons.

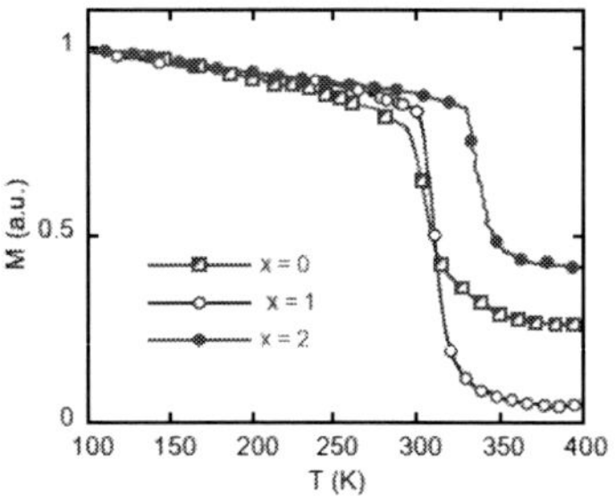

Fig. 5. Reduced thermomagnetization curves in an applied magnetic field of 100 Oe of $Pr_{2-x}Nd_xFe_{17}$ (x = 0, 1 and 2) ribbons after annealing at 900°C for 2 h.

Figure 6 displays the reduced thermomagnetization curves in an applied magnetic field of 100 Oe of $Pr_{2-x}Nd_xFe_{17}$ (x = 0, 1 and 2) annealed at different temperatures (900 - 1100°C) for various time (0.25 - 2 h). From obtained results, we see that the x = 1 sample annealed at 950°C for 2h has the sharp phase transition. The Curie temperature (T_C) of this sample is 310 K. In particular, its magnetization is almost zero after the first magnetic transition. This shows that this sample is almost single magnetic phase transition.

From figure 6b, it is apparent that the x = 1 sample annealed at 950°C for 2h has a quite strong magnetic phase transition, revealing the possibility of a large MCE. On the other hand, the T_C of this sample is around room temperature. Therefore, with the goal of finding magnetocaloric materials that possess magnetic transition temperature close to room temperature, we chose $NdPrFe_{17}$ ribbon annealed at 950°C for 2 h to investigate its MCE. We calculated magnetic entropy change (ΔS_m) basing on thermomagnetization curves of this ribbon at various magnetic fields ranging from 0.01 to 12 kOe (figure 7). From the thermomagnetization curves for the samples in various magnetic fields, we can deduce the magnetization versus magnetic field, M (H), at various temperatures (figure 8). According to previous results [26, 27], to check this derivation procedure, we compared data deduced from thermomagnetization curve with data of virgin magnetization ones and we found that the data obtained from the two different ways are coincided. Then, the magnetic entropy change, ΔS_m, is determined from M(H) data by using equation (1).

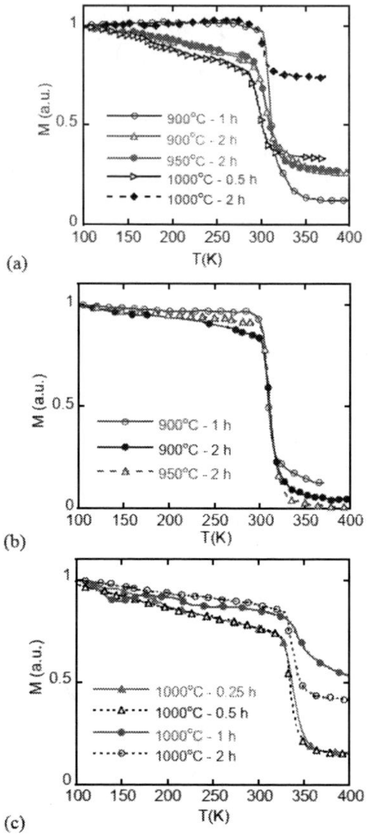

Fig. 6. Reduced thermomagnetization curves of Pr_2Fe_{17} (a), $NdPrFe_{17}$ (b) and Nd_2Fe_{17} (c) ribbons after annealing at different temperatures for various time.

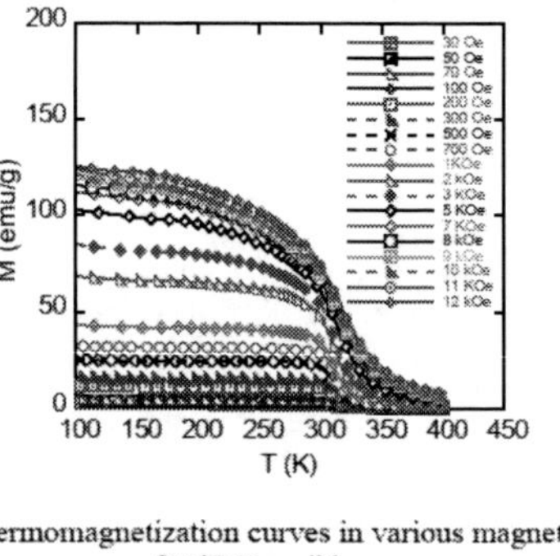

Fig. 7. Thermomagnetization curves in various magnetic field of $NdPrFe_{17}$ ribbon.

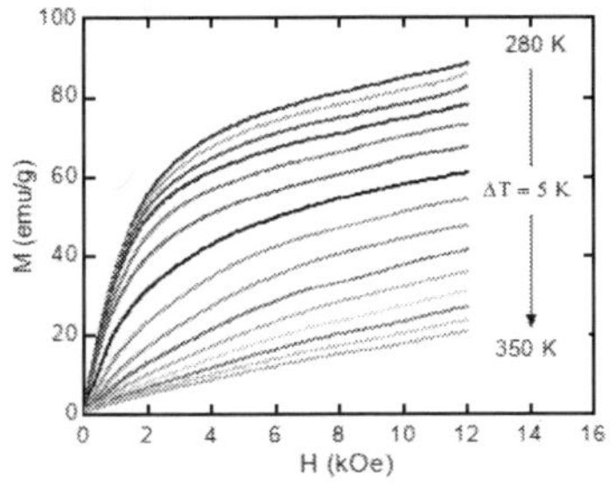

Fig. 8. Magnetization versus magnetic field at various temperatures deduced from thermomagnetization curves of $PrNdFe_{17}$ ribbon.

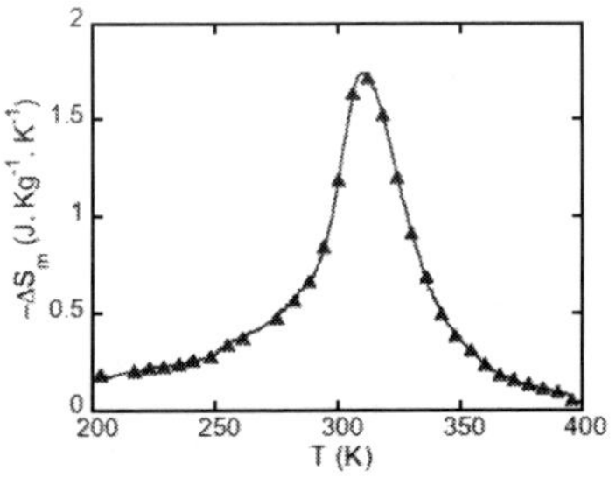

Fig. 9. - ΔS_m(T) curves (with $\Delta H = 12$ kOe) of $PrNdFe_{17}$ ribbon.

Figure 9 is the temperature dependence of - ΔS_m of $NdPrFe_{17}$ ribbon after annealing at 950°C for 2h in a magnetic field change of 0 – 12 kOe. The maximum magnetic entropy change, $|\Delta S_m|_{max}$, is 1.72 J.kg^{-1}.K^{-1} at 310 K. The refrigerant capacity (RC) of the samples, which is defined as product of maximum entropy change ($|\Delta S_m|_{max}$) and full width at half maximum (δT_{FWHM}) of entropy change peak, was also calculated. The δT_{FWHM} value is also mentioned as the working temperature range of magnetic refrigerant. The working temperature range of the alloys is about 40 K. The refrigerant capacity, which is determined for $NdPrFe_{17}$ ribbon, is 69 J/kg. This value is comparable to higher than that of Pr_2Fe_{17} and Nd_2Fe_{17} alloys [11, 18, 19].

4. Conclusion

The influence of Pr/Nd ratio and annealing process on structure, magnetic properties and magnetocaloric effect of (Pr,Nd)-Fe alloy ribbons has been investigated. The formation of $(Pr,Nd)_2Fe_{17}$ crystalline phase strongly depends on Pr/Nd ratio and annealing conditions. $(Pr,Nd)_2Fe_{17}$ phase is dominantly formed in the sample after annealing. Magnetic properties of the alloy changed markedly with annealing temperature and time. The sample with Pr/Nd ratio of 1 is relatively single phase after annealing. The maximum magnetic entropy change ($|\Delta S_m|_{max}$) of $NdPrFe_{17}$ ribbon was determined to be 1.72 J.kg^{-1}.K^{-1} at 310 K with magnetic field change $\Delta H = 12$ kOe.

This work was supported by Vietnam Academy of Science and Technology under grant No. VAST.HTQT.NGA.05/17-18 and Russian Foundation for Basic Research under grant No. 17-58-540002. A part of the work was done in Key Laboratory for Electronic Materials and Devices and Laboratory of Magnetism and Superconductivity, Institute of Materials Science, Vietnam.

References

1. X. Zhou, W. Li, H.P. Kunkel, G. Williams, J. Phys.: Condens. Matter **16**, L39 (2004).
2. X. Zhang, B. Zhang, S. Yu, Z. Liu, W. Xu, G. Liu, J. Chen, Z. Cao, G. Wu, Phys. Rev. B: Condens. Matter **76**, 132403 (2007).
3. W. Y. Tian, B. H. Yang, P. M. Xiang, Z. D. Qian, Wang Wei Hua, Sci. China. Ser. G-Phys. Mech. Astron **51**, 337 (2008).
4. L. Min, B. Yu, J. Cent. South Univ. Technol. **16**, 1 (2009).
5. V.K. Sharma, M.K. Chattopadhyay, J. Phys. D: Appl. Phys. **43**, 225001 (2010).
6. E. Yüzüak, B. Emre, Y. Elerman, A. Yüce, Chinese Phys. B, **19**, 057501 (2010).
7. M. Klimczak, E. Talik, J. Phys.: Conf. Ser. **200**, 092009 (2010).
8. S.P. Mathew, S.N. Kaul, A.K. Nigam, A.C. Probst, R. Birringer, J. Phys.: Conf. Ser. **200**, 072047 (2010).
9. J.J. Croat, J. Appl. Phys. **52**, 2509 (1981).
10. J.L. Sánchez Llamazares, M.J. Pérez, P. Aslvarez, J.D. Santos, M.L. Sasnchez, B. Hernando, J.A. Blanco, J. Sasnchez Marcos, P. Gorria, J. Alloys Compd. **483**, 682 (2009).
11. P. Alvarez, J. Sánchez-Marcos, J.L Sanchez Llamazares, V. Franco, M. Reiffers, J.A. Blanco and P. Gorria, 14th Czech and Slovak Conference on Magnetism, Košice, Slovakia, July 6–9 (2010).
12. H. Chen, Y. Zhang, J. Han, H. Du, C. Wang, Y.C. Yang, J. Magn. Magn. Mat. **320**, 1382 (2008).
13. Z.G. Zheng, X.C. Zhong, H.Y. Yu, V. Franco, Z.W. Liu, J. Appl. Phys. **111**, 07A922 (2012).
14. X.C. Zhong, P.F. Tang, Z.W. Liu, D.C. Zeng, Z.G. Zheng, J. Appl. Phys. **111**, 07A919 (2012).
15. Y. K. Fang, C.H. Lai, C.C. Hsieh, X.G. Zhao, H.W. Chang, W.C. Chang and W. Li, J. Phys. Conf. Series **266**, 012002 (2011).
16. K.H.J. Buschow, Rep. Prog. Phys. **40**, 1179 (1977).
17. F. Weitzer, K. Hiebl, P.J. Rogl, Appl. Phys. **65**, 4963 (1989).
18. S. Yu. Dan'kov, V.V. Ivtchenko, A.M. Tishin, Jr.K.A. Gschneidner, V.K. Pecharsky, Adv. Cryog. Eng. **46**, 397 (2000).
19. K. Mandal, A. Yan, P. Kerschl, A. Handstein, O. Gutfleisch, K.H. Müller, J. Phys. D **37**, 2628 (2004).
20. O. Tegus, E. Bruck, K.H.J. Buschow, F.R. de Boer, Nature **415**, 150 (2002).
21. V. Provenzano, A.J. Shapiro, R.D. Shull, Nature **429**, 853 (2004).
22. A. Fujita, S. Fujieda, Y. Hasegawa, K. Fukamichi, Phys. Rev. B **67**, 104416 (2003).
23. A.M. Tishin, Y.I. Spichkin, *The magnetocaloric effect and its applications* (Bristol: IOP Publishing, 2003).
24. Gorria P., J.L. Sánchez Llamazares, P. Álvarez, M.J. Pérez, J Sánchez Marcos, J.A. Blanco, J. Phys. D: Appl. Phys. **41**, 192003 (2008).
25. V. Franco, Annu. Rev. Mater. Res. **42**, 305 (2012).
26. N.H. Yen, P.T. Thanh, N.H. Duc, T.D. Thanh, T.L. Phan, S.C. Yu, N.H. Dan, Adv. Nat. Sci.: Nanosci. Nanotechnol. **4**, 025018 (2013).
27. N.H. Dan, N.H. Duc, T.D. Thanh, N.H. Yen, P.T. Thanh, N.A. Bang, D.T.K. Anh, T.L. Pham, S.C. Yu, J. Korean Phys. Soc. **62**, 1715 (2013).

Double magnetocaloric peak feature observed in quaternary Ni-Mn-In based Heusler alloys

I. D. Rodionov[1,*], *I. S.* Dubenko[2], *V. N.* Prudnikov[1], *A. B.* Granovsky[1], *I. S.* Titov[1,3], and *A. M.* Saletsky[1]

[1] *Faculty of Physics, Lomonosov Moscow State University, Moscow, Russia*
[2] *Department of Physics, Southern Illinois University, Carbondale, USA*
[3] *Physics and Materials Science Research Unit, University of Luxembourg, Luxembourg, Luxembourg*

Abstract. We present the results of the studies of magnetic and magnetocaloric (MCE) properties of quaternary Heusler alloys $Ni_{48.5}Mn_{35}In_{15}Co_{1.5}$ and $Ni_{50}Mn_{35}In_{13.5}Al_{1.5}$ in temperature range between 80 K and 400 K. While doping initial ternary Ni-Mn-In alloy with 1.5 at.% Al does not change T_C^A as well as M_S, doping with 1.5 at.% Co significantly increases both T_C^A and M_S. The direct magnetocaloric measurements around martensitic transition and austenite Curie temperature reveal the competition between the contributions of different signs. The latter is related to the vicinity of transition temperatures of austenite and martensite, resulting in mixed state, where both phases undergo magnetic order-disorder transformation in the same temperature region. Additionally temperature dependencies of adiabatic temperature change ΔT_{ad} and magnetic entropy ΔS_M of the system exhibit additional peak feature which we attribute to intermartensitic transition occurring in these alloys.

1 Introduction

Study of magnetocaloric effect (MCE) in different multifunctional materials is very attractive due to huge potential of it's practical applications [1]. Using magnetocaloric material as working body of magnetic refrigerator [2] leads to significant decrease of harmful waste release, reduction of consumed electricity and increasing of efficiency up to 30 % [3]. Additionally, MCE may be used in medicine for destruction of caner cells applying hyperthermia treatment [4, 5]. Heusler alloys are one of the most promising MCE materials with large amount of ΔT_{ad} at applied magnetic field [6, 7].

Off-stoichiometric Heusler alloys based on Ni-Mn-In undergo martensitic transition (MT) - structural diffusionless transition from high temperature austenite $L2_1$ phase to low temperature $L1_0$ phase. This transition reveals strong dependence between structural and magnetic subsystems - MT may be induced both by temperature change or by applying magnetic field. Doping by fourth element in low concentration (no more then 5 at.%) one can drastically change martensitic transition parameters (martensite start M_S and finish temperatures M_F, width of temperature hysteresis, saturation magnetization per unit cell in martensite and austenite states) [6].

MT proceeds with increase of volume of unit cell. Due to that reason inner mechanical stress appears and it may accompanied by formation of modulated structures [8–10]. Lowering temperature modulated structure may transform into non-modulated martensite phase and such transformation is called intermartensitic transition (IMT). In general IMT means sequence of phase transitions between modulated and non-modulated states of martensite phase. Similarly to MT, the presence of IMT may be detected by several methods. The most reliable way to find the IMT it is XRD analysis in vicinity of phase transition temperature range [11]. The temperature dependence of magnetization exhibits step-like behavior at hysteresis loop [12]. IMT also can be found from temperature dependencies of resistivity [11].

In this publication we present the results of magnetic and MCE study of quaternary Ni-Mn-In based Heusler alloys doped by Al and Co which reveal step-like behavior of magnetization, $\Delta T_{ad}(T)$ and $\Delta S_M(T)$ in vicinity of MT.

2 Experiment

Polycrystalline bulk samples were prepared by arc melting 4N purity metals under argon atmosphere followed by annealing in high vacuum (~ 10^{-5} Torr) for 24 h at 1123 K. The chemical composition of as-prepared samples was measured by energy-dispersive X-ray spectroscopy (EDS) and its relative deviations were less than 1 at.%. The room temperature structural analysis was carried out with conventional x-ray diffraction. Magnetic measurements were carried out with VSM Lake Shore in ZFC and FC regimes in a temperature range between 80 – 400 K in high (1.6 T) and low (5 mT) magnetic fields. Indirect determination of the magnetic entropy change ΔS_M were performed by calculation of ΔS_M via Maxwell equation from isothermal

*e-mail: rodionov@physics.msu.ru

magnetization curves $M(H)$:

$$\Delta S_{\mathrm{M}}(T, \Delta H) = \int_0^H \frac{\partial M(T, H)}{\partial T} dH \qquad (1)$$

Direct measurements of the adiabatic temperature change ΔT_{ad} in changing field $\mu_0 \Delta H = 1.8$ T were carried out using an adiabatic magnetocalorimeter (MagEq MMS 801) in a temperature range between 250 – 350 K.

3 Results and discussion

Temperature dependencies of magnetization at low and high magnetic fields for $Ni_{50}Mn_{35}In_{13.5}Al_{1.5}$ are presented on fig. 1 (a) and (c) . According this data the sample undergoes magnetic order-disorder transition at $T_C^A = 324$ K. Cooling to $M_S = 294$ K starts forward MT which finishes at $M_F = 250$ K. Abrupt change of magnetization between two phases are equal to $\Delta M = 33$ emu/g at $H = 1.6$ T. For $Ni_{48.5}Mn_{35}In_{15}Co_{1.5}$ $M(T)$ dependencies are presented on fig. 1 (b) and (d). Magnetic transition from ferromagnetic to paramagnetic state occurs at $T_C^A = 360$ K. MT start and finish temperatures are respectively $M_S = 342$ K and $M_F = 308$ K. Magnetization drop during MT $\Delta M = 54$ emu/g. Low temperature hysteresis was detected between 80 K and 140 K (fig. 1 (b)). The width of this hysteresis increases with increasing of external magnetic field.

It is worth mentioning that magnetization in area of martensitic transition reveals step-like $M(T)$ dependence on cooling and heating both in case of Al and Co-doped alloys. For Co-doped alloys such behavior clearly seen in vicinity of $M_{IM} = 338$ K (fig. 1 (b)). Also such non monotonous behavior of magnetization take place in case of $M(T)$ plotted from $M(H)$ data in wide magnetic field range from 5m T to 1.6 T.

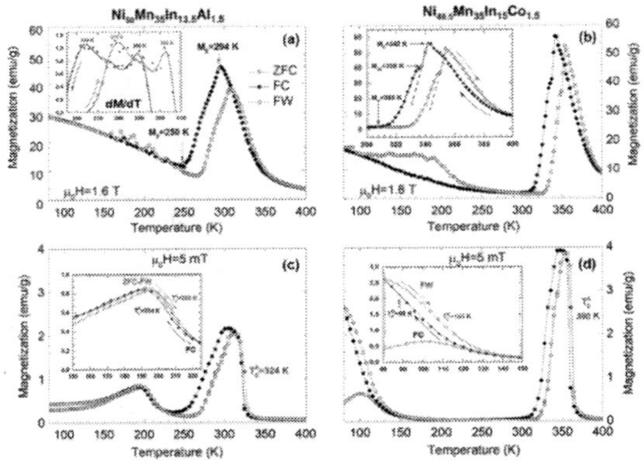

Figure 1. Temperature dependencies of magnetization for $Ni_{50}Mn_{35}In_{13.5}Al_{1.5}$ and $Ni_{48.5}Mn_{35}In_{15}Co_{1.5}$ samples measured in 1.6 T ((a) and (b)) and 5 mT ((c) and (d)), respectively.

Fig. 2 (a) and (b) show temperature dependencies of adiabatic temperature change ΔT_{ad} on applying 1.8 T for

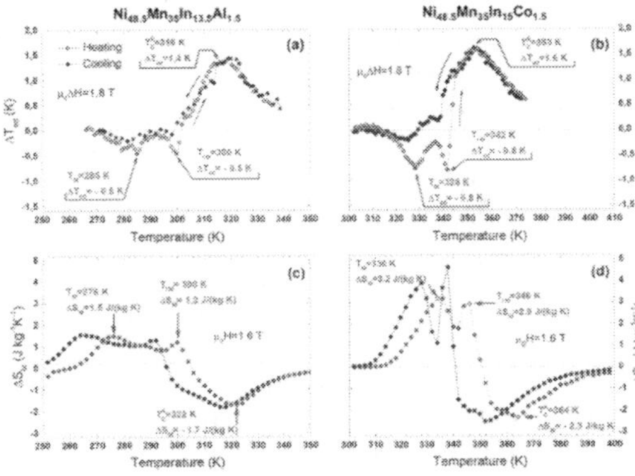

Figure 2. Temperature dependencies of adiabatic temperature change on applying 1.8 T ΔT_{ad} and magnetic entropy change ΔS_M in 1.6 T around the martensitic transition for $Ni_{50}Mn_{35}In_{13.5}Al_{1.5}$ and $Ni_{48.5}Mn_{35}In_{15}Co_{1.5}$ samples ((a) and (c)) and ((b) and (d)), respectively. Characteristic transition temperatures are marked with arrows.

$Ni_{50}Mn_{35}In_{13.5}Al_{1.5}$ and $Ni_{48.5}Mn_{35}In_{15}Co_{1.5}$ samples, respectively. Both samples exhibit maximum of ΔT_{ad} of about 1.5 K in the vicinity of austenite Curie temperature T_C^A: $Ni_{50}Mn_{35}In_{13.5}Al_{1.5}$ at 316 K and $Ni_{48.5}Mn_{35}In_{15}Co_{1.5}$ at 353 K, respectively. Al-doped sample shows two inverse MCE peaks in the region of the martensitic transition on heating: at $T_M = 285$ K 26 K wide and at $T_{IM} = 300$ K 7 K wide. Both peaks show the same magnitude of about 0.5 K. Co-doped sample also shows similar ΔT_{ad} dependence on heating with double peak feature: at $T_M = 328$ K 6 K wide and at $T_{IM} = 342$ K 24 K wide. Both peaks show the same magnitude of about 0.8 K.

Fig. 2 (c) and (d) show temperature dependencies of magnetic contribution to total entropy of the system ΔS_M for $Ni_{50}Mn_{35}In_{13.5}Al_{1.5}$ and $Ni_{48.5}Mn_{35}In_{15}Co_{1.5}$ samples, respectively. ΔS_M was calculated following the Maxwell equation (see eq. 1) from the series of isothermal magnetization curves measured on cooling and heating with step of 2 K in temperature with subsequent return martensite (austenite) state on heating (cooling).

For the Al-doped sample the maximum entropy change $|\Delta S_M| = 1.7$ J/kg K is observed around $T_C^A = 322$ K. In the region of the martensitic transition similar double peak feature is present: $|\Delta S_M| = 1.5$ J/kg K at $T_M = 276$ K and $|\Delta S_M| = 1.2$ J/kg K at $T_{IM} = 300$ K. For the Co-doped sample the maximum entropy change $|\Delta S_M| = 2.3$ J/kg K is observed around $T_C^A = 364$ K, two peaks are present during martensitic transition: $|\Delta S_M| = 3.2$ J/kg K at $T_M = 336$ K and $|\Delta S_M| = 2.9$ J/kg K at $T_{IM} = 346$ K. The difference between $\Delta S_M(T)$ curves on heating and cooling for Co-doped sample comes into agreement with $M(T)$ for this sample.

In summary, doping initial ternary $Ni_{50}Mn_{35}In_{15}$ alloy with 1.5 at.% Al does not change T_C^A as well as M_S, doping with 1.5 at.% Co significantly increases both T_C^A and M_S[13]. However, both Al- and Co-doped samples exhibit step-like behavior in temperature dependence of magnetization $M(T)$ as well as double peak feature in both magnetic entropy change ΔS_M and adiabatic temperature change ΔT_{ad}, these features are not observed in initial ternary alloy. One of the reasons for this peculiar behavior could be the intermartensitic transformation happening during the conventional martensitic transition. Such transitions between $L1_0$ and modulated martensite phases were reported by several groups, see[6, 12]. The origin and features of such behavior will be the subject of further temperature-dependent x-ray diffraction and differential scanning calorimetry studies.

We acknowledge financial support from the Russian Foundation for Basic Research (RFBR): grant No. 15-02-01976 A and 16-32-00460 mol a.

References

[1] O. Gutfleisch et al., Advanced Materials **23 (7)**, p. 841-842 (2011)

[2] K. Gschneidner et al., International Journal of Refrigeration **31 (6)**, pp. 945–961 (2008)

[3] C. Zimm et al., Advances in Cryogenic Engineering **43**, pp. 1759–1766 (1998)

[4] A. M. Tishin et al., Recent Patents on Anti-Cancer Drug Discovery **11 (4)**, pp. 360–375 (2016)

[5] A. M. Tishin et al., International Journal of Refrigeration **68**, pp. 177-186 (2016)

[6] J. Liu et al., Nature Materials **11 (7)**, p. 620-6 (2012)

[7] A. Quetz et al., Journal of Alloys and Compounds **683**, p. 139-142 (2016)

[8] K. Otsuka et al., Scripta Metallurgica et Materiala **29 (10)**, pp. 1359–1364 (1993)

[9] P. J. Brown et al., Journal of Physics: Condensed Matter **18 (7)**, p. 2249 (2006)

[10] S. Kaufmann et al., Physical Review Letters **104 (14)** (2010)

[11] A.Cakir et al., Journal of Applied Physics **114 (18)** (2013)

[12] S. Y. Dong et al., Scientific Reports **6** (2016)

[13] I. D. Rodionov et al., JETP Letters **311 (6)**, pp. 385–389 (2015)

Visualization of magnetostructural transition in Heusler alloys by Magnetic Force Microscopy

Pavel Geydt[1,*], *Igor D.* Rodionov[2], *Alexander B.* Granovsky[2], *Ekaterina* Soboleva[1], *Egor* Fadeev[1], *Ivan S.* Titov[2,3], *Igor* Dubenko[4], and *Erkki* Lähderanta[1]

[1]School of Engineering Science, Lappeenranta University of Technology (LUT), 53850 Lappeenranta, Finland
[2]Faculty of Physics, Lomonosov Moscow State University (MSU), 119991 Moscow, Russia
[3]Physics and Materials Science Research Unit, University of Luxembourg, 4365 Luxembourg, Luxembourg
[4]Department of Physics, Southern Illinois University, Carbondale, IL 62901, USA

Abstract. Magnetostructural transition was observed in Ni-Mn-In-Cr Heusler alloy with help of Magnetic Force Microscopy (MFM). The crystal structure of a sample and characteristic temperatures of the phase transition were controlled by roentgenostructural phase analysis and magnetometry, respectively. It appeared prominently important to prepare the surface of the sample until the nanometer level of surface roughness. Magnetic study performed with scanning probe microscope revealed existence of magnetic domains, which were spread across the surface evenly. Further studies revealed that intensity of magnetic signal decreases as fading out of the contrast of the MFM images. It was found that location of domains shifted after the heating/cooling cycle above Curie temperature for the studied alloy. Location of new domain walls appeared correlating with surface scrapings and defects, whilst it became independent from those after heating until just 70°C. The mechanism behind the observed transition is proposed.

1 Introduction and Motivation

Industrial and social needs are precursors of social and economic development, while scientific interest and creativity drive the state of technological advancement. Recently, sustainability paradigm requires more and more ecological, affordable and definitely non-toxic solutions for human scale development needs [1]. Latest three decades represented intensive growth of computerization and spreading of so-called smart technologies across human households. Unfortunately, even chemicals that are used inside coolers of typical refrigerators can hardly be considered as safe for life.

As a response to these challenges, one can observe a strong interest and investments oriented toward basic research of magnetic Heusler alloys. Heusler alloys represent a class of ternary intermetallic compounds where ferromagnetism emerges even in absence of 'typical' ferromagnetic materials [2] such as Co, Fe, Ni etc. Since the discovery of this phenomenon in 1903, Ni-Mn-In became the most widely acclaimed type of Heusler alloy. Exactly Ni-Mn-In is considered as the most probable solution for having the magnetostructural phase transition near room temperature, while the transition is sharp and the difference in magnetization density between martensitic and austenitic phases is also significant. It was shown that addition of the fourth component can significantly affect properties of the alloy, including the characteristic temperatures [3]. Addition of chromium was considered since it is an antiferromagnetic material and can improve the magnetocaloric output and decrease the electrical resistivity of the material.

Typically, Heusler alloys are studied in bulk form by SQUID magnetometry [4], optical magnetic methods [5], calorimetry [5], volumetry [6] and standard structural approaches such as XRD [7]. However, neither of the abovementioned methods can reveal micro- and nanoscale details of a mechanism of magnetostructural transition. It is yet unclear how the austenitic phase emerges in elevated temperatures under magnetic field, while martensitic phase experience paramagnetic transition in equivalent temperatures. Magnetic memory effect, i.e. hysteresis of magnetization, inhomogeneity of chemical composition within the samples and dissimilarity between thermodynamic characteristics of the surface and bulk material veil the definitive mechanism of transformation, which is nowadays far from being well understood.

Scanning Probe Microscopy (SPM) is exactly an accurate and spatially local family of methods, but surprisingly, we noticed a strong deficit of works published with its results concerned with Heusler alloys. Thus, we decided to perform a Magnetic force Microscopy (MFM) study of Ni-Mn-In alloy under variable temperature with strive to answer on how the magnetostructural transition occurs within the sample.

2 Sample preparation and Examination

* Corresponding author: pavel.geydt@gmail.com

Polycrystalline samples of $Ni_{48}Mn_{37}In_{13}Cr_2$ were fabricated with arc melting technique. Melting was repeated at least 7 times in argon atmosphere, followed by annealing of ~3 g ingots in vacuum 10^{-4} torr under heating T=850°C and cooling in water. After cutting and grinding performed with water protection from extensive reheat, the sample was polished by abrasive paste containing ~250 nm inclusions of diamond crystals on station EcoMet 250 (Buehler, USA). For details of the growth and polishing techniques, one can access [8].

Final verification of properties was acquired by energy-dispersive spectroscopy (EDS) and differential scanning calorimetry (DSC) (both not shown), and also X-ray diffraction (XRD), atomic force microscopy (AFM) and SQUID magnetometry. The roentgenostructural phase analysis (Fig. 1) have shown that both martensitic and austenitic phases are represented in the material at room temperature, which correlates well with previous studies of a similar system.

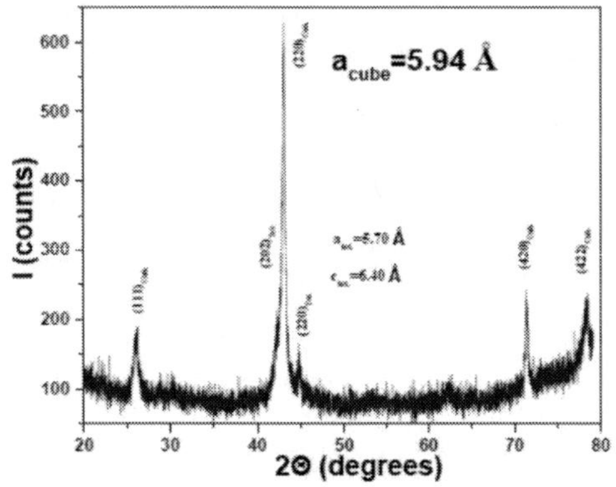

Fig. 1. XRD spectra for the Ni-Mn-In-Cr alloy.

Dependences of magnetization from temperature were studied with SQUID (Cryogenic R-S700X, UK). As it is visible from Figs. 2 and 3, the martensitic phase transition begins approximately at 320 K. It was possible to observe the difference between the gram-weight sample and approximately ten-milligram piece of sample after polishing (Fig. 2). It was visible that the temperature of martensitic transition for a smaller sample was ~10 K lower than for the initial one.

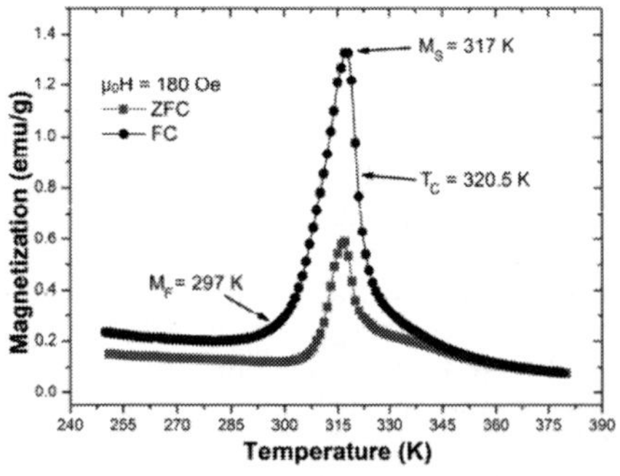

Fig. 2. Magnetization vs. temperature dependence for the Ni-Mn-In-Cr alloy under 180 Oe for milligram-weight sample.

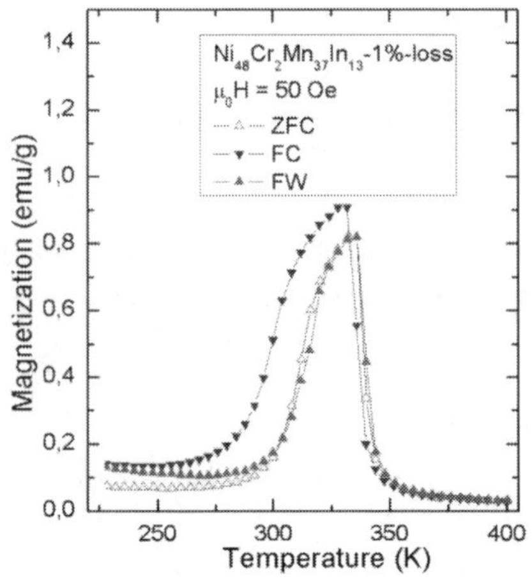

Fig. 3. Magnetization vs. temperature dependence for the Ni-Mn-In-Cr alloy under 50 Oe for gram-weight sample.

For the small sample, the entire material reaches the austenitic phase later than in Fig. 3, so that it appears that almost all austenitic phase is already paramagnetic.

Interesting accomplishments were attained from SPM study performed on Multimode 8 station (Bruker, USA), while they became possible solely due to the high-end polishing of the sample. In fact, more than two years and specific sample preparation were required to visualize the magnetic morphology with MFM. While previously samples could hardly be visualized (even to observe the topology), the polishing step permitted revealing separate magnetic domains within the smooth near-surface layer of the alloy. Therefore, desirable roughness is justified in the level of few nanometers per 10x10 μm scan area.

3 MFM during thermal cycling

Magnetic domains were visualized on the flat surface of the sample of Ni-Mn-In-Cr Heusler alloy. Approximately 1 μm wide, they were homogeneously spread on the entire surface, which was registered in 100x100 μm images taken in room temperature (partly represented in Fig. 4). Size of the domains was not limited by imaging resolution, because the tip radius of the microscopic probe MFMV (Bruker, USA) was ~40 nm and lift scan distance was ~40 nm.

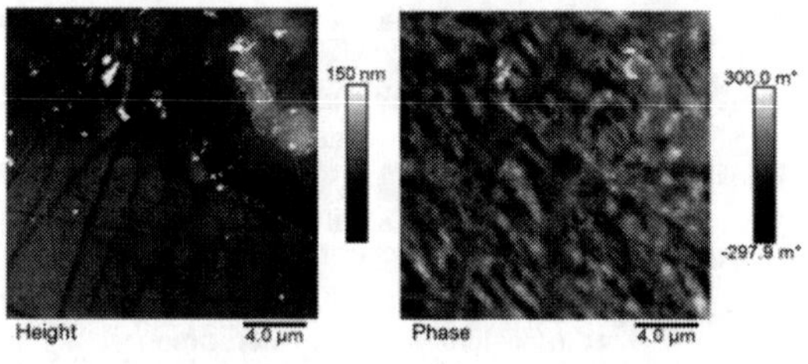

Fig. 4. Topology (*Height*) and Magnetic map (*Phase*) of the Ni-Mn-In-Cr sample for scan area 20x20 μm.

We need to admit that imaging of magnetic morphology with MFM is based on resonant oscillations of the probe during each scanning/tapping pattern. Typically, position of the resonance peak shifts to the lower values under heating, which is associated with decrease of elastic strength of the probe's material. Furthermore, intensity of resonant oscillations Arb decline, which is usually visible as decrease in height of the resonance peak. This effect is also associated with decline of stiffness of the probe and is to limited extent compensated by strengthening of the oscillations caused by decreased density of the heated ambient air. We observed that change of the peak followed the criterion -60 Hz/K and -1.5% Arb/K (Fig. 5). All measurements were carried out in gas cell, containing room air. The scanned area was separated from atmosphere as a specialized Gas cell (Bruker, USA), where this chamber was heated by the preinstalled thermal application controller CLS200 (Watlow, USA).

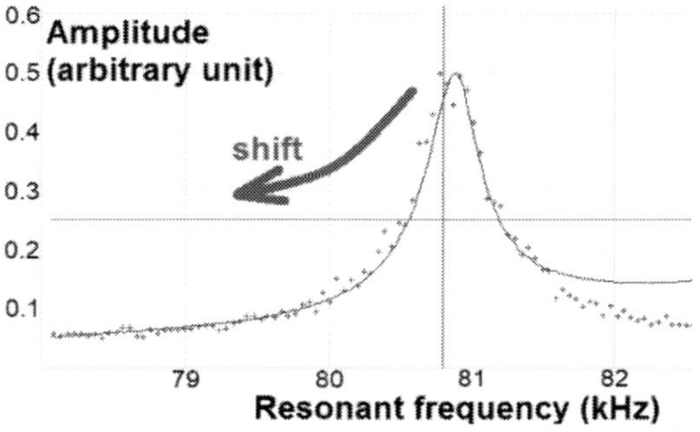

Fig. 5. Schematics of the change of resonance peak of SPM probe observed during thermal cycle.

Such control of the environment lead to accurate registration of the same 7x7 μm area of the sample, though thermal drift was compensated/controlled by enlarged scan sizes of all subsequent images. We performed the heating cycle with following temperatures 25-35-45-50-55-60-70-50-35 °C, so that peculiar changes of the domain structures becomes clear from selective temperatures in Fig. 6. Comparing with the SQUID results, it was possible to highlight that surface of the sample experienced a phase transition ~5 °C lower than the bulk material, while such phenomenon was reported by another research group previously [9]. Furthermore, it was possible to register the change in position of the domains with heating (see consecutive temperatures in Fig. 6). We suggest that mechanical stresses during the grinding/polishing have persisted within the solid material and were concentrated in close proximity to the scrapings appeared from the diamond particles in polishing suspension. We assume that crystal structure also experienced change near these edges. Since these centers have captured additional tension energy, they tended to release it, which is associated with a lower temperature of phase transition observed previously by other groups and us. Domains vanished above the Curie temperature, while they appeared again when sample was cooled in external field of ~180 Oe.

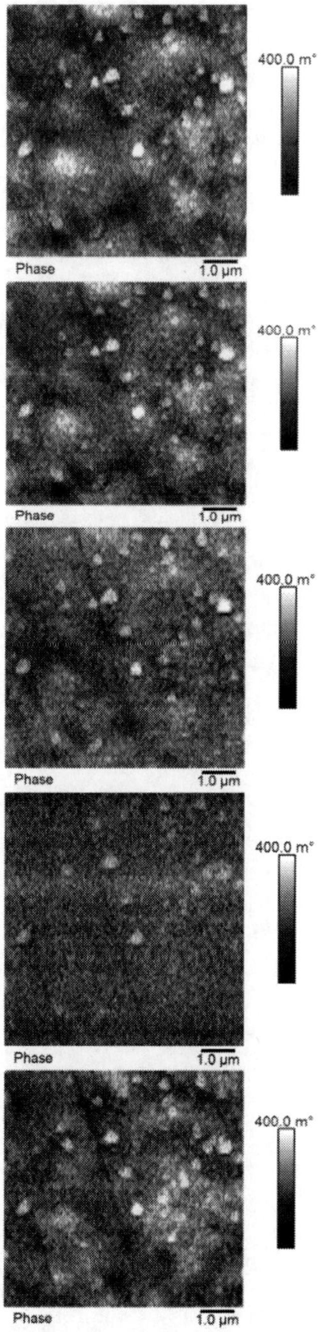

Fig. 6. Sequence of thermal MFM study represented as 25-35-55-70-35 °C. All MFM Phase images are presented for the same scanned area of 7x7 μm.

Remarkably, the boundaries of studied magnetic domains emerged in new locations. Since these magnetic domains are expected to represent mixture of martensitic and austenitic phases, then we can highlight that exactly the magnetostructural transition has been visualized by MFM for the first time.

References

1. D. Griggs, M. Stafford-Smith, O. Gaffney, J. Rockström et al., Nature **495** (2013)
2. A. Herdt, *Exploring the electronic properties of novel spintronic materials by photoelectron spectroscopy* (Dissertation, Jülich, 2012)
3. A. Zhukov, *Novel Functional Magnetic Materials* (Springer, 2016)
4. A. Quetz, Y. S. Koshkid'ko, I. Titov I., I. Rodionov et al., J. All. Comp. **683** (2016)
5. A. Novikov, A. Sokolov, E. A. Gan'shina, I. S. Dubenko, JMMM **432** (2017)
6. P. I. Polyakov, V. V. Slyusarev, V. V. Kokorin, S. M. Konoplyuk et al., J. Mater. Eng. Perform. **23** (2014)
7. S. Pandey, A. Quetz, A. Aryal, I. Dubenko et al., Magnetochemistry **3** (2017)
8. I. Titov, *Influence of magnetism on the martensitic transition and related magnetocaloric effect in NiMn-based Heusler alloys* (Dissertation, Duisburg-Essen, 2015)
9. K. Horakova, V. Chab, O. Heczko, V. Drchal et al., J. Appl. Phys **120** (2016)

Investigation of structural and magnetic properties of Fe-Rh-(Z) (Z = Co, Pt) alloys by first principles method

Oksana Pavlukhina[1,*], *Vladimir* Sokolovskiy[1,2], *Mikhail* Zagrebin[1], *Vasily* Buchelnikov[1]

[1] Chelyabinsk State University, Condensed Matter Physics Department, 454001 Chelyabinsk, Russia
[2] National University of Science and Technology "MIS&S", 119049 Moscow, Russia

Abstract. In this work, we present theoretical investigations of the structural and magnetic properties Fe (Rh, Co) and Fe (Rh, Pt) alloys using the density functional theory. The energy calculations were performed for the 16-atom supercell ($Fe_8Rh_{8-x}Z_x$) with different initial spin configurations. It is shown that a small variation of Pt or Co content leads to change the type of magnetic ordering. It is shown that the ferromagnetic configuration of $Fe_8Rh_{8-x}Co_x$ (x=2, 3) is more energetically favorable as compared with other configurations in austenite. The antiferromagnetic configuration is more energetically favorable for $Fe_8Rh_7Co_1$ alloy. For the $Fe_8Rh_{8-x}Pt_x$ system, the checkerboard-like antiferromagnetic configuration was found to be more energetically favorable. Besides, the addition of Pt into Fe-Rh system slightly changes the optimized lattice parameter and stimulates the martensitic phase transformation.

1 Introduction

Fe-Rh-based alloys showing a metamagnetic phase transition were studied experimentally and theoretically during the last decade [1-7]. This alloys have attracted a lot of attention because of their possible application in magnetic cooling, thermally assisted magnetic recording and spintronic devices [1-3]. Magnetic cooling attracts the attention of scientists around the world. At the present time, a large number of the intermetallic alloys and compounds including Fe-Rh-Z, Mn-Ga-(N, C), La-Ca-Ba-MnO_3 and Ni–Mn–(Ga, In, Sn, Sb) are also promising candidates in magnetic cooling technology [1-9]. The magnetocaloric effect reaches the maximal value at the temperatures of magnetic or magnetostructural phase transitions. Fe-Rh alloys exhibit a metamagnetic phase transition (AFM-FM). The metamagnetic phase transition in Fe-Rh succeeds also the large change in magnetization, which is responsible for a giant magnetocaloric effect (MCE) upon variation of a magnetic field. For $Fe_{49}Rh_{51}$, Nikitin et al. [9] reported for the first time the giant MCE about -13 K at magnetic field change of 2 T using a direct method of measurements. The similar values of MCE for the same composition were successfully repeated in Refs. [4,.5].

It well known, that the magnetic order in FeRh compounds depends strongly on the concentration. Therefore, it is important to study the effect of adding a third element on the magnetic and structural properties of the material. A. Jezierski et. al. [6] study the influence of dopants Co, Pd, Ru and Pl on the magnetic moment and the density of states at the Fermi energy in FeRh alloys. In this article the electronic and magnetic properties of FeRh alloys doped with Co, Pt, Pd Ir and Ru are studied by the TB LMTO-CPA method. The content of the third element in the $FeRh_{1-x}(Z)_x$ alloys was small and amounted to x = 0-0.05. The article says that a significant change of the magnetic moment and the density of states at the Fermi level during the substitution metals.

In this work, we present theoretical investigations of the structural and magnetic properties Fe- Rh-(Z) (Z = Co, Pt) alloys.

2 Computational details

In this work, the structural and magnetic properties of Co- and Pt-doped Fe-Rh alloys are investigated by using the density functional theory calculations as implemented in the Vienna *Ab initio* Simulation (VASP) package [10]. The ab initio calculations have been carried out by using the 16-atom supercell approach with different initial spin configurations. The generalized gradient approximation for the exchange correlation functional in the formulation of Perdew, Burke and Ernzerhof (PBE) was taken into calculations. The performed calculations were semirelativistic and the spin polarization was taken into account for all the cases. In the calculations the automatically generated uniform grid of k-point as in Monkhorst-Pack grids was taken into account. The k-points in the Brillouin zone for self-consistent field cycles were generated with 12^3 meshes for the lattice relaxation calculations and tetragonal distortion.

* Corresponding author: pavluhinaoo@mail.ru

 https://doi.org/10.1051/epjconf/201818505005

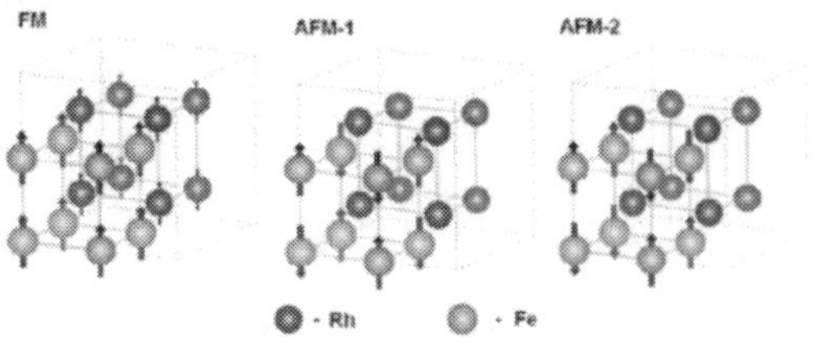

Fig. 1. Spin configurations taken into account in ab initio calculations.

The energy calculations were performed for the L2$_1$ supercell ($Fe_8Rh_{8-x}Z_x$). In detail, original cell has atomic coordinates of (0; 0; 0); (1/2; 0; 0); (1/2; 1/2; 1/2) for Rh atoms and those of (1/4; 1/4; 1/4); (3/4; 3/4; 3/4) for Fe atoms in the supercell. As the result, the supercell contains eight Fe and Rh atoms. Calculations were carried out for ferromagnetic (FM), and two kinds of antiferromagnetic states (AFM-1, AFM-2) with are shown in figures 1. Here, we considered that in the case of FM order the Rh atoms have small magnetic moment (< 1 μB) while it takes zero value for an AFM order.

3 Results of calculations

The variations of the total energy of the 16-atom supercells for $Fe_8Rh_{8-x}Co_x$ (x = 0, 1, 2 and 3) system calculated for different spin configurations as functions of the lattice parameter are presented in figure 2 (a, b, c, d).

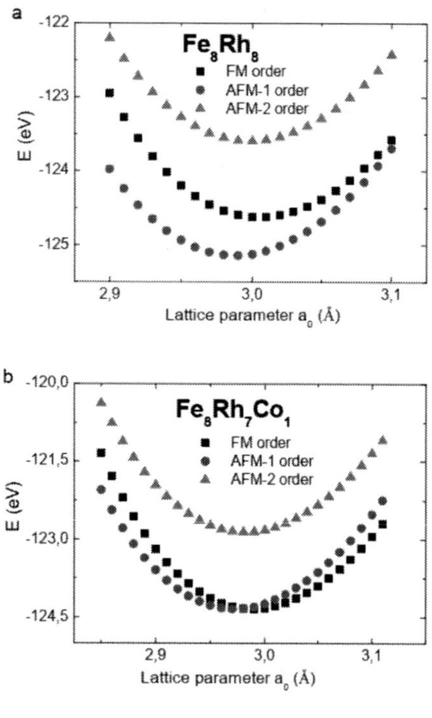

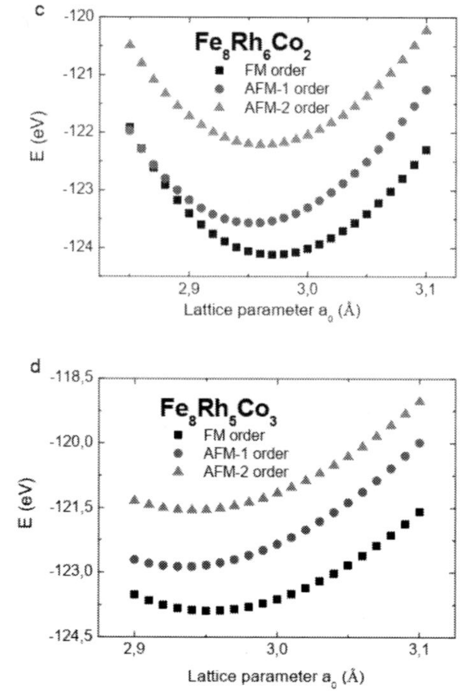

Fig. 2. The total energies of the 16-atom supercells for Fe_8Rh8-xCo_x system calculated for different spin configurations as functions of the lattice parameter.

We found that the AFM-1 (checkerboard-like) spin configuration in a cubic cell is energetically favorable compared to other AFM and FM configurations from Fe-Rh compound. In this case, a total magnetic moment is found to be of 0 μ_B/f.u. We note that the calculated optimized lattice parameter for Fe-Rh alloy is in a good agreement with experimental and other theoretical values [2, 7]. It is worth noting that for the alloy $Fe_8Rh_7Co_1$ of the energy of the FM and AFM-1 phases are close. However, the antiferromagnetic configuration is more energetically favorable for $Fe_8Rh_7Co_1$ alloy. As can be seen from Figure 2, when cobalt is added, the FM phase becomes energetically favorable. With an increase in the content of cobalt, the FM configuration becomes increasingly more energetically favorable. The total energies of the 16-atom supercells for $Fe_8Rh_{8-x}Pt_x$ system were also calculated for different spin configurations as functions of the lattice parameter. We found that the AFM-1 (checkerboard-like) spin configuration in a cubic cell is energetically favorable compared to other AFM and FM configurations for $Fe_8Rh_{8-x}Pt_x$. In table 1 present the total magnetic moments per formula unit for $Fe_8Rh_{8-x}Pt_x$ (x = 1 – 3) and $Fe_8Rh_{8-x}Co_x$ (x = 1 – 3) at equilibrium lattice parameter for all magnetic configurations. As it can be seen from table 1, the magnetic moments for $Fe_8Rh_{8-x}Pt_x$ (x = 1 – 3) and $Fe_8Rh_{8-x}Co_x$ (x = 1 – 3) are close to 4.0 and 4.2 μ_B/f.u. respectively. Further, we investigated the effect of the addition of the third element on the lattice parameter.

https://doi.org/10.1051/epjconf/201818505005

Table 1. The total magnetic moments per formula unit for Fe-Rh-(Z) (Z = Co, Pt) alloys (FM configurations).

x	Total magnetic moments (μ_B/f. u.) for $Fe_8Rh_{8-x}Co_x$	Total magnetic moments (μ_B/f. u.) for $Fe_8Rh_{8-x}Pt_x$
1	4.251	3.914
2	4.267	4.028
3	4.291	4.128

In Fig. 3 we present the total energy for $Fe_8Rh_{8-x}Co_x$ (x = 1 – 3) system with more energetically favorable spin configuration (AFM-1 for $Fe_8Rh_7Co_1$ and FM for $Fe_8Rh_{8-x}Co_x$ (x = 2 – 3)) and $Fe_8Rh_{8-x}Pt_x$ (x = 1 – 3) system with AFM-1 spin configuration calculated as a function of lattice parameter.'

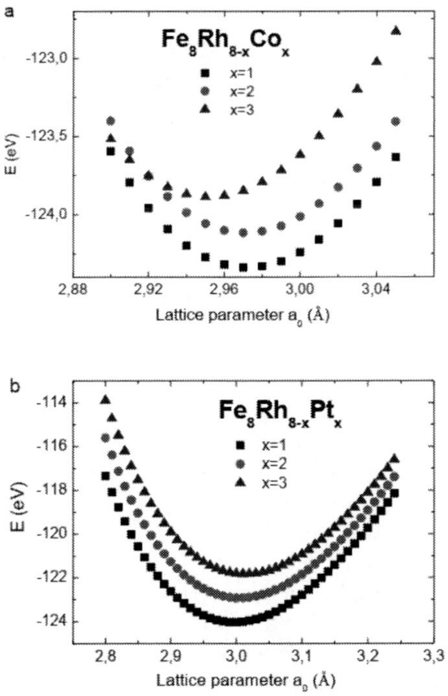

Fig. 3. The total energy curves of the 16-atom supercells in dependence on the lattice parameter for (a) $Fe_8Rh_{8-x}Co_x$ (x = 1 – 3) compositions and (b) for $Fe_8Rh_{8-x}Pt_x$ (x = 1 – 3) system with AFM-1 spin configuration.

Table 2 shows the equilibrium lattice parameters for $Fe_8Rh_{8-x}Pt_x$ (x = 1 – 3) and $Fe_8Rh_{8-x}Co_x$ (x = 1 – 3). Let us consider in more detail the influence of the third element. From the data presented in the table 2, it can be concluded that the addition of Pt atoms leads to an increase in the lattice equilibrium parameter due to the larger atomic radius of Pt compared to the lower Rh value. For cobalt, the reverse is observed. With increasing cobalt content, the equilibrium lattice parameter decreases. This can be explained by the fact that the atomic radius of cobalt is smaller in comparison with rhodium.

Table 2. The lattice parameters depending on the Pt and Co concentration.

x	Lattice parameter (in Å) for $Fe_8Rh_{8-x}Co_x$	Lattice parameter (in Å) for $Fe_8Rh_{8-x}Pt_x$
1	2.971	2.999
2	2.970	3.00
3	2.951	3.011

Further, the possibility of martensitic transformation in these alloys depending on the third element will be discussed.

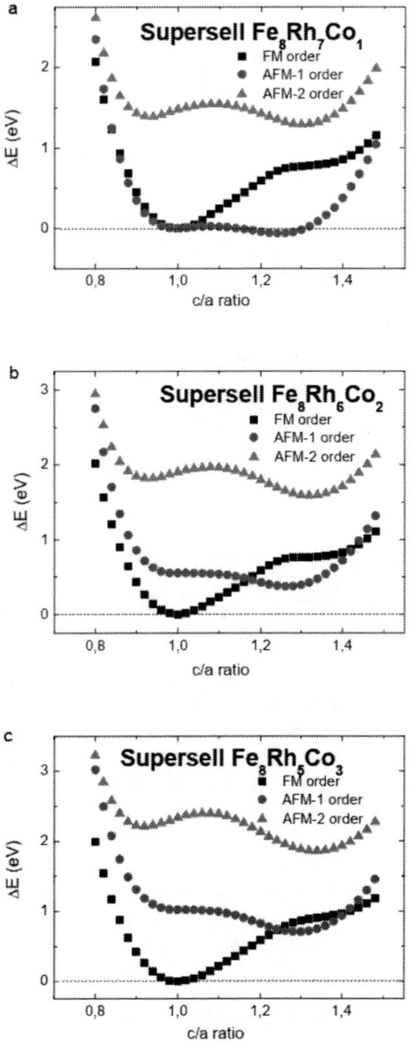

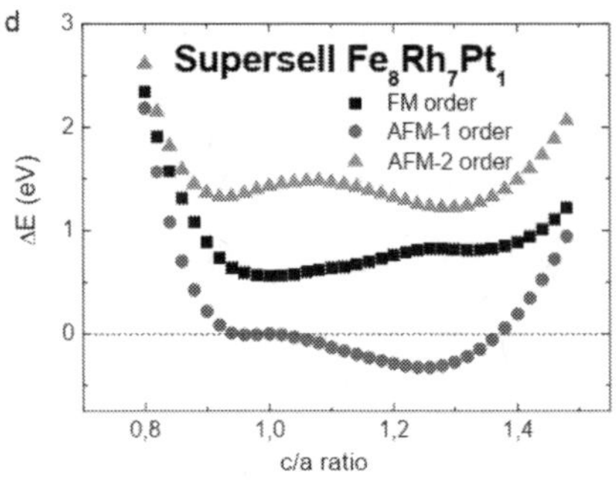

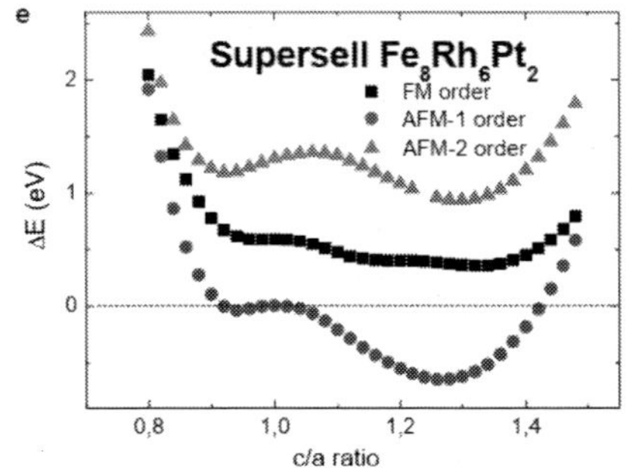

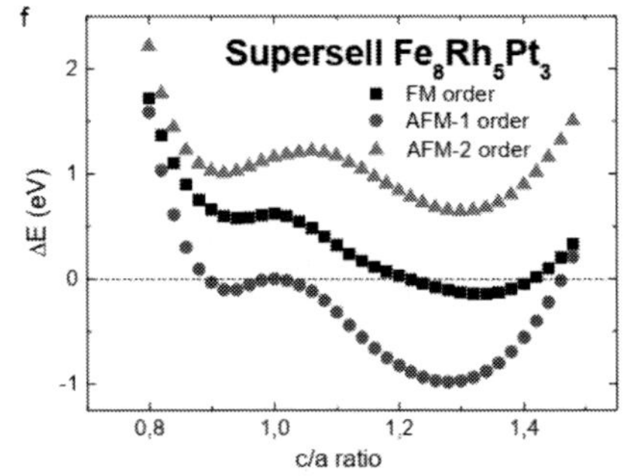

Fig. 4. The variation of the total energy as a function of tetragonal distortion c/a for (a, b, c) $Fe_8Rh_{8-x}Co_x$ (x = 1 – 3) system with FM spin configuration and (d, e, f) $Fe_8Rh_{8-x}Pt_x$ (x = 1 – 3) system with AFM-1 spin configuration.

The calculation of the total energy for the tetragonal distortion of the cubic structure along the z axis is performed in this paper. To accomplish this, we fixed the volume of a supercell as $V_0 = a_0^3 \approx a^2c$. The variation of the total energy as a function of tetragonal distortion c/a for $Fe_8Rh_{8-x}Co_x$ (x = 1 – 3) system and $Fe_8Rh_{8-x}Pt_x$ (x = 1 – 3) one with different spin configuration are shown in Fig. 4 (a-f). In this case, the zero value of ΔE corresponds to the austenitic phase for each compound. Let's consider more in detail the variation of the total energy as a function of tetragonal distortion c/a for $Fe_8Rh_{8-x}Pt_x$ (x = 1 – 3) system with FM, AFM-1 and AFM-2 spin configuration. In the case of $Fe_8Rh_{8-x}Pt_x$, the checkerboard-like antiferromagnetic configuration solution is favorable for martensite states. A tetragonal ratio for martensitic state increases slightly from c/a = 1.26 up to 1.28 with increasing Pt content. Moreover, an increase in Pt content leads to more deeper energy minimum at c/a ratio. It is important to note that the similar trend was observed experimentally by Yuasa et al. [11]. In the case of $Fe_8Rh_7Co_1$ the checkerboard-like antiferromagnetic configuration solution is favorable for both austenite and martensite states, and the structural transformation occurs without a magnetic phase transition.

4 Conclusion

In summary, the structural and magnetic properties of Co- and Pt-doped Fe-Rh alloys are investigated by using the density functional theory calculations as implemented in the VASP. In a cubic cell the antiferromagnetic configuration is more energetically favorable for $Fe_8Rh_7Co_1$ alloy. It is shown that the ferromagnetic configuration of $Fe_8Rh_{8-x}Co_x$ (x= 2, 3) is more energetically favorable as compared with other configurations in austenite. For the $Fe_8Rh_{8-x}Pt_x$ system, the checkerboard-like antiferromagnetic configuration was found to be more energetically favorable. The addition of Pt into Fe-Rh system slightly changes the optimized lattice parameter and stimulates the martensitic phase transformation with increasing doping concentration Pt due to an appearance of local minima in energy curves.

This work was supported by RSF-Russian Science Foundation No. 17-72-20022\17.

References

1. S. Cumpson, P. Hidding, and R. Coehoorn, IEEE Trans. Magn. **36**, 2271 (2000)
2. J.-U. Thiele, S. Maat, and E. E. Fullerton, Appl. Phys. Lett. **82**, 2859 (2003)
3. A. X. Gray, D.W. Cooke, P. Kruger, Phys. Rev. Lett. **108**, 257208 (2012)
4. M.P. Annaorazov, K.A. Asatryan, G. Myalikgulyev et al. Cryog. **32**, 867 (1992)
5. A. Chirkova, K.P. Skokov, L. Schultz et al. Acta Mater. **106**, 15 (2016)
6. A. Jezierski, G. Borstel. J. Magnet. Magnetic Mater. **140-144**, 81 (2005)
7. O. Pavlukhinaa, V. Sokolovskiy, V. Buchelnikov, Materials Today: Proceedings. **4**. 4642 (2017)
8. O. Pavlukhinaa, V. Sokolovskiy, V. Buchelnikov, Phys. Status Solidi A 213. **2**, 390 (2016)
9. S.A. Nikitin, G. Myalikgulyev, A.M. Tishin, et al., Phys. Lett A. **148**, 363 (1990).
10. G. Kresse and J. Furthmuller, Phys. Rev. B. **54**, 11169 (1996)
11. S. Yuasa, H. Miyajima, Y. Otani, J. Phys. Soc. Jpn. **63**, 3129 (1994)

The transformation of the magnetostructural phase transition with Ti addition in $Gd_5Si_2Ge_2$

Sergey Nikitin[1], Andrey Smirnov[1,*], Anatoly Bogdanov[1], and Ioulia Ovchenkova[1]

[1] Faculty of Physics, Lomonosov Moscow State University, Leninskie Gory, 119991, Moscow, Russia

Abstract. The magnetic, thermal expansion and magnetostriction measurements of $Gd_{5-x}Ti_xSi_2Ge_2$ (x=0, 0.05, 0.1) compounds have been carried out in wide range of fields and temperatures. It is found that Ti addition leads to the disappearance of the structural transformation, the change of the order of the phase transition and the decrease of the hysteresis. The large positive magnetostriction associated with the structural phase transition in $Gd_5Si_2Ge_2$ changes to the small negative one associated with the magnetic phase transition in $Gd_{4.9}Ti_{0.1}Si_2Ge_2$. The linear dependence of volume magnetostriction on the square of magnetization obtained for all investigated compounds is explained using the thermodynamic theory.

1 Introduction

The $Gd_5Si_2Ge_2$ compound has attracted the considerable attention since it exhibits the giant magnetocaloric effect (MCE) [1,2]. The structural and magnetic transformations occurring simultaneously at Curie temperature lead to the first order phase transition from paramagnetic monoclinic state to ferromagnetic orthorhombic state. At the same time, the large magnetic and thermal hysteresis typical to the first order phase transitions leads to the difficulties in possible practical application of this compound. Therefore the great interest presents the search of optimal substitution in p- or 4f-sublattice for reduction of hysteresis. The influence of small Sn and In substitutions on the magnetostructural phase transitions have been studied in detail [3-5]. These studies have shown that the Curie temperature slightly shifts by not more than 10 K, the first order phase transition is preserved and MCE slightly increases with small p-element addition. However, the hysteresis losses also increase. In this work the magnetic, thermal expansion and magnetostriction measurements of $Gd_{5-x}Ti_xSi_2Ge_2$ (x=0, 0.05, 0.1) compounds have been carried out in order to determine the influence of the temperature and magnetic field on the magnetoelastic properties of these alloys.

2 Experimental details

The $Gd_{5-x}Ti_xSi_2Ge_2$ (x=0, 0.05, 0.1) samples were prepared by arc-melting in an electric arc furnace under an argon atmosphere using a non-consumable tungsten electrode and a water-cooled copper tray. The ingots were remelted three times to insure homogeneity. The arc-melted samples were annealed at 970 K for 300 h in an argon atmosphere and quenched in ice-cold water. The quality of the samples was evaluated using X-ray diffraction and electron microprobe analysis.

The field and temperature dependencies of magnetization M were measured in applied magnetic fields up to 12 kOe in the temperature range 80–340 K both upon heating and upon cooling. The thermal expansion and magnetostriction measurements were carried out using the strain gauge method in static magnetic fields up to 12 kOe in the temperature range 80–330 K. The magnetostriction measurements were carried out under isothermal conditions.

3 Results

X-ray diffraction at room temperature and electron microprobe analysis reveal that all compounds contain a small amount of impurity GdGe-type phase. The main phase in $Gd_5Si_2Ge_2$ is monoclinic $Gd_5Si_2Ge_2$-type phase (space group $P112_1/a$) whereas in $Gd_{4.9}Ti_{0.1}Si_2Ge_2$ it is orthorhombic Gd_5Si_4-type phase (space group Pnma). $Gd_{4.95}Ti_{0.05}Si_2Ge_2$ contains both $Gd_5Si_2Ge_2$-type and Gd_5Si_4-type phases. The lattice parameters and unit cell volume are listed in Table 1.

According to the temperature dependences of magnetization, all investigated compounds demonstrate ferromagnetic behavior below Curie temperature T_C. The Curie temperatures determined from the maxima on $|dM/dT|$ curves are given in Table 2. For $Gd_5Si_2Ge_2$ the obtained Curie temperature is in a good agreement with

* Corresponding author: andrey07_91@mail.ru

the literature data [2]. The Ti addition shifts T_C to higher temperatures.

Table 1. The lattice parameters of monoclinic and orthorhombic unit cells for $Gd_{5-x}Ti_xSi_2Ge_2$ compounds.

x	a (Å)	b (Å)	c (Å)	γ (°)	V (Å^3)
Monoclinic structure					
0	7.571 (1)	14.773 (3)	7.765 (1)	93.15 (1)	867.18
0.05	7.546 (6)	14.73 (1)	7.759 (5)	93.05 (7)	861.21
Orhtorhombic structure					
0.05	7.490 (4)	14.735 (6)	7.782 (3)	90	858.86
0.1	7.4965 (8)	14.758 (2)	7.7930 (9)	90	862.14

The Belov-Arrott plots (*H/M* vs. M^2) obtained from the magnetization curves are linear near T_C for compounds with x=0.05 and 0.1 which indicates that the phase transition is second order for substituted compounds. Whereas the *H/M* vs. M^2 curves are nonlinear for $Gd_5Si_2Ge_2$ since the first order phase transition occurs for this compound. With changing of the order of phase transition the value of the thermal hysteresis also decreases from 17 K for $Gd_5Si_2Ge_2$ to 5-7 K for substituted compounds.

Table 2. The magnetic and magnetoelastic properties of $Gd_{5-x}Ti_xSi_2Ge_2$ compounds. T_C is Curie temperature, ω_{max} and ω_{min} are the magnitudes of the positive and negative peaks on the temperature dependence of volume magnetostriction

x	cooling			heating		
	T_C (K)	ω_{max} (10^{-4})	ω_{min} (10^{-4})	T_C (K)	ω_{max} (10^{-4})	ω_{min} (10^{-4})
0	271	9.4		288	2.5	
0.05	296	2.0	-0.8	301	0.1	-1.2
0.1	292		-0.5	299		-0.4

Fig. 1 shows the results of thermal expansion measurements. Above T_C the thermal expansion is observed to increase almost linearly. $Gd_5Si_2Ge_2$ demonstrates the change in the thermal expansion of about 10^{-3} near T_C. It also exhibits the large thermal hysteresis of about 10 K. For $Gd_{4.95}Ti_{0.05}Si_2Ge_2$, the change in the thermal expansion is observed near 270 K upon cooling and near 285 K upon heating, which is considerably lower than T_C values obtained from magnetization data. The observed change is significantly lower in magnitude than that for $Gd_5Si_2Ge_2$. For $Gd_{4.9}Ti_{0.1}Si_2Ge_2$, the thermal expansion increases almost linearly with temperature in the entire measured temperature range, which indicates an insignificant contribution from the magnetic subsystem compared with the phonon contribution. The hysteresis is almost absent for this compound.

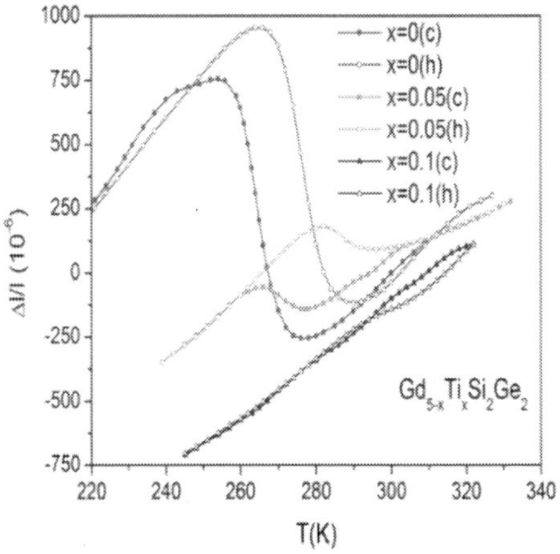

Fig. 1. The linear thermal expansion of $Gd_{5-x}Ti_xSi_2Ge_2$ compounds. Data obtained on continuous cooling (full symbols) and heating (open symbols).

Magnetostriction measurements revealed that the values of longitudinal $\lambda_{\parallel}$ and transverse magnetostriction $\lambda_{\perp}$ are almost the same. The volume magnetostriction was obtained as $\omega = \lambda_{\parallel} + 2\lambda_{\perp}$. The temperature dependences of volume magnetostriction in the applied field of 11.8 kOe are presented in fig. 2. The $Gd_5Si_2Ge_2$ compound possesses the large positive magnetostriction with the maximum near the Curie temperature. The ω values obtained upon cooling and upon heating are very different for this compound. The volume magnetostriction reaches 9.4×10^{-4} upon cooling and does not exceed 2.5×10^{-4} upon heating. For $Gd_{4.95}Ti_{0.05}Si_2Ge_2$, $\omega(T)$ has the positive peak near 280 K and the negative peak near 295 K. Below 290 K the hysteresis is very large while above 290 K the ω values obtained upon cooling and upon heating are rather close. The magnitude of positive peak is larger than that of negative peak but significantly lower than the maximal value for $Gd_5Si_2Ge_2$.

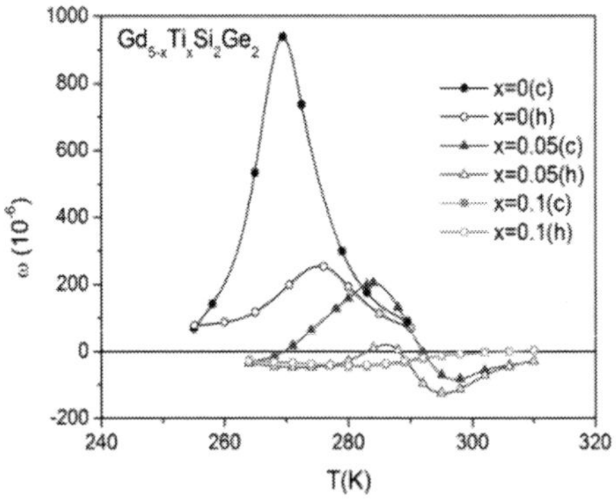

Fig. 2. The temperature dependences of volume magnetostriction for $Gd_{5-x}Ti_xSi_2Ge_2$ compounds in the applied field of 11.8 kOe. Data obtained on continuous cooling (full symbols) and heating (open symbols).

For $Gd_{4.9}Ti_{0.1}Si_2Ge_2$ the maghetostriction becomes negative. The small negative peak is observed near 280 K on $\omega(T)$ for this compound. The hysteresis is almost absent for $Gd_{4.9}Ti_{0.1}Si_2Ge_2$. The magnitude of the peak is much lower than that of negative peak for $Gd_{4.95}Ti_{0.05}Si_2Ge_2$.

4 Discussion

The obtained volume magnetostriction data points out that the maximal value of ω is observed in the vicinity of the magnetostructural phase transition in $Gd_5Si_2Ge_2$. This large positive magnetostriction is due to magnetostructural phase transition from paramagnetic monoclinic state to ferromagnetic orthorhombic state under the action of magnetic field. According to the thermodynamic relation [6,7]

$$\left(\frac{\partial\omega}{\partial H}\right)_P = -\left(\frac{\partial M}{\partial P}\right)_H, \tag{1}$$

where P is the applied pressure. The thermal expansion measurements revealed that the transition from paramagnetic monoclinic state to ferromagnetic orthorhombic state is accompanied by the abrupt increase in the sample size (see fig. 1). This effect can be considered as the negative pressure acting on the sample. So, the negative dP leads to the positive dM on cooling and conversely the positive dP leads to the negative dM on heating. Anyway, $(\partial M/\partial P)_H < 0$, thus $(\partial\omega/\partial H)_P > 0$. So, at the region of the structural phase transition the value of volume magnetostriction ω should increase with the applied magnetic field, which is observed in fig. 2.

For the ferromagnet in the magnetic field H and under the applied pressure P the thermodynamic potential G can be written as [6,7]

$$G = aM^2 + bM^4 + cP^2 - MH, \tag{2}$$

where a, b and c are the thermodynamic coefficients. According to Nikitin et al. [7]

$$a = a'(T - T_C) + \gamma P, \tag{3}$$

where γ is the magnetoelastic coefficient and a' is the thermodynamic coefficient at $P = 0$. The volume change ω can be found directly from the thermodynamic potential

$$\omega = \frac{\partial G}{\partial P} = \gamma M^2 + 2cP. \tag{4}$$

The equation (4) is valid for the second order phase transitions. The linear ω vs. M^2 plots obtained for $Gd_{4.9}Ti_{0.1}Si_2Ge_2$ (fig. 3b) confirm that the phase transition is second order for this compound. The volume magnetostriction also depends linearly on the square of magnetization for two other investigated compounds (fig. 3a). However, unlike $Gd_{4.9}Ti_{0.1}Si_2Ge_2$, for which ω vs. M^2 curves are almost parallel, in $Gd_5Si_2Ge_2$ and $Gd_{4.95}Ti_{0.05}Si_2Ge_2$ compounds the slope of the lines changes with temperature. It points to the change of the magnetoelastic coefficient γ, which is associated with the change of the compressibility of the sample due to the magnetostructural transition. Some deviation of ω vs. M^2 curves from the linear behavior is due to the contributions to the volume magnetostriction from terms with fourth and higher powers of M in the expansion of G into a power series of M.

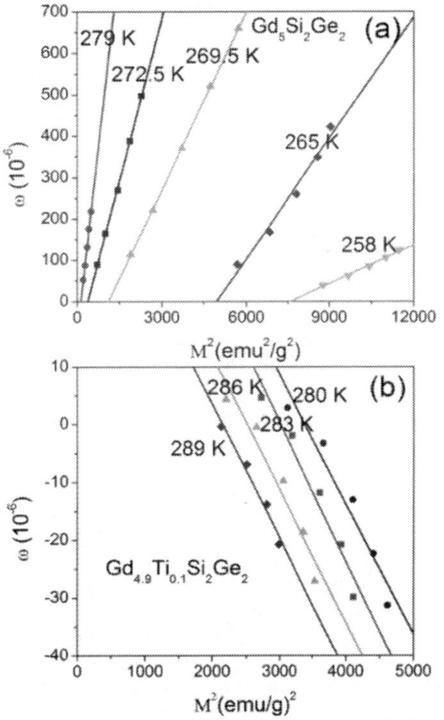

Fig. 3. The dependence of the volume magnetostriction on the square of magnetization for $Gd_5Si_2Ge_2$ (a) and $Gd_{4.9}Ti_{0.1}Si_2Ge_2$ (b)

The magnetostriction observed for the investigated compounds can be considered as the kind of the forced magnetostriction, which has the contributions from magnetostuctural phase transition and from the change of magnetization inside the magnetic domains under the action of field. As has been shown by Belov [6], the forced magnetostriction is connected to the exchange integral by relation

$$\frac{\partial\omega}{\partial H} = -\frac{T}{\kappa}\frac{1}{A}\frac{\partial A}{\partial\omega}\frac{\partial M_S}{\partial T}, \tag{5}$$

where κ is bulk modulus, A is the exchange integral, M_S is the spontaneous magnetization. The change of the sign of the volume magnetostriction of $Gd_{5-x}Ti_xSi_2Ge_2$ when x changes from 0 to 0.1 indicates the radical transformation of the magnetostructural state with Ti addition which is related with change of crystal structure from monoclinic to orthorhombic.

In $Gd_5Si_2Ge_2$ the temperature induced magnetostriction deformation due to the magnetostructural phase transition is added to the phonon part of the thermal expansion. For $Gd_{4.9}Ti_{0.1}Si_2Ge_2$, the structural phase transition is absent, therefore no noticeable anomalies are observed near the Curie temperature on the thermal expansion curve. In $Gd_{4.95}Ti_{0.05}Si_2Ge_2$ the orthorhombic and monoclinic phases are likely to coexist in the vicinity of the phase transition, which is also observed on the X-ray diffraction pattern for this compound. The comparison of the temperature dependences of magnetization and thermal expansion indicates that magnetic and structural phase transitions are likely to occur at different temperatures. The similar decoupling of magnetic and structural phase transitions have been observed by neutron diffraction in $Tb_5Si_2Ge_2$ [8]. It also seems to occur in $Gd_5Si_{2-x}Ge_{2-x}In_{2x}$ [3] according to magnetization, MCE and magnetostriction data. At first the magnetic ordering occurs in $Gd_{4.95}Ti_{0.05}Si_2Ge_2$ upon cooling. The negative magnetostriction is observed in the vicinity of this phase transition. Then the structural transformation from monoclinic to orthorhombic structure takes place and the positive maximum is observed on $\omega(T)$. In $Gd_5Si_2Ge_2$ the contribution of the structural phase transition is prevailing, therefore the positive magnetostriction is observed. $Gd_{4.9}Ti_{0.1}Si_2Ge_2$ initially crystalizes in orthorhombic structure therefore the structural phase transition is absent in this compound and only the negative magnetostriction due to the orientation of the magnetic moments in magnetic field is observed.

5 Conclusion

With small Ti addition in $Gd_5Si_2Ge_2$, the Curie temperature increases, the thermal hysteresis decreases and the magnetostriction is significantly reduced in magnitude and changes the sign. It is due to the transformation of the magnetostructural phase transition with Ti addition. For $Gd_{4.95}Ti_{0.05}Si_2Ge_2$, the decoupling of magnetic and structural phase transitions is found. At the same time only the magnetic phase transition is observed in $Gd_{4.9}Ti_{0.1}Si_2Ge_2$ since this compound initially crystalizes in orthorhombic structure.

The authors thank A.V. Morozkin for samples provided for the study. The work has been supported by RFBR Grants N 16-02-00472 A and N 16-52-00223 A.

References

1. V. K. Pecharsky, K. A. Gschneidner Jr., Phys. Rev. Lett. **78**, 4494 (1997)
2. A. O. Pecharsky, V. K. Pecharsky, K. A. Gschneidner Jr., J. Appl. Phys. **93**, 4722 (2003)
3. S.A. Nikitin, A.E. Bogdanov, I.A. Ovchenkova, E.A. Ovchenkov, A.V. Smirnov, A.V. Morozkin, Solid State Phenomena **233-234**, 208 (2015)
4. J.Q. Li, W.A. Sun, Y.X. Jian, Y.H. Zhuang, W.D. Huang, J.K. Liang, J. Appl. Phys. **100**, 073904 (2006)
5. A.M.G. Carvalho, J.C.G. Tedesco, M.J.M. Pires, M.E. Soffner, A.O. Guimarães, A.M. Mansanares, A.A. Coelho, Appl. Phys. Lett. **102**, 192410 (2013)
6. K.P. Belov, *Elastic, Thermal and Electric Phenomena in Ferromagnets* (Gostekhtizdat, Moscow, 1957) (in Russian)
7. S.A. Nikitin, N.Yu. Pankratov, A.I. Smarzhevskaya, G.A. Politova, Yu.G. Pastushenkov, K.P. Skokov, A. del Moral, J. Appl. Phys. **117**, 193908 (2015)
8. L. Morellon, C. Ritter, C. Magen, P.A. Algarabel, M.R. Ibarra, Phys. Rev. B **68**, 024417 (2003)

STRUCTURAL AND MAGNETIC PROPERTIES OF HEUSLER ALLOYS Pd_2MnZ (Z=Ga, Ge, As): *AB INITIO* STUDY

Olga Miroshkina[1,*], *Mikhail* Zagrebin[1,2], *Vladimir* Sokolovskiy[1,3], and *Vasiliy* Buchelnikov[1]

[1]Chelyabinsk State University, 454001 Chelyabinsk, Russia
[2]National Research South Ural State University, 454080 Chelyabinsk, Russia
[3]National University of Science and Technology "MIS&S", 119991 Moscow, Russia

Abstract. In this work, we report results of *ab initio* and Monte Carlo investigations of structural and magnetic properties in a series of Heusler compositions Pd_2MnZ (Z = Ga, Ge, As). It was found that for Pd_2MnGa and Pd_2MnAs, the stable martensitic state is realized on the contrast with Pd_2MnGe. The equilibrium lattice parameters for the series of Pd_2MnZ (Z = Ga, Ge, As) compounds increase with increasing the number of valence electrons per atom (*e/a* ratio). Having calculated total magnetic moments and magnetic exchange parameters from *ab initio* methods, the Curie temperature for Pd_2Mn-based alloys has been estimated in the framework of Monte Carlo simulations of Heisenberg model.

1 Introduction

Heusler alloys are of great interest to the scientific community due to their unique properties such as shape memory effect, effects of superelasticity and superplasticity, giant magnetocaloric effect, giant magnetoresistance and magnetostrain, etc. These effects are associated with martensitic transformations [1-4].

Full-Heusler alloys are ternary intermetallic systems with $L2_1$ crystal structure and are derived from a parent compound with the generic formula X_2YZ. Here, X and Y are 3*d* transition metals, and Z are *sp* elements of III–V group of periodic table of elements.

At present, a small number of works is devoted to the study of Pd_2Mn-based alloys. For instance, Natera et al. found that Pd_2MnGe shows a preferentially disordered $L2_1$ structure with 0.98 Pd + 0.02 Mn on the A- and C-sites and 0.96 Mn + 0.04 Pd on the B-sites [5]. Yin and Nash measured enthalpies of formation of Pd_2MnGa (-54.5 ± 1.7 kJ/mol) [6].

The objective of this paper is the theoretical investigation of the structural and magnetic properties of Pd_2MnZ (*Z* = Ga, Ge, As) Heusler alloys and comparison with available experimental data.

2 Computational details

In this study, we used density functional formalism (DFT) as implemented in the Vienna ab-initio simulation program (VASP) [7, 8] and spin-polarized relativistic Korringa-Kohn-Rostoker (SPR-KKR) package [9] jointly with the classical Monte Carlo (MC) simulations. The generalized gradient approximation (GGA) in Perdew-Burke-Ernzerhoff parameterization [10] was considered in both VASP and SPR-KKR calculations.

In the first step, the structural ground state calculations were performed using the VASP package. In order to determine the presence of a stable martensitic state, calculations of the total energy as a function of the tetragonal distortion (c/a, where *a* and *c* are the crystallographic axis) of a cubic crystal are performed. Subsequently, the optimized structural configurations from the VASP calculations were used for the calculations of magnetic exchange constants by means of the SPR-KKR package.

Computational parameters used for the VASP package are following. The pseudopotentials were taken with the electronic configurations as follows: Pd ($4p^6 4d^9 5s^1$), Mn ($3p^6 3d^6 4s^1$), Ga ($3d^1 4s^2 4p^1$), Ge ($3d^1 4s^2 4p^2$), and As ($4s^2 4p^3$). The interaction between ions and electrons was described by the projector augmented wave (PAW) method [7, 8]. For the kinetic energy cut-off was 400 eV and the kinetic energy cut-off for the augmentation charges was 800 eV. Brillouin zone sampling was performed according to the Monkhorst-Pack grid [10]. The *k*-points for self-consistent field cycles in the Brillouin zone were generated with 12^3 meshes. All considered structures were relaxed using the conjugate gradient algorithm and both the atomic position and the volume were optimized.

As for the SPR-KKR package, the calculations of exchange parameters were performed using of the spin-polarized scalar-relativistic mode.

The Heusler compounds Pd_2MnZ (*Z* = Ga, Ge, As) possess the $L2_1$ type cubic structure with space group *Fm-3m* (No. 225). For present calculations, the conventional cell consisting of 4 atoms with a cell

* Corresponding author: miroshkina.on@yandex.ru

formula of Pd_2MnZ is used. The Ga, Ge, and As atoms locate on the $4a$ Wyckoff position with coordinate (0, 0, 0), the Pd atoms occupy $8c$ Wyckoff position with (1/4, 1/4, 1/4), and Mn atom possess $4b$ Wyckoff position with coordinate (1/2, 1/2, 1/2).

To obtain the temperature dependences of the magnetization of the Pd_2MnZ (Z = Ga, Ge, As) compounds, well-working and well-tested Heisenberg model was applied [11, 12, 13, 14]. With the use of long-range exchange interactions, it allows obtaining Curie temperatures close to the experiment. We carried out MC simulations of the three-dimensional Heisenberg model in the absence of anisotropy and magnetic fields

$$H_{mag} = -\sum_{\langle ij \rangle} J_{ij} \mathbf{S}_i \mathbf{S}_j \qquad (1)$$

where $\mathbf{S}_i$ is the spin of the unit length ($|\mathbf{S}_i| = 1$) placed at the lattice site, and J_{ij} are the exchange coupling parameters calculated using an expression proposed by Liechtenstein *et al.* [15]. The magnetic exchange parameters clearly show the damping oscillatory behavior, changing the type of interaction from ferromagnetic (FM) to antiferromagnetic (AFM) and vice versa. We included the exchange interactions up to eight coordination shell for all interaction pairs. The MC simulations were performed for a simulation cell with periodic boundary conditions using the Metropolis algorithm [16]. Therefore, for the Pd_2MnZ (Z = Ga, Ge, As) alloys, we used the simulation cell containing 1098 Pd, 1099 Mn, and 1728 Z atoms. Changes in the independent spin variables $\{S_i^x, S_i^y, S_i^z\}$ are accepted or rejected in accordance with the single-site transition probability, $W = min\{1; exp(-\Delta H/k_BT)\}$. As a time unit, we used one MC step consisting of N attempts to change the spin variables. For each temperature, the properties (internal energy of the system H and magnetic order parameter m) were analyzed, allowing 5×10^5 MC steps and 10^4 thermalization steps.

To calculate the total average magnetization M at a certain temperature point, we used the following equation:

$$M = 2\mu_{Pd} m_{Pd} + \mu_{Mn} m_{Mn}, \qquad (2)$$

where μ_{Pd} and μ_{Mn} are the magnetic moments of Pd and Mn atoms, which were obtained from *ab initio* calculations; m_{Pd} and m_{Mn} are the normalized magnetizations calculated as follows:

$$m_X = \frac{1}{N}\sum_i \sqrt{\left(S_i^x\right)^2 + \left(S_i^y\right)^2 + \left(S_i^z\right)^2}. \qquad (3)$$

3 Results and discussion

3.1 Structural properties

Initially, we determined the equilibrium lattice constant of a crystal of Pd_2MnZ compounds using the geometry optimization in VASP, focusing on the equilibrium volume of the cell and the tetragonal distortion parameterized by the c/a ratio. In this case, a negligibly small change in volume between the cubic and tetragonal phases was assumed: $V_0 = a_0^3 = a^2c$.

The calculated equilibrium lattice parameters and bulk modulus of the Pd_2MnZ austenitic compounds are summarized in Table 1. Noted parameters were estimated according to the Birch-Murnagahan equation of state. As one can see from Table 1, the equilibrium lattice parameters for a series of Pd_2MnZ (Z = Ga, Ge, As) compounds increase with increasing e/a due to increasing Z element's radius. Experimental values of Pd_2MnGa and Pd_2MnGe are 6.180 Å and 6.174 Å, respectively [17].

Table 1. Calculated lattice constants a_0 (in Å) and bulk modulus B (in GPa) for Pd_2MnZ alloys

Compound	e/a	a_0	B
Pd_2MnGa	7.50	6.212	142.7
Pd_2MnGe	7.75	6.237	127.8
Pd_2MnAs	8.00	6.301	129.2

In order to investigate the possibility of the martensitic transformation in alloys studied, we performed calculations of the total energy relative to the cubic phase as a function of the tetragonality ratio c/a. The total energy differences between the tetragonally distorted and the cubic phases as functions of c/a ratio for Pd_2MnZ (Z = Ga, Ge, As) alloys are shown in Fig. 1. In this case, ΔE = 0 corresponds to the equilibrium austenitic phase for each compound.

It is seen that for Pd_2MnGa and Pd_2MnAs, the stable martensite can be realized at c/a ratio of 1.30 and 1.35, respectively. While in the case of the stoichiometric Pd_2MnGe alloy, the austenite–martensite transformation does not occur due to the absence of any energy minima in the $E(c/a)$ curve for $c/a \neq 1$, as shown in Fig. 1.

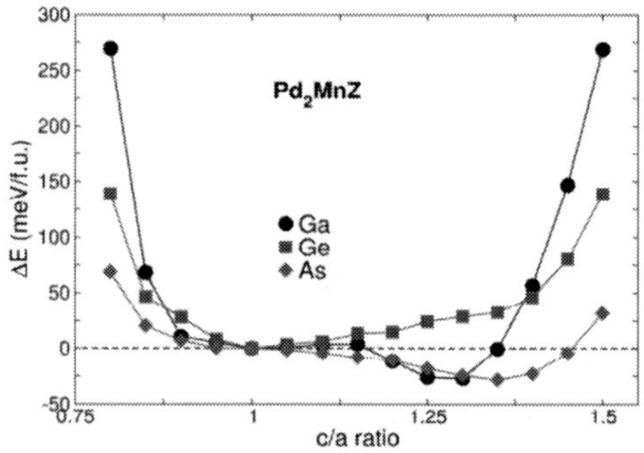

Fig. 1. The total energy differences with respect to equilibrium value for a series of Pd_2MnZ (Z = Ga, Ge, As) alloys as functions of tetragonal distortion, c/a.

3.2 Electronic and magnetic properties

The total densities of the electronic states are shown in Fig. 2(a-c). The analysis of the densities of electron states shows the relative shift of the Fermi level to higher energy regions in the Ga-Ge-As sequence, which arises from the increase in the number of valence electrons in this series of alloys. The character of the densities of electronic states is similar to those obtained for other Pd_2Mn-containing alloys [18, 19].

Magnetic properties of Pd_2Mn-based series of alloys are also of significant interest, because they allow us to predict compositional trends of the Curie temperature.

Firstly, the total magnetic moment of Pd_2MnZ (Z = Ga, Ge, As) compounds was calculated using the SPR-KKR package. We found that Pd_2MnGa possesses the lowest magnetic moment (4.12 μB/f.u.) and it increases slightly with the increasing number of valence electrons per atom *e/a*. Whereat we evaluated the magnetic exchange parameters.

The exchange coupling constants, J_{ij}, for Pd_2MnZ as a function of the distance between interacting atoms i and j are shown in Fig. 2 (d-f).

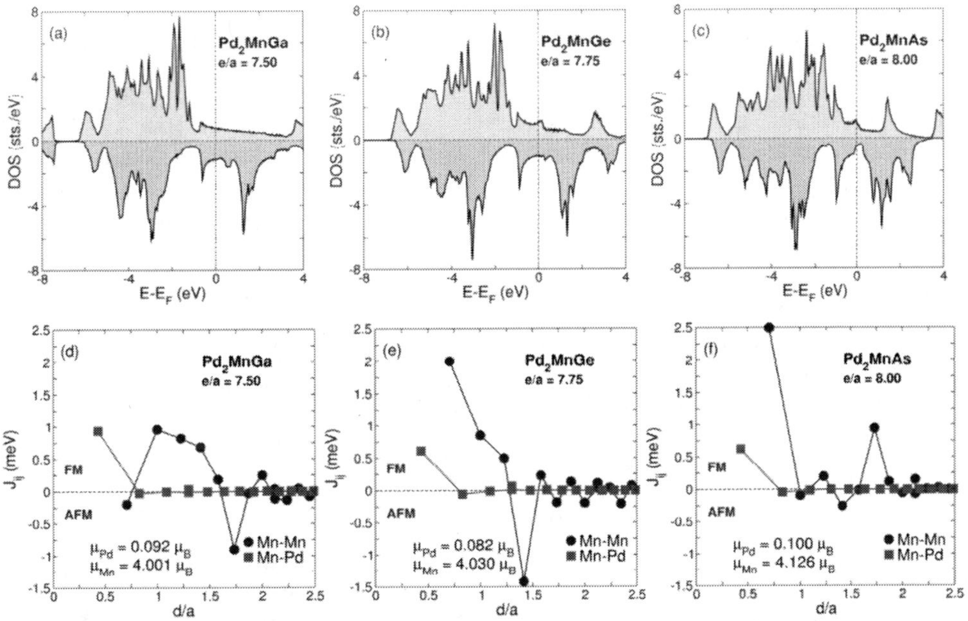

Fig. 2. The total densities of the electronic states and the exchange coupling parameters for (a) Pd_2MnGa, (b) Pd_2MnGe, and (c) Pd_2MnAs alloys as a function of the distance (*d/a*) between atoms i and j. Here a is the optimized lattice parameter.

It is worth mentioning that positive values of J_{ij} correspond to the FM exchange, whereas the negative ones mark the AFM exchange. It should be noticed that exchange interactions between Pd-Z (Z = Ga, Ge, As) and Pd-Pd atoms are very small due to the lowest magnetic moments and omitted from the figures. It is seen that for all compounds studied the Mn-Pd pair exhibits a similar behavior, where the nearest-neighbor interactions provide the maximum contribution to the exchange energy than subsequent interactions. The latter ones are close to zero.

As for the Mn-Mn exchange interactions, they have an AFM character between nearest-neighbor atoms within the first coordination shell for Pd_2MnGa, then it becomes FM and increases slightly with increasing *e/a* ratio. The magnetic exchange parameters clearly show the damping oscillatory behavior for all compositions, which can be associated with various exchange mechanisms such as the magnetic interaction between localized d and collectivized s electrons, magnetic exchange interaction between localized and collectivized d electrons, and super-exchange interaction via *sp*-conduction electrons contributing to the RKKY-type indirect exchange interaction [18]. In addition, for Pd_2MnGa and Pd_2MnGe, the AFM contributions to the exchange energy are accounted for Mn-Mn interactions within the sixth and fourth coordination shells, respectively. On the other side, for Pd_2MnAs, the AFM contributions are significantly less than for other alloys of studying series.

Obtained exchange coupling parameters and partial magnetic moments for Pd_2MnZ (Z = Ga, Ge, As) alloys were used for further MC simulations of the thermomagnetization curves to obtain corresponding Curie temperatures. The set of magnetization curves as a function of temperature for alloys studied is presented in Fig. 3. As can be seen from Fig. 3, for all cases, the ferromagnetic – paramagnetic phase transition in austenite occurs. In a case of Pd_2MnGa, the highest Curie temperature can be observed due to Mn-Mn FM interactions within second, third, fourth, and fifth coordination shells jointly with nearest-neighbor Mn-Pd FM interaction. On the other hand, for Pd_2MnGe, the lowest value of Curie temperature is found because of the largest AFM contribution between Mn atoms interacting within the fourth shell. We note that the experimental value of Curie temperature of Pd_2MnGe is 170 K [17].

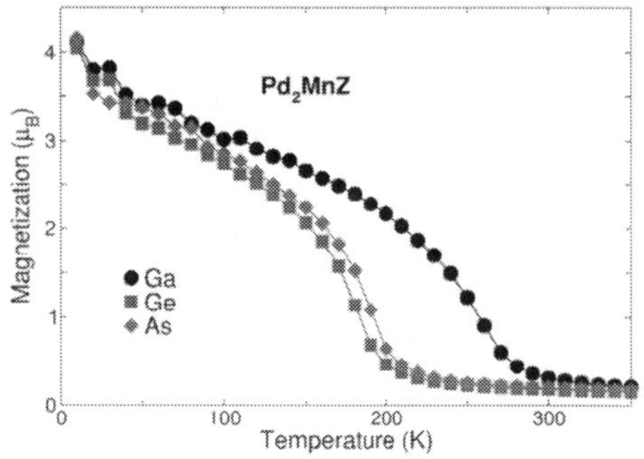

Fig. 3. Magnetization curves of Pd_2MnZ (Z = Ga, Ge, As) as a function of temperature.

The calculated total magnetic moments and Curie temperatures for alloys studied are summarized in Table 2.

Table 2. Calculated total magnetic moment μ_{tot} (in μ_B/f.u.) and Curie temperature T_C (in K) for Pd_2MnZ (Z=Ga, Ge, As) alloys

Compound	*e/a*	μ_{tot}	T_C
Pd_2MnGa	7.50	4.12	255
Pd_2MnGe	7.75	4.14	183
Pd_2MnAs	8.00	4.27	198

4 Conclusion

The optimized lattice parameters for austenite and martensite, total magnetic moments as well as thermomagnetization curves and Curie temperatures for a series of Pd_2MnZ (Z = Ga, Ge, As) compounds are successfully obtained using the first-principles ground state calculations in association with the finite-temperature Monte Carlo simulations. *Ab initio* calculations were performed using VASP and SPR-KKR packages.

It was shown that the optimized lattice parameters of considered alloys increase along the sequence of $Pd_2MnGa \rightarrow Pd_2MnGe \rightarrow Pd_2MnAs$. This has to do with the fact that Z element's radius increases along the respective sequence. The lattice energy calculations for the distorted cubic structure have shown that the stable martensitic phase can be realized only for Pd_2MnGa and Pd_2MnAs compounds with c/a ratio of 1.3 and 1.35, respectively. The total magnetic moments for the cubic austenitic state of Pd_2MnZ (Z = Ga, Ge, As) were found to be increasing with the increasing e/a ratio. Using the Heisenberg Hamiltonian and exchange coupling constants, the Curie temperatures for compounds studied have been obtained. It was shown that Pd_2MnGa possesses the Curie temperature close to room temperature, while for remaining compositions, it found to be around 190 K. We consider that discussed alloys could be interesting for further investigations.

Acknowledgments

This work was supported by Russian Science Foundation No. 17-72-20022\17 (VASP calculations) and Russian Science Foundation No. 14-22-00279 (SPR-KKR calculations).

References

1. P. Entel, V.D. Buchelnikov, V.V. Khovailo et al, J. Phys. D: Appl. Phys., **39**, 865 (2006)
2. V.D. Buchelnikov, A.N. Vasiliev, V.V. Koledov et al, Phys.-Usp., **49**, 871 (2006)
3. P. Entel, M. E. Gruner, A. Dannenberg, Mat. Sci. Forum, **635**, 3 (2010)
4. P.J. Webster, K.R.A. Ziebeck, S.L. Town, M.S. Peak, Philos. Mag. B, **49**, 295 (1984)
5. M.G. Natera, M.R.L.N. Murthy, R.J. Begum, and N.S. Satya Murthy, Phys. Status Solidi A, **3**, 959 (1970)
6. M. Yin, P. Nash, Intermetallics, **58**, 15 (2015)
7. G. Kresse, J. Furthmuller, Phys. Rev. B, **54**, 11169 (1996)
8. G. Kresse, D. Joubert, Phys. Rev. B, **59**, 1758 (1999)
9. H. Ebert, D. Kodderitzsch, J. Minar, Rep. Prog. Phys., **74**, 096501 (2011)
10. J.P. Perdew, K. Burke, M. Enzerhof, Phys. Rev. Lett., **77**, 3865 (1996)
11. P. Yu, X.F. Jin, J. Kudrnovský, D.S. Wang et al, Phys. Rev. B, **77**, 054431 (2008)
12. D. Comtesse, H.C. Herper, A. Hucht and P. Entel, Eur. Phys. J. B, **85**, 10 (2012)
13. D. Böttcher, A. Ernst, J. Henk, J. Magn. Magn. Mater., **324**, 610 (2012)
14. V.V. Sokolovskiy, V.D. Buchelnikov, M.A. Zagrebin et al, Phys. Rev. B, **86**, 134418 (2012)
15. A. Liechtenstein, M. Katsnelson, V. Antropov et al, J. Magn. Magn. Mater., **67**, 65 (1987)
16. D.P. Landau, K. Binder, *A Guide to Monte Carlo Simulations in Statistical Physics* (Cambridge University Press, Cambridge, 2005)
17. P.J. Webster *The Magnetic and Chemical Structures of the Heusler Alloys* (University of Sheffield, Sheffield, 1968)
18. E. Şaşıoğlu, L.M. Sandratskii, P. Bruno, Phys. Rev. B, **77**, 064417 (2008)
19. J. Kübler, A.R. William, C.B. Sommers, Phys. Rev. B, **28**, 1745 (1983)

Peculiarities of the magnetocaloric effect in FeRh-based alloys in the vicinity of the first order magnetic phase transition

Radel Gimaev[1,2,*], *Vladimir* Zverev[3], *Yury* Spichkin[1,2], *Alexander* Tishin[1,2,3] and *Takafumi* Miyanaga[4]

[1]Advanced Magnetic Technologies and Consulting LLC, 142191 Troitsk, Moscow, Russia
[2]National Research Centre "Kurchatov Institute", 123182, Moscow, Russia
[3]Faculty of Physics, M.V. Lomonosov Moscow State University, 119991 Moscow, Russia
[4]Department of Mathematics and Physics, Hirosaki University, Hirosaki, Aomori, 036-8561, Japan

Abstract. Medical applications of magnetocaloric effect (MCE) require possibility for precision shift of a temperature of the magnetic phase transition at the same MCE value and minimize irreversibility. Thus, detail dynamic MCE investigation of such alloys with non-toxic biocompatible dopants need to be done. In present work, the giant magnetocaloric effect, which is observed in the whole family of Fe-Rh alloys, has been investigated in Pd-doped samples in slowly cycled magnetic fields of up to 1.8 T in magnitude for a range of temperatures, $250\,K < T < 350\,K$. The shift of the ferromagnetic/antiferromagnetic transition temperature down towards room temperature and the decrease in the MCE have been observed in these alloys in comparison with a quasi-equiatomic FeRh alloy. The measurements have also shown an asymmetric behaviour of the first order magnetic phase transition with respect to whether the transition is traversed by heating from lower temperatures or cooling from above. These peculiarities have been explained in the framework of the ab-initio density functional theory-based disordered local moment theory of the MCE. The results have been compared with the those for the non-doped FeRh alloy. Thus features of the first order magnetic phase transition that these alloys have in common have been revealed which enable some predictions to be made appropriate for practical applications.

1 Introduction

Conventional nearly equiatomic stoichiometric Fe-Rh alloys exhibit a giant magnetocaloric effect (MCE) but have some limitations for widespread use in magnetic refrigeration technology. Among these are the sharp ferromagnetic/antiferromagnetic transition, the relatively high transition temperature and, of course, the high cost of Rh as a constituent of the working body of a refrigerator [1–4]. The obvious idea is to dope the pure alloy having CsCl-type structure with some elements so that the crucial parameters are adjusted to more suitable values. First attempts in making indirect ΔS measurements during the 1st order magnetic phase transition in Fe-Rh alloys doped with Pd, Pt and Ir have been performed in [5,6]. The authors notice the large contribution of the magnetic entropy change (~ 10 J/kg K) to the full ΔS change (~ 14 J/kg K) during the transition at magnetic field up to 1.2 T.

Doping Fe-Rh alloys with Pd, Cu, Co, Ni etc. leads to the shift of the phase transition towards low temperatures [7–11] which is important for medical purposes for example. The idea of this work is to move phase transition temperature around of human body temperature region at saving MCE value. Increase of Pd content is the reason of the lattice parameter increase [11]. For practical applications the shift of the MCE maximum during magnetization and demagnetization processes should be considered [12].

The importance of doping is to significantly increase the Refrigerant capacity (RC) value [10,13]. Adding some 3d or 4d transition metals changes the chemical order in the alloy drastically which leads to a significant broadening of the phase transition transformation range. Although doping decreases the MCE value, the peak broadens which leads to an overall increase of RC.

Nowadays there are just a few promising materials which undergo a 1st order antiferromagnetic to ferromagnetic (AFM-FM) transition and only FeRh has the transition occurring close to room temperature [14]. The idea to use this material in magnetic refrigerators was first proposed by [15]. Later on discussion about the creation of a magnetic refrigerator which employs the 1st order AFM-FM transition were published [16,17].

The high cost of Rh, however, drives research into other areas for application of this material where the cost problem is not so crucial. A good example is medicine where polycrystalline materials might be suitable. Additionally, Fe-Rh alloys and Pd are non-toxic biocompatible materials.

* Corresponding author: gimaev@amtc.org

https://doi.org/10.1051/epjconf/201818505008

2 Materials and methods

The $Fe_{49.7}Rh_{47.4}Pd_{2.9}$ and $Fe_{48.9}Rh_{46.8}Pd_{4.9}$ alloys were prepared using plasma arc melting in Ar atmosphere (pressure was 0.09 MPa). Samples were annealed in vacuum at 1273 K during 48 h and were quenched by cooling into water.

Direct magnetocaloric effect (MCE) measurements were performed using the MagEq MMS 801 (AMT&C LLC) system which has been described in [18–20]. Direct measurements were conducted in the same way as reported ones previously [20].

The theoretical description of MCE properties of $Fe_{49.7}Rh_{47.4}Pd_{2.9}$ and $Fe_{48.9}Rh_{46.8}Pd_{4.9}$ alloys was made using the ab-initio disordered local moment theory [21,22].

In an ideal stoichiometrically ordered Fe-Rh alloy with bcc structure, each iron atom is surrounded by 8 rhodium atoms as nearest neighbors and 6 iron atoms as second-order neighbors. Interactions between Fe-Fe atoms in the bcc structure of Fe-Rh alloys favor the onset of AFM ordering, while the FM ordering is caused by interaction with the nearest neighbors, i.e. Rh atoms. The ideal order in the bcc structure of Fe-Rh-based alloys is not achieved in real alloys. The phase analysis of the $Fe_{50}Rh_{50}$ alloy shows [20, 22] that during the annealing and subsequent cooling 1-2% of rhodium atoms are replaced by iron atoms and vice versa, which can not be eliminated by subsequent heat treatment. Thus, an insignificant disordering is introduced into the system: some of Rh atoms are replaced by Fe atoms and the nearest neighbors of these Fe atoms are Fe atoms. Such regions in which the nearest neighbors are two Fe atoms have a strong FM interaction, which is manifested in the bcc structure of iron. Thus, the appearance of a small number of nearest Fe-Fe neighbors (1-2%) leads to a strong effect on the AFM-FM magnetic transition.

Theoretical consideration of substitution defects is based on the model described in detail in [22]. Within the framework of this model, two sublattices of Fe and Rh atoms are considered in the crystalline bcc lattice of Fe-Rh alloys with close to equiatomic compositions. The first sublattice (sublattice A) has local locations for (a) Fe atoms, but Rh atoms occupy a small percentage (x) of these locations (real substitution occurs in real alloys) [22]. The second sublattice (sublattice B) is a sublattice of the arrangement of (b) Rh atoms, in which a small percentage (y) of Fe atoms also lies. When considering the equiatomic Fe-Rh alloy in the model of two sublattices (in the sublattice A, the main atom is Fe with the presence of a small number of x Rh atoms, in the sublattice B the main atom is Rh with the presence of a small number of y atoms Fe), the Fe-Rh alloy can be written in the form $Fe_{100-x}Rh_xRh_{100-y}Fe_y$. In this notation, the concentration of Fe atoms in the alloy can be written down as $c = \frac{100-x+y}{2}$, and long order parameter as $= \frac{100-y}{100}$.

For the composition of the Fe-Rh alloy, with certain values of x and y, the free energy of the alloy can be written as [22]:

$$F(c,y,H,T) = U(c,y,m_f,m_a) - T\left(S_{\text{Mar}}(m_f,m_a,T) + S_{\text{pem}}(T)\right) - Hm_f \quad (1)$$

where U is the internal energy, S_{mag} is the magnetic part of the entropy, S_{latt} is the entropy of the vibrations of the crystal lattice, H is the magnetic field, m_f is the PM ordering parameter (proportional to the magnetization), m_a is the AFM ordering parameter. The parameters m_a and m_f give the degree of FM and AFM ordering in the alloy, so for the pure AFM ordering the parameters $m_a = 1$, $m_f = 0$, for pure FM ordering $m_a = 0$, $m_f = 1$. In the high-temperature paramagnetic region, both parameters m_a and m_f are equal to zero.

3 Results and discussion

The magnetocaloric properties of Pd-doped Fe-Rh alloys $Fe_{49.7}Rh_{47.4}Pd_{2.9}$, and $Fe_{48.9}Rh_{46.8}Pd_{4.9}$ are studied. It is known that adding Pd leads to a decrease of the FM/AFM transition temperature bringing it closer to room temperature [5,6,9,11,13,23,24]. Thus, the study of such materials is promising from the practical point of view as it allows materials to be prepared which have desirable transition temperature values. In addition, doping the alloys with Pd does not lead to any significant MCE decrease. To ascertain whether the irreversibility effect is also present in these alloys we prepared two samples with different amounts of Pd. As with the Fe-Rh sample we measured the MCE on both heating and cooling through the phase transition. It should be noted, that $\Delta T(H)$ measurements were performed at magnetic field change rates of 1 T/s – 5 T/s but since they were practically identical here we present the ones obtained at 1 T/s.

The suppression of the transition temperature, T_{tr}, can be understood from the DLM-DFT (disordered local moments density functional theory) based model. As described in [22] an increase of the number of Fe-Fe nearest neighbors in Fe-Rh sharply depresses T_{tr}. This happens if Fe-Rh is made marginally iron-rich and/or if the CsCl chemical order is incomplete, i.e. a small number of positions on the Rh sub-lattice in the CsCl crystal structure are occupied by Fe atoms. Our DFT calculations reported in [20] based on model presented in [21,22] showed that addition of Pd enhances this process. They indicate that Pd atoms are liable to substitute for Fe and Rh atoms with roughly equal tendency. The net outcome is that Pd-doped Fe-Rh alloys have an increased likelihood of some Fe atoms residing on the 'Rh' sub-lattice as a consequence of the added Pd. This is consistent with the reported decrease in T_{tr}. Incidentally calculations for Pt-doped Fe-Rh alloys showed the reverse - Pt atoms preferentially displaced Rh atoms and reduced the tendency of the iron atoms to migrate to the 'Rh' sub-lattice. T_{tr} in this case would expected to increase with Pt-doping [13].

Fig. 1 shows the ΔT temperature dependences for (a) $Fe_{49.7}Rh_{47.4}Pd_{2.9}$ and (b) $Fe_{48.9}Rh_{46.8}Pd_{4.9}$ (magnetic field changed up to 1.8 T). The shift of the AFM-FM transition towards lower temperatures than that of $Fe_{50.4}Rh_{49.6}$ [20,25] was 25 K and 30 K for $Fe_{49.7}Rh_{47.4}Pd_{2.9}$ and $Fe_{48.9}Rh_{46.8}Pd_{4.9}$, respectively. In the present work we obtained transition temperatures 298 K and 293 K for $Fe_{49.7}Rh_{47.4}Pd_{2.9}$ and $Fe_{48.3}Rh_{46.8}Pd_{4.9}$, respectively, which were very close to room temperature values required in magnetic refrigeration technology. The shift of the transition temperature was in good agreement with previous results [5,6,9,11,13,23,24]. The transition in the Pd-doped samples spread over a wider range in comparison with $Fe_{50.4}Rh_{49.6}$ [20,25]: full width at half maximum (FWHM) 10 K for $Fe_{50.4}Rh_{49.6}$ and 12.4 K for $Fe_{49.7}Rh_{47.4}Pd_{2.9}$ and $Fe_{48.9}Rh_{46.8}Pd_{4.9}$ (Fig. 2). There was a decrease of the MCE values which was approximately -6 K at the maximum value. The same peculiarities were observed during the FM-AFM transition on cooling as with $Fe_{50.4}Rh_{49.6}$ [20,25] but with smaller MCE values distributed with a narrower peak in comparison with the results for the AFM-FM transition.

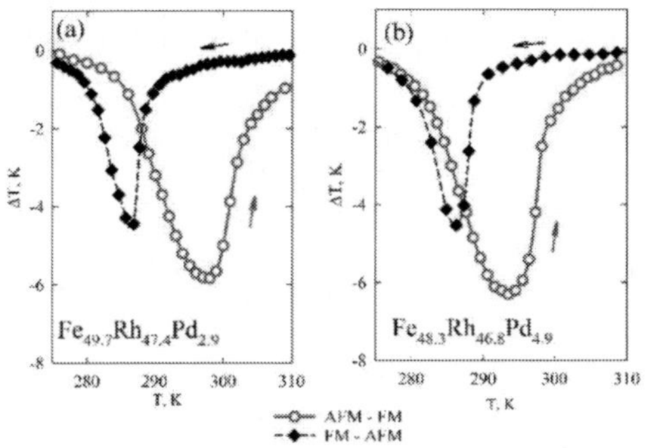

Fig. 1. $\Delta T(T)$ dependences for the bulk Pd-doped FeRh alloy samples $Fe_{49.7}Rh_{47.4}Pd_{2.9}$ (a) and $Fe_{48.9}Rh_{46.8}Pd_{4.9}$ (b) on heating (open red circles) and cooling (filled black diamonds). Magnetic field changed up to 1.8 T. The arrows show the direction of the temperature change.

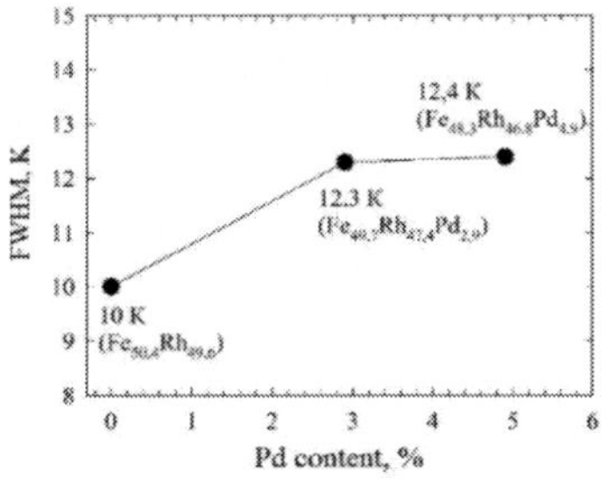

Fig. 2. FWHM for different Pd content in FeRh alloy.

The variation of ΔT as the field was cycled in $Fe_{49.7}Rh_{47.4}Pd_{2.9}$ and $Fe_{48.9}Rh_{46.8}Pd_{4.9}$ displayed similar behavior to that of the FeRh alloy [20,25]. The results of such dynamic ΔT measurements for one full cycle of the magnetic field in the range -1.8 T - +1.8 T for the alloy $Fe_{48.9}Rh_{46.8}Pd_{4.9}$ are presented in Fig. 3. We have obtained $\Delta T(H)$ curves for 40 different temperatures in the range 270 K – 310 K. Here we present three of them: where MCE reached its maximum (Fig. 3 (b) and (e)), and at temperatures with dominating AFM (Fig. 3 (a) and (d)) and FM phases, respectively (Fig. 3 (c) and (f)). The analogous $\Delta T(H)$ curves for $Fe_{49.7}Rh_{47.4}Pd_{2.9}$ showed the same features and thus are not shown.

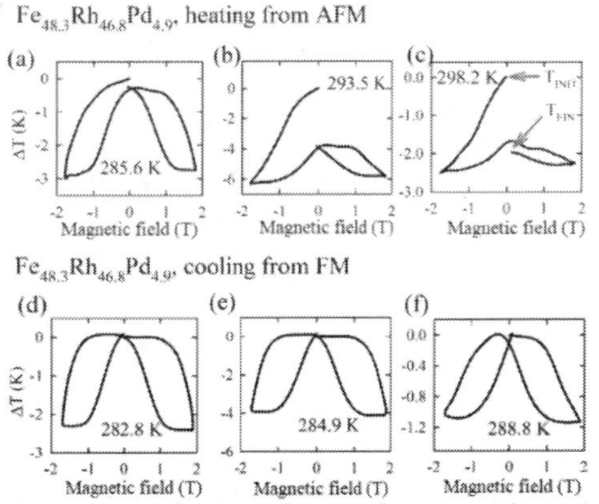

Fig. 3. ΔT(H) dependences for $Fe_{48.9}Rh_{46.8}Pd_{4.9}$ obtained at three different temperature values in the vicinity of the phase transition on heating from AFM phase (a)-(c) and cooling from FM one (d)-(f). The arrows in (c) show the initial T_{INIT} and final T_{FIN} temperatures of the sample during the full magnetic field cycle.

The analysis showed that the observed features in the experimental dependences ΔT (H) and ΔT (T) can be due to the presence of minor crystal replacement defects in which up to 2% of Fe atoms are replaced by Rh atoms and vice versa.

It is important to note that the change in chemical composition and addition of Pd did not alter the 'irreversibility' features observed in the FeRh sample. The difference between initial and final temperatures after one full cycle of magnetic field ($T_{FIN} - T_{INIT}$) was observed when heating through the AFM-FM transition (Fig. 3 (a)-(c)) but was not present when cooling through the FM-AFM transition (Fig. 3 (d)-(f)). The value of ($T_{FIN} - T_{INIT}$) depended on temperature and this dependence was similar to the one presented in Fig. 4 in ref. [20].

4 Conclusion

Direct MCE temperature and field measurements in FeRh-based alloys $Fe_{49.7}Rh_{47.4}Pd_{2.9}$ and $Fe_{48.9}Rh_{46.8}Pd_{4.9}$ showed common features of their MCE in the vicinity of the 1st order magnetic phase transition which each alloy possesses. Each material showed a non-symmetric character of its magnetocaloric properties on heating and cooling which was demonstrated by the presence/absence of the 'irreversibility' of the sample's temperature to its initial value during a full cycle of the magnetic field on heating/cooling. Doping samples with Pd did not lead to the disappearance of this effect. Increasing the amount of Pd decreased the MCE value and shifted the phase transition temperature down towards room temperature (down to 293 K in $Fe_{48.9}Rh_{46.8}Pd_{4.9}$ for the AFM-FM transition on heating).

At the same time it was shown that doping samples with Pd leads to increasing of FWHM of $\Delta T(T)$ dependence (from 10 K for FeRh alloy up to 12.4 K for Pd doped samples) and the relative cooling power (RCP) value is not changes. Thus, the Pd doping allows tuning of phase transition temperature without decreasing of RCP. This finding enables the appropriate positioning of the maximum MCE value for applications. Unlike ferromagnetic materials such as Gd, the MCE completely disappears on both sides of the phase transition point for temperatures more than 15-20 K away from the transition temperature. Consequently, in addition to their cost limitations these materials cannot be considered for conventional magnetic refrigeration technology over a wide enough range of temperatures. They are, however, promising choices for use in medical applications [26].

The authors thankful to Prof. Julie B. Staunton and Prof. Alexander Shtil for useful discussions. The authors thank the Ministry of Science and Education of Russia for financial support of the work (Federal target program "Research and development in priority areas of development of Russia's scientific and technological complex for 2014-2020", grant identification RFMEFI60417X0197). Work in Advanced Magnetic Technologies and Consulting, LLC is supported by Skolkovo Foundation, Russia.

References

1. S. A. Nikitin, G. Myalikgulyev, A. M. Tishin, M. P. Annaorazov, K. A. Asatryan, and A. L. Tyurin, Phys. Lett. A 148, 363 (1990).

2. M. P. Annaorazov, K. A. Asatryan, S. A. Nikitin, A. M. Tishin, and A. L. Tyurin, Pisma Zh. Tekh. Phys 12, (1991).

3. M. P. Annaorazov, K. A. Asatryan, G. Myalikgulyev, S. A. Nikitin, A. M. Tishin, and A. L. Tyurin, Cryogenics 32, 867 (1992).

4. M. P. Annaorazov, H. M. Güven, and K. Bärner, J. Alloys Compd. 397, 26 (2005).

5. J. S. Kouvel and C. C. Hartelius, J. Appl. Phys. 33, 1343 (1962).

6. J. S. Kouvel, J. Appl. Phys. 37, 1257 (1966).

7. N. V. Baranov, S. V. Zemlyanski, and K. Kamenev, in Itinerant Electron Magn. Fluct. Eff., edited by D. Wagner, W. Brauneck, and A. Solontsov (Springer Netherlands, 1998), pp. 345–351.

8. N. V. Baranov and E. A. Barabanova, J. Alloys Compd. 219, 139 (1995).

9. A. M. Chirkova, A. S. Volegov, D. S. Neznakhin, E. A. Stepanova, and N. V. Baranov, Solid State Phenom. 190, 299 (2012).

10. M. Manekar and S. B. Roy, J. Phys. Appl. Phys. 44, 242001 (2011).

11. K. Nishimura, Y. Nakazawa, L. Li, and K. Mori, Mater. Trans. 49, 1753 (2008).

12. Y. S. Koshkid'ko, J. Ćwik, T. I. Ivanova, S. A. Nikitin, M. Miller, and K. Rogacki, J. Magn. Magn. Mater. Complete, 234 (2017).

13. R. Barua, F. Jiménez-Villacorta, and L. H. Lewis, J. Appl. Phys. 115, 17A903 (2014).

14. J.-U. Thiele, T. Hauet, and O. Hellwig, Appl. Phys. Lett. 92, 242502 (2008).

15. S. A. Nikitin, G. Myalikgulyev, A. M. Tishin, M. P. Annaorazov, K. A. Asatryan, and A. L. Tyurin, Avtorskoe svidetel'stvo-№ SU 1 746 162 A1 (in Russia) (7 July 1992).

16. M. P. Annaorazov, M. Ünal, S. A. Nikitin, A. L. Tyurin, and K. A. Asatryan, J. Magn. Magn. Mater. 251, 61 (2002).

17. M. P. Annaorazov, S. A. Nikitin, A. L. Tyurin, S. A. Akopyan, and R. W. Myndyev, Int. J. Refrig. 25, 1034 (2002).

18. Y. I. Spichkin, A. M. Tishin, D. B. Kopeliovich, and A. Y. Malyshev, in (Third IIF-IIR International Conference on Magnetic Refrigeration at Room Temperature (Thermag III), Des Moines, Iowa, USA, 2009), pp. 175–180.

19. Y. I. Spichkin and R. R. Gimaev, Int. J. Refrig. 37, 230 (2014).

20. V. I. Zverev, A. M. Saletsky, R. R. Gimaev, A. M. Tishin, T. Miyanaga, and J. B. Staunton, Appl. Phys. Lett. 108, 192405 (2016).

21. B. L. Gyorffy, A. J. Pindor, J. Staunton, G. M. Stocks, and H. Winter, J. Phys. F Met. Phys. 15, 1337 (1985).

22. J. B. Staunton, R. Banerjee, M. dos S. Dias, A. Deak, and L. Szunyogh, Phys. Rev. B 89, 054427 (2014).

23. S. Yuasa, Y. Otani, H. Miyajima, and A. Sakuma, IEEE Transl. J. Magn. Jpn. 9, 202 (1994).

24. T. Zhou, M. K. Cher, L. Shen, J. F. Hu, and Z. M. Yuan, Phys. Lett. A 377, 3052 (2013).

25. R. R. Gimaev, V. I. Zverev, A. M. Saletsky, A. M. Tishin, T. Miyanaga, and J. B. Staunton, in Proc 7th Int C Magn Refrig Room Temp IIFIIR (International Institute of Refrigeration (IIR), 2016), pp. 279–286.

26. A. M. Tishin, Y. I. Spichkin, V. I. Zverev, and P. W. Egolf, Int. J. Refrig. 68, 177 (2016).

Magnetocaloric properties of Gd($Co_{1-x}Fe_x$)$_2$ compounds, with x ≤ 0.60

Maksim Anikin*, *Evgeniy* Tarasov, *Nikolay* Kudrevatykh, and *Aleksander* Zinin

Ural Federal University, Mira str. 19, 620002 Ekaterinburg, Russia

Abstract. In this paper the results of specific magnetization and magnetocaloric effect (MCE) measurements for Gd($Co_{1-x}Fe_x$)$_2$ system upon the Co substitution by Fe for the x = 0 ÷ 0.60 range are presented. Phase composition was controlled by X-ray diffraction analysis. MCE has been studied within the temperature range of 300-850 K in magnetic fields up to 17 kOe by the magnetic entropy change calculation (ΔS_m). It was found that in contrast to the previously studied R(Co-Fe)$_2$ compounds where R = Dy, Ho, Er, an ordinary symmetrical peak of $\Delta S_m(T)$ in the vicinity of T_C is observed for presented samples. Additionally, the MCE comparison of Gd($Co_{0.88}Fe_{0.12}$)$_2$ with that for the isostructural Gd($Ni_{0.88}Fe_{0.12}$)$_2$ compound having a plateau-like ΔS_m temperature dependence is given. The obtained results are discussed.

1 Introduction

In the last two decades intensive research has been devoted to the search for and development of new magnetocaloric materials for the purpose of their application in energy-efficient and environment-friendly systems of magnetic cooling. Different types of compounds with large magnetocaloric effect (MCE) in a narrow temperature interval around the magnetic phase transition temperature were pointed out: RCo_2 (R – heavy rare-earth metal), $Gd_5Ge_2Si_2$, MnFe($P_{1-x}As_x$), La($Fe_{13-x}Si_x$), Ni-Mn-Ga, et al. [1]. Nevertheless, the study of the magnetic properties and the MCE in the R($Me_{1-x}Fe_x$)$_2$ (R = Tb, Ho; Me = Al, Ni, Co) quasi-binary systems shown that a partial substitution of Me-element by Fe leads to Curie temperature (T_C) increase, and the emergence of significant MCE in a wide temperature range below its T_C [2-5]. The latter is an important material property when it is used for the manufacture of magnetic refrigerator working bodies.

Our recent MCE measurements for some Dy($Co_{1-x}Fe_x$)$_2$ [6], Ho($Co_{1-x}Fe_x$)$_2$, Er($Co_{1-x}Fe_x$)$_2$ [7], and Gd($Ni_{1-x}Fe_x$)$_2$ [8] compounds with Ni and Co substitution by Fe confirmed that results and allowed to suggest the reasons of MCE peak widening in the temperature range below their T_c. In general, the results obtained demonstrate a qualitatively identical picture, as in compounds of the R(Co-Fe)$_2$ system with R = Tb, Dy, Ho, Er, and R(Ni-Fe)$_2$ compounds with R = Gd, Tb, Ho . Thus it allows us to lean in favor of the "weak magnetic sublattice" MCE model [9], giving a qualitative interpretation of the magnetocaloric phenomena in the R(Me-Fe)$_2$ intermetallic compounds with Laves phases structure.

These data prompted us to undertake the magnetic and magnetocaloric properties studies of similar to mentioned - Gd($Co_{1-x}Fe_x$)$_2$ system with Fe content (x) from 0 to 0.60. These compounds, unlike those described above with iron concentration absence do not have a Curie point close to room temperature. However, they are of scientific interest from the point of view of the influence of cobalt replacement by iron on the Curie temperature, the magnitude and MCE peak width and refrigerating capacity (q) of these compounds.

2 Experimental details

Gd($Co_{1-x}Fe_x$)$_2$ alloys were melted in an electric arc furnace under a pure helium protection. An excess of rare earths metal (~ 3 wt. %) was added to the starting compositions the formation preventing of Co-rich phases. A homogenizing annealing of alloys was made in a vacuum furnace at 800 K during 24 hours. The samples structure was determined by X-ray diffraction technique (D8 Advance, Bruker) with Cu K_α radiation source. Diffraction patterns were analyzed by Rietveld method using the "Fullprof" software [10].

Magnetic susceptibility temperature dependence measurements of the samples were carried out in an alternating magnetic field of 50 Oe in the temperature range from 25 to 800° C.

Their magnetization was measured using vibrating sample magnetometer (7407, Lake Shore Cryotronics) in the temperature range from 300 K to 850 K under a magnetic field up to 17 kOe.

3 Results

3.1 Properties of Gd(Co-Fe)$_2$

* Corresponding author: maksim.anikin@urfu.ru

https://doi.org/10.1051/epjconf/201818505009

Analysis of the X-ray diffraction data at room temperatures showed that all samples contain mainly the 1:2 type Laves phase. Their crystal lattices belong to the *Fd*-3m space group. For example, figure 1 shows the X-ray diffraction pattern at room temperatures for $GdCo_2$.

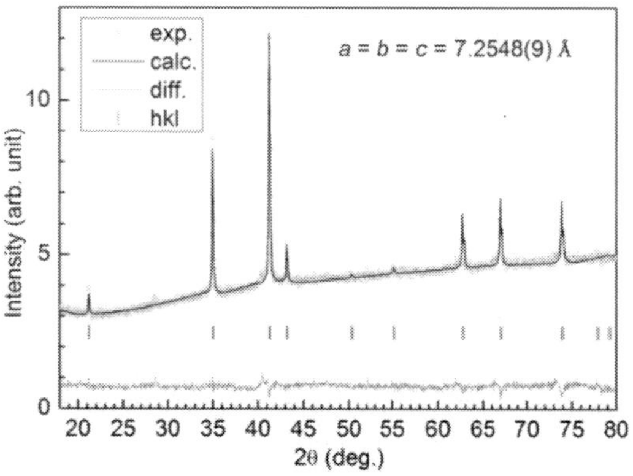

Fig. 1. The X-ray diffraction pattern of $GdCo_2$.

Also, samples were confirmed to contain only 1:2-phase by measurements of susceptibility temperature dependences (χ(T)) at external magnetic fields of H = 50 Oe. Curie temperatures (T_C) were determined from the positions of the dM/dT peaks on the magnetization (M) temperature axis, taken from the specific magnetization temperature dependencies (M(T)) in the magnetic field of 900 Oe (see Fig. 2).

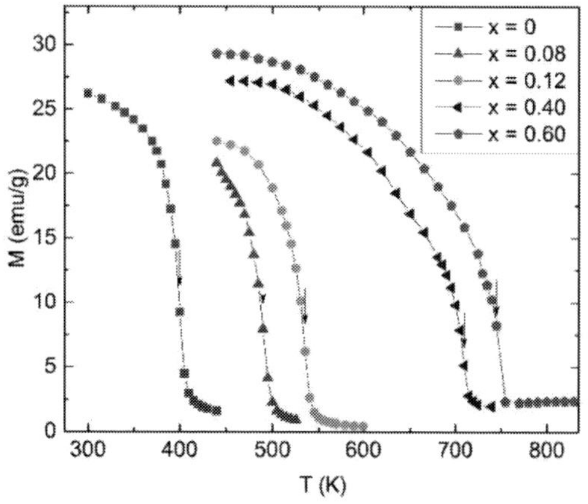

Fig. 2. Specific magnetization temperature dependences of $Gd(Co_{1-x}Fe_x)_2$ intermetallic compounds in external magnetic field of 900 Oe. Arrow indicates the T_C point.

Figure 3 shows the field dependences of the specific magnetization at different temperatures for the $Gd(Co_{0.60}Fe_{0.40})_2$ compound. Similar isotherms are observed for the samples of other compounds. It can be seen that even at T = 485 K, that is only of 225 K below T_C, the magnetization curve practically reaches the saturation. This indirectly indicates that in this temperature range the compound has a low magnetocrystalline anisotropy (MCA) energy and the relatively high degree of magnetic order is realized in Gd- sublattice. Consequently, the probability of the MCE appearance caused by the suppression of Gd atomic magnetic moments disorientation by an external magnetic field is small.

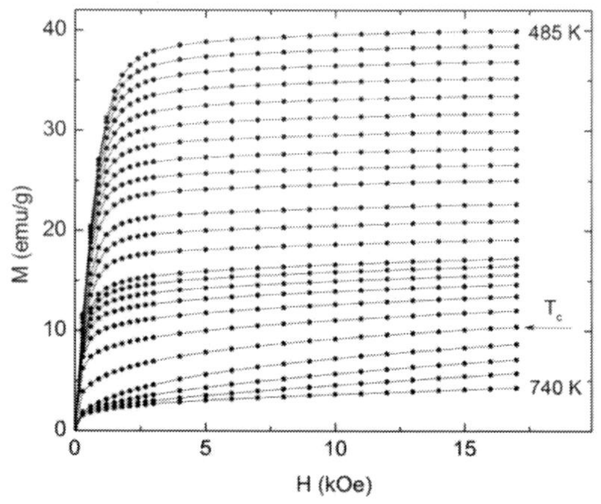

Fig. 3. Magnetization isotherms at various temperatures for the studied $Gd(Co_{0.60}Fe_{0.40})_2$ compound.

For the calculation of isothermal magnetic entropy change (ΔS_m), from the field magnetization M(H) isotherms (Fig. 3), we used the following equation [11]:

$$\Delta S_m(H,T) = \int_0^H \left(\frac{\partial M(H,T)}{\partial T}\right)_H dH \qquad (1)$$

In the case of a discrete temperature numbers and magnetic field values it takes the form:

$$\Delta S_m(H,T) = \sum_i \frac{M_{i+1}(T_{i+1},H) - M_i(T_i,H)}{T_{i+1} - T_i} \Delta H. \qquad (2)$$

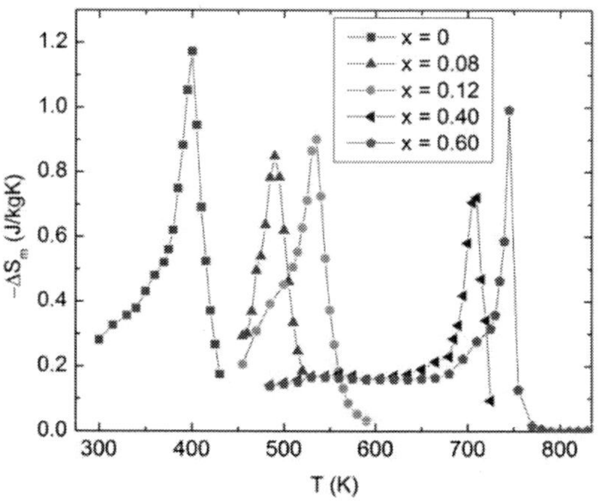

Fig. 4. Temperature dependences of magnetic entropy change $\Delta S_m(T)$ in magnetic fields (0-17) kOe for the studied $Gd(Co_{1-x}Fe_x)_2$ compounds.

Figure 4 shows the calculated values of ΔS_m when the change of magnetic field ΔH = 17 kOe. It can be seen that, in contrast to the previously studied R(Co-Fe)$_2$ compounds, where R = Dy, Ho, Er [6, 7], for the samples with R = Gd, an ordinary symmetrical peak of $\Delta S_m(T)$ in T_C neighborhoods is observed.

Using the $\Delta S_m(T)$ data, we calculated the refrigerant capacity (q) [11]:

$$q = -\int_{T_{cold}}^{T_{hot}} \Delta S_m(T) dT. \quad (3)$$

This parameter is used for comparison of the different magnetocaloric materials from the practical applications point of view. Values of q are presented in Table 1. Also the values of ΔT_{FWHM}, characterizing the distance between the highest and the lowest temperatures at the half maximum of the -ΔS(T) peak, are presented in this Table as well.

Table 1. The data on the temperature Curie (T_C), the refrigerating capacity parameter (q), the difference between the maximum and minimum temperatures at the half of ΔS_m maximum height (ΔT_{FWHM}) and maximum magnetic entropy change $-\Delta S_{max}$, of the studied Gd(Co$_{1-x}$Fe$_x$)$_2$ samples at 15 kOe.

Gd(Co$_{1-x}$Fe$_x$)$_2$	T_C, K	q, J/kg	ΔT_{FWHM}, K	-ΔS, J/kgK
x = 0	400	30	35	1.06
x = 0.08	490	23	35	0.78
x = 0.12	535	41	45	0.82
x = 0.40	710	18	30	0.72
x = 0.60	745	15	20	0.89

Consideration of the studied Gd(Co$_{1-x}$Fe$_x$)$_2$ compounds under the practical perspective does not cause much enthusiasm: they have small values of ΔS-effect – 1 vs 5 J/kgK for Gd (at 17 and 20 kOe, respectively), and hardly can be considered as candidate materials for the manufacture of the working bodies in the magnetic refrigeration systems.

3.2 Comparison of Gd(Co-Fe)$_2$ with Gd(Ni-Fe)$_2$

Figure 5 presents the relative ΔS_m temperature dependence of Gd(Me$_{0.88}$Fe$_{0.12}$)$_2$ (Me = Co, Ni) samples, where the ΔT_{FWHM} value characterizes the MCE peak width [1]. The grey and shaded areas (Fig. 5) correspond to the region of integration at q calculation. These data are 220 J/kg and 41 J/kg at ΔH = 15 kOe for compounds with Ni and Co respectively. It can be seen from the figure that the compound with Ni has a plate-like $\Delta S_m(T)$ dependence while the compound with Co has a MCE peak in the vicinity of T_C typical for many ferromagnets. Hence this large difference in the MCE temperature dependences for the Gd(Me$_{0.88}$Fe$_{0.12}$)$_2$ compounds with different Me is a consequence of a strong exchange interactions change in these compounds when Ni is replaced by Co.

Figure 6 shows the relative temperature dependences of the parapossess susceptibility - χ(T) of the studied compounds. The χ values were calculated from the linear parts of the M(H) curves at the 10÷15 kOe magnetic fields range (χ = dM/dH). It can be seen that the χ (T) dependence for the compound with Ni in the below T_C region lies significantly higher than for the compound with Co (the difference between the curves is shaded); For the sample with Co the χ(T) growth with T rise is much faster.. These differences in the χ(T) variation is an indirect confirmation of the much higher disorder degree in Gd –sublattice for the samples with Ni due to the weaker exchange interaction between Gd and 3d ions sublattices.

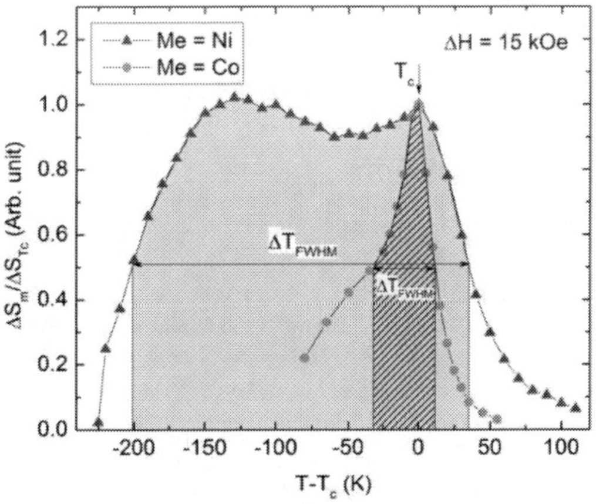

Fig. 5. Relative MCE temperature dependence of Gd(Me$_{0.88}$Fe$_{0.12}$)$_2$, were Me = Co, Ni.

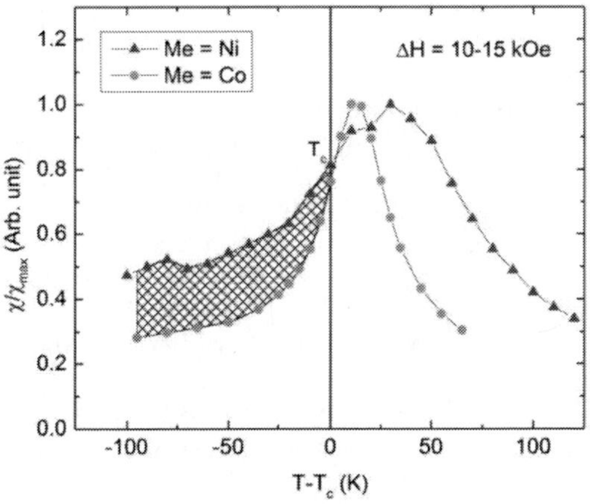

Fig. 6. Temperature dependencies of the relative para possess susceptibility - χ(T) on the Me-type in Gd(Me$_{0.88}$Fe$_{0.12}$)$_2$ compounds, were Me = Co, Ni.

4 Conclusions

Thus, the magnetic and magnetocaloric properties study of the Gd(Co$_{1-x}$Fe$_x$)$_2$ (x = 0, 0.08, 0.12, 0.40, 0.60) compounds revealed the following:

In these compounds no broadening of the MCE peak observed contrary to that for such compounds with heavier R = Tb, Dy, Ho, Er [5, 6];

In contrast to the isostructural $Gd(Ni\text{-}Fe)_2$ intermetallics, where the magnetic entropy $-\Delta S_m(T)$ and paraprocess susceptibility $-\chi(T)$ maxima broadenings are observed, in the given compounds, apparently much stronger internal and inter-sublattice exchange interactions keep the magnetic order in both sublattices in the higher degree almost for the whole 0 K ÷ T_C temperature interval. So such narrow peaks on $\Delta S_m(T)$ and $\chi(T)$ dependences appear only in the vicinity of T_C point.

The obtained results are important for understanding the nature of magnetic and magnetocaloric properties of the large $R(Co_{1-x}Fe_x)_2$ Laves phases group as a whole.

This work has been supported by the State contracts No. 3.6121.2017/8.9 between UrFU and the Ministry of Education and Science of Russian Federation.

References

1. K. A. Gschneidner Jr., V. K. Pecharsky, A.O. Tsokol, Rep. Prog.Phys. **68**, 1479-1539 (2005)
2. Niraj K. Singh, S. Agarwal, K.G. Suresh, R. Nirmala, A.K. Nigam, S.K. Malik, Phys. Rev. B: Condens. Matter, **72**, 014452 (2005)
3. Niraj K. Singh, K. G. Suresh, D. S. Rana, A. K. Nigam and S. K. Malik, J. Phys. Condens. Matter, **18**, 10775 (2006)
4. S. Mican, D. Benea, R. Tetean, J. Alloys Compd., **549**, 64–69 (2013)
5. Madhumita Halder, S. M. Yusuf, M. D. Mukadam, K. Shashikala, Phys. Rev. B: Condens. Matter, **81**, 174402 (2010)
6. M. Anikin, E. Tarasov, N. Kudrevatykh, A. Inishev, M. Semkin, A. V., A. Zinin, J. Magn. Magn. Mater., **418**, 181–187 (2016)
7. M.S. Anikin, E.N. Tarasov, N.V. Kudrevatykh, M.A. Semkin, A.S. Volegov, A.A. Inishev, A.V. Zinin, Refr. Sci. T., 236-239 (2016)
8. M. Anikin, E. Tarasov, N. Kudrevatykh, A. Inishev, M. Semkin, A. Volegov, A. Zinin, J. Magn. Magn. Mater., **449**, 353–359 (2018)
9. K. P. Belov, Adv. in Phys. Sci. **166**, 669-681 (1996) (In Russian)
10. J. Rodriguez-Corvajal, Physica B **192**, 55–69 (1993)
11. K.A. Gschneidner. Jr & V.K. Pecharsky, Annu. Rev. Mater. Sci. **30**, 387–429 (2000)

Giant magnetocaloric effect in composites based on polymeric matrix and manganese arsenide

Nikolay Pankratov[1,*], *Victor* Mitsiuk[2], *Alexander* Vetcher[2], *Andrei* Krokhotin[1], *Gennadiy* Govor[2], and *Sergey* Nikitin[1]

[1] *Lomonosov Moscow State University, Physics faculty, Leninskie Gory, 119991 Moscow, Russia*
[2] *Scientific-Practical Material Research Centre of the National Academy of Science of Belarus, P. Brovki 19, 220072 Minsk, Belarus*

Abstract. A field and/or temperature hysteresis is more than just an interesting incomprehensibility that occurs in materials with a first-order magnetic transition. Indeed, the reversibility of the magnetocaloric effect (MCE), being essential for magnetic heat pumps, strongly depends on the width of the thermal hysteresis and, therefore, it is necessary to find solutions to minimize losses associated with thermal hysteresis in order to maximize the efficiency of magnetic cooling devices. In this work, the polymer matrix composites with MnAs powder as reinforcing material were obtained and its MCE by direct method was studied. The influence of composite preparation process on the MCE was investigated. It was found that the MCE of bulk MnAs shows strong temperature hysteresis caused by magnetostructural transition from paramagnetic to ferromagnetic state, which also leads to an irreversible MCE at cooling at first magnetization. The reversible MCE of bulk MnAs in field 12.5 kOe is $\Delta T \sim 0.4$ K. It was shown that Mn-based polymeric composite cured in aligned magnetic field presents a reinforcement of the MCE temperature hysteresis. Thus, the MCE value in aligned composite is less than the value of the MCE in the bulk MnAs. It was shown that the decreasing of linear size of MnAs grains in the polymeric composite leads to sharp decreasing of the MCE to value of 0.05 K. It was established that the optimal properties, such as giant MCE and low temperature hysteresis, are found for composite hardened under pressure of 10 kBar. In this composite an increase of reversible MCE (compared to the bulk MnAs) till to the value of $\Delta T \sim 1.2$ K was observed.

1 Introduction

The composite material which is made of metallic and polymeric material by certain processes, can retain the advantages of the original components, overcome some shortcomings and show some new properties. A composite material is a multi-phase system consisted of matrix material and reinforcing material. Matrix material is a continuous phase, and reinforcing material is a dispersed phase, such as fiber, sheet or particle. The composite not only retains the key feature of the original component materials, but also gets the performances that are not depicted by the original components through the combined effects [1].

Various functional materials can be produced by compounding different materials of function. At present, electromagnetic functional materials develop most rapidly and are used most commonly as the functional composite materials. These composite materials can be processed into magnetic products with the required shape and the certain mechanical properties by the general forming methods (such as compression molding, laminating, etc.) of polymeric materials. In order to obtain a good overall performance, we must increase the amount of filler or enhanced material as much as possible, control the interface of composite materials and make composite materials have better process ability [2].

* e-mail: pankratov@phys.msu.ru

Magnetic refrigeration technology is a rapidly developing technology that is assumed to be capable of competing with and hopefully surpassing traditional vapour-compression refrigeration in terms of efficiency, device volume and ecological impact in the near future [3]. From this point of view, magnetic compounds exhibiting large magnetocaloric effects (MCE) in the temperature range 270-320 K (under magnetic field of 10-20 kOe) are attracting much attention due to their potential application in room temperature magnetic refrigeration [4–8]. The search for new magnetocaloric materials as well as the optimization of known materials is a growing field of interest. The new magnetic materials showing a large MCE with narrow hysteresis, particularly around room temperature, are the subject of intensive research activity.

During the past years, materials undergoing the first-order magnetic transition (FOMT) have attracted considerable attention. In these compounds, many fascinating functional properties, such as MCE [9], magnetic-field-induced shape recovery [10] and magnetoresistance (MR) [11], were gained near FOMT. It is know, FOMTs can be divided into two categories [12]. One is magnetoelastic transition, in which no crystal symmetry evolution but the change of lattice parameters is observed on both sides of the phase transition, such as $MnFeP_{1-x}As_x$ and $La(Fe,Si)_{13}$ [13–15]. The other is magnetostructural transition (MST), in which magnetic and crystallographic transitions occur simultaneously, such as $Gd_5(Si_2Ge_2)$, Mn-Ni-Ga(Fe) [16, 17] and MnAs [18–20]. In these cases, relatively smaller thermal/magnetic hysteresis than those possessing magnetostructural transition are generally revealed. In numerous materials presenting FOMT, Mn-based intermetallic compounds, such as MnFePAs(Si/Ge) [13, 18], MnAs [20, 21], Ni-Mn-based [9–11], and Mn(Co,Ni)Ge systems [12, 14], have extensively been investigated owing to the low cost and higher magnetic moment of Mn atoms, sometimes even 4 μ_B per Mn atom [18].

The large MCE and giant MR have to observe for Mn-based alloys, because such alloys often demonstrate the transitions with abrupt change of magnetisation or conductivity. It take notice, that magnetic properties in Mn-based alloys connect with interatomic Mn-Mn distance, thereby the contemporary investigation give its up to regulate the Mn-Mn interatomic distance and attain the steep change of magnetisation respectively, as for instance at partial atom replacement [12, 17] or by external pressure [22]. So, the alloys and composites based on MnAs have attracted attention during several decade.

The MnAs compound is a ferromagnetic with magnetic moment equal to $3.4\mu_B$ per Mn atom [23]. It displays one of the highest MCE neighboring room temperature. At the Curie temperature (T_C) MnAs demonstrates the MST from low-temperature ferromagnetic hexagonal NiAs-type state to a high-temperature paramagnetic state with orthorhombic MnP-type crystal structure [19, 24]. These MST and their concomitant entropy changes can be triggered by applying an external magnetic field [19, 24] and by an external pressure [25] giving rise to the MCE. It was shown that both ferro- and paramagnetic metastable phase coexist in MnAs in the region of phase transition, and so the MCE is caused by field induced magnetic phase transition from paramagnetic to ferromagnetic state [19].

The aim of this paper is to investigate the influence of both the polymeric matrix type and the preparation of the composites on the MCE temperature behaviour and finding the MCE dependence on unstable magnetic condition produced by thermal and magnetic history of sample. The submicron particles composite was also investigated. The MCE study was performed by direct measurements of temperature shift at an adiabatic change of magnetic field.

2 Sample preparation and experimental technique

Single crystals of equiatomic manganese arsenide were grown using the Bridgman-Stockbarger technique from a melt of the corresponding composition in evacuated quartz ampoules with a conical tip. The ampoule was dropped from the melting zone along the crystallization zone at a rate of 2-3 mm/h.

The crystallographic quality of the alloys was determined using a powder X-ray diffraction method (XRD) using Cu K_α radiation. The indexing procedure was applied to the obtained diffraction patterns in order to extract relevant crystallographic parameters. The X-ray emission analysis of the grown single crystals demonstrates that the deviation from the stoichiometric composition did not exceed $\Delta x < 0.01$ at %. The main phase was identified to be a hexagonal NiAs-type crystal structure with P6/mmm symmetry. The lattice parameters were closes to the ones reported for pure MnAs, e.g. a = 3.722 Å and c = 5.699 Å. The microstructure was examined using metallography.

The composites were prepared from microsrystalline and submicrosrystalline metallic powder fixed in polymeric matrix; the volume of polymer does not exceed 8%. The grain sizes was examined by electronic microscope. There were obtained both aligned by magnetic field and isotropic composites hardened under hydrostatic pressure up to 20 kbar.

The MCE was measured by direct method by detecting the temperature as function of the magnetic field change. The temperature was detected by the copper-constantan thermocouple. The MCE was studied in the temperature range of 100-350 K in magnetic field up to H = 12.5 kOe.

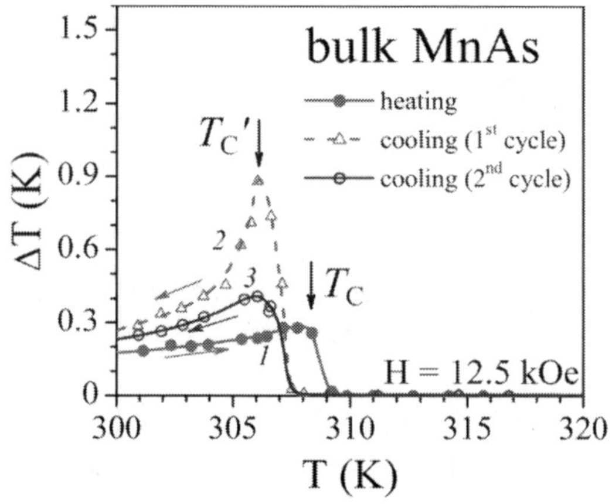

Figure 1. The MCE temperature dependencies of MnAs compound in field 12.5 kOe at heating (1) and cooling (2 and 3). Curve 2 is the irreversible MCE at first magnetization.

3 Results and discussion

The influence of composite preparation process on the MCE was studied. In this frame we choice three ways to manage the MCE behaviour of composites. One is to study the MCE dependence on the magnetic texture of composite. In this case the viscous mixture of MnAs powder and liquid polymer was hardened in uniform magnetic field produced by two permanent magnets. Second an influence of the magnetoactive material grain sizes on the MCE of composite was studied. The composite with micro- and submicrosizes grains were obtained and its MCE was studied. Other is the effect of the hydrostatic pressure applied at the composite's moulding process.

Fig. 1 shows the MCE temperature dependencies in bulk MnAs obtained at adiabatic field excursion from zero to the value of 12.5 kOe in the region of magnetic phase transition. Curve 1 relates to the case where the sample was heated up from ferromagnetic state (T < 300 K)

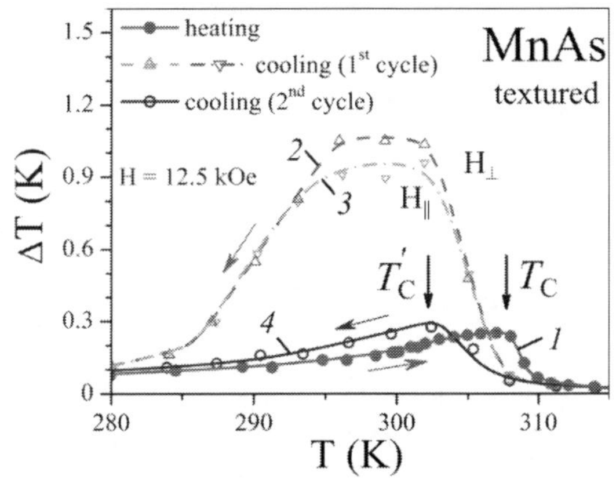

Figure 2. The MCE temperature dependencies of field aligned MnAs-based composite in magnetic field of 12.5 kOe at heating (1) and cooling (2,3, and 4). Curves 2 and 3 demonstrate the irreversible MCE at first magnetization for two different sample orientation in the external magnetic field.

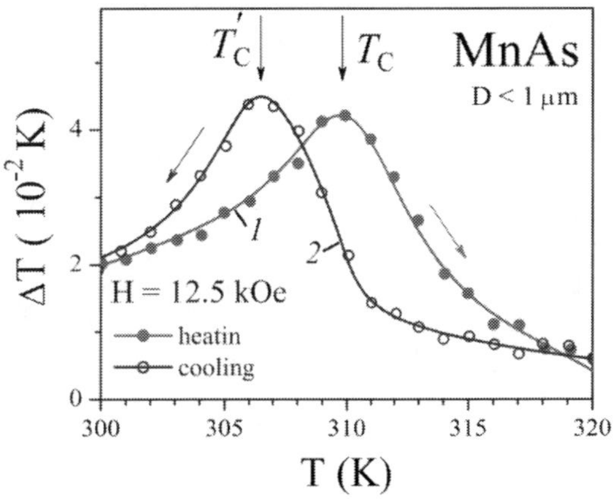

Figure 3. The MCE temperature dependencies of submicron's MnAs-based composite in field of 12.5kOe at heating (1) and cooling (2).

and above the temperature of phase transition the sample reaches the paramagnetic state. It was found that in bulk MnAs on the heating the MCE reaches maximum value of $\Delta T = 0.28$ K (in field of 12.5 kOe) at temperature $T = 308$ K.

In second case (curve 2) the sample is cooled from paramagnetic state ($T > 320$ K) to room temperature, the MCE shows maximum value of $\Delta T = 0.88$ K at temperature $T = 306$ K. Our data point to the fact that there are two metastable phase caused by transition from ferromagnetic to paramagnetic state [19]. As far as relative ratio of ferromagnetic and paramagnetic phases in the samples at heating and cooling is distinct, the MCE has different values depending on thermal history of the sample. This conclusion also is supported by the curve 3 on fig. 1 displayed the MCE behaviour of the preliminary magnetised sample. A first magnetic field switch on causes large MCE (curve 2) and changes relation between ferromagnetic and paramagnetic phases in comparison with the initial state. Obviously at the second magnetization the magnitude of MCE is less than at first one, because the MCE is caused by the transition from paramagnetic to ferromagnetic state and the amount of paramagnetic phase decreases after the first switching field.

Fig. 2 shows the MCE temperature dependencies in magnetic field change of 12.5 kOe in textured composite. This composite properties are texture-specific because there is an axis of preferred orientation of spontaneous magnetization of the grains. By this reason the MCE temperature dependencies at cooling and at heating were studied under the two orientation of magnetic field, along and perpendicular to texture axis, respectively. The maxima value of MCE at the heating is $\Delta T = 0.3$ K at temperature of $T_C = 309$ K. There is no any anisotropy of the MCE at heating of the field oriented composite.

It is found that at cooling the MCE temperature dependence demonstrates a broad maximum in temperature range 296-302 K with maximum value of $\Delta T = 1$ K, however it is irreversible MCE caused by magnetostructural transition from paramagnetic to ferromagnetic state. The MCE of the preliminary magnetized sample is quit less and comparable with MCE at heating (see curves 1 and 4). The temperature of maximum of MCE at cooling is found of $T'_C = 302$ K. It is noticed, that at cooling there is an influence of magnetic anisotropy of the MCE. In case of the field applied perpendicular to textured axis the MCE value is a bit large than in case where the direction of magnetic field coincide with textured axis. Here, $\Delta T_\perp$ is equal to 1.05 K and $\Delta T_\parallel = 0.95$ K. At the magnetization of preliminary magnetized sample the anisotropy of MCE is disappear. It can be conclude that the composite aligned by magnetic field demonstrates an increasing a MCE temperature hysteresis compared with bulk MnAs, while the value of reversible MCE at heating and cooling go up in this composite.

Fig. 3 shows the MCE temperature dependencies of the MnAs-based composite with submicrosizes powder in cases of heating and cooling. On both dependencies the MCE value does not exceed of $\Delta T = 0.05$ K. So, it is seen that the decreasing of the magnetoactive material grain size in the MnAs-based composite to the value less than 1 μm leads to strong decreasing of the MCE compared with bulk MnAs. The temperatures of the MCE maximum $T'_C = 306$ K and $T_C = 309$ K on the cooling and the heating, respectively. So, the temperature hysteresis of MCE preserves in submicron composite.

Fig. 4 displays temperature dependence of the MCE of the composite obtained under hydrostatic pressure of 10 kbar. It is seen that this composite shows a quasi-reversible MCE behaviour. The maximum of MCE ($\Delta T = 1.3$ K) is reached at $T_C = 309$ K for both cooling and heating. The MCE value in this composite is exceed the MCE in bulk MnAs at the same magnetic field [19].

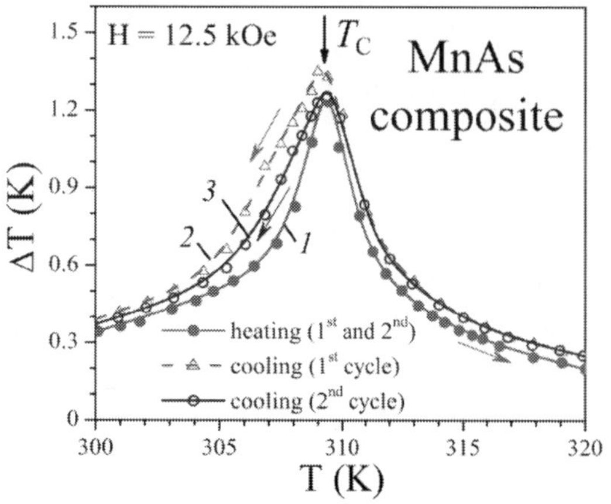

Figure 4. The MCE temperature dependencies of MnAs-based polymer composite cured on pressure 10 kbar at heating (1) and cooling (2 and 3) in field 12.5 kOe. Curve 2 is the irreversible MCE at first magnetization.

It is quite possible that this behaviour is result of residual pressure of polymeric matrix pushed to MnAs grains [25].

4 Summary

It was shown, that the decreasing of grain size under 1 μk reduces the MCE in submicron's MnAs-based composites. It was found, that both the MCE hysteresis and the MCE maximum value are increased in textured composites compared with bulk MnAs.

It was established that the optimal properties, such as giant MCE and low temperature hysteresis, are found in composite with particle's size of $\sim 1\mu$m and polyvinyl acetate matrix hardened at pressure of 10 kbar. The maximum of MCE in this composite is $\Delta T \sim 1.3$ K (at 309 K) in field of 12.5 kOe. At the same time the temperature hysteresis of MCE is foes down in this composite. These properties are caused by the residual pressure of the polymer mould on the grains.

The work is supported by RFBR grants # 16-52-00223, # 1602-00472 and BRFBR grant #T16P-170.

References

[1] F.C. Campbell, *Manufacturing Processes for Advanced Composites* (Elsevier Science, Amsterdam, 2004), ISBN 978-1-85617-415-2

[2] R.M. Wang, S.R. Zheng, Y.P. Zheng, *Polymer Matrix Composites and Technology*, Woodhead Publishing Series in Composites Science and Engineering (Woodhead Publishing, 2011), ISBN 978-0-85709-221-2

[3] C. Zimm, A. Jastrab, A. Sternberg, V. Pecharsky, K. Gschneidner, M. Osborne, I. Anderson, *Description and performance of a near-room temperature magnetic refrigerator* (Boston, MA: Springer US, 1998), pp. 1759–1766

[4] A.M. Tishin, Y.I. Spichkin, *The magnetocaloric effect and its applications*, Series in condensed matter physics (Institute of Physics Pub., Bristol and Philadelphia, 2003), ISBN 0-750-30922-9

[5] O. Gutfleisch, M.A. Willard, E. Brück, C.H. Chen, S.G. Sankar, J.P. Liu, Adv. Mater. **23**, 821 (2011)

[6] E. Brück, *Magnetocaloric refrigeration at ambient temperature* (Elsevier, Amsterdam, The Netherlands, 2007), chap. 4, pp. 235–291

[7] K.A. Gschneidner, V.K. Pecharsky, Int. J. Refrig. **31**, 945 (2008)

[8] K.A. Gschneidner, V.K. Pecharsky, Annu. Rev. Mater. Sci. **30**, 387 (2000)

[9] J. Liu, T. Gottschall, K.P. Skokov, J.D. Moore, O. Gutfleisch, Nat. Mater. **11**, 620 (2012)

[10] R. Kainuma, Y. Imano, W. Ito, Y. Sutou, H. Morito, S. Okamoto, O. Kitakami, K. Oikawa, A. Fujita, T. Kanomota et al., Nature (London) **439**, 957 (2006)

[11] H.C. Xuan, L.J. Shen, T. Tang, Q.Q. Cao, D.H. Wang, Y.W. Du, Appl. Phys. Lett. **100**, 172410 (2012)

[12] N.T. Trung, V. Biharie, L. Zhang, L. Caron, K.H.J. Buschow, E. Brück, Appl. Phys. Lett. **96**, 162507 (2010)

[13] O. Tegus, E. Brück, K.H.J. Buschow, F.R. de Boer, Nature (London) **415**, 150 (2002)

[14] F.X. Hu, B.G. Shen, J.R. Sun, Z.H. Cheng, G.H. Rao, X.X. Zhang, Appl. Phys. Lett. **78**, 3675 (2001)

[15] I.A. Radulov, K.P. Skokov, D.Y. Karpenkov, T. Gottschall, O. Gutfleisch, J. Magn. Magn. Mater. **396**, 228 (2015)

[16] V.K. Pecharsky, J. K. A. Gschneidner, Phys. Rev. Lett. **78**, 4494 (1997)

[17] E.K. Liu, W.H. Wang, L. Feng, W. Zhu, G.J. Li, J.L. Chen, H.W. Zhang, G.H. Wu, C.B. Jiang, H.B. Xu et al., Nat. Commun. **3**, 873 (2012)

[18] E. Brück, O. Tegus, D.T.C. Thanh, N.T. Trung, K.H.J. Buschow, Int. J. Refrig. **31**, 763 (2008)

[19] V.I. Mitsiuk, N.Y. Pankratov, G.A. Govor, S.A. Nikitin, A.I. Smarzhevskaya, Physics of the Solid State **54**, 1988 (2012)

[20] N.Y. Pankratov, V.I. Mitsiuk, A.I. Krokhotin, A.I. Smarzhevskaya, G.A. Govor, S.A. Nikitin, V.M. Ryzhkovskii, Solid State Phenomena **190**, 343 (2012)

[21] P.F. Xu, S.H. Nie, K.K. Meng, S.L. Wang, L. Chen, J.H. Zhao, Appl. Phys. Lett. **97**, 042502 (2010)

[22] S. Niziol, R. Zach, J.P. Senateur, J. Beille, J. Magn. Magn. Mater. **79**, 333 (1989)

[23] H. Wada, Y. Tanabe, Appl. Phys. Lett. **79**, 3302 (2001)

[24] G.A. Govor, in *Double Exchange in Heusler Alloys and Related Materials*, edited by K. Baerner (2007), chap. 2, pp. 29–62

[25] S. Gama, A.A. Coelho, A. de Campos, A. Magnus, G. Carvalho, F.C.G. Gandra, P.J. Ranke, N.A. Oliveira, Phys. Rev. Lett. **93**, 237202(4) (2004)

Magnetic phase diagrams of Gd-H, Tb-H, Dy-H systems

Victoria Chzhan[1], *Irina* Tereshina [2,*], *Gennady* Burkhanov[1], *Galina* Politova[1] and *Henryk* Drulis[3]

[1]Baikov Institute of Metallurgy and Materials Science RAS, 119334 Moscow, Russia
[2] Lomonosov Moscow State University, Faculty of Physics, 119991 Moscow, Russia
[3]Institute of Low Temperature and Structural Research, 50-422 Wroclaw, Poland

Abstract. We study the impact of hydrogen impurity on the magnetic phase transition temperatures of the high purity rare-earth metals Gd, Tb and Dy. Prior to hydrogenation, the rare earths were purified by the vacuum distillation method. Hydrogenation was carried out using a Sievert-type apparatus. Magnetic phase diagrams were constructed based on the thermomagnetic analysis.

1 Introduction

Characterization of physicochemical properties of the rare earth (R) metals should be performed on purified [1] samples due to their high chemical activity and strong affinity for gas-forming impurities. Literature survey revealed a large number of works dedicated to the description of the *R*'s properties [2-6]. However, electrical and magnetic properties studies were performed on single- and polycrystalline samples of various degrees of purity [7]. This unfortunately disables a direct comparison of the results obtained by different authors. It also makes it difficult to reveal intrinsic physical processes occurring in the metals. It is therefore important to use high-purity metals for the studies. Controlled doping (by e.g. gas-forming impurities) of the previously purified metal allows us to determine the influence of a particular impurity on the properties of a material under study. In this paper, a systematic study of the hydrogenation effect on magnetic properties of a distilled gadolinium, terbium and dysprosium is carried out. (The interest to Gd is due to its high magnetocaloric properties near room temperature that allow its use as a working body in magnetic refrigerators [8-10].) *R*'s interaction with hydrogen is rather extensively studied. Phase diagrams of *R*-H for heavy rare-earth metals and yttrium have common features (see Fig. 1a) [11]. Special attention in this paper is devoted to the study of the impact of hydrogenation on the phase transitions temperatures for small hydrogen concentrations ($x \leq 1.2$).

2 Materials synthesis and experimental details

Rare-earth metals Gd, Tb and Dy were purified by vacuum distillation [1]. High purity rare earths are characterized by a reduced content of gas-forming impurities (10^{-2} -10^{-3} wt. %). The metals purity with respect to the content of ~70 impurity elements are 99.92-99.96 wt. %. The amount of absorbed hydrogen was determined by measuring the pressure change in the reactor chamber after completing the reaction. The accuracy of the hydrogen concentration determination is 0.02 hydrogen atoms per formula unit (at.H/f.u.). The samples for hydrogenation and further studies were cut out of the central portions of the distillate [12]. Appearance of a rare-earth distillate is shown in Fig. 1b.

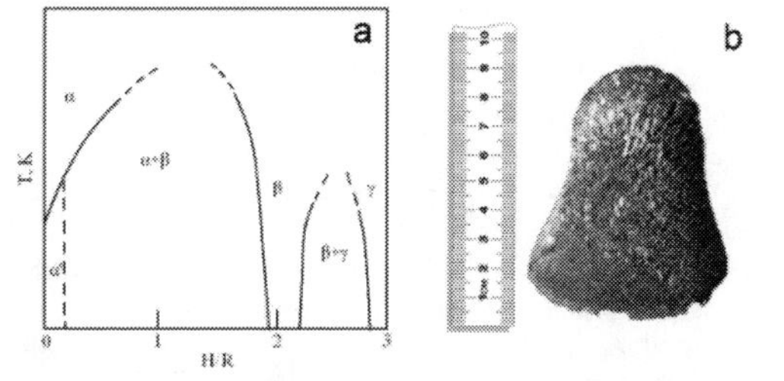

Fig. 1. Typical phase diagram of a hydrogenated rare earth (a) [11]. Appearance of a rare-earth distillate (b).

Purified metals were hydrogenated using a Sievert-type apparatus with hydrogen (impurities content < 10^{-3} – 10^{-4} vol. %) pressures up to 0.1 MPa. The samples of GdH_x (x = 0, 0.02, 0.05, 0.09, 0.12, 0.13, 0.16, 0.2), DyH_x (x = 0, 0.1, 0.2, 0.5, 0.7, 0.9, 1.2) and TbH_x (x = 0, 0.1, 0.2, 0.3, 0.5, 0.7, 0.9, 1.2) were obtained and investigated. Magnetic properties of the samples were studied over wide temperature range using a vibrating sample magnetometer equipped with a step motor [13]. Thermomagnetic analysis was carried out in a magnetic field of 0.02 T in the temperature range 4.2-300 K. The relative accuracy of data acquisition did not exceed 0.5%. The Curie temperatures of the parent and hydrided Gd were also determined by means of the Arrott-Belov plot method based on the thermodynamic theory [14].

* Corresponding author: irina_tereshina@mail.ru

 https://doi.org/10.1051/epjconf/201818505011

3 Results and discussion

The *R*-H phase diagrams (Fig. 1a) are characterized by a region of existing solid-solutions in the *R* phase (α-RH_x) having the structure of the parent *R*, as well as by the presence of β-RH_2 and γ-RH_3 phases with a variable composition. The β-RH_2 dihydride phase has a *fcc* CaF_2-type lattice and the γ-RH_3 trihydride phase crystallizes into a hexagonal HoH_3-type structure [11,15].

All the parent and hydrogenated samples of Gd, Tb and Dy were investigated by the X-ray diffraction analysis from the surface. The X-ray analysis indicated that a distilled Gd, and all obtained α-GdH_x solid-solutions are single-phase and have an *hcp* structure (space group *P6₃/mcc*) [12,14]. A similar crystalline structure is also present in the distilled Tb and Dy. The obtained structural characteristics are in good agreement with the literature [16]. It was found that all of the TbH_x and DyH_x samples contain two phases. In addition to the main α-RH_x solid-solution phase, a cubic β-RH_2 phase was found. The amount of the second phase increases linearly with the increasing of absorbed hydrogen content. For example, for the sample $DyH_{1.2}$, the amount of β-DyH_2 is 66 %. The Néel temperatures T_N and magnetic moments μ of the RH_2 dihydrides are listed in Table 1.

Table 1. Magnetic characteristics of the RH_2 dihydrides [19].

R	T_N, K	μ, μ_B
Gd	21	6.5
Tb	17	7.4
Dy	3.5	6.4

No data on the possible maximum hydrogen concentration in terbium and dysprosium solid solutions at room temperature is available in literature.

Experimental values of phase transition temperatures can vary depending on the method used for the determination [7]. The use of magnetic field typically leads to higher values of magnetic phase transition temperatures. The application of magnetic fields exceeding critical values for the particular compounds can also lead to the magnetic order change. In this work, the experiments were unified, i.e. we performed our measurements under the same conditions and the samples for the study had the same shape, thus enabling [17] precise determination of the changes in the magnetic state upon hydrogen absorption.

Previously the Curie temperatures of Gd were determined from the Arrott-Belov plots as 291 K [14]. The magnetic phase transition temperatures of the parent distilled Gd, Tb and Dy, and the same samples after hydrogenation were determined by using a thermomagnetic analysis (μ_0H = 0.02 T). The critical temperatures were found from the peak of derivative of magnetization dM(T)/dT. Two pronounced extrema were observed on the dM(T)/dT curve in the distilled Gd. The one at T_C = 292 K is associated with the transition from the ferromagnetic (FM) to the paramagnetic (PM) state in the distilled Gd and the other one at T_{SR}=223 K - with a spin-reorientation transition (SRT). According to Ref. 18, during the SRT the magnetic moments of Gd aligned along the *c*-axis at higher temperatures begin to deviate from the *c*-axis. The angle reaches 70° at 170 K, and then it decreases to ~30° at lower temperatures. It was found that α-GdH_x solid solutions below the Curie temperature retain the ferromagnetic properties. Figure 2a shows the temperature dependencies of magnetization for α-$GdH_{0.1}$. In the solid solution α-$GdH_{0.1}$ hydrogenation is found to increase the transition temperatures to T_C = 294 K and T_{SR} = 234 K (see inset into Fig. 2a). Note that introduction of 0.1 H/Gd increases the T_C by 2 K while the T_{SR} is increased by as much as ~ 10 K.

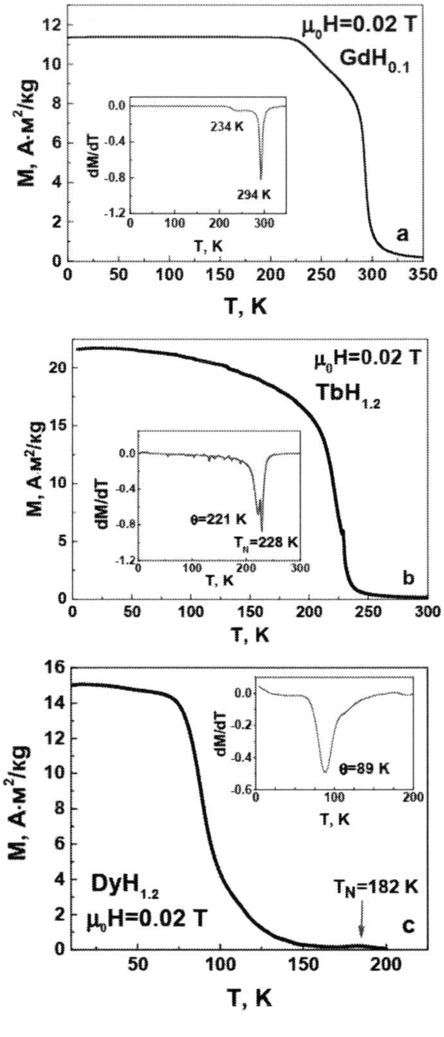

Fig. 2. The temperature dependencies of magnetization for α-$GdH_{0.1}$ (a), $TbH_{1.2}$ (b), $DyH_{1.2}$ (c). Insets show the temperature dependencies of the derivative dM/dT for the mentioned samples.

In contrast to ferromagnetic Gd, Tb and Dy are antiferromagnetic (AFM) below the T_N and have a second transition from the AFM to the FM state at a temperature θ. In Dy, however, the AFM region exists in a much wider (~90 K) temperature range than that in Tb (~10 K). The presence of the β-TbH_2 and β-DyH_2 antiferromagnetic phase does not have a significant effect on the M(T) dependence behavior because of their low ordering temperatures (see Table 1). As an example, Fig. 2 shows the magnetization curves of $TbH_{1.2}$ (2b) and $DyH_{1.2}$ (2c). As can be seen from the M(T) curves, the sharp decrease in the magnetization occurs at the temperature θ thus implying a change of the magnetic state. At higher temperatures, a maximum observed on the M(T) curves corresponds to the Néel temperature of the samples. Hydrogen addition in Tb and Dy up to the concentration of x = 1.2 H/R does not influence the M(T) behavior. Also two magnetic phase transitions are preserved while the temperatures T_N and θ decrease. Interestingly, similar magnetic behavior with two magnetic phase transitions was also demonstrated by rare-earth-Fe alloys e.g. R_2Fe_{17} (R = Ce [20] and Lu [21]). The temperatures of magnetic phase transitions and the magnetic structures are extremely sensitive to the interstitial impurity atoms (hydrogen, in particular) [22,23], to substitutional atoms [24] and high pressure [25]. The magnetic phase diagrams of such compositions have been obtained and studied extensively in literature [26].

We also constructed magnetic phase diagrams for α-GdH_x, TbH_x, and DyH_x. The effect of hydrogen addition on the T_C's of Gd is demonstrated in Fig. 3. As it is seen from the $T_C(x)$ dependence, the Curie temperature increases by about 5 K when x increases from 0 to 0.2 H/Gd. High pressure experiments give an opposite trend for T_C [27] and therefore the volume effect is very important. The main mechanisms of the Curie temperature increase are discussed in Ref. 15. Figures 4 and 5 show the concentration dependencies of the Néel temperature and the transition temperature from the AFM to the FM state (θ) on the hydrogen content (x) in TbH_x and DyH_x, respectively. Change in the phase transition temperatures is within 1 K in TbH_x when the hydrogen concentration (x) varies from 0 to 1.2 H/Tb. The Néel temperatures of DyH_x decrease from 184 to 182 K, and the temperature θ decreases from 94 to 89 K when x varies between 0 and 1.2 H/Dy. Although the variations in the R phase transition temperatures ranging between 1 and 5 degrees were also recorded previously [28-30], they cannot be attributed to a particular impurity but rather to a mixture of impurities absorbed by a metal. In this study, we were able to identify the real magnetic phase transitions temperatures and their variation as a function of hydrogen impurity content in Gd, Tb and Dy by using parent high purity samples and then those with a controlled impurity composition.

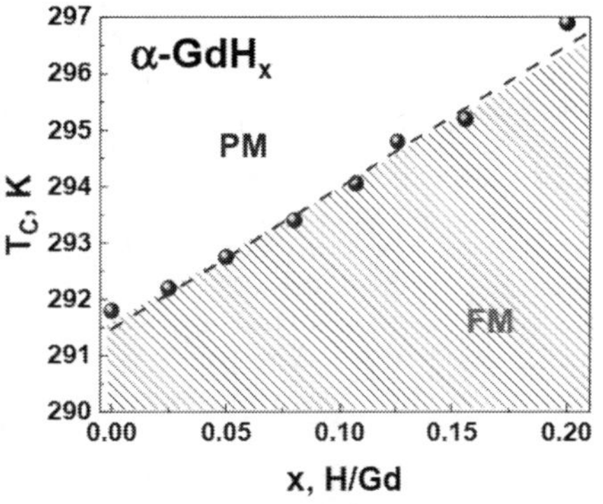

Fig. 3. The Curie temperature T_C *vs.* hydrogen content x in the α-GdH_x solid solutions.

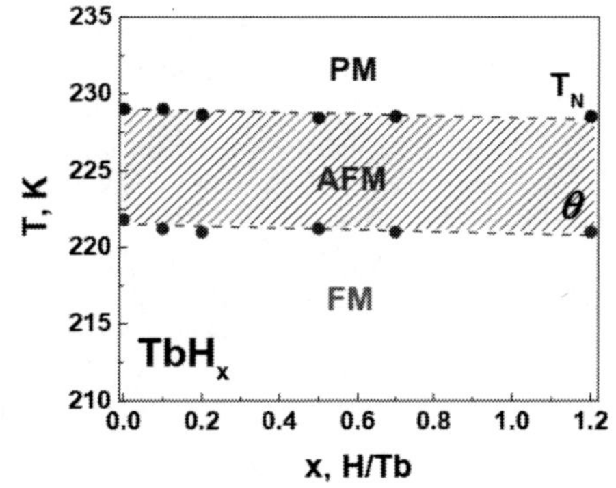

Fig. 4. Concentration dependencies of the Néel temperature T_N and the temperature θ on the hydrogen content x in TbH_x obtained in the magnetic field of 0.02 T.

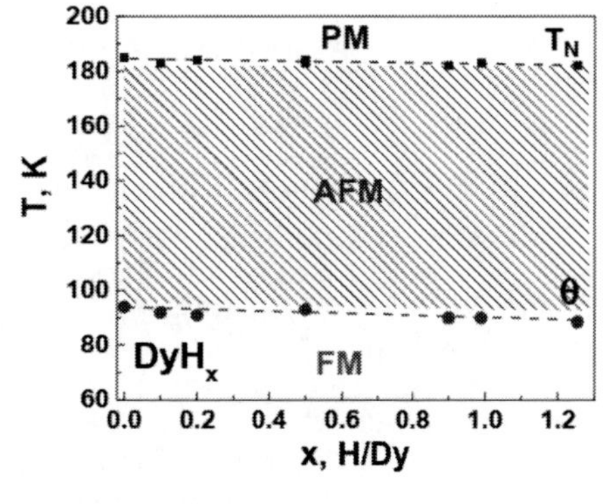

Fig. 5. Concentration dependencies of the Néel temperature T_N and the temperature θ on the hydrogen content x in DyH_x obtained in the magnetic field of 0.02 T.

4 Conclusions

The effect of hydrogenation on the magnetic phase transition temperatures for the three rare earths Gd, Tb, Dy having the highest magnetic ordering temperatures among *R*s is studied. Gadolinium has a Curie temperature T_C = 292 K. Tb and Dy have complex magnetic structures and display two magnetic phase transitions (FM-AFM and AFM-PM). It is found that hydrogen introduced into the Gd crystal lattice increases the Curie temperature in the range of existence of solid solutions. On the contrary Tb and Dy retain the values of magnetic phase transition temperatures or even have it slightly decreased under the influence of hydrogen. This is due to the presence of the second dihydride phase. As a main achievement, the refined magnetic characteristics of Gd, Tb and Dy after hydrogen treatment are obtained. It is important for the materials working in a hydrogen-containing environment [31].

This work was supported by the Russian Foundation for Basic Research, project no. 16-03-00612-a.

References

1. G.G. Devyatykh, G.S. Burkhanov, *High-Purity Refractory and Rare-Earth Metals* (Cambridge: Int. Sci. Publ., 1997)
2. S.A. Nikitin, *Magnetic Properties of Rare-Earth Metals and Compounds* (Moscow, Publishing House MSU, 1989) (in Russian)
3. V.I. Zverev, A.M. Tishin, A.S. Chernyshov, Ya. Mudryk, K.A. Gschneidner Jr., V.K. Pecharsky, J. Phys.: Condens. Matter **26**, 066001 (2014)
4. A.S. Chernyshov, A.O. Tsokol, A.M. Tishin, K.A. Gschneidner, Jr., V.K. Pecharsky, Phys. Rev. B. **71**, 184410 (2005)
5. M. Mito, K. Matsumoto, Yu. Komorida, H. Deguchi, S. Takagi, T. Tajiri, T. Iwamoto, T. Kawae, M. Tokita, K. Takeda. J. Phys. Chem. Solids **70**, 1290 (2009)
6. D. Michels, C. E. Krill, R. Birringer, J. Magn. Magn. Mater. **250**, 203 (2002)
7. S.Yu. Dan'kov, A.M. Tishin, V.K. Pecharsky, K.A. Gschneidner, Jr., Phys. Rev. B **57**, 3478 (1998)
8. A.M. Aliev, A.B. Batdalov, L.N. Khanov, V.V. Koledov, V.G. Shavrov, I.S. Tereshina, S.V. Taskaev, J. Alloys Compounds **676**, 601 (2016)
9. S.V. Taskaev, M.D. Kuz'min, K.P. Skokov, D.Yu. Karpenkov, A.P. Pellenen, V.D. Buchelnikov, O. Gutfleisch, J. Magn. Magn. Mater. **331**, 33 (2013)
10. Yu.S. Koshkid'ko, J. Cwik, T.I. Ivanova, S.A. Nikitin, M. Miller, K. Rogacki, J. Magn. Magn. Mater. **433**, 234 (2017)
11. P. Vajda, *Handbook on the Physics and Chemistry of Rare Earths* **20**, 207 (Amsterdam: North-Holland, 1995)
12. G.S. Burkhanov, N.B. Kolchugina, E.A. Tereshina, I.S. Tereshina, G.A. Politova, V.B. Chzhan, D. Badurski, O.D. Chistyakov, M. Paukov, H. Drulis, L. Havela, Appl. Phys. Lett. **104**, 242402 (2014)
13. V.I. Nizhankovskii, L.B. Lugansky. Meas. Sci. Technol. **18**, 1533 (2007)
14. E. Tereshina, S. Khmelevskyi, G. Politova, T. Kaminskaya, H. Drulis, I. Tereshina, Sci. Rep. **6**, 22553 (2016)
15. Y. Fukai, *The Metal-Hydrogen System. Basic bulk properties* (Berlin: Springer, 2005)
16. F.H. Spedding, B.J. Beaudry, J. Less Comm. Metals. **25**, 61 (1971)
17. V.I. Zverev, R.R. Gimaev, A.M. Tishin, Ya. Mudryk, K.A. Gschneidner, Jr., V.K. Pecharsky, J. Magn. Magn. Mater. **323**, 2453 (2011).
18. H.E. Nigh, S. Legvold, F. Spedding, Phys. Rev. **132**, 1092 (1963)
19. W.G. Bos, K.H. Gayer, J. Nucl. Mater. **18**, 1 (1966)
20. A.G. Kuchin, N.V. Mushnikov, M.I. Bartashevich, O. Prokhnenko, V.I. Khrabrov, T.P. Lapina, J. Magn. Magn. Mater. **313**, 1 (2007)
21. S.A. Nikitin, I.S. Tereshina, N.Yu. Pankratov, E.A. Tereshina, Yu.V. Skourski, K.P. Skokov, Yu.G. Pastushenkov, Phys. Solid State **43**, 1720 (2001)
22. E.A. Tereshina, A.V. Andreev, J. Kamarád, O. Isnard, J. Appl. Phys. 105, 07A747 (2009)
23. I.S. Tereshina, S.A. Nikitin, K.P. Skokov, T. Palewski, V.V. Zubenko, I.V. Telegina, V.N. Verbetsky, A.A. Salamova, J. Alloys Compounds, **350**, 264 (2003)
24. E.A. Tereshina, O. Isnard, A. Smekhova, A.V. Andreev, A. Rogalev, S. Khmelevskyi, Phys. Rev. B **89**, 094420 (2014)
25. E.A. Tereshina, A.V. Andreev, J. Kamarad, H. Drulis, J. Alloys Compounds **492**, 1 (2010)
26. W. Iwasieczko, A.G. Kuchin, H. Drulis, J. Alloys Compounds **392**, 44 (2005)
27. H. Bartholin, D. Bloch, Phys. Rev. **188**, 845 (1969)
28. G.S. Burkhanov, V.B. Chzhan, G.A. Politova, J. Cwik, N.B. Kolchugina, I.S. Tereshina, Doklady Physics **61**, 168 (2016)
29. F.H. Spedding, J. Less Comm. Metals. **25**, 61 (1971)
30. K.A. Gschneidner, J. Alloys Compd. **193**, 1 (1993)
31. A.P. Kamantsev, V.V. Koledov, V.G. Shavrov, I.S. Tereshina, Solid State Phenomena **215**, 113 (2014)

EPJ Web of Conferences **185**, 05012 (2018)
MISM 2017
https://doi.org/10.1051/epjconf/201818505012

Ternary phase diagram of Ni-Mn-Ga: insights from *ab initio* calculations

Yulia Sokolovskaya[1], *Mikhail* Zagrebin[1,2,⋆], *Vasiliy* Buchelnikov[1], and *Alexey* Zayak[3]

[1] *Chelyabinsk State University, 454001 Chelyabinsk, Russia*
[2] *National Reserach South Ural State University, 454080 Chelyabinsk, Russia*
[3] *Bowling Green State University, OH-43403 Bowling Green, USA*

Abstract. In this work we perform a wide-range systematic study of the family off-stoichiometric Ni-Mn-Ga alloys by using the supercell approach in the framework of density functional theory. Our goal is to explore the compositional variations of the structural stability and magnetic properties of Ni-Mn-Ga compositions. As a result equilibrium lattice parameters, bulk moduli, total magnetic moments, and formation energies of a wide range of Heusler alloys have been mapped on compositional ternary diagrams.

1 Introduction

Heusler alloys represent a large family of ternary metallic compounds that became known at first as magnetic shape-memory effect materials. Sometime later, more Heusler alloys became known as half-metals, promising for realizing spintronic technologies. Another subgroup of Heusler alloys is explored with respect to the magnetocaloric effect (magnetic refrigeration). A large number of Heusler family members and the richness of properties exhibited by them make this class of materials well suitable for emerging materials dependent technologies [1–3]. At present, Ni-Mn-Ga Heusler systems are the most studied among many ferromagnetic (FM) shape memory alloys. Such high interest is related to the unique properties exhibited by the prototypical stoichiometric shape-memory compound Ni_2MnGa [1–3]. A variety of different off-stoichiometric Ni-Mn-Ga compositions, which have been synthesized and characterized in the past twenty years are mapped on the ternary diagram. However, the most popular compositions are localized in the middle part of a ternary diagram, close to the stoichiometric system. The majority of well-studied compounds belong to the families of $Ni_{2+x}Mn_{1-x}Ga$ (x = 0.0 − 0.39) and $Ni_2Mn_{1+x}Ga_{1-x}$ (x = 0.0 − 0.40), showing a strong coupling between the magnetic and structural subsystems [1, 4].

In the previous studies, we presented ternary compositional phase diagrams that showed the variability of several major physical properties as a function of the chemical composition of Ni-Mn-Ga [5–7]. Those diagrams are compositional maps that show the most stable alloy compositions, or compositions that exhibit the largest magnetic moment, or those compositions that represent the hardest crystal structure. That was a proof-of-principle step, which showed that even compositions of Heusler alloys, which are far from the stoichiometric one, could have significant potential for exploring. However, those calculations had been done with the KKR-CPA method, which is great for studying random compounds, however, may lack the detailed chemical accuracy of the local chemical bonding, due to the mean-field approximation used in this method.

Our current study is reproducing of the ternary diagrams described above, but using the more accurate supercell approach, which takes into account local chemical effects without the mean-field approximation.

In this work, we perform a wide-range systematic study of the off-stoichiometric Ni-Mn-Ga alloys using the supercell approach in the framework of density functional theory.

2 *Ab initio* computational details

All calculations were performed using the density functional theory as part of VASP (Vienna *ab initio* simulation program) [8, 9]. This code is based on the projector augmented wave (PAW) method. The generalized gradient approximation (GGA) according to the Perdew-Burke-Ernzerhof (PBE) parametrization [10] was used to take into account the exchange correlation effects. Ground state calculations were performed with the help of the supercell approach with 16-atom supercell. With respect to the pseudopotentials in VASP calculations we used the next electronic configurations: Ni($3p^63d^84s^2$), Mn($3p^63d^64s^1$), Ga($3d^{10}4s^24p^1$), respectively. The kinetic energy cut-off was 400 eV and the kinetic energy cut-off for the augmentation charges was 800 eV. The k-points in the Brillouin zone for self-consistent field cycles were generated with 8^3 meshes for lattice relaxation calculations. To estimate the optimized lattice constants for all alloys under study, we fitted the dependences of total energy on a cell volume according to the Birch-Murnagahan equation of state. To investigate the phase stability of the

⋆e-mail: miczag@mail.ru

Ni-Mn-Ga compounds, we calculate their formation energies for the cubic phase. It should be noted that the formation energy is defined as a difference between the total energy of the alloy and the concentration weighted average of the pure atomic species, as follows:

$$E_{form} = E_{tot} - \left(x \times E^{\mathrm{Ni}} + y \times E^{\mathrm{Mn}} + z \times E^{\mathrm{Ga}}\right), \quad (1)$$

where, E_{tot} is the total energy of alloy, x, y, z and $E^{\mathrm{Ni(Mn,Ga)}}$ are the concentrations and pure element total energy of Ni, Mn, Ga atoms, respectively. The negative value of formation energy indicates the phase stability.

In this study, we considered 105 compositions presented in Fig. 1. Three of them are stoichiometric (Ni_2MnGa, Mn_2NiGa, and Ga_2MnNi).

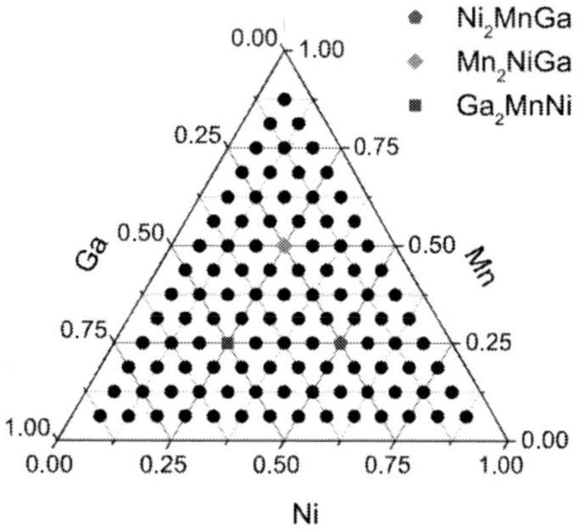

Figure 1. Ternary phase diagram of Ni-Mn-Ga alloys. Symbols indicate the compositions for which the calculations were performed. Color symbols indicate stoichiometric compositions: Ni_2MnGa, Mn_2NiGa, and Ga_2MnNi.

We used four types of crystal lattices to create off-stoichiometric compositions. The first one is the #225 ($Fm\bar{3}m$) symmetry group. In this case, for all compositions we fixed Ni, Mn, and Ga atoms at $8c$, $4b$, and $4a$ Wyckoff positions, respectively. The second one is the same as #225 ($Fm\bar{3}m$) symmetry group, and we used this type in a case there are no less than 8 Mn atoms in the supercell. In this case the Mn atoms are placed in $8c$ Wyckoff positions, Ni and Ga atoms in $4a$, and $4b$ positions, respectively. This type was denoted as #225'. In the third type (which we denoted as #225") Ga atoms are placed in $8c$ Wyckoff positions, Mn and Ni atoms in $4a$, and $4b$ positions, respectively. This case is used for compounds containing 8 and more Ga atoms. The last crystal structure is of #216 ($F\bar{4}3m$) symmetry group. In this case Mn atoms are placed in $4c$ and $4a$ positions, Ga and Ni atoms in $4b$, and $4d$ positions, respectively. Positions of Ni, Mn and Ga atoms for all considered types of crystal lattices are listed in Table 1.

As known, an antiferromagnetic (AFM) interactions between Mn atoms located in different sublattices are realized in nonstoichiometric alloys Ni-Mn-Ga. It is necessary to consider the different magnetic configurations of the Mn atoms. For compositions with the Ni and Ga excess, we considered three magnetic configurations refereed as FM (spins of Mn located at the Ni, Mn, and Ga sites are parallel) and two ferrimagnetic (FIM) ones: FIM-1 and FIM-2. For compositions with Mn excess the one FM and seven FIM magnetic configurations were considered. Sign of Mn atoms in different sublattices are listed in Table 2.

Table 1. Positions of the Ni, Mn and Ga atoms in the different crystal structures used in our calculations.

Wyckoff position		#225	#225'	#225"	#216
4a	$(0,0,0)$	Ga	Ga	Ni	Mn
4b	$\left(\frac{1}{2},\frac{1}{2},\frac{1}{2}\right)$	Mn	Ni	Mn	Ni
8c*	$\left(\frac{1}{4},\frac{1}{4},\frac{1}{4}\right)$	Ni	Mn	Ga	Mn
	$\left(\frac{3}{4},\frac{3}{4},\frac{3}{4}\right)$	Ni	Mn	Ga	Ga

* For #216 crystal structure **8c** Wyckoff position split into two positions: **4c**(1/4,1/4,1/4) and **4d** (3/4,3/4,3/4).

Table 2. The sign of the magnetic moment of Mn atoms placed at various positions of different crystal structures used in our calculations.

	$(0,0,0)$	$\left(\frac{1}{4},\frac{1}{4},\frac{1}{4}\right)$	$\left(\frac{1}{2},\frac{1}{2},\frac{1}{2}\right)$	$\left(\frac{3}{4},\frac{3}{4},\frac{3}{4}\right)$
FM	>0	>0	>0	>0
FIM-1	>0	>0	<0	<0
FIM-2	<0	>0	>0	>0
FIM-3	<0	>0	>0	>0
FIM-4	>0	>0	<0	>0
FIM-5	>0	>0	>0	<0
FIM-6	>0	>0	<0	<0
FIM-7	<0	>0	<0	>0
FIM-8	<0	>0	>0	<0
FIM-9	<0	>0	<0	<0

3 Results of *ab initio* calculations

Experimental studies of the crystal structure of the investigated alloys shows, that depending on the concentration of excess Mn, Ni, or Ga atoms (above 50%), alloys can crystallize into cubic structures of the symmetry groups #225 and #216 [11]. The research was started with the systematic calculations of the equilibrium energies and magnetic ordering of the mentioned above structures with allowance for different atomic placement. Current research results are shown in Figs. 2-4. In the Fig. 2 the distribution of equilibrium lattice parameter together with regions of stable crystal structure types is shown. These calculations show that only 3 types of crystal structures are realized in the ternary phase diagram. These are #225, #225" (#225 symmetry group with excess Ga in $8c$ Wyckoff positions) and #216. It should be noted that #225' crystal type (#225 symmetry group with excess of Mn atoms in 8*c* Wyckoff positions) is unstable on this diagram.

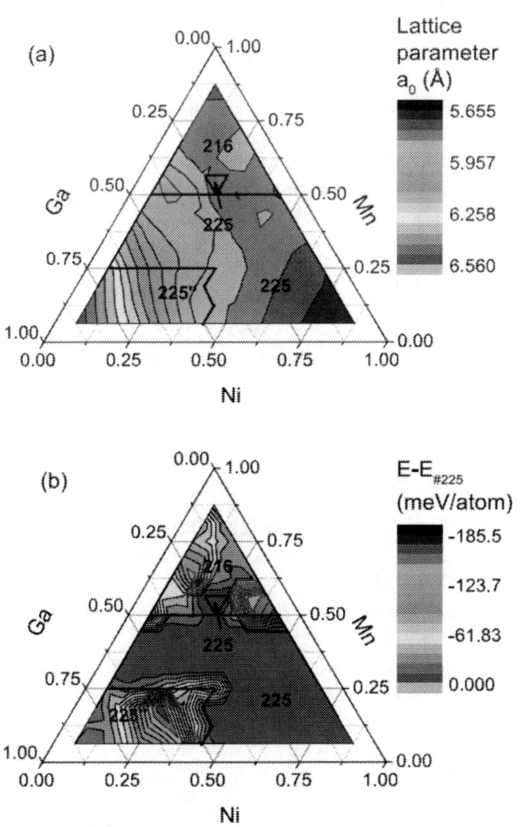

Figure 2. The distribution of (a) the equilibrium lattice parameter and (b) the equilibrium total energy relative to the total energy of #225 crystal lattice mapped on the ternary diagram of Ni-Mn-Ga.

As one can see from Fig. 2(a), the largest lattice parameter ($\approx$ 6.5 Å) is observed at the bottom - left corner of the ternary diagram indicating the compositions with the highest Ga content. Areas with the minimal lattice parameter ($\approx$ 5.7 Å) located at the top and right corners of the diagram corresponding to compositions with highest Mn and Ni concentrations, respectively. The largest lattice parameters for compounds with the excess of Ga is caused by the greater atomic radius of Ga as compared to Mn and Ni whose radii are approximately equal to each other. This result agrees with the previous KKR-CPA calculations [7]. Such agreement shows that, after all, KKR-CPA gives accurate predictions about disordered materials. In the Fig. 2(b) regions of stable crystal structure types together with the distribution of the equilibrium total energy relative to the total energy of #225 crystal lattice together are shown. Calculated energy difference allows us to estimate temperature of phase transitions between phases using the following expression $\Delta E \approx k_B T$ (where k_B is the Boltzman constant).

In Fig. 3(a) we illustrate the mapping of the total magnetic moment on the ternary diagram of Ni-Mn-Ga together with stable magnetic configurations. It can be seen from this distribution, that the largest magnetic moment is observed for alloys with significant Ni excess and with a small content of Ga atoms ($\approx$ 6.5 μ_B). At the same time, areas with the smallest ($\approx$ 0.15 μ_B) values of a magnetization are observed for the compositions with the Mn and Ga atoms excess. Fig. 3(b) shows regions of stable magnetic configurations together with the distribution of the equilibrium total energy relative to the total energy of FM state.

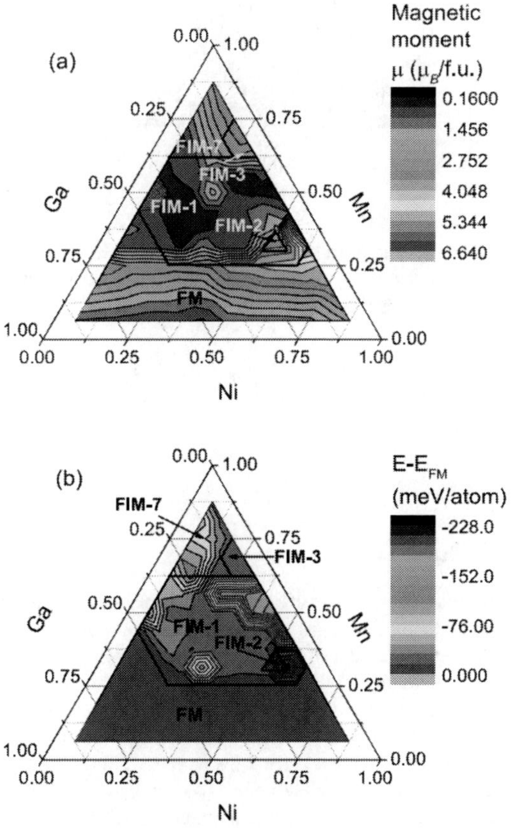

Figure 3. The distribution of (a) the total magnetic moment and (b) the equilibrium total energy relative to the total energy of FM configuration mapped on the ternary diagram of Ni-Mn-Ga.

The calculations also showed that for compositions with Mn excess, FIM ordering type (FIM-3 and FIM-7, see Table 2) is energetically favorable. As for compositions with an excess of Ni and Ga atoms (almost 2/3 of the diagram) an energetically favorable FM ordering except for a small area of compositions with an FIM-1 and FIM-2 ordering. The FIM ordering appears due to AFM interactions between Mn atoms located at the shortest distance in the cubic lattices of the symmetry groups #225 and #216.

Our calculations show that the most of FIM spin configurations are energetically unfavorable for all the cubic structures of Ni-Mn-Ga.

The formation energies for the cubic lattice of Ni-Mn-Ga alloys mapped on the ternary diagram are shown in Fig. 4.

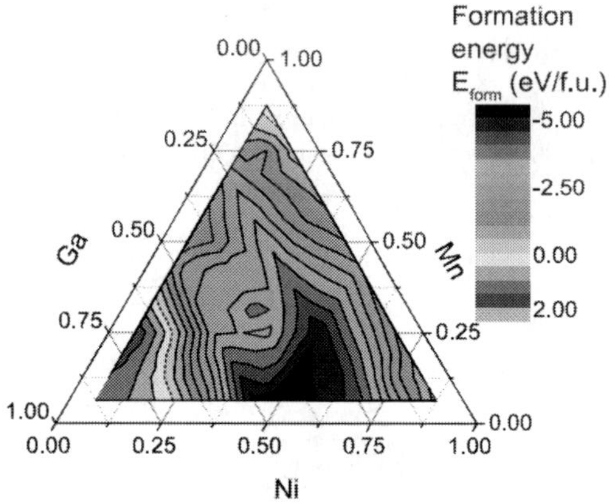

Figure 4. The distribution of the formation energy mapped on the ternary diagram of Ni-Mn-Ga. Dashed lines on diagram correspond to zero formation energy.

It is evident from the figure that the formation energies for the compounds with Ni and Mn excesses atoms have negative values while the positive values of the formation energies indicate that those structures are unstable for alloys with a high Ga content .

4 Conclusions

In this paper the results of investigations of the magnetic and structural properties of Ni-Mn-Ga Heusler alloys with the help of the density functional theory are presented.

Calculations of the ground-state energy of the crystal lattice, the equilibrium lattice parameter, the total magnetic moment, and formation energy were performed for the austenitic (cubic) structure of type $L2_1$ with the help of the VASP *ab initio* code. As an exchange correlation potential, the generalized gradient approximation was considered within the framework of the Perdew-Burke-Ernzerhof potential.

A systematic study of the magnetic and structural properties of Ni-Mn-Ga alloys have been carried out using 16-atoms supercell approach for 105 compositions containing three stoichiometric compositions (Ni_2MnGa, Mn_2NiGa, and Ga_2MnNi) and compositions with the excess of Ni, Mn, and Ga atoms. The calculated equilibrium lattice parameter, total magnetic moment, and formation energy of the cubic phase have been plotted for a ternary diagram of Ni-Mn-Ga alloys. Calculations have shown that a region with Ga atoms excess has the largest lattice parameter and the smallest magnetic moment, whereas regions with an excess of Ni atoms have the smallest lattice parameter. The region with an excess of Mn atoms is also characterized by a small lattice parameter and a small magnetic moment. Crystal formation energy calculations showed that compositions with a significant excess of Ga atoms are unstable in the cubic phase.

Acknowledgments

This work was supported by RSF-Russian Science Foundation No. 17-72-20022/17 (Section 3) and RFBR-Russian Foundation for Basic Research No. 18-32-00507 (Section 2).

References

[1] P. Entel, M.E. Gruner, A. Dannenberg, M. Siewert, S.K. Nayak, H.C. Herper, V.D. Buchelnikov, Mat. Sci. Forum **635**, 3 (2010).

[2] V.D. Buchelnikov, V.V. Sokolovskiy, Phys. Met. Metallogr., **112**, 633 (2011).

[3] V.V. Sokolovskiy, M.A. Zagrebin, V.D. Buchelnikov, Mat. Sci. Foundations, **81**, 38 (2015).

[4] P. Entel, V.D. Buchelnikov, V.V. Khovailo, A.T. Zayak, W.A. Adeagbo, M.E. Gruner, H.C. Herper, E.F. Wassermann, J. Phys. D: Appl. Phys., **39**, 865 (2006).

[5] V. Sokolovskiy, Yu. Sokolovskaya, V.D. Buchelnikov, M. Zagrebin and A.T. Zayak, MATEC Web of Conferences, **33**, 05005 (2015).

[6] V.V. Sokolovskiy, Yu.A. Sokolovskaya, M.A. Zagrebin, V.D. Buchelnikov, A.T. Zayak, Mat. Sci. Forum **845**, 130 (2016).

[7] Yu.A. Sokolovskaya, V.V. Sokolovskiy, M.A. Zagrebin, V.D. Buchelnikov, A.T. Zayak, J. Exp. Theor. Phys. **125**, 104 (2017).

[8] G. Kresse, J. Furthmüller, Phys. Rev. B **54**, 11169 (1996).

[9] G. Kresse, D. Joubert, Phys. Rev. B **59**, 1758 (1999).

[10] J.P. Perdew, K. Burke and M. Enzerhof, Phys. Rev. Lett. **77**, 3865 (1996).

[11] T. Graf, C. Felser, and S.S.P. Parkin, Progr. Sol. St. Chem., **39**, 1 (2011).

 https://doi.org/10.1051/epjconf/201818506001

Shubnikov-de Haas oscillations in diluted magnetic semiconductors $(Cd_{1-x-y}Zn_xMn_y)_3As_2$

Vasilii S. Zakhvalinskii[1*], *Tatyana B.* Nikulicheva[1], *Erkki* Lähderanta[2], *Aleksey V.* Kochura[3,4], *Ekaterina A.* Nikitovskaia[1], *Evgeny A.* Pilyk[1,4], and *Mikhail A.* Pogrebneak[1]

[1]Belgorod National Research University, 85 Pobedy St, Belgorod, 308015, Russia
[2]Department of Mathematics and Physics, Lappeenranta University of Technology, PO Box 20, FIN-53852 Lappeenranta, Finland
[3]SouthWest State University, 94 50 Let Oktjabrja str., Kursk, 305040, Russia
[4]P.N. Lebedev Physical Institute of the RAS, 53 Leninskiyprosp., 119991, Moscow, Russia

Abstract. Single crystals of a diluted magnetic semiconductor $(Cd_{1-x-y}Zn_xMn_y)_3As_2$ (CZMA) ($x + y = 0.4$;y=0.04 and 0.08) obtained by Bridgman method were used. The Shubnikov-de Haas (SdH) effect was observed within studying of the dependence of the resistivity on the magnetic field in CZMA solid solutions. The values of the cyclotron mass m_c, Hall and Shubnikov carrier concentrations were calculated.

1 Introduction

Diluted magnetic semiconductors (DMS) are the semiconductors in which a part of the cations is replaced by transition or rare-earth elements with partially filled d- or f-shells, respectively. These elements are randomly located in the semiconductor matrix, what leads to several types of interactions: the first, *sp-d* (*sp-f*) exchange interaction between the band carriers and the localized magnetic moments of the atoms of transition (or rare-earth) elements; the second, *d-d*(*f-f*) exchange interaction directly between the atoms of these elements.

The most known and studied DMS are Mn based solutions of semiconductors synthesized by different methods including crystallization from a melt and a molecular beam epitaxy [1, 2].At a sufficiently high concentration of magnetic ions, a ferromagnetic order appears in such DMS. Using materials with high Curie temperature (T_C) extends the capabilities of modern microelectronics and gives an impetus to develop spintronics [3]. By now, DMS structures with T_C ~200 K based on p-$(Ga_{1-x}Mn_x)As$ have been obtained [4]. In some materials such as p-(Ga, Mn)P, p- and n-(Ga, Mn)N, ferromagnetism was also detected at temperatures exceeding the room temperature [5].

On the other hand, a group of DMS based on the 3D Dirac semimetal Cd_3As_2 [6] is of particular interest. Recently evidences of an existence of a topological states in $(Cd_{1-x}Zn_x)As_2$ solid solutions have been observed [7, 8]. The research of magnetic solid solutions based on Cd_3As_2allowsusto trace the evolution of topological properties and will create prerequisites for a practical application of this unique material. The purpose of this article was to investigate the properties of solid solutions of a diluted magnetic semiconductor $(Cd_{1-x-y}Zn_xMn_y)_3As_2$.

2 Experimental details, results and discussion

A modified Bridgeman method was used to obtain single crystals of$(Cd_{1-x-y}Zn_xMn_y)_3As_2$ (CZMA) ($x + y = 0.4$; y=0.04 and 0.08). The composition and homogeneity of the samples were analyzed by X-ray powder diffraction and energy-dispersive X-ray spectroscopy (EDX) methods. All the investigated samples $(Cd_{1-x-y}Zn_xMn_y)_3As_2$ had a tetragonal crystal system (α''-Cd_3As_2), space group $P4_2/nmc$ [9].

The Shubnikov-de Haas (SdH) effect was observed within studying the dependence of the resistivity on the magnetic field in CZMA solid solutions. Measurements were made on the pulsed magnetic field facility in the temperature range from 1.6 K to 300 K and the magnetic fields from 0 to 25 T. The SdH oscillations observed in the CZMA samples are shown in Figure.

Increasing of an amplitude of the SdH oscillations in the investigated samples $(Cd_{0.6}Zn_{0.36}Mn_{0.04})_3As_2$ and $(Cd_{0.6}Zn_{0.32}Mn_{0.08})_3As_2$ with at increasing a magnetic field is similar to that observed earlier for other CZMA compositions [10, 11].

The amplitude of the SdH oscillations can be written as [2]:

$$A \sim B^{-1/2} z / \mathrm{sh}(z) \exp\left[-2\pi^2 m_c k_B T_D / (\eta e B)\right] \cos(\pi \nu) \quad (1)$$

* Corresponding author: zakhvalinskii@bsu.edu.ru

where $z = 2\pi^2 m_c k_B T/(\hbar eB)$, T_D is the Dingle temperature, $\nu = gm_c/(2m_0)$, m_c is the cyclotron mass, m_0, $\hbar$, e and k_B are the physical constant.

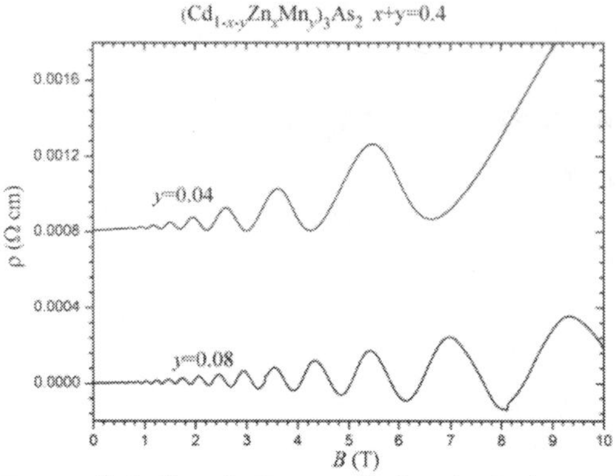

Figure. Shubnikov-de Haas oscillations in single crystals $(Cd_{1-x-y}Zn_xMn_y)_3As_2$($x + y = 0.4$;$y = 0.04$ and 0.08 at $T = 2.4$ K).

According to the results of [10], in our samples the g-factor should not depend on a temperature. So, an analysis of the amplitudes of the Shubnikov-de Haas oscillations using Eq. (1) and the values of the SdH amplitudes measured at two different temperatures at a fixed value of the magnetic field, as in conventional semiconductors, makes it possible to determine a number of parameters, for example, the cyclotron mass [9].The comparison of charge carrier concentrations obtained from Hall and Shubnikov measurements was carried out (Table). The Hall carrier concentration was calculated as follows [13]:

$$n_R = 1/(eR_H) \qquad (2)$$

where n_R is the concentration of charge carriers,e is electron charge and R_H is Hall coefficient.

Table. Parameters found from the SdH oscillations of CZMA samples ($x + y = 0.4$; $y = 0.04$ and 0.08) and for undoped Cd_3As_2 [15, 16].

y(Mn)	0.04	0.08	Cd_3As_2
n_R/n_{SdH}	0.97	1.04	1.2 [15]
$\mu_H \cdot 10^{-4}$, $cm^2V^{-1}s^{-1}$	2.28	1.53	2.9 [16]
P_{SdH}, T^{-1}	0.061	0.025	0.02 [16]
$m_c(0)/m_0$	0.0409	0.0435	0.043 [16]

Shubnikov carrier concentration was calculated according to the equation [14]:

$$n_{SdH} = \frac{1}{3}\pi^2\left(\frac{2e}{\hbar P_{SdH}}\right)^{3/2} \qquad (3)$$

where P_{SdH} is the period of the SdH oscillations, the ratio n_R/n_{SdH} is close to unity, in accordance with results obtained previously for CZMA [10,11,13]. When considering the properties of solid solutions of $(Cd_{1-x-y}Zn_xMn_y)_3As_2$, we were based on the similarity of the band structure of Cd_3As_2 and InSb [16].

3 Conclusions

In this paper we performed an investigation of the Shubnikov de Haas effect in single crystals of solid solutions of diluted magnetic semiconductors $Cd_{1-x-y}Zn_xMn_y)_3As_2$($x + y = 0.4$; y=0.04 and 0.08), obtained by Bridgman method based on 3D Dirac semimetals Cd_3As_2. For both samples the value of the cyclotron effective mass of charge carriers m_c, Shubnikov n_{SdH} and Hall n_R concentration, mobility, μ_H, of charge carriers and period of oscillations P_{SdH}, were defined (Table). The obtained parameters do not conflict with similar results obtained previously for diluted magnetic semiconductors CZMA (x+y = 0.2; x+y = 0.3) [10, 11, 13].

This work was supported by the Russian Science Foundation (grant №17-12-01345) in the part of crystal growth and structure characterization.

References

1. J. Cisowski.Phys. Stat. Sol. B**200**, 311 (1997).
2. J. K. Furdyna and J. Kossut.*Diluted Magnetic Semiconductors*, (Ed.Semicond. AndSemimet., Vol. 25, Academic Press, Boston, 1988).
3. I.Zutic, J.Fabian, S.D.Sarma.Reviews of Modern Physics **76**, 2, 323 (2004).
4. L.Chen, X.Yang, F.Yang, J.Zhao, J.Misuraca,P.Xiong, S.von Molnar.Nano.Lett. **11**, 2584(2011).
5. K.A. Kikoin and V.N. Fleurov.*Transition Metal Impurities in Semiconductors*(WorldScientific, Singapore, 1994).
6. Z. Wang, H. Weng, Q. Wu, X. Dai, Z. Fang.Phys. Rev. B**88**,125427(2013).
7. A.V. Galeeva, I.V. Krylov, K.A. Drozdov, A.F. Knjazev, A.V. Kochura, A.P. Kuzmenko, V.S. Zakhvalinskii, S.N. Danilov, L.I. Ryabova, D.R. Khokhlov. Beilstain J. Nanotechnol. **8**, 167 (2017).
8. H. Lu, X. Zhang, Y. Bian, Sh. Jia. Scientific Reports **7**, 3148 (2017).
9. ICSD Database, Version 2009-1, Ref. code 23245.
10. R.Laiho, K.G.Lisunov, V.N.Stamovand V.S. Zahvalinskii.J. Phys. Chem. Solids**57**, 1, 1, (1996).
11. R.Laiho, E.Lahderanta,K.G.Lisunov, V.N. Stamov and V.S. Zahvalinskii. J. Phys. Chem. Solids. **58**, 5, 717 (1997).
12. I.M. Tsidilkovsky.*Electrons and Holes in Semiconductors*,(Nauka, Moscow,1972).
13. R.Laiho, K.G.Lisunov, M.L. Shubnikov, V.N.Stamovand V.S. Zahvalinskii.Phys. Stat. Sol. B**198**, 135(1996).
14. B.I. Shklovskii and A.L Efros. Electronic Properties of Doped Semiconductors, Springer-Verlag, Berlin (1984).
15. .M. Tsidilkovsky. Gapless semiconductors – a New Class of Materials, Academie, Berlin (1988).
16. E.O. Kane. Band structure of indium antimonide. J. Phys. Chem. Sol. **1**, 249 (1957).

Effect of ageing on magnetic and electrical properties of Ti–substituted La-Sr manganites

A. Badelin[1,*], Z. Datskaya[1], S. Estemirova[2], V. Karpasyuk[1], D. Merkulov[1]

[1]Astrakhan State University, 414056 Astrakhan, Russia
[2]Institute for Metallurgy UB RAS, 620016 Yekaterinburg, Russia

Abstract. The changes of structural, magnetic and electrical parameters of ceramic $La_{0.8-x}Sr_{0.2+x}Mn_{1-x}Ti_xO_{3+\gamma}$ ($0.025 \leq x \leq 0.150$) manganites during long-term storage at room temperature and after thermal cycling were investigated. The unit cell volume (V) of all manganites decreased over time. Most considerable decrease of V and broadening of X-ray diffraction lines width were observed in the samples with high quantity of titanium. During the storage period magnetization (σ) and Curie point (T_c) of manganites with relatively low Ti content decreased, while the compositions with high values of "x" exhibited some trend of σ and T_c increase. The resistance of the samples with $x < 0.100$ increased over time. Fluctuations of magnetization of all thermal-aged samples were within the error of measurement, and Curie point of manganites with $x = 0.150$ showed the rise. From the data obtained it might be inferred that ageing phenomena in Ti-substituted manganites can be associated with oxidation processes, the transfer of Ti^{4+} to vacant sites in rare-earth sublattice and formation of microinhomogeneities.

1 Introduction

The development of new perspective materials for magnetoelectronics and spintronics, such as ferrites, manganites exhibiting colossal magnetoresistance or manganite electroceramics for thermistors, involves the achieving a number of valuable properties, among which stability of functional parameters in relation to thermal and prolonged storage ageing is of special importance [1-3]. However, the data concerning the ageing processes in bulk ceramic perovskite-like manganites are practically absent in the literature. The long-term stability of resistive and magnetic properties was investigated in nanocompounds p-$La_{0.8}Mn_{1.04}O_{3.5}$ [4] and in nanostructured La-Sr manganite thin films [5]. At the same time, one may specify ferrites [1, 6] and manganites with spinel structure for thermistor [3, 7] as types of polycrystalline oxide materials in which various effects of ageing have been extensively investigated. Generally, in considered materials the drift of their characteristics during prolonged storage in the absence of thermodynamic equilibrium under specified conditions may be due to various processes: interaction with the oxygen in surrounding atmosphere leading to the changes in the states of variable valency ions and concentration of nonstoichiometry defects; ordering or redistribution of cations and vacancies on the sublattices and in the volume of the crystals; relaxation of elastic stresses or grain boundaries, etc. The introduction of micro-components in the complex non-stoichiometric crystals may have different effects on the solid-state reactions and on exchange of oxygen with the ambient atmosphere depending on the nature of impurities [1,8-10].

Here, we present the experimental investigations on the in-time and thermal ageing of Ti-substituted manganites of specifically designed system $La^{3+}_{0.8-x}Sr^{2+}_{0.2+x}Mn^{3+}_{0.8-x-2\gamma}Mn^{4+}_{0.2+2\gamma}Ti^{4+}_xO_{3+\gamma}$ ($0.025 \leq x \leq 0.150$), in which Mn^{4+} content is independent of "x". These manganites under synthesis conditions in our experiments have excess oxygen content ($\gamma > 0$) [9-11], and after cooling they are not in a state of thermodynamic equilibrium. The transition into energetically more favorable state is a general cause of the ageing processes.

2 Experimental

Experimental samples of polycrystalline manganites have been prepared by traditional ceramic techniques in air, including operations of double ground with addition of ethanol, preliminary burning, introduction the binder (aqueous solution of polyvinyl alcohol), pressing the pelletes, final sintering at 1473 K for 10 h followed by cooling of the samples in the furnace. Procedures of synthesis are described in more detail in Ref. [9, 11].

In-time ageing was studied by periodical determination of the properties of manganites during long-term (up to 33 months) storage of initial samples at room temperature in air. Thermal ageing phenomena were investigated using five cycles "annealing for 6 hours at 573 K-cooling".

X-ray studies were carried out on the diffractometer Shimadzu XRD-7000 in CuK_α radiation.

* Corresponding author: alexey_badelin@mail.ru

Magnetization per mass unit (σ) was measured in magnetic field of 5.6 kOe with accuracy to 1.5-2%. Curie point (T_c) was determined as the temperature corresponding to the maximum of $|d\mu(T)/dT|$, where $\mu(T)$ is magnetic permeability as a function of temperature. The measurement error depended on the width of ferromagnetic-paramagnetic transition was not more than 5 K.

Measurements of dc electrical characteristics and magnetoresistance (MR=(R(H)-R(0))/R(H)) were made using copper contacts sputter-deposited onto opposite planes of pellets (thickness of about 4 mm). Magnetic field (H = 9.2 kOe) was parallel to the current direction.

3 Results and discussion

All samples were found to be single phased and rhombohedral. The unit cell volume (V) of all manganites decreased over time (Table 1). Most considerable decrease of V was observed in the samples with high quantity of titanium, which must reduce overstoichiometric oxygen concentration in sintered manganites [9-11].

Table 1. Unit cell volume of manganites in initial state and after 33 months.

Chemical composition	V, Å^3	
	Initial samples	Aged samples
$La_{0.750}Sr_{0.250}Mn_{0.950}Ti_{0.050}O_{3+\gamma}$	352.9	352.6
$La_{0.700}Sr_{0.300}Mn_{0.900}Ti_{0.100}O_{3+\gamma}$	353.2	352.9
$La_{0.650}Sr_{0.350}Mn_{0.850}Ti_{0.150}O_{3+\gamma}$	353.5	352.9

Variations of diffraction reflexes width during the storage of manganites (Table. 2) were dependent on their composition.

Table 2. Full width (W) at half maximum of diffraction reflexes of $La_{0.8-x}Sr_{0.2+x}Mn_{1-x}Ti_xO_{3+\gamma}$ manganites.

x	(HKL)	Initial samples		After 33 months	
		2 Θ,°	W,°	2 Θ,°	W,°
0.050	$(10\bar{2})$	22.832	0.154	22.856	0.153
	(110)	32.405	0.152	32.424	0.150
	(116)	52.743	0.170	52.763	0.166
	(214)	57.977	0.178	57.993	0.174
	(208)	68.433	0.205	68.452	0.196
0.100	$(10\bar{2})$	22.828	0.191	22.846	0.187
	(110)	32.416	0.198	32.432	0.191
	(116)	52.712	0.228	52.733	0.238
	(214)	57.975	0.238	57.996	0.252
	(208)	68.380	0.263	68.407	0.300
0.150	$(10\bar{2})$	22.832	0.178	22.854	0.191
	(110)	32.423	0.200	32.447	0.209
	(116)	52.689	0.258	52.716	0.278
	(214)	57.972	0.273	58.000	0.299
	(208)	68.346	0.316	68.375	0.345

At low titanium content (x = 0.050) all reflexes narrowed over time and at high content (x = 0.150) they broadened. At x = 0.100 the width of reflexes in the range of small angles decreased and in large angles it increased.

The values of magnetic parameters of manganites measured at certain intervals of time after beginning of the storage are presented on Figure 1. The following values of symbol parameter τ given on the x-axis of the figure are: τ = 0 – start of timing, it fit the parameters of the initial samples; τ = 1 means 1 month of storage, τ = 2 – 7 months, τ = 3 – 18 months, τ = 4 – 33 months.

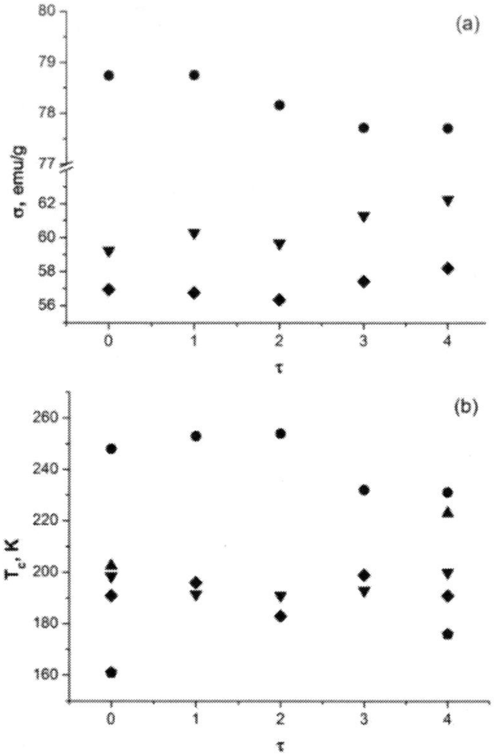

Fig. 1. Magnetization (a) and Curie point (b) of manganites $La_{0.8-x}Sr_{0.2+x}Mn_{1-x}Ti_xO_{3+\gamma}$ at different points in storage time: ● – x = 0.050; ▲ – x = 0.075; ▼ – x = 0.100; ◆ – x = 0.125; ⬟ – x = 0.150.

As it can be seen from the Figure 1, magnetization and Curie point of manganites with relatively low Ti content decreased during the storage period, while the compositions with high values of "x" exhibited some trend of σ and T_c increase. As an example of one conclusion concerning T_c, temperature dependencies of magnetic permeability of two samples are cited on Figure 2(a).

The resistance of all samples, with the exception of compositions with x ≥ 0.100, increased during storage period. Figure 2(b) illustrates the qualitatively different changes of the resistance of manganites with x = 0.025 and x = 0.100.

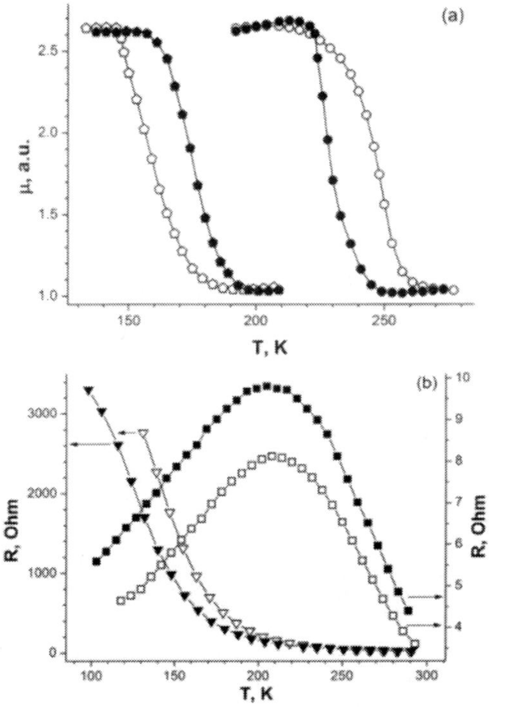

Fig. 2. Temperature dependencies of magnetic permeability (a) and electrical resistance (b) of manganites in initial state (light symbols) and in the end of storage period (black symbols): (○,●) – x = 0.050; (⬠,⬟) – x = 0.150; (□,■) – x = 0.025; (▽,▼) – x = 0.100

The curves on Figure 2(b) for x = 0.025 demonstrate that in manganite of this composition "metal-semiconductor" transition temperature lowered as a result of in-time ageing.

Magnetoresistance at different temperature intervals changed irregularly during the storage period. Table 3 shows maximumal values of |MR| and corresponding to them temperatures (T_{MR}) for initial and aged samples.

Table 3. Maximal absolute values of magnetoresistance at corresponding temperatures.

x	$\|MR\|_{max}$,%		T_{MR}, K	
	Initial samples	Aged samples	Initial samples	Aged samples
0.025	14	17	118	130
0.050	13	15	130	96
0.075	20	23	116	100
0.100	35	22	140	111
0.150	36	23	110	105

T_{MR} of the majority of the samples decreased during storage time, with the exception of manganite contained 0.025 formula units of Ti. Maximal |MR| values of manganites with low titanium content did not almost changed, while $|MR|_{max}$ of compositions with x ≥ 0.100 decreased slightly.

Variations of magnetic characteristics of manganites in the process of frequentative thermal cycling are presented on the Figure 3 (a,b). One can see that magnetization fluctuations of all samples were within the error of measurement, and manganite with x = 0.150 exhibited the rise of Curie point.

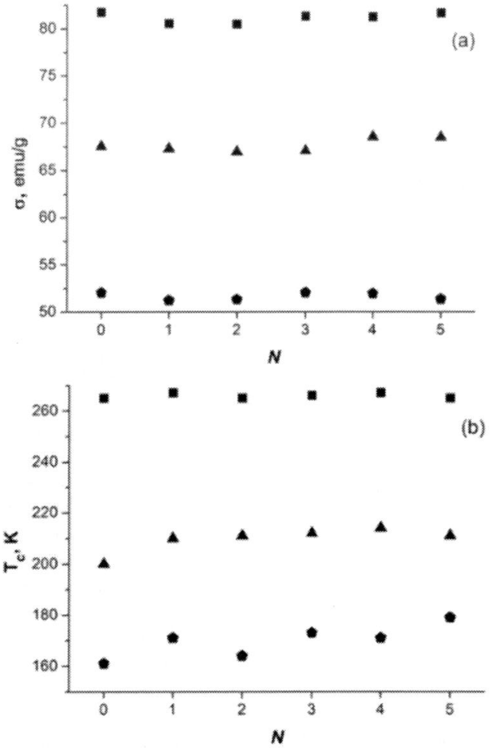

Fig. 3. Magnetization (a) and Curie point (b) of thermal-cycled manganites $La_{0.8-x}Sr_{0.2+x}Mn_{1-x}Ti_xO_{3+\gamma}$: ■ – *x* = 0.025; ▲ – *x* = 0.075; ⬟ – *x* = 0.150; the values of *N* denote the numbers of cycles (*N* = 0 corresponds to the initial sample).

Totality of obtained experimental data seems to agree with following notions.

From the viewpoint of thermodynamics, at the temperature lower than sintering one manganites must absorb the oxygen. Oxidation process causes $Mn^{3+} \rightarrow Mn^{4+}$ transition and formation of cation vacancies. As a result, unit cell volume decreases, and concentration of (Mn^{4+},Mn^{3+}) pairs bonded by double exchange interaction increases, which should lead to higher ferromagnetic characteristics and conductivity. At the same time, the decrease of magnetic moment of the manganese ions and the dilution of octahedral sublattice by cation vacancies are competing factors. Difference of radii and charges of the ions, Coulomb interaction cause the formation of microinhomogeneities and microstresses [8, 9, 12], broadening X-ray diffraction reflexes. At room temperature the contribution of entropy to free energy (-TS term) is less than at thermal processing, and separation of inhomogeneities becomes more evident, which can be promoted by the presence of Ti impurity.

Narrowing of diffraction reflexes over time in manganites with low titanium content requires further research.

Another important factor affecting the properties of the manganites is the transfer of Ti^{4+} to vacant sites in rare-earth sublattice. For this ion, due to the large electric charge, higher coordination may be favorable. The possibility of transition of the cations from octahedral sublattice to sublattice with coordination number of 12 was considered in the works [13, 14]. As a result of the transfer of Ti^{4+} ions unit cell volume decreases, diamagnetic dilution of Mn sublattice reduces, double exchange bonds recover, and eventually Curie point and conduction raise.

However, due to the oxidation of copper electrodes by the oxygen from manganites, resistance value of the contact zones may increase.

4 Conclusion

Analysis of adduced data permitted establishing the following regularities:

- during the storage period unit cell volume of all samples decreased, magnetization and Curie point of manganites with relatively low Ti content lowered, while the compositions with high values of "x" exhibited some trend of σ and T_c increase;
- resistance of the samples with $x < 0.100$ increased and with $x \geq 0.100$ decreased during storage period;
- maximal absolute values of magnetoresistance changed insignificantly in course of time;
- at low titanium content all X-ray reflexes narrowed over time and at high content they broadened;
- in the process of frequentative thermal cycling variations of magnetic characteristics of manganites were within the error of measurement, with the exception of compositions with $x = 0.150$ exhibited the rise of Curie point.

On the whole, it can be concluded that in-time and thermal ageing of manganites leads to the changes of their magnetic parameters within no more than 10 %.

Changes in electrical characteristics are more significant, but most likely they are due to oxidation processes in manganite-electrode interface.

Further work will be required to investigate more circumstantially the mechanisms of variations of diffraction reflexes width, electrical characteristics taking into consideration degradation of the contact zones, and to elucidate the role of the changes in microstructure.

This research was supported by the Ministry of Education and Science of RF (project 334).

References

1. A. Goldman, *Modern Ferrite Technology* (2nd Ed. – Pittsburg, PA, USA: Springer, 2006)
2. I.A. Abdel-Latif, J. Phys. **1(3)**, 15 (2012)
3. M. Vakiv, I. Hadzaman, O. Shpotyuk, O. Mrooz, J. Plewa, H. Altenburg, H. Uphoff, O. Bodak, P. Demchenko, J. Eur. Ceram. Soc. **24**, 1277 (2004).
4. A.I. Rykova, A.V. Terekhov, A.S. Cherny, E.N. Khatsko, A.V. Yeremenko, A. Zaleski, Low Temp. Phys. **38(6)**, 529 (2012).
5. L. Žurauskaitė, S. Balevičius, N. Žurauskienė, S. Keršulis, V. Stankevič, Č. Šimkevičius, J. Novickij, S. Tolvaišienė, Lithuanian J. Phys. **52(3)**, 224 (2012).
6. S. Krupićka, *Physik der ferrite und der verwandten magnetischen oxide* (Prag.: Academia, 1973)
7. Dao-lai Fang, Cui-hong Zheng, Chu-sheng Chen, A.J.A. Winnubst, J. Electroceram. **22(4)**, 421 (2009)
8. Z.R. Musaeva, N.A. Vybornov, V.K. Karpasyuk, A.M. Smirnov, L.S. Uspenskaya, S.Kh. Yazenkov, Journal of Surface Investigation. X-ray, Synchrotron and Neutron Techniques **1(4)**, 423 (2007)
9. A. Badelin, Z. Datskaya, I. Epifanova, S. Estemirova, V. Karpasyuk, A. Smirnov, Europ. Phys. J. (EPJ Web of Conf.) **40**, 15004 (2013)
10. O.Z. Yanchevskii, O.I. V'yunov, A.G. Belous, A.I. Tovstolytkin, V.P. Kravchik, Phys. Sol. State **48(4)**, 709 (2006)
11. D. Merkulov, A. Badelin, S. Estemirova, V. Karpasyuk, Acta Phys. Pol. A **127(2)**, 248 (2015)
12. P.G. Radaelli, R.M. Ibberson, D.N. Argyriou, H. Casalta, K.H. Andersen, S.W. Cheong, J.F. Mitchell, Phys. Rev. B **63**, 172419 (2001)
13. J. Mizusaki, N. Mori, H. Takai, Y. Yonemura, H. Minamiue, H. Tagawa, M. Dokiya, H. Inaba, K. Naraya, T. Sasamoto, T. Hashimoto, Solid State Ionics **129**, 163 (2000)
14. R.A. De Souza, M.S. Islam, E.Ivers-Tiffee, J. Mater. Chem. **9**, 1621 (1999)

Magnetotransport properties of InSb-MnSb nanostructured films

Aleksey Kochura[1,*], *Sergey* Marenkin[2], *Oleg* Novodvorsky[3], *Vladimir* Mikhalevsky[3], *Alexander* Davidov[4], *Mikhail* Shakhov[5,6], *Erkki* Lahderanta[6] *Vasilii* Zakhvalinskii[7], and *Alexander* Kuzmenko[1]

[1]SouthWest State University, 94 50 Let Oktjabrja str., Kursk, 305040, Russia
[2]Kurnakov Institute of the general and inorganic chemistry of the RAS, 31 Leninskii prosp., 119991, Moscow, Russia
[3] Institute on Laser and Information Technologies, RAS, 1 Svyatoozerskaya ul., Shatura, Moscow region, 140700 Russia
[4] P.N. Lebedev Physical Institute of the RAS, 53 Leninskiy prosp., 119991, Moscow, Russia
[5]A. F. Ioffe Physico-Technical Institute, 26 Politekhnicheskaya, St Petersburg 194021, Russian Federation
[6]Department of Mathematics and Physics, Lappeenranta University of Technology, PO Box 20, FIN-53852 Lappeenranta, Finland
[7]Belgorod National Research University, 85 Pobedy St, Belgorod, 308015, Russia

Abstract. Hybrid nanostructured InSb - MnSb films were obtained by the pulsed laser deposition using the mechanical droplet separation. Films structure was characterized by different methods (electron diffraction, scanning electron microscopy, atomic and magnetic force microscopy). The negative magnetoresistance (nMR) takes place below 100 K. This temperature is several times more than the temperature at which the nMR occurs in homogenous $In_{1-x}Mn_xSb$ films. At low temperatures the spin-dependent scattering of the holes by the localized Mn^{2+} moments prevails. When the temperature rises, the low nMR is observed due to the weak spin-dependent scattering on magnetic inclusions.

1 Introduction

Incorporating of local magnetic moments into a semiconductor matrix is one of suitable ways for a realization of manipulating by electron and spin subsystems [1]. The highest Curie temperature (T_C) in a dilute magnetic semiconductor is about 200 K, as found for $Ga_{1-x}Mn_xAs$ [2]. Ferromagnetic nanoclusters embedding in a nonmagnetic or paramagnetic semiconductor can help to rich the room temperature ferromagnetism. Such hybrid systems with coexisting of a ferromagnetic phase and a semiconducting matrix demonstrate interesting properties and can be used to fabricate spintronic devices [3, 4]. Incorporating of ferromagnetic nanoscale precipitations into III-V semiconductors with different dimensions: nanowires (1D), layers (2D) and bulk samples (3D) can substantially alter the observed magnetic characteristics and galvano-magnetic effects [3 - 7]. MnAs and MnSb are high temperatures ferromagnetic semimetals and formed nanoclusters with suitable structure compatibilities to III-V compounds [8]. It is shown that MnAs/III-V structures with the huge positive magnetoresistance and the robust tunnel megnetoresistance effect may find an application in high sensitivity magnetic sensors and other magneto-electric devices [3, 9]. The critical temperature of MnAs T_C = 317 K is slightly higher than the room temperature. So, MnSb with T_C = 585 K is a more available candidate for applications with operating temperatures of traditional electronic materials. GaSb/MnSb nanolayers were grown earlier by the molecular beam epitaxy [10 - 12], the pulsed laser deposition (PLD) [13] and the magnetron co-sputtering [14] methods. They demonstrate ferromagnetism and spin- depended scattering above the room temperature. An influence of the ferromagnetic MnSb inclusions on magnetic and electrical properties of InSb/MnSb nanostructures was observed both in bulk [15-17] and in films [18, 19].

Here we report the results about structural, magnetic, electrical and magnetotransport properties of magnetic eutectic InSb-MnSb nanostructured films grown by PLD method.

2 Results and discussion

2.1 Experimental results

Thin hybrid InSb - MnSb films were obtained by PLD. The mechanical droplet separation was used to avoid the droplet deposition. In details the growth procedure is described in Ref. 13.

The InSb-MnSb system has an eutectic near 6.5 mol% MnSb [20]. At first we grew by the Bridgman method the eutectic composition consisting of InSb single crystal matrix and faceted single-crystal MnSb inclusions and used it as a starting material to prepare the target.

Polished and oriented [0001] Leuco sapphire L-Al_2O_3 with dimensions 4 × 5 mm^2 have been used as substrates.

Structural properties and compositions of films were examined by the Tecnai G2 20F S-T (FEI)

* Corresponding author: akochura@gmail.com

scanning transmission electron microscope (STEM) with the high angle annular dark field (HAADF) detector and the scanning electron microscope (SEM) JSM-6610LV (Jeol) with the attachment for the energy dispersive X-ray spectroscopy (EDXS) X-MaxN (Oxford Instruments). The morphology and magnetic properties of films surface were determined using an AIST NT Smart SPM high resolution scanning probe microscope. Magnetization of the samples was measured in fields up to B = 5 T at temperatures T = 3 - 310 K using a superconducting quantum interference device (SQUID) magnetometer (model S600, Cryogenics ltd.). The electrical and magnetotransport properties were investigated between 4.2 K and 320 K using the standard six-point geometry in pulsed magnetic fields up to 20 T. The resistance was measured with a high accuracy (the error was at most 1 % of the measured value). The films had conductivity of p-type and linear current-voltage characteristics in the range of the measurements. The films thicknesses were about 50 nm.

2.2 Analysis of the results

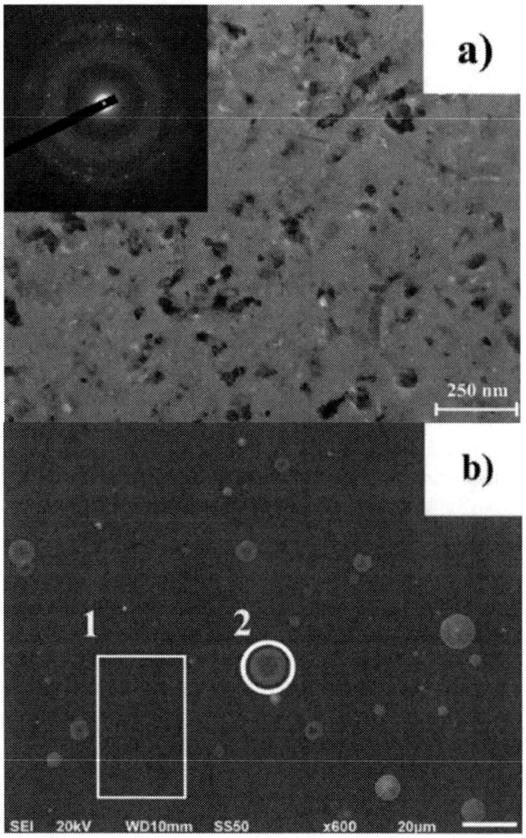

Fig. 1. Typical STEM (a) and SEM (b) images of InSb - MnSb films. The inset a) shows a ring structure of an electron diffraction pattern. EDXS analysis was performed for two different areas (b) labeled as 1 and 2

The microstructure observation showed a non homogeneous structure of InSb - MnSb film in different size scales (Fig. 1). The film is polycrystalline (Fig. 1a, inset) and nanostructured (Fig. 1a) with 30 – 100 nm size of granules and nanoinclusions. Less than 10 % of the surface is covered by drops with diameters from 1 to 15 mkm. All films surface contains In, Sb, Mn with the elements ratio close to a composition of the target and homogeneous distribution within EDXS size accuracy (1 mkm). MnSb content is slightly lower only in drops (Table).

Table. A surface elemental composition of InSb-MnSb film in different areas labelled on Fig.1b

Area	At. percent		
	In	Sb	Mn
1	44.5	49.8	5.7
2	45.7	49.4	4.9

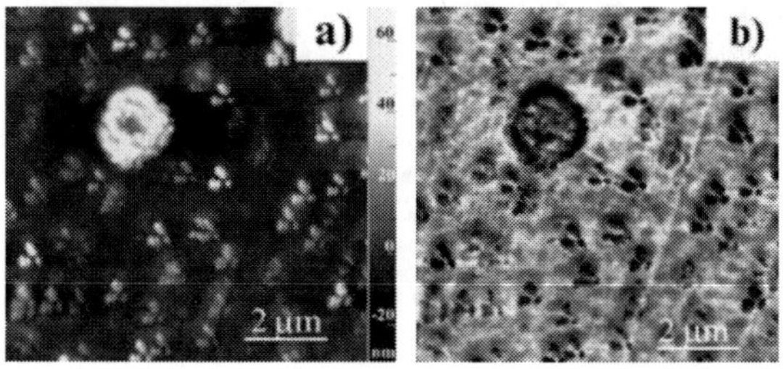

Fig. 2. Images of a site of the surface for the InSb-MnSb film obtained by the SPM in the mode of atomic force (a) and magnetic force (b) microscopy at 313 K.

The typical images of the InSb-MnSb film surface obtained by the scanning probe microscopy (SPM) at the room temperature are given in Fig. 2. The presence of ferromagnetic inclusions is well visible. The sizes of ferromagnetic granules measured by the SPM in the magnetic force regime (30 – 200 nm) were comparable with sizes obtained from TEM (Fig. 1a).

The electrical resistivity of the film had a semiconductor character and decreased with increasing a temperature in the range from 1.6 to 300 K. (Fig. 3). At a low temperature range the temperature dependence on resistivity is $\rho(T) = \rho_0 \exp[E_A/(kT)]$, where ρ_0 – pre-exponential factor, E_A – activation energy, k – Boltzmann constant. [21] So the local activation energy $E_A \equiv d\ln\rho/d(kT)^{-1} = 0.30$ meV we can found from low-temperature linear segment of the plots $\ln\rho$ versus $(kT)^{-1}$ (Fig 3, Inset). It is in a good agreement with $E_A = 0.28$ meV observed earlier in [22] for p – InSb(Mn) monocrystals at $N_{Mn} = 1.6 \cdot 10^{17}$ cm^{-3}.

The field dependences of magnetoresistance ($\Delta\rho/\rho = (\rho_B - \rho_0)/\rho_0$) of the InSb-MnSb film at different temperatures are shown in Fig. 4. The maximal negative magnetoresistance $\Delta\rho/\rho \approx -20\%$ was observed at $T = 4.2$ K. At $T_n < 100$ K magnetoresistance consisted of a negative component (nMR) prevailing up to 20 T, increased with increasing of a magnetic field and tended to the saturation at low temperatures. Note that T_n is

 https://doi.org/10.1051/epjconf/201818506003

several times more than the temperature at which the nMR is occur in homogenous $In_{1-x}Mn_xSb$ films [23].

The scattering of spin-polarized carriers on isolated Mn^{2+} ions at temperatures as below as above T_C leads to decreasing resistivity. When $T < T_C$ ferromagnetic ordered regions increase as T decreases but some of fractions of the spins can still remain outside ferromagnetic network and serve as a source of magnetic scattering [23]. In this case magnetoresistance is proportional both to the spin polarization P arising from the ferromagnetic block and to magnetization of the isolated paramagnetic moments $\Delta\rho/\rho \sim -4P\,M(B)$ [23].

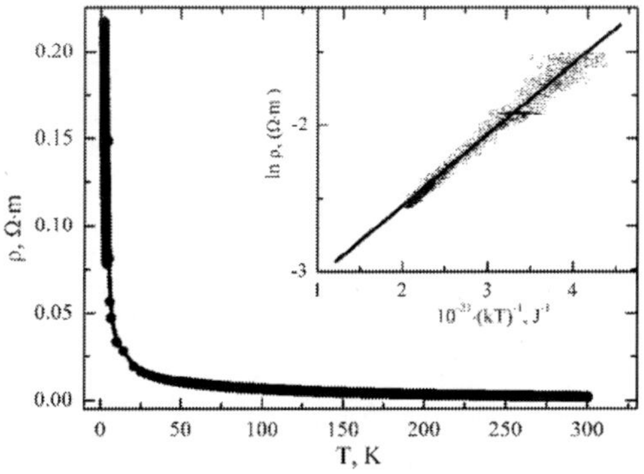

Fig. 3. A temperature dependence of the resistivity of InSb-MnSb film. Inset: linear segment of the plots of $\ln\rho$ versus $(kT)^{-1}$

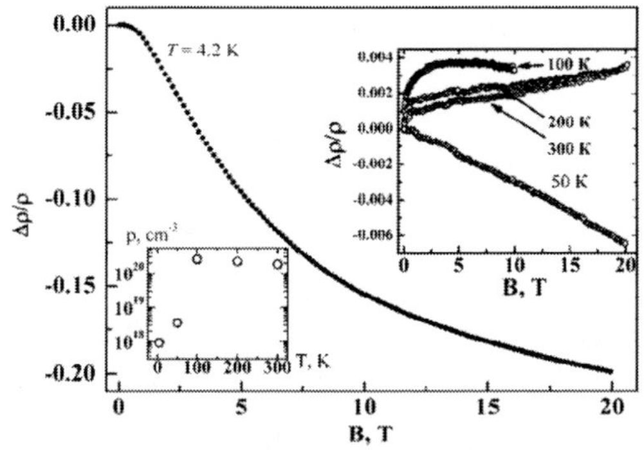

Fig. 4. A magnetic field dependence of the magnetoresistance for InSb-MnSb film at T = 4.2 K. Left inset: The plots of the hole concentration versus temperature. Right inset: Magnetoresistance as a function of magnetic field for different temperatures.

The second approach to a description of the nMR in our compounds is similar to one for granular paramagnetic-ferromagnetic GaAs:Mn/MnAs hybrid layers with nano-scaled MnAs clusters embedded in GaAs:Mn matrix [24]. Total magnetoresistance in these systems is formed by competition of negative and positive terms and paramagnetism of matrix play the sufficient role. At low temperatures in the external magnetic field B_{ext}=0 the holes are trapped near cluster interface due to the local giant Zeeman splitting and form a bound magnetic polarons (it is accompanied by decreasing hole concentration with decreasing of a temperature (fig. 4, the left inset). When $B_{ext} = 0$ giant Zeeman splitting spreads into paramagnetic GaAs:Mn matrix resulting in aligning of carrier spins and occurring of the negative magnetoresistance. A source of the positive magnetoresistance (pMR) in granular paramagnetic-ferromagnetic systems is a difference of a spin orientation of clusters and majority carriers [25]. The thermal disorder destroys the preferential spin orientation of the Mn ions and the positive magnetoresistance decreases for higher temperatures.

As temperature increases the magnetoresistance follows the second scenario: negative magnetoresistance decreases, becomes positive, reaches the maximum and then slowly decreases (Fig 4, right inset). The same behavior with smaller pMR was observed for (In,Mn)Sb crystals containing microsized MnSb clusters [26] and with the same and bigger pMR for GaAs:Mn/MnAs [26, 27]. At the same time pMR depended on both a size and a distribution of clusters. For clusters lining up perpendiculary to the current pMR reached 800 % and increased with decreasing clusters volumes [27, 28].

This work was supported by the Ministry of Education and Science of the Russian Federation (№ 16.2814.2017/PCh) and Russian Foundation for Basic Research (15-07-06081-a).

References

1. T. Dietl, H. Ohno, Rev. Mod. Phys., **86**, 187 (2014).
2. L. Chen, X. Yang, F. Yang, J. Zhao, J. Misuraca, P. Xiong, S. von Molnar, Nano Lett., **11**, 2584 (2011).
3. M.T. Elm, S. Hara, Advanced Materials, **26**, 8079 (2014).
4. S. Hara, K. Komagata, Physica Status Solidi (b), **252**, 1925 (2015).
5. K. Kabamoto, R. Kodaira, S. Hara, Journal of Crystal Growth, **464**, 80 (2017).
6. S.F. Marenkin, A.D. Izotov, I.V. Fedorchenko, V.M. Novotortsev, Russian Journal of Inorganic Chemistry, **60**, 295 (2015).
7. M. Khalid, S. Prucnal, M.O. Liedke, K. Gao, S. Facsko, W. Skorupa, M. Helm, S. Zhou, Materials Research Express, **1**, 026105 (2014).
8. T. Dietl, K. Sato, T. Fukushima, A. Bonnani, M. Jamet, A. Barski, S. Kuroda, M. Tanaka, P. N. Hai, H. Katayama-Yoshida, Rev. Mod. Phys., **87**, 1311 (2015).
9. P.N. Hai, Sh. Ohya, M. Tanaka, S.E. Barnes, S. Maekawa, Nature, **458**, 489 (2014).
10. J. Choi, J. Choi, S. Choi, S.C. Hong, S. Cho, M.H. Sohn, Y. Park, K.-W. Lee, H.-M. Park, J.H. Song, J.B. Ketterson. Journal of the Korean Physical Society, **47**, S497 (2005).
11. E. Dynowska, J. Bak-Misiuk, P. Romanowski, J.Z. Domagala, J. Sadowski, T. Wojciechowski, S. Kret, B. Kurowska, A. Kwiatkowski, W. Caliebe. Radiation Physics and Chemistry, **80**, 1051 (2011).

12. W. Braun, A. Trampert, V.M. Kaganer, B. Jenichen, D.K. Satapathy, K.H. Ploog. Journal of Crystal Growth, **301-302**, 20 (2007).
13. S.F. Marenkin, O.A. Novodvorsky, A.V. Shorokhova, A.B. Dasvydov, B.A. Aronzon, A.V. Kochura, I.V. Fedorchenko, O.D. Khramova, A.V. Timofeev, Inorganic Materials, **50**, 973 (2014).
14. J.A. Calderon, F. Mesa, A. Dussan, Applied Surface Science, **396**, 1113 (2017).
15. K. Ganesan, S. Mariyappan, H.L. Bhat, Sol. St. Com., **143**, 272 (2007).
16. A.V. Kochura, B.A. Aronzon, K.G. lisunov, A.V. Lashkul, A.A. Sidorenko, R. De Renzi, S.F. Marenkin, M. Alam, A.P. Kuzmenko, E. Lahderanta. J. Appl. Phys., **113** 083905 (2013).
17. E. Lahderanta, A.V. Lashkul, A.V. Kochura, K.G. Lisunov, B.A. Aronzon, M.A. Shakhov, Phys. Status Solidi A **211**, 991 (2014).
18. L. Lari, S. Lea, C. Feeser, B.W. Wessels, V.K. Lasarov, J. Appl. Phys., **111**, 07C311 (2012).
19. J. Liu, M.P. Hanson, J.A. Peters, B.W. Wessels, ACS Applied Materials and Interfaces, **7**, 24159 (2015).
20. V.M. Novotortsev, A.V. Kochura, S.F. Marenkin, I.V. Fedorchenko, S.V. Drogunov, A. Lashkul, E. Lahderanta, Russian Journal of Inorganic Chemistry, **56**, 1951 (2011).
21. B. I. Shklovskii, A. L. Efros. *Electronic Properties of Doped Semiconductors* (Berlin: Springer. 1984).
22. S.A. Obukhov, S.W. Tozer, W.A. Coniglio. Scientific Reports, **5**, 13451 (2015).
23. M. Csontos, T. Wojtowicz, X. Liu, M. Dobrovolska, B. Janko, J.K. Furduna, G. Mihaly Phys.Rew.Lett., **95**, 227203 (2005).
24. G. Zarand, C.P. Moca, B. Janko, Phys.Rev.Lett., **94**, 247202 (2005).
25. A. Milner, A. Gerber, B. Groisman, M. Karpovsky, A. Gladkikh. Phys.Rev.Lett., **76**, 475 (1996).
26. K. Kuzmina, B.A. Aronzon, A.V. Kochura, A.V. Lashkul, K.G. Lisunov, E. Lahderanta, M.A. Shakhov. EPJ Web of Conferencies, **75**, 05014 (2014)
27. W. Heimbrodt, P.J. Klar, S. Ye, M. Lampalzer, C.Michel, S.D. Baranovskii, P. Thomas, W. Stolz, J.Supercond, **18**, 315 (2005).
28. C. Michel, M.T. Elm, B. Goldlucke, S.D. Baranovskii, P. Thomas, W. Heimbrodt, P.J. Klar. Appl.Phys.Lett., **92**, 223119 (2008).

Effect of oxygen nonstoichiometry on magnetic phase transitions in frustrated cobaltites $YBaCo_4O_{7+x}$ ($x = 0, 0.1, 0.2$)

Zoya Kazei[1,*], *Vyacheslav* Snegirev[1], *Ludmila* Kozeeva[2], *Margarita* Kameneva[2], and *Alexander* Lavrov[2]

[1]Lomonosov Moscow State University, Leninskie Gory, 119991 Moscow, Russia
[2]Nikolaev Institute of Inorganic Chemistry, 630090 Novosibirsk, Russia

Abstract. Experimental studies were carried out for structural and elastic properties of cobaltites $YBaCo_4O_{7+x}$ (x = 0, 0.1, 0.2) obtained by various technologies and distinguished by oxygen content. Correlation was revealed between the structure distortion at room temperature and a hysteresis in a $\Delta E(T)/T_0$ curve in the temperature range 80 – 280 K, as well as the anomaly value in the region of T_N. It was found, that small deviations from stoichiometry (x = 0.1, 0.2) result in suppression of structural transition, reduce a hysteresis, quickly smooth out and reduce anomalies of the Young's modulus $\Delta E(T)/T_0$ in the range of magnetic phase transition. This behavior indicates that structural and magnetic transitions are suppressed in nonstoichiometric samples and only short range correlations of order parameters are retained

1 Introduction

Rare-earth cobaltites $RBaCo_4O_7$, which are characterized by geometrical frustration of exchange interactions, mixed valence and coupled 3d- and 4f- subsystems, have attracted considerable interest due to their remarkable magnetic, electronic and magnetoelectric properties [1 – 5]. Frustrations result in a degenerate ground state and in absence of a long-range magnetic order even at significant constants of exchange interaction. A small orthorhombic distortion $\Delta a/a \approx 10^{-2}$ of the structure at the phase transition [6, 7] removes frustration of exchange bonds and leads to occurrence of the distinct magnetic order in the Co subsystem.

The data on the magnetic configuration, ground state, temperatures and peculiarities of structural and magnetic phase transitions for cobaltites of various preparation procedure are often incoincident, that allows to assume essential influence of technology of preparation and oxygen contents on their properties. In the present work effect of oxygen nonstoichiometry on structural and elastic characteristic as well as on phase transition is investigated in cobaltite $YBaCo_4O_{7+x}$ with only magnetic cobalt subsystem.

2 Experimental details

Polycrystalline samples $YBaCo_4O_{7+x}$ (x = 0, 0.1, 0.2) for studies have been synthesized by standard ceramic technology with three-step annealing in air and quenching sample from 900–950°C after each stage [8]; the oxygen content after this procedure corresponds to x ~ 0.05–0.10. Further ceramic samples were exposed to additional heat treatment by using the $x(t)$ dependence from annealing time. Samples with x = 0 were obtained by annealing in vacuum at 500°C, and samples with x = 0.10 and x = 0.20 were obtained by annealing at T = 210°C within several days. The oxygen content in the samples was measured by change of weight of a tablet after heat treatment with a resolution $\Delta x \approx 0.001$ and absolute accuracy ± 0.01. For comparison, we also investigated samples Q_1 and Q_2, which were quenched from 900 - 950°C and not exposed to additional heat treatment, the sample Q_2 being exposed to one more additional step of trituration, pressing and annealing.

X-ray diffraction (XRD) measurements were performed on a Stoe powder diffractometer using monochromated $CoK_{\alpha 1}$ radiation. All lines in the XRD patterns of the $YBaCo_4O_{7+x}$ samples measured at room temperature were indexed in the frameworks of a hexagonal or weakly distorted hexagonal structures. A full-profile analysis of XRD patterns in the 18–120° range by Full_Prof software allowed to determine the change of structure at a small deviation from the stoichiometry. The Young's modulus E and internal friction coefficient q^{-1} were measured by the composite resonator method at frequency ~100–200 kHz in the temperature range 4.2–300 K on polycrystalline samples $YBaCo_4O_{7+x}$ in the steady-state regime.

3 Results and discussion

3.1 XRD analysis

Influence of oxygen nonstoichiometry on structure was investigated at room temperature for the series of the samples $YBaCo_4O_{7+x}$ with oxygen excess x = 0, 0.1, 0.2 and also for quenched samples Q_1 and Q_2. Three types of XRD patterns were observed for samples with various deviation from stoichiometry. Sample with x = 0.1 and quenched sample Q_1 have undistorted hexagonal

* Corresponding author: kazei@plms.phys.msu.ru

structure and don't reveal structural transition down to helium temperatures. On the XRD pattern of the stoichiometric sample, appreciable difference is observed for a number of peaks sensitive to the orthorhombic distortion and its structure is characterized by a small orthorhombic distortion $\varepsilon_o = (a_o - b_o/\sqrt{3})/a_o \approx 3{\cdot}10^{-3}$, where the relation between the hexagonal and orthorhombic lattice parameters is as follows $a_o = a_h$, $b_o = \sqrt{3}a_h$. Fig. 1 shows the part of the XRD pattern, containing peaks sensitive to orthorhombic distortion and/or presence of the two-phase state of a sample.

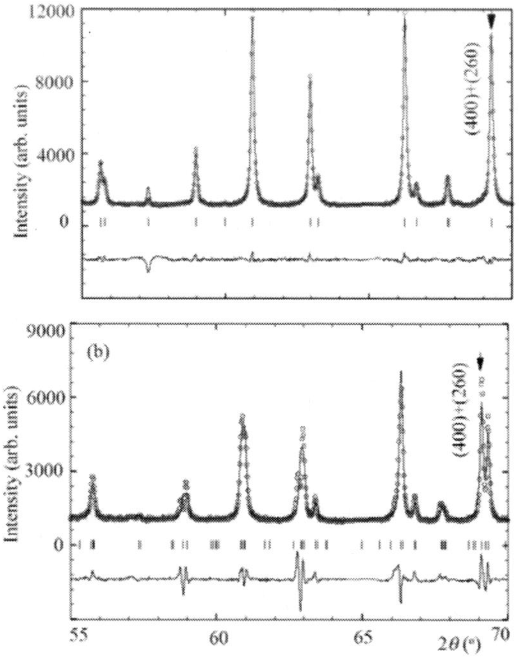

Fig. 1. Experimental (lines) and calculated (points) XRD patterns and their difference (lines below) for the series of the $YBaCo_4O_{7+x}$ samples with $x = 0.1$ (a) and $x = 0$ (b) on $CoK_{\alpha 1}$ radiation. Bars show positions of Bragg peaks according to space group.

This difference comes to light most precisely for the peak (400) + (260), having the Bragg's angle $2\theta = 69°$. The small orthorhombic distortion for the sample with $x = 0$ results in the peak splitting into two lines (Fig. 1 b), the ratio of the intensities $I_1/I_2 = 2:1$ for the split lines corresponding to statistical value for such type of distortion. Unsplit line is observed for the sample with $x = 0.1$, since $d_1 = d_2$ at $a_o = b_o/\sqrt{3}$ (Fig. 1 a).

According to full-profile analysis, the quenched sample Q_2 is a mix of two phases with hexagonal/quasi hexagonal structures. The difference of the cell parameters $\Delta a/a$ ($\Delta a = a_{h1} - a_{h2}$) for two phases is of the same value, as the value of the orthorhombic distortion ε_0 for the sample with $x = 0$, and the content of the minor phase is ~20%. This tendency of cobaltites to form a two-phase state at the oxygen deviation from the stoichiometry was discussed earlier for system $YbBaCo_4O_{7+x}$ [9]. The character and value of splitting reflexes on the XRD pattern for the sample with the oxygen excess $x = 0.2$ qualitatively differs from those observed for the samples with $x = 0$ and 0.1.

Thus, the development of two scenario is possible in ceramic samples at a small oxygen deviation from stoichiometry for x near to 0.1 value. It depends both on oxygen excess and synthesis procedure of ceramics (porosity of ceramics, temperatures and time of annealing, etc.). In the sample with $x = 0.1$ and quenched sample Q_1 the distortion of the hexagonal structures is absent, apparently because of irregular distribution of excess oxygen ions in the lattice. The structure becomes defect, that prevents the correlated ordering of local structural distortions thought the sample. Under other conditions, it is also possible a separation of the sample into two phases with different oxygen contents.

3.2 Young's modulus and internal friction

Due to the distortion of the crystal structure in the stoichiometric samples, frustrations are removed and subsequently the long-range magnetic order appears in the cobalt subsystem. In this case, one should expect the occurrence of the well defined magnetic phase transitions accompanied by the distinct anomalies of elastic properties of the samples. For this purpose, in this work the temperature dependences of Young's modulus $E(T)$ and internal friction coefficient $q^{-1}(T)$ were investigated for the series of $YBaCo_4O_{7+x}$ ($x = 0$, 0.1, 0.2) compounds distinguished by the oxygen content and value of the structure distortion. In figures relative

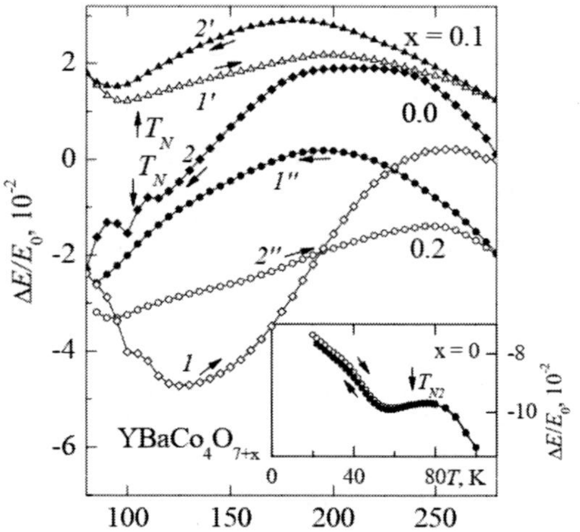

Fig. 2. Temperature dependence of the relative change $\Delta E/E_0$ of the Young's modulus for the series of samples $YBaCo_4O_{7+x}$ measured on heating (open symbols) and cooling (closed symbols; curves for $x = 0.1$, 0.2 are shifted along the vertical axis). The insert shows dependence $\Delta E/E_0(T)$ in the range of the spin reorientation transition for the sample with $x = 0$.

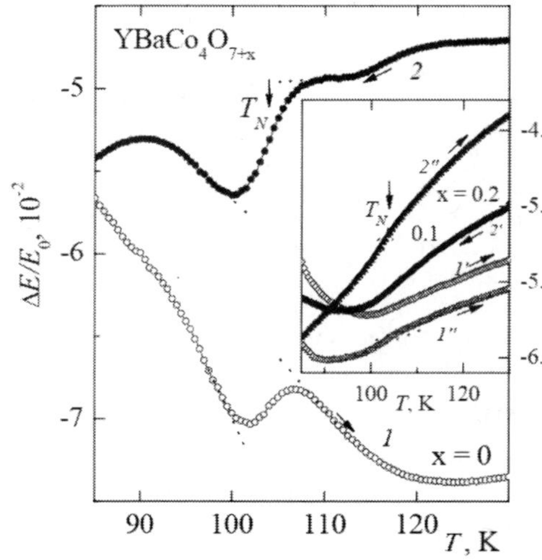

Fig. 3. Temperature dependence of the relative change $\Delta E/E_0$ of the Young's modulus for the series of the $YBaCo_4O_{7+x}$ samples measured on heating (open symbols) and cooling (closed symbols) in the range of the magnetic phase transition temperature T_N. (curves for various x are shifted along the vertical axis; the dotted line shows extrapolation of the temperature trend outside of anomaly).

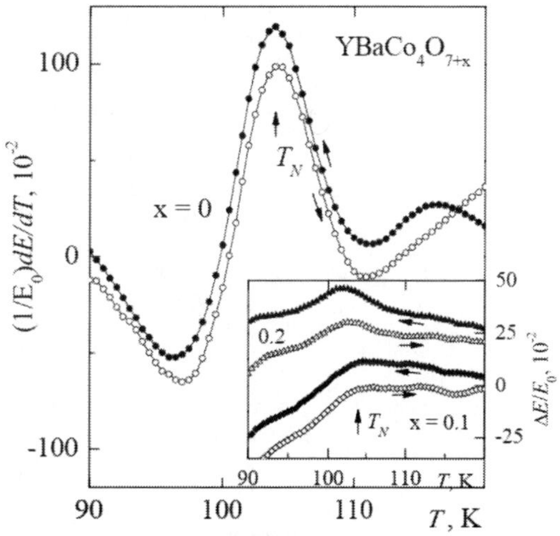

Fig. 4. Temperature dependence of the temperature derivative of the of Young's modulus $(1/E_0)\mathrm{d}E/\mathrm{d}T$ for the series of the $YBaCo_4O_{7+x}$ samples measured on heating (open symbols) and cooling (closed symbols) in the range of the magnetic phase transition temperature T_N (curves for various x are shifted along the vertical axis).

changes of the modulus $\Delta E(T)/E_0$, normalized on its value E_0 at $T = 280$ K, are shown ($\Delta E(T) = E(T) - E_0$; $E_0 = E(T = 280\text{ K})$).

The structural phase transition at temperature T_s in RE cobaltites, as was found previously, is accompanied by a softening (decrease) of the Young's modulus with decreasing temperature, which begins considerably above T_s, and then by its back jump at T_s and subsequent increase. Similarly, the magnetic phase transition is also accompanied by a softening of the Young's modulus, which begins above T_N. Thus, in the interval between two temperatures T_s and T_N, a complicated behavior may be observed due to two opposite tendencies to increase and decrease of the Young's modulus [7].

For the series of the samples with different x, correlation is traced between the structure distortion and the deviation from stoichiometry on the one side and features in the curves $\Delta E(T)/E_0$, such as the value of the hysteresis, full softening of the module in the temperature range (80 – 280) K, the value of the anomaly in the range of T_N on the other side (compare curves for samples with $x = 0$, 0.1, 0.2 in Fig. 2, 3). The $\Delta E(T)/E_0$ curve for the quenched sample Q_1 is similar to that observed for the sample with $x = 0.1$, only softening begins with lower temperatures ~130 K and is a little bit less in value. On the contrary, the behavior of the quenched sample Q_2 is close to that observed for the stoichiometric sample with $x = 0$.

For the stoichiometric and distorted sample with $x = 0$, a jump of the Young's modulus is observed (Fig. 3) in the range of the magnetic phase transition temperature $T_N \sim 105$ K. In view of the temperature variation in the curves $\Delta E(T)/E_0$ on the background of monotonous change of $\Delta E/E_0$ (the dotted line shows extrapolation of the temperature trend in the range of T_N), the values of jump at cooling and heating practically coincide and are $\delta E(T_N)/E_0 \approx 1\cdot10^{-2}$. The phase transition temperature T_N corresponds to the jump in the $\Delta E(T)/E_0$ curve or to the maximum of the temperature derivative of Young's modulus (Fig. 4). The similar anomaly of the modulus (comparable in character and value) is also observed in the range of T_N for the quenched sample Q_2. These anomalies are caused by the magnetic phase transition of the second order, the transition temperature coinciding well with the literary data. For the stoichiometric compound with $x = 0$ (as well as for quenched sample Q_2), similar anomalies are observed in the range of the second magnetic phase transition with T_{N2} ~70 K (insert in Fig. 2).

The jumps in temperature dependences of Young's modulus $\Delta E(T)/E_0$ at T_N decrease very sharply, smooth out and become practically unobservable at oxygen increase in the nonstoichiometric samples $YBaCo_4O_{7.1}$ and $YBaCo_4O_{7.2}$ (Fig. 3). Their traces can only be found out in the temperature derivative $(E_0^{-1})(\mathrm{d}E/\mathrm{d}T)$ (Fig. 4), having a maximum in the range of T_N. The transition temperature, determined by the maximum in the curve $(E_0^{-1})(\mathrm{d}E/\mathrm{d}T)$, is $T_N = (104.0 \pm 0.2)$ K, and does not appreciably vary at the change of the stoichiometry.

Thus, the distortion of the structure at structural phase transition for the stoichiometric $YBaCo_4O_7$ samples removes frustration of exchange bonds and promotes to a formation of the long-range magnetic order as the distinct magnetic phase transition. This phase transition is accompanied by the anomaly taking place both at heating and cooling, and reproducing at repeated thermo cycling. Structures for the nonstoichiometric samples remain undistorted, and frustrations in system are kept. As consequence, the short-range magnetic order develops gradually at temperature lowing, but the correlation length for magnetic order does not reach the sizes of crystallites.

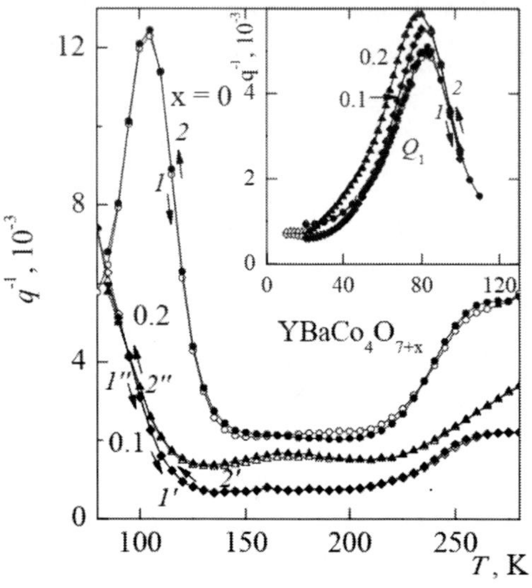

Fig. 5. Temperature dependence of the internal friction coefficient q^{-1} for the series of the $YBaCo_4O_{7+x}$ samples measured on heating (open symbols) and cooling (closed symbols). The insert shows comparison of absorption maxima of $q^{-1}(T)$ for the samples with $x = 0.1$, 0.2 and quenched sample Q_1.

Similarly, one can trace regularity in the behavior of the internal friction coefficient depending on the oxygen contents and distortion of the structure (Fig. 5). For the stoichiometric and distorted sample with $x = 0$ there is the large maximum consisting of two adjoined peaks close in temperature whereas for the nonstoichiometric samples with $x = 0.1$, 0.2, one low-temperature maximum at $T \sim 80$ K is only kept. The value of the $q^{-1}(T)$ maxima in the range of $T_N \sim 105$ K decreases for the quenched sample Q_2 which was not exposed to additional heat treatment in air. For the quenched sample Q_1, the behavior of the absorption maxima of the $q^{-1}(T)$ in the range of $T \sim 80$ K is very close to that observed for the nonstoichiometric samples with $x = 0.1$, 0.2 (insert in Fig. 5).

4 Summary

In this work, we performed experimental investigations of the structural and elastic properties for the series of the cobaltites $YBaCo_4O_{7+x}$ ($x = 0$, 0.1, 0.2) obtained by various technology and distinguished by oxygen content. Correlation was revealed between the structure distortion at room temperature and the hysteresis in $\Delta E(T)/T_0$ curve in the temperature range (80–280) K, as well as the anomaly value in the region of T_N. These compounds are found to exhibit significant hysteresis and irreversible elastic effects over a wide temperature range of (80–280) K between the temperatures of the structural and magnetic phase transitions. For the stoichiometric and distorted sample $YBaCo_4O_7$, the magnetic phase transition is accompanied by weak and distinct anomaly of the elastic properties, that is according to low dimension of exchange interactions in the Co subsystem. The small deviations from the stoichiometry for $x = 0.1$, 0.2 result in suppression of the structural transition, reduce the hysteresis, quickly smooth out and reduce the anomaly of the Young's modulus $\Delta E(T)/T_0$ in the range of the magnetic phase transition. This behavior indicates that structural and magnetic transitions are suppressed in nonstoichiometric samples and only short-range correlations of order parameters are retained.

References

1. L.C. Chapon, P.G. Radaelli, H. Zheng et al., Phys. Rev. B **74**, 172401 (2006)
2. P. Manuel, L.C. Chapon, P.G. Radaelli et al., Phys. Rev. Lett. **103**, 037202 (2009)
3. W. Schweika, M. Valldor, and P. Lemmens, Phys. Rev. Lett. **98**, 067201 (2007)
4. V. Caignaert, A. Maignan, K. Singh et al., Phys. Rev. B **88**, 174403 (2013)
5. Z.A. Kazei, V.V. Snegirev, G. P. Vorob'ev et al., J. Exp. Theor. Phys. **123**, 1025 (2016)
6. V. Caignaert, A. Maignan, V. Pralong et al., Solid State Sci. **8**, 1160 (2006)
7. Z.A. Kazei, V.V. Snegirev, A.S. Andreenko et al., J. Exp. Theor. Phys. **113**, 245 (2011)
8. L.P. Kozeeva, M.Yu. Kameneva, A.N. Lavrov et al., Inorgan. Mater. **49**, 668 (2013)
9. A. Huq, J.F. Mitchell, H. Zheng et al., J. Solid State Chem. **179**, 1136 (2006)

Vibronic interaction as main reason of magnetic ordering in insulating manganites $R_{1-x}A_xMnO_3$

Liudmila Gonchar[1,2*], *Anatoliy* Nikiforov[2]

[1]Ural State University of Railway Transport, 620034 Yekaterinburg, Russia
[2]Ural Federal University, 620002 Yekaterinburg, Russia

Abstract. The model of orbitally dependent magnetic structure of charge ordered insulated manganites is proposed. The model is semi-phenomenological. It allows using a few parameters to describe possible magnetic structures of compounds. The experimental crystal structure of compounds also could be taken into account. The compounds $LaMnO_3$, $La_{1/2}Ca_{1/2}MnO_3$, $La_{1/3}Ca_{2/3}MnO_3$, $BiMnO_3$ are considered.

1 Introduction

The manganites compounds are interesting because of strong correlation between crystal, charge, orbital, and magnetic subsystems. The general formula of compounds under consideration is $R_{1-x}A_xMnO_3$, where $R^{3+}=La^{3+}$ or Bi^{3+}, $A^{2+}=Ca^{2+}$, x=0, 1/2, 2/3. A field of correlation is a sublattice of Mn^{3+} ions. Mn^{3+} ion in octahedral oxygen neighborhood is a Jahn-Teller (JT) ion. The orbital states of Mn^{3+} ions form a regular structure, called orbital ordering (OO). In the case of non-isovalent doping, the charge carriers could be localized on Mn ions. This effect leads to charge ordering (CO). It means the regular structure of Mn^{3+}/Mn^{4+} ions. The magnetic structures (MS) of manganies are various [1–6]. There are A-type, E-type, CE-type, and F-type structures. OO is dependent upon crystal structure and charge distribution [7–11]. The interrelation between orbital and magnetic subsystems is also discussed [8–11]. In OO description, a choice of driving mechanism of cooperative ordering is discussed. The *ab initio* calculations require a complicated account of correlations [12, 13]. The most popular Kugel-Khomskii model [8] usually is used within the approximation of comparable vibronic and superexchange interaction. Nevertheless, the main vibronic term is estimated as ~1 eV [14], and the exchange parameters are ~1 meV [9, 10]. The rich set of MS makes difficult to create a uniform model. The main rules for estimation of the signs and relative values of superexchange (SE) parameters are Goodenough-Kanamory rules [1]. These rules use OO for MS explanation. The absence of precise values makes the MS description impossible for magnetically frustrated compounds such as $La_{1/3}Ca_{2/3}MnO_3$ or $BiMnO_3$. For SE parameters description, the *ab initio* calculations [12], Kugel-Khomskii model [8], orbitally-dependent phenomenological model [4, 9, 10, 17], next-nearest neighbor model [4, 7, 15], double-exchange model [7, 11] are applied. The *ab initio* calculations require preliminary assumption of MS [12, 13]. The double-exchange model is not applicable in suggestion of CO. We use previously proposed orbitally-dependent SE model [9, 10] to find possible MS for different orbital mixing within the framework of crystal symmetry. The aim of the paper is the description of orbital-magnetic correlation in insulating magnanites including frustrated cases.

2 Orbital state of Mn^{3+} ion

The ground state of Mn^{3+} ion's orbital subsystem with total spin S=2 in undistorted octahedral ligand neighborhood is ^5E multiplet. This state is twofold degenerated, thus some distortions of the neighborhood lead to removing of degeneration. We can use *e*-type orbital operators acting on many-electron eigenfunctions of ^5E ($|\theta\rangle$,$|\varepsilon\rangle$) in a linear combination as effective Hamiltonian

$$H_{eff}=\alpha_1 X_\theta+\alpha_2 X_\varepsilon, \quad (1)$$

where operators in matrix form are

$$X_\theta=\begin{pmatrix}-1 & 0\\ 0 & 1\end{pmatrix}\begin{matrix}\theta\\ \varepsilon\end{matrix};\; X_\varepsilon=\begin{pmatrix}0 & 1\\ 1 & 0\end{pmatrix}\begin{matrix}\theta\\ \varepsilon\end{matrix}. \quad (2)$$

The α_1 and α_2 coefficients are mainly phenomenological and could be described by linear vibronic interaction, non-linear vibronic interaction or crystal field models [9, 10]. Taking into account the effective interaction (1), the ground state function is considered as a linear combination of ^5E eigenfunctions:

$$\Psi=C_1|\theta\rangle+C_2|\varepsilon\rangle, \quad (3)$$

where $C_1^2+C_2^2=1$.

More convenient are polar coordinates:

$$\alpha_1=\rho\cos\Theta,\ \alpha_2=\rho\sin\Theta,\ \Theta\in[0;2\pi). \quad (4)$$

Thus, the ground state energy and the energy gap are

$$E=-\rho,\ \Delta E=2\rho. \quad (5)$$

The ground state wave function is

* Corresponding author: l.e.gonchar@urfu.ru

$$\Psi = \left|\cos\frac{\Theta}{2}\right|\left|\theta\right\rangle \pm \left|\sin\frac{\Theta}{2}\right|\left|\varepsilon\right\rangle, \quad (6)$$

where "+" is for $\alpha_2<0$, and visa versa. For magnetic properties prediction, the mean values of X-operators are needed.

$$\langle X_\theta \rangle = -\cos\Theta; \langle X_e \rangle = -\sin\Theta \quad (7)$$

In the crystal, there is a sublattice of Mn ions. The orbital wave functions and orbital operators must be indexed by Mn^{3+} ion number. Thus, there are two sets of orbital interaction characteristics for n^{th} Mn^{3+} ion: 1) $\alpha_{1,n}$ and $\alpha_{2,n}$; 2) ρ_n, Θ_n. For orbital state wave function (6) only one of characteristics– Θ_n –is sufficient.

3 Superexchange interaction

The exchange interaction of two Mn ions in octahedral neighborhood could be described by general expression

$$H_{ex} = \sum_{i>j} \hat{J}_{ij}\left(\mathbf{S}_i \cdot \mathbf{S}_j\right), \quad (8)$$

where SE operators are dependent upon orbital operators (2) of interacting ions:

$$\hat{J}_{ij} = \hat{J}_{ij}\left(X^i_\theta, X^i_\varepsilon, X^j_\theta, X^j_\varepsilon\right) \quad (9)$$

Let us review the different kinds of orbitally-dependent 180^0-SE interaction in insulating perovskite manganties. Finding the mean values of orbital operators on the orbital state wave functions (6) of interacting ions and taking into account the Mn–O– Mn bond configuration [9, 10], one could write the SE parameters in pseudoperovskite axes as a function of orbital angles (Θ_i, Θ_j), Mn–O–Mn bond angle φ_{ij} and mean Mn–O distance r_{ij}.

1) Mn^{3+}–Mn^{3+} (x, y, z):

$$J^z_{ij} = \frac{J^1_0 \cos^2\varphi_{ij}}{r^{10}_{ij}}\left(1-\alpha\left(\cos\Theta_i + \cos\Theta_j\right) + \beta\cos\Theta_i\cos\Theta_j\right),$$

$$J^{x,y}_{ij} = \frac{J^1_0 \cos^2\varphi_{ij}}{r^{10}_{ij}} \times$$

$$\times\left(1+\frac{\alpha}{2}\left(\cos\Theta_i \pm \sqrt{3}\sin\Theta_i + \cos\Theta_j \pm \sqrt{3}\sin\Theta_j\right) + \right.$$

$$\left. + \frac{\beta}{4}\left(\cos\Theta_i \pm \sqrt{3}\sin\Theta_i\right)\left(\cos\Theta_j \pm \sqrt{3}\sin\Theta_j\right)\right).$$

2) Mn^{3+}–Mn^{4+} (x, y, z):

$$J^z_{ij} = \frac{J^2_0 \cos^2\varphi_{ij}}{r^{10}_{ij}}\left(1-\alpha'\cos\Theta_i\right),$$

$$J^{x,y}_{ij} = \frac{J^2_0 \cos^2\varphi_{ij}}{r^{10}_{ij}}\left(1+\alpha'/2\left(\cos\Theta_i \pm \sqrt{3}\sin\Theta_i\right)\right)$$

3) Mn^{4+} – Mn^{4+}: $J_{ij} = \dfrac{J^3_0\cos^2\varphi_{ij}}{r^{10}_{ij}}$.

The values of parameters have been found in previous paper [9]: $J_0^1=1.69\cdot10^4$ K×Å^{10}, $\alpha=1.0$, $\beta=4.5$; $J_0^2=-4.90\cdot10^3$ K×Å^{10}, $\alpha'=5.1$; $J_0^3=1.12\cdot10^3$ K×Å^{10} Thus, we can determine the main component of magnetic interaction.

4 Results and discussion

Let us apply the theory above to certain manganite crystals. This application was performed sometimes earlier in previous papers [9, 10, 17], but using another characteristics of OO, i.e. ϕ angle instead of Θ. It helps to remove the equivalence with MS description used in Kugel-Khomskii model [8]

4.1 $LaMnO_3$

The space symmetry of the compound is *Pnma* [1]. The OO of $LaMnO_3$ (Fig. 1) is caused by vibronic interaction of Mn^{3+} ion with local distortions of oxygen octahedra. These distortions are described by symmetrized coordinates $Q_{\theta,n}$ and $Q_{\varepsilon,n}$. The coefficients of effective Hamiltonian (1) are:

$$\alpha_{1,n} = -|V_e|Q_{\theta,n}; \alpha_{2,n} = -|V_e|Q_{\varepsilon,n}, \quad (10)$$

where $-|V_e|Q_{\theta,n}$=0.11 eV, $-|V_e|Q_{\varepsilon,n}$=±0.34 eV.

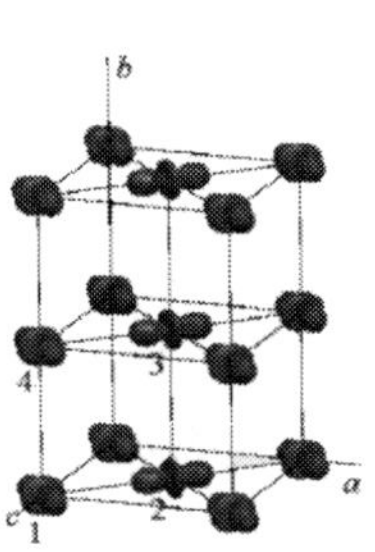

Fig.1. OO of $LaMnO_3$. Mn^{3+} ions as mixed single-electron e_g- functions are drawn. Here and on other figures, oxygen and lanthanum ions are omitted. Numbers denote Mn^{3+} ions of $4a$ position.

In terms of $\alpha_{1,n}$ and $\alpha_{2,n}$ constants, the wave function is

$$\psi_n = \left|\cos\tfrac{\Theta_n}{2}\right|\left|\theta\right\rangle_n \pm \left|\sin\tfrac{\Theta_n}{2}\right|\left|\varepsilon\right\rangle_n. \quad (11)$$

All Mn^{3+} ions are divided into four sublattices due to symmetry considerations (see Fig. 1). The correlation between angles is:

$$\Theta_1 = 2\pi - \Theta_2 = 2\pi - \Theta_3 = \Theta_4 = \Theta, \Theta = 73° \quad (12)$$

The SE interaction is of type 1. The SE dependences upon Θ angle are drawn on Fig. 2. In orthorhombic *Pnma* notations, $J_b=J^z$, $J_{ac}=J^{x,y}$.

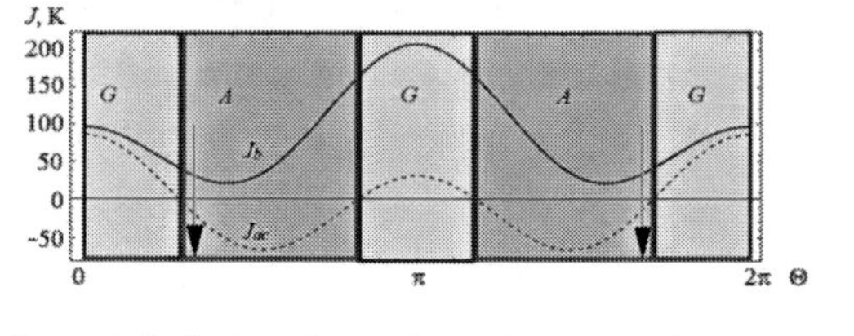

Fig. 2. Orbitally-dependent SE interaction. Arrows denote $LaMnO_3$ orbital states of Mn^{3+} ions. Letters *A* and *G* denote MS types.

As one can see at Fig. 2, the SE interaction allows only two types of AFM structure – *A* and *G*. The experimentally determined MS of $LaMnO_3$ is *A*-type AFM structure [1]. Some OO allow one-dimensional MS ($\Theta=52^0$, 149^0, 211^0, 308^0). The *G*-type MS is not observed in manganites.

4.2 $La_{1/2}Ca_{1/2}MnO_3$

The space symmetry of the compound is $P2_1/m$ [5]. At low temperature, there is CO state of the compound. The CO and OO are presented on Fig. 3.

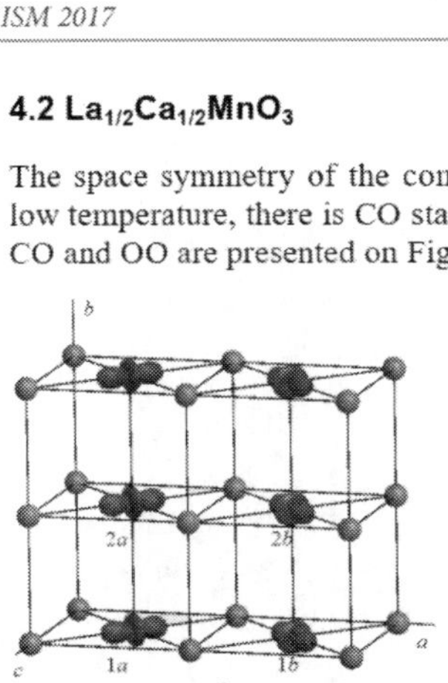

Fig.3. CO and OO of $La_{1/2}Ca_{1/2}MnO_3$. Mn^{3+} ions as mixed single-electron e_g-functions and Mn^{4+} ions as spheres are drawn. Ions of $2a$ and $2b$ positions are numbered

The OO of Mn^{3+} sublattice of $La_{1/2}Ca_{1/2}MnO_3$ is caused mainly by linear vibronic interaction of Mn^{3+} ions with local distortions of oxygen octahedra. The coefficients of vibronic Hamiltonian (1) of Mn^{3+} ions have the same form (10) as for $LaMnO_3$. The coefficients of effective Hamiltonian (1) are:

$$\alpha_{1,n} = -|V_e|Q_{\theta,n};\ \alpha_{2,n} = -|V_e|Q_{\varepsilon,n}, \qquad (13)$$

where $-|V_e|Q_{\theta,n}$=0.09 eV, $-|V_e|Q_{\varepsilon,a}$=–0.23 eV, $-|V_e|Q_{\varepsilon,b}$=0.21 eV.
The wave functions of the Mn^{3+} sublattice is of the same form as in Eqs. (11). The OO is described by correlation:

$$\Theta_{1a} = \Theta_{2a}\Theta_a, \Theta_{1b} = \Theta_{2b} = \Theta_b, \Theta_b \approx 2\pi - \Theta_a = 56°. \quad (14)$$

The SE interaction is of types 1–3. The second type of SE interaction defines planar interaction (J^x, J^y), and 1, 3 types are interplanar (J^z). The dependences of SE parameters upon Θ angle are drawn on Fig. 4. In monoclinic $P2_1/m$ notations, J_b=J^z, J_{ac}=$J^{x,y}$.SE parameter of Mn^{4+}–Mn^{4+} interaction is J=15 K [9].

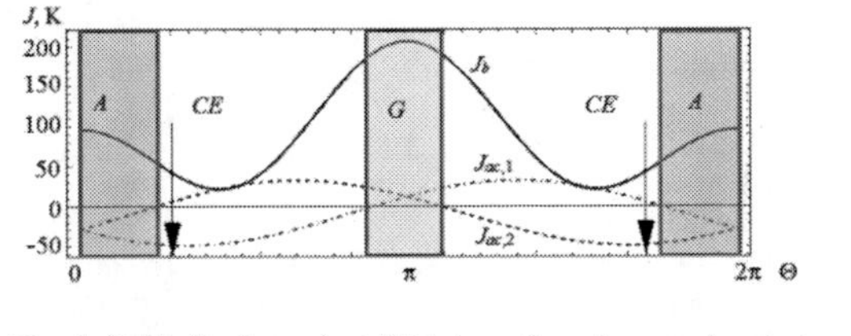

Fig. 4. Orbitally-dependent SE interaction. Arrows denote two states of Mn^{3+} sublattice in $La_{1/2}Ca_{1/2}MnO_3$. Letters *A*, *G*, and *CE* denote MS types.

The magnetic cell is described by doubled crystal cell.
The changes of OO may cause the change in the MS. The experimental MS of $La_{1/2}Ca_{1/2}MnO_3$ is *CE* structure [5]. It consists of FM zigzag chains along *a* axis with AFM order between chains. The *A*-type of MS is the nearest structure on the diagram.

4.3 $La_{1/3}Ca_{2/3}MnO_3$

The space symmetry of the compound is *Pnma* [21] that is tripled compared with $LaMnO_3$. At low temperature, the CO state is also proposed [17, 18]. The CO and OO are presented on Fig. 5. The OO of Mn^{3+} sublattice of $La_{1/3}Ca_{2/3}MnO_3$ is caused mainly by vibronic interaction of Mn^{3+} ion with local distortions of oxygen octahedra. The coefficients of vibronic Hamiltonian (1) of Mn^{3+} ions have the same form (11) as for $LaMnO_3$. They must be slightly corrected, because it helps describing MS [17].

$$\alpha_{1,n} = -|V_e|Q_{\theta,n} + \alpha_{\theta,n};\ \alpha_{2,n} = -|V_e|Q_{\varepsilon,n} + \alpha_{\varepsilon,n}, \quad (15)$$

where $-|V_e|Q_{\theta,n}$=0.24 eV, $-|V_e|Q_{\varepsilon,n}$=±0.31 eV.

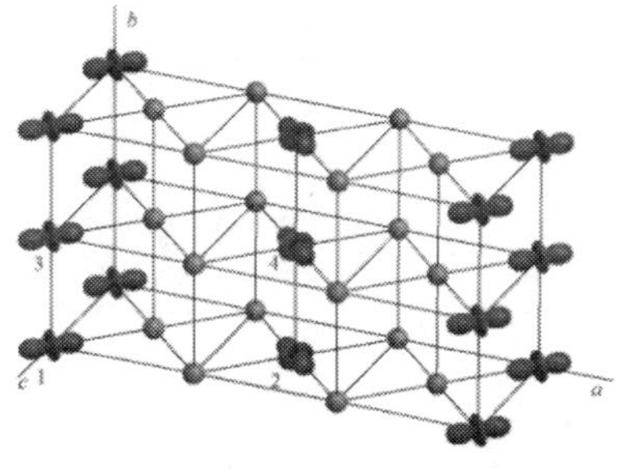

Fig. 5. CO and OO of $La_{1/3}Ca_{2/3}MnO_3$. Mn^{3+} ions as mixed single-electron e_g- functions and Mn^{4+} ions as spheres are drawn. Numbers denote Mn^{3+} ions of $4a$ position.

The wave function of the Mn^{3+} sublattice is of the same form as in Eq. (11). The OO is described by correlation:

$$\Theta_1 = 2\pi - \Theta_2 = 2\pi - \Theta_3 = \Theta_4 = \Theta, \Theta = 53° \quad (16)$$

The SE interaction is of types 1–3. The SE parameters dependent upon Θ angle are the same as in $La_{1/2}Ca_{1/2}MnO_3$ and are drawn on Fig. 4. SE parameter of Mn^{4+}–Mn^{4+} interplanar interaction is J_b=15 K [9], and J_{ac}=15 K [17] in the *ac*-plane. Taking into account the frustration of MS in Mn^{3+}–Mn^{4+} or Mn^{4+}–Mn^{4+} pairs within the *ac*-plane, one can suppose three possible types of MS. The magnetic moments in two structures form FM trimers along elongated O–Mn^{3+}–O bond. They are differently located relative to each other. The structure 1 (Fig 6a) has Mn^{3+}–Mn^{4+} FM ordered bond with AFM SE interaction. It does not double the crystal unit cell. On Fig. 6b, the FM Mn^{4+}–Mn^{3+}–Mn^{4+} trimers are ordered as stripes, which are perpendicular to each other. Fig. 6b structure has doubled crystal unit cell along *c*-axis.

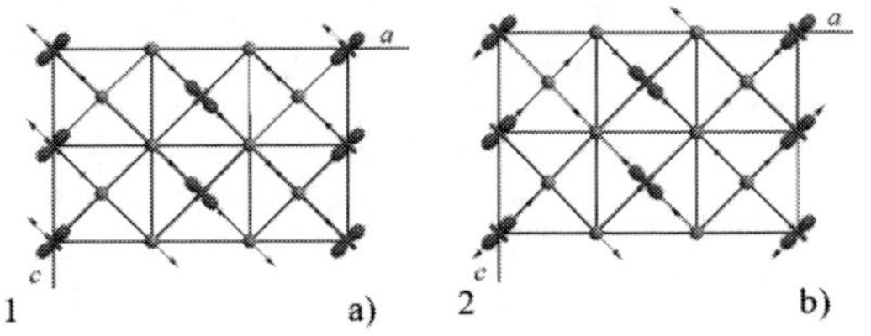

Fig. 6. Frustrated MS of 1 and 2 types of $La_{1/3}Ca_{2/3}MnO_3$ in one of *ac*-planes. In neighbor plane the spin moments are directed opposite.

The experimental investigation [18] has determined MS of type 2. However, the local oxygen octahedral distortions account in OO lead to type 1 structure. Taking into account possible terms $\alpha_{\theta,n}$, $\alpha_{\varepsilon,n}$ in vibronic interaction (15), Θ may change slightly. On Fig. 7, the region of mixing angle is drawn as wide arrows. The possible values of Θ angles supportingthe experimental MS are $\Theta \lesssim 305^0$, $\Theta \gtrsim 55^0$ The additional parts of Hamiltonian's coefficients could be estimated as $|\alpha_{\theta,n}|$, $|\alpha_{\varepsilon,n}|$~0.01 eV/Å. The edge of the region is established qualitatively, taking into account the Neel temperature of the compound.

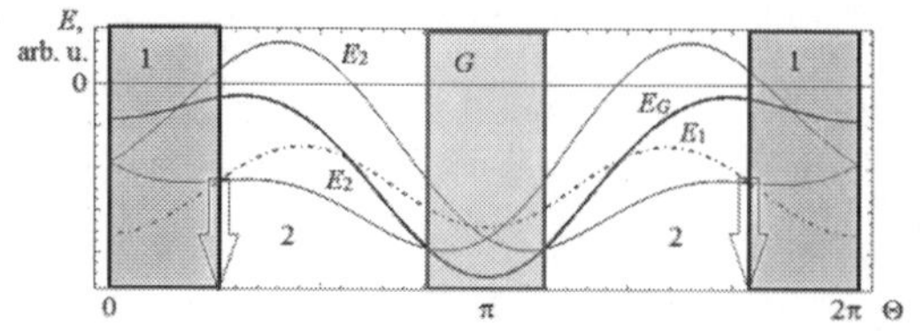

Fig. 7. Orbitally-dependent energy of possible MS of $La_{1/3}Ca_{2/3}MnO_3$. The white and colored stripes mean the regions of orbital-mixing angles corresponding to different MS.

4.4 $BiMnO_3$

The space symmetry of the compound is *C*2/*c* [6] that is unusual for manganites. The possible OO is presented on Fig. 8 [10]. The OO of $BiMnO_3$ is caused not only by vibronic interaction of Mn^{3+} ion with local distortions of oxygen octahedra [10]. The non-local and non-linear vibronic terms could be taken into account as parameters:

$$\alpha_{1,n} = -|V_e|Q_{\theta,n} - \alpha_{\theta,n};\ \alpha_{2,n} = -|V_e|Q_{\varepsilon,n} - \alpha_{\varepsilon,n},\quad (17)$$

where n is Mn^{3+} sublattice number (n=1*e*, 2*e*, 1*d*, 2*d*).

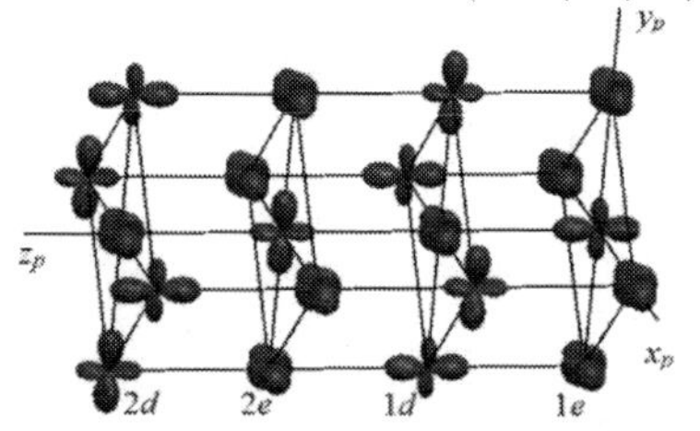

Fig.8. OO of $BiMnO_3$ in pseudoperovskite axes. Mn^{3+} ions as mixed single-electron e_g- functions are drawn. OO is drawn with respect of additional terms (16) in vibronic Hamiltonian (1) [10].

The main parts of coefficients could be estimated as $-|V_e|Q_{\theta,e}$=0.18 eV, $-|V_e|Q_{\varepsilon,e}$=–0.32 eV, $-|V_e|Q_{\theta,1d}$=0.10 eV, $-|V_e|Q_{\varepsilon,1d}$=0.32 eV, $-|V_e|Q_{\theta,2d}$=–0.36 eV, $-|V_e|Q_{\varepsilon,2d}$= –0.005 eV. The wave function of the Mn^{3+} sublattice is of the same form as in Eq. (11). There are two Mn^{3+} positions– 4*e* and 4*d*– with different local symmetry. OO with local symmetry account is described by correlation:

$$\Theta_1 = \Theta_2 = \Theta_e, \Theta_3 \approx -2\pi/3 - \Theta_4 = \Theta_d. \quad (18)$$

If $\alpha_{\theta,n}$=$\alpha_{\varepsilon,n}$ =0, it means only local linear vibronic interaction's influence. The orbital mixing angles are $\Theta_e = 300°, \Theta_d = 67°$. These values are inconsistent with found MS. However, the account of non-linear vibronic terms drastically changes the orbital states of *d*-position Mn^{3+} ions [10]. The superexchange interaction is of 1 type (Mn^{3+}–Mn^{3+}). As opposed to $LaMnO_3$, there are six different values of SE interaction.

The diagram of possible MS is drawn on Fig. 9. OO, caused by linear vibronic interaction (dark arrows on Fig. 9), leads to MS frustration in different directions (y_p or z_p). The possible structures are: *G*-type AFM without frustration, *F*-type FM structure, *C*-type AFM with FM chains approximately along x_p direction, A_1-type AFM with FM planes parallel to x_py_p-planes, A_2-type AFM with FM planes parallel to x_pz_p-planes. The regions of orbital mixing angles supporting the experimental *F*-type MS approximately are: $\Theta_e \approx 300^0$, $90 \le \Theta_d \le 160$.

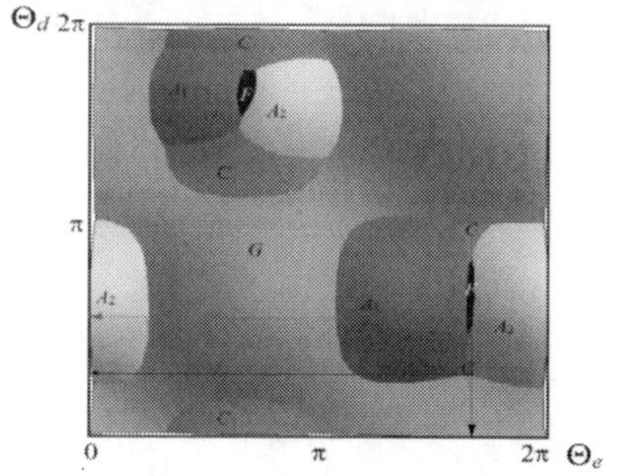

Fig. 9. Phase diagram of possible MS depending upon orbital mixing angles for *C*2/*c* space group. Black arrows denote OO of Mn^{3+} sublattices in $BiMnO_3$ due to local linear vibronic interaction [10]. Red arrow shows possible orbital state for experimental [6] MS. A_1, A_2, *C*, *G*, *F* denote types of MS.

5 Conclusions

The uniform approach to description of orbital and magnetic structure of insulating manganite crystals is proposed. The model of orbital structure uses 2 parameters per Mn^{3+} position. For magnetic structure description – only 1 parameter per JT position is needed.

The support by Act 211 Government of the Russian Federation, contract № 02.A03.21.0006 is acknowledged.

References

1. J.B. Goodenough, Phys. Rev. **100**, 564 (1955); E.O Wollan, W.C. Koehler, Phys. Rev. **100**, 545 (1955).
2. T. Kimura *et al.*, Phys. Rev. B, **68**, 060403 (2003)
3. Y. Yamada *et al.*, Phys. Rev. Lett., **77**, 904 (1996).
4. M. Mochizuki, N. Furukawa, Phys. Rev B, **80**, 134416 (2008).
5. P. G. Radaelli *et al.*, Phys. Rev. B, **55**, 3015 (1997).
6. D.P. Kozlenko *et al.*, Phys. Rev. B, **82**, 014401 (2010).
7. I.V. Soloviev, S.A. Nikolaev, Phys. Rev. B **90**, 184425 (2014)
8. D.I. Khomskii *et al.*, JETP **122**, 484 (2016)
9. L.E. Gontchar, A.E. Nikiforov, Phys. Rev. B **66**, 014437 (2002)
10. L.E. Gonchar, A.E. Nikiforov, Phys. Rev B., **88**, 094401 (2013)
11. I.V. Soloviev, Phys. Rev. Lett., **91**, 177201 (2003)
12. V.I. Anisimov, AIP Conference Proc. **1297**, 3 (2010).
13. D. Nazipov et al., J of Phys.: Conf. Ser., **833**, 012006 (2017).
14. A.E. Nikiforov, S.E. Popov, Appl. Phys. A, **74**, 1743 (2002).
15. R. Kováčik *et al.*, Phys. Rev. B, **93**, 075139 (2016).
16. Beom Hyun Kim, B.I. Min, Phys. Rev. B, **80**, 064416 (2009).
17. L.E. Gontchar, A.E. Nikiforov, J. Magn. Magn. Mater., **300**, 167 (2006).
18. P.G. Radaelli *et al.*, Phys. Rev. B, **59**, 14440 (1999).

Strain effect on resistivity of $La_{0.7}Ba_{0.3}MnO_3$ thin film

Timur A. Shaikhulov[1], *Valery A.* Shakhunov[1], *Victor V.* Demidov[1], *Gennady A.* Ovsyannikov[1a], *Nikolay V.* Andreev[2], *Anna E.* Pestun[2], *Vladimir L.* Preobrazhensky[3]

[1] Kotel'nikov Institute of Radio Engineering and Electronics, RAS, Moscow, 125009, Russia
[2]Moscow Technological University, MISIS, Moscow, 119936, Russia
[3]The Wave Research Center, Prokhorov General Physics Institute RAS, Moscow, 119991, Russia

Abstract. A complex study of the dc and magnetic characteristics of epitaxial manganite films $La_{0.7}Ba_{0.3}MnO_3$ (LBMO) was carried out under conditions of the crystal structure tension caused by a mismatch between the parameters of the LBMO crystal and the substrate. The epitaxial thin films with the thickness 150 nm were grown by laser ablation at T=700-800C under pure oxygen pressure 0.3-1 mbar. The substrates (001)$LaAlO_3$, (001)$SrTiO_3$, (110)$NdGaO_3$, (001)[$(LaAlO_3)_{0.3}$+$(Sr_2AlTaO_6)_{0.7}$] (LSAT) are used. It is shown that the temperature dependence of the film resistance in the low-temperature region does not depend on the film strain and it is in good agreement with the calculation that takes into account the interaction of carriers with magnetic excitations in the presence of strongly correlated electron states. The study of the of ferromagnetic resonance indicated the inhomogeneity of the ferromagnetic phase in LBMO films and an increase in the width of the FMR line with temperature decreasing.

1 Introduction

The parameters of epitaxial films of rare-earth manganite perovskites with the structure $Re_{1-x}A_xMnO_3$, where Re are the rare-earth materials (La or Nd) and A are the alkaline earth metals (Sr, Ca, Ba), very often differ substantially from the parameters of the crystals of the same composition. As it was shown in the references [1 - 3] the reason of the change in electric and magnetic properties is the strain in the films, caused by the lattices mismatch of the film with the substrate on which the films are deposited. A correlation was observed between the level of mechanical stresses in manganite films and the temperature of meta-insulator transition (T_{MI}) and the magnitude of the magnetoresistance. It was shown that the three-dimensional compression of the crystal lattice increases the hopping probability amplitude in the double exchange model, which leads to an increase in the Curie temperature (T_C), while the biaxial distortions of the Jahn-Teller type cause an increase the localization of electrons and decrease T_C [4-8].

The ferromagnetic phase transition for a single-phase $La_{0.7}Ba_{0.3}MnO_3$ (LBMO) crystal occurs at T_C = 345 K [9]. In the LBMO single crystal the giant magnetostriction (up to 4 10^{-4}) was observed at a temperature equal to Curie temperature [10]. In the LBMO films when the film tension changes multiple resistive states arise. They are modulated by the magnetic field over a wide temperature range [11]. This phenomenon is caused by the strong coupling of spin and charge degrees of freedom [10-13].

[a]Corresponding author: shcaihulov@hitech.cplire.ru

This paper presents the results of an experimental studies of the electronic transport and microwave magnetic parameters of epitaxial LBMO films deposited on various substrates.

2 Samples and experimental technique

Epitaxial LBMO films 150 nm thick were deposited on (001)$LaAlO_3$ (LAO), (001)$SrTiO_3$ (STO), (110)$NdGaO_3$ (NGO), (001) [$(LaAlO_3)_{0.3}$+$(Sr_2AlTaO_6)_{0.7}$] (LSAT) with a size of 5x5x0.5 mm^3 using laser ablation in off-axis geometry at the temperature of 700-800 C and the oxygen pressure of 0.2-0.5 mbar. The targets were made from a mixture of La_2O_3, $BaCO_3$ и $MnCO_3$ powders according to the ceramic technology. The stoichiometric composition of the target was used. The resistance of the films was studied by a four-point technique. Contact areas were produced by sputtering of Pt through a metal mask. The temperature of the metal-insulator transition T_{MI}, determined from the maximum resistance, varied in proportion to T_C.

The crystallographic parameters of the films and substrates were determined by means of a 4-circle x-ray diffractometer, by measuring in $2\Theta/\omega$ and φ scan mode. The magnetic characteristics were measured by magnetic resonance. The magnetic resonance spectra were recorded using ER-200 EPR Bruker spectrometer operating in a 3-cm electromagnetic wave band (frequency 9.76 GHz). When recording the signals a synchronous detection was used at the modulation frequency 100 kHz of an external magnetic field. The investigated samples were located in the microwave cavity of the spectrometer. The plane of the sample was always parallel to the direction of the constant external magnetic field and the magnetic component of the microwave field (parallel orientation). The parallelism of the plane of the samples to a constant magnetic field was controlled by getting minimum of the resonance value of the ferromagnetic resonance field (FMR). This arrangement of the samples eliminated the_contribution in the magnetic resonance spectra due form-factor of the sample.

3 Lattice parameters

The c-axis of the grown LBMO films are oriented along the normal of the plane of the substrate. The parameter of the interplanar distance in LBMO films along the normal to the substrate $a_\perp$ and the substrate plane a_s were determined from x-ray diffraction 2Θ/ω scan.

The 2Θ/ω scan for LBMO/LAO film is presented on Fig.1. The out of plane parameter $a_\perp$ for LBMO films strongly depends on a_s for the substrate. The in plane film parameters $a_{//}$ determined from asymmetric diffraction peaks are close to the a_s for the substrate crystals. These measurements indicate that a fast relaxation of the LBMO parameter of the film occurs at the substrate/film interface. Consequently, the grown manganite films are under the influence of mechanical stresses caused by the substrate during the film growth.

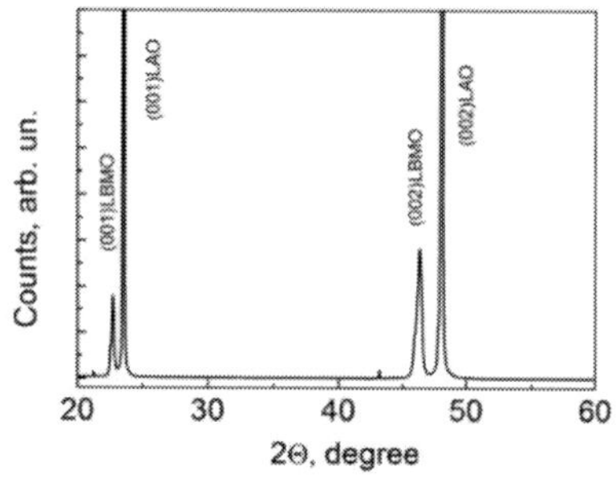

Fig.1. 2Θ/ω scan for LBMO film deposited on LAO substrate

The values of the strains are arisen in the plane of substrate $\varepsilon_{||}=(a_{||}-a_{LBMO})/a_{LBMO}$ and in the perpendicular direction $\varepsilon_\perp=(a_\perp-a_{LBMO})/a_{LBMO}$ are presented in Table 1. a_{LBMO} =0.390 nm obtained from XRD data for target is used. The temperatures of metal-insulator transition T_{MI} measured at max of temperature dependence of resistance R(T) (see Fig.2) are also shown. It can be seen that the compressive stresses in the plane of the substrate for LSMO films deposited on the NGO, LAO, and LSAT substrates reduce T_{MI}, and hence the Curie temperature.

We note that the unit cell volume of the crystal lattice of the LBMO film (V_{eff}) is not conserved when the substrate is changed. At small lattice distortions the Curie temperature (T_C) and magnetic transport parameters of the manganites depend on two parameters: the relative change in the V_{eff} : $\varepsilon_b=2\varepsilon_{||}+\varepsilon_\perp$, and the biaxial distortion $\varepsilon^*=(\varepsilon_\perp-\varepsilon_{||})\sqrt{2/3}$ [7, 14]. The distortions of unit cell volume lead to an increase or decrease in Tc dependently on the sign of the distortion, but the biaxial ones always decrease T_C [7, 13, 14].

Table 1. Crystallographic parameters and the temperature of metal-insulator transition

Subst	a_s, nm	$a_\perp$, nm	V_{eff} nm^3	ε_b, %	ε^*, %	T_{MI}, K
LSAT	0.387	0.396	0.0593	0.37	1.87	256
NGO	0.386	0.391	0.0583	-1.88	0.52	259
LAO	0.379	0.393	0.0564	-4.89	1.71	278
STO	0.391	0.393	0.060	3.02	-0.28	287

4 Conductivity

A temperature peak of resistance typical for manganites was observed at $T = T_{MI}$ for all the films studied (Fig. 2). The decrease in resistance at $T \leq T_M$ is most likely caused by an increase in the area of the conducting ferromagnetic regions. At low temperatures (T <100K), the resistance of all film tends to the asymptotic value which is determined by electron scattering on impurities, defects, grain boundaries, spin waves and domain walls [12, 15, 16].

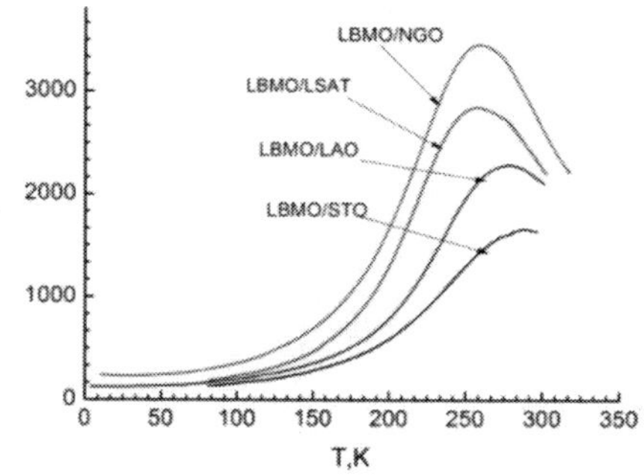

Fig. 2. Temperature dependences of the resistances of 150 nm thick LBMO films deposited on STO, NGO, LSAT, LAO substrates.

The highest value of T_{MI} = 283 K was observed for LBMO grown on STO substrates. However, even in the case of a small misalignment of the substrate and the film crystal parameters (1.12%) the T_{MI} film is several degrees lower than the T_{MI} for stoichiometric LBMO single crystal (345K). Perhaps the decrease in T_{MI} is due to the relatively low concentration of carriers in LBMO films due to a decrease in the Ba content [13]. Indeed, as was shown in the reference [17] for $La_{1-x}Ba_xMnO_3$ crystal with a x≤0.2 T_{MI} is for several degrees lower then in optimal case (x=0.3).

Curie temperature T_C depends on the strain according to the work of Millis as following [7]

https://doi.org/10.1051/epjconf/201818506006

$$T_c(\varepsilon)=T_c(0)(1-\alpha\varepsilon_b-\delta\varepsilon^{*2}/2), \quad (1)$$

where $\alpha=(1/T_c)dT_c/d\varepsilon_b$ and $\delta=(1/T_c)d^2T_c/d\varepsilon^{*2}$. Assuming that T_C and T_{MI} vary in the same way as the change in the film strain, we obtain that for $T_{MI}(0)=345K$ alpha changes within tens of units, and the gamma of units is hundreds. For example, for a pair of substrates with close crystallographic parameters NGO and LSAT, substrates we have $\alpha= 12$, $\delta=1200$.

For $T<T_{MI}$ the resistivity of manganite films depends both on scattering of carriers by impurities and defects, and also on the intensity of electron-electron and electron-magnon interactions. According to [18-23] the temperature dependence of the resistivity of a manganite film can be represented in the the following form

$$\rho=\rho_0+\rho_1 T^2+\rho_2 T^{4.5}, \quad (2)$$

where ρ_0 is the contribution to the resistivity of structural defects, ρ_1 and ρ_2 are determined by electron-electron scattering and the scattering of electrons by magnons correspondingly.

The smallest $\rho_0=2{,}3\ 10^{-3}\ \Omega$ cm is observed for LBMO films grown on STO substrates having the smallest mismatch between the crystal lattice of the substrate and the film. The obtained ρ_o for the film are several times greater than the resistivity for single crystals (less then $10^{-3}\ \Omega$ cm) [21]. In the LBMO film on LAO increasing of the ρ_0 could be due to twinning of the substrate. The dependence $\rho\infty T^{4.5}$ was indeed observed for manganite LBMO films sputtered on the (001) LSAT substrate [22]. At the same time, the dominance of the T^2 term in $\rho(T)$ was observed for $La_{0.67}Sr_{0.33}MnO_3$ (LSMO) films [3]. A small increase in the R (T) at T <50 K can be caused by a contribution to the resistance proportional to $T^{1/2}$ caused by the Coulomb interaction of carriers [12]. Detail analysis of the temperature dependences of the resistivity of the LBMO/NGO films shown that $\rho(T)$ is described with good accuracy by the expression (2) for $\rho_1=9\ 10^{-9}\ \Omega$ cm K^{-2} , $\rho_2 = 4\ 10^{-13}\ \Omega$ cm $K^{-4.5}$ and $\rho_0=2{,}3\ 10^{-3}\ \Omega$ cm in the temperature range (100-250 K). At T=100K the relation $\rho_0>\rho_2 T^{4.5}>\rho_1 T^2$ indicates the structural defects dominated over electron-electron and electron – magnon interactions.

5 Ferromagnetic resonance

Ferromagnetic resonance spectra (FMR) were obtained both for the purpose of determining the transition temperature of a paramagnet-ferromagnet and for studying the magnetization anisotropy in the plane of the films.

Shapes of all the FMR curves at nearly room temperature (T=280K) can not be described by a single Lorenz line (see Fig.3). This fact unambiguously indicates the inhomogeneity of the ferromagnetic phase in the films. The strain of the films can gradually decrease with increasing of their thickness. The difference in the values of the resonant fields H_0 for different parts of the inhomogeneous film did not exceed the width of the FMR line $\Delta_{1/2}$. The nonmagnetic layer, which according to [24] is at film- substrate interface,

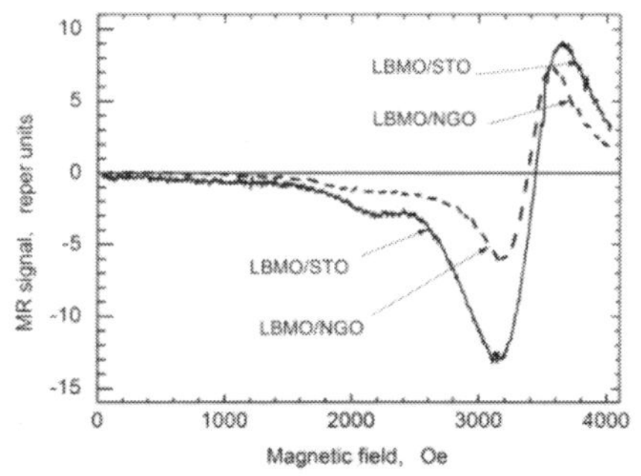

Fig. 3. Ferromagnetic resonance spectra for the two LBMO films measured at nearly room temperature (T=280K) in parallel orientation

does not have a noticeable effect on FMR. The magnetization values M_0 =30 Oe and uniaxial anisotropy fields H_u = 33 Oe and cubic anisotropy H_c = 3.5 Oe

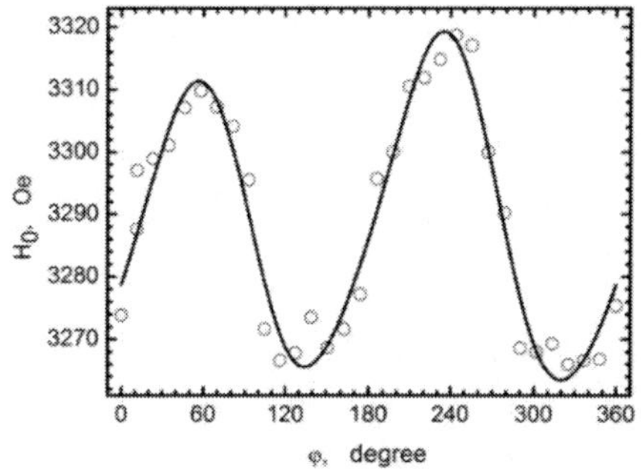

Fig.4. The angular dependence of the resonance magnetic field H_0 for LBMO/NGO for sample rotation around the normal to the film plane. The solid line describes the planar magnetic anisotropy with adjustable parameters given in the text.

were obtained from the fitting the solution of the Landau-Lifshitz equation for the evolution of the magnetization M in an external constant magnetic field H (see eq.1 ref.[25])

Fig. 5 shows the temperature dependences of the resistance the values of H_0 for the LBMO/NGO film and the width of the FMR line $\Delta_{1/2}$. The value of H_0 decreases due to the increase in the magnetization of the LBMO film. The Curie temperature could be determined from the inflection point of the H_0 (T) dependence. At the same time, as the temperature decreases the width of the FMR line increases. Usually in manganites the width of the FMR line decreases with temperature [25]. Possible cause of increase the line width is strong interaction of carries with magnons, which is absent in LSMO/NGO films [25 - 27].

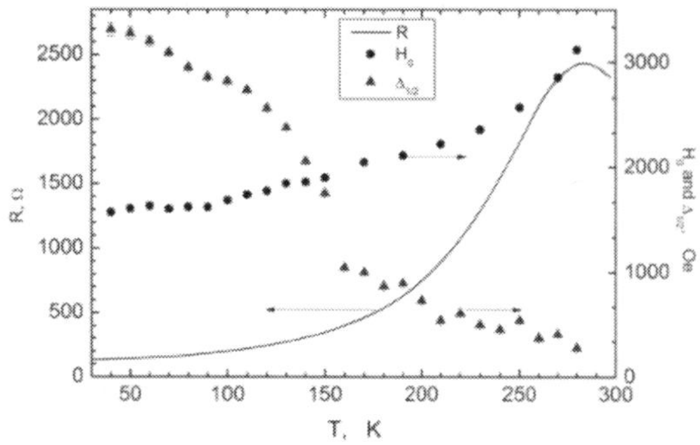

Fig.5. The temperature dependences of the resistance (solid line), the resonance magnetic field (circles) and the width of the FMR line (triangular) for LBMO/NGO film.

6 Conclusion

Epitaxial manganite lanthanum-barium films were deposited on 4 types of substrates that provide strain of the crystal lattice of films. The temperature dependence of the resistivity of all films at low temperatures is well described by a power polynomial that takes into account the effect of scattering on defects and impurities as well as electron-electron and electron-magnon interactions. The line of ferromagnetic resonance of the films is not described by a single Lorenz line, which indicates on the inhomogeneity of the ferromagnetic phase in the films. With temperature decreasing the widening of the FMR line is observed.

Supports by the International Associated Laboratory LEMAC-LICS, RFBR grants 17-02-00145 and 16-29-14022 are acknowledged

References

1. A.-M. Haghiri-Cosnet and J.P.Renard. J. Phys D:Appl. Phys. **36**, R127 (2003).
2. Zh. Huang, G.Y. Gao, Zh.Zh. Yin, X.X. Feng, Y.Zh. Chen, X.R. Zhao, J.R. Sun, W.B. Wu, J. Appl. Phys. **105**, 113919 (2009).
3. G.A. Ovsyannikov, A.M. Petrzhik, I.V. Borisenko, A.A. Klimov, Yu.A. Ignatov, V.V. Demidov, S.A. Nikitov, JETP, **108**, 48 (2009).
4. Y.Lu, J. Klein, C. Hofener, B. Wiedenhorst, J.B. Philipp, F. Herbstritt, A. Marx, L. Alff, R. Gross. Phys. Rev. B **62**, 15 806 (2000).
5. Y. Chen, B.G. Ueland , J.W. Lynn, G.L. Bychkov and S.N. Barilo, Y.M. Mukovskii, Phys. Rev. B 78, 212301(2008).
6. F. Tsui, M.C. Smoak, T.K. Nath, C.B. Eom, Appl. Phys. Lett. **76**, 2421 (2000).
7. A.J. Millis, T.Darling, A. Migliori, J. Appl. Phys. **83**, 1588 (1998).
8. M. Bibes, S. Valencia, L. Balcells, B Martínez, Phys. Rev. B **66**, 134416 (2002).
9. H.L. Ju, J. Gopalakrishnan, J.L. Peng, Qi Li, G.C. Xiong,T. Venkatesan, R.L. Greene, Phys. Rev. B **51**, 6143 (1995).
10. R.V. Demin, L.I. Koroleva, A.Z. Muminov, Ya.M. Mukovski, Physics of Solid States **48**, 322 (2006).
11. W. Zhou, Y. Xiong, Z. Zhang, D. Wang, W. Tan, Q. Cao, Z. Qian, and Yi Du. ACS Applied Materials & Interfaces, **8**, 5424 (2016).
12. P. Orgiani, C. Adamo, C. Barone, A. Galdi, S. Pagano, A. Yu. Petrov, O. Quaranta, C. Aruta, R. Ciancio, M. Polichetti, D. Zola, and L. Maritato, J. Appl. Phys. **103**, 093902 (2008).
13. Yu.A. Boikov, T. Claeson, Technical Physics, **59**, 1027 (2014).
14. H.Y. Hwang, T.T.M. Palstra, S-W. Cheong, B. Batlogg, Phys. Rev. B 52, 15046, (1995).
15. Y.Suzuki, H.Y. Hwang, S-W. Cheong et al., J. Appl. Phys. **83**, 7064 (1998).
16. G.J. Snyder, R. Hiskes, S. DiCarolis, M.R. Beasley, T.H. Geballe. Phys. Rev. B **53**, 14434 (1996).
17. J. Zhang, H. Tanaka,T. Kanki, J.-H. Choi, T.Kawai, Phys.Rev. B. **64.** 184 404 (2001).
18. K. Kubo, N. Ohata, J. Phys. Soc. Jpn. **33**, 21 (1972).
19. Yu.A. Boikov, I.T. Serenkov, V.I. Sakharov T. Claeson, A. Kalabukhov, V.V. Afrosimov, Physics of Solid States **55**, 2043 (2013).
20. S. V. Pietambaram, D. Kumar, R. V. Singh, C. B. Lee, and V. S.Kaushik, J. Appl. Phys. **86**, 3317 (1999).
21. W. Jiang, X. Z. Zhou, G. Williams, Y. Mukovskii, and K. Glazyrin, Phys. Rev. B 77, 064424 (2008).
22. Yu.A. Boikov, V.A. Danilov, Physics of Solid State **50**, 451 (2008).
23. N.G. Bebenin, R.I. Zainullina, V.V. Mashkautsan, V. V. Ustinov, Ya. M. Mukovskii, Phys. Rev. B. **69**, 104434 (2004).
24. Yan Wu, Y. Suzuki, U. Rudiger, J. Yu, and A. D. Kent T. K. Nath and C. B. Eom, Appl. Phys. Lett. **75**, 2295 (1999).
25. V.V. Demidov, G.A. Ovsyannikov, A.M. Petrzhik, I.V. Borisenko, A.V. Shadrin, and R. Gunnarsson, J. Appl. Phys. **113,** 163909 (2013).
26. R. von Helmolt, J. Wecker, B. Holzapfel, L. Schultz, and K. Samwer, Phys. Rev. Lett. **71**, 2331 (1993).
27. S. E. Lofland, S. M. Bhagat, H. L. Ju, G. C. Xiong, T. Venkatesan, and R. L. Greene, S. Tyagi. J. Appl. Phys. **79**, 5166 (1996).

Quantum effects in magnetotransport of InGaAs quantum wells with remote Mn impurities

Leonid Oveshnikov[1,2,⋆], *Leonid* Morgun[2,3,4], *Elena* Nekhaeva[1], *Vladimir* Kulbachinskii[1,5], and *Boris* Aronzon[1,2]

[1]*National Research Center "Kurchatov Institute", 123182 Moscow, Russia*
[2]*P.N. Lebedev Physical Institute RAS, 119991 Moscow, Russia*
[3]*Moscow Institute of Physics and Technology, 141700 Dolgoprudnyi, Moscow Region, Russia*
[4]*National Research University Higher School of Economics, 101000 Moscow, Russia*
[5]*M.V. Lomonosov Moscow State University, 119991 Moscow, Russia*

Abstract. We have studied magnetoresistance and Hall effect of GaAs/$In_xGa_{1-x}As$ quantum wells with remote Mn impurity. Temperature and magnetic field dependencies of samples resistivity indicate several effects related to the magnetic subsystem. Shubnikov - de Haas oscillations indicate the presence of several types of regions in conduction channel with significantly different hole mobilities. We discussed the impact of magnetic impurities on quantum corrections to conductivity by comparing our results with the data for similar non-magnetic structures. Our results suggest that the presence of Mn atoms leads to the damping of quantum corrections in in the investigated structures.

1 Introduction

Two-dimensional (2D) semiconductor structures with magnetic impurities are considered to be one of the most promising as a basic system of next-generation spintronics [1]. Substantial part of related publications are focused primarily on unique effects driven by magnetic subsystem (such as anomalous Hall effect (AHE)) [1–4], while the impact of magnetic impurities on more conventional properties of different systems, such as "metal-insulator" transition and quantum oscillations, is much less discussed in literature.

Introducing magnetic impurities into semiconductor matrix at concentrations high enough for the related effects (e.g. ferromagnetism) to be observed usually turns systems into strongly disordered media, for which quantum corrections should be significant. Conventional theory predicts that ferromagnetic phase tends to destroy quantum interference correction, while experimental data for such systems can be rather ambiguous [5–7]. Moreover, electron-electron (*e-e*) interaction in the presence of ferromagnetic subsystem is relatively poorly studied.

Here we have studied magnetotransport properties of GaAs/$In_xGa_{1-x}As$ quantum wells with separated by spacer Mn impurity. Spatial separation of acceptor Mn atoms and conducting holes leads to high mobility values of the latter along with feasibility of magnetic interactions between Mn ions. One of the reasons for interest to these studies is the recent experimental investigation of analogous systems with hopping conductivity, for which the indications of "topological" Hall effect contribution have been observed [2].

As far as we know, quantum effects in magnetotransport of systems with spatially separated conducting and magnetic subsystems are poorly studied.

2 Samples

The investigated structures were synthesized by MOCVD method on semi-insulating GaAs substrates (schematic structure of the similar samples is published in Ref's [2–4]). Quantum well (QW) $In_xGa_{1-x}As$ with $x \approx 0.2$ was sandwiched between carbon (lower) and manganese (upper) impurity layers separated by GaAs spacers. Samples with different Mn content in the range 0.4 to 0.5 monolayers (ML) were mesa etched into standard Hall bars. Temperature and magnetic field dependencies of transverse R_{xx} and Hall R_{xy} were measured in the temperature interval 2K $< T <$ 77K and in magnetic field B up to 9T. Additionally, the magnetoresistance of 0.5ML sample was studied for B = ±1T at temperatures down to 0.3K.

Previous studies of analogous samples showed that Mn atoms strongly diffuse into nearby layers forming comparatively thick (3-5 nm) (Ga,Mn)As layer [8]. Mn atoms in GaAs crystal matrix form electrically charged substitutional and interstitial defects. Inhomogeneous distribution of Mn in the impurity layer gives rise to the fluctuation potential (FP). FP forms additional potential wells in the conduction channel that leads to the "metal-insulator" transition for structures with sufficiently low Fermi energies. Theory predicts that in insulating state conduction channel in such structure consists of metallic droplets in-

⋆e-mail: Oveshln@gmail.com

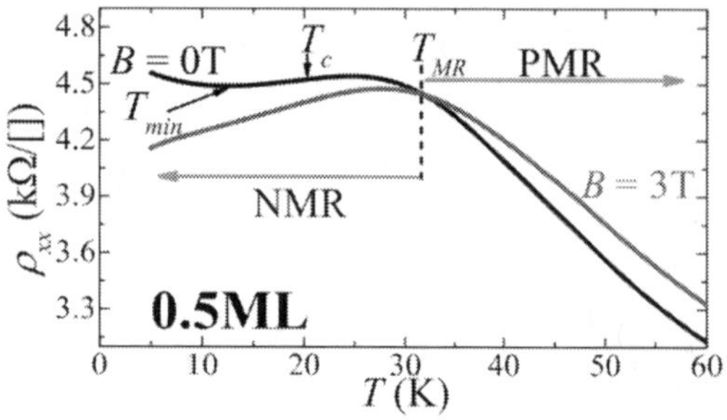

Figure 1. (colour online) Temperature dependencies of resistivity for 0.5ML sample at $B = 0$ and 3T. Temperature ranges of positive and negative MR are marked, along with several characteristic temperatures (see text).

serted into insulating matrix [9]. Significant difference of Hall mobility and estimated from Shubnikov - de Haas (SdH) oscillations mobility was experimentally observed for these structures [10]. Moreover, SdH oscillations were observed even for samples with hopping conductivity. It suggests the presence of at least two types of consecutively connected areas in QW with different mobilities, which is in good agreement with theory. It is important to note that mobility difference for these two types of areas remains significant even for samples with drift conductivity (including samples studied here).

3 Results and discussion

Due to relatively weak exchange interaction of conducting holes with Mn impurities, samples under study can be considered as a conventional QW perturbed by magnetic layer. Thus, the comparison of our data with results for conventional QW with 2D hole gas should help us to distinguish phenomena related to the magnetic nature of Mn impurity.

One of the most common feature of diluted magnetic semiconductor systems is the appearance of local maximum (peak) on the temperature dependence of resistivity near T_c, which is related to substantial growth of critical fluctuation density in the vicinity of transition to long-range ordered state. Observation of this peak for the investigated samples at $T \approx 26$K (Fig. 1) indicates the presence of significant hole scattering on the magnetic subsystem. Actual value of $T_c \approx 20$K can be estimated from the maximum on the temperature dependence of $d\rho/dT$ [11]. The mismatch of T_c and $\rho_{xx}(T)$ peak position is common for inhomogeneous systems.

One can see in Fig. 1 that below $T_{min} \approx 12.5$K the value of ρ_{xx} begin to grow upon cooling, which is related to quantum corrections to conductivity. Comparing $\rho_{xx}(T)$ curves for $B = 0$ and 3T, we can distinguish ranges of positive and negative magnetoresistance (MR), separated by $T_{MR} = 31.5$K.

It is important to note, that positive MR is quadratic, while low-temperature MR (Fig. 2a) is a superposition of negative MR in the whole range of magnetic fields and SdH oscillations that occur in high field region. Sharp MR in low field region is due to weak localization (WL). MR

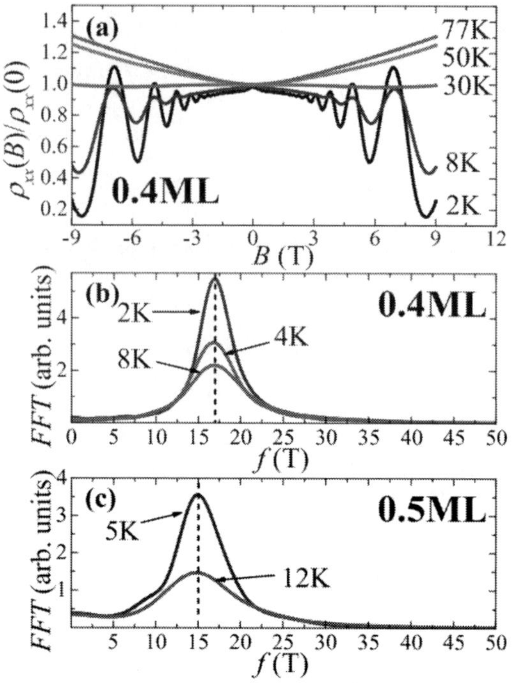

Figure 2. (colour online) (a) Magnetoresistance of 0.4ML sample at different temperatures. Fast Fourier Transform of SdH oscillation (after subtracting background MR) for 0.4ML (b) and 0.5ML (c) samples at different temperatures.

in the intermediate region and non-oscillating part of high-field MR have rather complicated functional form, which, however, cannot be described well by negative quadratic contribution. Thus, the spin-dependent scattering can be considered as the main reason of negative MR for $B > 1$T at $T < 30$K. This assumption is in a good agreement with the aforementioned observation of peak in $\rho_{xx}(T)$. For similar non-magnetic structures negative quadratic MR was ascribed to diffusive *h-h* interaction correction [12].

Observed SdH oscillations were treated in the same way as in Ref's [13, 14]. SdH oscillation frequencies derived by Fast Fourier Transform (FFT) are temperature-independent inside the experimental errors margins for both samples in the investigated temperature interval as it is shown in Fig. 2b and 2c. Assuming the hole spectrum of structures is spin-degenerate, we obtained carrier concentrations $p_{SdH} = 0.82$ and $0.72 \cdot 10^{12}$ cm^{-2} for 0.4ML and 0.5ML samples, correspondingly. Comparing these with values obtained from normal Hall effect $p_{Hall} = 0.96$ and $0.72 \cdot 10^{12}$ cm^{-2} (0.4ML and 0.5ML) one can see, that for 0.5ML sample there is an excellent agreement, while 15% discrepancy for 0.4ML sample indicates the occupation of second subband, that didn't reveal itself in SdH oscillations. As it was said earlier, Hall mobilities μ_{Hall} = 2100 and 1900 cm^2/(V·s) for 0.4ML and 0.5ML samples are significantly lower than values needed to fulfil $\mu B \geq 1$ condition for the observed SdH oscillations [10]. Thus, Hall effect includes average parameters of conduction channel, while SdH effect describes high-mobility regions. By exploring the temperature dependence of SdH

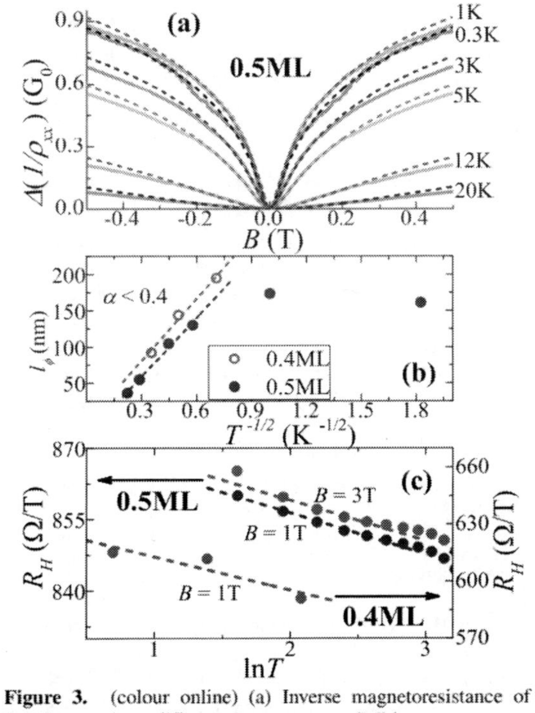

Figure 3. (colour online) (a) Inverse magnetoresistance of 0.5ML sample at different temperatures. Solid curves - experimental data, dashed lines - fitting by HLN formula (1). (b) Temperature dependence of the dephasing length l_ϕ. Dashed lines - linear fitting of $T > 2$K range. (c) Temperature dependencies of Hall coefficient R_H derived from different field ranges. Dashed lines - linear fit in low temperature region.

oscillation amplitudes, we calculated the effective mass of carriers $m^* \approx 0.15\, m_0$ (m_0 - free electron mass). This value is rather close to that reported for non-magnetic QW [12].

Low-field MR plotted in Fig. 3a has rather distinctive functional form for both samples, which is usually related to the damping of WL correction by magnetic field. It is important to note, that characteristic parameter of WL correction $B_{tr} = \hbar/(2el^2)$ (with $\hbar$ - Plank constant, e - electron charge, l - mean free path) for samples under study is approximately equal to 0.3T and 0.45T for 0.4ML and 0.5ML samples correspondingly for Hall parameters presented earlier. It should be mentioned, that for these samples we didn't observe any traces of weak antilocalization (WAL) effect, that was clearly seen for similar nonmagnetic structures with close parameters [15].

In order to investigate WL correction parameters in our samples we used common Hikami-Larkin-Nagaoki (HLN) formula [16] for 2D systems as a fitting function of low-field MR. In diffusive regime corresponding MR can be described in the following way:

$$\frac{\Delta(1/\rho_{xx}(B))}{G_0} = \alpha\left[\psi\left(\frac{1}{2} + \frac{\hbar}{4el_\phi^2 B}\right) - \ln\left(\frac{\hbar}{4el_\phi^2 B}\right)\right], \quad (1)$$

where ψ - digamma function, $G_0 = e^2/(2\pi^2\hbar)$, l_ϕ - dephasing length, α - prefactor. One can see that fitting curves are in a good agreement with experimental data in the field range of $B \leq \pm 0.3 B_{tr}$ (Fig. 3a). Obtained $l_\phi(T)$ dependencies are plotted in Fig. 3b. It is clearly seen that $l_\phi \sim T^{-1/2}$ at $T > 2$K for both samples. This behaviour usually indicates that inelastic e-e (h-h) scattering is the main dephasing mechanism [17], which is common for 2D systems. But below 2K the dependence $l_\phi(T)$ demonstrates saturation, which can be attributed both to conventional disorder and to Mn-related effects. However, the absence of this well pronounced saturation in nonmagnetic samples (see Fig. 4 in Ref. [15]), where $\tau_\phi \sim l_\phi^2 \sim 1/T$ behavior persists down to 0.4K for all samples, favours the latter assumption. Another peculiarity of the observed WL is prefactor values. Theory predicts $\alpha = 1$, as a mark that whole group of carriers contribute to WL [17, 18]. Also there are several effects, that can be responsible for lower experimental α values [18]. But none of these effects can decisively explain $\alpha < 0.4$ observed for our samples. Therefore, it can be related to the presence of Mn-layer. This assumption is partially supported by the absence of WAL effect.

It is commonly known that in 2D systems e-e interaction (or h-h interaction in our case) can be significant and the relation between interference (WL and WAL) and interaction corrections should be derived from the experiment in every particular case [17]. The contribution of h-h interaction in the diffusive channel is given by

$$\delta\sigma^{h-h} = K_{h-h} G_0 \ln\left(\frac{k_B T \tau}{\hbar}\right), \quad (2)$$

where k_B - Boltzman constant, T - temperature, τ - transport relaxation time, and K_{h-h} is the prefactor value, which is usually close to unity. A diffusive contribution should dominate in the interaction correction (favouring the drop of conductivity upon cooling) at $k_B T\tau/\hbar \ll 1$, which is for our samples is equal to $T \ll 45$K, for given Hall parameters. There are three basic approaches to determine K_{h-h} value from the experimental data. One of them is to investigate the logarithmical part of $\sigma_{xx}(T)$ dependence, assuming that σ_D (Drude conductivity) in studied region is temperature independent and changes of σ_{xx} are associated solely with the sum of interference and interaction correction. This approach is not applicable in our case because of the presence of a resistivity peak related to the critical fluctuations of magnetic subsystem. This peak can be very broad in inhomogeneous systems leading to additional temperature dependence of conductivity down to lowest temperatures. Indeed, using obtained parameters of WL we calculated the corresponding correction, taking it into account remaining part of conductivity changes derivative sign, i.e. below T_c conductivity increases upon cooling without upturns of well pronounced saturation region. In frame of the second approach of K_{h-h} determination one should examine negative quadratic MR related to h-h interaction. However, as it was said earlier, negative MR in our case is not quadratic and ascribed to spin dependent scattering. As far as we know the most suitable way to determine K_{h-h} in our case is to examine $R_H(T)$ dependencies. The diffusive h-h interaction does not give any contribution to σ_{xy}. Using standard relations between conductivity and resistivity tensors one can see that in this case interaction correction can be observed in Hall coefficient R_H [19]:

$$\frac{\delta R_H}{R_H} = -2\frac{\delta\sigma^{h-h}}{\sigma_D}, \quad (3)$$

Thus, in 2D system $R_H(T)$ dependence should demonstrate logarithmical increase upon cooling. In Fig. 3c corresponding dependencies are shown. One can see that for 0.5ML sample there is a difference of $R_H(T)$ derived by linear approximation of ± 1T and ± 3T field intervals of R_{xy}, which can be related to the presence of AHE contribution, while logarithmical slope of low-temperature region is the same for both curves. It should be noted that the positive value $R_H(3T) - R_H(1T) > 0$ indicates the presence of negative AHE contribution, though to investigate it's actual dependence on B and T an additional data processing should be applied (see e.g. [2]). Logarithmic slope of presented dependencies was estimated in the same temperature range, for which SdH oscillations were investigated. Using the SdH frequency temperature independence we can assume that logarithmic change of $R_H(T)$ is related to h-h correction. Thus, applying relations (2) and (3) we have obtained prefactor values $K_{h-h} \approx 0.4$ and 0.1 for 0.4ML and 0.5ML samples, correspondingly. The value of $K_{h-h} \approx 0.2$-0.3 was reported for similar non-magnetic structures in Ref. [12], if one uses the relation between K_{h-h} and Fermi liquid parameter F_0^{σ}. Although R_H data presented in that work wasn't discussed one can easily see that applying the same relations (2) and (3) yields sufficiently larger K_{h-h} value (by a factor of 2 under our estimations). Thereby the exact impact of Mn layer on h-h interaction in QW cannot be formulated from the current data and needs further investigation. The discrepancy of K_{h-h} for studied samples can be attributed to the occupation of second subband in 0.4ML sample and presumably different states of magnetic subsystems at different Mn concentrations.

Finally, we can compare different length scales observed for samples under study. For simplicity we will use data for 0.5ML sample. Magnetic length corresponding to SdH oscillations onset at 5K is about 17nm [10], which is rough estimation for minimum size of high mobility regions (metallic droplet in QW). Even for Hall parameters mean free path is about 26nm, while for parameters of high-conductive droplet it can be even larger. Thus, at zero field these droplets can be considered as areas of "ballistic" transport, which can affect different properties of system, e.g. interaction correction. For example, the logarithmic growth of $R_H(T)$ starts to be significant below 12K, while corresponding $L_T = (D\hbar/k_B T)^{1/2}$ (D - diffusion coefficient) at 12K is equal to 52nm. Thus, rather low temperature of interaction correction onset and low values of K_{h-h} can be related to the contribution of these ballistic areas (for which $k_B T\tau/\hbar > 1$) that have an opposite sigh compared to the diffusive one. Also, if we assume that saturation of $l_\phi(T)$ dependence is related to the spin-dependent scattering becoming main dephasing mechanism, then saturation value of $l_\phi \approx 170$ nm gives us estimation of low-field correlation length of magnetic subsystem.

4 Conclusions

We have studied magnetoresistance and Hall effect of $GaAs/In_xGa_{1-x}As$ QWs with remote Mn impurity. The presence of resistivity peak in $\rho_{xx}(T)$ and negative MR for $B > 1$T below 30K indicates the significant spin-dependent scattering of conducting holes on magnetic subsystem, which is partially supported by the observed AHE contribution. Observation of SdH oscillations indicates the presence of several types of regions in conduction channel with significantly different hole mobilities. The absence of pronounced WAL effect along with abnormally low prefactors for WL and h-h interaction contribution can be attributed to the damping of quantum corrections by magnetic impurities. The saturation of the dephasing length upon cooling can be also related to the presence of Mn atoms. Although for determination of the exact impact of Mn impurities on quantum corrections to conductivity in studied systems further investigations are needed.

This work was partially supported by Russian Foundation for Basic Research (grant #16-32-00446).

References

[1] A.M. Nazmul *et. al.*, Phys. Rev. Lett. **95**, 017201 (2005)

[2] L.N. Oveshnikov *et. al.*, Sci. Rep. **5**, 17158 (2015)

[3] L.N. Oveshnikov *et al.*, JETP Lett. **100**, 570 (2014)

[4] V.A. Kulbachinskii *et.al.*, Low Temp. Phys. **33**, 174 (2007)

[5] D. Neumaier *et. al.*, Phys. Rev. Lett. **99**, 116803 (2007)

[6] V.K. Dugaev *et. al.*, Phys. Rev. Lett. **101**, 129701 (2008)

[7] D. Neumaier *et. al.*, Phys. Rev. Lett. **101**, 129702 (2008)

[8] B.A. Aronzon *et al.*, J. Appl. Phys. **107**, 023905 (2010)

[9] V. Tripathi *et al.*, Phys. Rev. B **84**, 075305 (2011)

[10] L.N.Oveshnikov *et al.*, Bull. Lebedev Phys. Inst. **44**, 143 (2017)

[11] M. Wang *et al.*, App. Phys. Lett. **104**, 132406 (2014)

[12] G.M. Minkov *et al.*, Phys. Rev. B **72**, 165325 (2005)

[13] V.A. Kulbachinskii *et al.*, Semiconductors **49**, 921 (2015)

[14] V.A. Kulbachinskii *et al.*, Semiconductors **49**, 199 (2015)

[15] G.M. Minkov *et al.*, Phys. Rev. B **71**, 165312 (2005)

[16] S. Hikami *et al.*, Prog. Theor. Phys. **63**, 707 (1980)

[17] B.L. Altshuler, A.G. Aronov, *Electron-electron interaction in disordered conductors* (Elsevier, Amsterdam, 1985)

[18] G.M. Minkov *et al.*, Phys. Rev. B **70**, 245423 (2004)

[19] A.Yu. Kuntsevich *et al.*, Phys. Rev. B **94**, 235401 (2016)

Ab-initio study of electronic and magneto-optical properties of InAs:Mn

Elena Gan'shina[1,*], *Erkin* Kulatov[2,4], *Leonard* Golik[3], *Zoya* Kun'kova[3], *Yurii* Uspenskii[4], *Georgy* Zykov[1], *Ye* Yuan[5], and*Shengqiang*Zhou[5]

[1]Department of Physics, Moscow State University, Moscow 119991, Russia
[2]Prokhorov General Physics Institute, RAS, Moscow 119991, Russia
[3]Institute of Radioengineering and Electronics, RAS, Fryazino 141190, Russia
[4]Lebedev Physical Institute, RAS, Moscow 119991, Russia
[5]Helmholtz–Zentrum Dresden–Rossendorf, Institute of Ion Beam Physics and Materials Research, 01328 Dresden, Germany

Abstract. Energy difference between the ferromagnetic and antiferromagnetic collinear orderings has been calculated for the uniform and dimer Mn-pair geometries in order to find the ground state distribution of the Mn atoms in InAs host. We find the preference of the dimer ferromagnetic configuration of Mn dopants and an importance of optimizing the atomic site positions. The frequency-dependent optical and magneto-optical properties, namely the dielectric tensor (on- and off-diagonal components), the electron energy loss spectra, and the transversal Kerr effect (TKE), are calculated. Calculated TKE resonance in $In_{1-x}Mn_xAs$ (x=0.0625) is found to be in good agreement with corresponding experimental magneto-optical spectra. The origin of the large TKE is discussed.

1Introduction

Diluted magnetic semiconductors (DMS) are considered as promising materials for spintronics. However the Curie temperature (T_C) of the magnetic semiconductors is below room temperature and this fact limits their practical application. To increase T_C one should increase concentration of magnetic impurity in host semiconductor. In DMS the concentration is limited by low solubility of magnetic impurities in these materials. Exchange interaction between neighbouring Mn ions would be stronger in semiconductor based structures with inhomogeneous distribution of Mn ions, and achieved Curie temperature in such structures would be significantly higher than in DMS with uniform distribution of magnetic dopants. In this paper we study electronic and magneto-optical properties of InAs:Mn DMSregarding its possible application in spintronics.

2 Calculational method

Our calculations are performed using the relativistic Augmented Plane Wave + local orbital (APW + lo) method as implemented in WIEN2k code [1]. The exchange-correlation effects are described within the Generalized Gradient Approximation (GGA)of density functional theory [2]. Density-functional calculations presented here are performed within the supercell (2x2x2 cubic zinc blende unit cell), corresponding to 64 atoms. One or two Mn atoms are included in the supercell. We consider substitutional doping where In host atoms are replaced by Mn ones. Thus, the Mn concentration in our calculations varies in the experimentally relevant range (x_{Mn}=0.0312 and 0.0625). Experimental lattice parameter of InAs host (a_0= 6.058 Å) [3] is also used for InAs:Mn DMS's. The 4d states of In and 3d states of As are included in the APW basis set and treated as the valence orbitals. Converged results are obtained at the cut-off energy for the interstitial plane-wave expansion of 125 eV. Relaxations of all atomic positions have been included in our calculations. The total energy difference between ferromagnetic (FM) and antiferromagnetic (AFM) collinear orderings has been calculated for the uniform and dimer Mn-pair geometries in order to find the ground state distribution of the Mn atoms in InAs host. The magneto-optical transversal Kerr effect of $In_{1-x}Mn_xAs$ (x=0.0625) is calculated in the framework of the relativistic WIEN2k method [1], which takes into account the effect of spin-orbit coupling in the second variational procedure.

3 Results and Discussion

Optimization of the positions of all atoms in a supercell reveals that in InAs:Mn the bond length Mn-As shrinks by 2.5% in comparison with the unrelaxed case. Relaxation effects for the bonds Mn-In were found to be negligible (contraction by 0.1%).The calculated relativistic density of states (DOS) for an energetically

Corresponding author eagan@mail.ru Страница 1

important Mn dimmer configuration is shown in Fig. 1 for InAs:MnMn in the FM dimer and separate configurations and with Mn concentration equal to 0.0625. It is seen that InAs: Mn in both configurations is close to the half-metallic state. The main feature of shown DOS's is the large exchange splitting of the 3d-states of Mn (~4 eV).We find the preference of the dimer configuration of Mn dopants, i.e. the configuration

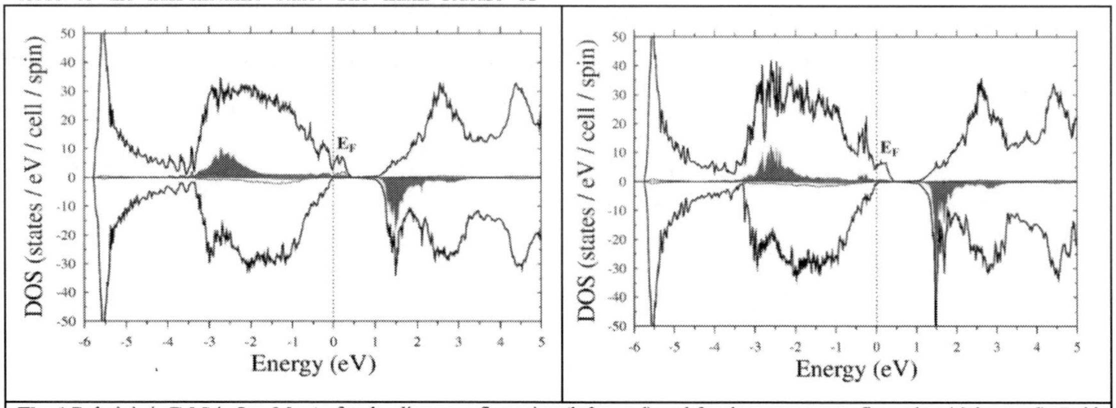

Fig. 1. Relativistic DOS in $In_{1-x}Mn_xAs$ for the dimer configuration (left panel) and for the separate configuration (right panel). Bold line – total DOS, shaded red region – 3d Mn partial DOS, thin line – 4p As partial DOS, E_F means the Fermi level.

Table 1. The AFM and FM energy differences ($\Delta(E_{AFM}$-$E_{FM})$) , total and local (in parentheses) magnetic moments (M_{FM}, M_{AFM}) in the FM and AFM phases,Δ(FM) and Δ(AFM) are the energies of the FM and AFM phases with respect to the energy of the FM dimer.

Configuration of Mn atoms	$\Delta(E_{AFM}$-$E_{FM})$, meV/Mn	M_{FM}, μ_B/cell (/Mn)	M_{AFM}, μ_B/cell (/Mn)	Δ(FM) Δ(AFM) meV/Mn
close (dimer)	+233	8.08 (+4.08)	0.002 (±4.11)	0 +233
intermediate	+218	8.08 (+4.09)	0.003 (±4.13)	+120 +338
separate	+ 55	8.13 (+4.11)	0 (±4.14)	+125 +180

where the Mn atoms share the same neighbouring As site is always found to be energetically most favorable (see the 5-th column of Table 1). So our calculations show that clustering effect is energetically favorable in agreement with the experimental results [4]. Also we found that the FM order is the most stable magnetic state (Table 1, the2-nd column).An interesting feature of DOS is a narrow unoccupied part of the valence band, which is placed just above the Fermi level. This area of 0.4 eV width is separated from the conducting band by the quasigap of 0.9 eV width. As is shown below, electron transitions from occupied states to this unoccupied part of the valence band and to the conduction band (across the quasigap) form important spectral features in the dielectric tensor and energy loss spectra.

It is known [5] that TKE is determined by both magneto-optical and optical parameters of a medium. The TKE value, $\delta_p = (I(H)\text{-}I_0)/I_0$, is the relative change in intensity of the reflected light caused by the magnetization of a sample in an external magnetic field whose direction is transverse to the plane of the light incidence. TKE observed at the angle of incidence φ can be expressed by the off- and on-diagonal components of the dielectric function as

$$\delta_p = 2\sin(2\varphi)(A/(A^2 + B^2))\cdot\varepsilon'_1 + (B/(A^2 + B^2))\cdot\varepsilon'_2 = a\cdot\varepsilon'_1 + b\cdot\varepsilon'_2\,, \quad (1)$$

where ε'_1 and ε'_2 are the dispersive and absorptive parts of the off-diagonal dielectric tensor, respectively, while $A = \varepsilon_2\cdot(2\varepsilon_1\cdot\cos^2\varphi - 1)$ and $B = (\varepsilon^2_2 - \varepsilon^2_1)\cdot\cos^2\varphi + \varepsilon_1 - \sin^2\varphi$ are the functions of the on-diagonal components of the dielectric tensor ε_1 and ε_2. Fig. 2 displays calculated TKE spectra, which show marked features in the infrared region. The energy position of our calculated MO resonance in $In_{1-x}Mn_xAs$ (x=0.0625) are found to be in good agreement with the corresponding experimental TKE spectra (obtained by us in[6]) at ~0.7 eV.

Spectra of calculated and off- and on-diagonal components of the dielectric tensor and electron energy loss spectrum are shown in Figs. 3, 4.Although the MO property is a rather complicated function of the on-diagonal and off-diagonal components of the optical conductivity tensor, our calculation provides a clear insight about its origin(see Figs. 2-4). The special feature of DOS noted above forms intensive electron transitions with ħω<0.4 eV (to the unoccupied part of the valence band) and with ħω>1.3 eV (to the conduction gap) and a peak of ε_2 at 2.2 eV (the left panel of Fig. 4).Large TKE observed in InAs:Mn can be understood as a combined effect of the on- and off-diagonal components, nearly maximal exchange splitting of the Mn3dstates and large spin-orbit coupling. Additional enhancement of the 0.7 eV TKE peak comes from the plasma resonance at about 0.65 eV, which occurs in our calculated electron energy loss spectrum (EELS) (the right panel of Fig. 4) near the minimum of ε_2.

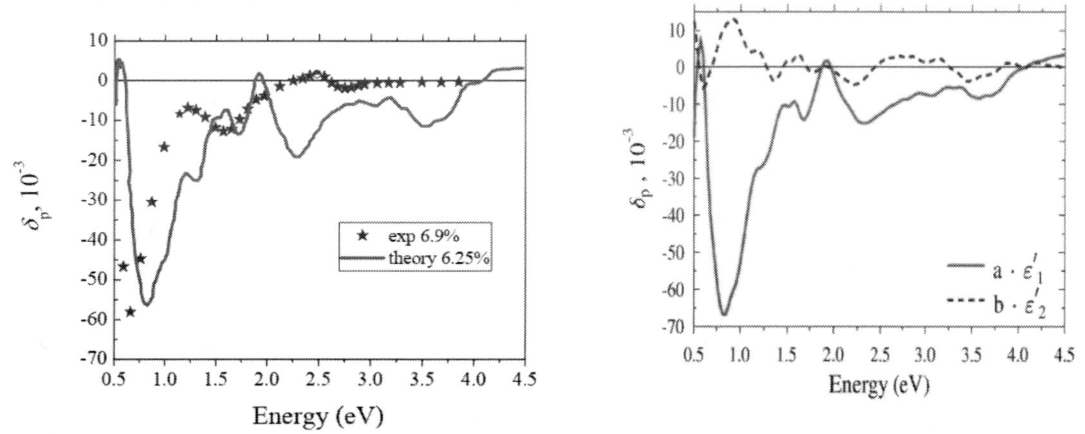

Fig. 2. Left panel:calculated TKE in InAs:Mnfor the dimer configuration. Symbols are the experimental TKE data for InAs: Mn (x = 0.069) from[6]. Angle of light incidence φ = 66°. Right panel shows decomposition of δ_p into the sum of products (see Eq. (1)).

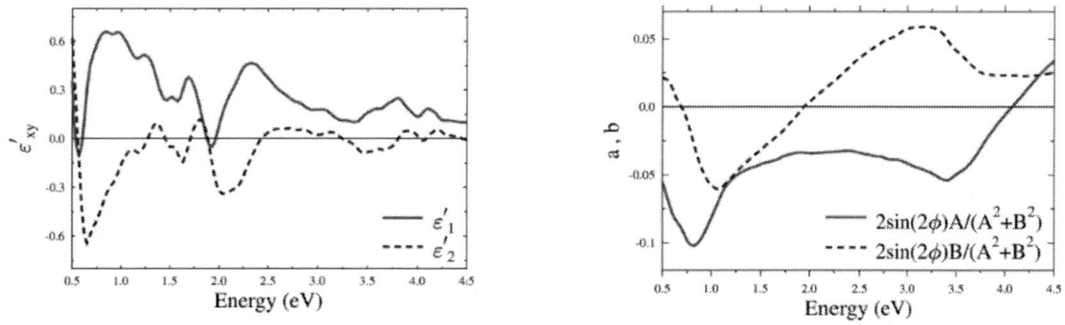

Fig. 3. Calculated off-diagonal components of the dielectric tensor ε'_{xy}=ε'₁+iε'₂(left panel) and **a**, **b** factors (see Eq. (1)) (right panel).

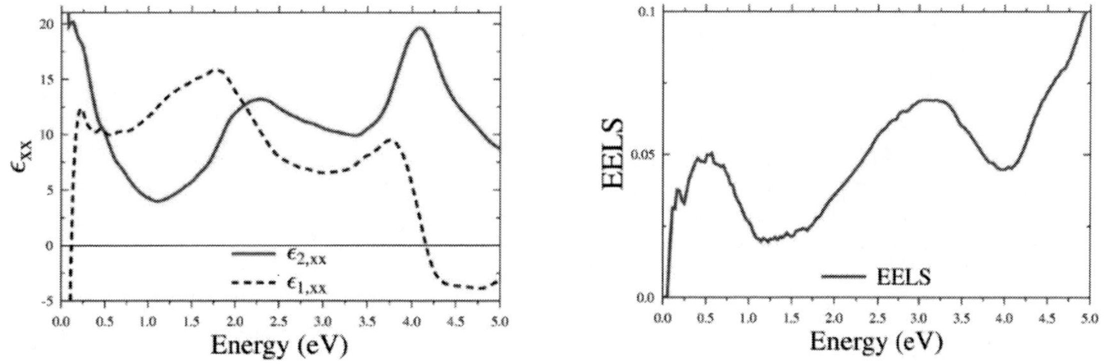

Fig. 4. Calculatedon-diagonal components of the dielectric tensor$\varepsilon_{xx}=\varepsilon_1+i\varepsilon_2$ (left panel) and electron energy loss spectrum (right panel).

Conclusion

Based on the calculation of the relativistic density of states for InAs:Mn in the dimer and separate configurations is found that InAs:Mn in both configurations is close to the half-metallic state. But the dimer ferromagnetic configuration of Mn dopants is always found to be energetically most favorable. Also found that the FM order is the most stable magnetic state.

The frequency-dependent optical and magneto-optical properties, the electron energy loss spectra, on- and off-diagonal conductivity tensor components and the transversal Kerr effect (TKE) are calculated. The modeled TKE resonance in $In_{1-x}Mn_xAs$ (x=0.0625) in near IK range is found to be in good agreement with the experimental magneto-optical spectrum. The additional enhancement of the 0.7 eV TKE peak comes from the plasma resonance at about 0.65 eV, which occurs in the calculated electron energy loss spectrum.

Support by Russian Foundation for Basic Research No 15-02-02077, 16-02-00612-a, 16-02-00024-a, 16-07-00798, and Program of Russian Academy of Sciences "Strongly correlated electrons in solids and structures" and "Basic investigations of nanotechnologies and nanomaterials" is acknowledged.

References

1. P. Blaha, K. Schwarz, G.K.H. Madsen, D. Kvasnicka, J. Luitz, Wien2k, An Augmented Plane Wave+Local Orbitals Program for Calculating Crystal Properties (Karlheinz Schwarz, Tech. Universitat Wien, Austria, 2001).
2. J.P. Perdew, S. Burke, M. Ernzerhof Phys. Rev. Lett. **77,** 3865(1996).
3. NIMS Materials Database, National Institute for Materials Science, Tsukuba, Japan,

   http://crystaldb.nims.go.jp/crystdb.
4. N.D. Parashar, N. Rangaraju, V.K. Lazarov *el al* Phys. Rev. B **81,** 115321(2010).
5. G.S. Krinchik and G.M Nurmukhamedov, Sov. Phys. JETP, **21,** 22(1965).
6. Elena Gan'shina, Leonard Golik, Zoya Kun'kova, IgorBykov, Andrey Novikov, Alexander Rukovishnikov, Ye Yuan, GeorgyZykov, Roman Böttger, and Shengqiang Zhou, Japanese Journal of Applied Physics, **55,**07MF02-1 (2016), http://doi.org/10.7567/JJAP.55.07MF02.

Superparamagnetic behavior of MOCVD grown ZnO:Co films

Vladimir G. Kytin[1,*], *Olga V. Maximova*[1], *Vladimir A. Kulbachinskii*[1,2], *Lidia I. Burova*[1], *Andrey R. Kaul*[1], *Sudipta Bandyopadhya*[3], *Azaharuddin Ahmed*[3], *Aritra Banerjee*[3]

[1]MV Lomonosov MSU, Moscow, 119991, GSP-1, Moscow, Russia
[2]National Research Nuclear University MEPhI, Moscow 115409, Russia
[3]University of Calcutta, Calcutta, India, 92, Acharya Prafulla Chandra Road, Kolkata-700 009, India

Abstract. Temperature and field dependences of magnetization have been measured for Co-doped ZnO films with different Co content grown by MOCVD on sapphire substrates. Measured field dependences of magnetization show ferromagnetic-like hysteresis loops and paramagnetic contribution. Coercive field decreases with an increase of Co content and temperature. The difference between temperature dependences of magnetization measured in zero field cooling and field cooling conditions was observed in the temperature range from 2 K to 300 K indicating superparamagnetic behaviour of the films. The distribution of blocking temperatures extracted from the temperature dependences of magnetization is broad and shifts to the low blocking temperatures with an increase of Co content

1 Introduction

Due to their specific transport and magnetic properties diluted magnetic semiconductors such as ZnO doped with transition metal ions attract a considerable interest of scientific society. In particular ZnO and ZnO:Co materials reveal weak ferromagnetic-like behavior persisting up to room temperatures [1-2]. This behavior remains the subject of discussion until now due to irregular dependence of magnetic properties on preparation method and on concentration of cobalt [1-7]. This irregularity points to the significant role of paramagnetic defects in the magnetic properties of undoped and Co-doped ZnO.

In this work we present the results of the investigation of magnetic properties of ZnO thin films (180-590 nm) doped with Co (1.5-7.2 at%) grown by MOCVD on crystalline sapphire substrates.

2. Experimental

The method and conditions of films preparation were described in our previous work where the conducting properties of the films have been reported [8]. In brief the films were grown on single crystalline sapphire (R-Al_2O_3) substrates. Volatile β_diketonates of metals (anhydrous acetylacetonates $Zn(C_5H_7O_2)_2$ and $Co(C_5H_7O_2)_2$ were used as precursors in the MOCVD process. The deposition temperature was 600°C. The composition of the films and their thickness were determined by X-ray structural microanalysis. Thickness and Co-content of investigated films are listed in table 1.

According to X-ray diffraction data investigated films crystallize in the hexagonal wurtzite structure. No second phases were detected up to the maximal concentration of Co.

Table 1. Thickness and Co content in investigated films

Film thickness, nm	Co content, at. %
180	1.5
380	6.3
590	7.1

The preferred orientation of ZnO (110) is observed, but the X-ray diffraction patterns of the films contain additional reflections corresponding to other orientations of wurtzite ZnO. The number and intensity of these reflections increase with the cobalt concentration and film thickness.

Substitution of Zn^{2+} by Co^{2+} in wurtzite structure was confirmed by X-ray absorption spectroscopy (XAS) using synchrotron radiation (the synchrotron radiation center was BESSY II, RGBL, Berlin). However it does not completely eliminate a possibility of embedding the cobalt ions in octahedral interstices of the wurtzite structure, which is implicitly supported by the slight increase in the wurtzite lattice parameter observed upon an increase of the cobalt concentration [9].

Field and temperature dependences of magnetization were measured using Squid magnetometer VSM (Quantum Design, USA, Model: SVSM-050). Field dependences of magnetizations were recorded with a sweep rate 50 Oe/s. The temperature sweep rate was 4 K/min during the measurements of temperature dependences of magnetization.

* Corresponding author: kytin@mig.phys.msu.ru

3. Results

Field dependences of magnetization measured for the film with 1.5 at. % and 6.3 at. % Co is shown in figures 1 and 2. Diamagnetic susceptibility of substrates was determined from the slope of magnetization dependence on magnetic field at 300 K in magnetic fields above 8 kOe. Then diamagnetic contribution to the magnetization was subtracted from measured magnetization. Magnetization of the films was calculated per formula unit (ZnO)

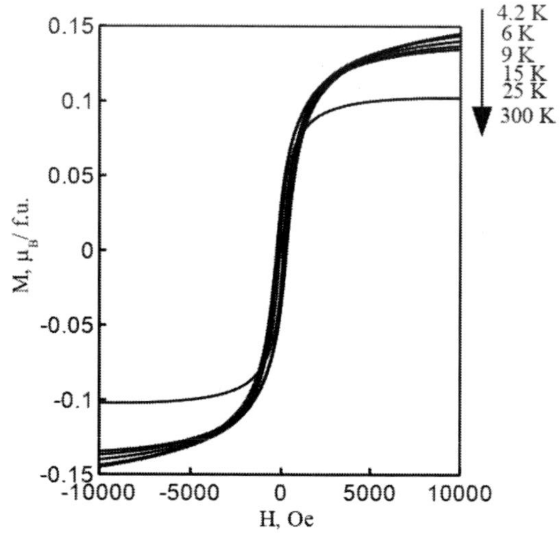

Fig. 1. Magnetization vs. magnetic field for the film with 1.5 at. % Co content

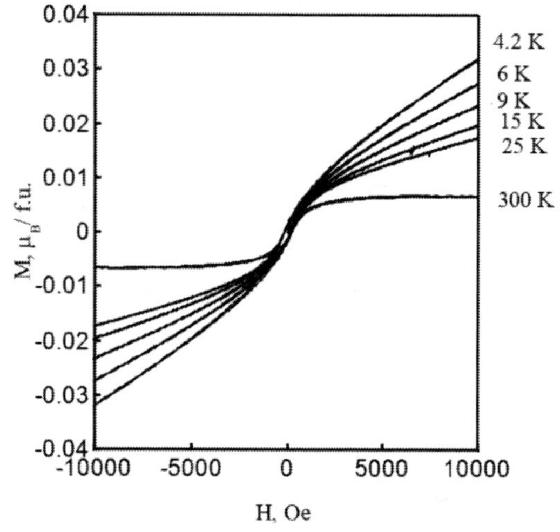

Fig. 2. Magnetization vs. magnetic field for the film with 6.3 at. % Co content

For all films clearly seen hysteresis loop is observed in the temperature range from 4.2 to 300 K. The coercive field decreases with an increase Co concentration. The magnitude of the hysteresis loop is significantly larger for the film with at. 1.5 % Co content than for the films with 6.3 at. % and 7.1 at. % Co.

At temperatures below 25 K magnetization increases nearly linearly with an increase of applied field in magnetic fields from 6 kOe to 10 kOe for the films with 6.3 at % Co and 7.1 at % Co, correspondingly. In this fields and temperature range hysteresis was not observed for these films. For the film with 1.5 at. % Co field dependences of magnetization are close to linear in magnetic field from 8 kOe to 10 kOe.

Temperature dependences of magnetization measured in magnetic field 200 Oe are shown in figures 3 and 4.

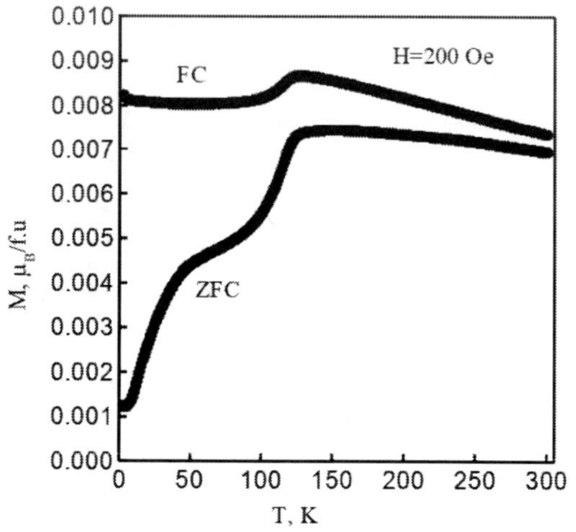

Fig. 3. Magnetization vs. temperature for the film with 1.5 at. % Co content

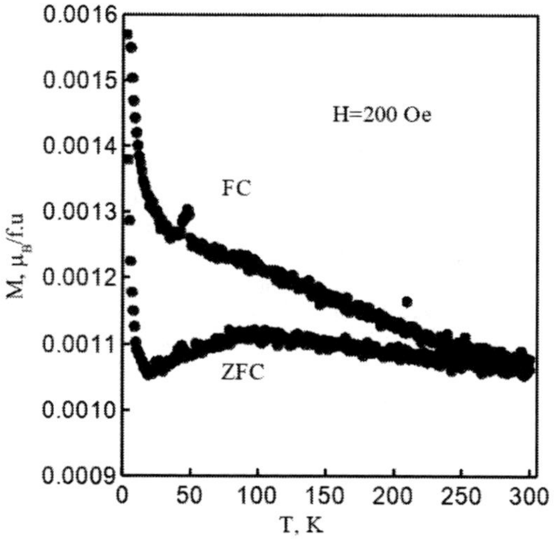

Fig. 4. Magnetization vs. temperature for the film with 6.3 at. % Co content

The temperature dependencies of magnetization measured in zero field cooling (ZFC) and filed cooling (FC) regimes differs for all investigated films at temperatures up to 300 K. The most pronounced difference was observed for the film with 1.5 at. % cobalt.

4. Discussion

Saturation magnetization at 300 K for the film with 1.5 at. % Co content significantly exceeds the maximum possible saturation magnetization originating from Co^{2+} ions. Thus observed hysteresis loop for this film should at least partially originate from paramagnetic defects present in ZnO. This is consistent with the observation of ferromagnetic-like hysteresis loops in MOCVD grown on $MgAlO_2$ substrates undoped ZnO films [10]. For the films with 6.3 % and 7.1 % Co content saturation magnetization measured at 300 K is several times smaller than for the film with 1.5 % Co and than the maximum saturation magnetization of Co^{2+} ions This suggests that Co impurity affects paramagnetic defects in ZnO.

The observation of hysteresis loops and the difference between magnetization values measured in ZFC and FC conditions points to the super paramagnetic contribution to the magnetization [11-13]. The properties of superparamagnetic materials are essentially determined by blocking temperature T_B. This temperature is directly related to the activation energy of the reorientation of magnetic moments of magnetically ordered nanoparticles or nanoclusters. Typically different nanoclusters have different blocking temperature. The following expression was derived for the calculation of blocking temperature distribution $p(T_B)$ from the temperature dependence of magnetization [12,13]:

$$p(T_B) = k\left[\frac{1}{T_B}\frac{d}{dT}\left(M_{ZFC}(T) - M_{FC}(T)\right)\right]_{T=T_B}$$

where k is the temperature independent coefficient, $M_{ZFC}(T)$, $M_{FC}(T)$ is the magnetization measured in ZFC and FC conditions correspondingly.

In figure 5 we plot the derivative of the difference between $M_{ZFC}(T)$ and $M_{FC}(T)$ divided by temperature as a function of temperature. For the film with 1.5 at. % Co content two maxima at 8 K and 110 K are clearly seen in figure 5. For the films with 6.3 at. % Co and at. 7.1 % Co the distribution of blocking temperatures is shifted to lower temperatures. No maxima of $p(T_B)$ can be seen for these films in the temperature range from 2 K to 300 K. The two maxima in the distribution of blocking temperature for the sample with at. 1.5 % Co content may indicate the presence of two different types of magnetic clusters in this film. The peculiarity near 100 K in the temperature dependences of magnetization is seen for both FC and ZFC measurement condition. Therefore it can be manifestation of intrinsic transition within magnetic clusters. For instance charge state of defects responsible for magnetism can vary with temperature due to thermal activation of electrons to the states with higher energy. The strong reduction of electrical conductivity with an increase of Co content was reported earlier for these films [8]. This typically means the change of Fermi level position and charge state of defects and thus their magnetic properties. According to X-ray diffraction data the structure of the investigated films changes significantly with an increase of Co content. Therefore concentration of paramagnetic defects in the films should vary significantly even with small variation of cobalt concentration. This can explain disappearance of the clusters with long blocking time in the samples with high cobalt concentration. Additionally interaction between Co^{2+} ions and paramagnetic defects can decrease blocking temperature.

The variation of coercive field H_c with temperature is shown in figure 6. The coercive field decreases with an increase of temperature for all films at temperatures above 15 K. For the films with 6.3 at. % and 7.1 at % of Co maximum in $H_c(T)$ dependence is observed below 15 K. This maximum most probably originates from linear paramagnetic contribution to the field dependence of magnetization.

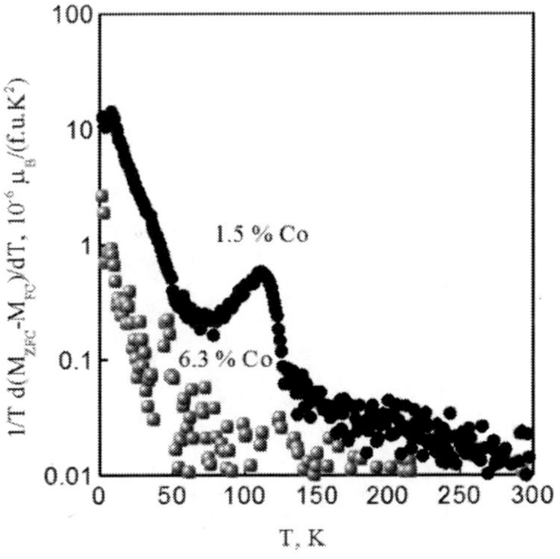

Fig. 5. Blocking temperature distribution ($1/Td/dT(M_{ZFC}-M_{FC})$) for the films with 1.5 % and 6.3 % Co

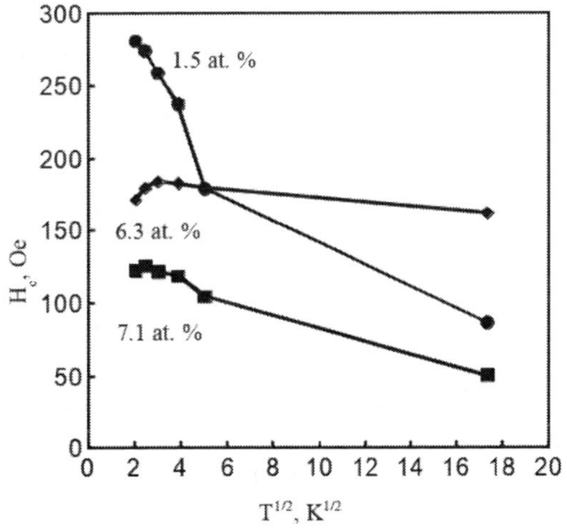

Fig. 6. Temperature dependence of coercive force for investigated films.

For ensemble of non-interacting magnetic nanoparticles coercive field decreases with temperature increase as square root approaching zero value at blocking temperature [13,14]. For investigated films coercive field decreases with temperature slower than it would be expected for distribution of blocking times (fig. 5). In particular a finite value of coercive field was obtained for all investigated films at 300 K. At this temperature the difference between magnetization measured in ZFC and FC conditions is zero within the accuracy of the measurements for the films with 6.3 at % and 7.1 at. % of cobalt concentration. Therefore ferromagnetic contribution can present in the measured magnetization especially in the films with high cobalt concentration.

In figure 7 the inverse magnetic susceptibility $(dM/dH)^{-1}$ calculated for field dependence of magnetization in magnetic field 10 kOe.

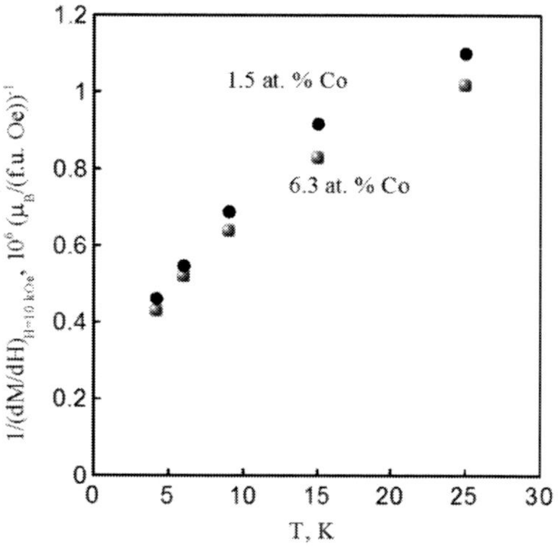

Fig. 7. Temperature dependence of the inverse magnetic susceptibility $(dM/dH)^{-1}$ calculated from the field dependence of magnetization in magnetic field 10 kOe.

At temperatures from 4,2 K to 15 K the dependence is close to linear. The corresponding straight line crosses temperature axis at -5.5 K for the film with 1.5 at % Co and – 5.9 K for the film with 6.3 at. % Co. The deviation from linear dependence at higher temperatures can originate from the rest of diamagnetic contribution. The magnitude of susceptibility calculated at 10 kOe is larger for the films with 6.3 at % Co and 7.1 at % Co than for the film with 1.5 at. % Co. However the difference is not proportional to the cobalt concentration. Therefore we associate the susceptibility at 10 kOe with the paramagnetic contribution to the magnetization then this paramagnetic contribution cannot be directly explained by the presence of additional Co^{2+} ions. Defect states in ZnO should also contribute to the paramagnetic properties of investigated films.

5. Summary

The analysis of field and temperature dependence of the magnetization of investigated films shows that measured magnetization contains superpamagnetic, paramagnetic and probably ferromagnetic contributions.

Defects of ZnO structure play significant role in magnetic properties of investigated films especially for the film with 1.5 at. % of Co content. These defects contribute to paramagnetic and superparamagnetic properties of investigated films.

Cobalt impurity affects the magnetic properties of the films essentially indirectly by variation of the defects concentration and charge state and probably by interaction of Co^{2+} ions with defect states.

References

1. T. Dietl, H. Ohno, Rev. Mod. Phys., **86**, 187 (2014)
2. Ü. Özgür,□ Ya.I. Alivov, C. Liu, A. Teke,□ M.A. Reshchikov, S. Doğan,□ V. Avrutin,, S.-J. Cho, H. Morkoç J. Appl. Phys. **98**, 041301 2005
3. K. Potzger, S. Zhou, Q. Xu, A. Shalimov, R. Groetzschel, H. Schmidt, A. Mücklich, M. Helm, J. Fassbender, Appl. Phys. Lett. **93**, 232504 (2008)
4. Z. H. Wang,ı D. Y. Geng, S. Guo, W. J. Hu, Z. D. Zhang, Appl. Phys. Lett. **92**, 242505 (2008)
5. M. Shatnawi, A.M. Alsmadi, I. Bsoul, B. Salameh, G.A. Alna'washi, Journal of Alloys and Compounds **655**, 244 (2016)
6. A. Franco Jr,, H.V.S. Pessonia, P.R.T. Ribeirob, F.L.A. Machado, F. Al-Dweri, F. El Akkad, Journal of Magnetism and Magnetic Materials **426**, 347 (2017)
7. B. Martínez, F. Sandiumenge, L.l. Balcells, J. Arbiol, F. Sibieude, C. Monty, Phys Rev. B **72**, 165202 (2005)
8. V. G. Kytin, V. A. Kul'bachinskii, D. S. Glebov, L. I. Burova, A. R. Kaul, O. V. Reukova, JETP,**111**, 225 (2010)
9. L.I. Burova, S.V. Samoilenkov, M.Fonin, E. Biegger, Yu. Dedkov, E.A. Ganshina, O.Yu. Gorbenko, U. Rüdiger, A.R. Kaul, Thin Solid Films **515,** 8490 (2007)
10. L.I. Burova, N.S. Perov, A.S. Semisalova, V.A. Kulbachinskii, V.G. Kytin, V.V. Roddatis, A.L. Vasiliev, A.R. Kaul, Thin Solid Films **520,** 4580 (2012)
11. Q. Li, Y. Wang, L. Fan, J. Liu, W. Konga, B. Ye Scripta Materialia **69**, 694 (2013)
12. R. Lard, E. Talbot, Ph. Pareige, H. Bieber, G. Schmerber, S. Colis, V. Pierron-Bohnes, A Dinia, J. Am. Chem. Soc., **133**, 1451 (2011)
13. M. Knobel, W.C. Nunes, L.M. Socolovsky, E. De Biasi, J.M. Vargas,J. C. Denardin, J. Nanosci. Nanotechnol. **8**, 2836 (2008)
14. B. Ghosh, M. Sardar, S. Banerjee, J. Phys.: Condens. Matter **24**, 366002 (2012)

Magnetic properties of $Pb_{1-x-y}Sn_xV_yTe$ alloys

Evgeny Skipetrov[1,*], *Nikita* Konstantinov[1], *Liudmila* Skipetrova[1], *Alexander* Knotko[2], and *Vasily* Slynko[3]

[1]M.V. Lomonosov Moscow State University, Faculty of Physics, 119991 Moscow, Russia
[2]M.V. Lomonosov Moscow State University, Faculty of Chemistry, 119991 Moscow, Russia
[3]I.M. Frantsevich Institute of Materials Science Problems, NAS of Ukraine, 58001 Chernivtsi, Ukraine

Abstract. We study the field and temperature dependences of the magnetization ($B\leq 7.5$ T, T=2.0-75 K) of samples from a single crystal $Pb_{1-x-y}Sn_xV_yTe$ (x=0.08, y=0.01) ingot synthesized by the Bridgman method. It is established that the magnetization of the samples contains two main contributions: the Brillouin-type paramagnetism of vanadium ions and the diamagnetism of the crystal lattice. Experimental field and temperature dependences of the magnetization are discussed in the framework of a theoretical model of electronic structure rearrangement in $Pb_{1-x-y}Sn_xV_yTe$ with doping and approximated as the sums of two terms based on the modified Brillouin functions for vanadium ions in the V^{3+} and V^{2+} states. The concentrations of magnetically active vanadium ions in two different charge states with increasing impurity concentration along the ingot are determined.

1 Introduction

It is well known that doping of PbTe with vanadium leads to appearance of a deep impurity level E_V under the bottom of the conduction band ($E_V\approx E_c-20$ meV) [1, 2]. When the concentration of tin in $Pb_{1-x}Sn_xTe$ alloys rises, the vanadium level slowly moves away from the bottom of the conduction band, remaining inside the forbidden gap for $x<0.20$ [3, 4].

As substituting ions in the metal sublattice, vanadium ions initially must be in the electrically neutral V^{2+} states. However, since undoped $Pb_{1-x}Sn_xTe$ alloys are usually p-type crystals, the Fermi level is located much lower than the impurity level of vanadium. Therefore, when the concentration of the impurity increases, the self-ionization of the impurity ions takes place ($V^{2+}\rightarrow V^{3+} + e_{band}$) due to the flow of electrons from the impurity level to the valence band [5]. Then, after filling the valence band with electrons, the Fermi level stabilizes inside the impurity band of vanadium, and the concentrations of filled and empty states in this band should become equal to the concentrations of the impurity ions in two different charge and magnetic states: $V^{2+}(3d^3)$ and $V^{3+}(3d^2)$, respectively. Thus, the magnetic properties of diluted magnetic semiconductors $Pb_{1-x-y}Sn_xV_yTe$ should be determined not only by the impurity concentration, but also by the ratio of the concentrations of vanadium ions in two different magnetic states.

Only the magnetic properties of lead telluride doped with vanadium were investigated previously [6-8]. It was shown that at low temperatures, the field dependences of the magnetization are linear at $B<0.5$ T and that the temperature dependences of the magnetic susceptibility follows the Curie–Weiss law. The effective magnetic moment per vanadium atom does not correspond to any of the possible charge states of vanadium ions and decreases with the increase of vanadium content.

In the present work, we investigate the field and temperature dependences of the magnetization ($B\leq 7.5$ T, $2.0\leq T\leq 75$ K) of samples from a $Pb_{1-x-y}Sn_xV_yTe$ (x=0.08, y=0.01) single-crystal ingot with the variation of the concentrations of tin and vanadium along the ingot. Our main goals are to determine the kinetics of changes in concentration of magnetic ions in two different magnetic states along the ingot and to establish a correlation between the magnetic and galvanomagnetic properties of alloys in the framework of the model of rearrangement of the electronic structure of $Pb_{1-x-y}Sn_xV_yTe$ upon doping.

2 Experimental details

A single-crystal $Pb_{1-x-y}Sn_xV_yTe$ ingot with a nominal tin and vanadium concentrations x=0.08, y=0.01 was grown by the vertical Bridgman method [3]. The resulting ingot with its axis roughly corresponding to the <111> crystallographic axis, was sliced using a wire saw into 26 wafers approximately 1.5 mm thick (wafer serial numbers are referred below as sample numbers). To determine the composition and to study the magnetic properties of alloys, samples weighing 40 to 80 mg were cleaved from the wafers at the liquid nitrogen temperature.

The phase and elemental composition of the samples were determined using a LEO SUPRA 50VP (Zeiss, Germany) scanning electron microscope equipped with an INCA Energy+ (Oxford Instruments, England) X-ray fluorescence microanalysis system. Figure 1 shows a

* Corresponding author: skip@mig.phys.msu.ru

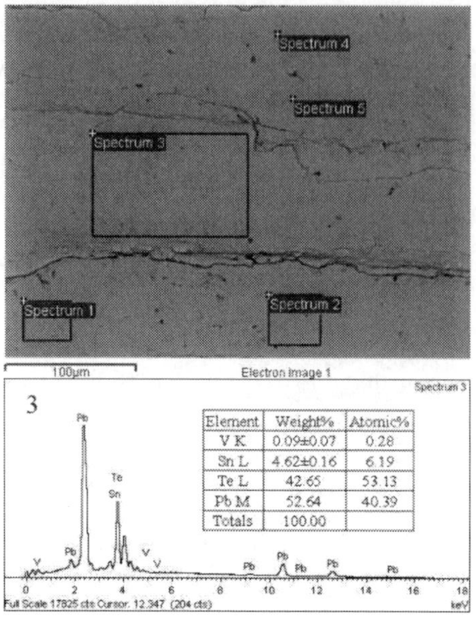

Fig. 1. Upper panel: a photograph of the cleavage surface of the sample 3. Lower panel: X-ray emission spectrum for one of the highlighted areas and the elemental composition of the sample.

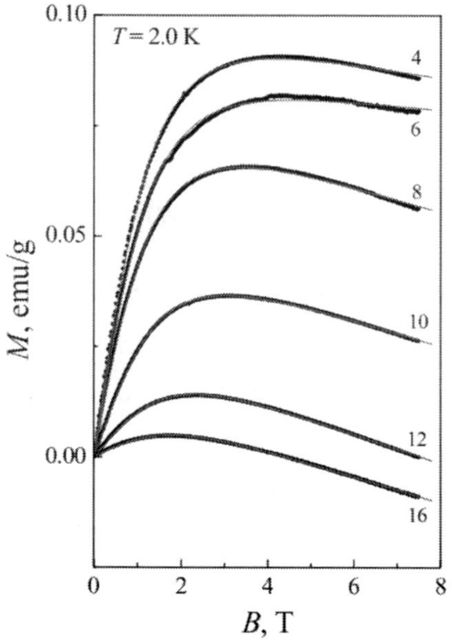

Fig. 2. Field dependences of the magnetization for series of investigated $Pb_{1-x-y}Sn_xV_yTe$ samples at T=2.0 K. Solid lines are theoretical curves calculated according to (2)-(4).

microphotograph of the cleavage surface and the X-ray emission spectrum for the sample 3, which has Sn content close to the maximum value. We found that all the wafers are rather homogeneous and contain no microscopic inclusions of second phases. The tin concentration increases along the ingot (from sample 26 to sample 2), covering a wide range of x from 0.05 to 0.18 [3]. The concentration of the vanadium impurity also increases along the ingot but no more than twice, reaching approximately 1.5 mol. %. However, the error in determining the concentration of vanadium is quite large and in the majority of samples comparable with the concentration of vanadium [3].

The magnetization of the samples was studied using a PPMS-9 (Quantum Design, USA) vibrating sample magnetometer. The field dependences of magnetization were measured at several temperatures from 2.0 K to 70 K in fields up to 7.5 T, while the temperature dependences of magnetization – in a constant magnetic field of 1 T and temperatures from 2.0 K to 75 K.

3 Results and discussion

The field dependences of the magnetization $M(B)$ for a series of investigated samples at T=2.0 K are shown in Fig. 2. The analysis of these dependences allow us to conclude that, in contrast to the previously investigated alloys $Pb_{1-y}Fe_yTe$ [9], in which the impurity level lies in the valence band, the magnetization of $Pb_{1-x-y}Sn_xV_yTe$ alloys contains only two main contributions. The initial linear increase of the magnetization and its subsequent non-linearity in the region of low temperatures are clearly associated with a Brillouin-type paramagnetic contribution of magnetic V^{2+} and V^{3+} ions. At the same time, the linear decrease of the magnetization at low temperatures and fields B>5 T can be associated with the diamagnetic contribution of the host crystal lattice under conditions of saturation of the paramagnetic contribution of vanadium ions. An increase of vanadium content along the ingot (from sample 16 to sample 4) leads to a significant increase of the paramagnetic contribution, while the diamagnetic contribution of the host lattice remains almost unchanged (see Fig. 2). An increase in temperature leads to a shift of the maximum in the dependence $M(B)$ towards higher fields and to a linearization of the field dependence of the magnetization typical for the Brillouin-type paramagnetism.

The temperature dependences of the magnetization for series of investigated $Pb_{1-x-y}Sn_xV_yTe$ samples at B=1 T are presented in Fig. 3. In general, they are of a standard Curie-Weiss type:

$$M(T) = M_0 + \frac{C}{T-\theta} \qquad (1)$$

where M_0 is mainly the temperature-independent diamagnetic contribution of the crystal lattice, C is the Curie constant and θ is the Curie temperature.

It is clearly seen that the magnetization at T=2.0 K monotonically increases by more than an order of magnitude with increase of vanadium concentration along the ingot (from sample 20 to sample 2).

Experimental field and temperature dependences of the magnetization are discussed in the framework of the model of the electronic structure rearrangement with doping proposed for $Pb_{1-x-y}Sn_xV_yTe$ alloys earlier [2, 10]. According to this model, doping leads to the appearance of vanadium ions only in the $V^{3+}(3d^2)$ state until the

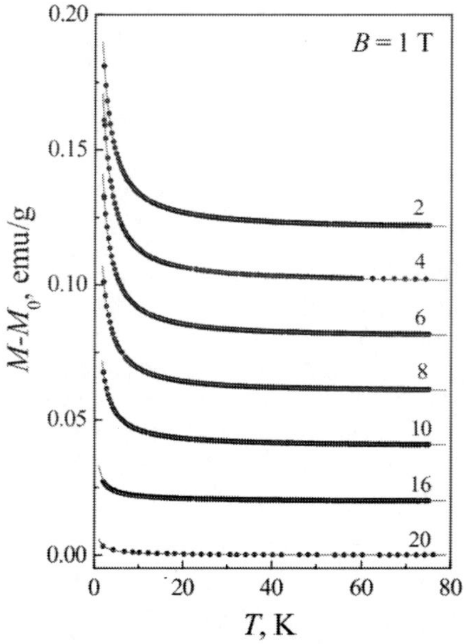

Fig. 3. Temperature dependences of the magnetization for series of investigated $Pb_{1-x-y}Sn_xV_yTe$ samples after subtraction of the temperature-independent contribution M_0. For clarity of presentation, curves 16-2 were consistently shifted upwards with respect to the curve for the sample 20.

pinning of the Fermi level by the vanadium deep level situated under the bottom of conduction band. Only when the Fermi level reaches the impurity level, the self-ionization of V^{2+} ions stops and the concentration of vanadium ions in the V^{3+} state saturates. Further doping should lead only to the appearance of vanadium ions in the $V^{2+}(3d^3)$ state and to a monotonous growth of their concentration.

Our previous results of a galvanomagnetic study of samples from the same $Pb_{1-x-y}Sn_xV_yTe$ ingot [3] indicate that the *p-n*-inversion of the conductivity type and pinning of the Fermi level by the vanadium level upon doping occur in the sample 10. So, when constructing the theoretical field dependences of the magnetization in the samples from 20 to 12, we consider the existence of vanadium ions only in the V^{3+} state, while in the samples from 10 to 2 – the existence of vanadium ions in two different charge and magnetic states (V^{3+} and V^{2+}).

To determine the parameters of V^{2+} and V^{3+} ions we approximate the *M(B)* and *M(T)* dependences as sums of two terms based on the modified Brillouin functions [11-13] for two types of impurity ions and temperature-independent contributions linear in magnetic field:

$$M(B,T) = \frac{1}{\rho}\sum_{i=2+,3+} N_{Vi} g S_i \mu_B B_{Si}(\xi) + \chi_0 B \quad (2)$$

$$B_{Si}(\xi) = \frac{2S_i+1}{2S_i}\coth\left(\frac{2S_i+1}{2S_i}\xi\right) - \frac{1}{2S_i}\coth\left(\frac{1}{2S_i}\xi\right) \quad (3)$$

$$\xi = \frac{g S_i \mu_B B}{k_B(T-T_0)} \quad (4)$$

Here ρ is the density of $Pb_{1-x}Sn_xTe$, which decreases linearly with the tin content in the alloy from 8.16 g/cm^3 in PbTe to 6.48 g/cm^3 in SnTe, $B_{Si}(\xi)$ and N_{Vi} are the modified Brillouin function and the concentration of the magnetic vanadium ions in $V^{2+}(3d^3)$ and $V^{3+}(3d^2)$ states with spins S=3/2 and S=1, respectively. The orbital angular momentum is considered "frozen" and g factor is equal to 2, χ_0 is the diamagnetic susceptibility of the host lattice, T_0 is an adjustable parameter that takes into account the exchange interaction of isolated impurity ions, μ_B is the Bohr magneton, and k_B is the Boltzmann constant.

We calculate the theoretical dependences *M(B)* and *M(T)* by treating concentrations of ions N_{V3+}, N_{V2+}, diamagnetic susceptibility χ_0 and temperature T_0 as fit parameters. The results of approximation of the experimental field and temperature dependences of the magnetization by the sums of two terms based on the modified Brillouin functions (2)-(4) are shown by solid lines in Figs. 2 and 3. The concentration of vanadium ions in V^{3+} and V^{2+} states along the investigated $Pb_{1-x-y}Sn_xV_yTe$ ingot determined in this way from the temperature dependences of the magnetization are presented in Fig. 4. Analysis of the obtained results shows that the theoretical and experimental curves virtually coincide for all the samples, and that the data obtained from the field and temperature dependences of the magnetization agree satisfactorily with each other. Diamagnetic susceptibility χ_0 slightly reduces along the ingot and in the majority of samples $\chi_0 = -(2\text{-}4)\cdot 10^{-7}$ emu g^{-1}Oe^{-1}, which is in good agreement with the theoretical value $\chi_0 = -3.0\cdot 10^{-7}$ emu g^{-1}Oe^{-1} for undoped PbTe [14]. In the investigated samples, adjustable

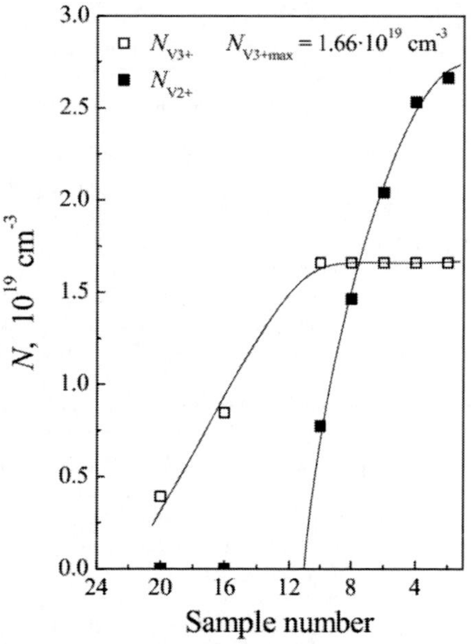

Fig. 4. Kinetics of changes in concentration of vanadium ions in V^{3+} and V^{2+} states along the investigated $Pb_{1-x-y}Sn_xV_yTe$ ingot, determined from the temperature dependences of the magnetization.

temperature T_0 varies in the range from –0.1 K to 0.6 K. When moving along the ingot, the concentration of V^{3+} ions increases and then saturates at $N_{V3+max} \approx 1.7 \cdot 10^{19}$ cm^{-3} (see Fig. 4). At the same time, the concentration of V^{2+} ions increases monotonically from sample 10 to sample 2 up to $N_{V2+} \approx 2.7 \cdot 10^{19}$ cm^{-3}.

It is interesting to note that the total concentration of the magnetic vanadium ions $N_{V3+}+N_{V2+}$, which monotonously increases along the ingot, reaches approximately $4.3 \cdot 10^{19}$ cm^{-3}, i.e. only about 0.3 mol. % at the end of the ingot (in the sample 2). Moreover, our estimates show that the integral concentration of magnetic vanadium ions in the ingot, dissolving, apparently, in the metal sublattice, is about 10% of the concentration of vanadium impurities introduced into the crystal (y=0.01). This means that most of the impurity introduced into the ingot is in the magnetically neutral state (pushed to the very end of the ingot during growth, forms nonmagnetic microscopic inclusions with tellurium or clusters, etc.).

4 Conclusions

The nature of the field and temperature dependences of the magnetization of the investigated $Pb_{1-x-y}Sn_xV_yTe$ alloys with varying matrix composition and concentration of vanadium impurity, allows us to conclude that the magnetization contains two main contributions: the Brillouin-type paramagnetism of vanadium impurity ions and the diamagnetism of the host crystal lattice. We analyze our experimental results in the framework of the model of electronic structure rearrangement in $Pb_{1-x-y}Sn_xV_yTe$ alloys upon vanadium doping. An important assumption of this model is that the concentration of V^{3+} ions in the alloys increases only until the pinning of the Fermi level by the vanadium impurity level lying under the bottom of the conduction band, after which the concentration of V^{3+} ions saturates and the concentration of V^{2+} ions starts to increase in its turn.

The experimental field and temperature dependences of the magnetization are successfully approximated by sums of two terms based on the modified Brillouin functions for two different types of impurity ions and temperature-independent contributions, linear in magnetic field. The kinetics of changes in the concentration of magnetic vanadium ions in V^{3+} and V^{2+} charge states along the investigated $Pb_{1-x-y}Sn_xV_yTe$ ingot is determined. It is shown that despite the monotonic increase in the total concentration of the magnetic vanadium ions $N_{V3+}+N_{V2+}$ along the ingot, it reaches only about 0.3 mol. % at the end of the ingot. The integral concentration of magnetic ions of vanadium in the ingot is approximately 0.1 mol. %, which is an order of magnitude smaller than the nominal concentration of vanadium introduced in the ingot.

The authors are grateful to M.M. Markina and K.V. Zakharov (M.V. Lomonosov Moscow State University, Faculty of Physics) for assistance in conducting measurements. The work was supported by the Russian Foundation for Basic Research (Project No. 16-02-00865).

References

1. A.A. Vinokurov, A.I. Artamkin, S.G. Dorofeev, T.A. Kuznetsova, V.P. Zlomanov, Inorg. Mater. **44**, 576 (2008)
2. E.P. Skipetrov, A.N. Golovanov, E.I. Slynko, V.E. Slynko, Low Temp. Phys. **39**, 76 (2013)
3. E.P. Skipetrov, A.N. Golovanov, A.V. Knotko, E.I. Slyn'ko, V.E. Slyn'ko, Semiconductors **46**, 741 (2012)
4. E.P. Skipetrov, A.N. Golovanov, B.B. Kovalev, A.V. Knotko, E.I. Slyn'ko, V.E. Slyn'ko, Semicond. Sci. Technol. **27**, 015019 (2012)
5. T. Story, E. Grodzicka, B. Witkowska, J. Gorecka, W. Dobrowolski, Acta Phys. Pol. A **82**, 879 (1992)
6. E.P. Skipetrov, E.A. Zvereva, A.N. Golovanov, N.A. Pichugin, A.E. Primenko, O.A. Savelieva, V.P. Zlomanov, A.A. Vinokurov, Solid State Phenom. **152-153**, 291 (2009)
7. E.P. Skipetrov, E.A. Zvereva, A.E. Primenko, O.A. Savelieva, N.A. Pichugin, A.N. Golovanov, V.V. Gorbachev, V.P. Zlomanov, A.A. Vinokurov, Mold. J. Phys. Sci. **8**, 63 (2009)
8. A.I. Artamkin, A.A. Dobrovolsky, A.A. Vinokurov, V.P. Zlomanov, S.Y. Gavrilkin, O.M. Ivanenko, K.V. Mitzen, L.I. Ryabova, D.R. Khokhlov, Semiconductors **44**, 1543 (2010)
9. E.P. Skipetrov, A.A. Solovev, A.V. Knotko, V.E. Slynko, Low Temp. Phys. **43**, 466 (2017)
10. E.P. Skipetrov, A.V. Knotko, E.I. Slynko, V.E. Slynko, Low Temp. Phys. **41**, 141 (2015)
11. J.A. Gaj, R. Planel, G. Fishman, Solid State Commun. **29**, 435 (1979)
12. J.R. Anderson, G. Kido, Y. Nishina, M. Gorska, L. Kowalczyk, Z. Golacki, Phys. Rev. B **41**, 1014 (1990)
13. M. Gorska, J.R. Anderson, G. Kido, S.M. Green, Z. Golacki, Phys. Rev. B **45**, 11702 (1992)
14. G.A. Bain, J.F. Berry, J. Chem. Educ. **85**, 532 (2008)

Colossal magnetoresistance and anomalous Hall effect in nonmagnetic semiconductors

Sergei Obukhov[1], and *Tatiana* Panysheva[2]

[1] Ioffe PTI of the RAS, 194021, St. Petersburg, Russian Federation
[2] Peter the Great St. Petersburg Polytechnic University, 194064, St. Petersburg, Russian Federation

Abstract. Colossal Magnetoresistance (CMR) in nonmagnetic semiconductors and magnetic materials has been investigated as a function of magnetic field, charge carriers concentration and temperature. Both types of materials demonstrated qualitative and quantitative coincidence of CMR dependence on magnetic field, charge carriers concentration and temperature. The findings support the CMR interpretation in the framework of the Excitonic Insulator (EI) model and transition of an insulating EI phase to conducting state under magnetic field for all types of materials under study. It is suggested that Jahn-Teller distortion caused by magnetic ions and external uniaxial stress could initiate EI phase formation

1 Introduction

The effect of specific resistance decrease in magnetic field, known as Negative Magnetoresistance (NM), was first observed by Lord Kelvin in ferromagnetic metals in 1856 [1]. Still the numerous attempts to reveal the nature of the NM effect in most types of solid state materials, like semiconductors, superconductors, perovskites, graphite remain debatable. Nowadays NM phenomenon exploration in the above materials has been boosted by the practical interest as their characteristics meet the demands of hard disk driver technology, spintronics and magnetoelectronics [2].

Special attention was attracted to NM effect in Magnetic Semiconductors (MS) like n-$Gd_{3-x}v_xS_4$ [3], Diluted Magnetic Semiconductor (DMS) $Hg_{1-x}Mn_xTe$ [4] and Manganite Perovskite like $Pr_{0,7}Ca_{0,3}MnO_3$ [5], all of which demonstrated colossal resistivity decrease in external magnetic field, the effect known as Colossal Magnetoresistance (CMR).

The physical models describing CMR, i.e. Magnetic Polaron and Phase Separation models [6,7] are based on the suggestion that MS and Manganite Perovskite are magnetic materials, where charge carriers transport is dependent on the interaction with the magnetic moments of the host magnetic ions.

However, this approach absolutely ignores the fact that CMR has been also experimentally observed in nonmagnetic semiconductors (NMS), i.e. p-Ge [8,9] and p-InSb [8,10], where resistivity decrease of $10^2 \div 10^5$ times was demonstrated in magnetic field B~3T and temperature T~1,5K, thus at the experimental conditions similar to the conditions applied to MS, DMS and Manganite Perovskite in CMR studies.

In this paper we aim to conduct the comparative study based on our magnetotransport experimental findings revealed for nonmagnetic semiconductors and the results of magnetic CMR materials studies described in scientific literature for the similar experimental conditions. This approach can help to expand CMR understanding and establish links between these two groups of solid state materials.

2 CMR comparison in nonmagnetic and magnetic materials

In this section we compare our experimental results of the CMR effect studies in NMS as the function of magnetic field, charge carrier concentration and temperature with the data adopted from literature for DMS, MS and Manganite Perovskite [3-5,12].

2.1 Experimental

p-InSb(Ge), p-InSb(Mn) and p-Ge(Ga) crystals were grown by Czochralski method. Magnetoresistance, the Hall effect and dc conductivity were measured at Hall bar samples 7×1.5×1 mm^3. The materials were studied under the low and high magnetic field (B=0-15T), temperature T= 40mK - 4K and charge carrier concentration n= $10^{16} \div 10^{18}$ cm^{-3}. Measurements at super-low temperatures were carried out in ^3He-^4He dilution refrigerator with 15T Oxford Instruments super-conductive magnet at NHMFL, Florida. Measurements in the temperature range T = 300 ÷ 1,5 K were carried out on 15T Bitter magnet of HMFL in Technical University Braunschweig, Germany. ^3He cryostate at T= 4 ÷ 0,3K with 7T superconductive magnet was used for

* Corresponding author: author@e-mail.org

magnetotransport measurements at The Institute of Physics,The University of Sao Paulo, Brasil.

2.2 CMR dependence on magnetic field

The dependence of CMR value on magnetic field strength in NMS, DMS, MS and Manganite Perovskite is demonstrated in figure 1. The curves represent data for the following CMR materials:

a. (NMS) uniaxially stressed InSb crystal doped with Ge at 1,84K; InSb crystal doped with Mn at 40mK and Ge crystal doped with Mercury at T=35K, adopted from [9]
b. (DMS) p-$Hg_{1-x}Mn_xTe$ at 1,7K [4]
c. (MS) n-$Gd_{3-x}v_xS_4$ at 4,2K [3]
d. Manganite Perovskite $Pr_{0,7}Ca_{0,3}MnO_3$ at 100K [5]

It can be seen from figure 1 that $\rho_B/\rho_{B=0}$ values in all observed nonmagnetic and magnetic CMR materials are comparable and demonstrate similar tendency to decrease in a low magnetic field (B<1T).

It is important to note that all observed nonmagnetic semiconductors demonstrate CMR at relatively low temperatures below 4K with the exception of p-Ge(Hg) demonstrating the same phenomenon at T=35K. It also should be stressed that all CMR nonmagnetic semiconductors are of p-type whereas magnetic semiconductors are both of p- and n-type.

In high magnetic field (B>4T) p-$Hg_{1-x}Mn_xTe$ and p-InSb(Mn) demonstrate the increase of resistivity with magnetic field increase, i.e. Positive Magnetoresistance.

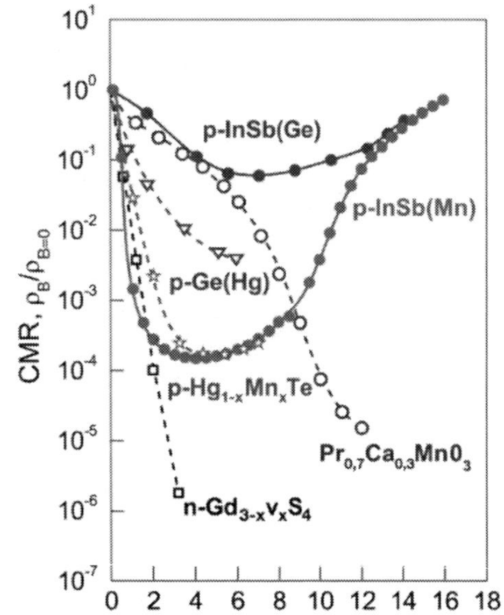

Fig. 1. CMR dependence on magnetic field in nonmagnetic semiconductors and magnetic materials. Dots on solid lines indicate our experimental results in nonmagnetic semiconductors p-InSb(Ge) and p-InSb(Mn). Dashed lines are results for magnetic materials adopted from Ref. [3-5, 9].

2.3 Charge carrier concentration influence on CMR

Figure 2 demonstrates the dependence of CMR on charge carrier concentration (n) in the following materials: NMS – Ge(Ga), InSb(Ge) and InSb(Mn); DMS - p-$Hg_{1-x}Mn_xTe$ [4]; MS - n-$Gd_{3-x}v_xS_4$ [3] and EuSe[11]. It can be seen from figure 2 that all CMR materials we study demonstrate similar behavior being the function of charge carrier concentration. At charge carrier concentration below the critical concentration of metal-insulator transition ($n<n_{cr}$) the relation $\rho_B/\rho_{B=0}$ first declines with the decrease of charge carrier concentration, reaches its minimum and finally increases with the decrease of charge carrier concentration.

Figure 2 also shows piezoresistance $\rho_\chi/\rho_{\chi=0}$ as a function of charge carrier concentration for the following p-type NMS materials, i.e. p-Ge(Ga), p-InSb(Ge) and p-InSb(Mn). Here the maximal piezoresistance value $\rho_\chi/\rho_{\chi=0}$ coincides with the minimal value of $\rho_B/\rho_{B=0}=f(n)$ for all observed semiconductors.

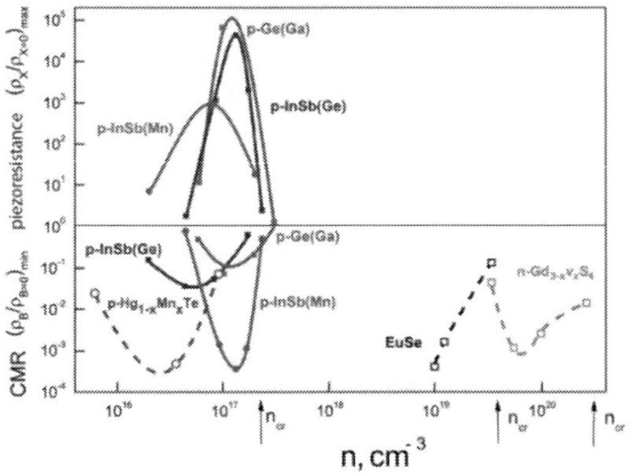

Fig. 2. CMR ($\rho_B/\rho_{B=0}$) dependence on charge carrier concentration in magnetic materials, nonmagnetic semiconductors and piezoresistance ($\rho_\chi/\rho_{\chi=0}$) vs n in p-Ge and p-InSb crystals. Both $\rho_B/\rho_{B=0}$ and $\rho_\chi/\rho_{\chi=0}$ dependencies on concentration (n) for p-Ge and p-InSb represent our experimental results. Dashed lines are CMR results adopted from Ref. [3,4,11].

2.4 CMR temperature dependence

CMR temperature dependence in nonmagnetic and magnetic materials is demonstrated in figure 3. It can be seen that $\rho_B/\rho_{B=0}$ exponentially declines with materials cooling in all studied types of semiconductors. The difference is that in nonmagnetic semiconductor InSb(Mn) at manganese concentration $N_{Mn}=1,7\times10^{17}cm^{-3}$ the minimal value $\rho_B/\rho_{B=0}=10^{-4}$ is reached within the millikelvine temperature range, whereas in diluted magnetic semiconductor p-$Hg_{1-x}Mn_xTe$ [4] the same CMR effect is observed at higher temperature level which is above 2K.

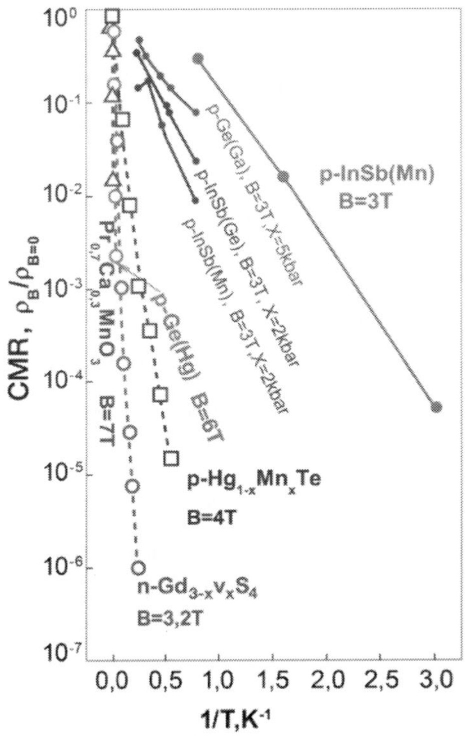

Fig. 3. CMR temperature dependence for nonmagnetic p-InSb and p-Ge crystal semiconductors (dots on solid lines -our experimental results) and magnetic materials (dashed lines-results adopted from the Ref. [3-5, 9]).

2.5 The Hall effect

To get information concerning the charge carrier contribution into the CMR effect we compare the Hall effect in NMS and DMS. Figure 4 depicts the Hall constant dependence on magnetic field in nonmagnetic InSb crystal doped with Mn at $N_{Mn}=1,5\times 10^{17}cm^{-3}$ and Diluted Magnetic Semiconductor p-$Hg_{1-x}Mn_xTe$ [12], where the concentration of Mn is 10^4 higher. It is clearly seen that both semiconductors demonstrate similar features in magnetic field, i.e. negative Hall constant in low magnetic field and positive Hall constant within the high magnetic field range. The difference between the observed NMS and DMS is that in InSb(Mn) crystal the Hall constant sign inversion occurs at B~1T but in p-$Hg_{1-x}Mn_xTe$ the inversion is revealed at B~5T.

Such Hall constant behavior demonstrates the same tendency of the Hall constant sign inversion and occurs in both types of semiconductors. Here the inversion under different levels of magnetic field could be explained by a large difference in effective masses of electrons and holes which is supported by the fact that InSb crystals have large hole/electron masses ratio ($m_h/m_e = 0{,}5/0{,}01 \sim 50$) [14].

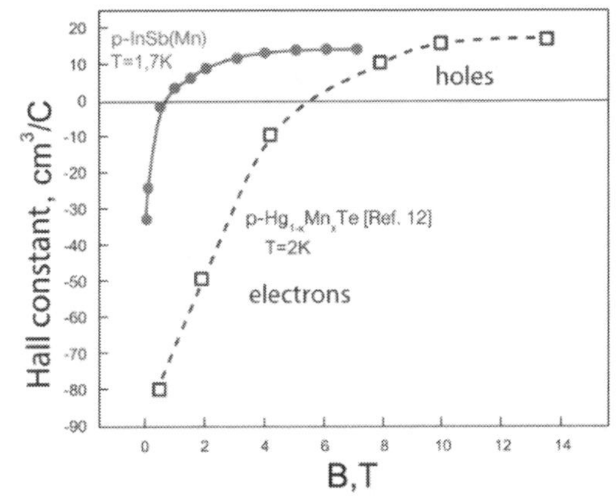

Fig. 4. The Hall constant vs magnetic field in nonmagnetic p-InSb(Mn) crystal (solid line- represents our results) and DMS p-$Hg_{1-x}Mn_xTe$ (dashed line-results adopted from [12]).

Discussion

Having summarized the revealed above qualitative and quantitative CMR and the Hall constant coincidences in nonmagnetic semiconductors and magnetic materials we have come to the following conclusion:

1 It can be suggested that the content of magnetic ions is not the only factor which influences CMR which can be seen from CMR comparison of nonmagnetic InSb doped with manganese at $N_{Mn} \sim 10^{17}cm^{-3}$ and p-$Hg_{1-x}Mn_xTe$ or Magnetic Perovskite $Pr_{0,7}Ca_{0,3}MnO_3$, where manganese concentration is 10^5-10^6 times higher.

2 It is also questionable whether the magnetic type of the CMR material, i.e. ferromagnetic or antiferromagnetic, could be considered the factor which influences NM value.

3 The fact that NM value increases on the insulator side of MIT supports the key role of the charge carrier concentration and the interaction of these charge carriers in producing NM effect.

In this view, the concept of excitonic insulator model (EI model) seems universal for the description of the CMR phenomenon in both magnetic and nonmagnetic materials. It has been suggested [15] that if electron-hole binding energy E_B in semiconductors exceeds the gap energy E_g, a new phase, which is called excitonic insulator, is formed. First the EI model was introduced for the interpretation of resistivity anomalies in $TmSe_{1-x}Te_x$ [16]. The peak of resistivity with increasing pressure in this material was explained as the sequence of pressure-induced transitions from normal semiconductor to excitonic insulator, and then to normal semimetal. We can suggest that under external magnetic field excitons in CMR materials dissociate and we observe transition from EI to normal semiconductor i.e. the CMR effect.

We propose that in the framework of EI model the EI phase is formed in CMR materials as the result of Jahn-Teller distortion. In DMS and MS Jahn-Teller (JT) distortion is caused by the host magnetic ions, whereas in NMS we see several possibilities to introduce JT distortion. It can be supposed that in p-InSb(Mn) crystals JT distortion can be caused by Manganese impurity ions while in p-Ge(Hg) the JT distortion could be induced by the deep impurity center of Mercury. As for uniaxially stressed p-InSb(Ge) and p-Ge(Ga), i.e. semiconductors doped with shallow impurities, JT distortion could appear when we apply uniaxial stress to crystals [17].

In spite of conflicting theories concerning the nature of excitonic insulator its application to the CMR phenomenon description seems to be the important point in the explanation of this phenomenon in both magnetic and nonmagnetic materials.

4 Conclusion

Colossal Magnetoresistance in nonmagnetic semiconductors and magnetic materials observed under magnetic field depends on charge carriers concentration and temperature. The revealed qualitative and quantitative coincidence of CMR behaviour in magnetic and nonmagnetic materials highlights the common nature of the CMR phenomenon in these types of materials. The interpretation of insulator-metal transition under magnetic field in the framework of the Excitonic Insulator phase model provides with the possibility to integrate the description of the conductivity transition from EI phase to conducting state for both magnetic and nonmagnetic materials into one model.

The authors acknowledgethe support of National Magnetic Field Laboratory, National Science Foundation Cooperative Agreement No. DMR- 0654118, the State of FL the support by Alexander von Humboldt Foundation, Germany and the support by FAPESP, Brasil. We are grateful to S.W. Tozer and W.A. Coniglio for help with the ^{3}He–^{4}He dilution refrigerator measurements. We are also grateful A.B. Henriques and N.F. Oliveira for help with ^{3}He cryostat measurements and Prof. W.Gey for help with Bitter magnet measurements.

References

1. Lord Kelvin, *Proc. of the Royal Soc. of London.* **8**, 546 (1857)
2. D. Chiba, M. Sawicki, Y. Nishitani, Y. Nakatani, F.Matsukura , H. Ohno, Nature. **455**, 515 (2008)
3. T. Penney, F. Holtzberg, L.I. Tao, S.von Molnar, *AIP Conf. Proc. Magnetizm and Magnetic Materials* (ed. C.D. Graham and J.J. Rhyne). **№18**, 908 (1973)
4. T. Wojtowicz, A. Mycielski, Acta Phys. Polon. **A67**, 363(1985)
5. Y. Tokura, Physics Today. **56**, 50 (2003)
6. E.L. Nagaev, *Colossal Magnetoresistance and Phase Separation in Magnetic Semiconductors* (Imperial College Press, London, 2002)
7. A. Moreo, E. Dagotto, Science. **283**, 2034 (1999)
8. S.A. Obukhov. Phys. Status Solidi. **B223**, 535 (2001)
9. S.D. Ganichev, H. Ketterl, W. Prettl, I.A. Merkulov, V.I. Perel, I.N. Yassievich, A.V. Malyshev, Phys. Rev. **B63**, 201204(2001)
10. S.A. Obukhov, S.W. Tozer, W.A. Coniglio, Sci. Rep. **5**, 13451(2015)
11. S.A. Obukhov, Phys. Status Solidi. **B242**, 1298 (2005)
12. A.V. Germanenko, L.P. Zverev, V.V. Kruzhaev, G.M. Min'kov, O.E. Rut', Soviet.Phys.Techn. Semicond. **20**, 80 (1984)
13. Y. Shapira, S. Foner, N.F. Oliveira jr, T.B. Reed, Phys. Rev. **B10**, 4765 (1974)
14. D. Bimberg, *Semiconductors* (Springer, Berlin 1982)
15. B.I. Halperin and T.M. Rice. Rev. Mod. Phys. **40**, 755 (1968)
16. J. Neueschwander, P. Wachter, Phys. Rev. **B41**, 12693 (1990)
17. G.L. Bir, JETP. **51**, 556 (1966)

Electronic and magneto-optical properties of ZnO:Co

Erkin Kulatov[1,5,*], *Oleg* Novodvorskii[2], *Andrey* Lotin[2], *Vladimir* Rylkov[3], *Elena* Gan'shina[4], *Dmitrii* Shevyakov[4], *Victor* Tugushev[3], and *Yurii* Uspenskii[5]

[1]Prokhorov General Physics Institute, RAS, Moscow 119991, Russia
[2] Institute on Laser and Information Technologies, RAS, Shatura 140700 , Russia
[3] National Research Centre "Kurchatov Institute", Moscow 123182, Russia
[4]Department of Physics, Moscow State University, Moscow 119991, Russia
[5]Lebedev Physical Institute, RAS, Moscow 119991, Russia

Abstract. High-quality ZnO:Co thin films were synthesized on the sapphire (0001) substrates by pulsed laser deposition. XRD result reveals that the $Zn_{1-x}Co_xO$ films are of wurtzite-type crystal structure with x up to x = 0.35. The structural, optical and magnetic properties of ZnO:Co thin films are studied by experiments and theoretical calculations. Electronic structure of ZnO:Co with native defects are studied by the first-principles density functional theory. The experimental data and theoretical data have demonstrated quite good agreement. The optical and magneto-optical properties, namely, on- and off-diagonal conductivity tensor and the transversal Kerr effect (TKE), are calculated for comparison with available experimental data. The origin of observed TKE peaks is discussed..

1 Introduction

It is widely expected that new and novel functionalities for electronics and photonics can be derived from the injection, transfer and detection of carrier spin at room temperature. Diluted magnetic semiconductors (DMSs) semiconductors containing magnetic ions as impurities, are considered as ideal systems for spintronics. The investigation of magnetic, magnetotransport, and magnetooptical properties of DMSs based on magnetic oxide semiconductors is important for semiconductor spintronic and photonic devices. Zinc oxide (ZnO) due to the wide direct band gap of 3.37 eV at room temperature is an important host material in many optoelectronic and spintronic devices. The observation of ferromagnetism in undoped ZnO has opened up the possibility to use this compound in magneto-optics and magnetoelectronics. ZnO doped with transition metals has been predicted theoretically to be very good candidates for room temperature ferromagnetism, but the investigation results are rather controversial. To increase T_C one should increase concentration of magnetic impurity in host semiconductor and consider an influence of various native defects. In DMS the concentration is limited by low solubility of magnetic impurities in these materials. In this paper we study *ab initio* electronic and magneto-optical properties of ZnO:Co co-doped by native defects and compare them with available experimental data [1, 2].

2 Calculational method

Our calculations are performed using the relativistic APW + lo method as implemented in WIEN2k code [3]. The exchange-correlation effects are described within the GGA local-density approximation [4] (standard DFT). Density-functional calculations presented here are performed within the supercell (2x2x2 hexagonal wurtzite-type unit cell), corresponding to 32 atoms. Two or four Co atoms are included in the supercell which contains also Zn vacancy as a double acceptor. We consider substitutional doping where Zn host atoms are replaced by Co ones (Fig. 1).Thus, both Co concentrations in our calculations vary in the experimentally relevant range (x_{Co}= 0.14 and 0.28). Experimental lattice parameters of ZnO host (a_0= 3.26 Å, c_0= 5.22 Å, and u_O=0.384) [5] are also used for ZnO:Co. The 3d states of Zn are included in the APW basis set and treated as the valence orbitals. Converged results are obtained at the cut-off energy for the interstitial plane-wave expansion of 260 eV. Relaxations of all atomic positions have been included in our calculations by using different computational scheme which employs the plane wave pseudopotential method VASP [6] and is significantly faster than that of the WIEN2k. Relaxed atomic positions first were found by the VASP, then were used as input in the WIEN2k computation for the final atomic relaxations and the following calculation of spectral and magnetic properties. To verify relaxed geometries, the DOS and the local spin moments of Co

* Corresponding author: kulatov@td.lpi.ru

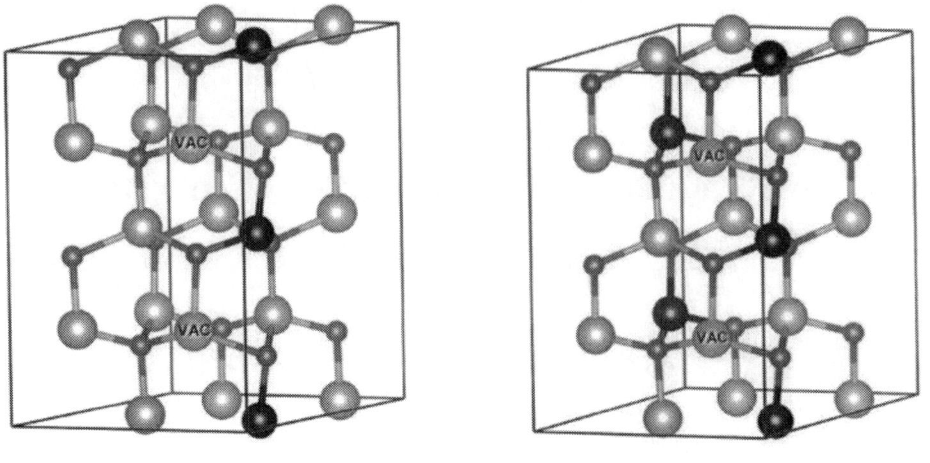

Fig. 1. 2x2x2 supercells of $Zn_{1-x}Co_xO$ for x=0.14 (left panel) and for x=0.28 (right panel). Grey balls are Zn, red – O, dark blue – Co atoms, green – Zn vacancies. Both structures are relaxed.

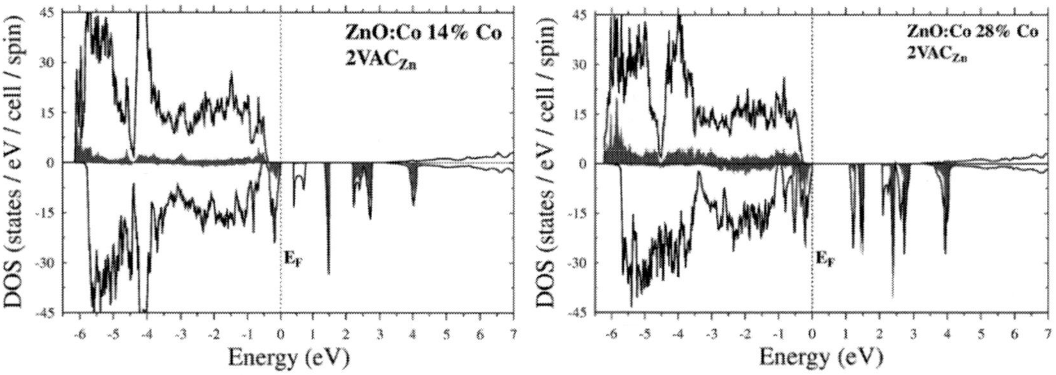

Fig. 2. Relativistic MBJ DOS with two Zn vacancies in $Zn_{1-x}Co_xO$ for x=0.14 (left panel) and for x=0.28 (right panel). Bold line – total DOS, shaded red region – 3d Co partial DOS. E_F means the Fermi level.

atoms, a series of complete WIEN2k calculations were made. The results of the two types of computations (the combined VASP + WIEN2k approach and the WIEN2k method alone) were nearly identical. This agreement indicates that the computational parameters of both methods were properly fitted.

The total energy difference between ferromagnetic (FM) and antiferromagnetic (AFM) collinear orderings has been calculated by standard DFT for both x_{Co} in order to find the ground state in ZnO:Co. Since the correct gap width is highly important for the prediction of transport and optical properties, particular attention was given to the exchange potential acting on electrons. For this we used the modified Becke-Johnson (MBJ) potential [7], which mimics very well the behaviour of orbital-dependent potential. The MBJ potential depends locally on electron density and its use in combination with L(S)DA-correlation provides the band gaps of semiconductors and insulators, which agree well with the experiment and are typically as accurate as the results of a very sophisticated and CPU time expensive *GW* method and hybrid functionals. Unless mentioned otherwise, MBJ+SOC method is used in all our spectral calculations. The magneto-optical transversal Kerr effect in $Zn_{1-x}Co_xO$ (x=0.14 and 0.28) with Zn vacancies is calculated in the framework of the relativistic WIEN2k method [1] which takes into account the effect of spin-orbit coupling in the second variational procedure and the MBJ exchange-correlation potential.

3 Results and Discussion

Optimization of the positions of all atoms in a supercell reveals that in $Zn_{1-x}Co_xO$ (x=0.14 and 0.28) the bond length Co-O shrinks by ~10% in z-direction and by ~9% in xy-plane in comparison with the unrelaxed case. Relaxation effects for the bonds Co-Zn were found to be small (max elongation by ~2 %). The calculated relativistic MBJ density of states (DOS) for $Zn_{1-x}Co_xO$ (x=0.14 and 0.28) is shown in Fig. 2. It is seen that ZnO:Co at both Co concentrations is a semiconductor

Table 1. The AFM and FM energy differences ($\Delta(E_{AFM}-E_{FM})$) , total (M_{FM}, M_{AFM}) and local (in parentheses) spin magnetic moments in FM and AFM phases, m_{orb} is the orbital magnetic moment in FM phase.

Concentration of Co atoms	$\Delta(E_{AFM}-E_{FM})$, meV/cell	M_{FM}, μ_B/Co	M_{AFM}, μ_B/Co	m_{orb}, μ_B/Co
14%	+30	5.0 (+2.92)	0.0 (±2.91)	+0.035
28%	+24	4.0 (+2.82)	0.0 (±2.90)	+0.043

with the energy gap of 0.4 (x=0.14) and 1.1 (x=0.28) eV. The main characteristics of shown DOS's is a full occupation of the majority 3d-states of Co, while the minority 3d-states of Co are partially occupied and are located inside the ZnO host gap (E_g(ZnO)~3 eV).

We find that the FM order is the stable magnetic state in both concentration of Co atoms (Table 1, 2-nd column). Magnetic spin moment on Co^{2+} site is ~3 μ_B while the Co orbital moment is almost quenched and has a positive sign in accordance with Hund's rule. It is known [8] that the TKE is determined by both magneto-optical and optical parameters of a medium. The TKE value, $\delta_p = (I(H)-I_0)/I_0$, is the relative change in intensity of the reflected light caused by the magnetization of a sample in an external magnetic field whose direction is transverse to the plane of the light incidence. The TKE observed at the angle of incidence φ can be expressed by the off- and on-diagonal components of the dielectric function as [8]:

$$\delta_p = 2\sin(2\varphi)(A/(A^2 + B^2))\cdot\varepsilon'_1 + (B/(A^2 + B^2))\cdot\varepsilon'_2 = a\cdot\varepsilon'_1 + b\cdot\varepsilon'_2 , \quad (1)$$

where ε'_1 and ε'_2 are the dispersive and absorptive parts of the off-diagonal dielectric tensor $i\varepsilon'_{xy}$, respectively, while $A = \varepsilon_2\cdot(2\varepsilon_1\cdot\cos^2\varphi-1)$ and $B = (\varepsilon^2_2-\varepsilon^2_1)\cdot\cos^2\varphi + \varepsilon_1-\sin^2\varphi$ are the functions of the on-diagonal components of the dielectric tensor ε_1 and ε_2. Our calculated TKE spectra show marked features in the visible optical region (Fig. 3). The energy positions of calculated MO resonances in $Zn_{1-x}Co_xO$ with Zn vacancies are found to be in quite good agreement with corresponding experimental TKE spectra [1]. Although the MO property is a rather complicated function of the on-diagonal and off-diagonal components of the optical conductivity tensor, our calculation provides a clear insight about its origin. Large TKE observed in ZnO:Co can be understood as a combined effect of the on- and off-diagonal components, an occurrence of minority Co 3d states inside the energy gap and the large spin-orbit coupling (Figs. 4, 5). Additional enhancement of 1.5, 2.0 and 3.0 eV TKE peaks comes from the plasma resonances at same energies which occur in our calculated electron energy loss spectrum (EELS).

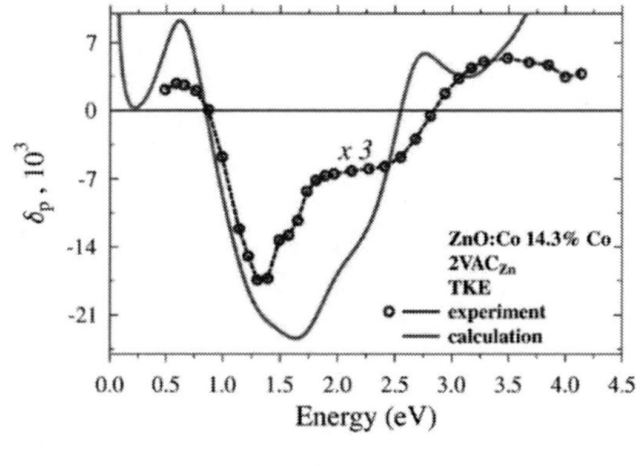

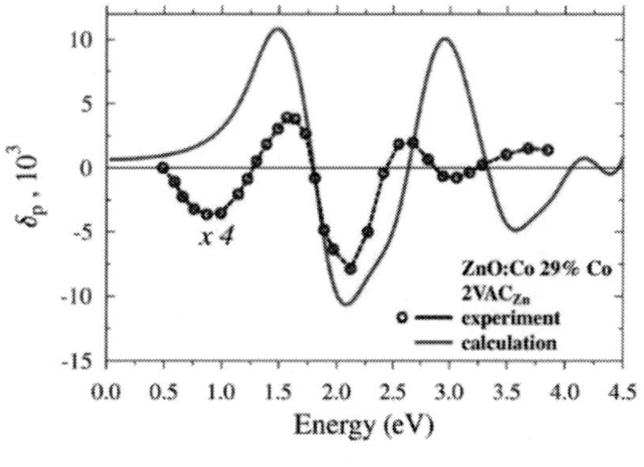

Fig. 3. Calculated (red line) and experimental (blue circles + broken line [1]) TKE in ZnO:Co for two Co concentration Left panel - 14% Co, right panel – 28% Co

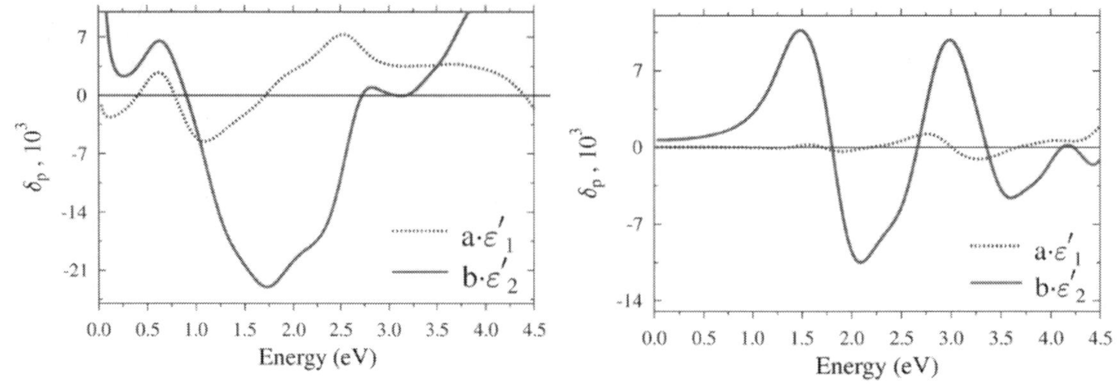

Fig. 4. Left (14% Co) and right (28% Co) panels show decomposition of δ_p into the sum of products (see Eq. (1)). Angle of light incidence $\varphi = 68^{\circ}$.

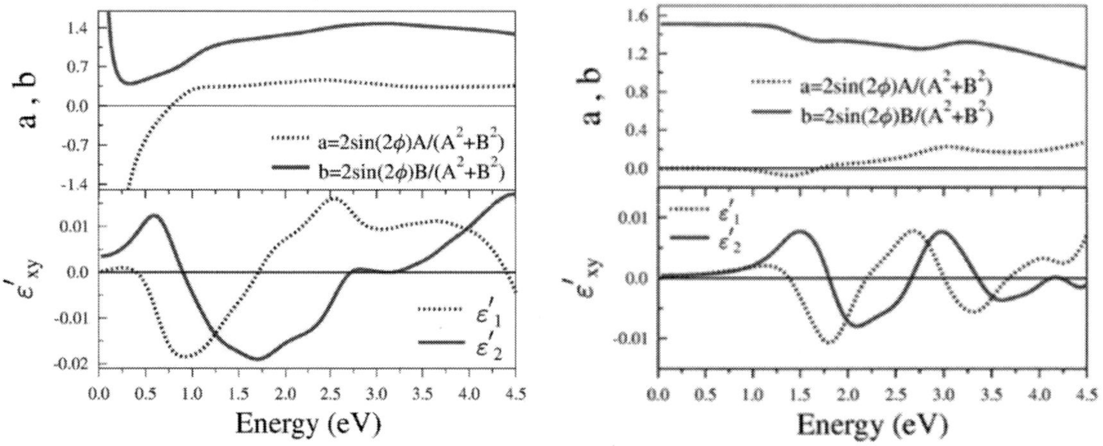

Fig. 5. Calculated **a, b** factors and off-diagonal components of the dielectric tensor $i\varepsilon'_{xy}$ (see Eq. (1)). Left panel – 14% Co, right panel – 28% Co.

Support by the Russian Foundation for Basic Research (RFBR No 15-02-02077, 16-02-00612-a, 16-02-00024-a, 16-07-00798), the Russian Science Foundation (Grant No 16-19-10233), and Programs of Russian Academy of Sciences "Strongly correlated electrons in solids and structures" and "Basic investigations of nanotechnologies and nanomaterials" is acknowledged. Calculations were performed at the Joint Supercomputer Center of RAS and "Lomonosov-1, Lomonosov-2" supercomputers of the Lomonosov Moscow State University.

References

1. A.S. Kuz'mina, A.A. Lotin, O.A. Novodvorsky, *et al* Materials Chemistry and Physics **198**, 291 (2017).
2. A.A. Lotin, O.A. Novodvorsky, V.V. Rylkov, *et al* Semiconductors **48**, 538 (2014).
3. P. Blaha, K. Schwarz, G.K.H. Madsen, D. Kvasnicka, J. Luitz, Wien2k, An Augmented Plane Wave+Local Orbitals Program for Calculating Crystal Properties (Karlheinz Schwarz, Techn. Universitat Wien, Austria, 2001).
4. J.P. Perdew, S. Burke, M. Ernzerhof Phys. Rev. Lett. **77**, 3865(1996).
5. NIMS Materials Database, National Institute for Materials Science, Tsukuba, Japan, http://crystaldb.nims.go.jp/crystdb.
6. G. Kresse and J. Furthmuller Phys. Rev. B **54**, 11169 (1996).
7. F. Tran and P. Blaha Phys. Rev. Lett. **102**, 226401 (2009).
8. G.S. Krinchik and G.M. Nurmukhamedov, Sov. Phys. JETP **21**, 22 (1965).

Influence of oxygen vacancies on magnetic and transport properties of thin $Zn_{1-x}Co_xO_y$ films

Andrey A. Lotin[1*], *Alina S.* Kuz'mina[2], *Oleg A.* Novodvorsky[1], *Liubov S.* Parshina[1], *Olga D.* Khramova[1], *Vladimir A.* Mikhalevsky[1], *Elena A.* Cherebilo[1], *Alexander G.* Shneider[2], *Nikolay S.* Perov[3], *Liudmila A.* Makarova[3]

[1]Federal Scientific Research Center "Crystallography and Photonics" RAS, 140700, Shatura, Russia
[2]Irkutsk National Research Technical University, Irkutsk, Russia
[3]Lomonosov Moscow State University, Moscow, Russia

Abstract. The features of the structural, transport and magnetic properties of thin $Zn_{1-x}Co_xO_y$ films (x=0-0.45), fabricated on *c*-sapphire substrates by the pulsed laser deposition method are studied. It is found that the transport and ferromagnetic properties of the wurtzite $Zn_{1-x}Co_xO_y$ films nonmonotonously depend on Co concentration at room temperature. For the $Zn_{0.87}Co_{0.13}Oy$ film, the strongest ferromagnetic signal is observed that is caused by formation of the greatest number of metallic Co clusters. A further increase of doping impurity concentration in the films leads to the oxidation of metallic Co and formation of the paramagnetic Co_3O_4 phase, in consequence of which the ferromagnetic signal subsides.

1 Introduction

An opportunity to control the charge carrier and its spin provided rapid evolution of spintronics and development of new materials with semiconducting and magnetic properties, and, as a result, will lead to development of new information technologies. Zinc oxide is one of the most promising materials thanks to its unique properties. Being the direct gap semiconductor with a band gap width of 3.3 eV and *n*-type conductivity zinc oxide is of great interest to applications in optoelectronics, solar energy, sensors and others [1]. Currently available theoretical estimations show that zinc oxide doped with 3d transition metals can be ferromagnetic, with a Curie temperature above room temperature, which makes the material attractive for semiconductor spintronics [1,2]. However, the experimental results were rather contradictory. Some authors [3,4] watched gain of a ferromagnetic signal in $Zn_{1-x}Co_xO$ films at increasing of cobalt concentration (x) while other authors [5] report about nonmonotonous dependence. At present there is no single idea about the nature of a high-temperature ferromagnetism in thin $Zn_{1-x}Co_xO_y$ films.
Early we have already demonstrated the ferromagnetic properties of thin $Zn_{1-x}Co_xO_y$ films at room temperature [6]. The purpose of present paper is investigation of oxygen deficiency effect on structural, transport and magnetic properties of thin $Zn_{1-x}Co_xO_y$ films in the range of cobalt concentration from 0 to 45 at. %.

2 Experimental details

2.1 The synthesis of samples

Thin $Zn_{1-x}Co_xO_y$ films (~ 200 nm) were synthesized from high-purity $ZnO-Co_3O_4$ ceramic targets by pulsed laser deposition method (PLD). Synthesis of films was carried out on *c*-sapphire substrates at 450 °C temperature [6]. The residual pressure in the vacuum chamber was $P \sim 5*10^{-6}$ Torr that provided an oxygen deficit in films up to 10% (0.9<y<1) and, therefore, the electron conductivity [7] in the range of cobalt concentration x = 0÷0.45.

2.2 Methods of sample research

The transport properties of thin films have been investigated using Hall effect at the room temperature by the HSM 3000 measure system in the field of a permanent magnet of 1 T. Study of crystal structure and phase compositions of thin $Zn_{1-x}Co_xO_y$ films have been conducted by the X-ray XRD 7000 diffractometer (Shimadzu) with Cu Kα a radiation source (λ=1.541 Å). In order to verify the element composition the synthesized films were examined by EDS (Oxford Instruments X-Max 80, UK) and XPS (Physical Electronics PHI 5600, USA) using Al Ka radiation ($h\nu$ = 1486.6 eV). Magnetic measurements were carried out by means of the vibrating sample magnetometer Lake Shore at the room temperature in the range of fields of ±1.6 T. The film magnetization was defined by subtraction of a magnetization signal of a clear substrate from the total magnetization signal of a sample (substrate + film).
The thickness of the films was measured with a modified microinterferometer MII-4 (LOMO, Russia). All measurements were carried out at the room temperature.

3 Results and discussion

* Corresponding author: lotin_82@mail.ru

Fig. 1 illustrates the X-ray diffraction patterns of thin $Zn_{1-x}Co_xO_y$ films in semilogarithmic scale in the range of 2θ-angles 30°-50°. It is shown that the thin films retain the main phase of the wurtzite structure (*P6₃/mc*) indicated by the peaks at 34.4; 72.1; 123.8° 2θ angles which correspond to the diffraction reflection orders from the plane (001) where n = 2; 4; 6 [8].

Starting with the $Zn_{0.9}Co_{0.1}O_y$ sample, an increase in Co concentration is accompanied by the formation of metallic cobalt phase [9] with hexagonal structure (*a* = 2.5031 Å, *c* = 4.0605 Å) with the Curie temperature of T_C = 848 K. The diffraction peak of metallic cobalt phase (44.24 °2θ) demonstrates the rigorous orientation of clusters along axis of film growth. The intensity of the peak diminishes as the Co content exceeds 13 at. %. This phenomenon results from the transition of the cobalt phase with hexagonal structure to the Co_3O_4 phase with cubic structure (*a* = 8.0837 Å). The Co clusters oxidize gradually: a small amount of metallic cobalt phase is maintained up to 30 at. %.

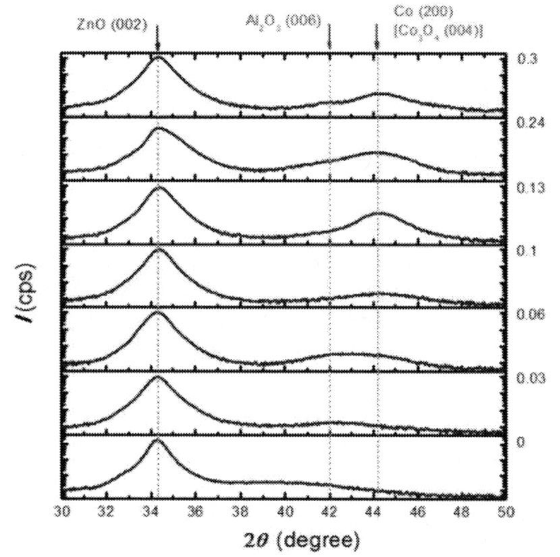

Fig. 1. The X-ray diffraction patterns of thin $Zn_{1-x}Co_xO_y$ films (x = 0 ÷ 0.3).

It is found that an increase in Co concentration in $Zn_{1-x}Co_xO_y$ thin film leads to reduction of the *c*-parameter of the crystal lattice. This fact is related to the tension of crystal lattice in the film plane which is caused by the different radii of substituted Zn and Co ions (r[Zn^{2+}] = 0.074 nm; r[Co^{2+}] = 0.072 nm) [10].

The previous studies of transport properties [11] have shown that $Zn_{1-x}Co_xO_y$ films (x=0.2) have *n*-type conductivity, and electron concentration significantly decreases with growth of oxygen pressure. The figure 2*a* represents the dependence of charge carrier concentration *n* on the cobalt content x in thin $Zn_{1-x}Co_xO_y$ films.

The approximation of experimental values of electron concentration has shown the exponential reduction of n with growth of the cobalt content in $Zn_{1-x}Co_xO_y$ films. Such dependence is related to increase in ionization energy of the donor state due to growth of energy of exchange interaction between electrons of the donor states and electrons of a *d*-state of cobalt ions [12].

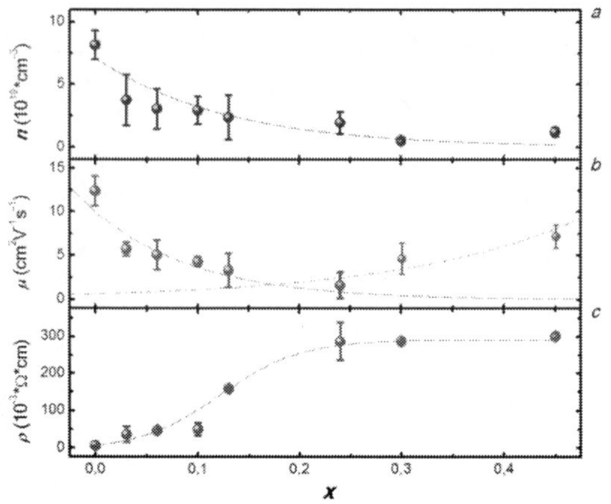

Fig. 2. The dependences of electron concentration (*a*), mobility (*b*) and resistivity (*c*) in thin $Zn_{1-x}Co_xO_y$ on cobalt content x.

It is obvious that high concentration of electrons in the samples synthesized in the oxygen depleted conditions is caused by high density of oxygen vacancies of V_O [1] oxygen. The increasing of oxygen pressure in the chamber leads to essential reduction of vacancy concentration and to stoichiometric growth of $Zn_{1-x}Co_xO_y$ films. Theoretical estimates [13,14] showed that the energy levels of Co lie deeply in the band gap of zinc oxide therefore they can't be thermally activated and, consequently, participate in conductivity.

It is known what wurtzite wide gap semiconductors demonstrate low mobility of charge carriers. High density of oxygen vacancies in nonstoichiometric $Zn_{1-x}Co_xO_y$ films form the additional intracrystalline fields leading to reduction of charge carrier mobility. The dependence of charge carrier mobility μ on cobalt concentration x in $Zn_{1-x}Co_xO_y$ films is provided in a figure 2b. The mobility exponentially decreases at increasing in cobalt concentration up to 24 at.% that is caused by degradation of crystalline lattice. The increasing of cobalt content to 30 at.% the growth of mobility is observed. Probably, this phenomenon is associated with enhancement of anions in the films due to increasing of oxygen in the targets consisted from mixed powders ZnO and Co_3O_4. For example, in target contained 45 at.% of cobalt the part of oxygen is 10% higher, than in a pure target of ZnO.

The field dependences of magnetization of $Zn_{1-x}Co_xO_y$ thin films were obtained at room temperature in the range of magnetic field ±1.6 T. Fig. 3 illustrates the field dependences of magnetization of $Zn_{1-x}Co_xO_y$ thin films. It is seen that the magnetic properties of $Zn_{1-x}Co_xO_y$ thin films change non-monotonically with the increase of the dopant concentration [15,16]. It has been revealed that $Zn_{1-x}Co_xO_y$ samples (x=0; 0.03; 0.1) are paramagnetics, while the other samples possess a weak ferromagnetism at room temperature. Although cobalt clusters start forming after cobalt reaches 10 at.%, their quantity is not sufficient for the onset of ferromagnetic interaction. This factor explains paramagnetism of the $Zn_{0.9}Co_{0.1}O_y$ film.

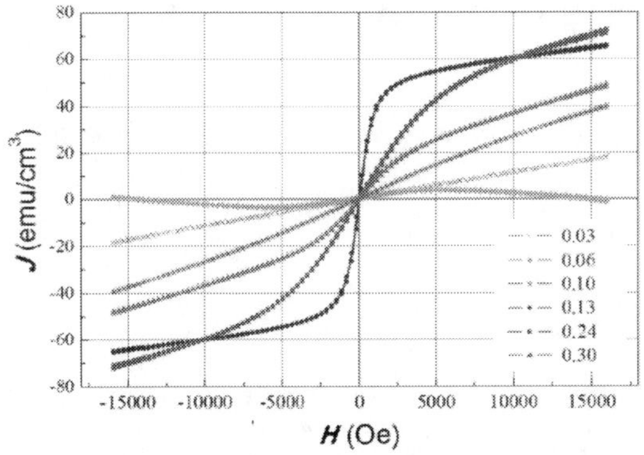

Fig. 3. The dependences of electron concentration (*a*), mobility (*b*) and resistivity (*c*) in thin $Zn_{1-x}Co_xO_y$ on cobalt content x.

$Zn_{0.87}Co_{0.13}O_y$ thin film is shown to have the highest value of magnetic moment signal. The saturation magnetization for this sample is J_s=54.5 emu/cm^3, and the coercive field equals to H_c=24.1±12 Oe. Reduced saturation magnetization with cobalt concentration exceeding 13 at. % (Fig.3) is attributed to the combined ferromagnetic and paramagnetic contribution in the samples. The ferromagnetic response is explained by metallic cobalt clusters while paramagnetism is associated with the formation of cobalt oxide Co_3O_4 during Co oxidation. The increase in Co_3O_4 paramagnetic phase in thin films leads to the decrease in the ferromagnetic signal.

The structural, transport and magnetic properties of thin $Zn_{1-x}Co_xO_y$ films (x=0-0.45), fabricated on *c*-sapphire substrates by the pulsed laser deposition method are studied. It is found that the transport and ferromagnetic properties of the wurtzite $Zn_{1-x}Co_xO_y$ films nonmonotonously depend on Co concentration at room temperature. For the $Zn_{0.87}Co_{0.13}Oy$ film, the strongest ferromagnetic signal is observed that is caused by formation of the greatest number of metallic Co clusters. It is shown that the oxygen deficiency significantly influence on transport and magnetic properties of thin $Zn_{1-x}Co_xO_y$ films.

The work has been supported by RFBR grants: 15-07-03331, 15-29-01171, 15-38-20369, 15-07-03580, 16-29-05385, 16-07-00842, 16-29-11719, 17-07-00615.

References

1. H.J. Ko, T. Yao, Y.F. Chen, S.K. Hong, Appl. Phys. **92**, 4354 (2002)
2. T. Dietl, H. Ohno, F. Matsukura, Science. **287**, 1019 (2000)
3. M. Ivill, S.J. Pearton, S. Rawal, et al., New J. Phys. **10**, 065002 (2008)
4. Y.Z. Peng, T. Liew, T.C. Chong, J. Appl. Phys. **98**, 114909 (2005)
5. D. Chakraborti, S. Ramachandran, G. Trichy, et al., J. Appl. Phys. **101**, 053918 (2007)
6. A.S. Kuz'mina, A.A. Lotin, O.A. Novodvorsky, N.S. Perov, E.A. Gan'shina, L.A. Makarova, A.S. Semisalova, A.G. Shneider, M.P. Kuz'min and S.S. Kolesnikov, Mater. Chem. Phys. **198**, 291 (2017)
7. Q. Liu, C.L. Yuan, C.L. Gan, G.C. Han, J. Appl. Phys. **101**, 073902 (2007)
8. A. Mecke, I. Lee, J.R. Baker jr., M.M. Banaszak Holl, B.G. Orr, Eur. Phys. J. E **14**, 7 (2004)
9. Y.Z. Peng, T. Liew, T.C. Chong, W.D. Song, H.L. Li, W. Liu, J. Appl. Phys. **98**, 114909 (2005)
10. N. Jedrecy, H. J. von Bardeleben, and D. Demaille, Phys. Rev. B. **80**, 205204 (2009)
11. A.A. Lotin, O.A. Novodvorsky, V.V. Rylkov, D.A. Zuev, O.D. Khramova, M.A. Pankov, B.A. Aronzon, A.S. Semisalova, N.S. Perov, A. Lashkul, E. Lahderanta, V.Ya. Panchenko, Semicond., **48**, 556 (2014)
12. V.G. Kytin, V.A. Kulbachinskii, D.S. Glebov, L.I. Burova, A.R. Kaul, O.V. Reukova Semicond., **44**, 155 (2010)
13. A. Zunger, Sol. St. Phys., **39**, 275 (1986)
14. J. Langer, C. Delerue, M. Lannoo, H. Heinrich, Phys. Rev. B. **38**, 7723 (1988)
15. D. Chakraborti, S. Ramachandran, G. Trichy, J. Narayan, J.T. Prater, J. Appl. Phys. **101**, 053918, (2007)
16. R. Al-Gaashani, S. Radiman, A.R. Daud, N. Tabet, Y. Al-Douri, Ceram. Int. **39**, 2283 (2013)

Connection of thermopower, magnetothermopower with resistivity and magnetoresistance in manganites with Nd and Sm

Luidmila Koroleva[1,*], *Ivan* Batashev[1], *Artem* Morozov[1], *Anatolii* Balbashov[2], *Henryk* Szymczak[3] and *Anna* Slavska-Wanniewska[3]

[1]Moscow State University, 19991 Moscow, Russia

[2]Moscow Power Engineering Institute, 19991 Moscow, Russia

[3]Institute of Physics Polish Academy of Sciences, Warsaw, Poland

Abstract. Thermopower, magnetothermopower resistivity and magnetoresistance of single crystal samples $Re_{(1-x)}Sr_xMnO_3$ ($0 \le x \le 0.3$, Re = Nd, Sm) were studied in wide temperature interval, included the Curie temperature T_C. Giant maxima of resistivity and thermopower, very big negative magnetothermopower and magnetoresistance were found in T_C region. So, a giant maximum of thermopower in T_C was found, which suppressed by magnetic field. Simultaneous magnetic field destroys magneto-impurity states – ferrons as evidenced the maxima resistivity and magnetoresistance in T_C Hence it follows that giant thermopower in manganites are produced by ferrons and its value is set by the quantity of impurities, that is the concentration of impurities and the volume of sample.

In our work, thermopower S, magnetothermopower $\Delta S/S$, electrical resistivity ρ, magnetoresistance Δρ/ρ and magnetization M of the single crustal samples $Re_{(1-x)}Sr_xMnO_3$ ($0 \le x \le 0.3$; Re = Nd, Sm) were studied. The temperature dependence of the magnetization $M(T)$ in different magnetic fields was measured by SQUID-magnetometer. The obtained results show, that the transition from ferromagnetic to paramagnetic state is heavily blurred. From this, we conclude that the samples we studied are magnetically heterogeneous. They consist of ferromagnetic clusters located in A-type antiferromagnetic matrix. Curie temperature T_C was defined as the temperature of minimum on $dM(T)/dT$ curve, measured in magnetic field 0.1 kOe. In this work the giant negative magnetoresistance was observed in these compounds near T_C. This indicates that these ferromagnetic clusters are of ferron type [1]. A sharp increase of S and module of negative $\Delta S/S$ were discovered near T_C, where $\Delta S/S_1$ reaches giant values up to 94.5% in magnetic field 13.2 kOe in case of compound with Re = Sm, x=0.3. In Fig. 1 the temperature dependence of thermopower $S(T)$ in magnetic field is shown for Nd compound with $x = 0.3$.

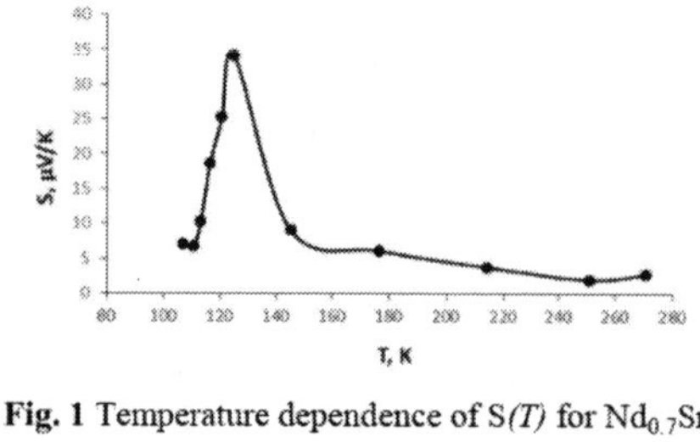

Fig. 1 Temperature dependence of $S(T)$ for $Nd_{0.7}Sr_{0.3}MnO_3$.

Figure 2 shows the temperature dependence of magnetothermopower $\{\Delta S/S\}(T)$ for this sample in different magnetic fields.

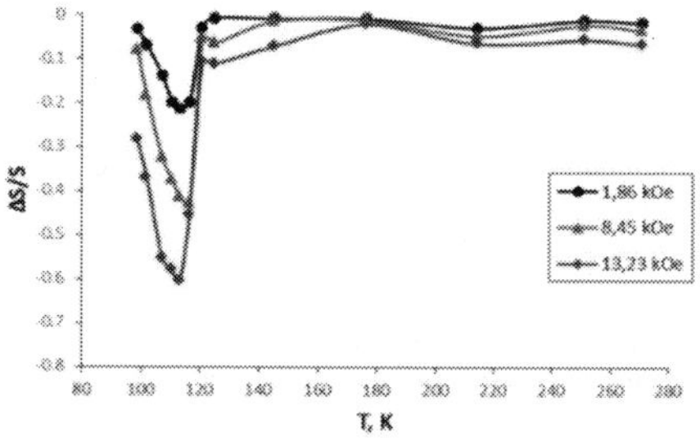

Fig. 2. Temperature dependence of magnetothermopower $\{\Delta S/S\}(T)$ in different magnetic fields for $Nd_{0.7}Sr_{0.3}MnO_3$

It's clear that magnetothermopower is negative and its absolute value near T_C reaches a giant value of 60% in a magnetic field 13.23 kOe. A similar behavior of the thermopower and magnetothermopower was observed in all compounds we researched. The curves presented above are typical for all examined compounds. Maximum values for all samples are shown in table 1. Also in table 1 maximum values of specified resistivity ρ and Δρ/ρ are shown. It should be noted that in compounds with $x = 0$ the value of S is very small. This means, that thermopower is caused by presence of ferrons, as the value of the thermopower falls with the destruction of ferrons caused by magnetic field or heating to temperatures over T_C. Inside ferrons the crystalline lattice is compressed as discussed in [2, 3], so it is possible to assume that the value of thermopower inside is different from thermopower in antiferromagnetic matrix. In this case thermopower S_1 appears on every single ferron under the influence of

* Corresponding author: koroleva@phys.msu.ru

local temperature gradient ΔT_1, caused by total temperature gradient ΔT, as shown in Fig. 3. Each such microregion contributes $(S_1\text{-}S_2)$ to the thermopower of the whole sample. Where S_2 is the thermopower of AFM matrix without ferrons.

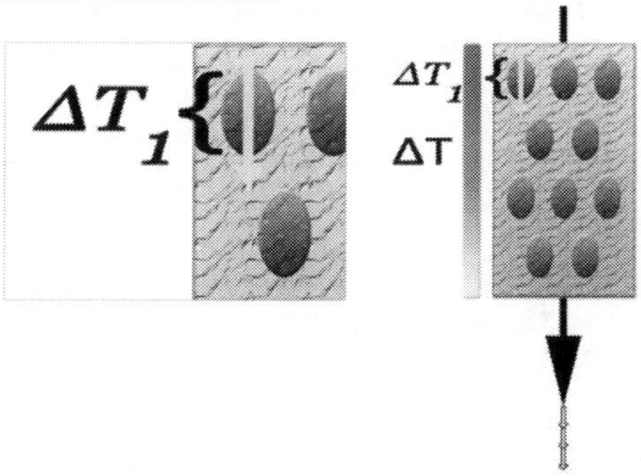

Fig. 3 Schematic representation of ferrons (ellipsoids) and their role in creation of thermopower in magnetic semiconductors.

So, in doped magnetic semiconductors thermopower is determined by number of ferrons and, therefore, by level of doping and volume of the sample. This was confirmed by changing the volume of sample. After dividing the sample in two parts thermopower of each part decreased proportionally the volume of the part. Thus, in doped magnetic semiconductors the value of thermopower can be considerably increased by enlarging the volume of sample, the level of doping and can be controlled by magnetic field.

Thus, in doped magnetic semiconductors, which include manganites, the value of thermopower can be significantly increased in comparison with undoped one, by increasing the concentration of impurities and the volume of the sample. Furthermore, thermomaterials based on magnetic semiconductors should be characterized by specific thermopower, that is the value of thermopower per unit volume. Exactly these values are shown in table 1, which are significantly different from those shown in the Fig. 1.

Table 1. Thermopower S, magnetothermopower $\Delta S/S$, electroresistivity ρ, magnetoresistance $\Delta\rho/\rho$ and Curie temperature for manganites $Nd_{(1\text{-}x)}Sr_xMnO_3$ and $Sm_{(1\text{-}x)}Sr_xMnO_3$.

Compound	T_C, K	S, μV/(K·cm^3)	$\Delta S/S$, %	ρ, kΩ·cm	$\Delta\rho/\rho$, %
$Sm_{0.85}Sr_{0.15}MnO_3$	80	525	54	168	70
$Sm_{0.8}Sr_{0.2}MnO_3$	90	575	37	208	33
$Sm_{0.75}Sr_{0.25}MnO_3$	95	900	64	260	51
$Sm_{0.7}Sr_{0.3}MnO_3$	87	210	94	67.7	76
$Nd_{0.7}Sr_{0.3}MnO_3$	110	86	15	25	34
$Nd_{0.8}Sr_{0.2}MnO_3$	155	240	42	70	45
$Nd_{0.85}Sr_{0.15}MnO_3$	120	1042	34	60	25
$NdMnO_3$	-	≤ 0,2 μV/K	-	-	-

References.

1. E. L. Nagaev. Phys. Rep. **346**, 387-531, (2001)
2. T. Kasuya, A. Yanase. Rev. Mod. Phys. **40**, 684-696, (1968).
3. T. Kasuya, A. Yanase. J. Phys. Soc. Jap. **25**, 1025-1042, (1968)

Diluted Magnetic Semiconductors InFeSb Prepared by Laser Ablation: Spectroscopic and Microscopic Investigations

Elena Gan'shina[1], *Leonard* Golik[2], *Zoya* Kun'kova[2], *Georgy* Zykov[1,*], *Alexander* Rukovishnikov[2], *Marina* Temiryazeva[2], *Vladimir* Lesnikov[3], *Yuri* Markin[2]

[1]Lomonosov Moscow State University, Department of Physics, 119991 Moscow Leninskie Gory, Russia
[2]RAS, Kotel'nikov Institute of Radioengineering and Electronics (Fryazino Branch), 141190 Fryazino, Russia
[3]RAS, Lobachevsky Institute of Physics and Technology, 603950 Nizhny Novgorod, Russia

Abstract. We report optical and magneto-optical results as well as atomic force microscopy (AFM) and magnetic force microscopy (MFM) results for InFeSb samples prepared by laser ablation. AFM and MFM studies have revealed the presence of magnetic particles on the samples surface, whose sizes depend on the Fe content and substrate temperature. It has found that both optical and magneto-optical spectra are superposition of spectra from the doped InFeSb layers and particles on their surface.

1 Introduction

Spintronic devices suggested in theoretical studies may possess characteristics that are unreachable for traditional electronics. However, the implementation of such devices is restrained by the lack of ferromagnetic semiconducting (FMS) thin films, whose Curie temperature (Tc) is above room one. Recently, sets of FMS GaFeSb [1] and InFeSb [2] thin films obtained using molecular beam epitaxy and laser ablation with T_C higher than 300 K have been reported. Magneto-optical spectroscopy is a powerful tool for probing the electronic structures of diluted magnetic semiconductors (DMS). It can provide unique information about such systems because MOKE spectra are sensitive to electronic and magnetic properties. We can distinguish features in the MO spectra associated with transitions in a doped semiconductor matrix (intrinsic ferromagnetism) and features associated with the presence of magnetic inclusions, since they are observed at different energies. In this work we study InFeSb thin films, which have been prepared by laser ablation [2], using optical (ellipsometry) and magneto-optical (transversal Kerr effect, TKE) spectroscopy as well as atomic and magnetic force microscopy (AFM and MFM, respectively).

1.1 Samples and experimental specifications

1.1.1 Samples

InFeSb samples are ≈ 40 nm-thick films, which were fabricated on i-GaAs substrates by pulsed laser deposition. Solid InSb- and Fe-targets were periodically sputtered by Nd:YAG laser in a vacuum chamber. The ratio of the sputtering times, $t_{Fe}/(t_{InSb}+t_{Fe})$ (Y_{Fe}), was varied in the range of 0.08 – 0.17. The substrate temperature, T_S, was 20, 150, 200, 250°C and 300°C. The samples characteristics are displayed in Table 1.

Table 1. InFeSb samples sputtered on GaAs substrate.

Sample number	Technology parameter, Y_{Fe}	Substrate temperature, T_s,°C
N1	0.08	250
N2	0.12	20
N3	0.12	150
N4	0.12	200
N5	0.12	250
N6	0.17	250
N7	0.17	300

1.1.2 Experimental methods

Magneto-optical properties were studied in the transversal Kerr effect (TKE) geometry. TKE is an intensity change of reflected linear p-polarized light during a sample magnetization. Optical properties were measured using spectral ellipsometry. The TKE measurement parameters were E = 0.5 - 4.0 eV, T = 20 – 460 K; H up to 3 kOe. Ellipsometry spectra were

* Corresponding author: V0tum-Separatum@yandex.ru

obtained in the energy range of E = 1.2 – 4.4 eV at room temperature.

2. Results

The AFM and MFM images (Fig. 1, left and right sides, respectively) obtained at room temperature show complex surface topography of the samples. There are particles, whose sizes depend on the Fe quantity and substrate temperature, on the samples surface.

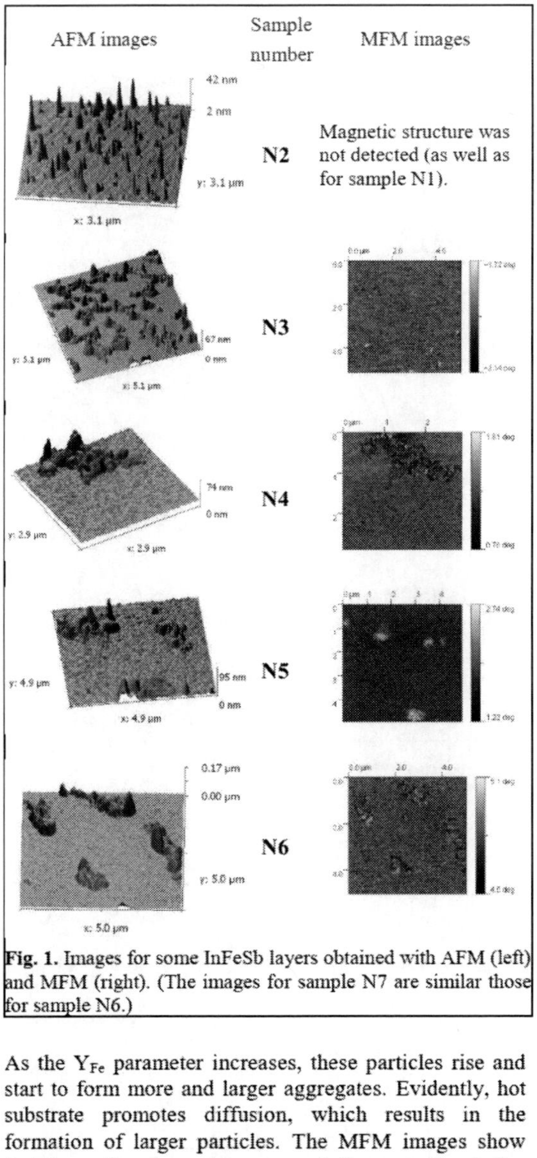

Fig. 1. Images for some InFeSb layers obtained with AFM (left) and MFM (right). (The images for sample N7 are similar those for sample N6.)

As the Y_{Fe} parameter increases, these particles rise and start to form more and larger aggregates. Evidently, hot substrate promotes diffusion, which results in the formation of larger particles. The MFM images show that the surface formations are partially magnetic. A fine magnetic structure (light spots), which expands as the T_S and Y_{Fe} values increase, is observed on the surface particles. Apparently, the surface formations contain inclusions of iron, iron oxides and some FeSb variations.

Figure 2 depicts spectral dependences of the real, $<\varepsilon_1>(E)$, and imaginary, $<\varepsilon_2>(E)$, parts of the samples pseudo-dielectric function calculated using the ellipsometry data. Peaks corresponding to the optical transitions near the L critical point of the Brillouin zone in the InSb parent semiconductor are observed in the $\varepsilon_2(E)$ spectra of the InFeSb layers. It confirms that the crystal structure of the parent semiconductor is conserved in the layers under study.

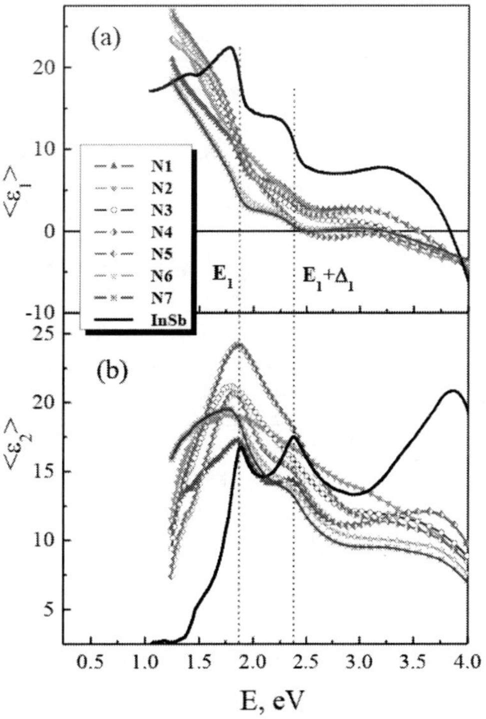

Fig. 2. Spectral dependences of the real, $<\varepsilon_1>(E)$, (a) and imaginary, $<\varepsilon_2>(E)$, (b) parts of pseudo-dielectric function for the investigated samples. Reported energies of the E_1 and $E_1+\Delta_1$ transitions at the critical L point in InSb are shown by vertical dashed lines [3].

An increase of matrix imperfection with the Fe doping and the surface particles contribution are possible reasons for smearing the doublet ($E \approx 1.85$–2.36 eV) and increasing the maxima in the $\varepsilon_2(E)$ spectra as well as for rise in the ε_1 values in the region $E < 1.75$ eV.

TKE room and low temperature spectra are displayed in Fig. 3(a,b). For the samples, whose the growth parameter is $Y_{Fe} \geq 0.12$, we observe large magneto-optical response even at room temperature. In the TKE spectra, two bands around 1.2 and 1.6 eV are distinguished. Some peculiarities (kinks) are seen near E_1 and $E_1+\Delta_1$ transitions in the TKE spectra of the samples with high Fe content (N6 and N7). The TKE values significantly decrease when the Fe amount, Y_{Fe}, and/or growth temperature, T_s, gets smaller (samples N1 and N3). At room temperature the TKE signal from sample N2 is not detected. The weak TKE response of

 https://doi.org/10.1051/epjconf/201818506015

sample N5 is apparently associated with a lower density of the surface particles.

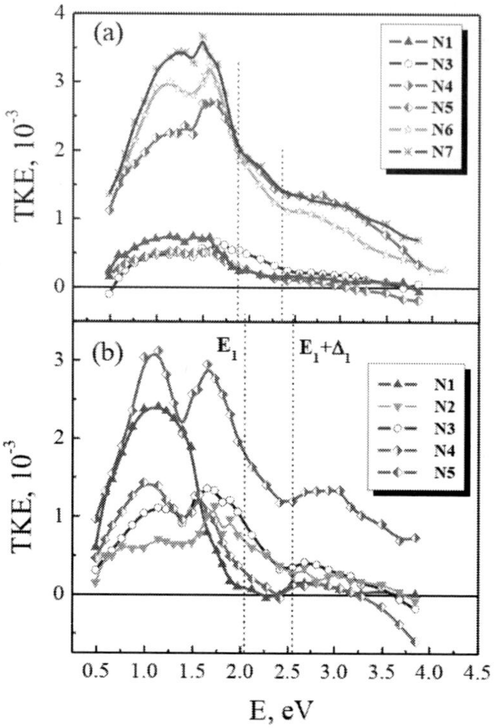

Fig. 3. Spectral dependences of TKE at room temperature (a) and at T=22 K (b) for the investigated samples at H=2.5kOe. Reported energies of the E_1 and $E_1+\Delta_1$ transitions at the critical L point in InSb are shown by vertical dashed lines for both room and low temperatures [3].

All the samples exhibit TKE signal increase as the temperature decreases. In the spectrum of sample N1 with the lowest Fe content even at the low temperature T=22 K only one band is present with the maximum around 1.2 eV. Similar band is observed in the spectra of other samples. There is a distinct peculiarity (minimum) around the $E_1+\Delta_1$ transition region in the low temperature TKE spectra and its location depends on the Fe content and growth temperature. We do not observe any peculiarities near the E_1 transition, however, a strong band is present around 1.7 eV. Most likely, the TKE spectra represent superpositions of contributions from InFeSb ferromagnetic matrix and magnetic surface particles containing Fe oxides.

Studies of the temperature and field TKE dependences (Fig. 4) have shown that the InFeSb layers with Y_{Fe} above 0.08 and $T_S \geq 200°C$ exhibit ferromagnetic behavior even at room temperature. At the same time, saturation in the fields used (3 kOe) was observed only for sample N7 with the maximum Fe content. The TKE temperature dependences (Fig. 4a) display that the samples with the lower Y_{Fe} values give more TKE contribution generated by a low-temperature ferromagnetic phase. It leads to the presence of well pronounced peculiarities in the TKE spectra related to this phase at temperatures lower than 100 K. In the spectra of the samples prepared at bigger Y_{Fe} values and T_S higher than 200°C, these peculiarities become less noticeable, since other magnetic phases make larger contribution to the effect.

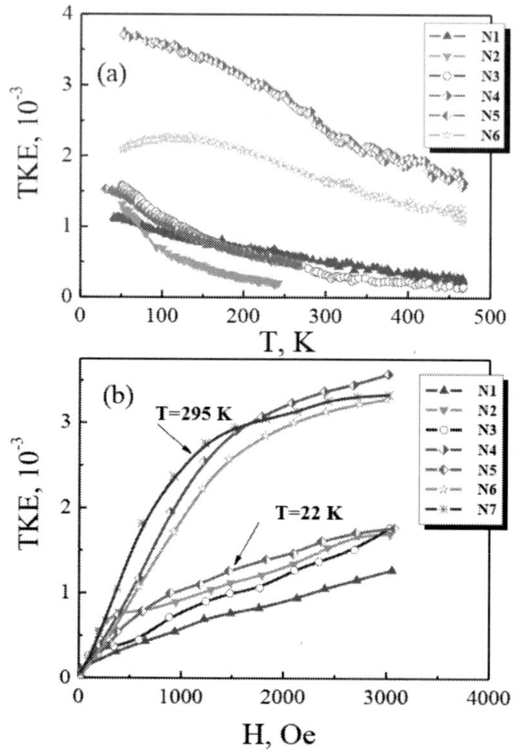

Fig. 4. (a) Temperature TKE dependences (E=1.73eV for N1, N2 and N4; E=1.65 eV for N3, N5 and N6, H=2.5kOe for all samples) and (b) field TKE dependences.

Figure 5 shows the normalized field TKE dependences, $TKE(H)/TKE(H_{max}) \sim M(H)/M(H_{max})$, for the samples with the lower Fe content and substrate temperatures (N1 and N2).

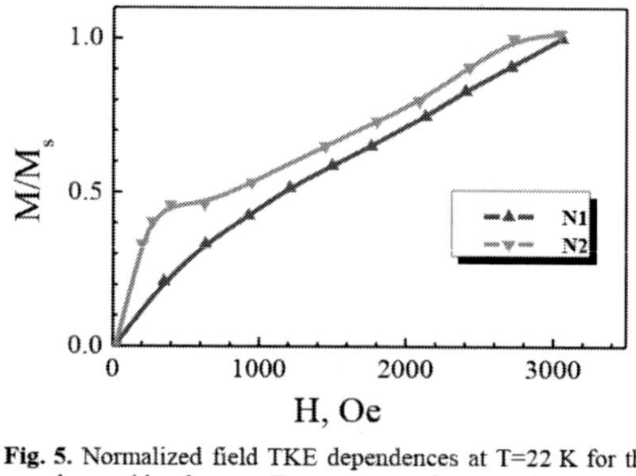

Fig. 5. Normalized field TKE dependences at T=22 K for the samples with lower Fe concentrations and substrate temperatures.

For sample N1, we observe approximately linear dependence on the magnetic field in the all temperature range. At the same time, a superposition of ferromagnetic and paramagnetic contributions is observed at low temperatures for sample N2.

3 Conclusions

Samples with Y_{Fe} >0.08 and $T_S \geq 200°C$ not only exhibit ferromagnetic behavior, but also have high Curie temperature. Even at T=194°C the TKE signal is also relatively strong. However, one should check, whether these properties are related to the bulk material or the surface particles, since the experimental data leads to a conclusion that these samples are multiphase. Most likely, the surface particles are compounds such as FeO, Fe_3O_4, Fe_2O_3 (since the small particles being on the surface would inevitably oxidize), which are mixed with FeSb variations. Despite the fact that at room temperature a distinguishable signal is not detected from sample N2, it shows a low-temperature ferromagnetic response. However, without the substrate heating the overall effect is lower that may be related to less iron oxides content. Matrix doping in the sample with the Fe lowest content (Y_{Fe} =0.08, N1) is apparently insufficient to create ferromagnetism. Further improvement of the growth technology aimed at the fabrication of single-phase InFeSb layers or those with negligibly small content of undesirable phases, as well as the studies continuation are required to confirm the intrinsic high-temperature ferromagnetism of the InFeSb layers obtained by laser ablation.

This research was supported by the Russian Foundation for Basic Researches №15-02-02077

References

1. N. T. Tu, P. N. Hai, L. D. Anh, and M. Tanaka, Appl. Phys. Lett. **108**, 192401 (2016).
2. A.V. Kudrin, Yu. A. Danilov et al. Trudy XI Mezhd. Symp. «Nanofizika i Nanoelectronika», **v.1**, 195 (2017).
3. V.I. Gavrilenko, A.M. Grekhov, D.V. Korbutyak, V.G. Litovchenko, Spravochnik "Optical properties of semiconductors", Kiev, Naukova Dumka, p.208 (1987).

Electroinduced magnetic bubble domain nucleation

Daria *Kulikova**, Timur *Gareev*, Elena *Nikolaeva, and* Alexander *Pyatakov*

Physics Department, M.V. Lomonosov Moscow State University, Moscow, Russia

Abstract. The magnetoelectric properties of micromagnetic structures in iron garnet films manifest themselves in electrically induced displacement of domain walls and magnetic bubble domains nucleation. In this paper we found the condition of the electrically induced bubble domain nucleation in terms of critical electric field, magnetic bias and temperature. The collapse of magnetic bubble domain under the action of electric field pulse is also demonstrated.

1 Introduction

The development of information technologies requires continuous search, synthesis and research of new materials allowing new physical principles for storing and recording information. Since the beginning of this century, interest for magnetoelectric materials and multiferroics [1, 2] has been increasing. Magnetoelectric properties can be manifested both in materials with a certain type of crystallographic symmetry and in materials with spatial modulation of the magnetization vector [2]. An example of a magnetic structure with magnetization modulation possessing electric polarization is a Neel-type domain wall [3].

The displacement of the domain wall in an iron garnet film with a crystallographic orientation (210) under the action of an inhomogeneous electric field of the tip electrode was experimentally demonstrated in [4]. This phenomenon is explained by the presence of electrical polarization of the domain wall: the effect is of magnetoelectric nature.

Later it was experimentally shown [4, 5] that the displacement of the domain wall in an inhomogeneous electric field can be increased by applying an in-plane magnetic field.

At the same time I.E. Dzyaloshinskii theoretically predicted the generation of a micromagnetic inhomogeneity from a single domain state by a non-uniform electric field [6]. Later the effect of an electroinduced nucleation of magnetic bubble domain in iron garnet film was observed [7].

In this paper the nucleation of the magnetic bubble domains in iron garnet films depending on external conditions is studied. We also present results on the dynamic collapse of bubble domains under the electric field action.

2 Experiment

2.1. Scheme of experiment

For the experiment the sample with strong magnetoelectric effect previously observed in [1] was chosen. It was $(BiLu)_3(FeGa)_5O_{12}$ garnet film grown by liquid-phase epitaxy on (210) $Gd_3Ga_5O_{12}$ substrate. In this sample the magnetization direction does not coincide with the normal to the film. The thickness of the substrate is 0.5 mm. The saturation magnetization is $4\pi Ms = 62$ Gs. A stripe domain structure is formed along the $[\bar{1}20]$ direction in the zero external magnetic fields. The period of the domain structure is about 29 μm.

The general scheme of the experiment is shown in Fig.1(a). To create electric field pulses we used a molybdenum wire as a tip electrode with a diameter of about 10 μm touching the surface of the film. The voltage was applied to the electrode by a high-voltage source.

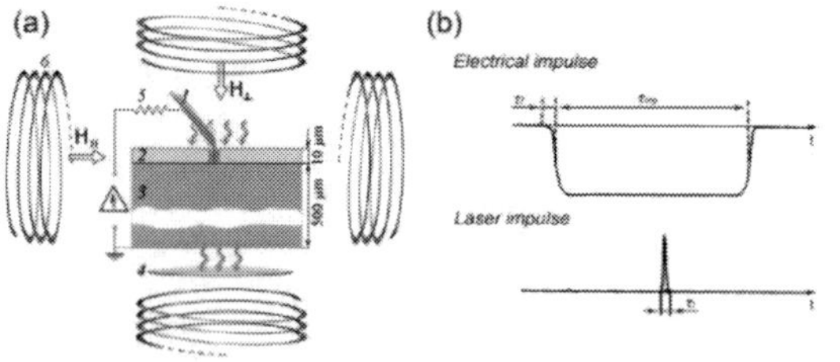

Fig.1. (a) The scheme of the experiment. 1 – tip electrode, 2 – iron garnet film, 3 – substrate, 4 – optical system of the microscope, 5 – ballast resistor, 6 – the coil system. (b) Negative electrical impulse from the generator and the laser impulse.

By supplying a voltage of 500 V to the wire we can obtain electric field strength up to 1 MV / cm. At the same time the voltage jump is enough sharp to nucleate bubble domain (time of charge is about 100 μs). To limit

* Corresponding author: dp.kulikova@physics.msu.ru

the effects of currents of charging, discharging and leakage we used a ballast resistor of 10 MOm.

According to the Fig.1, the sample was placed between two pairs of coils which created external magnetic bias fields: applied both in the plane of the iron garnet film ($H_{\parallel}$) and perpendicular ($H_{\perp}$) to it.

The visualization of the micromagnetic structure was realized with the magneto-optical Faraday effect. The image was recorded using a CCD camera.

To study the dynamic characteristics of the bubble domain generation the high-speed photography technique was used.

The signal of a sequence of rectangular pulses of the electrical voltage was applied to the electrode from the generator. Pulse time is τ_{imp} = 800 ns and a rise time is τ_f = 50 ns (Fig.1(b)).

The laser pulse duration is τ_l = 10 ns (Fig.1(b)), its start is synchronized with the start of the electric pulse generator. Varying the time delay between field and laser pulses enabled us to observe the consecutive position of domain and thus investigate its dynamics.

2.2 Results of the experiment

In [7] we first demonstrated the generation of the magnetic bubble domain under the action of non-uniform electric field.

We observed the bubble domain generation for two directions of the in-plane magnetic field: along the $[\bar{1}20]$ and [001] directions. The experimental results described below are presented for the in-plane magnetic field along the $[\bar{1}20]$ direction.

The collapse of the stripe domain structure of the iron garnet film occurs at magnetic field about $H_{\parallel}$ = 50 Oe. For the further increase of the field $H_{\parallel}$ we use the magnetic field $H_{\perp}$ which is oriented perpendicularly to the film surface. (Fig.2).

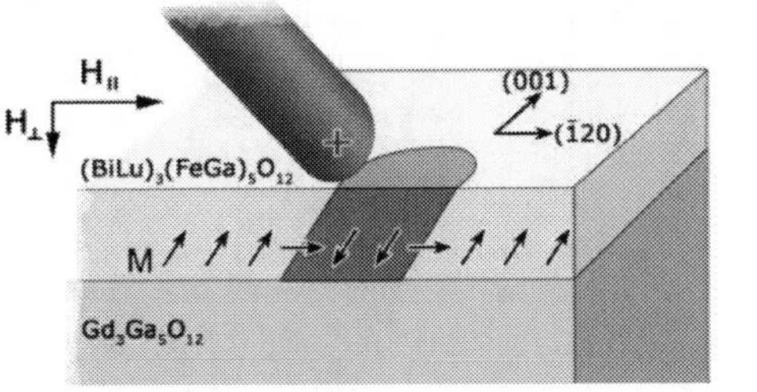

Fig.2. The schematic distribution of the magnetization in the iron garnet film with crystallographic orientation (210).

Magnetic fields of certain values were applied in order the sample was in the single domain state near the transition boundary into a state with a stripe domain structure. When the in-plane magnetic field is oriented along the $[\bar{1}20]$ direction, the values of $H_{\parallel}$ and $H_{\perp}$ were equal to 200 and 70 Oe, respectively. When a positive voltage was applied to the tip electrode over a critical value, the bubble domain nucleation occurred.

The magnetooptical images of the domain structure illustrating the process of the bubble domain nucleation are shown in Fig 3. Initially the sample was homogeneously magnetized (Fig.3(a)). When the positive voltage was applied to the tip, the bubble domain was blowed by it (Fig.3(b)). After the voltage was switched off the domain detached from the electrode and decreased in size (Fig.3(c)).

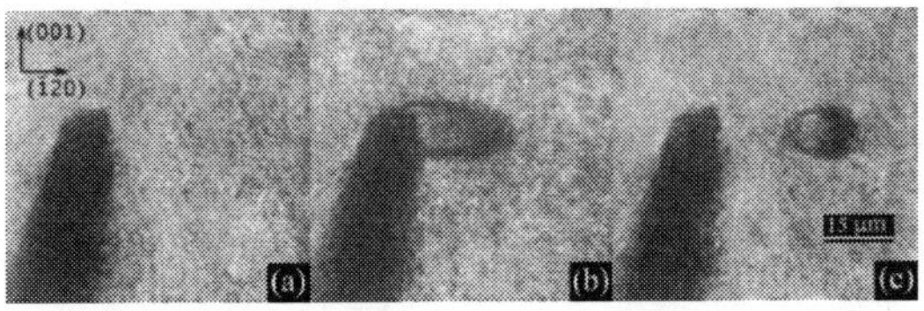

Fig.3. Magnetooptical images of the nucleation of the magnetic bubble domain at $H_{\parallel}$ = 200 Oe, $H_{\perp}$ = 70 Oe. (a) Tip electrode touches the surface of the film, which is in single domain state. (b) The voltage is switched on, the nucleation of the domain occurs. (c) The bubble domain detaches from the tip when the voltage is switched off.

The nucleation of magnetic bubble domains occurs only in electric fields above some critical value. In its turn the critical voltage depends on the magnitude of the bias magnetic field. The established dependence of the critical voltage on the electrode on the strength of the magnetic bias field is shown in the Fig 4. In the green area of the magnetic and electric fields (below the curve) the sample is homogeneously magnetized and no nucleation of the domain occurs. In the white area of fields on the graph (above the curve) there is observed both a homogeneously magnetized state and the stripe structure of the sample and also it is possible to generate magnetic bubble domain by electric field. The measurements were carried out at a fixed temperature of the system (t = 47°C).

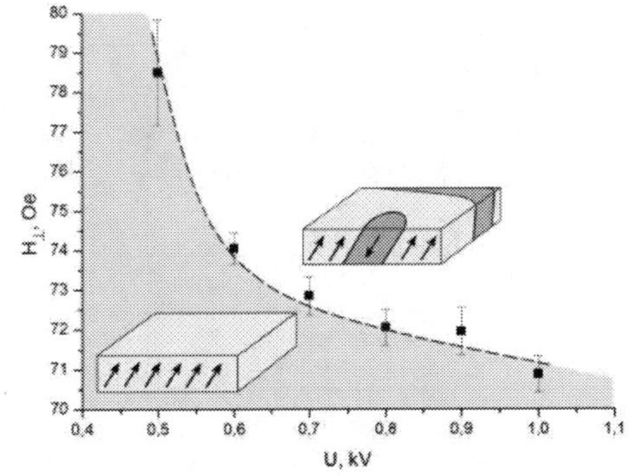

Fig.4. Dependence of the strength of the magnetic bias field on the critical voltage on the electrode.

With an increase of the temperature of the system the minimal magnetic field (which is necessary for the nucleation of the domain) decreases. In Fig.5 it is shown the dependence of the critical magnetic field on the temperature at fixed values of electric voltage on the electrode U = 700 V and of magnetic in-plane field $H_{||}$ = 200 Oe. Above the curve, there is a possibility of nucleation of bubble domains, below the curve - no. This indicates that the phase diagram in Fig.4 goes "down" when the temperature rises.

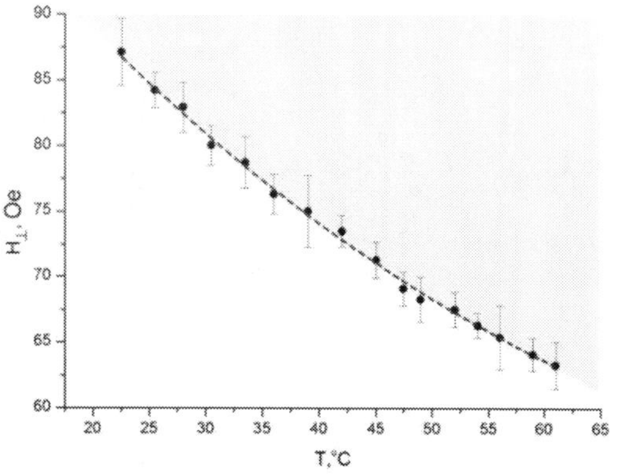

Fig.5. Dependence the critical magnetic bias field on the temperature of the system. $H_{||}$ = 200 Oe, U = 700 V.

We also investigated the collapse of the bubble domains under the action of a negative electric field pulse by the method of high-speed photography.

The magnetic bubble domain is generated at the rear front of the negative voltage pulse. Because the repetition period of electric pulses exceeds the width of the front, the domain nucleation process is completed before the next potential change at the electrode occurs. Using this technique we got magnetooptical images of the domain collapse on the forefront of the negative voltage pulse from the generator at different moments of time (Fig.6). The size of the domain decreases with time. At about 350 ns it collapses and the sample is turning to a homogeneously magnetized state again.

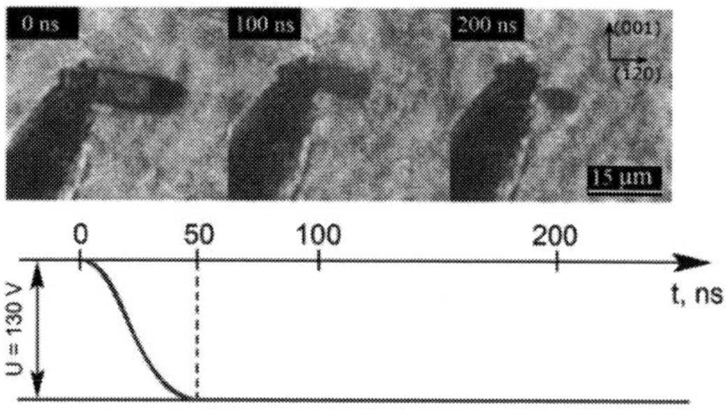

Fig.6. Magnetic bubble domain collapse dynamics in negative electric pulse with an amplitude U = 130V.

In Fig.7 it is presented the graph of the dependence of the bubble domain major semiaxis on time. Using the linear fitting it was determined the velocity of the bubble domain walls under the action of electric field. The value of it is about 50 m / s.

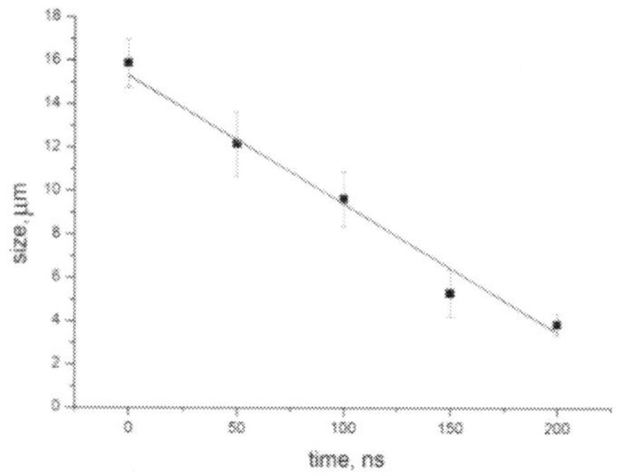

Fig.7. Bubble domain size on time dependence. $H_{||}$ = 200 Oe, $H_{\perp}$ = 70 Oe.

Conclusions

We have determined the conditions for bubble domain generation (critical values of external magnetic fields, electric field and their dependence on temperature). It was experimentally demonstrated that bubble domain can be both generated and collapsed under the action of an electric field.

Work was supported by RFBR grants ## 16-02-00494, and 16-29-14037_ofi_m.

References

1. W. Eerenstein, N.D. Mathur, J.F. Scott, Nature, **442**, 759 (2006)
2. A.P. Pyatakov, A.K. Zvezdin, Phys. Usp., **55**, 557 (2012)
3. V.G. Bar'yakhtar, V.A. L'vov, D.A. Yablonskii, JETP Letters, **37**, 673 (1983)
4. A.P. Pyatakov, A.S. Sergeev, E.P. Nikolaeva, T.B. Kosykh, A.V. Nikolaev, K.A. Zvezdin, A.K. Zvezdin, Phys.Usp., **58**, 981 (2015)
5. A.P. Pyatakov, D.A. Sechin, A.S. Sergeev, A.V Nikolaev, E.P. Nikolaeva, A.S. Logginov, A.K. Zvezdin. Europhysics Letters, **93**, 17001 (2011)
6. I. Dzyaloshinskii, EPL, **83**, 67001 (2008)
7. D.P. Kulikova, A.P. Pyatakov, E.P. Nikolaeva, A.S. Sergeev, T.B. Kosykh, Z.A. Pyatakova, A.V. Nikolaev, A.K. Zvezdin, JETP Lett., **104**, 197 (2016)

Optical spectroscopy of $ErFe_3(BO_3)_4$: detection of phase transitions and crystal-field levels of the Er^{3+} ground multiplet

Dmitry Erofeev[1,2,*], *Artjoms* Jablunovskis[2], and *Elena* Chukalina[1]

[1]Institute of spectroscopy RAS, Fizicheskaya Str., 5, Troitsk, Moscow, 108840, Russia
[2]Moscow Institute of Physics and Technology, 9 Institutskiy per., Dolgoprudny, 141701, Russia

Abstract. High-resolution spectra of oriented $ErFe_3(BO_3)_4$ single crystals are registered in the spectral range 500 – 8000 cm^{-1} at various temperatures (4 – 470 K). The temperature behaviour of the phonon mode at 976 cm^{-1} allowed us to register a structural phase transition at T_S=431 K. The energies of crystal-field levels of the ground $^4I_{15/2}$ and the first excited $^4I_{13/2}$ multiplets of the Er^{3+} ion are determined in the paramagnetic state of $ErFe_3(BO_3)_4$. The exchange splitting of the ground Kramers doublet in the magnetically ordered state was found to be Δ_0=6.3±1 cm^{-1}. The interference occurring due to birefringence in the single crystals was also registered. The temperature dependence of the position of the maximum of the interference band demonstrates two anomalies, at T_N and T_S, associated with the phase transitions in $ErFe_3(BO_3)_4$.

1 Introduction

$ErFe_3(BO_3)_4$ belongs to the family of rare-earth (RE) iron borates $RFe_3(BO_3)_4$ (R = Y, La – Lu) that have the same structure as a natural mineral huntite (*R32* space symmetry group). These compounds possess a number of fascinating properties. First of all, a magnetoelectric (ME) effect was discovered, which made it possible to classify them as multiferroics. Among the iron borates, the highest values of the electric polarization induced by a magnetic field were observed in $NdFe_3(BO_3)_4$ and $SmFe_3(BO_3)_4$, 400 and 500 μC/m^2, respectively [1, 2]. Below the Neel temperature, both these compounds order antiferromagnetically into an easy-plane structure. $ErFe_3(BO_3)_4$ also is an easy-plane antiferromagnet, therefore it was expected to have a large value of the induced polarization as well. In fact, its value is negligible [3]. The magnetic properties of iron borates are interesting also due to the presence of two magnetic subsystems, *f*- and *d*- ones, and their mutual interaction. The differences between the magnetic and ME properties of RE iron borates result from differences in crystal-field (CF) levels of the RE ion and RE-Fe exchange interaction. The data on CF levels and exchange interactions can be obtained from spectroscopic measurements and further theoretical calculations based on them. In this work, we present the first results on the temperature dependences of polarized optical spectra of $ErFe_3(BO_3)_4$ crystals. We demonstrate spectral signatures of the phase transitions in $ErFe_3(BO_3)_4$ and report the values of CF energies of the ground $^4I_{15/2}$ and the first excited $^4I_{13/2}$ multiplets of the Er^{3+} ion.

2 Structure and properties

In the trigonal crystal structure *R32*, the main elements are MO_6 octahedra forming quasi-one-dimensional helicoidal chains along the crystallographic axis *c*. Distorted ErO_6 prisms and triangular BO_3 groups unite the chains into a three-dimensional structure. Raman spectra measurements for $RFe_3(BO_3)_4$ revealed a structural phase transition at T_S=155, 198, 340, and 350 K for R = Gd, Tb, Er, and Y, respectively [4]. According to a single-crystal x-ray structural study of $GdFe_3(BO_3)_4$ [5], this transition corresponds to the crystal symmetry lowering from the *R32* space group to the $P3_121$ one. The first order structural phase transition to lower symmetry occurred in $ErFe_3(BO_3)_4$ was determined by excess heat capacity in the work [6]. The value T_S = 433 K was registered during cooling measurements, and 439 K – during heating measurements. Also in the work [6] the research provided absorption, magnetic circular dichroism and natural circular dichroism spectra of $ErFe_3(BO_3)_4$ as a function of temperature in the range of 90-293 K. As far as we know, there are no other works devoted to the spectroscopic study of $ErFe_3(BO_3)_4$ single crystals.

The specific heat data for $ErFe_3(BO_3)_4$ exhibit a lambda-type anomaly at 38 K, which matches closely the Neel temperature, and the Schottky anomaly [7]. A modeling of the latter allowed the authors of Ref. [7] to calculate the ground-state exchange splitting of Er^{3+} ions: 7.5 K [7]. The magnetic ordering as a second-order phase transition into an easy-plane magnetic structure at T_N = 39 K was detected by the erbium spectroscopic probe method [8]. The work also provided experimental data on the splitting of the Er^{3+} ground Kramers doublet, which equals 7.3 cm^{-1} at 5 K. The temperature dependences of magnetic susceptibility and

* Corresponding author: kinson@mail.ru

magnetization of $ErFe_3(BO_3)_4$ along and perpendicular to the c-axis at low temperatures demonstrate a contribution of the ground Kramers doublet of the Er^{3+} ion, which possesses a significant anisotropy of the g-tensor [3]. The compound also shows a spontaneous magnetic polarization in the magnetically ordered state [3].

3 Experimental methods

Transparent green single crystals of $ErFe_3(BO_3)_4$ of good optical quality were grown by a solution-melt method with a flux based on $Bi_2Mo_3O_{12}$ in the Kirensky Institute of Physics, Krasnoyarsk, Russia [9]. The growth details will be given elsewhere. The single crystals were oriented (by habitus and polarization method using a polarization microscope Olympus SZX7), cut, and polished. The result was several samples of different thickness, cut along (π- and σ-polarizations) and perpendicular (α-polarization) to the crystallographic axis *c*. Transmission spectra were recorded on a high-resolution Fourier spectrometer Bruker IFS 125HR in the spectral range 500-8000 cm^{-1} with a spectral resolution up to 0.1 cm^{-1}. Temperature measurements were carried out in the range 4–470 K using a Cryomech ST403 closed-cycle cryostat and a Bruker 2216e heating device.

To study the birefringence, a sample cut parallel to the crystallographic axis *c* of the crystal was placed into a cryostat between two crossed identical polarizers, so that the plane of polarization was at an angle of 45° to the *c* axis. As a result, interference of the ordinary and extraordinary rays was observed.

4 Results and discussion

4.1 Structural phase transition

Structural changes in a crystal are unambiguously detected in the temperature behavior of phonon modes. We were able to measure a phonon mode at 976 cm^{-1}, which is the E mode corresponding to internal oscillations ν_4 of the BO_3 group [4]. The mode appears only in the low-symmetry *$P3_121$* phase of $RFe_3(BO_3)_4$ [4]. The temperature dependence of the mode's integral intensity allowed us to determine the temperature of the transition $T_S = 431$ K (fig.1). The influence of the phase transition on the spectral lines corresponding to the *f-f* optical transitions of Er^{3+} is difficult to interpret, due to significant broadening of these lines at high temperatures. The difference in the values of T_S obtained by different research groups [4, 6] is associated with the entrance of impurities from the flux into the crystal in the course of growth, which depends on technology [10]. The powder samples synthesized by the solid-phase synthesis method do not contain such impurities. However, there are no temperature data on T_S for the $ErFe_3(BO_3)_4$ powder samples. But the approximation of the dependence of T_S on the ionic radius of the RE ion provides a value of about 430 K [11]. The fact that the value of T_S obtained in the present work is close to the one for powder samples proves a good quality of our samples. This gives us hope that the data we obtained are reliable.

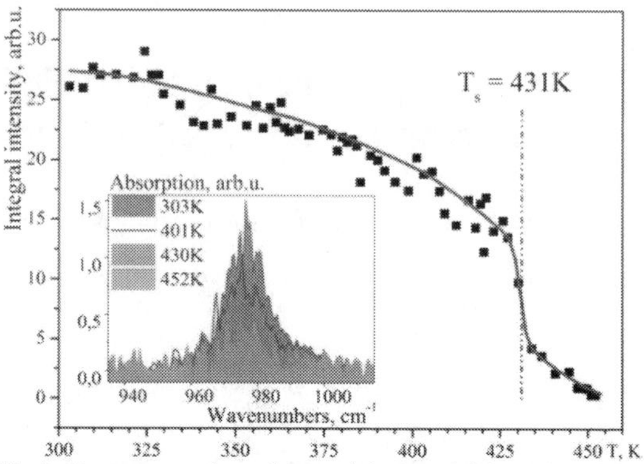

Fig.1. Temperature dependence of the integral intensity of the 976 cm^{-1} phonon appearing below structural phase transition temperature T_S in $ErFe_3(BO_3)_4$. The inset shows this phonon in the π-polarized absorption spectra at different temperatures.

4.2 Crystal-field levels and the exchange splittings

In the high-temperature structural phase (*R32*) of $ErFe_3(BO_3)_4$, the CF levels of Er^{3+} ion are the Γ_4 and Γ_{56} Kramers doublets of the D_3 point symmetry group. In the low-temperature phase, all of them are described by the Γ_{34} irreducible representations of the C_2 point group. At temperatures below the Neel temperature, the Er^{3+} Kramers doublets split in two in the effective magnetic field created by the ordered iron magnetic subsystem, and the corresponding spectral lines split generally into four components (fig.2).

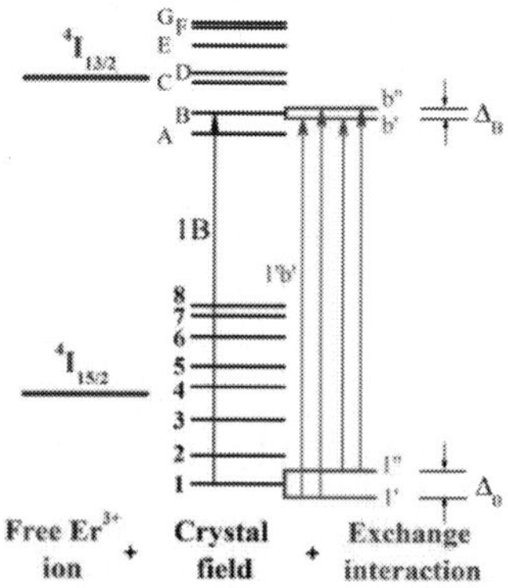

Fig.2. CF and exchange splittings of energy levels of the free Er^{3+} ion caused by the crystal-field and exchange interations in the magnetically ordered state of $ErFe_3(BO_3)_4$.

Fig.3 represents the spectra illustrated by the scheme on fig.2. A comparison of the spectra at different temperatures and polarizations allowed us to determine the energies of CF levels of the ground $^4I_{15/2}$ and the excited $^4I_{13/2}$ multiplets of Er^{3+} ion in $ErFe_3(BO_3)_4$. More lines appear at higher temperatures, due to population of higher levels according to the Boltzmann distribution. Seven lines that remain at low temperatures represent the structure of the CF levels of the excited multiplet $^4I_{13/2}$ (fig.3a). Further identification of the CF levels with energies higher than that of the level 3 (see fig.2) over high-temperature spectra is not possible. That is due to substantial broadening of the spectral lines at temperatures above 150 K. To increase the absorption of the spectral lines corresponding to the transitions from the excited levels of the ground multiplet, we measured the spectra of a much thicker sample (fig.3b). The analysis of these spectra at different temperatures allowed us to determine the energies of all the eight levels of the ground multiplet $^4I_{15/2}$. Table 1 presents the data on the CF levels of the two lowest multiplets of Er^{3+} in $ErFe_3(BO_3)_4$.

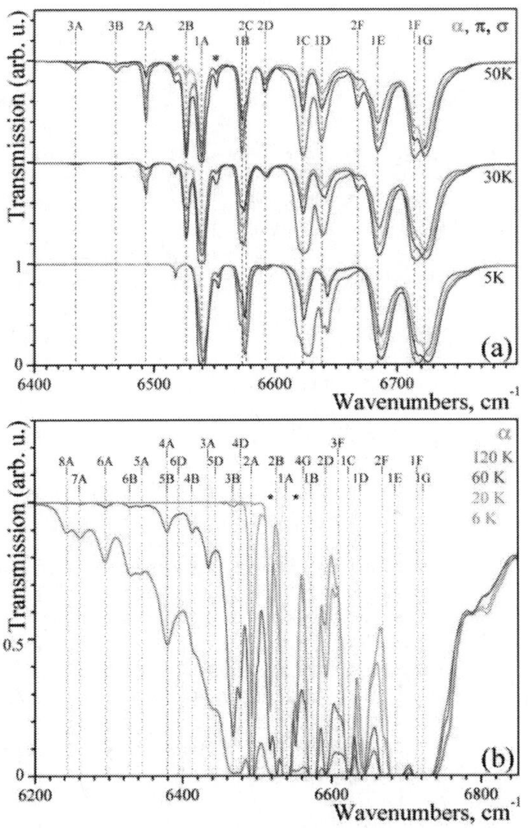

Fig.3. Transmisson spectra of $ErFe_3(BO_3)_4$ in the region of the $^4I_{15/2}\rightarrow{}^4I_{13/2}$ optical transition at different temperatures and polarizations for a sample with the thickness d=0.15mm (a) and for an α-polarized sample with d=2.016mm (b). The symbol * denotes the spectral lines related to the absorption of an impurity.

Table 1. Energies of CF levels of the Er^{3+} ion in the paramagnetic state of $ErFe_3(BO_3)_4$ (cm^{-1}).

$^4I_{15/2}$	0	46	105	160	194	244	279	296
$^4I_{13/2}$	6539	6573	6623	6639	6684	6714	6723	

The fact that in the low-temperature phase a polarization dependence is still present in the spectra, indicates that deviations from the D_3 local symmetry are small. At temperatures below T_N, the spectral lines split. The splitting of the spectral line 1B into four components (fig.4) allowed us to determine the exchange splitting of the ground level, according to the scheme of fig.2: $\Delta_0 = 6.3\pm1$ cm^{-1}. The obtained data within the error limit agree with the heat capacity data [7] and the estimate given in Ref. [8] from the spectral data on nonoriented $ErFe_3(BO_3)_4$ samples of a poor quality.

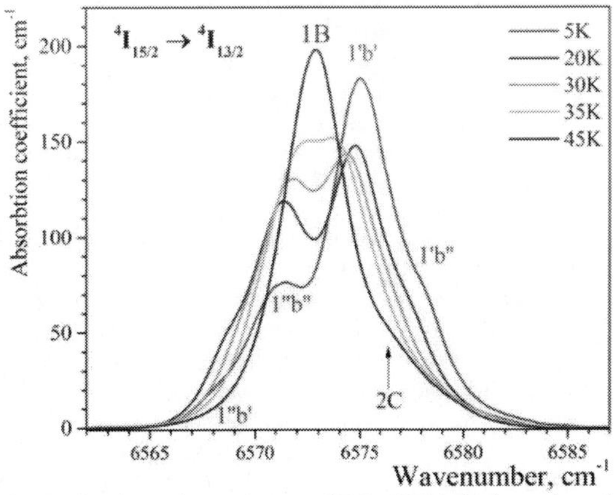

Fig.4. The absorption spectrum of $ErFe_3(BO_3)_4$ in the region of the 1B spectral line of the $^4I_{15/2}\rightarrow{}^4I_{13/2}$ optical transition at several temperatures.

The exchange splitting of the excited level B of the multiplet $^4I_{13/2}$ at T = 5 K is $\Delta_B = 2.7 \pm 1$ cm^{-1} (fig.4). Through the analysis of the temperature dependence of the transmission spectra of the thick sample (fig.3b), it was possible to determine the splitting of the first excited Kramers doublet of the ground multiplet $^4I_{15/2}$ (level 2 on fig.2): $\Delta_2 = 1.7 \pm 1$ cm^{-1} at T = 5 K. This value was established under the assumption that the nature of the temperature dependences of the exchange splittings is similar for the spectral lines 1B (fig.4), 2A and 2B fig.3b). The obtained data can be used for the explanation of the anomalies in the temperature dependences of the specific heat and magnetic susceptibility of $ErFe_3(BO_3)_4$.

4.3 Temperature dependences of the birefringence

The temperature dependence of the position of the maximum in the interference pattern $\nu_{max}(T)$ reveal two anomalies (see fig.5). The first one is a step-like shift at the temperature of the structural phase transition T_S. A further smooth decrease of ν_{max} with lowering the temperature changes to its growth at about 100 K with a saturation at the lowest temperatures. The low-temperature part of the dependence has an inflection point near T_N. This behavior is evidently associated with a magnetic contribution. An analogous dependence for $ErAl_3(BO_3)_4$, known to have no phase transitions in the studied temperature range, has a practically linear character without pronounced anomalies.

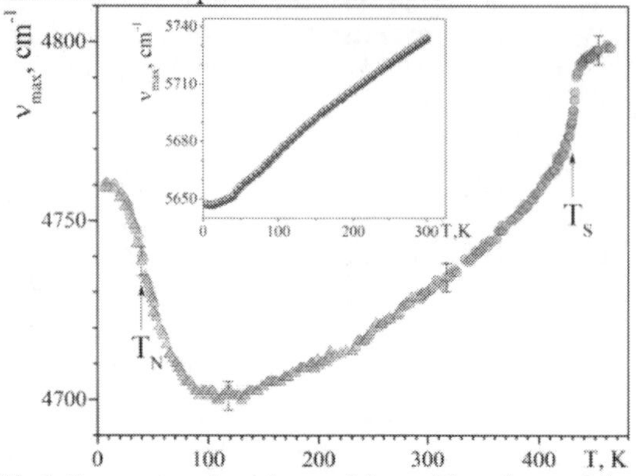

Fig.5. Temperature dependence of the position of one of the maxima in the interference between ordinary and extraordinary rays in $ErFe_3(BO_3)_4$ and $ErAl_3(BO_3)_4$ (inset).

The shift of the interference band is caused both by the temperature change in the refractive index (dielectric constant) of the compound and by its thermal expansion. The work [3] presents data on thermal expansion $\Delta L_a(T)/L_a$ for $RFe_3(BO_3)_4$ crystals (R = Eu, Gd, Tb). For $ErFe_3(BO_3)_4$ these data are not available. We assume that for isostructural compounds the behavior of $\Delta L_a(T)/L_a$ is similar. The maximum change in $\Delta L_a(T)/L_a$ caused by phase transitions is of the order of 10^{-4}. This is two orders of magnitude smaller than the observed relative shift of the maximum of the interference band $\nu_{max}(T)$. Thus, the temperature shift of the interference band is determined mainly by the change in the difference of refractive indices $\Delta n(T) = n_o(T) - n_e(T)$. The change in $\Delta n(T)$ is due to magnetic and lattice contributions. The magnetic contribution is proportional to the average value of the squared magnetization of the sublattices [12]. The lattice contribution consists of a change in birefringence due to the regular thermal expansion and to spontaneous magnetostriction. The spontaneous polarization in $ErFe_3(BO_3)_4$ is also providing a contribution into $\Delta n(T)$ below T_N. The structural phase transition in $RFe_3(BO_3)_4$ is accompanied by anomalies of the dielectric properties associated with the antiferroelectric phase transition. They are naturally related to the anomalies observed in fig.5.

Conclusion

In this work, we have presented the data on the energies of the CF levels of the ground and the first excited multiplets of the Er^{3+} ion in the paramagnetic state of $ErFe_3(BO_3)_4$. The exchange splitting of the ground Kramers doublet was found to be $\Delta_0 = 6.3\pm1$ cm^{-1}. The obtained results are important for the interpretation of thermodynamic and magnetic properties of $ErFe_3(BO_3)_4$. The temperature dependence of the interference occurring due to birefringence in the $ErFe_3(BO_3)_4$ single crystals was investigated. It demonstrates two anomalies near the temperatures of the phase transitions T_N and T_S. Possible contributions into the anomalies were discussed.

Acknowledgements

The authors are thankful to I.A. Gudim from the Kirensky Institute of Physics, Krasnoyarsk, Russia, for providing good optical quality single crystals of $ErFe_3(BO_3)_4$ and $ErAl_3(BO_3)_4$.

References

1. A.K. Zvezdin, G.P. Vorob'ev, A.M. Kadomtseva, Yu.F. Popov, A.P. Pyatakov, L.N. Bezmaternykh, A.V. Kuvardin, and E.A. Popova, JETP Lett. **83**, 509 (2006).
2. Yu.F. Popov, A.P. Pyatakov, A.M. Kadomtseva, G.P. Vorob'ev, A.K. Zvezdin, A.A. Mukhin, V.Yu. Ivanov, and I.A. Gudim, J. Exp. Theor. Phys. **111**, 199 (2010).
3. A.M. Kadomtseva, Yu.F. Popov, G.P. Vorob'ev, A.P. Pyatakov, S.S. Krotov, K.I. Kamilov, V.Yu. Ivanov, A.A. Mukhin, A.K. Zvezdin, A.M. Kuz'menko, L.N. Bezmaternykh, I.A. Gudim, and V.L. Temerov, Low Temp. Phys. **36**, 511 (2010).
4. D. Fausti, A. Nugroho, P. van Loosdrecht, S.A. Klimin, M.N. Popova, and L. N. Bezmaternykh, Phys. Rev. B **74**, 024403 (2006).
5. S.A. Klimin, D. Fausti, A. Meetsma, L.N. Bezmaternykh, P.H.M. van Loosdrecht, and T.T.M. Palstra, Acta Crystallogr. B **61**, 481 (2005).
6. A.V. Malakhovskii, A.L. Sukhachev, V.V. Sokolov, T.V. Kutsak, V.S. Bondarev, I.A. Gudim, J. Magn. Magn. Mater. **384**, 255–265 (2015).
7. A.N. Vasiliev, E.A. Popova, I.A. Gugim, L.N. Bezmaternykh, Z. Hiroi, J. Magn. Magn. Mater. **300**, e382–e384 (2006).
8. M.N. Popova, E.P. Chukalina, T.N. Stanislavchuk, L.N. Bezmaternykh, Journal of Magnetism and Magnetic Materials **300**, e440–e443 (2006).
9. I.A. Gudim, E.V. Eremin, V.L. Temerov, Journal of Crystal Growth **312**, pp. 2427–2430 (2010).
10. M.N. Popova, B.Z. Malkin, K.N. Boldyrev, T.N. Stanislavchuk, D.A. Erofeev, V.L. Temerov, and I.A. Gudim, Physical Review B **94**, 184418 (2016).
11. Y. Hinatsu, Y. Doi, K. Ito, M. Wakeshima, and A. Alemi, J. Solid State Chem. **172**, 438 (2003).
12. G.A. Smolenskii, R.V. Pisarev, I.G. Sinii, Sov. Phys. Usp. **18**, 410–429 (1975).

Polarizability of electrically induced magnetic vortex "atoms"

P.I. Karpov[1,*] and *S.I.* Mukhin[1]

[1] *National University of Science and Technology "MISiS", Leninsky avenue 4, 119991, Moscow, Russia*

Abstract. Electric field control of magnetic structures, particularly topological defects in magnetoelectric materials, draws a great attention, which has led to experimental success in creation and manipulation of single magnetic defects, such as skyrmions and domain walls. In this work we explore a scenario of electric field creation of another type of topological defects – magnetic vortices and antivortices. Because of interaction of magnetic and electric subsystems each magnetic vortex (antivortex) in magnetoelectric materials possesses quantized magnetic charge, responsible for interaction between vortices, and electric charge that couples them to electric field. This property of magnetic vortices makes possible their creation by electric fields. We show that the electric field, created by a cantilever tip, produces a "magnetic atom" with a localized spot of ordered vortices ("nucleus" of the atom) surrounded by antivortices ("electronic shells"). We analytically find the vortex density distribution profile and temperature dependence of polarizability of this structure and confirm it numerically by Monte Carlo simulation.

1 Introduction

Electric field control of magnetization and magnetic topological defects is very promising for creation of new types of magnetic memory devices. Magnetoelectric materials, such as type-II multiferroics [1] are particularly appealing for this purpose. The use of domain walls and other topological defects in multiferroics [2] for magnetic storage are very promising since they provide a mechanism of very dense packing of information, while topological nature protects information from loss under the influence of external perturbations such as heating or mechanical action. Among the bright examples of such topological defects are skyrmion lattices [3], as well as individual skyrmions observed in magnetoelectric materials [4], magnetic vortices [5, 6], and domain walls [7, 8].

In this paper we report our recent results [9]. We consider theoretically a phenomenological model developed by Mostovoy [10] for a type-II multiferroics and apply it to a thin-film material with easy-plane symmetry. We show both analytically and numerically that in this case a strong enough locally concentrated electric field may cause formation of a "magnetic atom" that consists of a vortex "nucleus" surrounded by antivortex "shells". This idea allows us to propose an experimental realization for 2D Coulomb plasma in a trap and make predictions about its electric polarizability.

2 The model

In order to describe a layered material with coupled magnetic and electric subsystems we use the following phenomenological model [9]. We write the total free energy density as a sum of parts arising from electric polarization, magnetization, and magnetoelectric coupling [10]:

$$w = \frac{P^2}{2\chi_e} - \mathbf{EP} - \frac{\mathbf{E}^2}{8\pi} - \gamma\mathbf{P}((\mathbf{M}\nabla)\mathbf{M} - \mathbf{M}(\nabla\mathbf{M})) + \\ + \frac{\alpha}{2}\left[((\partial\mathbf{M}/\partial x)^2 + (\partial\mathbf{M}/\partial y)^2\right]. \quad (1)$$

We assume that **M** is the primary order parameter, which can induce **P**. Thus we can find polarization by minimizing (1) with respect to **P**:

$$\mathbf{P} = \gamma\chi_e((\mathbf{M}\nabla)\mathbf{M} - \mathbf{M}(\nabla\mathbf{M})) + \chi_e\mathbf{E}. \quad (2)$$

Because of the thin-film geometry of the sample, magnetization tends to be parallel to the film plane, otherwise the demagnetizing field energy would greatly increase; also, material can possess an easy-plane type of symmetry. Hence we restrict the magnetization to lie in the plane

$$\mathbf{M}(\mathbf{r}) = M_0(\cos\phi(\mathbf{r}), \sin\phi(\mathbf{r}), 0). \quad (3)$$

Using the angle ϕ, the polarization can be written as

$$\mathbf{P} = \gamma\chi_e M_0^2\left(-\partial_y\phi,\, \partial_x\phi,\, 0\right) + \chi_e\mathbf{E}. \quad (4)$$

Substituting (3) and (4) into (1), we find that the electric, magnetic, and magnetoelectric parts of the energy combine to

$$w = \frac{1}{2}\left(\alpha M_0^2 - \chi_e\gamma^2 M_0^4\right)(\nabla\phi)^2 - \\ -\chi_e\gamma M_0^2\left(\partial_x\phi E_y - \partial_y\phi E_x\right) - \frac{1}{2}\chi_e E^2. \quad (5)$$

The expression for energy $W = \int w dV$ is similar to that of the XY model: $W_{XY} = \frac{1}{2}\rho_s\int(\nabla\phi)^2 dS$ with an effective spin-wave stiffness $\rho_s = (\alpha M_0^2 - \chi_e\gamma^2 M_0^4)h$, where h

*e-mail: karpov.petr@gmail.com

is the film thickness. The magnetoelectric coupling just renormalizes the spin stiffness of the XY model. For typical values of parameters [9], αM_0^2 is greater than $\chi_e\gamma^2 M_0^4$ by several orders of magnitude; therefore, ρ_s remains almost unchanged by magnetoelectric interaction and we use $\rho_s \approx \alpha M_0^2 h$. Despite this, some interesting features appear. It is known that in the XY and related models the magnetic vortices play the crucial role [11]. Vortex placed at the origin has the form $\phi = k \arctan \frac{y}{x} + \phi_0$, where k is the winding number of the vortex. Because of magnetoelectric interaction (2), magnetic vortex acquires an electric charge kq_e [10], where

$$q_e = 2\pi\gamma\chi_e M_0^2 h. \tag{6}$$

Neglecting small electrostatic interaction of vortices we consider only their magnetic interaction with 2D Coulomb potential $W = \pm 2q_m^2 \ln(r/a)$ [11], where a is the short-distance cut-off of order the lattice spacing and q_m is the "magnetic charge":

$$q_m^2 = \pi\rho_s \approx \alpha M_0^2 h. \tag{7}$$

Here and further we put for simplicity vorticities to be $k = \pm 1$, since vortices with higher winding numbers are energetically unstable.

The magnetic energy of a neutral system of N vortices and N antivortices with distances r_{ij} between i-th and j-th (anti-) vortex cores and vorticities $k_i = \pm 1$ is [12]

$$W_m = -2q_m^2 \sum_{i<j}^{2N} k_i k_j \ln \frac{r_{ij}}{a}. \tag{8}$$

3 Vortex - antivortex pairs creation by electric field

In this section we explore the basic idea of the paper, that a strong non-uniform electric field can break the "magnetic vacuum" of magnetoelectric material via creation of vortex-antivortex pairs. Fig.1 shows the typical geometry

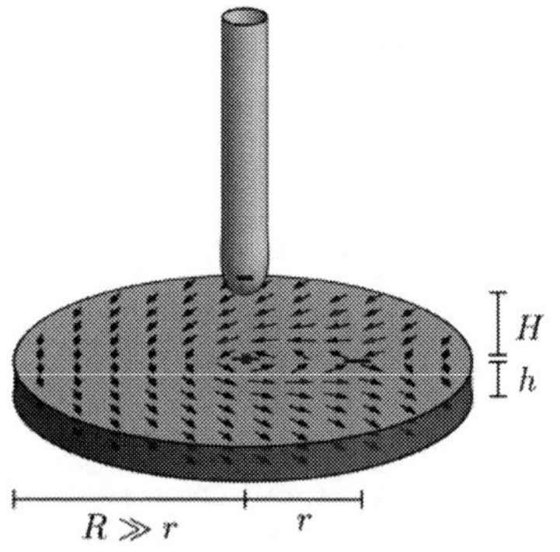

Figure 1. Geometry of the system. Sample is a disk with radius R and thickness h. Above its center at the height H the tip of the cantilever with a negative charge $-Q_{tip}$ is placed.

of the experiment that we keep in mind (which was proposed in Refs. [6, 13]). An electric field is produced by the tip of a cantilever with a fixed point charge $-Q_{tip} < 0$. Because of the magnetoelectric interaction, strong enough electric field can create magnetic vortices. Here we work in the limit of low temperatures, where there are no thermally activated vortex-antivortex (VA) pairs.

The electric potential produced by the tip is

$$\varphi(r) = -\frac{Q_{tip}}{\sqrt{r^2 + H^2}} \tag{9}$$

where H is the distance from the tip to the sample, and r is the polar radius. We assume that the sample is thin: its thickness $h \ll H$, so we neglect the variation of the electric field in the transverse to the film direction inside the film.

Now we want to find the critical value of the electric field that creates the first vortex-antivortex pair at $T = 0$ in a sample of big enough size to accommodate the pair as a whole. Estimates for the case of small samples were made in [13].

Let the electric field produces one VA pair with the vortex situated at the center of the disk and antivortex at the distance r from its center (for big samples $r \ll R$). The energy of the pair as a function of r consists of the magnetic energy (8) of the VA interaction and the magnetoelectric part due to electric charge (6) and electrostatic potential (9):

$$W = 2q_m^2 \ln \frac{r}{a} + Q_{tip}q_e \left(\frac{1}{\sqrt{r^2 + H^2}} - \frac{1}{H} \right). \tag{10}$$

Minimizing the energy with respect to the distance between the vortex and the antivortex r we obtain the following optimal distance:

$$r \approx \frac{Q_{tip}q_e}{2q_m^2}. \tag{11}$$

Note that for in the small sample case $r > R$ the antivortex leaves the sample and only vortex remains (which was analyzed in [13]). Substituting (11) into (10) we find the critical charge Q_{tip}^{crit}, above which the first VA pair forms:

$$Q_{tip}^{crit} = \frac{2q_m^2 H}{q_e} \ln \frac{H}{a}. \tag{12}$$

We also define the critical voltage of the tip as voltage produced at the nearest to the tip point of the sample:

$$\varphi_0^{crit} = \frac{Q_{tip}^{crit}}{H} = \frac{2q_m^2}{q_e} \ln \frac{H}{a} \tag{13}$$

Higher values of tip voltages can lead to systems with several vortex-antivortex pairs. Estimates for the prototypic type-I multiferroic $BiFeO_3$ give $\varphi_0^{crit} \simeq 20\,\mathrm{V} \times \ln(H/a)$ and for type-II multiferroic TbMnO3: $\varphi_0^{crit} \simeq 0.3\,\mathrm{V} \times \ln(H/a)$ (see [9] for the details).

4 Vortex distribution in continuous approximation

In this section we construct an analytical theory for vortex distribution in the continuous approximation. It proves that even such a crude approximation captures many effects and allows us to find the vortex distribution profile and the number of vortices for given Q_{tip}.

The number of vortices in the system is not fixed, but is rather determined by the condition of the energy minimum. Here we calculate self-consistently the number of vortices N created by the tip. We work in the continuoum limit [which means that at least $N \gg 1$, so $Q_{tip} \gg Q_{tip}^{crit}$; see (12)]. Let $n_v(\mathbf{r})$ and $n_a(\mathbf{r})$ be concentrations of vortices and antivortices, respectively; and $n(\mathbf{r}) = n_v(\mathbf{r}) - n_a(\mathbf{r})$.

Consider a ring from r to $r + dr$. In the ground state there is a balance between the electric and magnetic interactions. The magnetic interaction pushes vortices outside while the external electric force pulls them inside the ring (and vice versa for antivortices). From (9) external electric force acting on a given area element with electric charge $q_e dN = q_e n dS$ is

$$F_{el} = q_e dN \frac{d\varphi}{dr} = Q_{tip} q_e n dS \frac{r}{(r^2 + H^2)^{3/2}}. \quad (14)$$

From (8) the magnetic force between two equal charges magnetic q_m is $2q_m^2/r$. We can now calculate the magnetic force acting at the considered area element dS with magnetic charge $dq_m = q_m n dS$. Applying Gauss's theorem to magnetic charges we get the magnetic force:

$$F_{magn} = \frac{2dq_m q_m^{inside}}{r}, \quad (15)$$

where $q_m^{inside} = q_m \int n(\mathbf{r}) d^2r = 2\pi q_m \int_0^r n(r) r dr$ is a net magnetic charge inside the ring. Equating (14) and (15) we obtain the vortex distribution density profile:

$$n(r) = \frac{Q_{tip} q_e}{4\pi q_m^2} \frac{2H^2 - r^2}{(r^2 + H^2)^{5/2}} \quad (16)$$

Fig.2 shows plot for $rn(r)$, which is proportional to the number of vortices at distance r. Vortices are concentrated inside a small circle of radius $r_0 = H\sqrt{2}$, antivortices are smeared over the sample outside r_0 with their density decaying as $rn(r) \sim -1/r^2$.

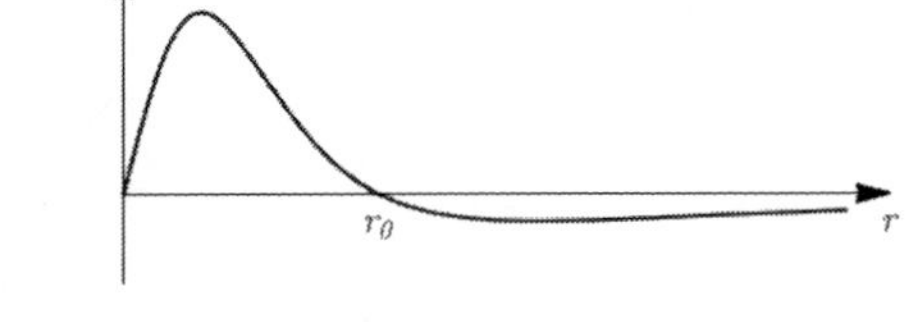

Figure 2. Schematic plot for $rn(r)$ vs r; $n(r)$ is given by (16).

From (16) we get the total topological charge of vortices and antivortices inside a circle of radius r:

$$N(r) = 2\pi \int_0^r n(r) r dr = \frac{Q_{tip} q_e}{2q_m^2} \frac{r^2}{(r^2 + H^2)^{3/2}} \quad (17)$$

Now we can calculate the total number of vortices. Border between vortices and antivortices lies at $r_0 = H\sqrt{2}$, so

$$N = \frac{1}{3^{3/2}} \frac{Q_{tip} q_e / H}{q_m^2}. \quad (18)$$

5 Numerical simulation

We modeled a system of pointlike vortices and antivortices, which interact between themselves (with the magnetic energy (8)) and with the external electric field created by the cantilever's tip (the electrostatic potential is given by (9)). The system shows a "glassy" behavior; i.e., there are many local minima that are close to the global one. In order to find a state corresponding to such a local energy minimum we performed simulated annealing using the Monte Carlo (MC) method. Three types of MC steps were used: movement of a vortex or an antivortex, creation of the VA pair, and destruction of the VA pair. We cooled the system from high temperatures $T_{start} \simeq q_m^2 > T_{BKT} \simeq q_m^2/2$ (which allowed the system to find the optimal number of vortices) down to as low as $T_{stop} = 10^{-6} q_m^2$, depending on the numerical experiment.

5.1 Number of vortices and the critical field

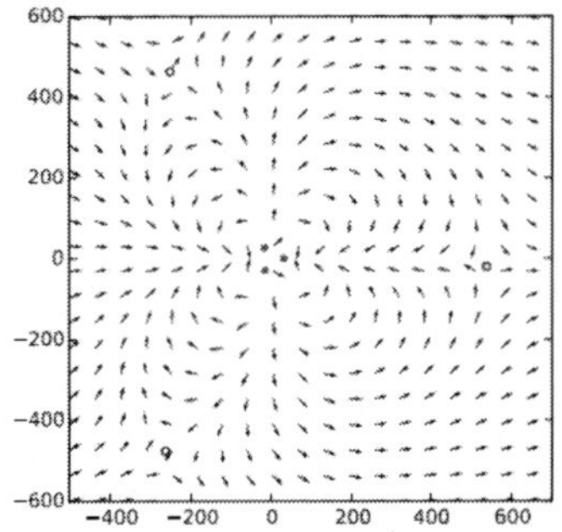

Figure 3. Configuration of a vortex-antivortex system for $Q_{tip} q_e / H = 23 q_m^2$ at $T = 0.001 q_m^2$. Vortex (filled circles) and antivortex cores (open circles) are shown; arrows show magnetization. Coordinates are given in the units of lattice spacing a.

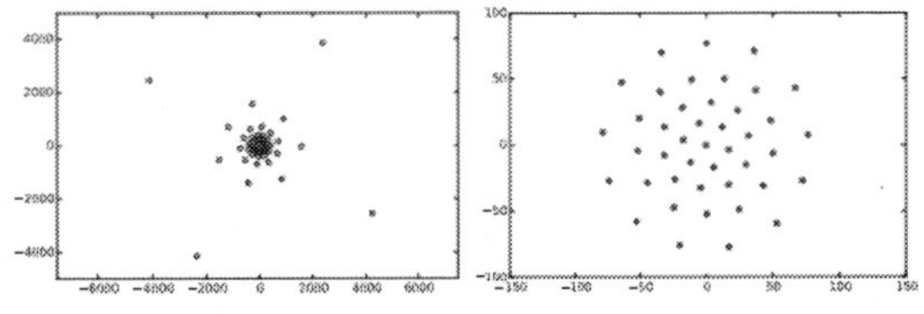

(a) (b)

Figure 4. Configuration of vortex-antivortex system for $Q_{tip} q_e / H = 240 q_m^2$ at $T = 10^{-6} q_m^2$. Vortex (filled circles) and antivortex cores (open circles) are shown. (a) Antivortex subsystem; vortex "nucleus" is not resolved. (b) Vortex subsystem.

As we described in Secs.3 and 4, the charged tip produces magnetic patterns ("magnetic atoms") in the system. Fig.3 shows a typical low-temperature distribution of magnetization in the system for moderate values of tip voltages, for which $N = 3$ vortices and 3 antivortices are created.

Fig.4 shows a typical low-temperature configuration of the system for higher values of the tip's voltage. Vortices and antivortices are arranged in a highly regular structure that is reminiscent of the structure of an atom: negatively

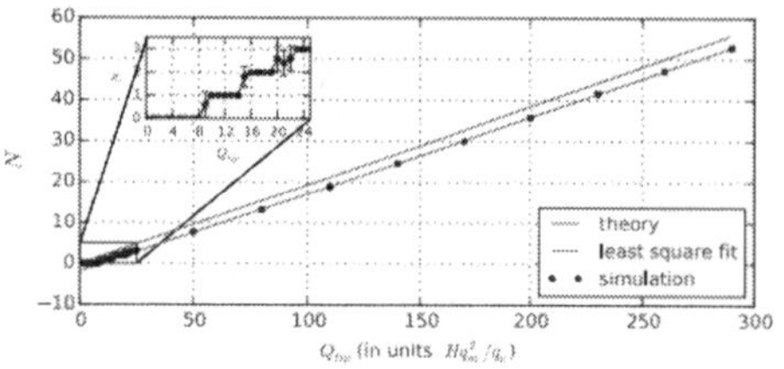

Figure 5. Plot for N vs Q_{tip}. The inset shows zoom of the plot for range $Q_{tip} = 0 \div 24Hq_m^2/q_e$, from which $Q_{tip}^{crit} \approx 9Hq_m^2/q_e$.

charged antivortex outer concentric circular "shells" surround the positive vortex "nucleus". Fig.4b depicts only this vortex "nucleus" of the "atom".

Fig.5 shows $N(Q_{tip})$ dependence for the ground states of the system with varying Q_{tip}. The inset shows zoom for small Q_{tip}: step-like dependence is due to the fact that the number of vortices is always integer. For higher Q_{tip} the dependence is step-like as well, but the steps are not resolved, and only linear monotonous part of the dependence can be seen. Analytical result (18) and the numerical modeling are in a good agreement.

5.2 Polarizability of big samples

Here we numerically find the temperature dependence of the polarizability for big samples. We performed cooling at a fixed value of electric field E, and then independently repeated this for different values of E. Fig.6 shows the temperature dependence of the polarizability in log-log coordinates, which in the limit of small β $(= 1/T)$ is fitted by $\ln\alpha \approx 0.74\ln(\beta q_m^2) + \text{const}$. This dependence is consistent with our theoretical estimate $\alpha \sim 1/T^{1-\eta}$, with $\eta \approx 0.26$.

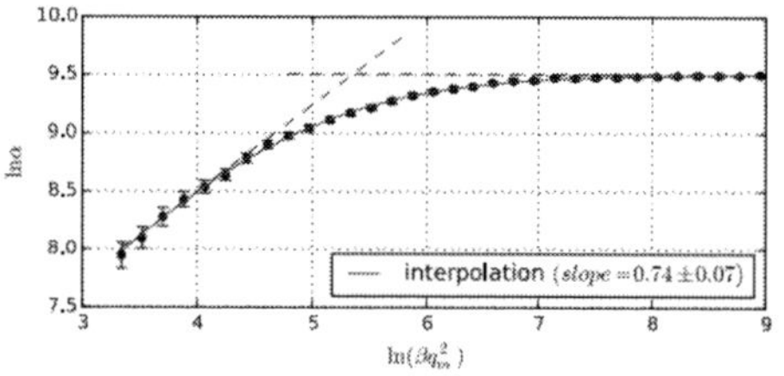

Figure 6. $\ln\alpha$ vs $\ln(\beta q_m^2)$ plot (for $Q_{tip}q_e/H = 200q_m^2$, $N = 36$). Symbols show data with their standard deviation; solid line shows fitting. Slope of the dashed inclined line is 0.74 ± 0.07.

6 Conclusions

Interaction of magnetic and electric subsystems in magnetoelectric materials makes possible the creation of magnetic structures by external electric field. In this paper we have presented analytical and numerical analysis of such quasi-2D materials, where "atom"-like patterns of magnetic vortices and antivortices are produced by a strong electric field, and studied the structure and electric polarizability of these magnetic "atoms".

When the cantilever's electric field exceeds some critical value, the first vortex-antivortex pair forms. Increasing electric field produces more and more vortices and antivortices. We found analytically and numerically the critical voltage required to create the first vortex-antivortex pair, and found the dependence of the number of vortex pairs on the applied voltage. We have shown that a "magnetic atom" forms: a "nucleus" consists of vortices arranged in a lattice-like structure with a short-range order; it is surrounded by antivortices forming outer concentric circular "shells". We obtained the vortex density profile in a continuous approximation.

In the type-II multiferroic-like materials studied here, vortices carry a positive and antivortices a negative electric charge; therefore an additional weak in-plane electric field pulls them in opposite directions, thus creating a polarization. In this work we have studied the temperature dependence of electric polarizability of such "magnetic atoms". By Monte Carlo simulation we deduced that polarizability α behaves at low temperatures like $1/T^{1-\eta}$, with $\eta \approx 0.26$.

7 Acknowledgements

We acknowledge the financial support of the Ministry of Education and Science of the Russian Federation in the framework of Increase Competitiveness Program of NUST MISiS (No. K2-2016-067). Figs.1, 3, 4 are reprinted with permission from P.I. Karpov and S.I. Mukhin, *Phys.Rev. B* **95**, 195136 (2017) doi.org/10.1103/PhysRevB.95.195136.

References

[1] D.I. Khomskii, *Physics* **2**, 20 (2009).
[2] J. Seidel, chapter in "Multiferroic Materials: Properties, Techniques, and Applications", ed. J. Wang (CRC Press, 2016), p.315.
[3] S. Seki et.al., *Science* **336** 198 (2012).
[4] P.-J. Hsu et.al., *Nat. Nanotechnol.* **12**, 123 (2017).
[5] K.T. Delaney et.al., *Phys.Rev.Lett.* **102**, 157203 (2009).
[6] A.P. Pyatakov et.al., *JMMM.* **324**, 3551 (2012).
[7] A.S. Logginov et.al., *Appl. Phys. Lett* **93**, 182510 (2008).
[8] A.P. Pyatakov et.al, *EPL* **93**, 17001 (2011).
[9] P.I. Karpov, S.I. Mukhin, *Phys. Rev. B* **95**, 195136 (2017).
[10] M.V. Mostovoy, *Ferroelectricity in spiral magnets. Phys.Rev.Lett.* **96**, 067601 (2006).
[11] J.M. Kosterlitz and D.J. Thouless, *J.Phys.C* **6**, 1181 (1973).
[12] P.M. Chaikin and T.C. Lubensky, *Principles of condensed matter physics*, CUP, (1995).
[13] A.P. Pyatakov et.al., *MSU Phys. Bull.* **65**, 329-331 (2010).

Temperature characteristics of the nonlinear magnetoelectric effect

Burdin Dmitry[*], *Ekonomov Nikolay*, *Chashin Dmitry* and *Fetisov Yury*

Moscow Technological University (MIREA), 119454, Vernadskogo prosp., 78, Moscow, Russian Federation

Abstract. The temperature characteristics of the nonresonant nonlinear magnetoelectric effect in the two-layer LGT-Metglas structure at a large excitation field amplitude are studied. It is shown that the harmonics of the signal have a different course when the temperature is varied in the range from 200 K to 400 K, which is explained by a change in the shape of the magnetization curve. Amplitude of alternating magnetic field affects the form of each ME voltage harmonic.

1 Introduction

Magnetoelectric (ME) effect in composite structures opens the prospect of creating a whole series of new devices of solid state electronics [1-2]. For the application of the composite ME structures it is important to investigate their temperature properties. Temperature characteristics of the linear ME effect in structures containg layers of PZT and amorphous alloy Metglas has been studied [3-6]. It was shown that temperature strongly affect the ME coefficient, which was explained by the concurrent variations with temperature of piezoelectric modulus of the PE layer, piezomagnetic modulus of the FM layer and mechanical coupling. Ye et al [3] also experimentally established relation between specimen size and ME signal temperature dependence.

Last few years a great interest is caused by the nonlinear ME effect in composite structures. Theory of nonlinear ME effect was developed [7, 8], and new devices based on it were suggested. In particular, new high sensitive ac and dc magnetic field sensors were described [9, 10]. In this connection, it is interesting to investigate the temperature characteristics of the nonlinear ME effect, which has so far not been practically studied.

Due to the nonlinear field dependence of the ferromagnet magnetostriction (Fig.1), when the ME structure is excited by an alternating magnetic field with frequency f, various voltage harmonics with frequencies f, $2f$, $3f$, etc. are generated. The magnitude of the harmonics depends on the amplitude of the alternating magnetic field h and the constant magnetic field H.

The aim of this work is to investigate the temperature dependences of the ME voltage harmonics generated by the piezoelectric-ferromagnetic (PE-FM) composite structure for different values of the amplitude h. For producing magnetostrictive layer an amorphous ferromagnetic alloys Metglas was chosen due to it's low saturation field and high magnetostriction. For piezoelectric layer most suitable material is lanthanum gallium tantalate due to its high temperature stability and rather high d/ε ratio [11].

2 Experiment

2.1 Samples and measurements

A two-layer structure with in-plane dimensions of 20 mm x 5 mm containing a 0.5 mm thick layer of monocrystalline lanthanum gallium tantalate (LGT) and a 20 μm thick layer of amorphous magnetostrictive alloy Metglas 2605S3A with composition FeBSiC was used in measurements. The surfaces of the LGT layer have been metalized and the layers were connected by a low-temperature soldering.

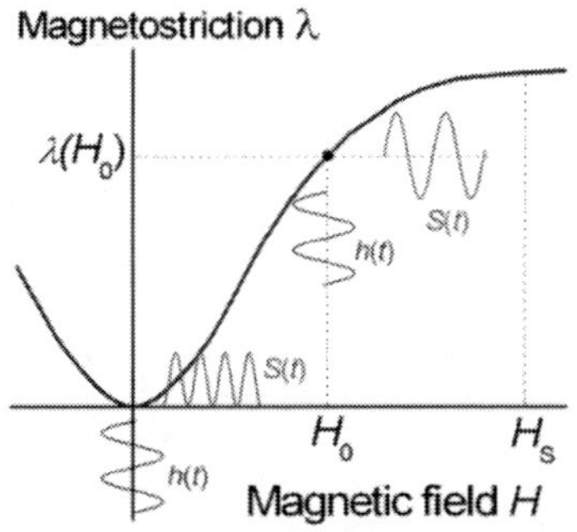

Fig. 1. Typical dependence of the FM layer magnetostriction λ on magnetic field H.

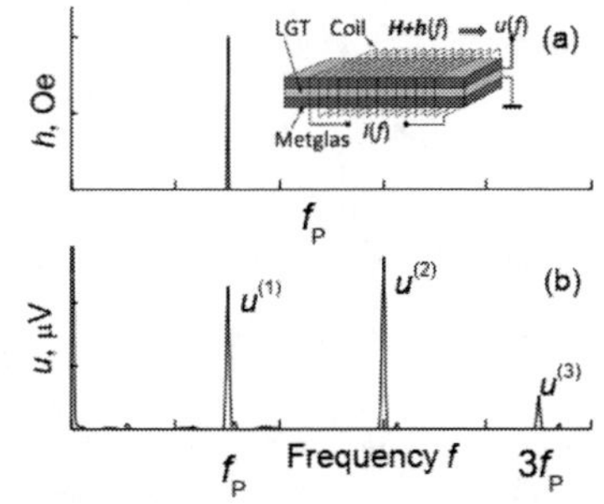

Fig. 2. Frequency spectra: (a) for pumping magnetic field and (b) for ME voltage generated by the LGT-Metglas structure.

[*] D.A. Burdin: phantastic@mail.ru

The sample was placed in a thermocell located inside magnetic coils which were fed from the signal generator Agilent 33210a. The amplitude of the pumping field h was varied between 0 ÷ 20 Oe and its frequency was 120 Hz. A permanent magnetic field H up to 100 Oe of the same direction was produced by the Helmholtz coils. The sample was blown with a stream of the nitrogen gas, whose temperature was adjusted in the range from 150 K to 400 K with an accuracy of 0.1 degree.

The ME signal was taken across the electrodes of the PE layer and applied to the input of the spectrum analyzer SR770. This allowed for simultaneous measuring of all harmonics of the ME voltage.

Figure 2 shows a typical voltage spectrum generated by the LGT-Metglas structure due to nonlinear ME effect at bias field $H = 4$ Oe and pumping field $h = 5$ Oe. Spectral components with amplitudes $u^{(1)}$, $u^{(2)}$ and $u^{(3)}$ are clearly visible. Based on the measured spectra, the dependences of the harmonic amplitudes on temperature were plotted for different amplitudes.

2.2 Measurement results

Figure 3 shows the temperature dependences of the amplitudes for the first three harmonics of the signal at bias field H=4 Oe and two pumping fields h=5 Oe and h=15 Oe. One should note the different character of the harmonics amplitudes temperature dependences. Thus, for h=5 Oe $u^{(1)}$ first grows with temperature, reaches a maximum of 6 mV at T≈220 K, and then decreases again. The amplitude of the 2nd harmonic grows from $u^{(2)}$=7 mV, reaching a maximum at T≈270 K, while $u^{(3)}$ gradually increases from 2.5 mV at T=150 K to 5 mV at T=300 K.

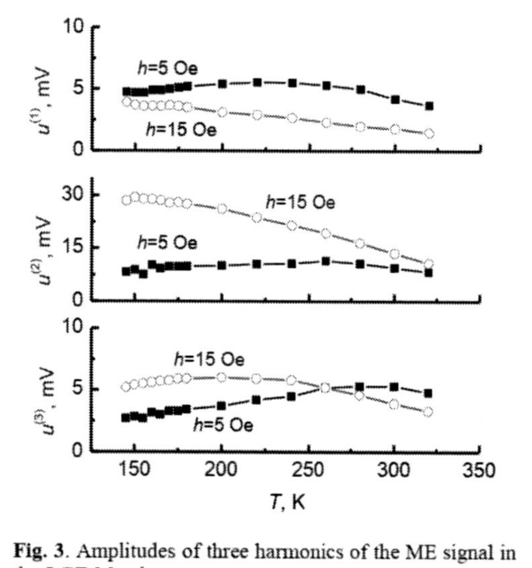

Fig. 3. Amplitudes of three harmonics of the ME signal in the LGT-Metglas structure versus temeprature at H=4 Oe and pumping fields h=5 Oe and h=15 Oe.

An increase in the pumping field up to 15 Oe has led to a change in the temperature dependences of the harmonics. The harmonics $u^{(1)}$ and $u^{(2)}$ decrease monotonically by a factor of about 2 at heating from 150 K to 350 K, and $u^{(3)}$ first increases with temperature, reaches a maximum at T≈200 K, and then falls again.

3 Discussion

The form of the temperature dependence of the given ME harmonic can be expressed by the relation [12]:

$$u^{(n)}(T) \sim \frac{d_{11}(T)q^{(n)}}{\varepsilon(T)}, \qquad (1)$$

where d_{11} is the piezoelectric coefficient of the PE layer, ε is the releative permittivity of the PE layer, $q^{(n)}$ is the piezomagnetic coeffictient for the n-th harmonic, whose magnitude is determined by the field dependence of the magnetostriction, the ac field amplitude h, and the dc field H. It follows that the differences in the character of the temperature dependences of the harmonics are caused by the behavior of the corresponding piezomagnetic coefficient. It has been experimentally established [13], that in the investigated temperature range the ratio d_{11}/ε for LGT is almost constant.

For the case of small h for a given biasing H=H_0 the piezomagnetic coefficient is simply the n-th derivative of the magnetostriction with respect to the field [8]:

$$q^{(n)} = \left.\frac{d^n\lambda}{dH^n}\right|_{H=H_0}, \qquad (2)$$

where λ is the magnetostriction of the FM layer.

In the case of large pumping, expression (2) is inapplicable and piezomagnetic coefficients should be considered as a function of several variables $q^{(n)}(h, H, T)$. Nevertheless, it is obvious that the quantity $q^{(n)}$ in any case is determined by the form of the dependence $\lambda(H)$.

With the help of the LakeShore 7400 vibrating sample magnetometer (Department of Magnetism, Moscow State University), the magnetization curves for the Metglas plate at temperatures from 140 K to 350 K have been measured (see Fig. 4).

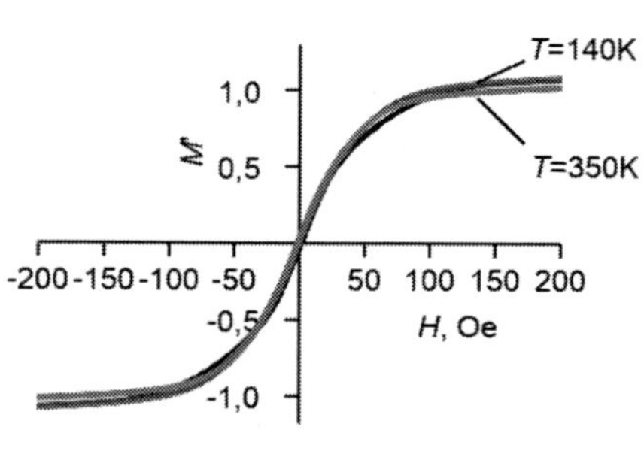

Fig. 4. The magnetization curves of the Metglas plate at different temperatures T.

One can see that saturation magnetization decreases by no more than 5% and some change in the shape of the curve $M(H)$ take place as the temperature increases up to 350 K. In order to characterize more accurately the observed change of the magnetization curve, the graphs of the first three derivatives of magnetostriction λ over the field H for different T were plotted, assuming that $\lambda \sim M^2$ [14] (Fig. 5).

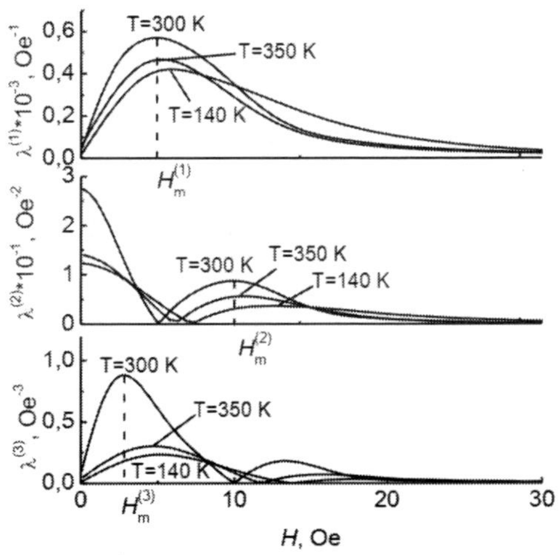

Fig. 5. Field dependences of the derivatives of M^2 over H at different temperatures.

It is seen that the 1-st derivative increases from zero with increasing H, reaches a maximum value $\lambda_m^{(1)}$ at $H_m^{(1)}$ $\approx$5 Oe, and then monotonically drops to zero. The 2nd derivative has the largest value $\lambda_m^{(2)}$ in the zero field, falls with increase of H, reaches zero and then has a second maximum at the field $H_m^{(2)}$. The 3-d derivative grows from zero with increase of H, reaches a maximum value $\lambda_m^{(3)}$ for the field $H_m^{(3)}$, then drops to zero and has a second maximum.

It can be noted, that all three derivatives change both their maximum values $\lambda_m^{(n)}$ and optimal fields under the influence of temperature

Let's denote the ratio of the maximum value of the n-th derivative at a given temperature T to the corresponding value at T=300 K as:

$$\beta^{(n)} = \frac{\lambda_m^{(n)}(T)}{\lambda_m^{(n)}(T=300K)}. \quad (3)$$

And we will characterize the change in the optimal filed $H_m^{(n)}(T)$ by the ratio:

$$\gamma^{(n)} = \frac{H_m^{(n)}(T=300K)}{H_m^{(n)}(T)}. \quad (4)$$

Figure 6 shows the dependences $\beta^{(1)}(T)$, $\beta^{(2)}(T)$, and $\beta^{(3)}(T)$, drawn using the data of Fig. 5 It is seen, that the coefficients $\beta^{(1)}(T)$, $\beta^{(2)}(T)$, and $\beta^{(3)}(T)$ for the Metglas layer posses similar temperature dependences and have a characteristic maximum in visinity of $T\approx$250 K. Wherein, $\beta^{(1)}(T)$ varies by 25%, $\beta^{(2)}(T)$ - by 50%, and $\beta^{(3)}(T)$ - by 75% in the investigated temperature range

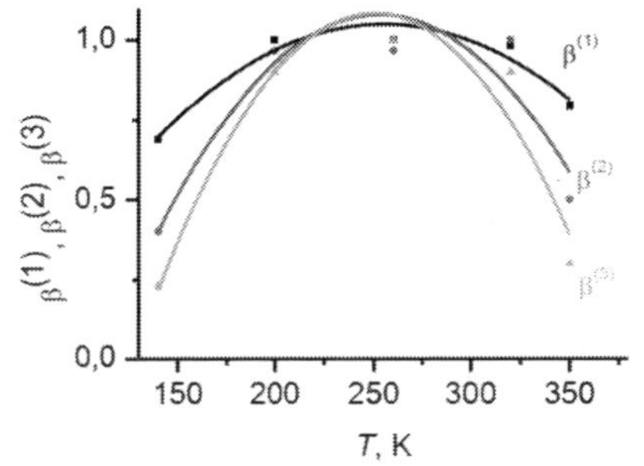

Fig. 6. Temperature dependences of the normalized derivatives of magnetostriction for the Metglas layer. Solid curves are approximations by a 2nd order polynoms.

Figure 7 shows the graphs $\gamma^{(1)}(T)$, $\gamma^{(2)}(T)$, and $\gamma^{(3)}(T)$. It can be seen that optimum fields for all derivatives change almost identically when the temperature is increased from 140 K to 350 K: γ first increases monotonically from 0.6-0.7 to 1, reaching a maximum at T=270-300 K, and then decreases again to a value of 0.7.

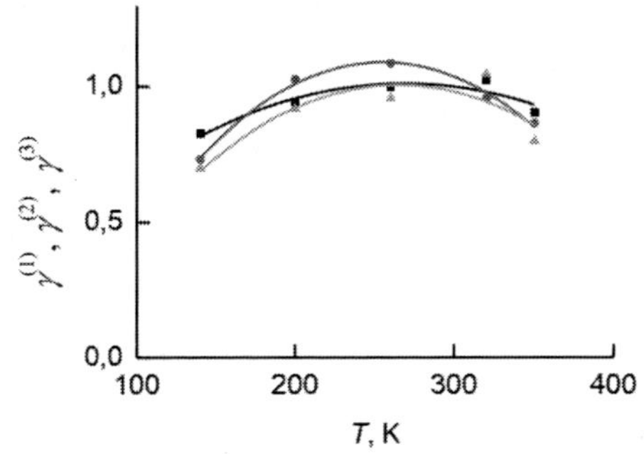

Fig. 7. Temperature dependences of the relative change in the optimal fields for the Metglas layer. . Solid curves are approximations by a 2nd order polynoms.

Let's suggest, that magnetization loop deformation can be expressed as x-axis stretching, which is equivalent to an effective magnetic field change: $H_{eff}=H\cdot\gamma(T)$. Indeed, if we replace H with H_{eff} in expression for the piezomagnetic coefficients (2), we obtain:

$$q^{(1)}(T,H) = \gamma(T) \cdot \frac{d\lambda}{dH}\Big|_{(H_{eff},\ T=300K)} \quad (5)$$

$$q^{(2)}(T,H) = \gamma^2(T) \cdot \frac{d^2\lambda}{dH^2}\Big|_{(H_{eff},\ T=300K)} \quad (6)$$

$$q^{(3)}(T,H) = \gamma^3(T) \cdot \frac{d^3\lambda}{dH^3}\Big|_{(H_{eff},\ T=300K)} \quad (7)$$

So, the higher the number of the harmonic, the more sensitive it is to the temperature "compression" of the magnetostriction curve, which is well illustrated by Fig. 5. Data on Fig.5 and Fig.6 are in a good agreement, which tells that deformation of the magnetization loop and, consequently, deformation of the field dependence of magnetostriction with temperature, can be represented as a compression-stretching along the field axis H.

In case of a large pumping amplitude, when the amplitude of the higher harmonics of the ME voltage becomes comparable with the amplitude of the first harmonic, then the relations (5-7) for $q^{(n)}(T)$ are no longer satisfied. But they can illustrate how temperature can affect amplitudes of ME signal harmonics in different way.

In other words, effective field change can be presented as equivalent movement left-right on $u(H)$ and $u(h)$ curves, which should be taken in an experiment. As $\gamma(T)$ goes down, our effective field decrease and we move left on $u(H)$, $u(h)$ curves, $\gamma(T)$ increase tends to right-hand movement. As a result nonmonotonic shape of $\gamma(T)$ on Fig.7 dependence can explain the same shape of $u(T)$ curves on Fig.3.

4 Conclusion

Thus, it has been shown experimentally that temperature significantly affects the nonlinear ME effect in the two-layer structure of the LGT-Metglas composition, and the character of the temperature characteristics for different ME voltage harmonics differs and depends on the magnitude of the excitation field. The last circumstance introduces a new factor, which should be taken into account when analyzing temperature properties of the ME effect. As far as authors know in all previous studies devoted to this problem influence of ac magnetic field amplitude was totally ignored [4-7]. We suppose that by selecting the appropriate amplitude of the alternating field h, it is possible to attain a decrease in the temperature coefficient for a given ME signal, which can be used in the development of new ME devices.

The work was supported by the Russian Science Foundation (agreement 17-12-01435).

References:

1. C.-W. Nan, M.I. Bichurin, S. Dong, D. Viehland, G. Srinivasan, J. Appl. Phys., **103,** p. 031101 (2008).
2. H. Palneedi, V. Annapureddy, S. Priya, J. Ryu, Actuators, **5,** No 9, p. 1-31 (2016).
3. J.X. Ye, J.N. Ma, J.M. Hu, Z. Li, M. Feng, Q.M. Zhang and C.-W. Nan C, J. App. Phys. **116**, 074103 (2014).
4. F. Fang, Y.T. Xu, W. Yang, J. Appl. Phys. **111**, 023906 (2012).
5. Y. Shen, J. Gao, Y. Wang, J. Li , and D. Viehland, Appl. Phys. Lett. **100**, 173505 (2012).
6. D. A. Burdin, D.V. Chashin, N.A. Ekonomov, Y.K. Fetisov, Solid State Phenomena pp.233-234.357-359 (2015).
7. V.M. Laletin, D.A. Filippov, T.O. Firsova, Tech. Phys. Lett. **40**, 3, pp 237-240.
8. D.A. Burdin, D.V. Chashin, N.A. Ekonomov, L.Y. Fetisov, Y.K. Fetisov, G. Srinivasan, G. Sreeniasulu, JMMM, **358- 359**, p. 98-104 (2014).
9. Y.K. Fetisov, D.D. Chashin, N.A. Ekonomov, D.A. Burdin, L.Y. Fetisov, IEEE Trans. Magn., **51** (11) 1–3 (2015).
10. D.A. Burdin, D.V. Chashin, N.A. Ekonomov., Y.K. Fetisov, A.A. Stashkevich, JMMM, **405,** p. 244-248 (2016).
11. G. Sreenivasulu, L.Y. Fetisov, Y.K. Fetisov and G. Srinivasan, Appl. Phys.Lett., **100**, 05290 (2012).
12. M. Vopsaroiu, J. Blackburn, M.G. Cain, J. Phys. D: Appl. Phys., **40**, p. 5027-5033 (2007).
13. P. Davulis, B.T. Sturtevant, S.L. Duy, M. Pereira, Proc. IEEE Ultrasonics Symposium, p. 1397–1400 (2007).
14. Y.K. Fetisov, D.V. Chashin, N.A. Ekonomov, D.A. Burdin, L.Y. Fetisov, IEEE Trans. on Magn., **51**, 11 (2015).

MAGNETOELECTRIC EFFECT IN FERROMAGNETIC-SEMICONDUCTOR LAYERED COMPOSITE STRUCTURES

Leonid Fetisov[1,*], Dmitri Chashin[1], Dmitri Saveliev[1], Daria Plekhanova[1], Ludmila Makarova[2], and Alexandr Stognii[3]

[1]Moscow Technological University, Moscow, Russia
[2]Moscow State University, Moscow, Russia
[3]Scientific and Practical Materials Research Centre of NAS of Belarus, Minsk, Belarus

Abstract. The results of magnetoelectric effect experimental studies in two different structures based on piezoelectric semiconductor gallium arsenide are presented. The monolithic structure consisted of a gallium arsenide substrate with deposited nickel layer (GaAs-Ni), and the composite structure contained a semiconductor substrate with an amorphous magnetic alloy (GaAs-Metglas) ribbon glued on one side. A quality factor $Q \approx 23500$ and magnetoelectric coefficient of 316 V/Oe·cm were achieved at the frequency of planar acoustic oscillations for GaAs-Ni structure at room temperature.

1 Introduction

Magnetoelectric (ME) effect manifests itself as polarization of the sample in the external magnetic field and observed in layered composite magnetostrictive-piezoelectric structures due to mechanical coupling between the layers [1]. Magnetic materials with large magnetostriction λ (Ni, Co, alloys FeCo, FeGa, terfenol, FeSiCB etc.) are used as magnetic layers in such structures. Most commonly used piezoelectric materials are various lead-based ceramics (PZT, PMN-PT), which posses good electric and piezoelectric properties.

There are number of other materials with high piezomodule to dielectric permittivity ratio d_{31}/ε, that is required for effective ME interaction, which are used in such composite structures along with ceramic piezoelectric materials: piezoelectric single crystals (langatate and quartz) [2] and flexible piezopolymers, like PVDF [3].

It is also known that some of semiconductor materials demonstrate piezoeffect, such as GaAs, ZnS, and CdS. Piezoelectric moduli of these materials lay in the range $2 \cdot 10^{-12}$ - $14 \cdot 10^{-12}$ C/N, and resistivity exceeds 10^6 Ω·cm. Magnetoelectric effect was recently experimentally discovered in composite structure based on GaAs semiconductor substrate [4-5]. Application of semiconductor materials in such composite structures will expand the functionality of ME devices. Standard planar technologies can be used for ME semiconductor-based devices production, which, for sure, will facilitate the integration of future magnetoelectric devices into modern electronic components.

In this paper authors investigated frequency and field characteristics of ME effect in composite structures based on semiconductor gallium arsenide substrate with most commonly used magnetostrictive materials such as nickel and amorphous magnetic alloy.

2 Structures and set-up

The measurements were carried out on two samples: monolithic structure GaAs-Ni and composite structure GaAs-Metglas. Commercially available semiconductor GaAs plates of lateral dimensions 13.5 mm x 5 mm and thickness 0.4 mm, cut along natural edges of the crystal with orientation (100) were used for structures fabrication. Gold electrodes with ~1 µm thickness were deposited on the surface of the plates. Monolithic structure GaAs-Ni was fabricated by the ion beam sputtering method by depositing ~1.4 µm thickness Ni

* Corresponding author: fetisovl@yandex.ru

layer on one side of the GaAs substrate [6]. Electronic image of this structure was obtained at FIB station FIE 450 in the Nanocentre of MIREA (see Fig. 1). The structure is characterized by the clear plane-parallel film-substrate and film-surface interface boundaries. The surface of the nickel film is uniform and the dimensions of the inhomogeneity's are close to the monomodal distribution at the submicron level. The composite structure was fabricated by gluing a ~30 µm thick ribbon of an amorphous alloy FeBSiC (Metglas) to one side of the GaAs plate. The structures were fastened with thin wires in the center in such a way that only longitudinal acoustic vibrations could be excited.

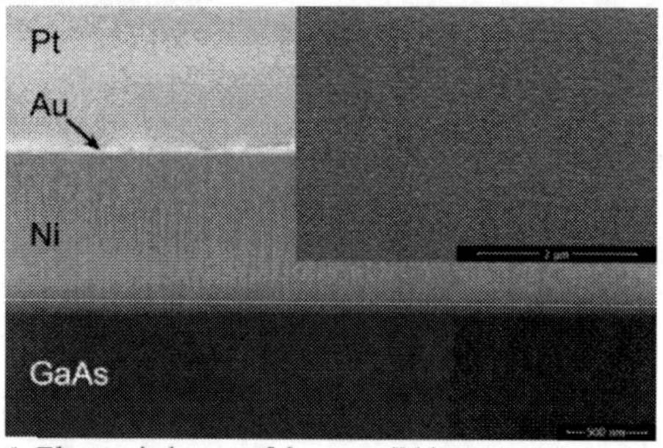

Fig. 1. Electronic image of the monolithic GaAs-Ni structure.

The samples were placed into external permanent magnetic field H and excited by the alternating magnetic field $h\cos(2\pi ft)$, both directed parallel to the long side of the structure. The dependencies of the voltage U generated across the electrodes of the GaAs layer were recorded when the field H was varied in the range 0-120 Oe, frequency f of the ac field was varied in the range 0-200 kHz, and amplitude h of the ac field changed from 0 to 10 Oe. The coefficient of ME interaction was calculated using the formula $\alpha = U/(h \cdot b)$ where U is the voltage amplitude, h is the ac field amplitude, and b is the thickness of the semiconductor layer. The dielectric properties were measured with RLC meter Actacom AM3026. The magnetic properties of a monolithic sample were measured using a LakeShore 7404 vibrating sample magnetometer when the structure was magnetized along its long side.

3 Results and discussion

Frequency dependencies of the dielectric permittivity and resistance for GaAs plates are shown in Fig. 2. It is seen that the resistance of the semiconductor decreases by a factor of 50, from 7.2 MΩ to 0.145 MΩ, as the frequency increases. The dielectric constant ε of the semiconductor, calculated using the measured capacitance of the structure, decreases sharply with increasing frequency from ~ 85 at a frequency f = 5 Hz to ~ 16 at frequencies f > 1 kHz.

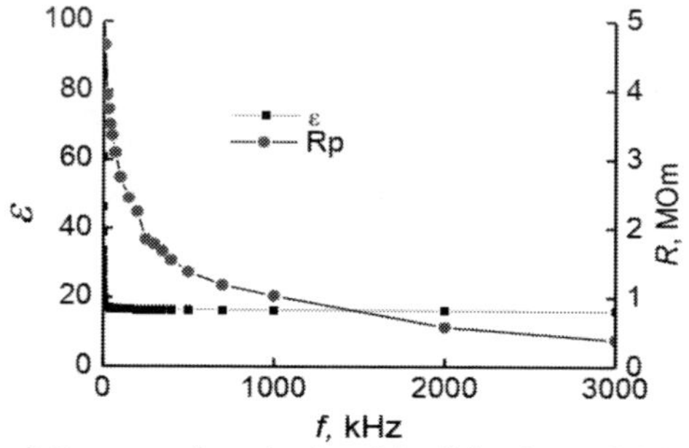

Fig. 2. Frequency dependencies of the dielectric permittivity and resistance for GaAs plate.

Hysteresis of the magnetic moment M of the monolithic GaAs-Ni structure is shown in Fig. 3. As it follows from the figure, magnetic layer of the structure saturates in the fields H_s ~60 Oe. The coercive field of the structure is H_c ~11 Oe.

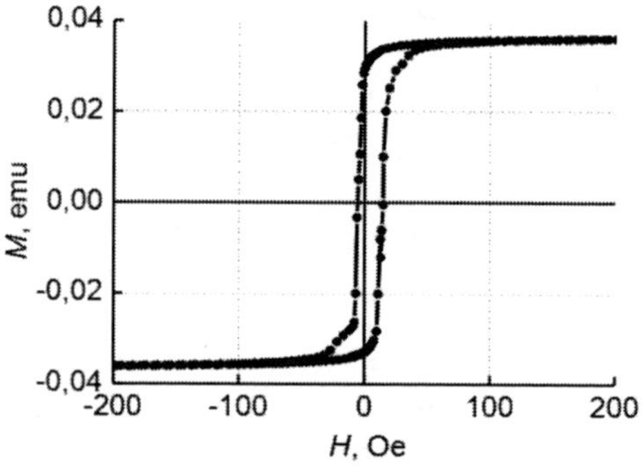

Fig. 3. Hysteresis loop for GaAs-Ni sample.

The dependencies of ME voltage, generated by the structures, on the excitation field frequency f measured during the experiment are shown in Fig. 4. Measurements were carried out at H = 20 Oe, h = 0.01 Oe for the GaAs-Ni sample and at H = 25 Oe, h = 0.01 Oe for the GaAs-Metglas sample. The monolithic structure GaAs-Ni demonstrates maximum of ME voltage U_1 = 127 mV at the planar resonance frequency f_1 = 155.17 kHz. The quality factor of the monolithic structure is $Q_1 \approx 23.5 \cdot 10^3$. Magnetoelectric coefficient at resonance frequency is $\alpha_1 \approx 316$ V/Oe·cm. The composite GaAs-Metglas structure generates ME voltage with maximum amplitude U_2 = 63.9 mV at the planar resonance frequency f_2 = 174.34 kHz. The quality factor of the composite structure $Q_2 \approx 1850$ is an order of magnitude smaller than that one of monolithic structure. Magnetoelectric coefficient for the composite structure is $\alpha_2 \approx 160$ V/Oe·cm at resonance.

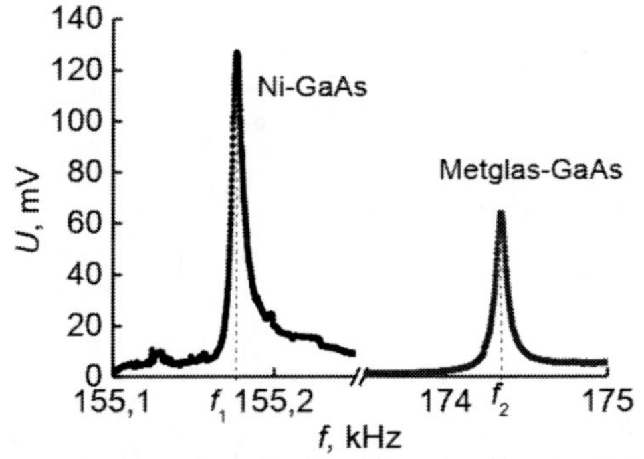

Fig. 4. Frequency dependencies of the ME voltage for GaAs-Ni and GaAs-Metglas samples.

Thus, the efficiency of the ME interaction in the monolithic structure GaAs-Ni under resonance turned out to be ~2 times higher than that one of the composite GaAs-Metglas structure, despite the fact that the thickness of the nickel layer is much smaller than the thickness of the Metglas layer. Apparently, the sputtering provides better mechanical contact between the layers than the gluing. Due to low thickness of the Ni layer the quality factor of the monolithic structure turned out to be ~ 13 times higher, which also leads to an increase in the ME interaction efficiency.

The dependencies of the ME voltage, quality factor and frequency of the resonance on the constant magnetic field for both structures are shown in Figs. 5 and 6. Measurements of the ME voltages and quality factors were carried out at frequencies f_1 and f_2 under excitation of the structure by ac magnetic field with amplitude h = 0.01 Oe. The ME voltage U(H) for the monolithic sample with Ni layer (see Fig. 5a) reaches maximum in small fields H ~3 Oe and voltage vs. magnetic field dependence demonstrates hysteretic behavior while the field decreases. The ME voltage vs. magnetic field dependence for the composite structure with Metglas layer (see Fig. 6a) demonstrates classical shape with maximum value in the field H ~28 Oe, which corresponds to the maximum of the piezomagnetic coefficient for the magnetostrictive layer.

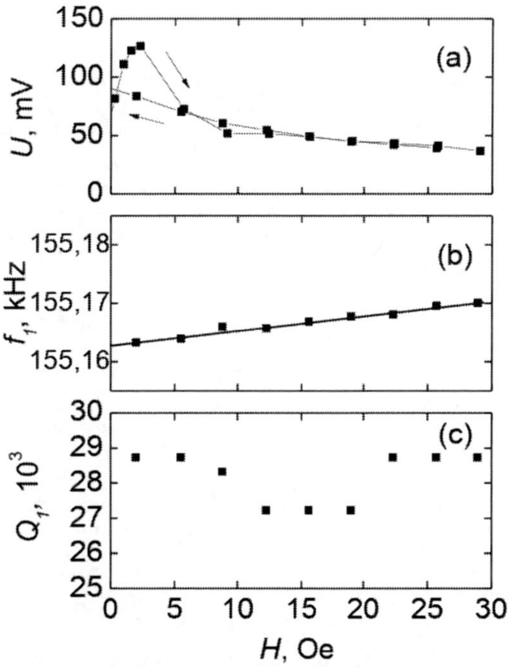

Fig. 5. Dependencies of the ME voltage U_1, resonance frequency f_1 and quality factor Q_1 on the field H for monolithic structure GaAs-Ni.

Resonance frequencies for both structures (see Fig. 5b and Fig. 6b) changed no more than on 0.005% with the permanent field increase. However, due to the bigger quality factor of the monolithic structure, even that small shift of the frequency results in significant change of the ME voltage measured near the resonance. From Fig. 5c and 6c it follows that when the constant field H is increased the acoustic quality factor for the monolithic structure changes no more than on 6%, whereas the quality factor of the composite structure changes on ~ 20%. Resonance frequency and quality factor dependences on the magnetic field may be due to the Young's modulus and dielectric losses magnetic field dependencies.

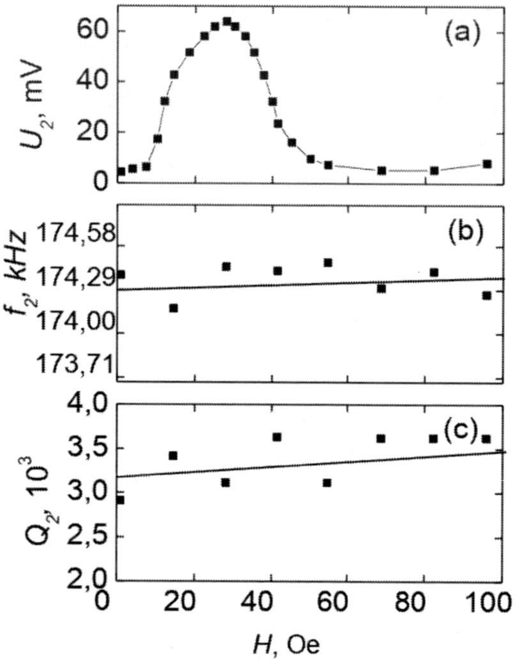

Fig. 6. Dependencies of the ME voltage U_2, resonance frequency f_2 and quality factor Q_2 on the field H for the composite structure GaAs-Metglas.

Thus, the resonant ME effect in composite structures based on the piezoelectric GaAs semiconductor was investigated. The ME coefficient at the acoustic resonance frequency was $\alpha_1 \approx 316$ V/Oe·cm for the monolithic GaAs-Ni structure and $\alpha_2 \approx 160$ V/Oe·cm for the composite GaAs-Metglas structure. The efficiency of ME interaction in a monolithic sample is higher due to better mechanical coupling between layers and smaller acoustic losses in a thin Ni layer.

The work was supported by the Russian Science Foundation (agreement 17-12-01435).

References

1. C.-W. Nan, M.I. Bichurin, and S. Dong, J. Appl. Phys. 103 031101 (2008)
2. G. Sreenivasulu, L.Y. Fetisov, G. Srinivasan, and Y.K. Fetisov, Appl. Phys. Let. V. 100 №. 052901 (2012)
3. P. Martins and S. Lanceros-Méndez, Adv. Funct. Mater. 23 3371 (2013)
4. D.A. Filippov, T.O. Firsova, V.M. Laletin, N.N. Poddubnaya, Письма в ЖТФ Vol. 43, №. 6, pp. 72-77. (2017)
5. V.M. Laletin, A.I. Stognii, N.N. Novitskii, N.N. Poddubnaya, Письма в ЖТФ Vol. 40, №. 21, pp. 71-78 (2014)
6. A. I. Stognij, N. N. Novitskii, V. A. Ketsko, S. A. Sharko, N. N. Poddubnaya, V. M. Laletin, A. V. Bespalov, O. L. Golikova, M. N. Smirnova, L. Yu. Fetisov, and A. O. Titova, Inorganic Materials Vol. 52 № 10 pp. 1070-1076 (2016)

SUPPRESSION OF THE MAGNETOELECTRIC HYSTERESIS IN A COMPOSITE LANGATATE-METGLAS STRUCTURE

Burdin D.A., Chashin D.V., Ekonomov N.A. Fetisov L.Y., and Fetisov Y.K.

Moscow Technological University (MIREA), 119454, Prosp. Vernadskogo, 78, Moscow, Russian Federation

Abstract. The suppression of the magnetoelectric effect hysteresis in a composite planar structure containing mechanically coupled layers of piezoelectric langatate and magnetostrictive amorphous alloy Metglas was observed and investigated. As the pumping magnetic field with frequency ~84 kHz was increased from 0.6 Oe up to 20 Oe, the coercive fields for the 1st and 3rd voltage harmonics generated by the structure were decreased from ~0.7 Oe up to zero. The suppression of the magnetoelectric hysteresis will allow increasing of absolute accuracy of the magnetoelectric magnetic field sensors by ~2 orders of magnitude.

1 INTRODUCTION

Magnetoelectric (ME) effect is observed in composite planar structures consisting of ferromagnetic (FM) and piezoelectric (PE) layers. The effect exhbits itself as generation of electrical voltage by the structure under simultaneous action of a permanent H and an alternating h magnetic fields [1,2]. The magnetic field produces magnetostrictive deformation of the FM layer, this deformation is transmitted to the PE layer, and it generates electrical voltage due to piezoelectricity. The ME effect is considered as one of the most prospective tools for design of high-sensitivity magnetic field sensors operating at room temperature [3,4].

However, it is well known that most of the ferromagnetic materials posses hysteretic behavior, that is the magnetization (as well as the magnetostrictive deformation) is an ambiguous function of the applied magnetic field. For materials with high magnetostriction, which are commonly used in composite ME structures, the coercive fields are H_C~1 Oe for amorphous alloy Metglas, ~ 20 Oe for nickel, and more that 50 Oe for rare-earth alloy Terfenol-D. The hysteretic phenomena result in big absolute measurement errors of ME magnetic field sensors or make necessary demagnetizing of the structure before each measurement, that strongly complicates design of the sensors. To the authors knowledge, these drawbacks of the ME magnetic field sensors have never been discussed up to now.

In this paper it is shown for the first time that hysteresis of ME effect in composite structures can be effectively suppressed by using pumping magnetic field of high amplitude. As it is known, an increase in the pumping field gives rise to a generation of higher harmonics of ME voltage [5]. So, the effect of hysteresis suppression was studied for the 1st and 3rd voltage harmonics, which can be used to design high-sensitivity magnetic field sensors [6,7]. This property of ME effect opens a possibility to elaborate high absolute accuracy censors for measuring permanent and alternating magnetic fields.

2 EXPERIMENT

A bilayer structure with dimensions of 20 x 5 mm^2, containing a layer of x-cut single-crystal lanthanum-gallium tantalite (LGT) of thickness 0.5 mm with metalized surfaces and a layer of amorphous alloy Metglas 2605S3A of thickness 20 μm was used in measurements. The layers were mechanically connected to each other by low-temperature soldering. The Metglas was chosen because it is the most prospective material for the sensor fabrication, it has the smallest coercive field H_C~1 Oe, rather high magnetostriction $\lambda_S \approx 20\cdot 10^{-6}$, and saturates in low magnetic fields H_S~100 Oe.

The sample was placed in an electromagnetic coil which was fed by the tuned oscillator Agilent 33210a and created a pumping magnetic field $h\cos(2\pi ft)$ with amplitude h up to 20 Oe and frequency f = 0-100 kHz. The permanent bias magnetic field H = 0-50 Oe was produced by the Helmholtz coils and was directed parallel to h along the longest side of the structure. The magnetic fields were measured with a Lake Shore gauss meter with absolute accuracy ~0.01 Oe. Magnetoelectric voltage u generated across electrodes of the LGT layer was measured with a lock-in amplifier SR 850.

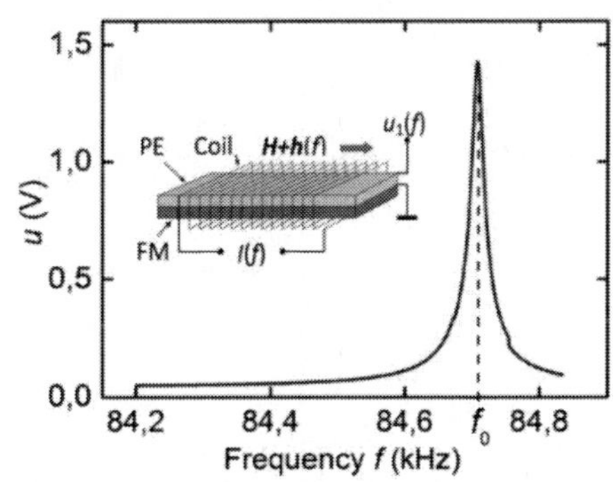

FIG. 1 Frequency responce of ME effect in a langatate-Metglas structure. The insert shows a schematic view of the composite structure.

* Y.K. Fetisov: fetisov@mirea.ru

The dependences of u as functions of excitation field frequency f at fixed H and h, and dependences of u as functions of bias magnetic field H for various pumping field amplitudes h have been measured.

3 RESULTS AND DISCUSSION

Figure 1 shows frequency response of the structure, that is dependence of generated voltage u on the excitation magnetic field frequency f for fixed H=1 Oe and alternating field h =0.08 Oe. One can see a peak with height of 1.42 V and quality factor of $Q \approx 5100$ in the vicinity of the frequency f_0 = 84.71 kHz. The peak corresponds to excitation of longitudinal acoustic oscillations in the structure which results in an increase in the deformation and "amplification" of the ME effect. For the given structure the ME coefficient $\alpha_E = u/(bh)$, where b is the FE layer thickness, was equal to ~12.5 V/(cm·Oe) far away of the resonance and ~355 V/(cm·Oe) at the resonance frequency, that coincide with data of [8].

In order to figure out the value of the hysteretic field for the described structure, the magnetization loop was measured (at Department of Magnetism, MSU) using the Lake Shore vibration sample magnetometer. The loop is shown in Fig. 2. One can see that the loop is symmetric with respect to the origin of coordinates and the sample saturates in the fields H_S ~100 Oe. The insert of Fig.2 shows initial part of the loop in expanded scale. The coercive field for Metglas is $H_C \approx 1$ Oe. As in the low field region the relation $\lambda \sim M^2$ is true, the coercive field for the dependence $\lambda(H)$ is also equal to 1 Oe.

In order to observe the hysteresis, the dependences of generated voltage u on the field H for fixed signal frequency f and different amplitudes of the pumping field h have been measured. To avoid changes in the generated voltage due to a shift in the resonance frequency caused by H variation (so called ΔE effect), all the measurements were carried out on the slope of the resonance curve (see Fig.1) at the frequency 84.6 kHz.

First, the magnetic field dependences for the 1st voltage harmonic have been measured. The structure was excited by magnetic field with a frequency 84.6 kHz and the signal was registered at the same frequency. Figure 3 shows dependences of the 1st harmonic amplitude u_1 on the field H for pumping field amplitudes h = 0.6 Oe and

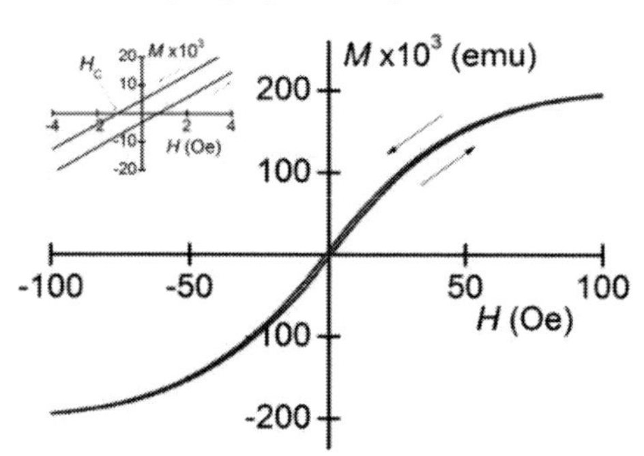

FIG. 2 Magnetization curve of the Metglas layer. The insert shows initial part of the curve in expanded scale.

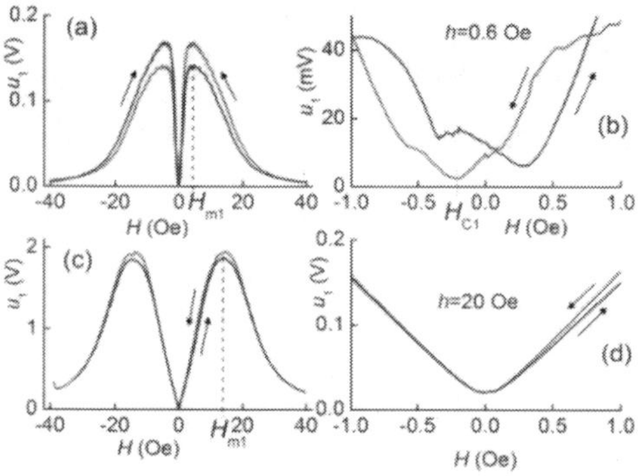

FIG. 3 Dependence of the 1st harmonic of the ME voltage u_1 on permanent magnetic field H for different amplitudes of the pumping field h = 0.6 Oe and h=20 Oe.

h=20 Oe. It is seen, that for low pumping field h=0.6 Oe (Fig.3a), u_1 first increases with the field increase, reaches a maximum at $H_{m1} \approx 4.2$ Oe, and then drops to zero. For high pumping field amplitude h=20 Oe (Fig.3c), the dependence has similar shape, but the maximum takes place in bigger field $H_{m1} \approx 13.7$ Oe. It is seen in expanded scale (Figs. 3b and 3d), that for low pumping field the hysteresis on the dependence $u_1(H)$ takes place for increasing and decreasing fields. The value of the coercive field is $H_{C1} \approx 0.25$ Oe. For high pumping field h=20 Oe (see Fig. 3d) the hysteresis vanishes. Note, that u_1 is not equal to zero at $H = 0$ for the 1st harmonic. This is due to the effect of direct electromagnetic interference at a large amplitude of the pumping field.

After that similar measurements have been carried out for the 3rd ME voltage harmonic. The structure was excited by magnetic field with frequency of 28.2 kHz and generated signal at the frequency 84.6 kHz was registered. Results of the measurements are presented in Fig. 4.

It is seen from Fig.4, that the dependence $u_3(H)$ has more complicated shape and contains three peaks as the field H increases from zero up to 40 Oe. For low pumping field amplitude h=4 Oe (Fig. 4a) the first maximum is reached in the field $H_{m3} \approx 1.3$ Oe, while for high pumping field (Fig. 4c) it is reached in the field H_{m3}

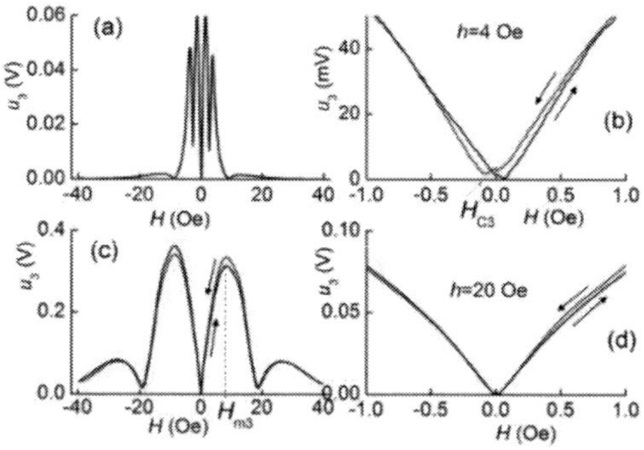

FIG. 4 Dependence of the 3rd harmonic of the ME voltage u_3 on permanent magnetic field H for different amplitudes of the pumping field h = 4 Oe and h=20 Oe.

≈ 8.2 Oe. It is seen in expended scale (Figs. 4b and 4d), that the dependence $u_3(H)$ also has a hysteresis. For low pumping field h=4 Oe, the coercive field is $H_{C3} \approx 0.07$ Oe. For high pumping field h=20 Oe the hysteresis is completely vanishes and u_3=0 in the field absence H =0. The direct electromagnetic interference with pumping frequency does not contribute to the amplitude of the 3rd harmonic.

Using measurements similar to those shown in Fig. 3 and Fig. 4, one can plot the dependences of characteristic fields H_{m1} and H_{m3}, corresponding to the first maximum of the 1st and the 3rd harmonics, on the pumping field h. It follows from Fig. 5, that these characteristic fields remain constant in the pumping field region h < 4 Oe, and then increase as the pumping field growths. Thus, the operation range of the ME magnetic field sensor expands with an increase of h, which coincides with conclusions of [7].

Figure 5 shows dependences of coercive fields H_{C1} and H_{C3} for the 1st and 3rd harmonics on the pumping field h, drawn using the data which are similar to those given in Fig. 3 and Fig. 4. It is seen, that the coercive field for both harmonics falls down with an increase of h

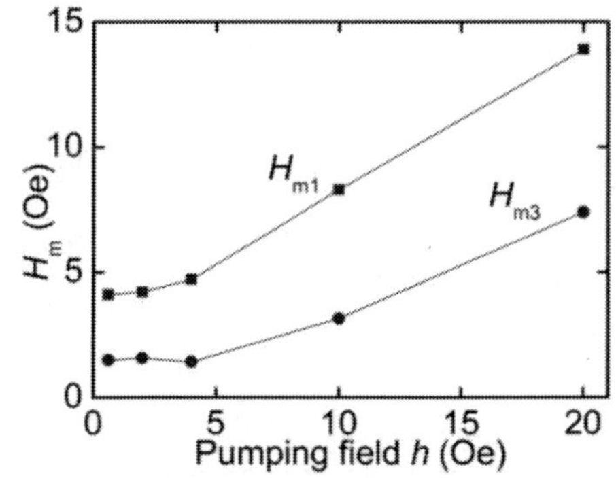

FIG. 5 Dependences of the characteristic fields H_{m1} and H_{m3}, corresponding to maxima of the 1st and 3rd ME voltage harmonics, on the pumping field h.

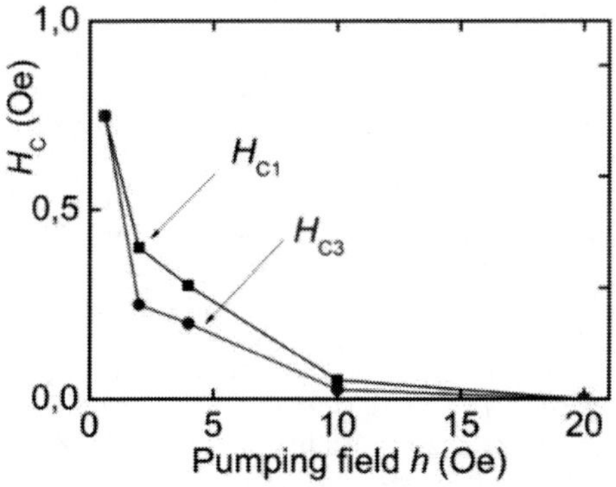

FIG. 6 Dependences of the coercive fields H_{C1} and H_{C3} for the 1st and 3rd ME voltage harmonics on the pumping field h.

and it turns out to zero at h ≈20 Oe. Inaccuracy in determining minima of curves in Fig. 3 and Fig. 4 does not allow to draw these curves more accurately and find an analytical formula for the hysteresis suppression.

Theoretical simulation showed that the effect of ME hysteresis suppression is due to a decrease in the relative influence of the constant field on the asymmetry of the minor magnetostriction loop of a ferromagnetic with increasing amplitude of the alternating field. Complete suppression of the hysteresis is achieved when amplitude of the pumping field turns out comparable with the saturation field of the ferromagnetic layer of the structure.

4 CONCLUSION

Thus, the effect of the hysteresis suppression in the field dependence of the ME voltage harmonics generated by the langatate-Metglas structure was experimentally observed. The effect will allow a ~2 order of magnitude enhancement in the accuracy of the absolute measurements of the ME magnetic field sensors.

The research was supported by the Russian Foundation for Basic Research (grant No 17-12-01435).

References:

1. C.-W. Nan, M.I. Bichurin, S. Dong, D. Viehland, G. Srinivasan, J. Appl. Phys., **103,** p. 031101 (2008).
2. H. Palneedi, V. Annapureddy, S. Priya, J. Ryu, Actuators, **5,** No 9, p. 1-31 (2016).
3. M.M. Vopson, Critical Rev. in Solid States and Mater. Sci. **40**, p. 1-28 (2014).
4. Y. Wang, J. Le, D. Viehland, Materials Today, 17, p. 269-275 (2014).
5. K.E. Kamentsev, Y.K. Fetisov, and G. Srinivasan, APL, **89**, p. 142510 (2006).
6. D.A. Burdin, D.V. Chashin, N.A. Ekonomov, Y.K. Fetisov, A.A. Stashkevich, JMMM, **405,** p. 244-248 (2016).
7. D.A. Burdin, D.V. Chashin, N.A. Ekonomov, L.Y. Fetisov, Y.K. Fetisov, M. Chamonin, J. Phys. D: Appl. Phys., 49, p. 375002 (2016).
8. G. Sreenevasulu, L.Y. Fetisov, Y.K. Fetisov, and G. Srinivasan, APL, 100, p. 052901 (2012).

Influence of Magnetic Order Parameter Symmetry and Spin Fluctuations on Forming of Ferroelectric Order in Mn_2O_5 Oxides

Vladimir Men`shenin[1]

[1]M.N. Miheev Institute of Metal Physics of Ural Branch of RAS, 620108 Yekaterinburg , Russia

Abstract. The magnetic phase transitions from paramagnetic to magnetic phases have studied in RMn_2O_5 oxides. Using a symmetry analysis and renormalization group method it was shown that an appearance of the electric polarization or lack whereof in the magnetic phases are determined the symmetry of these phases and spin fluctuations in the vicinity of the phase transitions.

The compounds were investigated intensively in the last twenty years in which the orderings of the different nature coexist. The RMn_2O_5 oxides are belonged to such systems. The long range magnetic order and ferroelectric order coexist in these oxides in the certain temperature interval. There is the strong interaction between the magnetic and ferroelectric subsystems in these compounds because the electric polarization arises only in the magnetic phase. The important difference RMn_2O_5 oxides from the manganites is that electric polarization achieves the biggest quantity in the commensurable magnetic phase whereas the polarization coexists with the long range magnetic order only in the incommensurable magnetic phase of the manganites. It was shown experimentally that electric polarization exists in the RMn_2O_5 (R = rare earth, Y, Bi) compounds where the rare earth element is heavier than neodymium. The ferroelectric order appears in the one direction incommensurable magnetic phase for the compounds with the rare earth element and Y. The symmetry of the crystal lattice is not violated at low temperature when electric polarization appears since where are only nuclear Bragg peaks on the neutron diffraction patterns which have existed in the paramagnetic phase. The authors of the some papers have supposed that the additional peaks were absent due to the extremely small atomic displacements. These displacements cannot be measured experimentally.

Let us note that chemical bondings have the ion-covalent character with a significant shear charge of the electrons, which delocalize between the ions. These electrons give the essential contribution to the electric polarization. In this case we can go over from the determination of the electric polarization through the charge distribution to its definition through the variation of the electric polarization in the process of the measurement for example when the electric field is turned on. In this case, the lattice of the values for the electric polarization can include the nonzero polarization vector. Besides the space inversion is absent as the element of the exact symmetry in the incommensurable magnetic phase.

The situation is never simple if the phase transition occurs from the paramagnetic phase (PMP) to the commensurable structure. For example, the transition takes place to magnetic phase with the propagation vector **k**= $\{1/2, 0, 1/2\}$ in the oxide $BiMn_2O_5$ and transition happens to phase with **k**= $\{1/2, 0, 0\}$ in the oxide $PrMn_2O_5$. The electric polarization has been found in the $BiMn_2O_5$ but it has been absent in the $PrMn_2O_5$ although the space group representations coincide for these propagation vectors. I would like to show the reasons of such behavior analyzing the magnetic phase transitions in these compounds.

Let us consider the appearance of the electric polarization for the phase transition from the PMP to the one direction incommensurable magnetic phase in the RMn_2O_5 oxides where R is the rare earth element. The propagation vector of the incommensurable magnetic structure is $\mathbf{k} = (1/2, 0, \nu)$, $\nu \neq 1/2$. The components of the order parameter transform under the full irreducible representation (IR) of the space group Pbam (D_{2h}^9). This IR is described by the matrices of fourth order. In this case, the order parameter has four components. The thermodynamic potential contains the usual terms of second and fourth order by the order parameter components, wherein a one of these terms is the product of four order parameter components, Lishitz invariant with first derivative with respect to the coordinate z of the order parameter components, two Dzjaloshinskii invariants and the heterogeneity exchange energy [1]. The W_1, W_2 values are coefficients in front of Dzjaloshinskii invariants respectively. Let us denote the order parameter components as$(\eta_1, \eta_2, \xi_1, \xi_2)$. There are the relations between these components

$$\eta_1 + \xi_2 = \rho_1 \exp(i\phi_1),\ \eta_1 - \xi_2 = \rho_1 \exp(-i\phi_1),$$

$$\xi_1 - \eta_2 = \rho_2 \exp(i\phi_2),\ \xi_1 + \eta_2 = \rho_2 \exp(-i\phi_2),$$

* Corresponding author: menshenin@imp.uran.ru

(1)

where values ρ_1,ρ_2 are the modules of the order parameter components and values ϕ_1,ϕ_2 are the phases of these components. The analysis of the thermodynamic potential leads to the following results. There is the transition to the incommensurable magnetic phase with $\varphi_2 = k_0 z,\ \ k_0 = |\gamma_1|/\gamma_{ex}$ if ρ_1,ρ_2 are changing slower than phases and the equation is performed for the phases $\varphi_1 = \varphi_2 + (2m+1)\pi/2$, m is an arbitrary integer. The γ_1 value is a coefficient in front of Lifshitz invariance and a γ_{ex} value is a coefficient in front of the heterogeneity exchange energy. The electric polarization P_y , which can be determined from the condition $\partial\Phi/\partial P_y$ =0, is $P_y = (-1)^{m+1}\kappa\alpha_1\rho_1\rho_2$. In this case, the polarization depends slowly from the coordinates and is the small value in vicinity of the phase transition. The polarization is the concomitant macroscopic order parameter.

The decrease of the temperature results in the creation of the higher harmonics. In this case, we derive the nonlinear differential equations for the phases. These equations transform to the equality $\partial^2\varphi_2/\partial z^2 = 4v\sin 4\varphi_2$, where v is equal $v = (4W_1 - u_2)\rho^2/16\gamma_{ex}$ if we assume that $\rho = \rho_1 = \rho_2$ and $\varphi_1 = \varphi_2 + 3\pi/2$. We can arrive at the conclusion on the base of the equation for the φ_2 value that the soliton lattice is formed in the system. The soliton lattice becomes the energetically unfavorable state under the certain conditions and the first order transition takes place to the commensurable antiferromagnetic structure. The electric polarization achieves the biggest value in this phase. The further decrease of the temperature leads to the transition to the incommensurable magnetic phase in two directions. The electric polarization is decreased significantly in this phase.

The renormalization group analysis have been performed of the transition from the PMP to the one direction incommensurable magnetic phase. We supposed that the relativistic contribution to the energy associated with the magnetic heterogeneities is small in the vicinity of this transition where magnetic fluctuations of the exchange nature are appeared. Thus, Lifshitz invariant does not play the role in this case. There are the three critical fixed points of the renormalization group transformations then the all interaction parameters are not equal zero in the approximation lineal in ε. The steadiness is analyzed of these points on the base of the recursive relations for the parameters of the effective Hamiltonian [2]. We consider first of all the point where the values of the g_i parameters are

$$g_1^* = \frac{\varepsilon}{80K_4}, g_2^* = \frac{3\varepsilon}{40K_4}, g_3^* = \frac{\varepsilon}{10K_4}, g_4^* = \frac{\varepsilon}{40K_4}. \tag{2}$$

The values Δg_i are the functions of the ξ parameter. The system of the equations for these values have the form

$$\frac{d(\Delta g_i)}{d\xi} = \sum_j A_j\left(g_1^*, g_2^*, g_3^*, g_4^*\right)\Delta g_j + R_i,$$

$$i = (1,..,4), \xi = \ln b. \tag{3}$$

where b is the parameter which describes the change of the spatial scale under the Kadanoff transformations.

In this case there are two different negative real eigenvalues and two different positive real eigenvalues of the A matrix. These results imply that trajectories of the system motion in the phase space $(g_1,.,g_4,r)$ with variable parameter ξ, which approach arbitrary close to point found above but do not pass through this point go away from this point to infinity. However where is the trajectory passing through the fixed point for which the system returns to this point if the one shifts from the fixed point. This behavior of the system allows one to conclude that this fixed point is a stable critical point. In this point where is the second order phase transition. The analysis of the system symmetry in this point leads to the following result. The terms with the coefficients g_1, g_2, g_4 in the Hamiltonian [2] are transformed to the expression $(\rho_1^2 + \rho_2^2)^2$ in the notation of the equality (1). These terms have the O(2) symmetry. The term with the g_3 coefficient does not this symmetry since the product fourth order parameter components has the orthorhombic symmetry. Hence, the system symmetry does not increase in the order point. The spin fluctuations do not happen to the disappearing of the macroscopic order parameter in the form of the electric polarization in this case.

Let us consider the situations then the magnetic transitions come about from the PMP to the commensurable structure in the RMn_2O_5 compounds. We compare the $BiMn_2O_5$ oxide and $PrMn_2O_5$ oxide. The magnetic phase transition occurs in the antiferromagnetic structure with the propagation vector **k** $=(1/2, 0, 1/2)$ in the $BiMn_2O_5$ oxide and in the one with **k**$=(1/2,0,0)$ in the $PrMn_2O_5$ compound. This comparison helps us to understand the influence of the spin fluctuations at the forming electric polarization in these oxides. We note first of all that the space group of the symmetry has the same irreducible representations for these propagation vectors. Where are the two representations. We have found that magnetic structure written above could be realized only for the ν_2 representation. The effective Hamiltonian of the system, which allows one to analyze the critical behavior of the $PrMn_2O_5$ oxide in the immediate vicinity of the transition from the paramagnetic to the antiferromagnetic phase, is given by

$$\begin{aligned} H_{eff} = \int d^d x \Bigg\{ & \frac{r}{2}\left(\varsigma_1^2(x) + \varsigma_2^2(x)\right) + \\ & + u_1\left[\varsigma_1^4(x) + \varsigma_2^4(x)\right] + \\ & + u_2\varsigma_1^2(x)\varsigma_2^2(x) + \\ & + \frac{1}{2}\left[\frac{\partial\varsigma_1}{\partial x_i}\frac{\partial\varsigma_1}{\partial x_i} + \frac{\partial\varsigma_2}{\partial x_i}\frac{\partial\varsigma_2}{\partial x_i}\right]\Bigg\}. \end{aligned} \tag{4}$$

In this equation $\varsigma_i (i = 1,2)$- are the order parameter components and the r, u_1, u_2 parameters have the same meaning as those in the expression for the thermodynamic potential that describes phase transition in Landau theory [3]. At first sight, the $\int d^d x \lambda P_y(\varsigma_1^2 - \varsigma_2^2)$ term must be included at the effective Hamiltonian due to its invariance under the space group transformations. However the expression for the electric polarization is equal zero within the framework of the Landau theory for the transition in which the both order parameter components obtain. The order parameter components are equal in the absolute value for this transition. The above expression in the Hamiltonian vanishes in the case of the transition where the one order parameter component is nonzero below the Neel's temperature. The disappearing of this term happens due to spin fluctuations as we show below.

Let us introduce a parameter $s = u_2/u_1$. Where are the three fixed points of the renormalization group transformations [4] s=0, s =2, s =6. The point s =2 is the stable critical point and the points s =0, s =6 are unstable critical points. The transition is realized in the system with $\zeta_1 = 0, \zeta_2 \neq 0$. The condition $u_2 = 2u_1$ is satisfied in the stable critical point. An O_2 symmetry arises in the system at the phase transition point. Formally, it is rather easy to understand the emergence of symmetry. The above condition implies that, instead of two fourth-order invariance, a significant invariant $(\zeta_1^2 + \zeta_2^2)^2$ arises in the effective Hamiltonian. Then, considering the order parameter as a two-component vector we can see that the first invariant $\zeta_1^2 + \zeta_2^2$ is a scalar product and, $(\zeta_1^2 + \zeta_2^2)^2$-the square of this product [4]. It is clear, that these invariants are invariant under the rotation of two-component vector at an arbitrary angle. However, the expression $\int d^d x \lambda P_y(\varsigma_1^2 - \varsigma_2^2)$ is not invariant under the rotations at an arbitrary angle and is equal zero in this point. Thus, the appearance of the electric polarization is suppressed of the spin fluctuations in the $PrMn_2O_5$ oxides.

Let us consider the $BiMn_2O_5$ compound. The magnetic susceptibility at low temperature increases upon the cooling and has a broad peak with two characteristic temperatures T=39K and T=36K, indicated the occurrence of the long-range magnetic order. The existence of the second transition indicates a slight delay in the ordering of magnetic moments associated with one position of manganese relative to other. It is found [5] that the sets of the spin orientations for the positions 4h (Mn^{3+}) and (Mn^{4+}) is included in the sets of the basis functions irreducible representation ν_1 space group. The effective Hamiltonian that allows to describe the two experimentally observed phase transitions has the form

$$\begin{aligned} H_{eff} = \int d^d x \Big\{ & \Big[\frac{r_1}{2}(\eta_1^2 + \eta_2^2) + \frac{1}{2}\Big(\frac{\partial \eta_1}{\partial x_i}\frac{\partial \eta_1}{\partial x_i} + \frac{\partial \eta_2}{\partial x_i}\frac{\partial \eta_2}{\partial x_i}\Big) + u_1(\eta_1^4 + \eta_2^4) + u_2(\eta_1^2\eta_2^2) + \frac{1}{2\kappa}P_y^2 \\ & + \lambda_1 P_y(\eta_1^2 - \eta_2^2) \Big] \\ & + \Big[\frac{r_2}{2}(\varsigma_1^2 + \varsigma_2^2) + \frac{1}{2}\Big(\frac{\partial \varsigma_1}{\partial x_i}\frac{\partial \varsigma_1}{\partial x_i} + \frac{\partial \varsigma_2}{\partial x_i}\frac{\partial \varsigma_2}{\partial x_i}\Big) + u_3(\varsigma_1^4 + \varsigma_2^4) + u_4(\varsigma_1^2\varsigma_2^2) \\ & + \lambda_2 P_y(\varsigma_1^2 - \varsigma_2^2) \Big] + u_5(\eta_1^2\varsigma_1^2 + \eta_2^2\varsigma_2^2) + u_6\eta_1\eta_2\varsigma_1\varsigma_2 + u_7(\eta_1^2\varsigma_2^2 + \eta_2^2\varsigma_1^2) \Big\} \end{aligned} \quad (5)$$

Here d=4-ε is the spatial dimension, κ is the dielectric susceptibility, the last three terms describe the interaction of the order parameters for two transitions with different temperatures, the $u_1, .., u_7$ parameters have the same meaning as in Landau theory. The effective Hamiltonian contains the terms including the electric polarization P_y oriented along the y-axis of the oxide.

We believe that the fluctuations of the order parameters are significant for the first and second transitions because $r_1 \approx r_2$ in this region. Let us consider the method of the renormalization group. The system has the following fixed points then the all u_i parameters differ from zero

$$x_1^* = \frac{1}{160}, x_2^* = \frac{1}{80}, x_3 = \frac{1}{160}, x_4^* = \frac{1}{80}, x_5^* = \frac{1}{80}, x_6^* = \frac{1}{20}, x_7^* = \frac{3}{80},$$

$$x_1^* = \frac{1}{160}, x_2^* = \frac{1}{80}, x_3^* = \frac{1}{160}, x_4^* = \frac{1}{80}, x_5^* = \frac{3}{80}, x_6^* = \frac{1}{20}, x_7^* = \frac{1}{80}, \quad (6)$$

where $x_i = u_i K_4/\varepsilon, K_4 = 1/8\pi^2$, the value ε is a small parameter. The analysis of the stable of these points leads to the following results. The first two fixed points are the unstable points of the renormalization group transformations. The first of these points has the critical surface $\{u_1, u_5\}$. The all trajectories belonging to this surface lead to stable fixed point. The second order phase transition is realized on these trajectories. Thus, the first transition occurs as second order if the trajectory associated with this transition belongs to this surface whereas the second transition is the first order. In general case, the both transitions are the first order. The u_6 parameter is the essential parameter while u_7 is the intermediate parameter. The spin fluctuations lead to the growth in the symmetry in the other fixed points where the u_6 parameter is zero and result in the disappearance of the invariances containing the electric polarization from the effective Hamiltonian in the point of the phase transitions. The above analysis implies that only the first unstable fixed point at which the u_6 parameter is significant materializes in the system.

[1] V.V. Men`shenin, JETP, **135**, 236 (2009).
[2] V.V. Men`shenin, JETP, **116**, 980 (2013).
[3] V.V. Men`shenin, JETP, **120**, 1019 (2015).
[4] K.G. Wilson, M. Fisher, Phys. Rev., Lett., **28**, 240(1972).
[5] V.V. Men`shenin, JETP, **122**, 145 (2016).

Composite multiferroic materials consisting of NdFeB and PZT particles embedded in elastic matrix: the appearance of electrical polarization in a constant magnetic field

Liudmila Makarova[1,*], *Yuliya* Alekhina[1], *Elena* Kramarenko[1], *Alexander* Omelyanchik[2,3], *Valeria* Rodionova[2,3], *Olga* Malyshkina[4] and *Nikolai* Perov[1,2]

[1]Lomonosov Moscow State University, Faculty of Physics, 119991Moscow, Leninskie gory 1-2, Russia
[2]Immanuel Kant Baltic Federal University, 236004 Kaliningrad, Nevskogo 14, Russia
[3] Center for Functionalized Magnetic Materials (FunMagMa), Immanuel Kant Baltic Federal University, 236041, Kaliningrad, Russia
[4]Tver State University, 170100, Tver, Russia

Abstract. New composite materials consisting of polymer matrix with PZT and NdFeB microparticles were prepared and investigated in the work. It was found that magnetic properties such as saturation magnetization, coercivity, permeability, depend on mass concentration of the ferromagnetic particles in the samples. Also it was found that all samples had electrical polarization in DC external electric field. The electric properties such as coercivity, remanent polarization, the maximum polarization value, had changes in the external constant magnetic field 1.1 kOe. These changes depended on both concentrations of ferromagnetic and ferroelectric particles. This type of magnetoelectric transformation allows us to classify new materials as multiferroic materials. These new composite materials can easily be prepared of any shape, the final materials are flexible and resistant to external chemical influences. The area of application of new multiferroic materials varies from sensors to autonomous energy sources.

1 Introduction

Smart materials are materials which convert one form of energy to another [1]. Usual ferromagnetic or ferroelectric materials are also smart. Magnetostrictive effect which is observed in ferromagnetic materials is the example of changes of shape and sizes in the external magnetic field. Piezoelectric, magnetostrictive and ferroelastic materials have been already used in sensors, actuators and controllers. The area of application of smart materials includes different types of sensors, microphones, injectors, energy harvesting and well logging [2,3]. Multiferroic materials are smart structures due to combination of magnetostriction and piezoelectric properties in one material [4,5].

Magnetorheological materials such as fluids, foams, elastomers are smart structures. Well known examples of energy transformation are magnetorheological effect, magnetodeformation effect, Payne effect, magnetodielectric effect [6,7].

New type of composite multifunctional material with both ferromagnetic, ferroelectric and ferroelastic properties was prepared. Unlike the existing types of composite multiferroics, such materials are easy to prepare, they are flexible and chemoresistive.

Previously it has been shown that magnetodielectric effect was observed in magnetic elastomers, namely the dielectric permittivity increase in the external magnetic field [8]. It is supposed that magnetic moments of all ferromagnetic particles rotate to the field direction and particles tend to shift forming chain-like structures. The internal stresses arise in the elastic matrix and these stresses can result in rotation or displacement of ferroelectric particles or prevent it what results in the changes of electric properties. In this work we investigated the influence of external magnetic field on the electric properties of new three-component composite material. Also we investigated the effect of ferroelectric particles on the magnetic properties of the samples. The direct magnetoelectric transformation in new three-component samples was found. The opposite effect can be observed: the displacement of ferroelectric particles leads to changes in magnetic properties [9].

2 Materials and methods

2.1 Materials

The preparation of ferromagnetic and ferroelectric particles was described in our previous work [9]. Filling particles were hard magnetic NdFeB particles (sizes varied from 5 to 30 μm) and ferroelectric particles PZT-26 (size was about 5 μm). There were prepared 7 samples, based on silicone polymer, the sample compositions are presented in Table 1. The components of polymer compound with total volume 1 ml were mixed with powders. The amount of each type of powder was calculated from mass concentration and density of each material. The correlations between volume and mass concentration are also presented in Table 1. The total volume concentrations of filling powder varied from 20% to 40%. The volume ratios between NdFeB and PZT particles were 1:1, 2:1, 1:3 and 1:6. Also the elastomer with 20 vol% of PZT particles was prepared.

*Corresponding author: la.loginova@physics.msu.ru

Table 1. Samples composition.

Sample No.	NdFeB:PZT (vol)	Total vol. %	NdFeB mass %	PZT mass %
0-1	0:1	20	-	50.5
1	1:6	20	12.0	41.0
2	1:3	20	20.3	34.5
3	1:1	20	37.3	21.2
4	2:1	20	47.2	13.4
5	1:1	30	45.1	25.6
6	1:1	40	50.4	28.6

2.2 Methods

Magnetic properties were investigated by vibrating sample magnetometer (VSM) Lake Shore 7400 in the field range 16 kOe at room temperature. The sample for measurement with sizes 3mm*3mm*1mm was cut from the material and fixed at the holder, the plane of the sample was parallel to the direction of the external magnetic field.

Also using VSM the magnetic relaxation properties were investigated. The time dependences of the magnetic moment of the sample were measured either with switching on the external magnetic field or after switching off the external magnetic field. The sample was placed between the copper plates (using a non-conductive gasket between the samples and the plate to avoid accidental breakdown currents) and fixed at the holder at the same orientation (plane was parallel to the direction of magnetic field). Plates were connected to the high voltage source. Magnetic relaxation properties were also measured with switching on the external electric field with the voltage 5 kV.

The electric properties, namely the dependence of electric polarization on the electric field, were investigated by Soyer-Tauer method. The ferroelectric hysteresis loop was observed at the screen of the oscilloscope. The AC voltage from the generator is transmitted to the series-connected sample and the reference capacitor. The input Y of the oscilloscope is supplied with a voltage from the reference capacitor. The input X is supplied with a voltage from the ferroelectric sample.

The sample with average sizes 1.5cm*1.5cm*2.5mm was covered by liquid silver solution at both sides and placed between conducting clips. The polarization *P* of the sample is calculated from the formula:

$$P = \frac{C_{ref} * U_{out}}{S} \quad (1)$$

Where $C_{ref} = 1\mu F$ – reference capacitor, U_{out} – value of voltage of output signal (ordinate axes at the screen of oscilloscope), S – square of the sample's electrodes.

The electric field $E=U*d$, where U – the amplitude of AC voltage (abscissa axes at the screen of oscilloscope), d – the distance between the plates with the sample, or the width of the sample.

The influence of external magnetic field on the electric properties was measured by placing the sample into the magnetic field of permanent calibrated magnet. The plane of the sample was perpendicular to the direction of magnetic field.

Phase transition properties were investigated with Differential Scanning Calorimeter NETZSCH 204 F1 Phoenix at the temperature range from 0 C to 450 C.

3 Results and discussion

3.1 Magnetic properties

The typical hysteresis loops of different samples are presented in Figure 1.

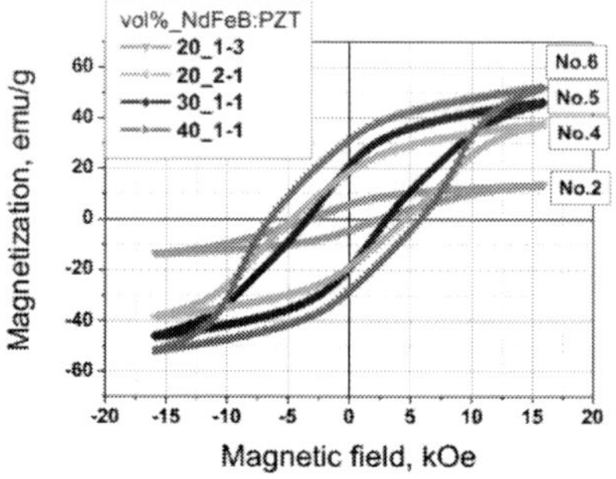

Fig. 1. The dependences of magnetization of the samples 2, 4, 5, 6 on magnetic field at room temperature.

All magnetic parameters such as maximum magnetization, coercivity, remanent magnetization depend on the sample composition. For example maximum magnetization increases with the increase of mass concentration of ferromagnetic particles, what is presented in Figure 2. As the saturating field of NdFeB material is about 25-30 kOe, we can not present the dependence of saturation magnetization on concentration of magnetic filler. In the field 16 kOe not all magnetic moments of the particles can be oriented in the direction of the external magnetic field. Moreover, in non-saturating field it is necessary to take the interaction between magnetic particles and the change of the internal field associated with this interaction into account. Therefore, the Figure 2 presents the nonlinear dependence of magnetization in the maximum field on the concentration of magnetic filler.

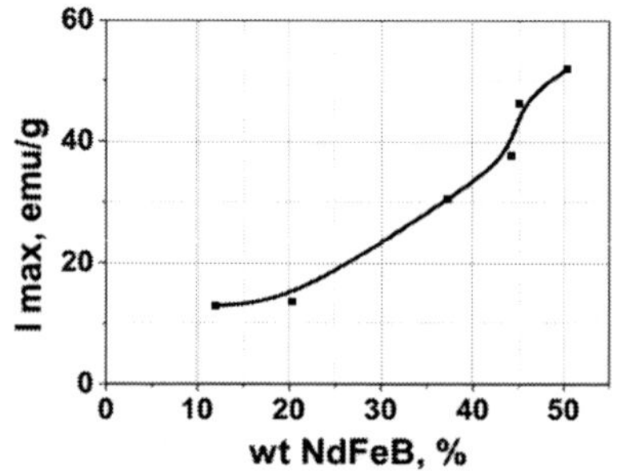

Fig.2. The dependence of maximum value of magnetization of the samples 1-6 on mass concentration of NdFeB particles.

The elastic properties depend on the relative concentration of PZT particles. Namely the Young's modulus decrease with increase of relative content of PZT particles (Fig.3).

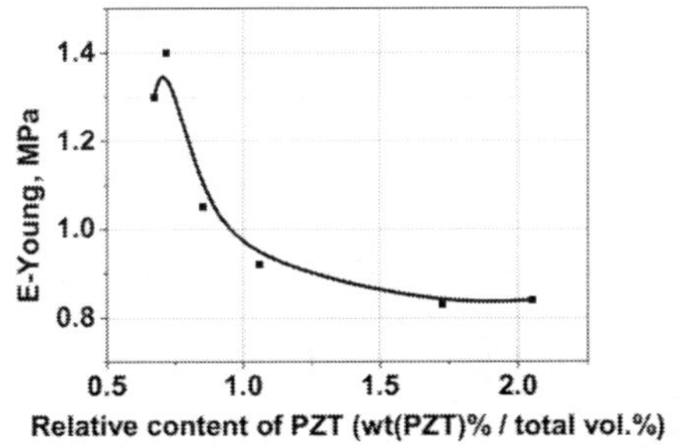

Fig.3. The dependence of Young's modulus of the samples 1-6 on the relative concentration of PZT particles.

The dependence of Young's modulus on PZT particles concentration is significant as the coercivity depends on elastic modulus (Fig.4a). It was shown earlier [10] that the remagnetization of the sample is associated with the rotation of the ferromagnetic particle in the elastic medium. In softer medium the elastic force acting on the particle is smaller, so the coercivity has a smaller value.

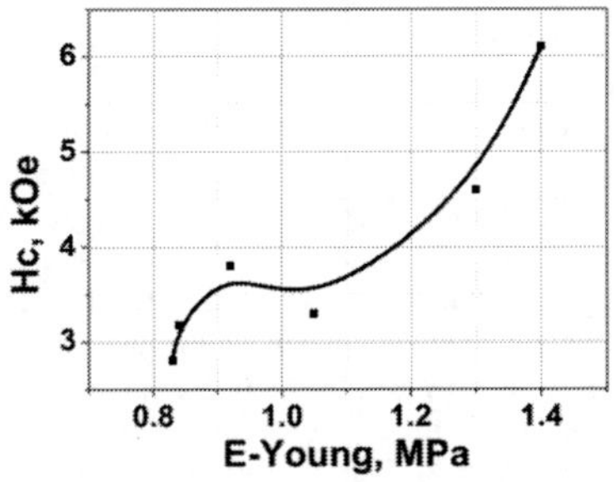

Fig.4(a). The dependence of coercivities of the samples 1-6 on the corresponding Young's modulus.

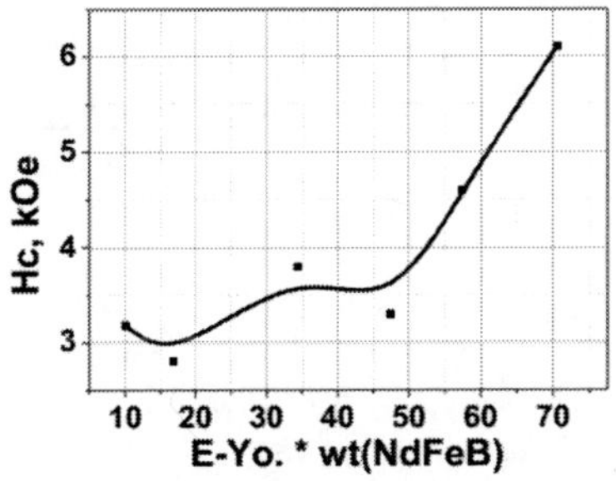

Fig.4(b). The dependence of coercivities of the samples 1-6 on both Young's modulus and mass concentration of NdFeB particles.

At the same time coercivity depends on concentration of ferromagnetic particles as interparticle interactions influence on the remagnetization process. The dependence of coercivity on both elastic properties and concentration of NdFeB particles is presented in Figure 4(b).

3.2 Ferroelectric properties

The dependences of polarization on electric field were measured for each sample with ferroelectric particles, including the sample with only ferroelectric particles No.0-1.

Also temperature measurements were carried out with Differential Scanning Calorimeter NETZSCH 204 F1 Phoenix. Curie temperature of PZT powder was 321.1 C. All electric measurements were carried out at room temperature, so we assume that samples revealed ferroelectric hysteresis properties.

It was confirmed that addition of PZT particles to the silicone matrix led to appearance of ferroelectric properties of the material (Fig. 5-10). For example, the dependence of electric polarization of the sample No.0-1 with PZT particles on electric field at frequencies 20 Hz and 50 Hz are presented in Figure 5. Moreover the loop at lower frequency 20 Hz has larger value of coercivity than the loop at higher frequency 50 Hz. It should be noted that at even higher frequencies (100 Hz for instance) the loop transformed to straight line. The loop is not saturated for several reasons: 1) the value of saturating electric field exceeds 1 kV/mm, that can not be achieved at the equipment; 2) the width of the sample was about 2 mm, that decreases the total value of electric field; using of thinner sample may lead to breakdown of capacitor.

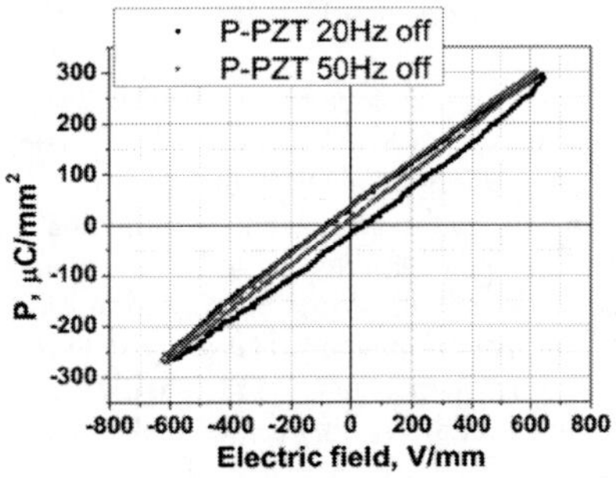

Fig.5. The dependences of polarization of the sample No.0-1 on electric field at different frequencies.

It was found that hysteresis loops of different samples were observed at different modes of generator – at different frequencies and at different amplitudes of the electric field. In Figure 6 the hysteresis loops of the samples No. 3, 5, 6 are presented. These loops were obtained at the extreme mode of generator, namely, at lower frequency or at the lower value of amplitude the loop was not observed.

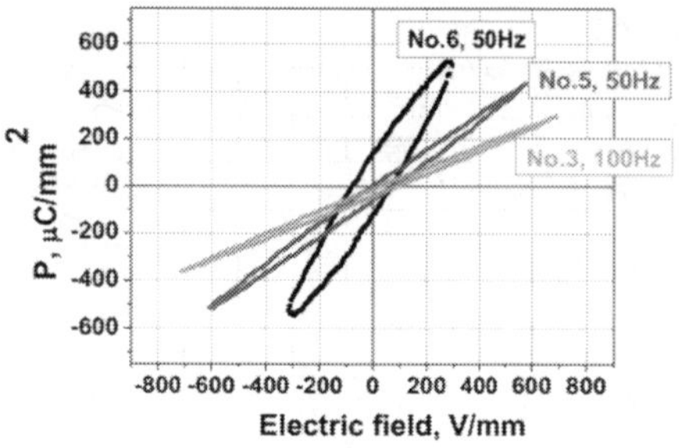

Fig.6. The dependences of polarization of the samples 3, 5, 6 on electric field.

Electric properties changed in the external magnetic field. The maximum value of polarization and coercive electric field increased in the magnetic field. For example, the change of the electric hysteresis loop of the sample No.6 in the external magnetic field is presented in Figure 7.

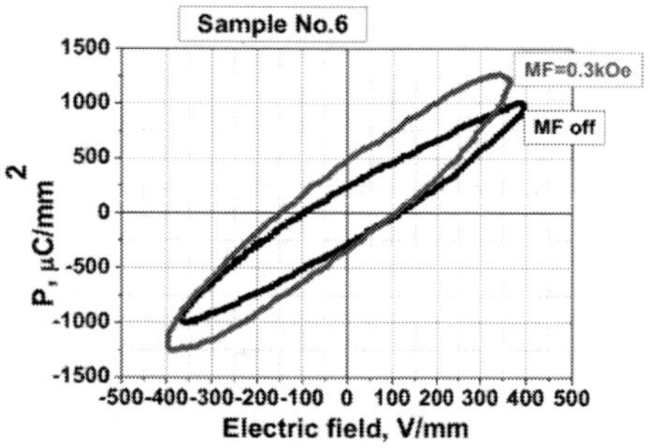

Fig.7. The dependences of polarization of the sample No.6 on the electric field in zero magnetic field (black) and in 300 Oe magnetic field (red).

The change of maximum value of polarization was about 25% for the sample No.6. Also the electric coercivity increased in magnetic field at 22 V/mm.

The coefficient of magnetoelectric transformation is the change of electric coercivity divided by value of magnetic field (α_E=ΔEc/H, V/(mm*Oe)), so the values of this coefficient were calculated for all samples (Table 2). It was not found any dependence of the coefficient α_E on the elastic, ferroelectric or ferromagnetic characteristics of materials.

Table 2. Young's modulus and the coefficient of electric transformation for the samples 1-6.

Sample No., (vol% NdFeB:PZT)	E-Young, MPa	α_E=ΔEc/H, V/(cm*Oe)
1 (20 1:6)	0.84	0.14
2 (20 1:3)	0.83	0.43
3 (20 1:1)	0.92	0.19
4 (20 2:1)	1.3	0.11
5 (30 1:1)	1.05	0.44
6 (40 1:1)	1.4	0.73

Also the anisotropy of the samples was found. Namely, the placing of the sample by different sides to the magnet leaded to different changes in the hysteresis loop. Firstly the hysteresis loop was measured for the sample without magnetic field, then the sample was placed by one side to the magnet (first direction), then the sample was turned by 180^0 and was placed by another side to the magnet (second direction). In Figure 8 the results of initial loop and loops in the magnetic field are presented.

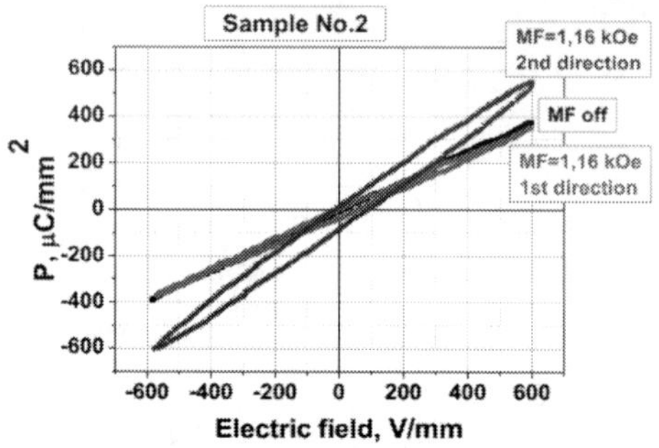

Fig.8. The dependences of polarization of the sample No.2 on electric field in zero magnetic field (black), in 1.16kOe magnetic field applied in first direction (red), in 1.16kOe magnetic field applied in second direction (blue).

If the sample was placed to the magnet by first side, the changes of electric properties were not observed. And the turning of the sample to the other side leaded to significant changes in the electric properties. This anisotropy could appear from the previous measurements in the external magnetic field [9]. This anisotropy is associated with the alignment of magnetic moments of magnetic particles. In the first case the value of magnetic field is too small, the direction on magnetic field is opposite to the direction of magnetic moments of the particles, and particles stay at their position. In the second case the particles can move in the elastic matrix, producing the internal stresses in the matrix, that can change the electric polarization of the sample.

At the smaller concentration of ferroelectric particles, the electric hysteresis loop was almost not observed (Figure 9, sample No.1). The electric hysteresis loop of the sample No.1 was not observed even in the external magnetic field, but after taking the sample out from the magnetic field, the explicit hysteresis loop appeared.

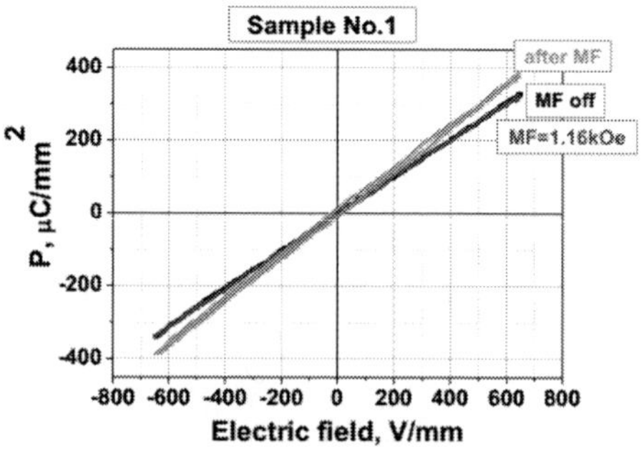

Fig.9. The dependences of polarization of the sample No.1 on electric field.

The other sample (No.5) showed the smooth decrease of the electrical parameters of the loop after the magnetic field switched off, which is presented in Figure 10(a,b). It is probably associated with magnetic relaxation properties of the samples.

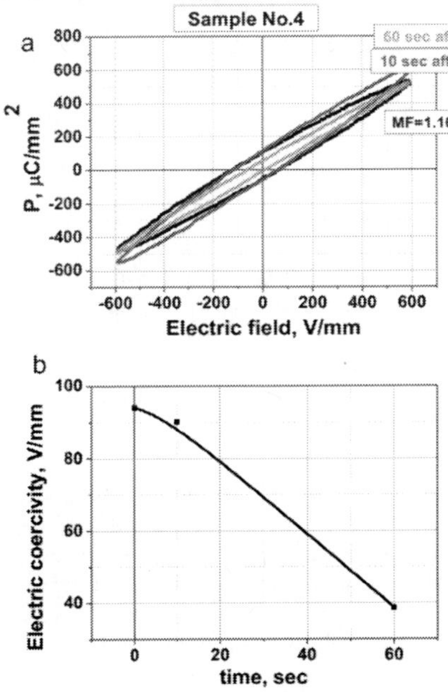

Fig.10(a,b).a) the dependences of polarization of the sample No.5 on electric field in magnetic field (black), 10 seconds after magnetic field switched off (red), 60 secondsafter magnetic field switched off (green); b) the dependence of electric coercivity of the sample No.5 on time after magnetic field switched off.

3.3 Magnetic relaxation properties with external electric field

The magnetic relaxation properties were measured in two modes: first mode was in the constant magnetic field 500 Oe, second mode was after switching magnetic field off. For example, the relaxation time dependence for the sample No.6 is presented in Figure 11 [9]. The relaxation time τ for each section was determined by approximating by the exponential function $e^{\pm t/\tau}$. The relaxation time in the first mode in the external electric field exceeds one without electric field. We proposed that the internal stresses, induced by displacements of ferroelectric particles in external electric field, prevent the rotation of ferromagnetic particles in the field direction. The relaxation times approximately coincide after the magnetic field is turned off (second mode).
Also it was found that it takes a long time to reach the constant value of magnetic moment after switching the magnetic field off. This magnetic relaxation influences the electric properties of the materials, namely, the dependence of electric coercivity on time after magnetic field is switched off (Fig.10b).

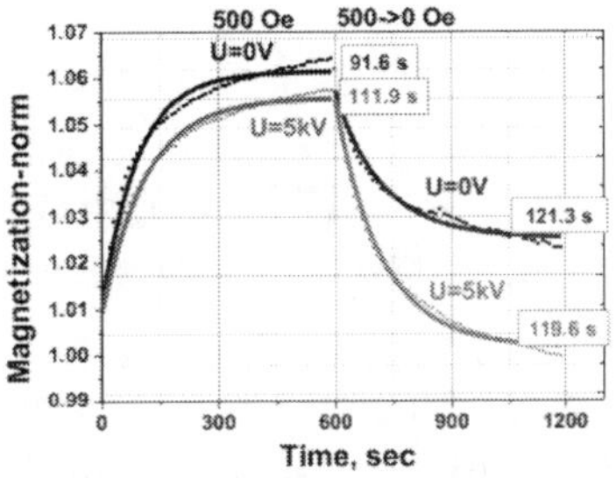

Fig.11. The dependence of normalized magnetization of the sample No.6 on time in different modes in zero and 5kV external electric voltage [9].

4 Conclusion

New composite materials with ferromagnetic, ferroelectric and ferroelastic properties were investigated in the work. It was found that ferroelectric particles influence the Young's modulus of the composite, indirectly affect the magnetic characteristics (coercivity). It was shown that materials change their electric properties in the external magnetic field. Maximum value of the coefficient of magnetoelectric transformation is 0.73 V/(cm*Oe). Magnetic relaxation properties were investigated and it was shown that magnetic relaxation influences the electric properties of the materials.

Acknowledgement

This work was supported by the Ministry of Education and Science of the Russian Federation in the framework of government assignments 3.9002.2017/6.7 and 3.4168.2017/4.6.

References

1. Y. Huang, et.al., Adv. Mat., 8344-8364 (2016)
2. R. Sabo, et.al., J. of Renewable Mat., **4**, 297 (2016)
3. Y. Yang, et.al., Mat. Sc. And Engineering R: Reports, **115**, 1-37 (2017)
4. Q. Zhu, et.al., J. Mater. Res., **29**, 657 (2014)
5. J. Zhang, et.al., Sensors and Actuators A, **214**, 149 (2014)
6. Y. Li, et.al., Smart Materials and Structures, **23 (12)**, 123001 (2014)
7. X.L. Gong, et.al., Int. J. Mod. Phys. B. **21 (28-29)**, 4875-7882 (2007)
8. A.S. Semisalova, et.al., Soft Matter, **9(47)**, 11318 (2013)
9. L.A. Makarova, et.al, IEEE Transactions on Magnetics, **53(11)**, 17-03-0349 (2017)
10. L.A. Makarova, et.al., JMMM, **440**, 10-32 (2017).

https://doi.org/10.1051/epjconf/201818507009

MFM Study of the Domain Structure of Permalloy Microparticles under Mechanical Stress

Dmitry Biziyaev[1], *Anastas* Bukharaev[1,2], *Niyaz* Nurgazizov[1,2*]

[1]Kazan E. K. Zavoisky Physical-Technical Institute, 420029 Kazan, Russia
[2]Kazan Federal University, Institute of Physics, 420008 Kazan, Russia

Abstract. The domain structure of planar permalloy (Py) microparticles was studied under mechanical stress. An array of Py particles was formed by electron beam evaporation of Py on flat and preliminarily bent glass substrates. After evaporation the substrate was unbent and the Py particles were compressed along one axis. The change of the domain structure of stressed particles in comparison with that of unstressed particles was studied. It was shown that the change of the domain structure of Py particles depends on their compression ratio.

1 Introduction

Recently the possibility of using the Villari effect (the change of magnetic properties of a body under mechanical impact) for the magnetic reversal processes of micro and nanostructures is intensively studied [1-4]. This is due to the possibility of a significant reduction in the energy required for re-writing a single bit of information due to the simultaneous use of a magnetic field and mechanical stress.

Piezoelectric substrates, on the surface of which nanoparticles are formed, are usually used to create stress. Such substrates can change their geometric sizes under the action of an electric potential for compression or tension [4,5]. Another method of creating stress in particles is the elastic bending of the substrate. The substrates can be also bent before the particles are formed and to create stresses in the particles by straightening the substrate [2,6-9].

In addition, the change in the magnetic structure of micro- and nanoparticles upon their compression can be used to detect mechanical stress in solids. In this case, the lateral resolution of this technique will be comparable with the linear sizes of the nanoparticles under study. In this work the change in the magnetic properties of permalloy (Py) particles under mechanical compression was studied.

2 Sample preparation

The studies were carried out on samples representing an array of the planar Py (Ni79%, Fe16%, Mo4%) particles (with the size of $25\times25\times0.03$ μm^3) located on a glass substrate. Before forming the particles, the glass substrate was coated with a thin copper film (~5 nm) for discharging the sample during measurements by the magnetic force microscope (MFM). The copper film and particles were fabricated by electron beam evaporation under ultrahigh vacuum conditions. The particles were deposited using a "Multiprobe P" device (Omicron). An array of identical particles was formed by sputtering through a metal grid placed on the surface of the substrate.

The optical image of obtained particles is shown in fig. 1a. The 3D-image of a single Py particle is shown in fig. 1b. Atomic force microscope (AFM) scanning of different parts of the sample made it possible to confirm the identity of all particles. The size of the used glass substrate was 12×4 mm^2, the thickness was 0.15 mm. The region occupied with Py particles was in the center of the substrate and had the sizes of about 6×4 mm.

To create stressed particles, the substrate was elastically bent by fixing in the holder before conducting evaporation. After the substrate was straightened, the stresses were created in the particles. A sample holder with the constant curvature radius (of about 70 mm) was used to create particles with the same degree of stress (fig. 2a). To create a sample with different compression ratios of particles along the long axis, an even holder was used, where a metal wire (with the diameter of 0.08 mm) was placed under the substrate perpendicular to the axis (fig. 3a). When using a wire with the diameter of about 0.1 mm (or more), the substrate was broken. This meant that the non-uniform stresses produced in the glass during the creation of particles are close to the critical values.

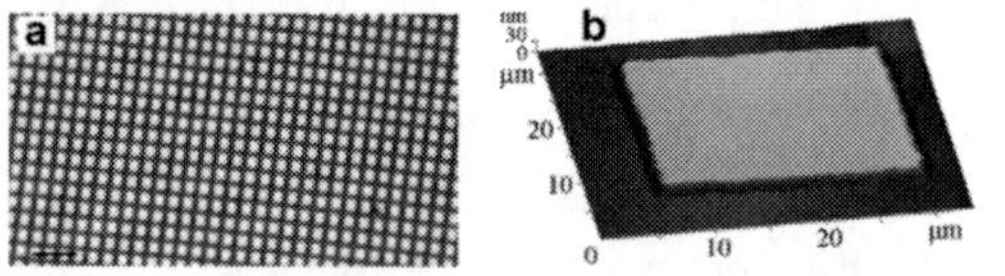

Fig. 1. Optical image of Py particles under study (a). The bar length is 0.1 mm. AFM image of a single Py particle (b).

* Corresponding author: niazn@mail.ru

3 MFM studies of the domain structure of Py particles

A Solver HV (NT MDT) device working in the MFM mode was used to study the domain structure of Py particles. The measurements were carried out in the one-pass technique. In this case, the MFM tip did not affect the magnetization of Py particles. The N18/CoCr (MikroScience) cantilevers were used. MFM scanning of the surface of the sample was carried out with a fixed step along the long axis of the sample (parallel to the compression direction of particles).

To detect the magnetization distribution in the particle, the MFM images (fig. 2b) were compared with results of computer simulation of its magnetic structure. These calculations were carried out by OOMMF [10] (fig. 2d). The following parameters were used: the exchange interaction constant of $1.3 \cdot 10^{-11}$ J/m, the unit cell size of 20 nm in the in-plane direction and 30 nm in the out-of-plane direction (equal to the thickness of the particle). The obtained magnetization distribution was used for simulation of the MFM image (fig. 2e) using the "Virtual MFM" software [11], which was compared with the experimental results. The coincidence of the model image with the experimental made it possible to make a conclusion about the correct calculation of the magnetization distribution.

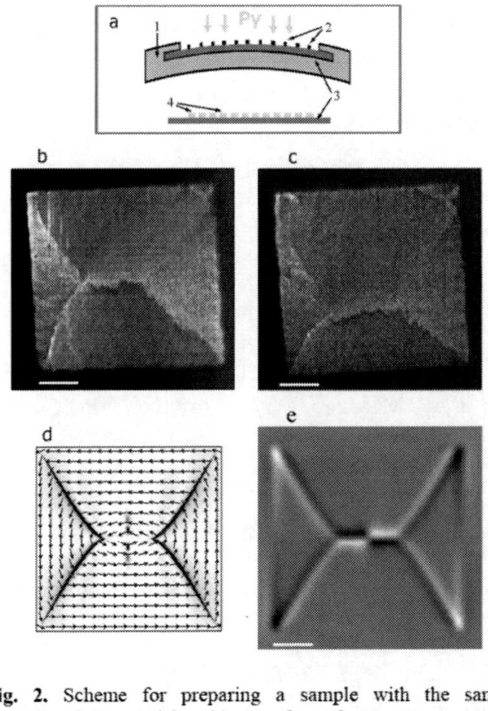

Fig. 2. Scheme for preparing a sample with the same compressed Py particles (a). Numbers denote: 1 – sample holder (with the constant curvature radius); 2 – metal grid; 3 – substrate; 4 – Py particles. MFM images of Py particles obtained in different regions of the sample (b,c). The particle magnetization simulated in OOMMF (d). The virtual MFM image calculated for such magnetization distribution (e). The bar length - 5 μm.

In order to achieve the coincidence of the model MFM images with the experimental ones (i.e., to find the magnetization distribution in the particle), two parameters were varied during the OOMMF calculation. The main varied parameter was the anisotropy coefficient (k_{eff}). The second varied parameter was the saturation magnetization value ($8.6 \div 9.6 \cdot 10^5$ A/m).

In our case, the anisotropy coefficient of the particle depends on the crystal anisotropy coefficient and the magnetoelastic anisotropy coefficient. The obtained Py particles had the polycrystalline structure therefore the anisotropy coefficient of the particle is determined by the magnetoelastic anisotropy coefficient. The anisotropy coefficient (k_{eff}) obtained by the calculations in the OOMMF program was used for the calculation of the effective anisotropy field (H_a). The following formula $H_a = 2k_{eff}/M_S$, where M_s is the saturation magnetization of the Py particle, was used. A similar approach was used in [3] to determine the magnetization distribution in Ni particles under mechanical stress.

MFM images obtained on uncompressed samples and in certain regions of a sample with differently compressed particles show that the uncompressed Py particles have the characteristic four-domain structure (fig. 3c). All domains have the same sizes and shape caused by the shape anisotropy. A small spread in the shape of the observed domains is due to the nonideal square shape of the prepared Py particles.

The change of domain structure was observed on samples with equally compressed Py particles (fig. 2b,c). Two domains with the magnetization direction parallel to the compression direction increase in size (fig. 2d). Domains with the magnetization direction perpendicular to the compression direction decrease in size. This is due to the negative sign of the Py magnetostrictive constant.

A characteristic bridge between two large domains was observed on MFM images. Its length is proportional to the ratio between the areas of domains with the magnetization direction parallel to the compression axis and those with the magnetization direction perpendicular to the compression axis. This bridge length can be easily found from the MFM image of the particle and used for the comparison of the compression ratio of particles. The bridge length may be also used for the comparison of the experimental and model MFM images (fig. 2b,c and e).

For samples with identically compressed particles the same MFM images were observed over the whole sample. Obtained MFM images confirmed that in such samples all particles undergo the same stress. The effective anisotropy field of the Py particles due to uniform bending of the substrate during sputtering was 0.94 mT.

Three typical states of MFM images were observed on the samples with the different compression ratios of particles. In the regions near clumps (fig. 3a) the bridge on MFM images is perpendicular to the compression direction (fig. 3b). This indicates that the particles in this region are stretched. In the middle of the sample the bridge on the MFM image has the maximal length and is parallel to the compression direction (fig. 3e). It means that the particles in the center of the sample are maximally compressed. Since the Py particles are stretched at the edges of the sample and compressed in the center, they are unstressed between these regions. The effective anisotropy field of the Py particles varies to 3.12 mT.

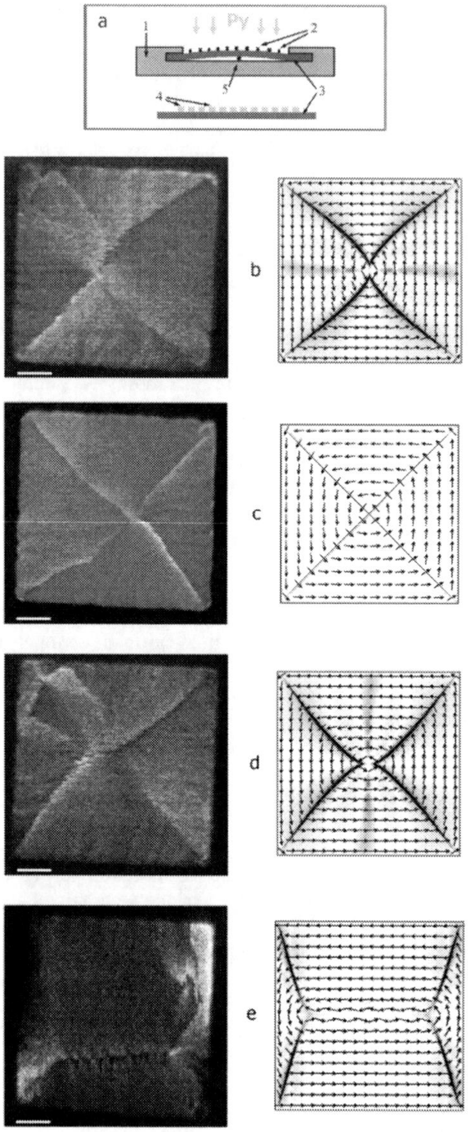

Fig. 3. Scheme of preparing of the sample with non-uniformly compressed Py particles (a). Numbers denote: 1 – sample holder (with the constant curvature radius); 2 – metal grid; 3 – substrate; 4 – Py particles; 5 – Me wire for the creation of the compression gradient of particles. MFM images of Py particles with the corresponding magnetization simulated in OOMMF obtained at different distances from the sample edge: 0.3 mm - (b); 0.9 mm - (c); 1.3 mm - (d); 2.9 mm - (e). The bar length is 5 μm.

According to the obtained MFM images, the "cross-tie" structure was observed in highly compressed particles (fig. 3e). Such structure is formed at computer simulation in the case of the increase in the saturation magnetization value (up to $9.6 \cdot 10^5$ A/m). Ferromagnetic resonance studies carried out by us and other authors [5] showed that the saturation magnetization of Py particles may change under mechanical stress. It is possible that the "cross-tie" structure is formed in uniformly compressed Py particles resulting in the disappearance of the magnetic vortex in the centre of the particle (fig. 2c). At the same time the saturation magnetization almost does not affect the bridge length.

4 Conclusions

It was shown that uncompressed Py particles have the classical four-domain structure with domains of the same size. The uniaxial compression of particles results in the increase in the size of domains with the magnetization parallel to the compression direction (this is due to the negative Py magnetostrictive constant). The effective anisotropy field of Py particles with the same compression ratio was 0.94 mT.

It was established that the anisotropy effective field of the Py particles on non-uniformly curved samples changes to 3.12 mT. The particles are stretched at the edge of the sample and compressed in the center of the sample. The bridge length between domains was proportional to the compression ratio.

The work is supported by the Russian Foundation for Basic Research 17-08-00915.

References

1. Y. Liu, Q. Zhan, B. Wang, H. Li et al., IEEE Transactions on Magnetics, **51**, 2501404 (2015)
2. G. Dai, Q. Zhan, H. Yang et al., J. Appl. Phys. **114**, 173913 (2013)
3. S. Finizio, M. Foerster, M. Buzzi et al., Phys. Rev. Applied, **1**, 021001 (2014)
4. T. Wu, A. Bur, J.L. Hockel et al., IEEE Magnetics Letters, **2**, 6000104 (2011).
5. A.I. Morozov, Physics of the Solid State **56,** 865–872 (2014)
6. A. Bur, T. Wu, J. Hockel et al., J. Appl. Phys., **109**, 123903 (2011)
7. Z. Hao, L. Yuan-Yuan, Y. Mei-Yin et al., *Chinese* Phys. B, **24**, 07750 (2015).
8. A.P. Chuklanov, N.I. Nurgazizov, D.A. Bizyaev et al., Journal of Physics: C 7**14,** 012006 (2016)
9. B.A. Belyaev, A.V. Izotov, Physics of the Solid State, **49**, 1731–1739 (2007)
10. M.J. Donahue, D.G. Porter, *OOMMF*, (http://math.nist.gov/oommf/).
11. D.V. Ovchinnikov, A.A. Bukharaev, *Technical Physics,* **46**, 1014–1019 (2001).

Temperature Mössbauer study of the spatial spin-modulated structure in the multiferroic $BiFeO_3$

Vyacheslav Rusakov[1,*], *Vyacheslav* Pokatilov[2], *Alexander* Sigov[2], *Mikhail* Matsnev[1], and *Alexander* Pyatakov[1]

[1]Lomonosov Moscow State University, 119991 Moscow, Russia
[2]Moscow Technological University (MIREA), 119454 Moscow, Russia

Abstract. ^{57}Fe Mössbauer detailed study of the spatial spin-modulated structure of the multiferroic $BiFeO_3$ was carried out in a wide temperature range including the temperature of magnetic phase transition. The Mossbauer spectra have been analysed by fitting in terms of the anharmonic spin cycloid mode. It is established that at temperatures below ~330 K a magnetic anisotropy of the "easy axis" type is realized and above is the magnetic anisotropy of the "easy plane" type. An explanation for the change in the type of magnetic anisotropy is proposed, based on taking into account the different temperature dependences of the two contributions to the effective uniaxial magnetic anisotropy constant: a crystal anisotropy of net antiferromagnet and a weak ferromagnetism.

1 Introduction

The bismuth ferrite $BiFeO_3$ is known to possess multiferroic properties, being both antiferromagnetic and ferroelectric. High antiferromagnetic Neel temperature (643 K) and ferroelectric Curie temperature (1103 K) of the ferrite are of great interest in many fields of science and technology [1]. Bismuth ferrite has a rhombohedral distorted perovskite crystal structure (space group R3c) [2] in which Fe^{3+} and Bi^{3+} cations are displaced along the hexagonal [001]-axis, leading to a spontaneous polarization in this direction. In accordance with neutron diffraction investigations [3], $BiFeO_3$ has a spatial spin-modulated structure (SSMS). It was established that there existed a magnetic cycloid spiral with a long period of 620 Å incommensurate with the lattice parameter, in which G-type antiferromagnetic ordered moments of Fe atoms rotate in the direction $[110]_{hex}$ of wave propagation in the plane containing hexagonal symmetry axis of the rhombohedral ferrite cell. The existence of SSMS in BiFeO3 was theoretically explained in [4, 5], where it was shown that spatial dependence of the angle ϑ between antiferromagnetism vector and symmetry axis on the distance x along the propagation direction, was described by the elliptic Jacobi sn-function.

The temperature-dependent evolution of the magnetic structure in BiFeO3 is an important subject for investigation with several currently unresolved issues. ^{57}Fe Mössbauer spectroscopy was used in studying $BiFeO_3$ in several papers [6 – 11] that were devoted to analysis of the magnetic structure features in terms of the cycloid type incommensurate anharmonic SSMS model. In analyzing the Mössbauer spectra, the dependence of the angle ϑ on the coordinate x corresponded to the magnetic anisotropy of the "easy plane" type, was used in [6]. Thus the values of the anharmonicity parameter $m = 0.5$ at RT and $m = 0.6$ at 90 K were obtained [6]. It was shown [7] that at low temperatures a magnetic anisotropy of the "easy axis" type is realized. In this case the anharmonicity parameter m obtained from the Mössbauer spectrum ($m = 0.26 \pm 0.06$ at 4.85 K) was close to the value obtained from the NMR spectrum ($m = 0.25 \pm 0.07$ at 4.2 K). In subsequent papers [8 – 11] it was assumed that magnetic anisotropy of the "easy axis" type is realized in the entire temperature range of SSMS existence. As a result of an analysis of the Mossbauer spectra, it was shown on this assumption that with increasing temperature, the anharmonicity parameter m remained virtually constant up to $T \sim 150$ K and then decreased, tending to zero at 350 – 400 K and until the system reaches the harmonic state ($m = 0$). It was noticed in [10, 11] that with increasing temperature in the region $T > 350$ K when the value of the anharmonicity parameter equal zero, description of the spectrum deteriorates and the Chi-square functional increases.

The purpose of this work was the detailed temperature Mössbauer study of the feature of the spatial spin-modulated structure of multiferroic $BiFeO_3$.

2 Methods

2.1 Synthesis

Polycrystalline sample of $BiFeO_3$ with relative content stable isotope ^{57}Fe at 10% was prepared by the solid ceramic technology method. The mixture of oxide powders of the components compound in the proper proportions pressed into tablets, which were annealed for one day at a temperature equal to 700-830°C in air (with

* Corresponding author: rusakov@phys.msu.ru

an intermediate triple milling and compression into tablets). At $T = 830°C$, final annealing was performed and then rapidly cooling the sample in air.

Phase composition of samples was characterized by X-ray diffraction using Panalytical Empyrean powder diffractometer in Bragg-Brentano geometry (step size – 0.026°, scan step time – 200 s) using nickel-filtered Cu(Kα) ($\lambda = 1.5406$Å) radiation. As shown X-ray measurements, the obtained single-phase sample $BiFeO_3$ have rhombohedral structure with parameters: $a = 3.963$ Å and $\alpha = 89.43°$, space group R3c.

2.2 ^{57}Fe Mössbauer spectroscopy

Mössbauer studies were carried out using a spectrometer operating in constant acceleration mode with a triangular shape changes in Doppler velocity of the source relative to the absorber. The source is ^{57}Co performed in the matrix Rh. Mössbauer spectrometer calibration was performed at room temperature using a standard absorber α-Fe. Measurements were carried out at temperatures in the range of 4.85 – 700 K using a closed cycle helium cryostat SHI-850-5, compact cryogenic complex produced by VNIIFTRI and Mössbauer furnace MBF-1100-TR.

For processing and analysis of the Mössbauer spectra we used methods of recovery hyperfine parameter distributions and fitting in the SSMS cycloid type model implemented in the program SpectrRelax [12, 13].

3 Result and discussions

3.1 Study of the spatial spin-modulated structure

According to papers [4, 5] the spatial dependence of angle $\vartheta(x)$ between the antiferromagnetism vector and symmetry axis in bismuth ferrite BiFeO3 is defined by the two possible solutions, depending on the sign of magnetic anisotropy constant K_u:

$$\cos\vartheta(x) = \mathrm{sn}\left(\frac{4K(m)}{\lambda}x, m\right) \text{ at } K_u > 0 \text{ ("easy axis")}, \quad (1)$$

$$\sin\vartheta(x) = \mathrm{sn}\left(\frac{4K(m)}{\lambda}x, m\right) \text{ at } K_u < 0 \text{ ("easy plane")}, \quad (2)$$

where x is the coordinate along the direction of wave propagation, λ is the anharmonic spin modulation wavelength, $0 \le m < 1$ is the (anharmonicity) parameter of Jacobi's elliptic function, and $K(m)$ is a complete elliptic integral of the first kind.

The results of the analysis of Mössbauer spectra of nuclei ^{57}Fe in $BiFeO_3$ at $T < T_N$ in the context of the cycloid type SSMS model are presented in Fig. 1. One can see a model envelope curve fitting well experimental Mössbauer spectrum (Fig. 1a). In Fig. 1b, the hyperfine field distribution $p(H_n)$, calculated in context of the cycloid type SSMS and the distribution obtained as a result of recovery are provided for comparison.

As we can see (Fig. 1b), the distributions $p(H_n)$ contain two distinct local maxima with different intensities with depressions between them. The partial spectra related to these local maxima have almost the same shifts, but different in the magnitude and sign of the quadrupole shift. For the multiferroic $BiFeO_3$ the maxima of the intensity of the distribution $p(H_n)$ with large field values correspond to the orientations of the magnetic moment of the iron atoms parallel to the axis of symmetry of the crystal ($H_\parallel$), and the maxima with lower field values – perpendicular to this axis ($H_\perp$) [7, 14].

The distribution profiles of $p(H_n)$ allow us to judge not only the existence of SSMS in the studied compound, but also on the degree of anharmonicity of SSMS. When $m \to 0$, the SSMS becomes harmonic (sinusoidal). This means that the $p(H_n)$ distribution will have a symmetrical shape with equal height peaks at the edges and a minimum in the middle. If $m \to 1$, then the anharmonicity of the SSMS increases, and the difference in the intensities of the maxima increases.

A comparison of the distributions $p(H_n)$ shown in Fig. 1, indicates a very important experimental fact, namely, as the temperature increases, the ratio of the intensities of the maxima of the distribution $p(H_n)$ changes. At temperatures of 80, and 300 K, the left maximum in the distribution $p(H_n)$ is of lower intensity, and the right-hand maximum is of higher intensity, while for the spectra measured at temperatures of 350 and 500 K, the right-hand maximum of $p(H_n)$ already has a smaller, and the left one – most intensity. This means that with increasing temperature the relative number of Fe atoms that have magnetic moments oriented perpendicular to the axis of symmetry of the crystal, increases, and the relative number of Fe atoms with the orientation of the moments along the axis of symmetry decreases.

The fitting of the spectra showed that for the temperature range 4.85–300 K the best description of the experimental spectrum was achieved using the equation (1), which corresponds to a uniaxial magnetic anisotropy ($K_u > 0$), and in the region 350 – 585 K – of the equation (2), which corresponds to a magnetic anisotropy of the easy plane type ($K_u < 0$). A change in the sign of the magnetic anisotropy constant with an increase in temperature means a change in the type of magnetic anisotropy with "easy axis" on the "easy plane".

The values of the anharmonicity parameters m of the cycloid type SSMS, obtained as a result of fitting the Mössbauer spectra of $BiFeO_3$, measured in the temperature range 4.85–585 K, are shown in Fig. 2. With increasing temperature in the region $T < 330$ K, in which a magnetic anisotropy of the "easy axis" type exists, the parameter m decreases, the ratio of the intensities of the maxima in the distribution $p(H_n)$ and SSMS becomes more harmonious. At $T \to 330$ K, the anharmonicity parameter $m \to 0$, which corresponds to a harmonic SSMS. With a further increase of the temperature in the region $T > 330$ K, in which there is a magnetic anisotropy of the "easy plane" type, the parameter m increases and the SSMS becomes more anharmonic.

We estimate the values of the magnetic anisotropy constant K_u as a function of temperature for the multiferroic $BiFeO_3$, using the experimental values of the parameter m obtained by us.

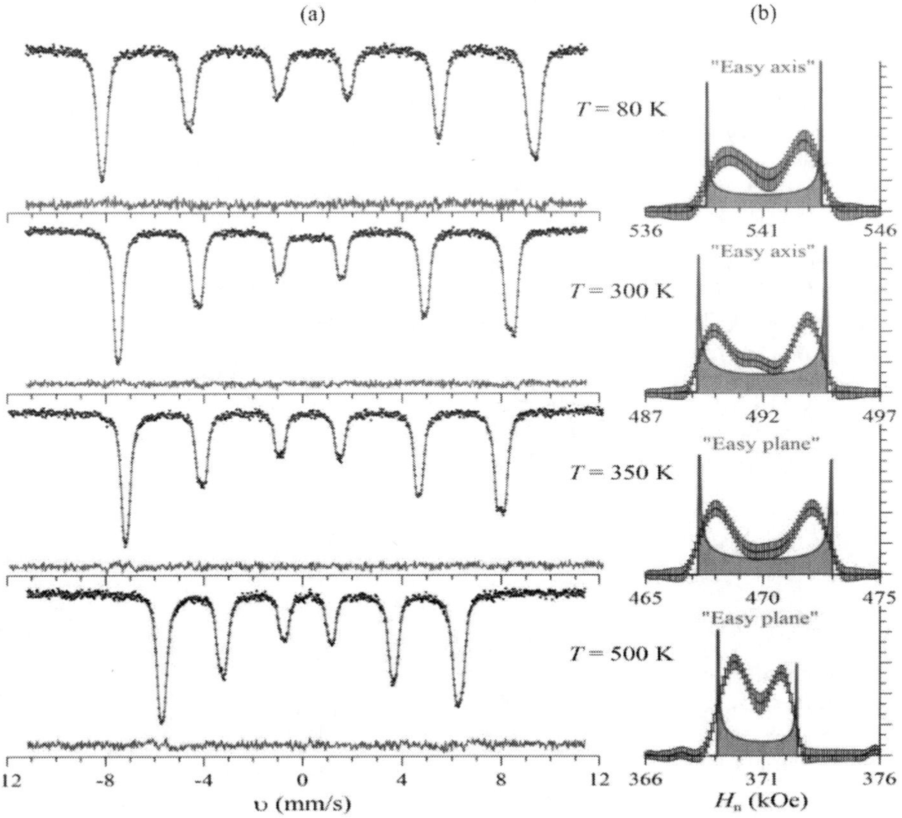

Fig. 1. ^{57}Fe Mössbauer spectra (a) and results of its processing using the method of the reconstruction of the hyperfine parameter distributions and within the model of the cycloidal type SSMS for $Bi^{57}Fe_{0.10}Fe_{0.90}O_3$ at various temperatures. (b) Distribution of the hyperfine magnetic field $p(H_n)$ (points with standard deviation are the result of distribution reconstruction; solid line with shaded area is $p(H$n) distribution calculated in the terms of cycloid type SSMS model).

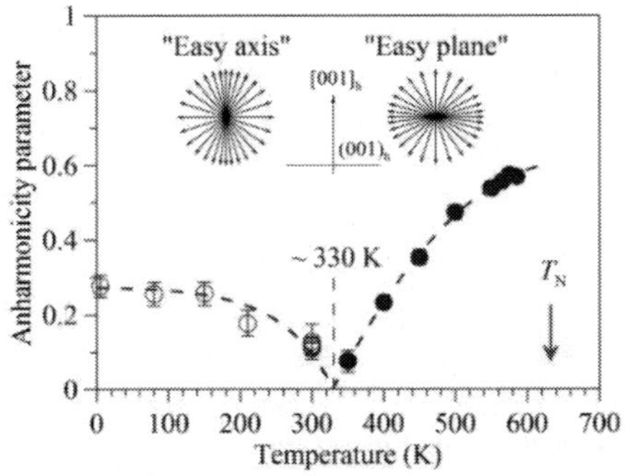

Fig. 2. The anharmonicity parameter m of the cycloid type SSMS as a function of temperature. The insets schematically show the spin spatial variation in the cycloids with "easy axis" and "easy plane" anisotropy.

For this purpose, we use the relations connecting the magnetic anisotropy constant K_u, the anharmonicity parameter m, the SSMS wavelength λ, the Néel temperature T_N and the distance between the nearest iron atoms $a_{Fe\text{-}Fe}$ [14].

When estimating the magnetic anisotropy constant K_u the length of the cycloid in $BiFeO_3$ was assumed to be $\lambda = 620$ Å (according to [3]) and only slightly change with temperature [15], the value of T_N was taken equal to 633.1 K [10], the distance $a_{Fe\text{-}Fe}$ was set to 4 Å [14].

As a result, it was found that at T = 4.85 K the anisotropy constant is $K_u \approx 1.0\cdot10^6$ erg/cm^3, at $T \to$ 330 K $K_u \to 0$, then changes sign and increases in absolute value to $3.0\cdot10^6$ erg/cm^3 at T = 580 K.

3.2 Mechanism of changing the type of magnetic anisotropy

The temperature induced switching of the magnetic anisotropy type from easy plane to easy axis one observed in the experiment can be explained by the competition of several contributions to the effective anisotropy of bismuth ferrite:

$$K_{u\_eff} = K_u - \chi_\perp \frac{H_{DM}^2}{2} - \chi_\perp \frac{H^2}{2} \qquad (3)$$

where K_u is the anisotropy constant of antiferromagnetic system, the second term is the contribution of weak ferromagnetism, the third one appears in external magnetic field H; $\chi_\perp = 4.7\cdot10^{-5}$ is magnetic susceptibility in the direction perpendicular to the moments in antiferromagnetic lattices, and H_{DM} is Dzyaloshinskii-

Moriya field responsible for the weak ferromagnetism. At low temperatures ($T = 4.2$ K) the constant values are: $K_u = 6.6 \cdot 10^5$ erg/cm^3, $H_{DM} = 120$ kOe [16] and the resultant anisotropy in the absence of external magnetic field is the easy axis one.

The high magnetic field experiments [16-18] show that at some critical field the spin cycloid state is suppressed and the phase transition occurs to the homogeneous magnetic state with the in-plane orientation of the magnetic moments.

Noteworthy that this critical field strongly depends on the temperature [18] decreasing from 18 Tesla to several Tesla in the vicinity of Neel temperature. Since the perpendicular susceptibility $\chi_\perp$ is temperature independent this fact denotes the different temperature dependence of the first and second terms in (3). At some temperature, the second term in (3) can outbalance the first term and the magnetic anisotropy becomes the easy-plane one even in the absence of the external magnetic field.

We can estimate this critical temperature using the graph from [18] and the condition of spin cycloid suppression [16]:

$$K_{u\_eff} < -\Delta W_{cycloid} \quad (4)$$

where $\Delta W_{cycloid}$ is the energy gain due to the cycloid formation.

At critical temperature, the first two terms in (3) compensate each other and the phase transition condition (4) takes the form:

$$\chi_\perp \left.\frac{H_c^2}{2}\right|_{T=T_c} = \Delta W_{cycloid}. \quad (5)$$

$\Delta W_{cycloid}$ can be estimated as ~ 500 kerg/cm^3 [19] that in accordance with (5) corresponds to the H_c ~ 15 Tesla. Using the graph in Fig. 3 [18] we obtain the corresponding critical temperature about 350 K that is close to the value 330 K for pure bismuth ferrite and lies within the range of critical temperatures for substituted compounds (300-400 K).

4 Conclusions

The temperature-dependent evolution of the spatial spin-modulated structure in multiferroic $BiFeO_3$ was detailed study by ^{57}Fe Mössbauer spectroscopy. The Mossbauer spectra have been analyzed by recovery of hyperfine parameter distributions and fitting in the SSMS cycloid type model methods. It is established that at temperature ~330 K, the type of magnetic anisotropy from the "easy axis" to the "easy plane" changes. With increasing temperature in the region $T < 330$ K, in which there is a magnetic anisotropy of the "easy axis" type, the anharmonicity parameter m decreases, and SSMS becomes more harmonious. At $T \rightarrow 330$ K, the anharmonicity parameter $m \rightarrow 0$, which corresponds to a harmonic SSMS. With a further increase of the temperature in the region $T > 330$ K, in which there exists a magnetic anisotropy of the "easy plane" type, the parameter m increases and the SSMS becomes more anharmonic. The magnetic anisotropy constant K_u as a function of temperature has been derived from the observed anharmonicity.

References

1. A.P. Pyatakov, and A.K. Zvezdin, Phys.-Usp., **55**, 557 (2012).
2. F. Kubel and H. Schmid, Acta Cryst. B, **46**, 698 (1990).
3. I. Sosnowska, T. Peterlin-Neumaier, and E. SteicheleJ, Phys. C: Solid State Phys., **15** 4835-4846 (1982).
4. I. Sosnowska, A.K. Zvezdin. JMMM, **140-144**, 167 (1995).
5. M.-M. Tehranchi, N.F. Kubrakov, and A.K. Zvezdin. Ferroelectrics, **204**, 181 (1997).
6. A. Palewicz, T. Szumiatab, R. Przeniosłoa, I. Sosnowskaa, I. Margiolakic. Solid State Communications, **140**, 359 (2006).
7. V.S. Rusakov, V.S. Pokatilov, A.S. Sigov, M.E. Matsnev, and T.V. Gubaidulina. JETP Letters, **100**, 463 (2014).
8. J. Landers, S. Salamon, M. Escobar Castillo, D. C. Lupascu, and H. Wende, Nano Lett., **14**, 6061 (2014).
9. A. Sobolev, I. Presniakov, V. Rusakov, A.Belik, M. Matsnev, D. Gorchakov, and I. Glazkova. AIP Conference Proceedings, **1622**, 104, (2014).
10. V. Rusakov, V. Pokatilov, A. Sigov, M. Matsnev, and T. Gubaidulina. Journal of Materials Science and Engineering B, **4,** 302 (2014).
11. V.S. Rusakov, V.S. Pokatilov, A.S. Sigov, M.E. Matsnev, and T.V. Gubaidulina. Bulletin of the Russian Academy of Sciences. Physics, **79**, 708 (2015).
12. M.E. Matsnev and V.S. Rusakov. AIP Conference Proceedings, **1489**, 178 (2012).
13. M.E. Matsnev and V.S. Rusakov. AIP Conference Proceedings, **1622**, 40 (2014).
14. A.V. Zalesskii, A.K. Zvezdin, A.A. Frolov, and A.A. Bush, JETP Lett. **71** (11), 465 (2000).
15. R. Przenioslo, A. Palewicz, M. Regulski, I. Sosnowska, R.M. Ibberson, R.S. Knight. J. Phys.: Condens. Matter **18**, 2069 (2006).
16. B. Ruette, S. Zvyagin, A. Pyatakov, et al. Physical Review B, **69**, 64114 (2004).
17. Y.F. Popov, A.K. Zvezdin, G.P. Vorob'ev, et al. JETP Letters, **57**, 69 (1993).
18. M. Tokunaga, M. Azuma, Y. Shimakawa. Journal of the Physical Society of Japan, **79**, 64713 (2010).
19. D. Sando, A. Agbelele, D. Rahmedov, et al. Nature Materials, **12**, 641 (2013).

This work was supported by the RFBR grant No. 14-02-01109a and No. 17-02-00911.

Isolation of proximity-induced triplet pairing channel in a superconductor/ferromagnet spin valve

Andrey Kamashev[1,*], *Aidar* Validov[1], *Nadir* Garif'yanov[1], *Yakov* Fominov[2,3], *Pavel* Leksin[4], *Joachim* Schumann[4], *Jürgen* Thomas[4], *Vladislav* Kataev[4], *Bernd* Büchner[4,5], and *Ilgiz* Garifullin[1]

[1]*Zavoisky Physical-Technical Institute, Russian Academy of Sciences, 420029 Kazan, Russia*
[2]*L.D. Landau Institute for Theoretical Physics, Russian Academy of Sciences, Chernogolovka, Russia*
[3]*Moscow Institute of Physics and Technology, 141700 Dolgoprudny, Russia*
[4]*Leibniz Institute for Solid State and Materials Research IFW Dresden, D-01069 Dresden, Germany*
[5]*Institute for Solid State Physics, Technical University Dresden, D-01062 Dresden, Germany*

Abstract. In the present work we have studied the proximity-induced superconducting triplet pairing in CoO_x/Py1/Cu/Py2/Cu/Pb spin-valve structure (where Py = $Ni_{0.81}Fe_{0.19}$). For CoO_x(3 nm)/Py(3 nm)/Cu(4 nm)/Py(0.6 nm)/Cu(2 nm)/Pb(70 nm) we have studied the dependence of the T_c on the angle α between the direction of the cooling field and the external field both applied in the plane of the sample. We obtained that the T_c does not change monotonically with the angle but passes through a minimum. To observe an "isolated" triplet spin-valve effect we exploited the oscillatory feature of the magnitude of the ordinary spin-valve effect ΔT_c in the dependence of the Py2-layer thickness d_{Py2}. We determined the value of d_{Py2} at which ΔT_c caused by the ordinary spin-valve effect is suppressed. This means that the difference in the T_c between the antiparallel and parallel mutual orientation of magnetizations of the Py1 and Py2 layers is zero. For such a sample a "pure" triplet spin-valve effect which causes the minimum in T_c at the orthogonal configuration of magnetizations has been observed.

1 Introduction

The superconducting spin-valve effect consists in the fact that the degrees of suppression of superconductivity in thin-film multilayer structures F1/F2/S or F1/S/F2 are different for the parallel (P) and antiparallel (AP) relative orientations of the magnetizations of the F1 and F2 ferromagnetic layers [1]. The operation of the superconducting spin valve based on the superconductor/ferromagnet (S/F) proximity effect provides a key to the understanding of many fundamental aspects of the relationship between superconductivity and magnetism. Spin valves have good prospects for use as passive elements of superconducting spintronic devices and should operate with a change of the direction of a weak external magnetic field. The manifestation of this effect in S/F systems with a good contact between the metallic ferromagnetic and superconducting layers prepared from ordinary metals and conventional ferromagnets was confirmed in a large number of experimental studies (see, for example, recent reviews [2–4] and references therein). Recent theoretical studies (see, for example,reviews [2–8]) have predicted the generation of a long-range triplet component of a superconducting condensate in S/F-structures. As follows from the theory proposed by Fominov et al. [9], the presence of a minimum of the superconducting transition temperature T_c in the vicinity of the orthogonal configuration of the magnetizations clearly indicates the generation of the long-range triplet component. We have experimentally confirmed the existence of the long-range triplet component for the CoO_x/Fe1/Cu/Fe2/Pb superconducting spin-valve structure [10]. The observed angular dependence of the superconducting transition temperature T_c is caused by a combination of the singlet and triplet components of the superconducting condensate. The existence of the triplet contribution to the superconducting spin-valve effect has also been shown in [11–15]. An important question for basic and applied research in the field of spintronics is whether there is the possibility of observing the "isolated" triplet contribution to the superconducting spin-valve effect. At first glance, this seems to be impossible, because the long-range triplet components arise from the singlet components and cannot exist without them. However, in this study, we have experimentally observed the "isolated" triplet effect caused by the oscillatory behavior of the dependence $\Delta T_c(d_{F2})$, which effectively suppresses the conventional superconducting spin-valve effect with an appropriate choice of the thickness of the F2 d_{F2} layer.

2 Samples

To investigate the proximity-induced triplet pairing we measured the magnitude of the spin-valve effect in

*e-mail: kamandi@mail.ru

the dependence of the Py2-layer thickness for the CoO_x/Py1/Cu/Py2/Cu/Pb multilayer grown on the MgO (001) substrate. Here Py denotes permalloy $Ni_{0.81}Fe_{0.19}$. Optimization of the preparation conditions of the samples and their characterization are described in [16]. The optimal thickness of the Pb layer d_{Pb} = 70 nm was determined from the $T_c(d_{Pb})$ curve measured at a constant d_{Py1} = 5 nm, which is much larger than the penetration depth ξ_h of Cooper pairs into ferromagnetic Py. Basing on our data on $T_c(d_{Py})$ at fixed d_{Pb} we estimate this value as $\xi_h \sim 1.1$ nm.

Earlier we revealed that the F1-layer thickness at a fixed d_{F2} does not significantly influence ΔT_c for the sample set CoO_x(2.5 nm)/Fe1(d_{Fe1})/Cu(4 nm)/Fe2(d_{Fe2})/Cu(1.2 nm)/Pb(60 nm) and that a thin Cu(1.2 nm) interlayer between F2- and S- layers is completely transparent for the Cooper pairs [17].

3 Experimental results

We investigated the transport properties of samples. We analyzed the dependence of the superconducting transition temperature T_c on the angle α between the magnetizations of the Py1 and Py2 ferromagnetic layers in the magnetic field H_0 = +100 Oe applied in the plane of the sample. The most interesting results on the dependence $T_c(\alpha)$ were obtained for the structures CoO_x/Py(3 nm)/Cu(4 nm)/Py(0.6 nm)/Cu(2 nm)/Pb(70 nm) (sample 1) and CoO_x/Py(3 nm)/Cu(4 nm)/Py(3 nm)/Cu(2 nm)/Pb(70 nm) (sample 2) (see Fig 1). As can be seen from Fig. 1, the angular dependences of the superconducting transition temperatures T_c of these samples have a nonmonotonic behavior. The dependence $T_c(\alpha)$ passes through a minimum in the range between the parallel ($\alpha = 0^o$) and antiparallel ($\alpha = 180^o$) orientations in the vicinity of the orthogonal configuration of the magnetizations. For sample 1, the magnitude of the superconducting spin-valve effect ΔT_c exceeds the width of the superconducting transition with a change in the relative orientation of the magnetizations from the antiparallel to orthogonal orientation. Consequently, in this structure, there is the possibility of the complete turn-on/turn-off switching of the superconducting current. In this case, we observed the complete turn-on/turn-off switching of the superconducting current due to a combination of the conventional and triplet spin-valve effects.

In the samples under investigation, the long-range triplet component of the superconducting condensate contributes to the superconducting spin-valve effect. However, the triplet contribution to the spin-valve effect coexists with the contribution from the conventional component of the condensate. As was shown previously, the amplitude of the conventional superconducting spin-valve effect can be suppressed to zero for a particular thickness of the second ferromagnet layer due to the oscillatory behavior of the dependence $T_c(d_{F2})$ [18, 19]. These oscillations are caused by the interference of the pair wave function at the F2/S interface with the wave function incident on it and reflected from the F1/F2 interface.

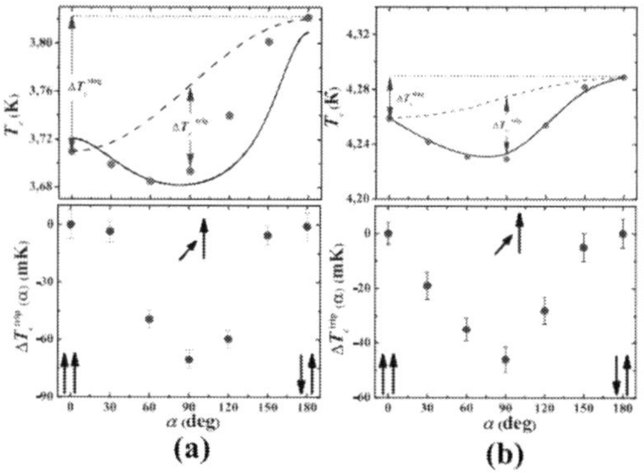

Figure 1. Dependences of the superconducting transition temperatures on the angle between the magnetizations of the ferromagnetic layers in samples (a) 1 and (b) 2. The upper panels show the angular dependences of the temperature T_c measured in the external magnetic field H_0 = +100 Oe (circles). The dashed lines are the reference curves (see formula (1)). The solid lines are the theoretical curves calculated according to the theory proposed by Fominov and his colleagues (see [10]). The lower panels show the angular dependences of the quantity ΔT_c^{trip} , i.e., the difference between the actual value of the temperature T_c and the reference curve for the corresponding samples.

For the CoO_x/Fe1/Cu/Fe2/Cu/Pb structure, the superconducting spin-valve effect ΔT_c vanishes in the range of thicknesses $d_{Fe2} \approx 0.8$–1.0 nm [17, 20]. From the analysis of the data on the dependence $T_c(d_{Py})$, it was concluded that the depth of penetration of the Cooper pairs into the permalloy layer is two times greater than the value obtained for iron. This means that the equality $\Delta T_c = 0$ for the samples with permalloy should be achieved for layer thicknesses in the range $d_{Py2} \approx 1.5$–1.7 nm. Indeed, for the CoO_x/Py(3 nm)/Cu(4 nm)/Py(1.7 nm)/Cu(2 nm)/Pb(70 nm) structure (sample 3), we observed an isolated triplet contribution to the superconducting spin-valve effect (see Fig. 2). The angular dependence $T_c(\alpha)$ for this sample has the form of a curve in which the critical temperatures T_c at $\alpha = 0^o$ and $\alpha = 180^o$ coincide with each other (the conventional superconducting spin-valve effect $\Delta T_c = \Delta T_c^{AP}$ - $\Delta T_c^{AP} = 0$). At the same time, when the configuration of the magnetizations of the Py1 and Py2 layers is noncollinear, the dependence $T_c(\alpha)$ has a minimum. A comparison of these data with the theory for the same parameters as for the description of the data obtained for the samples in Fig. 1 demonstrated a satisfactory agreement (see the solid line in Fig. 2). The only difference was in the values of d_{Py2}, which was taken to be equal to 1.7 nm. Thus, it was demonstrated that there is the possibility of the isolation (separate observation) of the triplet contribution to the superconducting spin-valve effect. The creation of an S/F heterostructure, where the contribution from the singlet component to the superconducting spin-valve effect is ineffective in terms of the influence on the critical temperature T_c due to the interference of pair wave functions, is an interesting problem.

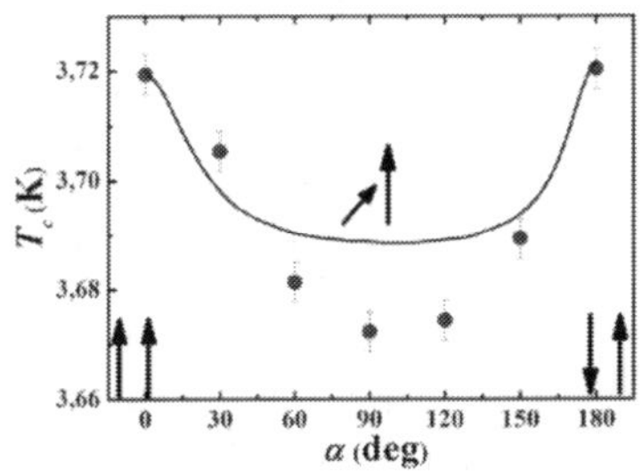

Figure 2. Sample of the CoO_x/Py(3 nm)/Cu(4 nm)/Py(3 nm)/Cu(2 nm)/Pb(70 nm) structure (sample 3) with zero conventional spin-valve effect (ΔT_c^P=ΔT_c^{AP}). The circles show the angular dependence of the superconducting transition temperature $T_c(\alpha)$, which is caused by the long-range triplet component of the superconducting condensate. The solid line is the theoretical curve.

4 Discussion of experimental results

As can be seen from Fig. 1, the magnitude of the conventional superconducting spin-valve effect is equal to ΔT_c = 110 mK for sample 1 and ΔT_c = 30 mK for sample 2. This difference in magnitude of the spin-valve effect ΔT_c is associated with the fact that, for the thickness of the permalloy layer d_{Py2} = 3 nm in sample 2, the wave function of superconductor Cooper pairs almost completely decays in the Py2 ferromagnetic layer. Therefore, the layer Py1 has little influence on the superconductivity and, consequently, on the magnitude of the superconducting spin-valve effect ΔT_c. This is consistent with the fact that, in accordance with the theory, a decrease in the thickness of the permalloy layer d_{Py2} leads to an increase in the superconducting spin-valve effect ΔT_c. It can also be seen from Fig. 1 that, when the relative orientation of the magnetizations changes from the parallel (P) ($\alpha = 0^o$) to antiparallel (AP) ($\alpha = 180^o$) orientation, the superconducting transition temperature T_c varies monotonically, passing through the minimum for the noncollinear orientation. According to the theory (see [10]), the characteristic minimum in the dependence $T_c(\alpha)$, which is most pronounced in the vicinity of the angle $\alpha = 90^o$, clearly indicates the generation of the long-range triplet component in the superconducting condensate. By assuming that there is no triplet component (even though, according to the theory, the appearance of such a component is inevitable), it can be expected that the dependence $T_c(\alpha)$ will be monotonic. From the general considerations, the superconducting transition temperature T_c should be a function of the quantitiesα^2 and $(\pi - \alpha)^2$ for the angle α varying from 0 to π. Thus, the dependence $T_c(\alpha)$ can be expressed in terms of the temperatures and as follows:

$$T_c^{ref}(\alpha) = T_c^P cos^2(\alpha/2) + T_c^{AP} sin^2(\alpha/2). \quad (1)$$

This imaginary curve is represented by the dashed line in Fig. 1 for the dependences $T_c(\alpha)$. This curve will be referred to as the reference curve. The deviation of the actual temperature T_c from the reference curve is shown in Fig. 1 for the dependences $\Delta T_c^{trip}(\alpha)$. The value of this deviation demonstrates the contribution from the long-range triplet component to the magnitude of the superconducting spin-valve effect. Let us designate this magnitude as ΔT_c^{trip} and find that the difference in the temperatures T_c between the antiparallel and perpendicular orientations of the magnetizations already is equal to 130 mK for sample 1 and 60 mK for sample 2. This means that the long-range triplet component makes a significant contribution to the superconducting spin-valve effect.

In all previous studies of our group (see [10, 17–20]), the experimental data on the critical temperature T_c were compared with the parameter of the theory W, which determines the degree of suppression of superconductivity by the ferromagnetic layer (see [7]). The effective boundary condition for the real component of the anomalous Green's function can be written as

$$\xi_S \frac{df_0}{dx} = W f_0. \quad (2)$$

This analysis demonstrated that there is a qualitative agreement between the theory (parameter W) and the experiment (T_c) without calculating the critical temperature itself. Here, for the first time, we carried out a direct comparative analysis of the theoretically calculated and experimentally measured values of the superconducting transition temperature T_c. For the theoretical calculation of the dependence $T_c(\alpha)$, we used the theory proposed by Fominov and his colleagues (see [9]) for the case of an arbitrary transparency of the F2/S interface (see [17, 20]). The result of this calculation is represented by the solid line in Fig. 1. The theoretical calculation was performed with the following parameters: the coherence lengths for the S- and F-layers were taken as ξ_S = 41 nm and ξ_F = 13 nm, respectively; the thickness of the superconducting layer was d_{Pb} = 73.5 nm; the temperature of the transition to the superconducting state of bulk lead was ΔT_c^{Pb} = 7.2 K; the transparency parameters at the Py2/Pb interface were γ = 0.734 and γ_b = 1.8; and the exchange field in the ferromagnet was h = 0.3 eV. In the theoretical calculations, it was found that, according to the theory, the thickness of the Py2 layer d_{Py2} should be two times less than the actual value of the layer thickness. This is likely associated with the fact that the Py2 layer is surrounded on both sides by copper layers, and the effective thickness of the Py2 layer decreases due to the mutual diffusion of the copper atoms and permalloy. As can be seen from Figs. 1 and 2, the theory reproduces the main features of the experimental dependences. The significant discrepancy between theory and experiment can be caused by two factors. In the theory proposed by Fominov et al. [9], the F-layers were represented by weak ferromagnets, whereas permalloy is not ferromagnetic. In addition, we ignored the limitation of the transparency of the Py1/Cu/Py2 interface, as was done in the theoretical studies of Deminov et al. [21, 22], because, in our case, this parameter would be free. In contrast to the parameters characterizing the transparency of the Py2/Pb interface, we cannot estimate this quantity from the experiment.

5 Conclusions

Thus, the main results obtained in the investigations presented in this paper can be summarized as follows. First, the long-range triplet component in CoO_x/Py1/Cu/Py2/Cu/Pb superconducting spin valve samples has been studied experimentally. It has been revealed that there is a complete switching between the normal and superconducting states. The switching is caused by a combination of the conventional and triplet spin-valve effects when the relative orientation of the magnetizations of the Py1 and Py2 layers changes from the antiparallel to perpendicular orientation. Second, a direct comparison of the theoretically calculated and experimentally obtained values of the superconducting transition temperature T_c has been performed for the first time.

This work was partially supported by Russian Foundation for Basic Research (No. 17-02-00229) and the Program of the Russian Academy of Sciences.

References

[1] S. Oh, D. Youm, M.R. Beasley, Appl. Phys. Lett. **71**, 2376 (1997)
[2] M. Eschrig, Physics Today **64**, 43 (2011)
[3] M.G. Blamire, J.W.A. Robinson, J. Phys: Condens. Matter **26**, 453201 (2014)
[4] J. Linder, J.W.A. Robinson, Nat. Phys. **11**, 307 (2015)
[5] F.S. Bergeret, A.F. Volkov, K.B. Efetov, Rev. Mod. Phys. **77**, 1321 (2005)
[6] A.I. Buzdin, Rev. Mod. Phys. **77**, 935 (2005)
[7] K.B. Efetov, I.A. Garifullin, A.F. Volkov, K. Westerholt, *Magnetic Heterostructures. Advances and Perspectives in Spinstructures and Spintransport*, Vol. 227 of *Springer Tracts in Modern Physics* (Springer Berlin Heidelberg, 2007)
[8] K.B. Efetov, I.A. Garifullin, A.F. Volkov, K. Westerholt, *Magnetic Nanostructures. Spin Dynamic and SpinTransport.*, Vol. 246 of *Springer Tracts in Modern Physics* (Springer Berlin Heidelberg, 2013)
[9] Y.V. Fominov, A.A. Golubov, T.Y. Karminskaya, M.Y. Kupriyanov, R.G. Deminov, L.R. Tagirov, JETP Letters **91**, 308 (2010)
[10] P.V. Leksin, N.N. Garif'yanov, I.A. Garifullin, Y.V. Fominov, J. Schumann, Y. Krupskaya, V. Kataev, O.G. Schmidt, B. Büchner, Phys. Rev. Lett. **109**, 057005 (2012)
[11] V.I. Zdravkov, J. Kehrle, G. Obermeier, D. Lenk, H.A. Krug von Nidda, C. Müller, M.Y. Kupriyanov, A.S. Sidorenko, S. Horn, R. Tidecks et al., Phys. Rev. B **87**, 144504 (2013)
[12] A.A. Jara, C. Safranski, I.N. Krivorotov, C.T. Wu, A.N. Malmi-Kakkada, O.T. Valls, K. Halterman, Phys. Rev. B **89**, 184502 (2014)
[13] X.L. Wang, A. Di Bernardo, N. Banerjee, A. Wells, F.S. Bergeret, M.G. Blamire, J.W.A. Robinson, Phys. Rev. B **89**, 140508(R) (2014)
[14] N. Banerjee, C.B. Smiet, R.G.J. Smits, A. Ozaeta, F.S. Bergeret, M.G. Blamire, J.W.A. Robinson, Nat.Commun. **5**, 3048 (2014)
[15] M.G. Flokstra, T.C. Cunningham, J. Kim, N. Satchell, G. Burnell, P.J. Curran, S.J. Bending, C.J. Kinane, J.F.K. Cooper, S. Langridge et al., Phys. Rev. B **91**, 060501 (2015)
[16] P.V. Leksin, A.A. Kamashev, J. Schumann, V. Kataev, J. Thomas, B. Büchner, I.A. Garifullin, Nano Research **9**, 1005 (2016)
[17] P.V. Leksin, N.N. Garif'yanov, A.A. Kamashev, Y.V. Fominov, J. Schumann, C. Hess, V. Kataev, B. Büchner, I.A. Garifullin, Physical Review B **91**, 214508 (2015)
[18] P. Leksin, N. Garif'yanov, I. Garifullin, J. Schumann, V. Kataev, O. Schmidt, B. Büchner, Phys. Rev. Lett. **106**, 067005 (2011)
[19] P.V. Leksin, N.N. Garif'yanov, I.A. Garifullin, J. Schumann, V. Kataev, O.G. Schmidt, B. Büchner, Phys. Rev. B **85**, 024502 (2012)
[20] P.V. Leksin, A.A. Kamashev, N.N. Garif'yanov, I.A. Garifullin, Y.V. Fominov, J. Schumann, C. Hess, V. Kataev, B. Büchner, JETP **97**, 478 (2013)
[21] R. Deminov, L. Tagirov, R. Gaifullin, T. Karminskay, M. Kupriyanov, Y. Fominov, A. Golubov, Journal of Magnetism and Magnetic Materials **373**, 16 (2015)
[22] R. Deminov, L. Tagirov, R. Gaifullin, T. Karminskay, M. Kupriyanov, Y. Fominov, A. Golubov, Solid State Phenomena **233–234**, 745 (2015)

Superconducting mesa-structures with a spin filtering manganite interlayer

Karen Constantinian*, *Gennady* Ovsyannikov, *Yulii* Kislinskii, *Victor* Demidov, *Anton* Shadrin, *Andrey* Petrzhik

Kotel'nikov Institute of Radio Engineering and Electronics RAS, Moscow, Russia

Abstract. We report on current transport and magnetism in hybrid superconducting mesa-structures prepared from epitaxial superconductor $YBa_2Cu_3O_{7-\delta}$ and manganite $LaMnO_3$ thin films. Superconducting Au-Nb bilayer was the top electrode. Measurements of magnetic resonance in heterostructure with the only Au top electrode (without Nb) showed ferromagnetic state close to Curie temperature of $LaMnO_3$, 150 K. A spin filtering was revealed by measurements of tunneling magneto-resistance. Mesa-structures had no superconducting current even at reduced temperature T=300 mK and the smallest manganite interlayer thickness, d_M=5.6 nm. Varying the external magnetic field $|H|$<10 Oe and dc bias current $|I|$<0.5 mA the microwave generation with power P <1 pW with a line-width of order 50 MHz was observed in GHz frequency band. The frequency of generation could be tuned by the biasing current with a rate 10^{13} Hz/A. The impact of spin polarized current on microwave generation is discussed taking into account the asymmetry of oscillation amplitudes over the applied magnetic field.

1 Introduction

Recently, superconducting structures with a spin-polarized current have attracted increasing interest in view of the prospects for their use in spintronics and novel elements of microwave electronics. Interaction of superconductivity and magnetism was studied in hybrid structures comprised from superconductors (S) and ferromagnetic materials (F), tunnel junctions with a ferromagnetic barrier [1, 2] and etc. Tunnel current through a ferromagnetic barrier becomes spin-polarized due to the splitting of the energy band into the spin-up and spin-down levels [3, 4]. The precession of magnetization excited by a spin-polarized current was theoretically considered in [5-7]. Experimentally, the excitation of precession of the magnetization with a sufficiently high current density (above 10^6 A/cm^2) was observed in multilayer structures made of ferromagnets and metals [8-10] and in the tunnel-type structure [11]. At the same time, the dynamics of the spin-polarized current in superconducting structures with a ferromagnetic interlayer remains less studied.

2 Experimental

Hybrid mesa-structures were prepared from epitaxial films of superconducting cuprate $YBa_2Cu_3O_x$ (YBCO) and manganite $LaMnO_3$ (LMO) - materials having good chemical and crystalline compatibility [12]. The top superconducting electrode was thin film bilayer consisting of Nb and Au film, providing the contact with LMO manganite. The superconducting YBCO film was produced by laser ablation at temperature of 700-800° C on the substrate (110) of $NdGaO_3$ (NGO). A thin film of LMO interlayer d_M = 5-20 nm thick was grown epitaxially in the same vacuum chamber *in situ* at high temperature, and after cooling to room temperature was covered with an Au layer 20-30 nm thick. The critical temperature of the YBCO film was T_C = 88-89 K. The subsequent layer of Nb was deposited by magnetron sputtering. The Au film was used to reduce the diffusion of oxygen from manganite to Nb. The proximity effect between Nb and Au films provided a superconducting critical temperature of the bilayer T_C^*= 8.5-9 K, close to the critical temperature of the Nb film (9.2 K). Superconducting mesa-structures with interlayer of LMO and in-plane dimensions from 10x10 to 50x50 μm^2 (see Fig.1) were formed by means of photolithography and plasma-chemical and ion-beam etching. According to results [13, 14], the LMO manganite of stoichiometric composition has the properties of an insulator and is an antiferromagnet at low temperatures. But a small change in the oxygen content leads to the appearance of ferromagnetism with the Curie temperature of order 140-150 K. A ferromagnetism for the LMO films was reported in [15] and confirmed by measurements of temperature dependence of ferromagnetic resonance magnetic field $H_0(T)$ [16]. Fig.2 shows temperature dependences $\rho(T)$ of resistivity of LMO film and the ferromagnetic resonance dependence. The investigation of the microwave properties of mesa-structures was carried out using apparatus schematically shown in Fig.1. Low temperature measurements were carried out

* Corresponding author: karen@hitech.cplire.ru

 https://doi.org/10.1051/epjconf/201818508002

using cooled HEMT amplifier operating in frequency band 1 - 2 GHz with the noise temperature T_{N1}= 8 ± 2 K and gain K_1 = 20 dB at T = 4.2 K. The balance circuit of the input stage of the amplifier provided stable operation in a wide range of load resistance 10-100 Ω.

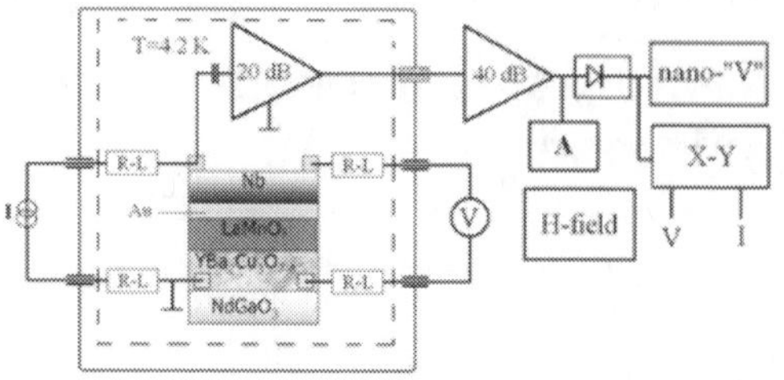

Fig.1. Experimental set-up for microwave measurements. The mesa-structure (the cross section is shown in the middle of the figure) and the cold amplifier were placed inside the magnetic field coil (dashed contour) and the screen (solid-line contour) at T = 4.2 K. The (H-field) coil had bipolar dc feed. The dc biasing of mesa-structure (I) and voltage measurement unit (V) as well the power supply of the cold amplifier were set through the filtering (R-L) circuits and through the built-in capacitors. The amplified microwave signal from mesa-structure was monitored either by the spectrum analyzer (A), or the quadratic detector. The output signal level was registered by the nano-voltmeter (nano-"V") and recorded by (X-Y) plotter.

The noise temperature of the "room" amplifier was T_{N2}=130 K and gain K_2 = 40 dB. The voltage signal, proportional to the power P of microwaves emitted by mesa-structure, was taken from the output of the quadratic detector. Simultaneously, the spectrum analyzer monitored the amplitude-frequency response at the output of the "room" amplifier. The magnetic field at microwave measurements was set perpendicular to the plane of the substrate and, according to the calculations [17], had efficiency close to the parallel field for the case of a sandwich tunnel junction with superconducting electrodes with a square shape.

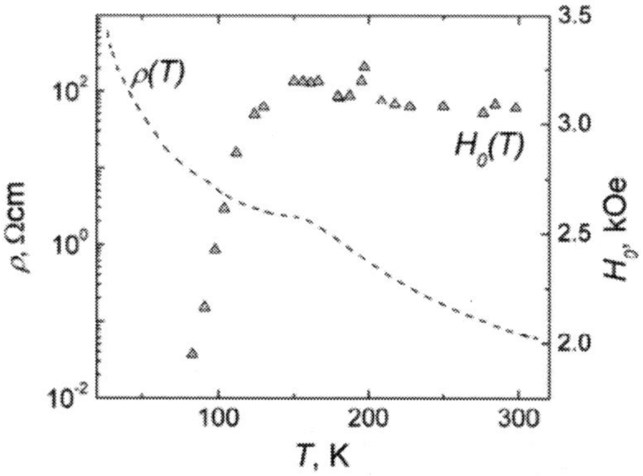

Fig.2. Temperature dependence $\rho(T)$ of resistivity of an autonomous LMO film (dashed curve). Temperature dependence of the magnetic resonance field H_0 for the LMO film (triangles).

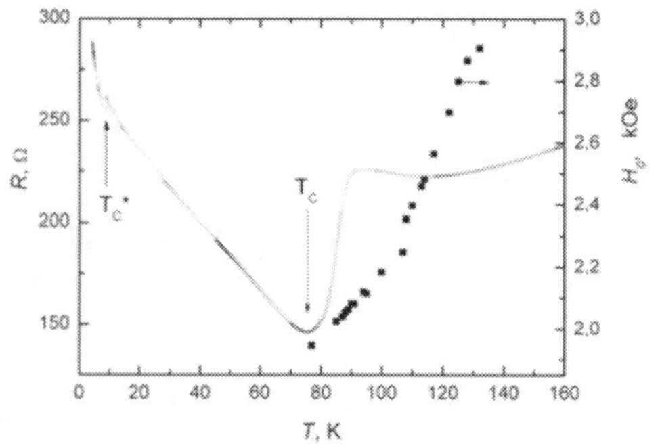

Fig.3. Temperature dependence of the resistance of mesa-structure with area of 400 μm² and thickness d_M = 17 nm of the LMO interlayer. The superconducting transition temperature is T_C= 78 K for YBCO and T_C^* = 8.7 K for Nb–Au. Squares show the temperature dependence of the ferromagnetic resonance field of Au/LMO/YBCO heterostructure with d_M = 16 nm.

The differential resistance R_d=dV/dI was measured separately using lock-in amplifier with the synchronous detector and the built-in modulating unit of signal generator. Modulating signal was applied to the mesa-structure through the dc feed noise filtering circuit. Fig.3. shows temperature dependence of the resistance of mesa-structure along with the ferromagnetic resonance experimental function of Au/LMO/YBCO heterostructure, made without superconducting Nb film. The magnetic uniaxial anisotropy Hs = 375 Oe, and magnetization M_0 = 1.75μB/Mn were obtained.

The case when magnetic field was applied to mesa-structure is shown in Fig.4. The temperature dependence of differential conductivity $G = 1/R_d$ is presented in Fig.4a when no magnetic field was applied. Fig.4b shows the $G(V)$ dependences under the action of a magnetic field. An asymmetry of $G(V)$ function for positive and negative biasing V is seen, which could be caused by the shape of the tunnelling barrier and the difference of work functions of contacting to the LMO, the YBCO and Au films. An other feature is the change of $G(V)$ under the influence of magnetic field. This change of $G(V)$ occurs only within the voltage range at the mesa-structure up to the voltage of niobium energy gap. It could not be attributed to the Zeeman splitting of the density of states of the electrodes [13], since the applied magnetic field H is much smaller than the Zeeman splitting field. After the transition to the ferromagnetic state of the LMO layer at temperature T < 140 K, a spin-polarized current appears in the mesa-structure [3] and magnetic field affects the tunneling conductivity due to spin polarization [14]. The observed value of magnetoresistance of mesa-structure is typical for manganites [12] and took place only at temperatures below the critical temperature of bilayer Au-Nb. With an increase of dc bias voltage above 3 mV (about twice higher than the niobium energy gap), the conductivity of the mesa-structure no longer depends on magnetic field [16]. The magnetoresistance measured at wider magnetic field sweeping range is shown in Fig.5. The hysteresis, typical for ferromagnets caused by the change in magnetic field direction takes place. Upon an increase of dc bias voltage up to 10 mV, the differential resistance of mesa-structure becomes independent of the magnetic field (curve 2 in Fig.5).

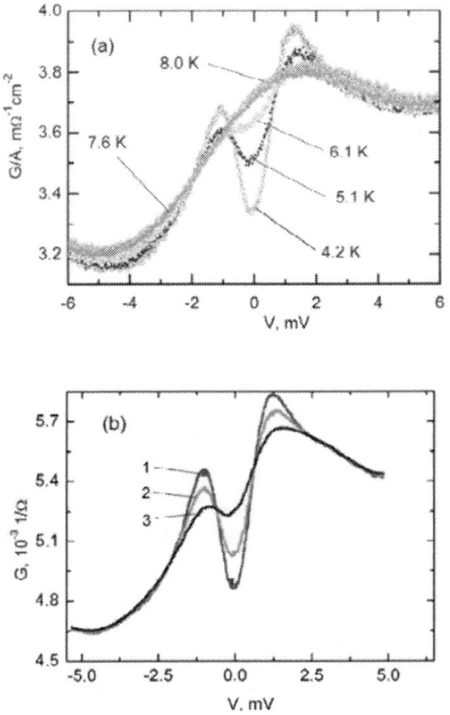

Fig.4. (color online) Family of differential conductivity G vs. voltage V for mesa-structure with d_M=5.6 nm, A = 140 μm². (a) Characteristic conductivity G/A measured at fixed temperatures in the range 4.2 – 8 K for H = 0; (b) for fixed magnetic field directed parallel to the substrate plane, 1-H = 0 Oe, 2-H = 263 Oe, 3-H = 526 Oe. Ambient temperature T = 4.2 K.

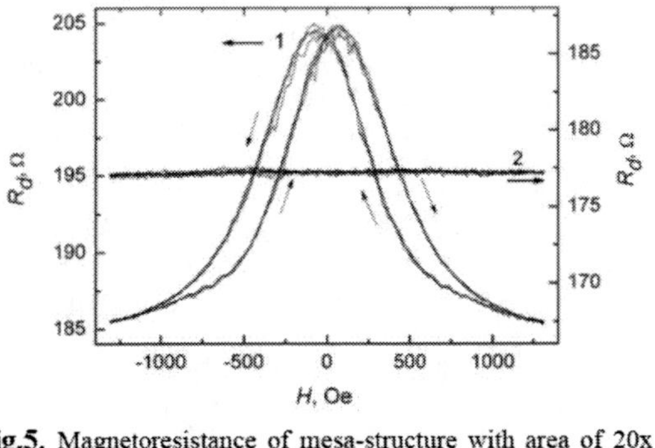

Fig.5. Magnetoresistance of mesa-structure with area of 20x 20 μm² at T = 4.2 K measured using a few runs of the external magnetic field applied in parallel to the plane of the substrate. Arrows indicate the direction of magnetic field change. Family of curves 1 was obtained for zero bias voltage (V = 0); family 2 was obtained for fixed V = 10 mV.

Results of microwave measurements are shown in Fig.6. Fig.6a demonstrates a family of $P(V)$ dependences of the microwave power emitted by mesa-structure vs. bias voltage, obtained for several values of the external magnetic field at T = 4.2 K. As the temperature rises above the critical temperature of superconducting Nb-Au bilayer, the microwave generation disappears. The smooth growth of P with increasing V is most likely caused by the shot noise of the mesa-structure. Symmetrically located relative to V= 0 the excess signal in $P(V)$ dependence, significantly exceeding the noise level, are observed. At the microwave spectrum analyzer, the appearance of excess P signal was manifested in the form of a sharp generation peak with a power of order of fraction of pW and with a linewidth of ~ 50 MHz. The central frequency of the peak was tuned within the whole experimental frequency range f = 1 - 2 GHz. Power P of excess signal showed tendency to grow with increasing the bias current I (or voltage V): the highest peaks were observed at larger voltages V. Note, that in the structures discussed in [8, 9], on the contrary, the generation frequency decreased with increasing bias current I. The dependence of the generation power P on the magnetic field H (for a fixed V) had an oscillating form with an average oscillation period of order 0.5 Oe. Fig.6a also indicates that an average range of bias voltages ΔV where microwave generation was observed is amounted of ~ 8 mV. An estimate of the frequency/current coefficient of frequency tuning $\Delta f/\Delta I$ gives a value of 7.5 10^{12} Hz/A, where $\Delta I=\Delta V/R_d$. Note, R_d was of order 60 Ω and weakly depends on V. In our experiment the generation frequency f and the corresponding voltage V differs from the Josephson ratio 483.6 GHz/mV by three orders in magnitude (h is the Planck constant, e is the electron charge). Taking into account the absence of superconducting current, it is impossible to attribute such microwave generation to the non-stationary Josephson effect. Fig.6b shows the magnetic field dependence of the maximum values of the generation amplitudes P. Despite the fact that the amplitude of the detected signal is rugged and irregular in shape, the values of the maxima corresponding to both polarities of the bias voltage coincided well (see Fig. 6b). On one hand, the occurrence of microwave generation could be explained by the trivial mechanism of the coherent motion of magnetic vortices. However, on the other hand, the stability of the regular motion of vortices in superconducting structures is difficult to achieve, and it easily disturbed by external microwave influence [18]. An application to the mesa-structure of a monochromatic signal with a power up to 0.5 mW at a frequency f = 70 GHz was resulted in a slightly (about 20%) suppression the generation amplitude without changing in its shape. Moreover, the repeated measurements of dependences of $P(V)$ at fixed H both with the external microwave signal of 70 GHz, and without it, completely coincided, repeating the ruggedness of $P(V)$. Consequently, neither the Josephson effect, nor the mechanism of coherent motion of magnetic vortices could explain the nature of the observed microwave generation. At the same time, it was

https://doi.org/10.1051/epjconf/201818508002

theoretically shown [19] that the controlled rotation of spin vortices (skyrmions [20]) under the action of a constant spin-polarized current can excite stable narrow-band microwave generation and, importantly, without the necessary condition for exceeding the threshold value of the electric current. Another coincidence of the results of the experiment and the model [19] is the increase of generation frequency with increasing of bias current through the mesa-structure. The strong spin-orbit interaction, which is relatively weak in the autonomous manganite LMO [21], and which is necessary for the formation of spin vortices, can be enhanced due to Jahn-Teller distortion in the mesa-structure due to the influence of the lower YBCO film grown on the NGO substrate [22], and also by the presence of an interlayer of gold, the spin-orbit interaction in which was analyzed in [23].

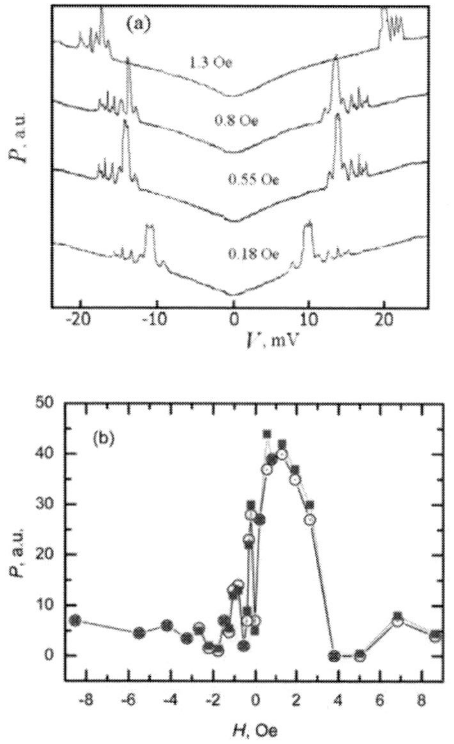

Fig.6. (a) The family of dependences of the signal P on the bias voltage V (shifted along the ordinate axis) obtained from the output of the quadratic detector at magnetic field changed in the range 0.18-1.3 Oe. Mesa-structure had the thickness of the LMO interlayer $d_M = 5.6$ nm, the resistance $R = 60\ \Omega$, and size $A = 1000\ \mu m^2$, (b) (color online) Dependence of the maximum microwave generation P vs. magnetic field. Open circles denote peaks obtained with negative bias V, filled squares – at the positive bias.

In conclusion, the YBCO/LMO/Au-Nb mesa-structure with the spin-filter properties demonstrated microwave generation in GHz frequency band with a linewidth of order 50 MHz under the influence of external weak magnetic field $|H| < 10$ Oe and the bias current. Generation frequency could be tuned by changing the bias current with a ratio of ~ 10^{13} Hz/A, and its amplitude could be controlled by a weak magnetic field $H \sim 0.5$ Oe.

The authors are grateful to I.V. Borisenko for a fruitful discussions of the results. The work was supported by the Russian Academy of Sciences, the RFBR project 16-29-14022 and the Scientific School NSh-8168.2016.2.

References

1. F. S. Bergeret, A. F. Volkov and K. B. Efetov, Rev. Mod. Phys., **77**, 1321 (2005).
2. S. Kawabata, Y. Tanaka, A. A. Golubov, et al. J. Magn. Magn. Mater., **324**, 3467 (2012).
3. R. Meservey, P.M. Tedrow, Phys. Rep. **238**, 173 (1994).
4. T.S. Santos, J.S. Moodera, K.V. Raman, et al., Phys. Rev. Lett. **101**, 147201 (2008).
5. E. Zhao, T. Löfwander, J.A. Sauls, Phys. Rev. B**70**, 134510 (2004).
6. J.C. Slonczewski, J. Magn. Magn. Mater.**159**, L1 (1996).
7. A. Manchon and S. Zhang, Phys. Rev. B**79**, 094422. (2009).
8. S.I. Kiselev, J.C. Sankey, I.N. Krivorotov, et al. Nature (London), **425**, 380 (2003).
9. W.H. Rippard, M.R. Pufall, et al. Phys. Rev. Lett. **92**, 027201. (2004).
10. V.S. Pribiag, I.N. Krivorotov, G.D. Fuchs, et al. Nature Physics, **3**, 498 (2007).
11. Luqiao Liu, Chi-Feng Pai, et al., Phys. Rev. Lett., **109**, 186602. (2012).
12. A.M. Petrzhik, G.A. Ovsyannikov, A.V. Shadrin, et al., JETP, **112**, 1042. (2011).
13. J.S. Moodera, T.S. Santos, T. Nagahama, J. Phys. Cond. Matter. **19**, 165202. (2007).
14. X. Hao, J.S. Moodera, R. Meservey, Phys. Rev. B**42**, 8235 (1990).
15. X. Renshaw Wang, C. J. Li,W. M. Lü, et al. Science, **349,** Issue 6249, 717 (2015).
16. G.A. Ovsyannikov, Yu.V. Kislinskii, K.Y. Constantinian, JETP, **124**, 628 (2017).
17. R. Monaco, M. Aaroe, J. Mygind, V.P. Koshelets, Phys. Rev. B**79**, 144521. (2009).
18. K.I. Konstantinyan, G.A. Ovsyannikov, L.E. Amatuni, Z.G. Ivanov, Sov. Phys. JETP, **72**, 376 (1991).
19. F. Garcia-Sanchez, J Sampaio, et al., New J. Phys. **18**, 075011 (2016).
20. A. Fert, V. Cros and J. Sampaio, Nature Nanotechnology, **8**, 152 (2013).
21. M. Snamina, A.M. Ole´s, Phys. Rev. B**94**, 214426 (2016).
22. I.V. Borisenko, M.A. Karpov, G. A. Ovsyannikov, Tech. Phys. Lett. **39**, 1027 (2013).
23. A. Dal Corso, A. M. Conte, Phys. Rev. B**75**, 115106 (2005).

Interplay between magnetism and superconductivity in metallic hydrogen and hydrides at high pressure

Lev Mazov[1,*]

[1]Dept.of Phys.of Supercond., Inst.for Phys.of Microstruct. Russian Academy of Sciences, 603600 Nizhny Novgorod, Russia

Abstract. The detailed analysis of resistive and magnetic measurements with sulfur hydrides at high pressure is performed. The hydrogen system at high pressure can exhibit ferromagnetic (FM) as well as superconducting (SC) properties with layered structure. The onset temperature of resistive transition at 200 K in metallic sulfur hydrides corresponds to magnetic (AF SDW) phase transition rather than SC one. SC transition in these sulfur hydrides occurs only when magnetic (AF SDW) phase transition is over (~ 40 K). The SC mechanism in metallic sulfur hydrides is not conventional but corresponds to Keldysh-Kopaev model characteristic for systems with coexistence of dielectric and SC pairing.

1 Introduction

According to modern viewpoint, hydrogen at high pressure must pass from the molecular to the metallic phase, which can be SC at relatively high temperatures [1]. However, the experimental implementation of such a transition in laboratory conditions still encounters a number of problems related to the stability of this state. On the other hand, as is known, such pressures are achievable for a number of stars and planets, a significant part of whose composition is hydrogen. Moreover, such celestial bodies, as follows from observational data, have a significant magnetic field. In particular, this is indicated by the polarization of the radio emission observed from "white dwarfs" (WD) [2]. Among the planets in our solar system, as is known, Jupiter and Saturn have the largest magnetic fields, having a large share of hydrogen in their structure (gas giants) (see, e.g. [3]). A recent comparison of theoretical and observational data for magnetic WD indicates that large magnetic field of astrophysical objects is due to monoatomic hydrogen in metallic, FM-ordered state [4].

2 Hydrogen phase change with increasing pressure

2.1 Molecular and metallic hydrogen

As known, liquid-metal hydrogen in laboratory was obtained under dynamic compression, at a certain density, P = 140 GPa and T ~ 2600 K (see, e.g. [5,6]). These states were obtained repeating shock compression generated by a shock wave in liquid H_2 at T = 20 K, enclosed between two sapphire anvils. Multiple shock pulse was induced by impact on the cryogenic sample holder by means of either an Al or Cu plate accelerated to a certain speed with two-stage light-and-gas gun. The pressures generated were in the range of 90-180 GPa.

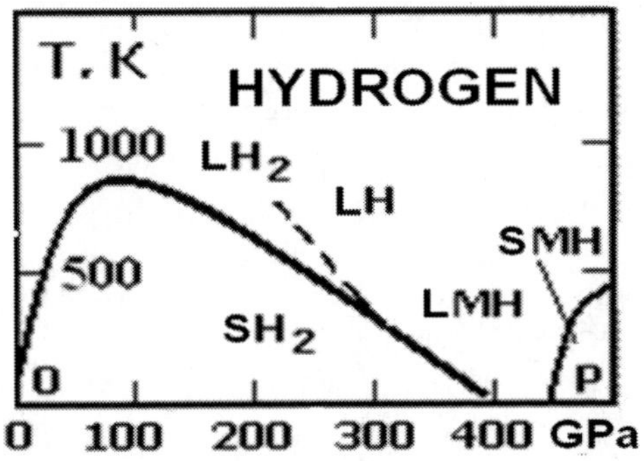

Fig. 1. Phase T-P diagram of hydrogen (see, e.g. [5,6]). (Here, symbols are: S – solid, L – liquid, M – metallic, H – hydrogen)

The electrical conductivity increased by almost four orders of magnitude between 90 and 140 GPa, in which range liquid H at T > 1000 K is a semiconductor with conduction electrons thermally excited through the energy gap. It closes with a density of up to ~ k_BT by ~ 0.64 mol/cm^3. At 140 GPa to 180 GPa thermal excitation is quenched and liquid H is a bad metal [5,6]. Measured conductivity 2000 Ohm^{-1}cm^{-1} characteristic for liquid metals and highly disordered alloys is in good agreement with the theoretical calculations (see, e.g. [5,6]).
It is interesting that later, at a static pressure P > 220 GPa, a layer-like lattice, called Phase IV, was detected.

* Corresponding author: mazov@ipm.sci-nnov.ru

 https://doi.org/10.1051/epjconf/201818508003

Phase IV is insulating and, possibly, semi-metallic up to P ~ 360 GPa, above which, as predicted, it can become metallic (see, e.g. [5,6]). Since the layer-like fcc phase of H_2 is similar to graphite in terms of "very low electron charge density in the ground state," Ashcroft [1,7] considered the optical properties of graphite to obtain an understanding of the nature of solid H_2 at very high pressures. It was concluded that solid semimetallic H_2 at high pressures probably has a layered, crystalline structure of the fcc type and that the paired-H zone in this structure will support this semimetallic phase at very high pressures, thereby preventing a transition to metallic phase at a pressure higher than the pressure in the lamellar phase.

2.2 Ferromagnetism in metallic hydrogen

There were put forward two concepts indicating that metallic hydrogen can be FM in nature at high pressure [3,4]. The nature of magnetic transition from a paramagnetic (PM) to a FM phase is in fact in a competition between the spin-orientation dependent exchange processes which favor a spin-aligned FM state and the contributions of kinetic energy which favor a PM state [4]. In the ground state a FM solid is formed due to the predominance of exchange effects between parallel spins. In one of above approaches [4], the phase diagram for dense metallic hydrogen can be essentially characterized by two dimensionless parameters

$$r_s = me^2 a/\hbar^2 \text{ and } \theta = 2mk_B T/\hbar^2 (3\pi^2 n)^{2/3},$$

corresponding to density and temperature, respectively. Here m, n, and k_B refer to the mass of a fermion, the number density of the fermions, and the Boltzmann constant, respectively; $a = (4\pi n/3)^{-1/3}$.

In a FM state the Fermi liquid then induces a magnetic field with the strength $B_M = 4\pi\mu_M n$, where μ_M denotes the magnetic moment of a fermion, and for an electron, μ_M = 9.28477 x 10^{-21} (erg / G).

The free energy for the strongly coupled (r_s > 20 and Γ > 10) PM electron liquids can been expressed in a parametrized form as [4]

$$f_{PM}(r_s, \theta) = f_0(\theta) + f_1(\theta, \Gamma, \Theta), \quad (1)$$

where $\Gamma = e^2/ak_B T$. First term $f_0(\theta)$ corresponds to the Helmholtz free energy per particle of the ideal-gas electrons in the PM state in units of the thermal energy. The ideal-gas contribution to the free energy is expressed as a balance between those of the chemical potential and of the pressure as

$$f_o(\theta) = \alpha_0(\theta) - p_0(\theta) \quad (2)$$

The free-energy formulas for the strongly coupled FM electron fluids in the finite temperature regime, were constructed analogously to the way Eq. (1), where parameter θ should be then naturally replaced to

$$\theta_F = 2m_e k_B T/\hbar^2 (6\pi^2 n)^{2/3} \text{ [4].}$$

The resulting magnetic phase diagram for metallic hydrogen was constructed in [4] (see, Fig.2). In this diagram, PM and FM phases correspond to the PM and FM fluids, respectively; S-phase corresponds to bcc crystalline solid. Points T_{spf} and C_{mag} correspond to the triple point and the critical point for the FM transition

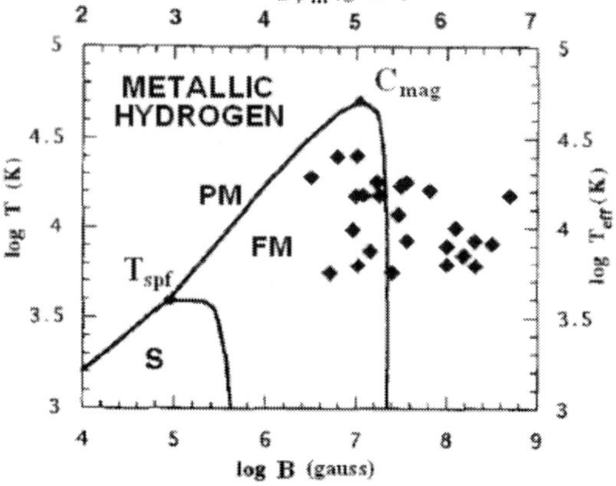

Fig. 2. Magnetic phase H-T diagram of metallic hydrogen [4]. The diamonds represent the data for magnetic white dwarfs estimated from observational data [4] (for legend, see text).

Table 1. Physical parameters at the triple point T_{spf} and the critical point C_{mag} for the metallic hydrogen ($\rho_m = mn$ is the mass density)

-- --	T_{spf}	C_{mag}
n (cm^{-3})	3.2 x 10^{17}	6.0 x 10^{28}
ρ_m (g/cm^3)	8.3 x 10^2	1.0 x 10^{-5}
B_M(G)	8.9 x 10^4	1.1 x 10^7
r_s, R_s	183	43
T(K)	3.9 x 10^3	5.1 x 10^4
θ, Θ	0.281	0.141

In this Table, $R_s = (m_H e^2 a/\hbar^2)\exp(-0.85K_s)$, $\theta_F = 2m_H k_B T/\hbar^2 (3\pi^2 n)^{2/3}$, n and m_H refer to the number density and the mass of hydrogen (after [4]).

The data in Fig.2 elucidate physical mechanisms for the origin of strong magnetization in the magnetic WD [4]. For the nuclear FM to be of relevance to the stellar magnetism, the state of matter should fall in the FM domain. Solidification of metallic hydrogen, on the other hand, may not take place in such WD, since each of the T_{eff} values exceeds the temperature at the triple point [4]. The FM phase transition is a concept which may provide a relevant account for an origin of the strong magnetization e.g. in degenerate magnetic stars.

3 Analysis of experimental data

3.1 Metallic hydrogen

The above work [5] was of great interest and excitation in the metallization of hydrogen at an unprecedentedly high pressure, reporting experimental Raman spectra, phase transition, metallization, and reflection coefficients observed visually. Unfortunately, the values of the electrical resistances obtained from their measured electrical resistances and the reduced dimensions of the sample was in order of magnitude larger than for the metal, and it was not clear to what extent hydrogen caused these observations [6].

In the beginning of this year, it appears an interesting paper [8] reporting the observation of the phenomenon of a phase transition, which indicates the production of metallic hydrogen. Atomic hydrogen from the liquid phase passed to a solid at a pressure up to 495 GPa (see Fig.1) at a temperature of 5.5 K in a diamond anvil between the tips of artificial diamonds. Hydrogen from a state of transparent, like glass, transformed into a state of shiny metal, like copper or gold reflecting light (see Fig.3). The transition to the solid phase was confirmed by spectroscopic analysis. Then in [8] it was predicted that after removal of pressure, metallic hydrogen in the new experiments would remain metastable.

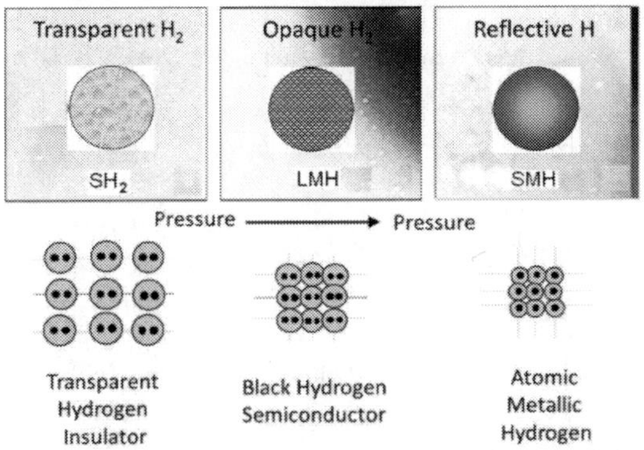

Fig. 3. Schematic illustration of reflectance in hydrogen [8].

However, a month later the first sample of metallic hydrogen was lost. Also in literature it was reported that now authors of [8] will be preparing another experiment to see if they can repeat high pressures in the first experiment and again make metallic hydrogen.

3.2 Metallic sulfur hydrides

As can be seen from the phase diagram for hydrogen (Fig.1), very high static pressures (~ 500 GPa) are required to obtain a solid metallic phase. Conditions soften for a number of hydrides [7]. In [9], an attempt was made to realize these effects in a system with a hydrogen compound H_2S, which is more stable. According to their measurements, the resistive curve in the normal state is linear in temperature, which is typical for metals, and at T ~ 200 K the electrical resistance begins to decrease - a resistive transition is observed. In their work, the temperature of the sharp drop of the electrical resistivity was identified with the onset of the transition to the SC state (Fig.4). However, the shape of the resistive curve with an inflection in the transition region indicated the possibility of an additional transition

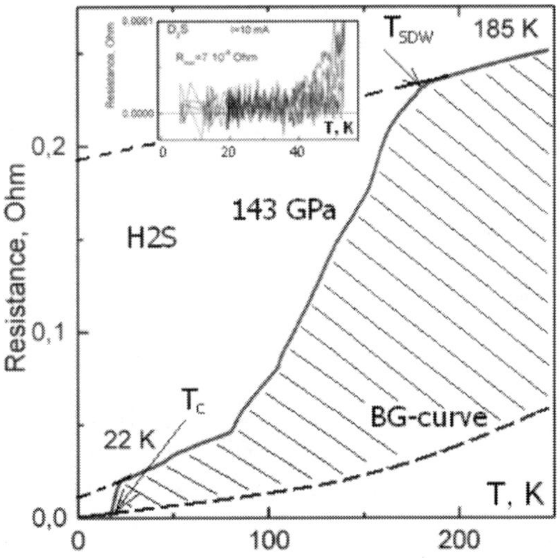

Fig. 4. T-dependence of resistance in H_2S (see [9,10]).

not directly related to SC. In our paper [10], it was suggested that the upper part of the resistive transition is associated with a magnetic (AF SDW-like) phase transition in the system (shaded area in Fig.4), while the SC transition occurs at much lower temperatures. The magnetic resistivity $\rho_m(T)$ can be estimated from the total resistivity $\rho_{tot}(T)$ via subtracting phonon contribution $\rho_{ph}(T)$ (Bloch-Gruneisen (BG) curve in Fig.4) and residual part ρ_{res} (given in Fig.4 by the point of intersection of dashed lines with ordinate axis). This is reminiscent the picture observed in copper oxides and pnictides (selenides) of iron. While pure iron is known to be a FM, iron atoms in pnictides (selenides) are AF-ordered, what allows SC coexist with dynamical AF-ordering in the whole temperature (or pressure) range of interest (cf. [10,16]]). The same situation, apparently, is observed in metallic sulfur hydride. Moreover, for the solid-state phase of hydrogen, a layered structure is characteristic [1,7], as in the known HTSC compounds (see also [11]). In [11] the study of the evolution of the magnetic field within cavity in sulfur hydride at high pressures was performed as a function of temperature by means of synchrotron radiation (Fig.5), causing resonant nuclear scattering (NRS) in the system. As a magnetic field sensor, a strip of Sn foil was placed inside a cell with diamond anvil (DAC) filled with H_2S. As it can be seen from Figs.5 there is essential anisotropy in a NRS spectrum relative to the magnetic field direction what can be evidence of magnetic nature of the sample. The similar picture follows from measurements of magnetic field $H_{Sn}(T)$ within the sulfur hydride sample (Fig.6). In accordance with [10] it can be proposed that upper part of drop of $H_{Sn}(T)$ curve can be due to drop of magnetic susceptibility $\chi^{//}_{AF}$ characteristic for AF along AF-axis; SC drop of $H_{Sn}(T)$ should occur only below.

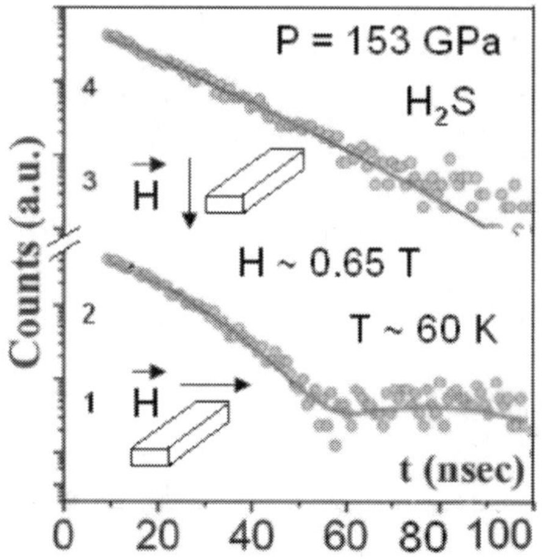

Fig. 5. Schematic anisotropy of NRS spectrum in H_2S; in insert – geometry of experiment (the data are from [11]).

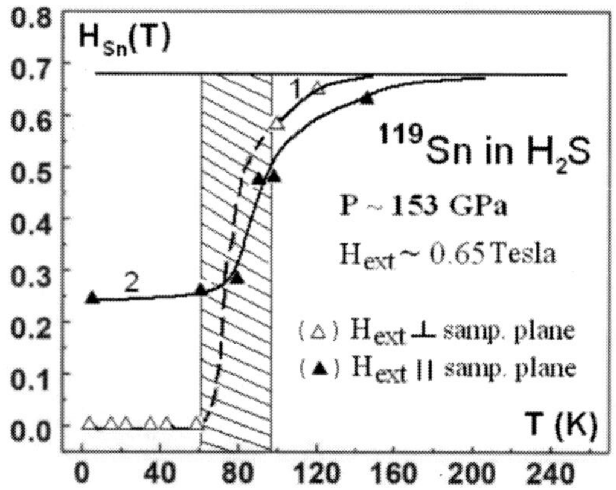

Fig. 6. Possible nonmonotonic disappearance (dashed curve [9]) of magnetic field in cavity within sulfur hydride; triangles are the experimental data from [11].

A detailed analysis of resonant nuclear scattering data performed in [12] demonstrates that the purely exponential nature of the spectrum corresponding to the SC state of the H_2S sample is observed only in the helium temperature range T < 59 K, and not up to 140 K, as discussed in [11]. Since detailed data are not given in their work in the range 60 K <T <100 K (shaded area in Fig.6), it can be assumed that the phase transition (dashed curve in Fig.6) has a complex (two-step character (cf. [9]) and its upper part can correspond to the magnetic (AF SDW-like) phase transition [10] and only its lower part relates to genuine SC transition.

4 Discussion and conclusion

So, the decrease in the resistance observed in [9,11] at T ~ 200 K can be related to the appearance in the hydrogen subsystem of a planar, modulated magnetic structure (AF SDW-like) (cf. [10,12]). The magnetic character of the transition, beginning at T ~ 200 K, is indicated also by the anisotropy of NRS in orthogonal directions of the magnetic field [11]. As for drop in magnetization [9] at ~ 200 K, then "anomalous diamagnetism" was observed in CuCl [14] and in YCaBaCuFeO [15] before, however, SC was absent in these works.

This behavior is rather typical for systems with coexistence of dielectric and SC pairing (see, e.g. [10,12]). In such a system, with a decrease in T, first a dielectric (AF SDW) transition takes place, and only then a SC transition occurs at lower temperatures. (The thermodynamics of dielectric (SDW) phase transition here is the same as for superconductor [13]). Such high Tc is a natural consequence of (e-e)-pairing at the background of high density of states which singularity arises in the narrow energy range near dielectric-(SDW)-gap edges due to removing of electronic states from the energy region of dielectric (SDW) gap (already formed in the normal state). However, formation of the dielectric (SDW) gap at the Fermi surface leads to decreasing of energy region for (e-e)-pairing what, in its turn, leads, on the contrary, to decreasing of Tc.

Of course, to study the magnetic nature of normal-state dielectric phase transition in H_2S the additional efforts are necessary, and such more detailed search with "deploying a sensitive magnetic technique" is now in progress [17].

References

1. N.W.Ashcroft // Phys.Rev.Lett. V.21, 1748 (1968).
2. V.L.Ginzburg, V. V.Zheleznyakov, V.V. Zaitsev // Nature v. 220, 355 (1968).
3. J.E.Hirsch // Phys.Lett. A v.141, 191 (1989).
4. S.Ichimaru // Phys.Lett. A v. 258, 354 (1999).
5. M. I. Eremets & I. A. Troyan // Nat.Mater. 10, 927 (2011).
6. W.J.Nellis et al. arXiv: cond-mat/1201.0407
7. N.W.Ashcroft // Phys.Rev.Lett. v.92, 187002 (2004).
8. R.P.Dias, I.F.Silvera // Science DOI: 10.1126 / science.aal1579 (2017).
9. A.P.Drozdov et al. // Nature v.525, 73 (2015).
10. L.S.Mazov // arXiv: cond-mat-1510.00123 (2015).
11. I.Troyan et al. // Science v.351, 1303 (2016).
12. L.S.Mazov // 14 conf. "Strongly correlated electronic systems and quantum critical phenomena "(HPPI RAS, Troitsk-Moscow, June, 2016), Conf.Abstr., P.17.
13. L.V. Keldysh, Yu.V. Kopaev // Fiz.Tv.Tela v.6, 2791 (1964).
14. N.B.Brandt et al. // JETP Lett. v.27, 37 (1978).
15. N.S.Ovanesyan et al. // Progress in HTSC, v.32 (World Scientific, Singapore, 1972), p.316.
16. Li Xiang et al. // Phys.Rev B 96, 024511 (2017).
17. C.W.Chu // High-Tc Copper Oxide Superconductors and Related Novel Materials (Eds. A.Bussmann-Holder et al., Springer Ser. in Mater. Sci. 255, Springer Int. Publ. AG 2017), p.15.

Inhomogeneity and irreversibility field of superconducting Nb-Ti tapes

Valentin Guryev*, *Sergey* Shavkin and *Vitaly* Kruglov

National Research Centre "Kurchatov Institute", Kurchatov sq. 1, 12318, Moscow, Russia

Abstract. Magnetizations and current-voltage characteristics of superconducting tapes from Nb-50% wt.Ti alloy are studied. In the region of external magnetic fields close to the upper critical field, there is a so called irreversibility field, which is often associated with the phase transition of vortex matter. However, in this study it is shown that the whole volume of experimental data can be described if we abandon the hypothesis of a phase transition and go over to the hypothesis of macroscopic inhomogeneity. This hypothesis implies the existence of two superconducting components with different upper critical fields, which naturally arise as a result of the production process.

1 Introduction

One of the most puzzling questions in the modern thermodynamics of vortex matter in type-II superconductors is the nature of the so-called irreversibility line in the field-temperature (H-T) phase diagram. At a fixed temperature, the irreversibility field $H^*(T)$ can be defined as the field above which there is no hysteresis (irreversibility) of the magnetization loop $M(H)$, but nevertheless the superconducting state is conserved. Usually, in high-T_c oxide superconductors, simultaneously with the appearance of reversibility, the critical current density reduces to zero, so $H^*(T)$ is also an important parameter for applications [1].

A number of models have been proposed to explain the existence of the irreversibility line in high-T_c oxide superconductors [2]. Regardless of the mechanisms, it seemed that consensus was reached on what is the main cause of this phenomenon: a large anisotropy in electronic properties (layered structure) and a high impact of thermal fluctuations. However it was later discovered that the irreversibility field is not unique to high-T_c oxides, but also exists in conventional superconductors such as Nb [3], Nb_3Sn [4], Nb-Ta [5] and Nb-Ti [4-6], which are supposed to be isotropic and weakly subject to thermal fluctuations due to low operating temperatures. Attempts to apply existing models of the irreversibility field to low-T_c superconductors have led to the fact that the most adequate phenomenological description is given by the model of the flux lattice melting [3, 4], although the validity of this model gives rise to doubt, as due to high pinning in commercial superconductors, the vortex matter should be in a glass state below $H^*(T)$ [7].

In this study it is shown that in the Nb-Ti tape the hypothesis of a phase transition of a vortex matter to a liquid state is not consistent. Moreover, in order to describe the phenomenon of the irreversibility field, it is not necessary to resort this hypotheses, it is sufficient to take into account the inhomogeneity that naturally arises in the fabrication process. Such a model of the irreversibility field fits concordantly into modern concepts of pinning mechanisms in conventional cold-worked (rolled or drawn) superconducting alloys.

1.1 Flux pinning mechanism and electrodynamics in high magnetic fields of cold-rolled Nb-Ti tape

It is well known that in single-phase cold-rolled superconducting alloys at high degree of deformation, the main mechanism of the pinning is a pinning at the grain boundaries [1]. At the microscopic level, the main reason for pinning is the decrease in the mean free path of conduction electrons near the grain boundary, which leads to a local decrease in the coherence length. Thus, the size of vortex core is effectively reduced, self-energy of the vortex is decreased and it is energetically favorable for vortex to remain near the grain boundary. This model allows one to unequivocally describe the main experimental facts in a wide range of magnetic fields, namely, the anisotropy of the pinning force in cold-rolled tapes, the guided motion of the vortexes, the growth of the critical current with increasing field in small magnetic fields and etc. [8]. Such a local decrease in the coherence length leads to the fact that the upper critical field is locally increased there, as the relation between these quantities is [1]:

$$\mu_0 H_{c2}(T) = \Phi_0 / 2\pi\xi(T)^2 \tag{1}$$

where Φ_0 - flux quantum, ξ - coherence length, H_{c2} - upper critical field. Thus, in a range of sufficiently high

* Corresponding author: GuryevVV@mail.ru

magnetic fields, it is to be expected that the grain bodies will transform into a normal state, while the grain boundary will still be superconducting. Indeed, this division into two superconducting components with different H_{c2} has been observed experimentally earlier: through the features of the appearance of the satellite electric field [9], and through the apparent anisotropy of the upper critical field [10].

2 Methods and samples

The irreversibility field has been investigated on a cold-rolled Nb-Ti tape. Electrodynamics and the structure of this tape have been studied in details by our group earlier [8-10]. This is a single-phase tape of the Nb-50%wtTi alloy, with the pinning due to the reduction of the mean free path of the electron at the grain boundaries. Besides, the tape has been investigated after additional heat treatment in high vacuum at 385 °C during 25 hours. As a result, 6% by volume of α-Ti have been precipitated. This has changed the pinning mechanism sensibly, bringing it closer to the mechanism which is implemented in commercial NbTi wires.

The magnetization loops were measured in a shunted superconducting magnet up to 13 T in liquid helium, by a couple of Hall sensors, under one of which a sample was placed. Thus, the magnetization was determined from the difference of the signals. The magnetic field was oriented orthogonal to the plane of the Nb-Ti tape. For the purpose of amplification the wanted signal, the tested samples were stacks of 10 tapes cut in the form of a square 6 mm x 6 mm. The field sweep rate was controlled by the sweep rate of solenoid current and was approximately 0,008 T/s.

Transport measurements were carried out in the same superconducting solenoid. Flat samples, adhered to dielectric corundum substrate, with bridge geometry along the rolling direction were made by laser cutting (the bridge was 1.0 mm in width and 6.7 mm in length) by analogy with [10]. In all experiments, the field was perpendicular to the plane of the tapes, so the pinning force was determined as a simple product of the field and the critical current density, determined by the criterion 1 μV/cm.

3 Results

Figure 1 shows experimental data from the magnetization, the pinning force density and resistivity as a function of applied field for the cold-rolled NbTi tape. A linear background signal, which is present above H_{c2}, has been subtracted from the magnetization data (fig. 1a), and hence the curve becomes flat at fields beyond H_{c2}. After the collapse of the $M(H)$ loop at the field H^{*}~10.6, one can see a weak but reproducible inverse (ferromagnetic) hysteresis whose nature is difficult to explain. The hysteresis completely disappears in fields above ~11.3 T. Since the value of this hysteresis is only about 2 times greater than the noise, the

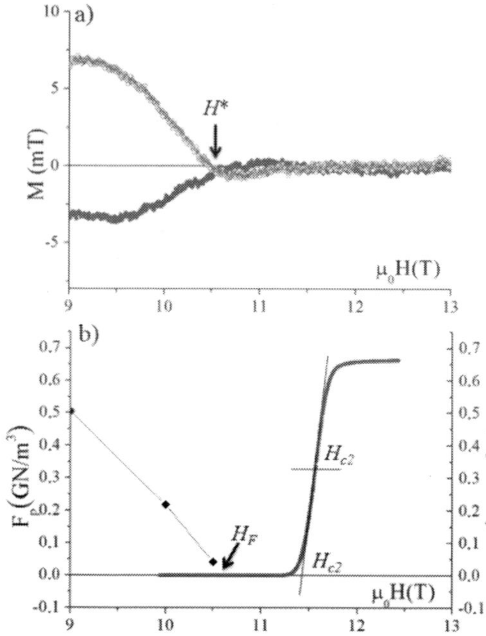

Fig. 1. Experimental magnetization, pinning force, and resistivity data for cold-rolled NbTi tape as a function of applied field. An extricating current of 5 mA was used for resistivity measurements

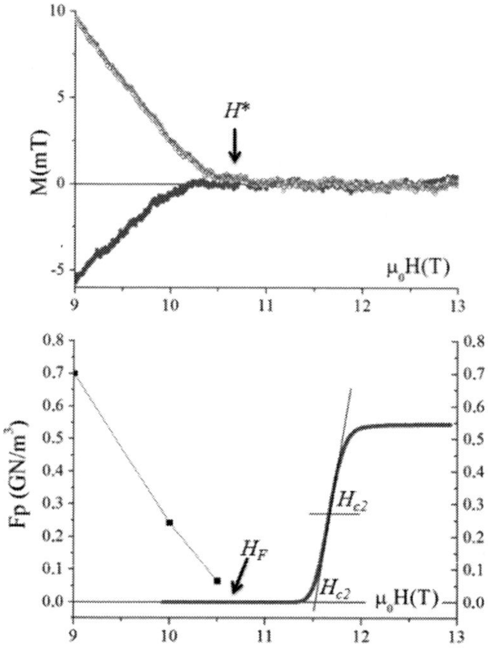

Fig. 2. Experimental magnetization, pinning force, and resistivity data for heat-treated NbTi tape as a function of applied field. An extricating current of 5 mA was used for resistivity measurements.

magnetization can be regarded as reversible. No pronounce changes in the slope of reversible magnetization around H_{c2} have been observed. This is possibly related to roughness of used method.

The irreversibility field has a clear correlation with the extrapolation of the pinning force to zero (H_F ~ 10.6T) (fig. 1b). The upper critical field (H_{c2} = 11.6T) has been determined from measurements of the resistivity with an exciting current of 5 mA. It was shown in [10] that a decrease in the current below does not lead to a significant shift in H_{c2} determined in the middle of the transition. An alternative way to determine upper critical field [6] by the intersection of the linear extrapolation of the resistive transition through the baseline voltage gives value H_{c2}=11.4 T. Thus, the difference between H^* and H_{c2} is about 0.9 T.

Experimental data for heat-treated sample are shown on figure 2. After heat treatment, the normal resistance has decreased by 20% and pinning in the 9 T field has increased by 40%. As in the case of cold-rolled, no changes in the magnetization slope near H_{c2} (11.5T or 11.7T, depending on the criteria) are observed, there is a clear correlation between H_F (10.6T) and H^* (10.6T) and the difference between H^* and H_{c2} is about 1T.

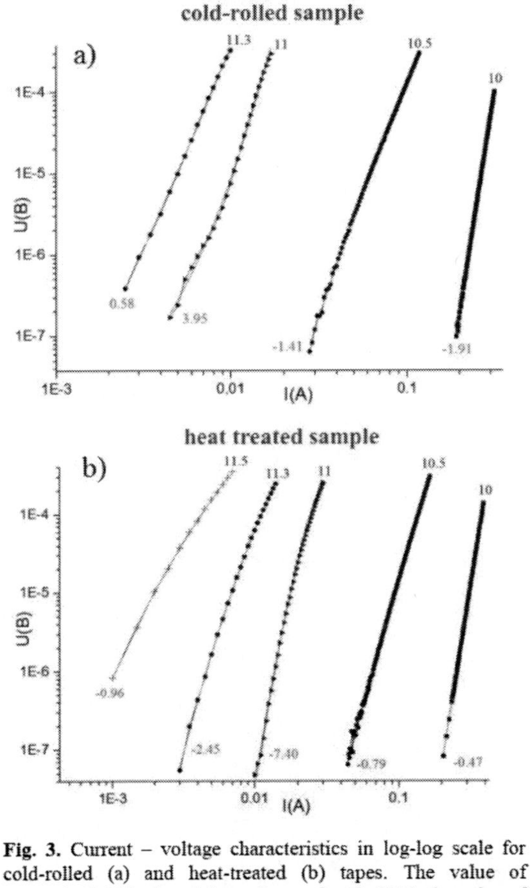

Fig. 3. Current – voltage characteristics in log-log scale for cold-rolled (a) and heat-treated (b) tapes. The value of magnetic field is signed from above of each CVC; the value of the coefficient in front of the quadratic term of approximation is signed from below each CVC.

The irreversibility field is often associated with a vortex matter phase transition. A common practice of the identification such a transition is to analyze the current – voltage characteristics (CVCs) plotted in log-log space. The transition identified as the field at which the CVC curvature changes sign. Figure 3 shows the CVCs in double logarithmic scales for different values of an external magnetic field (signed from above in Tesla), and also approximations (thin red lines) by quadratic fit. The values of the coefficient of the second order of this approximation are signed below each CVC.

4 Discussions

The obtained data validate the existence of the irreversibility field both in cold-rolled and heat-treated Nb-Ti tapes. The hypothesis of the phase transition of vortex matter from glass state to liquid state, as the cause of the irreversibility field, clearly fails, as can be seen from Fig. 3. Indeed, in accordance with modern concepts, one of the necessary conditions for the phase transition of the vortex matter is the change in the curvature of CVCs at the irreversibility field [11]. At that, in the fields below the irreversibility field, all CVCs have a negative curvature; in fields above the irreversibility field they have a positive curvature. For the cold-rolled tape, some experimental CVCs change the curvature several times, which contradicts the model of the vortex matter phase transition, in which the curvature should be either positive or negative. For the heat-treated tape, all the CVCs have a negative curvature up to the upper critical field.

On the other hand, in the statistical inhomogeneity model, the changes in the sign of the CVC curvature do not have a deep physical meaning, but are determined by the specific character of the inhomogeneity of the sample [12].

The fact that the irreversibility field has a clear correlation with the extrapolation of the volume pinning force, what was also observed in [6], hints at the fact that both characteristic fields are a manifestation of a single physical phenomenon. It was shown in [9, 10] that in the fields above H_F only the grain boundaries remain in superconducting state, while the grain bodies become normal. This mechanism gives a qualitative explanation for such dramatic decrease in the critical current and the appearance of reversible magnetization, since superconducting currents are mainly caused not by pinning, but by reversible surface currents.

As a result of the rolling, the grains of NbTi are flattened; therefore in fields above H^* the cold-rolled NbTi tape becomes a strongly anisotropic superconductor, which can be roughly represented as a stack of superconducting films (grain boundaries) [10]. This fact makes it possible to draw a formal analogy with high-T_c layered superconductors with a large parameter of anisotropy. As a result of heat treatment, α-Ti particles are precipitated at the grain boundaries, disrupting their flatness and decreasing the anisotropy [10], the pinning force is enhanced substantially, but the irreversibility field remains unchanged. Thus, the irreversibility field is not directly related to the pinning force in low-T_c superconductors, as it is usually assumed in models which attach to thermal fluctuations a decisive role in the mechanisms of the phenomenon.

5 Conclusion

The field of irreversibility of the cold-rolled and additionally heat-treated NbTi tapes has been investigated. Based on the analysis and generalization of the experimental facts obtained in this work, as well as those obtained earlier in our group, it is concluded that a dramatic decrease in the critical current in fields above the irreversibility field of NbTi tapes is a direct consequence of the inhomogeneity created during production process, in order to achieve high current carrying capacity in lower fields.

References

1. T. Matsushita, *Flux pinning in superconductors* (Springer, 2014)
2. M. Tinkham, IEEE Trans. On Mag. **27** (1991)
3. M. F. Schmidt, N. E. Israeloff, A. M. Goldman Phys. Rev. Lett., **70** (1993)
4. M. Suenaga, A. K. Ghosh, Y. Xu, D. O. Welch, Phys. Rev. Lett., **66** (1991)
5. D. N. Zheng, N. J. C. Ingle, A. M. Cambell, Phys. Rev. B, **61** (2000)
6. M. T. Naus, R. W. Heussner, A. A. Squitieri, D. C. Larbalestier, IEEE Trans. on App. Super. **7** (1997)
7. T. Matsushita, Phys. C **214** (1993)
8. E. Yu. Klimenko, S. V. Shavkin, P. V. Volkov, Zh. Eksp. Teor. Fiz **112** (1997)
9. E. Yu. Klimenko, S. V. Shavkin, P. V. Volkov, Phys. Met. Metallorg. **92** (2001)
10. V. V. Guryev, S. V. Shavkin, V. S. Kruglov, P. V. Volkov, A. L. Vasiliev, A. V. Ovcharov, I. A. Likhachev, E. M. Pashaev, R. D. Svetogorov, Y. V. Zubavichus, J. of Phys.: Conf. Ser. **747** (2016)
11. D.R. Strachan, M.C. Sullivan, P. Fournier, S.P. Pai, T. Venkatesan, C.J. Lobb, Phys. Rev. Lett. **87** (2001)
12. E. Yu. Klimenko, *Superconductivity Theory and Applications* ed. by A. M. Luiz (InTech, 2011)

EPJ Web of Conferences **185**, 08005 (2018) https://doi.org/10.1051/epjconf/201818508005
MISM 2017

Nonlinear Response of a Superconductor of the Bi-Sr-Ca-Cu-O System to an Alternating Magnetic Field in the Superconducting Transition Temperature Range

Alexander Sergeev[1,*], *Igor* Golev[2], *Victoria* Gvozdevskaya[1] and *Anastasia* Barkalova[1]

[1]Voronezh State Technical University, 394026 Moscow Avenue 14, Voronezh, Russia
[2]The Military Educational and Training Center of Air Forces «The Air Force Academy, named after Prof. N.E. Zhukovsky and Y.A. Gagarin», 394064 ul. Old Bolsheviks 54 "A", Voronezh, Russia

Abstract. The nonlinear response of the superconductor of the Bi-Sr-Ca-Cu-O system in the temperature range of the superconducting transition under the action of a harmonic alternating magnetic field is experimentally studied. For multiphase superconductors having in their volume regions with distinct critical temperatures, the effect of odd harmonics in the response signal is observed. The contribution of crystallites and the system of weak bonds between the crystallites in the nonlinear response is singled out. It was found that the nonlinear properties of the investigated samples in the resistive state are determined mainly by the nonlinear current-voltage characteristics of the system of weak bonds between the crystallites.

1 Introduction

The study of the behavior of high-temperature superconductors (HTSC) under the influence of alternating magnetic fields of different amplitudes still raises an increased practical interest. The results obtained can be a source of additional information on physical processes occurring in high-temperature superconductors with a complex real crystal structure, including regions with different chemical composition, crystallites, and intercrystalline media [1-4].

This paper presents the results of experimental studies of the nonlinear properties of ceramic HTSC Bi-Sr-Ca-Cu-O systems during their transition from the normal to the superconducting state, when the volume of the superconductor has an electrical resistance to a constant electric current.

2 Samples and experimental procedure

Samples of the Bi-Sr-Ca-Cu-O system measuring $3.2 \times 3.2 \times 20$ mm^3, obtained by solid-phase synthesis, were studied. The X-ray photograph is shown in fig.1. The fig. 2 presents the photograph of the sample surface microstructure, obtained using an electronic scanning microscope with a detector of inversely scattered electrons (BEI). The phase composition of samples was determined by the Rietveld method. The samples consisted of two phases: Bi-2223 – 47.6 % (a = 5.411 Å, b = 5.411 Å, c = 37,21 Å) and Bi-2212 – 52.3 % (a = 5.397 Å, b = 5.4 Å, c = 30.783 Å). The temperature of the onset of the superconducting transition of the sample is T = 103 K, the temperature of the end of the superconducting transition is T = 94 K, the sample density was 4.1 g / cm^3.

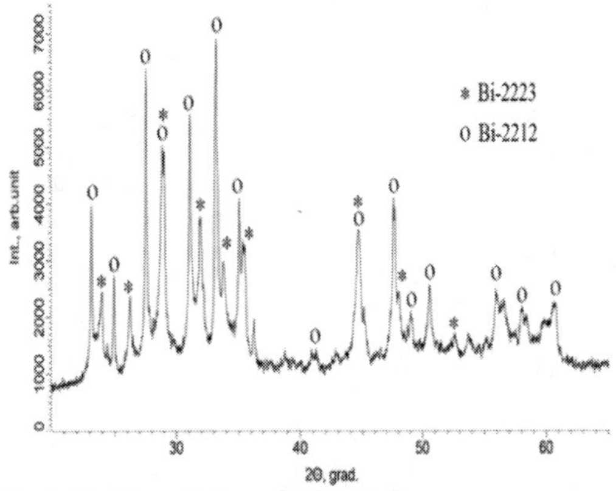

Fig. 1. The X-ray photograph of a sample

In the experiments we measured the voltages of odd (n = 1, 3, 5, 7) harmonics from a measuring single-layer coil wound around the sample under the action of a low-frequency alternating magnetic field. The alternating magnetic field was directed along the principal axis of symmetry of the superconductor sample. Harmonic components from the response signal of a superconductor sample were isolated using a selective microvoltmeter. In Fig. 3 is a schematic diagram of the measuring stand.

* Corresponding author: sergeev-av@bk.ru

Fig. 2. The microstructure of a sample split

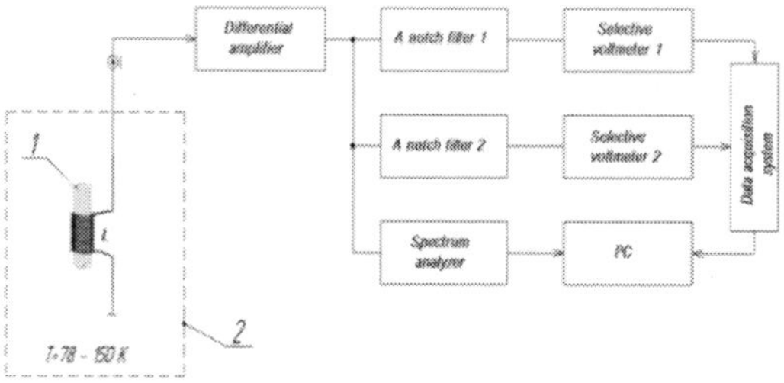

1 – sample, 2 – thermo-stabilized volume
Fig. 3. The circuit diagram of the measuring unit

3 Results and discussion

In Fig. 4 shows the temperature dependences of the first harmonic of the response signal of a superconductor on an alternating magnetic field and the resistivity. In Fig. 5 shows the temperature dependences of the odd harmonics (third, fifth, seventh) of the sample response signal to an alternating magnetic field.

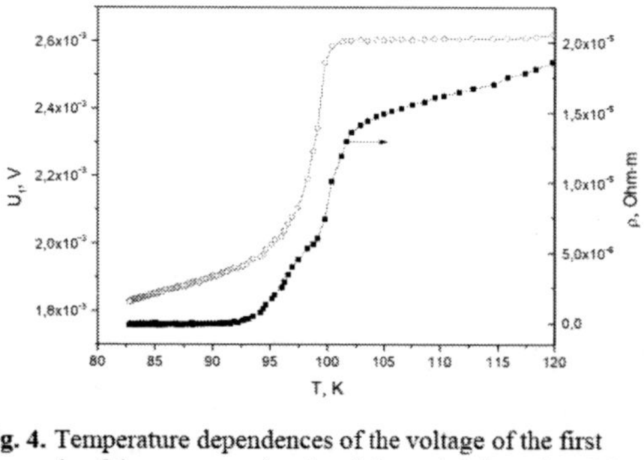

Fig. 4. Temperature dependences of the voltage of the first harmonic of the response signal and the resistivity at b = 5 G; f = 1000 Hz

As can be seen from the results of the experiments (Fig. 5), maxima are observed in the superconducting transition temperature range, with a specific electric resistance $\rho > 0$ on the dependences $U_3(T)$, $U_5(T)$, $U_7(T)$. In addition, the maximum value of the harmonic voltage ($U_3(T)$, $U_5(T)$, $U_7(T)$) decreases with decreasing ordinal number of the harmonic, and the temperature of realization of the maximum does not change.

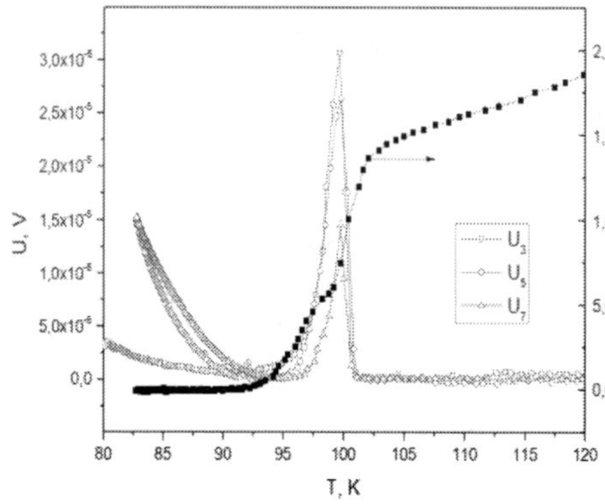

Fig. 5. Temperature dependences of the voltages of the third, fifth, seventh harmonics of the response signal at b = 5 G; f = 1000 Hz

In the future, only $U_3(T)$ dependences will be presented, characterizing the nonlinearity of the superconductor response to an alternating magnetic field. Even harmonics were not detected under the action of an alternating magnetic field of different amplitudes [5].

The effect of the amplitude of the alternating magnetic field b on the temperature dependence of the third harmonic of the response signal was studied. As seen from the experimental data, the temperature region T = 120 ÷ 102 K, the third harmonic U_3 has a zero value, i.e. The sample responds linearly to the alternating magnetic field (Fig. 6).

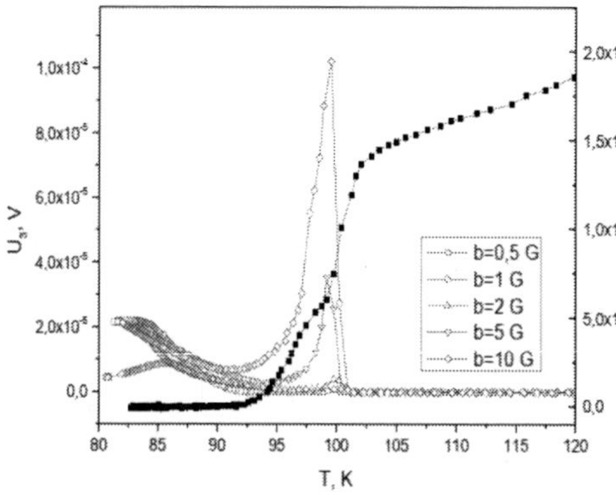

Fig. 6. Influence of the amplitude of the alternating magnetic field on the dependence $U_3(T)$ at f = 1000 Hz

https://doi.org/10.1051/epjconf/201818508005

The investigated superconductors of the Bi-Sr-Ca-Cu-O system have a complex structure consisting of Bi-2223 and Bi-2212 phase crystallites separated by weak bonds with weaker critical parameters (J_c, T_c). An electric current induced by an alternating magnetic field with density J transforms the weak links into a resistive state and their current-voltage characteristic becomes nonlinear. It can be described by the following power-law dependences [6]

$$U = AJ^N, \quad (1)$$

or

$$U = (J - J_c)^N, \quad (2)$$

where $U(J_c)$ is the voltage drop in the elementary volume of the superconductor; J_c is the critical current; A, N are constants determined experimentally. The quantity $U(J_c)$ depends nonlinearly on the current density, which determines the nonlinear properties of the superconductor and, accordingly, the signal of its response to the alternating magnetic field. If N is an odd number, which is observed in our experiments, then the odd harmonics are present in the response signal of the superconductor. (see Fig. 5).

Let us single out the contributions of the nonlinear response of crystallites and the system of weak bonds between the crystallites. To do this, we analyze the dependencies obtained for the powder obtained from the sample.

The results of measurements of the temperature dependence of the third harmonic of the response signal to an alternating magnetic field for a powder are shown in Fig. 7. The value of $U_3(T)$ was normalized per unit volume, taking into account the sample density.

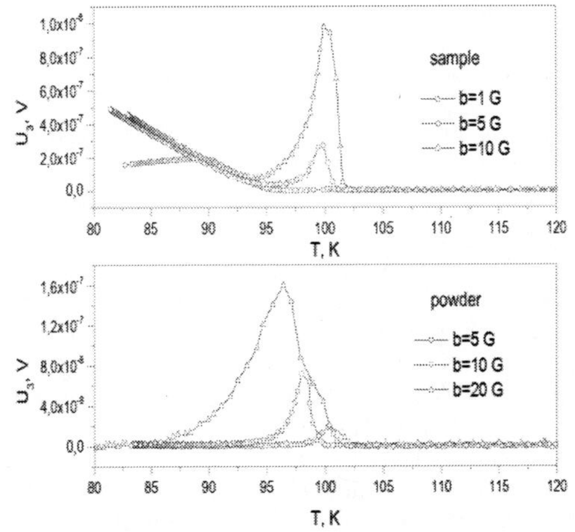

Fig. 7. Influence of the amplitude of the alternating magnetic field on the dependence $U_3(T)$ at f = 1000 Hz

As can be seen from Fig. 7, for a powder obtained from a sample, in the absence of a percolation superconducting network, only one maximum is observed on the $U_3(T)$ curve, the mechanism of which is analogous to the third harmonic generation mechanism in the integral samples described in [3]. The contribution of the nonlinear response in the volume of crystallites of the Bi-2223 phase is approximately 100 times less than in the system of weak bonds of the Bi-2223 phase.

4 Conclusions

The nonlinear properties of the investigated samples in the resistive state are determined mainly by the nonlinear current-voltage characteristics of the system of weak bonds between the crystallites. The observed effect of the generation of odd harmonics in the response signal to an alternating magnetic field can be used as a tool for investigating the dissipative processes of multiphase or inhomogeneous superconductors.

References

1. V. Ya. Pokrovskii, I. G. Gorlova, and S. G. Zybtsev, Physica C, **4**, 2077 (1997).
2. N. V. Dalakova, B. I. Belevtsev, E. Yu. Belyaev, and A.S. Panfilov, Low Temp. Phys. **40** (5), 397 (2014).
3. A. I. Golovashkin, N. D. Kuz'michev, and V. V. Slavkin, Tech. Phys. **53** (10), 1314 (2008).
4. I. M. Golev, A. V. Sergeev, V. G. Kadmenskii, and O.V.Kalyadin, Vestn. Voronezh. Gos. Tekh. Univ. **11** (5), 129 (2015).
5. I. M. Golev, A. V. Sergeev, and O.V.Kalyadin, Phys. Solid State, **59** (1), 19 (2017).
6. Babic E., Prester M., Biskup N. Phys. Rev.B, **41** (10A), 6278 (1990).

Unconventional phase transitions in strongly anisotropic 2D (pseudo)spin systems

Vitaly Konev[1,*], *Evgeny* Vasinovich[1], *Vasily* Ulitko[1], *Yury* Panov[1], and *Alexander* Moskvin[1]

[1] *Ural Federal University, Ekaterinburg, 620083, Russia*

Abstract. We have applied a generalized mean-field approach and quantum Monte-Carlo technique for the model 2D $S = 1$ (pseudo)spin system to find the ground state phase with its evolution under application of the (pseudo)magnetic field. The comparison of the two methods allows us to clearly demonstrate the role of quantum effects. Special attention is given to the role played by an effective single-ion anisotropy ("on-site correlation").

1 Introduction

These days spin algebra and spin Hamiltonians are used not only in the traditional fields of spin magnetism and magnetic resonance but in so-called pseudospin lattice systems with the on-site occupation constraint (see, e.g., Ref. [1]). Standard pseudospin formalism represents a variant of the equivalent operators technique widely known in different physical problems from classical and quantum lattice gases, binary alloys, (anti)ferroelectrics,.. to neural networks, however, usually for simplest $s = 1/2$ pseudospin value. At variance with quantum $s = 1/2$ systems the Hamiltonian of $S = 1$ spin lattices in general is characterized by several additional terms such as a single ion anisotropy, biquadratic isotropic and anisotropic exchange couplings, that results in their rich phase diagrams and novel types of the order such as quantum paramagnet and spin-nematic order.

Typical $S = 1$ spin Hamiltonian with an uniaxial on-site and exchange anisotropies reads as follows:

$$\hat{H} = \sum_{i>j} J_{ij}(S_{ix}S_{jx} + S_{iy}S_{jy} + \lambda S_{iz}S_{jz}) + \\ + \sum_i DS_{iz}^2 - \sum_i hS_{iz}. \quad (1)$$

The spectrum of the spin Hamiltonian (1) in the absence of external magnetic field changes drastically as D varies from very small to very large positive or negative values. A strong "easy-plane" anisotropy for large positive $D > 0$ favors a singlet phase where spins are in the $S_z = 0$ ground state. This "quadrupole" phase has no magnetic order, and is aptly referred to as a quantum paramagnetic phase (QPM), which is separated from the "ordered" state by a quantum critical point (QCP) at some $D = D_C$. This is a quadrupole state with no magnetic order, so that all linear order parameters vanish and only a quadrupole (spin-nematic) order parameter such as $Q_{zz} = \left\langle S_z^2 - \frac{2}{3}\right\rangle$ is nonzero. The QPM phase consists of a unique ground state with total spin $S_z^{total} = 0$, separated by a gap from the first excited states, which lie in the sectors $S_z^{total} = \pm 1$. This is purely a quantum phase without a counterpart in the classical models. The QPM order differs in principle from the conventional paramagnetic state, because for $S = 1$ in the classical paramagnetic state $\langle S_x^2\rangle = \langle S_y^2\rangle = \langle S_z^2\rangle = 2/3$, while in the quantum paramagnetic state $\langle S_z^2\rangle = 0$, $\langle S_x^2\rangle = \langle S_y^2\rangle = 1$.

Quite a few compounds can be mentioned with the quantum paramagnetic ground state of the considered type which are described by the Hamiltonian (1), these are several Ni^{2+} compounds with a real $S = 1$ ground state and Fe^{2+} compounds $CsFeBr_3$, $CsFeCl_3$ with a ground state described by the fictitious S=1 spin (pseudospin) due to strong spin orbit coupling (see, e.g., Ref. [2] and references therein).

However, studying anisotropic $S = 1$ spin and pseudospin systems is of a great importance not only for quantum magnets but for different bosonic-like systems with the on-site Hilbert space truncated to the three lowest occupation states $n = 0, 1, 2$, in particular, for so-called semi-hard core bosons. Furthermore, the $S = 1$ pseudospin formalism was suggested to describe the charge degree of freedom in a model high-T_c cuprate with the on-site Hilbert space reduced to the only triplet of the three effective valence centers $[CuO_4]^{7-,6-,5-}$ (nominally $Cu^{1+;2+;3+}$), where the electronic and lattice degrees of freedom get strongly locked together [3–6]. Indeed, recently [3] it was argued that an unique property of high-T_c cuprates is related with a dual nature of the Mott insulating state of the parent compounds that manifests itself in two distinct energy scales for the charge transfer (CT) reaction: $Cu^{2+} + Cu^{2+} \rightarrow Cu^{1+} + Cu^{3+}$. The d-d CT gap as derived from the optical measurements in parent cuprates such as La_2CuO_4 is 1.5-2.0 eV while the true (thermal) d-d CT

*e-mail: konefvitbka53@gmail.com

gap, or effective correlation parameter U_d, appears to be as small as 0.4-0.5 eV. It means the cuprates should be addressed to be *d-d* CT unstable systems whose description implies accounting of the three many-electron valence states, or nominally the $Cu^{1+,2+,3+}$ centers, on an equal footing as a well-defined charge triplet. Effective $S = 1$ pseudospin Hamiltonian for such a cuprate does incorporate all the on-site and inter-site couplings, three types of correlated one-particle (fermionic) transport and a two-particle (bosonic) charge transfer with a charge density constraint. With the exception of the three-spin terms the model becomes equivalent to a generalized strongly anisotropic $S = 1$ quantum magnet in an external magnetic field with a constraint on the magnetization.

The Hamiltonian implies possible simplifications, in particular, so-called negative-U ($U = 2D$) model, which in the limit of large negative U becomes equivalent to the model of hard-core bosons. For the model we made use of a special algorithm for CUDA architecture for NVIDIA graphics cards, a nonlinear conjugate-gradient method to minimize energy functional, and Monte-Carlo technique for large square lattices 256×256 to directly observe the forming of the ground state configuration with lowering the temperature and its evolution under deviation from half-filling [7]. The technique allowed us to examine earlier implications and uncover novel features of the phase transitions, in particular, look upon the nucleation of different metastable topological structures, in particular, odd domain structure with emergence of filamentary superconductivity nucleated at the antiphase domain walls in the insulating charge-ordered phase.

Here, in the paper, we have applied a generalized mean-field approach and quantum Monte-Carlo (QMC) technique for a simplified model 2D $S = 1$ (pseudo)spin system to find the ground state phase with its evolution under application of the (pseudo)magnetic field. Despite the generalized MFA implies a quantum-mechanical treatment of the $S = 1$ sites in a molecular field the more comprehensive QMC technique allowed us to clearly uncover the role of quantum effects.

2 $S = 1$ (pseudo)spin Hamiltonian

The $S = 1$ spin algebra includes the eight nontrivial independent spin operators: spin-dipole moment **S** and five spin-quadrupole operators $Q_{ij} = \left(\frac{1}{2}\{S_i, S_j\} - \frac{2}{3}\delta_{ij}\right)$ whose mean values define so-called spin-nematic order. Spin operators $S_\pm$ and $T_\pm = \{S_z, S_\pm\}$ change the pseudospin projection (and occupation number!) by ±1, while $S_\pm^2$ changes the pseudospin projection by ±2.

Hereafter in the paper we will focus on a simplified 2D $S = 1$ (pseudo)spin Hamiltonian for bosonic-like systems with the on-site Hilbert space truncated to the three lowest occupation states n = 0, 1, 2, in particular, for so-called semi-hard core bosons, or for model high-T_c cuprate with the on-site Hilbert space reduced to the only triplet of the three effective valence centers $[CuO_4]^{7-,6-,5-}$ (nominally $Cu^{1+;2+;3+}$). We restrict ourselves by the nearest neighbour coupling and the only two-particle transport term (inter-site biquadratic anisotropy) so the (pseudo)spin Hamiltonian takes the form as follows:

$$\hat{H} = \sum_i (\Delta S_{iz}^2 - \mu S_{iz}) + V \sum_{\langle ij \rangle} S_{iz} S_{jz} - t \sum_{\langle ij \rangle} (S_{i+}^2 S_{j-}^2 + S_{i-}^2 S_{j+}^2), \quad (2)$$

where $V > 0$, $t > 0$. The first single-site term in $\hat{H}$ describes the effects of a bare pseudo-spin splitting and relates with the on-site density-density interactions, or correlations: $\Delta = U/2$. The second term, or a pseudospin Zeeman coupling may be related with a pseudo-magnetic field $\parallel z$ which acts as a chemical potential μ for boson systems with a boson density constraint:

$$\frac{1}{N} \sum_i \langle S_{iz} \rangle = n \,, \quad (3)$$

where n is the deviation from a half-filling ($n = 0$).

The third (Ising) term in $\hat{H}$ describes the effects of the short- and long-range inter-site density-density interactions. The last term in $\hat{H}$ describes the two-particle inter-site hopping. In the strong on-site attraction limit of the model (large easy-axis pseudospin on-site anisotropy) we arrive at the Hamiltonian of the hard-core, or local, bosons which was earlier considered to be a starting point for explanation of the cuprate high-T_c superconductivity [8]. The spin counterpart of $\hat{H}$ corresponds to an anisotropic $S = 1$ magnet with a single ion (on-site) and two-ion (bilinear and biquadratic) symmetric anisotropy in an external magnetic field. It describes an interplay of the Zeeman, single-ion and two-ion anisotropic terms giving rise to a competition of an (anti)ferromagnetic order along Z-axis with an in-plane XY spin-nematic order. A remarkable feature of the Hamiltonian (2) is that the on-site pseudospin states $M = 0$ and $|M| = 1$ do not mix under the inter-site coupling. The model allows us to directly study a continuous transformation of the semi-hard-core bosons to the effective hard-core bosons formed by boson pairs under driving the correlation parameters $\Delta = U/2$ to large negative values ("negative-U model"). The simplified model can be directly applied to a description of bosonic systems with suppressed one-particle hopping.

3 Mean-field approximation

To analyse the simplified model we start with a mean-field approximation (MFA) for 2D square lattice, however, at variance with a conventional classical MFA we made use of more correct approach that takes into account the quantum nature of the $S = 1$ (pseudo)spin states [9]. First we introduce a set of the on-site $S = 1$ coherent states

$$|\mathbf{c}\rangle = c_{-1}|-1\rangle + c_0|0\rangle + c_{+1}|+1\rangle \,, \quad (4)$$

where the c_M coefficients can be represented as follows

$$c_1 = \sin\frac{\theta}{2}\cos\frac{\phi}{2}e^{-i\frac{\alpha}{2}},\, c_0 = \cos\frac{\theta}{2}e^{i\frac{\beta}{2}},\, c_{-1} = \sin\frac{\theta}{2}\sin\frac{\phi}{2}e^{i\frac{\alpha}{2}} \quad (5)$$

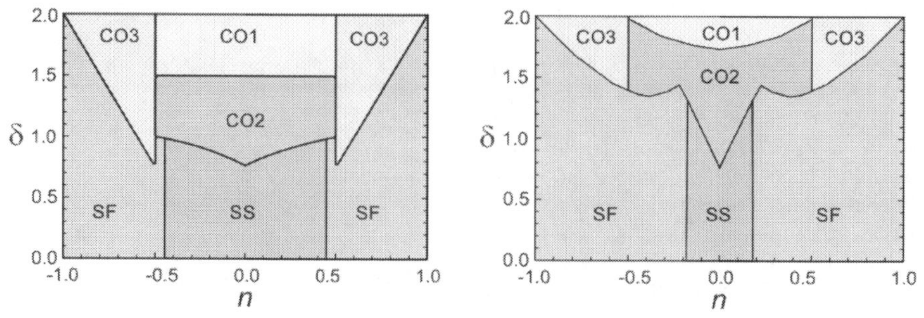

Figure 1. (Color online) The δ - n ground state phase diagrams for the model system given $V/t = 0.75$ (left panel, the MFA results, right panel, the QMC results).

with $\theta, \phi, \alpha, \beta$ to be parameters defined by the minimization of the energy. The MFA energy can be written as follows

$$E = \frac{\Delta}{2}\sum_i(1-\cos\theta_i) - \frac{\mu}{2}\sum_i(1-\cos\theta_i)\cos\phi_i +$$

$$+\frac{V}{4}\sum_{\langle ij\rangle}(1-\cos\theta_i)(1-\cos\theta_j)\cos\phi_i\cos\phi_j -$$

$$-\frac{t}{8}\sum_{\langle ij\rangle}(1-\cos\theta_i)(1-\cos\theta_j)\sin\phi_i\sin\phi_j\cos(\alpha_i-\alpha_j). \quad (6)$$

It is worth noting that the energy does not depend on β. Below we denote $\delta = \Delta/t$ and $v = V/t$. In a two-sublattice A-B model which seems to be a reasonable approximation for the system with the nearest-neighbor coupling we arrive at a high-temperature non-ordered (NO) phase and the five MFA uniform phases, two phases with nonzero local superfluid order parameter, or pseudospin nematic order $\langle S^2_{A,B\pm}\rangle \neq 0$ and three charge ordered phases with $\langle S^2_{A,B\pm}\rangle = 0$ but different types of the sublattice occupation (pseudospin S_z components):

Superfluid (SF) phase: $\langle S_{A,Bz}\rangle = n$, $\langle S^2_{A,Bz}\rangle = 1$, $\langle S^2_{A,B\pm}\rangle = \frac{\zeta}{2}\sqrt{1-n^2}e^{\pm i\alpha}$, uncertain factor $\zeta = \pm 1$.

Supersolid (SS) phase: $\langle S^2_{A,Bz}\rangle = 1$,

$$\langle S_{A,Bz}\rangle = n \mp \sqrt{1+n^2-\frac{4|n|v}{\sqrt{4v^2-1}}},$$

$$\langle S^2_{A,B\pm}\rangle = \frac{\zeta}{2}e^{\pm i\alpha}\times$$

$$\times\left(\sqrt{|n|\sqrt{\frac{2v+1}{2v-1}}-n^2} \pm \operatorname{sgn} n\sqrt{|n|\sqrt{\frac{2v-1}{2v+1}}-n^2}\right).$$

Charge ordered CO1 phase: $\langle S_{Az}\rangle = 0$, $\langle S^2_{Az}\rangle = 0$, $\langle S_{Bz}\rangle = 2n$, $\langle S^2_{Bz}\rangle = 2|n|$, $(|n| \le 0.5)$.

Charge ordered CO2 phase: $\langle S_{Az}\rangle = 2n - \operatorname{sgn} n$, $\langle S^2_{Az}\rangle = 1-2|n|$, $\langle S_{Bz}\rangle = \operatorname{sgn} n$, $\langle S^2_{Bz}\rangle = 1$, $(|n| \le 0.5)$.

Charge ordered CO3 phase: $\langle S_{Az}\rangle = \operatorname{sgn} n$, $\langle S^2_{Az}\rangle = 1$, $\langle S_{Bz}\rangle = 2n - \operatorname{sgn} n$, $\langle S^2_{Bz}\rangle = 2|n| - 1$, $(|n| \ge 0.5)$.

Table 1. The energies of the MFA ground-state phases.

Phase	$E/(tN)$
SF	$\delta - 1 + n^2(2v+1)$
SS	$\delta - 2v + 2\|n\|\sqrt{4v^2-1}$
CO1	$\|n\|\delta$
CO2	$(1-\|n\|)\delta + 4\left(\|n\|-\frac{1}{2}\right)v$
CO3	$\|n\|\delta + 4\left(\|n\|-\frac{1}{2}\right)v$

The energies of the MFA ground state phases are listed in Table 1. Note that the energy of CO1 phase does not depend on inter-site interaction parameter v.

The density of superfluid component is related to helicity modulus [10]. This allows us to find an expression for the superfluid density ρ in terms of the local superfluid order parameters in the two-sublattice MFA:

$$\rho = \operatorname{Re}\left(\langle S^2_{A+}\rangle\langle S^2_{B-}\rangle\right). \quad (7)$$

The superfluid density in the SF phase, $\rho = (1-n^2)/4$, has a maximum value at $n = 0$ and does not depend on the the inter-site density-density interaction parameter v. In the SS phase, the superfluid density $\rho = |n|/(2\sqrt{4v^2-1})$ decreases with rising of v.

Interestingly, all the local order parameters do not depend on the correlation parameter Δ, while this parameter governs the energy of different phases. Taking into account the on-site correlations we arrive at very rich and intricate phase diagrams for the model system as compared with relatively simple phase diagrams for hard-core bosons [8, 11]. In Fig. 1 (left panel) we present an example of the MFA δ - n phase diagrams calculated given $v = 0.75$. At half-filling $n = 0$ the positive values of the correlation parameter δ stabilize a limiting CO1 phase with $\langle S_{A,Bz}\rangle = \langle S^2_{A,Bz}\rangle = 0$, or a "parent Cu^{2+} phase" for a model cuprate, while positive values of v stabilize a limiting CO2 phase with $\langle S_{A,Bz}\rangle = \pm 1$; $\langle S^2_{A,Bz}\rangle = 1$, or a checkerboard "antiferromagnetic" order of pseudospins along z-axis, or a disproportionated Cu^{1+}-Cu^{3+} phase for a model cuprate. As a result of the competition between the on-site and inter-site correlations we arrive at a "starting" CO1 phase for

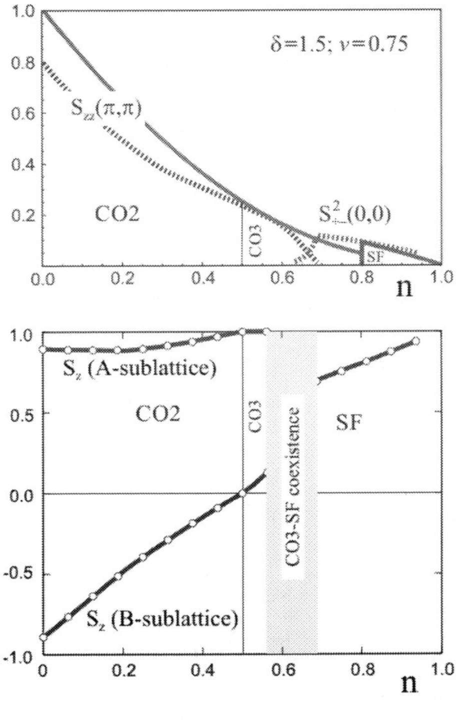

Figure 2. (Color online) Top panel: Correlation functions for the model $S = 1$ pseudospin system given $\Delta/t = 1.5$, $V/t = 0.75$, solid lines are the MFA results, dotted lines are the QMC results. Bottom panel: QMC data for the sublattice S_z-components as functions of the deviation from the half-filling. Filling points to a CO3-SF coexistence phase typical for the first kind phase transition.

$\delta > 2v$ or CO2 phase for $\delta \leq 2v$. At $n = 0.5$ we see a transformation of the CO1 and CO2 phases into the CO3 phase. The line of the first order phase transition CO3-SF in Fig. 1 corresponds to the equality of the respective energies. It is worth to note that the critical concentration n for the SS-SF, CO1, CO2-CO3 transitions does not depend on the correlation parameter δ. In Fig. 2 (top panel, solid lines) we present the n-dependence of the correlation functions $S_{zz}(\pi,\pi) = \langle S_z, S_z\rangle$ (static structure factor) and $S^2_{+-}(0,0) = \langle S^2_+, S^2_-\rangle$ at $\delta = 1.5$, $v = 0.75$, determining the long-range CO and SF orders, respectively, given $\delta = 1.5$, that is in an immediate closeness to CO2-CO1 phase transition for small n.

4 Quantum Monte-Carlo calculations

We have performed QMC calculations for our model Hamiltonian (2). In our QMC calculations, we use the D. Galanakis open source program [12]. This program is an implementation of the Stochastic Green Function algorithm for the simulation of bosonic Hamiltonians [13]. During the simulation, we performed 5×10^6 Monte-Carlo steps for equilibration and 10×10^6 Monte-Carlo steps for measurement the values under investigation in the square lattice 12×12 with periodical boundary conditions. In Fig. 1 (right panel) we compare the ground state δ - n phase diagram of our model 2D system calculated on square lattice 12×12 given $v = 0.75$ with that of calculated within a generalized MFA approach (left panel). First of all, it should be noted that the QMC calculations do not "obey" the MFA two-sublattice restrictions and at $n \neq 0$ all the QMC phases are inhomogeneous (disordered) phases confined by the concentration constraint. Nevertheless, the configuration averaging does lead to results similar those for the CO1, CO2, and CO3 phases of the two-sublattice MFA model.

As for simple hard-core counterpart [8, 11], despite some qualitative agreement, we see rather large quantitative difference between two phase diagrams in Fig. 1. In particular, it concerns a clearly larger volume of the quantum SF phase that might be related with a sizeable suppression of quantum fluctuations within MFA approach. In Fig. 2 (top panel, two dotted lines) we present the QMC calculated static structure factor $S_{zz}(\pi,\pi)$ and the superfluid (pseudospin nematic) correlation function $S^2_{+-}(0,0)$. It is worth to note a semiquantitative agreement with the MFA data. Smaller value of the quantum structure factor $S_{zz}(\pi,\pi)$ at $n = 0$ is believed to be a result of the pseudospin reduction due to quantum fluctuations. Bottom panel in Fig. 2 shows the n-dependence of the mean sublattice S_z values, S_{Az} and S_{Bz}, that clearly demonstrates the pseudospin quantum reduction effect within CO2 phase and specific features of the sublattice occupation, or "pseudomagnetization" under CO2-CO3-SF transformation.

5 Conclusions

A simplified 2D $S = 1$ pseudospin Hamiltonian with a two-particle transport term (pseudospin nematic coupling) was analyzed within a generalized MFA and QMC technique. We have obtained the ground-state phase diagrams and correlation functions given different values of the coupling parameters with a focus on the role of the on-site correlation effect (single-ion anisotropy). The comparison of the two methods allows us to uncover fundamental shortcomings of the MFA technique and clearly demonstrate the role of quantum effects.

The research was supported by the Government of the Russian Federation, Program 02.A03.21.0006 and by the Ministry of Education and Science of the Russian Federation, projects Nos. 2277 and 5719.

References

[1] C.D. Batista, G. Ortiz, Adv. Phys. **53**, 1 (2004)
[2] A.V. Sizanov and A.V. Syromyatnikov, Phys. Rev. B **84**, 054445 (2011)
[3] A.S. Moskvin, Phys. Rev. B **84**, 075116 (2011)
[4] A.S. Moskvin, J. Phys.:Condens. Matter. **25**, 085601 (2013)
[5] A.S. Moskvin, JETP. **121**, 477 (2015)
[6] A.S. Moskvin, J. Phys.: Conf. Ser. **592**, 012076 (2015)

[7] A.S. Moskvin, Yu.D. Panov, F.N. Rybakov, A.B. Borisov, J. Supercond. Nov. Magn. **30**, 43 (2017).
[8] R. Micnas, J. Ranninger, S. Robaszkiewicz, Rev. Mod. Phys. **62**, 113 (1990)
[9] N.A. Mikushina, A.S. Moskvin, Phys. Lett. A **302**, 8 (2002)
[10] M.E. Fisher, M.N. Barber, D. Jasnow. Phys. Rev. A. **8**, 1111–1124 (1973).
[11] G. Schmid, S. Todo, M. Troyer, A. Dorneich, Phys. Rev. Lett. **88**, 167208 (2002)
[12] https://github.com/galanakis/QMCSGF
[13] V.G. Rousseau, Phys. Rev. E **78**, 056707 (2008)

Features of microstructure and magnetic flux dynamics in superconducting Nb-Ti tapes with strong anisotropic pinning

Sergey Shavkin*, *Valentin* Guryev, *Vitaly* Kruglov, *Alexey* Ovcharov, *Igor* Likhachev, *Alexandre* Vasiliev, *Alexey* Veligzhanin, and *Yan* Zubavichus

National Research Centre "Kurchatov Institute", Kurchatov sq. 1, 12318, Moscow, Russia

Abstract. Results of micro-structural, phase and texture analysis of Nb-Ti tapes and results of magnetic flux dynamics fixed by Magneto-Optical Imaging (MOI) method are presented in order to investigate origin and consequences of the anisotropic pinning model proposed, and implemented earlier experimental results including Guided Vortex Motion (GVM) phenomenon.

1 Introduction

All practically used low-temperature (conventional) superconducting materials including them with the cubic lattice structure (Nb-Ti alloy, Nb_3Sn compound, etc.) demonstrate pronounced anisotropy of the electromagnetic superconducting properties, caused by the directional nature of manufacturing processes (rolling, drawing, etc.) which lead to formation of the non-isotropic field of natural or artificial pinning centers and hence to anisotropic current-carrying capacity [1, 2].

The macroscopic electrodynamics of such superconductors with anisotropic (i.e. depending on the directions of a magnetic field and Lorentz force) pinning was proposed phenomenologically, on the basis of empirical studying of critical Lorentz forces. Interaction of the vortex structure (magnetic flux) with the anisotropic complex of pinning centers in the superconductor is described by a global potential well which depth depends on the direction of a magnetic induction and described by the tensor field. Force counteraction of system of the pinning centers to driving of a magnetic flux is described by a derivative of this tensor field on coordinates. It is also the tensor field. The proposed macroscopic approach allowed to avoid a problem of summation of the individual pinning forces.

For thin Nb-50%wt.Ti cold-rolled tapes this formalism allowed to calculate the value of critical Lorentz force for the arbitrary orientation of a magnetic field and transport current, measured by 2D Voltage Current Characteristic method [2], and to explain quantitatively the measured value of the angle between magnetic flux movement direction (and corresponding direction of the electric field induced by this movement) and the direction of the transport current (this is so-called "guided vortex motion" phenomenon - GVM) [3]. It is important that combination of critical currents measurements and GVM angle measurements give a way to restore a shape of pinning surface cross section. The angle between electric field and current usually is very large (around some dozens of degrees) in high-anisotropy pinning superconductors.

In this paper we presented micro-structural, phase and texture analysis of Nb-Ti tapes (both cold-rolled and heat-treated) and results of magnetic flux dynamics fixed by Magneto-Optical Imaging (MOI) method in order to investigate origin and interesting consequences of such a anisotropic pinning.

2 Samples and methods

Two kinds of samples were investigated by structural methods. First series of samples were fabricated from long-length 80-mm width and 10-μm thick cold-rolled tape of the Nb-50wt.%Ti, tape manufactured at the metallurgic factory in 1978 [2]. Second series of samples (heat-treated) were initially the same tape, and additionally they had been annealed in high vacuum at 385°C during 25 hours to precipitate non-superconducting α-Ti nano-particles. We expected that such heat treatment should change pinning system significantly.

Detailed TEM and SEM investigation of Nb-Ti tapes microstructure were done using Titan S-TWIN 80-300 (FEI) microscope with resolution up to 0.08 nm.

Detailed phase analysis by wide angle X-ray diffraction and Small angle X-ray scattering (SAXS) measurements were carried out at Kurchatov Synchrotron Radiation Source (STA and DIKSI stations) in transmission geometry using 2D detectors. The texture analysis was performed using conventional Bruker D8 Discover diffractometer with Cu-K_α radiation.

MOI experiments were performed in University of Oslo, Norway. Small (up to 2 mm in length) rectangular slices of cold-rolled Nb-Ti tape were mounted in the optical cryostat and the temperature were ~ 5.5-6K in an perpendicular external field up to 60 mT, generated by a pair of non-superconducting electromagnets outside the

* Corresponding author: shavkin@mail.ru

cryostat A Bi-doped YIG film with in-plane anisotropy was used as the Faraday active indicator for MOI experiments [4].

3 Results and discussion

3.1 Structural and textural studies

Typical SEM images of cross-sections of cold-rolled and heat-treated tapes are shown in Figure 1. The Nb-Ti grains look like stacks of thin long ribbons extended along tape rolling direction (RD). The main grains sizes in all directions were determined by averaging of statistical data from several dozen of images. Calculated main grains sizes were: 65 nm in the direction of normal to the plane of the tape; 440 nm in the direction perpendicular to the RD; and more than 1 μm in the rolling direction. Pronounced grain boundaries were observed only in parallel to the plane. Black spots are clearly α-Ti precipitates. Distributions of Nb-Ti grain sizes were similar both for cold-rolled and heat-treated tapes.

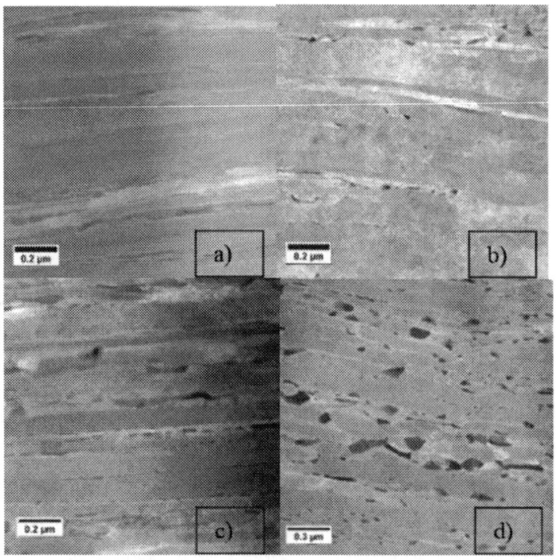

Fig. 1. SEM cross-section images for Nb-Ti samples: a) cold-rolled, RD is horizontal; b) cold-rolled, RD is perpendicular to image plane; c) heat-treated, RD is horizontal; d) heat-treated, RD is perpendicular to image plane.

X-ray diffraction measurements were performed at the «Structural Materials Science» beamline at the Kurchatov Synchrotron Radiation Source using the following parameters:

- Transmission (Debye-Scherrer) geometry,
- Beam size 200x200 mkm,
- X-ray wavelength λ=0.688862 Å,
- Si(111) monochromator to $\Delta E/E \sim 2 \cdot 10^{-4}$,
- FujiFilm ImagingPlate 2D detector.

Figure 2 shows diffraction patterns of cold-rolled and heat-treated tapes for two typical quadrants emphasizing features of diffraction peaks in rolling direction (RD) and in perpendicular to RD. It is clear that significant texture exists in both samples.

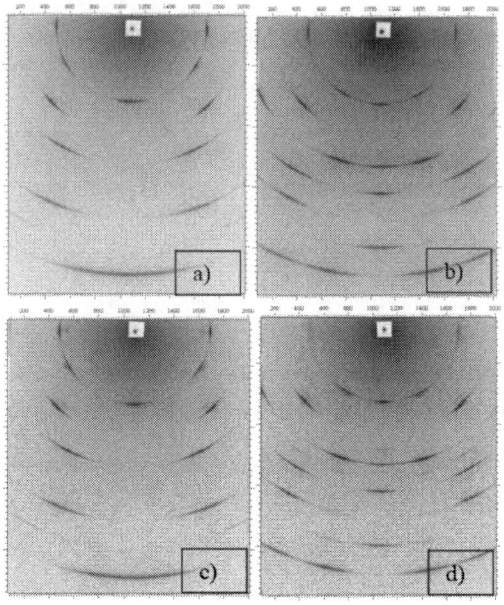

Fig. 2. Synchrotron XRD 2D images for Nb-Ti samples: a) cold-rolled, RD is vertical; b) cold-rolled, RD is horizontal; c) heat-treated, RD is vertical; d) heat-treated, RD is horizontal.

All diffraction lines in Figure 2 can be indexed in terms of cubic β-phase of NbTi (main phase) and small amount of α-Ti phase. Volume fracture of α-Ti phase can be estimated as about 5% for heat treated sample and less than 1% for initial cold-rolled sample.

It should be noted that α-Ti precipitates which mainly are located nearby grain boundaries (see Figure 1) are also high textured and moreover the texture in α-Ti succeeds the texture of main cubic NbTi phase. Possibly it is due to diffusion nature of α-Ti precipitates formation mechanism during heat-treatment of NbTi alloy.

Small angle X-ray scattering (SAXS) 2D images for both types of samples are presented in Figure 3.

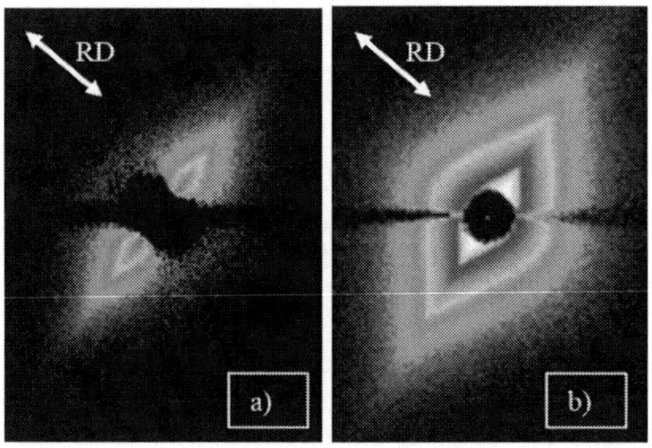

Fig. 3. Small angle X-ray scattering (SAXS) 2D images for Nb-Ti samples: a) cold-rolled, b) heat-treated. Rolling direction (RD) is shown by arrows.

One can see "eye-shape" of the SAXS images which points at anisotropic form of precipitates along RD and in perpendicular to RD. For heat-treated Nb-Ti tape the brightness of SAXS image is much higher (more than one order of magnitude) in contrast with cold-rolled tape. This can be explained by significant increasing of α-Ti precipitates volume in heat-treated Nb-Ti tape.

Figure 4 shows the pole figures of Nb-Ti phase registered by standard X-ray diffraction for the reflections (110), (200) and (222) in the cold-rolled Nb-Ti tape [5]. Analysis of ODF showed almost uniform distribution of the orientations from {100} <110> to {112} <110>, which is typical for the rolling of Nb-Ti alloys [6]. Heat treatment does not change texture of the main Nb-Ti phase.

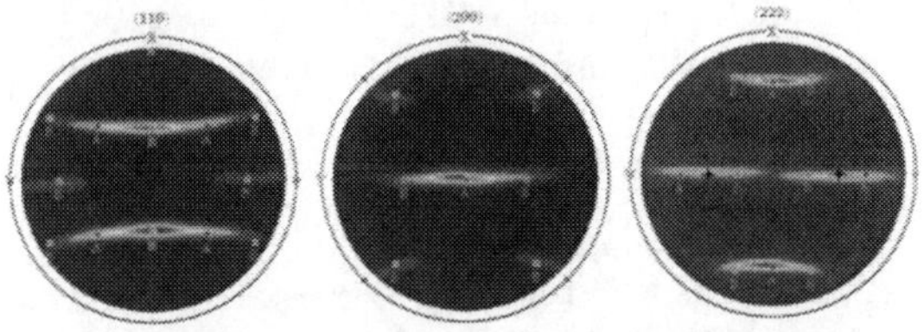

Fig. 4. The direct pole figure for the (200), (110), (222) reflections for β-phase of NbTi tape. Rolling direction (RD) is vertical.

3.2 Magneto-optical study

Features of magnetic flux penetration (field is normal to sample plane) into cold-rolled Nb-Ti tape were observed in MOI experiment with optical cryostat in polarized light. The penetration picture in general corresponds to model of a critical state taking into account anisotropy of a pinning in cases when current flow along RD and in perpendicular to RD. Due to small inhomogeneity of a superconductor the path of a magnetic flux movement can be observed that allowed to estimate the angle between the direction of flux movement and the Lorentz force originated this driving which is normal to sample edge.

In Figure 5 the evolution of the magnetic flux penetration into Nb-Ti sample with an angle of 10 degrees to rolling direction is shown. Some interesting peculiarities can be marked on these images.

- the flux penetration corresponds to a model of a critical state in perpendicular geometry;
- it is visible, that in case of a full flux penetration (in field more than 45 mT) the anisotropy factor of the critical current density is about 2 (see caption of a the Figure 5 for details). The similar value of anisotropy factor was obtained from direct transport measurements (see Figure 6) at low field end [2].
- it is visible, that the flux starts to penetrate into superconductor as "branches". It can be explained by some heterogeneity of a material.
- it is visible, that the direction of such a "branches" does not coincide with tape rolling direction and/or perpendicular to sample edge. For example, it is about 73 deg. for MO image at 30 mT in Figure 5, which corresponds to GVM angle of 90-73=17 deg. This angle is very close to calculated GVM angle for anisotropy factor of 2 derived from anisotropic pinning model [2] (see Figure 7).

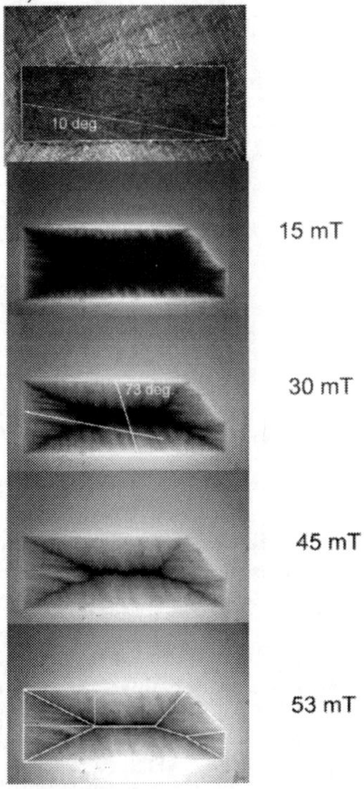

Fig. 5. MOI for various magnetic field. At the top is a general view of the sample and the rolling direction (white line). At the bottom the method for estimation of the critical current density anisotropy factor is shown: The height of the left triangle is approximately twice bigger than the half of the sample's width. This means that the currents flowing in transverse to sample direction occupy an area approximately two times larger than the currents flowing along the sample. Consequently, the current density in the transverse direction is proportionally smaller than in the longitudinal direction.

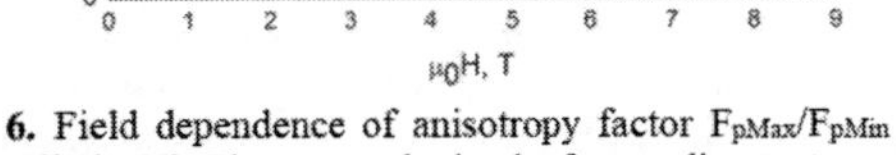

Fig. 6. Field dependence of anisotropy factor F_{pMax}/F_{pMin} for cold-rolled Nb-Ti tape obtained from direct transport measurements [2].

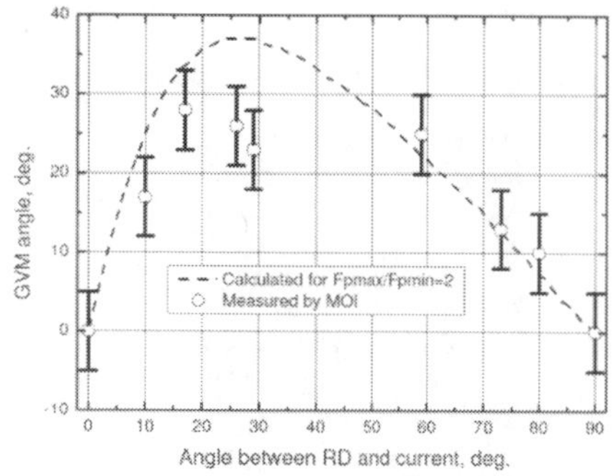

Fig. 7. GVM angle for cold-rolled Nb-Ti tape measured (open circles) by MOI in comparison with calculated (dash line) by anisotropy pinning model [2] for the anisotropy factor of $F_{pMax}/F_{pMin}=2$.

4 Discussion and conclusions

In previous papers [1, 2] the macroscopic electrodynamics of hard superconductors with anisotropic pinning was proposed phenomenologically, on the basis of empirical studying of critical Lorentz forces. For Nb-50%wt.Ti tapes this formalism allowed to calculate the value of critical Lorentz force for the arbitrary orientation of a magnetic field and transport current, and to define the direction of magnetic flux movement. GVM phenomenon was experimentally investigated by analysis of 2D Voltage-Current Characteristics.

In this paper we present a number of various modern structural analytical methods (including SEM, wide angle and small angle X-ray diffraction) and texture analysis in connection with the same Nb-Ti material. In cold-rolled Nb-Ti tape the pinning force anisotropy can be explained mainly by anisotropy of grain boundaries density. Calculated main grains sizes were ~65 nm in the direction of normal to the plane of the tape; ~0.4 μm in the transverse direction; and more than 1 μm in the rolling direction. Pronounced grain boundaries observed only in the tape plane and this results to very high pinning force if magnetic field lies in the tape plane. In contrary when magnetic field is perpendicular to the tape plane the pinning force drops dramatically [2]. In-plane anisotropy factor of pinning force can be explained by difference of main grain size (and correspondent grain boundaries density) in rolling direction and in perpendicular to rolling direction. Volume fracture of α-Ti precipitates in cold-rolled tapes is negligible (less than 1%) so precipitates do not influence on global pinning system significantly. After annealing in high vacuum at 385°C during 25 hours the volume fracture of precipitates increases up to 5% and the precipitates mainly are located nearby grain boundaries. It is obvious that critical current anisotropy in Nb-Ti tape after such a heat-treatment should change significantly. Detailed analysis of transport and magnetic measurements of heat-treated Nb-Ti samples will be published additionally.

Magneto-Optical Imaging (MOI) method was implemented to verify if the model of pining anisotropy can be applied at low magnetic field (less than 0.1 T) and in perpendicular geometry (taking into account known limitations of the method). Good quantitative agreements with previous transport measurements [2] were obtained.

The authors are very grateful to Prof. E. Yu. Klimenko for discussions of the results and to Prof. T. H. Johansen for arranging of MOI investigations. The SEM and X-ray measurements structural and texture were performed on equipment of the Resource Center of Probe and Electron Microscopy, Resource Center of X-ray methods (Kurchatov Complex of NBICS-Technologies, NRC "Kurchatov institute") and on Kurchatov Synchrotron Radiation Source.

References

1. E. Yu. Klimenko, *Superconductivity Theory and Applications* ed. by A. M. Luiz. InTech (2011)
2. E. Yu. Klimenko, S. V. Shavkin, P. V. Volkov, Zh. Eksp. Teor. Fiz **112** (1997)
3. A. K. Niessen, C. H. Weijsenfeld, J. Appl. Phys. **40** (1969)
4. T. H. Johansen, M. Baziljevich, H. Bratsberg, Y. Galperin, P. E. Lindelof, Y. Shen and P. Vase, Phys. Rev. B, **54**, N22 (1996)
5. V. V. Guryev, S. V. Shavkin, V. S. Kruglov, P. V. Volkov, A. L. Vasiliev, A. V. Ovcharov, I. A. Likhachev, E. M. Pashaev, R. D. Svetogorov, Y. V. Zubavichus, J. of Phys.: Conf. Ser. **747** (2016)
6. M. Gepreel, *Recent Developments in the Study of Recrystallization* ed. P Wilson, chapter 4, InTech (2013)

Thermomagnetic convection in a cylindrical enclosure. Part 1: Zero gravity

Mikhail S. Krakov[1,*], *Ivan V.* Nikiforov[2]

[1]Belarusian National Technical University, 65 Independence Ave., 220013 Minsk, Republic of Belarus
[2]Belarusian State University, 4 Independence Sq., 220050 Minsk, Republic of Belarus

Abstract. The effects of a uniform external magnetic field on the thermomagnetic convection of magnetic fluid in a cylindrical enclosure are numerically studied. It was found that uniform magnetic field can be cause the convection in the non-uniformly heated magnetic fluid in a cylindrical enclosure even in the case of zero gravity. There are two threshold values of fluid magnetization and temperature gradient. Above the first threshold the convective flow is realized as circular flow, above the second one there are two cells in the enclosure. The angle between a magnetic field strength and a temperature gradient is found to be significant factor influencing structure of a convective flow and heat transfer control.

1 Introduction

Natural convection of magnetic fluids has attracted much research over the years [1 – 6]. The additional body force $\mu_0 M \nabla H$ gives an opportunity to control the natural convection under magnetic field. This ponderomotive Kelvin force drives the stronger magnetized colder fluid to the regions with a stronger magnetic field. Such a motion is referred to as thermomagnetic convection. The thermomagnetic convection has a great potential for many engineering applications, such as electronic cooling devices, heat exchanger, and pure magnetic convection in space engineering.

In the case of uniform magnetic field the Kelvin force is equal to zero and magnetic field can not to effect convection. But there are two factors which even in the case of uniform external magnetic field give an opportunity to control the thermoconvection in an enclosure.

The first one has been founded by Finlayson [7] and studied in details in [4, 8] for infinite flat layers. It is shown that uniform transversal external magnetic field becomes nonuniform in the infinite layer of a heated magnetic fluid and influence a convection. Magneto-gravitational convection in a vertical layer of magnetic fluid in an oblique magnetic field is studied in [9].

The second factor is the geometry of a magnetic fluid-filled enclosure. Any finite enclosure such as square or cube has the corners and magnetic field becomes nonuniform near them because of boundary conditions for magnetic field. The effect of the external uniform magnetic field on the thermomagnetic convection in square and cubic enclosures was studied in [10 – 13].

It is interesting to exclude the factor of geometry and study the convection in the situation when the external magnetic field remains uniform in an enclosure. Such enclosure with finite dimension would be very convenient both for theoretical and experimental investigation.

It's well known that uniform external magnetic field remains to be uniform in the magnetic enclosures formed by the surfaces of the second order: the cylinder magnetized across the axes, ellipsoid or sphere. Experimental study of the thermomagnetic convection in the sphere is presented in [14]. We study the effect of the uniform external magnetic field on 2D-convection in the cylindrical enclosure in order to understand the results of [14] too.

2 Physical model

Cylindrical enclosure heated from below is one of the classical and the most considered configuration for the study of natural convection because of its relative simplicity and practical importance. Fig. 1 shows the schematic view of the physical model considered in this paper. The cylindrical enclosure with the radius R filled with magnetic fluid is surrounded with the solid massive with the radius R_∞. The temperature at the external boundary is fixed as

$$T = -\frac{T_b - T_t}{2R} y + \frac{T_b + T_t}{2} \quad (1)$$

so, the temperature gradient at the "infinity" is oriented vertically downwards and equal to

$$\gamma_0 = -\frac{T_b - T_t}{2R} \quad (2)$$

If the thermal conductivity of the fluid and the solid massive are equal then T_b is the temperature at the bottom point of the cylinder ($x = 0$, $y = -R$) and T_t – at the top one, otherwise the temperature gradient for the regime of conductivity is constant and equal to [16]

* Corresponding author: mskrakov@gmail.com

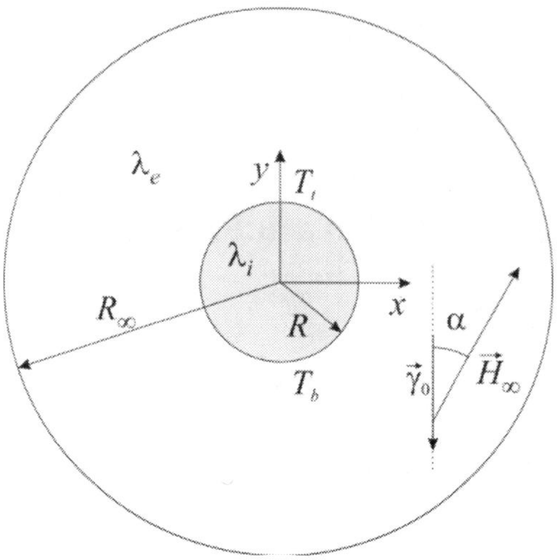

Fig. 1. Problem geometry.

$$\gamma = \frac{2\gamma_0}{1+\lambda_i/\lambda_e} \quad (3)$$

The magnetic field is supposed to be uniform at the "infinity" R_∞: $\mathbf{H}_\infty = \{H_{\infty x}, H_{\infty y}\}$. If the magnetic permeability of the fluid is μ and the solid massive is nonmagnetic than, in the isothermal case, the magnetic field in the cylinder is uniform and equal to [17]

$$\mathbf{H} = \frac{2\mathbf{H}_\infty}{1+\mu(H)/\mu_0} \quad (4)$$

where μ_0 is the permeability of vacuum.

In our computations we use the properties of magnetic fluid from the experimental paper [14]:

Density, ρ	1370 kg/m^3
Dynamic viscosity, η	0.069 Pa·s
Kinematic viscosity $\nu = \eta/\rho$	$5.04 \cdot 10^{-5}$ m^2/s
Thermal conductivity, λ_i	0.15 W/(m·K)
Magnetization saturation, M_S	44.9 kA/m

The heat capacity of the magnetic fluid could be calculated taking into account the additivity law. As the heat capacities of the surfactant (oleic acid) and the carrier fluid (transformer oil) are very close and equal to c_p^{fluid} = 1.75 kJ/(kg·K) then the heat capacity of the magnetic fluid can be calculated as

$$c_p = c_p^{\text{solid}}\varphi_s + c_p^{\text{fluid}}\varphi_f \quad (5)$$

where φ_s is the volume concentration of the solid phase and φ_f is the volume concentration of liquid phase (oleic acid and transformer oil). The volume concentration of solid phase can be calculated through the density of magnetic fluid and its components. As for transformer oil ρ_f = 817 kg/m^3 and for magnetite ρ_s = 5000 kg/m^3, the volume concentration of the solid phase can be estimated as

$$\varphi_s = \frac{\rho-\rho_f}{\rho_s-\rho_f} = 0.132 \quad (6)$$

Heat capacity of the magnetite is c_p^{solid} = 0.52 kJ/(kg·K) and in according to (5) the heat capacity of the magnetic fluid is c_p = 1.58 kJ/(kg·K). Then the thermal diffusivity is $\kappa_{\text{mf}} = \lambda_i/\rho c_p = 6.9 \cdot 10^{-8}$ m^2/s.

We assume, as in [14], that the cylindrical enclosure is surrounded by Plexiglas massive with the properties: ρ = 1190 kg/m^3, λ = 0.18 W/(m·K), c_p = 1.27 kJ/(kg·K), and thermal diffusivity $\kappa_e = 1.2 \cdot 10^{-7}$ m^2/s.

The dependence of the magnetization of a magnetic fluid with monodisperse particles on a magnetic field is described by the well-known Langevin function. But real magnetic fluids contain particles with the different sizes and this dependence is much more complicated. So, we use empiric relationship which rather correctly describes the experimental dependences of the magnetization on the magnetic field [15]

$$M(H) = M_S \frac{\chi_0 H/M_S}{1+\chi_0 H/M_S} = M_S \frac{\chi_0 \hat{H}}{1+\chi_0 \hat{H}} \quad (7)$$

Here χ_0 is the initial magnetic susceptibility, the dimensionless value marked by "^", $\hat{H} = H/M_S$. In futher computations we used χ_0 = 2 which corresponds to real magnetic fluids. Taking into account equation (7), the equation (4) has an analytical solution: the magnetic field in the cylindrical enclosure filled with a magnetic fluid is related with the external uniform magnetic field by the equation

$$\hat{H} = \frac{(2\chi_0\hat{H}_\infty - 2 - \chi_0) + \sqrt{(2\chi_0\hat{H}_\infty - 2 - \chi_0)^2 + 16\chi_0\hat{H}_\infty}}{4\chi_0} \quad (8)$$

3 Governing equations

3.1 Momentum equation

The momentum equation including gravity buoyancy force and magnetizing force is as follows [3]

$$\rho\left[\frac{\partial \mathbf{u}}{\partial t} + (\mathbf{u}\cdot\nabla)\mathbf{u}\right] = -\nabla p + \eta\Delta\mathbf{u} + \rho\mathbf{g} + \mu_0 M\nabla H. \quad (9)$$

The fluid is assumed to be an incompressible fluid

$$\text{div}\mathbf{u} = 0. \quad (10)$$

For the thermal convection with low temperature difference, the classical Boussinesq approximation that the thermophysical properties of the fluid are assumed to be constant except for the density and the magnetization difference for buoyancy force is used.

There are two reasons for dependence of a fluid magnetization on the temperature: change of the particles magnetic moment and thermal expansion of the fluid [4]. As the Curie temperature for the particles of magnetite is 858 K, around a reference state T_0 change of the magnetic moment is negligible and, if magnetic field is not very small, only thermal expansion is the reason of magnetization change. Then we can write [4]

$$\left(\frac{\partial M}{\partial T}\right)_0 = -M_0\beta_{mf} = -M_0[\varphi_s\beta_s + (1-\varphi_s)\beta_f] \quad (11)$$

where M_0 is determined with Eq. (7), β_s is the coefficient of thermal expansion of solid particles, and β_f – of fluid. As for magnetite $\beta_s = 2\cdot10^{-4}\,\text{K}^{-1}$, and for transformer oil $\beta_f = 6.9\cdot10^{-4}\,\text{K}^{-1}$, $\beta_{mf} = 6.25\cdot10^{-4}\,\text{K}^{-1}$.

Using Eq. (7) $(\partial M/\partial H)_0$ can be written as

$$\left(\frac{\partial M}{\partial H}\right)_0 = \frac{\chi_0}{(1+\chi_0\hat{H}_0)^2} \quad (12)$$

Thus, the equation of state finally can be written as

$$M(H,T) = M_0 + \frac{\chi_0}{\left(1+\chi_0\hat{H}\right)^2}(H-H_0) - M_0\beta_{mf}(T-T_0)$$
$$= M_0 + a(H)(H-H_0) + b(H)(T-T_0) \quad (13)$$

Then the momentum equation can be written as follows:

$$\rho\left[\frac{\partial \mathbf{u}}{\partial t}+(\mathbf{u}\cdot\nabla)\mathbf{u}\right]=-\nabla(p)+\eta\Delta\mathbf{u}-\rho_0\beta_{mf}(T-T_0)\mathbf{g}+\mu_0\begin{bmatrix}a(H)(T-T_0)\nabla H+b(H)(H-H_0)\nabla H\\ +M_0\nabla(H-H_0)\end{bmatrix} \quad (14)$$

We will use a stream function ψ and a vortex ω as follows:

$$u_x=\frac{\partial\psi}{\partial y},\ u_y=-\frac{\partial\psi}{\partial x},\omega=\frac{\partial u_y}{\partial x}-\frac{\partial u_x}{\partial y}=-\Delta\psi \quad (15)$$

After an operation curl applied to Eq. (15) and taking into account that $\nabla\times\mathbf{H}=0$, the momentum equation becomes

$$\frac{\partial\omega}{\partial t}+(\mathbf{u}\cdot\nabla)\omega=\nu\Delta\omega+\beta_{mf}g\frac{\partial T}{\partial x}+\frac{\mu_0 M_S\beta_{mf}}{\rho}\frac{\chi_0\hat{H}_0}{1+\chi_0\hat{H}_0}\left(\frac{\partial H}{\partial x}\frac{\partial T}{\partial y}-\frac{\partial H}{\partial y}\frac{\partial T}{\partial x}\right) \quad (16)$$

3.2 Energy equation

This equation has a standard form:

$$\rho c_p\left[\frac{\partial T}{\partial t}+(\mathbf{u}\cdot\nabla)T\right]=\lambda\Delta T \quad (17)$$

3.3 Magnetic field

A magnetic field both in the enclosure and in the external space is described with Maxwell's equations for nonconducting fluids

$$\nabla\times\mathbf{H}=0,\quad \nabla\cdot\mathbf{B}=0,\quad \mathbf{B}=\mu_0(\mathbf{H}+\mathbf{M})$$

As for the magnetic fluids with magnetite particles the vectors of magnetic field strength and magnetization are parallel for flow velocities typical for convection, i.e. $\mathbf{M}=\chi(H,T)\mathbf{H}$, then, using (7), (11) and (12), we get

$$\mathbf{B}=\mu_0(1+\chi(H,T))\mathbf{H}=\mu_0\left(1+\frac{\chi_0}{1+\chi_0\hat{H}}\left[1-\beta_{mf}(T-T_0)\right]\right)\mathbf{H}=\mu_0\hat{\mu}(H,T)\mathbf{H}$$

As $\nabla\times\mathbf{H}=0$ the magnetic field strength can be expressed as the gradient of a scalar potential F

$$\mathbf{H}=\nabla F \quad (18)$$

and finally the magnetic field is described with the nonlinear equation

$$\nabla\cdot(\hat{\mu}(|\nabla F|,T)\nabla F)=0 \quad (19)$$

where

$$\hat{\mu}(\nabla F,T)=1+\frac{\chi_0}{1+\chi_0|\nabla F|/M_S}\left[1-\beta_{mf}(T-T_0)\right] \quad (20)$$

3.4 Boundary conditions

On the walls of the enclosure, no-slip conditions for the velocity are applied: the stream function is equal to zero on the walls of the enclosure; the vortex is calculated from the Eq. (15 taking into account the curvature of the walls.

The temperature at the "infinity" R_∞ is determined by means of the Eq. (1) and the conditions of continuity must be fulfilled at the boundary of the enclosure

$$T_i=T_e,\quad \lambda_i\frac{\partial T}{\partial n}=\lambda_e\frac{\partial T}{\partial n} \quad (21)$$

The external magnetic field is uniform and at the "infinity" R_∞ is equal to $\mathbf{H}_\infty=\{H_{x\infty},H_{y\infty}\}$, i.e. the scalar potential is determined at the external boundary as

$$F=H_{x\infty}x+H_{y\infty}y \quad (22)$$

The conditions $(\mathbf{B}_i-\mathbf{B}_e)\cdot\mathbf{n}$, $(\mathbf{H}_i-\mathbf{H}_e)\cdot\boldsymbol{\tau}=0$ must be fulfilled at the boundary of the enclosure, i.e.

$$\hat{\mu}\frac{\partial F_i}{\partial n}=\frac{\partial F_e}{\partial n},\quad \frac{\partial F_i}{\partial\tau}=\frac{\partial F_e}{\partial\tau} \quad (23)$$

where $\mathbf{n}$ is the unit vector normal to the boundary and $\boldsymbol{\tau}$ is the unit vector tangential to the boundary.

3.5 Dimensionless form of the equations

We use as the scales: $[t]=R^2/\nu_i$, $[x,y]=R$, $[T]=\gamma_0 R$, $[u]=\kappa_i/R$, $[H]=M_S$, $[F]=M_S R$, $[\psi]=\kappa_i$, $[\omega]=\kappa_i/R^2$. The system of equations can be written in dimensionless form as follows:

$$\nabla\cdot\left(\mu(|\nabla\hat{F}_i|,\theta)\nabla\hat{F}_i\right)=0,\quad \Delta\hat{F}_e=0 \quad (24)$$

$$\frac{\partial\hat{\omega}}{\partial t}+\frac{1}{\mathrm{Pr}}(\mathbf{u}\cdot\nabla)\hat{\omega}=\Delta\hat{\omega}+\mathrm{Ra}\frac{\partial\theta}{\partial x}+\mathrm{Ra}_m\frac{\chi_0\hat{H}_0}{1+\chi_0\hat{H}_0}\left(\frac{\partial\hat{H}}{\partial x}\frac{\partial\theta}{\partial y}-\frac{\partial\hat{H}}{\partial y}\frac{\partial\theta}{\partial x}\right) \quad (25)$$

$$\Delta\hat{\psi}=-\hat{\omega} \quad (26)$$

$$\mathrm{Pr}\frac{\partial\theta_i}{\partial t}+(\mathbf{u}\cdot\nabla)\theta_i=\Delta\theta_i \quad (27)$$

$$\mathrm{Pr}\frac{\partial\theta_e}{\partial t}=\frac{\kappa_e}{\kappa_i}\Delta\theta_e \quad (28)$$

Here $\hat{F}=F/RM_S,\hat{H}=H/M_S,\hat{\omega}=\omega R^2/\kappa_i,\hat{\psi}=\psi/\kappa_i$, $\theta=(T-T_0)/\Delta T,\Delta T=2\gamma_0 R$, the magnetic permeability is equal to 1 for external space and is given by (20) for the fluid in the cylinder.

The Rayleigh number, the magnetic Rayleigh number and the Prandtl number are:

$$\mathrm{Ra}=\frac{g\beta_{mf}\Delta TR^3}{\nu\kappa},\mathrm{Ra}_m=\frac{\mu_0M_S^2\beta_{mf}\Delta TR^2}{\rho\nu\kappa},\mathrm{Pr}=\frac{\nu}{\kappa} \quad (29)$$

4 Numerical method

The problem was solved numerically by finite elements method with triangle elements. The linear interpolation functions are used for the magnetic field scalar potential and the stream function

$$F=A_Fx+B_Fy+C_F,\quad \psi=A_\psi x+B_\psi y+C_\psi$$

and the exponential functions [18] for the temperature and the vortex are used:

$$\omega=A_\omega\exp\left[\frac{U_{\mathrm{el}}}{\mathrm{Pr}}(X-X_{\max})\right]+B_\omega Y+C_\omega$$

$$T=A_T\exp[U_{\mathrm{el}}(X-X_{\max})]+B_TY+C_T$$

where the coefficients A, B, C are calculated from the function values in the nodes of the triangle elements, a local coordinate system X, Y is oriented at each element so that the axis X directed along the local velocity $\mathbf{U}_{\mathrm{el}}$ in this element.

The vortex at the solid boundary of the enclosure is calculated with taking into account the curvature of the boundary according to the method supposed in [19].

5 Results and Discussion

The Rayleigh number is only one governing parameter in the case of the thermo-gravitational convection. In our case there are three independent dimensionless parameters: the numbers Ra and Ra_m and the parameter $\beta_{mf}\Delta T$. The last one specifies the magnetic field gradient in the enclosure through the equations (20) and (24).

 https://doi.org/10.1051/epjconf/201818509001

In order to exclude the influence of the gravity and to highlight the influence of the uniform external magnetic field on the thermal convection in the pure state we put the Rayleigh number equal to zero (Ra = 0) in this study. As both parameters Ra_m and $\beta_{mf}\Delta T$ depend on the temperature difference ΔT, we use in the study ΔT and the magnetization M_0 as the independent physical parameters. All the further computations are fulfilled for the fluid with properties presented above and corresponding to the magnetic fluid used in the experiments [14]. The dimensionless radius of the external boundary of the calculation domain is $R_\infty = 11$, the radius of the enclosure, as in the [14], is 8 mm.

The heat flux across the boundary of the cylinder can be characterized by the Nusselt number which was calculated as

$$\mathrm{Nu} = \frac{1}{\mathrm{Nu}_0}\int_{\vartheta_1}^{\vartheta_2} \frac{\partial \theta}{\partial n} d\vartheta$$

where Nu_0 is the Nusselt number in the case of conductivity regime, ϑ_1 and ϑ_2 are the angles for which $\partial\theta/\partial n = 0$. This integral was calculated both from ϑ_1 to ϑ_2 and from ϑ_2 to ϑ_1. The difference between the results was less than 0.01%.

To validate the quality of the numerical results, the problem of convective stability in the field of gravity in a cylindrical enclosure without magnetic field ($Ra_m = 0$) is tested and results are compared with the linear theory solution [16]. Table 1 shows the studied grid systems:

Table No 1. Grid parameters.

No of the grid	1	2	3	4	5
Number of nodes	9091	16201	25381	36541	49771
Number of elements	18000	32160	50400	72720	99120

Table 2 illustrates the mesh independency analysis. The grid No 4 is used in all computations presented below.

For the case of zero gravity (Ra = 0, Ram ≠ 0) it was found that there are two types of convective flow

Table 2.
Comparison of the present results of critical Rayleigh number for different grid resolution with linear theory [16].

λ_i/λ_e	Galerkin method [16]	Grid No 1	Grid No 2	Grid No 3	Grid No 4	Grid No 5
0.01	**408.2**	421.52	417.63	415.32	413.8	413.01
0.2	**373.6**	384.33	382.06	380.7	379.83	379.46
0.5	**372.3**	379.65	378.36	377.62	377.14	377.15
1	**412.4**	415.36	414.74	414.44	414.26	414.31
2	**527.1**	525.88	525.97	526.11	526.30	526.40

structures in a cylindrical enclosure (Fig. 2). In contrast to the thermo-gravitational convection, for which the only one structure (a single cell) can be realized, two structures (single cell and two cells) are possible for the thermomagnetic convection in the case of zero gravity.

To check the dependence on the Prandtl number, the problem for the maximum value of the parameters (Ra = 0, $Ra_m = 2383686$, $H_\infty = 20M_S$, $\Delta T = 120$ K) is solved. Table 3 shows that the results are the Prandtl number almost independent. The value Pr = 709 and $\lambda_i/\lambda_e = 0.833$ corresponding to the above-stated properties of the magnetic fluid are used in all computations.

Table 3. The Prandtl number dependency test.

Pr	ψ_{min}	ψ_{max}	Nu
10	-5.13317	5.1306	1.6393
100	-5.13528	5.1358	1.6426
1000	-5.1356	5.1366	1.6430

The distribution of the temperature and magnetic field potential in the calculation domain equal to the external boundary conditions and $\psi = 0$, $\omega = 0$ are used as initial conditions.

There is no convection for small values of ΔT and H_∞ (remember Ra = 0). Above some critical values of the fluid magnetization and the temperature difference the circulation flow arises (Fig. 2a). Fig. 3 shows the

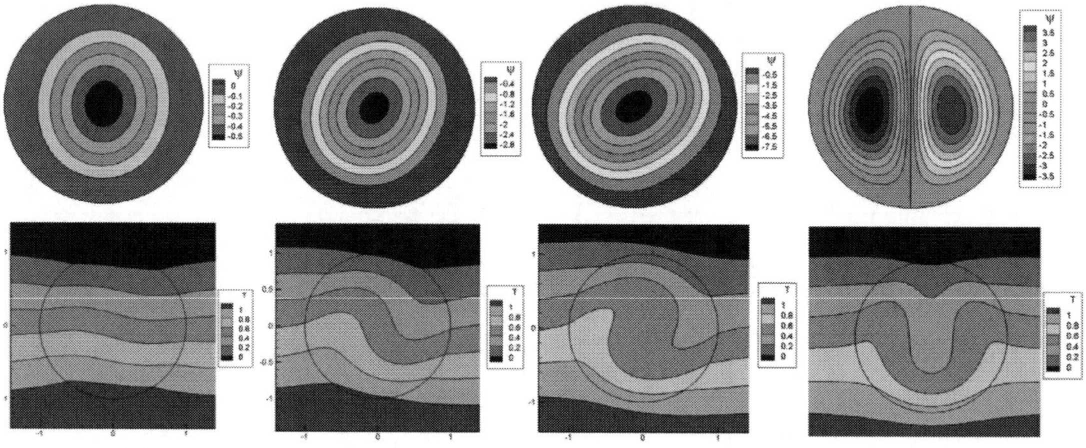

a) $\Delta T = 23$ K b) $\Delta T = 40$ K c) $\Delta T = 56$ K d) $\Delta T = 57$ K

Fig. 2. Stream function and temperature fields. $\hat{H}_\infty = 20$, $\alpha = 0$.

correspondence the dependence of the maximum stream

function on the temperature difference to the square root law of the Landau $\psi \sim (\Delta T - \Delta T^*)^{0.5}$ where ΔT^* is the critical value of the temperature difference. The same character of dependence we can see for the Nusselt number (Fig. 4).

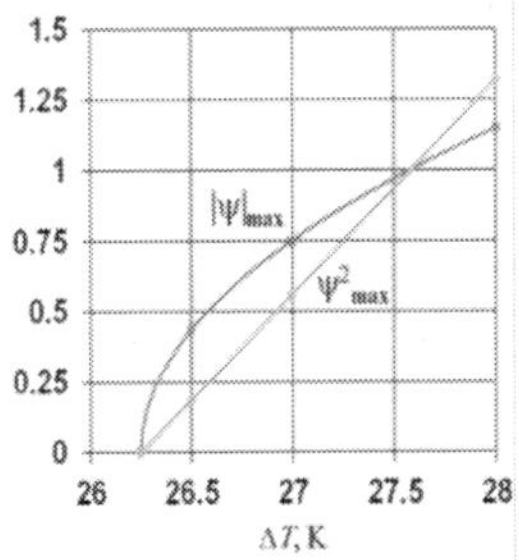

Fig. 3. The maximum of stream function and the square of maximum stream function in the cell vs temperature difference. $\widehat{H}_\infty = 2, \alpha = 0$.

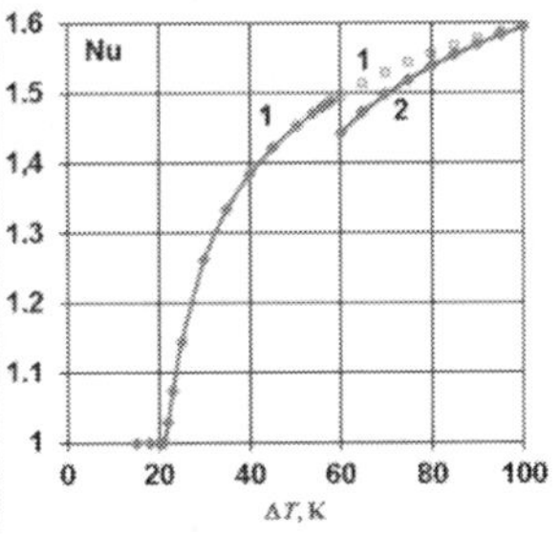

Fig. 4. The effect of temperature difference on Nusselt number. $\widehat{H}_\infty = 20$, $\alpha = 0$. Zero initial conditions for the flow give solid line, result for using of previous solution as initial condition is presented by □. 1 – one cell, 2 – two cells.

As in the case of thermo-gravitational convection, the circulation flow takes the form of an oval with the increasing of ΔT and the inclination of the major axis of the oval increases too (Figs. 2b, 2c). The direction of rotation in the case of parallelism of the magnetic field and the temperature gradient is arbitrary and it's possible both clockwise, and counter-clockwise.

The investigations of a linear stability of the basic state in the case of thermo-gravitational convection show [16] that the onset of the convection is possible in one of the four different perturbations: circular flow, two cells oriented vertically and horizontally, and four cells. However, numerous experimental studies and computer simulations found that only one cell flow can be realized. Fig. 2d shows that for sufficient value of the magnetization and the temperature difference two cells flow can be realized for thermo-magnetic convection. Above this second threshold the Nusselt number significantly decreases in comparison with one cell, but the rate of the Nusselt number increasing with the ΔT growth is much more (Fig. 4).

Fig. 4 shows that, for the given set of the parameters, the convective flow can be realized as two cells if the initial condition is $\psi = 0$, $\omega = 0$ and $\Delta T > 56$ K. But if the initial condition is the one sell solution for $\Delta T < 56$ K then the convective flow is one cell for $\Delta T > 56$ K too. The corresponding values of the Nusselt numbers are shown with the marker □. It depends on the way of the fluid heating in the enclosure what structure can be realized in the experiment. The circular flow, formed at the small temperature differences, is stable in all range of temperatures if the heating is slow. In the case of quick heating (it corresponds to initial conditions is $\psi = 0$, $\omega = 0$ and the Prandtl number Pr = 709) two cell flow is formed for sufficiently high temperature difference. This critical ΔT depends on the fluid magnetization. Fig. 5 shows two neutral curves which divide the space of M - ΔT on three zones: zone of stable state without convection, zone of the convection with one cell flow and zone of the two cell flow convection. The connection between the magnetization M_0 in Fig. 5 and the external magnetic field H_∞ is given by the equations (7), (8).

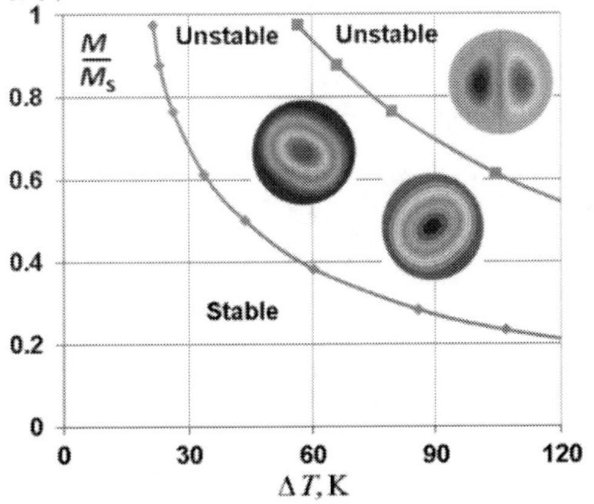

Fig. 5. Neutral curves for $\alpha = 0$. The direction of rotation in the case of one cell is arbitrary. M and $\widehat{H}_\infty$ are connected through Eqs. (7) and (8).

Till now the magnetic field parallel to the temperature gradient is studied. As the magnetization gradient is formed by the temperature gradient, they are parallel always and, it seems, the direction of the magnetic field vector has not to influence the convective flow. But after the onset of the convection the convective flow distorts the temperature field and, as consequence, distorts the magnetic field too. As a result, the verticality deviation of the magnetic field leads to very complicated interaction with the field of the temperature and can influence the convective flow.

The numerical simulation shows that the maximum absolute value of the stream function for ΔT = 50 K (only one cell flow exists for this temperature difference) increases with the increasing of the angle of vertically deviation up to ±30° and then decreases (Fig. 6a). The dependence of the Nusselt number on the angle of vertically deviation has a similar character (Fig. 6b). The variation of the stream function ψ and the Nusselt number Nu not very large (less than 2%) in this case but it exists. Thus, we can confirm the influence of the uniform magnetic field direction on the convective flow.

For ΔT = 100 K and $\alpha = 0$ the convective flow has a structure of two cells. In this case the influence of the external magnetic field inclination is more significant (Fig. 7a, b). The flow almost isn't changed in the range of the angles $-28° \leq \alpha \leq 31°$: the maximum stream function varies between 6.79 and 6.64, i.e. within the range of 2%; the Nusselt number varies even less than ψ : in the range of 1.596 – 1.598. It should be mentioned that if the angle of the field inclination increases in positive direction then the intensity of the flow in the right cell increases and in the left cell decreases. In the case of counter-clockwise inclination of the field we see an opposite situation.

For $76^{\circ} > |\alpha| > 31^{\circ}$ the cell with the maximum absolute value of the stream function becomes to dominate and only one cell convective flow is possible in all range of the parameters. The direction of the circulation flow is counter-clockwise for positive α and clockwise for negative angles.

For $90^{\circ} > |\alpha| > 76^{\circ}$ we can see two cell structure again.

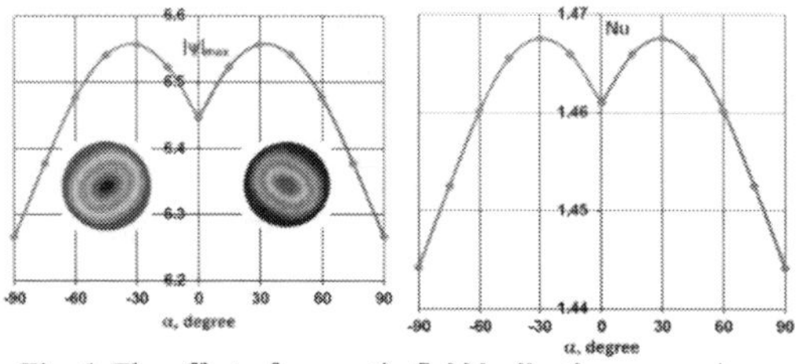

Fig. 6. The effect of magnetic field inclination on maximum stream function (6a) and the Nusselt number (6b). $\Delta T = 50$ K, $\hat{H}_{\infty} = 20$.

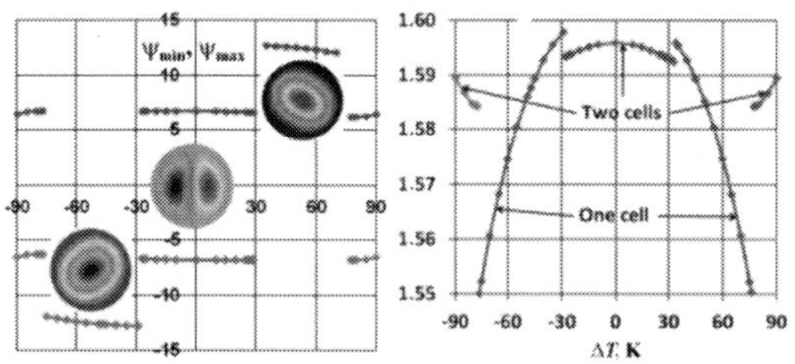

Fig. 7. The effect of magnetic field inclination on maximum stream function (7a) and the Nusselt number (7b). $\Delta T = 100$ K, $\hat{H}_{\infty} = 20$.

6 Conclusions

The buoyancy-induced thermal convection under external uniform magnetic field is numerically studied in a cylindrical enclosure. The effect of temperature, magnetization, direction of magnetic field and temperature gradient on thermomagnetic convection inside the enclosure is discussed in details. The main important conclusions can be summarized as follows:

1. Convection in a non-uniformly heated cylindrical enclosure can be induced by the external uniform magnetic field even in the case of zero gravity.
2. There are two threshold values of temperature gradient and fluid magnetization.
3. Above the first threshold values the convective flow is realized as circular flow.
4. There are the second threshold values of the parameters of the problem for angles near 0° and 90° between magnetic field and temperature gradient. Flow realized above these values is two cells.
5. The direction of rotation in one cell structure is arbitrary for the external magnetic field and temperature gradient parallelism. It is counter-clockwise for positive and clockwise for negative angles.

All these conclusions are right if zero initial conditions for stream function and vortex are used. If the convective structure formed for some parameters (for example, for $\Delta T = 100$ K, $M_0 = 0.96$, $\alpha = 0^{\circ}$ – two cells) is used in computer simulation as initial condition for another set of parameters (for example, $\alpha = 40^{\circ}$ or 70°), restructuring of the convective flow doesn't occur. Thus, it's possible to state that formed structures have some margin of stability.

References

1. J.R. Carruthers, R. Wolfe, J. Appl. Phys. **39,** 5718–5722, (1968)
2. D. Braithwaite, E. Beaugnon, R. Tournier, Nature, **354,** 134–136 (1991)
3. R.E. Rosensweig, *Ferrohydrodynamics* (Cambridge University Press, 1985)
4. V.G. Bashtovoy, B.M. Berkovsky, A. N. Vislovich, *Introduction to thermomechanics of magnetic fluids.* (Hemisphere, Washington, DC, 1988)
5. B.M. Berkovsky, V.F. Medvedev, M.S. Krakov, *Magnetic Fluids: Engineering Applications* (Oxford University Press Inc., NY, 1993)
6. E. Blūms, A.O. Cebers, M.M. Maĭorov, *Magnetic Fluids* (Walter de Gruyter, 1997)
7. B.A. Finlayson, J. of Fluid Mech. **40** 753 – 767 (1970)
8. V.G.Bashtovoy, J. of Appl. Mech. and Technical Physics. No 3, 126 – 133 (1979) (in Russian.)
9. H. Rahman, S.A. Suslov, J. of Fluid Mech. **795** 847-875 (2016)
10. M.S. Krakov, I.V. Nikiforov J. Magn. Magn. Mater. **252** 209–211 (2002)
11. M.S. Krakov, I.V. Nikiforov, J. Magn. Magn. Mater. **289** 278-280 (2005)
12. S. Kamiyama, M.S. Krakov, I.V. Nikiforov, Magnetohydrodynamics, **40** 335-345 (2004)
13. M.S. Krakov, A.G. Reks, I.V. Nikiforov, J. of Magn. Magn. Mater. **289** 272-274 (2005)
14. M.T. Krauzina, A.A. Bozhko, P.V. Krauzin, S.A. Suslov. J. of Magn. Magn. Mater. **431** 241 – 244 (2017)
15. A.N. Vislovich, Magnetohydrodynamics **26** 178–183 (1990)
16. G.Z. Gershuni, E.M. Zhukhovitskii. *Convective Stability of Incompressible Fluids.* (I.P.S.T. 1976)
17. L.D.Landau and L.P.Pitaevskii. *Electrodynamics of Continuous Media, Second Edition: Volume 8 (Course of Theoretical Physics S).* (Butterworth-Heinemann 1984)
18. S. Patankar, *Numerical Heat Transfer and Fluid Flow. Hemisphere Series on Computational Methods in Mechanics and Thermal Science* (Taylor and Fransis, London, 1980).
19. M.S. Krakov, Numer. Heat Transfer, Part B Fundamentals **21/2** 125 – 145 (1992)

Effects of magnetic fields on capillary-gravity waves in the presence of magnetic surfactants

Alexander V. Zhukov [1,*]

[1]Institute of Mechanics, Moscow State University , 117192 Moscow, Russia

Abstract. The stability of capillary-gravity wave motion on horizontal free surface of viscous non-compressible fluid in the presence of magnetic surfactant in an external magnetic field was studied. It is shown that for normal as well as for tangential external magnetic field the horizontal free liquid surface is unstable for field strength exceeding some critical value that does not depend on the elastic constant of the surfactant film. However, for oblique external magnetic field the stability of the free surface depends not only on the field value but also on the surfactant elastic constant.

1 Governing equations

Recently, new surfactant molecules with magnetic properties have been synthesized [1]. This makes it possible the magnetic control over liquid surface properties by sufficiently strong (0.4...1 T) external magnetic fields. The surface tension tensor for these media is anisotropic and depends on magnetic field strength. The analogous dependence was found for magnetic fluid — water interfaces subjected to moderate-intensity (60...110 G) magnetic fields [2]. We study the stability of capillary-gravity wave motion on horizontal free surface of viscous non-compressible fluid in the presence of magnetic surfactant in external magnetic field.

Let D^- be the domain occupied by liquid, D^+ be the external domain and Σ be the interfacial boundary, defined by the equation $z \equiv x^3 = \eta(x^a), a = 1,2$, where $x^i, i = 1,2,3$ are the Cartesian coordinates: $D^-(t): z < \eta(x^a,t), D^+(t): z > \eta(x^a,t), \Sigma(t): z = \eta(x^a,t)$

In equilibrium $\eta = 0$. The z axis is directed upwards. The normal vector $\mathbf{n}$ on the boundary is defined as the external normal to D^-.

Inside the domain D^- the Navier-Stokes equations are valid:

$$\partial_k v^k = 0, \rho\left(\partial_t v^i + v^k \partial_k v^i\right) = \partial_k p^{ik(f)} + \rho g^i;$$
$$\mathbf{g} = (0,0,-g); p_{ik}^{(f)} \equiv -p\delta_{ik} + \mu\left(\nabla_i v_k + \nabla_k v_i\right) \quad (1)$$

Here ρ is the density and μ is the dynamic viscosity of the liquid, v^i are the velocity components, p is the pressure, $p_{ik}^{(f)}$ is the stress tensor (excluding the magnetic field stress) and $\mathbf{g}$ is the acceleration of gravity. Inside D^+ $p = p_0 = \mathrm{const}$.

The following conditions are valid at the interface [2]:

$$\partial_t \eta - v^3 + v^a \partial_a \eta = 0;$$
$$[H_{ta}] = -4\pi\lambda_n \nabla_a \langle B_n \rangle, [B_n] = -4\pi\lambda_t \nabla_a \langle H_t^a \rangle;$$
$$\left[p_b^i\right] n_i - \nabla_a \sigma_b^a = 0, \left[p^{ik} n_i n_k\right] - \sigma^{ab} b_{ab} = 0; \quad (2)$$
$$p_{ik} = p_{ik}^{(f)} - \frac{H^2}{8\pi}\delta_{ik} + \frac{H_i H_k}{4\pi}.$$

Here $\lambda_n = \Gamma b_n, \lambda_t = \Gamma b_t$, where b_n, b_t are (generally different) constants defining surface magnetization of surfactant film in normal and tangential directions: $\mathbf{m}_n = -\lambda_n \langle \mathbf{B}_n \rangle, \mathbf{m}_t = -\lambda_t \langle \mathbf{H}_t \rangle$ [2]. Γ is the surface surfactant density, $[f] \equiv f_+ - f_-, \langle f \rangle \equiv (f_+ + f_-)/2$; $\mathbf{H}, \mathbf{B}$ are the vectors representing magnetic field and magnetic flux density, respectively, p_{ik} are the components of total stress tensor, b_{ab} are the components of the second quadratic form of the boundary and σ_{ab} are the components of surface tension tensor

$$\sigma_{ab} = \left(\gamma_0 - \beta\Gamma - \frac{1}{2}\lambda_n \langle B_n \rangle^2 + \frac{1}{2}\lambda_t h_c h^c\right) a_{ab} - \lambda_t h_a h_b, \quad (3)$$
$$h_a \equiv \langle H_{ta} \rangle$$

The constant β characterizes the dependence of surface tension on surfactant concentration [4]. The dependence of σ_{ab} on magnetic field strength is taken from [2]. The following stability conditions are necessary: $\lambda_n \geq 0, \lambda_t \leq 0$ [3].

* Corresponding author: az@imec.msu.ru

https://doi.org/10.1051/epjconf/201818509002

The dynamics of surfactant surface density is described by the following equations:

$$\partial_t \Gamma + \nabla_a\left(\Gamma u_\Sigma^a\right) + \frac{1}{2}\Gamma a^{ab}\partial_t a_{ab} = 0; u_\Sigma^a = v_t^a \qquad (4)$$

Here the surfactant diffusion effects and mass transfer effects to or from the dividing surface are neglected and the surfactant surface velocity u_Σ^a is taken equal to the tangential fluid velocity [4]; a_{ab} are the components of first quadratic form on the surface.

The Maxwell equations for magnetic field in non-magnetizable media are

$$\operatorname{div}\mathbf{H} = 0, \operatorname{rot}\mathbf{H} = 0, \mathbf{B} = \mathbf{H} \qquad (5)$$

The conditions at the infinity take the form

$$z \to -\infty : v^k \to 0; z \to \pm\infty : \mathbf{H} \to (H_t, 0, H_n) \qquad (6)$$

Here H_t, H_n are the given values of external magnetic field strength in tangential and normal directions to the horizontal unperturbed surface.

2 Linearization and dispersion equation

To study capillary-gravity waves we perform the linearization of the equations (1)-(6) with respect to the equilibrium solution

$$\begin{aligned} &z > 0 : p = p_0, \mathbf{H} = (H_t, 0, H_n); \\ &z < 0 : p = p_0 - \rho g z, v^i = 0, \mathbf{H} = (H_t, 0, H_n); \eta = 0 \end{aligned} \qquad (7)$$

Each of the perturbations may be represented as a superposition of normal modes, for example:

$$p' = p - p_{(eq)} = \int \exp\left(st + ik_\alpha x^\alpha\right)\hat{p}\left(k_\alpha, s, z\right)d^2k, \alpha = 1,2$$

Here $s = s\left(k_1, k_2\right)$ is the solution of the dispersion equation and the (eq) subscript denotes values taken from the equilibrium solution (7).

The linearization of equations (1)-(6) give the following expressions for Fourier coefficients of the perturbations:

$$\hat{v}_3 = A\exp(kz) + C\exp(lz); l^2 = k^2 + \frac{s}{\nu},$$

$$\nu = \frac{\mu}{\rho}, \operatorname{Re}(l) > 0; \hat{\eta} = \frac{A + C}{s};$$

$$\hat{v}_\alpha = \frac{ik_\alpha}{k^2}\left(kA\exp(kz) + lC\exp(lz)\right) + \varepsilon_{\alpha\beta}k^\beta B\exp(lz),$$

$$\hat{p} = -\frac{\rho sA}{k}\exp(kz); \hat{\Gamma} = \frac{\Gamma}{s}(kA + lC);$$

$$\hat{H}_a = ik_a\hat{h}^\pm\exp(\mp kz), \hat{H}_3 = \mp k\hat{h}^\pm\exp(\mp kz); k = |\mathbf{k}|;$$

$$\hat{h}^+ + \hat{h}^- = \frac{4\pi b_t\Gamma H_3 k\hat{\eta}}{1 - 2\pi b_t\Gamma k}, \hat{h}^+ - \hat{h}^- = -\frac{4\pi i b_n\Gamma H_a k^\alpha\hat{\eta}}{1 - 2\pi b_n\Gamma k}$$

Using these expressions we obtain the dispersion equation

$$\begin{aligned} &-(\omega_0^2 + 2\nu kls)(2\nu k^2 s + \varepsilon k^3 - i\theta k^3) + \\ &(s^2 + \omega_0^2 + 2\nu k^2 s)(s^2 + 2\nu k^2 s + \varepsilon k^2 l - i\theta k^3) = 0 \end{aligned} \qquad (8)$$

where functions $l(k,s), \omega_0^2(\mathbf{k}), \theta(\mathbf{k})$ and the elastic constant of surfactant ε [4] are defined as follows:

$$\varepsilon \equiv \frac{\Gamma\beta}{\rho}; l^2 = k^2 + \frac{s}{\nu}, \nu = \frac{\mu}{\rho}, \operatorname{Re}(l) > 0;$$

$$\theta(\mathbf{k}) \equiv \frac{1}{\rho}\left(\frac{b_n\Gamma}{1 + 2\pi b_n\Gamma k} - \frac{b_t\Gamma}{1 - 2\pi b_t\Gamma k}\right)\frac{H_\alpha k^\alpha}{k}H_3$$

$$\frac{\rho\omega_0^2}{k} = \rho g + \left(\gamma_0 - \beta\Gamma + \frac{\Gamma b_t}{2}H_t^2 - \frac{\Gamma b_n}{2}H_n^2\right)k^2 + \left(\frac{\Gamma b_n}{1 + 2\pi k\Gamma b_n} - \Gamma b_t\right)\left(H_\alpha k^\alpha\right)^2 + \frac{\Gamma b_t H_n^2 k^2}{1 - 2\pi k\Gamma b_t}$$

3 Stability conditions

If $H_n H_t = 0$ then $\theta = 0$ and the equation (8) is analogous to the dispersion equation for capillary-gravity waves in the presence of surfactant but in the absence of magnetic fields [4]; the only difference is the expression for ω_0^2, that may be negative even for positive surface tension in the presence of magnetic fields. The stability condition in this case is

$$\forall k_1, k_2 : \omega_0^2(\mathbf{k}) > 0 \qquad (9)$$

The condition (9) is equivalent to one of the more simple conditions $H_n < H_*$ or $H_t < H_*$, where H_* is the critical value of magnetic field strength depending on other parameters of the problem. For $H_n \neq 0, H_t = 0$ and three different values of $\delta = -\lambda_t / \lambda_n$ the dimensionless $q^* = \lambda_n H_*^2 / 2(\gamma_0 - \beta\Gamma)$ is shown on Fig.1.

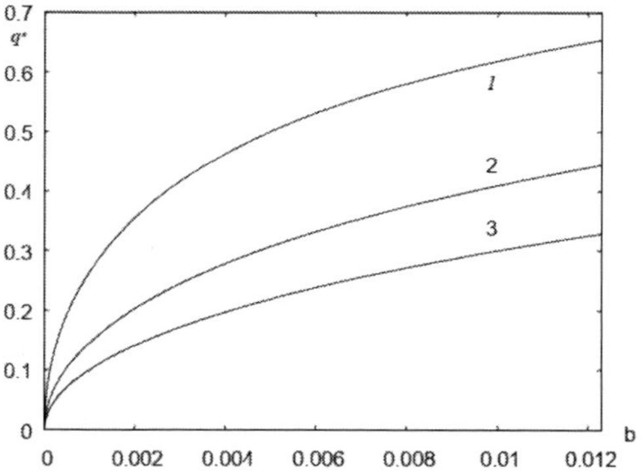

Fig.1. The dependence of critical dimensionless magnetic field strength $q^* = \lambda_n H_*^2 / 2(\gamma_0 - \beta\Gamma)$ on parameter $b = (\gamma_0 - \beta\Gamma) / \rho g\left(2\pi\lambda_t\right)^2$ at $\delta = 0.1$ (1),0.2(2),0.3(3).

The function $q^*(b)$ is defined by the following parametric expressions:

$$\delta q^* = t + 2t^2 + t^3, b(1 - q^*) = t^2 + 2t^3, t > 0$$

For $H_n = 0, H_t \neq 0$ the critical value of magnetic field strength is defined by the equation

$$\gamma_0 - \beta\Gamma + \Gamma b_t H_*^2 / 2 = 0$$

Note then the stability condition requires $b_t < 0$ [3].The value of critical field may be estimated if we use the experimental data [1] qualitatively and suppose that for the magnetic field value 0.5 T the surface tension changes by 20%. In this case the critical field equals $0.5\sqrt{5} \approx 1.118$ T.

If $H_n H_t \neq 0$, the condition (9) is insufficient for the stability of the liquid surface. In that case the stability condition is equivalent to $\varepsilon > \varepsilon_* (H_t / H_n)$, where the critical value ε_* depends on other parameters of the problem (see Fig.2).

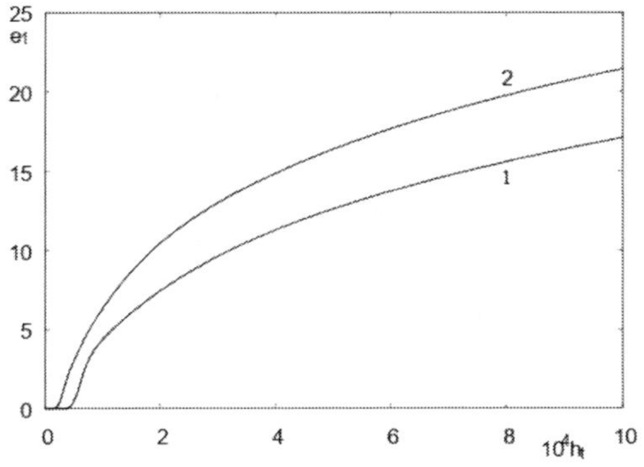

Fig.2. The dependence of $e_1 = \varepsilon_* \rho / \sigma_*$ on dimensionless tangential component of external magnetic field $h_t (h_t^2 = \lambda_n H_t^2 / \sigma_*)$, where $\sigma_* = \gamma_0 - \beta\Gamma - \lambda_n H_n^2 / 2 + \lambda_t H_t^2 / 2$, for the following parameters: $\lambda_n H_n^2 / \sigma_* = 1, \nu^2 \rho / \sigma_* \lambda_n = 0.1, \rho g \lambda_n^2 / \sigma_* = 10^{-6}$ and $\delta = 0.1(1), \delta = 1(2)$.

4 The damping decrement

The damping decrement of capillary-gravity waves depends significally on elastic constant. Let $s = s_*(\mathbf{k})$ be the solution of the dispersion equation with maximal $\mathrm{Re}(s)$, i.e. with minimal damping decrement. The dependence of $\mathrm{Re}(a) \equiv \mathrm{Re}(s_*) / \omega_0(\mathbf{k})$ on dimensionless wave number $m = \nu k^2 / \omega_0(\mathbf{k})$ for several values of problem parameters is shown on Fig.3.

For small values of $e_1 \neq 0$ $\mathrm{Re}(a) << 1$ and $\mathrm{Re}(a) \to 0$ if $e_1 \to 0$ but $e_1 \neq 0$ and $k \neq 0$. Then, for sufficiently long wavelengths, the value of $\mathrm{Re}(a)$ increases with growing e_1 for sufficiently small $e_1 < e_1^*(k)$ but decreases for $e_1 > e_1^*(k)$ where $e_1^* \approx 0.3$.

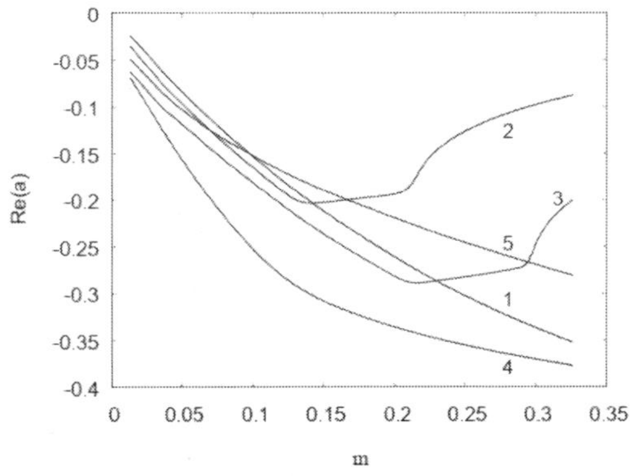

Fig.3 The dependence of damping decrement on dimensionless wave number for different values of dimensionless elastic constant $e_1 = \varepsilon_* \rho / \sigma_*$: $e_1 = 0(1); 0.05(2); 0.1(3); 0.3(4); 1(5)$. The main problem parameters are $\lambda_t / \lambda_n = -0.1, \lambda_n H_n^2 / (\gamma_0 - \beta\Gamma - \lambda_n H_n^2 / 2) = 1, H_t = 0$.

5 Conclusions

The stability of capillary-gravity wave motion on horizontal free surface of viscous non-compressible fluid in the presence of magnetic surfactant in an external magnetic field was studied taking into account the anisotropy of surface tension tensor and the dependence of surface tension on surfactant surface density and on magnetic field strength. It is shown that for normal as well as for tangential external magnetic field the horizontal free liquid surface is unstable for field strength exceeding some critical value that does not depend on the elastic constant of the surfactant film. However, for oblique external magnetic field the stability of the free surface depends not only on the field value but also on the surfactant elastic constant.

The damping decrement of capillary-gravity waves depends significally on surfactant elastic constant.

Support by the Russian Foundation for Basic Research (projects Nos. 16-01-00157 and 17-01-00037) is acknowledged.

References

1. Brown P., Hatton T.A., Eastoe J. Current Opinion in Colloid & Interface Science, **20**, 140 (2015).
2. Golubyatnikov A. N., Subhankulov G. I. Magnetohydrodynamics (Engl. Transl.), **22**, 62 (1986).
3. Golubyatnikov A. N., Uspekhi Mekhaniki, **4**, 3(2006) (in Russian).
4. Levich V. G. *Physicochemical hydrodynamics*, Prentice-Hall, 1962, 700 pp.

EPJ Web of Conferences **185**, 09003 (2018) https://doi.org/10.1051/epjconf/201818509003
MISM 2017

Frequency dependence of magnetothermal properties for magnetic fluid and magnetically functionalized implants

Salakhova R.T.[1], *Vylegzhanin A. G.*[1], *Kashtanov E.A.*[1], *Zverev V.I.*[1], *Müller R.*[2], *F.D. de Sena Pereira*[3], *Parfenov V.A.*[3], *V.A. Mironov V.A.*[3], *Kritskaya E.A.*[4], *Pyatakov A.P.*[1], *Markov V.K.*[5], *Malyshev A.Yu*[5], *Kamilov K.I*[1], *and Tishin A.M*[1,5]

[1] M.V. Lomonosov Moscow State University, Leninskie gori street 1, Moscow, Russia
[2] Institute of Photonic Technology, Albert-Einstein-Straße 9, Jena, Germany,
[3] 3D Bioprinting Solutions Lab, 68/2, Kashirskoe highway, Moscow, Russia
[4] LLC Shell, Novinskii boulevard 31, Moscow, Russia
[5] AMT&C Group, Promishlenaya street 4, Troitsk, Russia,

Heating of the magnetic nanoparticles in AC magnetic field is the effect promising for application in medicine. The mechanisms of heating in AC-magnetic field implies nontrivial dependence of the power dissipated by magnetic nanoparticles on frequency. With the use of a reconfigurable experimental setup, this frequency-dependent magnetic heating was measured on two characteristic examples: the magnetite nanoparticles conventionally used in medicine and polymer coating with micrometer sized magnetite particles. The saturation of the heating power with frequency is shown that is more pronounced for the second case of microparticles.

1. Introduction

Magnetic nanoparticles (MNp) have attracted attention in modern medicine due to their possibility of injecting and accumulating in the target tissue. Heating of MNp in alternating (AC) magnetic field leads to the therapeutic effect either through direct thermal treatment [1,2] or due to the heat-assisted release of the drug carried by nanoparticles [3]. The magnetically functionalized implants, which consist of polymeric matrices and MNp, can also be used as effective method to prevent undesired effect of inflammation or cell proliferation.

There are three types of heating mechanisms of magnetic nanoparticles in AC-magnetic field: Brown relaxation, Néel relaxation, and hysteresis losses. Brown relaxation implies the nanoparticles have freedom of rotation in the fluid and their rotation in the external AC magnetic field is accompanied with energy dissipation due to the viscous damping. In contrast to the Brown relaxation, the Néel relaxation mechanism is based on the rotation of the magnetic moments of the nanoparticles in AC magnetic field with energy dissipated during the relaxation of the magnetic moments around the equilibrium position. The Néel relaxation prevails for immobilized magnetic nanoparticles or particles with the large time of Brown relaxation. Finally the hysteresis losses are dominant in multidomain particles and associated with the power dissipated during the cycle of magnetization reversal.

A common parameter for quantitative characterization of magnetic heating is the Specific Absorption Rate (SAR), i.e. the ratio of the power dissipated to the mass of the particles. This value is of clinical importance for calculation of particle concentration and field exposure time.

Since all of those three above mentioned mechanisms are related to some charactersitic relaxation time [4] or hystersis cycle period, SAR should be strongly dependent on the frequency of the AC magnetic field. Thus it is crucial to have the setup with easily changeable frequency range.

In this paper the special reconfigurable experimental setup was developed to study the frequency dependence of the magnetothermal properties on two characteristic examples: the magnetic fluid and magnetically functionalized polymer implants.

2. The experimental setup and measurements procedure

The experimental setup for calorimetric SAR measurements consists of an AC magnetic field module, a water cooling system and a PC data acquisition system (fig.1). The AC magnetic field module comprises the 400W power supply, the generator connected in series with the oscillating circuit that includes the reconfigurable capacitor system and the main part of the setup: the solenoid providing magnetic field 100 Oe (its stability is controlled in real time with a test coil) in the working volume with the inner diameter 7 mm. Due to the reconfigurable capacitor system there is a possibility to adjust the frequency in the vicinity of the oscillating circuit resonance frequencies corresponding to the set of values: 150, 200, 250, 300 kHz.

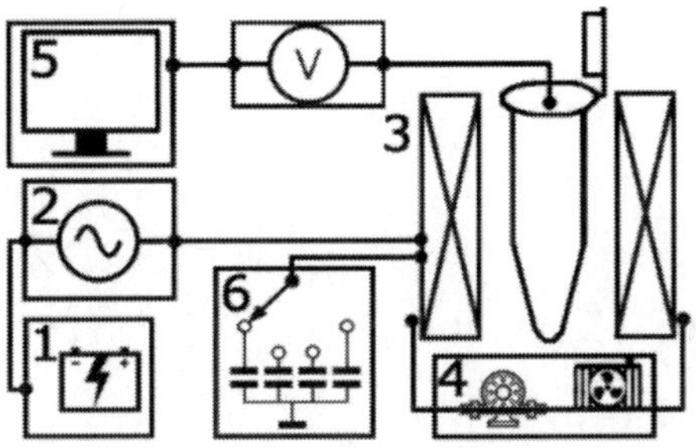

Fig. 1. Equipment for SAR measurement.1. Power supply; 2. generator; 3. solenoid; 4. water cooling system ; 5. data acquisition and monitoring system, 6. reconfigurable capacitor system.

During the measurements, the test tube with a sample and thermocouple was placed into homogeneous magnetic field region deep inside the coil.
SAR was calculated according to the definition:

$$SAR = C\frac{M}{m}\frac{dT}{dt} \qquad (1)$$

where C is water specific heat capacity, dT/dt is the rate of heating in the absence of heat losses, M/m is the ratio of the water mass to the mass of the sample (powder of nanoparticles or a polymer implant functionalized).

Due to the non-adiabatic measurement condition the actual slope of the heating curve differs from dT/dt in eq.(1). Generally speaking, the detailed thermodynamic model of the heat exchange mechanisms in experimental setup is needed to calculate the energy losses [5]. We used the empirical approach that is the variation of the "corrected slope" method [6]: to obtain the adiabatic value of dT/dt at any given temperature in the real environment one should sum the absolute values of slopes of the heating and cooling curves at this temperature.

To exclude the parasitic effects related to the heating of the coil and the inductive interference in thermocouple the control measurements with distilled water without samples were carried out.

3. Results

In this part we studied the dependence of heating dynamics on frequency for two samples: the magnetic fluid of magnetite nanoparticles conventionally used in medicine [7] and the polymer implant coating with magnetite microparticles embedded in it.

3.1 Suspension of magnetite@dextran nanoparticles

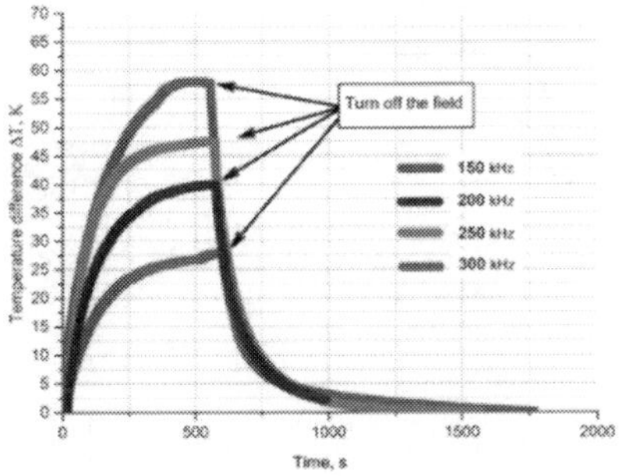

Fig. 2. The heating/cooling curves for suspension of magnetite nanoparticles at various frequencies of AC magnetic field.

The set of heating/cooling curves corresponding to various magnetic field frequencies for magnetite nanoparticles water suspension is shown in Fig 2. The magnetite nanoparticles with dextran coating (core diameter 15±2 nm, and hydrodynamical diameter about 100 nm) were prepared by wet chemical precipitation method followed by coating procedure with biocompatible carboxymethyldextran (for details, see [7]). The sample was prepared by stirring 20 mg of nanoparticles in 0.1 ml of distilled water. The amplitude of AC magnetic field was 100 Oe. One can see the growth of the ultimate temperature with the frequency of magnetic field.

3.2 Polymer matrix with embedded magnetite microparticles

The analogues heating/cooling curves for functionalized polymer implant is shown in fig.3. It was the electrospun polyurethane matrices embedded with magnetite particles (3% v/v). The particles were prepared by co-precipitation method from aqueous solution with a 2 Fe(III):1 Fe(II) mole ratio by hydrolyzing in 25% ammonium hydroxide solution. The average size of particles was 650 nm (size dispersion 350 nm). The sample was merged into 0.5 ml of water. The sample size, thickness and mass were 2x2 cm, 0.25 mm, and 60 mg, respectively. The amplitude of AC magnetic field was 100 Oe. The saturation of the ultimate temperature with frequency is clearly seen.

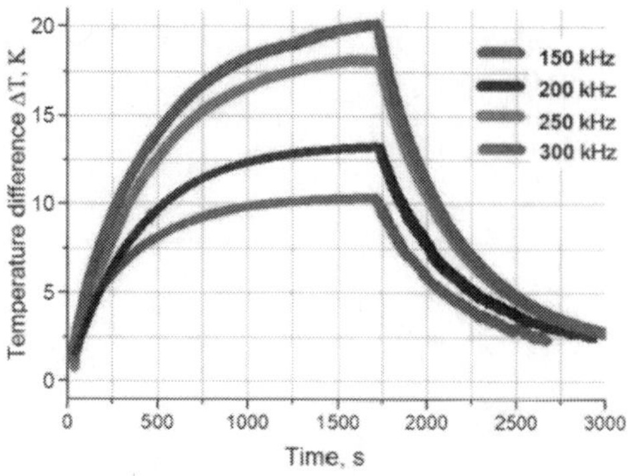

Fig. 3. The heating/cooling dynamics at various frequencies for magnetite microparticles in polymer implant coating.

4. Discussion

To characterize the heating of the samples quantitatively, the value of SAR according to eq.1 was calculated. The temperature increase rate *dT/dt* was obtained by following the procedure described in section 2.

We also approximate the heating/cooling curves with the following equation, corresponding to the time evolution of the temperature [1]:

$$\Delta \mathrm{T(t)} = \tau_R \left.\frac{dT}{dt}\right|_{t=0} \left(1-\exp\left(-\frac{t}{\tau_R}\right)\right), \quad (2)$$

where τ_R is a relaxation constant which depends on the heat capacity, the surface of the sample, and the heat transfer coefficient between the sample and the medium. The value *dT/dt* obtained from approximation of the experimental data with the dependence of eq. (2) gives the same results as the corrected slope method described in section 2. That means that we deal with conventional magnetic heating far from the blocking temperature, Curie point and other self-regulated mechanisms limiting the maximum temperature [2].

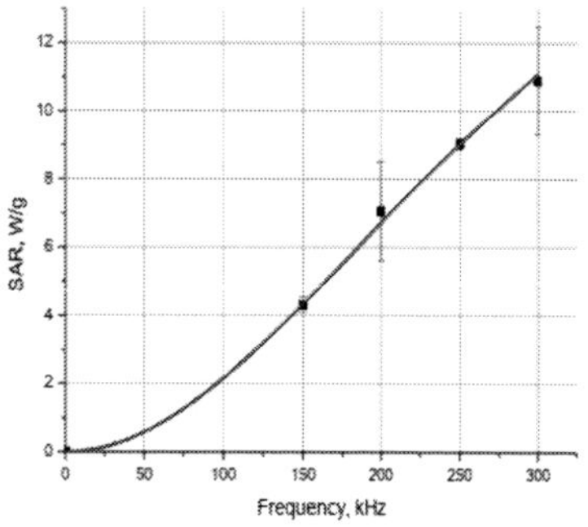

Fig. 4. The heating rate dependence on frequency of the AC field for suspension of magnetite nanoparticles.

The frequency dependences of SAR for the samples are shown in Fig.4 and 5.

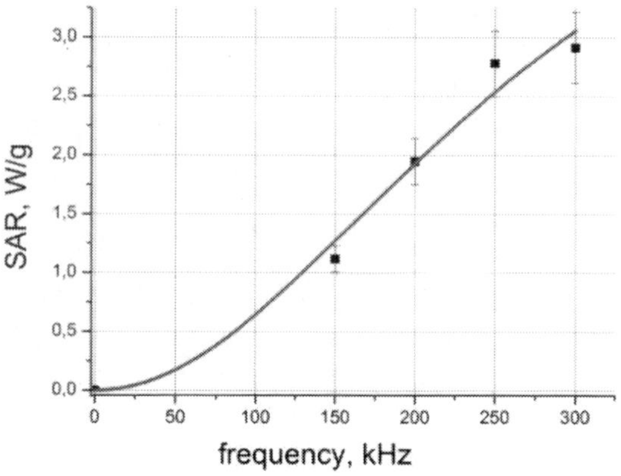

Fig. 5. The heating rate dependence on frequency for microparticles in polymer implant coating.

It should be noted that the values of SAR was calculated per unit of mass of the sample (magnetic part and the polymer coating) so they cannot be used for comparison of heating ability of bare magnetite nanoparticles. However we can compare the character of the frequency dependence (the asymptotics, the saturation etc). The SAR is increasing with the frequency for both of the samples, but the saturation of the dependence for the second sample is more prominent.

To gain insight into the origin of this saturation the frequency dependence of various heating mechanism should be analyzed. The naive approach to the problem supposes the linear dependence of SAR on frequency since a certain amount of heat is released in every cycle of AC magnetic field. However, the characteristic time of the reaction of the particles on external magnetic field should be taken into account. The more accurate approach implies the dependence saturating with the frequency [4]:

$$SAR = C\cdot\frac{(2\pi f\tau)^2}{1+(2\pi f\tau)^2} \quad (3)$$

where f is the AC magnetic field frequency, C is proportionality constant, and τ is characteristic time of the magnetization relaxation or hysteresis reversal process.

Approximation of the data in fig.4 by eq. (3) gives the value for relaxation constant 510±10 ns. The analogous approximation of the SAR frequency dependence gives somewhat higher value 600±50 ns and the approximation seems to be less accurate in this case (red curve in Fig.5).

The characteristic time strongly depends on the size of the particles, it also depends on the mechanism of heating. The Brown and Neel mechanism are predominant for magnetic nanoparticle suspension, while they are irrelevant for magnetic microparticles embedded into polymer. Since the particles are immobilized (that eliminates Brown heating) and too large for Neel relaxation the most probable mechanism is hysteresis losses, that can be the reason why the superparamagnetic-like dependence of eq.3 not fully describes the experimental data. Intriguingly, the magnetic heating rate is not just a linear function of the frequency. The number of remagnetization cycles per second is not the only factor that matters: the energy released in one hysteretic cycle is also the function of frequency.

5. Conclusions

Theory of magnetic heating in AC field predicts nontrivial dependence on frequency with quadratic dependence at low frequencies, linear law in the intermediate frequency range and saturation of heating at high frequencies. To explore this frequency dependence the reconfigurable experimental setup is constructed that enables to measure the magnetic heating in 100 Oe AC magnetic field at frequencies 150, 200, 250, and 300 kHz. The frequency dependence of magnetic heating shows the sublinear dependence of magnetic heating in high frequency range. The demonstrated saturation of heating with the frequency more pronounced for larger particles implies that the "naive" notion of magnetic heating rate proportional to the frequency is not valid both for Neel and hysteretic mechanism of magnetic heating.

The support of The Foundation for Assistance to Small Innovative Enterprises in Science and Technology (project #C1-21200) is acknowledged .

References

1. B. Thiesen, A. Jordan. Int. Journal of Hyperthermia.; **24(6)**, 467 (2008)
2. E. A. Périgo, G. Hemery, O. Sandre, D. Ortega, E. Garaio, F. Plazaola, and F. J. Teran, Appl. Phys. Rev. **2**, 041302 (2015)
3. Ch. S.S.R. Kumar, F. Mohammad, Advanced Drug Delivery Reviews, **63**, 789 (2011).
4. R. E. Rosensweig, JMMM, **252**, 370 (2002).
5. M. Coisson, G. Barrera, F. Celegato et al, Biochimica et Biophysica Acta, **1861**(6), 1545 (2017).
6. R.R. Wildeboer, P. Southern, Q.A. Pankhurst J. Phys. D: Appl. Phys. **47**, 495003 (2014).
7. S. Dutz, W. Andra, R. Hergt et al, JMMM, **311,** 51 (2007) .

Microstructure formation and macroscopic dynamics of ferrofluid emulsion in rotating magnetic field

Anastasia Kolesnikova[1], *Arthur* Zakinyan[1,*], and *Yuri* Dikansky[1]

[1]Department of General and Theoretical Physics, North Caucasus Federal University, 1 Pushkin Street, Stavropol 355009, Russia

Abstract. The ferrofluid-in-water emulsion has been produced and studied. The microstructure formation in a flat layer of such emulsion under the action of in-plane rotating magnetic field has been observed. Several microstructure types have been found such as chain-like aggregates, disc-like clusters, branching and space-occupying structures. The time evolution of the formation process has been analysed. The revealed microstructures have been computationally investigated by molecular dynamics simulations. The microstructure dynamics in ferrofluid emulsions manifests itself in macroscopic mechanical effects. We analysed the appearance of macroscopic torque on a ferrofluid emulsion sample in a rotating magnetic field. The value of the torque has been measured as a function of magnetic field strength and frequency.

1 Introduction

Processes of formation and growth of clusters from individual particles or components attract great interest in the context of investigation of complex heterogeneous systems. The aggregation process constitutes an important field of basic research and applications in various branches of physics and materials sciences. Many efforts have been undertaken to understand the aggregation dynamics of magnetic microparticles dispersed in a fluid (magnetorheological suspension), and subjected to constant and/or varying magnetic fields. The works [1–3] can be cited as a few examples of the research in this direction. The particles under the action of applied magnetic field undergo an induced, complex self-organization process. In the context of the present research the structurization processes under the rotating uniform magnetic field are of particular interest. In our case, the system under study is a coarse ferrofluid-in-water emulsion. The use of ferrofluid emulsion gives some new opportunities in the aggregation process studies such as the ability to produce the samples with different droplet sizes and the ability to produce the inverse systems of nonmagnetic droplets suspended in ferrofluid.

The microstructure formation dynamics in composite material can manifest itself in macroscopic mechanical effects. Thus the rotating clusters forming in a magnetically active suspension under the action of rotating magnetic field can produce a macroscopic torque on a suspension sample. Previously the effects of internal rotations have been studied with regard to the pure ferrofluids [4–7]. It can be expected that the internal rotation effect will be even more pronounced in magnetic suspensions (such as ferrofluid emulsion) than that in a pure ferrofluid.

2 Microstructure formation processes

The ferrofluid emulsion under study has been produced by dispersing a kerosene-based ferrofluid with magnetite nanoparticles in distilled water by means of ultrasonic disperser. Sodium dodecyl sulfate has been used as emulsifier. The emulsion with droplets sizes of a few tens of micrometers has been obtained. The microstructure formation in a flat horizontal layer of such emulsions under the action of in-plane rotating magnetic field has been observed and studied by means of an optical microscope. Uniform rotating magnetic field has been produced by two pairs of perpendicular Helmholtz coils.

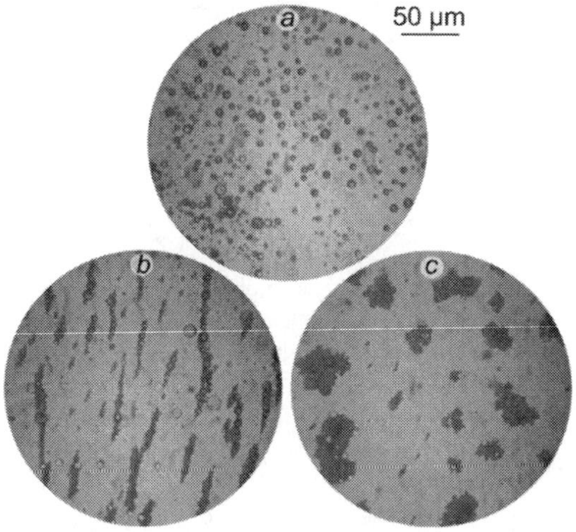

Fig. 1. Microstructures at low concentration. (a) Zero field. (b) Chain microstructure (f=1 Hz, H=9 kA/m). (c) Disc-like microstructure (f=7 Hz, H=7.8 kA/m).

* Corresponding author: zakinyan.a.r@mail.ru

It has been found several microstructure types. At comparatively low emulsion concentrations two types of clusters have been revealed: chain-like aggregates and disc-like clusters (Fig. 1). At low magnetic field frequencies and high magnetic field strengths the chine-like structures takes place. At higher frequencies and lower magnetic fields the droplet discs are formed. The structures evolve in time and their type can change. Thus the initial chains can transform into discs. The phase diagrams of structure types in frequency–strength coordinates at three time moments are shown in Fig. 2.

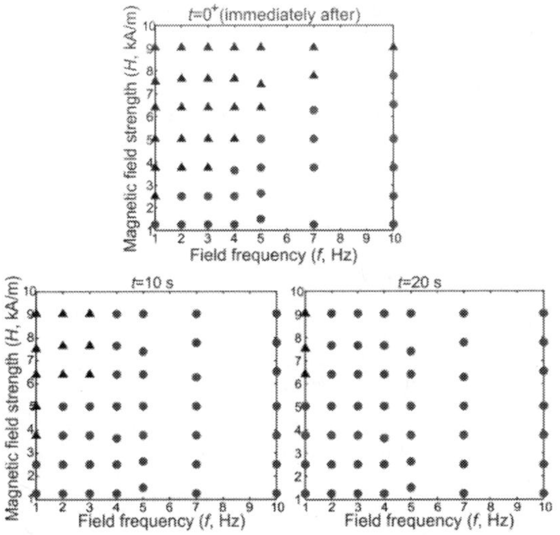

Fig. 2. Phase diagrams of structures at different moments after magnetic field application. Triangles (▲) correspond to chain-like aggregates; circles (●) correspond to disc-like clusters.

The microstructure time evolution has been analysed be measuring the relative total cluster number. Fig. 3 shows the time dependence of the relative total cluster number at different values of the rotating magnetic field strength. As is seen, this dependence tends to saturation after first ten seconds of magnetic field action. At higher fields the aggregation process is more intensive.

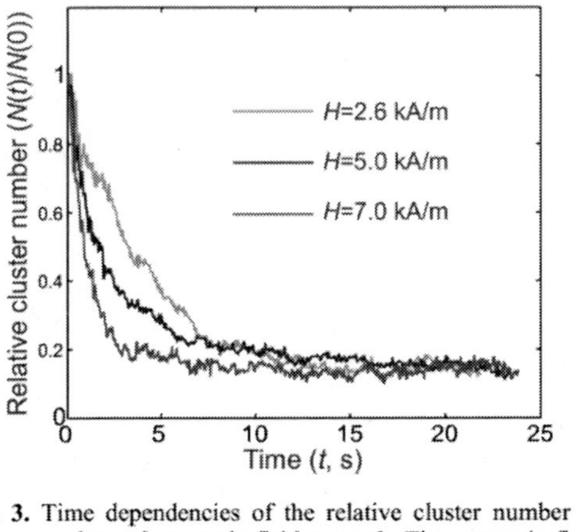

Fig. 3. Time dependencies of the relative cluster number at different values of magnetic field strength. The magnetic field frequency is f=5 Hz.

The emulsion concentration is also an essential factor for the microstructure formation. It has been found that at comparatively high concentrations the mesoscale structures of particle clusters are formed under the action of rotating magnetic field independently of its frequency and strength. Thus, Fig. 4a shows the branching structures formed at middle concentrations. At higher concentrations the space-occupying structures take place (Fig. 4b).

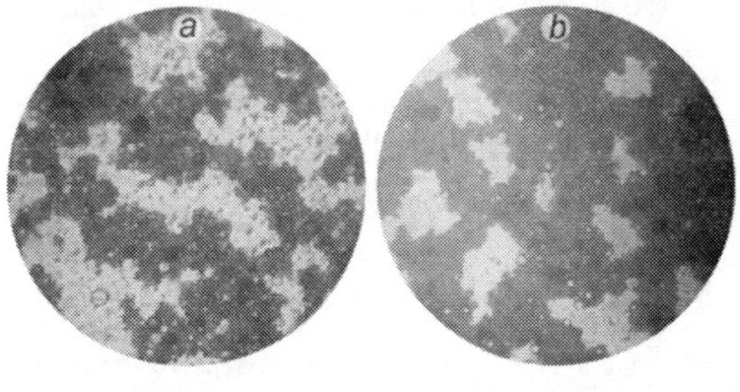

Fig. 4. Branching (a) and space-occupying (b) structures at high concentrations of ferrofluid emulsion.

The studied systems have been computationally investigated by molecular dynamics simulations similar to that presented in [8]. The main interactions that capture the general behavior of the droplets are the dipolar magnetic interaction, the drag force and the repulsive force avoiding droplets from overlapping. In contrast to [8], the local magnetic field was used to calculate the droplets magnetic moments instead of the external field; and Hadamard–Rybczynski formula was used instead of Stokes' formula for the drag force. The structures analogous to the experimentally observed have been obtained in the calculations with the use of experimental values of the system parameters. The results of calculations are shown in Fig. 5.

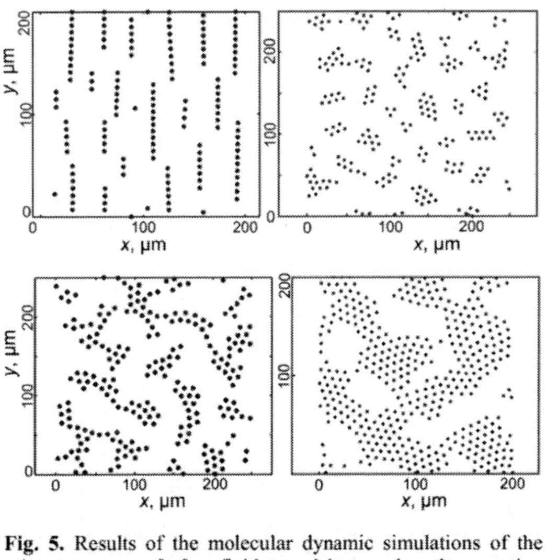

Fig. 5. Results of the molecular dynamic simulations of the microstructure of ferrofluid emulsion under the rotating magnetic field. The chain-like and disc-like aggregates, branching and space-occupying structures can be observed.

To characterize the emulsion microstructure the pair distribution function has been calculated based on the simulations results (Fig. 6).

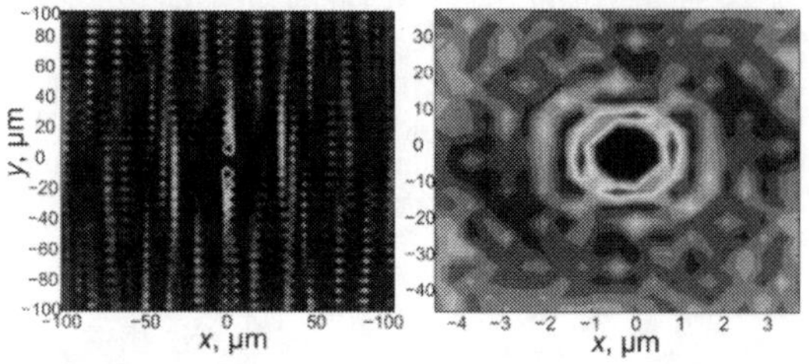

Fig. 6. Pair distribution functions calculated on the basis of simulations results. Left image corresponds to the chain microstructure; right image corresponds to the branching microstructure.

3 Macroscopic torque measurements

We have analysed the appearance of macroscopic torque on a ferrofluid emulsion sample in a rotating magnetic field. The internal rotations of formed microstructural elements produce torque acting on a whole emulsion sample due to the viscous coupling. To measure the torque the emulsion sample was placed into a hollow plastic sphere which is neutrally buoyanced in a surrounding liquid. Under the action of uniform rotating magnetic field the sphere rotation has been observed. The sphere rotation frequency has been measured from which the toque acting on sphere can be easily determined. The macroscopic torque generating inside the sphere in a rotating magnetic field due to the structurization processes is balanced by the viscous torque acting on the outside from the surrounding liquid in this case.

The macroscopic torque has been measured as a function of magnetic field strength and frequency. The results of measurements are presented in Fig. 7. As it can be seen, the torque grows with magnetic field strength and frequency.

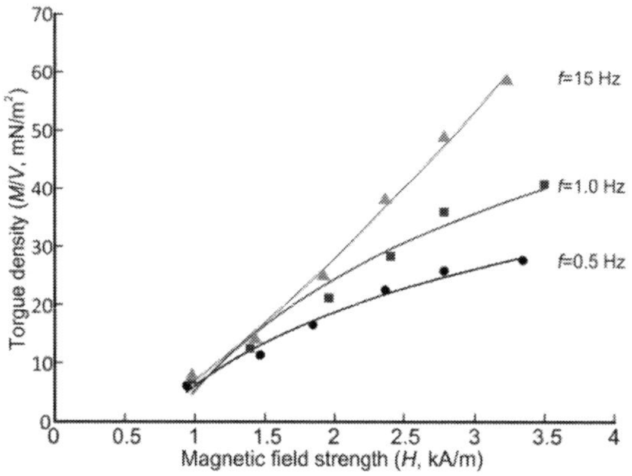

Fig. 7. Macroscopic torque density as a function of magnetic field strength at different values of magnetic field frequency.

4 Conclusions

Thus, in the present work the processes of structure formation in a ferrofluid emulsion have been studied experimentally. Different microstructure types have been observed and studied depending on the magnetic field strength, frequency, time and emulsion concentration. The chain-like, disc-like, branching and space-occupying structures have been analyzed. Analogous structures have been obtained in numerical simulations. The appearance of the macroscopic torque in a ferrofluid emulsion due to the internal rotations of formed microstructure elements has been found. It can be concluded that the ferrofluid emulsions can be regarded as controllable soft magnetic materials. The studied effects can be useful for actuation purposes.

This work was supported by the grant of the President of the Russian Federation No. MK-3169.2017.2 and also by the Ministry of Education and Science of the Russian Federation in the framework of the base part of governmental ordering for scientific research works (project № 3.5822.2017/8.9).

References

1. L.D. Yépez, J.L. Carrillo, F. Donado, J.M. Sausedo-Solorio, P. Miranda-Romagnoli, J. Magn. Magn. Mater. **408**, 321-329 (2016).
2. M. Llera, J. Codnia, G.A. Jorge, J. Magn. Magn. Mater. **384**, 93-100 (2015).
3. J. Černák, G. Helgesen, Phys. Rev. E **78**, 061401 (2008).
4. M.I. Shliomis, Sov. Phys. JETP 34, 1291-1294 (1972).
5. A.V. Lebedev, A.F. Pshenichnikov, Magnetohydrodynamics **27**, 4-8 (1991).
6. A. Chaves, M. Zahn, C. Rinaldi, Phys. Fluids **20**, 053102 (2008).
7. I. Torres-Diaz, A. Cortes, Y. Cedeño-Mattei, O. Perales-Perez, C. Rinaldi, Phys. Fluids **26**, 012004 (2014).
8. S. Melle, O.G. Calderón, G.G. Fuller, M.A. Rubio, J. Colloid Interf. Sci. **247**, 200-209 (2002).

MECHANICS OF MAGNETIC FLUID ACTIVE ELEMENT IN STRONG MAGNETIC FIELD

Polunin V.M.[1], *Ryapolov P.A.*[1], *Sheldeshova E.V.*[1]

[1] Southwest State University, Kursk, Russia

Abstract. Measurements and theoretical analysis of the processes of static displacement and oscillations of the magnetic fluid column confined by magnetic levitation in a strong magnetic field in a horizontally placed tube are carried out. The calculations of the saturation magnetization, made on the basis of the obtained results of the displacement and the oscillation frequency for the sample of the magnetic fluid under study, are in good agreement with the experimental data. The described technique is of interest when studying saturation magnetization, magnetophoresis, aggregation of nanoparticles and their temporal dependence in magnetic colloids.

1 INTRODUCTION

The design of many devices using magnetic fluids MF is based on the impact of inhomogeneous magnetic field, i.e., ponderomotive force, on the drop of magnetic colloid; this can be observed while the drop displaces from equilibrium position. Such devices can be called oscillatory systems, wherein MF acts as an inertial-viscous element [1-4]. The most common oscillation systems based on nanodispersed MFs, such as controlled shock absorbers and vibration damping complexes, acoustic dynamics with the use of MF, are based on linear vibrations of MF elastic elements on the basis. Therefore, of particular interest are measurements of the elastic-magnetic parameters of a magnetic fluid held by magnetic levitation, in which linear movements of the MF sample are tracked.

2 EXPERIMENTAL SETUP

To measure MF-column displacement in a strong magnetic field under hydrostatic pressure an experimental setup, the diagram of which is presented in figure 1 was designed.
In the experiment, we used a laboratory electromagnet FL-1 with a diameter of the pole tips equal to 100 mm. The construction of the electromagnet FL-1 and its specifications are given in [5]. The transparent tube 1 made of plexiglas with the internal diameter d = 12 mm is fixed between the pole tips. The tube axis passes horizontally through the centre of the pole gap and is parallel to the surface of the pole tips. When the magnetic field intensity is ≥ 100 kA/m, the tube is filled with a magnetic fluid 2, which is captured by the field and holds in the region of high magnetic field. The position of the tube is fixed with the holding device. In the initial state, MF-column is symmetric with respect to the pole centre. One end of the tube using a silicone tube 3 is connected with the U-shaped tube 4 filled with distilled water. The U-shaped tube elbows are of 800 mm, its internal diameter is 16 mm. The difference in water levels Δh in the tube elbows is determined using cathetometer.
To measure MF-column displacement under hydrostatic pressure – Δz on the same plane with the horizontal axial section of the tube the reference mark is made as it is shown dashed in figure 1; a camera was placed above the whole system to record data of MF-column boundary displacement. MF-column moves to equilibrium state due to the resultant action of ponderomotive force and hydrostatic pressure.
The diagram of the experimental setup for measuring MF-column oscillation frequency ν is shown in Figure 2. There are some new parts in this setup in comparison with the setup shown in Figure 1. The piston 3 is used to drive oscillations. There is a fine air gap between the piston and magnetic fluid. The oscillation indicator, the inductance coil 4 is mounted in the gap between the tube and the pole tip at the level of MF-column free surface. The inductance coil has 5,000 turns of copper wire with a diameter of 0.071 mm wound around the Plexiglas spool. The signal from the oscillation indicator comes to Selective Nanovoltmeter type 237 5. After amplification, the signal is transmitted to GwInstek GOS-72074 oscillograph 6, and then to a computer (not shown in Figure 2) for further processing.
The required value of the magnetic field intensity H in the pole gap of the electromagnet FL-1 is reached by means of two series-connected direct current sources – QJE QJ6030S and is registered by Aktakom ATE-8702 teslameter (not shown in Figures 1 and 2).

* Corresponding author: r-piter@yandex.ru

The experimental dependence of magnetic field intensity (to be more precise, its component Hx) on the z coordinate along the OZ axis extending from the centre of the pole gap parallel to the surface of the pole tips were measured. The measurements were carried out in the range of magnetic field intensity from 100 kA/m to 900 kA/m in increments of 50 kA/m. The teslameter probe moves over the distance of 90 mm using cathetometer. The probe position at the distances from 0 mm to 40 mm and from 70 mm to 90 mm is set at a pitch of 10 mm, and at the distance of 40 mm to 70 mm it is set at a pitch of 2 mm is fixed with an accuracy of 0.01 mm. The values of the magnetic field intensity peculiar to MF-column top and bottom are marked with the

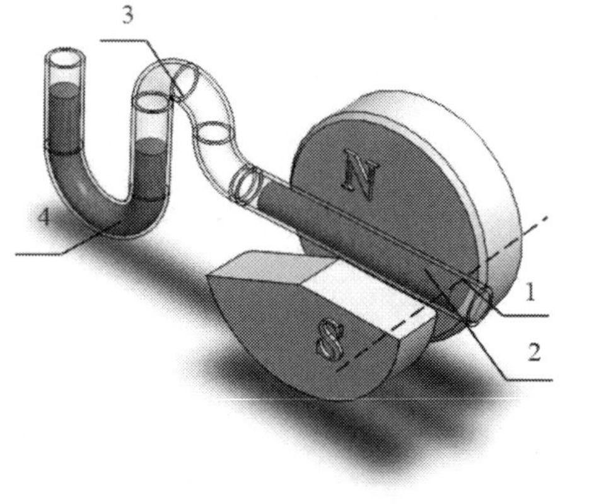

Fig. 1. Static experimental setup diagram

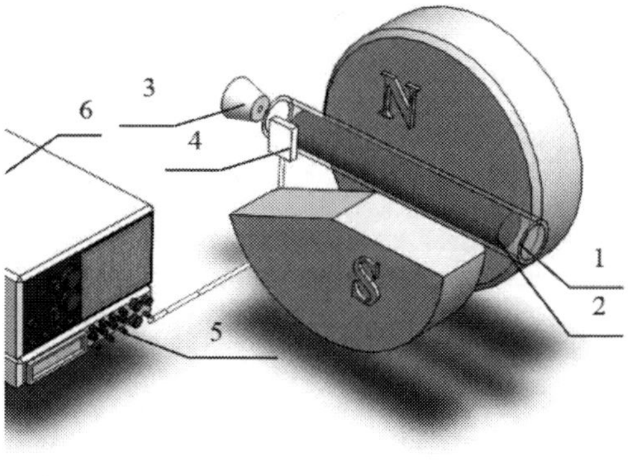

Fig. 2. Dynamic experimental setup diagram

dashed line. A significant feature of the obtained dependencies Hx(z) is the presence of a linear segment on the curves at the level of z = 57.5 mm, which allows us to take the magnetic field intensity gradient at this segment as follows: $\Delta H_x / \Delta z = const$

The measurements of physical quantities ν, Δh, and Δz are made at fixed magnetic field parameters. Table 1 shows the used combination of magnetic field parameters: H_0, which is magnetic field intensity in the centre between the poles of the electromagnet, H*, which is magnetic field intensity, and ΔHx/Δz, which is the magnetic field intensity gradient.

Table 1

H_0, kA/m	H_*, kA/m	$\Delta H_x/\Delta z$, MA/m^2
200	156	3.42
300	236	5.13
400	313	6.95
500	391	8.7
600	469	10.5
700	540	12.4
800	619	14.2
900	686	15.9

In a strong and inhomogeneous magnetic field, MF-column takes a form close to a cylindrical one; in the conducted experiment, the distance between the bases of the cylinder is 115 mm.

3 THEORY OF MF-COLUMN ELASTOMAGNETIC PROPERTIES

The computational pattern introduced for solving the task is shown in Figure 3. It is assumed that the MF left and right free surfaces are on the linear segments of the magnetic field decrease.

In the view of the assumptions concerning the tube 'thinness', magnetic field at the points of the MF free surface is tangential to it, i.e. it has only the tangential component Hx and the magnetic field intensity gradient is directed perpendicular to the surface along the OZ axis to its beginning. Under the action of the differential pressure on the bases of the MF-column 2 with the cross section S and the length b, a small mass center shift equal to δz occurs, which is accompanied by the S·δz volume gain of the fluid at the point with the coordinate z = b/2 and decrease of its volume at the same value at the point z=–b/2.

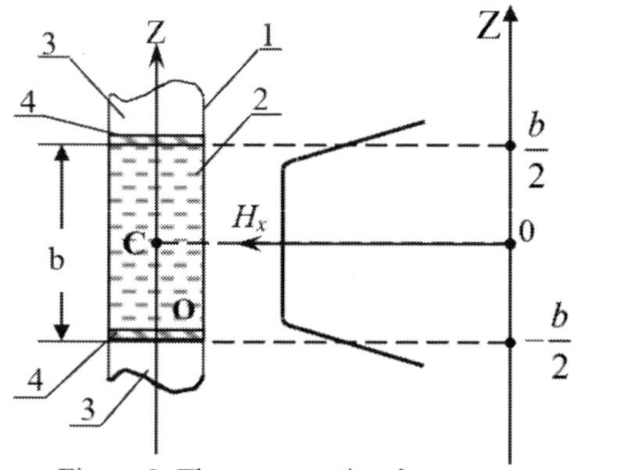

Figure 3. The computational pattern

For the elasticity coefficients of this system, the following values of the elasticity coefficients were obtained in [6]:

For the static experiment the hydrostatic coefficient determined by the water pressures in the U-shaped manometer:

$$k_s = \pi d^2 \rho_w g \Delta h / 4\Delta z, \quad (1)$$

where ρ_w is the water density.

The ponderomotive coefficient is determined by the magnetic fluid displacement from the equilibrium position:

$$k_p = \mu_0 \frac{\pi d^2}{2}\left(M_x \frac{\partial H_x}{\partial z}\right)_{z=b/2}, \quad (2)$$

For the dynamic experiment the coefficient is determined from the oscillation frequency:

$$k_{d\omega} = \omega^2 \pi \rho b d^2 / 4 \quad (3)$$

where ρ is the MF density.

The ponderomotive coefficient is determined by the magnetic fluid displacement from the equilibrium position with addition to viscous friction:

$$k_d = \mu_0 \frac{\pi d^2}{2}\left(M_x \frac{\partial H_x}{\partial z}\right)_{z=b/2} - \frac{\pi^2}{2\sqrt{2}} \cdot bd\sqrt{\omega^3 \eta \rho} \quad (4)$$

We studied a sample of finely dispersed magnetite-based magnetic fluid – Fe_3O_4 which is stabilized by a surfactant – oleic acid $C_8H_{17}CH{=}CH(CH_2)_7{-}COOH$. Undecane $C_{11}H_{24}$, an alkane hydrocarbon, was used as the dispersion medium – liquid carrier. The Undecane fluid MF-1 was synthesized in Ivanovo State Power Engineering University.

Figure 4 shows MF-column elastomagnetic parameters in the strong magnetic field which include oscillation frequency ν, elasticity coefficients k_s, k_d, $k_{d\omega}$, k_p and the correction for viscous fluid flow δ_η calculated using formulas (1), (2), (3), (4) depending on the magnetic field intensity gradient. The calculations were performed using the experimental results: by measuring the MF-column oscillation frequency; the data obtained by measuring the magnetic field intensity gradient at the MF-column base, static measurements of the water column level differences in the elbows of the U-tube Δh and Δz – MF-column displacement under hydrostatic pressure. In this case, the calculations of k_d were performed using the values of M_x belonging to the magnetization curve of the sample under study.

As it can be seen from the data shown in Figure 4, the elasticity coefficients k_p, k_s and k_d, $k_{d\omega}$ corrected for the viscous fluid flow are approximately equal to each other to each other. Thus, we got an experimental confirmation of the above assumptions concerning the presence of equality, and we can obtain the value of the magnetization from two experiments:

magnetostatic saturation magnetization, which is obtained from (2) for the static displacement of the magnetic fluid horizontal column from equilibrium position.

$$M_{SS} = \frac{\rho_w g}{2\mu_0} tg\alpha_s$$

magnetodynamic saturation magnetization, which is obtained from (3) for magnetic fluid dynamic oscillations, taking into account viscous magnetic fluid flow.

$$M_{Sd} = \frac{2}{\mu_0 \pi d^2} tg\alpha_d$$

To test the sensitivity of the proposed method to the change in the structure of the magnetic fluid, these fluids were modified. The sample MF-2 was partially evaporated from the sample MF-1. The physical parameters of the samples are shown in Table 2

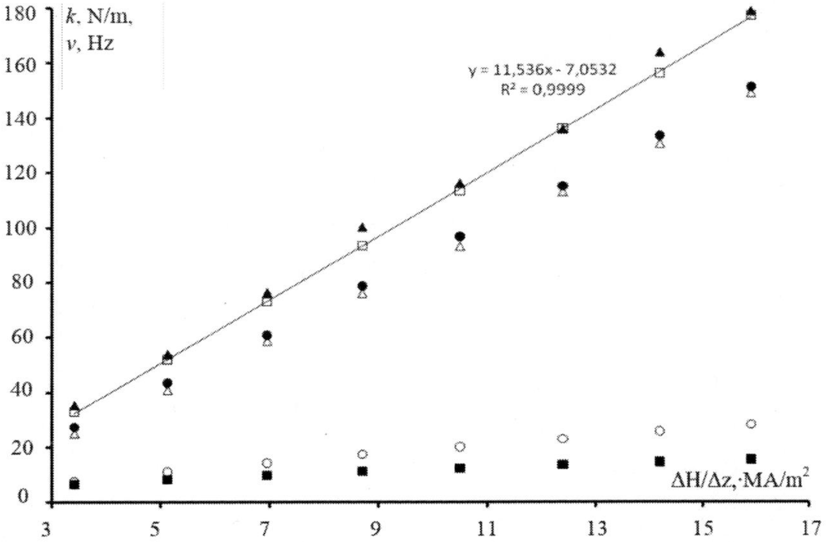

Figure 4. Parameters of oscillation system ■ - ν, ○ - $|\delta_\eta|$, ●- $k_{d\omega}$, Δ- k_d, □ - k_p, ▲- k_s depending on the magnetic field intensity gradient

Table 2

Sample	MF-1	MF-2
Carrier fluid	Undekan	
Modification		Evaporation of carrier fluid
ρ, kg/m3	1227	1240
φ, %	10,8	11,1
χ	1,6	2,1
Ms, kA/m	40,4	45,1
$M_{SS,}$ kA/m	40,9	46,2
M_{Sd}, kA/m	40,6	45,8

The magnetization curves obtained for liquids based on undecane are presented on the figure 5. The rhombus is a ballistic method, rectangles are the magnetodynamic method.

It can be seen that the data of the two methods are in good agreement. But for a structured magnetic fluid MF-2, the difference between the data of the two methods is greater. Data presented in Table 2 show that a small change in the nanoparticle volume fraction 'phi' (from 10.8 to 11.1%) seems to produce an unproportionnally strong increase in saturation magnetization (from about 41 to about 46 kA/m). This can be explained by the fact that when magnetic fluid is evaporated, a structure is formed in it.

4 CONCLUSIONS

The following results were obtained: the construction and validation of the model theory of elasticity formation of MF in a tube confined by magnetic levitation in a strong magnetic field were carried out, the described technique of complex measuring MF elasticomagnetic parameters in the strong magnetic field is of interest when studying magnetophoresis and aggregation of nanoparticles in magnetic colloids.

The research was carried out as a part of the project of state assignment of the Ministry of Education and Science of the Russian Federation. Project code: 3.2751.2017/PP.

References

1. Bogdan Sapinski. Of MR Fluids Recommended For Use In Shock Absorbers / Bogdan Sapinski, Wojciech Horak. Rheological Properties // acta mechanica et automatica, vol.7 no.2 (2013).p.107-110.
2. Bashtovoi, V.G. The effect of magnetophoresis and Brownian diffusion on the levitation of bodies in a magnetic fluid / Bashtovoi V.G., Polevikov V.K., Suprun A.E. // Magnetohdrodynamics. 2008. Vol. 44. №2. P. 121-126.
3. Boev M. L., Polunin V.M, P.A. Ryapolov, [et al.] Instability of the Flow of a Magnetic Liquid Pushing Down an Air Cavity // Russian Physics Journal. 2015. V. 57. №10. pp 1348-1355
4. Polunin V. Acoustics of nanodispersed magnetic fluids. New York- London: CRC Press, CISP, 2015. P. 472
5. Chechernikov V.I. Magnetic measurements. - M.: MGU, 1969. 387 p.
6. Polunin V. M., Ryapolov P. A., Platonov V. B., and Kuz'ko A. E. Free Oscillations of Magnetic Fluid in

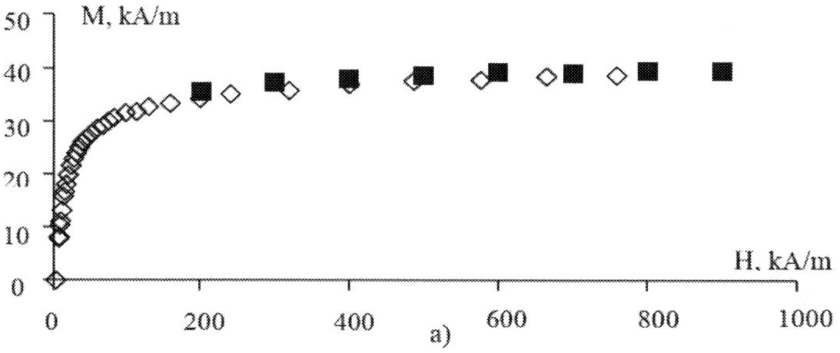

a)

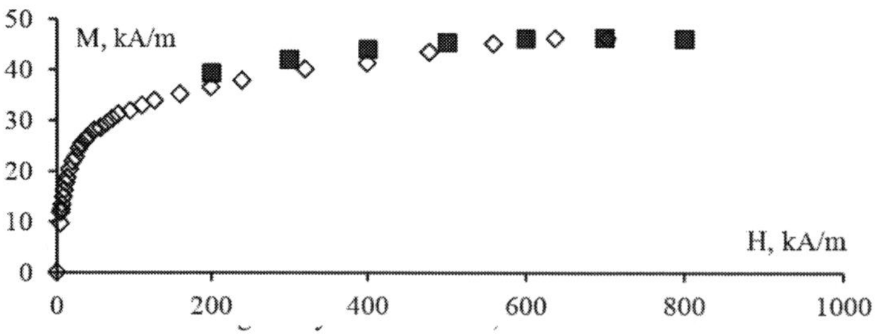

Strong Magnetic Field // Acoustical Physics, 2016, Vol. 62, No. 3, pp. 313–318.

Deformation of a magnetic liquid drop in an applied non-stationary uniform magnetic field

Alexander Tyatyushkin[1,*]

[1] *Institute of Mechanics, Moscow State University, Michurinskiy Ave., 1, Moscow 119192, Russia*

Abstract. Small steady-state deformational oscillations of a drop of magnetic liquid in a non-stationary uniform magnetic field are theoretically investigated. The drop is suspended in another magnetic liquid immiscible with the former. The Reynolds number is so small that the inertia can be neglected. The variation of the magnetic field is so slow that the quasi-stationary approximation for the magnetic field and the quasi-steady approximation for the flow may be used.

1 Introduction

A non-stationary magnetic field applied to a drop of a magnetic liquid suspended in an ordinary liquid or to a drop of an ordinary liquid suspended in a magnetic liquid causes a motion of the liquids inside and outside the drop. Investigation of this phenomenon is interesting from the point of view of application to the actuation of motion of liquids in various devices including microfluidic ones. Besides, the study of this phenomenon is interesting from the point of view of basic science.

Results of experimental investigation of the behavior of magnetic liquid drops are presented in a number of works (see [1] for the references). For the theoretical description of the behavior of a magnetic liquid drop, models based on rather strong assumptions about the shape of the drop and about the flow inside and outside it were used [1]. The goal of the present work is to solve the problem about the shape of the drop and flow inside and outside it without the use of similar assumptions. Within this approach, the shape of the drop is determined by equations obtained with the help of asymptotic methods from the system of equations and boundary conditions that determine the magnetic field and the flow. In order to provide the applicability of the asymptotic methods, the scope of the present theoretical investigation is restricted by the case of weakly deformed drops.

2 Setting of the problem

2.1 Object of investigation

Consider a drop of incompressible magnetic liquid in an applied non-stationary uniform magnetic field with intensity $\vec{H}_a = \vec{H}_a(t)$, where t is the time. The viscosity and magnetic permeability of the liquid inside the drop are η_i and μ_i. The radius of the drop in the undeformed state is a. The drop is suspended in an incompressible magnetic liquid with the viscosity η_e and magnetic permeability μ_e. The surface tension of the interface between the liquids is σ_s. The liquids are regarded as sufficiently viscous for the small Reynolds number approximation to be valid. The variation of the magnetic field is regarded as sufficiently slow and the conductivities of the liquids are regarded as sufficiently small for the magnetic field due to the variation of magnetic field and the electric current in the liquid can be neglected, i.e., for the approximation of ferrohydrodynamics to be valid (see [2]).

2.2 System of equations

In order to find the drop shape, magnetic field intensity, $\vec{H}$, velocity, $\vec{v}$, and pressure, p, as functions of the radius-vector, $\vec{r}$, and time, t, under the above made assumptions, the system of equations of ferrohydrodynamic written down for magnetic fields in the quasi-stationary approximation and for flows in the quasi-steady approximation is used. This system consists of the continuity equation for an incompressible liquid

$$\nabla \cdot \vec{v} = 0, \tag{1}$$

the motion equation in the small Reynolds number and quasi-steady flow approximations

$$\nabla \cdot \left(-p\hat{I} + \hat{\sigma}_m^T + \hat{\sigma}_v\right) = 0, \tag{2}$$

the Maxwell's equations in the ferrohyrodynamics and quasi-stationary field approximations

$$\nabla \cdot \vec{B} = 0, \qquad \nabla \times \vec{H} = 0, \tag{3}$$

*e-mail: tan@imec.msu.ru

and the constitutive relation

$$\vec{B} = \mu\vec{H}. \tag{4}$$

Here, the formulas are written down for Gaussian system of units, $\cdot$ and $\times$ denote the scalar and vector products, $\vec{B}$ is the magnetic induction, $\eta = \eta_{\rm i}$ and $\mu = \mu_{\rm i}$ inside the drop, $\eta = \eta_{\rm e}$ and $\mu = \mu_{\rm e}$ outside it, ∇ is the nabla operator, $\hat{I}$ is the identity tensor, $\hat{\sigma}_{\rm m}$ and $\hat{\sigma}_{\rm v}$ are the tensors of magnetic and viscous stresses expressed as follows

$$\hat{\sigma}_{\rm m} = \frac{1}{4\pi}\vec{H}\vec{B} - \frac{1}{8\pi}\vec{B}\cdot\vec{H}\hat{I}, \tag{5}$$

$$\hat{\sigma}_{\rm v} = 2\eta\left(\nabla\vec{v}\right)^{\rm S}, \tag{6}$$

where $\vec{A}\vec{B}$ denotes the dyadic product of the vectors $\vec{A}$ and $\vec{B}$, $\nabla\vec{f}$ denotes the dyadic product of the nabla operator and the vector field $\vec{f} = \vec{f}(\vec{r})$, $\hat{T}^{\rm T}$ denotes the transpose of the tensor $\hat{T}$, $\hat{T}^{\rm S} = \frac{1}{2}\hat{T} + \frac{1}{2}\hat{T}^{\rm T}$ denotes the symmetric part of the second order tensor $\hat{T}$.

Using the continuity equation Eq. (1) and Maxwell's equations Eq. (3) and taking into account that the magnetic permeability is uniform, the motion equation Eq. (2) can be rewritten (see, e.g., [2]) in the form of the Navier–Stokes equation in the small Reynolds number approximation

$$-\nabla p + \eta\Delta\vec{v} = 0, \tag{7}$$

where Δ is the Laplacian.

2.3 Boundary conditions

The boundary conditions on the interface between the liquids include the impenetrability condition

$$\vec{v}\big|_{\rm e}\cdot\vec{n} = \vec{v}\big|_{\rm i}\cdot\vec{n} = v_{sn}, \tag{8}$$

the no-slip condition

$$\left[\vec{v}\right]_{\rm s}\times\vec{n} = 0, \tag{9}$$

the conditions for the jumps of the normal and tangential components of the stress vector $\vec{\sigma}_n = \vec{n}\cdot\left(-p\hat{I} + \hat{\sigma}_{\rm m} + \hat{\sigma}_{\rm v}\right)$

$$\vec{n}\cdot\left[-p\hat{I} + \hat{\sigma}_{\rm m} + \hat{\sigma}_{\rm v}\right]_{\rm s}\cdot\vec{n} = -\sigma_{\rm s}\mathcal{H}, \tag{10}$$

$$\vec{n}\cdot\left[\hat{\sigma}_{\rm m} + \hat{\sigma}_{\rm v}\right]_{\rm s}\times\vec{n} = 0, \tag{11}$$

the continuity conditions for the tangential component of the magnetic field intensity

$$\left[\vec{H}\right]_{\rm s}\times\vec{n} = 0 \tag{12}$$

and for the normal component of the magnetic field induction

$$\left[\vec{B}\right]_{\rm s}\cdot\vec{n} = 0. \tag{13}$$

Here, $A\big|_{\rm i}$ and $A\big|_{\rm e}$ denote the values of the quantity A on the interface between the liquids approached from inside and outside the drop, respectively, $[A]_{\rm s} = A\big|_{\rm e} - A\big|_{\rm i}$ denotes the jump of the quantity A at the interface when moving from the inside to the outside, $\vec{n}$ is the external normal unit vector at a given point of the interface, v_{sn} is the normal component of the velocity of the surface of the drop at a given point, $\mathcal{H}$ is the mean curvature at a given point of the surface of the drop. Note that the definition of the mean curvature used in the present work is such that it takes negative values on the surface of a convex domain.

The continuity conditions for the normal component of the magnetic induction Eq. (13) and for the tangential component of the magnetic field intensity Eq. (12) allow determining the following scalar and vector fields defined on the surface of the drop

$$B_{sn} = \vec{B}\Big|_{\rm i}\cdot\vec{n} = \vec{B}\Big|_{\rm e}\cdot\vec{n}, \tag{14}$$

$$\vec{H}_{s\tau} = \vec{n}\times\left(\vec{H}\Big|_{\rm i}\times\vec{n}\right) = \vec{n}\times\left(\vec{H}\Big|_{\rm e}\times\vec{n}\right). \tag{15}$$

Using the boundary conditions Eq. (12) and Eq. (13), the constitutive relation Eq. (4), and the introduced notations Eq. (14) and Eq. (15), one can transform the conditions for the stress vector Eq. (10) and Eq. (11) into the form

$$-\left[p\right]_{\rm s} + \frac{1}{8\pi}\left(\frac{1}{\mu_{\rm e}} - \frac{1}{\mu_{\rm i}}\right)B_{sn}^2 - \frac{1}{8\pi}\left(\mu_{\rm e} - \mu_{\rm i}\right)H_{s\tau}^2 + 2\vec{n}\cdot\left[\eta\nabla\vec{v}\right]_{\rm s}\cdot\vec{n} = -\sigma_{\rm s}\mathcal{H}, \tag{16}$$

$$\vec{n}\times\left(\left[\eta\nabla\vec{v}\right]_{\rm s}\cdot\vec{n} + \vec{n}\cdot\left[\eta\nabla\vec{v}\right]_{\rm s}\right) = 0. \tag{17}$$

The boundary conditions at infinity have the form

$$\vec{v}\to 0 \quad \text{as} \quad r\to\infty, \tag{18}$$

$$p\to p_\infty \quad \text{as} \quad r\to\infty, \tag{19}$$

$$\vec{H}\to\vec{H}_{\rm a} \quad \text{as} \quad r\to\infty, \tag{20}$$

where p_∞ is the pressure at infinity.

Besides, $\vec{v}(\vec{r},t)$, $p(\vec{r},t)$, and $\vec{H}(\vec{r},t)$ should be bounded for all the bounded values of $\vec{r}$.

2.4 Deformation of the drop

Let the surface of the drop be given by the following equation

$$r = |\vec{r}| = a + h\left(\frac{\vec{r}}{r}, t\right). \tag{21}$$

The function $h\left(\vec{r}/r, t\right)$ can be represented in the form

$$h = h\left(\frac{\vec{r}}{r}, t\right) = \sum_{n=2}^{\infty}\hat{h}_n \overset{n}{\cdot} \frac{\vec{r}^{\,n}}{r^n}, \tag{22}$$

which excludes displacements of the drop by some vector without variation of its shape. Here, $\hat{h}_n = \hat{h}_n(t)$ $(n > 1)$ are some tensors each of which is an arbitrary irreducible tensor of nth order, i.e., a tensor of nth order symmetric with respect to any pair of indices and such that its contraction with the identity tensor over any pair of indices is equal to zero (see [3]). Here and in what follows, $\vec{b}^{\,n}$ denotes nth dyadic degree of the vector $\vec{b}$ (i.e., the $(n-1)$-multiple dyadic product of the vector $\vec{b}$ by itself), $\overset{n}{\cdot}$ denotes n-multiple contraction of tensors over the adjacent indices.

3 Solution

3.1 Magnetic field

The intensity of the magnetic field is sought for in the form

$$\vec{H} = \nabla\psi, \tag{23}$$

where

$$\psi = \begin{cases} \vec{H}_{\mathrm{a}} \cdot \vec{r} - \sum\limits_{n=1}^{\infty} \hat{M}_{en} \overset{n}{\cdot} \dfrac{\vec{r}^{\,n}}{r^{2n+1}}, & r > a+h, \\ \vec{H}_{\mathrm{a}} \cdot \vec{r} - \sum\limits_{n=1}^{\infty} \hat{M}_{in} \overset{n}{\cdot} \dfrac{\vec{r}^{\,n}}{a^{2n+1}}, & r \leqslant a+h. \end{cases} \tag{24}$$

Here, $\hat{M}_{en} = \hat{M}_{en}(t)$ and $\hat{M}_{in} = \hat{M}_{in}(t)$ are tensors which are either some vectors for $n = 1$ or, for $n > 1$, some irreducible tensors of nth order. For a field given in this form, Maxwell's equations Eq. (3), the constitutive relation Eq. (4), the condition at infinity Eq. (20), and the condition of boundedness of $\vec{H}$ are automatically satisfied.

3.2 Flow

The velocity and pressure in the flow are sought for in the form which can be constructed from general Lamb's solution (cf. [4], art. 336) for a quasi-steady flow

$$\vec{v} = \begin{cases} \vec{v}_{\mathrm{e}}, & r > a+h, \\ \vec{v}_{\mathrm{i}}, & r \leqslant a+h, \end{cases} \qquad p = \begin{cases} p_{\mathrm{e}}, & r > a+h, \\ p_{\mathrm{i}}, & r < a+h, \end{cases} \tag{25}$$

$$\begin{aligned} \vec{v}_{\mathrm{e}} = {} & \sum_{n=1}^{\infty} \left[(2n-1)r^2 - (2n+1)a^2\right] a^n \hat{B}_{en} \overset{n}{\cdot} \frac{\vec{r}^{\,n+1}}{r^{2n+3}} \\ & + \sum_{n=1}^{\infty} \left[(2n-1)nr^2 - (n-2)a^2\right] a^n \hat{C}_{en} \overset{n}{\cdot} \frac{\vec{r}^{\,n+1}}{r^{2n+3}} \\ & + \sum_{n=1}^{\infty} \left[(-n+2)r^2 + na^2\right] a^n \hat{B}_{en} \overset{n-1}{\cdot} \frac{\vec{r}^{\,n-1}}{r^{2n+1}} \\ & - \sum_{n=1}^{\infty} (n-2)n\left(r^2 - a^2\right) a^n \hat{C}_{en} \overset{n-1}{\cdot} \frac{\vec{r}^{\,n-1}}{r^{2n+1}} \\ & + \sum_{n=1}^{\infty} n \left[\hat{\Omega}_{en} \overset{n-1}{\cdot} \frac{\vec{r}^{\,n-1} a^{n+2}}{r^{2n+1}}\right] \times \vec{r}, \end{aligned} \tag{26}$$

$$\begin{aligned} \vec{v}_{\mathrm{i}} = {} & -\sum_{n=1}^{\infty} 2\hat{B}_{\mathrm{i}n} \overset{n}{\cdot} \frac{\vec{r}^{\,n+1}}{a^{n+1}} + \sum_{n=1}^{\infty} 2(n+1)\hat{C}_{\mathrm{i}n} \overset{n}{\cdot} \frac{\vec{r}^{\,n+1}}{a^{n+1}} \\ & + \sum_{n=1}^{\infty} \left[(n+3)r^2 - (n+1)a^2\right] \hat{B}_{\mathrm{i}n} \overset{n-1}{\cdot} \frac{\vec{r}^{\,n-1}}{a^{n+1}} \\ & - \sum_{n=1}^{\infty} (n+1)(n+3)\left(r^2 - a^2\right) \hat{C}_{\mathrm{i}n} \overset{n-1}{\cdot} \frac{\vec{r}^{\,n-1}}{a^{n+1}} \\ & + \sum_{n=1}^{\infty} n \left[\hat{\Omega}_{\mathrm{i}n} \overset{n-1}{\cdot} \frac{\vec{r}^{\,n-1}}{a^{n-1}}\right] \times \vec{r}, \end{aligned} \tag{27}$$

$$p_{\mathrm{e}} = \eta_{\mathrm{e}} \sum_{n=1}^{\infty} 2(2n-1)\left(\hat{B}_{en} + n\hat{C}_{en}\right) \overset{n}{\cdot} \frac{\vec{r}^{\,n} a^n}{r^{2n+1}} + p_\infty, \tag{28}$$

$$p_{\mathrm{i}} = \eta_{\mathrm{i}} \sum_{n=1}^{\infty} \frac{(2n+2)(2n+3)}{n} \times \left[\hat{B}_{\mathrm{i}n} - (n+1)\hat{C}_{\mathrm{i}n}\right] \overset{n}{\cdot} \frac{\vec{r}^{\,n}}{a^{n+1}} + p_0. \tag{29}$$

Here, $\hat{\Omega}_{en} = \hat{\Omega}_{en}(t)$, $\hat{B}_{en} = \hat{B}_{en}(t)$, $\hat{C}_{en} = \hat{C}_{en}(t)$, $\hat{\Omega}_{\mathrm{i}n} = \hat{\Omega}_{\mathrm{i}n}(t)$, $\hat{B}_{\mathrm{i}n} = \hat{B}_{\mathrm{i}n}(t)$, $\hat{C}_{\mathrm{i}n} = \hat{C}_{\mathrm{i}n}(t)$ are some tensors depending on the time each of which is either an arbitrary vector for $n = 1$ or an arbitrary irreducible tensor for $n > 1$, $p_0 = p_0(t)$ is the pressure at the center of the drop, which is to be found.

3.3 Asymptotic expansion

The coefficients are sought for in the form of the following asymptotic expansions over the parameter

$$\alpha = \frac{9a\mu_{\mathrm{e}} \max(H_{\mathrm{a}})^2}{32\pi\sigma_{\mathrm{s}}} \tag{30}$$

$$\begin{gathered} \hat{M}_{en} \sim \sum_{j=0}^{\infty} \alpha^j \hat{M}_{en,j}, \ \hat{\Omega}_{en} \sim \sum_{j=1}^{\infty} \alpha^j \hat{\Omega}_{en,j}, \\ \hat{B}_{en} \sim \sum_{j=1}^{\infty} \alpha^j \hat{B}_{en,j}, \ \hat{C}_{en} \sim \sum_{j=1}^{\infty} \alpha^j \hat{C}_{en,j}, \\ \hat{M}_{\mathrm{i}n} \sim \sum_{j=0}^{\infty} \alpha^j \hat{M}_{\mathrm{i}n,j}, \ \hat{\Omega}_{\mathrm{i}n} \sim \sum_{j=1}^{\infty} \alpha^j \hat{\Omega}_{\mathrm{i}n,j}, \\ \hat{B}_{\mathrm{i}n} \sim \sum_{j=1}^{\infty} \alpha^j \hat{B}_{\mathrm{i}n,j}, \ \hat{C}_{\mathrm{i}n} \sim \sum_{j=1}^{\infty} \alpha^j \hat{C}_{\mathrm{i}n,j}, \\ p_0 \sim \sum_{j=0}^{\infty} \alpha^j p_{0j}, \ \hat{h}_n \sim \sum_{j=1}^{\infty} \alpha^j \hat{h}_{n,j} \text{ as } \alpha \to 0. \end{gathered} \tag{31}$$

The found nonzero terms of the asymptotic expansions are as follows

$$\hat{M}_{\mathrm{e}1,0} = \vec{M}_{\mathrm{e}1,0} = \hat{M}_{\mathrm{i}1,0} = \vec{M}_{\mathrm{i}1,0} = \frac{\mu_{\mathrm{i}} - \mu_{\mathrm{e}}}{\mu_{\mathrm{i}} + 2\mu_{\mathrm{e}}} a^3 \vec{H}_{\mathrm{a}}, \tag{32}$$

$$p_{0,0} = p_\infty + \frac{2\sigma_{\mathrm{s}}}{a} + \frac{3\mu_{\mathrm{e}}}{8\pi} \frac{\mu_{\mathrm{e}} - \mu_{\mathrm{i}}}{\mu_{\mathrm{i}} + 2\mu_{\mathrm{e}}} H_{\mathrm{a}}^2, \tag{33}$$

$$\hat{B}_{\mathrm{e}2,1} = \hat{B}_{\mathrm{i}2,1} = \frac{3}{10} \frac{2\eta_{\mathrm{e}} + \eta_{\mathrm{i}}}{\eta_{\mathrm{e}} + \eta_{\mathrm{i}}} \frac{\mathrm{d}\hat{h}_{2,1}}{\mathrm{d}t}, \tag{34}$$

$$\hat{C}_{\mathrm{e}2,1} = \hat{C}_{\mathrm{i}2,1} = \frac{1}{6} \frac{\mathrm{d}\hat{h}_{2,1}}{\mathrm{d}t}, \tag{35}$$

$$\vec{M}_{\mathrm{e}1,1} = \frac{2}{5}(\mu_{\mathrm{e}} - \mu_{\mathrm{i}}) \frac{13\mu_{\mathrm{e}} + 11\mu_{\mathrm{i}}}{(\mu_{\mathrm{i}} + 2\mu_{\mathrm{e}})^2} a^2 \vec{H}_{\mathrm{a}} \cdot \hat{h}_{2,1}^*, \tag{36}$$

$$\vec{M}_{\mathrm{i}1,1} = \frac{4}{5}(\mu_{\mathrm{e}} - \mu_{\mathrm{i}}) \frac{7\mu_{\mathrm{i}} + 5\mu_{\mathrm{e}}}{(\mu_{\mathrm{i}} + 2\mu_{\mathrm{e}})^2} a^2 \vec{H}_{\mathrm{a}} \cdot \hat{h}_{2,1}^*, \tag{37}$$

$$\hat{M}_{\mathrm{e}3,1} = \frac{\mu_{\mathrm{i}} - \mu_{\mathrm{e}}}{\mu_{\mathrm{i}} + 2\mu_{\mathrm{e}}} \frac{5\mu_{\mathrm{i}} - 8\mu_{\mathrm{e}}}{3\mu_{\mathrm{i}} + 4\mu_{\mathrm{e}}} a^4 \left[\left(\vec{H}_{\mathrm{a}} \hat{h}_{2,1}^*\right)^{\mathrm{S}}\right]^{\mathrm{D}}, \tag{38}$$

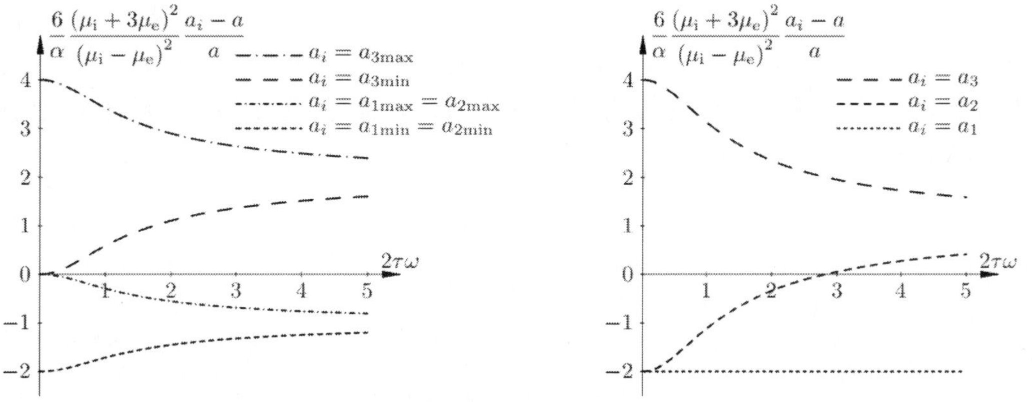

Figure 1. Steady-state oscillations of the drop: in an oscillating (left) and in a rotating (right) magnetic fields

$$\hat{M}_{i3,1} = -4\frac{\mu_i-\mu_e}{\mu_i+2\mu_e}\frac{\mu_i-\mu_e}{3\mu_i+4\mu_e}a^4\left[\left(\vec{H}_a\hat{h}^*_{2,1}\right)^S\right]^D, \tag{39}$$

where

$$\left[\left(\vec{H}_a\hat{h}^*_{2,1}\right)^S\right]^D = \left(\vec{H}_a\hat{h}^*_{2,1}\right)^S - \frac{2}{5}\left(\vec{H}_a\cdot\hat{h}^*_{2,1}\hat{I}\right)^S, \tag{40}$$

and $\hat{h}_{2,1}=\hat{h}_{2,1}(t)$ is a solution of the following ordinary differential equation

$$\frac{(16\eta_e+19\eta_i)(3\eta_e+2\eta_i)}{10(\eta_e+\eta_i)}\alpha\frac{d\hat{h}_{2,1}}{dt}+\frac{4\alpha\sigma_s}{a}\hat{h}_{2,1} = \frac{9a}{8\pi}\left(\frac{\mu_i-\mu_e}{\mu_i+2\mu_e}\right)^2\left(\vec{H}_a^2-\frac{1}{3}H_a^2\hat{I}\right). \tag{41}$$

3.4 Steady-state oscillations

With accuracy up to the terms of the first order, in an oscillating applied magnetic field with the intensity $\vec{H}_a = H_{am}\cos(\omega t)\vec{k}$, the drop is a prolate spheroid with the axis directed along $\vec{k}$ and with the semiaxes

$$a_1 = a_2 = a\left\{1-\frac{\alpha}{6}\left(\frac{\mu_i-\mu_e}{\mu_i+2\mu_e}\right)^2\left[1+\frac{\cos(2\omega t-2\phi)}{\sqrt{1+4\tau^2\omega^2}}\right]\right\}, \tag{42}$$

$$a_3 = a\left\{1+\frac{\alpha}{3}\left(\frac{\mu_i-\mu_e}{\mu_i+2\mu_e}\right)^2\left[1+\frac{\cos(2\omega t-2\phi)}{\sqrt{1+4\tau^2\omega^2}}\right]\right\} \tag{43}$$

where

$$\tau = \frac{(16\eta_e+19\eta_i)(3\eta_e+2\eta_i)}{10(\eta_e+\eta_i)^2}\frac{a(\eta_e+\eta_i)}{4\sigma_s}, \tag{44}$$

$$\phi = \frac{1}{2}\arctan(2\tau\omega). \tag{45}$$

Thus, the drop performs deformational oscillations with the angular frequency 2ω and the phase lag 2ϕ. As $\omega\to\infty$, the drop tends to take the shape of an unvarying prolate spheroid (see figure 1 left).

In a rotating applied magnetic field with the intensity $\vec{H}_a = H_{am}\left[\cos(\omega t)\vec{i}+\sin(\omega t)\vec{j}\right]$, the drop takes the shape of a tri-axial ellipsoid with the semiaxes

$$a_1 = a\left[1-\frac{\alpha}{3}\left(\frac{\mu_i-\mu_e}{\mu_i+2\mu_e}\right)^2\right], \tag{46}$$

$$a_2 = a\left[1+\frac{\alpha}{6}\left(\frac{\mu_i-\mu_e}{\mu_i+2\mu_e}\right)^2\left(1-\frac{3}{\sqrt{1+4\tau^2\omega^2}}\right)\right], \tag{47}$$

$$a_3 = a\left[1+\frac{\alpha}{6}\left(\frac{\mu_i-\mu_e}{\mu_i+2\mu_e}\right)^2\left(1+\frac{3}{\sqrt{1+4\tau^2\omega^2}}\right)\right]. \tag{48}$$

The ellipsoid rotates around its minor axis, directed along $\vec{k}=\vec{i}\times\vec{j}$, with the angular speed ω so that its major axis lags from $\vec{H}_a$ by the angle ϕ. Here, $\vec{i}$, $\vec{j}$, and $\vec{k}$ form a triple of orthonormal vectors. As $\omega\to\infty$, the drop tends to take the shape of an unvarying oblate spheroid (see figure 1 right).

The partial support by RFBR grants 16-01-00157 and 17-01-00037 is acknowledged.

References

[1] A. V. Lebedev, A. Engel, K. I. Morozov, and H. Bauke. Ferrofluid drops in rotating magnetic fields. New J. Phys., Vol. 5 (2003), pp. 57.1–57.20.

[2] R.E. Rosensweig, *Ferrohydrodynamics* (Cambridge University Press, 1985).

[3] L.D. Landau, E.M. Lifshits, *The classical theory of fields* (Pergamon Press, 1975).

[4] H. Lamb, *Hydrodynamics* (Cambridge University Press, 1932).

COMPLEX REFRACTIVE INDEX OF CONCENTRATED MAGNETIC FLUIDS

Constantine Yerin[1,*], *Victoria* Lykhmanova[1], and *Marina* Erina[1]

[1]North Caucasian Federal University, Stavropol, Russia

Abstract. The spectra of the real and imaginary parts of the refractive index of magnetic fluids based on kerosene with a solids phase concentration of 1% to 35% are experimentally measured. The comparison of calculations results have been made according to Maxwell-Garnett and Bruggeman's models of effective medium approximation. The influence of refractive index spectra on the extinction of light in magnetic fluids is shown.

1 Introduction

Magnetic fluids are colloidal systems of magnetic nanoparticles in a liquid medium with significant magneto-optical properties. In magnetic fluids birefringence, dichroism, anisotropic light scattering, and the change in transparency are observed under the action of external magnetic field. With these spectacular and useful magneto-optical properties, magnetic fluid show promise in the development of various devices, such as optical switches, tuneable gratings, filters, magnetic-field sensors, and static or dynamic displays, etc [1]. The refractive index of magnetic nanoparticles is one of the most important parameters that determine various optical effects in magnetic fluids. The complex refractive index of magnetite for the infrared region is usually assumed to be equal to $\tilde{n}$ =2.2-0.58i [2]. The other indexes in visible region of the spectrum, are known to differ in larger [3] as well as in smaller [4] extend. The influence of the magnetic field makes it possible to change the refractive index of a magnetic fluid and it gives prospects for the creation of magnetically tuneable optical devices. [5,6].

Magnetic emulsions based on magnetic fluid are the object of a large number of experimental and theoretical studies [7]. In such systems, a number of unusual optical effects is associated with the formation of structures from microdroplets in electric and magnetic fields, as well as with deformation of droplets are observed [8,9]. Data of the complex refractive index of magnetic microdrops from a concentrated magnetic fluid are necessary for creating the models of optical effects in magnetic emulsions.

In this paper, we present experimental results of complex refractive index in the visible region for kerosene based magnetic fluids with various concentrations of the magnetite. According to the experiment, the refractive index of magnetite nanoparticles is determined using known approximations of effective medium theory. The experiment was carried out in attempt to value the refractive index for magnetite nanoparticles extrapolation on the ground of common models.

2 Experiments

To determine the complex refractive index of magnetic fluids, we used the "Ellipse-1891" spectral ellipsometric complex, which made it possible to measure ellipsometric parameters in the wavelength range 350-1050 nm. 16 samples of a magnetic fluid with volume concentrations in the range of 1-35% were measured. The initial sample for all samples was a magnetic fluid with a concentration of 14%, from which all other samples were obtained by dilution or evaporation. The most concentrated magnetic fluids with a concentration of more than 28% were pasty-like or solid samples.

The method of ellipsometry consists in determining the refractive index of the surface from the measured ellipsometric parameters. Ellipsometry measures the complex reflectance ratio ρ of a system, which may be parametrized by the amplitude component ψ and the phase difference Δ. The polarization state of the light incident upon the sample may be decomposed into an *s*- and a *p*- component (the *s*- component is oscillating perpendicular to the plane of incidence and parallel to the sample surface, and the *p*- component is oscillating parallel to the plane of incidence, see. Fig. 1). The reflection coefficients of the *s*- and *p*- components of light are denoted by R_s and R_p, respectively. The relationship between the basic ellipsometric parameters (ψ, Δ) and the complex reflection coefficients (R_s, R_p) is given by the basic equation of ellipsometry [10]:

* Corresponding author: exiton@inbox.ru

$$\rho = \tan\psi \exp(i\Delta) = \frac{R_p}{R_s}. \qquad (1)$$

In case of reflection from a semi-infinite medium, equation (1) can be solved in an analytical form and the refractive index of the reflecting surface:

$$n_1 = n_0 \sin\varphi \sqrt{1 + \tan^2\psi\left(\frac{1-\rho}{1+\rho}\right)^2}, \qquad (2)$$

where φ is the angle of light incidence, n_0 is the refractive index of the second medium (air in our case).

Fig. 2 shows the spectral dependences of the real part of the complex refractive index for three samples of magnetic fluids of various concentrations. Fig. 3 and Fig. 4 depict the concentration dependences of real and imaginary parts of the refractive index for a wavelength of 633 nm.

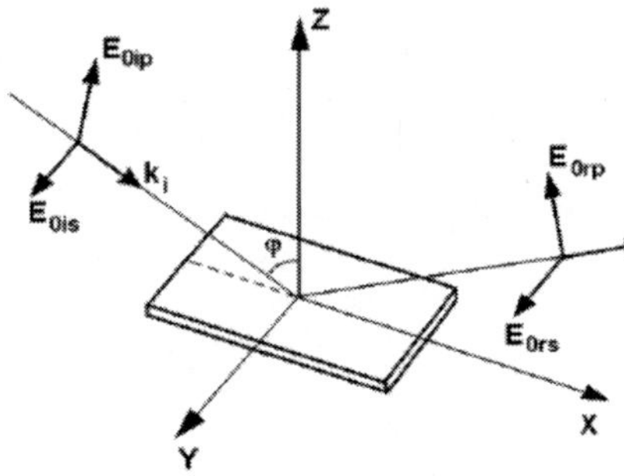

Fig. 1. Scheme of ellipsometric experiments.

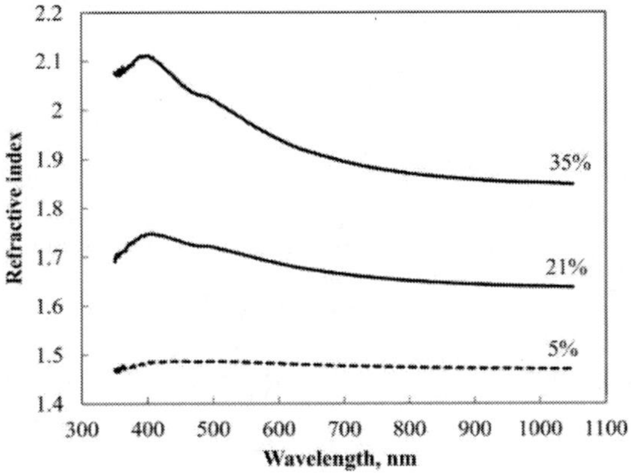

Fig. 2. Spectra of the real part of a complex refractive index for magnetic fluids various concentrations.

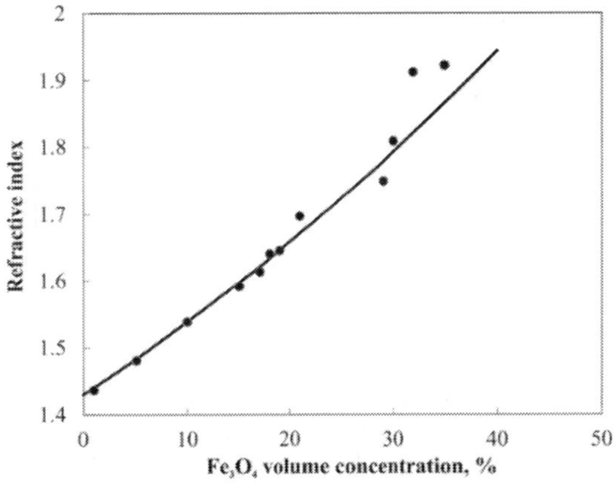

Fig. 3. Dependence of real part of refractive index on the concentration of the magnetite (dots – experimental data; line – Maxwell Garnett effective medium approximation).

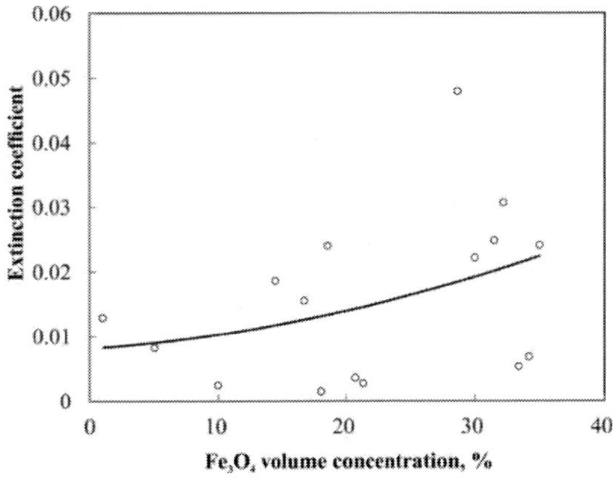

Fig. 4. Dependence of imaginary part of refractive index on the concentration of the magnetite.

3 Results and Discussion

For a theoretical estimate of the concentration dependence of magnetic fluid refractive index various approximations of the effective medium theory were used. One of the most famous and widely used models of an effective medium are the following models: Maxwell (known from Landau's book [11]), as well as the Maxwell-Garnett and Bruggeman's models [12]. To determine the effective dielectric constant at high particle concentrations, the Maxwell-Garnett (Wiener-Wagner) equation is used, which is valid for volume concentration f>1/3:

$$\frac{\varepsilon_{eff} - \varepsilon_0}{\varepsilon_{eff} + 2\varepsilon_0} = f\frac{\varepsilon_1 - \varepsilon_0}{\varepsilon_1 + 2\varepsilon_0}, \qquad (3)$$

Or Bruggeman's equation:

$$(1-f)\frac{\varepsilon_0 - \varepsilon_{eff}}{\varepsilon_0 + 2\varepsilon_{eff}} + f\frac{\varepsilon_1 - \varepsilon_{eff}}{\varepsilon_1 + 2\varepsilon_{eff}} = 0. \qquad (4)$$

Where ε_0 and ε_1 are dielectric constants of liquid medium and particle material (magnetite) respectively. The Maxwell-Garnett and Bruggeman's equations give the same results for f=0 ($\varepsilon_{eff} = \varepsilon_0$) and for f=1 ($\varepsilon_{eff} = \varepsilon_1$), but for intermediate values of f the curves are different. The relationship between the effective dielectric permittivity $\tilde{\varepsilon} = \varepsilon' + i\varepsilon''$ and the refractive index $\tilde{n} = n + ik$ of the system is given by the equations:

$$n = \sqrt{\frac{\varepsilon' + \sqrt{\varepsilon'^2 + \varepsilon''^2}}{2}}, \qquad k = \frac{\varepsilon''}{2n}. \tag{5}$$

We calculated the concentration dependence of the refractive index for the above-listed models and the values of the real and imaginary parts of the refractive index of magnetite particles, with the best coincidence between the experimental data and calculations. In Fig. 3 the results of calculations of the real part of the refractive index are shown. It is obvious that in the field of low concentrations (f <10%) all the approximations give approximately the same results. It should be noted that the value of the refractive index at which the experimental and calculated data are equal to $\tilde{n}$=2.9+0.015i is significantly different from the tabulated data for bulk magnetite [13] at the same wavelength (633 nm).

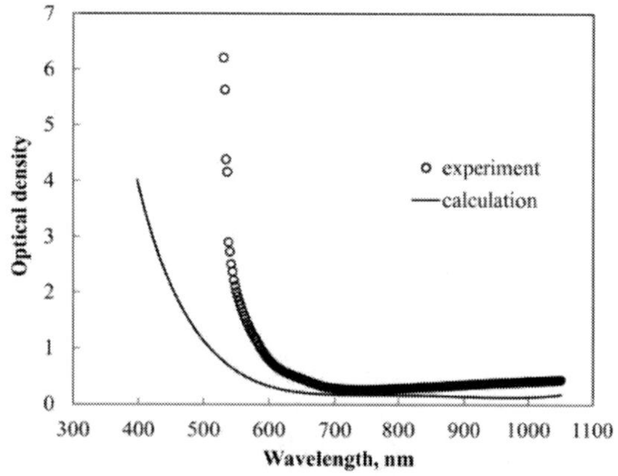

Fig. 5. Spectra of optical density of magnetic fluid (volume concentration of magnetite 5%, sample thickness is 0.04 mm).

Optical density spectra were calculated using the obtained spectral dependences of real and imaginary parts of the refractive index of a highly concentrated magnetic fluid. It is well known that [14], the extinction of light passing through the dispersion, as subject to single scattering of light (i.e., not a very large concentration of particles) is described by Bouguer-Lambert law:

$$I = I_0 \exp(-\sigma N l), \tag{6}$$

where σ is cross-section of light extinction, l is the radiation path length in a cell, N is numeric concentration of particles, and I_0 is the intensity of the incident light. By definition, the optical density of the medium is:

$$D = \log\frac{I_0}{I} = 0.43\sigma N l. \tag{7}$$

Let us to consider the contribution of light extinction and scattering by magnetic colloids in the spectral dependence of the optical properties. The cross section of the light extinction is made up of the scattering efficiency factor Q_{sca} and absorption efficiency factor Q_{abs}:

$$\sigma = \pi a^2 (Q_{sca} + Q_{abs}) \tag{8}$$

For small Rayleigh's particles expressions for the efficiency factors are in form [14]:

$$Q_{sca} = \frac{8}{3}\left(\frac{2\pi a}{\lambda}\right)^4 \left|\frac{m^2-1}{m^2+2}\right|^2. \tag{9}$$

$$Q_{abs} = \frac{8\pi a}{\lambda}\operatorname{Im}\left(\frac{m^2-1}{m^2+2}\right)^2. \tag{10}$$

Where a, m– are the size of particles and relative refractive index respectively. Measurements of the spectral dependencies of the optical density were performed using spectral ellipsometric apparatus "Ellipse 1891" in the mode of the spectral photometer. The weakening of the light transmitted through the sample was detected by a photodetector, and received data is automatically calculated dependence of $D(\lambda)$. Fig. 5 shows the calculation of the optical density spectrum of a magnetic fluid with a concentration of 5% by formulas (7-10) and the results of experimental measurement of this spectrum. The difference between experimental and calculated curves can be explained by the polydispersity of the sample.

4 Conclusions

Spectral dependences of real and imaginary parts of the refractive index of a magnetic fluid based on kerosene with magnetite nanoparticles are obtained. It is shown that the concentration dependence of the real part of the refractive index is properly described by the Bruggeman's equation in the effective medium theory. Extrapolation of the values $\mathrm{Re}(\tilde{n})$ for bulk magnetite gives a higher value than known tabulated data. Calculation of the spectral dependence of the extinction of thin films of a concentrated magnetic fluid shows that extinction is mainly determined by light absorption.

This work was supported by Russian Foundation for Basic Researches (project No.16-03-00054).

References

1. S.Y. Yang, J.J. Chieh, H.E. Horng Appl. Phys. Lett. **84**, 5204 (2004)
2. U. Buchenau, I. Muller, Solid State Comm. **11,** 1291 (1972)
3. S.Y. Yang, Y.F. Chen, H.E. Horng, Chin-Yih Hong, W.S. Tse et al. Appl. Phys. Lett. **81**, 4931 (2002)
4. E.Ya. Levitin, N.G. Kokodiy, V.A. Timanjuk, I.O. Vedernikova, T.M. Chan Inorganic Mater. **50**, 8 (2014)

5. H. E. Horng, Chin-Yih Hong, S.Y. Yang, H.C. Yang Appl. Phys. Lett. **82**, 2434 (2003)
6. Xiaopeng Fang, Yimin Xuan, Qiang Li Nanoscale Res. Lett. **6**, 237 (2011)
7. Yu.I. Dikansky, A.R. Zakinyan, A.N. Tyatyushkin Phys. Rev. E **84**, 031402 (2011)
8. K.T. Wu, Y.D. Yao, H.K. Huang J. Magn. Magn. Mater. **209**, 246 (2000)
9. K.V. Erin, S.A. Kunikin Optics and Spectroscopy **104**, 277 (2011)
10. H. Fujiwara *Spectroscopic Ellipsometry: Principles and Applications* (John Wiley & Sons, 2007)
11. L.D. Landau and E.M. Lifshitz *Electrodynamics of Continuous Media* (Pergamon Press, Oxford, 1960)
12. T.C. Choy *Effective Medium Theory* (Oxford: Clarendon Press, 1999)
13. M.R. Querry *Optical Constants* (Aberdeen, 1985)
14. C.F. Bohren, D. R. Huffman *Absorption and Scattering of Light by Small Particles* (Wiley, 2007)

Levitation and motion of a magnet partially immersed into a magnetic fluid

Daria Pelevina*, *Mikhail* Kobzev, *Sergey* Kalmykov, *Dmitrii* Merkulov and *Vera* Naletova

Lomonosov Moscow State University, 119991, Moscow, Russia

Abstract. The levitation of a permanent disc magnet partially immersed into a magnetic fluid volume on a horizontal plane is studied theoretically and experimentally. Motion of the magnet along a thin magnetic fluid layer is investigated experimentally. A mathematical model of the magnet dynamics is proposed.

1 Introduction

In [1] static levitation of a magnet inside a magnetic fluid (MF) in a cylindrical vessel on a horizontal plane is observed for the first time. The motion of a magnet along a thin MF layer is studied experimentally in [2, 3]. The mathematical model of the magnet motion is not proposed in [2, 3]. In the present paper, static levitation of a disc magnet partially immersed into a MF with free surface (without a vessel in contrast to [1]) on a horizontal plane is studied theoretically and experimentally. The motion of a disc-like magnet along a MF layer on a horizontal plane is studied experimentally. The influence of the layer thickness and length and of the MF volume around the magnet on the magnet trajectory is studied experimentally, in contrast to [2, 3]. A mathematical model of the magnet dynamics is proposed. The comparison of experimental and theoretical data allows us to estimate the friction coefficient and the height of the MF layer remaining behind the magnet.

2 Statics

2.1 Experiment

In the MF volume, a magnet can levitate in equilibrium, as it is shown, e.g., in [1], at a certain distance h from the plane (levitation height).

Fig. 1. Scheme of the magnet immersed into a MF volume.

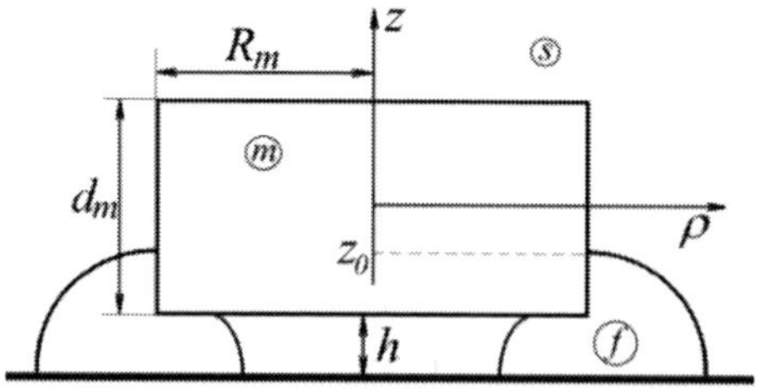

We introduce a cylindrical coordinate system with the centre at the magnet centre, and the z-axis is directed vertically upward (see Fig. 1). In the experiments, a cylindrical (disc-like) magnet made of NdFeB rare earth alloy (height $d_m = 2.04$ mm, radius $R_m = 3$ mm, $m_m = 0.4 \cdot 10^{-3}$ kg), see Fig. 1, is used. The magnetic field intensity H_0 of the magnet is measured at the distance of 1 mm from the magnet surface at center ($\rho = 0$, $z = d_m/2 + 1$ mm) by the Hall sensor $H_{0c} = 187$ kA/m. The MF (Ferrotec) based on a light hydrocarbon (initial susceptibility $\chi_{0f} = 1.89$ ($\mu_f = 1 + \chi_{0f}$), saturation magnetization $M_f = 31.8$ kA/m, density $\rho_f = 1210$ kg/m^3, viscosity $\eta_f = 6$ mPa·s) is chosen. The environment (air, NMF) is not magnetizable ($M_s = 0$) and its density is small in comparison with the MF density ($\rho_s = 1.2$ kg/m^3 $<< \rho_f$). Here and after, the parameters corresponding to the magnet, the MF and the NMF are noted by the indices m, f, s, respectively.

Fig. 2. The MF shape around the disc magnet:
1 – experiment, side view; 2 – experiment, bottom view;
3 – theory; a – $V = 1$ mm^3; b – $V = 2.9$ mm^3;
c – $V = 36.7$ mm^3.

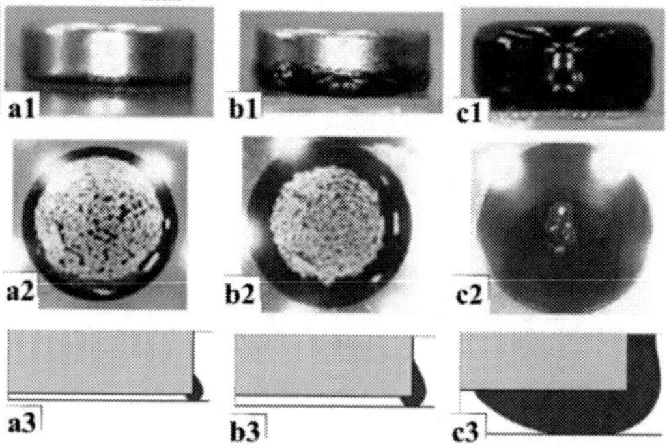

The magnet is placed in the MF volume on a horizontal plane. The magnet collects the MF around itself, so the MF has a symmetrical surface shape with respect to the z-axis, see Fig. 2, row 1. It is

* Corresponding author: pelevina.daria@gmail.com

experimentally observed that the magnet can levitate in the MF volume at the height h. Different MF volumes are considered. The limited MF volume ($V < 4.26$ mm^3) corresponds to the vertical coordinate of the contact between the MF surface and the lateral surface of magnet $z_0 < 0$. For a small MF volume $V < 3.2$ mm^3 ($z_0 < -0.33$ mm) the MF touches only the lower part of the magnet. For big volumes, the MF also touches the upper part of the magnet. There is a region not occupied by the MF under the magnet, see Fig. 2, row 2.

The levitation height h for different MF volumes is measured in the experiment, see Fig. 3, points. The levitation height h increases with the increasing of the MF volume.

Fig. 3. The dependence of the levitation height h on the MF volume V.

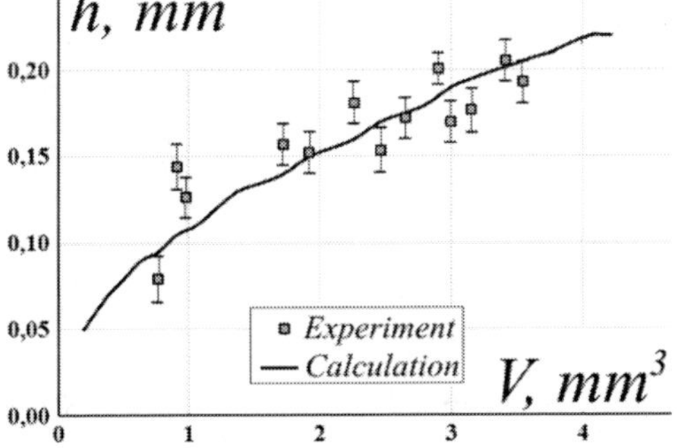

2.2 Theory

Let us consider the statics of a disc magnet in a small MF volume on a horizontal plane under the following assumptions: 1) Non-induction approximation: $\mu_f - 1 << 1$ ($\mu_f = const$). In this case, we can assume that the magnetic field is a magnet field in the vacuum $H \approx H_0$ (the distortion of the magnetic field by the MF shape is not considered). 2) There is no surface tension ($\sigma = 0$) at the boundary MF – NMF. 3) We neglect the influence of gravity on the MF shape. These assumptions are true near the magnet, since the magnetic force is much greater than the forces of gravity and of surface tension.

The stress tensor has the form ($k = f, s$; $\boldsymbol{B} = \mu_0\mu_f\boldsymbol{H}$):

$$p_{ij}^{(k)} = -p_k g_{ij} + \tau_{ij}^{(k)},\ \tau_{ij}^{(k)} = H_i B_j - \frac{HB}{2} g_{ij} \quad (1)$$

It follows from the equilibrium equation that the pressures in the MF and the NMF are constant: $p_s = const$, $p_f = const$. The dynamic boundary condition on the interface MF – NMF (which is described by the equation $z = f(\rho)$) in the assumptions described above has the following form:

$$0 = p_{nn}^{(s)} - p_{nn}^{(f)} \approx p_f - p_s + \mu_0(\mu_f - 1)\frac{H_{0f}^2}{2} \quad (2)$$

Here $\boldsymbol{n}$ is the external normal to the MF, H_{0f} is the magnetic field modulus on the MF surface. It follows from (2) that the MF surface coincides with the curves of constant magnetic field values: $H_{0f}(\rho, z) = const$. We can express the surface force acting on the magnet, taking into account $\nabla^j \tau_{ij}^{(k)} \mathbf{e}_i = -H^2 \mu_0 \nabla \mu_k / 2 = 0$ and that the magnetic field decreases as $o(1/r^2)$ ($r^2 = z^2+\rho^2$), as follows:

$$\mathbf{F} = \int_{S_m} p_{ij} n_m^j e^i ds = \mu_0 \frac{(\mu_f - 1)}{2} \int_{S_b} \left(H^2 - H_{0f}^2\right) ds \mathbf{e}_z \quad (3)$$

Here $\boldsymbol{n}_m$ is the external normal to the magnet, S_m is the surface of the magnet, S_b is the contact surface between the MF and the plane. The magnetic field H_0 is simulated using the ANSYS software package for parameters mentioned above.

The equilibrium equation for the magnet has the following form:

$$\mathbf{F} + \mathbf{P} + \mathbf{P}_A = 0,\ \mathbf{P} = m_m \mathbf{g},$$
$$\mathbf{P}_A = -(d_m + z_0) S_{mf} \rho_f \mathbf{g} \quad (4)$$

Here S_{mf} is the area of the magnet bottom wetted by the MF. For a given MF volume V from (2) and (4) the MF shape and the levitation height h can be determined by the following way. Let the MF surface $z = f(\rho)$ begins at some point with coordinates (ρ_0, z_0). We can calculate the magnetic field modulus at this point: $H_{0f} = H(\rho_0, z_0)$. From the equation $H(\rho, z) = H_{0f}$, using Maple software, we compute the MF shape $z = f(\rho)$ up to the intersection with the horizontal plane $z = -d_m/2 - h$. The value of the height h is chosen in a manner that the equilibrium condition (4) is satisfied. Then the MF volume V is calculated.

Calculated MF shapes, obtained for the experimental data, are shown in Fig. 2, row 3. Only the lower right quadrant is shown in Fig. 2, row 3 because of the symmetry. The curve in Fig. 3 shows the dependence of the levitation height h on the MF volume V. Fig. 3 shows that the results of numerical calculations and of experimental studies are in a good agreement.

It should be noted that in [2, 3] the study of the magnet statics in the MF is not carried out.

3 Dynamics

If the cylindrical magnet is immersed in the non-symmetric MF volume, horizontal force acting on a levitated magnet arises. Under the influence of this force, the magnet can move along the plane. The motion of a magnet, partially immersed in the MF, along a thin MF layer, see Fig. 4, is investigated experimentally and theoretically.

3.1 Experiment

The magnet is placed on the left edge of the MF layer (thickness d_l, length l_l, height $h_{l1} = 0.3$ mm), see Fig. 8. In [2, 3], the layer is a trace of the magnet sliding down the inclined plane. In contrast, here we fill a rectangular region of a given size by the MF using a syringe to make a layer. Parameters of the layer are denoted by the index l. The magnet levitates and moves to the right, collecting the MF. Thus, the MF non-symmetry with respect to the magnet centre remains until the magnet reaches the right layer edge: the MF amount on the right side of the magnet is greater than on the left side. For some parameters of the experiment, the magnet does not collect all the MF in the layer and the MF layer, which is thinner than the initial layer, remains on the left side of the magnet. Height of this layer is $h_{l2} < h_{l1}$. In this case, when the magnet reaches the right edge, there is a non-symmetry of the MF volume with respect to the magnet centre: the MF amount on the left side is more than on the right side. Therefore, the magnet moves to the left and can return to its initial position, see Fig. 4.

Fig. 4. The magnet motion in a MF layer in experiment (d_l = *4* mm, l_l = *160* mm), top view. Time between frames $t = 1/3$ s.

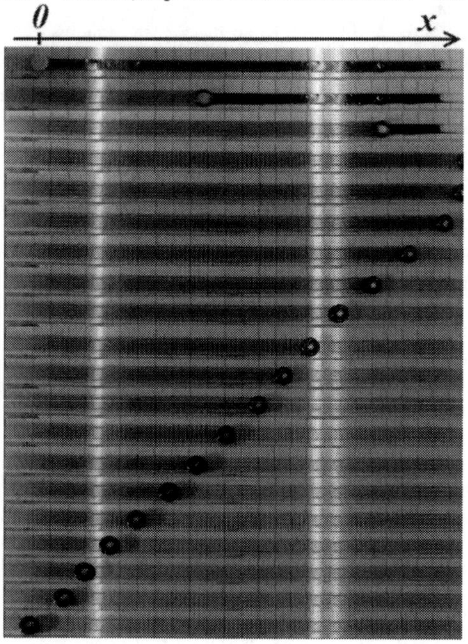

The dependencies of the position of the magnet on time, using numerical processing of experimental video, are plotted. Different thicknesses d_l, lengths l_l of the MF layer, and the MF volumes V on the magnet at the initial position ($t = 0$) are considered.

Fig. 5. The dependencies of the magnet position x on time t for d_l = *3* mm, h_{l1} = *0.3* mm, V = *14.5* mm^3 and for different layer lengths: 1 – l_l = *50* mm; 2 – l_l = *100* mm; 3 – l_l = *150* mm.

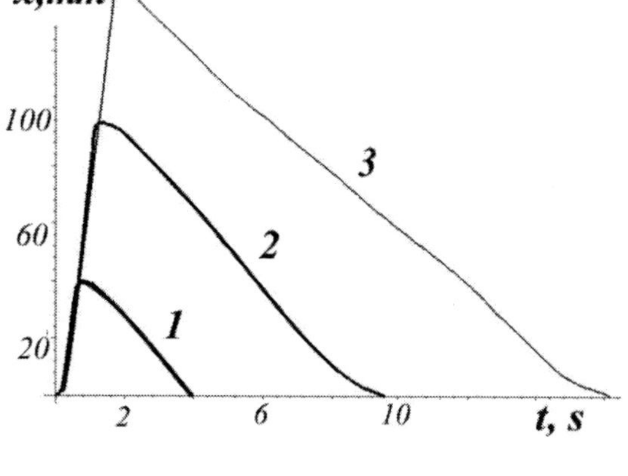

For example, the dependencies of the magnet centre position x on time t for different layer lengths l_l are shown in Fig. 5. The magnet is fully immersed into the MF, as in Fig. 2, c. It is observed that far from the layer edges, the magnet moves with the nearly constant velocity v. The velocity v_1, when magnet moves to the right, is greater than the velocity v_2, when it moves to the left (the return motion along a thin layer h_{l2}). Far from the layer edges, the constant velocities do not depend on the layer length l_l.

Fig. 6. The dependencies of the magnet centre position x on time t for l_l = 100 mm, V = *0* and for different layer thicknesses: 1 – d_l = *3* mm; 2 – d_l = *4* mm; 3 – d_l = *5* mm.

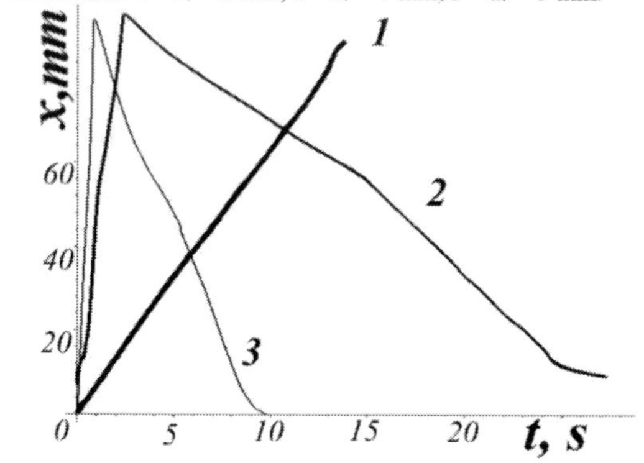

The dependencies of the position of the dry magnet (V = *0*) on time t for different layer thicknesses d_l are shown in Fig. 6. The dry magnet collects the MF from the layer, rises above the plane and moves along the layer. Thus, in the experiment, h varies significantly, for example, h varies from *0* to *0.2* mm for curve 1 in Fig. 6. For small thickness d_l = *3* mm, the magnet has a relatively small velocity v_1 and collects all the MF from the layer, so that it does not return to its initial position. For d_l = *4* mm and *5* mm the magnet velocity is higher, and the magnet moves back. The velocities v_1 and v_2 increase with the increasing of the layer thickness d_l due to the increase of the levitation height h (decrease of the viscosity force) and of the magnetic force, which is proportional to the area of the edge layer section.

Fig. 7. The dependencies of the magnet centre position x on time t for l_l = *100* mm, d_l = *3* mm and different MF volumes: 1 – V = *0*; 2 – V = *2.9* mm^3; 3 – V = *36.7* mm^3.

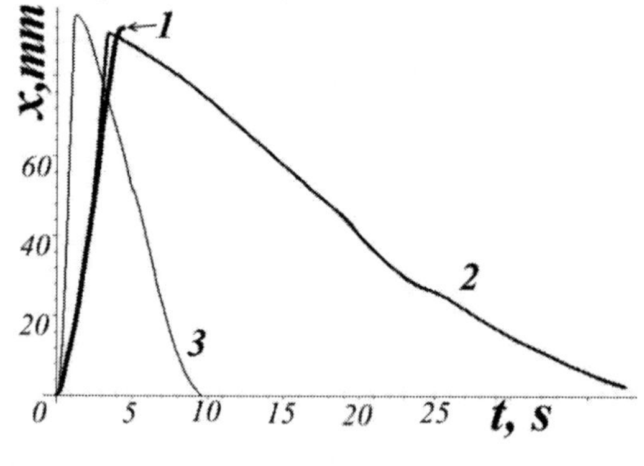

In Fig. 7, different MF volumes V on the magnet at the initial position are considered. The dry magnet ($V = 0$, curve 1) moves more slowly, collects all the MF from the layer and does not move back. A partially immersed magnet (as in Fig. 2, b) and a fully immersed magnet (as in Fig. 2, c) have a reverse motion. Motion velocities increase with the increasing of the MF volume V.

In [3] an experimental investigation of a magnet dynamics is carried out for various angles of plane inclinations. In contrast, in this paper a horizontal plane is considered, but the effect of the layer parameters and of the volume V on the problem is studied.

3.2 Theory

Let us consider the interaction between a magnet levitating in the MF volume V and a MF layer with the shape of a rectangular parallelepiped, see Fig. 8. The left edge of the layer is located at the distance r_f (r_f is the radius of the MF volume) from the magnet centre. We assume that the contact surface shape between the MF volume and the MF layer is close to the rectangular shape S_{lc}. We introduce the fixed x-axis horizontally along the layer. The beginning of the x-axis is in the magnet centre at its initial position. If the magnet completely collects all the MF from the layer, the expression for the horizontal component of the force acting on the magnet at rest from the MF layer with the surface S_l is written as following:

$$F_{x1} = -\int_{S_l} p_{xj} n_l^j \, ds = \mu_0 \frac{\mu_f - 1}{2} \left(\int_{S_{lc}} H^2 ds - \int_{S_{lf}} H^2 ds \right) > 0 \qquad (5)$$

Here S_{lc} and S_{lf} are the left and right edges of the layer (see Fig. 6). If the MF layer of the thickness h_{l2} remains to the left of the magnet, then the force $F_{x2} < 0$, acting from the left layer, is calculated in a similar way to (5).

Fig. 8. The interaction between the magnet and the MF layer.

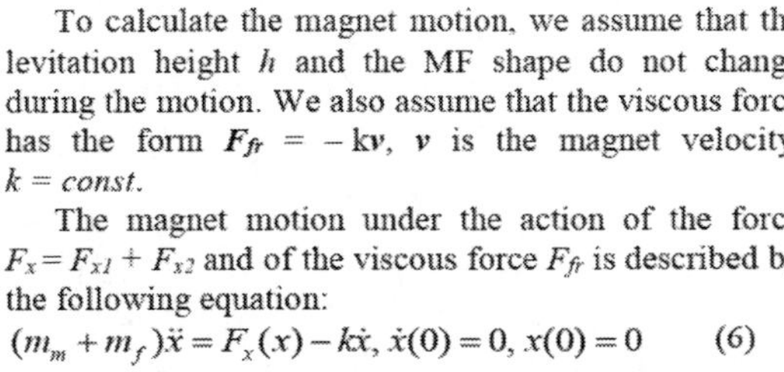

To calculate the magnet motion, we assume that the levitation height h and the MF shape do not change during the motion. We also assume that the viscous force has the form $\boldsymbol{F_{fr}} = -k\boldsymbol{v}$, $\boldsymbol{v}$ is the magnet velocity, $k = const$.

The magnet motion under the action of the force $F_x = F_{x1} + F_{x2}$ and of the viscous force F_{fr} is described by the following equation:

$$(m_m + m_f)\ddot{x} = F_x(x) - k\dot{x}, \; \dot{x}(0) = 0, \; x(0) = 0 \qquad (6)$$

Taking into account that far from the layer edges, in the non-inertial region during the motion to the right and to the left, the velocity is constant, we can write:

$$kv_1 = F_{x1,\max} + F_{x2,\max}, \quad kv_2 = F_{x2,\max} \qquad (7)$$

Here $F_{x1,max}$ and $F_{x2,max}$ are the maximum values of the magnetic force from right and left layers, v_1 is the velocity, when the magnet moves to the right, v_2 – to the left. The parameters of the MF right layer are measured in the experiment, so we can calculate the magnetic force F_{x1} and find the maximum force $F_{x1,max}$. The velocities v_1 and v_2 are known from the experiment. Under the assumption that $F_{xi,\ max} \sim h_{li}$ ($i = 1, 2$), from (7) we can find the coefficient k, the force $F_{x2,max}$ and the thickness of the layer h_{l2} remaining behind the magnet.

For the following parameter values, corresponding to the experiment, see Fig.5, curve 1: $V = 14.5$ mm^3, $h = 0.6$ mm, $x_0 = r_f = 3.8$ mm, $d_l = 3$ mm, $h_{l1} = 0.3$ mm, $l_l = 50$ mm, $v_1 = 88.8$ mm/s, $v_2 = -13.3$ mm/s, we obtain: $k = 46$ g/s, $|F_{x2max}|/|F_{x1max}| = 0.13$, $h_{l2} = 0.039$ mm.

4 Conclusion

The statics of a permanent disc-like magnet in a MF volume on a horizontal plane is experimentally and theoretically investigated. The analytical expression for the force acting on the magnet from the small surrounding MF volume is obtained in the static case in the non-inductive approximation. The magnetic field of the magnet is modelled in ANSYS. Various static shapes of the MF surface are calculated and obtained in the experiments. The dependence of the levitation height of the magnet on the MF volume is investigated. A good coincidence between the theoretical and experimental results is obtained.

The motion of a magnet along a MF layer on a horizontal plane is experimentally studied for various initial conditions (the MF layer geometry, the MF volume on the magnet). It is shown that the increase of the MF volume and of the layer thickness leads to the increase of the magnet velocity. It is found that for some parameters the magnet returns to its initial position. A mathematical model of the magnet dynamics along a thin MF layer is proposed. The expression for the force acting on the magnet from the MF layer is obtained analytically. In the non-inertial approximation, the estimations for the viscous force coefficient and the height of the layer to the left of the magnet are obtained from the comparison with the experiment.

This study can be used to create new types of actuators based on a MF.

This work was supported by the RFBR grant 16-31-60091 and partially supported by FASIE 8541ГУ/2015.

References

1. R. E. Rosensweig, Nature, 210, 5036, p. 613 (1966)
2. Hartung S., Rehberg I., Richter R., Abs. Book: 14th German Ferrofluid Workshop, 82 (2014)
3. Freundorfer A., Rehberg I., Richter R., Abs. Book: 15th German Ferrofluid Workshop, 74 (2015)

EPJ Web of Conferences **185**, 09009 (2018)
MISM 2017

https://doi.org/10.1051/epjconf/201818509009

Research of the deformation of a thin body with a magnetizable elastomer in the magnetic field of a coil

Dmitrii Merkulov[1,*]

[1]Lomonosov Moscow State University, 119991, Russia, Moscow, 1, Leninskiye Gory

Abstract. Deformation of a thin cylindrical body with a magnetizable elastomer based on silicone in the magnetic field of an electromagnetic coil is investigated experimentally and theoretically. It is observed that a bistability of the thin body equilibrium shape exists at some values of the coil current. The method of measuring elasticity coefficients of the magnetizable elastomer in the Mooney – Rivlin model is proposed.

1 Introduction

A magnetizable elastomer (ME) is a composite material consisting of elastic basis and of ferromagnetic nano- or microparticles. The ME small deformation in the applied uniform magnetic field is studied in [1-4], where only one stable equilibrium shape is found theoretically. In [5] the deformation of spherical bodies with ME in a uniform magnetic field is studied experimentally, and a unique dependence of the body length on the applied magnetic field intensity is obtained. In [6-7] the bistability of the ferrogel equilibrium shapes in a non-uniform magnetic field of an electromagnet is found experimentally. In [8-9] the possibility of bistable and of multistable equilibrium shapes of bodies with ME in different magnetic fields is shown theoretically.

In the present work, the possibility of bistable equilibrium shapes of the thin body with ME based on silicone (unlike the ferrogel in [6-7]) in the magnetic field of an electromagnetic coil is found experimentally. The method of measuring elasticity coefficients in the Mooney – Rivlin model is proposed. Bistability of the deformation is shown also by numerical computations.

2 Experiment

In the experiment, the axisymmetric cylindrical thin body with ME is investigated in the applied non-uniform magnetic field $\mathbf{H_0}$, which is created by an electromagnetic coil (see Fig. 1). The magnetic field $H(x)$ on the coil axis is measured by a Hall effect sensor. The body top end is fixed at the distance of *104* mm from the bottom face of the coil. The point $x = 0$ coincides with the point of the sample fixation. The x-axis is directed vertically downward. The experimental setup scheme is shown in Fig. 1.

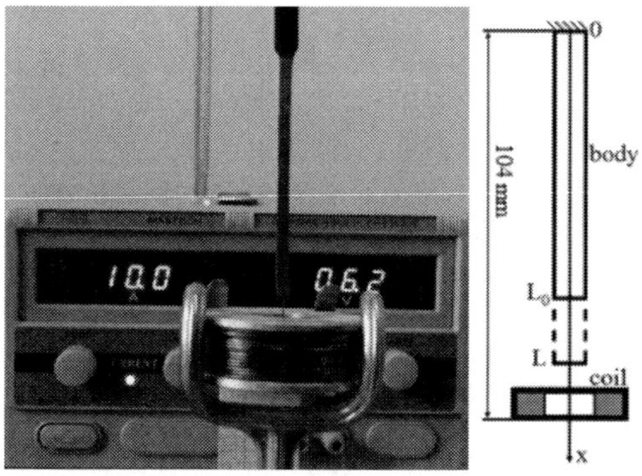

Fig. 1. Experimental setup.

In the experiment, the coil current I is varied and the measurement of the body length L is carried out ($L_0 = 71$ mm is the undeformed length of the body, the density of ME $\rho = 1960$ kg/m^3). The experimentally measured dependence of the relative body lengthening $(L - L_0)/L_0 * 100\%$ on the current I is shown in Fig. 2.

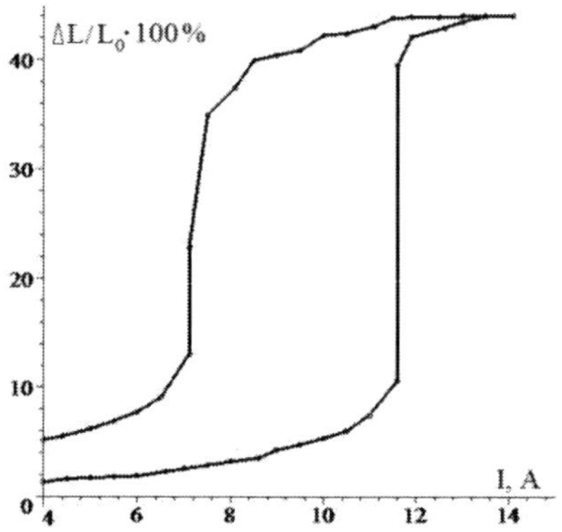

Fig. 2. The experimental dependence of the relative body lengthening on the current.

* Corresponding author: merkulovdima@mail.ru

It is seen that the shape hysteresis of the body with ME under cyclic increase and decrease of the current in the coil exists and that two stable equilibrium positions exist in some range of currents (bistability).

Measurement of the body elongation in a horizontal position upon application of forces to its ends allows to find the dependence $p_{11n}/(\lambda - \lambda^{-2})$ on λ^{-1} (p_{11n} is a component of the stress tensor acting on the undeformed cross-section S_0 of the body; here $\lambda = L/L_0$, because the horizontal deformation is uniform), see Fig. 3.

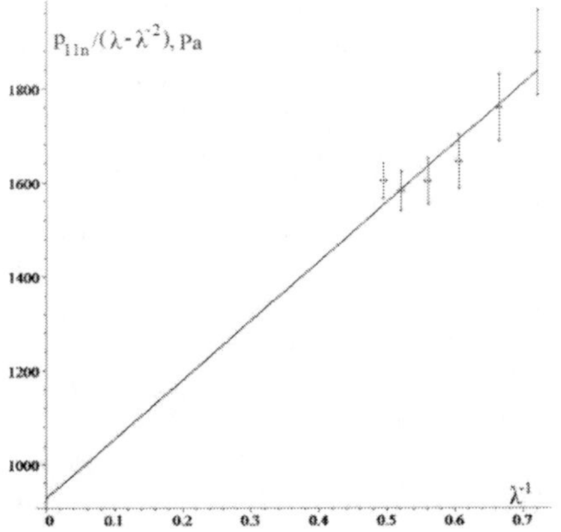

Fig. 3. The experimental dependence of $p_{11n}/(\lambda - \lambda^{-2})$ on λ^{-1} for **H** = **0**.

If the elastic properties of incompressible ME are described by the Mooney – Rivlin model, the equality $p_{11n} = 2(\lambda - \lambda^{-2})(C_1 + \lambda^{-1}C_2)$ is true [10]. The graph in Fig. 3 allows to find the parameters $C_1 = 463.8$ Pa and $C_2 = 631$ Pa using the linear least squares approximation. In [6-7] the Mooney – Rivlin model coefficients are determined experimentally using the formula (9) in [7]. However, when a sample is compressed or stretched along the gravitational force (the samples are deformed in such way in [7]), the deformation cannot be considered homogeneous, and one cannot use the formula (9) in [7]. Perhaps the fact that for some samples the coefficient $C_2 < 0$ in [6-7] is associated with the heterogeneity of the deformations. In the present work, unlike [7], the samples based on silicone are stretched horizontally in this experiment, so the deformation is homogeneous, and one can use the formula (9) in [7]. Thus, the Mooney – Rivlin model coefficients are determined, moreover, all the coefficients are positive.

The measurement of the magnetic force, acting on a spherical body with ME in the non-uniform magnetic field of the coil, allows to calculate the magnetic permeability of the material $\mu = 4.6$.

3 Theory

In [8] the mathematical model to calculate the length of a thin body with ME in the applied axisymmetric non-uniform magnetic field $\mathbf{H_0}$ is proposed for the non-inductive approximation. In the case of $\mathbf{M} = \chi\mathbf{H}$, $\chi = (\mu - 1)/(4\pi) = const$, the magnetic force $\mathbf{F_m} = M\,\mathbf{grad}(H)$ can be written as follows $\mathbf{F_m} = \mathbf{grad}(G(H))$, $G(H) = \chi H^2/2$. In the non-inductive approximation, the surface magnetic forces are small, and the magnetic force $\mathbf{F_m} = \mathbf{grad}(G(H_0))$ is a known function of coordinates. For thin enough samples, we can assume that the magnetic force inside the body is defined as $\mathbf{F_m} = \mathbf{F_m}(x, y = 0, z = 0)$. The x-axis is directed down along the axis of symmetry of the body. The point $x = 0$ coincides with the point of the body fixation. All body surfaces are free except the surface $x = 0$, so we can assume that the stress tensor components p_{22} and p_{33} are equal to zero on these surfaces, and the following boundary condition is true: $p_{11}(x = L) = 0$. The following equations are true inside the thin body:

$$\frac{\partial p_{11}}{\partial x} + \frac{\partial G}{\partial x} + \rho g = 0, \qquad p_{22} = p_{33} = 0; \tag{1}$$

$$p_{11} = 2\left(\lambda^2 - \lambda^{-1}\right)\left(C_1 + C_2\lambda^{-1}\right). \tag{2}$$

Here, $\lambda = \sqrt{g_{11}} = dx/dx_0 > 0$, g_{11} is the component of the metric tensor, x_0 is the Lagrangian coordinate.

Using the equations (1), (2) and the condition $p_{11}(L) = 0$, the following equation is obtained:

$$G(L) - G(x) + \rho g(L - x) = 2\left(\lambda^2 - \lambda^{-1}\right)\left(C_1 + C_2\lambda^{-1}\right). \tag{3}$$

The first integral (3) of the static equation (2) allows to calculate $\lambda(x, L)$.

The undeformed length of the body L_0 and its length L after deformation are related as follows:

$$L_0 = \int_0^L \frac{dx}{\lambda}. \tag{4}$$

The function $\lambda(x, L)$ is determined by the equation (3), and then we can find the length of the deformed body L using the equation (4).

It should be noted that in [7] the equation (20), which does not take into account the non-homogeneity of deformation, is used for the bistability theoretical research. The calculations in [7], made using the equation (20), cannot be true, because in any non-uniform magnetic field and in the gravitational field one cannot neglect the non-homogeneity of deformation. In [7] the equation (24), which describes non-homogeneous deformations in the Neo-Hooke model with the constant elasticity coefficient, is written out. In the present work, the equation (3) for statics of the elastic body in the more general Mooney – Rivlin model [10] with the constants C_1 and C_2 (which can be non-uniform in general) is obtained. In (3) in addition to the magnetic force, gravity force is also taken into account.

4 The numerical calculations

The potential $G(x)$ can be calculated by using the experimentally measured distribution of the magnetic field of the coil $H(x)$. The calculations using the experimentally measured parameters ($\mu = 4.6$, $\rho = 1960$ kg/m^3, $C_1 = 463.8$ Pa, $C_2 = 631$ Pa) show that for some values of the current in the electromagnetic coil, the dependence of L on L_0 is ambiguous. Three values of L correspond to one value L_0 (in some range of the current I, see Fig. 4).

It is possible to show that if $dL/dL_0 < 0$ the equilibrium state is unstable [9]. So, the middle of these three values L corresponds to the unstable equilibrium state because the $dL/dL_0 < 0$ at this branch of the solution. Thus, the existence of the bistable body deformation in some current range is shown theoretically. Some quantitative difference between theory and experiment may be due to the fact that in calculations the dependence of C_1 and C_2 on the magnitude of the magnetic field is not taken into account.

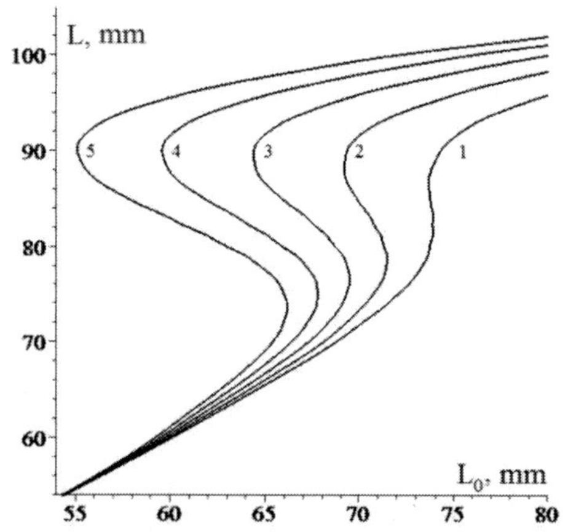

Fig. 4. The theoretical dependence of the deformed body length on its initial length at currents: 1 – 6 A, 2 – 7 A, 3 – 8 A, 4 – 9 A, 5 – 10 A.

5 Conclusion

The deformation of a thin cylindrical body with the magnetizable elastomer based on silicone in the applied non-uniform axisymmetric magnetic field of the electromagnetic coil is investigated experimentally and theoretically. The bistable deformation of such body in some range of currents in the coil is observed in the experiment. The magnetic permeability of the magnetizable elastomer is measured by the power method. The technique of measuring elasticity coefficients of the elastomer in the Mooney – Rivlin model is proposed. The numerical calculations with the experimental data show that in some current range the bistability of the body equilibrium shapes exists. Some quantitative difference between theory and experiment may be due to various reasons, for example, because the dependence of C_1 and C_2 on the magnitude of the magnetic field is not taken into account in calculations.

This work is supported by the Russian Foundation for Basic Research (16-51-12024), the Russian President grant MK-4541.2016.1 and the FASIE (8541ГУ/2015).

References

1. O.V. Stolbov, Yu.L. Raikher, Tech. Phys. Lett., **26**(2), 156-158 (2000)
2. G.V. Stepanov, Yu.L. Raikher, O.V. Stolbov, Tech. Phys., **53**(9), 1169-1176 (2008)
3. K. Morozov, M. Shliomis, H. Yamaguchi, Phys. Rev. E, **79**, 040801: 1-4 (2009)
4. D. Ivaneyko, V. Toshchevikov, M. Saphiannikova, G. Heinrich, Cond. Matt. Phys., **15**(3), 33601: 1-12 (2012)
5. L.V. Nikitin, K.G. Kornev, L.S. Mironova, *VIII Plyos international conference on magnetic fluids* (1998, in Russian), 95-96
6. M. Zrinyi, L. Barsi, D. Szabo, H.-G. Kilian, J. Chem. Phys., **106**, 5685-5692 (1997)
7. G. Filipcsei, I. Csetneki, A. Szilágyi, M. Zrínyi, Adv. Polym. Sci., **206**, 137-189 (2007)
8. V.A. Naletova, D.A. Pelevina, D.I. Merkulov, I. Zeidis, K. Zimmermann, Magnetohydrodynamics, **52**(3), 287-298 (2016)
9. D.I. Merkulov, V.A. Naletova, D.A. Pelevina, V.A. Turkov, JMMM, **431**, 123-125 (2017)
10. G.V. Vinogradov, A.Y. Malkin. Rheology of Polymers. Springer-Verlag Berlin Heidelberg, 1980, p. 468

Magnetic fluid bridge between two cones and a cylinder in the magnetic field of a current-carrying wire

Alexandra S. Vinogradova[1,*], *Vladimir A.* Turkov[1], and *Vera A.* Naletova[1,2]

[1]Lomonosov Moscow State University, Research Institute of Mechanics, 119192, Moscow, Michurinsky Pr., 1, Russia
[2]Lomonosov Moscow State University, Faculty of Mechanics and Mathematics, 119991, Moscow, Leninskiye Gory, 1, Russia

Abstract. A magnetic fluid (MF) changes its surface shape in the magnetic field of a current-carrying wire while the current is changing. In the present paper, we propose to study a MF bridge which can open or close the channel formed by two cones and a cylinder by imposing the magnetic field of a wire. Numerical modelling of the bridge behavior for different values of MF volumes and currents in the wire is done for two cases: when the MF wets and does not wet surrounding solid boundaries. It is shown that the presence of limiting cones allows the MF to sustain the pressure drop which is much higher in case of non-wetting than in case of wetting. In case of wetting, the MF cannot sustain any pressure drop at low currents, but in case of non-wetting, the MF can do it even at zero current. It is found that in case of non-wetting, spasmodic and hysteresis phenomena are possible for some values of MF volumes and currents in the wire. The use of a MF in valves, dispensers and pumps is one of possible actuation methods.

1 Introduction

A magnetic fluid (MF) is a suspension of monodomain ferromagnetic particles (its size is about 10 nm), coated with a surfactant, in a non-magnetic carrier fluid. Such superparamagnetic fluid is attracted strongly by magnetic fields, for example, by the magnetic field of a straight current-carrying wire. The spasmodic spreading of a MF drop along a wire is studied theoretically [1-3] and observed in the experiment [3]. A MF drop on a wire with limiting conical surfaces is studied in [4]. A MF bridge between coaxial cylinders (there is a straight current-carrying wire on the axis of these cylinders) cannot sustain any pressure drop [1, 5]. In the present paper, we propose to use the magnetic field created by a wire to operate a MF bridge which can open and close the gap of special geometry (formed by coaxial conical and cylindrical surfaces). A numerical technique for investigating a MF bridge which can open such gap by the varying magnetic field of a current-carrying wire is proposed in [6]. Such MF bridge is limited by two cones, so the bridge can serve as a MF valve which sustains a pressure drop.

A MF is widely used in microfluidics as an actuation method in valves, dispensers and pumps. In [7] it is shown that a level of pressure in the range of tens of millibar can be expected from a MF. Different magnetic fields could be used to actuate the MF in such devices: for example, external permanent magnets [8-13], a coil [13-14], an oscillating magnetic line dipole [15], an electromagnet [16], and a straight current-carrying wire as in the present paper.

2 Problem statement

We consider a heavy, incompressible, homogenous, isothermal MF of the volume V between a cylindrical surface of the radius R_c and two limiting right circular truncated conical surfaces with different apex angles α_1 and α_2 (Fig. 1). All these surfaces are coaxial, and a straight wire of the radius r_0 carrying the current I is located on its axis. The cones intersect in a circle of the wire radius r_0. In this geometry, the MF bridge can sustain some pressure drop $\Delta p = p_1 - p_2$. The pressure p_1 is maintained above the MF and the pressure p_2 is maintained under the MF. The MF is immersed in a non-magnetic liquid (its magnetization $M_l = 0$) with the same density (the case of hydroweightlessness). If the MF does not wet solid boundaries then $90^\circ < \theta_1, \theta_2, \theta_3 \leq 180^\circ$, where θ_1 is the wetting angle of the upper conical surface, θ_2 – of the lower conical surface, θ_3 – of the outer cylinder. If the MF wets solid boundaries then $0^\circ \leq \theta_1, \theta_2, \theta_3 \leq 90^\circ$ (the case $\theta_i > \alpha_i$, $i = 1, 2$ is only considered).

The MF has an axially symmetric surface $z = h(r)$, $r^2 = x^2 + y^2$ (the axis z is directed along the axis of the wire). In this geometry, the magnetic field of the wire $\boldsymbol{H}$ is not deformed by the MF and $|\boldsymbol{H}| = H$, $H(r) = 2I/(cr)$, where c is the speed of light in vacuum [17]. We consider that for a MF with a low concentration of the same ferromagnetic particles, its magnetization M_f can be described by the Langevin law [18]: $M_f(\xi) = M_S L(\xi)$, $L(\xi) = \coth \xi - 1/\xi$, $\xi = mH/(kT)$, $m = M_S/n$. Here M_S is the saturation magnetization of a MF, m is the magnetic moment of one ferromagnetic particle, n is the number of

* Corresponding author: vinogradova.msu@gmail.com

ferromagnetic particles per unit volume of a MF, T is the MF temperature, k is the Boltzmann constant, ξ is the Langevin parameter which is directly proportional to the current I in the wire.

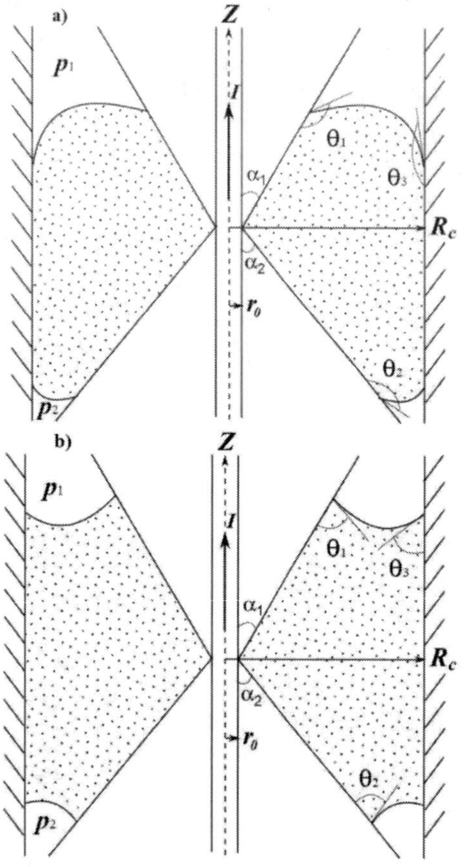

Fig. 1. MF bridge in case of (a) non-wetting and (b) wetting between two cones and a cylinder.

The hydrostatic equation and the boundary condition on the MF surface $h(r)$ [19] are used to get the general analytical solution for any axisymmetric shape of the MF surface $h(r)$ in the magnetic field created by a straight current-carrying wire [1-2, 4-6]. Let us introduce the following dimensionless parameters: $z^* = z/r_0$, $V^* = V/r_0^3$, $W = r_0\Delta p/(2\sigma)$, $\xi_0 = mH_0/(kT)$, $H_0 = 2I/(cr_0)$. Later, the signs "*" are omitted and parameters are considered as non-dimensional, unless otherwise specifically agreed.

Without loss of generality, we will further assume that $p_1 \geq p_2$. It should be noted that for $p_1 > p_2$, the MF bridge can take two different positions: to contact simultaneously the upper and the lower conical surfaces (Fig. 1) or to contact only the lower conical surface.

3 Numerical simulation

To simulate numerically the static shapes of MF surface, we fix the following problem parameters: $r_0 = 0.05$ cm, $R_c = 0.5$ cm, $T = 300$ K, $n = 1.9 \cdot 10^{17}$ cm^{-3}, $\alpha_1 = \alpha_2 = 5°$, $M_S = 56.6$ G, the coefficient of surface tension $\sigma = 20$ dyn/cm, $\theta_1 = \theta_2 = \theta_3 = 175°$ (non-wetting) and $\theta_1 = \theta_2 = \theta_3 = 70°$ (wetting).

3.1 Break-up of the magnetic fluid bridge

At each value of the current ξ_0 it is possible to calculate the MF surface shapes with the fixed volume V before reaching the value of current $\xi_0 = \xi_{break}$ or $\xi_0 = \xi_{cr}$. At some break value of current $\xi_0 = \xi_{break}$, the upper MF surface contacts the lower one, the MF volume becomes minimal to bridge the gap between conical and cylindrical surfaces ($V = V_0$) and the MF bridge breaks up and opens the channel predictably (at the same time, in case of non-wetting the MF surfaces touch simultaneously the cylinder at the same point). However, at some critical value of current $\xi_0 = \xi_{cr}$, the analytical solution [1-2, 4-6], which describes the static shape of MF surface, may stop existing earlier than the upper MF surface contacts the lower one. In this case, the MF surface has a vertical tangent at some point $r = r_{cr}$ and the MF bridge breaks up and opens the channel unpredictably.

3.1.1 Case of non-wetting

We consider the dependence of the value z_1 (z-coordinate of the upper contact line of three media: MF, non-magnetic liquid and upper/lower cone) on the current ξ_0 (Fig. 2) for different values of the MF volume V in case of non-wetting at the pressure drop $W = 1.27$.

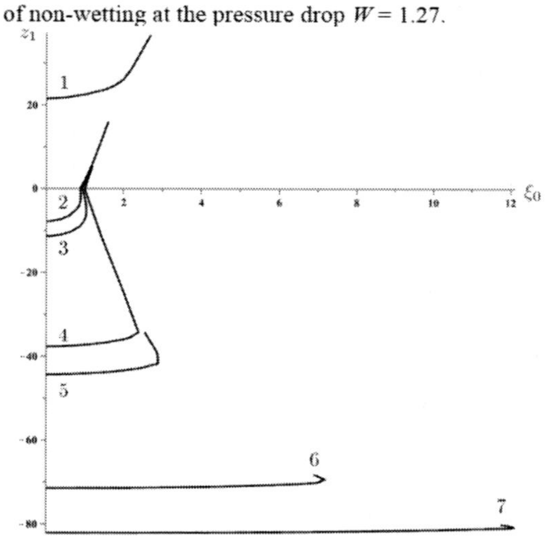

Fig. 2. Dependences $z_1 = z_1(\xi_0)$ for 1) $V = 27525$; 2) $V = 17770$; 3) $V = 16772$; 4) $V = 9580$; 5) $V = 7925$; 6) $V = 2460$; 7) $V = 1076$ in case of non-wetting at $W = 1.27$.

From Fig. 2 we can see that up to three different values of z_1 are possible for the fixed current ξ_0 and volume V (the higher MF equilibria for the curves 5, …, 7 and the middle one for the curves 3, 4 are unstable), so spasmodic and hysteresis phenomena are possible for some values of the MF volume V and of the current ξ_0 in the wire. The curves 1, …, 7 in Fig. 2

($0 < V \leq 27525$) come abruptly to an end when the MF bridge breaks up and opens the channel predictably at the current $\xi_0 = \xi_{break}$ (for the curves 4, 5 it happens twice at a lower and at a higher break current). The curves for $27525 < V < 40720$ (not shown in Fig. 2) are monotonically increasing functions like the curve 1 and it comes abruptly to an end when the MF bridge breaks up and opens the channel unpredictably at the current $\xi_0 = \xi_{cr}$. The volume $V = 40720$ corresponds to the maximal MF volume, which can be placed in the gap between conical and cylindrical surfaces.

3.1.2 Case of wetting

We consider the dependence of the value z_1 on the current ξ_0 (Fig. 3) for different values of the MF volume V in case of wetting at the pressure drop $W = 0.001$.

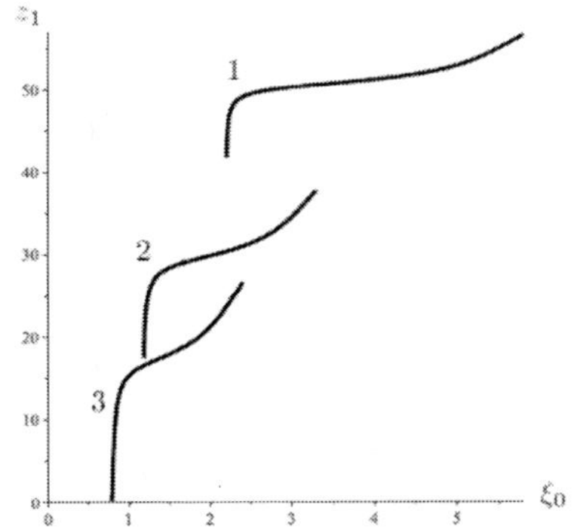

Fig. 3. Dependences $z_1 = z_1(\xi_0)$ for 1) $V = 27525$; 2) $V = 16772$; 3) $V = 9580$ in case of wetting at $W = 0.001$.

For these values of the problem parameters, the MF bridge can take only one position: to contact simultaneously the upper and the lower conical surface (Fig. 1). The curves in Fig. 3 come abruptly to an end on the right when the MF bridge breaks up and opens the channel either predictably at the current $\xi_0 = \xi_{break}$ for $V \leq 1367$ (the curves are not shown in Fig. 3) or unpredictably at the current $\xi_0 = \xi_{cr}$ for $1367 < V < 40720$ (the curves 1, …, 3 in Fig. 3). The same curves come abruptly to an end on the left at the minimal current, for which the pressure drop $W = 0.001$ can be sustained by the MF bridge of the fixed MF volume V. Fig. 3 shows that in case of wetting MF bridge cannot sustain the low pressure drop $W = 0.001$ at low currents in the wire, but in case of non-wetting the MF bridge can do it even at zero current.

3.2 Rebuilding of the magnetic fluid bridge

If the MF bridge breaks up and opens the channel predictably at the current $\xi_0 = \xi_{break}$, then in case of non-wetting the whole MF volume V turns into the MF drop on the cones of the thickness r_d and in case of wetting the whole MF volume V divides into two volumes: one volume on the cones and one on the cylinder. The case when the MF bridge breaks up and opens the channel unpredictably at the current $\xi_0 = \xi_{cr}$ is not considered.

3.2.1 Case of non-wetting

We consider the dependence of the minimal MF volume V_0, which can bridge the gap between conical and cylindrical surfaces (when the MF drop thickness r_d is equal to the cylinder radius R_c), on the current ξ_0 at the pressure drop $W = 0$ (Fig. 4).

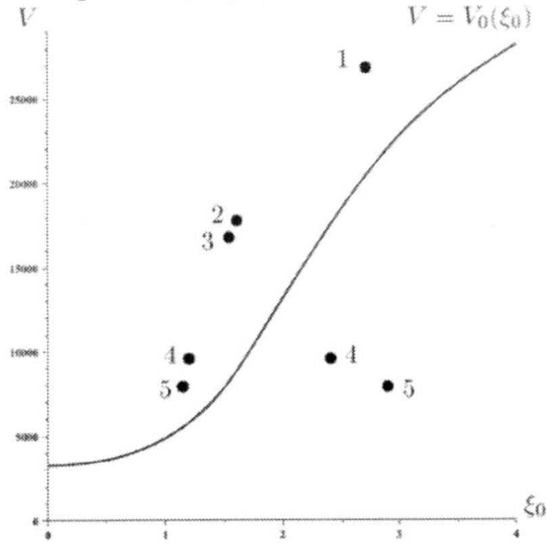

Fig. 4. The solid curve denotes the dependence $V = V_0(\xi_0)$ in case of non-wetting at $W = 0$.

Fig. 4 shows that the MF bridge cannot be rebuilt at any values of the MF volume, but only at $V \geq V_0(\xi_0 = 0) = 3428$. The points "•" of indices $i = 1, \ldots, 5$ in Fig. 4 correspond to the breaks-up of the MF bridge at the current $\xi_0 = \xi_{break}$ at the pressure drop $W = 1.27$ (see the curves in Fig. 2 of indices i which come abruptly to an end). The points "•" of indices $i = 4, 5$ are given twice in Fig. 4, because the "left" points correspond to the break-up of the MF bridge at a lower break current and the "right" points correspond to the break-up of the MF bridge at a higher break current. The points "•" of indices $i = 6, 7$ should be placed under the graph $V = V_0(\xi_0)$, but it is not depicted in Fig. 4, because it corresponds to $\xi_0 > 4$.

If the point "•" is placed above the graph $V = V_0(\xi_0)$ (for example, the "left" points of indices $i = 4, 5$ in Fig. 4), then the thickness r_d of the drop, formed on the cones because of the MF bridge break-up at $\xi_0 = \xi_{break}$, is greater than the radius R_c and the MF bridge is rebuilt. If the point "•" is placed under the graph $V = V_0(\xi_0)$, then the thickness r_d of such drop is less than the radius R_c and at the current $\xi_0 = \xi_{break}$ the MF bridge is not rebuilt. For example, points of indices $i = 6, 7$ are under the graph $V = V_0(\xi_0)$ at $\xi_0 > 4$, but the corresponding volumes $V = 2460$ and $V = 1076$ are not enough to bridge the gap at the pressure drop $W = 0$ even if the current ξ_0 decreases to zero. However, for big enough volumes $V > V_0(\xi_0 = 0) = 3428$, the MF bridge can be rebuilt through the decreasing of the current ξ_0. For example, the "right" points of indices $i = 4, 5$ in Fig. 4, which are placed under the graph $V = V_0(\xi_0)$, correspond to the volumes $V = 9580$ and $V = 7925$. If the current ξ_0 decreases from the value $\xi_0 = \xi_{break}$ to some value on the graph $V = V_0(\xi_0)$ corresponding to the volumes $V = 9580$ and $V = 7925$, the drop thickness r_d increases to the value of the radius R_c and the MF bridge is rebuilt.

3.2.2 Case of wetting

We consider the dependence of the minimal volume V_0, which can bridge the gap between cones and a cylinder (when the upper MF surface contacts the lower one), on the current ξ_0 at the pressure drop $W = 0$ (Fig. 5).

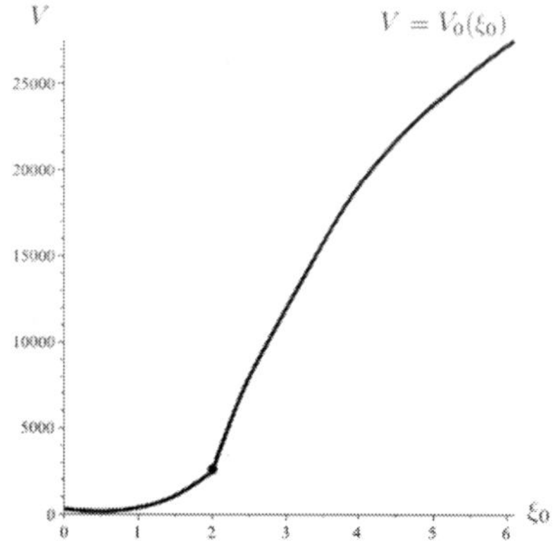

Fig. 5. Dependence $V = V_0(\xi_0)$ in case of wetting at $W = 0$.

In Fig. 5 the MF bridge of the volume $172 \leq V \leq 2460$ breaks up predictably at the current $0 \leq \xi_0 = \xi_{break} \leq 2$ (the part of the graph $V = V_0(\xi_0)$ to the left of the point "•"). The bridge of the volume $2460 < V < 40720$ breaks up unpredictably at $\xi_0 = \xi_{cr} > 2$ (the part of the graph $V = V_0(\xi_0)$ to the right of the point "•"). The graph $V = V_0(\xi_0)$ in Fig. 5 has a minimum $V = 172$. It means that for some range of volumes V and currents ξ_0, the bridge break-up can occur for both increase and decrease of the current. The volumes $172 \leq V \leq 2460$ are not enough to bridge the gap again even if the current ξ_0 decreases to zero. However, the bridge of big enough volume $2460 < V < 40720$ may be rebuilt after its break-up at the critical current.

4 Conclusion

The work demonstrates that the presence of limiting cones near a wire allows the MF bridge to sustain the pressure drop which is much higher in case of non-wetting than in case of wetting. In case of non-wetting, spasmodic and hysteresis phenomena may be presented for some fluid volumes and currents in a wire. In case of wetting, such phenomena are not found. In case of wetting, the bridge cannot sustain any pressure drop at low currents, but in case of non-wetting, the MF bridge can do it even at zero current.

The break-up and rebuilding of the MF bridge are investigated. The bridge breaks up either unpredictably at the critical current, for which the static shape of the MF surface stops existing and the MF surface has a vertical tangent, or predictably at the break current, for which the MF volume is minimal to bridge the gap between cones and a cylinder. In case of non-wetting, depending on the MF volume and the break value of current, the MF bridge can either be rebuilt or not. The bridge can be rebuilt only for big enough volumes. In case of wetting, the MF bridge of small volume cannot be rebuilt after its break-up at the break current, but the MF bridge of big enough volume may be rebuilt after its break-up at the critical current.

The results of the conducted research can be useful for the construction of different devices with the MF position controlled by the magnetic field, such as valves, dispensers and pumps.

Support by the Russian Foundation for Basic Research (project 16-51-12024) is acknowledged.

References

1. V.A. Naletova, V.A. Turkov, A.S. Vinogradova, Phys. Procedia, **9**, 68-73 (2010)
2. A.S. Vinogradova, V.A. Naletova, V.A. Turkov, A.G. Reks, Magnetohydrodynamics, **49**, 119-126 (2013)
3. J.-C. Bacri, C. Frenois, R. Perzynski, D. Salin, Rev. Phys. Appl., **23**, 1017-1022 (1988)
4. A.S. Vinogradova, V.A. Naletova, V.A. Turkov, A.G. Reks, Magnetohydrodynamics, **49**, 350-354 (2013)
5. A.S. Vinogradova, V.A. Naletova, V.A. Turkov, Solid State Phenomena, **233-234**, 335-338 (2015)
6. A.S. Vinogradova, V.A. Naletova, *Proc. 9th Pamir Int. Conf. Fund. and Applied MHD*, 297-301 (2014)
7. R. Pérez-Castillejos, J.A. Plaza, J. Esteve, P. Losantos, M.C. Acero, C. Cané, F. Serra-Mestres, Sens. Actuators A, **84**, 176-180 (2000)
8. H. Hartshorne, C.J. Backhouse, W.E. Lee, Sens. Actuators B, **99**, 592-600 (2004)
9. Y.-J. Chang, C.-Y. Hu, C.-H. Lin, *Proc. 4th Int. Conf. on Bioinformatics and Biomedical Engineering* (IEEE, 2010)
10. L.-M. Fu, W.-C. Fang, T.-F. Hong, C.-Y. Lee, Int. J. AUSMT, **4**, 77-82 (2014)
11. C. Yamahata, M. Chastellain, V.K. Parashar, A. Petri, H. Hofmann, M.A.M. Gijs, J. Microelectromech. Syst., **14**, 96-102 (2005)
12. A. Hatch, A.E. Kamholz, G. Holman, P. Yager, K.F. Böhringer, J. Microelectromech. Syst., **10**, 215-221 (2001)
13. J.-C. Bacri, J. Lenglet, R. Perzynski, J. Servais, J. Magn. Magn. Mater., **122**, 399-402 (1993)
14. T.G. Liu, J. Wu, C. Xia, Z.H. Qian, Adv. Mater. Res., **433-440**, 3767-3772 (2012)
15. K. Das, M. Sarkar, A. Mukhopadhyay, R. Ganguly, Magnetohydrodynamics, **49**, 499-504 (2013)
16. N.E. Greivell, B. Hannaford, IEEE Trans. Biomed. Eng., **44**, 129-135 (1997)
17. L.D. Landau, E.M. Lifshitz, *Electrodynamics of continuous media* (Pergamon, Oxford, 1960)
18. S.V. Vonsovskij, *Magnetism* (John Wiley & Sons, New York, 1974)
19. R.E. Rosensweig, *Ferrohydrodynamics* (Cambridge University Press, New York, 1985)

On the temperature dependence of ferrofluid susceptibility

Yuri Dikansky[1], *Anna* Ispiryan[1], *Stanislav* Kunikin[1,*]

[1] North-Caucasus Federal University, Institute of mathematics and natural science, 355009 1 Pushkina str., Stavropol, Russia

Abstract. The fulfillment of Curie–Weiss laws for a ferrofluid is critically discussed. The results of investigations of the magnetic susceptibility temperature dependences of ferrofluid and dried powder of magnetic particles are presented. The effective field parameter is computed in approximation of the effective field theory. It is indicated that the value of the effective field parameter is not constant and depends on the concentration of magnetic nanoparticles and the temperature of the sample. It was found that Curie-Weiss law with $\lambda = 4\pi/3$ is satisfied for ferrofluid with a limited range of volume concentration of magnetic particles.

1 Introduction

A magnetic colloid, also known as a ferrofluid (FF), is a colloidal suspension of single-domain magnetic particles, with typical dimensions of about 10 nm, dispersed in a liquid carrier [1–3]. The liquid carrier can be polar or nonpolar. Since the nineteen sixties, when these materials were initially synthesized, their technological applications did not stop to increase. As time went by, concurrent to the improvements in the liquids themselves was the discovery of numerous other practical applications for ferrofluids, some of which have gained high commercial importance. At present, ferrofluids are an important class of materials for industrial applications.

First models of ferrofluid magnetization based on the simple idea when magnetic nanoparticles are treated as an ideal paramagnetic gas and magnetic properties were described by Langevin function. At finite concentration, taking into account the magnetic interparticle interaction under the applied field, it is possible to use a mean field approximation, an effective field model, and the magnetization then is described as

$$H_{\text{eff}} = H + \lambda M , \qquad (1)$$

where λ is the effective field parameter. In the case of the classical dipolar system λ is Lorentz parameter that equals $4\pi/3$. With the help of Eq. (1) we express equation for initial susceptibility in a form similar to classical Curie-Weiss law:

$$\chi = \frac{C}{T - T_C}, \qquad (2)$$

here $C = \dfrac{nm^2}{3k}$, $T_C = \lambda C$

The previously conducted experimental studies [4] revealed the limited use of the formula for the effective field in a ferrofluid in the form of expression (2), since it turned out that the effective field parameter does not remain constant when the concentration of colloidal particles changes. At the same time, the temperature studies of the magnetic susceptibility of various samples of ferrofluid at a fixed concentration of the particles showed that at sufficiently high temperatures the Curie-Weis law could describe dependence of the susceptibility on temperature. This result indicates the possibility of using the effective field model in this temperature interval.

Actually, well known that some properties of the carrier and magnetic particles strongly depends on temperature. In addition, magnetization relaxation mechanism can switch from Néel to Brownian with temperature change. That means we have to use effective field model more careful. In this paper, we focus on what effects on the magnetic susceptibility has these properties of magnetic particles and carrier.

2 SAMPLES AND SETUP

Investigation of the magnetization processes in the ensemble of magnetic nanoparticles, taking into account their interaction, was carried out by an analysis of the temperature dependences of the magnetic susceptibility. The main sample for research was a kerosene based ferrofluid with a concentration of magnetite nanoparticles of 17.6 vol. %. We changed the concentration by diluting kerosene sample, or by evaporating carrier at 400 K. Additional studies were carried out on a powder containing magnetite nanoparticles. This sample was obtained by complete evaporation of kerosene and further grinding of the dry residue. In this work, the temperature dependences of the magnetic susceptibility obtained on three powder samples were analysed. The difference between these samples was the different size of magnetite nanoparticles.

Investigations of the temperature dependence of the magnetic susceptibility were carried out by the AC-bridge method using a setup described in [5]. The probing field frequency f was in the range from 20 Hz up to 10^4 Hz,

* Corresponding author: skunikin@yandex.ru

its magnitude 20 A/m. Experimental data processing take into account effects of carrier's thermal expansion and temperature dependence of the magnetite saturation magnetization..

Table 1. Setting Word's margins.

Sample	Carrier	Particles mean diameter d, nm
1	Kerosene	13.2
2	– / powder	13.2
3	– / powder	9.7
4	– / powder	7.8

3 RESULTS AND DISCUSSION

The features of the process of magnetization of ferrofluid are related to the mechanism of relaxation of the magnetic moment of a single-domain colloid particle. Well known that two magnetic relaxation mechanisms are actualized in ferrofluid. The first is the Brownian rotation of particles with a magnetic moment fixed relating to the particle. The second is the rotation of the magnetic moment of the particle, related to the thermal fluctuations the Néel relaxation. The implementation of a specific mechanism is due to both particle size and temperature. The polydispersity of ferrofluids allows one particle to be Brownian, the other to be Néel particles at the same temperature [6]. This leads to the possibility of the appearance of peculiarities in the temperature dependences of the magnetic susceptibility.

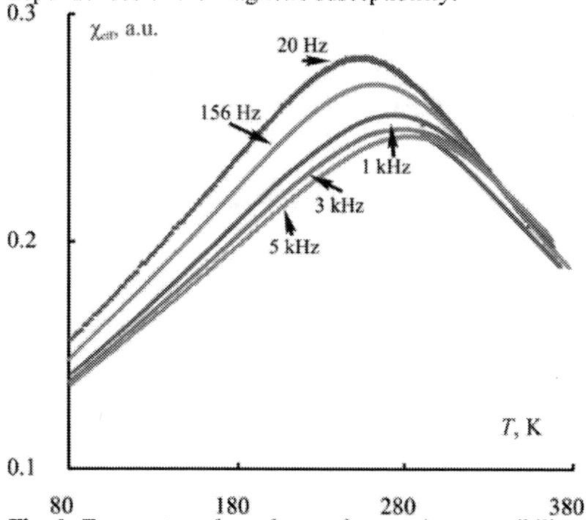

Fig. 1. Temperature dependence of magnetic susceptibility of sample No. 2 at different frequencies of probing field.

For Brownian particles, a change in the viscosity of the carrier can lead to a change in the behaviour of the functional dependencies of the dynamic magnetic susceptibility. For example, when the carrier solidifies, the Brownian degrees of freedom are blocked. With such a phase transition, the magnetic susceptibility should decrease.

An experimentally similar pattern of the dependence of the magnetic susceptibility is described in some papers, for example [7]. For the sample of a ferrofluid No. 1 the maximum of magnetic susceptibility observed at a temperature of 207 K close to the temperature of solidification of the carrier. Based on the analysis of the temperature dependence, we conclude the Brownian relaxation mechanism for magnetic moments of the particles is dominant for this sample.

The temperature dependence of the magnetic susceptibility of the powders also revealed the peak of the magnetic susceptibility temperature dependence at a specific temperature, depending on the size of the particles and the frequency of the probing field (Fig. 1). In magnetic powders, Brownian degrees of freedom are blocked, and therefore the presence of a maximum unexplainable in the same way as for the sample No. 1. A peak on the temperature dependence of magnetic susceptibility for this sample we associated with the transition of magnetic particles from the superparamagnetic state into ferromagnetic state related to the temperature change. The temperature corresponding to the maximum of the temperature dependence of magnetic susceptibility depends on the frequency of probing field (Fig. 2). Observation of the transition of the ensemble of magnetic particles to a superparamagnetic state is possible if the relaxation time of the ensemble correspond to the observation time (period of probing field) [9]. The relaxation time of the activation magnetization reversal is [10]:

$$\tau = \tau_0 \exp\left(\frac{\Delta E}{kT}\right), \qquad (3)$$

here $\tau_0 = 10^{-9} - 10^{-10}$ s, ΔE – energy barrier depending on magnetocrystalline anisotropy.

From the Eq. (3), one could obtain that the transition to a superparamagnetic state is observed when:

$$\Delta E \le \ln \frac{\tau_{meas}}{\tau_0} kT\,. \qquad (4)$$

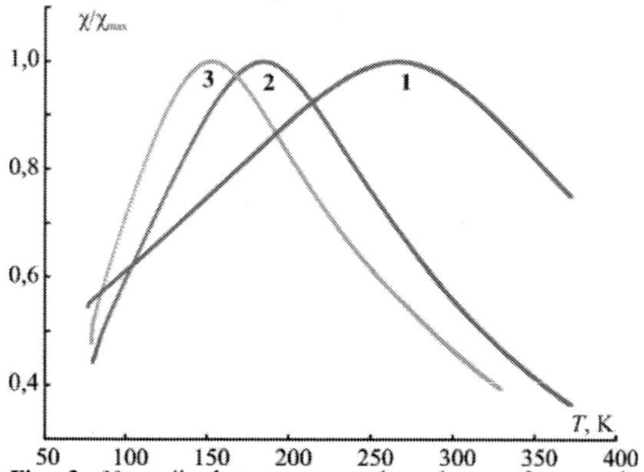

Fig. 2. Normalized temperature dependence of magnetic susceptibility of samples No. 2 (curve 1), No. 3 (curve 2), No. 4 (curve 3). Frequency of probing field 328 Hz.

Sample No. 2 contains larger particles than sample No. 3 and No 4. These leads to increasing of energy barrier magnitude therefore their blocking temperature is higher (Fig. 2). From (4) it appears that the transition of an ensemble of particles into a superparamagnetic state at a given temperature is also determined by the observation time (in the case of an AC field, by its period). An increase of the probing field frequency (a decrease in the observation time) leads to a blocking of the direction of the magnetic moment of the particle (transition to the ferromagnetic state) at higher temperatures. This explains the change in the maximum of the magnetic susceptibility temperature dependence with increasing of the probing field frequency.

Simple analysis lead us to figure out that at the temperature corresponding to the peak of the magnetic susceptibility temperature dependence we achieve a situation when $\tau_{meas} = \tau$. For each curve, we determined temperature corresponding to maxima and also found value $\tau \sim 1/f$. One could find that the dependence of relaxation time on the temperature of the sample well fitted by Vogel-Fulcher law (Fig. 3)

$$\tau = \tau_0 \exp\left(\frac{A}{T - T_0}\right). \quad (5)$$

Analysis of fitting parameters τ_0 , A and T_0 allows us to compute energy barrier height. For investigated sample, it was about 0.17 eV.

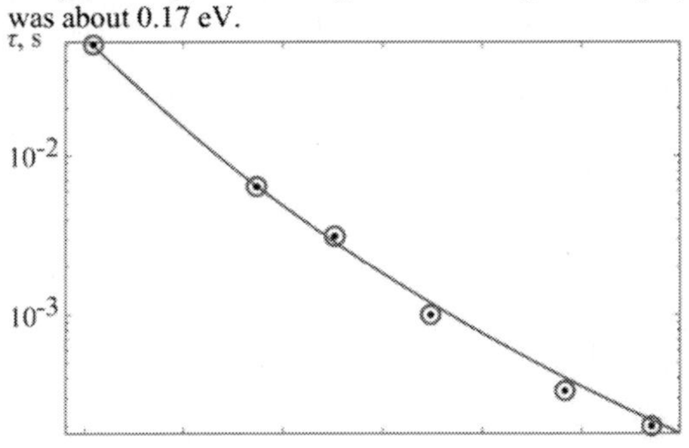

Fig. 3. Dependence of relaxation time computed by temperature dependencies of magnetic susceptibility of sample No. 3. (Circles – experiment, solid line Vogel-Fulcher law approximation)

Well known that this dependence describes relaxation properties of the Ising-like correlations systems. The results obtained from fitting of experimental data by Vogel-Fulcher law allows us to conclude that relaxation properties of investigated samples are intermediate between those of spin glasses and ensemble of noninteracting ferromagnetic particles. We have figure out that studied sample obeys the behavior similar to a progressive freezing of clusters [4 ,8].

Thus, investigating the possibility of describing the temperature dependence of the magnetic susceptibility by the Curie-Weiss law we need to exclude the influence of factors associated with different mechanisms of particle's magnetic moment relaxation. To achieve needed conditions we done measurements in the temperature range corresponding to the superparamagnetic state of particles. Superparamagnetic state of investigating samples corresponding to the temperature range higher than the temperature of the peak observed on temperature dependencies of susceptibility at Fig. 1 and Fig. 2. Therefore, studies of liquid sample No 1 were carried out at temperatures above 310 K, for the magnetic powder with large-sized particles (sample No. 2) – at temperatures above 320 K.

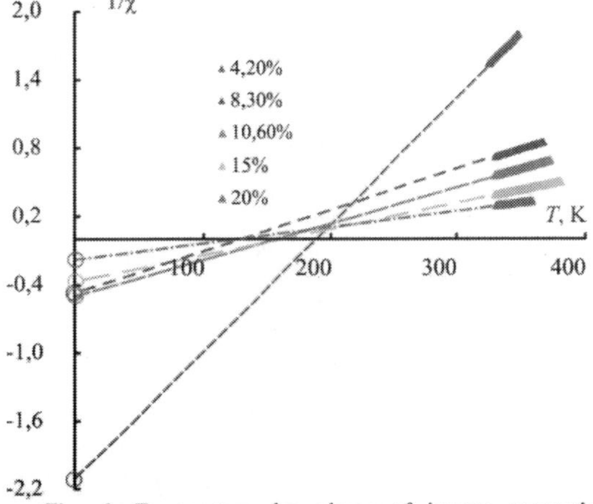

Fig. 4. Temperature dependence of inverse magnetic susceptibility of sample No. 1 at different concentrations of particles at temperatures higher than 310 K (Solid points – experiment, dash and dash-dotted lines – straight-line approximation).

We figured out that the magnetic susceptibility temperature dependence for all the investigated samples of ferrofluid at temperatures above 310 K is well described by behaviour similar to the classical Curie-Weiss law. This allows us to use the theory of the effective field to determine magnetic field acting on the magnetic particle. Figure 4 reveals the dependence of the inverse magnetic susceptibility on the temperature for the sample with different particle concentrations. In this figure, the fitting lines intercept segments of different magnitude on y-axis. Quantitatively the length of these segments are equal to effective field parameter λ determined by eq. (2). On the one hand for small concentrations (less than 8 vol. %), the magnitude of this parameter exceeds the same value for classical dipolar systems. On the other, an increase in concentration leads to a decrease of λ up to zero for more dense samples. This analysis allows us to compute magnitude of λ on concentration. Fig 5. illustrates the dependence of the effective field parameter on the volume concentration of the particles. From the slopes of the straight lines the magnetic moments of the particles were determined. These values are in good agreement with those calculated from the size determined by electron microscopy (Table 1) and a constant intrinsic magnetization.

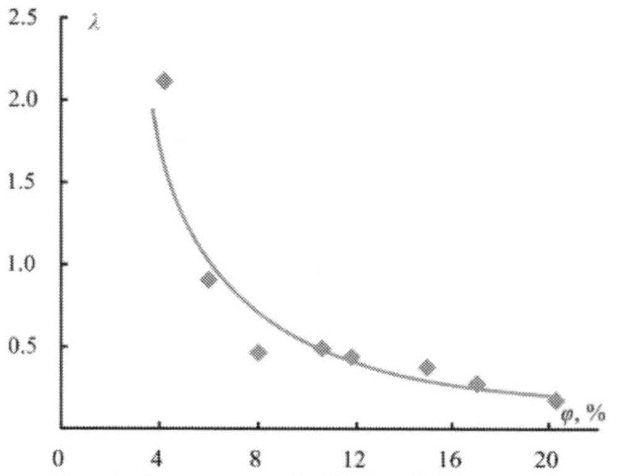

Fig. 5. Dependence of effective field parameter λ on concentration of magnetic particles.

With a decrease in the concentration of particles, the intensity of the particle interaction should decrease and for a diluted ferrofluid, the magnetic susceptibility could be expressed by behaviour similar to the Curie law. On the contrary, to predictions experimental results show different behavior. Dependence on Fig. 5 illustrates that while volume concentration of magnetic particles decrease effective field parameter λ increase (similar dependence for other samples of ferrofluid were expressed in [9]). This means that intensity of particle interaction increase We assumed that this is due to a violation of the homogeneity of the sample due to the formation of aggregates when the initial sample is diluted by the carrier. In this case, it is possible to form both nanoscale chain aggregates [10,11] and larger ones with non-zero magnetic moments [12–14]. The possibility of its formation was previously presented in [1]. We did not observe aggregates in vivo but found it presence qualitatively by means of the laser correlation spectroscopy and the Vogel-Fulcher behavior analysis. The decrease of λ with increasing in volume concentration (especially for the samples corresponding to dense state), is associated with an increase in the impact of the local fields produced by particle's neighbour and contributing for the orientation of the particle moment along the applied field direction.

Conclusion

Thus, we conclude that the temperature dependence of the magnetic susceptibility of such systems corresponds to the behaviour similars to the Curie–Weiss law at temperature range of 310–380 K. For samples with concentrations of magnetic particles exceeding 8 % the value of the parameter λ is always less $4\pi/3$, which corresponding to the effective field theory for dipolar systems. It decreases with increasing volume concentration of magnetic particles. With a further decrease in the concentration of particles, the λ value, calculated from the temperature dependences of the magnetic susceptibility, increases. It is assumed that this is the result of forming aggregates with uncompensated magnetic moment.

Acknowledgements

We thank Russian foundation for basic research (project 18-03-00279) and Ministry of Science and Education of Russian Federation for funding.

References

1. Y. I. Dikansky, D. V. Gladkikh, S. A. Kunikin, and A. A. Zolotukhin, Magnetohydrodynamics **48**, 493 (2012).
2. S. Yoon, J. Magn. **16**, 368 (2011).
3. Y. I. Dikanskii, V. A. Silaev, K. A. Balabanov, Y. M. Kozlov, and N. G. Polikhronidi, Magnetohydrodynamics **25**, 108 (1989).
4. J. L. Dormann, L. Bessais, and D. Fiorani, J. Phys. C Solid State Phys. **21**, 2015 (1988).
5. S. A. Kunikin and Y. I. Dikanskii, Tech. Phys. **55**, 866 (2010).
6. P. C. Fannin and S. W. Charles, J. Phys. D. Appl. Phys. **22**, 187 (1989).
7. A. F. Pshenichnikov, A. V. Lebedev, and K. I. Morozov, Magnetohydrodynamics **23**, 31 (1987).
8. V. Kumar, R. Kumar, K. Singh, S. K. Arora, I. V. Shvets, and R. Kumar, J. Appl. Phys. **116**, 73903 (2014).
9. Y. I. Dikanskii, A. G. Ispiryan, and S. A. Kunikin, Russ. Phys. J. **60**, 181 (2017).
10. A. O. Ivanov, Colloid J. **66**, 679 (2004).
11. L. M. Pop and S. Odenbach, J. Phys. Condens. Matter **18**, S2785 (2006).
12. D. Eberbeck, A. Lange, and M. Hentschel, J. Appl. Crystallogr. **36**, 1069 (2003).
13. A. O. Ivanov and A Y. Zubarev, Phys. Rev. E. Stat. Nonlin. Soft Matter Phys. **64**, 41403 (2001).
14. Y. I. Dikansky, K. A. Balabanov, O. V. Borisenko, and V. V. Kiselev, Magnetohydrodynamics **33**, 202 (1997).

EPJ Web of Conferences **185**, 10001 (2018) https://doi.org/10.1051/epjconf/201818510001
MISM 2017

Magnetic properties and giant magnetoimpedance of surface modified Co-based amorphous ribbons with carbon covering

Elizaveta V. Golubeva[1], *Anna* A. Chlenova[1], *Elena* A. Stepanova[1] and *Galina* V. Kurlyandskaya[1,2]

[1]UrFU Ural Federal University, Laboratory of Magnetic sensoric, 620002, Lenin Ave. 51, Ekaterinburg, Russia
[2] Universidad del País Vasco UPV-EHU, Depto Electricidad y Electrónica, 48940, Leioa, Spain

Abstract. $Fe_3Co_{66}Cr_3Si_{16}B_{12}$ and $Fe_6Co_{60}Ni_{10}Si_{14}B_{10}$ amorphous ribbons were surface modified in toluene at room temperature. Such a treatment resulted in deposition of thin carbon-based layer. As a result of the carbon covering deposition the stress distribution in the near-surface layers was changed due to partial compensation of the initial quenching stresses. Comparative analysis of magnetic and magnetoimpedance properties of as-quenched and surface modified ribbons confirms changes in effective magnetic anisotropy as a result of surface treatment. An increase of the corrosion resistance of the ribbons with carbon covering can be useful for the development amorphous ribbon based magnetic biosensor.

1 Introduction

Miniature compact analytical devices (biosensors) have been developed for a large range of applications [1]. Biosensors for detection of magnetic markers acting as biomolecular labels are promising for medical diagnostics, drugs delivery, tissue engineering and environmental control [2]. Various working principles for the magnetic particles detection can be mentioned: inductive coils, giant and anisotropic magnetoresistance, Hall effect, giant magnetoimpedance (GMI) [3, 4]. GMI effect consists in a large change of the impedance of a ferromagnetic conductor (Z) under application of the magnetic field [5]. Due to high sensitivity GMI effect is widely used for the development magnetic field sensors [6-7]. High frequency impedance depends on the classic skin penetration depth (δ). In magnetic materials, the application of a magnetic field changes the magnetic permeability, skin depth and the total impedance: $Z = Z(\delta)$. As the effect depends on the effective magnetic anisotropy, properties of a surface layer become very important. Amorphous ribbons are cheap materials with high GMI sensitivity, some of them having good corrosion resistance [4,8]. They were proposed as disposal sensitive elements for biosensors. However, surface functionalization of the metallic glass is difficult. Stability of the sensitive element in the biocorrosive environments is also important. It can be done by polymer covering [9] but good adhesion between the ribbon and polymer is not always a case.

Defected graphene-like carbon layer can be used to increase adhesion and to protect samples from the corrosion. Room temperature carbon deposition was discovered for the case of Fe and Ni nanoparticles and Fe and FeNi films [10]. Heterogeneous Scholl oxidation of the aromatic hydrocarbons was proposed as possible mechanism of the reaction with the first step being an oxidation of atoms of metals by oxygen dissolved in the solvent. No data on low temperature carbon deposition in aromatic solvents were reported for the case of amorphous materials. The main objective of this work was to study the possibility and particular features of the process of ambient conditions of carbon deposition onto the surface of amorphous ribbons with GMI responses.

2 Experimental methods

Rapidly quenched Co-based amorphous ribbons of two compositions were studied: S1 – $Fe_3Co_{66}Cr_3Si_{16}B_{12}$ (0.8 × 0.025 mm) and S2 – $Fe_6Co_{60}Ni_{10}Si_{14}B_{10}$ (1.3 × 0.025 mm). Surface modification of the samples was done in the toluene at ambient conditions for one or two months period. Toluene was used because the best results in previous experiments were observed with a use of it [10]. Table 1 lists all samples and their states. Scanning electron microscopy (SEM) and energy dispersive x-ray spectroscopy (EDX) were used for structural characterization and elemental analysis (AURIGA CrossBeam equipment). For mass losses determination ME 235S analytical balance was used. Relative weight loss is determined by follows: $\Delta m/m_0 = (m_0 - m_i)/m_0$. Where m_0 is a mass of the sample in initial state and m_i is a mass of the sample after i-days exposure in 0.9% NaCl solution.

Magnetic properties were measured by MMKS 100-05 installation. Visualization of domain structure was done with EVICO magnetooptical Kerr-microscope using the equatorial effect (MOKE). The complex impedance (Z) was measured as a function of the external magnetic field (H) applied in the direction of the flowing

* Corresponding author: anna.chlenova@urfu.ru

alternating current and parallel to the long side of the ribbon in a frequency range 1–50 MHz. Measurements were made in a "microstripe" line by standard methodology [11]. The GMI ratio was defined as follows: $\Delta Z/Z = [Z(H_o)-Z(H_{max})]/Z(H_{max})$, where H_{max} = 100 Oe is maximum applied magnetic field. Maximum values of the total impedance $((\Delta Z/Z)_{max})$ were also used for analysis.

Table 1. Designations of the samples used in this work.

Composition	Name	Time of exposure in toluene (months)
$Fe_3Co_{66}Cr_3Si_{16}B_{12}$	S1-0	0
	S1-1	1
	S1-2	2
$Fe_6Co_{60}Ni_{10}Si_{14}B_{10}$	S2-0	0
	S2-1	1
	S2-2	2

3 Results and discussions

The structure and elemental composition of the samples were analyzed in as-prepared state and after exposure in toluene for two months. Fig.1 and Table 2 show selected examples. The morphology of as-prepared ribbons was typical for this kind of materials. It was observed differences in the compositions before and after surface modification: carbon and oxygen content of the surface layer was significantly increased after modification. Moreover, an appearance of dark circle areas on SEM images of the sample after the treatment can be noticed on Fig.1 (b). Similar defects corresponding to areas an accumulation of carbon were also observed in previous experiments with thin films [10]. Fig.1(c) shows the effect of carbon re-distribution as a result of covering interaction with electron beam during SEM studies (black squares are previously scanned areas). Observed compositional changes are consistent with proposed mechanism of the low temperature carbon deposition [10].
The next step is to understand the role of the surface modification in the change of magnetic properties. Figure 2 shows that the general shape of the M(H) loops was not changed by the surface modification: only a small reduction of the coercivity takes place in the S2 ribbon. In all cases the effective magnetic anisotropy was clearly a longitudinal one with possible small contribution of a transverse anisotropy (slow saturation approach). Fig. 3-4 show changes of magnetic domains under application of the magnetic field. For analysis of the magnetization process we must take into account both measurements of bulk M(H) loops and MOKE data. In the initial state of sample S1 there is a surface contribution of the transverse magnetic anisotropy (perpendicular to the ribbon axis bar domain structure). After the surface modification transverse and longitudinal components begin to compete in close to surface layer. In the S2 case the elongated bar domains parallel to the ribbon axis confirm that the main contribution in the initial state is the longitudinal magnetic anisotropy. After the treatment domains become invisible due to the higher surface roughness. As those changes took place in a thin (≈20 nm) surface layer, mentioned effects are absent in the M(H) loops.

Table 2. Percentage elemental composition of S1 and S2 ribbons before and after modification in toluene for 2 months.

Element	S1-0	S1-2	Element	S2-0	S2-2
C	21.07	40.58	C	22.63	49.32
O	0.00	5.77	O	0.00	15.14
Si	14.01	10.37	Si	11.05	9.17
Cr	3.04	1.92	Fe	4.81	1.87
Fe	3.08	2.04	Co	52.21	20.83
Co	58.79	39.32	Ni	9.30	3.66

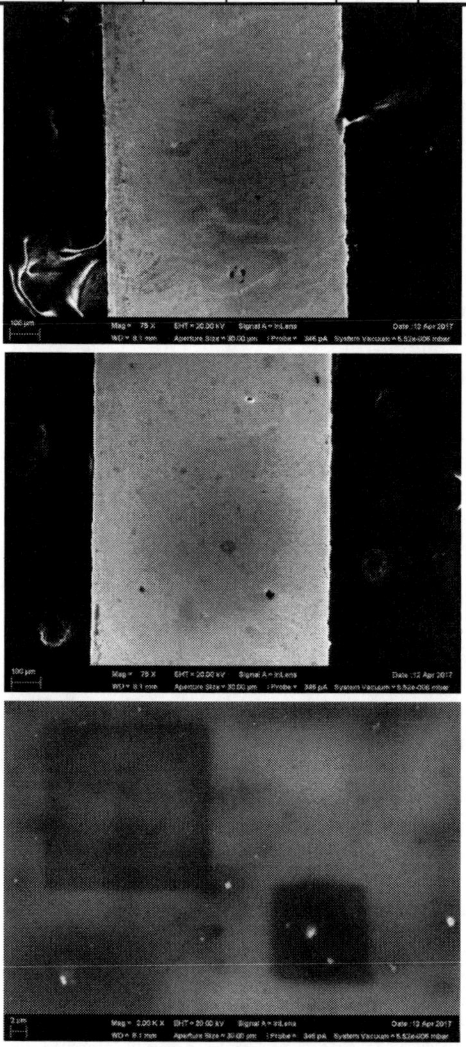

Fig. 1. SEM images of the FeCoCrSiB ribbon before (a) and after (b,c) surface modification in toluene for 2 months.

Frequency dependencies for $\Delta Z/Z_{max}$ ratio are shown in Figure 5. Decrease of the value of $\Delta Z/Z_{max}$ and its shift toward the higher frequencies as a result of surface modification were observed. Fig. 6 shows field dependences of the GMI ratios. For both samples the one-peak shape of the curves was not changed as a result of the surface modification, but fine changes are evident in the small fields. In the case of S1 ribbon both decrease of a distance between GMI peaks and «down» (decreasing) and «up» (increasing field) branches take place. In the S2 case (Fig. 6(c)) there is also a decrease of a distance between «down» and «up» branches.

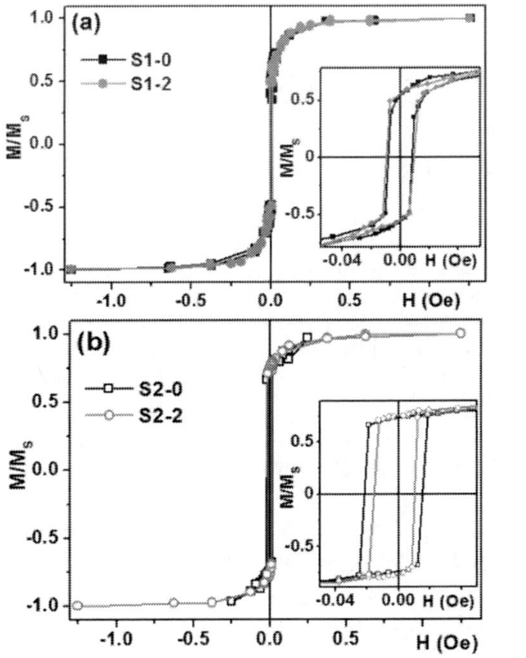

Fig. 2. Longitudinal hysteresis loops of FeCoCrSiB (a) and FeCoNiSiB (b) ribbons before and after their treatment in toluene. Insets show curves in a smaller field interval.

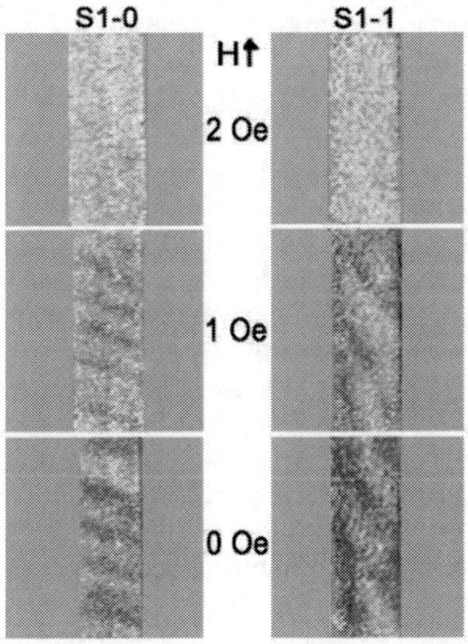

Fig. 3. MOKE images of domain structure of FeCoCrSiB ribbons before and after surface modification.

Changes of the shape of GMI curve of S1 sample indicate an increase of the contribution of longitudinal component of the effective magnetic anisotropy in a good agreement with MOKE studies. The existence of quenching stresses layers of as-cast quenched ribbons is known [12]. Surface modification in aromatic solvent seems to cause stress relaxation in the near-surface layer due to an appearance of chemical bonds between the carbon covering and a practically oxidized surface of the samples. As the changes of magnetic anisotropy of

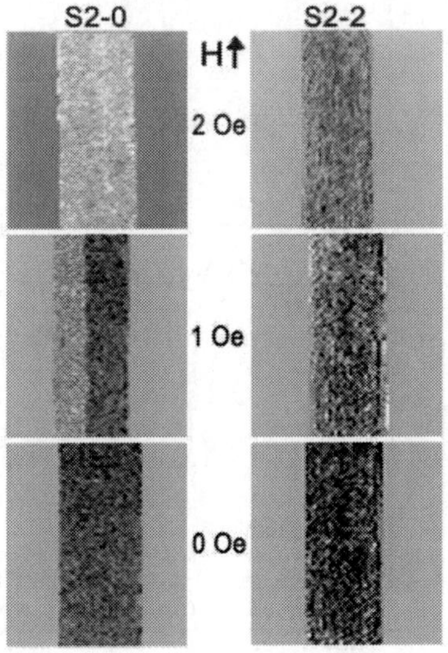

Fig. 4. MOKE images of domain structure of and FeCoNiSiB ribbons before and after surface modification.

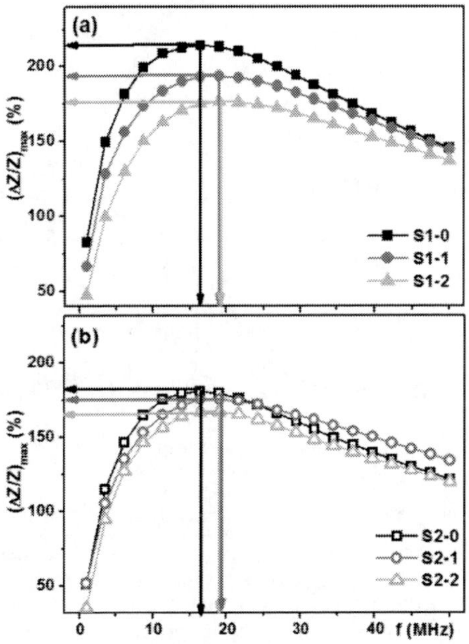

Fig. 5. Frequency dependencies of total impedance ratio for S1 (a) and S2 (b) ribbons before and after treatment in toluene. Arrows indicate positions of the maxima.

the sample S1 are more significant the decrease of the GMI ratio is much larger than in the S2 case. Corrosion stability of the samples was studied for 0.9%NaCl solution, they were placed into solution and mass loss was systematically measured (Fig. 7). The mass loss is a bit smaller in the case of the ribbon S2-1 with surface passivation by low temperature carbon layer deposition.

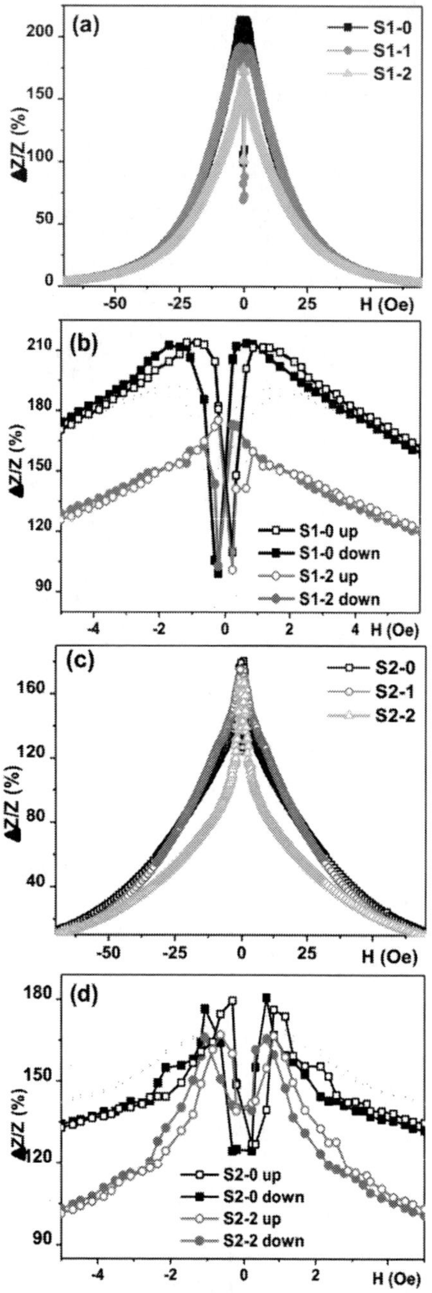

Fig. 6. Field dependences of GMI ratio for FeCoCrSiB (a) and FeCoNiSiB (b) ribbons before and after surface modification. Decreasing (down) and increasing (up) branches are shown.

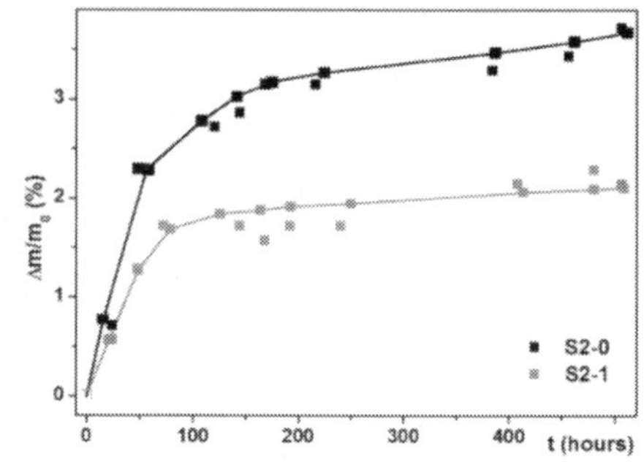

Fig. 7. Relative weight change of S2 ribbon in 0.9% NaCl solution before (S2-0) and after (S2-1) treatment in toluene.

4 Conclusion

Room temperature carbon deposition from toluene was confirmed for $Fe_3Co_{66}Cr_3Si_{16}B_{12}$ and $Fe_6Co_{60}Ni_{10}Si_{14}B_{10}$ of amorphous ribbons. Deposition process involves change of the surface magnetic anisotropy in the near-surface layers and modifies the GMI effect. An increase in corrosion resistance was observed for FeCoNiSiB ribbon in 0.9% NaCl solution. It might be useful as a method for biosensor surface passivation.

Acknowledgments

We thank UCSU «Modern nanotechnologies» and S.O. Volchkov for special support.

References

1. A.P.F. Turner, Science **290** 1315 (2000)
2. D.R. Baselt, G.U. Lee, M. Natesan et al., Biosens. Bioelectron. **13** 731 (1998)
3. G. Kurlyandskaya and V. Levit, Biosens. Bioelectron. **20** (8) 1611 (2005)
4. J. Devkota, et al, J. Appl. Phys. **113**, 104701 (2013)
5. R.S. Beach and A.E. Berkowitz, Appl. Phys. Lett. **64** 3652 (1994)
6. L. V. Panina, K. Mohri, T. Uchiyama, M. Noda, et al. IEEE Trans. Magn. **31** 1249 (1995)
7. A. S. Antonov, S. N. Gadetskii, A. B. Granovskii, et al., Phys. Met. Metal. **83** (6) 612 (1997)
8. X. Liang, X. Li, C. Chen, Z. Liang, at al., Int. J. of Electrochem. Sci. **10** 8395 (2015)
9. A.V. Semirov, M.S. Derevyanko, D.A. Bukreev, at al, J. Magn. Magn. Mater. **415** 97 (2016)
10. A.P. Safronov, G.V. Kurlyandskaya, A.A. Chlenova et al, Langmuir **30** 3243 (2014)
11. G.V. Kurlyandskaya, D. de Cos, S.O. Volchkov, Russian J. Nondestruct. Test. **45** 377 (2009)
12. A. Saad, J.A. Garcia, G.V. Kurlyandskaya., J.D. Santos, L. Elbaile, Jap. J. of Appl. Phys. **44** № 7A 4939 (2005)

Heterogeneity of population of microorganisms grown in presence of iron oxide maghemite nanoparticles

Tatyana P. Denisova[1,*], *Elena* V. Simonova [2], *Lubov* A. Kokorina [2], *Evgenia* N. Maximova [1], *Oleg* M. Samatov [3], *Alexander* P. Safronov [3,4], *Galina* V. Kurlyandskaya [4,5]

[1] Irkutsk State University, Nizhnyaya Embankment St., 6, Irkutsk, 664011, Russia
[2] FSBEI HE Irkutsk State Medical University MOH, Red Revolt str., 1, Irkutsk, 664003, Russia
[3] Institute of Electrophysics, Amundsen str., 106, Ekaterinburg, 620016, Russia
[4] Ural Federal University, Mira St. 19 Ekaterinburg, 620000, Russia
[5] Universidad del País Vasco UPV-EHU, Leioa, 48940 Spain

Abstract. In this work γ-Fe_2O_3 MNPs were obtained by laser target evaporation and water based suspensions were prepared. Maximum permissive dose of iron in water (MPD) is 0.3 mg/L. It was found that 100 MPD dose of iron induces formation of non-typical colonies after 72 or 96 hours exposition: against a background of small black colonies large white colonies appeared due to a disruption in tyrosine synthesis. Multiple re-cloning of the white colonies grown with MNPs showed that they retained their properties both under standard conditions (temperature of 24 °C) and at the temperatures up to 37° C. *E.nigrum* grown with MNPs demonstrated very scant extension of small colonies at the cultivation temperature of 24° C, their growth was completely blocked at 37°C. Significant changes in the structure of the population were noted. First of all, large cells with pronounced aggregation were observed among the black colonies. These aggregates consisted of large cells connected to each other by matrix. In the white colonies the appearance of very long threadlike cells connecting different groups of the cells establishing an intercellular communication was evident. Fe_2O_3 MNPs induce an increase in the heterogeneity of the population, expressed as a change in morpho-physiological states.

1 Introduction

The universal property of any microbial population showing its adaptive capability is the heterogeneity. The study of this phenomenon is of interest for understanding the development of the microbial population in conditions of stress arising from the unfavourable environmental factors, as well as the mechanism of "quorum sensing", which provides a way of the microorganisms behaviour and survival in the microbial community [1]. It is known that heterogeneity in microbial populations can be discrete and continuous, arising due to phenotypic or genotypic variability [2]. Heterogeneity in the structure of microbial population is determined by factors of interaction between the external and internal factors, and also due to the variability of their physiological and biochemical properties. Heterogeneity provides microbial communities chance to survive in changing living conditions.

Many living systems, including microbial populations strongly depend on iron in variety of life processes [3]. Therefore, consideration that variation of small amount of iron is an instrument of the study of heterogeneity in the biological samples can provide useful insight on the features of the development of microbial population in artificially created stress conditions.

Exophiala nigrum (*E. nigrum,* black unicelular yeast) of the Baykal Lake play an important role in the equilibrium of this unique ecosystem [4]. Black yeasts were previously used as a model system for the estimation of cytotoxicity, MNPs accumulation and inductive heat capacity studies of the biological samples grown in the presence of MNPs [5]. For fabrication of spherical iron oxide magnetic nanoparticles (MNPs) electrophysical technique of the laser target evaporation (LTE) was employed [6].

The aim of the work was to study the heterogeneity in the structure of the microbial population of *Exophilia nigrum* under the influence of controlled amount of iron provided by iron oxide γ-Fe_2O_3 nanoparticles in form of electrostatically stabilized water-based suspension.

2 Experimental

Iron oxide MNPs were synthesized by the laser target evaporation technique. In the LTE installation, the solid target (compacted commercial Fe_2O_3 powder from Alfa Aesar) is overheated by the high-energy pulse of a focused laser beam. Rapid evaporation in the laser spot area is a result of beam energy transfer into the kinetic energy of the vapours, ejected away from the surface of

* Corresponding author: denis_tp@inbox.ru

the target made of the. More details can be found elsewhere [6-7].

Structural studies of LTE MNPs were performed by transmission electron microscopy (TEM, JEOL JEM2100 microscope operating at 200 kV). The X-ray diffraction (XRD) studies were performed by DISCOVER D8 (Bruker) diffractometer using Cu-Kα radiation with wave lengths λ=1.5418 Å. The stoichiometric ratio of Fe^{2+}/Fe^{3+} in all samples of the iron oxide MNPs was determined based on the analysis of XRD diffractograms, and redox potentiometric titration using an automatic TitroLine titrator.

Electrostatically stabilized suspensions based on γ-Fe_2O_3 LTE MNPs were prepared by ultrasound treatment using sodium citrate solutions (5 mM) in distilled water. A Hermle Z383 centrifuge operated at 10.000 rpm for 10 min was used in order to remove large aggregates. Magnetic measurements were carried out on a MPMS XL-7 SQUID-magnetometer.

As a test culture, the strain R-11 *Exophiala nigrum*, belonging to the saprophyte forms of black yeast, was isolated from the natural population by streaking on a solid nutrient medium for yeast cultivation. Strain has a black colour due to the produced melanin. For experimental studies, a test culture was used in the exponential stage of growth. The R-11 strain was registered in the All-Russian Collection of Microorganisms of the Research Institute of Genetics No. Y 309. It can be used as a destructor of the main components of wastewater from the cellulose and paper industry (lignin, in particular) [8-9]. For both scanning (SEM) and transmission (TEM) electron microscopy cell cultures were prepared by standard procedure [10].

Maximum permissive dose of iron ions in water (MPD) is 0.3 mg/L. Electrostatically stabilized suspension of MNPs was used for preparation of solutions for biological part of the experiments on the basis of Saburo liquid nutrient medium in concentration 100 MPD of iron. First, a cell suspension of yeast was prepared, which was used for all studied cases including control. Next, the flasks were prepared with MNPs suspensions for studied concentrations (liquid medium Saburo + MNPs). In the control case Saburo medium had no MNPs. After this, the same volume of the same cell suspension of the test culture was added to the prepared liquid medium Saburo + MNPs, after which the mixture was vigorously shaken, and then the bite was sewn onto a solid nutrient medium. The control samples were grown with Saburo liquid medium and corresponding addition of a sterile physiological solution. The exposure was 72 and 96 hours at a temperature of 24° C. In the control version, instead of suspension of MNPs an equivalent saline volume was added. Exposed for 72, and 96 hours samples were used in order to determine the properties of the *E. nigrum* cultures. For this purpose, the yeast was sown to a solid nutrient medium. The smear preparations were also elaborated, staining them with gentianviolet. The heterogeneity of the microbial population was evaluated under the experimental conditions for the following indicators: growth of culture on a solid nutrient medium and change in the morphological features of cells that are specific for each microorganism species under standard conditions.

3 Results and discussion

Fig. 1 shows the as-prepared spherical LTE MNPs. XRD studies gave a mean crystallite size for MNPs of 19 ± 2 nm in a good agreement with TEM data. The chemical composition found by redox titration was close to the stoichiometric maghemite ($Fe_{2.72}O_4$).

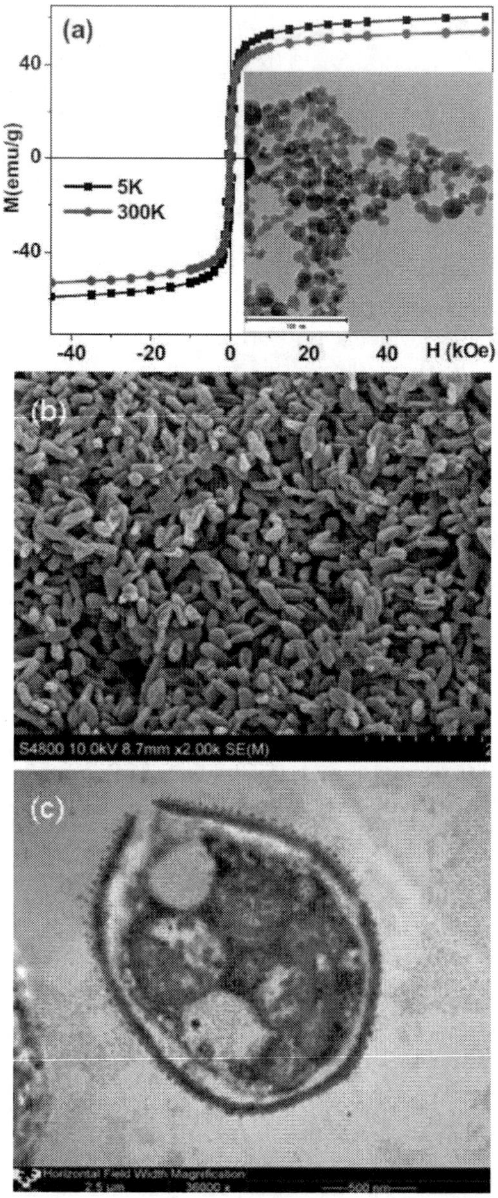

Fig. 1. Hysteresis loops of as-prepared spherical LTE MNPs of iron oxide (a) Inset shows TEM image of MNPs. *E. Nigrum* cell cultures grown without MNPs: SEM (b) and TEM (c).

One can see that hysteresis loops of MNPs show no saturation at both temperatures: the magnetization in the field of 65 kOe in both cases is close to 54 emu/g for 300 K and 61 emu/g for 5 K temperature. The comparative analysis of structural and magnetic parameters confirms consistency of obtained data about close to superparamegnetic behaviour for MNPs of this size.

Fig. 1(b) shows general view of *E. Nigrum* cell cultures grown in the presence of 0 MPD of iron: individual cell size and morphologies were typical for *E. nigrum*. It was found that culturing the E. nigrum test culture on Saburo solid nutrient medium without MNPs results in good grows of the culture with the formation of colonies typical for this strain. When MNPs were added in iron concentration of 100 MPD for 72 and 96 hours cultivation, a change in the properties of the cell cultures was observed. Namely, the large white colonies appeared on the background of very small *E. nigrum* colonies of black colour (Fig. 2).

White colonies grown without MNPs onto the Saburo nutrient medium did not alter their properties after several times re-cloning both under standard growth conditions (the cultivation temperature of 24 °C) and with an increase in the cultivation temperature up to 37 °C. At the same time, very low growing rate of the black colonies, grown up without MNPs was observed at the cultivation temperature of 24° C and their growth was completely blocked at 37° C temperature. White colony color is associated with a breakdown in the synthesis of melanin, and the root cause may be a violation of any stage of the biochemical cycle of tyrosine, from which the precursors of melanin are formed. The MNPs amount corresponding to 100 MPD of iron change biochemical properties of black yeasts leading to a disruption of the synthesis of melanin in the *E. nigrum* strain R-11 culture.

It is known that *E. nigrum* strain R-11, grown without MNPs is a compactly organized uniform cell mass. The degree of heterogeneity of the morphotype is poorly expressed. During the time the cells were in liquid medium Saburo + MNPs (1 min), no changes were found (taking into account the number of cells in 1 ml, the number of colonies grown latter onto a solid medium and the morphology of the colonies).

At the same time, significant changes in the structure of the microbial population were noted in culture grown with MNPs for the exposure of 72 or 96 hours. First of all, among the black colonies of the *E. nigrum* strain R-11, morphotypes (Fig. 2) of large cells were isolated with their pronounced aggregation and connected to each other by the matrix. Such a morphotype is typical feature of a static culture, for which the concentration of cells ceases to increase due to a decrease in the growth rate, depending on a number of factors, such as low partial pressure of oxygen or the content of toxic substances in the nutrient medium. In the culture of the microbial population of *E. nigrum*, which had grown in the form of white colonies, along with ordinary cells, very long threadlike cells appeared connecting different groups of cells in course of intercellular communication (Fig. 3). The presence of local aggregates of cells, interconnected by the intercellular matrix, was detected. The formation of such a morphotype is also associated with a violation of the development of microbial culture: the biomass of the cell increases and the processes of its division do not occur. This indicates that biofilm formation around the MNP is likely to occur as a result of the process of adaptation to unfavourable living conditions. Comparing the character of the change in the properties of cell culture and morphological properties of *E. nigrum* strain R-11 grown without MNPs, it should be noted that the observed degree of heterogeneity under experimental conditions is the result of its reserve of stability, as a result of which physiologically active cells die in unfavourable environmental conditions, whereas, physiologically inactive forms are able to survive.

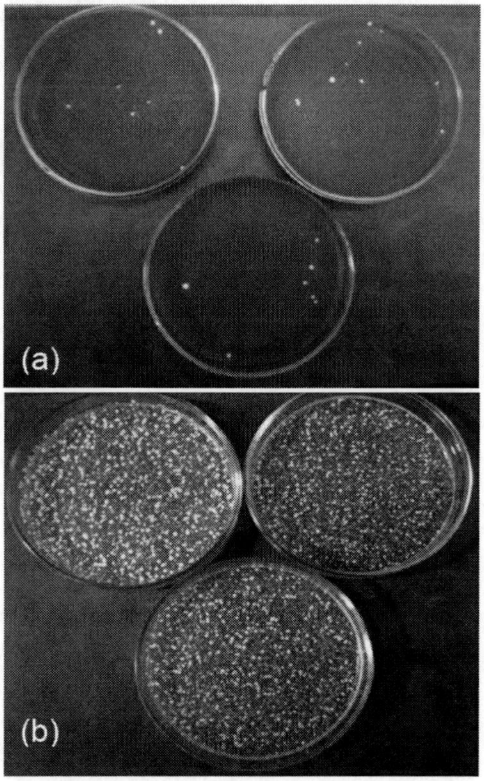

Fig. 2. Colonies of yeast of white colour in the experiment with 100 MPD addition of iron in the form of iron oxide MNPs: at exposures of 72 hours (a) and 96 hours (b).

Although careful magnetic characterization of iron oxide MNPs was discussed above at this study we did not provide a magnetic characterization of cell cultures grown in the presence of MNPs. At the same time obtained results on the heterogeneity of the microbial population and activation of the mechanism of a biofilm formation as a reaction on the presence of LTE MNPs in nutrient give new insight on our previous results [11], where total reflection x-ray fluorescence spectroscopy and magnetic measurements were used as power tools for evaluation of iron concentration in ferrofluids and yeast samples. Observed surprisingly low amount of iron

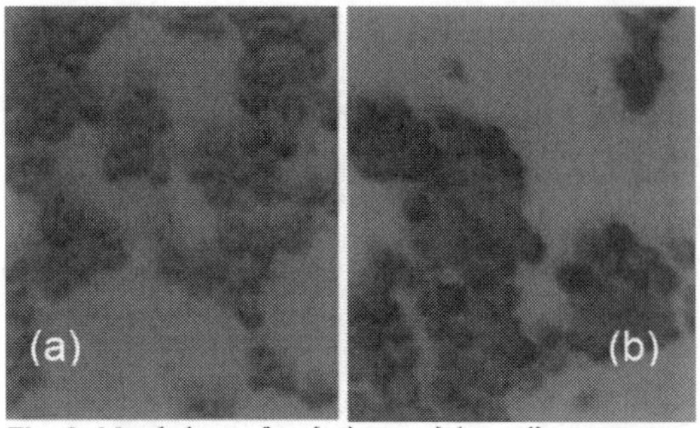

Fig. 3. Morphology of melanin-containing cells grown on a solid medium in the presence of LTE MNPc in concentration corresponding to 100 MPD of iron: (a) - 72 hours exposire, (b) - 96 hours exposure.

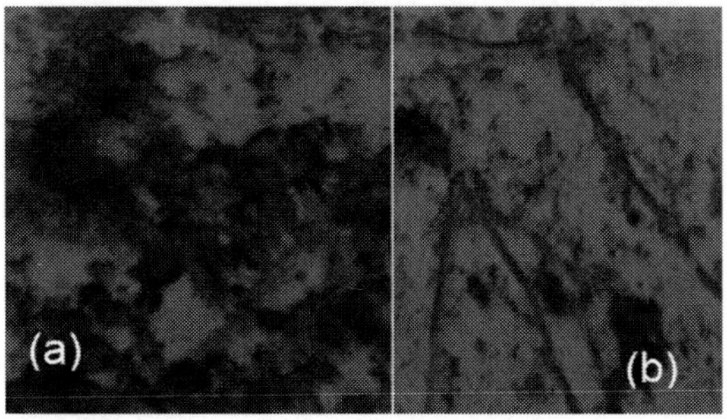

Fig. 4. Morphology of *E. nigrum*, cells grown on a solid medium. in the presence of LTE MNPc in concentration corresponding to 100 MPD of iron forming white colonies: (a) - 72 hours exposire, (b) - 96 hours exposure.

in the yeast samples grown in the presence of LTE MNPs was previously given a suggestion that membrane absorption can be proposed as a possible mechanism of MNPs incorporation. Here we observed that biofilm formation around the MNP is likely to occur. This additional barrier can also contribute to the low degree of iron MNPs incorporation.

The next step in the studies of the process of adaptation of *E. nigrum* to unfavourable living conditions and better understanding of the mechanisms of iron accumulation would be application of external magnetic field (both static and alternating) during cell cultivation and comparative studies of the iron content and magnetic responses. MNPs containing model samples with different concentration of magnetic nanoparticles are highly requested by a new interdisciplinary field – development of magnetic biosensors [12]. To be reliable the main features of the cell cultures with different amount of MNPs must be similar and therefore heterogeneity of population of microorganisms is important part of the development of this research line.

4 Conclusions

Exophiala nigrum strain R-11 unicelular organisms of the Baikal Lake were used for heterogeneity of population analysis. Yeasts were grown in the presence of 19 ± 2 nm maghemite MNPs produced by laser target evaporation. Electrostatically stabilized water based suspensions were prepared for biological experiments to study the ability of iron oxide nanoparticles at a concentration equivalent to 100 MPD of iron to induce a change in the structure of a microbial population.

It has been established that under the influence of the iron oxide nanoparticles at a concentration equivalent to 100 MPD of iron the heterogeneity of the population of saprophytic eukaryotic microorganisms was increased in comparison with the control case (grown without MNPs). A change in the morpho-physiological and biochemical characteristics of cells of *E. nigrum* strain R-11 induced by iron oxide nanoparticles was detected. Heterogeneity ensures the stability of the microbial population and allows it to remain viable using the mechanism of a biofilm formation.

The work was supported by the Ministry of Education and Science of the Russian Federation (project No. 3.6121.2017) and RFFI №16-34-50192 grants. We thank R. Andrade, Iu. Novoselova and I.V. Beketov for special support. Selected measurements were made at SGIKER services UPV/EHU.

References

1. M.B. Miller, B.L. Bassler Quorum sensing in bacteria. Annu Rev Microbiol. **55**, 199 (2001).
2. L.A. Magdanova, N.V. Golyasny. Microbiology. **82**, 3 (2013).
3. M. Ilbert and V. Bonnefoy, Biochimica et Biophysica Acta – Bioenergetics, **1827**, 161 (2013).
4. I.P. Babieva and I.Yu. Chernov, *Biology of yeasts* (M.V. Lomonosov MSU, Moscow, 2004).
5. J.P. Novoselova, A.P. Safronov, O. Samatov, I.V. Beketov, H. Khurshid, Z. Nemati, H. Srikanth, T. P. Denisova, R. Andrade, and G.V. Kurlyandskaya, IEEE Trans. Magn. **50,** 4600504 (2014).
6. A.P. Safronov, I.V. Beketov, S.V. Komogortsev, G.V. Kurlyandskaya, A.I. Medvedev, D.V. Leiman, A. Larrañaga, and S.M. Bhagat, AIP Adv., **3**, 052135 (2013).
7. V.V. Osipov, V.V. Platonov, M.A. Uimin, and A.V. Podkin, Tech. Phys. **57,** 543 (2012).
8. V. V. Pavlenko, V. M. Kharlamov, V. I. Chemerilova, S. S. Timofeeva. SU 1071637 A, C 12 N 15/00// C 02 F 3/34; C 12 R 1/00.
9. S.S.Pavar, V.V. Pavlenko. Messenger of IrGSHA. **60**, 67 (2014).
10. G.V. Kurlyandskaya, Iu.P. Novoselova, V.V. Schupletsova, R. Andrade, N.A. Dunec, L.S. Litvinova, A.P. Safronov, K.A.Yurova, N.A. Kulesh, A.N. Dzyuman, and I.A. Khlusov **431,** 249 (2017).
11. N.A. Kulesh, I.P. Novoselova, A.P. Safronov, I.V. Beketov, O.M. Samatov, G.V. Kurlyandskaya, M.V. Morozova, and T.P. Denisova, J. Magn. Magn. Mater. **415**, 39 (2016).
12. G.V. Kurlyandskaya, E. Fernandez, A.P. Safronov, A.V. Svalov, I.Beketov, A. Burgoa Beitia, A. Garcıa-Arribas, F.A. Blyakhman, Appl. Phys. Lett. **106** 193702 (2015).

Magnetic properties and giant magnetoimpedance effect for CoFeMoSiB surface modified amorphous ribbons covered by water based ferrofluid

Zahra Lotfollahi [1,*], *Ahmad* Amirabadizadeh [1], *Alexander* P. Safronov [2,3], *Galina* V. Kurlyandskaya [2,4]

[1] University of Birjand, 97175-615, Birjand, Iran
[2] Ural Federal University, Mira str., 19, Ekaterinburg, 620002, Russia
[3] Institute of Electrophysics, Amundsen str., 106, Ekaterinburg, 620016, Russia
[4] Universidad del País Vasco UPV-EHU, Depto Electricidad y Electrónica, Leioa, 48940, Spain

Abstract. Giant magnetoimpedance (GMI) effect is a powerful technique for magnetic label detection. Co-based amorphous ribbons are cheap materials showing high GMI effect at low operation frequencies for close to zero magnitostriction compositions. In this work magnetic properties and GMI were studied for CoFeMoSiB amorphous ribbons in as-quenched and surface modified states without and in the presence of water-based ferrofluid with electrostatic stabilization of γ-Fe_2O_3 nanoparticles. Surface modification by ultrasound treatment resulted in appearance of round defects with average diameter of about 150 micrometers. The GMI difference for as-quenched ribbons in absence and in the presence of ferrofluid was measured for the frequency range of 0.5 to 10 MHz. Although proposed surface modification by the ultrasound treatment did not improve the sensitivity limit for ferrofluid detection, it did not decrease it either. Observed changes of GMI are useful for understanding of functionality of GMI biosensors.

1 Introduction

Giant magnetoimpedance (GMI) effect in soft ferromagnets is promising for applications in the area of small magnetic field sensors including magnetic biosensors [1-4]. Magnetoimpedance phenomenon is a change of the total impedance of ferromagnetic conductor under application of an external magnetic field when high frequency alternating current flows through it [2]. Different kinds of materials were proposed for GMI biosensors [3, 5-7]. The idea of GMI biosensor based on amorphous ribbon is attractive because this kind of material is cheap and sensitive element can be designed as a disposable strip [5]. Two types of GMI-based biosensors are discussed in a literature. The first one is related to magnetic label detection [3] and the second is for label free detection process [8]. In the second case the rapidly quenched ribbon sensitive element slightly changes the mass during the sensing process. The surface layer properties of the rapidly quenched ribbon differ from the properties of the central part. Therefore, gradual removal of the surface layer causes the change of the effective magnetic anisotropy and GMI [8].

Such a parameter as surface roughness is important in the case of amorphous ribbon based biosensors. On one hand, it might be seen as an obstacle in the case of magnetic label detection influencing the sensitive element response (due to strong contribution of the surface anisotropy). On the other hand, one can expect the improvement of the magnetic flux closure for the rough surface with magnetic ferrofluid spread on it. The origin of sensitivity is still not fully understood and it depends on surface features, size and agglomeration state of magnetic labels and other parameters [6,9].

Mechanical processing of the surface modification of the amorphous materials is practically impossible due to their physical properties. For the magnetic detection of the MNPs they usually select Mo or Cr doped compositions with enhanced corrosion stability. Consequently, chemical processing of these kinds of ribbons (like lithographic process including chemical etching) is also not simple. There were attempts to use advanced technique of lithography for surface defects creation or even mechanical drilling [10, 11, 12] but employment of such an expensive technique makes whole device less adapted for applications. Search for new methods of creation of artificial surface defects for the ribbons with enhanced corrosion stability is therefore a challenge.

In this work, spherical magnetic nanoparticles were fabricated by the laser target evaporation. Electro-statically stabilized water-based suspensions were prepared on the basis of obtained MNPs. Their physical properties were studied by different techniques prior to GMI measurements without and with immersing the sensitive element into a magnetic suspension. CoFeMoSiB amorphous ribbon GMI sensitive element was either in as-quenched or in ultrasound treatment surface modified state.

* Corresponding author: lotfollahi@gmail.com

2 Experimental

Iron oxide nanoparticles (MNPs) were synthesized by the laser target evaporation (LTE) method using Ytterbium (Yb) fiber laser with 1.07 μm wavelength. More details of LTE technique can be found elsewhere [12-13]. Structural studies of as prepared MNPs were performed by transmission electron microscopy (TEM) using a JEOL JEM2100 microscope operating at 200 kV. The X-ray diffraction (XRD) studies of the MNPs and the ribbons were performed by DISCOVER D8 (Bruker) diffractometer using Cu-Kα radiation (wave lengths λ=1.5418 Å). Electrostatically stabilized ferrofluid (FF) was prepared by ultrasound treatment using sodium citrate solutions (5 mM) in distilled water. The final concentration of MNPs in ferrofluid was 5.0 %. The specific surface area of MNPs was measured by the low-temperature sorption of nitrogen (Brunauer–Emmett–Teller physical adsorption method, BET) using Micromeritics TriStar 3000 analyzer.

$Co_{68.6}Fe_{3.9}Mo_{3.0}Si_{12.0}B_{12.5}$ amorphous 0.7 mm wide and 20 μm thick ribbons were prepared by rapid quenching onto Cu weal technique (tangential velocity of ~30 m/s). Saturation magnetostriction of this ribbon was close to zero [14]. Surface modification of the amorphous ribbon was done in an ultrasonic bath with 5% H_3PO_4 acid concentration during 120 minutes treatment for creation of equidistantly distributed artificial surface defects. As-quenched ribbon was called as AQ, surface modified as SM.

Surface features of the ribbons were studied by the scanning electron microscopy (SEM) using by TM3000 HITACHI instrument. Magnetic measurements at room temperature were done by vibrating sample magnetometer (VSM, Lake Shore 7404). The total impedance (Z) was measured by the four-point technique for different frequencies of 0.5 to 10 MHz and of the driving current peak-to-peak intensity of 5 mA. Giant magnetoimpedance ratio for total impedance was defined as follows: $\Delta Z/Z = 100\times(Z(H)-Z(H_{max}))/Z(H_{max})$, where H_{max}= 200 Oe. For GMI measurements in presence of water-based ferrofluid with MNPs of the ferrofluid 3.5 cm long sample was placed into a plastic tube of 30 mm length and 1 mm in diameter. Ribbon was located in centre of a tube filled with ferrofluid.

3 Results and discussion

Fig. 1 shows XRD spectrum with evidence of amorphous structure of the ribbon (very wide peak near 45 ± 3° for 2θ angles). The SEM micrographs (Fig. 1, insets) show the surface morphology before and after ultrasound treatment in acid revealing sufficiently smooth surface of the bright side of AQ ribbon with typical for rapid quenching technique defects in the direction of the ribbon displacement during the solidification. SM ribbon has surface defects which can be described as round indentations with average diameter of about 150 micrometers with no anisotropy in the shape of the defects or their orientation. Surface defects were equidistantly separated from each other as a

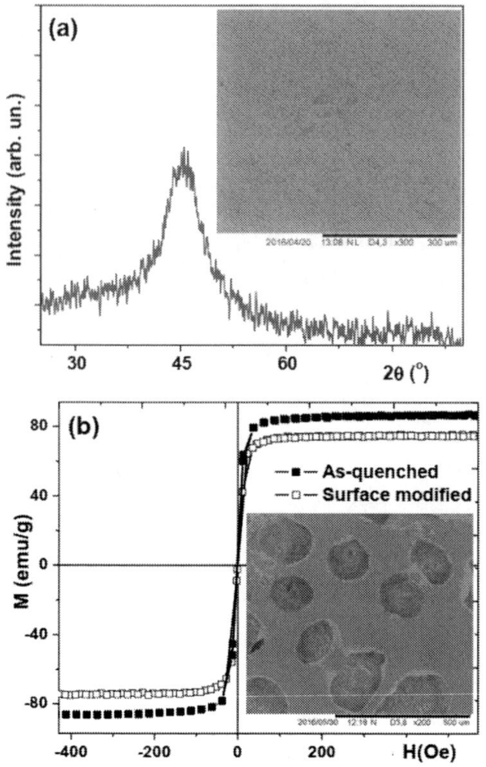

Fig. 1. XRD pattern for $Co_{68.6}Fe_{3.9}Mo_{3.0}Si_{12.0}B_{12.5}$ amorphous ribbon in as-quenched state. Inset shows SEM image of bright side of the AQ ribbon (a). VSM hysteresis loops for AQ and SM states, inset shows SEM image of bright side of the SM ribbon (b).

result of self assembling process and approximately 50% of the surface was covered by the defects. The surface layer of about 2-3 μm was most affected part of the ribbon in the course of the ultrasound treatment. Magnetic characterization for AQ and SM ribbons shows in both cases quite rapid saturation approach and round shoulder shape in the saturation approach indicating mixed anisotropy with domination of the longitudinal one. Part of the contribution (transverse anisotropy component) can come from the surface anisotropy.

Surface modification by the controlled ultrasound treatment results in a slight decrease of saturation magnetization from 87 to 76 emu/g and does not affect the coercivity. Saturation magnetization decrease is consistent with surface modification which includes the removal of the initial surface layer and formation of a passivation layer with lower saturation magnetization, comparing with the material of the AQ ribbon.

XRD spectra of MNPs gave a mean crystallite size for air-dried LTE MNPs of 19 ± 2 nm (log-normal distribution) in a good agreement with TEM data and specific surface area evaluation (Fig. 2). Although the experimental XRD data were well fitted by the magnetite database, it was impossible to distinguish the magnetite and maghemite on the basis of XRD studies solely. The chemical composition of LTE MNPs was determined by the combination of redox titration and the lattice period analysis provided by XRD: it was close to the stoichiometric maghemite ($Fe_{2.72}O_4$). Also we can see that the M_s of the MNPs is about 57 emu/g. This value is lower than for the bulk maghemite as to expect for the MNPs of the observed average size due to nanoscaling effects [12]. At the same time magnetic measurements confirm well the concentration of the MNPs of ferrofluid previously defined by chemical titration technique.

Fig 3. shows frequency dependences of the maximum GMI ratio $(\Delta Z/Z)_{max}$ for AQ and SM ribbons in the presence and absence of FF: until f= 5 MHz, GMI maximum increases and afterwards it changes very little.

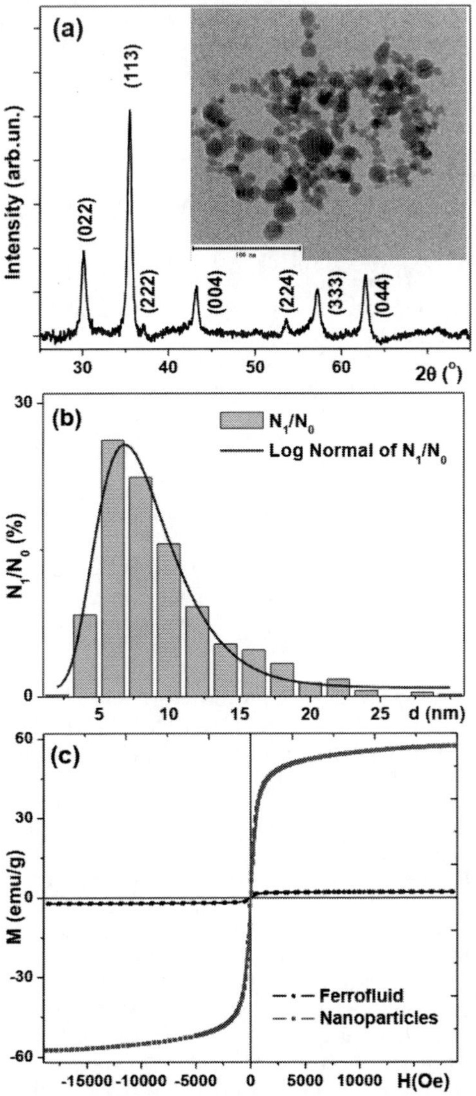

Fig. 2. XRD pattern for LTE MNPs from ferrofluid, inset shows TEM image of LTE MNPs (a). Histogram of particle size distribution and log-normal fits (blue line) (b). Hysteresis loops of as-prepared LTE MNPs and electrostatically stabilized water-based suspension of LTE MNPs (c).

This behavior is consistent with skin depth changes: the condition of strong skin effect appears for the frequencies above 4-5 MHz. In a frequency range 0.5-10 MHz GMI responses of the as-quenched and surface modified ribbons had very similar shapes but the GMI values were slightly higher for AQ ribbons. M(H) hysteresis loops are similar. We therefore assign GMI difference to the surface anisotropy contribution and dynamic magnetic permeability. An increase of the surface roughness and the partial removal of very thin surface layer during ultrasound treatment due to formation of the round defects can be a reason of $(\Delta Z/Z)_{max}$ value decrease. $(\Delta Z/Z)_{max}(f)$ curves were very

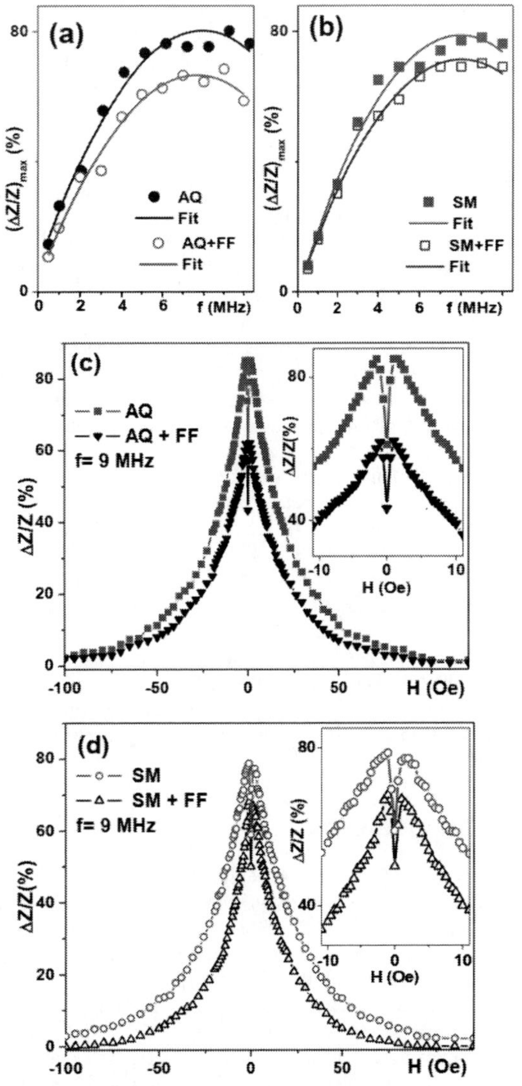

Fig. 3. Frequency dependences of the $(\Delta Z/Z)_{max}$ ratio in the presence and in absence of ferrofluid for as-quenched (a) and surface modified (b) and Field dependences of GMI ratio in the presence and in absence of ferrofluid for as-quenched (c) and surface modified (d) CoFeMoSiB amorphous ribbons.

much affected by the presence of FF in AQ and lesser in SM ribbons. For AQ ribbons covering by the FF resulted in a decrease of GMI ratio in comparison with response without FF. The change in the GMI ratio due to the presence of the MNPs can be explained by the effect of their fringe fields on the superposition of the applied direct current magnetic field and the induced transverse alternating current field created by the driving current.

Field dependences of GMI ratios (Fig.3) shows that the general shapes of the ΔZ/Z(H) in both cases (AQ and SM ribbons) correspond well to the longitudinal effective magnetic anisotropy (one-peak GMI curve) [13-14]. At the same time the appearance of double peak shape related to small contribution of the surface component [8] of transverse magnetic anisotropy is evident in a small field. Increase of the frequency from 3 to 9 MHz results in the change of the shape of ΔZ/Z(H) curves and double peak behaviour become more obvious: dip of the double-peak at zero field increases with the frequency increase. For the frequency of 3 MHz maximum GMI in presence of FF decreases from 65% to 46% and for surface modified ribbon decreases from 53% to 43%, i.e. GMI response of AQ ribbon offers more stable detection opportunities with better noise-to-signal characteristics. For frequency 9 MHz similar conclusion can be made as the maximum GMI in presence of FF for AQ ribbon decreases from 85% to 61% and for SM ribbon decreases from 78% to 68%.

Creation of round defects makes ribbon less sensitive to the presence of FF but the origin of this sensitivity is not trivial. The GMI sensing process counts with the magnetic field of the alternating current passing through the ribbon, the constant external magnetic field applied during GMI measurements for the change of magnetic permeability of the sensitive element, the stray fields created by MNPs and stray fields of round created defects. Surface modification changes demagnetizing fields: defects corresponding to rapid solidification become removed but new defects in the shape of round indentations are formed. The indentations are large, comparing with typical quenching defects but out of the border of the round defect the surface roughness become even smaller. We can suppose that the flux closing by the magnetic chains formed in FF [15] contributes to GMI.

4 Conclusions

Amorphous $Co_{68.6}Fe_{3.9}Mo_{3.0}Si_{12.0}B_{12.5}$ ribbons with good corrosion stability were prepared by rapid quenching technique. Magnetic and GMI properties were studied in as-prepared state and after creation of artificial surface defects by the ultrasound treatment in the acid.

GMI responses in as-quenched and surface modified state in the absence and in presence of LTE MNPs based ferrofluid model the magnetic biosensor functionality for magnetic labels detection. GMI curves were affected by the presence of ferrofluid in as-quenched and much less in the surface modified ribbon cases. Covering by the ferrofluid resulted in a decrease of GMI ratio in all frequency range. For the frequency 3 MHz maximum GMI in presence of FF decreases from 65% to 46% and for surface modified ribbon decreases from 53% to 43%. GMI response of AQ ribbon shows better noise-to-signal characteristics. The change in the GMI ratio with ferrofluid covering can be explained by the effect of fringe fields of MNPs and magnetic flux closure due to the MNPs chains formation. Obtained results can be useful for the development of the magnetic biosensor prototype with cheap disposal amorphous ribbon based GMI sensitive element operating at low frequencies of the order of 5 MHz.

The work was supported in part by the Government of the BC is acknowledged for financial support under the Elkartek Program, the Project Micro4Fab (KK-2016/00030) and the Ministry of Education and Science of the RF, Project N° 055, within the state job 3.6121.2017/8.9. We thank I.V. Beketov, A.I. Medvedev and A. Larranga for special support. Selected measurements were made at SGIKER services of UPV/EHU.

References

1. R.S. Beach and Berkowitz, A.E., *Appl. Phys. Lett.* **64**, 3652–3654 (1994).
2. L.V. Panina and K. Mohri, Sens. Act. A **81**, 71-77 (2000).
3. G.V. Kurlyandskaya and V.I. Levit, Mater. *Sci. Eng.* C **27** 495–503 (2007).
4. A. Antonov, S. Gadetsky, A. Granovsky, A. Diachkov,M. Sedova, N. Perov, T. Furmanova, A.Lagarkov, *Physica* A **241,** 414 (1997).
5. G.V. Kurlyandskaya, E. Fernandez, A.P. Safronov, A.V. Svalov, I.V. Beketov, A. Burgoa Beitia, A. Garcıa-Arribas, F.A. Blyakhman, *Appl. Phys. Lett.* **106**, 193702 (2015).
6. F. Blanc-Beguin, S. Nabily, J. Gieraltowski, A. Turzo, S. Querellou, P.Y. Salaun, *J. Magn. Magn. Mater.* **321**, 192-197 (2009).
7. H. Chiriac, M. Tibu, A.E. Moga, and D. D. Herea, *J. Magn. Magn. Mater.* **293**, 671 (2005).
8. G.V. Kurlyandskaya, V. Fal Miyar, V., A. Saad, E. Asua, and J. Rodriguez, J. *Appl. Phys.* **101(5)**, 054505 (2007).
9. J. Devkota, A. Ruiz, P. Mukherjee, H. Srikanth, and M. H. Phan, *IEEE Trans. Magn.* **49**, 4060 (2013).
10. S. Kumar Manna and V. Srinivas *J. Magn. Magn. Mater.* **418**, 62-67 (2016).
11. A. Garcia-Arribas, M. Goiriena-Goikoetxea, E. Fernández, and J.M. Barandiarán *IEEE Trans. Magn.* **53**, 400310 (2017).
12. J.P. Novoselova, A.P. Safronov, O. Samatov, I.V. Beketov, H. Khurshid, Z. Nemati, H. Srikanth, T. P. Denisova, R. Andrade, and G.V. Kurlyandskaya, *IEEE Trans. Magn.* **50,** 4600504 (2014).
13. L. Kraus, *Sens. Act. A* **106**, 187–194 (2003).
14. Z. Lotfollahi, A. García-Arribas, A. Amirabadizadeh, I. Orue, and G.V. Kurlyandskaya, *J. Alloys and Compounds*, **693**, 767-776 (2017).
15. A. Hubert and R. Schäfer, Magnetic Domains. Berlin, Germany: Springer (1998).

Selected features of morpho-functional reactions of eukaryotic microorganisms grown in the presence of maghemite iron oxide nanoparticles obtained by laser target evaporation

Elena V. Simonova [1,*], *Tatyana P. Denisova* [2], *Roman T. Khandukhanov* [2], *Grigory Yu. Melnikov* [3], *Anatoly I. Medvedev* [3,4], *Igor V. Beketov* [3,4], *Ricardo Andrade* [5]

[1] FSBEI HE Irkutsk State Medical University MOH, Red Revolt str., 1, Irkutsk, 664003, Russia
[2] Irkutsk State University, Nizhnyaya Embankment St., 6, Irkutsk, 664011, Russia
[3] Ural Federal University, Mira St. 19 Ekaterinburg, 620000, Russia
[4] Institute of Electrophysics, Amundsen str., 106, Ekaterinburg, 620016, Russia
[5] Universidad del País Vasco UPV-EHU, Leioa, 48940 Spain

Abstract. Biological activity of iron introduced into nutrient medium as a suspension of iron oxide nanoparticles (MNPs) synthesized by the laser target evaporation was investigated. *Exophiala nigrum* (*E. nigrum*) eukaryotes were grown either in the presence or in absence of MNPs. De-aggregated suspensions of $Fe_{2.75}O_4$ MNPs were added in concentrations of 1 to 10^4 maximum permissive dose (MPD, being 0.3 mg/L of Fe ions in water). Cells were exposed for 24 to 96 hours periods and then plated onto a solid medium. The effect of MNPs was evaluated by the change in the number of cells during exposure and the number and morphology of the colonies. For 1-10 MPD yeast showed unaltered characteristics. For 100 or 1000 MPD for 72 hours of exposure and above the number of cells increases up to 30 times in comparison with the control. A pronounced stimulating effect was revealed at 10^4 MPD of iron. A significant excess of the number of cells was observed for the first day. At exposures of 72 and 96 hours the differences in the number of cells in comparison with the control were 11 and 30 times, respectively. A change in the colonies morphology was observed at 100 MPD concentration.

1 Introduction

Recent focus on wide employment of magnetic nanoparticles (MNPs) in wide spectrum of biomedical applications [1-2] requires the study of the features of their interactions with biocomponents, in particular with eukaryotic organisms. Iron oxides such as magnetite (Fe_3O_4) and maghemite (γ-Fe_2O_3) are well known as materials with a high degree of biocompatibility, a sufficiently high saturation magnetization at room temperature and a high Curie temperature [3]. Physical and chemical properties of their nanosized analogues are less studied, especially in the case of assessments of their cytotoxicity and environmental impact [4-5]. Biological applications require the most complete attestation of MNPs by various methods [6-7], since even the most insignificant changes in the conditions of production can lead to a change in their functional properties. Nanoparticles of iron oxides with very small non-stoichiometry of the composition are able to act as catalysts and introduce additional peculiarities to the morpho-functional reactions of the living systems. In addition, for pharmaceutical products containing nanocomponents, there is a requirement for additional testing of samples of each lot just prior to application of the drug [7]. These circumstances are the reason that the toxic, cytotoxic and genetic effects of MNPs require further study and refinement. The understanding of the interaction between the living system and magnetic nanoparticles and possibility of incorporation of MNPs with reasonably large magnetic moment open new research and clinical opportunities [4,8].

Exophiala nigrum (*E. nigrum*, black unicelular eukaryote yeast) playing an important role in the equilibrium of ecosystem of the Baykal Lake [2] can be used as a model system biological experiments with MNPs.

This study was focused on the morpho-functional reactions of *E. nigrum* eukaryotic microorganisms in response to the effect of$Fe_{2.75}O_4$ iron oxide magnetic nanoparticles in form of electrostatically stabilized water-based suspension.

2 Experimental

For the fabrication of spherical iron oxide magnetic nanoparticles an electrophysical technique of the laser target evaporation (LTE) was employed [2]. Compacted commercial Fe_2O_3 powder target (Alfa Aesar) was used in the LTE installation designed and developed at the

* Corresponding author: denis_tp@inbox.ru

Laboratory of pulsed processes of the Institute of Electro physics UD RAS. The target was overheated by the high-energy pulse of focused laser beam of ytterbium-doped fibre laser (1.07 μm wavelength). More details on LTE technique can be found elsewhere [2,9].

Structure of LTE MNPs was studied by transmission electron microscopy (TEM, JEOL JEM2100 microscope operating at 200 kV) and the X-ray diffraction (XRD) (DISCOVER D8, Bruker diffractometer using Cu-Kα radiation). The stoichiometric ratio of Fe^{2+}/Fe^{3+} in the iron oxide MNPs was determined based on the XRD analysis and redox potentiometric titration using a TitroLine titrator. Suspensions of LTE MNPs in distilled water electrostatically stabilized by the sodium citrate (5 mM) were prepared by ultrasound treatment (ColePalmer CPX750). Hermle Z383 centrifuge operated at 10 000 rpm for 10 min was used to remove large aggregates. Magnetic measurements were carried out on a MPMS XL-7 SQUID-magnetometer.

As a test culture, melanin-containing microscopic fungi *Exophiala nigrum*, was isolated from the natural population of Baikal Lake. Toxic properties of a suspension of iron oxide MNPs synthesized by laser evaporation were studied for *E. nigrum*, grown in the presence or absence of (control samples) MNPs. *E. nigrum* has a black colour due to the produced melanin [10-11]. Primary concentration of MNPs in suspension was 5 % by weight. Suspensions of $Fe_{2.75}O_4$ MNPs solutions for biological experiments were prepared on the basis of Saburo liquid nutrient medium in concentrations 1 - 10^4 MPD. According to the sanitary and hygienic norms, 1 MPD is 0.3 mg/l of Fe ions in water, but the determination of the maximum allowable concentration refers to soluble forms of iron [11].

The control was Saburo liquid medium with the corresponding addition of a sterile physiological solution without MNPs. In all cases, 1 ml of E. nigrum cell suspension was added. Flasks with sowings were placed on a mechanical rocking chair. Exposure times were 24, 48, 72 or 96 hours. At the end of each exposure period, cultures were planted onto a solid nutrient medium (agar Saburo). Considered indicators were: the number of cells at the beginning of the experiment and after exposure (measured in the Goriaev's chamber), the number and morphology of the colonies grown on a solid nutrient medium [12].

For TEM studies of biological samples cell cultures were pre-fixed in 0.5% glutaraldehyde in Sörenson buffer and centrifuged. The supernatant was removed and 2% glutaraldehyde in Sörenson buffer was added to the pellet. Samples were fixed in 1% osmium tetroxide in Sörenson buffer, dehydrated, embedded in EponPolarbed resin and cut in ultrathin sections for TEM studies (Philips EM208S). More details on the technique can be found elsewhere [4].

3 Results and discussion

Figure 1(a) shows a TEM micrograph of the MNPs of iron oxide: the shape of the nanoparticles is close to spherical. For this batch the size distribution calculated

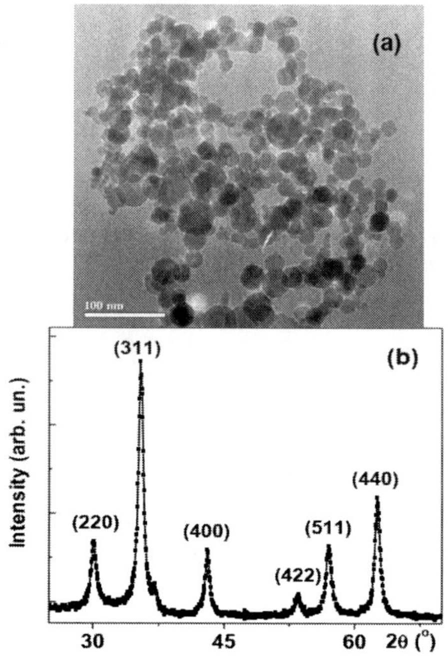

Fig. 1. TEM image of LTE MNPs (a). XRD patterns of iron oxide MNPs. Miller indexes is given in round brackets (b).

from the TEM images of the MNPs was lognormal. Figure 1(b) shows the X-ray spectrum of the MNPs of iron oxide. The experimental points for studied MNPs are fitted well by the magnetite XRD database. However, it is well known that the magnetite (Fe_3O_4) structure can not be distinguished from that of maghemite (γ -Fe_2O_3) based only on XRD data [3]. The crystalline cells of these oxides correspond to the inverse spinel structure (Fd3m space group). In the present study, the stoichiometric ratio Fe^{2+}/Fe^{3+}, determined from the analysis of the interplanar distances of the crystal lattice and the potentiometric titration data, was $Fe_{2.75}O_4$.

The number-average (D_N) and mass median (D_W) diameters [13] were measured using the graphical analysis of TEM images and the statistical sampling of not less than 2000 individual particles. It should be specially noted that the mean values calculated using various techniques (TEM, XRF) correlate very well with each other $D_{XRD} = 23 \pm 3$ nm (for XRD), $D_N = 23 \pm 3$ nm and $D_W = 34 \pm 4$ nm. The size of the coherent scattering region D_{XRD} was determined by the Scherrer method based on the XRD patterns [14].

The hysteresis loop of as-prepared MNPs (not shown here) indicates no saturation at room temperature and very low coercivity: the magnetization in the field of 65 kOe was close to 55 emu/g. The comparative analysis of structural features and magnetic parameters confirms consistency of obtained data about close to superparamegnetic behaviour for MNPs of this size. Reasonably high value of the magnetic moment of each superparamagnetic nanoparticle makes this kind of materials promising for such biomedical applications as magnetic hyperthermia or thermal ablation, magnetic

Table1. Number of cells in 1 ml of test solutions

Exposure, hours	Number of cells in 1 ml, × 10^6					
	0 MPD	1 MPD	10 MPD	100 MPD	10^3 MPD	10^4 MPD
0	2.0	2.2	1.6	2.0	2.6	2.0
24	10.6	14.6	18.0	15.2	13.2	546.0
48	13.4	17.0	13.2	15.2	15.0	584.0
72	10.6	11.4	6.8	30.0	43.0	116.0
96	7.6	15.4	9.4	35.0	230.0	228.0

Field assisted drugs delivery and magnetic biosensing [4,8].

The maximum iron oxide concentration of the electrostatically stabilized aqueous suspensions was 37 g/l with an average hydrodynamic aggregate size of 86 nm and an averaged value of the zeta potential (-42.5) mV. A detailed analysis of the structure of the MNPs and the parameters of aqueous suspensions allows accurate calculation of the iron content in media for the biological part of the studies. But we must mention the existing ambiguity in determining the maximum permissible dose in the case of MNPs. Nanoparticles and, in a broader sense, colloidal particles, strictly speaking, can not be considered a soluble form. The particles are a separate insoluble phase with an interface with the aqueous medium. Although some iron solubility on the boundary of the particle is, but most likely it does not exceed one MPD [15].

Figure 2 shows examples of TEM typical morphology of *Exophiala nigrum* cells grown without and with iron oxide MNPs. The pronounced inclusion of nanoparticles into the cells was not observed. Counting the number of cells in the Goriaev's chamber was carried out in 20 squares, and then the average value of the number of cells in 1 ml was determined from the known formulas [11]. The results are shown in Table 1. It is important to mention that at high iron concentrations of 1000 and 10000 MPDs intensive yeast growth was noted, which resulted in a very high cell density in 1 ml: from 228 to 584 (× 10^6). Figure 3 shows the change in the number of cells in 1 ml of test solutions, depending on the exposure period. Considering the high concentration of cells, the log of cell number per ml is shown on the ordinate axis. Thus, suspensions of iron oxide nanoparticles in the concentrations of 1 or 10 MPD do not have an active stimulating effect for *Exophiala nigrum* yeast cells, in contrast to 100, 1000 and 10000 MPD. It was found that the higher the concentration of MNPs in the suspension of nanoparticles, the stronger the stimulating effect, expressed in the high activity of yeast cell multiplication.

Additions of a suspension in iron concentrations of 100 and 1000 MPD provided by MNPs of the iron oxide starting from 72 hours of exposure resulted in the increase of the number of *Exophiala nigrum* cells from 3-5 to30 times in comparison with the control ($P < 0.05$). For a suspension of MNPs at a concentration of 10^4 MPD of iron a significant excess (52 times) of the number of cells in 1 ml compared with the control was observed in the first day of exposure. At exposures of 72 and 96 hours, the difference with the number of cells in the control variant is 11 and 30 times, respectively ($P < 0.05$).

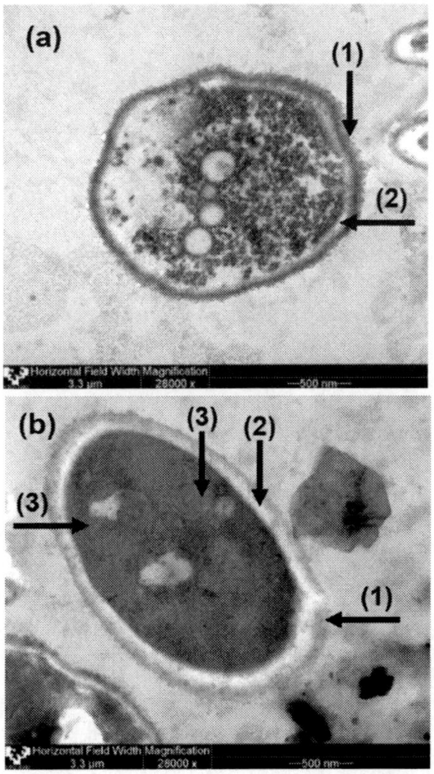

Fig. 2. TEM image *of E. Nigrum* cell grown without MNPs (control) (a) and in the presence of 10^0 MPD of iron provided by citric acid stabilized MNPs from suspension (b). Cell wall (1), a periplasmic space (2) and mitochondria (3) of black yeast.

Naturally, a high density of cells in 1 ml, led to the fact that a large number of colonies grew onto solid substrates in Petri dishes. Moreover, after 48 hours of exposure, counting colonies in Petri dishes proved to make no sense, because the growth of yeast reached the saturation level. Suspensions of MNPs in concentrations of 100 and 1000 MPD are moderately toxic, and at a concentration of 10000 MPC – are toxic.

Interesting results were obtained in an experiment with MNPs suspension in a concentration of 100 MPD for biological experiments. At the beginning of the exposure, as well as after exposures at 24 and 48 hours, all grown up colonies on a solid substrate visually had a typical morphology for *E. nigrum*: black color. After exhibiting the yeast in this suspension for 72 hours, on a solid substrate against a background of a set of typical black colonies of *E. nigrum* unusual single white colonies grew. After the exposure for 96 hours, colonies of white colour predominated on a solid substrate. It should be noted that very similar pattern was observed in all replicates of the same experiment.

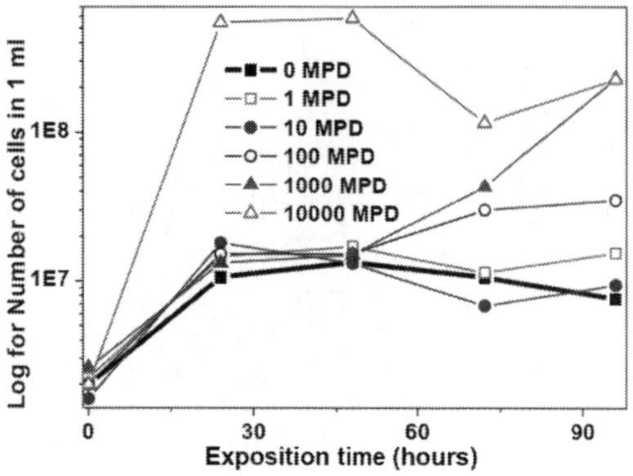

Fig. 3. Change in the concentration of *Exophiala nigrum* yeast cells in 1 ml, depending on the exposure time in different for different iron concentrations measured in maximum permissive doses.

The results obtained in the present work require further development for better understanding of the mechanism of stimulation of cell divisions of *E. nigrum* yeasts and revealing the nature of the heterogeneity of the population according to the morphological criteria induced by the presence of MNPs. We suggest that *E. nigrum* cages contain (or adhere on the surface of the membrain) iron oxide in the form of MNPs [2, 10,16], which makes yeast an attractive model object at the stage of development of suspensions for hyperthermia, thermal ablation and magnetic biosensing.

4 Conclusions

Magnetic maghemite nanoparticles ($Fe_{2.75}O_4$) were synthesized by laser target evaporation. They had an inverse spinel structure. The size of the coherent scattering region D_{XRD} determined by the Scherrer method based on the XRD patterns and the number average diameter calculated from the TEM data correlate well with each other: $D_{XRD} = 23 \pm 3$ nm and $D_N = 23 \pm 3$ nm. The synthesized MNPs were used to produce electrostatically stabilized aqueous suspensions with 5 mM sodium citrate. The maximum iron oxide concentration of was 37 g/l with an average hydrodynamic aggregate size of 86 nm and averaged value of the zeta potential (-42.5) mV.

Exophiala nigrum eukaryotes were grown either in the presence or in absence of MNPs. Strain R-11 *Exophiala nigrum* was isolated from a sample of Baikal water taken from the surface of the layer at a distance of 5 m from the shore in the Baikalsk area. De-aggregated suspensions of $Fe_{2.75}O_4$ MNPs were added in concentrations of 1 to 10^4 maximum permissive dose. Cells were exposed for 24 to 96 hours periods and then plated onto a solid medium. Suspensions of MNPs in the concentration range of 1 MPD to 10 MPC did not induce toxic effects. Additions of a MNPs suspension in iron concentrations of 100 and 1000 MPD starting from 72 hours of exposure caused a stimulating effect.

Very pronounced stimulating effect was revealed for a suspension of MNPs at a concentration of 10000 MPD of iron. A significant excess of the number of cells in 1 ml compared with the control was manifested in the first day of exposure. At exposures of 72 and 96 hours, the difference with the number of cells in the control was 11 and 30 times, respectively ($P < 0.05$). Suspensions of nanoparticles in concentrations of 100 MPC and 10^3 MPD are moderately toxic, but at a 10^4 MPD they become toxic.

The work was supported by MEC MAT2014-55049-C2-1-R, state work № 0389-2014-0002 and RFBR №16-34-50192 grants. We thank A.P. Safronov, Iu. Novoselova, A.M. Murzakaev and I.S. Tyukova for special support. Selected measurements were made at SGIKER services UPV/EHU.

References

1. H. Matusiewicz, Acta Biomater.**10**, 2379 (2014).
2. J.P. Novoselova, A.P. Safronov, O. Samatov, I.V. Beketov, H. Khurshid, Z. Nemati, H. Srikanth, T. P. Denisova, R. Andrade, and G.V. Kurlyandskaya, IEEE Trans. Magn. **50,** 4600504 (2014).
3. R. C. O'Handley, Modern Magnetic Materials (John Wiley & Sons, New York, 1972) p. 740.
4. G.V. Kurlyandskaya, Iu.P. Novoselova, V.V. Schupletsova, R. Andrade, N.A. Dunec, L.S. Litvinova, A.P. Safronov, K.A.Yurova, N.A. Kulesh, A.N. Dzyuman, and I.A. Khlusov **431**, 249 (2017).
5. I. S. Lyubutin, C. R. Lin, Yu. V. Korzhetskiy, T. V. Dmitrieva, R. K. Chiang, J. Appl. Phys. **106**, 034311 (2009)
6. Y. Hwang, J.-K. Lee, Y.-M. Jeong, S. Cheong, Y.-Ch. Ahn, and S. H. Kim, Powder Technology **186**, 145 (2008).
7. J. H. Grossman and S. E. McNeil, Physics Today, **38** (August, 2012).
8. G.V. Kurlyandskaya, E. Fernandez, A.P. Safronov, A.V. Svalov, I.Beketov, A. Burgoa Beitia, A. Garcıa-Arribas, F.A. Blyakhman, Appl. Phys. Lett. **106** 193702 (2015).
9. V.V. Osipov, V.V. Platonov, M.A. Uimin, and A.V. Podkin, Tech. Phys**. 57,** 543 (2012).
10. I.P. Babieva and I.Yu. Chernov, *Biology of yeasts* (M.V. Lomonosov MSU, Moscow, 2004).
11. A.I.Netrusov, M. A. Egorova, L. M. Zakharchuk et al. Workshop on a microbiology (Moscow, 2005).
12. N. V. Abramov, S. P. Turanska, A. P. Kusyak, A. L. Petranovska, P. P. Gorbyk, J. Nanostruct. Chem. **6**, 223 (2016).
13. I.V. Beketov, A.P. Safronov, A.I. Medvedev, A. I. Medvedev, J. Alonso, G. V. Kurlyandskaya, and S. M. Bhagat AIP Adv. **2,** 022154 (2012).
14. P. Scherrer, Nachr Ges Wiss G¨ottingen **26**, 98 (1918).
15. A.P. Safronov, private communication at Ural federal university (2017).
16. N.A. Kulesh, I.P. Novoselova, A.P. Safronov, I.V. Beketov, O.M. Samatov, G.V. Kurlyandskaya, M.V. Morozova, and T.P. Denisova, J. Magn. Magn. Mater. **415**, 39 (2016).

MODELLING OF THIN FILM MAGNETOIMPEDANCE SENSITIVE ELEMENT DESIGNED FOR BIODETECTION

Volchkov S.O.[1], Chlenova A.A.[1], Lepalovskij V.N.[1]

[1]Ural Federal University, Laboratory of Magnetic Sensors, Ekaterinburg, Russia

Abstract. Magnetic soft matter (ferrofluids or ferrogels) is one of the rapidly growing areas of research and applications including magnetic biosensing. Giant magnetoimpedance is the effect with proven capacity to magnetic label detection. In this work, we describe a universal model to simulate conditions of magnetic biodetection and to check its validity with giant magnetoimpedance sensitive element based on magnetic multilayer. Finite element method allows calculations of high-frequency current distribution using the Maxwell's equations taking into account the magnetodynamics of iron oxide water-based ferrofluid in small channels similar to the blood vessels. The modelling was realized with the licensed software Comsol©. The calculations were performed on a specialized engineering server based on four processors Intel Xeon E5 and 124 Gb RAM, adapted for parallel computations and suitable for description of individual layers with nanometer dimensions for the number of elements in the mesh structure above 10^6 cells. The designed model allows calculations of the current density, the outside magnetic flux, resistivity, etc. for each one of the created cells and total values by integration of sub-domains. One can quantitatively describe concentration of ferrofluid, velocity and pressure in the blood vessel. These changes affecting on the giant magnetoimpedance of the FeNi-based multilayer were both calculated and measured.

1 Introduction

Magnetic drug targeting (MDT) provide special opportunity to treat malignant tumours locally without systemic toxicity. An external magnetic field is focused on a particular region of interest to retain magnetic nanoparticles (MNPs) in form of water-based ferrofluids (FF) from the blood flow. MNPs can carry the chemotherapeutic agents to tumours for specific delivery. It is possible to use MNPs as a “carrier system” for a variety of anticancer agents (radionuclides, specific antibodies and genes) [1-2]. MDT is a complex process, where several phenomena concur: the pulsating blood flow that conveys the FF mix, its action on the vessels and muscular volume, and the interaction with an external magnetic field (due to the magnetization of the MNPs) aimed at prolonging the time of residence of the magnetizable drug carriers conveyed by the blood stream in the region of interest. Giant magnetoimpedance (GMI) is the effect which with proven capacity to quantify the presence of MNPs in the biological samples from model ferrofluids, ferrogels to in vitro grown cell cultures [3-4].

In this work we design a universal model to simulate conditions of magnetic biodetection in GMI FeNi/Cu/FeNi multilayers forming part of the “microstrip” transmission line with predetermined parameters, where impedance parameters are controlled by an external magnetic field. To confirm the model predictions, the real FeNi/Cu/FeNi GMI elements have been obtained and their impedance was measured in the installation mimicking the blood vessel conditions.

2 Modelling

The modelling and simulations were performed by finite element method (FEM) on licensed Comsol Multi-physics software version 5.0 The hardware implementation was realized with computer station having four CPU Intel Xeon E5, RAM 128 Gb and shared-memory parallelism architecture on Ubuntu 14.04 (Linux). The topology of the magnetic multilayered structure was developed in our previous experimental studies [5-7]. The GMI elements, the heterogeneous structures consisting of alternating layers of copper and permalloy onto the glass substrate: glass/ $[FeNi/Cu\_s]_4/FeNi/Cu\_c/[FeNi/Cu\_s]_4/FeNi$ (Fig. 1) were deposited by magnetron sputtering. The variation parameters were length, width and thickness of FeNi, Cu_c (copper central layer) and Cu_s (copper intermediate layers). The topological model of the multilayer was connected with the architecture model of the “microstrip” transmission line with SMA-adapters for calculations.

The topological model of “microstrip” line and FeNi-element was divided into sub-areas, which are tetrahedrons for 3D-configuration (Fig. 2(A)). The size of finite elements in the mesh partition depended on the wavelength of the electromagnetic excitation wave.

* Corresponding author: stanislav.volchkov@urfu.ru

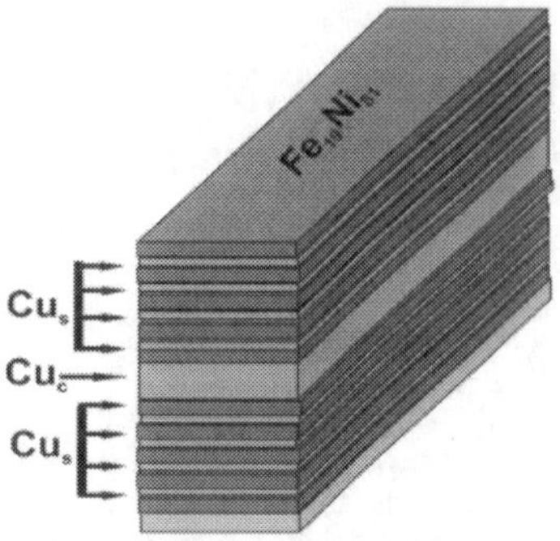

Figure 1. The topological model of the ferromagnetic multilayered film GMI element deposited onto glass substrate

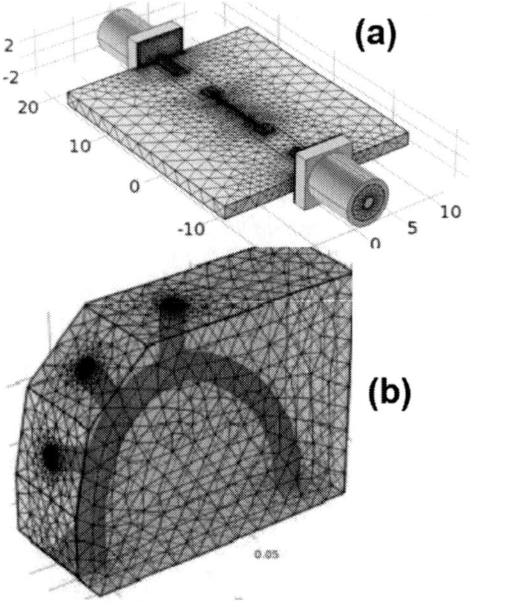

Figure 2. The mesh model of the GMI sensitive element placed in the microstip line with SMA adapters (a), blood vessel related part of the model (b).

A separate sub-grid of hexahedral elements was set for the boundary of the magnetic layer. This model used a transition boundary condition (TBC) programming function because of the need to incorporate the interface related to diffusion (1 nm) between the FeNi and Cu, layers and glass substrate as well as the propagation of electromagnetic waves for the layers with different dielectric and magnetic properties. For the numerical solution of FEM models we used frequency-dependent form of Maxwell's equations for waves with transmission electromagnetic mode (TEM) [7]:

$$\nabla \times \mu^{-1}(\nabla \times \vec{E}) - k_0(\varepsilon - \frac{j\sigma}{\omega\varepsilon_0})\vec{E} = 0 \quad , \quad (1)$$

where E - electric field vector, μ - relative permeability tensor, ω is the angular frequency, σ is the conductivity tensor, ε is the relative permittivity tensor, and k_0 is the free space wave number. Microwave losses in the sample are introduced as permittivity and permeability anisotropic tensors. The model assumes that the static magnetic bias field (H_0) is much stronger than the alternating magnetic field of the microwaves. So the quoted expressions are a linearization for a small-signal analysis around this operating point. Following parameters of the materials have been used in the model: ε_{Sub} = 3.38 - dielectric permittivity of sample holder; ε_{SMA} = 2.10- dielectric permittivity of polytetrafluoroethylene in SMA-adapter; ε_{GS} = 4.80 - dielectric permittivity of glass substrate, σ_{hold}= 5.998 × 10^7 (S/m) -electrical conductivity of Cu in SMA-adapter and "microstrip" line roads; σ_{Cu}= 2.5 × 10^7 (S/m) electrical conductivity of Cus; σ_{FeNi}=6.6 × 10^6 (S/m) electrical conductivity of FeNi. Remaining volume in the model is air. An important step in the design of this microwave device is matching the characteristic impedance of total circuit (transmission line and sensitive element) to 50 Ohm through the parameters of variation. Impedance matching is equivalent to minimizing the reflections back to the input port [9].

Fig. 2(B) shows an example of the modelling of a portion of vascular system (the upper part of the aorta). Both aorta and its ramified blood vessels are embedded into a biological matrix (cardiac muscle). The following material properties were used: Blood - density = 1060 kg/m³ - dynamic viscosity = 0.005 Ns/m²; Artery - density is equal to 960 kg/m³ - Neo-Hookean hyperelastic behavior: the coefficient μ is equal to 6.20 × 10^6 N/m², while the bulk modulus is equal to 20μ and corresponds to a value for Poisson's ratio (v = 0.45). An equivalent elastic modulus equals 1.0 × 10^7 N/m². Cardiac muscle - density = 1200 kg/m³ - Neo-Hookean hyperelastic behavior: the coefficient μ equals 7.20 × 10^6 N/m², while the bulk modulus is equal to 20μ and corresponds to a value for Poisson's ratio (v = 0.45). An equivalent elastic modulus equals 1.16 × 10^6 N/m².

3 Experimental

The experimental multilayers glass/[FeNi(100 nm)/Cus(3 nm)]$_4$/FeNi(100 nm)/Cuc/[FeNi(100 nm)/ Cus(3 nm)]$_4$/FeNi(100 nm) were designed after simulations. They were deposited by magnetron sputtering in a magnetic field of 100 Oe onto 0.2 mm thick Corning glass 2875-25 substrates with dielectric permittivity ε=4.8. The composition of magnetic layer was $Fe_{19}Ni_{81}$. The formation of the GMI elements was carried out by metallic masks. Magneto-optical Kerr effect measurements confirmed formation of the well defined uniaxial magnetic anisotropy with easy magnetization axis oriented in plane of the elongated stripe sample along the short side, i.e. in the direction of the application of a magnetic field during the sample deposition. Small coercivity of about 0.5 Oe and anisotropy field of about 6 Oe – are typical features of soft ferromagnets. The impedance parameters were measured in 50 Ohm "microstrip" line by Agilent Impedance Analyzer follow standard procedure [9].

Table 1. Types of GMI experimental studies

Name	Type of GMI measurements
S1	Film
S2	Film with water capillary
S3	Film with FF capillary (concentration 0.02% = 1 g/l)
S4	Film with FF capillary (concentration 0.04% = 2 g/l)

For experimental studies of the FF dynamic behaviour the capillary was a tube of silicone with an outer diameter of 5 mm and an inner diameter of 3 mm of 1 m length. A motor working with of 4.5 V voltage pushed liquid through the capillary mechanically. The capillary was located perpendicular to the long side of the film parallel to the film plane (in the centre) at a minimum distance above the film. It was fixed without direct contact and therefore did not press the GMI element surface. Additional study confirmed that the operation of the motor does not affect the created external magnetic field features: both GMI curves for motor switch on and switch of curves were identical with an error of 1% for the GMI ratio of the total impedance. Series of the types of experimental studies are described in Table 1.

Commercial EMG 807 ferrofluid was de-aggregated for 10 min. for obtaining the desired concentrations in distilled water using ultrasonic bath. Initial concentration of the nanoscale iron oxide was 2 volume % [10].

4 Result and Discussion

The pressure values were defined as the mean values over a heart beating cycle with in substitution the blood sample by diluted FF. During a cycle the pressure varies between a minimum and a maximum values which are calculated using a relative amplitude. For the time-dependent analysis with FEM simulations, a simple trigonometric function is used for varying the pressure distribution and velocity over time (Fig 3).

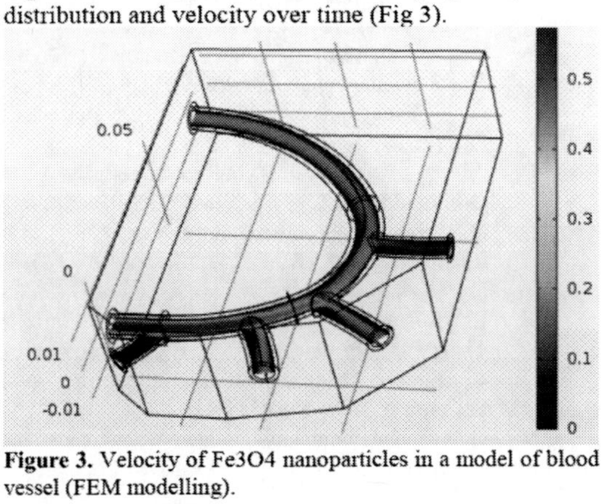

Figure 3. Velocity of Fe3O4 nanoparticles in a model of blood vessel (FEM modelling).

The transport and capture of MNPs in the channel under application of a magnetic field with different strength were calculated using the particle tracing module of finite element based COMSOL software. The dominant magnetic and fluidic drag forces are considered to predict the transport and capture of magnetic nanoparticles in a channel. Moreover, the diameter of tube and radius of MNPs are important parameters included in the model.

Fig. 5 shows FEM modelling results corresponding to different types of GMI experimental studies (S1-S4). One can see a good agreement for the frequency range below 180 MHz predicting the same GMI response for capillary with and without water. The highest difference between S1-S2 and S3, S4 responses was observed in a low frequency range below 110 MHz. Following tendency was clearly observed: the higher the MNPs concentration, the lower the total impedance value for all frequencies under consideration.

Figure 4. Magnetic flux distribution in a model system consisting of a blood vessel with FF and small magnet placed at a distance: thin film GMI element should be located in the gap between the magnet and the capillary.

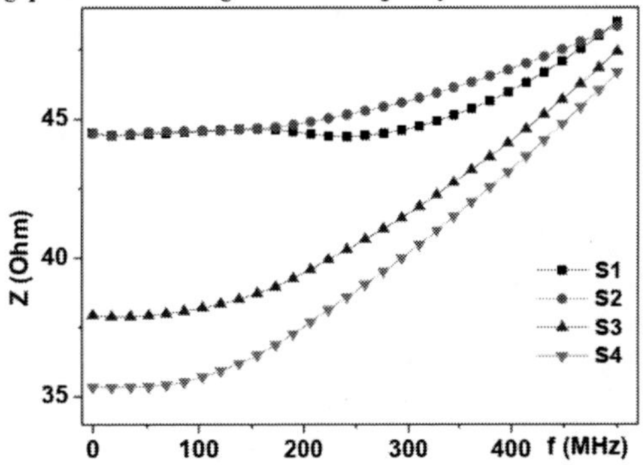

Figure 5. The modeling of frequency dependence of MI ratio of total impedance (a) and real part (b) for structure S1-S4

Let us now analyze experimental results. Fig. 6 shows the frequency dependences of the maximum of the total impedance Z for all types of the measurements. In the case of Fig.6(a) Z(f) dependences are represented with the type of the measurements as a parameter and in the case of Fig.6(b) Z(Sx) are given for selected frequencies. Generally speaking, in a frequency range of 80 to 400 MHz we observe the same tendency as the tendency observed in the case of the modelling (Fig. 5): the higher the MNPs concentration, the lower the maximum value of the total impedance.

We must mention that the experimental values of the total impedance were approximately 2-3 times lower in comparison with the obtained by the modelling. One can clearly see that the deviation comes from the modelling of thin film GMI structure. As it was discussed before

[11] thick cooper conductor may cause formation a thick interface decreasing an effective conductivity and

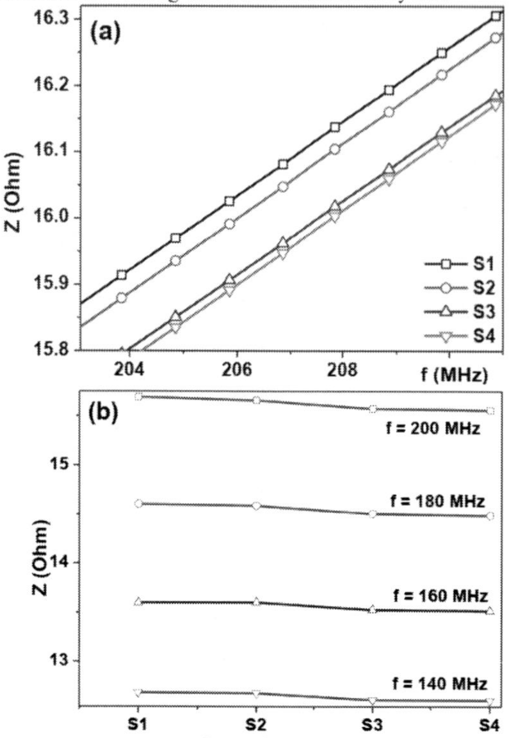

Figure 6. The experimental frequency dependence of the maximum of the total impedance Z of S1-S4 types of measurements. The summary of Z(Sx) given for selected frequencies (see also Table 1).

changing the current distribution. Although improvements of the GMI model are requested as the next step of the future research the general tendencies observed for the measurements of S1-S4 types shows reasonable agreement between the model and experimental results.

5 Conclusions

Universal model to simulate conditions of magnetic biodetection and to check its validity with GMI sensitive element based on magnetic multilayer was developed and tested. FEM allowed calculations of high-frequency current distribution using the Maxwell's equations taking into account the magnetodynamics of iron oxide water-based ferrofluid in small silicon channels mimicking the blood vessels. The calculations were performed on a specialized engineering server based suitable for description of individual layers of the GMI multilayered structure with nanometer dimensions for the number of elements in the mesh structure above 10^6 cells. Glass/[FeNi(100 nm)/Cu_s(3 nm)]$_4$/FeNi(100 nm)/Cu_c/[FeNi(100 nm)/Cu_s(3 nm)]$_4$/FeNi(100 nm) GMI multilayers were deposited by the magnetron sputtering. The designed model allowed calculations of the current density, the outside magnetic flux and resistivity for each one of the created cells and total values by integration of sub-domains. We were able to quantitatively describe behaviour of ferrofluids of different concentrations, flowing with certain velocity and under certain pressure in the blood vessel in the tube mimicking a capillary. These changes affecting the GMI of the FeNi-based multilayer were both calculated and measured showing the same tendency: the higher the MNPs concentration, the lower the total impedance value for all frequencies under consideration.

Acknowledgments

The research was supported by the Ministry of Education and Science of the Russian Federation Agreement no. 02.A03.21.0006. The some results were obtained within the framework of the state task of the Ministry of Education and Science of Russia 3.6121.2017/8.9. We thank Prof. F.A. Blyakhman for special support.

References

1. J. Llandro, J.J. Palfreyman, A. Ionescu, C.H.W. Barnes, Med. Biol. Eng. Comput. **48,** 977 (2010)
2. G.V. Kurlyandskaya, Iu.P. Novoselova, V.V. Schupletsova, R. Andrade, N.A. Dunec, L.S. Litvinova, A.P. Safronov, K.A. Yurova, N.A. Kulesh, A.N. Dzyuman, I.A. Khlusov, J. Magn. Magn. Mater. **431**, 249 (2017)
3. G.V. Kurlyandskaya, E. Fernandez, A.P. Safronov, A.V. Svalov, I.V. Beketov, A. Burgoa Beitia, A. Garcıa-Arribas, F.A. Blyakhman, Appl. Phys. Lett. **106,** 193702 (2015)
4. F. Blanc-Beguin, S. Nabily, J. Gieraltowski, A. Turzo, S. Querellou, P.Y. Salaun, J. Magn. Magn. Mater. **321**, 192 (2009)
5. G. V. Kurlyandskaya, A. García-Arribas, E. Fernández, and A. V. Svalov, IEEE Trans. Magn. **48(4)**, 1375 (2012).
6. A. Yuvchenko, V. N. Lepalovskii, V. O. Vas'kovskii, A. P. Safronov, S. O. Volchkov, G. V. Kurlyandskaya Technical Physics **59 (2)**, 230 (2014)
7. A. S. Antonov, S. N. Gadetskii, A. B. Granovskii, A.L. D'achkov, V.P. Paramonov, N.S. Perov, A.F. Prokoshin, N.A. Usov, A.N. Lagar'kov, Fiz. Met. Metalloved. **83 (6)**, 60 (1997).
8. L. D. Landau and E. M. Lifshitz, Course of Theoretical Physics, Vol. 8: Electrodynamics of Continuous Media (Nauka, Moscow, 1982; Pergamon, New York, 1984).
9. G.V. Kurlyandskaya, D. de Cos, S.O. Volchkov, Russ. J. Nondestruct. Test. **4,** 377 (2009)
10. https://ferrofluid.ferrotec.com
11. G.V. Kurlyandskaya, A.A. Chlenova, E. Fernández, K.J. Lodewijk J. Magn. Magn. Mater. **383**, 220 (2015)

Volumetric registration of magnetic nanoparticles for optimization of quantitative immunochromatographic assays for detection of small molecules

Natalia V. Guteneva[1,2,*], *Sergey* L. Znoyko[1], *Alexey* V. Orlov[1,2], *Maxim* P. Nikitin[2], and *Petr* I. Nikitin[1,3]

[1]Prokhorov General Physics Institute, Russian Academy of Sciences, 38 Vavilov St, 119991 Moscow, Russia
[2]Moscow Institute of Physics and Technology (State University), 9 Institutskii per., 141700 Dolgoprudny, Russia
[3]National Research Nuclear University MEPhI (Moscow Engineering Physics Institute), 115409 Moscow, Russia

Abstract. Precise quantitative and highly sensitive detection of small molecules (haptens) is highly demanded in medicine, food quality control, *in vitro* diagnostics, criminalistics, environmental monitoring, etc. In the present work, the magnetic method of particle quantification and the optical methods of spectral correlation and spectral phase interferometry complement each other for optimization of a quantitative assay for measuring concentrations of small molecules. The assay employs magnetic nanoparticles as labels in rapid immunochromatographic format. The approach was demonstrated with fluorescein as a model molecule. The interferometric label-free biosensors were employed for selection of optimal reagents that produced high specificity and sensitivity. The method of magnetic particle quantification counted the magnetic labels over the entire volume of the immunochromatographic membrane to provide their distribution along the test strip. Such distribution was used for optimization of such parameters as concentrations of the used reagents and of antibody immobilized on the labels, amount of the labels and conjugates of haptens with protein carriers to realize the advanced quantitative immunochromatographic assay.

1 Introduction

Small molecules (haptens), i.e., toxins, hormones, drugs, antibiotics and vitamins exhibit biological activity even in extremely low concentrations. Development of methods for ultrasensitive quantitative detection of the small molecules is among the high-priority research tasks in many fields of science and industry [1, 2].

Liquid chromatography and mass-spectrometry traditionally used for detection of small molecules [3, 4] offer very high sensitivity and specificity. However, long sample preparation, expensive equipment and reagents significantly limit their use.

Immunological methods are simpler and more cost-effective. This group includes sensitive and affordable label–free biosensors that employ microscope cover slips as single-used sensor chips [5], fluorescence polarization [6], lateral flow (LF) tests based on optical labels [7], smart polymer-based bioanalysis [8], etc. A more complicated enzyme-linked immunosorbent assay (ELISA) is the most commonly used test considered as the "gold standard" of immunological methods [9]. The quantitative ELISA-type methods for measuring hapten concentrations are relatively long (2-24 h) and labor-consuming, require re-calibrating for each measurement using standard probes. Some of the mentioned drawbacks can be overcome using the immunochromatographic (IC) approach realized with nitrocellulose LF test strips. The approach is simple, rapid, cost-effective, does not require specially trained personnel. The existing LF methods for hapten detection employ visual or instrumental registration of various optical labels such as quantum dots, fluorescent or color markers, and gold nanoparticles, which are the most popular [7]. Although numerous studies have reported quantitative readout of gold labels on the LF strips by different electronic detectors or scanners, in the majority of countries, such assays are still approved as the threshold "yes/no" tests only.

Magnetic nanoparticles (MP) employed as the labels instead of the optical markers offer substantial advantages because they can be counted on flat surfaces by magnetic field sensors built in microfluidic chips [10] or from the entire volume of 3D porous structures used as the solid phase. The latter solution was successfully applied for multiplex measuring in a wide dynamic range of concentrations of protein molecules [11].

Here we propose several quantitative measuring techniques as candidates for development of an analytical platform for hapten detection with magnetic labels in the competitive format. As small molecules possess only one antigen determinant, competitive binding of antibody (Ab) with either free hapten in the test sample or that conjugated with protein deposited on the IC membrane should be recorded. Therefore, the reproducibility of such assay strongly depends on

* Corresponding author: nguteneva@gmail.com

accessibility of the hapten antigen determinant in the conjugate, which varies according to the ways of formation. For quantitative monitoring of reproducibility of these conjugates and kinetics of binding with different Ab, the original optical biosensors have been modified. The fluorescein was used as a model small molecule for optimization of conjugate design for further application of the results for development of magnetic ultra-sensitive and quantitative LF assays.

2 Experimental

2.1 Reagents

The following reagents were used in the experiments: mouse monoclonal antibody against fluorescein (clone FCII) were purchased from the Russian Research Center of Molecular Diagnostic and Therapy (Moscow, Russia); fluorescein isothiocyanate (FITC) was purchased from Thermo Fisher Scientific; bovine serum albumin (BSA), phosphate buffered saline (PBS) tablets were purchased from Sigma-Aldrich (Germany); 198-nm Bio-Estapor microspheres were purchased from Estapor – Merck Millipore (Germany); nitrocellulose membrane NC-140 (255 μm thick and 100 μm backing) was kindly donated by Sartorius AG (Germany); absorbent sinks/wicking pads were purchased from Ahlstrom CytoSep (Finland); backing cards were obtained from Lohmann (USA); N-(3 - dimethylaminopropyl) - N′ - ethylcarbodiimide hydrochloride (EDC) was from (Fluka, Switzerland); deionized water, MilliQ grade, was from Millipore Corp. (USA). All other chemicals were of analytical grade.

2.2 Magnetic particle quantification

Magnetic nanolabels were counted along the LF strips by original readers based on the magnetic particle quantification (MPQ) method [11]. The MPQ employs nonlinear magnetization of MP by a magnetic field at two frequencies and records the response at combinatorial frequencies. The method is insensitive to linear dia- and paramagnetic materials such as glass, water, etc. The MPQ readers modified for these studies feature improved spatial resolution for various biophysical applications including precise quantification of MP along the LF strips. They have the limit of detection of 0.4 ng or 60 zmol of MP within the volume up to 0.2 mL and the 7-order linear dynamic range.

2.3 Immunochromatographic assay procedure

Standard IC test strips were used, each being composed of overlapping sample pad, nitrocellulose and absorbent membranes assembled on an adhesive plastic backing sheet (Fig. 1). The Sartorious UniSart Cn140 membrane of 255 μm thick characterized by the manufacturer was composed of cellulose nitrate polymers and exhibited capillary speed of 140 s/40 mm.

The conjugate of BSA with FITC at concentration of 2.5 mg/ml and goat Ab against mouse IgG (GAM Ab) at concentration of 1 mg/ml were used to form on the nitrocellulose membrane the test (TL) and control (CL) lines, respectively. Then conjugates of MP with different amounts of antibody were prepared according to the procedure described in [12]. The superparamagnetic 198-nm Bio-Estapor carboxyl-modified spheres were used as the detectable labels. These magnetic nanoparticles conjugated with antibody migrate well through the membrane delivering high and specific signal. The conjugates were added to the samples containing a known amount of fluorescein used in this study as a model and incubated for 5 min. The main parameters of the assay such as reaction time, amount of reagents and sample volume were optimized using the MPQ readers toward higher sensitivity.

The IC test strips were placed vertically into the prepared solutions for 20 min. Then the test strips were passed through the measuring coil of the MPQ reader to measure the signals. All the experiments were implemented at least thrice and statistically processes. In graphs, each value represents an average, and error bars show standard deviations.

2.4 Label-free monitoring of molecular bindings

For development of the sensitive and rapid assays, we carried out selection of optimal antibodies and conjugates of fluorescein with BSA by their binding kinetic characterization using modified label-free biosensors based on spectral-phase interferometry (SPI) [13] and spectral-correlation interferometry (SCI) [14] that provide real-time monitoring of thickness of a layer of biomolecular complexes bound on the sensor chip surface due to biochemical reactions. The techniques use microscopic cover glass slips as two-beam reflective interferometers and affordable sensor chips simultaneously. These biosensors showed high efficiency for optimization of Ab immobilization interfaces and development of protocols for sandwich magnetic immunoassays [14].

3 Results and discussion

The single-channel SPI and 3-channel SCI biosensors were used for real-time monitoring of immunochemical reactions of the reagents, as well as to control reproducibility of MP functionalization, of accessibility of antigen determinants in synthesized conjugates, etc.

The carboxylated glass cover slip was activated with EDC solution, and then FITC/BSA was covalently immobilized to it. To block the unreacted carboxyl groups on the surface, 1% glycine solution was pumped along the glass cover slip installed in the biosensor flow system followed by pumping of Ab solution. The examples of recorded sensograms, used to control reproducibility of antigen determinants on the conjugates and specific binding, are shown in Fig. 2. The sensograms were obtained using the label-free biosensors based on the interferometric techniques [13,14] briefly described in the "Experimental" section.

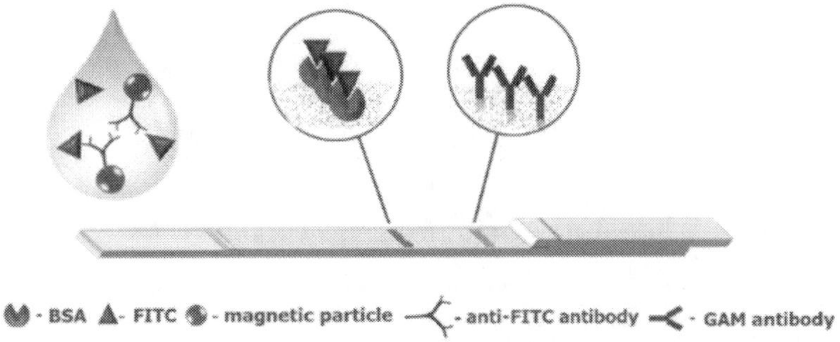

Fig. 1. Scheme of the competitive immunochromatographic assay procedure. Goat anti mouse antibody (GAM) is deposited onto the control line, the BSA/FITC conjugate deposited onto the test line competes with free antigen in sample for the limited amount of anti-FITC antibody binding sites.

The sensograms show the temporal dependence of Δd, which is a change in thickness of the biomolecular layer on the sensor chip surface averaged over the measuring area. Two test samples were used: i) without haptens so that free Ab binds to the conjugate and ii) with haptens that binds to all Ab in solution preventing binding with the conjugate. The same experiment was repeated for different concentrations of antibody solutions and immobilized FITC/BSA conjugates.

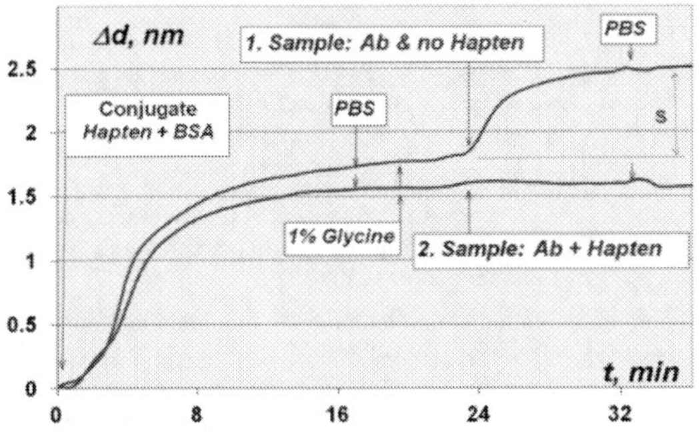

Fig. 2. Sensograms of competitive label-free assay in the presence and absence of haptens.

The observed difference S in the signal is traditionally used as assay result. It depends on availability and accessibility of antigen determinants on the conjugate. To get the results faster, the Ab were selected, which demonstrated the steepest slope of the sensograms.

New efficient quantitative methods of assay optimization have been proposed based on the developed technique of magnetic detection. One of them is based on local counting of MP over the entire depth of the strip and mapping of MP distribution (Fig. 3) along all the constituent components of lateral flow test strips (not merely on the membrane surface of TL and CL as it is typically done with optical labels). The MP distribution was used for optimization of assay parameters such as Ab immobilized on MP, pH of buffers, sample volume, amounts of MP and conjugates deposited onto the test lines to achieve highest sensitivity. According to the experiments, the optimal values were: pH - 7.5, MP amount - 3 μg, and sample volume - 75 μl.

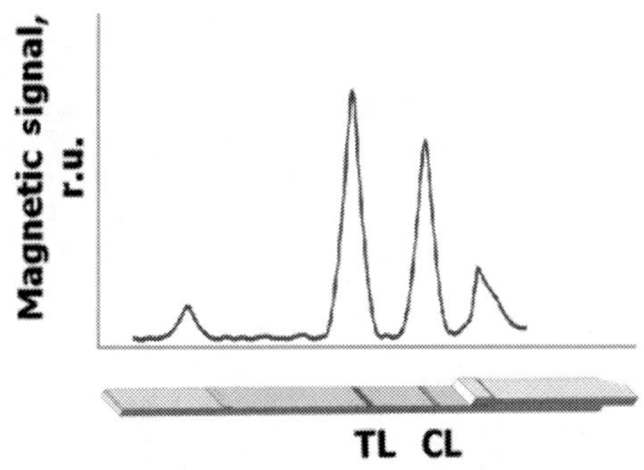

Fig. 3. Distribution of magnetic nanolabels along the test strip measured by the MPQ reader. The signal is proportional to the local amount of magnetic nanoparticles.

The optimization procedure is illustrated here for amount of Ab immobilized on MP, which is among the most important parameters that may strongly influence the assay sensitivity. As small molecules are detected here in the competitive format, if excessive amount of Ab is present in the test sample, then the antigen concentration does not practically affect the magnetic signal because MP bind to the test line anyway. On the contrary, insufficient amount of Ab conjugated with MP yields a weak signal on the test line due to low-effective binding of MP, and this significantly reduces the assay sensitivity. The examples of dependences of magnetic signal upon amount of Ab conjugated with MP obtained with the proposed method are shown in Fig. 4.

One may see from the Fig. 4 that the best ratio of the signal in presence and absence of fluorescein corresponds to 3 μg of Ab immobilized on MP. As expected, in case of insufficient amount of Ab on MP (0.3 μg), the magnetic signal registered on the TL is too weak for sensitive detection of antigen. Otherwise, when large amount of Ab on MP (30 μg) is used, the magnetic signals in presence and absence of fluorescein are almost equal, and that hinders detection of the small molecules.

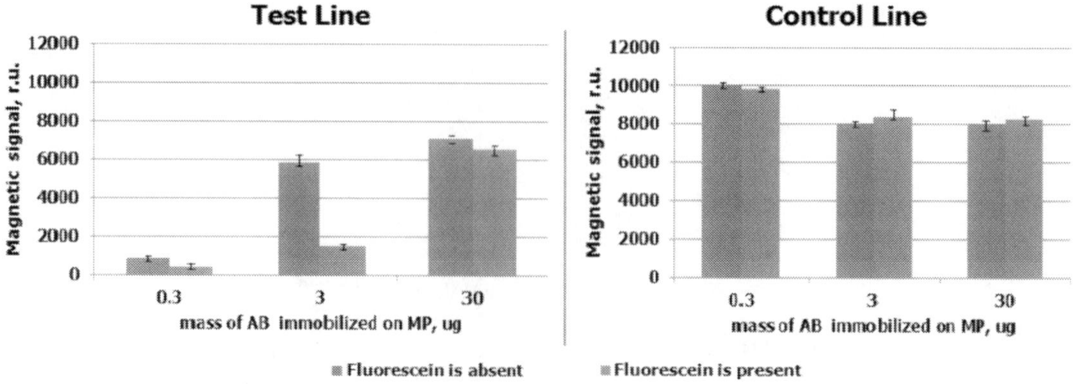

Fig. 4. Dependences of magnetic signal from MPQ reader on test and control lines upon amount of antibody conjugated with magnetic nanoparticles in the presence and absence of antigen.

Thus, the proposed method of MP distribution mapping provides a tool for rapid, simple and cost-efficient optimization of all stages of the IC assay without high consumption of reagents. The method has been also used for quantitative monitoring of total MP mass by determination of square under the curve of their distribution along the test strip. According to the experiments, this parameter did not depend on antigen concentration and remained constant for each batch of the test strips. The method for optimization of quantitative IC assays for detection of small molecules can be used for rapid and cost-effective development of highly sensitive express test systems for food quality control, in vitro diagnostics, criminalistics, environmental monitoring, etc.

4 Conclusion

The proposed method of recording quantitative distribution of magnetic nanoparticles along all constituent components of immunochromatographic lateral flow strips can be used for selection of proper concentrations of the used reagents such as amounts of Ab immobilized on MP, amounts of MP and conjugates deposited onto the test lines, etc.

The advanced label-free optical biosensors based on the SPI and SCI have allowed rapid selection of optimal reagents that feature high affinity and specificity. They also have shown high efficiency to control MP functionalization and synthesis of conjugates, interaction of conjugate with specific Ab, etc. These magnetic and optical methods are also promising for transformation of the previously reported magnetic LF sandwich assays for sensitive detection of small molecules in a wide dynamic range.

The experiments have demonstrated that the synergetic combination of original highly sensitive magnetic instruments having extraordinarily wide 7-order dynamic range with interferometric methods yields powerful tools for development of advanced analytical systems based on magnetic nanolabels.

Different aspects and parts of this multidisciplinary research were partially supported by the Russian Foundation for Basic Research (grants No. 16-33-60228, 15-04-09372, 15-32-70022, 15-04-99499, 15-02-07791, 17-54-560024, 17-02-01415) and the Competitiveness Programs of NRNU "MEPhI" (P.I.N.).

References

1. X. Wang and S. Wang, Sensors, **8**, 6045 (2008)
2. D.R. Shankaran, K.V. Gobi, N. Miura, Sensors and Actuators B: Chemical, **121**, 158 (2007)
3. E. González-Peñas, C. Leache, M. Viscarret, A. Pérez De Obanos, C. Araguás, A. López de Cerain, J. Chromatogr. A, **1025**, 163 (2004)
4. A.M. Evans, C.D. Dehaven, T. Barrett, M. Mitchell, E. Milgram, Anal. Chem. **81**, 6656 (2009)
5. A.V. Orlov, A.G. Burenin, N.G. Massarskaya, A.V. Betin, M.P. Nikitin, P.I. Nikitin, Sensors and Actuators B: Chemical, **246**, 1080 (2017)
6. S.C. Pei, Y.Y. Zhang, S.A. Eremin, W.J. Lee, Food Control **20**, 1080 (2009)
7. D. Kong, L. Liu, S. Song, S. Suryoprabowo, A. Li, .H. Kuang, L. Wang, C. Xu, Nanoscale, **8**, 5245 (2016)
8. A.E. Ivanov, V.P. Zubov, Russ. Chem. Rev. **85**, 565 (2016)
9. D. Wild, *The Immunoassay Handbook,* 4th edn. (Elsevier Science, 2013)
10. I. Giouroudi, G. Kokkinis, Nanomaterials, **7**, 171 (2017)
11. A.V. Orlov, S.L. Znoyko, V.R. Cherkasov, M.P. Nikitin, P.I. Nikitin, Anal. Chem. **88**, 10419 (2016)
12. G.T. Hermanson *Bioconjugate techniques*, 3rd edn. (Academic, New York, 2013)
13. P.I. Nikitin, B.G. Gorshkov, M.V Valeiko, S.I. Rogov, Quantum Electronics, **30**, 1099 (2000)
14. A.V. Orlov M.P. Nikitin, V.A. Bragina, S.L. Znoyko, M.N. Zaikina, T.I. Ksenevich, B.G. Gorshkov, P.I. Nikitin, J. Magn. Magn. Mater. **380**, 231 (2015)

Methodological aspects of small iron concentrations determination in black yeasts grown in the presence of iron oxide nanoparticles

Grigory Yu. Melnikov[1,*], *Tatyana* P. Denisova[2], *Alexander* P. Safronov[1,3], *Oleg* M. Samatov[3], *Roman* T. Khandukhanov[2], *Nikita* A. Kulesh[1], *Ricardo* Andrade[4], and *Galina* V. Kurlyandskaya [1,5]

[1] Ural Federal University, Mira str., 19, Ekaterinburg, 620002, Russia
[2] Irkutsk State University, Nizhnyaya Embankment St., 6, Irkutsk, 664011, Russia
[3] Institute of Electrophysics, Amundsen str., 106, Ekaterinburg, 620016, Russia
[4] Universidad del País Vasco UPV-EHU, SGIKER, Leioa, 48940, Spain
[5] Universidad del País Vasco UPV-EHU, Depto Electricidad y Electrónica, Leioa,48940, Spain

Abstract. Nonpathogenic *Exophiala nigrum* (black yeasts) unicelular organisms of the Baikal Lake were used as a model system for determination of small iron concentrations in the samples grown without or with controlled amount of maghemite nanoparticles (MNPs) in nutrient. MNPs were produced by the electrophysical laser target evaporation technique. Electrostatically stabilized suspensions were prepared using sodium citrate solutions in distilled water. We assumed that one maximum permissive dose of ionic iron in water 1 MPD is equal to 0.3 mg/L. For biological experiments Saburo liquid nutrient medium was prepared with iron concentrations of 0, 10^2, 10^3 and 10^4 MPD. One ml of *E. Nigrum* cell suspension was added to Saburo liquid nutrient for 24 hours exposure. Followed by sowing onto a solid agar Saburo for 30 days colonies grows. Biosamples for electron microscopy, magnetic and total reflection X-ray fluorescence spectroscopy measurements were collected simultaneously. We were able to comparatively analyze the trace concentrations of iron in the yeast of the order of 10 ppm for control group and 600 ppm for the group grown in the presence of 10^4 MPD of iron.

1 Introduction

Living systems strongly depend on iron in variety of life steps [1-2]. Therefore, precise determination of a small amount of iron in the biological and artificial samples mimicking natural tissues is a challenge. Magnetic nanoparticles (MNPs) can enter occasionally or to be on-purpose injected for therapeutic action into the human body. Magnetite (Fe_3O_4) and maghemite (γ-Fe_2O_3) are most studied magnetic biocompatible materials [3-4]. An accumulation of MNPs and their degradation products in the tissues/biofluids is a dangerous insufficiently understood health hazard [5]. Nonpathogenic *Exophiala nigrum* (unicelular black yeast) play an important role in the equilibrium of the Baykal Lake ecosystem [6]. *E. Nigrum* were previously used as a model system for the estimation of cytotoxicity, MNPs accumulation and inductive heat capacity studies of the samples grown in the presence of MNPs [4,7]. One of the main obstacles for rapid extension of biomedical applications of MNPs is the variation of the properties of MNPs from batch to batch [8]. Therefore, the fabrication techniques providing enhanced batches attract special attention. From this point of view electrophysical techniques of the electric explosion of wire ensuring fabrication of spherical MNPs [9-10] and laser target evaporation (LTE) with the production rates up to 50 g/h [11-12] are of special interest. Preparation of model samples with of MNPs is highly requested nowadays for the development of cheap compact analytical devices (magnetic biosensors) [13-15]. They are also requested for better understanding of mechanisms and elaboration of the treatment conditions for such cancer therapies as hyperthermia and thermal ablation [4, 16-17].

In this work, maghemite spherical magnetic nanoparticles were fabricated by LTE technique, and water-based suspensions were prepared on the basis of obtained MNPs. Their physical properties were carefully checked by a number of techniques prior to determination of trace concentrations of iron in the *Exophiala nigrum* samples grown without or with controlled amount of MNPs in nutrient medium.

2 Experimental

Iron oxide MNPs were synthesized by LTE method using Ytterbium (Yb) fiber laser with 1.07 μm wavelength. More details can be found elsewhere [4, 7]. Structural studies of air-dry MNPs were performed by transmission electron microscopy (TEM) using a JEOL

* Corresponding author: grisha2207@list.ru

JEM2100 microscope operating at 200 kV and scanning electron microscopy (Hitachi S-4800, SEM). The X-ray diffraction (XRD) studies were performed by DISCOVER D8 (Bruker) diffractometer using Cu-Kα radiation (wave lengths λ=1.5418 Å).

Electrostatically stabilized suspensions were prepared by ultrasound treatment using sodium citrate solutions (5 mM) in distilled water. De-aggregation was insured by ultrasound treatment on Cole-Parmer CPX-750 homogenizer operated at power output 300 W. During the ultrasound treatment the decrease of an average hydrodynamic diameter of aggregates in suspension was monitored by dynamic light scattering. A Hermle Z383 centrifuge operated at 10,000 rpm for 10 min was used to remove all large aggregates. The specific surface area of MNPs was measured by the low-temperature sorption of nitrogen (Brunauer–Emmett–Teller physical adsorption method, BET) using Micromeritics TriStar 3000 analyzer. More details on suspension preparation can be found elsewhere [4, 11].

Maximum permissive dose of iron in water (MPD) is determined as 0.3 mg/L for ionic iron [18]. The control was Saburo liquid medium with the corresponding addition of a sterile physiological solution without MNPs. As the first step of the biological experiments Saburo liquid nutrient medium was prepared with following iron concentrations (provided by MNPs from suspension): 10^2, 10^3 and 10^4 MPD. One ml of *E. Nigrum* cell suspension was added to Saburo liquid nutrient for 24 hours exposure in flasks using a mechanical rocking chair. At the end of the exposure period sowing onto a solid agar Saburo was done from which biosamples for TEM, magnetic measurements and total reflection X-ray fluorescence (TXRF) measurements were collected simultaneously after 30 days colonies grows. For TEM studies of biological samples cell cultures were pre-fixed in 0.5% glutaraldehyde in Sörenson buffer and centrifuged. The supernatant was removed and 2% glutaraldehyde in Sörenson buffer was added to the pellet. Sampleswere fixed in 1% Osmium Tetroxide in Sörenson buffer, dehydrated, embedded in EponPolarbed resin and cut in ultrathin sections for TEM studies (Philips EM208S). More details on LTE technique can be found elsewhere [15]. TXRF is a method of elemental analysis that can be applied to samples in the form of thin layer. Nanohunter spectrometer by Rigaku and previously elaborated protocol [7] were used. For both scanning electron microscopy (SEM) and TEM of *E. Nigrum* cell cultures were prepared by standard procedure [15]. Magnetic measurements were carried out on a MPMS XL-7 SQUID-magnetometer in polystyrene envelopes of known mass, contribution of which was carefully subtracted from the biological sample signal. Magnetic measurements were done at room temperature.

3 Results and discussion

XRD spectra gave a mean crystallite size for air-dried MNPs of 19 ± 2 nm in a good agreement with TEM and specific surface area evaluation. Fig. 1 shows the air-dry

Fig. 1. LTE iron oxide MNPs, TEM (a). Petri dish with *E. Nigrim* grown onto solid nutrient medium without MNPs (b). General view of *E. Nigrum* grown in the presence of 0 (c) or 10^4 MPD (d) of iron: MPD of iron provided by MNPs.

spherical LTE MNPs. The experimental XRD data were well fitted by the magnetite database. The chemical composition of LTE MNPs ($Fe_{2.72}O_4$) close to the stoichiometric maghemite was determined by the combination of red-ox titration and the lattice period analysis provided by XRD as in [9]. We were not able to observe significant difference in the average size/shape of the cells in all cultures under consideration grown in the presence or absence of MNPs (Fig. 1-2, Table 1).

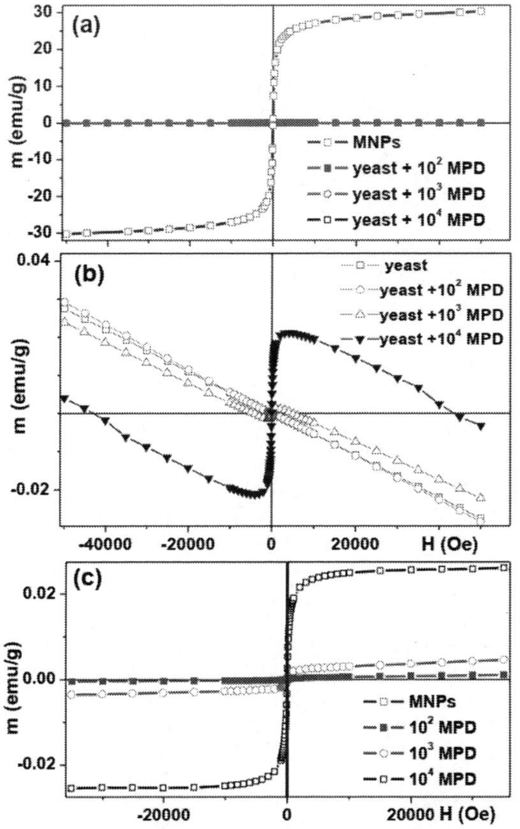

Fig. 2. Hysteresis loops of LTE MNPs from ferrofluid and *E. Nigrum* cell cultures grown in the presence of different amounts of MNPs (a). Hysteresis loops of all cell cultures (b). Recalculated hysteresis loops of all cell cultures grown in the presence of MNPs without contribution of yeast part (c).

Table 1. Iron concentration in *Exophiala Nigrum* cell cultures grown without or in the presence of MNPs. M_s^*- magnetic moment for 1 g of the biosample in the field of 35 kOe.

MPD in nutrient mg/L	M_s^* (emu/g) SQUID	ppm Fe by SQUID	ppm Fe by TXRF
0	0.0000	0 ± 10	9 ± 10
10^2	0.0009	21 ± 10	15 ± 10
10^3	0.0043	101± 20	44 ± 15
10^4	0.0255	603 ± 100	440 ± 50

Fig.2(a) shows results of the SQUID measurements of the air dried MNPs and yeast samples. The yeast samples were taken by sterile scriber and immediately placed into polystyrene envelopes of known mass, mass

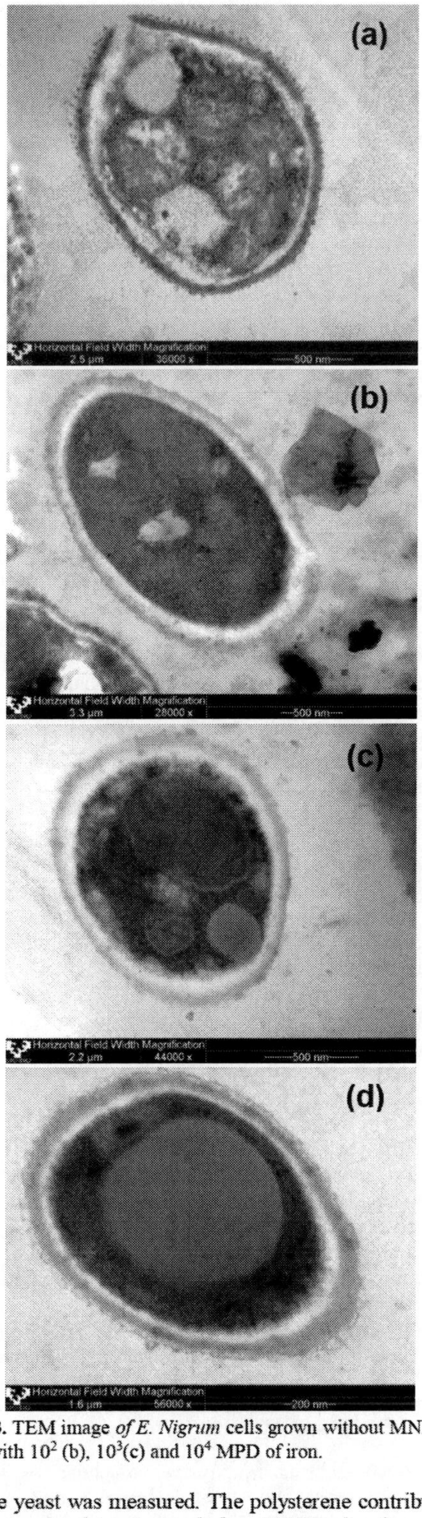

Fig. 3. TEM image *of E. Nigrum* cells grown without MNPs (a) and with 10^2 (b), 10^3(c) and 10^4 MPD of iron.

of the yeast was measured. The polysterene contribution was previously measured by SQUID in the same conditions. As to expect, the highest magnetic moment corresponds to the MNPs dried from ferrofluid due to the highest MNPs concentration. No saturation at room temperature was observed in the field of 50 kOe which is typical for superparamagnetic samples [4,14]. Observed magnetization value is consistent with the mean diameter of 19 ± 2 nm for MNPs [7, 9]. Although non-zero coercivity (≈ 50 Oe at 300 K) was observed it can be understood in the frame of the core-shell model invoking a ferromagnetic core and a spin-glass like surface layer in which the spins are frozen without long range magnetic order [11].

Magnetic moment (m) field dependences for all cases had strong diamagnetic contribution as to expect for water containing biosamples (Fig. 2(b)). In order to calculate the magnetic signal corresponding to MNPs per mass unit we used *E. Nigrum* cell culture grown without MNPs signal for calibration. For each field point we recalculated the hysteresis loops of all samples without diamagnetic contribution (Fig. 2(c). As the saturation was not reached even for high field all calculations were made for the field of H = 35 kOe wich was reasonably high: M_s^* is a magnetic moment of 1 g of biosample for the external field H = 35 kOe. Increase of the M_s^* value with MPD increase and "S" shapes of the hysteresis loops of biosamples after exposure to MNPs indicate MNPs accumulation with the increase of MPD of iron in the initial nutrient. Taking into account parameters of the hysteresis loop for MNPs we were able (Table 1) to re-calculate iron concentration in ppm for SQUD measurements and compare them whith ppm concentrations obtained by TXRF studies of dried samples: the higher MPD in nutrient the higher ppm of iron in the biological samples in both TXRF and SQUID measurements. For recalculation we considered that air-dry MNPs magnetic signal (H = 35 kOe) corresponds to 100 % of concentration. Therefore, simple proportion relations with air-dry signal as calibration gave numbers for all concentrations. One can see that SQUID measurements tend to give slightly higher value comparing with TXRF which is surprising because TXRF measures both ionic and MNPs contribution but SQUID signal is mostly corresponds to the MNPs. More studies are necessary in order to clarify the origin of such a discrepancy but it seems to be that *E. Nigrum* cells are rather large and special improved protocol is necessary for thin film on their basis preparation for TFRF. Trace iron concentrations indicated limited penetration of MNPs inside the cell or their adsorption by cell membranes (Fig. 3 shows some examples). One can clearly see thick cell membrane with typical for *E. Nigrum* morphology. Comparative estimation of the average size of the oval-shaped Exophiala Nigrum cells gives approximate values of 6 ± 1 μm × 3 ± 1 μm for SEM and of 5 ± 1 μm × 3 ± 1 μm for TEM studies for all cell cultures in a good agreement with each other.

4 Conclusions

Nonpathogenic *E. Nigrum* unicelular yeasts were used as a model system for determination of trace iron concentrations. They were grown in the presence of 19 ± 2 nm maghemite MNPs produced by laser target evaporation.Electrostatically stabilized water based suspensions were prepared: liquid nutrient medium was elaborated with different iron concentrations of 0 to 10^4 MPD provided by MNPs. One ml of *E. Nigrum* cell suspension was added to Saburo liquid nutrient for 24 hours exposure followed by sowing onto a solid agar Saburo for 30 days cell grows after which the biosamples for TEM, magnetic and TXRF measurements were collected simultaneously. In both TRXRF and SQUID measurements similar tendency was observed: the higher MDP in nutrient the higher ppm of iron in the biological samples. Trace iron concentrations of the order of 10 ppm for control group and 600 ppm for the group grown in the presence of 10^4 MPD of iron provided by MNPs indicated limited penetration of MNPs into the cell or their adsorption by membranes.

The work was supported in part by the Spanish MEC project MAT2014-55049-C2-1-R and RFFI №16-34-50192 grant. We thank Iu.P. Novoselova, I.V. Beketov, E.A. Stepanova, A. Larrañaga and A.S. Volegov for special support. Selected measurements were made at SGIKER services of UPV/EHU and URFU common services.

References

1. M. Ilberta and V. Bonnefoy, BBA – Bioenerg. **1827**, 161-175 (2013).
2. E. Vorburger Mielczarek and Sh. Bertsch McGrayne, *Iron, Nature's Universal Element,* (Rutgers University Press, 2000).
3. R. C. O'Handley, *Modern Magnetic Materials* (New York, NY, USA: Wiley, 1972).
4. J.P. Novoselova, A.P. Safronov, O. Samatov, *et al.*, IEEE Trans. Magn. **50,** 4600504 (2014).
5. H. Matusiewicz, Acta Biomater.**10**, 2379-2403 (2014).
6. I.P. Babieva and I.Yu. Chernov, *Biology of yeasts* (M.V. Lomonosov MSU, Moscow, 2004).
7. N.A. Kulesh, I.P. Novoselova, A.P. Safronov, *et al.*, J. Magn. Magn. Mater. **415**, 39-44 (2016).
8. J.H. Grossman and S.E. McNeil, Phys. Today, **65,** 38-42 (2012).
9. I.V. Beketov, A.P. Safronov, A.I. Medvedev, *et al.*, AIP Adv. **2**, 022154 (2012).
10. V. Romanova, G. Ivanenkov, A. Mingaleev, *et al.*, Plasma Phys. Rep. **41**, 617-636 (2015).
11. A.P. Safronov, I.V. Beketov, S.V. Komogortsev, *et al.* , AIP Adv. **3**, 052135 (2013).
12. V.V. Osipov, V.V. Platonov, M.A. Uimin, and A.V. Podkin, Tech. Phys**. 57**, 543-549 (2012).
13. J. Llandro, J.J. Palfreyman, A. Ionescu, *et al.,* Med. Biol. Eng. Comput. **48**, 977–998 (2010).
14. G.V. Kurlyandskaya, E. Fernandez, A.P. Safronov, *et al.*, Appl. Phys. Lett. **106,** 193702 (2015).
15. G.V. Kurlyandskaya, Iu.P. Novoselova, *et al.,* J. Magn. Magn. Mater. **431,** 249-254 (2017).
16. P. Moroz, S.K. Jones, and B.N. Gray, J. Surgical Oncol. **77**, 259–269, (2001).
17. M. Coïsson, G. Barrera, F. Celegato, *et al.*, J. Magn. Magn. Mater. **415,** 2-7 (2016).
18. https://ohranatruda.ru/ot_biblio/normativ/data_normativ/9/9742/SanPiN2.1.4.1074-01.

https://doi.org/10.1051/epjconf/201818510008

Possibility to use of the Fe_3O_4/Ta_2O_5 core-shell nanoparticles in radiotherapy

Kseniya Lukyanenko[1,*], Leonid Afremov[1], *Vladimir* Apanasevich[2], *Mariya* Shmykova[1], *Mikhail* Medkov[3], *Pavel* Lukyanov[4], *Olga* Tarakova[1], *Vladimir* Kustov[5]
[1]Far Eastern Federal University, 690950, 8 Sukhanova St., Russia
[2]Pacific State Medical University, 690002, 2 Ostryakova Avenue, Russia
[3]Institute of Chemistry FEB RAS, 690022, 159, Prospekt 100-let Vladivostoku, Russia
[4]G.B. Elyakov Pacific Institute of Bioorganic Chemistry FEB RAS, 690022, 159 Prospect 100 let Vladivostoku, Russia
[5]Vladivostok's Branch of the Customs Academy, 690034, 16v Strelkovaya st., Russia

Abstract. The study was carried out of the possibility of using magnetic core-shell nanoparticles Fe3O4/Ta2O5 as a radio-modifier. It is investigated the influence of the inhomogeneous magnetic field on the distribution of the nanoparticles in the region of its maximum inhomogeneity. The increase of the core-shell nanoparticles' concentration leads to the increase of the number of 511keV gamma-quanta. The absorption of gamma-quanta with lower energy (20-200 keV) increases with increase in concentration of nanoparticles.

1 Introduction

A development of the biomedical technologies including such as hyperthermia [1-4], targeted drug delivery [5-11] as well as a creation of biosensors and biochips [12] resulted in the heavy growth in the experimental investigations of the magnetic properties of the systems of different nanoparticles and nanofilms/nanotape. The possibility to manipulate living cells anchored with NPs via a magnetic field, ferromagnetic NPs affect hemopoietic /stromal cells and promote cellular adhesion and formation of cell-to-cell contacts along the magnetic field lines, NPs in regenerative medicine, thermal ablation, etc. are describe in [13,14].

Over the last years, the interest has increased in the study of magnetic nanoparticles for purposes of their use in the cancer diagnostics and treatment. The unique properties of the nanoparticles (NPs), ability to control their geometric (sizes, shapes) and surface characteristics determining the hydrophility and hydrophobicity of the matter as well as possibility to inject the nanoparticles to the biological media in a number of different ways make them perspective in many aspects of oncology.

Magnetic core-shell nanoparticles are important for applications because they allow to control them via an external magnetic field. Direct control of the motion and distribution of magnetic nanoparticles forms at the basis of their using for targeted drug delivery or for the delivery of magnetic nanoparticles to cancer tumors [7, 10] for the purpose of the subsequent hyperthermia process or to enhance the effect of radiation therapy due to scattering radiation by injected particles.

Various published data referred to as iron oxide toxicity [15] and its biocompatibility [16]. At the present time, the studies related to obtaining the nontoxic magnetic nanoparticles based on iron oxides covered, for example, with the tantalum oxide and designed to biomedical applications become more and more actual [17]. In addition, tantalum oxide nanoparticles have a long circulation period in the bloodstream and they are usually easily biodegradable and have low toxicity [18].

It should be noted that the core-shell Fe_3O_4/Ta_2O_5 nanoparticles can be used to construct magnetically controlled non-toxic X-ray contract drugs [19].

The objective of this paper is to study the possibility of using magnetic core-shell nanoparticles of Fe_3O_4/Ta_2O_5 to enhance the effectiveness of radiation therapy.

2 Synthesis of nanoparticles of Fe_3O_4/Ta_2O_5 and their properties

For achievement of the above-referenced objective, we have synthesized the core/shell nanoparticles the cores of which consist of the iron oxide while envelope of the tantalum oxide.

To achieve the above goal, we synthesized Fe3O4/Ta2O5 nanoparticles. An aqueous fluoride solution of tantalum was used as a precursor, and sodium oleate as a stabilizer of nanoparticles, in contrast to the widely used method of obtaining a core-shell structure [15]. To a solution containing 50 ml of ammonia was added 30 ml solution containing 0.3 g of tantalum in hydrofluoric acid with stirring. Further the pulp was allowed to settle, decanted some of the solution over the precipitate, and the remaining pulp was filtered. The precipitate was washed on the filter twice, after which the precipitate was pelletized in 100 ml of water. To the

* Corresponding author: ks.lukyanenko@gmail.com

received of the mixture was added successively 100 ml of a solution containing 0.5 g of $FeSO_4.7H_2O$ and 0.7 g of $Fe_2(SO_4)_3$, 0.2 ml of a 5% solution of sodium oleate and 50 ml of concentrated ammonia solution with stirring. After that, 15 ml of a solution of tantalum in hydrofluoric acid, 50 ml of ammonia and 0.2 ml of 5% solution of sodium oleate. The pH value in the stages of tantalum deposition and the iron should be not less than 10. Then the pulp was allowed to settle, poured decanting part of the solution over the precipitate, and the remaining pulp was filtered. The precipitate on the filter is washed twice, after which it is pulped precipitate in 100 ml of water. The resulting pulp was ultrasonically dispersed. The result was a magnetic fluid that was magnetically radiopaque agent. Advantages of the proposed method, we obtained as described in [20], are associated with fewer stages of the process, which significantly reduces the synthesis time of nanoparticles.

The geometric characteristics of nanoparticles and their size distribution were determined using the atomic-force microscope. The atomic-force microscopy allowed us to construct the size distribution (Fig.1). It is described by close to normal law of distribution with mean value d= 44 nm and variance σ= 3 nm.

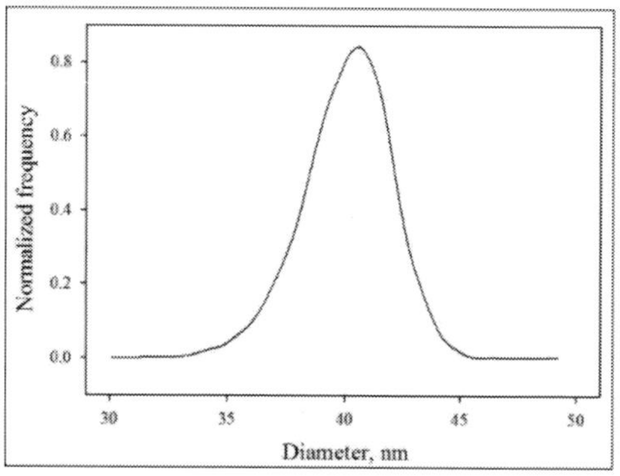

Fig. 1. The size distribution of Fe_3O_4/ Ta_2O_5 core-shell magnetic NPs with the use of AFM.

3 Nanoparticles distribution in the magnetic field

A modeling of the Fe_3O_4/Ta_2O_5 nanoparticles distribution in the magnetic field was performed in the sodium oleate solution. For this, the cuvette with the dissolved Fe_3O_4/Ta_2O_5 nanoparticles was placed into the inhomogeneous magnetic field H established by the coil with the pointed ferromagnetic core. Recording the image brightness was performed using the digital microscope and, afterwards, the mathematical processing of images obtained was accomplished.

Recording the image brightness was carried out on two solutions the concentrations of nanoparticles $\Delta n/n_0$ in which differ twice. We assume that the image brightness is proportional to relative concentration of nanoparticles $\Delta n/n_0$ in the solution. In Fig. 2, the image brightness of solution in different fields is shown.

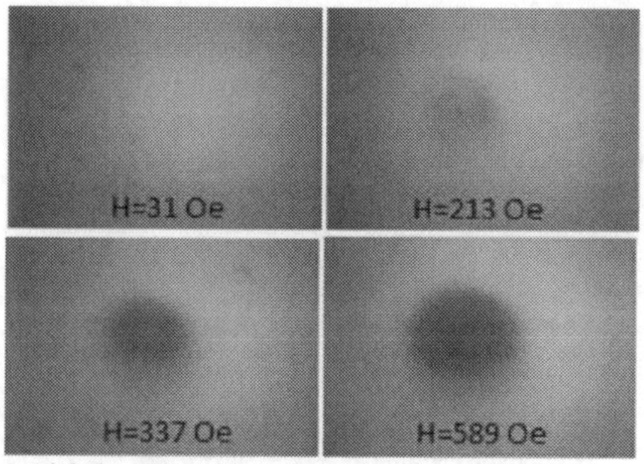

Fig. 2. Light intensity of the solution in different fields H.

The dependence of the distribution of relative nanoparticles concentration in the area of solution near the spot center on the magnetic field and initial concentration of particles is presented in Fig. 3. The effective spot size R_{eff} increases with magnetic field's increasing and decreases with decreasing of the nanoparticle concentration as expected.

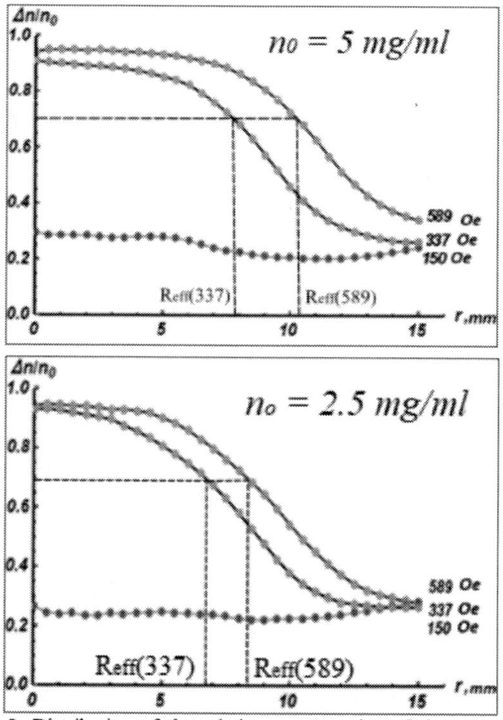

Fig. 3. Distribution of the relative concentration of the NPs $\Delta n/n_0$ on the magnetic field H (Oe) and solution density n_0. r – distance between the cloud center and measuring range.

From Fig. 4, it is evident that the relative concentration of nanoparticles at the spot center increases with increase in the magnetic field strength and reaches the saturation at its particular value. The initial concentration of nanoparticles n_0 in the solution determines the value of the magnetic field at which the curve of relative concentration dependence attains the saturation (Fig. 4): the lesser magnetic field corresponds to the larger initial concentration.

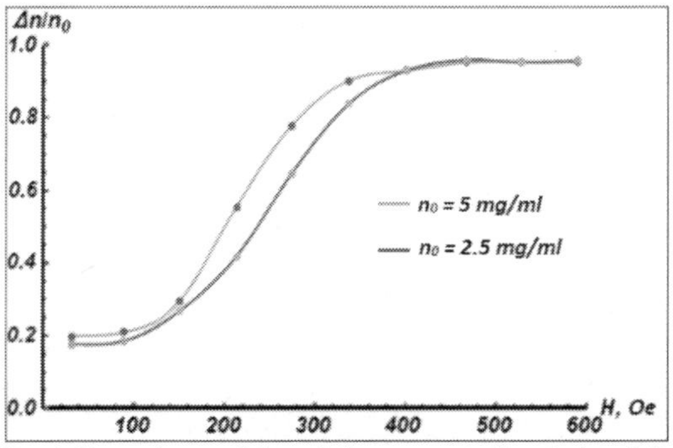

Fig. 4. Dependence of the relative concentration $\Delta n/n_0$ of the NPs in cloud center on magnetic field H

Therefore, for the nanoparticles under consideration, the major values of the magnetic field needed for their maximal concentration at the fixed point.

4 Energy spectrum of gamma-ray emission absorption

The major objective of the modern therapy of the oncology diseases is to defeat the tumor cells with minimum damage to the healthy tissues. As it's known, the maximum biological effect is exercised by the kilo-volt radiation (20-200 keV) [21].

One expects that the introduction of nanoparticles containing tantalum into the area of malignant tumor should result in increase of the number of the low-energy quanta which will contribute to the apoptosis of the tumor cells. In order to test this hypothesis, we conducted the experimental study of the influence of the Fe_3O_4/Ta_2O_5 nanoparticle concentration on the energy spectrum of gamma-ray radiation absorption. In the experiment in vitro, we used the gamma-ray spectrometer with the semiconductor detector GEM-15P4 and two gamma radiation sources ^{60}Co (sample spectrometry gamma source type) with the aggregate activity of about 2×10^5 Bq. The concentration of nanoparticles in the specified area was changed by use of magnetic field: before each radiation, the nanoparticles were subjected to impact of different magnetic fields, at that, the delay time for each measurement was 5 minutes. The curves of dependence of the gamma-quanta number $\Delta n=n(H)-n(H=0)$ on their number in the field H=0 are shown in Fig.5.

The increase in the magnetic field force and, thereby, concentration of nanoparticles Fe_3O_4/Ta_2O_5, result in growth in the number of quanta in the range of 511 keV. Most likely, such behavior of the absorption curve is related to the creation of electron-positron pairs [22]. As the substantiation of this statement, the fact that the energy of gamma-quant higher than $2mc^2$=1022 keV is needed for the creation of electron-positron pair can serve. While the energy of the ^{60}Co gamma-ray is slightly higher and is equal to 1.25 MeV. The electron and positron formed as a result of the interaction between the gamma-ray and substance are absorbed in the radiated tissues, in what connection, the positron annihilates, producing at this two secondary photons with energy of 511 keV each. The secondary gamma-quanta with lower, than the primary radiation, energy have the higher potential of interaction with the biological environment.

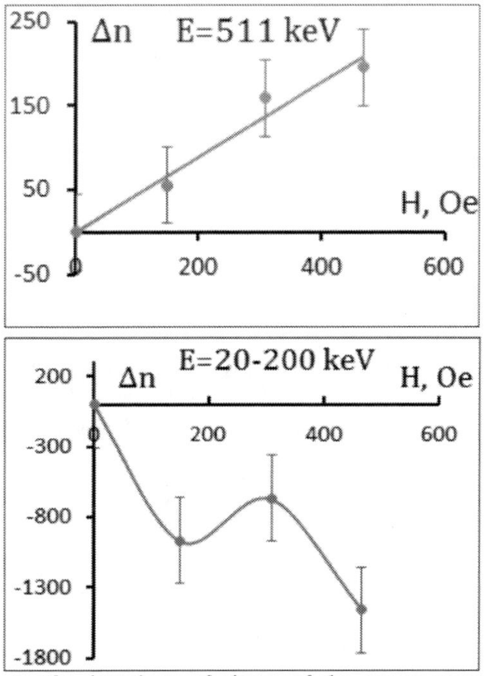

Fig. 5. The dependence of change of the gamma-quanta number Δn on the magnetic field H.

The number of absorbed gamma-quanta increases in the range of low energies (20-200 keV) with growth in the concentration of nanoparticles Fe_3O_4/Ta_2O_5.

5 Results

The core-shell nanoparticles Fe_3O_4/Ta_2O_5 were studied as a radio-modifier in the X-ray therapy. It is shown that the increase of the concentration of core-shell nanoparticles' leads to the increase of the number of 511 keV gamma-quanta and the increase of the absorption of the 20-200 keV gamma-rays.

Therefore, the above-described interaction of gamma-ray with nanoparticles Fe_3O_4/Ta_2O_5 results in formation of the low-energy secondary gamma quanta which can contribute to disruption of the tumor cells and do not disturb the healthy ones.

The work was supported by financial support in the framework of state task 3.7383.2017/8.9

References

1. Hergt, Rudolf, Silvio Dutz, and Michael Röder, Journal of Physics: Condensed Matter **20**, 385214 (2008)

2. Moroz, P., S. K. Jones, and B. N. Gray, International Journal of Hyperthermia **18**, 267-284 (2002)

3. Gneveckow, U., Jordan, A., Scholz, R., Eckelt, L., Maier-Hauff, K., Johannsen, M., and Wust, P., Biomed. Tech **50**, 92-3 (2005)

4. Johannsen, M., Gneveckow, U., Thiesen, B., Taymoorian, K., Cho, C. H., Waldöfner, N., ... and Wust, P., European urology **52**, 1653-1662 (2007)

5. Pankhurst, Q. A., Thanh, N. T. K., Jones, S. K., and Dobson, J., Journal of Physics D: Appl. Phys. **42**, 224001 (2009)

6. Laurent, S., Dutz, S., Häfeli, U. O., and Mahmoudi, M., Advances in colloid and interface science **166**, 8-23 (2011)

7. Murthy, Shashi K., International journal of nanomedicine **2**, 129 (2007)

8. Sinibaldi, E., Pensabene, V., Taccola, S., Palagi, S., Menciassi, A., Dario, P., and Mattoli, V., Journal of Nanotechnology in Engineering and Medicine **1**, 021008 (2010)

9. Ventrelli, L., Ricotti, L., Menciassi, A., Mazzolai, B., and Mattoli, V., Journal of Nanomaterials **2013**, 39 (2013)

10. Alexis, F., Pridgen, E. M., Langer, R., & Farokhzad, O. C., Drug delivery. Springer Berlin Heidelberg 55-86 (2010)

11. Ghosh Chaudhuri, Rajib, and Santanu Paria, Chemical reviews **112**, 2373-2433 (2011)

12. Varadan, Vijay K., Linfeng Chen, and Jining Xie, John Wiley & Sons (2008)

13. Kurlyandskaya, G. V., Novoselova, I. P., Schupletsova, V. V., Andrade, R., Dunec, N. A., Litvinova, L. S., ... and Khlusov, I. A., Journal of Magnetism and Magnetic Materials **431**, 249-254 (2017)

14. Khlusov, I. A., Zagrebin, L. V., Shestov, S. S., Itin, V. I., Sedoi, V. S., Feduschak, T. A., ... and Puchkovskaya, E. S., Bulletin of experimental biology and medicine **145**,151-157 (2008)

15. Lin, Hong-Yi, Wei-Yu Chen, and Yu-Chie Chen, Analytical and bioanalytical chemistry **394,** 2129-2136 (2009)

16. Novoselova, J. P., Safronov, A. P., Samatov, O. M., Beketov, I. V., Khurshid, H., Nemati, Z., ... and Kurlyandskaya, G. V. IEEE Transactions on Magnetics **50**, 1-4 (2014)

17. L. Larionov, M. Zuev, I. Strekalov, S. Sokovnin, Bulletin of the Ural Medical Academy of Science **3**, 37 (2011)

18. Bulte, Jeff WM, and Dara L. Kraitchman, NMR in Biomedicine **17**, 484-499 (2004)

19. Lei, F., Yan, B., Chen, H. H., Zhang, Q., & Zhao, J. T., Crystal Growth and Design, **9**, 3730-3736 (2009)

20. Steblevskaya N.I., Belobeletskaya M.V., et al. The application for the patent of the Russian Federation №2016132930 from 9.08.2016 (priority). Method of obtaining magnetoactive radiocontrast agent ;

21. I.V. Schegolkov, I.N. Sheyno, V.F. Hohlov, A.A. Lipengolts, Medical physics **4**, 12-16 (2010)

22. Kostylev VA Narkevitch BJ "Medical Physics". - M .: JSC "Publishing house" Medicine " (2008)

EPJ Web of Conferences **185**, 11001 (2018) https://doi.org/10.1051/epjconf/201818511001
MISM 2017

CRITICAL PROPERTIES OF *2D* DISORDERED 3-STATE ANTIFERROMAGNETIC POTTS MODEL ON TRIANGULAR LATTICE

A.K. Murtazaev[1,2], *A.B. Babaev*[1,3], *G.Y. Ataeva*[1,4]

[1]Institute of Physics, Dagestan Scientific Center, Russian Academy of Sciences, Makhachkala, 367003 Russia
[2]Dagestan State University, Makhachkala, 367025 Russia
[3] Dagestan State Pedagogical University, Makhachkala, 367003 Russia
[4] Dagestan State University of National Economy, Makhachkala, 367008 Russia
ataeva20102014@mail.ru

Abstract. By introducing a small amount of non-magnetic impurities into an antiferromagnetic (AF) two-dimensional (*2D*) Potts model on a triangular lattice it is that the impurities in spin systems described by this model result in the change of a first order to a second-order phase transition. The systems with linear sizes $L \times L = N$, L = 9-144 are considered. Investigations are performed using the standard Metropolis algorithm along with Monte-Carlo single-cluster Wolff algorithm. On the basis of the theory of finite-size scaling, critical exponents (CE) are calculated: the heat capacity α, the susceptibility γ, the order parameter β, and the CE of the correlation radius υ.

1 Introduction

The Monte Carlo investigations of thermal and critical properties of magnetic systems influenced by impurities and structure defects indicate the non-ideal features of real magnetic systems to effect on phase transitions (PT) and critical properties, in particular, non-magnetic impurities can change a phase transition order [1-3]. Two-dimensional statistic systems can act as models of real physical systems, for instance, the Potts model possessing a variety in critical behavior. The most interesting object, up to date, is the triangular antiferromagnetic *2D* Potts model. This stems from the ability of *2D* lattice models to describe a large class of real physical systems: layered magnetic materials, liquid helium films, superconducting films, etc. [4].

2 The 2*D*weakly diluted antiferromagnetic Potts model

In this work, we perform the Monte Carlo study of phase transitions and critical phenomena of impure triangular *2D* antiferromagnetic Potts model at a spin concentration p=0.90 with lattice linear sizes of $L \times L = N$, L=9-144, and periodic boundary conditions.

The Hamiltonian reads [5]

$$H = -\frac{1}{2} J \sum_{i,j} \rho_i \rho_j \cos\theta_{i,j}, \quad S_i = 1, 2, 3 \tag{1}$$

where J is the parameter of nearest neighbor antiferromagnetic exchange interaction (J<0); ρ_i=1, if the site i is occupied by a magnetic atom,ρ_i =0, if the site is occupied by a non-magnetic impurity; $\theta_{i,j}$ is the angle between interacting spins S_i- S_j.

3 Simulation method

To analyze a phase transition character and to define critical temperatures we use the method of fourth order Binder cumulants [6]

$$V_L(T, p) = 1 - \frac{\langle E^4 \rangle_L}{3\langle E^2 \rangle_L^2}, \tag{2}$$

$$U_L(T,p) = 1 - \frac{\langle m^4(T,p;L)\rangle_L}{3\langle m^2(T,p;L)\rangle_L^2}, \tag{3}$$

where E is the energy and $m=m_{AF}$- is the order parameter of a system with L. Equations 2 and 3 permit an accurate evaluation of $T_N(p)$ at first and second order transitions, respectively. This technique works well at detection of a transition order. One knows the distinctive features inherent in PT [7]: at a first order transition the averaged value $V_L(T, p)$ tends to some non-trivial value V^* according to the equation

$$V(T,p) = V^* + bL^{-d}, \tag{4}$$

at $L \to \infty$ and $T = T_c(L)$, where V^* is non 2/3; the minimum value $U_{L,\min}(T = T_{\min}, p)$ diverges to $U_{L,\min}(T = T_{\min}, p) \to -\infty$ at $L \to \infty$, at a second order transition the averaged value $V_L(T, p)$ at $T = T_c(L)$ will tends to 2/3, and temperature dependence curves of Binder cumulants $U_L(T\ p)$ will have a well-expressed cross point. A procedure of a PT order determination was detailed in works [8-12].

4 Results of simulation

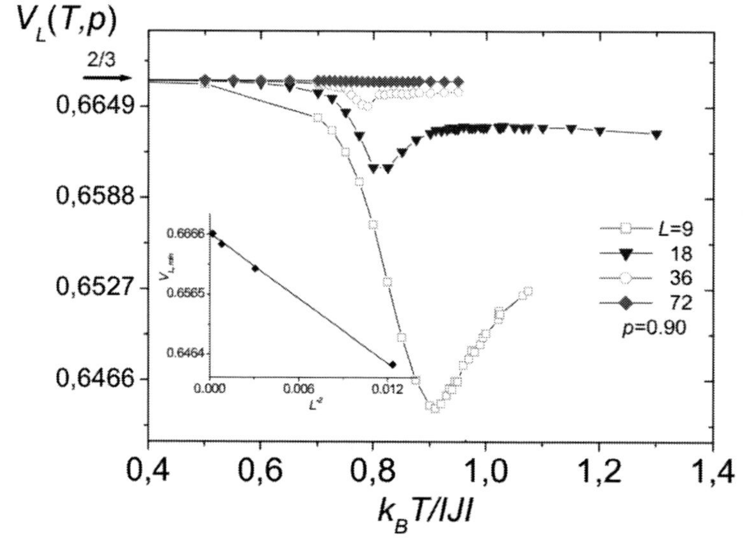

Fig. 1. The temperature dependence of Binder cumulants $V_L(T, p)$ for the 2D 3-state site-diluted AF Potts model at p=0.90.

Figure1 depicts the Binder cumulants $V_L(T, p)$ temperature dependences for systems with different L at p=0.90. It is apparent from Fig.1 that the $V_L(T, p)$ value tends to 2/3 at $L\to\infty$. Such a behavior is, as has been mentioned above, typical for a second order transition [7]. Phase transition temperature $T_l(p)$ determined for all considered systems by the Binder cumulants method in units of $|J|/k_B$ are equal: T_l (1.00)=0.940(1), T_l (0.90)=0.79(1), T_l (0.80)=0.65(2), T_l (0.70)=0.42(3), T_l (0.65)=0.35(4).

We calculate the static CE for the heat capacity α, susceptibility γ, and order parameter β by means of the finite-size scaling theory (FSST). The correlations of this theory indicate that for a sufficiently large system with periodic boundary conditions the order parameter, susceptibility, and parameter V_n for the evaluation of radius correlation (ν) critical exponents T=T_N obey the following analytical expressions [13, 14]:

$$\chi \sim L^{\gamma/\nu}, \tag{5}$$

$$m \sim L^{-\beta/\nu}. \tag{6}$$

$$V_n = L^{1/\nu} g_{V_n} \tag{7}$$

where g_{Vn} is some constant, and can act as V_n,

$$V_i = \frac{\langle m^i E \rangle}{\langle m^i \rangle} - \langle E \rangle, (i=1, 2, 3), \tag{8}$$

$$V_3 = \frac{dU_L}{d\beta} = \frac{1}{3\langle m^2 \rangle^2} \left[\langle m^4 \rangle \langle E \rangle - 2 \frac{\langle m^4 \rangle \langle m^2 E \rangle}{\langle m^2 \rangle^2} + \langle m^4 E \rangle \right], \tag{9}$$

where $\beta = 1/T$, T is the temperature.

The temperature dependence of the heat capacity on L is usually approximated using other expressions, for instance [7]

$$C_{\max}(L) = C_{\max}(L = \infty) - AL^{\alpha/\nu}, \tag{10}$$

where A is some coefficient.

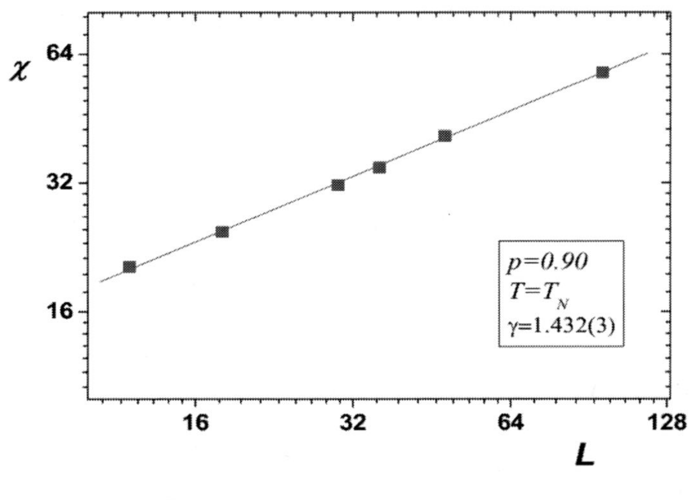

Fig. 2. The dependence of the susceptibility on L at $T=T_N$ and $p=0.90$.

To calculate critical exponents for α, β, γ, and ν we draw diagrams of the dependence of C, m, χ, and V_n on L. The non-linear least-square analysis allows calculation of α/ν, β/ν, γ/ν, and $1/\nu$ values. Next, using derived ν values, we determine α, β, γ exponents.

In Fig. 2, we draw in log-log scale a diagram of susceptibility dependence on L for the $2D$ disordered 3-state antiferromagnetic Potts model on triangular lattice. It should be noted that data obtained for the susceptibility don't deviate from the line even at small L. It is evident that the number of different initial configurations and sizes $L \geq 9$ used for averaging allow us to achieve the asymptotic critical mode. The most crucial moment is that the ν exponents is computed from results of the numerical experiment within the framework of this investigation, while in many other works this exponents is determined from different scaling correlations. Values of critical exponents obtained in this study are presented in Table 1.

Table 1. The finite-size scaling theory-based critical exponents of *2D* disordered 3-state antiferromagnetic Potts model on triangular lattice.

	Spin concentration	**Phase transition temperature T_l**	**Critical exponents**			
Our data	P	T_l	α	β	γ	υ
	1.00	0.940(1)	***First order phase transition***			
	0.90	0.79(1)	0.310(3)	0.102(2)	1.432(3)	0.88(1)

5 Conclusions

Most valuable result of performed investigations is evaluation of the critical exponents for *2D* disordered 3-state antiferromagnetic Potts model on triangular lattice using the finite-size scaling theory. The obtained values of critical exponents obey the Rashbrook relation $\alpha+2\beta+\gamma=2$. A disorder caused by non-magnetic impurities results in the change of a first order transition to a second order state. Such a behavior is associated with the suppression of the fluctuation instability caused by introduced impurities.

References

1. M. Aizenman, J. Wehr. Phys. Rev. Lett. **62**, 2503 (1989).
2. K. Hui, A.N. Berker. Phys. Rev. Lett. **62**, 2507 (1989).
3. A. K. Murtazaev, A. B. Babaev, G. Ya. Ataeva. J. Exp. Theor. Phys. **109** (3), 412 (2009).
4. R.J. Baxter, *Exactly Solved Models in Statistical Mechanics* (London: Academic Press, 1982).
5. F.Y. Wu. Rev. Mod. Phys. **54**, 235 (1982).
6. K. Eichhorn, K. Binder. J. Phys.: Cond. Matter **8**, 5209 (1996).
7. D. Loison, K.D. Schotte. Eur. Phys. J. B **5**, 735 (1998).
8. A.K. Murtazaev, A.B. Babaev, G.Y. Aznaurova. J. Exp. Theor. Phys. **115**, 6, 1042 (2012).
9. A.K. Murtazaev, A.B. Babaev. J. Magn. Magn. Mater. 324, 3870 (2012).
10. A.K.Murtazaev, A.B. Babaev, G.Ya. Aznaurova, Low Temperature physics **39**, 2, 147 (2013).
11. A.K. Murtazaev, A.B. BabaevJ. Exp. Theor. Phys. **116**, 1, 101 (2013).
12. A.K. Murtazaev, A.B. Babaev, JETP Lett **99,**535 (2014).
13. D. Loison. *Physics Letters A***257**, **83** (1999).
14. M.E. Fisher, M.N. Barber. *Phys. Rev. Lett.* **28**, 1516 (1972).

Frustrations and phase transitions in magnets of various dimensionality

Felix Kassan-Ogly[1,*] and *Alexey* Proshkin[1]

[1] *Institute of Metal Physics of Ural Branch of Russian Academy of Sciences, 620990, Ekaterinburg, Russia*

Abstract. We studied magnetic orderings, phase transitions, and frustrations in the Ising, 3-state Potts and standard 4-state Potts models on 1D, 2D, and 3D lattices: linear chain, square, triangular, kagome, honeycomb, and body-centered cubic. The main challenge was to find out the causes of frustrations phenomena and those features that distinguish frustrated system from not frustrated ones. The spins may interrelate with one another via the nearest-neighbor, the next-nearest-neighbor or higher-neighbor exchange interactions and via an external magnetic field that may be either competing or not. For problem solving we mainly calculated the entropy and specific heat using the rigorous analytical solutions for Kramers-Wannier transfer-matrix and exploiting computer simulation, par excellence, by Wang-Landau algorithm. Whether a system is ordered or frustrated is depend on the signs and values of exchange interactions. An external magnetic field may both favor the ordering of a system and create frustrations. With the help of calculations of the entropy, the specific heat and magnetic parameters, we obtained the points and ranges of frustrations, the frustration fields and the phase transition points. The results obtained also show that the same exchange interactions my either be competing or noncompeting which depends on the specific model and the lattice topology.

1 Methods and basic formulae

All considered models on all the lattices are described by the Hamiltonian:

$$\mathcal{H} = -\frac{J}{2}\sum_{i,\Delta}(\mathbf{s}_i \cdot \mathbf{s}_{i+\Delta}) - \frac{J'}{2}\sum_{i,\Delta'}(\mathbf{s}_i \cdot \mathbf{s}_{i+\Delta'}) - \sum_i(\mathbf{h} \cdot \mathbf{s}_i), \quad (1)$$

where J and J' mean exchange interactions between nearest-neighbor and next-nearest-neighbor spins; h is an external magnetic field; sums run over lattice sites i, nearest-neighbors Δ, next-nearest-neighbors Δ', vectors $\mathbf{s}$ (with a module equal to unity) take on two, three, or four values, admissible in the Ising, 3-state Potts and 4-state Potts models.

If the analytical solution for maximum eigenvalue λ of Kramers-Wannier transfer-matrix was found, the entropy S, specific heat C, magnetization M, etc. are only expressed in terms of λ:

$$S = \ln(\lambda) + \frac{T}{\lambda}\frac{\partial\lambda}{\partial T}, \; C = T\frac{\partial S}{\partial T}, \; M = \frac{T}{\lambda}\frac{\partial\lambda}{\partial h}, \quad (2)$$

In the absence of analytical solution, we performed the computer simulations by Monte-Carlo method, mainly in Wang-Landau algorithm [1].

2 Types of frustrations

Figures 1, 2 and 3 display several types of frustrations. The majority of papers in the 2D case are devoted to the Ising model with competing antiferromagnetic nearest and next-nearest interactions on a square lattice (see, for example, [2]), frustrations are found at one point $J'/J = 0.5$. The phase transition point and the zero-temperature entropy $S(T = 0)$ are equal to zero at this frustration point.

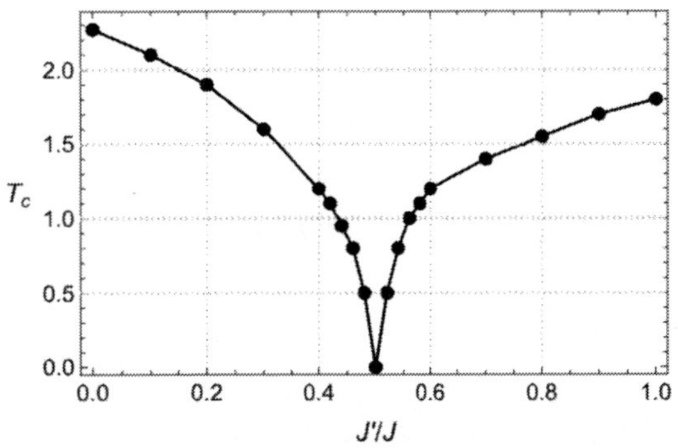

Figure 1. Phase transition temperature as a function of J'/J in Ising model on square lattice.

We studied Ising model with the same papameters on a body-centered cubic lattice [3] and obtain the phase diagram, shown in Fig. 2. The main difference of it from Fig. 1 is that the phase transition point has nonzero value, and this is another type of frustrations. We have discovered a novel (third) type of frustrations [4], shown in Fig. 3. First of all, in this type, a whole range of frustrations rather than a single frustration point is present. Second, despite the frustrations, the phase-transition temperature

[*] e-mail: felix.kassan-ogly@imp.uran.ru

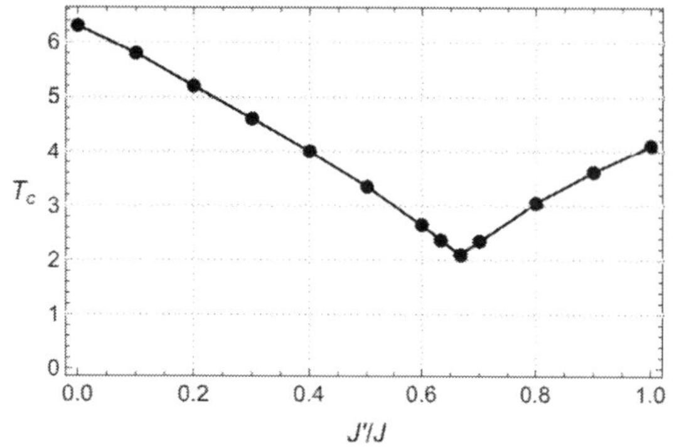

Figure 2. Phase transition temperature as a function of J'/J in Ising model on bcc lattice.

does not vanish at all. Third, the zero-temperature entropy $S(T = 0)$ is nonzero and has one and the same value over the whole frustration range.

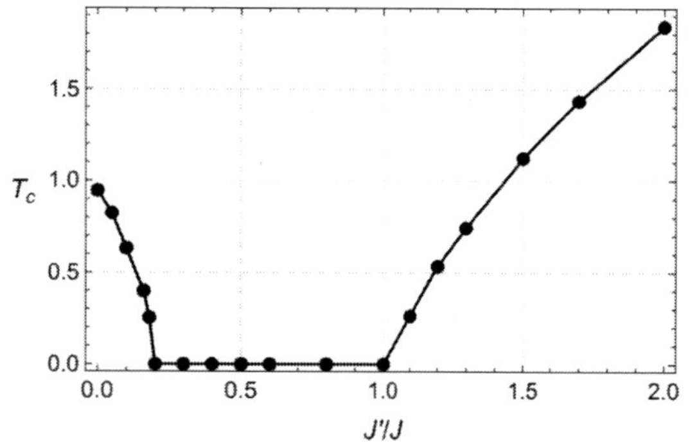

Figure 3. Phase transition temperature as a function of J'/J in 3-state Potts model on a triangular lattice [4].

3 Splitting of specific heat near frustration points

We have revealed that frustrations create a remarkable peculiarity inherent to any model and any lattice. In the vicinity of every point of frustration or frustration field, specific heat splits into a sharp peak and a broad dome-shaped peak. The phase-transition temperature decreases with approaching the frustration point or field. It may vanish, and the sharp peak disappears in this case at the frustration point or field. Otherwise, the broad dome-shaped peak disappears.

Using Onsager solution [5], keeping interaction along one side of a square J_x constant and varying J_y from −0.1 to +0.1, we will obtain the temperature of phase transition as a function of J_y/J_x (solid line in Fig. 4a), and the dependence of a broad peak maximum on J_y/J_x (circles in Fig. 4a). Figure 4b also shows specific heat at $J_y/J_x = 0.01$.

The same behavior of heat capacity splitting takes place in all considered models on all the lattices considered, in particular, linear chain, triangular [6], and kagome lattice [7]. A very similar temperature dependence of magnetic contribution into heat capacity is observed in several real 3D crystals. Figure 5 displays, as an example, such heat capacity splitting in ErBi crystal [8]. As is seen, the authors' calculation attempt did not give an adequate agreement with experiment, since they were not aware that the effect has been created by frustrations.

The splitting of specific heat and decreasing of the phase transition temperature with increasing magnetic field towards frustration point were also observed in some real 3D crystals, for example, in TmTe [9] and in $Eu_{0.95}Sr_{0.05}Te$ [10] crystals with magnetic face-centered cubic sublattices, but the frustration field had not been reached. Most probably, this was not done due to unawareness of the authors about such effect of frustrations.

4 Partial ordering

One more remarkable phenomenon, the so-called partial ordering (or ordering with lowered dimensionality) is observed in frustrated systems, which is closely related to the general properties of entropy. In the models considered, the entropy (per site) at $T \to 0$ equals to $\ln(n)/N$, where $N \to \infty$ is the number of lattice sites and n is the number of configurations with the lowest energy, i.e. those that survive at $T \to 0$. If n is finite, for example, in Ising model it is equal to 2. Then entropy is equal to zero, and we have some kind of ordering. If n is infinite and of the order of q^N (q is the number of permissible states at a site) then the zero-temperature entropy is nonzero, and we have no ordering (frustrations). Sometimes n can be computed, for example, in the standard antiferromagnetic 4-state Potts model on 1D chain it is equal to $4 \cdot 3^{N-1}$, and entropy is equal to ln(3).

The number n may be infinite, but not of the order of q^N. The first example is the Onsager's case considered in the previous section. At $J_y = 0$, and $J_x < 0$ the system consists of separate chains, ordered antiferromagnetically and unlinked to one another. The number n of such a system is equal to 2^{N_y}, where N_y is the number of sites along the y direction in a square lattice. The zero-temperature entropy is equal to $N_y \ln(2)/N_y N_x$, or $\ln(2)/N_x$ and consequently, vanishes in the thermodynamic limit ($N_x \to \infty$). As a whole, this 2D system has no finite transition point; nevertheless, in one direction it is ordered.

This phenomenon occurs very often in the systems with competing interactions, for example, in Ising model on a square lattice at the frustration-point value $J'/J = 0.5$, where J is the antiferromagnetic exchange interaction between nearest-neighbor spins and J' is also the antiferromagnetic interaction between next-nearest-neighbor spins. It is more convenient to consider the pattern of partial ordering from the region, where $J'/J > 0.5$. Hereinafter, open and solid circles denote opposite spin directions. In this case, the magnetic structure looks like in Fig. 6a and is often called super-anti-ferromagnetic (SAF). It consists of alternating ferromagnetic columns in antiferromagnetic

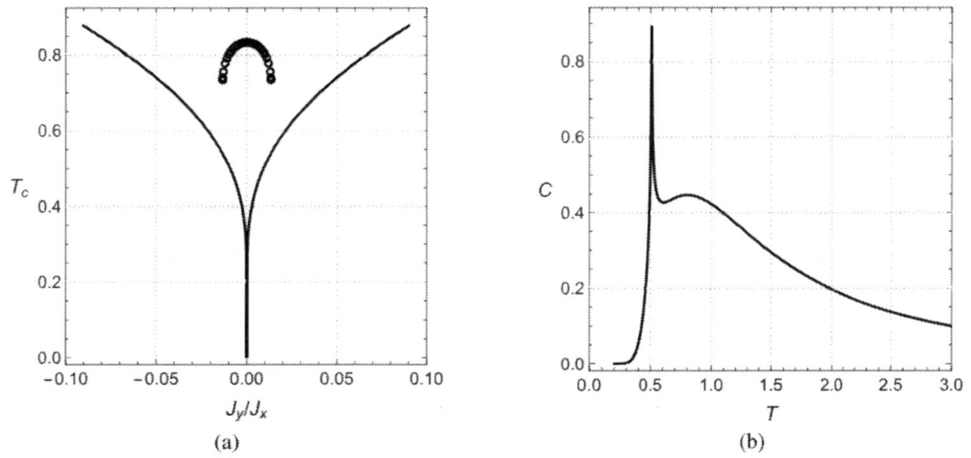

Figure 4. Phase transition temperature as a function of J_y/J_x (a) and temperature dependence of heat capacity at $J_y/J_x = 0.01$ (b) in Ising model on a square lattice.

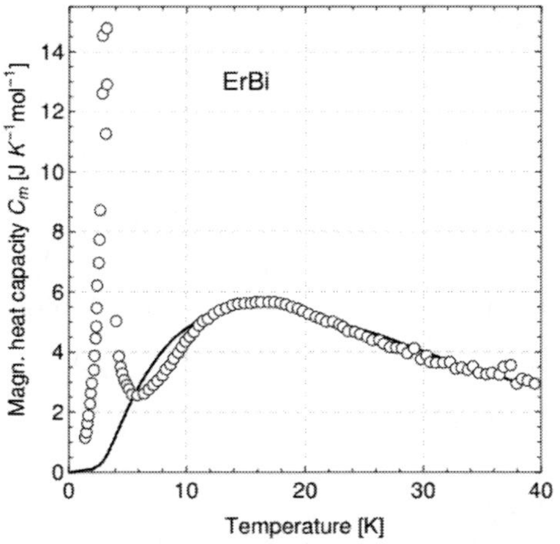

Figure 5. Magnetic heat capacity phase in ErBi. Circles — experiment. Solid line — calculation.

stacking. The partial ordering in this case is displayed in Fig. 6b. It consists of alternating ferromagnetic columns, but their stacking is violated. Calculating the internal energies of Néel structure, SAF structure, and the structure with stacking faults such as in Fig. 6b, we find out that all of them coincide.

In 3D systems, which consist of $N_xN_yN_z$ sites, the frustrations may create two possible kinds of partial ordering. The first is related to ordering in two dimensions that is the ordering within parallel planes and stacking faults of these planes. In this case, the zero-temperature entropy is equal to $\ln(q)/N_z$ and vanishes in the thermodynamic limit; the phase-transition temperature has non-zero value, and heat capacity has a form of sharp λ-shaped Onsager's peak. This is the partial ordering from 3D to 2D dimensionality.

The second kind of partial ordering corresponds to ordering in one dimension that is the ordering within parallel lines and stacking faults of these lines. In this case, the zero-temperature entropy is equal to $\ln(q)/N_xN_y$ and also vanishes in the thermodynamic limit. However, the phase-transition temperature vanishes, phase transition does not occur and heat capacity has gently sloping shape, as in all 1D lattices. This is the partial ordering from 3D to 1D dimensionality.

Our studies [3] in Ising model with competing antiferromagnetic J and J' interactions on bcc lattice, in which the frustration point is equal to $J'/J = 2/3$, meets all the features of first-kind partial ordering. The 2D ordering takes place within planes perpendicular to cube space-diagonals of (111) type. The second kind of partial ordering is not yet found.

5 Competing and noncompeting interactions

It is commonly accepted that competing interactions may be of two kinds, either $J < 0$ and $J' < 0$ or $J > 0$ and $J' < 0$, and it is really valid for Ising model in all dimensions, for 3-state Potts model on a triangular lattice, and many others. However, $J < 0$ and $J' < 0$ interactions appear to be favoring rather then competing, which strongly depends on the choice of a model and a lattice topology. The example of noncompeting $J < 0$ and $J' < 0$ interactions in 4-state Potts model on a triangular lattice is shown in Fig. 7. Hereinafter, numbers in circles denote four permissible spin directions corresponding to the 4-state standard Potts model. They are such that any unlike pair makes one and the same angle. Both interactions demand nonparallel arrangement for all pairs of nearest and next-nearest spins, and these demands are simultaneously fulfilled in the Fig. 7, resulting in the absence of frustrations at any values of interactions.

 https://doi.org/10.1051/epjconf/201818511002

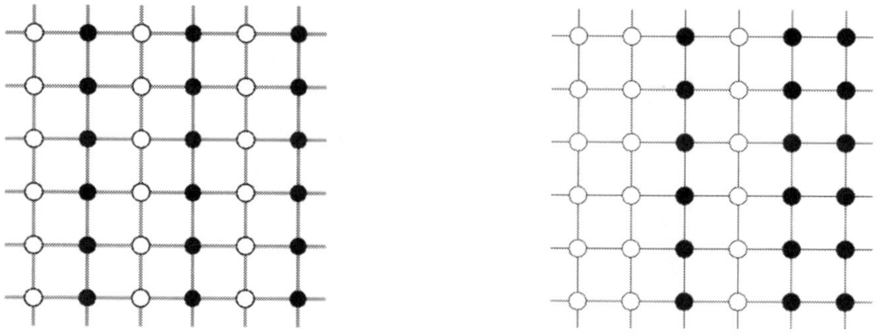

(a) (b)

Figure 6. SAF ordering, $J'/J > 0.5$ (a) and partial ordering, $J'/J = 0.5$ (b) in Ising model on a square lattice.

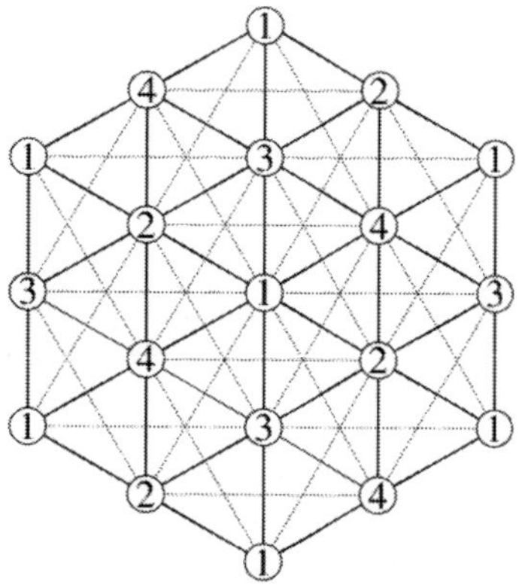

Figure 7. Noncompeting interactions on triangular lattice. 4-state Potts model.

6 Summary

We considered the Ising and Potts models with taking into account the nearest and next-nearest neighbor interactions on 1D, 2D, and 3D lattices: linear chain, square, triangular, kagome, honeycomb, and body-centered cubic. We have revealed many distinguishing features concerning magnetic orderings, phase transitions, and frustrations.

The competition between interactions and an external magnetic field may create either frustration points or phase transitions points, and sometimes even the ranges of frustrations. The interplay of competing interactions results sometime in the partial ordering (which means lowering of ordering dimensionality). This phenomenon occurs in Ising model on square, kagome, triangular, and body-centered cubic lattice.

In some cases the same interactions may either be competing or noncompeting. For example, in 4-state Potts model, the antiferromagnetic nearest and next-nearest interactions on 1D chain compete so strong that lead to frustrations at any values, so that only the third-neighbor interaction can create ordering. Quite the contrary, on a triangular or honeycomb lattice these interactions are noncompeting and lead to the phase transitions also at any values.

We predict that at appropriate choice of a model the magnetic properties of 1D, 2D, and 3D lattices will be similar despite that linear chain does not possess a phase transition while 2D and 3D lattices do. The rescaling of the temperature and magnetic field is the necessary condition thereto.

The most astonishing result is that in every model on any lattice, in the vicinity of frustration point or frustration field the specific heat splits into a sharp peak and a broad dome-shaped peak, both of which move with the field variation.

The research was carried out within the state assignment of FASO of Russia (theme «Quantum» no. 01201463332).

References

[1] F. Wang, D.P. Landau, Phys. Rev. E **64**, 056101 (2001)

[2] A. Kalz, A. Honecker, S. Fuchs, T. Pruschke, Eur. Phys. J. B **65**, 533 (2008)

[3] A.K. Murtazaev, M.K. Ramazanov, F.A. Kassan-Ogly, D.R. Kurbanova, JETP **120**, 127 (2015)

[4] A.B. Babaev, M.A. Magomedov, A.K. Murtazaev, F.A. Kassan-Ogly, A.I. Proshkin, JETP **122**, 310 (2016)

[5] L. Onsager, Phys. Rev. **65**, 117 (1944)

[6] R.M.F. Houtappel, Physica **16**, 425 (1950)

[7] K. Kanô, S. Naya, Prog. Theor. Phys. **10**, 158 (1953)

[8] H. Wada, H. Imai, M. Shiga, J. All. Comp. **218**, 73 (1995)

[9] T. Matsumara, H. Shida, T. Suzuki, Physica B, Condensed Matter **230–232**, 738 (1997)

[10] E. Scheer, J. Wosnitza, H.v. Löhneysen, Z. Phys. B **85**, 79 (1991)

 https://doi.org/10.1051/epjconf/201818511003

Laws of volume elasticity in physical processes of the effect of T, P, H parameters on the lattice modifications, phase transitions, conductivity and magnetism

Petr Polyakov
Institute for Physics of Mining Processes, 72 R. Luxemburg Str., 83114 Donetsk

Abstract. Analysis of the experimental results allows to single out the law of bulk elasticity that implies that the structural binding energy is formed by physical and chemical methods and transformed into elastic stresses that control lattice changes, formation of phase transitions, resistivity and magnetism in the course of physical processes under the effect of parameters. The laws of volume thermo-, baro- and magnetoelasticity are the determining factors in lattice modifications in the structure sites together with the composition variation.

1 Introduction

The laws of volume thermo-, baro- and magnetoelasticity are the determining factors in lattice modifications in the structure sites together with the composition variation. It is the binding energy transformed into the stresses due to thermoelastic compression that redistributes the state of electron-phonon bonds, the volume and the density. Besides, baro- and magnetoelasticity combined with magnetic uncompensation demonstrate "cooling" and "heating" effects due to additional energy of strains that allow redistribution of free and semi-free electrons into a phonon bond and adding to modification of the structure as a whole. The deformation factor accompanied by thermoelastic strains determining phase transitions and magnetic field uncompensation control regularities of conductivity jump in the form of colossal magnetoresistance. The presented work is aimed at settling the question of the presence of anomalies and regularities in view of the laws of the bulk elasticity.

2 Results and discussion

When analyzing a wide spectrum of the results of the experimental studies of physical processes under the thermodynamic parameters varied, a question arises, if an analogy in the effects should be singled out. A number of known scientists contributed to the investigation of the magnetic field effect as presented in the Kapitsa's works [1]. The experiments were based on the employment of temperature and high magnetic fields. The works of Bridgeman [2] dealt with a wide pressure range at a fixed temperature. In Vereschagin's works [3], a number of methods involving high pressure were presented. The works of A.A. Galkin scientific school [4] were very representative, because they employed sensitive methods of investigation and extensive results about the effect of three parameters were analyzed.

The main determining factor of the evolution in the form of deforming elastic stress is the elastic energy that is higher of the energies of other microprocesses. This fact requires understanding of the interaction laws within the structure, to control modifications of the density, volume, the number of the neighbors, alternation of the high symmetry to the low one (polymorphous transformations related to the reconstruction of the structure as a transition to a dense packed structure). Phase transformations are associated with a volume jump and insignificant structure transformation within the range of 1%, with the anisotropy of the elastic stresses participating. It should be noted that the energy of the elastic stresses is the basis separating the phase states of the physical processes [5].

Under the effect of H and P, energy changes are generated in the structure sites of the samples. The changes of the energy affect the state of the electron and phonon subsystems that are analogous to those caused by impurities or defects.

X-ray methods provided registered four-fold difference in the pressure effect on the structure compressibility that emphasizes the role of the anisotropy of elasticity [6]. The analysis of the effect of magnetoelastic and elastic energies allowed separation of combinations of the parameters of the elastic interactions. The results of the studies of magnet-containing semiconductors revealed a regularity in the T-H-P effect [7]. It was found that the effect of 6,2K corresponds to that of 2,7 kOe and 1 kbar [8]. This was an evidence of the causing role of the structure modifications.

* Corresponding author: poljakov@mail.fti.ac.donetsk.ua

Analogies and regularities of the phase state modifications were established in single crystal semiconductors and dielectrics. The presented results illustrate the causing role of the deforming elastic stresses that modifies the structure and the properties and forms the basis of the mechanics in the course of the structural transformations in a wide spectrum of solids.

Further, an attention is concentrated on the analysis of the studies of manganite-based polycrystal structures and cuprate-based monocrystals [5]. As examples, the measurements of conductivity [9] are presented; the temperature dependence under the effect of the hydrostatic pressure P and the magnetic field H is separated. Some details of the reduction of the resistance within the whole temperature range are observed. The metal part of the resistance reduction is explicitly emphasized to compare it with Fig.1. As a comparison, Fig.2 is presented [10] where the resistance of the $La_{1-x}Sr_xMnO_3$ samples is modified in the course of the processes of varied composition *x*. Thus, any structural changes are reflected in the properties and these are the laws of the volume elasticity that are the determining factors of the measurements in the physical processes of the parameter effects.

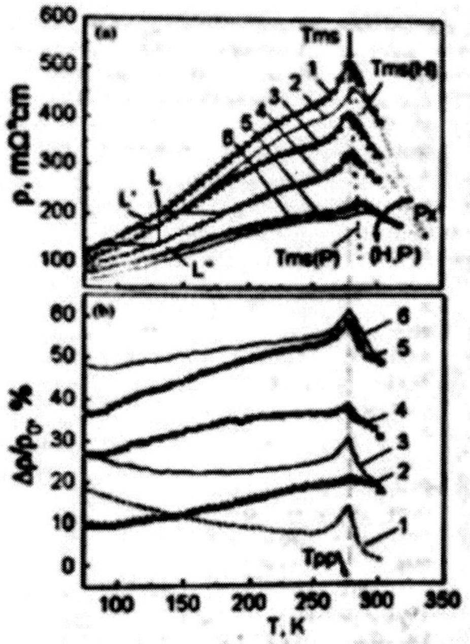

Fig.1. Temperature dependence of the resistivity (a) and magneto-, baro-, magnetobaroresistive effects (b) in $La_{0.7}Ca_{0.3}MnO_3$ under the effect of magneretic field and hydrostatic pressure: a) 1 – P=0, H=0; 2 – P=0,H=8kOe; 3 – P=6 kbar, H=0; 4 – P=12 kbar, H=0; P=17 kbar, H=0; b) 1 – P-0, H=8kOe; 2 – P=6 kbar, H=0; 3 – P=6 kbar, H=8 kOe; 4 – P=12 kbar, H=0; 5- P=17 kbar, H=0; 6 – P=17 kbar, H=8 kOe [5].

The introduced energy of the elastic stresses is transformed into the united energy that forms electron and phonon bonds. The energy of 0.1 eV corresponds to 10 kbar. In the course of the structure formation, the binding energy of the nucleus and the electrons of the closed internal shells is much higher of the binding energy of the stressed shells. These are the energies that play the major role in the formation of the site structure and redistribution of the electron levels in the phonon structure. In the course of gradual filling of the external shells, an additional energy of the elastic stresses related to the binding energy is about 0,1-1 eV.

The selected research area including the volume elasticity allows establishing the regularities of the elasticity anisotropy in a magnetodielectric [11]. The regularities of the first-order phase transition have been found in the frequency-field dependences in three parameters. The temperature of the phase transition at 9,8°K [12] has been demonstrated first.

Now we shall evaluate the levels of energy contribution in the course of consideration of critical phenomena and properties. This matter touches upon principal questions of the direct relation between thermodynamic parameters and structure evolution through the mechanisms of elastic stresses affecting the volume and other parameters controlling the structural phase transitions and the properties of phase states. We shall estimate the energy of Coulomb interactions as 1÷10 eV, the influence of the crystal field as 0.1÷1 eV, spin-orbit relationship as 10^{-2}÷10^{-1} eV, spin-spin (magnetic) bond as 10^{-4} eV, electron-nuclear bond as 10^{-4}÷10^{-5} eV. According to our estimations, the energy of elastic stresses with taking into account the coefficient of compressibility is 1÷10 eV.

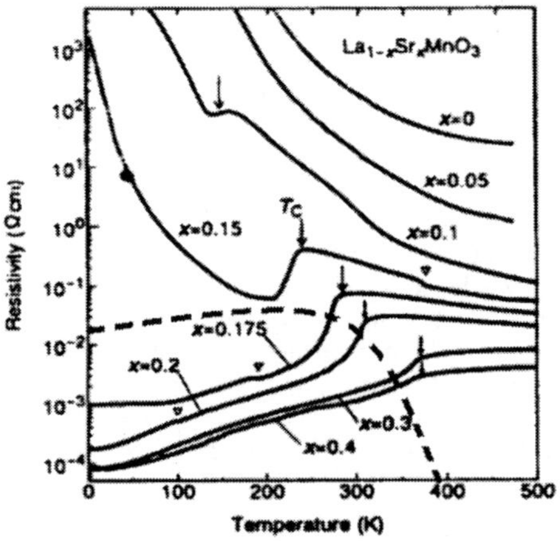

Fig.2. Temperature dependence of the resistivity for various single crystals of $La_{1-x}Sr_xMnO_3$. Arrows indicate the Curie temperature. The open triangles indicate anomalies due to structural transitions [10].

Unfortunately, methods of investigations under the hydrostatic pressure were far from being perfect at that time. Our result shows only a combined T-P-H effect (effect of temperature, pressure and magnetic field) through mechanisms of elastic deforming (ED) stresses. It provides an objective consideration of physical processes; otherwise, new approaches cannot be elaborated. The magnetism of magnet-containing structures can get the most effective consideration from the viewpoint of the experiment. There the real physical process is registered if the form of the evolution of the properties of a solid that testifies the regular bond of the changes of the crystal structure and the electronic and magnetic properties.

Magnetization is a consequence of the influence of the magnetic field on magnetically non-compensated constituent of a particle provided it is high enough to affect the structure significantly.

3 Conclusions

We should take into account that quantum mechanical forces forming the magnetism are of short range in fact, so deep understanding of the regularity if interaction is necessary for the estimation of this process. The elastic energy is an immediate energy of long range. These are elastic stresses in the structure that form the linearity of the magnetization regardless of important details of microscopic interactions. By definition, atoms and molecules have small magnetic moments of non-compensation. The structure is the totality of molecules and atoms brought into the sites of the structure (several atoms and molecules) and bound in the lattice by the compatibility of valent and free electrons. The state of the structure and its predisposition to paramagnetism is formed by high-temperature physical and chemical processes where the magnetic factor is absent by definition. In other words, the high-temperature area has no ability for magnetization.

The succeeding cooling and bulk compression result in the loss of heat energy transformed into the loss of the energy of the elastic stresses in the structure. This process is followed by the linear evolution of the parameter due to the mechanisms of bulk elasticity.

The object of the present paper is an analysis, drawing analogies and generalization of regularities of an evolution of structural phase transitions and properties in different multi-component magnet-containing structures. It is also aimed at the presentation of unambiguous T-P-H impact through mechanisms of elastic anisotropic deforming stresses and settling the question of the presence of anomalies and regularities in view of the laws of the bulk elasticity.

The presented results allow statement that the basis of magnetism is formed in the energetic bonds of the structure sites. Here magnetic non-compensation is manifested that appears in the course of electron-phonon interaction under the effect of the introduced energies to be transformed into the elastic stresses by the effect of temperature, pressure and magnetic field.

References

1. P.L. Kapitsa, The strong magnetic fields, Nauka, Moscow, 1988 (in Russian).
2. P. Bridgmen, Newest works in the high-pressure area, Nauka, Moscow, 1948 (in Russian).
3. L.F. Vereschagin, Solids under high pressure, Nauka, Moscow., 1981 (in Russian).
4. A.A. Galkin,, Physics of Solid State, Naukova dumka, Kiev, 1986 (in Russian).
5. P.I. Polyakov, T.A. Ryumshina, Magnetism and laws of bulk elasticity, Transworld research network, Kerala, 2009.
6. A.A. Galkin, S.V. Ivanova, P.I. Polyakov, V.I. Kamenev, Structural, magnetic and resonance properties of $CuCl_2$ $2D_2O$ Sol. State Phys., 8 (1979) 2580-2583
7. P.I. Polyakov, S.S. Kucherenko, Regularities in temperature, magnetic field and pressure effect on the resistive properties of magnetic semiconductors, JMMM 248, (2002) 396-401.
8. P.I. Polyakov, S.S. Kucherenko, Elastic deformation mechanisms of the effect of temperature, magnetic field and pressure on the resistive and magnetic properties of magnetic semiconductors, Fiz. Nizk.Temp. 28, (2002) 1041-1047.
9. P.I Polyakov, S.S. Kucherenko, Dynamics of structural phase transition and changes in properties under the influence of elastically anisotropic deforming stresses, regularities of critical lines and points in magnetic semiconductors and magnetic dielectrics, JMMM 278, (2004) 138-155.
10. E. Dagotto, T. Hotta, A.Moreo, Colossal magnetoresistant materials: the key role of phase separation, Phys. Rep. 344, (2001) 1-153.
11. A.A Galkin, S.V. Ivanova, V.I. Kamenev, Magnetyostriction effect on resonant properties of cuprpus chloride dehydrate under pressure, Proc. Of XX congress AMPERE, Tallinn, 1978.
12. P.I Polyakov, Materials science Feb.2015 arxiv: 1502.07473

CURRENT-DRIVEN QUANTUM DYNAMICS OF TOROIDAL MOMENT IN RARE-EARTH NANOCLUSTERS

D.I. Plokhov[1,⋆], *A.I.* Popov[2,3], and *A.K.* Zvezdin[1,2]

[1] *A.M. Prokhorov General Physics Institute, Russian Academy of Sciences, Moscow, Russia*
[2] *Moscow Institute of Physics and Technology, Dolgoprudny, Moscow Region, Russia*
[3] *National Research University of Electronic Technology, Zelenograd, Moscow, Russia*

Abstract. The current-driven quantum dynamics of toroidal moment in single molecule magnets with polygonal rare-earth ion core is investigated. The effects of an external current is considered both in equilibrium and in the frames of the Landau-Zener-Stückelberg tunneling model. It is shown that the toroidal moment is a suitable degree of freedom for using the systems in question as qubits in quantum computing.

1 Introduction

The idea of quantum computation began to be actively discussed in the world in 1982, after the publication of R. Feynman [1]. The avalanche-like growth in the number of publications on quantum computation was caused by Shor's article [2] on quantum (implemented on a quantum computer) algorithms. Further progress in the area of quantum computing mainly follows two lines, the development and improvement of quantum algorithms and the search for suitable physical implementations of a qubit. The basic requirements (so called DiVincenzo checklist) for physical systems claiming the role of qubits are formulated in ref. [3].

Over last two decades, several physical implementations of a qubit have been proposed. Some of them origin from quantum optics: cavity quantum electrodynamics [4], including linear optic quantum computation [5] with the so-called boson sampling [6]. Others deal with ultracold atoms: trapped ion quantum computer [7] and optical lattice one [8]. Very exciting are spin-based representations: Kane quantum computer [9], Loss–DiVincenzo quantum computer [10], and nitrogen-vacancy center quantum computer [11]. Quite elaborated are superconducting qubits: charge, flux, and phase [12]. There are also proposals [13, 14] on the use of magnetic particles in quantum computing.

There has been made a great progress on the way, but the «ideal» qubit has not been developed yet. Research efforts are focused mainly on the properties of the very physical systems. At the same time, the search for new implementations still remains topical. We propose to use a new degree of freedom, the toroidal (anapole) moment, which is natural characteristics of recently synthesized family of dysprosium based ring single molecule magnets (toroics). The toroidal moment is easily accessible by electric means (electric fields, which are more localizable than magnetic ones, or just an electric current). In this paper, we describe the macroscopic quantum dynamics of the toroidal moment coupled with an electric current. The results make it possible to conclude that toroidal moment is convenient to implement an anapole qubit. The compliance with the DiVincenzo criteria is discussed.

⋆e-mail: dmitry.plokhov@gmail.com

2 Single molecule toroics

We consider a spin ring, a system of N non-Kramers rare-earth ions located in the apices of a regular polygon. The ground state of rare-earth ions in compounds is formed mainly due to the influence of a crystal field and often has a strong anisotropy of the magnetic moment. For example, the ground state of Dy^{3+} ions in Dy_3 cluster is the Kramers doublet $|M_J = \pm 15/2\rangle$ [15] and responds only to the z_i local component of an external magnetic field (see fig. 1). The first excited states are close to $|M_J = \pm 13/2\rangle$ and separated from the ground one by the gap of 200 cm^{-1}. In the case of non-Kramers ions, asymmetrical perturbations remove degeneration, thus making the ground state be quasi-doublet, i.e. the two close singlets with splitting small compared with distance from the higher levels. The effective Hamiltonian of the system reads

$$\mathcal{H} = -\frac{j}{4}\sum_{i<k}^{N}\sigma_{iz}\sigma_{kz} - \frac{\Delta}{2}\sum_{i=1}^{N}\sigma_{ix} - g_J\mu_B\sum_{i=1}^{N}H_{z_i}J_{z_i}\sigma_{iz}, \quad (1)$$

where σ_x, σ_y, and σ_z stand for the Pauli matrices, j is the exchange interaction constant ($j \sim 10$ cm^{-1}), Δ is the energy gap between the singlet levels of an ion in the crystal field with respect to the local axes ($\Delta \sim 0.05$ cm^{-1}). It is shown [16] that eq. (1) is valid not only for non-Kramers ions, but for Kramers ions as well. In the latter case, the splitting Δ can be produced by an external magnetic field H, and $\Delta \to 0$ at $H \to 0$.

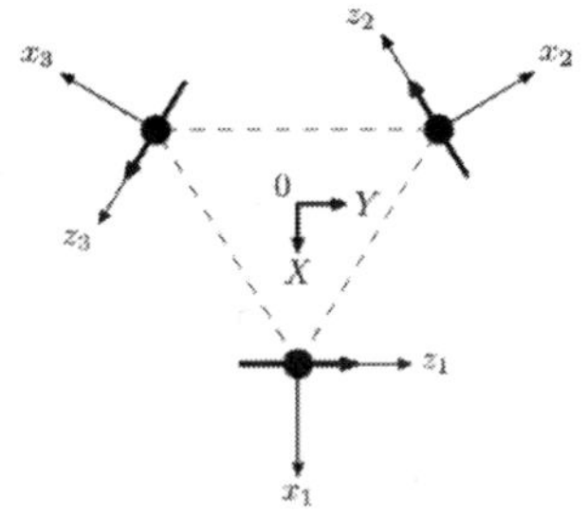

Figure 1. The spin structure of a triangular rare-earth molecular cluster and the local easy axes orientation in respect of the laboratory XYZ reference frame. The thicker arrows represent the spins of the rare-earth ions in the molecule in the state with toroidal moment $T_Z = +3T_0$ ($T_0 = \frac{1}{2}g_J\mu_B r_0 J$).

There are 2^N possible orderings of the ion spins in the rare-earth polygonal cluster. The spin orderings can be characterized in terms of spin chirality. It is clear that there are always pairs of states with spins inversely twisted, i.e. the states have opposite chirality. The natural physical quantity associated with spin chirality in this case is the T-odd polar vector of the anapole (toroidal) moment, so the systems in question can be called single molecule toroics (SMTs) by analogy of single molecule magnets (SMMs). We would remind that the toroidal moment operator in the case of localized magnetic ions can be defined as

$$\mathbf{T} = \frac{1}{2}g_J\mu_B \sum_{i=1}^{N} [\mathbf{r}_i \times \mathbf{J}_i],$$

where $\mathbf{r}_i$ is the radius-vector connecting the center of the N-gon with the i-th apex (all $r_i = r_0$). The toroidal moment of the system is just formally equal to the effective "magnetization", associated with the effective Hamiltonian in eq. (1). The values of the toroidal moments for conjugate states are different in sign.

3 The current driven dynamics

To describe the interaction between the toroidal moment and an external electric current it is necessary to put the term $\hat{V} = \frac{4\pi}{c}\mathbf{j}\hat{\mathbf{T}}$ into the Hamiltonian in eq. (1). The electric current $\mathbf{j}$ could be produced as the displacement current $\mathbf{j} = \frac{1}{4\pi}\frac{\partial \mathbf{E}}{\partial t}$. If the electric field strength $\mathbf{E}$ linearly depends on the time and is directed along the Z-axis, then $\hat{V} = \frac{v}{c} \cdot \hat{T}$, where $v = \partial E_Z/\partial t$. The energies of the two low-lying levels could be approximated in the vicinity of the avoided level crossing at $j_z = 0$ as (see ref. [16])

$$E_\pm(x, j_z) = E_0 \mp \sqrt{\left(\frac{4\pi}{c} \cdot 3T_0 \cdot j_z\right)^2 + \left(\frac{\delta}{2}\right)^2}, \qquad (2)$$

where the crossing energy is $E_0 = -\frac{3}{4}j$, the avoided level splitting is $\delta = \frac{3}{2}jx^3$, $x = \Delta/j$, and $T_0 = \frac{1}{2}g_J\mu_B r_0 J$.

The toroidal moment is a conjugate variable against the electric current density and can be found at zero temperature as

$$T(x) = \frac{c}{4\pi}\frac{\partial E_-(x, j_z)}{\partial j_z}, \qquad \tau(x) = \frac{T(x)}{T_0}. \qquad (3)$$

The plot of the relative (equilibrium) dependence is given in fig. 2 (curve 1). It is seen, that the current changes the direction of the spin twisting to the opposite, thus reversing the toroidal moment of the system in a way of the relatively sharp jump of $\Delta\tau = 6$. The anapole moment reversal, which could be called reanapolization, the reanapolization requires the currents of 10^7 A/cm^2.

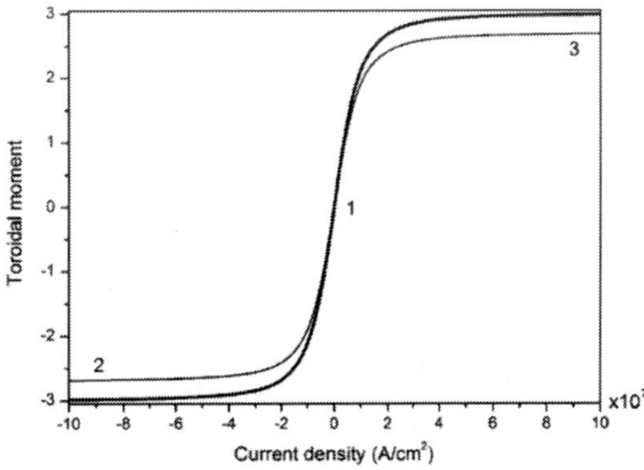

Figure 2. The zero temperature plot of toroidal moment projection onto the laboratory Z-axis vs the external current density: thicker curve (1) in equilibrium, the other curves in the case of the finite value of current sweeping rate 10^{13} A/(cm^2·s), (2) from $\tau = +3$ to $\tau = -2.7$ and (3) from $\tau = -3$ to $\tau = +2.7$.

The tunneling processes when sweeping the current j_z at a constant rate over an avoided energy level crossing can be treated in the frames of the Landau-Zener tunneling model. The probability P to change the state characterized by the quantum number of the toroidal moment at the avoided level crossing is given by the expression

$$P = 1 - \exp\left(-\frac{\pi\delta^2}{(4\pi/c)\hbar \cdot T_0 \cdot |\tau_1 - \tau_2| \cdot |dj_z/dt|}\right), \qquad (4)$$

where $\tau_{1,2} = \pm 3$ are the toroidal moment quantum numbers of the avoided level crossing with the splitting δ, and dj_z/dt is the current sweeping rate. For $dj_z/dt \sim 10^{13}$ A/(cm^2·s) and $\Delta/j = 0.005$ we have $P \approx 0.95$.

With the Landau-Zener-Stückelberg model in mind, we can now start to understand qualitatively the hysteresis in the system considered. If current j_z is sufficiently large negative, than, at very low temperature, all molecules are in the $|\tau = -3\rangle$ ground state. When the current is come down to zero, all molecules will stay in the $|\tau = -3\rangle$ ground state. When passing the current over the avoided level crossing region at $j_z \approx 0$, there is Landau-Zener probability P ($0 < P < 1$) to tunnel from the $|\tau = -3\rangle$ to the $|\tau = +3\rangle$ state. The dependence of the toroidal moment on the current then reads as follows

$$\tau(x, j_z) = \frac{c}{4\pi T_0}\left(P \cdot \frac{\partial E_+(x, j_z)}{\partial j_z} + (1 - P) \cdot \frac{\partial E_-(x, j_z)}{\partial j_z}\right) \qquad (5)$$

https://doi.org/10.1051/epjconf/201818511004

shown in fig. 2 (curve 3). The toroidal moment undergoes the jump of $\Delta\tau = 6P$. So the average value of toroidal moment τ becomes less than 3, namely $\tau = 2.7$ for $P = 0.95$. The relevant mixed quantum state relaxes to the equilibrium $|\tau = +3\rangle$ state due to interaction with environment (the $3 \to 1$ process in fig. 2).

Sweeping the current in the opposite direction from the $|\tau = +3\rangle$ state, we arrive to the mixed state with toroidal moment average value of -2.7, see curve 2 in fig. 2. The state relaxes as above to the equilibrium $|\tau = -3\rangle$ state (the $2 \to 1$ process in fig. 2). This way, we will come to the closed hysteresis loop.

4 Anapole qubit

In this section, we will briefly discuss the possibilities of using single-molecule toroics as qubits in quantum computing. It was shown, that the most important characteristic of such systems is the toroidal (anapole) moment, which is effectively coupled with electric current, therefore we will call such a qubit an anapole one.

1. The first requirement of creating systems of many ($N > 10^3$) qubits, each of which has two discrete energy levels, can easily be fulfilled. Indeed, all the molecules are obviously identical and can be synthesized in any necessary amount. What levels can be working ones in the qubit? To answer the question, suppose that the molecules are subjected to an electric current. The electric current, generally speaking, depends on time, which is necessary for controlling the particle states during initialization and when performing logical operations.

We will focus on the states $|T_Z = +3T_0\rangle$ and $|T_Z = -3T_0\rangle$, denoting them $|\uparrow\rangle$ and $|\downarrow\rangle$ accordingly. All other states have a much higher energy than those indicated if the electric current is not too strong. In this case the system can be considered a two-level system, and its Hamiltonian, written in the $\frac{1}{2}$-spin representation, has the simple form

$$\mathcal{H} = -\frac{\varepsilon}{2}\sigma_z - \frac{\delta}{2}\sigma_x = -\frac{E}{2}\left(\sigma_z \cdot \cos\varphi + \sigma_x \sin\varphi\right),$$

where $\varepsilon = \frac{4\pi}{c} \cdot 3T_0 \cdot j_z$, δ is the avoided level splitting for zero current (see the previous section), $E = \sqrt{\varepsilon^2 + \delta^2}$, and $\tan\varphi = \delta/\varepsilon$. The eigenstates of the Hamiltonian are the logical states of the qubit:

$$|0\rangle = \cos\frac{\varphi}{2}|\uparrow\rangle + \sin\frac{\varphi}{2}|\downarrow\rangle,$$

$$|1\rangle = -\sin\frac{\varphi}{2}|\uparrow\rangle + \cos\frac{\varphi}{2}|\downarrow\rangle.$$

2. The second requirement to the sterling qubit deals with the possibility of preparing qubits of a quantum computer in the ground state $|00\ldots0\rangle$. Obviously, initialization can be achieved only with such deep cooling of an array of qubits, in which thermal excitations can not cause transitions to the higher state. This implies the natural requirement for an interval of operating temperatures $T << j/k_B \sim 15$ K. Of course, achieving such low temperatures is not an easy task, and it should be noted that the vast majority of qubit implementations have this drawback, but the limitation on the maximum possible temperature in the case of single-molecule toroics can be weakened by using molecules with larger interionic exchange constant. At the same time, this operating temperature range is three orders wider than that of traditional magnetic particles using the magnetic moment as the degree of freedom. Thus, the anapole qubit has an indisputable advantage.

3. The next requirement concerns the possibility of overcoming the relaxation of the degrees of freedom used in the qubit (decoherence). During the execution of calculations (unitary transformation), the qubits of the computer are constantly exposed to the environment. As a result, the amplitude and phase of the state vector of each qubit experience unwanted changes thus randomness acquires after the decoherence time has elapsed. However, it is possible to achieve a stable quantum computing process for an arbitrarily long period of time if the methods of phase and amplitude error correction are used. It is proved [17] that with relatively low requirements for error-free execution of elementary operations (the probability of error is not more than 10^{-4} per computational step), quantum correction methods ensure stable operation.

The calculation of the decoherence time is a non-trivial task, the solution of which is far beyond the scope of this paper. However, here we can make some numerical estimates. As we shall see, the execution time of one step of computation is the order of $\Delta t = 10\,\mu$s. This means that the decoherence time required for a stable quantum computing process must be at least $\tau_{coh} = 10^4 \cdot \Delta t \sim 100$ ms. According to ref. [18], the dysprosium based molecular clusters demonstrate the slow relaxation behavior with comparable lifetime of the ground states limited by the macroscopic quantum tunneling at 5 K.

4. The fourth requirement concerns the feasibility of the basic logic elements (gates) on qubits of this type. It is proved [21] that a quantum computer can be constructed from gates of only two types: a single-qubit gate and the two-qubit CNOT (controlled NOT). It should be noted that almost any logically invertible two-qubit gate (the exception is the classical SWAP gate) in combination with one-qubit gates allows performing any predetermined computational algorithm. The selection of the CNOT element is owed to the simplicity of its realization.

Suppose that the anapole qubit is initially in a state far from the degeneracy point. In this case, the eigenstates $|0\rangle$ and $|1\rangle$ are almost the states $|\uparrow\rangle$ and $|\downarrow\rangle$, respectively. The sudden pulse of an electric current forces the system to arrive in the point of degeneracy. Then, a rotation occurs in the effective spin space under the influence of the operator

$$U_x(\alpha) = \exp\left(\frac{i\alpha}{2}\sigma_x\right), \quad \alpha = \frac{\delta\Delta t}{\hbar},$$

after which the backward pulse of the electric current dropping to zero removes the system from the degeneracy point. With this sequence of operations, the logical operation NOT corresponding to a rotation through angle $\alpha = \pi$ produced over a period of time $\Delta t = \frac{\pi\hbar}{\delta} \sim 10\,\mu$s.

The phase shift between two logical states can be achieved by varying the magnitude of the electric current for a certain time by a small amount. The corresponding operator has the form

$$U_z(\beta) = \exp\left(\frac{i\beta}{2}\sigma_z\right), \quad \beta = \frac{E_0\Delta t}{\hbar}.$$

The CNOT valve is executed by acting on two interacting qubits: by means of the interaction, one qubit controls the evolution of the other. There exists a weak interaction between the qubits. However, when carrying out calculations on a quantum computer, it is desirable to have a controlled coupling between different qubits. Therefore, the coupling must be organised artificially. Without specifying yet the nature of the coupling between the qubits, we point out that the Ising type interaction ($H_{int}^{(1,2)} = J^{(1,2)}\sigma_z^{(1)}\sigma_z^{(2)}$) during the following sequence of state transformations of the 1st (controlling) and 2nd (controlled) qubits leads to the execution of the CNOT gate:

$$H^{(2)} \cdot U_z^{(1)}\left(\frac{\pi}{2}\right) \cdot U_z^{(2)}\left(\frac{\pi}{2}\right) \cdot U^{(1,2)}\left(\frac{\pi}{2}\right) \cdot H^{(2)},$$

$$U^{(1,2)}(\phi) = \exp\left(-i\phi\sigma_z^{(1)}\sigma_z^{(2)}\right), \quad \phi = \frac{J\Delta t}{\hbar}.$$

In this sequence, the operator $H^{(2)}$ performs the Hadamard transformation,

$$H^{(2)} = U_x^{(2)}\left(\frac{\pi}{2}\right) \cdot U^{(2)}\left(\frac{\pi}{2}\right) \cdot U_x^{(2)}\left(\frac{\pi}{2}\right).$$

5. Finally, the problem of measuring the states of qubits at the end of the computational process emerges. This, in fact, is the weakest point in almost all known qubit implementations. Unfortunately, there are no mastered technologies for such measurements.

The problem is closely related to the problem of controlling their states, which can be solved by the aid of superconductive current loops surrounding SMT particles. Electromagnetic signals coming from the computer-controlled device change the current in the loops, thereby generating an unsteady electric fields resulting in the displacement currents necessary to perform single-qubit operations. Using an additional system of current loops equipped with Josephson switches, a controlled interaction between any two qubits can be organized. The same loops can be used to measure the states of magnetic particles at the end of the calculations.

This scheme is aided by the effective coupling between current and toroidal moment, and has a number of advantages. First, the current loops have a small value of the power of the dissipated heat energy, which obviously favors the maintenance of the quantum coherence of the qubit states in the calculations. Secondly, the Josephson switches available in the circuit are capable of switching in a very short time.

5 Conclusion

The analysis of the toroidal moment dynamics gives the opportunity to assert that the electric current is an effective means of controlling the states of the ring rare-earth molecular nanoclusters. Thus, the toroidal moment appears to be a promising versatile degree of freedom for quantum computing. The rare-earth single molecule toroics are considered to be a possible physical realization of a qubit as they meet the requirements of DiVincenzo.

We would like to acknowledge the financial support of Russian Foundation for Basic Research (projects 15-02-08509 and 16-29-14005).

References

[1] R. Feynman, Inter. Jour. Theor. Phys. **21**, 467 (1982).

[2] P. W. Shor, Proc. of the 35th Annual Symposium on the Foundations of Computer Science, 1994, p. 124.

[3] D. P. DiVincenzo, Fortschritte der Physik, **48**, 771 (2000), see also arXiv:quant-ph/0002077.

[4] H. Walther, B. T. H. Varcoe, B.-G. Englert, T. Becker, Rep. Prog. Phys. **69**, 1325 (2006).

[5] P. Kok, W. J. Munro, K. Nemoto et al., Rev. Mod. Phys. **79**, 135 (2007).

[6] S. Aaronson and A. Arkhipov, Theory of Computing **9**, 143 (2013).

[7] D. Kielpinski, C. Monroe, and D. J. Wineland, Nature **417**, 709 (2002).

[8] G. K. Brennen, C. Caves, P. S. Jessen, and I. H. Deutsch, Phys. Rev. Lett. **82**, 1060 (1999).

[9] B. E. Kane, Nature **393**, 133 (1998); J. O'Gorman et al., Nat. Phys. J. Quant. Inf. **2**, 15019 (2016).

[10] D. Loss and D. P. DiVincenzo, Phys. Rev. A **57**, 120 (1998).

[11] L. Gordon, J. R. Weber, J. B. Varley et al., MRS Bulletin **38**, 802 (2013).

[12] C. Rigetti, J. M. Gambetta, S. Poletto et al., Phys. Rev. B **86**, 100506 (2012).

[13] J. Tejada, E. M. Chudnovsky, E. del Barco et al., Nanotechnology **12**, 181 (2001).

[14] A. K. Zvezdin and D. I. Plokhov, Bulletin of the Lebedev Physics Institute, issue 12 (2003), in Russian.

[15] L.F. Chibotaru, L. Ungur, and A. Soncini, Angew. Chem. Int. Ed. **47**, 4126 (2008).

[16] D.I. Plokhov, A.K. Zvezdin, and A.I. Popov, Phys. Rev. B **83**, 184415 (2011).

[17] A. M. Steane, Rep. Progr. Phys. **61**, 117 (1998).

[18] P.-H. Guo, J.-L. Liu, Z.-M. Zhang et al., Inorg. Chem. **51**, 1233 (2012).

[19] A. Barenco, C. H. Bennett, C. Cleve et al., Phys. Rev. A **52**, 3457 (1995).

Non-equilibrium critical behavior of Heisenberg thin films

Maria A. Shlyakhtich[1,2,*] and *Pavel V.* Prudnikov[2,**]

[1] *Siberian federal university, 79 Svobodny Av., Krasnoyarsk, 660041, Russia*
[2] *Omsk State University, Mira prospekt 55-A, Omsk, 644077, Russia*

Abstract.
In this work we study the non-equilibrium properties of Heisenberg ferromagnetic films using Monte Carlo simulations by short-time dynamic method. By exploring the short-time scaling dynamics, we have found thickness dependency of critical exponents z, θ' and β/ν for ferromagnetic thin film. For calculating the critical exponents of ferromagnetic films we considered systems with linear size $L = 128$ and layers number $N = 2$; 4; 6; 10. Starting from initial configurations, the system was updated with Metropolis algorithm at the critical temperatures

1 Introduction

It is well known now the fundamental role of competing interactions in the emerging features of low-dimensional systems. Among a wide number of numerical and theoretical investigations on equilibrium and dynamical properties of several model Hamiltonian of low-dimensional magnets, Heisenberg-like models are one of the most widely used to approach real magnetic materials. In fact, anisotropic versions of the Heisenberg model describe several compounds as K_2NiF_4 [1], $BaCo_2(AsO_4)_2$ [2], $CoCl_2$ – GIC [3] and Rb_2CrCl_4 [4].

Magnetic order in thin ferromagnetic films is very complex due to a strong influence of the shape and the magnetocrystalline anisotropies of the sample. In the past decade, a considerable amount of experimental results on different aspects of magnetism in ultrathin films has appeared [5]. Nevertheless it is difficult to reach general conclusions even in seemingly basic things such as the kind of magnetic order at low temperatures. In view of this complexity, theoretical work on simplified models and computer simulations are essential for rationalizing and guiding new experimental work.

In the vicinity of the critical temperature T_c, the thermodynamic observables associated with statistical models display universal characteristics, which may be parametrized in terms of critical exponents. These quantity tend to zero or infinity at the transition and depend only on the spatial dimensionality of the system, the range of the interactions and the number of components of the order parameter.

*e-mail: mmed@mail.ru
**e-mail: prudnikp@univer.omsk.su

2 Model and Methods

We have performed Monte Carlo simulations of macroscopic spin system with crystalline structure of ultrathin film which is described by anisotropic Heisenberg model [6] with Hamiltonian:

$$H = -J\sum_{i,j}^{N_s}\left[(1-\Delta(N))\left(S_i^xS_j^x + S_i^yS_j^y\right) + S_i^zS_j^z\right], \quad (1)$$

where $\mathbf{S}_i = (S_i^x, S_i^y, S_i^z)$ is a unit vector in the direction of the classical magnetic moment at lattice site i; the sum is extended over nearest-neighbor pairs on the cubic lattice; $J > 0$ – ferromagnetic exchange constant; $N_s = L\times L\times N$ – size of film; Δ characterizes the amount of anisotropy; $\Delta = 0$ corresponds to the isotropic Heisenberg case; $\Delta = 1$ – the Ising case. Periodic and free boundary conditions are used for the in-plane and out-plane directions, respectively.

The effective anisotropy constant $\Delta(N)$ as a function of film thickness N was chosen from experimental studies of the Curie temperature T_C for thin films of $Ni(111)/W(110)$ [7] with different thicknesses of Ni film. It was proposed that Δ is proportional to the critical temperature for different film thicknesses. In approximation procedure it was used fact that Ni films with large number of layers demonstrate bulk critical properties of 3D isotropic Heisenberg magnets [8, 9]. The effective anisotropy constant $\Delta(N)$ as a function of film thickness N presented on figure 1.

According to the argument of Janssen et al. [10] obtained with the RG method and ε-expansion, one may anticipate a generalized scaling relation for the k-th moment of the magnetization

$$m^{(k)}(t,\tau,L,m_0) = b^{-k\beta/\nu}m^{(k)}(t/b^z, b^{1/\nu}\tau, L/b, b^{x_0}m_0). \quad (2)$$

And it is realized after a time scale $t_{\rm mic}$ which is large enough in a microscopic sense but still very small in a

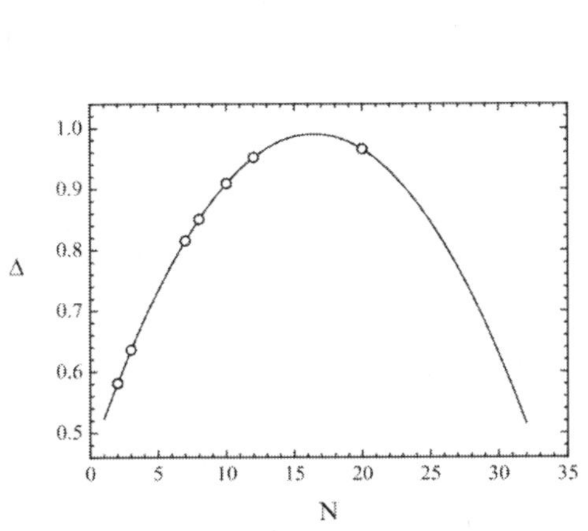

Figure 1. Anisotropy constant $\Delta(N)$ as a function of film thickness N

macroscopic sense. In (2) b is a spatial rescaling factor, β and ν are static critical exponents, z is the dynamic exponent, the new independent exponent x_0 is the scaling dimension of the initial magnetization m_0 and $\tau = (T - T_c)/T_c$ is the reduced temperature.

Since the system is in the early stage of the evolution the correlation length is still small and finite size problems are nearly absent. Therefore, we generally consider L large enough ($L = 128$) and skip this argument.

When system evolving from high-temperature state ($m_0 \ll 1$) we can choose the scaling factor $b = t^{1/z}$ and applying the scaling form (2) for $k = 1$ to the small quantity $t^{x_0/z}m_0$, one obtains

$$\begin{aligned} m(t,\tau,m_0) &\sim m_0 t^{\theta'} m(t^{1/\nu z}\tau, t^{x_0/z}m_0) = \\ &= m_0 t^{\theta'}(1 + a t^{1/\nu z}\tau) + O(\tau^2, m_0^2) \end{aligned} \quad (3)$$

where $\theta' = (x_0 - \beta/\nu)/z$ has been introduced. For $\tau = 0$ and small enough t and m_0, the scaling dependence for magnetization (3) takes the form $m(t) \sim m_0 t^{\theta'}$. For almost all statistical systems studied so far, the exponent θ' is positive. Time scale for this initial increase of magnetization is $t_0 \sim m_0^{-z/x_0}$. The increasing of the magnetization $m(t)$ from a high-temperature initial magnetization $m_0 \ll 1$ at short-time regime $t < t_{cr} \lesssim m_0^{-1/(\theta' + \beta/\nu z)}$ have the form $m(t) \sim t^{\theta'}$.

As the spatial correlation length in the beginning of the time evolution is small, for a finite system of dimension d with lattice size L the second moment $m^{(2)}(t,L) \sim L^d$. Combining this with the result of the scaling form in (3) for $\tau = 0$ and $b = t^{1/z}$, one obtains

$$m^{(2)}(t) \sim t^{-2\beta/\nu z} m^{(2)}(1, t^{-1/z}L) \sim t^{c_2}, \quad (4)$$

where $c_2 = (d - 2\beta/\nu)/z$. Until now, a completely disordered initial state has been considered as starting point, i.e., a state of very high temperature. The question arises how a completely ordered initial state evolves, when heated up suddenly to the critical temperature. In the scaling form (3), one can skip besides L, also the argument $m_0 = 1$,

$$m^{(k)}(t,\tau) = b^{-k\beta/\nu} m^{(k)}(b^{-z}t, b^{1/\nu}\tau). \quad (5)$$

The system is simulated numerically by starting with a completely ordered state, whose evaluation is measured at or near the critical temperature. The quantities measured are $m(t)$ and $m^{(2)}(t)$. With $b = t^{1/z}$, one avoids the main t dependence in $m^{(k)}(t)$ and for $k = 1$ one has

$$\begin{aligned} m(t,\tau) &= t^{-\beta/\nu z} m(1, t^{1/\nu z}\tau) = \\ &= t^{-\beta/\nu z}(1 + a t^{1/\nu z} + O(\tau^2)). \end{aligned} \quad (6)$$

In order to calculate the critical exponent z we have used the ratio $F_2(t) = m^{(2)}(t)|_{m_0=0}/m^2(t)|_{m_0=1} \sim t^{d/z}$ [11].

3 Results of Monte Carlo modeling

For calculating the critical exponents β/ν, z and θ' of ferromagnetic films we considered systems with linear size $L = 128$ and layers number $N = 2; 4; 6; 10$.

For ferromagnetic films the order parameter can be defined as magnetization

$$m^{(k)}(t) = \left\langle \left(\frac{1}{N_s} \left| \sum_{i=1}^{N_s} S_i(t) \right| \right)^k \right\rangle, \quad (7)$$

where angle brackets denote the statistical averaging. We have studied evolution of thin film both from high-temperature ($m_0 = 0.0001 \ll 1$) and from low-temperature ($m_0 = 1$) initial states. Starting from initial configurations, the system was updated with Metropolis algorithm at the critical temperatures $T_c = 1.03(8)$ for $N = 2$, $T_c = 1.25(5)$ for $N = 4$, $T_c = 1.35(4)$ for $N = 6$, $T_c = 1.57(8)$ for $N = 10$ [12, 13]. Simulations have been performed up to $t = 10000$ MCS/s. One would like to mention that measurements starting from a low-temperature initial state ($m_0 = 1$) are more favorable, since they are much less affected by fluctuations, because the quantities measured are rather big in contrast to those from a random start with zero or small initial magnetization ($m_0 \ll 1$).

In order to estimate values of critical exponents z, θ' and β/ν, we have measured the time evolution of the magnetization $m(t)$, the second moment $m^{(2)}(t)$ (figure 2*a*) from high-temperature and low-temperature initial states and autocorrelation function (figure 2*b*)

$$A(t) = \left\langle 1/N_s \sum_{i=1}^{N_s} S_i(t) S_i(0) \right\rangle - m(t)m(0) \sim t^{-c_a}, \quad (8)$$

where $c_a = d/z - \theta'$. In figure 3*a* the magnetization is plotted for a high-temperature initial state and a low-temperature initial state for different layer number $N = 2, 4, 6, 10$. It allowed to calculate the time-dependent cumulant $F_2(t)$ (figure 3*b*).

From slope of magnetization at high-temperature initial state on $t \in [710; 1800]$ we can estimate the value of critical non-equilibrium exponent θ'. The time interval $[t_{\rm left}; 1800]$ for $t_{\rm left} = 710$ gives the minimum of errors for exponent θ' for film thickness $N = 2, 4, 6, 10$ (figure 3*a*).

 https://doi.org/10.1051/epjconf/201818511005

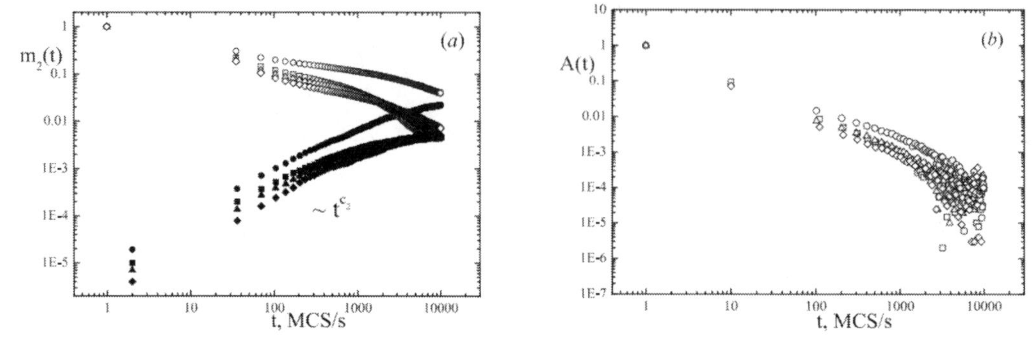

Figure 2. Time dependencies of second moment of magnetization $m^{(2)}(t)$ (a) for different initial state and autocorrelation function $A(t)$ (b) for film with $N = 2$ (○), $N = 4$ (□), $N = 6$ (△), $N = 10$ (◇)

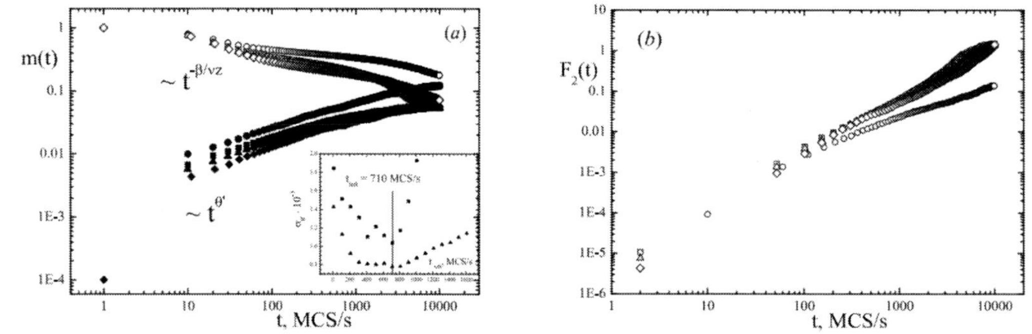

Figure 3. Time dependencies of magnetization $m(t)$ (a) for different initial state and cumulant $F_2(t)$ (b) for film with $N = 2$ (○), $N = 4$ (□), $N = 6$ (△), $N = 10$ (◇)

From the slope of time dependence of the cumulant $F_2(t)$ we can estimate the value of critical exponent d/z. Using the hyperscaling relation $\gamma/\nu + 2\beta/\nu = d$, the effective dimension of the system d_{eff} can be obtained [13]. This effective dimension $d_{\text{eff}} = 2.048$ for $N = 2$, $d_{\text{eff}} = 1.993$ for $N = 4$, $d_{\text{eff}} = 2.200$ for $N = 6$ and $d_{\text{eff}} = 2.199$ for $N = 10$ we use to estimate critical exponents z.

4 Conclusion

The dimensionality aspects of magnetic and structural phase transitions represent one of the key problems of ultrathin film [14, 15]. For magnetic systems, the spin dimensionality as well as the spatial extension determines the universality class, giving rise to a great number of ordering phenomena on different length scales. Furthermore, there are transition regions not represented by any universality class with corresponding critical exponents, but representing something in between. For example, the dependence of the critical exponents of thin magnetic films with thickness exhibit such a transition, in which the exponents are continuously changes with increasing film thickness of the layers, from two-dimensional (2D) Ising ($\beta = 0.125$) to three-dimensional (3D) Heisenberg ($\beta = 0.364$) [16] behaviour. The crossover from 2D to 3D critical exponents has been observed in thin films of Ni on W(110) [7] and of Co on Cu(111) [8] as their thickness is increased. Thus, studying the dimensional crossover of a system as its thickness is increased often provides a significant degree of insight.

In this paper we present result of Monte Carlo simulation of non-equilibrium critical behavior of Heisenberg thin film. We obtain values of $z = 2.187(16)$ for $N = 2$, $z = 2.197(33)$ for $N = 4$, $z = 2.300(26)$ for $N = 6$, $z = 2.108(21)$ for $N = 10$ and $\theta' = 0.248(5)$ for $N = 2$, $\theta' = 0.238(10)$ for $N = 4$, $\theta' = 0.225(8)$ for $N = 6$, $\theta' = 0.198(7)$ for $N = 10$. Comparing the calculated values z for $N = 2, 4$ with the results for 2D Ising model in various works $z = 0.191(1)$ [17], $z = 2.0842(39)$ [18], $z = 2.166(7)$ [19], $z = 2.1667(5)$ [20], $z = 2.14(2)$ [21],

$z = 2.12(6)$ [22] we see the very good agreement among them, i.e. thin ferromagnetic films belong to universality class non-equilibrium behaviour of 2D Ising model. Value of critical exponents $z = 2.300(26)$ for $N = 6$ and $z = 2.108(21)$ for $N = 10$ demonstrate the dimensional crossover to characteristic values for critical behavior of the 3D model.

5 Acknowledgments

The work was performed within the project of Russian Fund of Basic Research No. 16-32-00581 mol_a and within the grant No. MD-6024.2016.2 of Russian Federation President. The simulations were supported by the Supercomputing Center of Lomonosov Moscow State University, Moscow Joint Supercomputer Center and St. Petersburg Supercomputer Center of the Russian Academy of Sciences.

References

[1] R. J. Birgeneau, J. Skalyo, Jr. Shirane, G. Shirane, J. Appl. Phys. **41**, 1303 (1970).

[2] L. P. Regnault, P. Burlet, and J. Rossat-Mignod, Physica **86**, 660 (1977).

[3] D. G. Wiesler and H. Zabel, Phys. Rev. B **36**, 7303 (1987).

[4] L. M. Castro, J. A. Plascak, and A. S. T. Pires, J. Magn. Magn. Mater. **248**, 62 (2002).

[5] C. A. F. Vaz, J. A. C. Bland, and G. Lauhoff, Rep. Prog. Phys. **71**, 056501 (2008).

[6] K. Binder and D. P. Landau, Phys. Rev. B **13**, 1140 (1976).

[7] Y. Li and K. Baberschke, Phys. Rev. Lett. **68**, 1208 (1992).

[8] F. Huang, M. T. Kief, G. J. Mankey, and R. F. Willis, Phys. Rev. B **49**, 3962 (1994).

[9] A. Ney, A. Scherz, P. Poulopoulos, et al., Phys. Rev. B **65**, 024411 (2001).

[10] H.K. Janssen, B. Schaub and B. Schmittmann, Z. Phys. B **73**, 539 (1989).

[11] R. da Silva, N. A. Alves and J. R. Drugowich de Felecio, Phys. Lett. A **298**, 325 (2002).

[12] P.V. Prudnikov, V.V. Prudnikov, et al., JMMM **387**, 77 (2015).

[13] P.V. Prudnikov, V.V. Prudnikov, M.A. Medvedeva, JETP Letters **100**, 446 (2014).

[14] M. Parnaste, M. Marcellini, E. Holmstrom et al., J. Phys.: Condens. Matter **19**, 246213 (2007).

[15] E.T. Papaioannou, V. Kapaklis, A. Taroni et al., J. Phys.: Condens. Matter **22**, 236004 (2010).

[16] K. Chen, A. M. Ferrenberg, and D. P. Landau, Phys. Rev. B **48**, 3249 (1993).

[17] K. Okano, L. Schülke, K. Yamagishi and B. Zheng, Nucl. Phys. B **485**, 727 (1997).

[18] A. S. Krinitsyn, V. V. Prudnikov and P. V. Prudnikov, Theor. Math. Phys. **147**, 561 (2006).

[19] F.-G. Wang and C.-K. Hu, Phys. Rev. E **56**, 2310 (1997).

[20] M. P. Nightingale and H. W. Blöte, Phys. Rev. B **62**, 1089 (2000).

[21] C. Kalle, J. Phys. A **17**, 801 (1984).

[22] N. Jan, L. Leo Moseley and D. Stauffer, J. Stat. Phys. **33**, 1 (1983).

Unconventional phase separation in the model 2D spin-pseudospin system

K.S. Budrin[1,*], *Yu.D.* Panov[1,**], *A.S.* Moskvin[1], and *A.A.* Chikov[1]

[1] *Ural Federal University, Institute of Natural Sciences, 620002, 19 Mira street, Ekaterinburg, Russia*

Abstract. The competition of charge and spin orderings is a challenging problem for strongly correlated systems, in particular, for high-T_c cuprates. We addressed a simplified static 2D spin-pseudospin model which takes into account both conventional spin exchange coupling and the on-site and inter-site charge correlations. Classical Monte-Carlo calculations for large square lattices show that homogeneous ground state antiferromagnetic solutions found in a mean-field approximation are unstable with respect to phase separation into the charge and spin subsystems behaving like immiscible quantum liquids. In this case, with lowering of a temperature one can observe two sequential phase transitions: first, antiferromagnetic ordering in the spin subsystem diluted by randomly distributed charges, then, the charge condensation in the charge droplets. The inhomogeneous droplet phase reduces the energy of the system and changes the diagram of the ground states. On the other hand, the ground state energy of charge-ordered state in a mean-field approximation exactly matches the numerical Monte-Carlo calculations. The doped charges in this case are distributed randomly over a system in the whole temperature range. Various thermodynamic properties of the 2D spin-pseudospin system are studied by Monte-Carlo simulation.

1 Introduction

The models with competing or intertwining order parameters are popular in the condensed matter theory in connection with such real systems as, for example, multiferroics or high-T_c cuprate superconductors. In cuprates, the competition of static magnetic order, bulk superconductivity and charge-density waves has attracted a lot of attention over the years, but its nature remains a challenge [1]. Earlier we suggested a simplified static 2D spin-pseudospin model [2, 3] which takes into account both conventional Heisenberg spin exchange coupling and the on-site and inter-site charge correlations. A detailed qualitative and quantitative analysis of the spin-charge competition within the model with the ground state (GS) and temperature phase diagrams was done in the mean field approximation (MFA). Here, in the paper, we present the results of classical Monte-Carlo (MC) calculations in a "strong" exchange limit. An interactive visualization of the actual states of the system allowed us to observe qualitatively different behavior of doped charges in the charge ordered (CO) and antiferromagnetic (AFM) phases with a clear trend to a bulk spin-charge phase separation in the AFM phase and a random distribution of the doped charges in the CO phase.

2 The model

In our model approach [4] to copper oxides such as $La_{2-x}Sr_xCuO_4$ we assume that the on-site Hilbert space is reduced to only three effective charge states (nominally $Cu^{1+;2+;3+}$) of copper ions in the CuO_2 planes. These charge states are associated with components of the $S = 1$ pseudospin triplet with $M_S = -1, 0, +1$, respectively. The on-site states are characterized by different hole occupation: $n_h = 0, 1, 2$ for $Cu^{1+;2+;3+}$, respectively, and different conventional spin: $s = 1/2$ for Cu^{2+} and $s = 0$ for $Cu^{1+;3+}$. The doped hole concentration n are related to the pseudomagnetization: $nN = \sum_i \langle S_{iz} \rangle$. Conventional spin density for mixed valence superpositions can vary inbetween 0 and 1 in accordance with the weight of the Cu^{2+} in the on-site superposition.

Hereafter, in the paper we will consider only simplified spin-pseudospin Hamiltonian which takes into account the on-site and inter-site correlations, and conventional Heisenberg spin exchange coupling:

$$\mathcal{H} = \Delta \sum_{i=1}^{N} S_{iz}^2 - \mu \sum_{i=1}^{N} S_{iz} + V \sum_{\langle ij \rangle} S_{iz} S_{jz} + J \sum_{\langle ij \rangle} P_{0i} \vec{s}_i \vec{s}_j P_{0j}, \quad (1)$$

where the sums run over the sites of a two-dimensional square lattice, $\langle ij \rangle$ means the nearest neighbors, S_{iz} and $\vec{s}_i$ are the on-site pseudospin and conventional spin operators, respectively. The first on-site term with $\Delta = U/2$ relates to the on-site density-density interactions, the second term with chemical potential μ is needed to account for the charge density constraint, $n = const$, the third term with $V > 0$ describes the effects of the inter-site density-density interactions. The last term is the antiferromagnetic ($J > 0$) Cu^{2+}–Cu^{2+} Heisenberg spin exchange cou-

*e-mail: kbudrin@yandex.ru
**e-mail: yuri.panov@urfu.ru

GS phase	$\varepsilon = \langle\mathcal{H}\rangle/N$	$\langle S_z\rangle_j$	$\langle P_0\rangle_j$
COI	$\Delta - 2V(1-2\|n\|)$	$n+(-1)^j(1-\|n\|)$	0
COII	$\|n\|\Delta - 2V(1-2\|n\|)$	$n+(-1)^j(1-\|n\|)$	$(1-\|n\|)(1-(-1)^j \operatorname{sgn} n)$
COIII	$(1-\|n\|)\Delta - 2V(1-2\|n\|)$	$n+(-1)^j(1-\|n\|)$	$\|n\|-(-1)^j n$
FIM	$\|n\|\Delta - \frac{J}{2}(1-2\|n\|)$	$n+(-1)^j\|n\|$	$1-\|n\|-(-1)^j n$
AFM	$\|n\|\Delta - \frac{J}{2}(1-\|n\|)^2 + 2n^2V$	n	$1-\|n\|$

Table 1. The mean energy ε, on-site charge density $\langle S_z\rangle_j$ and on-site spin density $\langle P_0\rangle_j$ of the MFA GS phases for 2D spin-pseudospin system. The index $j = 0, 1$ distinguishes two checkerboard sublattices.

pling, where the projection operator $P_{0i} = 1 - S_{iz}^2$ takes into account the on-site occupation dependence. In order to study the competition of charge and spin orderings, we used in numerical calculations the Ising type spin-spin interaction with the same account of the on-site occupation dependence as in expression (1). The classical MC calculations in this case are comparable with MFA for the model (1). The most important limitation of the model is the static character of the charge and spin subsystems, since the Hamiltonian does not contain any transfer terms.

The MFA analysis [3] gives five GS solutions or phases of spin-pseudospin system. The energies and structural characteristics in terms of the on-site charge and spin densities for these phases are given in Table 1.

In a "weak" exchange limit, at $J/4 < V$, all the GS phases (COI, COII, COIII, FIM) correspond to the various types of charge ordering. The COI is a charge-ordered phase without spin centers. In the COII and COIII phases the charge ordering is diluted by the non-interacting spins. In the FIM phase charge and spin orderings coexist. In a "strong" exchange limit, at $J/4 > V$, there are only COI and AFM phases.

3 Numerical results

Here we present some results of classical MC calculations in a "strong" exchange limit ($J/4 > V$) with the heat-bath algorithm on the square lattice 256×256 under periodical boundary conditions. As an initial state, we choose the random distribution of pseudospins and spins with a fixed total z-component of pseudospins for a given value of n. We implemented high-performance parallel computing on NVIDIA graphics cards and an interactive visualization of the actual states of the system. This allows us to observe the relaxation of the system to the ground state in the process of calculation.

First, we address the on-site correlations $\Delta < 0$ when they stabilize the GS COI phase. In Fig.1 the temperature dependence of the specific heat $C(T)$ and snapshots of the real states of the system at some characteristic points are shown for $\Delta = -1.5$ and $n = 0.1$. The $C(T)$ dependence reveals the maximum near $T/J \approx 0.22$. The plateau at $T/J \approx 0.5$ is related to a freezing of the spin subsystem. A direct observation of the state of the system shows that

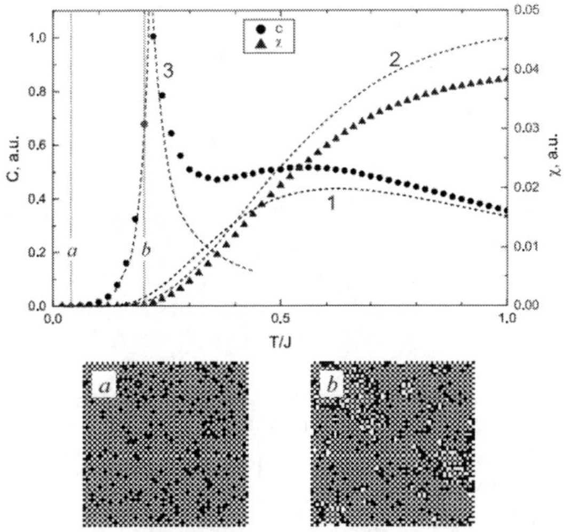

Figure 1. (colour online). The MC calculated temperature dependence of specific heat and snapshots of real states of 2D lattice for the CO phase at $n = 0.1$, $\Delta = -1.5$, $V = 0.1$, $J = 1$. The dashed lines 1 and 2 correspond to the "free doublets" approximation (2, 3) for $n = 0.1$ and $\Delta = -1.5$. The dashed line 3 corresponds to the Ising type dependence of specific heat for the charge subsystem given $V = 0.1$. In the snapshots of real states of the lattice in the CO phase the blue and white colors correspond to the on-site value of $\langle S_z\rangle = \pm 1$, respectively. Comparison of the snapshots at temperatures $a = 0.04J$ and $b = 0.2J$ points to a weak temperature dependence of the random character of the doped charge distribution over the CO matrix.

the spin excitations mostly disappear at $T/J \approx 0.5$ before the ordering in the spin subsystem occurs. Qualitatively this part of the $C(T)$ dependence can be described within a rough approximation of free charge and spin doublets. In this case $V = 0$ and $J = 0$, so taking into account the charge density constraint we come to the expressions for the specific heat and susceptibility as follows:

$$C = \frac{\Delta^2 e^{\Delta/T}(1-n^2)^2}{T^2 f\,(e^{\Delta/T}+f)^2}, \tag{2}$$

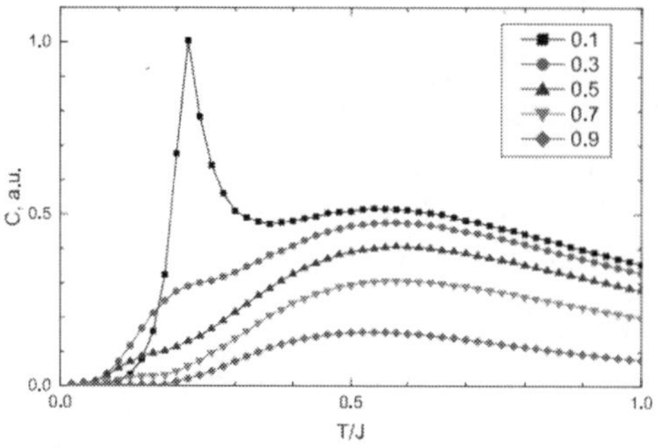

Figure 2. (colour online). The temperature dependence of the specific heat for the CO phase given different doping at $\Delta = -1.5$, $V = 0.1$, $J = 1$.

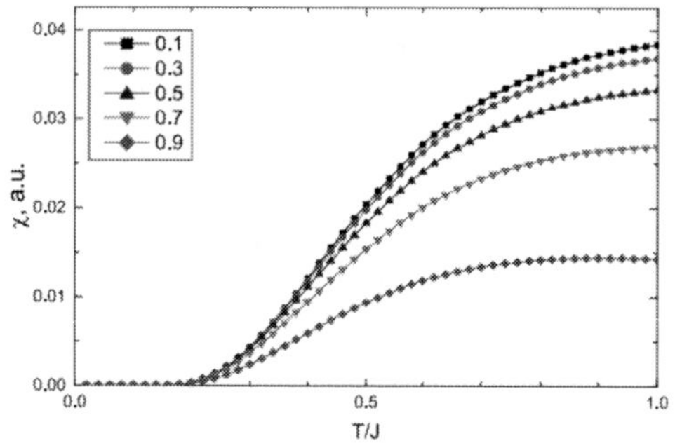

Figure 3. (colour online). The temperature dependence of the magnetic susceptibility for the CO phase given different doping at $\Delta = -1.5$, $V = 0.1$, $J = 1$.

$$\chi = \frac{1 - n^2}{4T\,(1 + e^{-\Delta/T} f)}, \tag{3}$$

where $f = \sqrt{1 - n^2 + n^2 e^{2\Delta/T}}$. The $C(T)$ and $\chi(T)$ dependencies (2) are shown in Fig.1 for $n = 0.1$ and $\Delta = -1.5$ by the dashed lines 1 and 2, respectively.

The $C(T)$ peak at $T/J \approx 0.22$ is related to the charge ordering. The snapshots *a,b* clearly demonstrate this feature. The temperature of the charge ordering nearly corresponds to the Ising value $T^* = 2V/\log(1 + \sqrt{2}) \approx 2.26\,V$. The distribution of doped charges over the CO matrix remains the random one with the temperature decrease so that the energy of the low-temperature state is exactly equal to the MFA GS energy. The concentration dependencies of the specific heat and susceptibility for the CO1 phase are shown in Fig.2 and Fig.3. With increasing of the charge doping the Ising type maximum of the specific heat is rapidly reduced and almost disappears above the $n \approx 0.3$. The concentration dependence of susceptibility is roughly following the expression (3).

One observes qualitatively different temperature behavior for the on-site correlations $\Delta > 0$ when they stabilize the GS AFM phase. Temperature dependence of the specific heat for the AFM phase at $n = 0.1$ is shown in Fig.4. The sharp maximum at $T/J \approx 0.45$ corresponds to the AFM ordering of the spin subsystem. The temperature of the ordering nearly corresponds to the value $T^* \approx 2.26\,(1 - n^*)\,J/4$, where $n^* \approx 0.2$ is a total concentration of the doped and the excited charge centers. When the temperature is lowered, the specific heat demonstrates second peak at $T/J \approx 0.08$. The snapshots *a* and *b* show that this puzzling peculiarity is related to a condensation of doped charges in the charge droplets. The concentration dependencies of specific heat and susceptibility for the AFM phase are shown in Fig.5 and Fig.6, respectively. Phase separation in the AFM phase exists in the whole

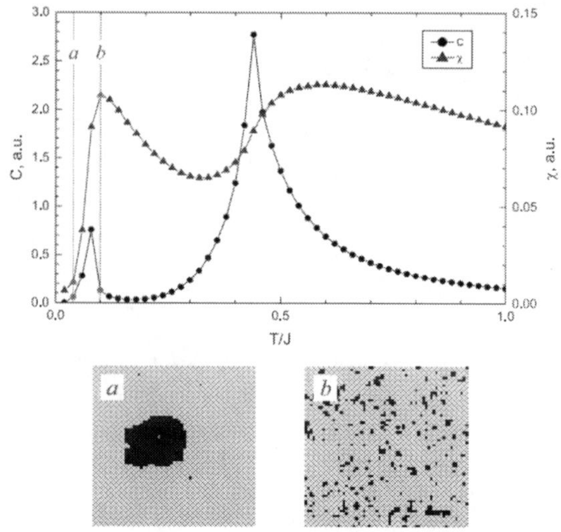

Figure 4. (colour online). The temperature dependence of the specific heat and the magnetic susceptibility for the AFM phase at $n = 0.1$, $\Delta = 0.5$, $V = 0.1$, $J = 1$. The Ising type high-temperature peak corresponds to the AFM ordering in spin subsystem at $T^* \approx 0.45$. The low-temperature peak corresponds to the charge droplet condensation shown in the snapshots of the real states of the 2D lattice at the temperatures $a = 0.04J$ and $b = 0.1J$. Blue color in the snapshots points to doped charge distribution, $\langle S_z \rangle = 1$, yellow and green colors correspond to the on-site spin values: $\langle s_z \rangle = \pm 1/2$, respectively.

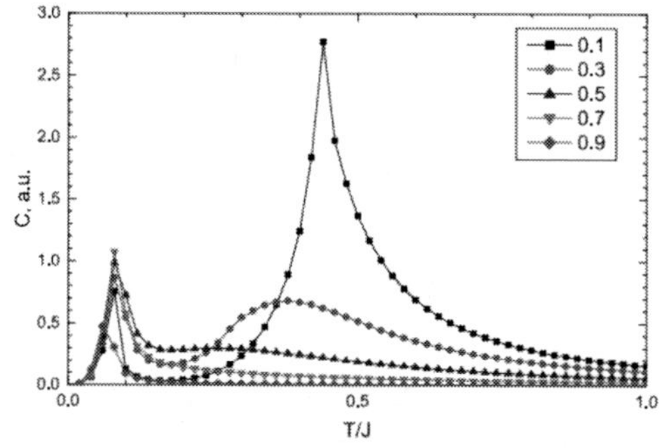

Figure 5. (colour online). The temperature dependence of the specific heat for the AFM phase given different doping at $\Delta = 0.5$, $V = 0.1$, $J = 1$.

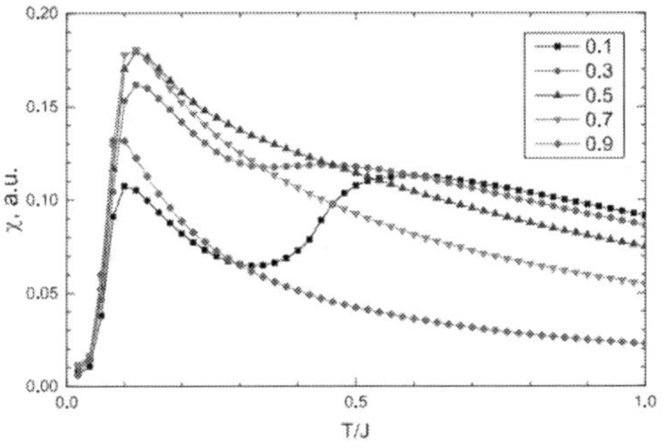

Figure 6. (colour online). The temperature dependence of the magnetic susceptibility for the AFM phase given different doping at $\Delta = 0.5$, $V = 0.1$, $J = 1$.

range of the doped charge concentrations except for $n = 0$ and $n = \pm 1$. The sharp peak on the specific heat curve at $n = 0.1$ that relates to the ordering of spin subsystem rapidly flattens with the charge doping. At the same time, the second maximum does not change its position and magnitude over a wide range of concentration of the doped charge. The variation with n of the susceptibility clearly indicates that the driving force of the spatial phase separation of charge and spin subsystems is related to a magnetic ordering. The calculated energy of the low temperature phase separated state is lower than the MFA GS energy of AFM phase, since the surface energy effects are omitted in MFA.

4 Conclusions

We have performed classical MC calculations for the 2D spin-pseudospin system with competing CO and spin AFM orderings in a "strong" exchange limit. The behavior of the system strongly depends on the sign and value of the on-site correlation parameter Δ either stabilizing CO ($\Delta < 0$) or AFM ($\Delta > 0$) phase, respectively. We show that homogeneous ground-state AFM solutions found in the MFA [1] are unstable with respect to phase separation with the charge and spin subsystems behaving like immiscible quantum liquids. The specific heat temperature dependence reveals two sequential phase transitions: first, antiferromagnetic ordering in the spin subsystem diluted by randomly distributed charges, then, the charge condensation in the charge droplets. The inhomogeneous droplet phase reduces the energy of the system and changes the diagram of the GS. Charge doping does suppress the long-range spin order, but the phase separation of doped charges and short-range spin order exists in a whole range of the charge doping. Specific heat for the system with the GS COI phase shows a feebly marked maximum due to a spin freezing at elevated temperatures with a low-temperature singular peak due to the charge ordering. The doped charges remain distributed randomly over the CO matrix up to $T = 0$ since for the nearest-neighbor interaction the energies of all possible distributions of extra charges over the CO matrix are equal. For this reason the GS energy of the COI MFA solutions exactly matches the energy of the low-temperature MC state and the entropy of the low-temperature state in the doped CO phase is higher than in the doped AFM phase.

The work is supported by Act 211 Government of the Russian Federation, agreement No 02.A03.21.0006 and by the Ministry of Education and Science, projects 2277 and 5719.

References

[1] E. Fradkin, S. A. Kivelson, J. M. Tranquada, Rev. Mod. Phys. **87**, 457 (2015). DOI: 10.1103/RevModPhys.87.457
[2] Yu. D. Panov, A. S. Moskvin, A. A. Chikov, I. L. Avvakumov, J. Supercond. Nov. Magn. **29**, 1077 (2016). DOI: 10.1007/s10948-016-3378-5
[3] Yu. D. Panov, A. S. Moskvin, A. A. Chikov, K. S. Budrin, J. Low Temp. Phys. **187**, 646 (2017). DOI: 10.1007/s10909-017-1743-9
[4] A. S. Moskvin, Physical Review B **84**, 075116 (2011). DOI: 10.1103/PhysRevB.84.075116
[5] A. S. Moskvin, JETP **148**, 549 (2015). DOI: 10.1134/S1063776115090095

Magnetic excitations of carbon nanotubes in chiral model of graphene

Yu.P. Rybakov[1,⋆]

[1] *RUDN University, Moscow, Russian Federation*

Abstract. Within the scope of the chiral model of graphene in 8-spinor realization, suggested earlier [1, 2], the combined spinor field Ψ is used as a new order parameter, which contains information on spin and quasi-spin excitations in graphene (8-spinor part) and also that on sp-hybridization effect of carbon valence electrons [$SU(2)$-matrix part]. This model admits two kinds of configurations describing, in the long-wave approximation, the graphene mono-atomic plane and the carbon nanotube respectively. The first one is given by the domain-wall (or kink-like) solution and the second one is described by the hedgehog ansatz with large topological charge (winding number n). We consider the infinite carbon nanotube in a static external magnetic field oriented along the tube and analyze the structure of the magnetic intensity on the outside. The diamagnetic behavior of the material is shown by the perturbation method which proves to be self-consistent for $n \gg 1$.

1 Introduction. Structure of the model

The discovery of carbon nanotubes [3, 4], which are endowed with very high electric and heat conductivities and also with an extraordinary tensile and bending strength, opened new possibilities in spintronics and in production of composite materials for their reinforcing. This fact stimulated a large number of theoretical studies on the electronic properties of carbon nanotubes. The magnetic properties of carbon nanotubes were studied by H. Ajiki and T. Ando [5, 6] within the scope of the effective-mass theory. These authors found that in a magnetic flux parallel to the tube axis the magnetic moment oscillated as a function of the flux. For a magnetic field perpendicular to the axis carbon nanonubes exhibited a strong diamagnetism. Large diamagnetic susceptibility was found experimentally for a magnetic field both perpendicular and parallel to the tube axis [7, 8].

We suggest here a phenomenological model which permits to calculate all necessary characteristics of both carbon nanotubes and other possible graphene-like configurations. The concept behind this research is the following. The carbon atom possesses of four valence electrons in the so-called hybridized sp^2-states. The one of them is "free" in the honeycomb graphene lattice and all others form the strong covalent bonds with the neighbors. It appears natural to introduce scalar a_0 and 3-vector $\vec{a}$ fields corresponding to the s-orbital and the p-orbital states of the "free" electron respectively. These two fields can be combined into the unitary matrix

$$V = a_0\tau_0 + \imath\vec{a}\cdot\vec{\tau}$$

considered as the order parameter of the model, the long-wave approximation being adopted, where τ_0 is the unit 2×2-matrix and $\vec{\tau}$ are the three Pauli matrices, with the $SU(2)$-condition

$$a_0^2 + \vec{a}^2 = 1$$

being imposed. The model admits the kink-like configuration

$$V = \exp[\imath\tau_3\Theta(z)]$$

of the graphene plane and the hedgehog one of the carbon nano-tube [1]:

$$V = \exp[\imath\sigma\Theta(r)], \quad \sigma = \tau_1\cos\phi + \tau_2\sin\phi, \quad \phi = n\varphi, n = 1, 2 \ldots, \tag{1}$$

with r, φ, z being polar coordinates and Θ designating the chiral angle. Here n stands for the topological charge Q of the degree type which is known as the helicity or the winding number:

$$Q = \deg\{S^2 \to S^2\} = \frac{1}{4\pi}\int\limits_{\mathbb{R}^2\cup\infty} d\phi\, d\Theta\,\sin\Theta = n.$$

In the sequel the helicity n will be chosen positive.

For the description of spin and quasi-spin excitations, the latter ones corresponding to the independent excitation modes of the two triangular sub-lattices, we introduce the two Dirac spinors ψ_1, ψ_2 and consider the combined spinor field Ψ as a new order parameter:

$$\Psi = \xi\otimes(\psi_1\oplus\psi_2), \tag{2}$$

where ξ stands for the first column of the matrix V. The Lagrangian density L of the model

$$L = \frac{I}{2}\overline{D_\mu\Psi}P\,D^\mu\Psi - \frac{\lambda^2}{2}\vec{a}^2 j_\mu j^\mu + \\ + \imath\mu_0\,\vec{a}^2\overline{\Psi}\sigma_{\mu\nu}F^{\mu\nu}\Psi - \frac{1}{16\pi}F_{\mu\nu}F^{\mu\nu} \tag{3}$$

⋆e-mail: rybakov_yup@pfur.ru

contains the projector $P = \gamma^\nu j_\nu$ on the positive energy states, where $j_\mu = \overline{\Psi}\gamma_\mu\Psi$, $\mu = 0, 1, 2, 3$, designates the Dirac current, $\overline{\Psi} = \Psi^+\gamma_0$, γ_μ stands for the Dirac matrices. Here standard denotations are introduced for the Pauli antisymmetric spin operator

$$\sigma_{\mu\nu} = [\gamma_\mu, \gamma_\nu]/4,$$

and for the electromagnetic field tensor $F_{\mu\nu} = \partial_\mu A_\nu - \partial_\nu A_\mu$, the constant $\mu_0 > 0$ denoting the Bohr magneton per lattice spacing cubed. The model contains the two constant parameters: the exchange energy I per lattice spacing a and some characteristic inverse length $\sqrt{\lambda}$. The interaction with the electromagnetic field is realized through the extension of the derivative:

$$D_\mu = \partial_\mu - \imath e_0 A_\mu \Gamma_e,$$

with $e_0 > 0$ being the coupling constant and $\Gamma_e = (1-\tau_3)/2$ being the charge operator chosen in accordance with the natural boundary condition at infinity: $a_0(\infty) = 1$. However, the additional interaction term of the Pauli type should be added to take into account the proper magnetic moments of the electrons.

The comparison of the Lagrangian (3) with that of the Landau—Lifshitz theory shows the proportionality of the parameter I to the exchange energy E_{exch}between atoms and also the following structure of the coupling constants e_0 and μ_0:

$$I = \frac{E_{exch}}{a}, \quad e_0 = \frac{e}{\hbar c}, \quad \mu_0 = \frac{e\,\hbar}{2m_e c a^3},$$

where the exchange energy is usually adopted as $E_{exch} = 2{,}9eV$ and the lattice spacing as $a = 3{,}56 \cdot 10^{-8} cm$, with e being the absolute value of the electron charge. Finally, one can find the following numerical values of the parameters in question:

$$e_0\, I = 2 \cdot 10^3 Gauss, \quad \mu_0 = 2 \cdot 10^2 Gauss.$$

Within the frame of the graphene chiral model we investigated earlier the magnetic excitations of the graphene plane. In particular, magnetic properties of the graphene plane appear to be highly anisotropic: for the case of an external magnetic field parallel to the plane the material reveals diamagnetic behavior, however, for the case of the orthogonal field orientation it reveals paramagnetic behavior [9]. It is worth-while to stress that similar behavior is predicted for the bilayer graphene [10, 11]. These results are in agreement with experiments [12].

We consider now an infinite carbon nanotube interacting with an external static magnetic field oriented along the tube. The axially-symmetric configuration being studied, we analyze the structure of the corresponding solution via the perturbation method. This method proves to be self-consistent under the supposition that the topological charge of the tube is large. Using the Green function method, we find the magnetic intensity of the tube and show the field to decrease in the central domain.

2 Axially-symmetric configuration

Let us describe first the static magnetic field through the azimuth component of the vector potential:

$$A_\varphi = A(r), \quad B_3 = B = \frac{1}{r}\partial_r(r A), \tag{4}$$

with the natural boundary condition at infinity:

$$A(r \to \infty) = \frac{1}{2}B_0\, r, \quad B_0 = \text{const}. \tag{5}$$

It is also convenient to use the tangential basis of the Dirac γ-matrices in 8-spinor realization:

$$\gamma_0 = \tau_0 \otimes \tau_1 \otimes \tau_0; \gamma_1 = -\imath\lambda_1 \otimes \tau_2 \otimes \tau_0;$$
$$\gamma_2 = -\imath\lambda_2 \otimes \tau_2 \otimes \tau_0; \gamma_3 = -\imath\lambda_3 \otimes \tau_2 \otimes \tau_0,$$

where

$$\lambda_1 = \tau_1 \cos\varphi + \tau_2 \sin\varphi; \lambda_2 = \tau_2 \cos\varphi - \tau_1 \sin\varphi; \lambda_3 = \tau_3.$$

Due to this choice of the basis, the spinor connection vanishes in the plane case and one finds

$$\sigma_{12} = \frac{1}{4}[\gamma_1, \gamma_2] = -\imath\frac{r}{2}\tau_3.$$

As can be seen from (3), the model admits the mirror symmetry

$$\Psi \Rightarrow \gamma_0\Psi$$

and also the quasi-spin symmetry

$$\psi_1 \Leftrightarrow \psi_2,$$

with the spinor invariant field being

$$\psi_1 = \psi_2 = \psi = \text{col}\,(w, w), \tag{6}$$

where w stands for the 2-spinor. The structure (6) indicates that the spatial Dirac current vanishes: $\vec{j} = \overline{\Psi}\vec{\gamma}\Psi = 0$. On the other hand, if one takes into account that the electron spin should be oriented contra magnetic field, the new combined symmetry group can be revealed:

$$G = \text{diag}\,[SO(2)_S \otimes U(1)],$$

where the rotations around z-axis are combined with the $U(1)$-transform: $w \Rightarrow \exp(\imath\beta)\,w$. Therefore, the G-invariance yields the following equation for w:

$$J_3 w = -\imath\partial_\varphi w + \frac{1}{2}\tau_3 w = -\frac{1}{2}w, \tag{7}$$

where J_3 stands for the rotation generator. We infer that the solution to the equation (7) takes the form

$$w = \text{col}\,[w_1(r)\exp(-\imath\varphi),\, w_2(r)], \tag{8}$$

where the correspondence with the scalar chiral model implies the boundary condition at $r \to \infty$: $j_\mu j^\mu(\infty) = 1$ or

$$w_1 \to 0, \quad |w_2|^2 \to \frac{1}{4}. \tag{9}$$

Inserting (8) into the Lagrangian (3) and taking into account the structure (1) of the carbon nanotube and the boundary condition (9), we confine our attention to the simplest possible configuration $w_1 = 0$, $w_2^* = w_2$. As a result, we obtain the following effective Lagrangian density:

$$L = -8I\Big[U^2(\partial_r\Theta)^2 + \frac{1}{4}(\partial_r U)^2 + \\ + \frac{1}{r^2}(n + e_0 rA)^2 U^2 \sin^2\Theta\Big] - \\ - 8\lambda^2 U^2 \sin^2\Theta + \frac{2\mu_0}{r} U \sin^2\Theta\partial_r(rA) - \\ - \frac{1}{8\pi r^2}(\partial_r(rA))^2, \quad (10)$$

where $U = w_2^2$ and the boundary conditions are imposed:

$$U(\infty) = \frac{1}{4}, \quad \Theta(\infty) = 0, \quad \Theta(-\infty) = \pi. \quad (11)$$

Changing the independent variable $r = \ell e^\eta$, where $\ell = \sqrt{I}/\lambda$, let us first consider the asymptotic domain $\eta \to \infty$ through the substitution:

$$U = \frac{1}{4} + \zeta, \quad A = \frac{1}{2}B_0 \ell e^\eta + \alpha, \quad \Theta = 2\mathrm{arctan}R, \quad (12)$$

with ζ, α, R vanishing at infinity. One derives from (10) and (12) the following equation for R:

$$R''(1 + R^2) - 2RR'^2 = R(1 - R^2)\times \\ \times\left[(n + e_0 B_0 \ell^2 e^{2\eta}/2)^2 + e^{2\eta}\left(1 - \mu_0 B_0 \ell^2/I\right)\right]. \quad (13)$$

By inspection that $n \gg 1$ one can find the approximate solution to the equation (13) of the form:

$$R \approx k \exp\left(-n\eta - m e^{2\eta}\right), \quad (14)$$

where $4m = e_0 B_0 \ell^2$, and the constant k determines the radius r_0 of the nanotube according to the formula $k^2 = r_0/\ell$. It is worth-while to notice that the helicity n has no any relation to the number N of walls. In fact, for $N > 1$ the substitution (12) for Θ should be replaced with $\Theta = 2N\mathrm{arctan}R$.

Let us study now the magnetic field structure through the perturbation method. To this end, we consider (14) as the source for the perturbation α of the vector potential satisfying the equation:

$$\alpha'' - \alpha = 2\pi\ell(\mu_0 - e_0 I)\left(\sin^2\Theta\right)' \equiv j(\eta). \quad (15)$$

The corresponding magnetic intensity $B = B_0 + b$ can be found from (15) by the Green's function method. In our case the Green's function reads:

$$G(\eta, \eta') = -\frac{1}{2}\begin{cases} e^{\eta'-\eta}, & \text{if } \eta' < \eta, \\ e^{\eta-\eta'}, & \text{if } \eta' > \eta, \end{cases}$$

and the solution to (15) acquires the form:

$$\alpha(\eta) = -\frac{1}{2}\int_{-\infty}^{\eta} d\eta'\, e^{\eta'-\eta} j(\eta') - \frac{1}{2}\int_{\eta}^{\infty} d\eta'\, e^{\eta-\eta'} j(\eta'). \quad (16)$$

Taking into account that $b = \ell^{-1}e^{-\eta}(\alpha + \alpha')$, one obtains from (15) and (16) the perturbation part of the magnetic intensity:

$$b = -\ell^{-1}\int_{\eta}^{\infty} d\eta'\, e^{-\eta'} j(\eta') = 2\pi(\mu_0 - e_0 I)\sin^2\Theta. \quad (17)$$

Numerical estimates, based on the correspondence with the Landau—Lifshitz theory, show that $e_0 I \gg \mu_0$. Thus, in virtue of (17), the magnetic intensity decreases inside the nanotube, with the susceptibility $\chi = b/B_0$ being proportional to the radius of the nanotube since $\sin^2\Theta \propto k^2 = r_0/\ell$. This result is confirmed experimentally [7].

3 Conclusion

We applied the chiral model of graphene in 8-spinor realization to the description of the infinite carbon nanotube interacting with the static magnetic field oriented along the tube. We construct the approximate solution to the model equations, which proves to be valid asymptotically for the case of the tube endowed with the topological charge $n \gg 1$. We find the magnetic intensity outside the tube by applying the perturbation scheme based on this solution and show the diamagnetic behavior of the material. As can be seen from (3) and the structure of the extended derivative D_μ, the predicted magnetically insulating properties of carbon nanotubes seem to be emerging from the superconductive behavior of carbon structures.

Acknowledgement

This work was financially supported by the Ministry of Education and Science of the Russian Federation (the Agreement number 02.a03.21.0008).

References

[1] Yu.P. Rybakov, Solid State Phenomena **190**, 59-62 (2012)
[2] Yu.P. Rybakov, Solid State Phenomena **233-234**, 16-19 (2015)
[3] S. Iijima, Nature **354**, 56-58 (1991)
[4] S. Iijima and T. Ichihashi, Nature **363**, 603-605 (1993)
[5] H. Ajiki and T. Ando, J. Phys. Soc. Jap. **62**, 2470-2480 (1993)
[6] H. Ajiki and T. Ando, J. Phys. Soc. Jap. **64**, 4382-4391 (1995)
[7] X.K. Wang et al., J. Mater. Res. **9**, 1578-1590 (1994)
[8] A.P. Ramirez et al., Science **265**, 84-86 (1994)
[9] Yu.P. Rybakov, M. Iskandar, A.B. Ahmed, RUDN J. of Math. Inform. Physics **25**, Issue 3, 266-275 (2017)
[10] Yu.P. Rybakov, A.B. Ahmed, Intern. J. of Appl. Math. and Theor. Physics **3**, Issue 3, 74-77 (2017)
[11] Yu.P. Rybakov, A.B. Ahmed, S. Massou and M. Iskandar, Far East J. Math. Sci. **102**, Issue 2, 399-407 (2017)
[12] M. Sepioni et al., Phys. Rev. Lett. **105**, 207205 (2010)

Phase transitions and thermodynamic properties of the antiferromagnetic Potts model on a face-centered cubic lattice

Albert Babaev[1,2*], *Akai* Murtazaev[1,3], *Felix* Kassan-Ogly[4], *Alexey* Proshkin[4]

[1]Institute of Physics of Daghestan Scientific Center of RAS, Makhachkala, Russia
[2]Daghestan State Pedagogical University, Makhachkala, Russia
[3]Daghestan State University, Makhachkala, Russia
[4]Institute of Metal Physics, Ural Branch, RAS, Yekaterinburg, Russia

Abstract. Using the Monte Carlo method we investigate the phase transitions and thermodynamic properties of magnetic structures with non-collinear directions of magnetic moments corresponded to antiferromagnetic q=4 Potts model on a face-centered cubic lattice. Monte Carlo simulations are performed on lattices with linear sizes L=20÷44. Thermodynamic parameters: the order parameter m_{AF}, susceptibility χ, internal energy U, and specific heat C are evaluated for all studied systems. By employing the fourth order Binder cumulant method, a first order transition is shown to be occurred in the model.

1 Introduction

There exists at present a rich body of the experimental material on magnetic substances with crystalline structure and properties of *AnX*-type compounds with the *NaCl* structure. These are so called monopnictides and monochalcogenides, where *An* is a rear-earth element or actinide, *X* is V or VI groups' element (*UN*, *UAs*, *USb*). They display most interesting properties. The evolution of the magnetic neuronography increases a number of experiments in determination of different compound magnetic structures. Rossat-Mignod *et. al.* [1] proved that a single wave vector component is insufficient to describe the experimental data for *UP*, *UAs*, and *USb* compounds. Therefore, multi-**k** structures with wave vectors comprised more than one component (2-**k** structures for two-component wave vector and 3-**k** structures for three-component vector) were proposed. The interpretation of *USb* neuronography is in a better agreement with experimental data of a 3-**k** structure than a collinear model. We notice that this 3-**k** structure corresponds precisely to the q=4 standard Potts model illustrated in Fig. 1.

The investigation of phase transitions (PTs) and thermodynamic properties of the 4-state antiferromagnetic (AF) Potts model is the main goal of this study.

* Corresponding author: b_albert78@mail.ru

2 Model and Method

The Hamiltonian of the antiferromagnetic q=4 standard Potts model on a face-centered cubic (FCC) lattice (Fig.1) reads

$$H = -J\sum_{i,j}\cos\theta_{i,j}\,, \tag{1}$$

where J is the parameter of antiferromagnetic exchange interaction (J<0), $\theta_{i,j}$ is the angle between interacting spins S_i - S_j. As is evident from Fig. 1, in the model spins occupying lattice sites can orient in hypertetrahedron q-symmetric directions in the dimensionality space q-1 retaining equal angles between any two different directions.

The cubic ferromagnetic q=4 Potts model with nearest neighbor ferromagnetic interactions were studied in works [2, 3]. Authors reported that a second order transition occurred in the models considered.

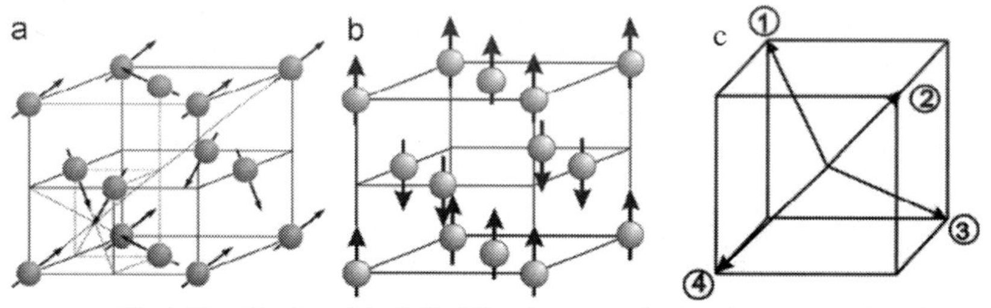

Fig. 1. Identification of the USb diffraction pattern in the 3-k structure (a) and in the collinear one (b). 4-State Potts model (c).

In our study we use the Metropolis algorithm along with the Wolff cluster algorithm [4]. We performe the Monte Carlo simulations for systems of linear sizes 4×(L×L×L) with periodic boundary conditions, where L=28-40. Initial configurations were given so that every spin inside each of sublattices is ordered. To bring the system to equilibrium state the relaxation time τ_0 for all systems with L is computed. Next, the averaging is performed along the Markov chain of length τ=150τ_0 and additional averaging is done by ten different initial spin configurations.

3 Simulation Results

The temperature behavior of the specific heat and susceptibility is evaluated from fluctuation correlations [5]

$$C = (NK^2)(\langle U^2\rangle - \langle U\rangle^2), \tag{2}$$

$$\chi = (NK)(\langle m^2\rangle - \langle m\rangle^2), \tag{3}$$

where $K=|J|/k_BT$, $N=4\times L^3$ is the number of sites, U is the intrinsic energy, m is the order parameter, angle brackets denotes the ensemble average. The order parameter for the antiferromagnetic Potts model is defined as [6]

$$m_{AF} = \left\langle \frac{4}{3}\sum_{\alpha\neq\beta\neq\gamma\neq\varphi}\left(\frac{(N_\alpha+N_\beta+N_\gamma+N_\varphi}{N}-\frac{1}{4}\right)\right\rangle^{1/2}, \tag{4}$$

where $N_\alpha = \{N_1, N_2, N_3, N_4\}$, N_1, N_2, N_3, and N_4 are the number of spins in q=1, q=2, q=3, and q=4, correspondingly, N_α, N_β, N_γ, N_φ are the number of spins in sublattices A, B, C, and D, correspondingly.

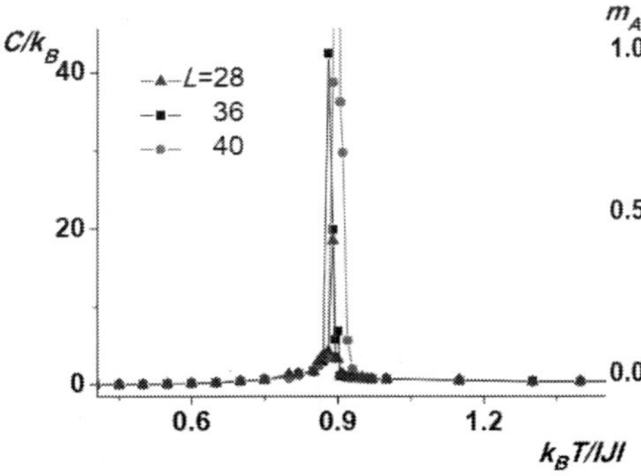

Fig. 2. The temperature dependence of the specific heat C for the 3D 4-state Potts model on the FCC lattice.

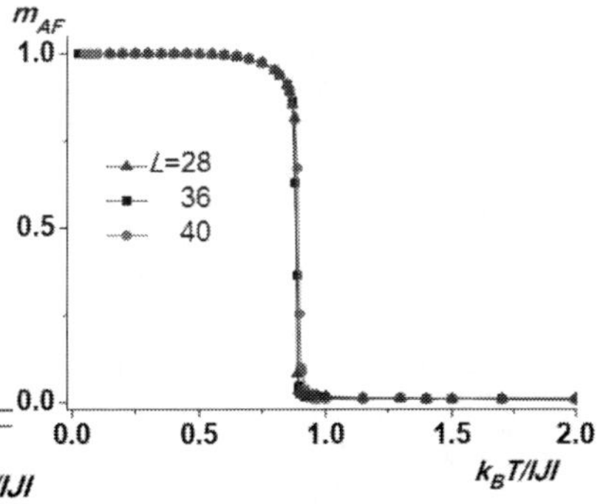

Fig. 3. The temperature dependence of the order parameter m_{AF} for the 3D 4-state Potts model on the FCC lattice.

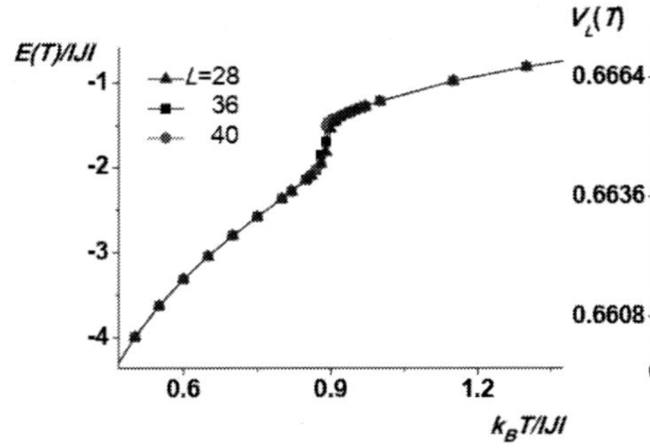

Fig. 4. The temperature dependence of the average energy for the 3D 4-state Potts model on the FCC lattice.

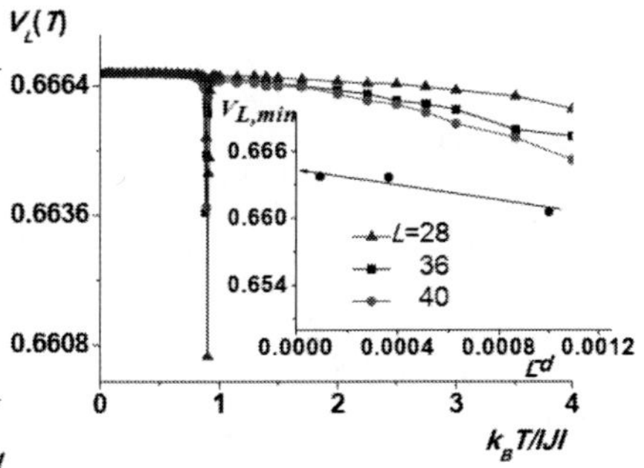

Fig. 5. Dependence of the Binder cumulants $V_L(T)$ on temperature for the 3D 4-state Potts model on the FCC lattice.

Figures 2, 3, 4 and 5 depict the temperature dependences of the specific heat C, order parameter m_{AF}, internal energy E and Binder cumulants V_L for 3D 4-state antiferromagnetic Potts model on a FCC lattice for systems with L=28; 36; 40. From now on an error of data in all figures doesn`t exceed symbol sizes used at plotting the diagrams. As is evident from Figs. 2, 3, and 4, the temperature dependences of the specific heat C, order parameter m_{AF}, internal energy E for all studied systems exhibit a behavior featured to a first order transition (a jump occurs at PT point). Beyond that, we investigate the dependence of the Binder's cumulants $V_L(T)$ on temperature T for this model (Fig. 5). A clearly marked jump at PT point indicates a first order transition. The fourth-order Binder cumulants method [7] has proved to be the most efficient method for the analysis of the character of the phase transition at $T = T_l(L)$.

$$V_L(T) = 1 - \frac{\langle E^4 \rangle_L}{3\langle E^2 \rangle_L^2}, \quad (5)$$

$$U_L(T) = 1 - \frac{\langle m^4(T,L) \rangle_L}{3\langle m^2(T,L) \rangle_L^2}, \quad (6)$$

where E is the internal energy and m is the order parameter of the system with L. This technique has proved to be successful for the detection of a PT order, a detail description of that was provided in [8-12]. The averaged value $V_L(T)$ tends to a certain non-trivial value V^* at the first order transition

$$V(T) = V^* + bL^{-d} \quad (7)$$

at $L\to\infty$ and $T=T_l(L)$, where V^* is distinct from 2/3 at $L\to\infty$, which is shown in the insert in Fig. 5 for the 3D q=4 antiferromagnetic Potts model on a FCC lattice.

Thus, in this study we evaluate within the unified technique the phase transitions and thermodynamic properties of magnetic structures with non-collinear directions of magnetic moments corresponded to the antiferromagnetic q=4 Potts model on a FCC lattice. The temperature dependences of thermodynamic parameters for the specific heat C, susceptibility χ, order parameter m_{AF}, and Binder cumulants $V_L(T)$ demonstrate the first order transition. The investigation of the antiferromagnetic q=4 Potts model on a FCC lattice permits to elucidate magnetic and thermodynamic properties of real crystals with multi- **k** structures.

References

1. J. Rossat-Mignod, *Neutron scattering. In: Methods of experimental physics* (New York: Academic Press, 1987)
2. A.K. Murtazaev, A.B. Babaev, G.Y. Aznaurova, Solid State Phenomena **168-169**, 357 (2011)
3. A.K. Murtazaev, A.B. Babaev, G.Y. Aznaurova, Low Temperature Physics **37,** 134 (2011)
4. U. Wolff. Phys. Lett. **62**, 361 (1989)
5. P. Peczac, A.M. Ferrenberg, D.P. Landau, Phys.Rev. B **43**, 6087 (1991)
6. Y. Saito, J. Phys. A **15**, 1885, (1982)
7. K. Eichhorn, K. Binder, J. Phys.: Condens. Matter **8**, 5209 (1996)
8. A.K. Murtazaev, A.B. Babaev, JETP Lett. **99**, 535 (2014)
9. A.B. Babaev, A.K. Murtazaev, Low temperature physics **41**, 608 (2015)
10. A.K. Murtazaev, A.B. Babaev, G.Ya. Ataeva, Low Temperature Physics **39,** 147 (2013)
11. A.K. Murtazaev, A.B. Babaev, G. Ya. Ataeva, Physics of the solid state **59**, 141 (2017)
12. A.B. Babaev, T.R. Rizvanova, A.K. Murtazaev, Physics of the solid state **59**, 2444 (2017)

Non-equilibrium critical dynamics of low-dimensional magnetics and multilayer structures

Pavel V. Prudnikov[1,*], *Vladimir V.* Prudnikov[1], *Alena Yu.* Danilova[1], *Vadim O.* Borzilov[1], and *Georgy G.* Baksheev[2]

[1] *Omsk State University, Pr. Mira 55A, Omsk 644077, Russia*
[2] *Novosibirsk State University, Pirogova str. 1, Novosibirsk 630090, Russia*

Abstract. The Monte Carlo simulation of the critical behavior of multilayer structures based on anisotropic Heisenberg model is performed. The influence of the uniaxial anisotropy on the critical behavior of the thin Heisenberg-like film is described. The investigation of non-equilibrium critical behavior of multilayer structure which correspond to the nanoscale superlattice Co/Cu demonstrates that the aging effects can be observed in a wider temperature range than for bulk magnetic systems.

1 Introduction

Magnetic order in the multilayers is complex due to a strong influence of the shape and the magnetocrystalline anisotropies of the sample. It is well known now the fundamental role of competing interactions in the emerging features of low-dimensional systems. Among a wide number of numerical and theoretical investigations on equilibrium and dynamical properties of several model Hamiltonian of low-dimensional magnets, Heisenberg-like models are one of the most widely used to approach real magnetic materials [1].

This study includes the Monte Carlo simulation of the critical behavior of thin films and multilayers based on anisotropic Heisenberg model [2].

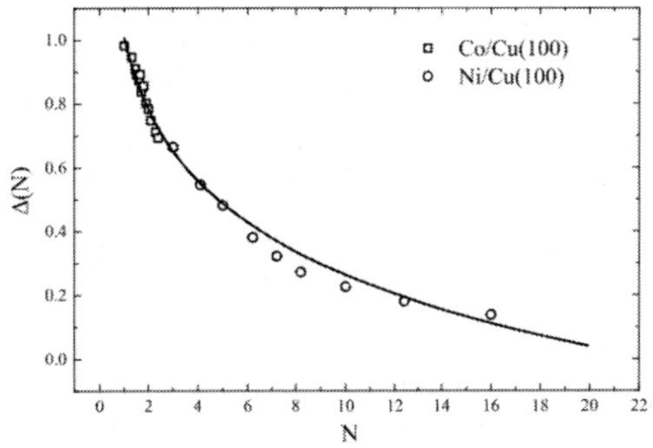

Figure 1. Anisotropy parameter $\Delta(N)$ versus the thickness of the film N. The dotes are corresponds to experimental data [3]

2 Model and methods

Magnetic film is constructed as a spin lattice with anisotropic Heisenberg Hamiltonian

$$H = -J\sum_{i,j}\left[S_iS_j - \Delta(N)S_i^zS_j^z\right] - A\sum_i (S_i^z)^2 - h\sum_i S_i^x, \tag{1}$$

where $\mathbf{S}_i$ is a three-dimensional spin in the lattice cite i, J is exchange constant $J/k_bT = 1$, A characterizes uniaxial anisotropy in out-of-plane direction, $\mathbf{h}$ is applied magnetic field.

Anisotropy parameter $\Delta(N)$ (Figure 1) characterizes the amount of anisotropy: $\Delta = 0$ corresponds to the isotropic Heisenberg case, $\Delta = 1$ – the case of XY-model. Microscopic nature of anisotropy in films of Fe, Co, Ni and it's dependence from film thicknesses N is determined by influence of crystalline field of substrate surface, magnetic single-ion anisotropy [2, 4].

*e-mail: prudnikovpv@omsu.ru

2.1 Influence of uniaxial anisotropy on critical behavior of thin Heisenberg films

We calculate the equilibrium characteristics of the thin Heisenberg films with the aim to to investigate the influence of uniaxial anisotropy on the critical behavior. We used values $A = 0.1$ and 1.0, $\Delta = 0$. For accurate determination of the critical temperatures, we consider films with different thicknesses $N = 1$ ML, 3 ML, 10 ML and linear sizes $L = 32, 48, 64$. Open boundary conditions were used for z axis, which perpendicular to plane of films and periodic boundary conditions w ere used in xy plane. The Metropolis algorithm was used for Monte Carlo simulations. Magnetic properties are calculated for these systems with values of $A = 0.1J$, $1.0J$, and $\Delta = 0$.

https://doi.org/10.1051/epjconf/201818511009

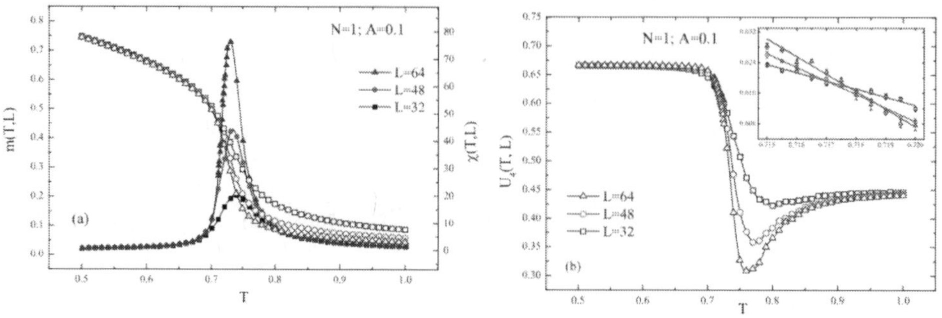

Figure 2. Temperature dependencies of the total magnetization $m(T, L)$ (open symbols), magnetic susceptibility $\chi(T, L)$ (full symbols) (a), and Binder cumulant $U_4(T, L)$ (b) for the film thickness $N = 1$ and anisotropy $A = 0.1$. The inlayer demonstrates the intersection region of the curves which determined the critical temperature $T_c(N)$.

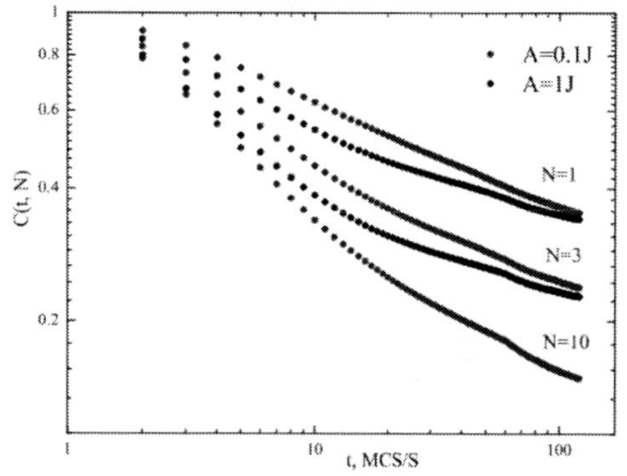

Figure 3. Non-equilibrium time dependencies of the autocorrelation function $C(t)$ for different values of $N = 1, 3, 10$ and $A = 0.1J, 1J$

We measured temperature dependencies of the total magnetization of film (Figure 2a)

$$m(N, T, L) = 1/NL^2 \sum_i \left(S_i^{x2} + S_i^{y2} + S_i^{z2}\right)^{1/2} \sim L^{-\beta/\nu}, \tag{2}$$

and magnetic susceptibility (Figure 2a)

$$\chi(N, T, L) = \frac{\langle m(N, T, L)^2\rangle - \langle m(N, T, L)\rangle^2}{TNL^2} \sim L^{\gamma/\nu}, \tag{3}$$

where $\langle \ldots \rangle$ – denotes the statistical averaging.

The temperature dependencies of the Binder cumulant (Figure 2b)

$$U_4(N, T, L) = 1/2\left(3 - \frac{\langle m(N, T, L)^4\rangle}{\langle m(N, T, L)^2\rangle^2}\right). \tag{4}$$

were calculated to clarify the critical temperatures of second-order phase transition in samples. The scaling dependence of the cumulant

$$\frac{dU_4}{dT} = uL^{1/\nu} \tag{5}$$

makes it possible to determine the critical temperature $T_c(N)$ from the coordinate of the intersection points of the curves specifying the temperature dependence $U_4(N, T, L)$ for different L.

We consider finite size scaling form of film for $m(L, N)$ and $\chi(L, N)$ and use this to extract the effective critical exponents β/ν and γ/ν. The scaling dependence of the cumulant allowed us to calculate values of the critical exponent ν. The calculated values of the critical temperatures $T_c(N)$ and critical exponents are presented in Table 1. We can conclude that the presence even small amount of anisotropy lead to the realization of the second order phase transition in 2D Heisenberg model. More large amount of "easy axis" anisotropy lead to the 2d ising-like critical behavior with the same critical exponents, especially for $N = 1$, but with different critical temperature.

The influence of the anisotropy "easy axis" on the non-equilibrium behavior can be revealed from time dependence of the autocorrelation function $C(t)$ =

Table 1. The values of the critical temperatures and exponents

Thin anisotropic films with $A = 0.1J$, $\Delta = 0$				
	T_c, J/k_b	ν	β	γ
$N = 1$	0.7178(2)	0.88(9)	0.143(34)	1.59(28)
$N = 3$	1.1842(3)	0.96(10)	0.159(26)	1.69(26)
$N = 10$	1.4285(20)	0.99(2)	0.184(14)	1.70(8)
Thin anisotropic films with $A = 1J$, $\Delta = 0$				
$N = 1$	1.005(5)	1.01(6)	0.134(15)	1.84(15)
$N = 3$	1.483(1)	0.96(7)	0.103(9)	1.56(15)
Multilayer structure with $\Delta(N = 3) = 0.5677$, $A = 0$				
	T_c, J/k_b	ν	β	γ
	0.784(7)	0.854(31)	0.626(32)	0.697(42)
Well-known systems				
2D Ising	2.269	1	0.125	1.75
2D XY	0.893	–	0.231	–
3D Ising	4.511	0.630	0.325	1.241
3D XY	2.202	0.672	–	1.318
3D Heis.	1.443	0.711	0.369	1.396

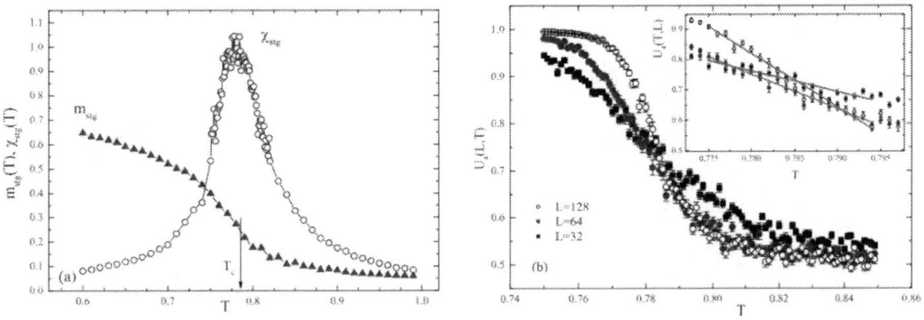

Figure 4. Temperature dependence of the of the staggered magnetization $m_{stg}(T, L)$ (full symbols), magnetic susceptibility $\chi(T, L)$ (open symbols) (a), and Binder cumulant $U_4(T, L)$ (b) for the trilayer structure

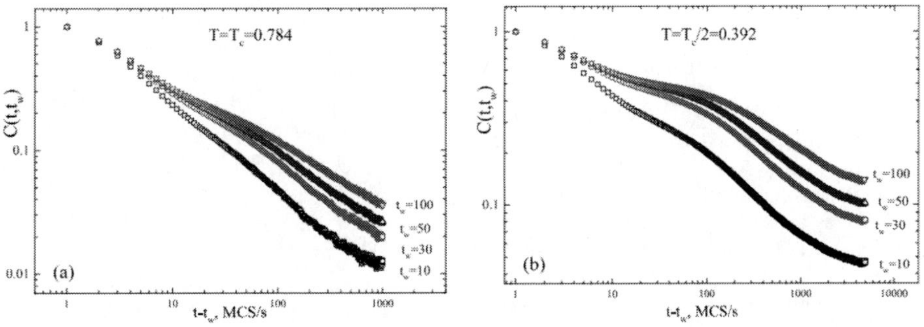

Figure 5. Autocorrelation function $C(t, t_w)$ versus the observation time $(t - t_w)$ at the evolution of the multilayer structure starting from initial state with $m^0_{stg} = 0$ for different temperatures $T = T_c = 0.784 J_1/k_b$ (a) and $T = T_c/2 = 0.392 J_1/k_b$ (b)

$\langle 1/NL^2 \sum \boldsymbol{S}(t)\boldsymbol{S}(0)\rangle - \boldsymbol{m}(t)\boldsymbol{m}(0)$ (Figure 3). We can see that the presence of the "strong" anisotropy effects on non-equilibrium behavior only on short time interval up to 100 MCS/s. The non-equilibrium relaxation of thin anisotropic films has a strong dependence from film thickness N.

2.2 Non-equilibrium critical behavior of trilayer magnetic structure

We simulate behavior modeling of the trilayer structure consisting of two ferromagnetic films with long-range RKKI interlayer interaction separated by nonmagnetic metal film. The exchange integral J_1 determine the interaction between the neighboring spins, whereas that for the interlayer interaction is $J_2 = -0.3J_1$. The sign of J_2 is negative because the thickness of nonmagnetic spacers in multilayer structures exhibiting the giant magnetoresistance effect is tuned such that both the long-range and oscillating RKKY exchange interactions between the ferromagnetic layers effectively provide antiferromagnetism.

We measured the staggered magnetization $\boldsymbol{m} = \boldsymbol{m}_1 - \boldsymbol{m}_2$, where $\boldsymbol{m}_1, \boldsymbol{m}_2$ are magnetizations of adjacent magnetic layers. The simulations were carried out for layers with sizes $L \times L \times N$ with the use of periodic and free boundary conditions for the in-plane and out-plane directions, respectively. N here is the thickness of ferromagnetic layers. Magnetic properties are calculated for this trilayer with $N = 3$, $\Delta = 0.5677$, $A = 0.0$. The equilibrium properties were simulated by the same way as in Section 2.1. The critical temperature $T_c = 0.784(7) J_1/k_b$ for trilayer system was calculated by Binder's cumulant method (Figure 4).

In investigations of influence of the initial states on characteristics of non-equilibrium critical behaviour we computed the two time-dependent autocorrelation function

$$C(t, t_w) = \left\langle \frac{1}{NL^2} \sum_{i=1}^{NL^2} \boldsymbol{S}_i(t)\boldsymbol{S}_i(t_w) \right\rangle - \boldsymbol{m}(t)\boldsymbol{m}(t_w), \quad (6)$$

The time variable t_w characterizes the age of a sample and is called waiting time. Autocorrelation function in nonequilibrium processes depend not only on the difference $t - t_w$ but also on each time variable individually at $t - t_w$ and t_w much smaller than the relaxation time of the system t_{rel}. The non-equilibrium evolution starts when the system after initial preparation is placed to the thermostat with $T = T_c$. At the waiting time t_w, we have begun to measure two-time correlation function within the $1 \ll t, t_w \ll t_{rel}$ range. The aging of a system is manifested in slowing down relaxation processes with the time passing from the preparation of a sample.

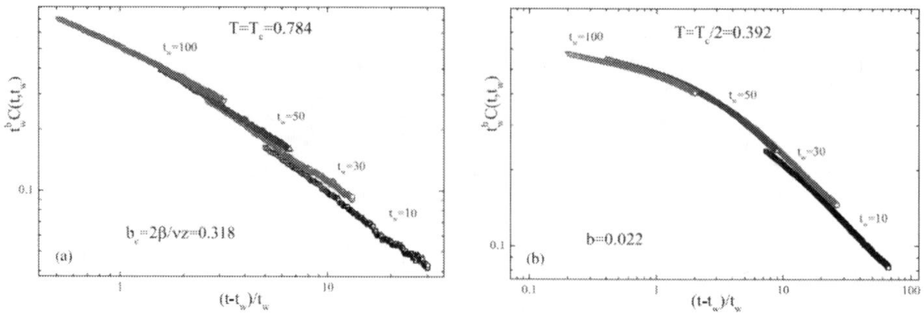

Figure 6. Scaling function $F(t/t_w) = t_w^b C(t, t_w)$ versus $(t - t_w)/t_w$ at $m_{stg}^0 = 0$ for $T = T_c = 0.784J_1/k_b$ (a) and $T = T_c/2 = 0.392J_1/k_b$ (b)

For the aging regime at $t - t_w \sim t_w \gg 1$, the two-time autocorrelation function has the scaling form [5]

$$C(t, t_w) \sim t_w^{-b} F_C(t/t_w), \tag{7}$$

where the exponent b at the T_c is expressed in terms of the critical exponents as $b = 2\beta/\nu z$. The scaling function $F_C(t/t_w)$ is a homogeneous function of its argument t/t_w.

The features of non-equilibrium critical dynamics [5] can be considered as basis for understanding and interpretation of the experimental data [6]. The next values of modelling parameters $\Delta(N = 3) = 0.5677$ and $A = 0$ are used for the multilayer structure, which correspond to the nanoscale superlattice [Co/Cu] [3]. The investigation of the autocorrelation function relaxation in the multilayer superlattice has revealed the magnetic aging phenomena (Figure 5). The realization of dynamical scaling which demonstrated in Figure 6 gives the possibility to determine the critical exponent $b_c = 2\beta/\nu z = 0.318$. Obtained before static critical exponents (Table 1) gives the possibility to estimate a effective system dimension $d_eff = \gamma/\nu + 2\beta/\nu = 2.28$. Effective exponents demonstrates dimensionality crossover from 2d XY-model to 3D isotropic Heisenberg model.

3 Conclusions

Therefore we have studied the magnetic properties of the trilayer structure consisting of two ferromagnetic films separated by nonmagnetic metal film. The study of aging effects in multilayered structures shows that these effects brightly appear in non-equilibrium relaxation from initial states.

The nanoscale periodicity creates in these magnetic multilayer structures mesoscopic spatial magnetic correlations with slow relaxation dynamics when quenching of the system into a non-equilibrium state is realized. In comparison with bulk magnetic systems where the slow dynamics and aging phenomenon are displayed close to critical point, the magnetic superlattices, structured on the nanoscale, give possibility to increase relaxation time due to the increasing in the characteristic spin-spin correlation length. The aging can be observed in the multilayer structures in a wider temperature range $T < Tc$ than for bulk magnetic systems.

Acknowledgments

This work was supported by Grant No. MD-6024.2016.2 of Russian Federation President and by Grant No. 17-02-00279 of Russian Foundation for Basic Research. The simulations were supported by the Laboratory of Applied Theoretical Physics and Parallel Computation OmSU (2×IBM P590), the Supercomputing Center of Lomonosov Moscow State University, Moscow and Saint Petersburg Joint Supercomputer Center of the Russian Academy of Sciences.

References

[1] C.A.F. Vaz, J.A.C. Bland, G. Lauhoff, Rep. Prog. Phys. **71**, 056501 (2008)
[2] P.V. Prudnikov, V.V. Prudnikov, M.A. Medvedeva, JETP Lett. **100**, 446 (2014)
[3] F. Huang, M.T. Kief, G.J. Mankey, R.F. Willis, Phys. Rev. B **49**, 3962 (1994)
[4] P. Prudnikov, V. Prudnikov, M. Menshikova, N. Piskunova, J. Magn. Magn. Mater. **387**, 77 (2015)
[5] V. Prudnikov, P. Prudnikov, E. Pospelov, J. Stat. Mech. **043303** (2016)
[6] T. Mukherjee, M. Pleimling, Ch. Binek, Phys. Rev. B **82**, 134425 (2010)

The Monte Carlo simulation of 2D ANNNI-model

Murtazaev Akai Kurbanovich[1,2, a], Ibaev Zhavrail Gadzhievich.[1,b*]

[1]Institute of Physics DSC RAS, Makhachkala, Yaragskogo 94, Russia

[2]Dagestan State University, Makhachkala, Gadzhieva 43a, Russia

[a]akai2005@mail.ru, [b]eibaev77@mail.ru

Keywords: condensed state, competing interaction, modulated ordering, Fourier analysis.

Abstract. In this, study we present the data for 2D Axial Next Nearest Neighbor Ising model (ANNNI-model) obtained from Monte Carlo (MC) simulations using the standard Metropolis algorithm. The temperature dependences of thermodynamic parameters for a cubic lattice with linear sizes L=32 at different values of the competing interaction parameter $|J_1/J|=0.1\div1.0$. Transition temperatures of ferromagnetic ordering to the paramagnetic state at $|J_1/J|<0.3$ and to the modulated state at $0.3<|J_1/J|<0.5$ are shown to shift towards low temperatures with an increase in a competing interaction parameter absolute value. Conversely, transition temperatures of the modulate state to the paramagnetic ordering grow. The modulated ordering in the 2D ANNNI-model appears in the temperature range $0.1<T<2.0$ at $0.2<|J_1/J|\leq1.0$. Modulated structure parameters are computed using a mathematic apparatus of Fourier transform spectral analysis. According to the Fourier analysis results, the wave number grows with an increase in the competing interaction parameter absolute value. Summarizing obtained results, we plot a phase diagram of 2D anisotropic Ising model with competing interactions.

Introduction

Successes achieved in study of quasi-1D and quasi-2D transition metal-based magnetic materials are due in large part to the application of simplified multi-electron approaches by means of the Heisenberg and Ising models [1, 2]. The Hamiltonians of these models have a fairly simple algebraic structure. This permits to prove a number of theorems concerning a character of precise energy spectrum and a spin character of the ground state, which have the considerable importance for testing the numerical simulation data [6-8]. However, as experience shows, main inferences about a magnetic structure deduced within the framework of these simplified models are, as a rule, reserved when passing to stricter approaches too, for instance, which take into account the repulsion of electrons located in non-neighboring crystal lattice sites [1, 6].

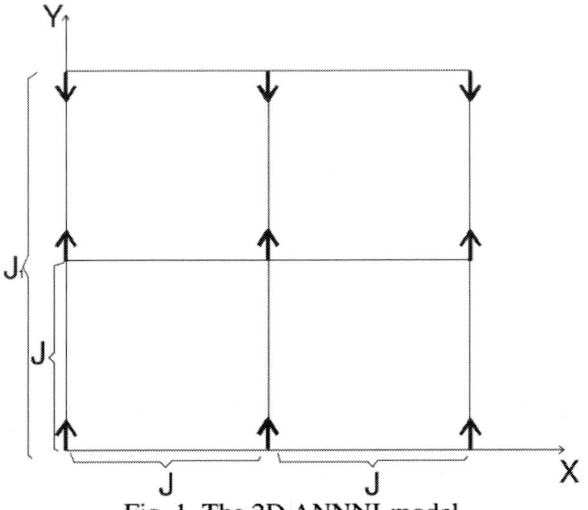

Fig. 1. The 2D ANNNI-model

The cubic Ising model with competing ferromagnetic and antiferromagnetic interactions (termed the ANNNI-model) was introduced to describe ordered magnetic phases in CeSb crystals [8]. A similar model is successfully used to evaluate the thermodynamics of oil microemulsions and is termed the Widom model [9]. In this study, we address 2D analog of such a model (Fig. 1), suitable for the description of microemulsion films on solid surfaces. It is assumed that similar models can be used to optimize extraction processes, for instance, in an oil recovery.

The Hamiltonian of the model is

$$H_{ANNNI} = -J\sum_{i,j} s_i s_j - J_1 \sum_i s_i s_{i+2}\ , \tag{1}$$

where $s_i=\pm 1$, $J>0$ is the parameter of nearest neighbor exchange interactions, $J_1<0$ is the parameter of next-nearest neighbor antiferromagnetic interaction along Y axis.

To define a phase behavior of the 2D ANNNI-model we employ approximate theoretical methods including high- and low-temperature splitting, the mean field theory, and other theoretical approximations.

According to literature data [8], when decreasing the temperature the ANNNI-model undergoes a second order transition from the paramagnetic to nearest ordered state, while the ferromagnetic to modulated phase transition, as we reported earlier, is of a first order [10].

Methods and Results

To address the thermodynamic and magnetic properties, we use the square systems of $L\times L$ sizes; $L=32$ with periodic boundary conditions. The number of spins in the simulated system N_{eff}=1024. The Monte Carlo simulations are performed by the Markov process of $\tau=1000\tau_0$ in length ($\tau_0=10^4$ MC steps per spin, the length of non-equilibrium part). Thermodynamic parameters of the system are calculated by the averaging over one thousand different initial configurations along this chain. The estimation of an error for results estimated doesn`t exceed sizes of symbols used in figures.

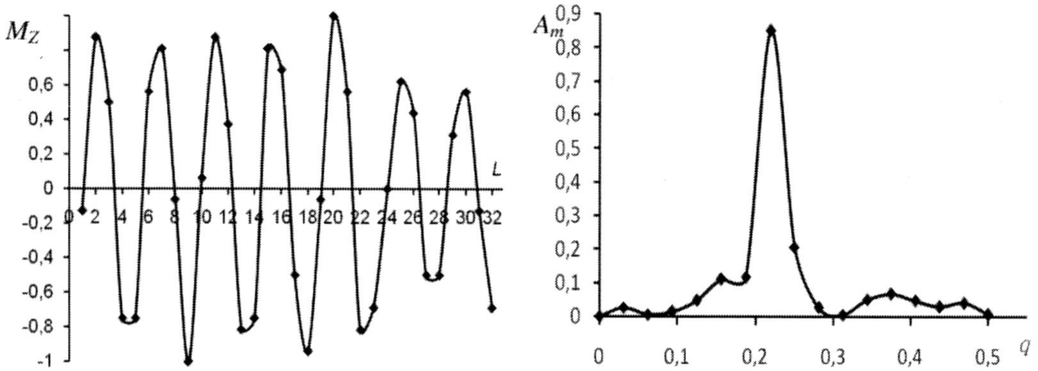

Fig. 2. The modulated structure of the 2D ANNNI-model at $L=32$, $|J_1/J|=1.0$, $t=1.95$ and the Fourier spectrum.

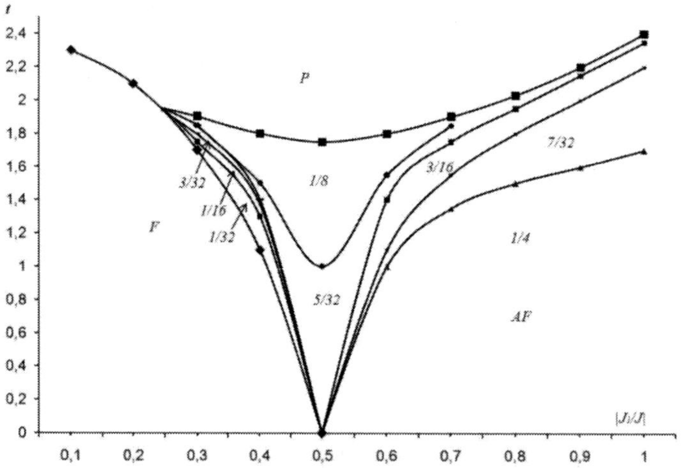

Fig. 3. The phase diagram of the 2D ANNNI-model (F is ferromagnetic, P is paramagnetic, and AF is antiferromagnetic phases with k=1/4, fractional numbers denote the values of the wave number k, $t=k_BT/J$ is the temperature).

Magnetic properties of the modulated region are evaluated by means of a value

$$M_Z = \frac{1}{L}\sum_{x}^{L} S_{x,y}, \tag{2}$$

which is the averaged magnetization of spin chains, perpendicular to Y axis. The averaged amplitude of the modulated structure is taken as the order parameter.

The parameters of modulated structures and transition temperatures between those are determined by a mathematic apparatus of Fourier spectral analysis

$$M_Z = \frac{a_0}{2} + \sum_{k=1}^{N}[a_k \cos(2\pi kz / L) + b_k \sin(2\pi kz / L), \tag{3}$$

where

$$a_0 = \frac{1}{L}\sum_{z=1}^{L} M_Z, \tag{4}$$

$$a_k = \frac{2}{L}\sum_{z=1}^{L} M_Z \cos(2\pi kz / L), \tag{5}$$

$$b_k = \frac{2}{L}\sum_{z=1}^{L} M_Z \sin(2\pi kz / L), \tag{6}$$

$$q = k / L, \tag{7}$$

k is the harmonic number, z is the layer number.

As the modulated structures are harmonic, the wave number q is defined by the value of k, at which the amplitude function $A_m = (a_k^2 + b_k^2)$ has a maximum.

Figure 2 depicts a typical ordering occurred in the 2D ANNNI-model at L=32, $|J_1/J|=1.0$, and temperature t=1.95 and a respective Fourier spectrum. As is easily seen from the figure, the model gains the modulated structure, which can be described by using the harmonic function

$$M_z = A\cos(2\pi qz + \varphi) \tag{8}$$

where A=0.749578 is the amplitude of the modulated structure, q=7/32 is the wave number, φ=3.14 is the initial phase.

In Fig. 3, we plot a phase diagram of two-dimensional anisotropic Ising model with competing interactions (Fig. 3) employing the Fourier analysis results of modulated structures occurred at L=32 and different temperatures and values of the competing interaction parameter. As is evident from the Fig.3, in the model, a region with the modulated ordering is observed at $0.2<|J_1/J|\leq 1.0$ in the temperature interval $0.1<T<2.0$.

Conclusion

It is known that real magnetic systems comprise different physical and chemical interactions as well one or several kinds of the anisotropy. The full account of all possible anisotropy varieties and interactions results in the appearance of several types of the magnetic ordering and, as a consequence, of multicritical phenomena, which complicate a behavior and properties of the magnetic system. The complete description of solids requires a correct consideration of such processes and phenomena.

In this work, we investigate the model system with competing exchange interaction of nearest and next-nearest spins. The presence of the competing interaction is shown to lead to the occurrence of the modulated ordering in the system and the complication of thermodynamic and magnetic properties.

These results can be used for interpretation of real magnetic systems, where the modulated ordering is possible. However, we fail, at present, to find the literature data on real magnetic systems obtained experimentally in considered temperature ranges and values of exchange interaction constants.

References

[1] A. A. Ovchinnikov, I. I. Ukrainskii, G. V. Kventsel', Theory of one-dimensional mott semiconductors and the electronic structure of long molecules having conjugated bonds, Physics-Uspekhi 15 (1973) 575–591
[2] S.R. White, Strongly correlated electron systems and the density matrix renormalization group, Phys. Reports 301 (1998) 187-204.
[3] I. Moreira, R. Dovesi, Periodic approach to the electronic structure and magnetic coupling in $KCuF_3$, K_2CuF_4, and $Sr_2CuO_2Cl_2$ low-dimensional magnetic systems, Int. J. Quant. Chem. 99 (2004) 805-823.
[4] Hitesh J. Changlani, Norm M. Tubman & Taylor L. Hughes, Charge density waves in disordered media circumventing the Imry-Ma argument, Scientific Reports 6 (2016) 31897-31900.
[5] Zeph Landau, Umesh Vazirani & Thomas Vidick, A polynomial time algorithm for the ground state of one-dimensional gapped local Hamiltonians, Nature Physics 11 (2015) 566–569.
[6] M. A. Shakirov, S. V. Tsibulsky, A. E. Antipov, Y. E. Shchadilova & A. N. Rubtsov, Modeling the metastable dynamics of correlated structures, Scientific Reports 5 (2015) 8005-8008.
[7] D.J. Klein, Ground-state features for Heisenberg models, Chem. Phys. 77 (1982) 3098-3100.
[8] R.J. Elliott, Phenomenological discussion of magnetic ordering in the heavy rare-earth metals, Phys. Rev. 124 (1961) 346-353.
[9] B. Widom, Lattice model of microemulsion, J. Chem. Phys. 84 (1986) 6943-6954.
[10] A.K. Murtazaev, J.G. Ibaev, Critical properties of an ANNNI-model in the neighborhood of multicritical Lifshitz point, Solid State Communications 152 (2012) 177-179.

Spin-Density Correlations and Short-Range Order in Metals above the Curie Temperature

Nikolai B. Melnikov[1,*], *Boris I.* Reser[2,**], and *Georgii V.* Paradezhenko[1,***]

[1] *Lomonosov Moscow State University, Moscow, Russia*
[2] *Mikheev Institute of Metal Physics, RAS, Ekaterinburg, Russia*

Abstract. In the dynamic spin-fluctuation theory, we calculate the effective and local magnetic moments and spatial spin-density correlator. Our theoretical results are demonstrated by the example of bcc Fe. The effective and local moments are found in good agrement with results of polarized neutron scattering experiment over a wide temperature range. The calculated short-range order is small (up to 5 Å) and slowly decreases with temperature.

1 Introduction

Polarized neutron scattering experiments [1, 2] point to the existence of large short-range order (SRO) of about 20 Å in metals above the Curie temperature, but the interpretation of the experiment based on the spin waves is controversial.

Static spin-fluctuation theories all support the existence of SRO, but there is no agreement about the spin correlation length (see, e.g., [3]). The dynamic theories based on the local approximation [4, 5] do not allow to estimate the SRO quantitatively.

The dynamic spin-fluctuation theory (DSFT) [6–8] takes into account both quantum nature and nonlocal character of the spin fluctuations. For selected temperatures, calculations of the spin-density correlator and effective moment in the dynamic approximations of spin-fluctuation theory were carried out by [6, 9].

We use the DSFT to calculate the spin correlator as a function of distance and temperature; calculate the effective moment as a function of wavevector and temperature and compare the results with experiment [10–12]. Our theoretical results are demonstrated in the example of bcc Fe.

2 Spin correlations in metals

From a theoretical point of view, we are interested in the spatial correlator $\langle s^\alpha(\mathbf{r})s^\alpha(0)\rangle$ of the spin density

$$s^\alpha(\mathbf{r}) = \sum_{i=1}^{N_e} s_i^\alpha \delta(\mathbf{r}-\mathbf{r}_i), \qquad \alpha = x, y, z,$$

where $\mathbf{s}_i$ is the spin of the ith electron and N_e is the number of electrons. In the experiment, the measured scattering is expressed in terms of the square of the effective moment

$$M^2(q) = \frac{1}{N}\langle \mathbf{s_q s_{-q}}\rangle g^2\mu_B^2, \qquad (1)$$

where N is the number of atoms and

$$s_{\mathbf{q}}^\alpha = \int s^\alpha(\mathbf{r})\,\mathrm{e}^{-\mathrm{i}\mathbf{qr}}\,\mathrm{d}\mathbf{r} \qquad (2)$$

is the Fourier transform of the spin-density $s^\alpha(\mathbf{r})$.

From the spin correlator we can obtain various magnetic characteristics. For instance, integration over the Wigner-Seitz cell of the volume Ω_{WS} gives the square of the local spin moment

$$s_{\mathrm{L}}^2 = \iint_{\mathrm{WS}} \langle \mathbf{s}(\mathbf{r})\mathbf{s}(\mathbf{r}')\rangle\,\mathrm{d}\mathbf{r}\,\mathrm{d}\mathbf{r}' \approx \Omega_{\mathrm{WS}}^2\langle \mathbf{s}(0)\mathbf{s}(0)\rangle. \qquad (3)$$

The correlator can be calculated by inverting the Fourier transformation (2):

$$\langle s^\alpha(\mathbf{r})s^\alpha(0)\rangle = \frac{1}{V}\sum_{\mathbf{q}}\langle s_{\mathbf{q}}^\alpha s_{-\mathbf{q}}^\alpha\rangle\,\mathrm{e}^{\mathrm{i}\mathbf{qr}},$$

where $V = N\Omega_{\mathrm{WS}}$ is the volume of the crystal. Replacing the sum over the Brillouin zone by the integral over a sphere of equal volume with the radius q_{B}, we obtain [12]

$$\langle s^\alpha(\mathbf{r})s^\alpha(0)\rangle = \frac{1}{2\pi^2 N\Omega_{\mathrm{WS}}}\int_0^{q_{\mathrm{B}}}\langle s_{\mathbf{q}}^\alpha s_{-\mathbf{q}}^\alpha\rangle\frac{q\sin(qr)}{r}\,\mathrm{d}q. \qquad (4)$$

Passing to the limit $r \to 0$ in expression (4) and substituting the result in (3), we come to

$$s_{\mathrm{L}}^2 = \frac{\Omega_{\mathrm{WS}}}{2\pi^2 N}\int_0^{q_{\mathrm{B}}}\langle \mathbf{s_q s_{-q}}\rangle q^2\,\mathrm{d}q. \qquad (5)$$

Using formula (1), we can write the square of the local magnetic moment $m_{\mathrm{L}} = s_{\mathrm{L}} g\mu_{\mathrm{B}}$ as

$$m_{\mathrm{L}}^2 = \frac{4\pi}{\Omega_{\mathrm{BZ}}}\int_0^{q_{\mathrm{B}}} q^2 M^2(q)\,\mathrm{d}q, \qquad (6)$$

*e-mail: melnikov@cs.msu.ru
**e-mail: reser@imp.uran.ru
***e-mail: gparadezhenko@cs.msu.ru

where $\Omega_{BZ} = (2\pi)^3/\Omega_{WS}$ is the volume of the Brillouin zone.

Calculation of $M^2(q)$ by formula (1) requires the spin correlator $\langle s^\alpha_{\mathbf{q}} s^\alpha_{-\mathbf{q}}\rangle$. We use the fluctuation-dissipation theorem:

$$\langle s^\alpha_{\mathbf{q}} s^\alpha_{-\mathbf{q}}\rangle = \frac{1}{2\pi}\int B(\varepsilon)\,\mathrm{Im}\,\chi^\alpha_{\mathbf{q}}(\varepsilon)\,\mathrm{d}\varepsilon, \qquad (7)$$

where

$$\chi^\alpha_{\mathbf{q}}(\omega) = \frac{2\mathrm{i}}{\hbar}\int_0^\infty \langle [s^\alpha_{\mathbf{q}}(t), s^\alpha_{-\mathbf{q}}]\rangle\,\mathrm{e}^{\mathrm{i}\omega t}\,\mathrm{d}t$$

is the dynamic susceptibility (in units of $\frac{1}{2}g^2\mu_B^2$) and $B(\varepsilon) = (\mathrm{e}^{\varepsilon/T} - 1)^{-1}$ is the Bose function (for details, see, e.g., [13]).

In the DSFT, the enhanced susceptibility is given by

$$\chi^\alpha_{\mathbf{q}}(\varepsilon) = \frac{\chi^{0\alpha}_{\mathbf{q}}(\varepsilon)}{1 - \tilde{u}\chi^{0\alpha}_{\mathbf{q}}(\varepsilon)},$$

where $\chi^{0\alpha}_{\mathbf{q}}(\varepsilon)$ is the unenhnaced susceptibility of the non-interacting electrons in the fluctuating field and $\tilde{u} = u/N$ is the Fourier transform of the effective interaction constant u. Since the Bose function $B(\varepsilon)$ is localized around $\varepsilon = 0$, the unenhnaced susceptibility $\chi^{0\alpha}_{\mathbf{q}}(\varepsilon)$ is replaced by its Taylor series in ε. Then

$$\langle s^\alpha_{\mathbf{q}} s^\alpha_{-\mathbf{q}}\rangle = \frac{T}{2\tilde{u}\lambda^\alpha_{\mathbf{q}}}\frac{2}{\pi}\arctan\frac{\tilde{u}\varphi^\alpha_{\mathbf{q}}\pi^2 T}{6\lambda^\alpha_{\mathbf{q}}},$$

where $\lambda^\alpha_{\mathbf{q}} = 1-\tilde{u}\chi^{0\alpha}_{\mathbf{q}}(0)$ and $\varphi^\alpha_{\mathbf{q}} = \mathrm{d}\chi^{0\alpha}_{\mathbf{q}}(0)/\mathrm{d}\varepsilon$. To take the mode-mode interaction into account, the static susceptibility $\chi^{0\alpha}_{\mathbf{q}}(0)$ is interpolated between the *homogeneous* $\chi^{0\alpha}_0(0)$ and *local* $\chi^{0\alpha}_L(0)$ susceptibilities. Thus,

$$\lambda^\alpha_{\mathbf{q}} = \lambda^\alpha_0 + \left(\lambda^\alpha_L - \lambda^\alpha_0\right)q^2/\overline{q^2},$$

where $\lambda^\alpha_L = 1 - u\chi^{0\alpha}_L(0)$ and $\overline{q^2}$ is the average over the Brillouin zone [7]. For simplicity, $\varphi^\alpha_{\mathbf{q}}$ is replaced by its mean value. Then the spin correlator is given by [12]

$$\langle s^\alpha_{\mathbf{q}} s^\alpha_{-\mathbf{q}}\rangle = \frac{NT}{2u\lambda^\alpha_L}\frac{1}{a^2_\alpha + b^2_\alpha(q/q_B)^2}\frac{2}{\pi}\arctan\frac{c_\alpha}{a^2_\alpha + b^2_\alpha(q/q_B)^2}, \qquad (8)$$

where the coefficients a_α, b_α, c_α and λ^α_L are calculated self-consistently at each T.

Now we consider high-temperatures ($T \gg T_C$). Approximating the Bose function $B(\varepsilon) \approx T/\varepsilon$, we write the fluctuation-dissipation theorem (7) as

$$\langle s^\alpha_{\mathbf{q}} s^\alpha_{-\mathbf{q}}\rangle = \frac{T}{2\pi}\int\frac{\mathrm{Im}\,\chi^\alpha_{\mathbf{q}}(\varepsilon)}{\varepsilon}\,\mathrm{d}\varepsilon.$$

From the Kramers-Kronig relation

$$\mathrm{Re}\,\chi^\alpha_{\mathbf{q}}(0) = \frac{1}{\pi}\int\frac{\mathrm{Im}\,\chi^\alpha_{\mathbf{q}}(\varepsilon)}{\varepsilon}\,\mathrm{d}\varepsilon$$

we obtain the high-temperature form of the fluctuation-dissipation theorem:

$$\langle s^\alpha_{\mathbf{q}} s^\alpha_{-\mathbf{q}}\rangle = \frac{T}{2}\chi^\alpha_{\mathbf{q}}(0). \qquad (9)$$

Using formula (1), we write relation (9) as

$$\frac{1}{N}\chi^\alpha_{\mathbf{q}}(0)\,\frac{1}{2}g^2\mu_B^2 = \frac{M^2(q)}{3T}.$$

Taking the latter at $\mathbf{q} = 0$ and comparing with the Curie-Weiss law

$$\chi^\alpha_{CW} = \frac{m^2_{\mathrm{eff}}}{3(T - \Theta_C)}, \qquad T > \Theta_C,$$

we see that

$$\chi^\alpha_{CW} = \frac{1}{N}\chi^\alpha_0(0)\frac{1}{2}g^2\mu_B^2, \qquad M(0) = m_{\mathrm{eff}}, \qquad (10)$$

for $T \gg \Theta_C$. In the DSFT, calculating the static susceptibility $\chi^\alpha_{\mathbf{q}}(0)$ on the right-hand side of (9), we obtain that the spin correlator is given by the Lorentzian function

$$\langle s^\alpha_{\mathbf{q}} s^\alpha_{-\mathbf{q}}\rangle = \frac{NT}{2u\lambda^\alpha_L}\frac{1}{a^2_\alpha + b^2_\alpha(q/q_B)^2}. \qquad (11)$$

This high-temperature formula differs from the exact one in the DSFT (8) by the factor with the arctangent.

Fig. 2 shows the square of the effective moment (1) for bcc Fe calculated in the DSFT. We see that the exact formula (8) and high-temperature formula (11) give qualitatively similar results.

Next, we consider the high-temperature approximation of the spatial correlator. At large distances $r > \pi/q_B$, we can extend the integration interval in (4) to infinity. Then the spin correlator reduces to the Ornstein-Zernike form

$$\langle s^\alpha(\mathbf{r})s^\alpha(0)\rangle = \frac{B}{q_B r}\,\mathrm{e}^{-r/r_c}.$$

Here $r_c = |b_\alpha|/(|a_\alpha|q_B)$ is the correlation radius, which can be considered as a measure of SRO. In the DSFT, it was shown theoretically that r_c increases to infinity as T goes to T_C, and r_c decreases to zero inversely with temperature: $r_c \sim 1/T$, at high temperatures $T \gg T_C$ [10].

3 Polarized magnetic neutron scattering

We compare the DSFT calculation results with the results of the polarized magnetic neutron scattering experiment. The magnetic contribution is given by

$$\left(\frac{\mathrm{d}^2\sigma^{\uparrow\downarrow}}{\mathrm{d}\Omega'\mathrm{d}E'}\right)_\parallel - \left(\frac{\mathrm{d}^2\sigma^{\uparrow\downarrow}}{\mathrm{d}\Omega'\mathrm{d}E'}\right)_\perp = \left(\frac{\gamma e^2}{m_e c^2}\right)^2\frac{1}{3}\frac{V}{\Omega_{WS}}\frac{k'}{k}S(\kappa,\omega),$$

where

$$S(\kappa,\omega) = \sum_\alpha\frac{1}{2\pi\hbar N}\int\langle s^\alpha_{\kappa}(t)s^\alpha_{-\kappa}\rangle\,\mathrm{e}^{\mathrm{i}\omega t}\,\mathrm{d}t \qquad (12)$$

is the scattering function and $\boldsymbol{\kappa} = \mathbf{k} - \mathbf{k}'$ is the scattering vector (for details, see [12]). In the experiment [1] the integral scattering over all energies is measured:

$$\left(\frac{\mathrm{d}\sigma^{\uparrow\downarrow}}{\mathrm{d}\Omega'}\right)_\parallel - \left(\frac{\mathrm{d}\sigma^{\uparrow\downarrow}}{\mathrm{d}\Omega'}\right)_\perp = \left(\frac{\gamma e^2}{m_e c^2}\right)^2\frac{1}{3g^2\mu_B^2}\frac{V}{\Omega_{WS}}\frac{k'}{k}M^2(\kappa),$$

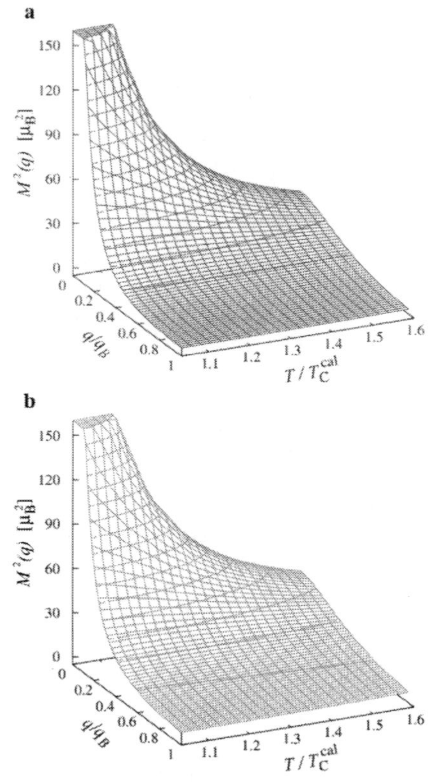

Figure 1. Square of the effective moment for bcc Fe calculated by the exact formula (**a**) and high-temperature formula (**b**) in the DSFT.

In the experiment [1], the energy cutoff is $\varepsilon_{max} = 50$ meV.

Comparison of the squared effective moment calculated with the energy cutoff shows good agreement with experiment over a wide range of temperatures above T_C (Fig. 2; for detailed results, see [11]). Results of the exact formula (8) and approximate high-temperature formula (11) can be viewed as a theoretical prediction.

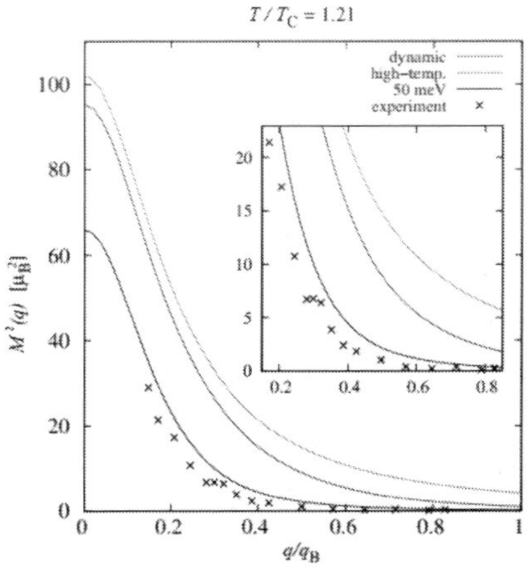

Figure 2. Square of the effective moment (1) for bcc Fe calculated by exact formula (8), approximate high-temperature formula (11) and with the energy cutoff $\varepsilon_{max} = 50$ meV compared with the experimental values [1]. The inset shows a region with the experimental points on a larger scale.

where $M^2(\kappa)$ is the square of the effective moment (1). It is related to the scattering function (12) by the formula

$$M^2(\kappa) = \left(\hbar \int S(\kappa, \omega)\, d\omega\right) g^2 \mu_B^2.$$

Comparing the theoretical results with experimental results, one should keep in mind three circumstances. The first one is the spatial spin distribution, which leads to the magnetic form-factor

$$F(\mathbf{q}, \mathbf{k}) \approx F(q) = \int |w(\mathbf{r})|^2 \, e^{-i\mathbf{qr}} \, d\mathbf{r},$$

where $w(\mathbf{r})$ is a Wannier state. For bcc Fe, $|F^2(q)| \approx 0.81$. The second is vibrations of the crystal lattice. This can be taken into account by the Debye-Waller factor e^{-2W}. Here

$$2W(\kappa) = \frac{1}{3}\kappa^2 \langle u^2 \rangle,$$

where $\langle u^2 \rangle$ is the mean-square atomic displacement. For bcc Fe, within the Brillouin zone $e^{-2W} \approx 1$ (for details, see [14]). The third is the energy cutoff $\varepsilon \leq \varepsilon_{max}$ in the scattering experiment. To be consistent with the experiment, we integrate over the same energy interval

$$\langle s_{\mathbf{q}}^{\alpha} s_{-\mathbf{q}}^{\alpha} \rangle = \frac{1}{2\pi} \int_0^{\varepsilon_{max}} B(\varepsilon) \, \mathrm{Im}\, \chi_{\mathbf{q}}^{\alpha}(\varepsilon) \, d\varepsilon. \qquad (13)$$

The square of the local magnetic moment m_L^2 is calculated by formula (6), where the spin correlator is given by the exact formula (8), high-temperature formula (11) and formula with the energy cutoff (13). The integrand of (6) is shown in Fig. 3. Results of the calculation for bcc Fe are presented in Table 1. The local moment calculated with the energy cutoff is in reasonable agreement with the experimental one $m_L^{exp} = 1.3\,\mu_B$ at $T/T_C^{exp} = 1.2$ [1]. The value m_L calculated with the energy cutoff decreases with increasing temperature (Table 1, last column). This implies that the energy window in the experiment should be wider to include all magnetic scattering, especially at higher temperatures. Calculations by the exact formula are in agreement with the ones using effective Hamiltonians with classical spins, e.g. $m_L \approx 2.2\,\mu_B$ at $T/T_C^{exp} = 1.1$–1.5 [15]. Comparing the second and third columns in Table 1, we see that, despite qualitative similarities, the exact and high-temperature formulae give different quantitative results.

In paper [1] the experiment was interpreted based on the spin wave theory. The idea is as follows. The normalized correlation function is given by

$$C(r, T) = \left(\int_0^{q_B} q^2 M^2(q)\, dq\right)^{-1} \int_0^{q_B} q^2 M^2(q) \frac{\sin(qr)}{qr}\, dq.$$

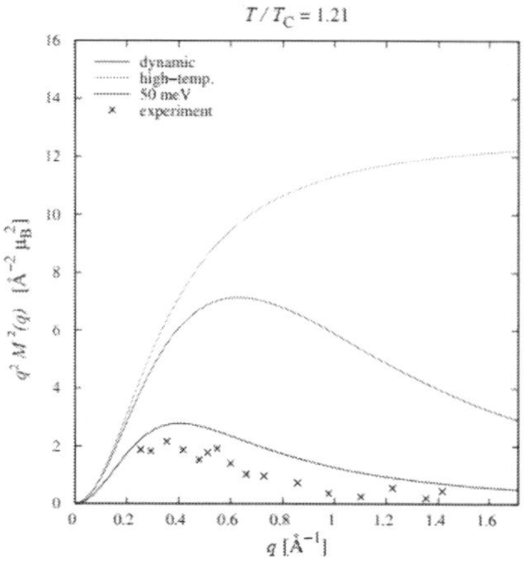

Figure 3. Values of $q^2M^2(q)$ for bcc Fe (the legend as in Fig. 2).

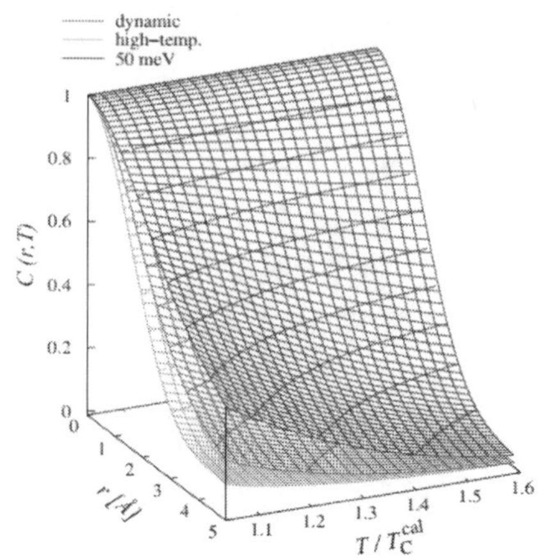

Figure 4. Correlation function $C(r, T)$ of bcc Fe calculated in the DSFT.

T/T_C^{cal}	exact, μ_B	high-temp., μ_B	50 meV, μ_B
1.226	2.214	3.060	1.120
1.322	2.218	3.080	1.116
1.418	2.222	3.097	1.047
1.514	2.228	3.114	0.993

Table 1. Local moment m_L of bcc Fe calculated in the DSFT.

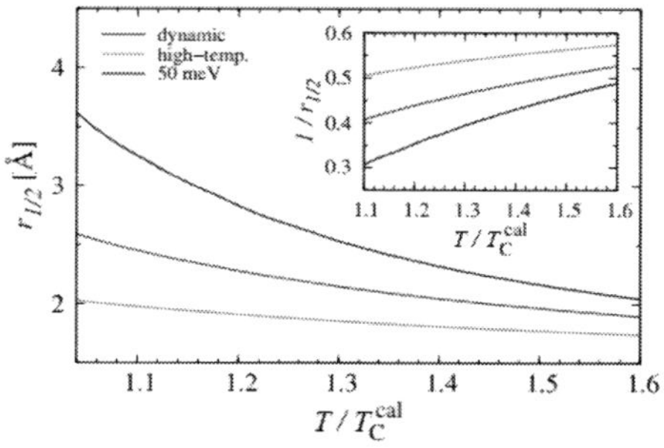

Figure 5. Halfwidth $r_{1/2}(T)$ of correlation function $C(r, T)$ for bcc Fe calculated in the DSFT. The inset shows the inverse halfwidth $r_{1/2}^{-1}(T)$.

Assuming that $q^2M^2(q) \propto \delta(q - q_0)$, we obtain a *spin wave* of the wavelength $\lambda_0 = 2\pi/q_0$:

$$C(r, T) = \int_0^{q_B} \delta(q - q_0)\,\frac{\sin(qr)}{qr}\,dq = \frac{\sin(q_0 r)}{q_0 r}.$$

In bcc Fe $\lambda_0 \approx 18$ Å at $T = 1.21T_C^{exp}$. The problem with this interpretation is that the peak is too wide: the height and width of the peak are about equal (see Fig. 3).

Instead of approximating the normalized correlation function by a delta-peak, we calculate it in the DSFT (Fig. 4). Since Ornstein-Zernike correlation radius r_c works only at large distances, we measure the SRO domain using the halfwidth $r_{1/2}(T)$ of the correlation function $C(r, T)$.

The value of the halfwidth $r_{1/2}(T)$ lies within 5 Å in all approximations (Fig. 5). The halfwidth decreases with temperature and comes down to 1–2 Å at $1.6\,T/T_C^{cal}$. From the inset of Fig. 5 we see that the reciprocal halfwidth has an almost linear temperature dependence similarly to the theoretical result for the correlation radius. Moreover, this linear dependence is already observed from about $1.1\,T/T_C^{cal}$.

4 Conclusions

In the DSFT, we have calculated the effective moment $M(q)$ as a function of q and T and spin correlator $\langle \mathbf{s}(\mathbf{r})\mathbf{s}(0)\rangle$ as a function of r and T. At high temperatures, $M(q)$ is given by the Lorentzian function, spin correlator has the Ornstein-Zernike form, and $M(0)$ tends to m_{eff} in the Curie-Weiss law. Calculated $M(q)$ and m_L of bcc Fe above T_C are in good agreement with results of the polarized neutron scattering experiment. Calculated spatial correlator in bcc Fe above T_C gives a small SRO domain (less than 5 Å). The SRO remains at high temperatures.

Acknowledgments

The research was carried out within the state assignment of the Federal Agency for Scientific Organizations of Russia (program "Electron" No. 01201463326) and was supported in part by the Program of Ural Branch of Russian Academy of Sciences (Project No. 15–8–2–10).

References

[1] P. Brown, H. Capellmann, J. Deportes, D. Givord, K. Ziebeck, J. Magn. Magn. Mater. **30**, 243 (1982)
[2] G. Shirane, P. Böni, J.P. Wicksted, Phys. Rev. B **33**, 1881 (1986)

[3] T. Moriya, *Spin Fluctuations in Itinerant Electron Magnetism* (Springer, Berlin, 1985)

[4] A.I. Lichtenstein, M.I. Katsnelson, G. Kotliar, Phys. Rev. Lett. **87**, 067205 (2001)

[5] Y. Kakehashi, M.A.R. Patoary, J. Phys. Soc. Jpn. **80**, 034706 (2011)

[6] V.I. Grebennikov, Phys. Solid State **40**, 79 (1998)

[7] B.I. Reser, V.I. Grebennikov, Phys. Met. Metallogr. **85**, 20 (1998)

[8] N.B. Melnikov, B.I. Reser, V.I. Grebennikov, J. Phys.: Condens. Matter **23**, 276003 (2011)

[9] V.I. Grebennikov, J. Magn. Magn. Mater. **84**, 59 (1990)

[10] N.B. Melnikov, B.I. Reser, J. Magn. Magn. Mater. **397**, 347 (2016)

[11] N.B. Melnikov, B.I. Reser, G.V. Paradezhenko, J. Magn. Magn. Mater. **411**, 133 (2016)

[12] N.B. Melnikov, G.V. Paradezhenko, B.I. Reser, Theor. Math. Phys. **191**, 602 (2017)

[13] N.B. Melnikov, B.I. Reser, Phys. Met. Metallogr. **117**, 1328 (2016)

[14] G.V. Paradezhenko, N.B. Melnikov, B.I. Reser, EPJ Web of Conferences (2017), this volume

[15] P.-W. Ma, S.L. Dudarev, Phys. Rev. B **86**, 054416 (2012)

Quasi-classical localized states in the 2D ferrimagnet

Yury Panov[1,*] and *Alexander* Moskvin[1]

[1] *Ural Federal University, 620002, Mira str. 19, Ekaterinburg, Russia*

Abstract. We consider highly anisotropic 2D quantum $s = 1/2$ (pseudo)magnetic system which is equivalent to the frequently used system of charged hard-core bosons on a square lattice. In the continuous quasi-classical approximation, the types of localized excitations are determined by asymptotic analysis and compared with numerical results. Depending on the homogeneous ground state, the excitations are the ferro and antiferro type vortices, the skyrmion-like topological excitations or linear domain walls.

1 Introduction

We construct a continuous two-sublattice approximation for the 2D system of charged hard-core (hc) bosons, that is equivalent to the highly anisotropic $s = 1/2$ 2D magnets with a constant magnetization. Quasi-classical continuous description of the 2D magnetic systems reveals their striking features, namely, the collective localized inhomogeneous states with nontrivial topology and finite excitation energy. These include topological solitons [5, 6], magnon drops [7], in- and out-of-plane vortex-antivortex pairs [8], and various spiral solutions [9–11]. Basically these solutions have been obtained for the isotropic and anisotropic ferromagnet. In the system of lattice bosons, these states correspond to an inhomogeneous distribution of the charge and the superfluid densities. We consider the asymptotic behavior of localized solutions which converge to homogeneous solutions at infinity and do find their phase diagram. The results are compared with numerical calculations of the ground-state energy in the quasi-classical approximation.

2 The model

Hereafter, in the paper we make use of a pseudospin formalism. For instance, the $S = 1$ pseudospin formalism [3, 4] developed for the copper oxides implies that the on-site Hilbert space described by three effective valence states of the copper ions, $Cu^{1+,2+,3+}$, or, strictly speaking, of the copper-oxygen clusters $CuO_4^{7-,6-,5-}$ in the CuO_2 plane. These states correspond to the hole occupation numbers $n_h = 0, 1, 2$ and can be considered as components of the pseudospin $S = 1$ triplet with projections $M_S = -1, 0, +1$, respectively. The full pseudospin Hamiltonian [4] includes the two-hole transfer and several types of one-hole transfer terms, but here we use a simplified pseudospin Hamiltonian that takes into account only the two-hole transport:

$$\mathcal{H} = \Delta\sum_i S_{iz}^2 + V\sum_{\langle ij\rangle} S_{iz}S_{jz} - t\sum_{\langle ij\rangle}(S_{i+}^2 S_{j-}^2 + S_{i-}^2 S_{j+}^2). \quad (1)$$

A pair of holes plays the role of one boson in the model, however, the condition $\Delta \to \infty$ eliminates the on-site state with $n_h = 1$, so in the limit of $\Delta \to -\infty$, it is equivalent to the Hamiltonian of charged hc bosons in terms of a pseudospin $\vec{\sigma}$, $\sigma_z = \pm 1$:

$$\mathcal{H} = -t\sum_{\langle ij\rangle}(\sigma_{i+}\sigma_{j-} + \sigma_{i-}\sigma_{j+}) + V\sum_{\langle ij\rangle}\sigma_{iz}\sigma_{jz}. \quad (2)$$

Here σ_α, $\alpha = x, y, z$ are Pauli matrices, $\sigma_\pm = (\sigma_x \pm i\sigma_y)/2$. The z-component of the pseudospin describes the local density of bosons, so that antiferromagnetic z-z exchange corresponds to the repulsive density-density interaction, while isotropic ferromagnetic planar exchange corresponds to the kinetic energy of bosons. The constant total number of bosons leads to the constraint on the total z-component of the pseudospin. We define the n, as the density of the total doped charge counted from the state with a zero total z-component, or parent Cu^{2+} state. Then n is the sum of z-components of the pseudospin: $\sum_i \sigma_{iz} = nN$, where N is the total number of sites. If ρ is the density of hc bosons, then n is the deviation from the half-filling: $\rho = (1+n)/2$.

The energy functional in a quasi-classical approximation

$$E = \langle\Psi|\mathcal{H}|\Psi\rangle, \quad |\Psi\rangle = \prod_i |\psi\rangle_i, \quad (3)$$
$$|\psi\rangle = \cos\frac{\theta}{2}\, e^{-i\frac{\phi}{2}}\, |{+1}\rangle + \sin\frac{\theta}{2}\, e^{i\frac{\phi}{2}}\, |{-1}\rangle,$$

where $|\pm 1\rangle$ are the eigenfunctions of the σ_z, takes the form

$$\varepsilon = -\sum_{\langle i,j\rangle}\sin\theta_i\,\sin\theta_j\,\cos(\phi_i - \phi_j) + \lambda\sum_{\langle i,j\rangle}\cos\theta_i\,\cos\theta_j - \xi\left(\sum_i \cos\theta_i - nN\right). \quad (4)$$

*e-mail: yuri.pano@urfu.ru

Here we define $\varepsilon = 2E/t$, $\lambda = 2V/t$, $\xi = 2\mu/t$, where the chemical potential μ takes into account the bosons density constraint, and $\langle i, j\rangle$ denotes summation over nearest neighbors in a square lattice. The θ_i and ϕ_i are the polar and azimuthal angles of the quasiclassical pseudospin vector at an i-th site.

Hereafter, we introduce two sublattices A and B with the checkerboard ordering. The energy functional (4) in the two-sublattices continuous approximation takes the form

$$\begin{aligned}\varepsilon = \frac{1}{2}\int\Big[&(\cos\theta_A\cos\theta_B\cos(\phi_A-\phi_B) -\\ &-\lambda\sin\theta_A\sin\theta_B)(\nabla\theta_A,\nabla\theta_B) +\\ &+\sin\theta_A\sin\theta_B\cos(\phi_A-\phi_B)(\nabla\phi_A,\nabla\phi_B) -\\ &-\sin\theta_A\cos\theta_B\sin(\phi_A-\phi_B)(\nabla\phi_A,\nabla\theta_B) +\\ &+\cos\theta_A\sin\theta_B\sin(\phi_A-\phi_B)(\nabla\theta_A,\nabla\phi_B) +\\ &+4(-\sin\theta_A\sin\theta_B\cos(\phi_A-\phi_B)+\lambda\cos\theta_A\cos\theta_B) -\\ &-\xi(\cos\theta_A+\cos\theta_B-2n)\Big]\,d\vec{r}, \end{aligned}\quad (5)$$

The Euler equations for this functional have rather complicated form. But we note that the first sum in the Exp.(4) has its lowest value if $\cos(\phi_i - \phi_j) = 1$. This allows us to make a simplifying assumption, that $\phi_A(\vec{r}) = \phi_B(\vec{r}) \equiv \phi(\vec{r})$. It is worth to note that this assumption is confirmed by the results of our numerical simulations. We define functions $u(\vec{r}) \equiv \theta_A(\vec{r})$, $v(\vec{r}) \equiv \theta_B(\vec{r})$, and their combinations $f = \cos u\cos v$, $F = -f + \lambda f_{uv}$, where subscripts u and v denote derivatives with respect to these quantities. Then the Euler equations in the continuous approximation take a compact form

$$\begin{cases} f_{uv}\Delta\phi - 2f_v\cdot(\nabla u,\nabla\phi) = 0,\\ f_{uv}\Delta\phi - 2f_u\cdot(\nabla v,\nabla\phi) = 0,\\ F\Delta u + F_u\cdot(\nabla u)^2 - f_u\cdot(\nabla\phi)^2 - 4F_u + \xi\sin v = 0,\\ F\Delta v + F_v\cdot(\nabla v)^2 - f_v\cdot(\nabla\phi)^2 - 4F_v + \xi\sin u = 0.\end{cases}\quad (6)$$

These equations need to add the bosons density constraint. With the relevant exchange constants, the equations (6) lead to the equations of [12, 13].

3 The asymptotic behavior of localized solutions

The system (6) along with the boson density constraint has constant (homogeneous) solutions, $\phi = \phi_0$, $u = u_0$, $v = v_0$, that determine the well-known ground-state phase diagram [14] of the hc bosons system in the mean-field approximation. Correspondence between phases of anisotropic magnet and hc bosons is given in table 1.

Given $\lambda < 1$ or $n^2 > (\lambda-1)/(\lambda+1)$ the ground state of the system is a ferromagnetic (FM) (superfluid (SF))

$$\cos u_0 = \cos v_0 = n, \quad \varepsilon_0 = -2 + 2(\lambda+1)n^2. \quad (7)$$

hc bosons	anisotropic magnet
superfluid (SF)	ferromagnetic (FM)
supersolid (SS)	ferrimagnetic (FIM)
charge-ordered (CO)	antiferromagnetic (AFM)

Table 1. The corresponding phases of hc bosons and anisotropic magnet.

Given $n^2 < (\lambda-1)/(\lambda+1)$ the ground state is a ferrimagnetic (FIM) (supersolid (SS)):

$$\cos u_0 = n + z, \quad \cos v_0 = n - z, \qquad \varepsilon_0 = -2\lambda + 4|n|\sqrt{\lambda^2-1}. \quad (8)$$

where $z^2 = 1 + n^2 - 2|n|\lambda/\sqrt{\lambda^2-1}$. In all phases, the value of ξ satisfies the regular expression $\xi_0 = \partial\varepsilon_0/\partial n$. When $\lambda > 1$ and $n = 0$ the FIM phase transforms into a conventional antiferromagnetic (AFM) (charge-ordered (CO)) phase with the checkerboard ordering.

The equations (6) in the case of $\lambda < 0$ have localized solutions with nonzero topological charge and finite energy [5–11]. However, in our case $\lambda > 0$, numerical calculations with the conjugate gradient method for minimizing of the energy functional (3) on the lattice 256×256 indicate the existence of similar solutions, at least, as metastable states. The results are shown in figure 1. The actual stability of these solutions in our calculation was different. The FM-phase solutions, similar to that of in figure 1, cases 1 and 2, quickly evolved to an uniform one. The FIM- and AFM-phase solutions, similar to that of in figure 1, cases 3 and 4, retain its form for more than 10^6 iterations.

We investigated the asymptotic behavior of localized solutions, suggesting that at $r \to \infty$ they have the form $\phi(\vec{r}) = \phi_0 + \tilde{\phi}(\vec{r})$, $u(\vec{r}) = u_0 + \tilde{u}(\vec{r})$, $v(\vec{r}) = v_0 + \tilde{v}(\vec{r})$, where $\tilde{\phi}, \tilde{u}, \tilde{v} \to 0$ at $r \to \infty$. Hereinafter, the index 0 means the corresponding values for constant solutions. The linearized system (6) for the functions $\tilde{\phi}, \tilde{u}, \tilde{v}$ takes the form

$$\begin{cases} f_{uv0}\Delta\tilde{\phi} = 0,\\ F_0\Delta\tilde{u} + 4F_0\tilde{u} + 4\left(-F_{uv0} + \dfrac{\xi_0}{4}\cos v_0\right)\tilde{v} = 0,\\ F_0\Delta\tilde{v} + 4F_0\tilde{v} + 4\left(-F_{uv0} + \dfrac{\xi_0}{4}\cos u_0\right)\tilde{u} = 0.\end{cases}\quad (9)$$

In the case of the FM and FIM phases, the solutions for the first equation can be written as

$$\tilde{\phi}(\vec{r}) = \sum_{m=1}^{\infty}\frac{c_m}{r^m}\cos m(\varphi - \varphi_m), \quad (10)$$

with c_m and φ_m determined by the boundary conditions. In the case of the AFM phase, the first equation reduces to an identity since $f_{uv0} = 0$.

In the case of the FM phase, the second and the third equations become independent Helmholtz equations for the ferro- and antiferro-type combinations $U = \tilde{u} + \tilde{v}$, $V = \tilde{u} - \tilde{v}$:

$$\Delta U + A_1 U = 0, \quad A_1 = 4\frac{(\lambda+1)(1-n^2)}{\lambda - (\lambda+1)n^2}; \quad (11)$$

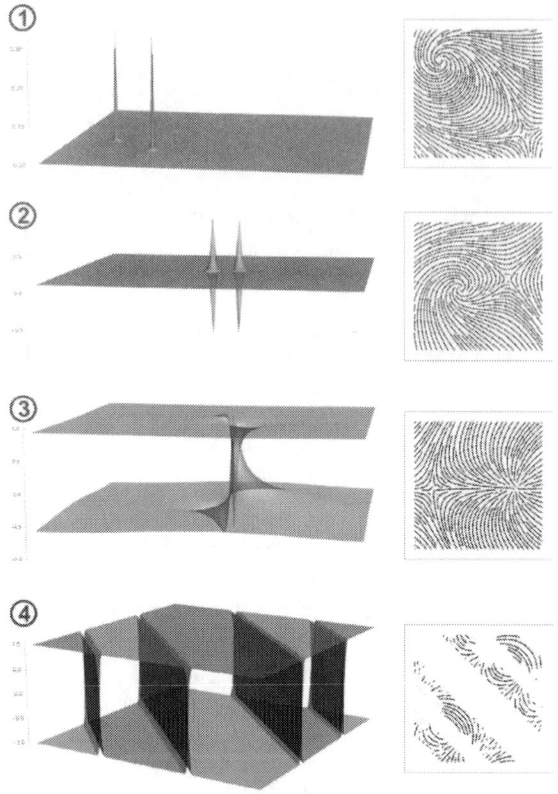

Figure 1. (color online) Inhomogeneous states in 2D system of charged hard-core bosons. The left panels show local charge density $n_i = \sigma_{zi} = \cos\theta_i$. The difference of sublattice states is clearly evident in the cases 2, 3 and 4. The right panels show phase flow of planar components of the pseudospin, σ_{xi} and σ_{yi}. The phase flow reveals the vortex-antivortex pair structure in the core of inhomogeneity of the local charge density in the cases 1, 2 and 3. The parameters of the model are: 1) $n = 0.2$, $\lambda = 0.5$ (FM phase); 2) $n = 0.1$, $\lambda = 0.9$ (FM phase); 3) $n = 0.2$, $\lambda = 1.5$ (FIM phase); 4) $n = 0.0$, $\lambda = 1.5$ (AFM phase). These sets of parameters are denoted with numbers 1-4 on the ground-state phase diagram in figure 2.

$$\Delta V + A_2 V = 0, \quad A_2 = 4\frac{\lambda - 1 - (\lambda+1)n^2}{\lambda - (\lambda+1)n^2}. \tag{12}$$

The solutions has the form

$$\Phi(\vec r, r_i) = \sum_{l=0}^{\infty} a_l K_l(r/r_i)\cos l(\varphi - \alpha_l), \tag{13}$$

$$\Psi(\vec r, r_i) = \sum_{l=0}^{\infty}\Big[b_{1l}J_l(r/r_i)\cos l(\varphi - \beta_{1l}) + b_{2l}Y_l(r/r_i)\cos l(\varphi - \beta_{2l})\Big], \tag{14}$$

where K_l are the Macdonald functions, J_l and Y_l are the Bessel functions of the first and second kind, and a_l, α_l, b_{kl}, β_{kl}, $k = 1,2$ are some constants. An analysis of the asymptotic behavior of solutions (13,14) and the requirement that the omitted nonlinear terms in equations (9) are small in comparison with the remaining linear terms determine that $\Psi = 0$. The account in the lowest order of the mixing with the function $\tilde\phi$ does not change $V(\vec r)$ and gives additional term in $U(\vec r)$ having asymptotic behavior:

$$U_1(\vec r) \approx -\frac{n c_m^2}{2(\lambda+1)\sqrt{1-n^2}}\frac{m^2}{r^{2m+2}}, \tag{15}$$

where m is the number that specifies first nonzero term in (10).

Line $n^2 = \lambda/(\lambda+1)$ is the boundary of areas of the FM-phase with a different behavior of the U and V functions

$$n^2 > \frac{\lambda}{\lambda+1}: \; U(\vec r) = \Phi(\vec r, r_1) + U_1(\vec r), \; V(\vec r) = 0; \tag{16}$$

$$n^2 < \frac{\lambda}{\lambda+1}: \; U(\vec r) = U_1(\vec r), \; V(\vec r) = \Phi(\vec r, r_2). \tag{17}$$

Here we define characteristic lengths $r_i^{-2} = |A_i|$.

In the case of the FIM phase, we need to define the ferri-type combinations: $\tilde U = A\tilde u + \tilde v$ and $\tilde V = A\tilde u - \tilde v$, where $A = -F_{uu0} + \frac{\varepsilon_0}{4}\cos u_0$. The equations (9) lead to Helmholtz equation for the $\tilde U$ function having solution $\Psi(\vec r, r_3)$, $r_3^{-2} = 8$. As in previous case we have to put $\Psi = 0$. The $\tilde V$ function obeys to Laplace equation. Taking into account the mixing in the lowest order with the function $\tilde\phi$ we come to the expressions

$$\tilde U(\vec r) = \frac{c_m^2}{8}(Af_{u0} + f_{v0})\frac{m^2}{r^{2m+2}}, \tag{18}$$

$$\tilde V(\vec r) = \sum_{l=1}^{\infty}\frac{C_l}{r^l}\cos l(\varphi - \gamma_l) + \frac{c_m^2}{4}(Af_{u0} - f_{v0})\frac{1}{r^{2m}}, \tag{19}$$

where m is the number that specifies first non-zero term in (10), and the expressions $f_{u0} = -\sin u_0\cos v_0$, $f_{v0} = -\cos u_0\sin v_0$ are determined by the expressions (8).

Similarly the case of the AFM phase, we obtain

$$\tilde\phi(\vec r) = 0, \quad U(\vec r) = 0, \quad V(\vec r) = \Phi(\vec r, r_4), \tag{20}$$

where $r_4^{-2} = 4(\lambda - 1)$.

4 Discussion

The analysis of the asymptotic behavior of the localized states reveals qualitative differences of the finite energy excitations in the FM, FIM, and AFM phases.

In the FM phase an asymptotic of the polar angle of the pseudospin vector is determined by the expressions (16, 17). When comparing these results with numerical calculations it is worth to note that the characteristic lengths obey to inequality $r_i < 1$ in the most part of the phase diagram in the SF phase except for the areas indicated shadowed in figure 2, so the function Φ goes to zero value very fast with increasing of r. On the contrary, the asymptotic behavior of the azimuthal angle of the pseudospin (10) has no characteristic scale. This means that in the FM phase the main excitations are almost in-plane vortex-antivortex pairs. They have well localized out-of-plane core of the ferro type, with $\sigma_{zA} = \sigma_{zB}$, as shown in figure 1, and become the pure in-plane ones at $n = 0$ in accordance with

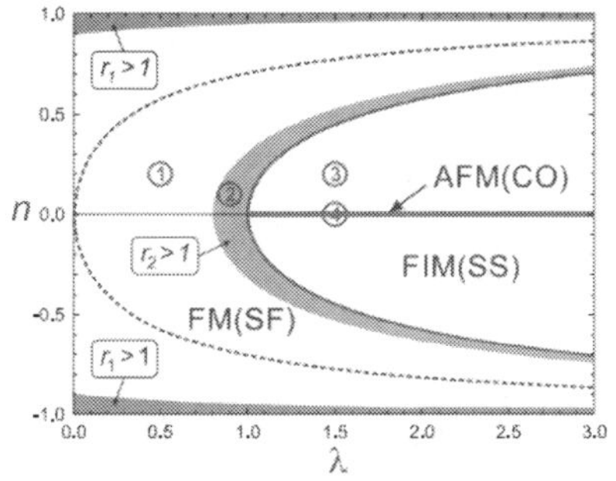

Figure 2. (color online) The ground state phase diagram of the hc bosons in the mean-field approximation. The solid line corresponds to SF-SS phase boundary, $n^2 = (\lambda - 1)/(\lambda + 1)$. The thick line at $\lambda > 1$, $n = 0$ shows CO phase. The dotted line, $n^2 = \lambda/(\lambda + 1)$, separates the two types of the asymptotic behavior in accordance with the expressions (16) and (17). In the shaded areas inside the SF phase region the characteristic lengths satisfy to inequalities $r_i > 1$. The numbers 1-4 in the circles correspond to the parameters sets in figure 1.

expression (15). The same type of localized solutions was found by the authors of [8]. For the hc bosons, the polar angle is related with the density of bosons, while the azimuthal angle is responsible for the superfluid density, hence these states correspond to the excitation of the superfluid component with highly localized heterogeneity of bosons density in the foci of the vortex-antivortex pairs. In the shaded region in the FM phase near the border of the FM-FIM phases in figure 2, the antiferro type vortices, with $\sigma_{zA} \neq \sigma_{zB}$, begin to dominate, their inflation is preceded by a change of the homogeneous ground state from FM to FIM phase.

The AFM phase has no linear excitation of $\tilde{\phi}$. The characteristic lengths of the azimuthal excitations (20) are small except the region near $\lambda = 1$. This results in a high stability of the homogeneous AFM phase. A typical picture of the inhomogeneous state shown in figure 1 is represented by linear domains of the AFM phase. The non-zero values of the FM(SF) order parameter are realized within the domain walls, thus giving rise to appearance of a filamentary superfluidity in the hc bosons.

In the FIM phase the asymptotic behavior of the polar and azimuthal excitations is qualitatively the same without characteristic scales. Hence in this case there are skyrmion-like excitations as shown in figure 1. For the hc bosons these coherent states include both the excitations of the superfluid component and the boson density. In a center of skyrmion the difference $\sigma_{zA} - \sigma_{zB}$ has maximal magnitude, that corresponds to AFM(CO) phase, and near there is a region where $\sigma_{zA} - \sigma_{zB} = 0$, that corresponds to FM(SF) phase. So, the skyrmion-like excitations in the FIM(SS) phase generate the topological phase separation. Note, that another type of instability in FIM(SS) phase were also found by Quantum Monte Carlo calculations [15].

The work supported by Act 211 Government of the Russian Federation, agreement No 02.A03.21.0006 and by the Ministry of Education and Science, projects 2277 and 5719.

References

[1] T. Matsubara, H. Matsuda, Progr. Theor. Phys. **16**, 569 (1956)
[2] E. Kim, M.H.W. Chan, Nature **427**, 225 (2004)
[3] A.S. Moskvin, Phys. Rev. B **84**, 075116 (2011)
[4] A.S. Moskvin, JETP **121**, 477 (2015)
[5] A.A. Belavin, A.M. Polyakov, JETP Letters, **22**, 245 (1975)
[6] V.P. Voronov, B.A. Ivanov, A. Kosevich, Sov. Phys.: JETP **57**, 1303 (1983)
[7] B.A. Ivanov, A.M. Kosevich, Sov. Phys.: JETP **45**, 1050 (1977)
[8] M.E. Gouvea, G.M. Wysin, A.R. Bishop, F.G. Mertens, Phys. Rev. B **39**, 11840 (1989)
[9] A.B. Borisov, JETP Letters **73**, 242 (2001)
[10] I.G. Bostrem, A.S. Ovchinnikov, JETP Letters **76**, 716 (2002)
[11] A.B. Borisov, I.G. Bostrem, A.S. Ovchinnikov, JETP Letters **80**, 103 (2004)
[12] I. Bostrem, A. Ovchinnikov, R. Egorov, Physics Letters A **279**, 33 (2001)
[13] R. Egorov, I. Bostrem, A. Ovchinnikov, Physics Letters A **292**, 325 (2002)
[14] R. Micnas, J. Ranninger, S. Robaszkiewicz, Rev. Mod. Phys. **62**, 113 (1990)
[15] G.G. Batrouni, R.T. Scalettar, Phys. Rev. Letters **84**, 1599 (2000)

AUTHOR INDEX

Afremov Leonid 567
Ahmad Haseeb 290
Ahmed Azaharuddin 405
Akai Kurbanovich Murtazaev 605
Alekhina Yuliya 455
Alexander Kamantsev 324
Alexey Mashirov 324
Alla Novakova 279
Amirabadizadeh Ahmad 547
Andrade Ricardo 551
Andreev Nikolay V. 393
Andrianov Timofey 78
Anikin Maksim 306, 359
Annenkov Alexander Yu. 106
Antonov Anatoliy S. 119
Apanasevich Vladimir 567
Arnold Zdenek 275
Aronov Aleksey 66
Aronzon Boris 397
Asadullin Fanur 98, 145
Ataeva G.Y. 571
Atul Thakur 320
Averyanov Dmitry V. 35
Babaev A.B. 571
Babaev Albert 597
Babaizev Georgy 1, 39
Badelin A. 377
Baksheev Georgy G. 601
Balbashov Anatolii 424
Bandyopadhya Sudipta 405
Banerjee Aritra 405
Barkalova Anastasia 483
Basiladze Georgy 130
Bataronov Igor 271
Batashev Ivan 424
Bazhanov Andrey 86
Beketov Igor V. 302, 551
Beketova Anna I. 302
Beklemisheva Anna V. 320
Berzhansky Vladimir 130, 134, 141
Beznosikov Dmitry 102
Biziyaev Dmitry 460
Blinov Ilya 20
Blinov Mikhail 47
Bobrovskii Stanislav 298
Bobrovskii Stanislav Y. 90, 151
Bogdanov Anatoly 347
Bolshakov A.P. 232
Bondarev Alexey 271
Borisenko Igor V. 23, 109
Borzilov Vadim O. 601
Buchelnikov Vasiliy 256, 343, 351, 371
Buchner Bernd 162, 467
Budrin K.S. 590
Bukharaev Anastas 460
Bukunov Kirill 1
Buravtsova Victoria 200
Burdin 441
Burdin D.A. 449
Burkhanov Gennady 367
Burova Lidia I. 405
Buznikov Nikita A. 119
Chashin D.V. 449
Chashin Dmitri 445
Chechenin Nikolay 1, 39
Chepkov Grigory 12
Cherebilo Elena A. 421
Cherebylo Elena 66
Cherepkov Maksim 306
Chikov A.A. 590
Chinenkov Maksim 8
Chistotina E.A. 232
Chistyakov V.V. 5
Chlenova A.A. 555
Chlenova Anna A. 539
Chukalina Elena 433
Chzhan Anatoly V. 197
Chzhan Victoria 367
Constantinian Karen 471
Constantinian Karen Y. 23
Cwik Jacek 264
Danilova Alena Yu. 601
Datskaya Z. 377
Davidov Alexander 381
Davydova M.D. 16
de Sena Pereira F.D. 504
Demidov V. 23, 109, 393, 471
Demin Gleb 74
Demishev Sergey 94
Demokritov Sergej O. 170
Denisova Tatyana P. 543, 551, 563
Dikansky Yuri 508, 535
Dinh Quang 324
Djuzhev Nikolay 8
Dmitriev Alexei 314
Dmitry 441
Dmitry Chashin 441
Dolgov Aleksandr 130
Domozhirova A.N. 5
Drulis Henryk 367
Dubenko I. S. 336
Dubenko Igor 339
Edelman Irina 193
Eisterer M. 5
Ekonomov N.A. 449
Erina Marina 520
Erofeev Dmitry 433
Estemirova S. 377
Evstigneeva Svetlana 290
Fadeev Egor 339

AUTHOR INDEX

Fetisov L.Y. ... 449
Fetisov Leonid ... 445
Fetisov Y.K. ... 449
Filippov Dmitry ... 134
Fominov Yakov ... 467
Gaidukova Irina ... 239
Gan'shina Elena ... 115, 200, 215, 401, 417, 426
Garanov Victor A. ... 90
Gareev Timur ... 430
Garif'yanov Nadir ... 467
Garifullin Ilgiz ... 467
Garshin Vladimir ... 215
Gaviko Vasily ... 286
Gerasimov Evgenii.G. ... 235
Gerasimov Evgeny ... 286
Gerus Sergey V. ... 106
Geydt Pavel ... 211, 339
Gilmanov Marat ... 94
Gimaev Radel ... 355
Golev Igor ... 483
Golik Leonard ... 401, 426
Golubeva Elizaveta V. ... 539
Gonchar Liudmila ... 389
Govor Gennadiy ... 363
Granovsky A. B. ... 47, 155, 336, 339
Granovsky Nikita ... 115
Granovsky Sergey ... 239
Grebennikov Vladimir I. ... 235
Grigoriev Evgenii ... 226
Grishin Alexander S. ... 23
Gunbin Anton ... 317
Guryev Valentin ... 479, 491
Guteneva Natalia V. ... 559
Gvozdevskaya Victoria ... 483
Harin Eugene ... 222, 298
Hashim H. ... 260
Hashim Hisham ... 290
Hirazawa H. ... 243
Huang J.C.A. ... 5, 16
Ilyushin Alexander ... 243
Inishev Alexander ... 286
Inyushkin A.V. ... 232
Ispiryan Anna ... 535
Iurov Aleksey ... 8
Iusipova Iulia ... 27, 43
Ivanova Oxana ... 193
Ivanova Tatiana ... 264
Ivantsov Ruslan ... 193
Jablunovskis Artjoms ... 433
Kabanov Vladislav ... 243
Kalinin Yu.E. ... 59, 155
Kalinin Yuri ... 200
Kalmykov Sergey ... 524
Kamantsev Alexander ... 328, 332
Kamarád Jiri ... 275
Kamashev Andrey ... 467
Kameneva Margarita ... 385
Kamilov K.I ... 504
Kapranov Mikhail ... 189
Karateev Igor A. ... 35
Karavainikov Andrey ... 130, 141
Karpasyuk V. ... 377
Karpenkov Alexey ... 310
Karpenkov Dmitriy ... 310
Karpov P.I. ... 437
Kashtanov E.A. ... 504
Kassan-Ogly Felix ... 166, 575, 597
Kataev Vasiliy ... 211
Kataev Vladislav ... 467
Katkova Olga ... 189
Kaul Andrey R. ... 405
Kazakevich I.S. ... 248
Kazei Zoya ... 385
Kevraletin Alexandr ... 219
Khandukhanov Roman T. ... 551, 563
Kharlamova Anna ... 185
Khramova Olga ... 66
Khramova Olga D. ... 421
Kieu Hau ... 328, 332
Kirman Marina ... 267
Kirpan Veronika ... 51
Kiryukhantsev-Korneev Philipp ... 298
Kiseleva Tatiana ... 243
Kislinskii Yulii ... 471
Kliava Janis ... 193
Klimin Sergei ... 162
Klimov Alexey A. ... 23
Klingeler Rudiger ... 162
Knotko Alexander ... 409
Kobzev Mikhail ... 524
Kochura Aleksey ... 66, 381
Kochura Aleksey V. ... 375
Koksharov Yuri ... 222
Koledov Victor ... 123, 328, 332
Kolesnikova Anastasia ... 508
Kolodkin Denis ... 294
Kolotov Oleg ... 113
Komlev Anton ... 252
Kondratev Oleg A. ... 35
Konev Vitaly ... 486
Konstantinov Nikita ... 409
Korolev Alexander ... 294
Korolev Alexander V. ... 170
Koroleva Luidmila ... 424
Koshkid'ko Yurii ... 264
Kotov Leonid ... 98, 102, 127, 137, 145
Kozabaranov Roman ... 158
Kozeeva Ludmila ... 385
Krakov Mikhail S. ... 495
Kramarenko Elena ... 455

AUTHOR INDEX

Krekhno Roman V. 302
Krinitsina Tatiana 20
Kritskaya E.A. 504
Krokhotin Andrei 363
Kruglov Vitaly 479, 491
Kuchin Anatoly 275, 294
Kudasov Yuri 158
Kudrevatykh Nikolay 306, 359
Kulatov Erkin 401, 417
Kulbachinskii Vladimir 397
Kulbachinskii Vladimir A. 405
Kulesh Nikita A. 563
Kulikova Daria 430
Kun'kova Zoya 401, 426
Kunikin Stanislav 535
Kurenkov Alexander 39
Kurlyandskaya Galina 185
Kurlyandskaya Galina V. 539, 543, 547, 563
Kustov Vladimir 567
Kuz'mina Alina S. 421
Kuzmenko Alexander 381
Kuznetsova Tatyana V. 235
Kuznetsova Yulia 310
Kytin Vladimir G. 405
Lähderanta Erkki 47, 211, 252, 339, 375, 381
Larin Vladimir 290
Lavrov Alexander 385
Leksin Pavel 467
Lepalovskij V.N. 555
Lepalovskij Vladimir 211
Leshchiner Dmitry 12
Lesnikov Vladimir 426
Likhachev Igor 491
Lileev Alexey 317
Lin J.G. 16
Lock Edwin H. 106
Lotfollahi Zahra 547
Lotin Andrey 66, 417
Lotin Andrey A. 421
Lubov A. Kokorina 543
Lugovskoy Nazar 134
Lukyanenko Kseniya 567
Lukyanov Pavel 567
Lyakhova Marina 310
Lyapilin Igor 82
Lyashko Sergey 141
Lykhmanova Victoria 520
Makagonov V.A. 155
Makarov Andrey 185
Makarov Pavel 127, 137
Makarova L. 421, 445, 455
Maltceva Lidiya 127, 137
Malyshev A.Yu 504
Malyshkina Olga 455
Mamonova Marina 177
Mamonova Marina V. 31, 181
Marat Khairullin 279
Marchenkov V.V. 5
Marenkin Sergey 66, 381
Markin Yuri 426
Markov Gennadiy 243
Markov V.K. 504
Mashirov Alexey 328, 332
Matsnev Mikhail 463
Matyunin Andrey 113
Matyunina Mariya 256
Maximova Evgenia N. 543
Maximova Olga V. 405
Mazov Lev 475
Mazurkin Nikita 8
Medkov Mikhail 567
Medvedev Anatoly I. 551
Melnikov Grigory Yu. 551, 563
Melnikov Nikolai B. 609
Men'shenin Vladimir 452
Menshikh Ivan 166
Menushenkov Vladimir 282
Merkulov D. 377
Merkulov Dmitrii 524, 528
Mikhailova Tatyana 141
Mikhalevsky Vladimir 66, 381
Mikhalevsky Vladimir A. 421
Mikhalitsyna Evgeniya 211
Milyaev Mikhail 20
Minin A.S. 174
Mironov V.A. V.A. 504
Miroshkina Olga 351
Mitsiuk Victor 363
Miyanaga Takafumi 355
Mokhov E.N. 232
Morchenko Alexander 290
Morchenko Alexander T. 320
Morgun Leonid 397
Morgunov Roman 267
Morozov Artem 424
Moskalev Vladimir 306
Moskvin A.S. 590
Moskvin Alexander 486, 614
Mostovshchikova E.V. 174
Mukhin S.I. 437
Müller R. 504
Murtazaev A.K. 571
Murtazaev Akai 597
Mushnikov Nikolai.V. 235
Mushnikov Nikolay 286
Naboko Andrei S. 90
Naletova Vera 524
Naletova Vera A. 531
Nekhaeva Elena 397
Nematov Makhsudsho 290

AUTHOR INDEX

Nematov Makhsudsho G. 320
Nepochatykh Yuriy 123
Neznakhin Dmitriy 294
Nguyen An 332
Nguyen Dan 324, 328, 332
Nguyen Duong 324
Nguyen Ha 332
Nguyen Lam 324
Nguyen Mai 328
Nguyen Ngoc 324
Nguyen Yen 324, 328, 332
Nikiforov Anatoliy 389
Nikiforov Ivan V. 495
Nikitin Maxim P. 559
Nikitin Petr I. 559
Nikitin Sergey 264, 347, 363
Nikitovskaia Ekaterina A. 375
Nikoladze Georgiy 113
Nikolaeva Elena 430
Nikolay Ekonomov 441
Nikulicheva Tatyana B. 375
Novodvorskii Oleg 417
Novodvorsky Oleg 66, 381
Novodvorsky Oleg A. 421
Nurgazizov Niyaz 460
Obukhov Sergei 413
Okorokov Mikhail 82
Okulov Vsevolod 64
Omelyanchik Alexander 455
Orekhov Anton 226
Orlov Alexey V. 559
Osipov Aleksey V. 151
Osipov Alexei V. 90
Ostrovskaya Natalia 27
Ovcharov Alexey 491
Ovchenkova Ioulia 347
Oveshnikov Leonid 226, 397
Ovsyannikov Gennady 471
Ovsyannikov Gennady A. 23, 109, 393
P Panaetov V. 230
Pamyatnykh Evgeny 64
Panchenko Vladislav 66
Panina L. 260, 290, 320
Pankratov Nikolay 363
Panov Yu.D. 590
Panov Yury 486, 614
Panysheva Tatiana 413
Paradezhenko Georgii V. 609
Parfenov Oleg E. 35
Parfenov V.A. 504
Parshina Liubov 66
Parshina Liubov S. 421
Patrin Gennady S. 197
Pavlukhina Oksana 343
Pelevina Daria 524
Perevozchikova Y.A. 5
Perlo Pietro 12
Pernod Phillipe 23
Perov Dmitry V. 170
Perov Nikolai 455
Perov Nikolay S. 421
Perova Natalia 215
Pestun Anna E. 393
Petrakovskaja Eleonora 193
Petrov Dmitry A. 151
Petrov P.N. 16
Petrzhik Andrey 471
Pham Thanh 324, 328, 332
Pilyk Evgeny A. 375
Pinkas Victoria 317
Platonov Sergey 275, 294
Plekhanova Daria 445
Pleshev Dmitriy 98, 145
Plokhov D.I. 582
Podgornaya S.V. 260
Podorozhnyak Sergey A. 197
Pogrebneak Mikhail A. 375
Poimanov Vladislav 123
Pokatilov Vyacheslav 463
Poleshchikov Sergey 98, 145
Politova Galina 367
Polunin V.M. 511
Polyakov Petr 113, 579
Ponomareva Tatyana 166
Popkov Anatoly 12, 74
Popov A.I. 582
Popov Vladimir 20
Popov Zakhar 204
Popova Elena 162
Popova Marina 162
Preobrazhensky Vladimir L. 23, 393
Prokopov Anatoly 130, 134, 141
Proshkin Alexey 166, 575, 597
Proskurnina Natal'ya 294
Prudnikov Pavel V. 31, 181, 586, 601
Prudnikov V. N. 336
Prudnikov Valerii 47
Prudnikov Vladimir 177
Prudnikov Vladimir V. 31, 181, 601
Pudonin F.A. 260
Purtov Alexandr N. 181
Pushkina Marina 8
Pyatakov A.P. 504
Pyatakov Alexander 430, 463
Ralchenko V.G. 232
Repina Nataliya 208
Reser Boris I. 609
Ricardo Andrade 563
Rinkevich Anatoly B. 170
Ritter Clemens 239

AUTHOR INDEX

Rodionov I. 47, 336, 339
Rodionova Valeria 455
Rogacki Krzysztof 264
Romanovskii Dmitry E. 31
Rozanov Konstantin 298
Rozanov Konstantin N. 90, 151
Rukovishnikov Alexander 426
Rusakov Vyacheslav 463
Ryapolov P.A. 511
Rybakov Yu.P. 594
Rylkov Vladimir 47, 417
Safin Ansar 189
Safronov Alexander P. 302, 543, 547, 563
Salakhova R.T. 504
Saletsky A. M. 336
Saletsky Alexander 69
Samarin Alexander 94
Samatov Oleg M. 543, 563
Samoilovich Mikhail I. 170
Sampath Vedamanickam 123
Sanga Deleg 243
Savchenko Alexander 282
Saveliev Dmitri 445
Schumann Joachim 467
Semeno Alexey 94
Semenova Elena 310
Semkin Mikhail 306
Seong Yu 328
Sergeev Alexander 483
Sergei Chvalun 226
Severin Pavel 102
Shadrin Anton 471
Shaikhulov Timur A. 393
Shakhov Mikhail 381
Shakhunov Valery A. 393
Shalygina Elena 185
Shaposhnikov Alexander 141
Shavkin Sergey 479, 491
Shavrov Vladimir 98, 123, 145
Shcaihulov Timur A. 109
Shcheglov Vladimir 98, 127, 137, 145
Shchetinin Igor 282
Sheftel Elena 222, 298
Sheldeshova E.V. 511
Sherstnev I.A. 260
Shevchenko Evgeniy 252
Shevyakov Dmitrii 417
Shiryaev Artem O. 151
Shkar Vladiimir 123
Shkliaev Alexey A. 119
Shlyakhtich Maria A. 586
Shmykova Mariya 567
Shneider Alexander G. 421
Shpetnyy I. 260
Shuyskyy Alexandr 134
Sigov Alexander 463
Simonova Elena V. 543, 551
Singh S.P. 260
Sitnikov A. 47, 51, 59, 155, 200
Skidanov Vladimir 27
Skipetrov Evgeny 409
Skipetrova Liudmila 409
Skirdkov P.N. 16
Skokov Konstantin 310
Skvortsov Maxims 27
Slavska-Wanniewska Anna 424
Slynko Vasily 409
Smelova Ekaterina 69
Smirnov Andrey 347
Snegirev Vyacheslav 385
Soboleva Ekaterina 211, 339
Sokolov Alexey 115
Sokolovskaya Yulia 371
Sokolovskiy Vladimir 256, 343, 351
Spichkin Yury 355
Starostenko Sergey N. 151
Stepanova Elena A. 539
Stognei O.V. 59
Stognei Oleg 51
Stognii Alexandr 445
Storchak Vyacheslav G. 35
Streltsov Dmitry 226
Sultan-Zade Timur T. 119
Svalov Andrey 185
Syr'ev Nikolaj 215, 219
Szymczak Henryk 424
Taldenkov A.N. 35, 232
Tarakova Olga 567
Tarasov Evgeniy 306, 359
Tarasov Ivan 204
Tedzhetov Valentin 222, 298
Temiryazeva Marina 426
Terentev Pavel 286
Tereshina Irina 367
Thomas Jürgen 467
Tiercelin Nicolas 23
Tigran Baluyan 279
Tishin A.M 504
Tishin Alexander 355
Titov I. S. 336, 339
Tokmachev Andrey M. 35
Tomilin Sergey 141
Tran Thanh 328, 332
Tregubova Tatyana 51
Trifonov Aleksey 8
Trukhanov Alex V. 248
Trukhanov Alexey 290
Trukhanov Sergei V. 248
Tsysar Kseniya 69
Tugushev Victor 417

AUTHOR INDEX

Turchenko Vitalii 248
Turkov Vladimir A. 531
Tyatyushkin Alexander 516
Udalov Nikolay 189
Uimin M.A. 174
Ulitko Vasily 486
Urzhumtsev Andrey 306
Useinov Niazbeck 55
Uspenskii Yurii 401, 417
Ustinov Vladimir 20
Validov Aidar 467
Varnakov Sergey 204
Vasiliev A. 162, 226, 491
Vasinovich Evgeny 486
Vdovichenko Artem 226
Vedyayev Anatoly 78
Veligzhanin Alexey 491
Vetcher Alexander 363
Victor Koledov 148, 324
Vinogradova Alexandra S. 531
Visotin Maxim 204
Vladimir Shavrov 148
Vladyslav Poimanov 148
Vlasov Vladimir 98, 102, 145
Volchkov S.O. 555
Volegov Alexey 294
Volochaev Mikhail N. 197
Volodina Natalia 290
Vorob'ev-Desyatovskii Nikolay 252
Voronchihina Kristina 317
Voronin Vladimir 294
Vu Quang 328
Vylegzhanin A. G. 504
Vyzulin Sergei 215, 219
Wiafe-Akenteng Dennis 282
Yakovlev Ivan 204
Yakubtsev Rodion A. 320
Yan Zubavichus 226
Yerin Constantine 520
Yermakov A.Ye. 174
Yu Seong 332
Yuan Ye 401
Yudanov Nicolay A. 320
Yudanov Nikolay 290
Yurasov Alexey 115
Yurii Nepochatykh 148
Yury Fetisov 441
Zagrebin Mikhail 256, 343, 351, 371
Zakharchuk Ivan 47, 211
Zakhvalinskii Vasilii 381
Zakhvalinskii Vasilii S. 375
Zakinyan Arthur 508
Zarubin Alexander 166
Zavyalov Sergei 226
Zayak Alexey 371
Zazymkina Daria 115
Zezyulina Polina 298
Zhavrail Gadzhievich Ibaev 605
Zhou Shengqiang 401
Zhukov Alexander V. 501
Zinin Aleksander 306, 359
Znoyko Sergey L. 559
Zubavichus Yan 491
Zverev V.I. 504
Zverev Vladimir 355
Zvezdin A.K. 582
Zvezdin K.A. 16
Zvezdin Konstantin 12
Zvonov Alexey 264
Zykov Georgy 401, 426